The Ulnar Nerve

Giuliano Gentili • Mario Di Napoli

The Ulnar Nerve

Sensory and Motor Conduction Studies

Springer

Giuliano Gentili
Neurological Service
S. Camillo de' Lellis General Hospital
Rieti
Italy

Mario Di Napoli
Neurological Service
S. Camillo de' Lellis General Hospital
Rieti
Italy

ISBN 978-3-319-43291-5 ISBN 978-3-319-43292-2 (eBook)
DOI 10.1007/978-3-319-43292-2

Library of Congress Control Number: 2016957746

Printed on acid-free paper

This Springer imprint is published by Springer Nature
The registered company is Springer International Publishing AG Switzerland
The registered company address is: Gewerbestrasse 11, 6330 Cham, Switzerland

To my wife Maria Carla,
who believed in me: day after day, wave after wave.
Thank you for everything you have given me: white sands, hidden treasures, and some shells.
I love you, Ami, ever since you taught me to listen to the sound of the sea.

To my son Giulio,
from one world traveler to another.

Giuliano

"We choose our next world through what we learn in this one. Learn nothing, and the next world is the same as this one, all the same limitations and lead weights to overcome."

Jonathan Livingston Seagull, Richard Bach

Foreword

Clinical neurophysiology is of great clinical value as a functional counterpart to the emerging biochemical and imaging techniques. Over the years, we see continuous progress such as new methods, modified methods, reference values, and the consolidation of useful databases. This book about the ulnar nerve gives a historic overview of the nerve conduction studies and reference values from the 1950s until present days. The fact that it covers about 70 different methods or modifications is a sign that no single method is really so good that it clearly supersedes all others. This development is based on continuously better understanding of physiological mechanisms, better equipment, and improved methods for signal analysis. There is also an obvious need for quantitation of results, which automatically require reference values to be useful. This has been successively more clear for examiners using neurophysiological techniques in clinical routine and also for equipment manufacturers that must provide means for post processing of data.

With this book, the authors have made an impressive work of collecting detailed information about the majority of methodological descriptions and available reference values. The book is a summary of efforts put into improving diagnostic and follow-up techniques and is a collection of methods that have been or that are presently used for patient care. It also points to methods that have helped understand some of the underlying factors that explain nerve-related symptoms such as weakness, fatigue, and sensibility loss. The early methods also revealed new information about anatomy, anomalous innervation, frequency, and degree of nerve involvement for a given diagnosis and often gave unique supplementary information to the neurologist, physiotherapist, or surgeon who helped to define treatment strategies in the days before other methods became available.

The content is presented in a standardized easily accessible form: the original settings, the recording and stimulation details, and many very effective and well-conceived figures for each technique. They have summarized the main information from literature, but also given their own comments and interpretations when pertinent. In the literature, there is very little description of sensitivity, specificity, and reproducibility when comparing different neurophysiological parameters (ranking). This type of information is extracted from reports of comparisons between the neurography results and other data such as clinical, biochemical, or data from imaging techniques.

The authors have also performed their own recordings by reproducing all methods being described and presented signals of high quality despite the old techniques that can no longer be exactly reproduced due to lack of the description about electrodes and specific equipment. Many articles do not provide pictures of signals, so that a detailed comparison cannot be done. In addition, many publications have small material and poor statistics and lack information about correlation to factors such as age, height, and temperature.

Which readership is this book meant for? One group of users is certainly those who just came up with an excellent idea, but want to check how much is previously published in that field. This will help to avoid duplication of old methods. It saves a lot of time and reduces large frustration. Another group is the EMG laboratory that is just going to implement a new method, not used routinely in this laboratory. Help and advice as well as the important information of reference values, at least for comparison with the new values that the laboratory is collecting as a methodological check, is needed. A third application is the use as a textbook, both for physicians and technicians performing electroneurography.

The book will be of great importance to give ideas for future sophistication of neurography, in this case, for conditions that include the ulnar nerve, a very common referral reason. The book is an impressive historical document not only as a review of methods for studies of the ulnar nerve, but it also gives indication how methods have been developed or modified based on evidence markers. Some have stood the test of time; others have slowly (too slowly) been taken out of service. Only relatively recently have reference materials been of sufficient quality for better description of findings, prompted by the obvious need for the user and for the equipment manufacturers. It will give ideas how to improve the neurographical testing. The modern way of making multivariate analysis, or making interpretation from a constellation of data (various indices), will require the highest possible signal quality and good reference databases.

The authors help us understand this; they provide advice and show excellent examples of recordings and helpful anatomy pictures with superimposed drawings of electrode placements. I congratulate them on these results and thank them for undertaking this task, helpful to all of us involved in clinical neurophysiology.

Uppsala, Sweden Erik Stålberg

Contents

List of Abbreviations

BMI	Body mass index
CIDP	Chronic inflammatory demyelinating polyradiculoneuropathy
CMAP	Compound muscle action potential
CMT	Charcot–Marie–Tooth disease
CTS	Carpal tunnel syndrome
DML	Distal motor latency
EMG	Electromyography
ENG	Electroneurography
GBS	Guillain–Barrè syndrome
MMN	Multifocal motor neuropathy
MNCV	Motor nerve conduction velocity
PML	Proximal motor latency
SNAP	Sensory nerve action potential
SNCV	Sensory nerve conduction velocity
UNE	Ulnar neuropathy at the elbow
UNW	Ulnar neuropathy at the wrist

Units of Measurement

cm	Centimeter
Hz	Hertz
kHz	Kilohertz
mA	Milliampere
ms	Millisecond
m/s	Meter per second
mV	Millivolt
µs	Microsecond
µV	Microvolt

Introduction

The ulnar nerve is one of the most studied nerve during the daily routine activities in each clinical neurophysiopathological laboratory not only for the frequency of the pathologies of which it is affected like the ulnar neuropathy at the elbow (UNE) but also for the simplicity to investigate that makes it a preferred nerve in multiple disease study. With the first studies of Dawson and Scott (1949) and Dawson (1956) through the propagation of electrical impulses along the median and ulnar nerves and the recording of the peripheral nervous responses by means of electrodes applied to the skin, the knowledge process about nerve conduction in man kicked off in the following years a number of validation studies of the electroneurographic method allowing to determine normal and pathological ranges. The electroneurography (ENG) was born in those years as a diagnostic method. It was initially used in the diagnosis of compressive diseases such as the UNE or in the definition of diabetic neuropathy, and its utility was gradually extended to diagnose many other diseases.

This volume represents a chronological path that allows the reader to move from pilot studies on peripheral nerve conduction in humans during the 1950s to the most recent scientific evidence. It is a journey through the first sensory and motor conduction studies of the 1960s, the various comparative orthodromic and antidromic studies of the sensory conduction of the ulnar–median nerves and the F-wave studies introduced in the 1970s, the first recordings of the responses from the dorsal sensory branch of the ulnar nerve and the comparison of the distal motor latency and amplitude of the *first dorsal interosseous* (FDI) and the *abductor digiti minimi* (ADM) or *abductor pollicis brevis* (APB) muscles described in the 1980s, the various observations about changes in the nerve conduction velocity (NCV) due to the elbow position performed in the 1990s, and the more recent mixed and motor conduction studies described since the early 2000s to today.

The main attention has been given to techniques that have shown higher sensitivity and specificity in the diagnosis of UNE and, for this reason, are widely used in clinical practice, such as the short-segment incremental study (SSIS).

About 70 techniques are described in this text-atlas including methods from the original rationale and variants and organized according to practical criteria for easy reference; all of the tests are presented by facilitated pictures in order to make it easy to understand and follow a reproduction of every method. There have been reports of the parameters and settings used by the authors and the normative and pathological values cited in the entries and present in subsequent articles in order to enrich and update the evidence. With particular attention, it has been reproduced in every test in the laboratory so as to present iconographically the signals acquired in the normal subject; an effective diagnostic utility of each test was then evaluated in several diseases and in varying degrees of the same pathology as in the UNE. All traces acquired, both in normal and pathological subjects, are shown with different settings to adapt easily and be more accessible to the different situations that may arise in clinical practice.

The execution of the techniques described in the manuals of electroneurography on the market is not always able to provide a complete answer to the many clinical questions.

The nonuse of the correct technique because not known incorrect use of a technique, and the absence of normative and pathological values represent all factors that can alter the diagnostic power of the electroneurographic examination.

From this premise to the volume was born a systematic monograph of electroneurographic techniques focusing on the ulnar nerve, with the aim to validate the various methods described in the literature since the early articles of the 1950s to the present to define the feasibility, reproducibility, and actual usefulness in the diagnostic field. Some of these techniques, abandoned over the years, still show their worth and deserve to be known in the clinical setting, still not having found the right recognition in any of the manuals on the market. This text-atlas, which is the result of over five years of hard work, comes from our daily neurophysiopathological experience; we have described in detail the technique, standardization, and normal and pathological values derived from the original articles and the subsequent literature as never seen in any other work before.

We, therefore, hope that this text, which is useful to both the novice and the experienced specialist, is always in a laboratory of electromyography as a valid means of reference for study on the field.

Synoptic Table

Sensory Conduction Studies

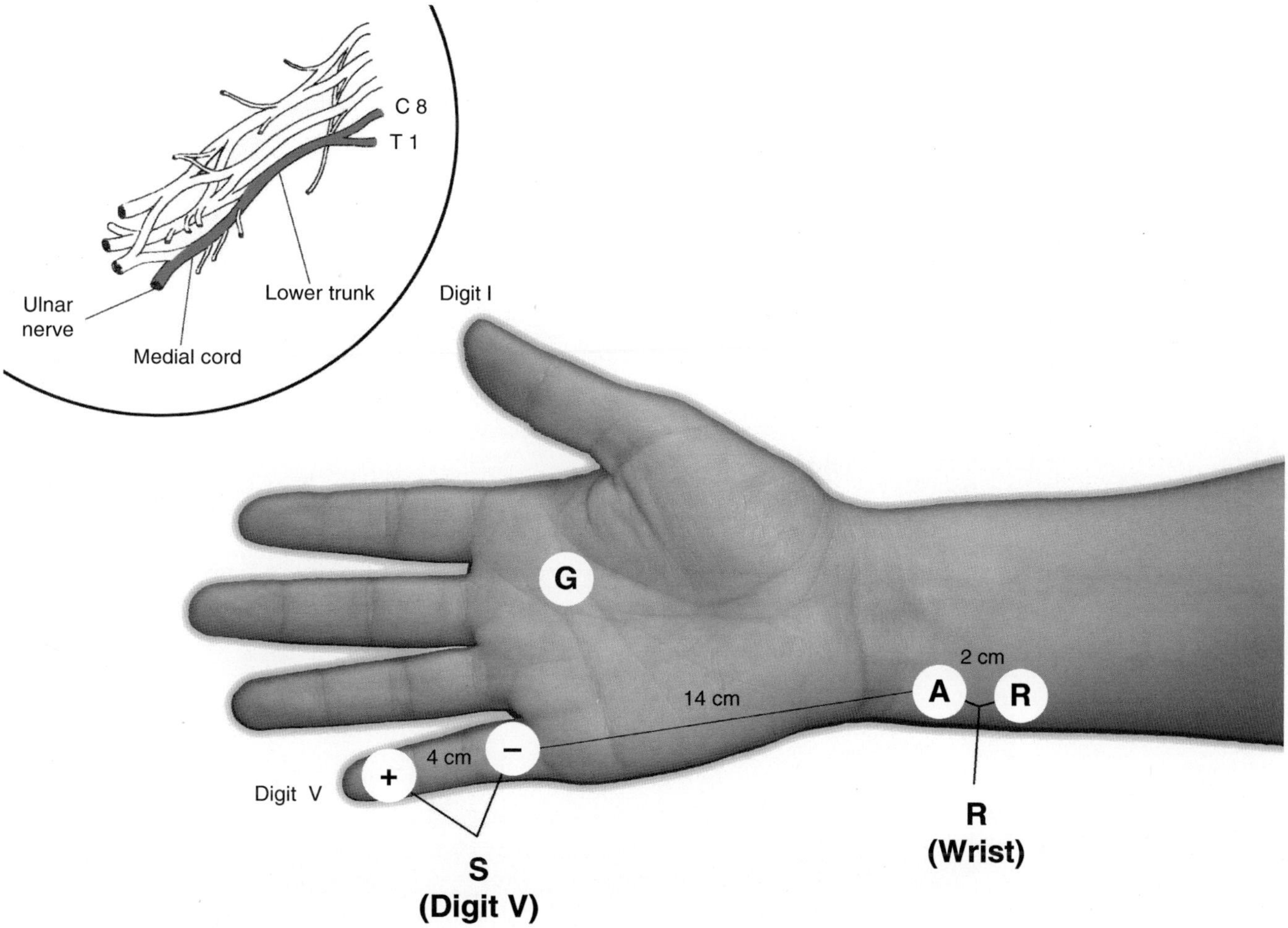

(A)	Active Electrode
(R)	Reference Electrode
(G)	Ground Electrode
(−)	Cathode
(+)	Anode
[R]	Recording Site
[S]	Stimulating Site

Fig. 1 Sensory conduction study at the wrist; stimulation of digit V; stimulation (*S*)

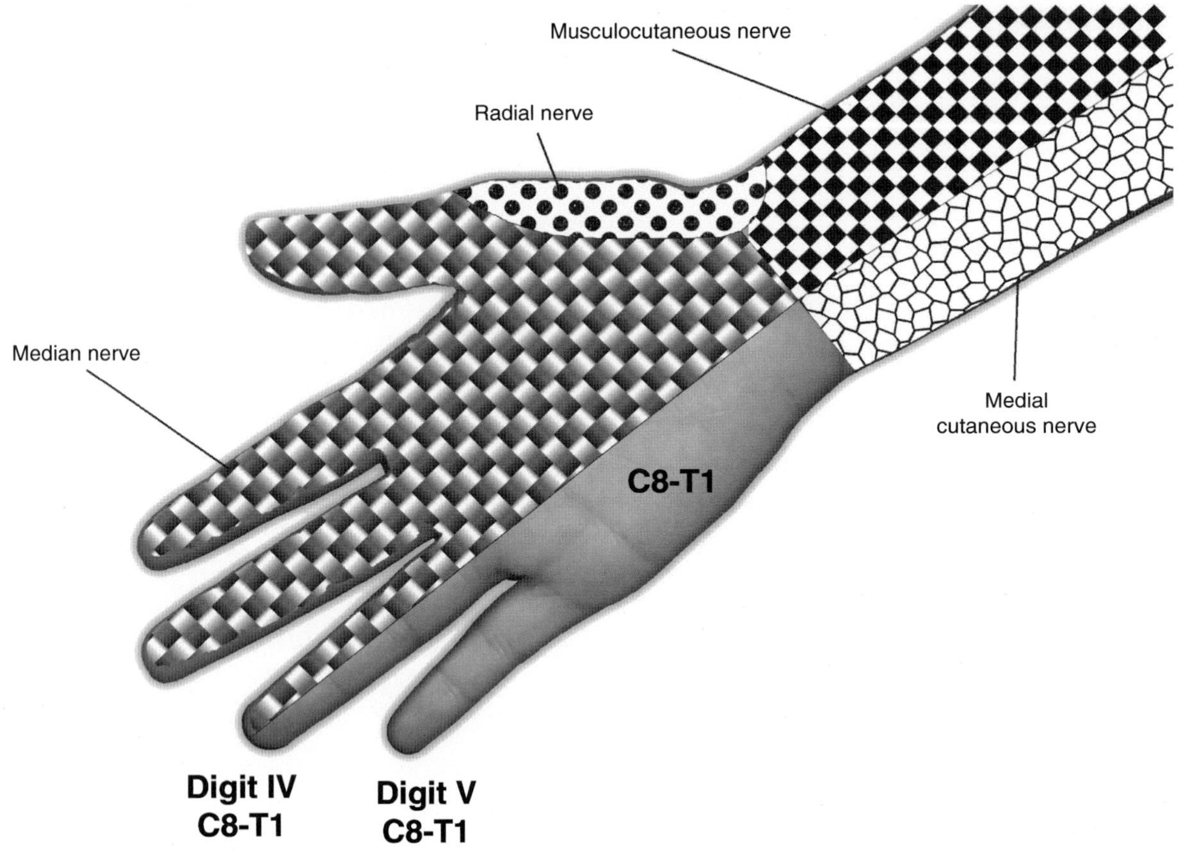

	AREA	BRANCH	CORD	TRUNK	ROOT
UPCN	Hypothenar eminence	Ulnar palmar cutaneous nerve	Medial	Lower	C8-T1
DIV	Digit IV (lateral ring finger)	Digital	Medial	Lower	C8-T1
DV	Digit V (little finger)	Digital	Medial	Lower	C8-T1

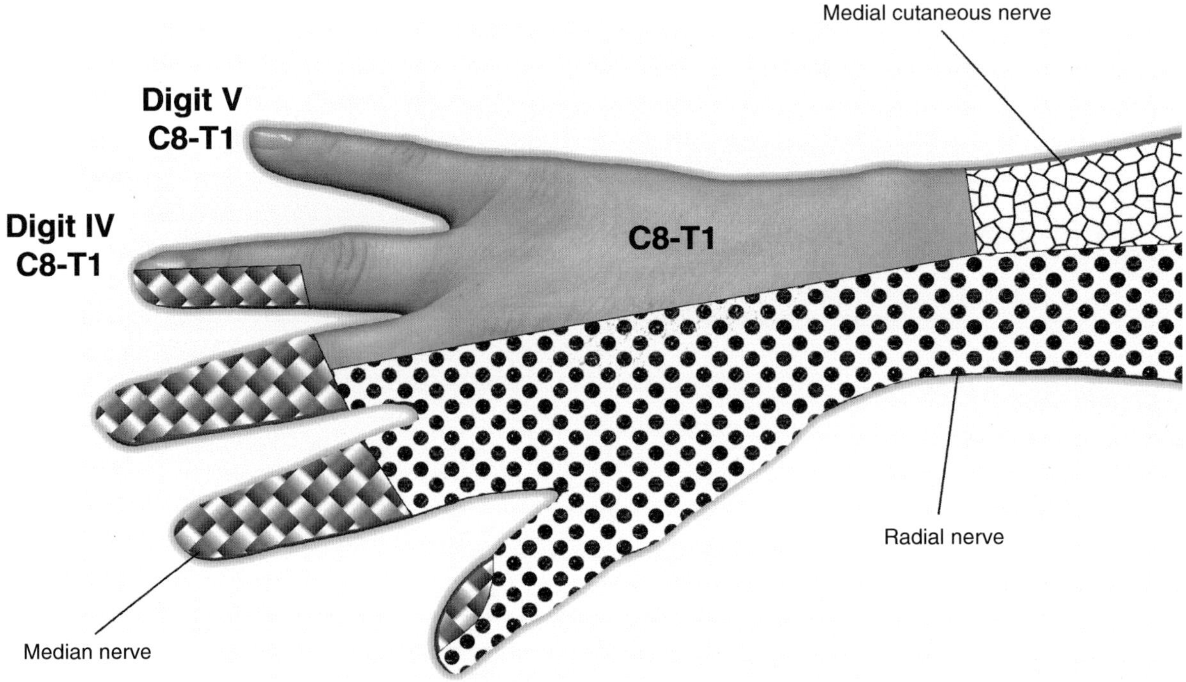

Fig. 2 Hand: typical sensory territory of the ulnar nerve (palmar and dorsal side)

Motor Conduction Studies

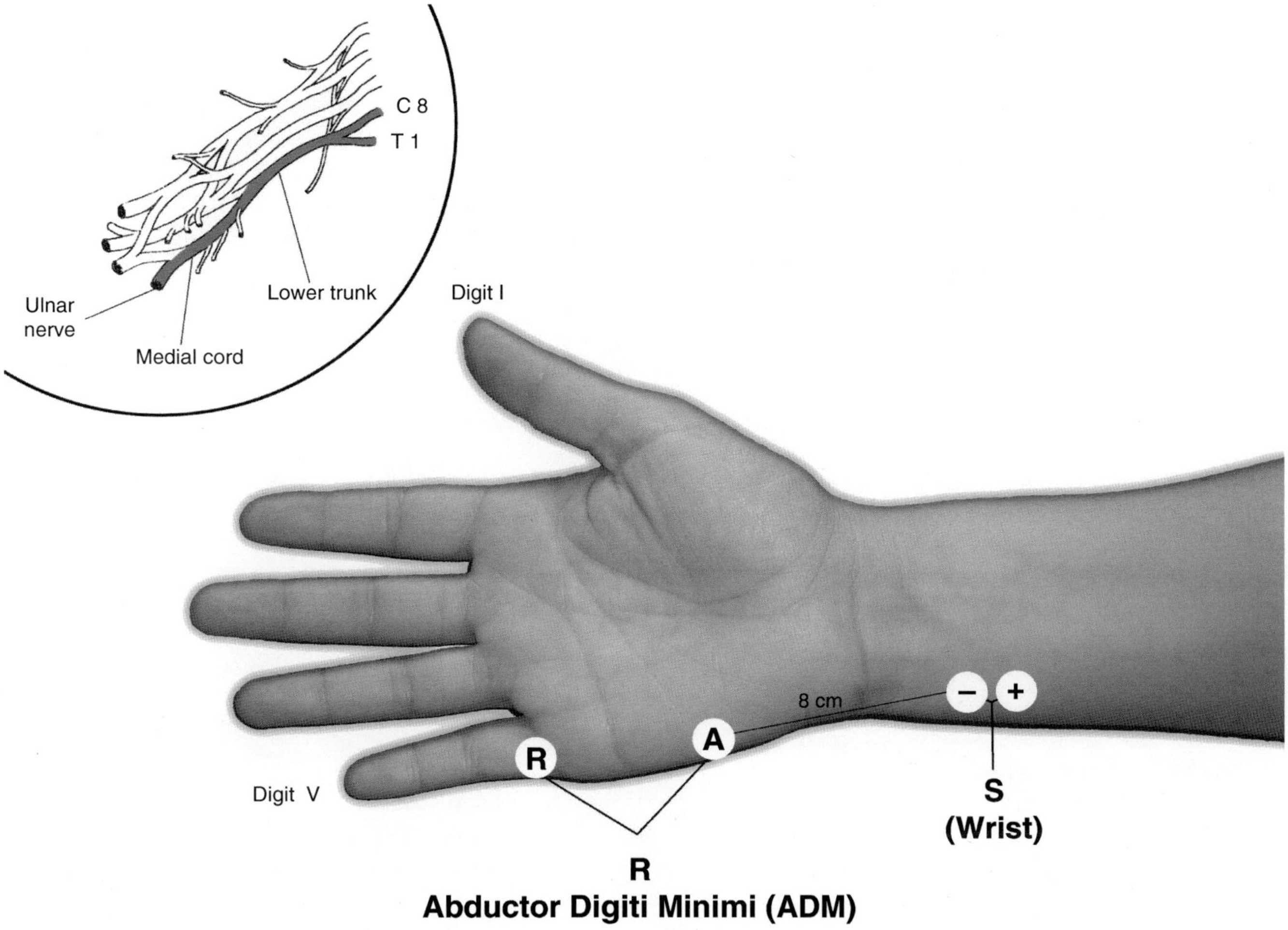

(A)	Active Electrode
(R)	Reference Electrode
(G)	Ground Electrode
(−)	Cathode
(+)	Anode
[R]	Recording Site
[S]	Stimulating Site

Fig. 3 Motor conduction study to the *abductor digiti minimi* (ADM) muscle; stimulation of the wrist (S)

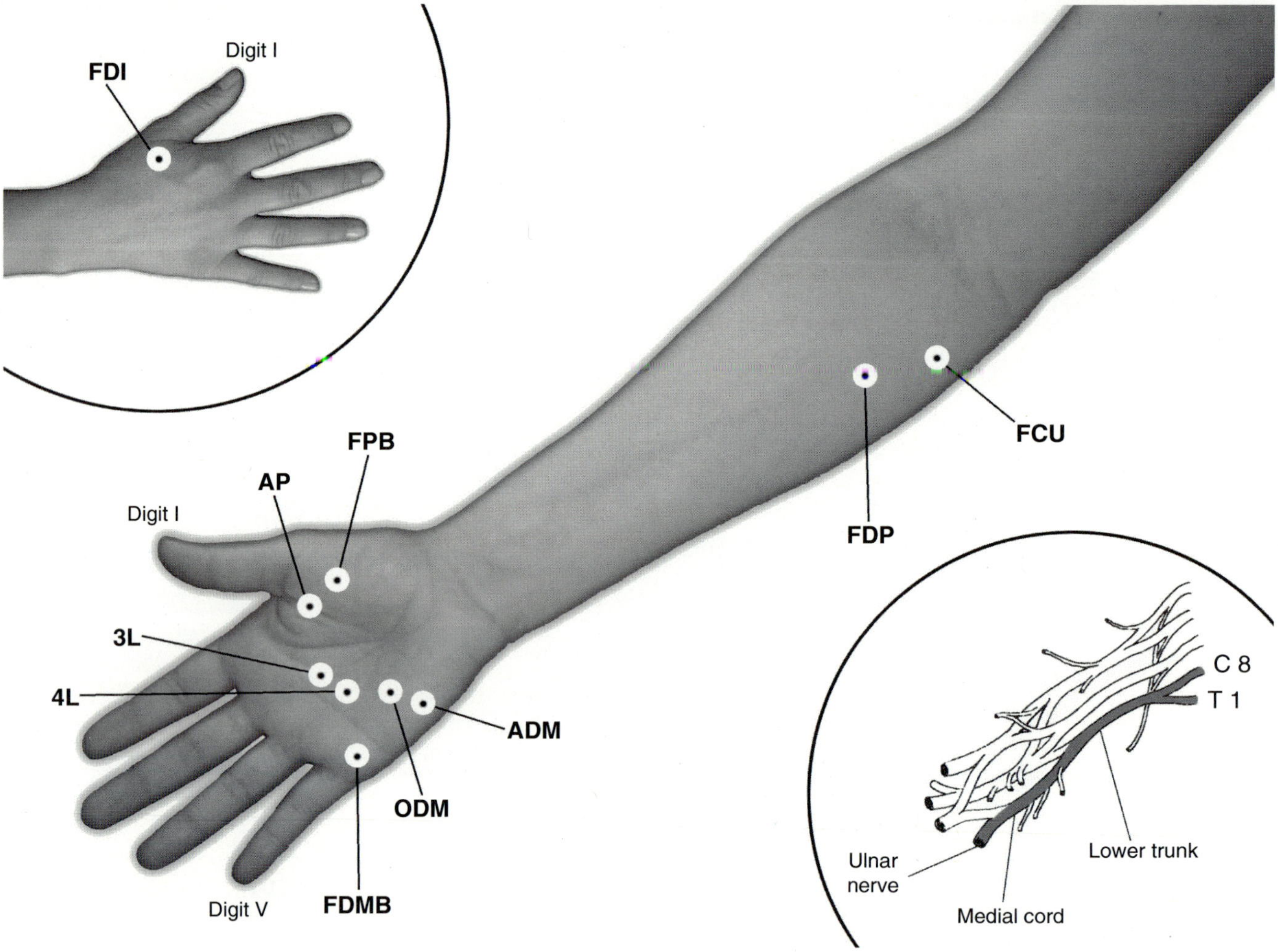

	MUSCLE	BRANCH	CORD	TRUNK	ROOT
FCU	Flexor Carpi Ulnaris	Ulnar	Medial	Lower	C8-T1
FDP	Flexor Digitorum Profundus	Ulnar	Medial	Lower	C8-T1
ADM	Abductor Digiti Minimi	Ulnar Deep Branch	Medial	Lower	C8-T1
ODM	Opponens Digiti Minimi	Ulnar Deep Branch	Medial	Lower	C8-T1
FDMB	Flexor Digiti Minimi Brevis	Ulnar Deep Branch	Medial	Lower	C8-T1
3L	Third Lumbrical	Ulnar Deep Branch	Medial	Lower	C8-T1
4L	Fourth Lumbrical	Ulnar Deep Branch	Medial	Lower	C8-T1
FPB	Flexor Pollicis Brevis	Ulnar Deep Branch	Medial	Lower	C8-T1
AP	Adductor Pollicis	Ulnar Deep Branch	Medial	Lower	C8-T1
FDI	First Dorsal Interosseous	Ulnar Deep Branch	Medial	Lower	C8-T1

Fig. 4 Muscles innervated by the ulnar nerve (forearm and hand)

Sensory Conduction Studies Index (by Practical Criteria)

Orthodromic Studies

S1 – Orthodromic technique at the wrist and at the elbow; stimulation of digit V
S2 – Orthodromic technique at the wrist and above elbow; stimulation of digit V. Mixed technique at the elbow; stimulation of the wrist. Mixed technique at the axilla; stimulation at the elbow
S3 – Orthodromic technique at the wrist, at the elbow, and at the axilla; stimulation of digit V. Mixed technique at the elbow and at the axilla; stimulation of the wrist
S4 – Orthodromic technique at the wrist and at the elbow; stimulation of digit IV and V. Orthodromic technique at the wrist; stimulation of digit IV and V
S6 – Orthodromic technique at the wrist, below elbow, and above elbow; stimulation of digit V
S7 – Orthodromic technique at the wrist; stimulation of digit II and V
S10 – Orthodromic technique at the palm, wrist, and elbow; stimulation of digit V
Mixed technique at the elbow and at the axilla; stimulation of the wrist
S11 – Orthodromic technique at the wrist; stimulation of digit I, II, III, and digit V
S14 – Antidromic technique to the dorsal hand; stimulation of the forearm. Orthodromic technique at the forearm; stimulation of the dorsal hand
S17 – Mixed technique to the arm and axilla; stimulation of the wrist. Mixed technique to the axilla; stimulation below elbow. Orthodromic technique to the arm; stimulation of digit V
S20 – Orthodromic technique to the distal forearm and elbow; stimulation of the palm
S21 – Antidromic technique to digit V; stimulation of the wrist, below elbow, and above elbow. Orthodromic technique at the wrist, below elbow, and above elbow; stimulation of digit V. Mixed technique below elbow and above elbow; stimulation of the wrist

Antidromic Studies

S5 – Antidromic technique to digit IV and V; stimulation of the wrist and elbow. Antidromic technique to digit IV and V; stimulation of the wrist
S8 – Antidromic technique to digit V; stimulation of the wrist and elbow
S9 – Antidromic technique to digit II and V; stimulation of the wrist
S12 – Antidromic technique to digit II and V; stimulation of the wrist and palm
S13 – Antidromic technique to the dorsal hand; stimulation of the wrist
S14 – Antidromic technique to the dorsal hand; stimulation of the forearm. Orthodromic technique at the forearm; stimulation of the dorsal hand
S15 – Antidromic technique to digit V; stimulation of the wrist, below elbow, above elbow, and of the upper arm
S16 – Antidromic technique to digit V; stimulation of the wrist, below elbow, above elbow, and of the axilla
S18 – Antidromic technique to the dorsal hand; stimulation of the forearm
S19 – Antidromic technique to digit V; stimulation of the wrist, below elbow, and above elbow

S21 – Antidromic technique to digit V; stimulation of the wrist, below elbow, and above elbow. Orthodromic technique at the wrist, below elbow, and above elbow; stimulation of digit V. Mixed technique below elbow and above elbow; stimulation of the wrist
S24 – Antidromic technique to digit V; stimulation of the wrist, below elbow, and above elbow

Digit–Wrist (Long Distance) Studies

S4 – Orthodromic technique at the wrist and at the elbow; stimulation of digit IV and V. Orthodromic technique at the wrist; stimulation of digit IV and V
S5 – Antidromic technique to digit IV and V; stimulation of the wrist and of the elbow. Antidromic technique to digit IV and V; stimulation of the wrist
S7 – Orthodromic technique at the wrist; stimulation of digit II and V
S9 – Antidromic technique to digit II and V; stimulation of the wrist
S11 – Orthodromic technique at the wrist; stimulation of digit I, II, III, and V

Palm–Wrist (Short Distance) Studies

S10 – Orthodromic technique at the palm, wrist, and at the elbow; stimulation of digit V. Mixed technique at the elbow and at the axilla; stimulation of the wrist
S12 – Antidromic technique to digit II and V; stimulation of the wrist and palm

Forearm Velocity (Wrist–Elbow) Studies

S1 – Orthodromic technique at the wrist and at the elbow; stimulation of digit V
S3 – Orthodromic technique at the wrist, at the elbow, and at the axilla; stimulation of digit V. Mixed technique at the elbow and at the axilla; stimulation of the wrist
S6 – Orthodromic technique at the wrist, below elbow, and above elbow; stimulation of digit V
S10 – Orthodromic technique at the palm, wrist, and elbow; stimulation of digit V. Mixed technique at the elbow and at the axilla; stimulation of the wrist
S15 – Antidromic technique to digit V; stimulation of the wrist, below elbow, above elbow, and of the upper arm
S16 – Antidromic technique to digit V; stimulation of the wrist, below elbow, above elbow, and of the axilla
S20 – Orthodromic technique to the distal forearm and elbow; stimulation of the palm
S21 – Antidromic technique to digit V; stimulation of the wrist, below elbow, and above elbow. Orthodromic technique at the wrist, below elbow, and above elbow; stimulation of digit V. Mixed technique below elbow and above elbow; stimulation of the wrist

Forearm Velocity (Palm–Elbow) Studies

S21 – Antidromic technique to digit V; stimulation of the wrist, below elbow, and above elbow. Orthodromic technique at the wrist, below elbow, and above elbow; stimulation of digit V. Mixed technique below elbow and above elbow; stimulation of the wrist

Arm Velocity (Elbow–Axilla) Studies

S3 – Orthodromic technique at the wrist, at the elbow, and at the axilla; stimulation of digit V. Mixed technique at the elbow and at the axilla; stimulation of the wrist

S10 – Orthodromic technique at the palm, wrist, and elbow; stimulation of digit V. Mixed technique at the elbow and at the axilla; stimulation of the wrist

S16 – Antidromic technique to digit V; stimulation of the wrist, below elbow, above elbow, and of the axilla

S17 – Mixed technique to the arm and axilla; stimulation of the wrist. Mixed technique to the axilla; stimulation below elbow. Orthodromic technique to the arm; stimulation of digit V

Amplitude Ratio Studies

S7 – Orthodromic technique at the wrist; stimulation of digit II and V

S9 – Antidromic technique to digit II and V; stimulation of the wrist

S11 – Orthodromic technique at the wrist; stimulation of digit I, II, III, and V

S21 – Antidromic technique to digit V; stimulation of the wrist, below elbow, and above elbow. Orthodromic technique at the wrist, below elbow, and above elbow; stimulation of digit V. Mixed technique below elbow and above elbow; stimulation of the wrist

Median–Ulnar Comparative Studies

S7 – Orthodromic technique at the wrist; stimulation of digit II and V

S9 – Antidromic technique to digit II and V; stimulation of the wrist

S11 – Orthodromic technique at the wrist; stimulation of digit I, II, III, and V

S12 – Antidromic technique to digit II and V; stimulation of the wrist and palm

Dorsal Cutaneous Branch (DCU) Studies

S13 – Antidromic technique to the dorsal hand; stimulation of the wrist

S14 – Antidromic technique to the dorsal hand; stimulation of the forearm. Orthodromic technique at the forearm; stimulation of the dorsal hand

S18 – Antidromic technique to the dorsal hand; stimulation of the forearm

Subdermal Recording Studies

S6 – Orthodromic technique at the wrist, below elbow, and above elbow; stimulation of digit V

Residual Latency (RL) Studies

S8 – Antidromic technique to digit V; stimulation of the wrist and elbow

Nerve Conduction Studies with the Elbow Fully Extended

S1 – Orthodromic technique at the wrist and at the elbow; stimulation of digit V

S2 – Orthodromic technique at the wrist and above elbow; stimulation of digit V. Mixed technique at the elbow; stimulation of the wrist. Mixed technique at the axilla; stimulation at the elbow

S3 – Orthodromic technique at the wrist, at the elbow, and at the axilla; stimulation of digit V. Mixed technique at the elbow and at the axilla; stimulation of the wrist

S4 – Orthodromic technique at the wrist and at the elbow; stimulation of digit IV and V. Orthodromic technique at the wrist; stimulation of digit IV and V

S5 – Antidromic technique to digit IV and V; stimulation of the wrist and of the elbow. Antidromic technique to digit IV and V; stimulation of the wrist

S6 – Orthodromic technique at the wrist, below elbow, and above elbow; stimulation of digit V

S8 – Antidromic technique to digit V; stimulation of the wrist and elbow

S10 – Orthodromic technique at the palm, wrist, and elbow; stimulation of digit V. Mixed technique at the elbow and at the axilla; stimulation of the wrist

S15 – Antidromic technique to digit V; stimulation of the wrist, below elbow, above elbow, and of the upper arm

S17 – Mixed technique to the arm and axilla; stimulation of the wrist. Mixed technique to the axilla; stimulation below elbow. Orthodromic technique to the arm; stimulation of digit V

S21 – Antidromic technique to digit V; stimulation of the wrist, below elbow, and above elbow. Orthodromic technique at the wrist, below elbow, and above elbow; stimulation of digit V. Mixed technique below elbow and above elbow; stimulation of the wrist

S22 – Mixed technique above elbow; stimulation of the wrist

S23 – Mixed technique to the elbow; stimulation of the wrist. Mixed technique to the axilla; stimulation of the elbow

Nerve Conduction Studies with the Elbow Flexed (90°)

S19 – Antidromic technique to digit V; stimulation of the wrist, below elbow, and above elbow

S21 – Antidromic technique to digit V; stimulation of the wrist, below elbow, and above elbow. Orthodromic technique at the wrist, below elbow, and above elbow; stimulation of digit V. Mixed technique below elbow and above elbow; stimulation of the wrist

S22 – Mixed technique above elbow; stimulation of the wrist

S24 – Antidromic technique to digit V; stimulation of the wrist, below elbow, and above elbow

Nerve Conduction Studies with the Elbow Flexed (135°)

S16 – Antidromic technique to digit V; stimulation of the wrist, below elbow, above elbow, and of the axilla

Motor Conduction Studies Index (by Practical Criteria)

Distal Latency Studies

M1 – Technique from the *abductor digiti minimi* muscle; stimulation of the wrist, elbow, and axilla
M2 – Technique from the *abductor digiti minimi* muscle; stimulation of the elbow and of the wrist
M3 – Technique from *abductor digiti minimi* and *first dorsal interosseous* muscles; stimulation of the wrist, below the elbow, above the elbow, at the mid-arm, and of the axilla
M5 – Technique from *abductor digiti minimi* and *first dorsal interosseous* muscles; stimulation of the wrist, above the elbow, and of the axilla
M9 – Technique from the *abductor digiti minimi* muscle; stimulation of the axilla, above the elbow, and of the wrist
M14 – Technique from *abductor digiti minimi* and *first dorsal interosseous* muscles; stimulation of the axilla, above the sulcus, below the sulcus, and of the wrist. Technique from the *flexor carpi ulnaris* muscle; stimulation above the sulcus
M17 – Technique from the *abductor digiti minimi* muscle; stimulation of the axilla, above the elbow, below the elbow, and of the wrist
M18 – Technique from the *abductor digiti minimi* muscle; stimulation of the wrist, elbow, above the elbow, and at Erb's point
M24 – Technique from the *abductor digiti minimi* muscle; stimulation of the wrist, elbow, axilla, and at Erb's point

Wrist–Palm Studies

M23 – Technique from the *abductor digiti minimi* muscle; stimulation of the elbow and wrist and on the palm
M37 – Technique from the *first dorsal interosseous* muscle; stimulation on the palm, of the wrist, above the elbow, of the axilla, and at Erb's point
M38 – Technique from the *first dorsal interosseous* muscle; stimulation on the palm, of the wrist, below the elbow, and above the elbow; stimulation on the palm and of the wrist

Proximal Velocity Studies

M12 – Technique from the *abductor digiti minimi* muscle; stimulation of the wrist, below the elbow, at the elbow, above the elbow, and at Erb's point
M15 – Technique from the *abductor digiti minimi* muscle; stimulation of the wrist, elbow, and axilla
M17 – Technique from the *abductor digiti minimi* muscle; stimulation of the axilla, above the elbow, below the elbow, and of the wrist

M18 – Technique from the *abductor digiti minimi* muscle; stimulation of the wrist, of the elbow, above the elbow, and at Erb's point
M20 – Technique from the *abductor digiti minimi* muscle; stimulation of the axilla, above the elbow, below the elbow, and of the wrist
M24 – Technique from the *abductor digiti minimi* muscle; stimulation of the wrist, elbow, and axilla and at Erb's point
M30 – Technique from the *abductor digiti minimi* muscle; stimulation of the wrist, below the elbow, above the elbow, and of the axilla

Forearm Velocity Studies

M4 – Technique from *abductor digiti minimi* and *first dorsal interosseous* muscles; stimulation of the wrist and above the elbow
M9 – Technique from the *abductor digiti minimi* muscle; stimulation of the axilla, above the elbow, and of the wrist
M10 – Technique from the *abductor digiti minimi* muscle; stimulation of the wrist and of the elbow
M14 – Technique from *abductor digiti minimi* and *first dorsal interosseous* muscles; stimulation of the axilla, above the sulcus, below the sulcus, and of the wrist. Technique from the *flexor carpi ulnaris* muscle; stimulation above the sulcus
M15 – Technique from the *abductor digiti minimi* muscle; stimulation of the wrist, elbow, and arm
M17 – Technique from the *abductor digiti minimi* muscle; stimulation of the axilla, above the elbow, below the elbow, and of the wrist
M20 – Technique from the *abductor digiti minimi* muscle; stimulation of the axilla, above the elbow, below the elbow, and of the wrist
M24 – Technique from the *abductor digiti minimi* muscle; stimulation of the wrist, elbow, and axilla and at Erb's point
M30 – Technique from the *abductor digiti minimi* muscle; stimulation of the wrist, below the elbow, above the elbow, and of the axilla

Arm/Forearm Velocity Ratio Studies

M7 – Technique from the *abductor digiti minimi* muscle; stimulation of the axilla, elbow, and wrist
M25 – Technique from the *abductor digiti minimi* muscle; stimulation above the sulcus, below the sulcus, and of the wrist

Across Elbow Velocity and Conduction Time Studies

M11 – Technique from *abductor digiti minimi* and *adductor pollicis* muscles; stimulation of the wrist, below elbow, above elbow, and of the axilla
M14 – Technique from *abductor digiti minimi* and *first dorsal interosseous* muscles; stimulation of the axilla, above the sulcus, below the sulcus, and of the wrist. Technique from the *flexor carpi ulnaris* muscle; stimulation above the sulcus
M15 – Technique from the *abductor digiti minimi* muscle; stimulation of the wrist, elbow, and arm
M17 – Technique from the *abductor digiti minimi* muscle; stimulation of the axilla, above the elbow, below the elbow, and of the wrist
M20 – Technique from the *abductor digiti minimi* muscle; stimulation of the axilla, above the elbow, below the elbow, and of the wrist

M30 – Technique from the *abductor digiti minimi* muscle; stimulation of the wrist, below the elbow, above the elbow, and of the axilla
M43 – Technique from the *abductor digiti minimi* muscle; stimulation below the elbow and above the elbow

Ulnar–Ulnar Comparative Studies

M13 – Technique from *abductor digiti minimi* and *first dorsal interosseous* muscles; stimulation of the wrist
M28 – Technique from *abductor digiti minimi* and *first dorsal interosseous* muscles; stimulation of the wrist
M40 – Technique from *abductor digiti minimi* and *first dorsal interosseous* muscles; stimulation of the wrist
M41 – Technique from *abductor digiti minimi* and *first dorsal interosseous* muscles; stimulation of the wrist, below the elbow, and above the elbow
M42 – Technique from *abductor digiti minimi* and *first dorsal interosseous* muscles; stimulation of the wrist, below the elbow, and above the elbow

Median–Ulnar Comparative Studies

M22 – Technique from *abductor pollicis brevis* and *abductor digiti minimi* muscles; stimulation of the wrist
M31 – Technique from *abductor pollicis brevis* and *first dorsal interosseous* muscles; stimulation of the wrist
M32 – Technique from *abductor pollicis brevis* and *abductor digiti minimi* muscles; stimulation of the wrist
M44 – Technique from *abductor pollicis brevis* and *abductor digiti minimi* muscles; stimulation of the wrist

F-Wave Studies

M16 – Technique from the *abductor digiti minimi* muscle; stimulation of the wrist, elbow, and axilla
M21 – Technique from the *abductor digiti minimi* muscle; stimulation of the wrist and of the elbow
M44 – Technique from *abductor pollicis brevis* and *abductor digiti minimi* muscles; stimulation of the wrist

Residual Latency (RL) Studies

M6 – Technique from the *abductor digiti minimi* muscle; stimulation of the axilla, above the elbow, at the mid-forearm, and of the wrist
M8 – Technique from the *abductor digiti minimi* muscle; stimulation of the wrist, lower forearm, upper forearm, elbow, and arm
M19 – Technique from the *abductor digiti minimi* muscle; stimulation of the wrist and of the elbow
M23 – Technique from the *abductor digiti minimi* muscle; stimulation of the elbow and wrist and on the palm

Needle Recording Studies

M3 – Technique from *abductor digiti minimi* and *first dorsal interosseous* muscles; stimulation of the wrist, below the elbow, above the elbow, at the mid-arm, and of the axilla
M4 – Technique from *abductor digiti minimi* and *first dorsal interosseous* muscles; stimulation of the wrist and of the elbow
M5 – Technique from *abductor digiti minimi* and *first dorsal interosseous* muscles; stimulation of the wrist, above the elbow, and of the axilla
M12 – Technique from the *abductor digiti minimi* muscle; stimulation of the wrist, below the elbow, of the elbow, above the elbow, and at Erb's point
M14 – Technique from *abductor digiti minimi* and *first dorsal interosseous* muscles; stimulation of the axilla, above the sulcus, below the sulcus, and of the wrist. Technique from the *flexor carpi ulnaris* muscle; stimulation above the sulcus
M20 – Technique from the *abductor digiti minimi* muscle; stimulation of the axilla, above the elbow, below the elbow, and of the wrist
M25 – Technique from the *abductor digiti minimi* muscle; stimulation above the sulcus, below the sulcus, and of the wrist
M27 – Technique from the abductor digiti minimi and flexor carpi ulnaris muscles; stimulation of the wrist, below the elbow, above the elbow, and upper arm; stimulation of the upper arm, and above the elbow
M32 – Technique from *abductor pollicis brevis* and *abductor digiti minimi* muscles; stimulation of the wrist

Short-Segment Incremental (SSIS) Studies

M26 – Technique from the *abductor digiti minimi* muscle; stimulation at the axilla, above the elbow, below the elbow, and of the wrist; short-segment incremental stimulations from above-elbow to below-elbow sites
M33 – Technique from the *abductor digiti minimi* muscle; short-segment incremental stimulations from below-elbow to above-elbow sites
M34 – Technique from the *abductor digiti minimi* muscle; short-segment incremental stimulations from below-elbow to above-elbow sites
M36 – Technique from the *first dorsal interosseous* muscle; short-segment incremental stimulations from above wrist to below wrist sites
M39 – Technique from the *flexor carpi ulnaris* and *abductor digiti minimi* muscles; short-segment incremental stimulations from below-elbow to above-elbow sites

Forearm: Flexor Carpi Ulnaris (FCU) Muscle Studies

M27 – Technique from the abductor digiti minimi and flexor carpi ulnaris muscles; stimulation of the wrist, below the elbow, above the elbow, and upper arm; stimulation of the upper arm, and above the elbow
M29 – Technique from *abductor digiti minimi*, *flexor carpi ulnaris*, and *flexor digitorum profundus* muscles; stimulation of the wrist, below the elbow, and above the elbow; stimulation below the elbow and above the elbow
M35 – Technique from *flexor carpi ulnaris*, *abductor digiti minimi*, and *first dorsal interosseous* muscles; stimulation below the elbow and above the elbow; stimulation of the wrist, below the elbow, and above the elbow
M39 – Technique from the *flexor carpi ulnaris* muscle; short-segment incremental stimulations from below-elbow to above-elbow sites

Forearm: Flexor Digitorum Profundus (FDP) Muscle Studies

M29 – Technique from *abductor digiti minimi, flexor carpi ulnaris,* and *flexor digitorum profundus* muscles; stimulation of the wrist, below the elbow, and above the elbow; stimulation below the elbow and above the elbow

Hand: Abductor Digiti Minimi (ADM) Muscle Studies

M1 – Technique from the *abductor digiti minimi* muscle; stimulation of the wrist, elbow, and axilla
M2 – Technique from the *abductor digiti minimi* muscle; stimulation of the elbow and of the wrist
M3 – Technique from *abductor digiti minimi* and *first dorsal interosseous* muscles; stimulation of the wrist, below the elbow, above the elbow, at the mid-arm, and of the axilla
M4 – Technique from *abductor digiti minimi* and *first dorsal interosseous* muscles; stimulation of the wrist and of the elbow
M5 – Technique from *abductor digiti minimi* and *first dorsal interosseous* muscles; stimulation of the wrist, above the elbow, and of the axilla
M6 – Technique from the *abductor digiti minimi* muscle; stimulation of the axilla, above the elbow, at the mid-forearm, and of the wrist
M7 – Technique from the *abductor digiti minimi* muscle; stimulation of the axilla, elbow, and wrist
M8 – Technique from the *abductor digiti minimi* muscle; stimulation of the wrist, lower forearm, upper forearm, elbow, and arm
M9 – Technique from the *abductor digiti minimi* muscle; stimulation of the axilla, above the elbow, and of the wrist
M10 – Technique from the *abductor digiti minimi* muscle; stimulation of the wrist and of the elbow
M11 – Technique from *abductor digiti minimi* and *adductor pollicis* muscles; stimulation of the wrist, below elbow, above elbow, and of the axilla
M12 – Technique from the *abductor digiti minimi* muscle; stimulation of the wrist, below the elbow, of the elbow, above the elbow, and at Erb's point
M13 – Technique from *abductor digiti minimi* and *first dorsal interosseous* muscles; stimulation of the wrist
M14 – Technique from *abductor digiti minimi* and *first dorsal interosseous* muscles; stimulation of the axilla, above the sulcus, below the sulcus, and of the wrist. Technique from the *flexor carpi ulnaris* muscle; stimulation above the sulcus
M15 – Technique from the *abductor digiti minimi* muscle; stimulation of the wrist, elbow, and arm
M16 – Technique from the *abductor digiti minimi* muscle; stimulation of the wrist, elbow, and axilla
M17 – Technique from the *abductor digiti minimi* muscle; stimulation of the axilla, above the elbow, below the elbow, and of the wrist
M18 – Technique from the *abductor digiti minimi* muscle; stimulation of the wrist, of the elbow, above the elbow, and at Erb's point
M19 – Technique from the *abductor digiti minimi* muscle; stimulation of the wrist and of the elbow
M20 – Technique from the *abductor digiti minimi* muscle; stimulation of the axilla, above the elbow, below the elbow, and of the wrist

M21 – Technique from the *abductor digiti minimi* muscle; stimulation of the wrist and of the elbow

M22 – Technique from *abductor pollicis brevis* and *abductor digiti minimi* muscles; stimulation of the wrist

M23 – Technique from the *abductor digiti minimi* muscle; stimulation of the elbow, wrist, and on the palm

M24 – Technique from the *abductor digiti minimi* muscle; stimulation of the wrist, elbow, and axilla and at Erb's point

M25 – Technique from the *abductor digiti minimi* muscle; stimulation above the sulcus, below the sulcus, and of the wrist

M26 – Technique from the *abductor digiti minimi* muscle; stimulation at the axilla, above the elbow, below the elbow, and of the wrist; short-segment incremental stimulations from above-elbow to below-elbow sites

M28 – Technique from *abductor digiti minimi* and *first dorsal interosseous* muscles; stimulation of the wrist

M29 – Technique from *abductor digiti minimi, flexor carpi ulnaris,* and *flexor digitorum profundus* muscles; stimulation of the wrist, elbow, and above the elbow; stimulation elbow and above the elbow

M30 – Technique from the *abductor digiti minimi* muscle; stimulation of the wrist, below the elbow, above the elbow, and of the axilla

M32 – Technique from *abductor pollicis brevis* and *abductor digiti minimi* muscles; stimulation of the wrist

M33 – Technique from the *abductor digiti minimi* muscle; short-segment incremental stimulations from below-elbow to above-elbow sites

M34 – Technique from the *abductor digiti minimi* muscle; short-segment incremental stimulations from below-elbow to above-elbow sites

M35 – Technique from *flexor carpi ulnaris*, *abductor digiti minimi*, and *first dorsal interosseous* muscles; stimulation below the elbow and above the elbow; stimulation of the wrist, below the elbow, and above the elbow

M40 – Technique from *abductor digiti minimi* and *first dorsal interosseous* muscles; stimulation of the wrist

M41 – Technique from *abductor digiti minimi* and *first dorsal interosseous* muscles; stimulation of the wrist, below the elbow, and above the elbow

M42 – Technique from *abductor digiti minimi* and *first dorsal interosseous* muscles; stimulation of the wrist, below the elbow, and above the elbow

M44 – Technique from *abductor pollicis brevis* and *abductor digiti minimi* muscles; stimulation of the wrist

Hand: Adductor Pollicis (AP) Muscle Studies

M11 – Technique from *abductor digiti minimi* and *adductor pollicis* muscles; stimulation of the wrist, below elbow, above elbow, and of the axilla

Hand: First Dorsal Interosseous (FDI) Muscle Studies

M3 – Technique from *abductor digiti minimi* and *first dorsal interosseous* muscles; stimulation of the wrist, below the elbow, above the elbow, at the mid-arm, and of the axilla

M4 – Technique from *abductor digiti minimi* and *first dorsal interosseous* muscles; stimulation of the wrist and of the elbow

M5 – Technique from *abductor digiti minimi* and *first dorsal interosseous* muscles; stimulation of the wrist, above the elbow, and of the axilla

M13 – Technique from *abductor digiti minimi* and *first dorsal interosseous* muscles; stimulation of the wrist

M14 – Technique from *abductor digiti minimi* and *first dorsal interosseous* muscles; stimulation of the axilla, above the sulcus, below the sulcus, and of the wrist. Technique from the *flexor carpi ulnaris* muscle; stimulation above the sulcus

M28 – Technique from *abductor digiti minimi* and *first dorsal interosseous* muscles; stimulation of the wrist

M31 – Technique from *abductor pollicis brevis* and *first dorsal interosseous* muscles; stimulation of the wrist

M35 – Technique from *flexor carpi ulnaris*, *abductor digiti minimi*, and *first dorsal interosseous* muscles; stimulation below the elbow and above the elbow; stimulation of the wrist, below the elbow, and above the elbow

M36 – Technique from the *first dorsal interosseous* muscle; short-segment incremental stimulations from above wrist to below wrist sites

M37 – Technique from the *first dorsal interosseous* muscle; stimulation on the palm, of the wrist, above the elbow, of the axilla, and at Erb's point

M38 – Technique from the *first dorsal interosseous* muscle; stimulation on the palm, of the wrist, below the elbow, and above the elbow; stimulation on the palm and of the wrist

M40 – Technique from *abductor digiti minimi* and *first dorsal interosseous* muscles; stimulation of the wrist

M41 – Technique from *abductor digiti minimi* and *first dorsal interosseous* muscles; stimulation of the wrist, below the elbow, and above the elbow

M42 – Technique from *abductor digiti minimi* and *first dorsal interosseous* muscles; stimulation of the wrist, below the elbow, and above the elbow

Nerve Conduction Studies with the Elbow Fully Extended

M1 – Technique from the *abductor digiti minimi* muscle; stimulation of the wrist, elbow, and axilla

M2 – Technique from the *abductor digiti minimi* muscle; stimulation of the elbow and of the wrist

M3 – Technique from *abductor digiti minimi* and *first dorsal interosseous* muscles; stimulation of the wrist, below the elbow, above the elbow, at the mid-arm, and of the axilla

M4 – Technique from *abductor digiti minimi* and *first dorsal interosseous* muscles; stimulation of the wrist and of the elbow

M5 – Technique from *abductor digiti minimi* and *first dorsal interosseous* muscles; stimulation of the wrist, above the elbow, and of the axilla

M7 – Technique from the *abductor digiti minimi* muscle; stimulation of the axilla, elbow, and wrist

M8 – Technique from the *abductor digiti minimi* muscle; stimulation of the wrist, lower forearm, upper forearm, elbow, and arm

M9 – Technique from the *abductor digiti minimi* muscle; stimulation of the axilla, above the elbow, and of the wrist

M10 – Technique from the *abductor digiti minimi* muscle; stimulation of the wrist and of the elbow

M11 – Technique from *abductor digiti minimi* and *adductor pollicis* muscles; stimulation of the wrist, below elbow, above elbow, and of the axilla

M12 – Technique from the *abductor digiti minimi* muscle; stimulation of the wrist, below the elbow, of the elbow, above the elbow, and at Erb's point

M14 – Technique from *abductor digiti minimi* and *first dorsal interosseous* muscles; stimulation of the axilla, above the sulcus, below the sulcus, and of the wrist. Technique from the *flexor carpi ulnaris* muscle; stimulation above the sulcus

M16 – Technique from the *abductor digiti minimi* muscle; stimulation of the wrist, elbow, and axilla

M17 – Technique from the *abductor digiti minimi* muscle; stimulation of the axilla, above the elbow, below the elbow, and of the wrist

M18 – Technique from the *abductor digiti minimi* muscle; stimulation of the wrist and elbow, above the elbow, and at Erb's point

M19 – Technique from the *abductor digiti minimi* muscle; stimulation of the wrist and of the elbow

M21 – Technique from the *abductor digiti minimi* muscle; stimulation of the wrist and of the elbow

M23 – Technique from the *abductor digiti minimi* muscle; stimulation of the elbow, wrist, and on the palm

M24 – Technique from the *abductor digiti minimi* muscle; stimulation of the wrist, elbow, axilla and at Erb's point

M25 – Technique from the *abductor digiti minimi* muscle; stimulation above the sulcus, below the sulcus, and of the wrist

M26 – Technique from the *abductor digiti minimi* muscle; stimulation at the axilla, above the elbow, below the elbow, and of the wrist; short-segment incremental stimulations from above-elbow to below-elbow sites

M27 – Technique from the abductor digiti minimi and flexor carpi ulnaris muscles; stimulation of the wrist, below the elbow, above the elbow, and upper arm; stimulation of the upper arm, and above the elbow

M37 – Technique from the *first dorsal interosseous* muscle; stimulation on the palm, of the wrist, above the elbow, of the axilla, and at Erb's point

M44 – Technique from *abductor pollicis brevis* and *abductor digiti minimi* muscles; stimulation of the wrist

Nerve Conduction Studies with the Elbow Flexed (35°)

M20 – Technique from the *abductor digiti minimi* muscle; stimulation of the axilla, above the elbow, below the elbow, and of the wrist

Nerve Conduction Studies with the Elbow Flexed (70°)

M6 – Technique from the *abductor digiti minimi* muscle; stimulation of the axilla, above the elbow, at the mid-forearm, and of the wrist

M15 – Technique from the *abductor digiti minimi* muscle; stimulation of the wrist, elbow, and arm

M29 – Technique from *abductor digiti minimi*, *flexor carpi ulnaris,* and *flexor digitorum profundus* muscles; stimulation of the wrist, below the elbow, and above the elbow; stimulation below the elbow and above the elbow

M34 – Technique from the *abductor digiti minimi* muscle; short-segment incremental stimulations from below-elbow to above-elbow sites

M39 – Technique from the *flexor carpi ulnaris* muscle; short-segment incremental stimulations from below-elbow to above-elbow sites

M42 – Technique from *abductor digiti minimi* and *first dorsal interosseous* muscles; stimulation of the wrist, below the elbow, and above the elbow

Nerve Conduction Studies with the Elbow Flexed (90°)

M15 – Technique from the *abductor digiti minimi* muscle; stimulation of the wrist, elbow, and arm

M28 – Technique from *abductor digiti minimi* and *first dorsal interosseous* muscles; stimulation of the wrist

M33 – Technique from the *abductor digiti minimi* muscle; short-segment incremental stimulations from below-elbow to above-elbow sites

M34 – Technique from the *abductor digiti minimi* muscle; short-segment incremental stimulations from below-elbow to above-elbow sites

M35 – Technique from *flexor carpi ulnaris*, *abductor digiti minimi*, and *first dorsal interosseous* muscles; stimulation below the elbow and above the elbow; stimulation of the wrist, below the elbow, and above the elbow

M38 – Technique from the *first dorsal interosseous* muscle; stimulation on the palm, of the wrist, below the elbow, and above the elbow; stimulation on the palm and of the wrist

M43 – Technique from the *abductor digiti minimi* muscle; stimulation below the elbow and above the elbow

Nerve Conduction Studies with the Elbow Flexed (135°)

M30 – Technique from the *abductor digiti minimi* muscle; stimulation of the wrist, below the elbow, above the elbow, and of the axilla

M41 – Technique from *abductor digiti minimi* and *first dorsal interosseous* muscles; stimulation of the wrist, below the elbow, and above the elbow

Part I

Sensory Conduction Studies

Digit V – Wrist, Elbow

S1

Surface Recording Technique, Orthodromic Study

Original Settings Sensitivity was 20 μV/division; low-frequency filter, high-frequency filter, sweep speed, duration of pulse, rate of pulse, and the machine used were not specified.

Position This study was performed in the supine position, with the subject lying on a comfortable couch with the arm slightly abducted and supported at three or four places by broad webbing straps hung from the beam.

Recording Following the orthodromic method, signals were recorded using surface electrodes (chloride-coated silver plates 16 by 24 mm, held on with adhesive tape and making contact through electrode paste) at the wrist and just above the elbow and on the course of the ulnar nerve (Fig. 1). For recording at the wrist (R), the active (A) electrode was placed proximally to the distal crease at the wrist; the reference (R) electrode was placed proximally to the active electrode [1]. For recording at the elbow (R), the active (A) electrode was placed proximally to the wrist; the reference (R) electrode was placed proximally to the active electrode. Distance between recording and stimulating electrodes was not fixed. Ground (G) electrode was placed on the dorsum of the hand; the figure shows the ground electrode placed on the forearm.

Stimulation The ulnar nerve was stimulated distally to digit V (S). For stimulation, the author used silver strips (10 mm wide) coated with silver chloride, covered with lint, and soaked with brine. They were bent to encircle the finger and stimulate the digital nerves. The active stimulating electrode (cathode, -) was placed near the metacarpophalangeal joint (at the base of the digit, proximal to the recording site); the anode (+) was positioned distally in the region of the terminal interphalangeal joint. The author used a different intensity of stimulation to study the threshold of excitability of nerve fibers, where supramaximal stimulation was mainly used. He showed that the more excitable the sensory afferent fibers were from the fingers, it had a lower threshold to electrical stimulation than that of the motor fibers to the small muscles of the hand.

Measurements Onset latency (ms) was measured from the stimulus artifact to the initial deflection of the evoked sensory nerve action potential (SNAP). Peak latency (ms) was measured from the stimulus artifact to the peak of the negative deflection of the SNAP. The author adjusted the recording electrode positions until the nerve action potentials had nearly as much of the same shape, whether recorded at the wrist or elbow. Afferent conduction times were taken as the difference in time between corresponding points on the action potentials and recorded at the wrist and elbow. No control of limb temperature was carried out. The author [1] studied ten ulnar nerves from healthy (Table 1) subjects; sample data were not reported.

Comment

Dawson [1] observed a difference between the conduction times in the wrist–elbow segment measured by the onset of latencies and peak latencies. The measurement of sensory conduction time made between the peaks (peak latencies) of the action potentials gave on the average a longer time than the measurements between the start (onset latencies). This longer time was evidently due to broadening of the action potentials recorded near the elbow caused by dispersion of impulses in fibers with different conduction velocities and over the longer conduction distance. For the author, the conduction time measured using peak-to-peak latencies probably represents the activation of the fastest and slower fibers, while the conduction time of the fastest fibers can be measured using the onset latencies. He also observed the sensory afferent nerve fibers from the fingers to have a higher maximum conduction velocity than that of the motor fibers to the small muscles of the hand.

G. Gentili, M. Di Napoli, *The Ulnar Nerve*, DOI 10.1007/978-3-319-43292-2_1

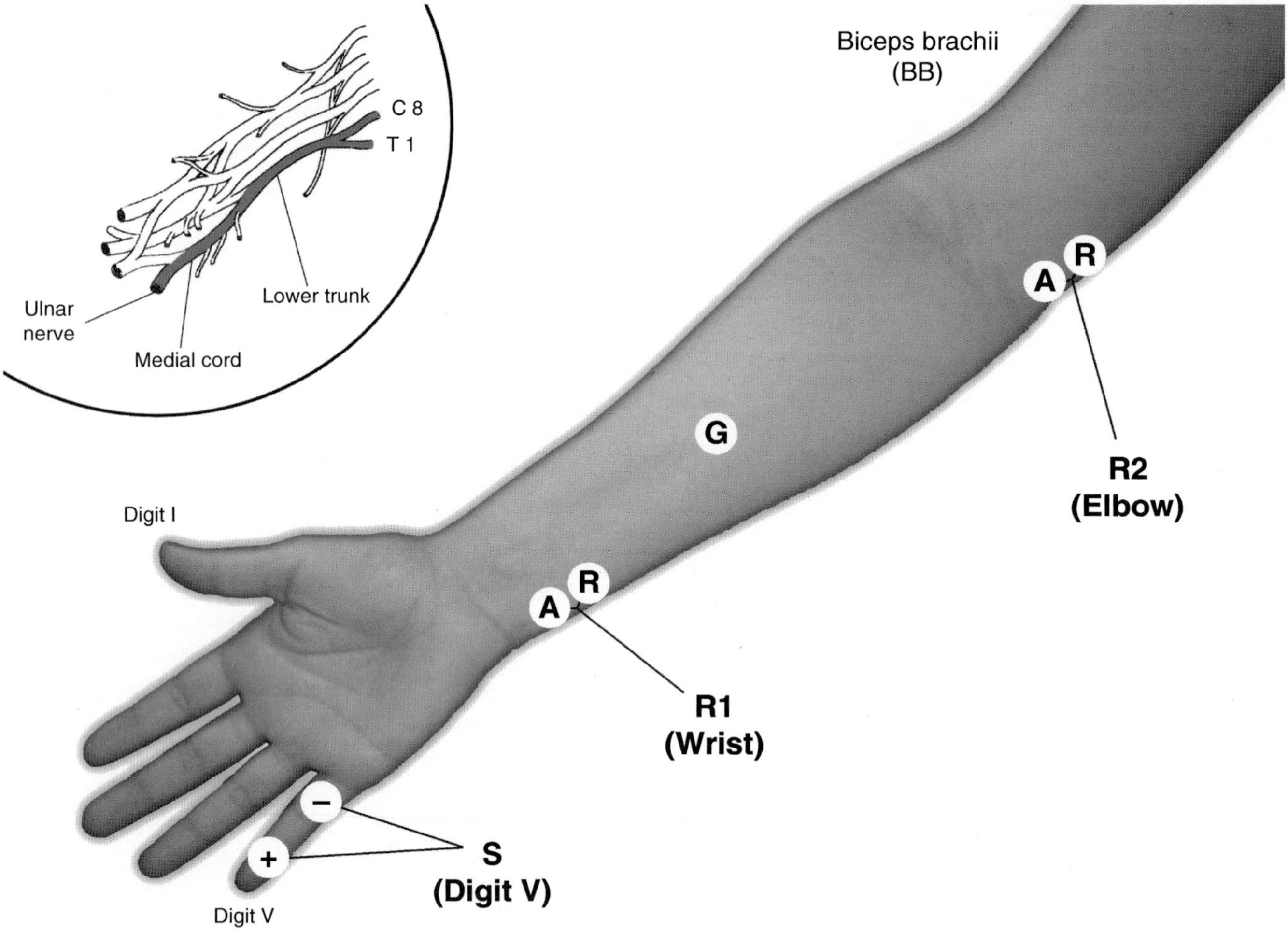

Typical waveform (digit V – wrist, elbow):

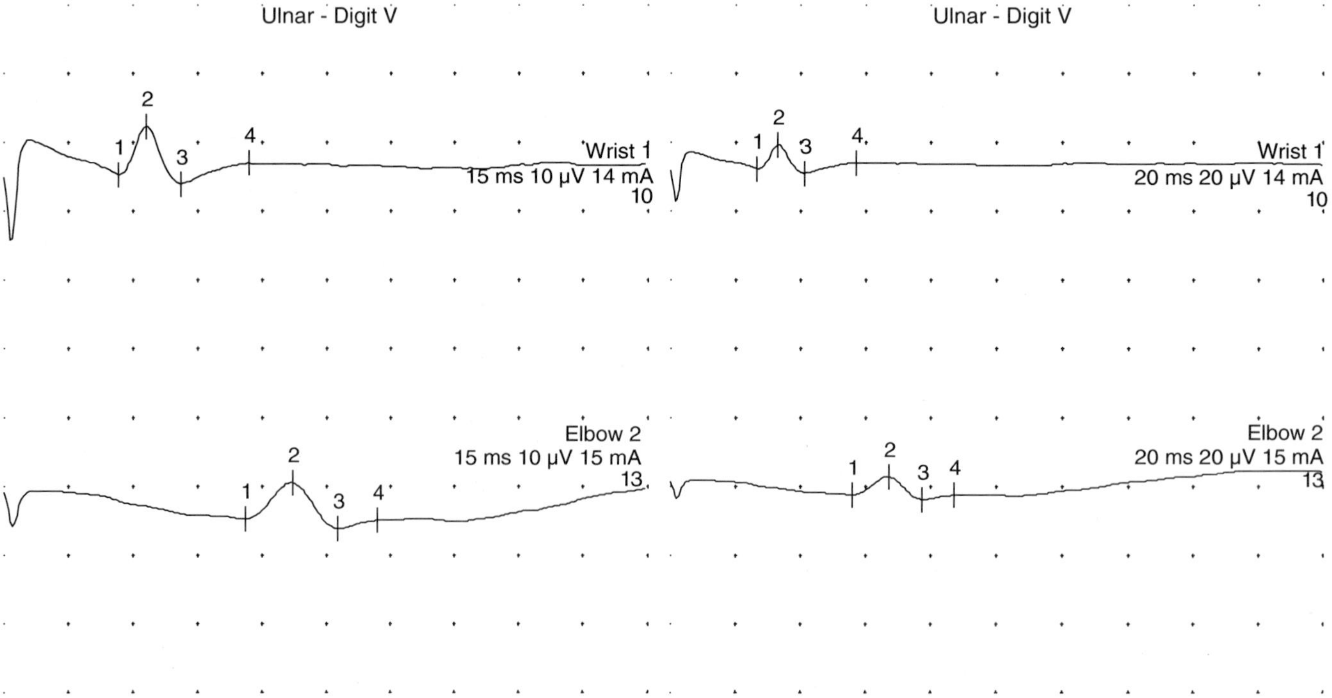

Sensitivity 10 µV/div, sweep speed 1.5 ms/div

Sensitivity 20 µV/div, sweep speed 2 ms/div

Fig. 1 Orthodromic sensory nerve action potentials (*SNAPs*) recorded at the wrist (*upper trace*) and at the elbow (*lower trace*), stimulation of digit V

Table 1 Normal values [1]

	Mean
Wrist–elbow, onset latency (ms)	4.96
Wrist–elbow, peak latency (ms)	5.15

Reference

1. Dawson GD (1956) The relative excitability and conduction velocity of sensory and motor nerve fibres in man. J Physiol 131:431–451

Digit V – Wrist; Digit V – Above Elbow; Wrist – Above Elbow; Elbow – Axilla

S2

Surface Recording Technique, Orthodromic Study

Original Settings Sensitivity was 20 μV/division; sweep speed was 1 ms/division; low-frequency filter, high-frequency filter, duration of pulse, rate of pulse, and the machine used were not specified.

Position This study was performed in the supine position, with the subject lying on a couch and covered with blankets.

Recording Following the orthodromic method, signals were recorded at the wrist and just above the elbow, on the course of the ulnar nerve [1]. For recording at the wrist (R), the active (A) electrode was placed proximally to the distal crease at the wrist; the reference (R) electrode was placed 3 cm proximally to the active electrode (Fig. 1). For recording at the elbow (R), the active (A) electrode was placed proximally to the ulnar groove; the reference (R) electrode was placed 3 cm proximally to the active electrode (Figs. 2 and 3). Distances between recording and stimulating electrodes were not fixed; the authors mapped out the course of the ulnar nerve, before applying the recording electrodes, by stimulation and observation of the motor response. In a few patients, the course of the nerve could not be determined in this way; in such a case, recording electrodes were placed over the expected position of the nerve trunk and adjusted to give the largest response to sensory stimulation. Ground (G) electrode position was not specified in the report; the figure shows the ground electrode placed on the palm.

Stimulation The ulnar nerve was stimulated distally to the digit V (recording at the wrist and above the elbow) and proximally at the wrist (recording just above the elbow). For stimulation of digit V (S), the authors used silver strips (2–4 mm wide) covered by lint moistened in saline and firmly wrapped around the finger for stimulating the sensory fibers of the ulnar nerve. The active stimulating electrode (cathode, −) was placed near the metacarpophalangeal joint (at the base of the digit, proximal to the recording site); the anode (+) was positioned distally in the region of the terminal interphalangeal joint (distal to the recording site). For stimulation at the wrist (S), the same electrodes used for the stimulation to digit V were used, firmly wrapped on the skin at the wrist level. Supramaximal stimulation was mainly used, a few subjects were unable to tolerate the repetitive stimuli required to obtain superimposed records, and in such cases, single sweeps were photographed.

Measurements Distal peak latency (ms) was measured from the stimulus artifact to the peak of the negative deflection of the evoked sensory nerve action potential (SNAP). Peak amplitude (μV) was measured from negative to positive peak. All tests were carried out in a warm room with the subject lying on a couch and covered with blankets. Before examining people with cold hands, the arms were immersed in hot water for 5 or 10 min before the session, but no other method of controlling temperature was attempted. The authors [1] studied 39 normal ulnar nerves (Table 1) in 12 patients with suspected ulnar nerve lesions at the elbow (Table 2), referred by members of the hospital staff for routine electrodiagnosis. Values from stimulation of the wrist and recording at the elbow were not reported.

Comment

For Gilliatt and Sears [1], the SNAPs recorded from the elbow region were always small and in some obese, but otherwise normal, subjects were absent altogether. For this reason, recording at the wrist level or just above it was preferred as a standard arrangement, the stimulus being applied to digit V when testing the ulnar nerve. Stimulation of the ulnar trunk at the wrist was used with recording electrodes just above the elbow; with this arrangement, the nerve action potentials recorded were large and easy to record, providing useful supplementary information. Twelve patients with suspected ulnar nerve lesions at the elbow were enrolled for the purpose of the study; nine patients showed wasting of the ulnar-supplied muscles, and weakness without

G. Gentili, M. Di Napoli, *The Ulnar Nerve*, DOI 10.1007/978-3-319-43292-2_2

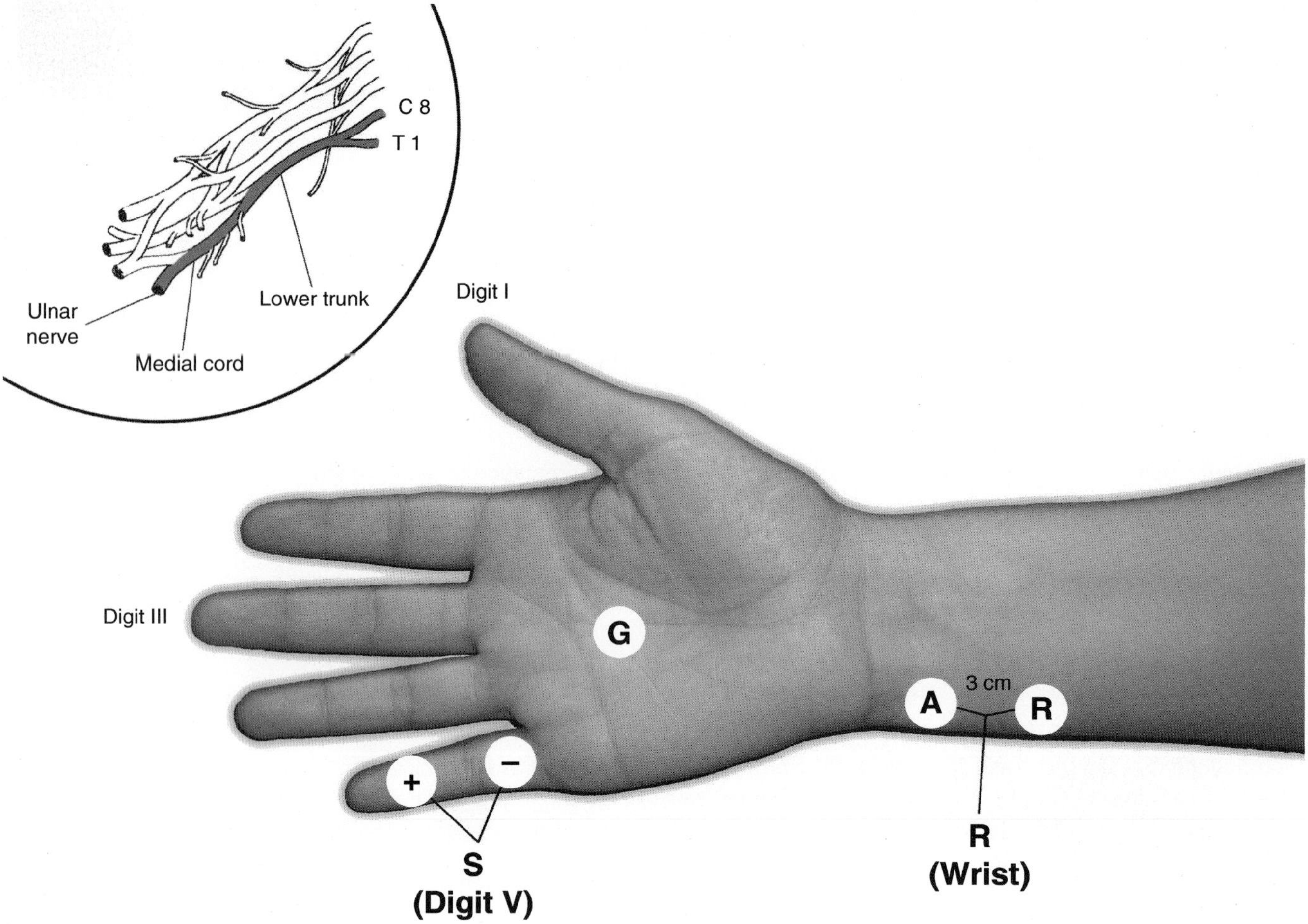

Typical waveform (digit V – wrist):

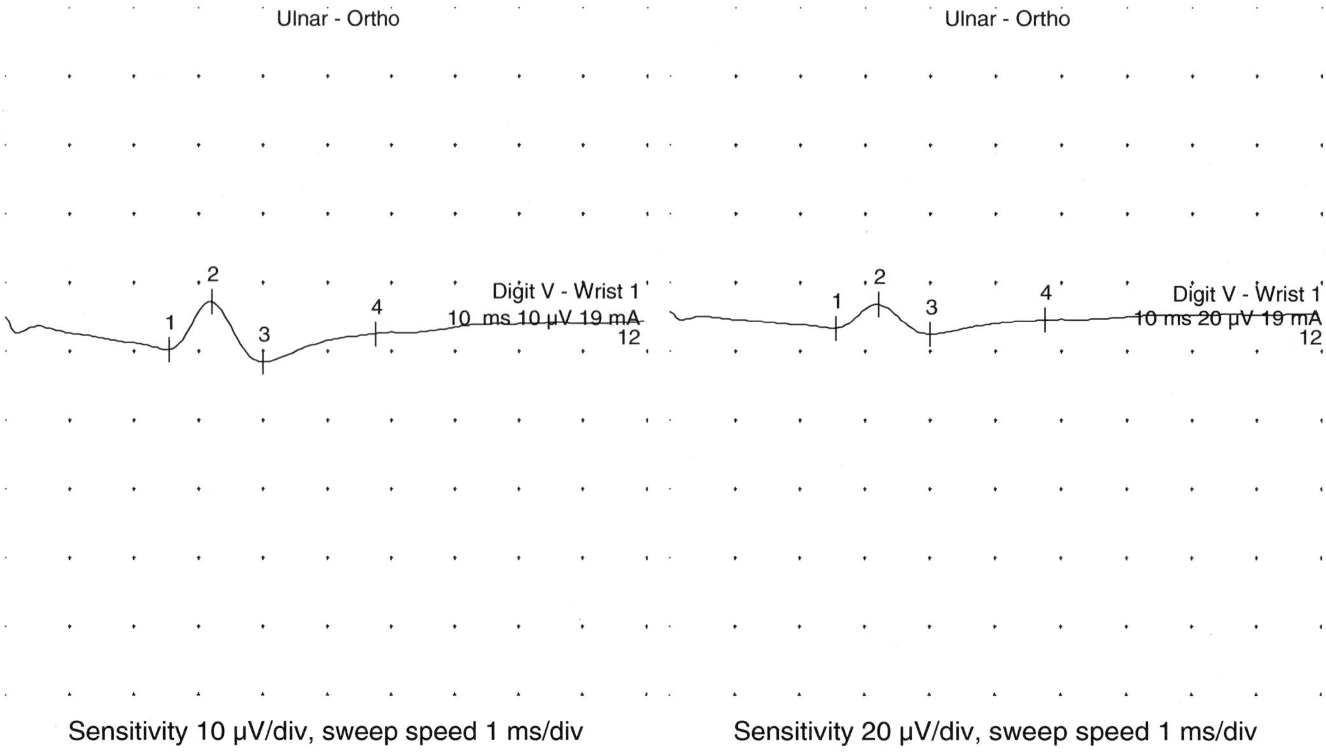

Fig. 1 Orthodromic sensory nerve action potential (*SNAP*) recorded at the wrist, stimulation of digit V

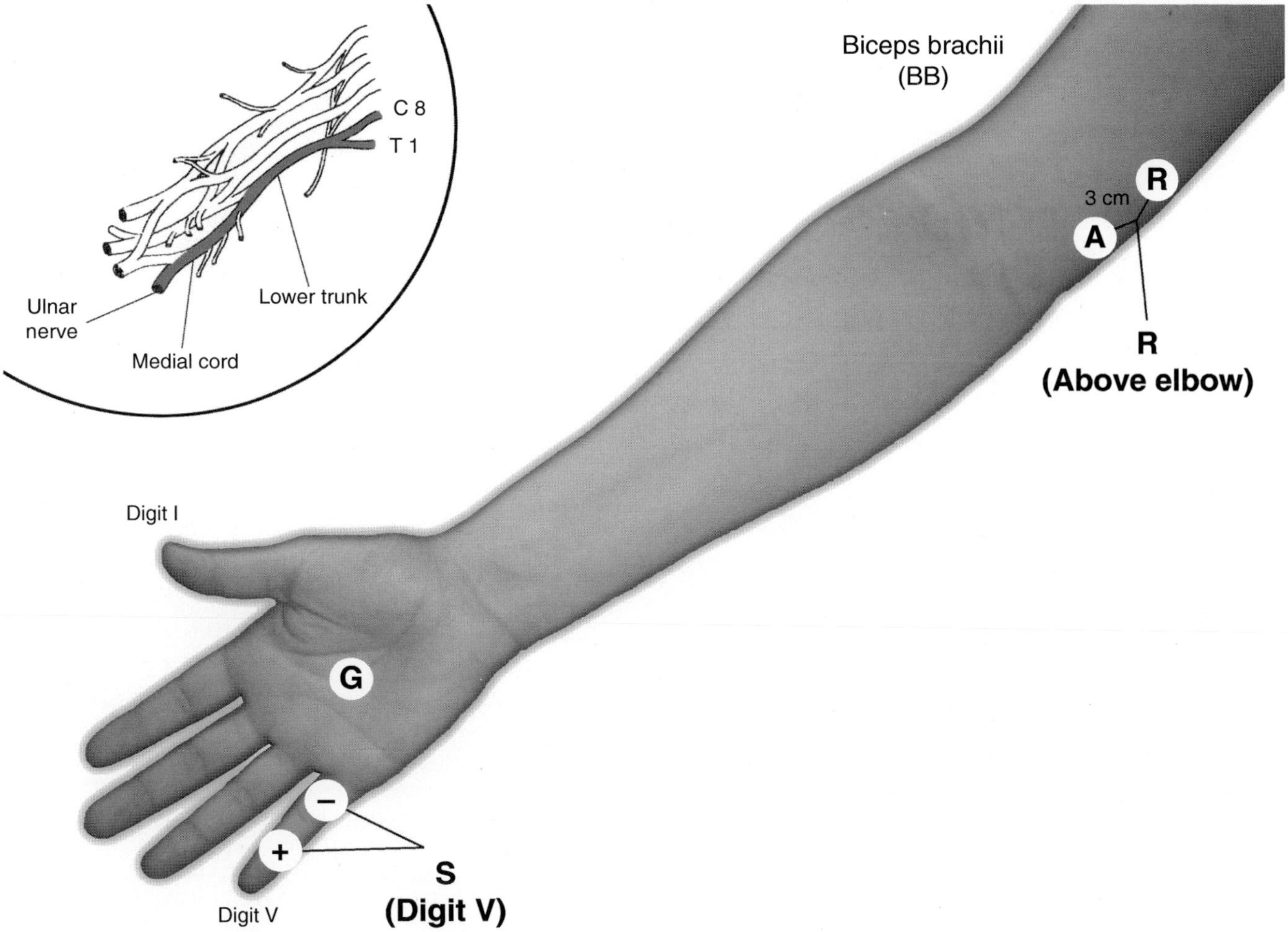

Typical waveform (digit V – above elbow):

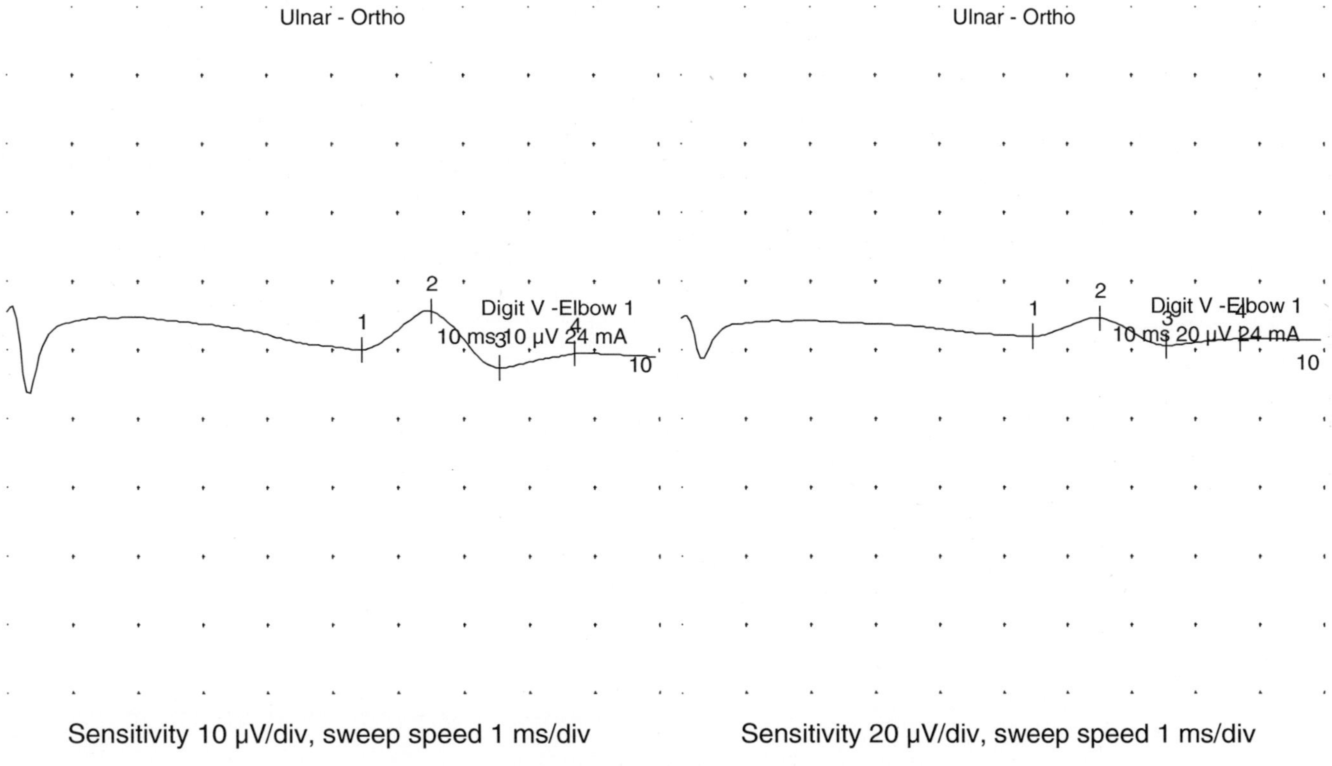

Fig. 2 Orthodromic sensory nerve action potential (*SNAP*) recorded above the elbow, stimulation of digit V

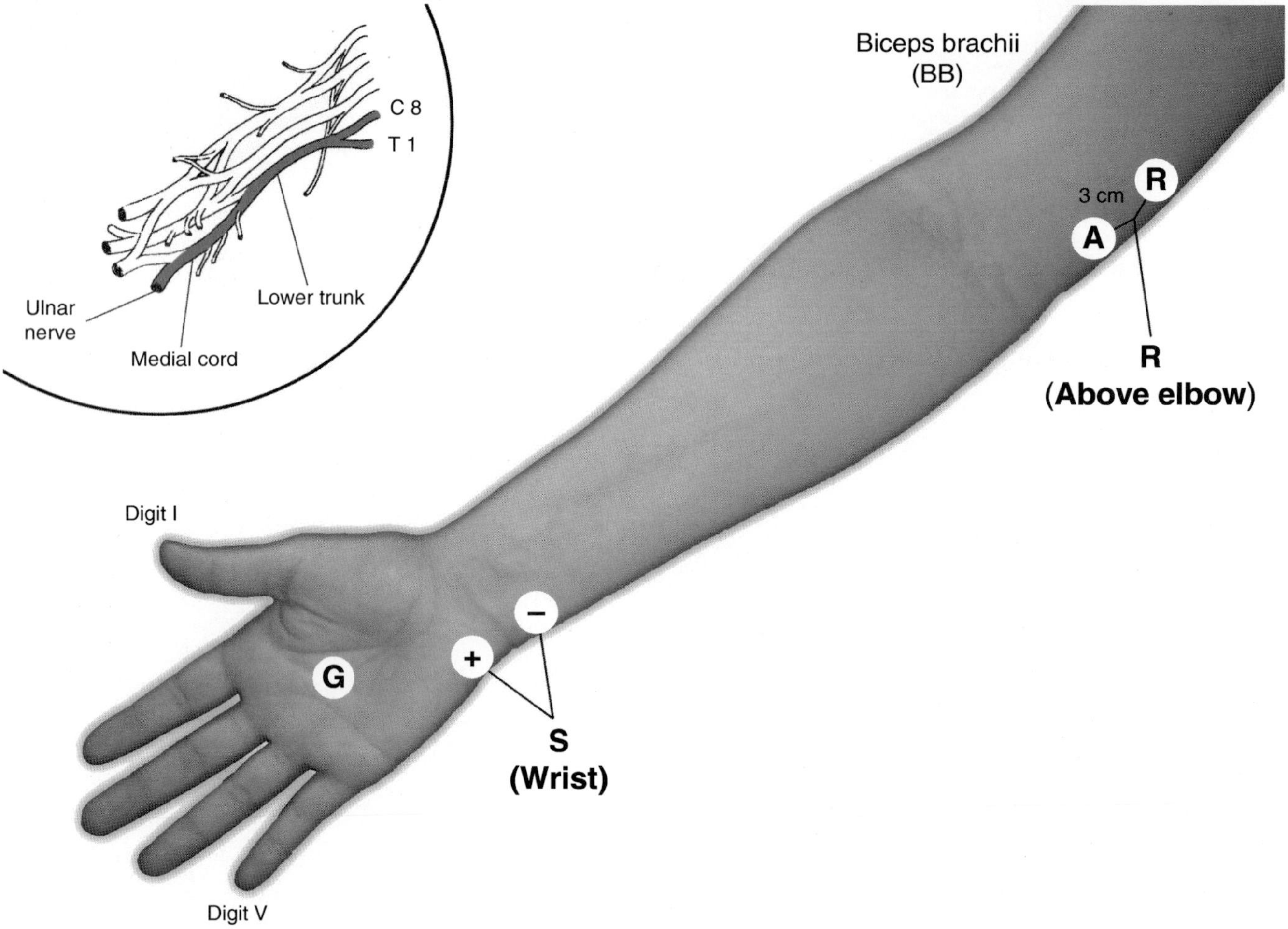

Typical waveform (wrist – above elbow):

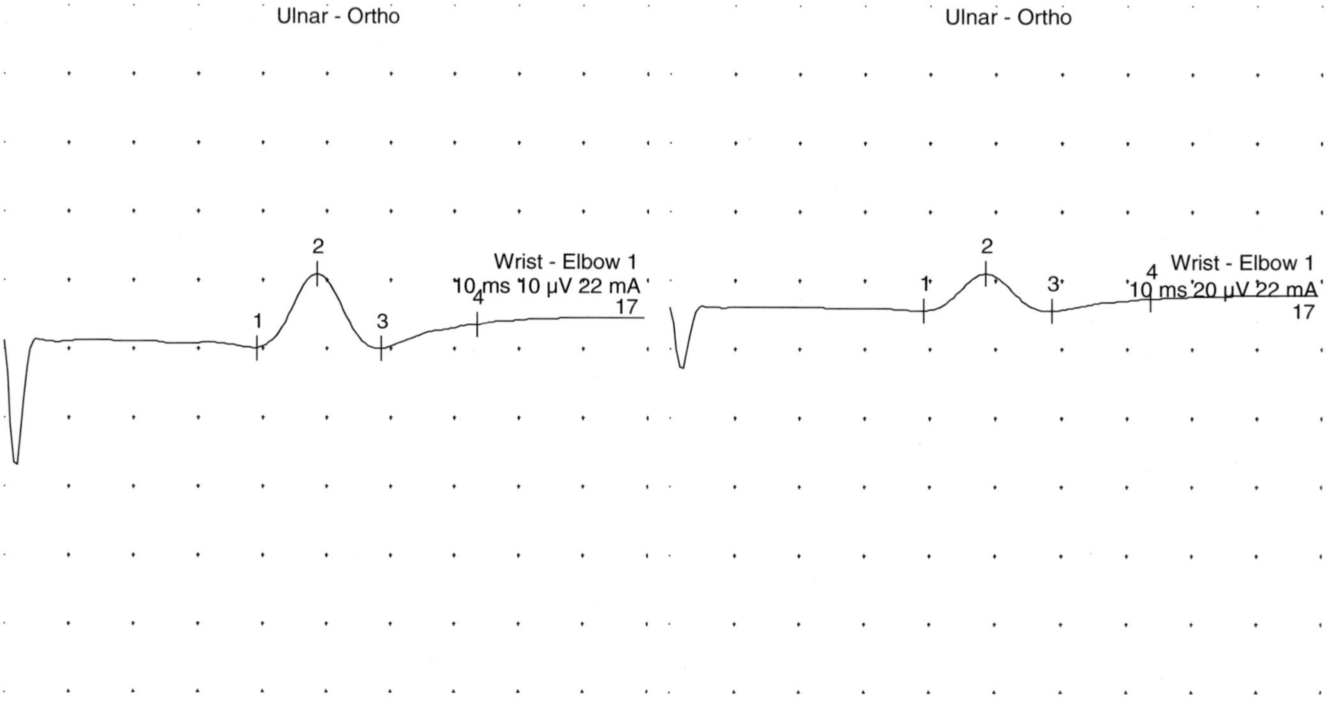

Sensitivity 10 μV/div, sweep speed 1 ms/div

Sensitivity 20 μV/div, sweep speed 1 ms/div

Fig. 3 Orthodromic sensory nerve action potential (*SNAP*) recorded above the elbow, stimulation of the wrist

Table 1 Normal values [1]

	Range
Digit V–wrist, peak latency (ms)	2.2–3.4
Digit V–wrist, peak amplitude (μV)	8–28

Table 2 Pathological values [1]

	Mean
Digit V–wrist, peak amplitude (μV)	0–14

wasting was observed in two others. Eleven patients referred subjective impairment of slight sensation in the ulnar territory. Only amplitude sensory values were reported, varying in normal subjects considerably, so little significance has been placed to small variations in actual potential size in pathological material. Amplitude measurements showed markedly reduced values at the wrist level in the patients with ulnar neuritis at the elbow, and authors obtained more information by stimulating the ulnar nerve trunk at the wrist and recording the motor volley from the ulnar nerve just above the elbow. In a patient with a history of numbness and tingling in the fourth and fifth fingers of the left hand, they found a rather small ulnar SNAP (6 μV) compared with the median SNAP (18 μV) while recording at the wrist, and when they recorded sensory responses just above the elbow, a small delayed ulnar SNAP than normal median SNAP was found.

Gilliatt and Thomas [2], following the method by Gilliatt and Sears [1], recorded ulnar SNAPs at the wrist (digit V stimulation) and just above the elbow (wrist stimulation). In healthy subjects, they observed that stimulation to digit V with recording just above the elbow provided very small potentials, and they preferred to stimulate the ulnar nerve at the wrist level when examining the ulnar nerve conduction over the forearm segment. The authors also performed nerve conduction study over a more proximal segment stimulating the ulnar nerve just above the elbow and recording SNAPs at the axilla (Figs. 4 and 5). In order to examine ulnar nerve conduction above the suspected level of the lesion, they studied the elbow–axilla segment; the authors suggested to shave the axilla and place the recording electrodes as high as possible with the arm abducted to a right angle. With the recording electrodes in the apex of the axilla (3 cm fixed interelectrode distance), they recorded potentials from the nerve with equal facility; great care was necessary when placing the stimulating electrodes on the ulnar nerve just above the elbow to avoid cross stimulation of the median nerve. Control observations were made in 15 healthy subjects (age range 18–62 years), and supramaximal stimulation was applied at each stimulation site, such that a further increase of stimulus intensity produced little change in the size of the sensory or mixed response. The authors observed that nerve action potential amplitude values for the forearm and elbow segment varied from 33 to 100 μV. The relationship between ulnar nerve action potential latency (measured to the negative peak) and conduction distance (measured between the stimulating cathode to the recording electrode) showed values from 4.3 to 6.2 (wrist–above-elbow segment) and from 2.7 to 4.3 ms (above-elbow–axilla segment). Gilliatt and Thomas [2] performed sensory and mixed nerve conduction studies on 14 patients with ulnar neuropathy at the elbow (UNE) clinically diagnosed (sample data were not reported). In all patients, they detected no action potentials by surface electrodes placed on the skin over the ulnar nerve at the wrist and just above the elbow, and when they stimulated the ulnar nerve above the elbow, a mixed nerve action potential was successfully recorded from the axilla (amplitude range 8–92 μV, peak latency range 2.5–3.7 ms, conduction distance range 13.5–20.5 cm) in each patient. In two patients the axillary potentials were small when compared with those obtained in the control subjects suggesting that conduction may be mildly disturbed above the level of a chronic lesion.

Comment

For Gilliatt and Thomas [2], with the electrode arrangement shown in Figs. 1 and 2, the potentials recorded at the wrist and just above the elbow were derived solely from impulses traveling in afferent digital sensory fibers, whereas with the electrode arrangements shown in Figs. 3, 4, and 5, antidromic impulses in motor fibers also contribute to the recorded action potentials (mixed potentials). The recording of mixed nerve action potentials from the axilla with stimulation above the elbow has been introduced to define the upper level of the lesion in cases without a clear-cut abnormality of motor conduction in the elbow segment. The proximal evaluation of the ulnar nerve conduction showed somewhat technical difficulty, and it was more time-consuming than the other procedures, but in each patient, they were able to demonstrate an ascending volley in the upper arm, whereas none could be detected below this level.

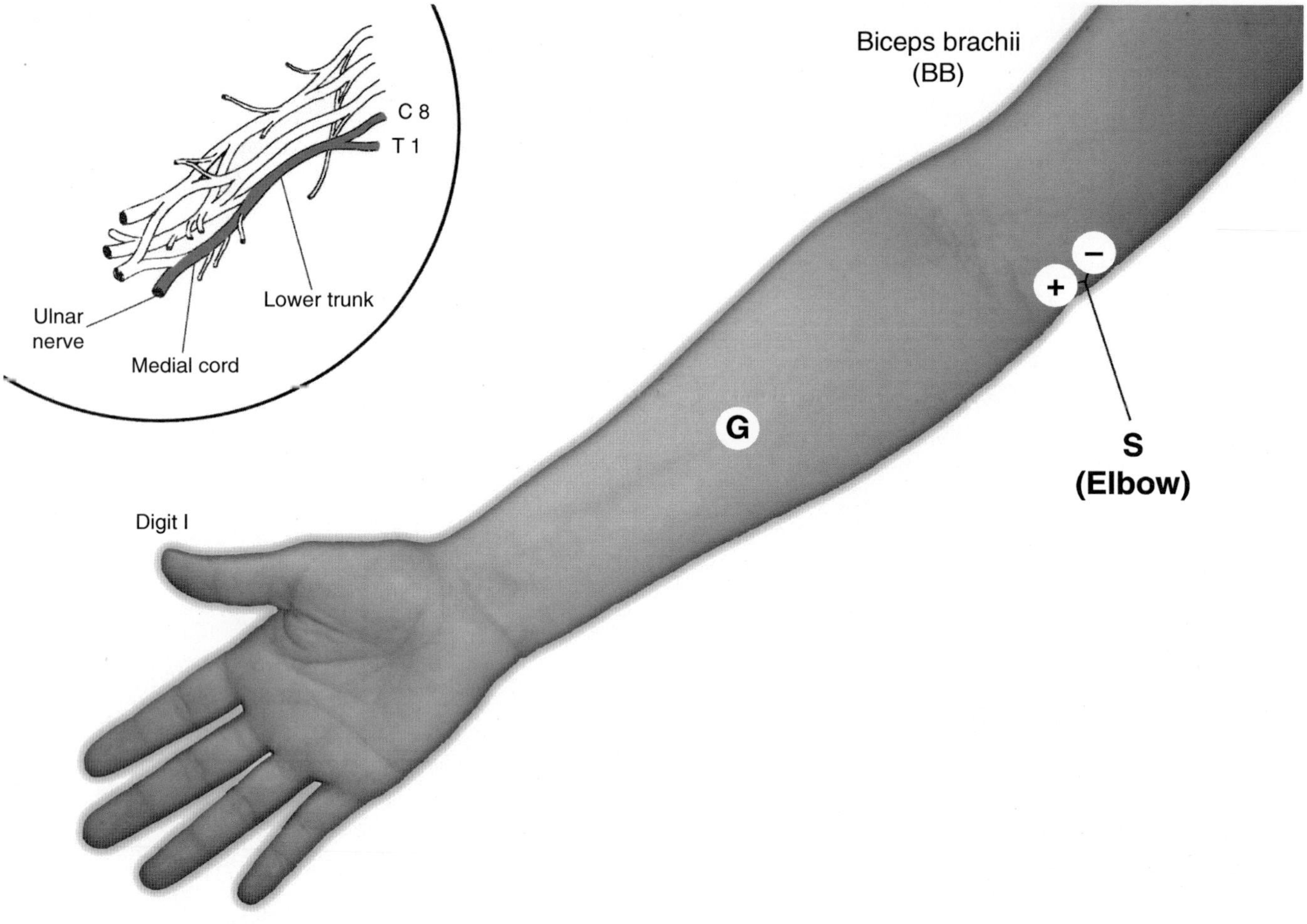

Typical waveform (elbow–axilla):

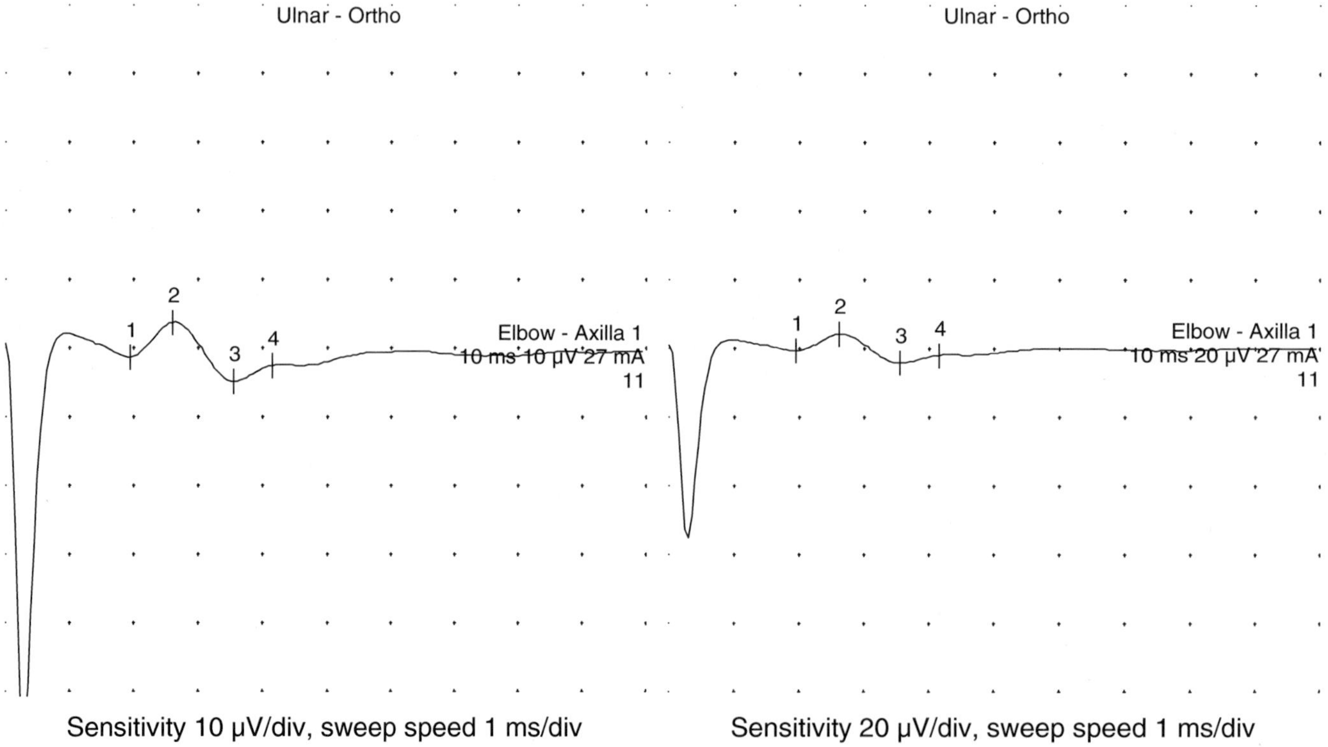

Fig. 4 Orthodromic sensory nerve action potential (*SNAP*) recorded at the axilla, stimulation of the elbow

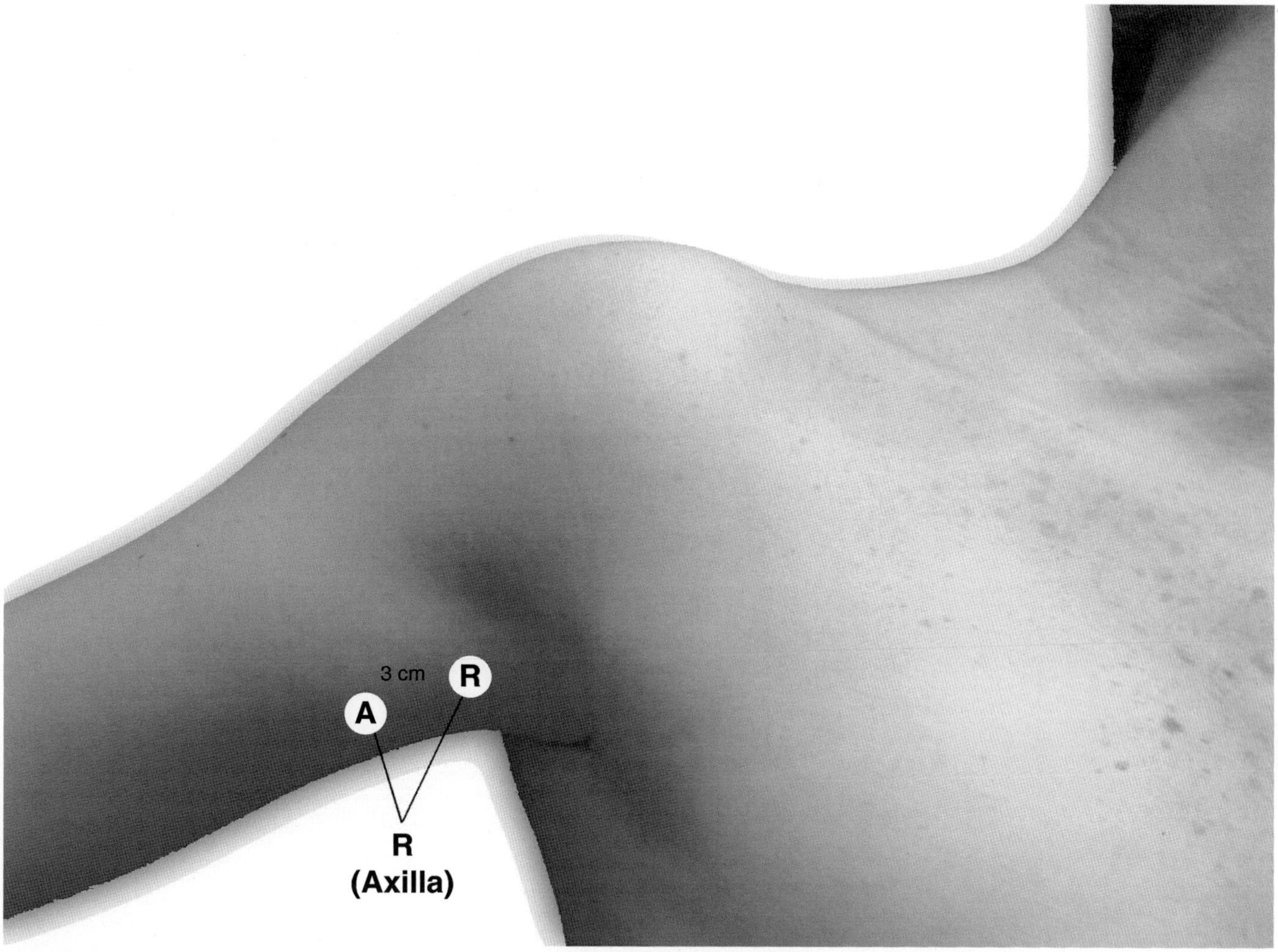

Fig. 5 Ulnar proximal recording (axilla), stimulation of the elbow

Table 3 Normal values [3]

	Mean ± SD	Range
Subject 1 (33 years): wrist–elbow, conduction velocity (m/s)	52.2 ± 1.8	48.7–54.2
Subject 2 (25 years): wrist–elbow, conduction velocity (m/s)	54.6 ± 1.5	52.0–56.4
Subject 3 (24 years): wrist–elbow, conduction velocity (m/s)	58.5 ± 1.8	57.8– 61.4
Subject 4 (33 years): wrist–elbow, conduction velocity (m/s)	56.2 ± 0.9	53.0–56.9
Subject 5 (22 years): wrist–elbow, conduction velocity (m/s)	53.7 ± 1.5	51.5–56.2
All subjects (27 years): wrist–elbow, conduction velocity (m/s)	55.0 ± 2.6	48.7–61.4

McQuillen and Gorin [3] stimulated 5 left ulnar nerves of 5 healthy (Table 3) subjects (age range 22–33 years, mean age 27 years). Mixed potentials were recorded using surface disk electrodes (DISA 13 K 60) placed at the elbow (Fig. 6). The authors [2] used a DISA 14 A 20 amplifier, and mixed potentials were displayed on a storage oscilloscope (Tektronix RM564). Following the Dawson's method [1], maximal stimulation was applied at the wrist by bipolar, saline-moistened pad, stimulating electrode (DISA 13 K 62). All measurements were obtained in an air-conditioned laboratory, and the room temperature was maintained at a temperature of 23–24 °C. The nerve conduction distance was measured from the center of the stimulating cathode (–) at the wrist to the center of the recording electrode (A) at the elbow. Mixed conduction velocity was derived by dividing conduction distance by conduction time, and it was measured in meter per second (m/s). The mixed conduction time was measured from stimulus at the wrist to the initial negative peak (peak latency) of the nerve action potential recorded at the elbow.

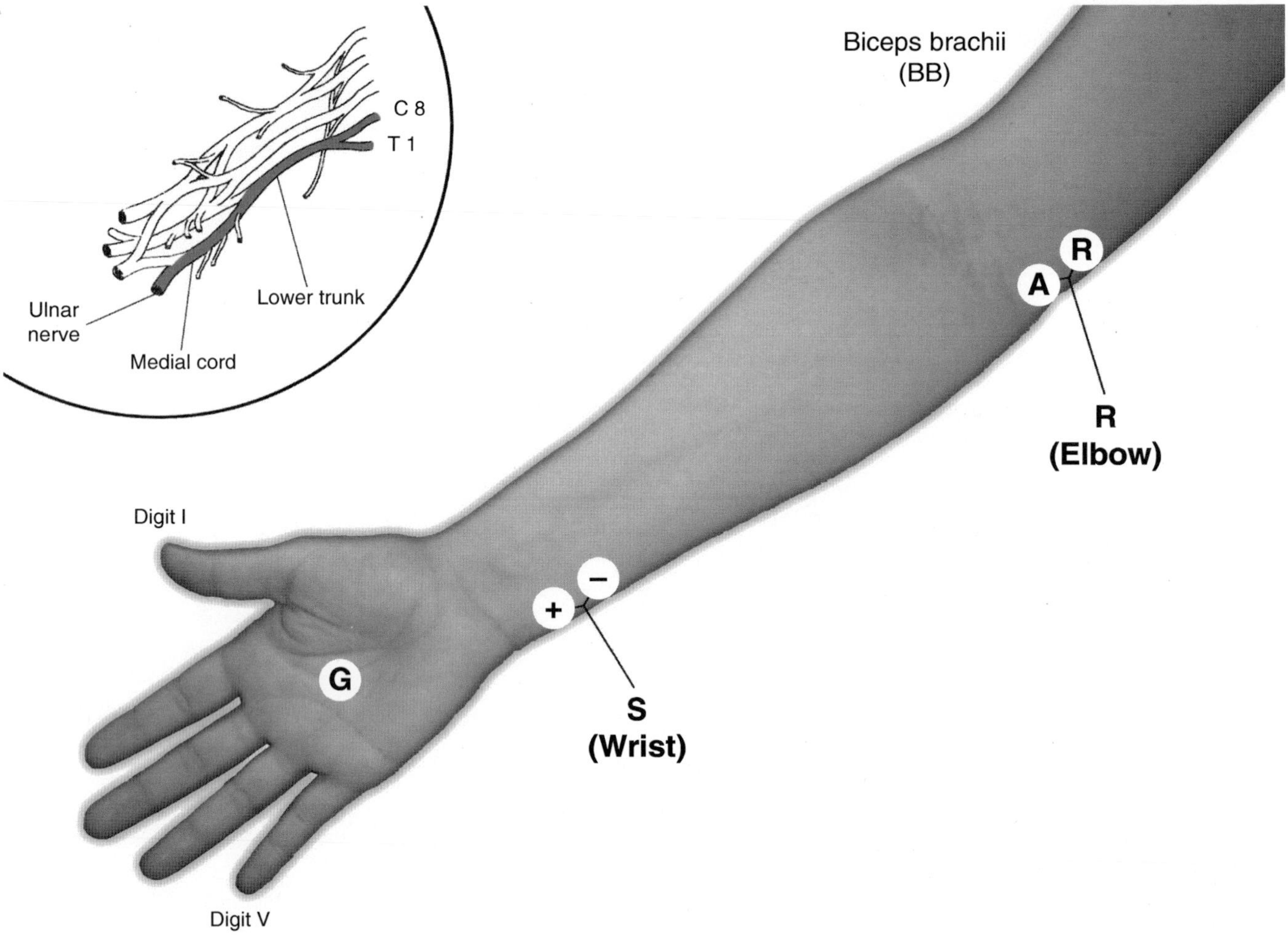

Fig. 6 Ulnar proximal recording (elbow), stimulation of the wrist

Comment

In order to assess the ulnar nerve function in the forearm, McQuillen and Gorin [3] expressed the conduction velocity in the mixed nerve (Vm) as the velocity derived from stimulation of both motor and sensory nerve fibers at the wrist using the latency of the negative peak of the mixed response recorded at the elbow. Since Vm was derived from the peak of the nerve action potential, it was a modal velocity, as the motor conduction velocity to the peak (Vp) of the compound motor potential (CMAP). The authors compared Vm with Vp, and they found no statistical differences between Vm and Vp. They observed that only a quarter of the mixed nerve action potential was derived from the sensory fibers, and three-quarters was contributed by motor nerve [4]; for this reason, Vm and Vp were similar in adult subjects.

Sedal et al. [5], following the method by Gilliatt and Sears [1], recorded ulnar SNAPs at the wrist (digit V stimulation) in 34 control (mean age 47 years, age range 18–77 years) subjects (Table 4) and in 234 patients (Table 5) with clinical-confirmed carpal tunnel syndrome (CTS) – mean age 54 years and age range 19–87 years. The authors used electrical stimulus of 0.2 ms duration derived from a DISA Ministim; the recording electrodes were connected to a Tektronix FM122 preamplifier, and the SNAPs were displayed on a Tektronix 502A oscilloscope. The temperature was measured on all control subjects and a significant number of patients with CTS.

Table 4 Normal values [5]

	Mean ± SD	Range
Digit V–wrist, peak latency (ms)	2.6–0.3	2.0–3.4
Digit V–wrist, peak amplitude (μV)	13.7–6.4	6–36

Table 5 Pathological values [5]

	Mean ± SD	Range
Digit V–wrist, peak latency (ms)	3.0–0.6	1.9–5.0
Digit V–wrist, peak amplitude (μV)	7.9–4.3	0–18

Comment

Sedal et al. [5] observed that at all ages, the regression line for the patients with CTS has a lower value for ulnar sensory amplitude than for the control subjects. The mean latency (ms) and amplitude (μV) of the ulnar SNAP in the patients were 3.0 ± 0.6 ms and 7.9 ± 4.3 μV, which may be compared with those of 2.6 ± 0.3 ms and 13.7 ± 6.4 μV in the controls. Both the differences are highly significant if compared to the normal limit values, respectively: in 39.3 % of CTS cases, the amplitude of the SNAP was less than 7.0 μV, and in 4.8 % of CTS cases, the latency of the SNAP was greater than 4.0 ms. The authors found a statistically significant correlation between the decreasing of the median motor conduction velocity and amplitude of the ulnar SNAPs; they suggested that factors outside the carpal tunnel itself may be important in the pathogenesis of the CTS syndrome. This finding lends support to the idea of an underlying subclinical neuropathy which predisposes them to entrapment neuropathies [6] or a superimposed ulnar neuropathy at the wrist.

Eisen [7], following the method by Gilliatt and Sears [1], recorded orthodromic ulnar SNAPs at the wrist (digit V stimulation) in 48 nerves of healthy (Table 6) subjects (mean age 43.7 years, age range 19–78 years). The author preferred near-nerve technique to surface technique, and stainless steel needle recording electrodes (0.5 mm diameter, and 0.3 mm bare tip) were placed at the wrist, on the course of the ulnar nerve. The anode was placed 2 cm proximally to the active recording electrode. Pathological values were recorded from 100 ulnar nerves of patients divided into two groups (Group 1, severe ulnar lesions, and Group 2, mild ulnar lesions): Group 1–34 ulnar nerves from patients (Table 7) showing clear-cut clinical evidence of an ulnar nerve compression at the elbow (mean age 52.5 years, age range 21–77 years). All patients showed atrophy and significant weakness of the *first dorsal interosseous* (FDI) and *abductor digiti minimi* (ADM) muscles on the involved side and sensory deficit in an appropriate distribution; Group 2 – 56 ulnar nerves from patients (Table 8) referred because of subjective complaints of numbness and tingling limited to the digits IV and V on the involved side and because of a presumptive diagnosis of ulnar palsy (mean age 44.1 years, age range 17–74 years).

Table 6 Normal values [7]

	Mean ± SD	Range
Digit V–wrist, onset latency (ms)	2.3 ± 0.45	2.0–3.5

Table 7 Pathological values [7]

Group 1	Mean ± SD	Range
Digit V–wrist, onset latency (ms)	3.7 ± 1.0	2.4–5.2

Table 8 Pathological values [7]

Group 2	Mean ± SD	Range
Digit V–wrist, onset latency (ms)	2.5 ± 0.4	2.0–5.2

Comment

Eisen [7] recorded SNAPs at the wrist in patients with severe ulnar nerve lesions (Group 1–34 nerves) and mild ulnar nerve lesions (Group 2–56 nerves) at the elbow. In Group 1, in 27 patients (78.8 %) it was not possible to obtain a sensory evoked response even after averaging 128 signals. In Group 2, in 18 patients (32.1 %) a sensory evoked response could not be obtained from the wrist, even after averaging 128 responses. In contrast, SNAPs were easily recorded at the wrist from all the control subjects. The above-elbow mixed nerve action potential recording was not performed because, in agreement with Gilliatt and Thomas [2], when the SNAP is absent at the wrist, the mixed nerve action potential above the elbow also will be absent. Pathological waveform and values in a case of Guyon's canal syndrome are reported (Fig. 7).

Pathological waveform (digit V – wrist):

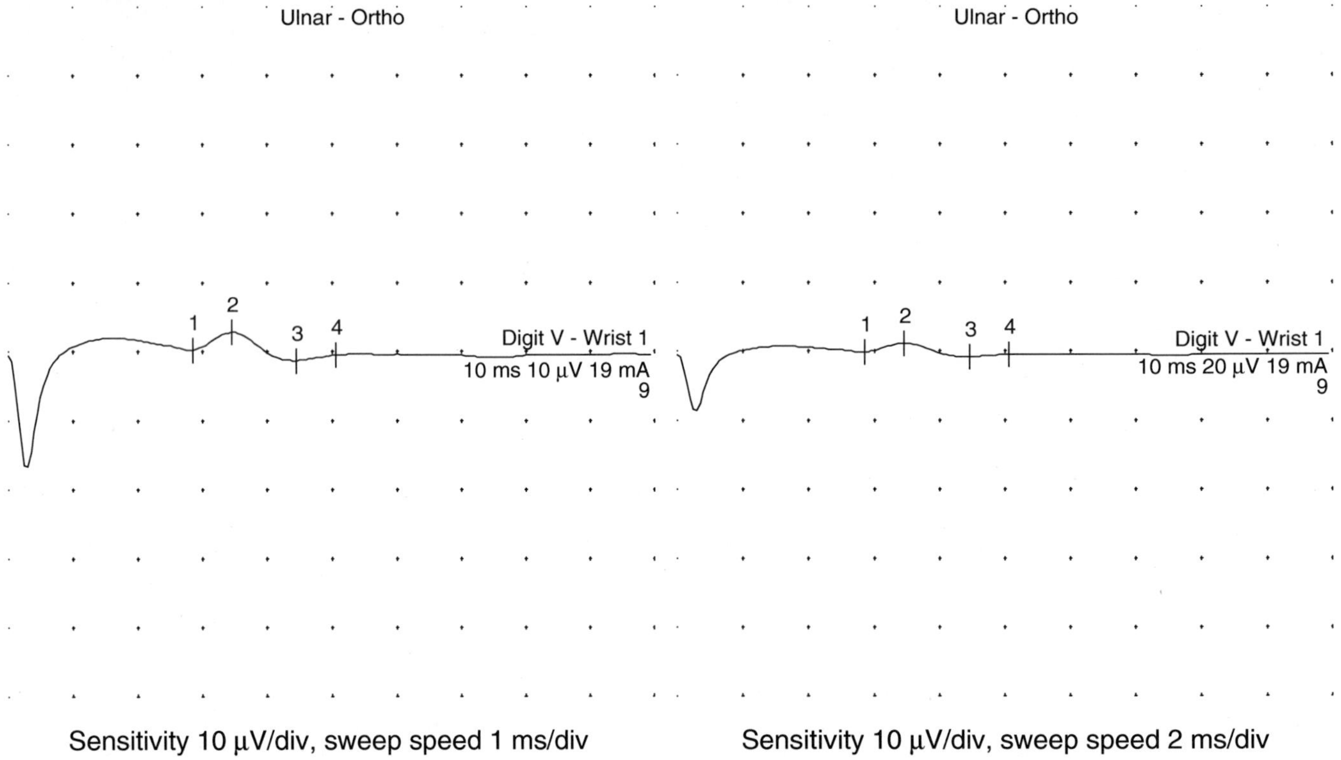

Onset latency: 2.90 ms; **Peak latency**: 3.45 ms; **Onset to peak amplitude**: 2.5 μV; **Peak to peak amplitude**: 4.2 μV; **SNCV**: 41.4 m/s (12 cm distance)

Fig. 7 Orthodromic sensory nerve action potential (*SNAP*) recorded at the wrist in a Guyon's canal syndrome, stimulation of digit V

References

1. Gilliatt RW, Sears TA (1958) Sensory nerve action potentials in patients with peripheral nerve lesions. J Neurol Neurosurg Psychiatry 21:109–118
2. Gilliatt RW, Thomas PK (1960) Changes in nerve conduction with ulnar lesions at the elbow. J Neurol Neurosurg Psychiatry 23:312–320
3. McQuillen MP, Gorin FJ (1969) Serial ulnar nerve conduction velocity measurements in normal subjects. J Neurol Neurosurg Psychiatry 32:144–148
4. McQuillen MP, Johns RJ (1967) The nature of the defect in the Eaton-Lambert syndrome. Neurology 17:527–536
5. Sedal L, McLeod JG, Walsh JC (1973) Ulnar nerve lesions associated with the carpal tunnel syndrome. J Neurol Neurosurg Psychiatry 36:118–123
6. Gilliatt RW, Willison RG (1962) Peripheral nerve conduction in diabetic neuropathy. J Neurol Neurosurg Psychiatry 25:11–18
7. Eisen A (1974) Early diagnosis of ulnar nerve palsy. Neurology 24:256–262

Digit V – Wrist, Elbow, Axilla; Wrist – Elbow, Axilla

S3

Surface Recording Technique, Orthodromic Study

Original Settings Sensitivity was 40 μV/division, and sweep speed was 1 ms/division; low-frequency filter, high-frequency filter, duration of pulse, rate of pulse, and the machine used were not specified.

Position This study was performed in the supine position.

Recording Following the orthodromic method [1], sensory nerve action potentials (SNAPs) were recorded on the course of the ulnar nerve (Fig. 1) above the wrist (R1), at the elbow (R2), and in the axilla (R3). For recording at the wrist (R1), the active (A) electrode was placed proximally to the distal crease at the wrist; the reference (R) electrode was placed 3 cm proximally to the active electrode. For recording at the elbow (R2), the active (A) electrode was placed proximally to the ulnar groove; the reference (R) electrode was placed proximally to the active electrode. For recording in the axilla (R3), electrodes were placed medially on the course of the ulnar nerve, with the active (A) electrode placed proximally to the elbow and the reference (R) electrode placed proximally to the active electrode. The authors also performed an orthodromic method to record the mixed nerve action potentials (MNAPs) at the elbow (R1) and the axilla (R2) recording sites, stimulating the ulnar nerve at the wrist (Fig. 2). Ulnar proximal recording at the axilla, for both the orthodromic SNAPs recording (Fig.1) and mixed potentials recording (Fig. 2) is showed in Fig. 3. Distances between the recording and stimulating electrodes were not fixed. Bipolar silver electrodes applied to the skin were used for recording. In cases in which the nerve action potentials were extremely small, the author used needle electrodes for recording. In these cases, the needles were inserted through the skin and placed in the vicinity of the ulnar nerve with an interelectrode distance of 3–4 cm. With needle electrodes, the amplitude of the SNAP was always greater than that obtained with surface electrodes, but the latency from stimulus to response remained the same. The ground (G) electrode position was not specified in the report; the figures (Figs. 1 and 2) show the ground electrode placed on the palm.

Stimulation The ulnar nerve was stimulated distally to digit V (S) using surface silver electrodes. The active stimulating electrode (cathode, −) was placed near the metacarpophalangeal joint (at the base of the digit); the anode (+) was positioned distally in the region of the terminal interphalangeal joint. The author used supramaximal electric stimulus as reported by Dawson [2]. SNAPs recorded at the elbow (R2) and in the axilla (R3) were extremely small; however, when the ulnar nerve was stimulated at the wrist, a large potential (mixed) could be orthodromically recorded at the elbow and in the axilla.

Measurements Onset latency (ms) was measured from the onset of the stimulus artifact to the junction of the negative inflection potential and the baseline; this was converted to sensory nerve conduction velocity (SNCV) and measured in meter per second (m/s), reflecting the conduction in the largest afferent sensory fibers. The temperature was maintained between 26 and 30 °C. The temperature of the extremities ranged between 33 and 36 °C, surface measurements (34–37 °C, intramuscular measurements). If the extremities were cooler than this, they were warmed with a heating pad during the procedure. Mayer [1] recorded SNAPs at the wrist, at the elbow, and in the axilla in 64 healthy (Table 1) subjects (age range 10–35 years, 30 cases; age range 36–50 years, 16 cases; age range 51–80 years, 18 cases); he also evaluated 41 patients (Table 2) with diabetes mellitus without clinical evidence of peripheral neuropathy (age range 10–35 years, 14 cases; age range 36–50 years, 19 cases; age range 51–80 years, 8 cases) and 64 patients (Table 3) with diabetes mellitus with clinical evidence of peripheral neuropathy (age range 10–35 years, 9 cases; age range 36–50 years, 22 cases; age range 51–80 years, 33 cases).

G. Gentili, M. Di Napoli, *The Ulnar Nerve*, DOI 10.1007/978-3-319-43292-2_3

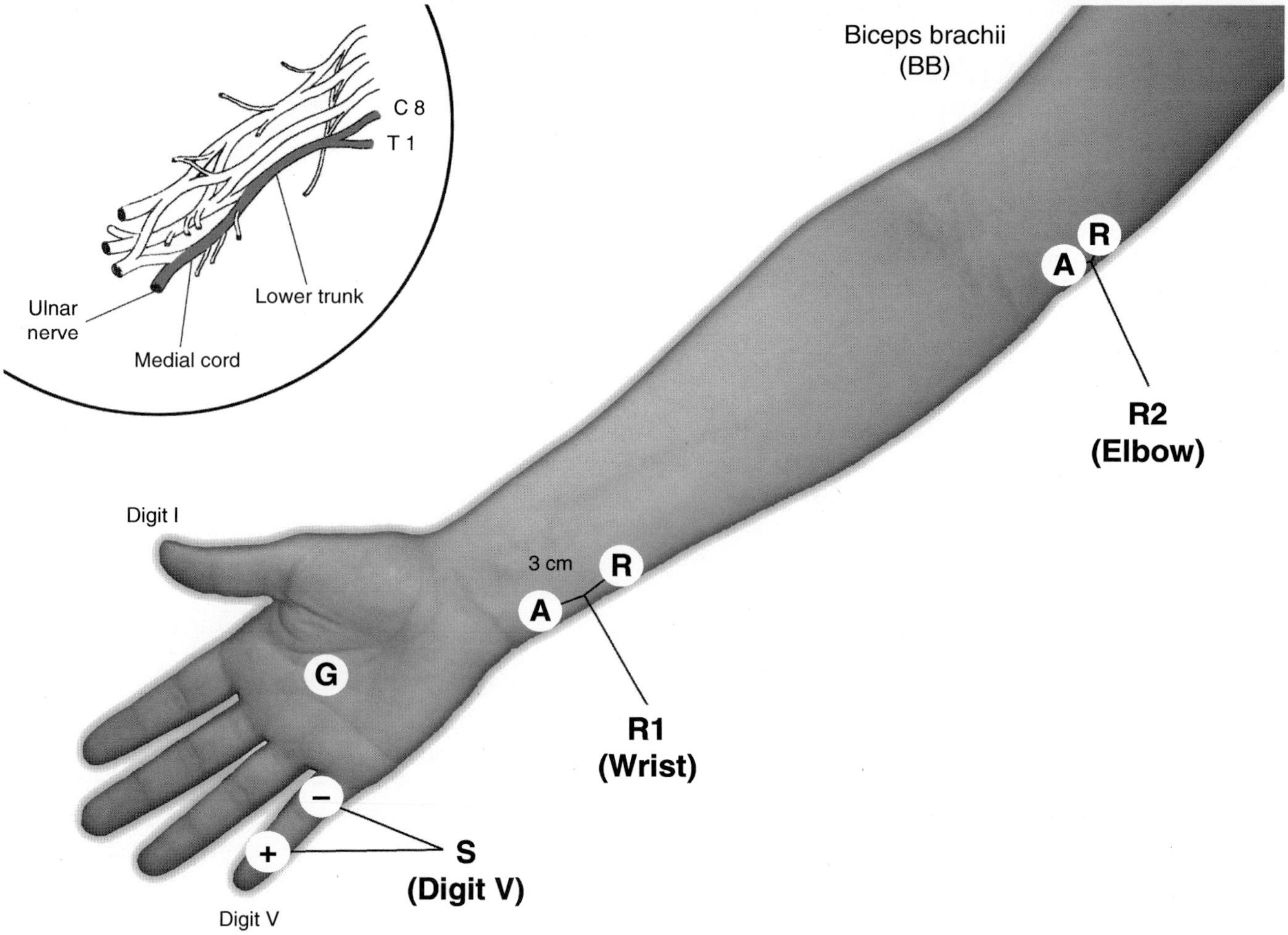

Typical waveform (digit V – wrist, elbow, axilla):

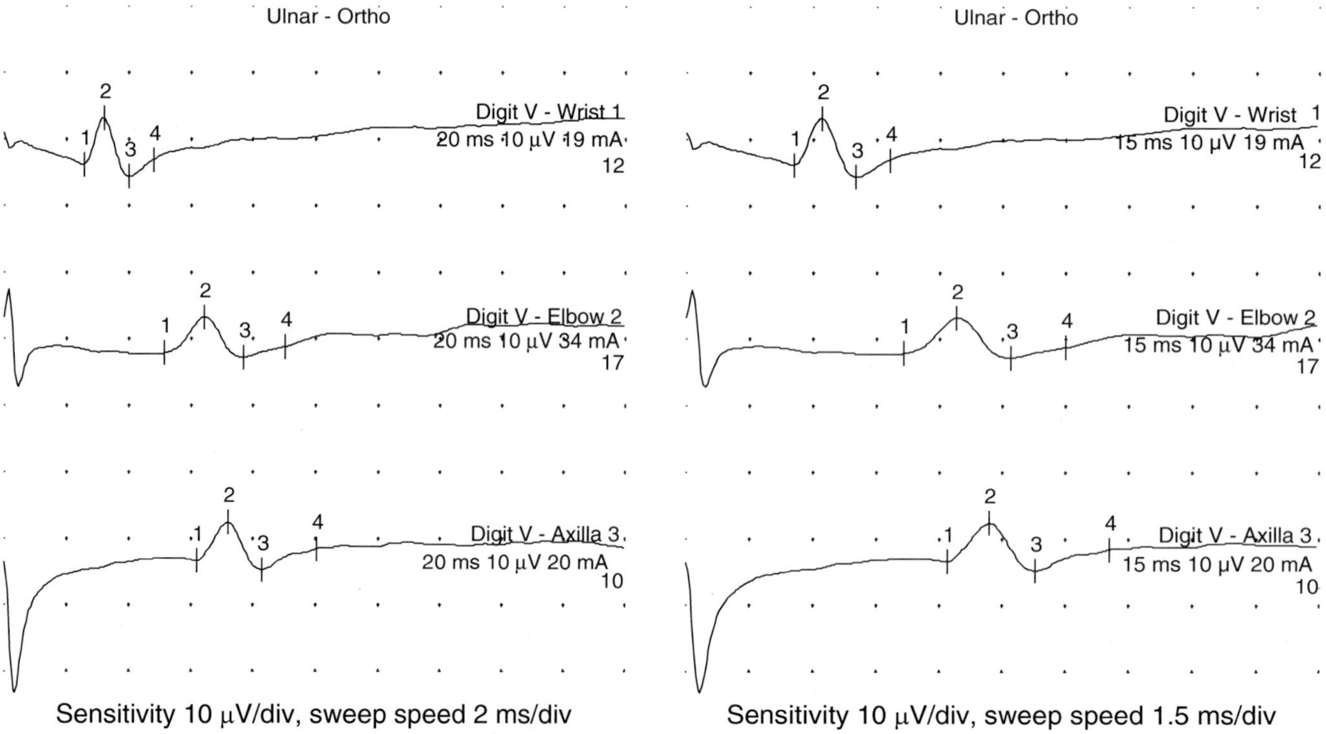

Fig. 1 Orthodromic sensory nerve action potentials (*SNAPs*) recorded at the wrist (*upper trace*), at the elbow (*middle trace*), and at the axilla (*lower trace*), stimulation to the digit V

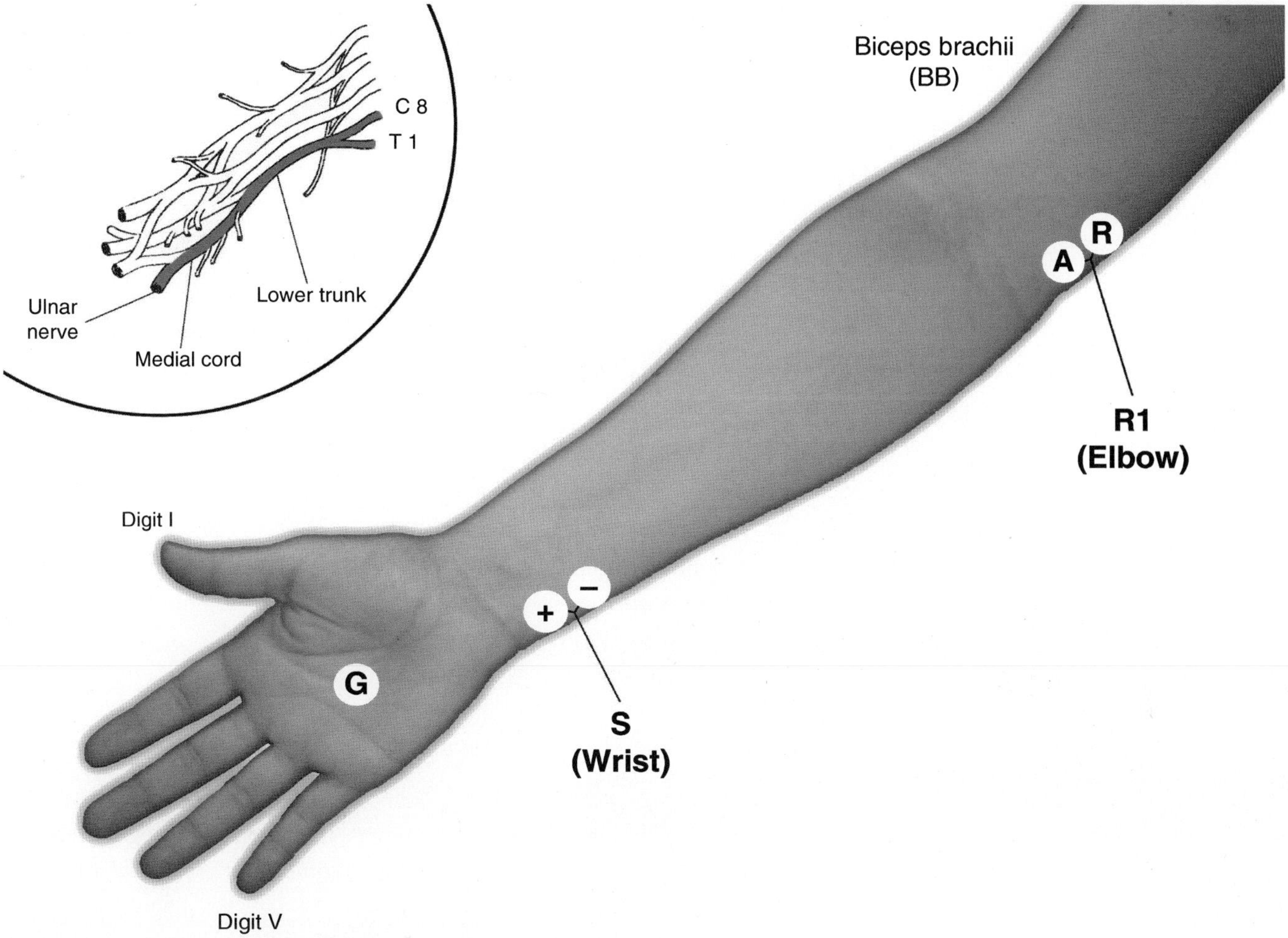

Typical waveform (wrist – elbow, axilla):

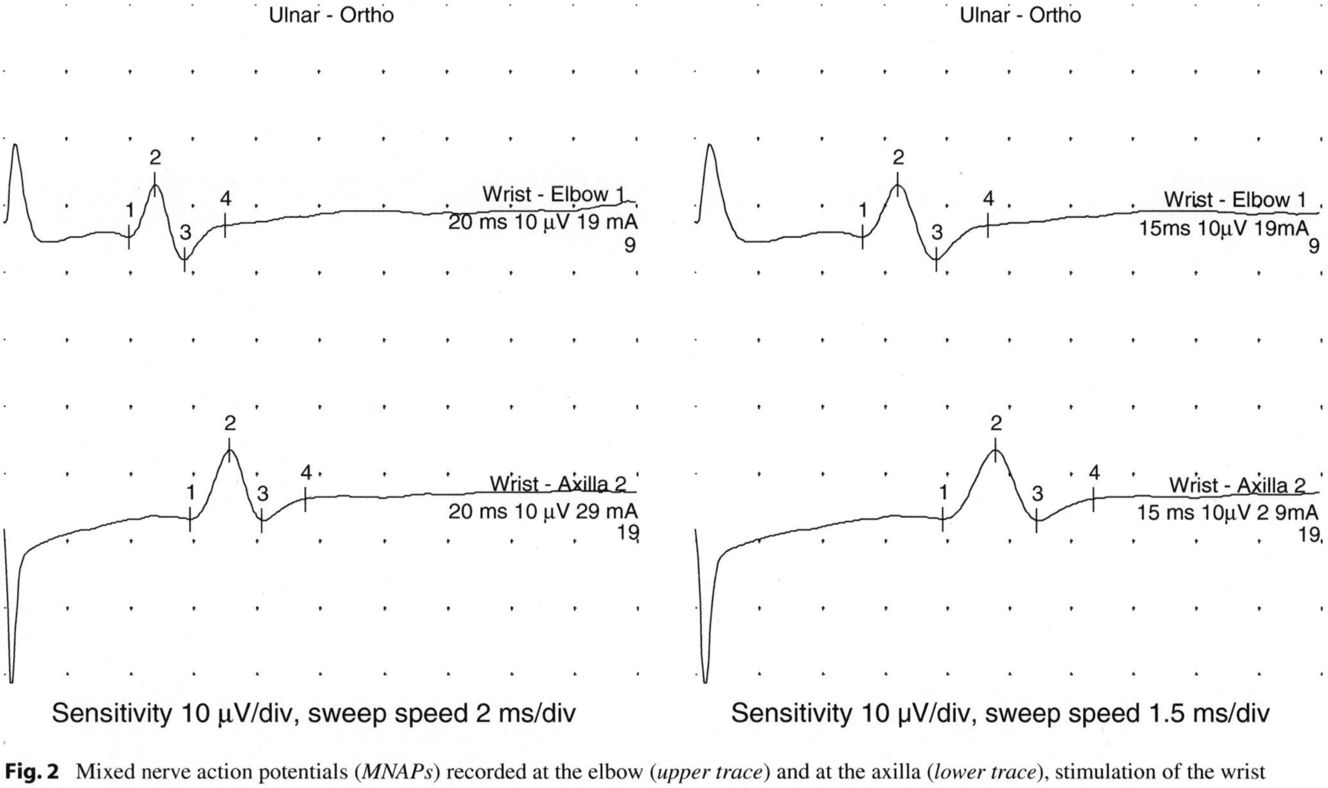

Fig. 2 Mixed nerve action potentials (*MNAPs*) recorded at the elbow (*upper trace*) and at the axilla (*lower trace*), stimulation of the wrist

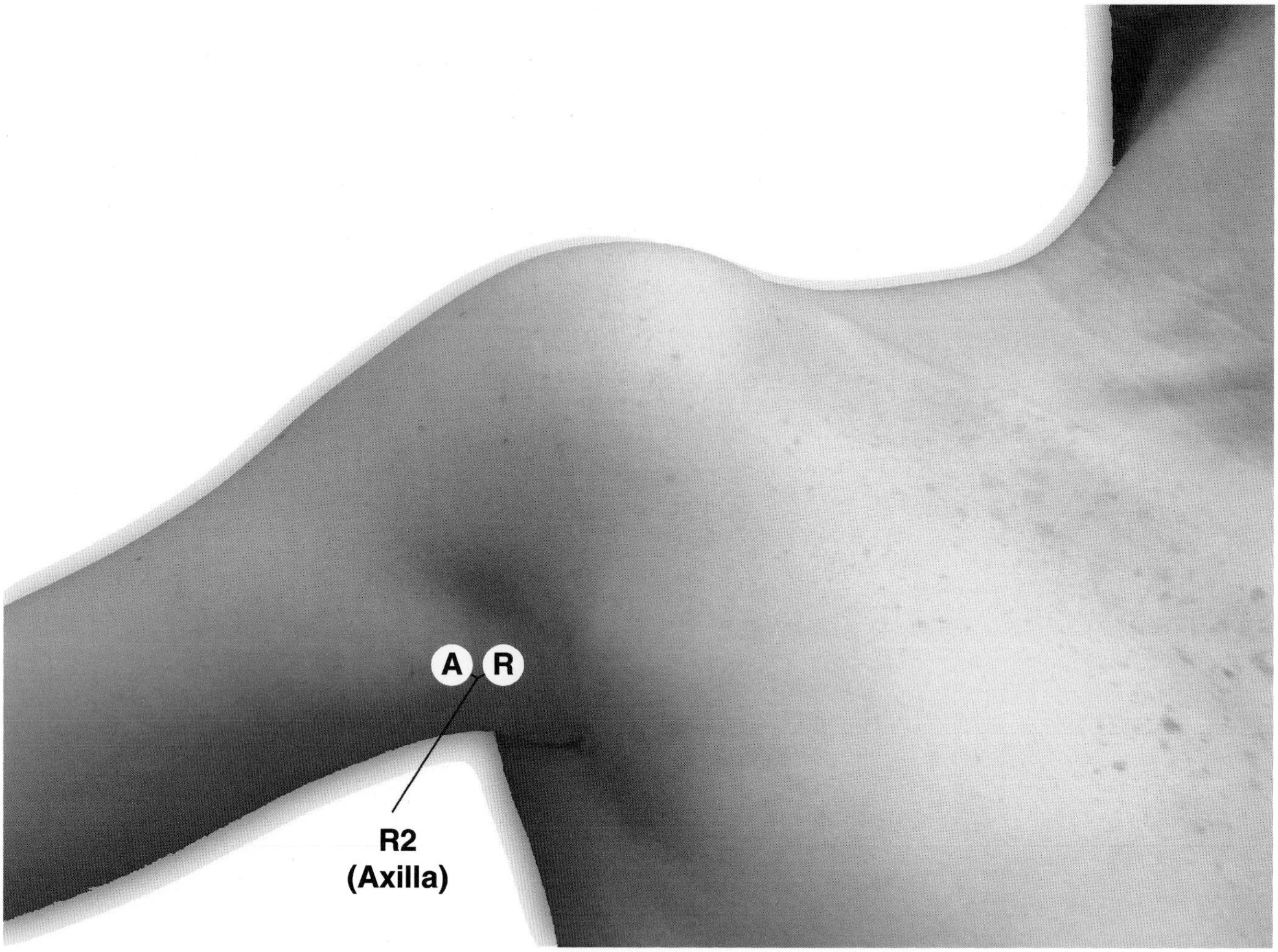

Fig. 3 Ulnar proximal recording (axilla), stimulation of the wrist

Table 1 Normal values [1]

	Mean ± SD
Digit V–wrist, SNCV (m/s) – age range 10–35 years	64.7 ± 3.9
Digit V–wrist, SNCV (m/s) – age range 36–50 years	66.5 ± 3.4
Digit V–wrist, SNCV (m/s) – age range 51–80 years	57.5 ± 6.6
Wrist–elbow, SNCV (m/s) – age range 10–35 years	64.8 ± 3.8
Wrist–elbow, SNCV (m/s) – age range 36–50 years	67.1 ± 4.7
Wrist–elbow, SNCV (m/s) – age range 51–80 years	56.7 ± 3.7
Elbow–axilla, SNCV (m/s) – age range 10–35 years	69.1 ± 4.3’
Elbow–axilla, SNCV (m/s) – age range 36–50 years	70.6 ± 2.4
Elbow–axilla, SNCV (m/s) – age range 51–80 years	64.4 ± 3.0

Table 2 Pathological values [1]

	Mean ± SD
Digit V–wrist, SNCV (m/s) – age range 10–35 years	62.1 ± 8.3
Digit V–wrist, SNCV (m/s) – age range 36–50 years	59.6 ± 5.1
Digit V–wrist, SNCV (m/s) – age range 51–80 years	55.7 ± 8.6
Wrist–elbow, SNCV (m/s) – age range 10–35 years	59.2 ± 3.9
Wrist–elbow, SNCV (m/s) – age range 36–50 years	60.6 ± 6.0
Wrist–elbow, SNCV (m/s) – age range 51–80 years	56.1 ± 5.3
Elbow–axilla, SNCV (m/s) – age range 10–35 years	64.4 ± 3.7
Elbow–axilla, SNCV (m/s) – age range 36–50 years	63.9 ± 3.5
Elbow–axilla, SNCV (m/s) – age range 51–80 years	68.7 ± 4.4

Table 3 Pathological values [1]

	Mean ± SD
Digit V–wrist, SNCV (m/s) – age range 10–35 years	53.5 ± 13.0
Digit V–wrist, SNCV (m/s) – age range 36–50 years	52.9 ± 8.1
Digit V–wrist, SNCV (m/s) – age range 51–80 years	51.7 ± 8.0
Wrist–elbow, SNCV (m/s) – age range 10–35 years	57.8 ± 8.8
Wrist–elbow, SNCV (m/s) – age range 36–50 years	50.1 ± 6.6
Wrist–elbow, SNCV (m/s) – age range 51–80 years	53.0 ± 6.2
Elbow–axilla, SNCV (m/s) – age range 10–35 years	65.6 ± 6.0
Elbow–axilla, SNCV (m/s) – age range 36–50 years	65.0 ± 6.0
Elbow–axilla, SNCV (m/s) – age range 51–80 years	62.7 ± 6.9

Comment

For Mayer [1], the rate in the proximal segment (axilla–elbow) was significantly faster than that in the distal segments, which were similar. There was no significant change in conduction velocity in various age groups until over the age of 50, when there was a slowing of the sensory fibers. The slowing was present in all the segments but appeared more prominent in the wrist–elbow segment. In 41 patients with diabetes mellitus without clinical evidence of peripheral neuropathy, there was slowing of conduction in the afferent fibers in all segments, more in the region of the elbow, but this change was not statistically significant. After the age of 50, the slowing did not differ significantly from the normal. Any alteration of ankle jerk or vibration sense was accepted as clinical evidence of peripheral neuropathy. The author also studied 64 patients with diabetes mellitus with clinical evidence of peripheral neuropathy. These patients were older (52 % over the age of 50) than the diabetics without peripheral neuropathy (20 % over the age of 50). Under the age of 35, the conduction velocity was reduced in the afferent fibers and to a greater degree than in the diabetics without peripheral neuropathy (an average of 9 m/s compared to 6 m/s). The slowing was more prominent in the below-elbow than above-elbow segments. Pathological waveform and values in a case (39 years old male) of diabetes mellitus with peripheral neuropathy are reported (Fig. 4).

Pathological waveform (wrist – elbow, axilla):

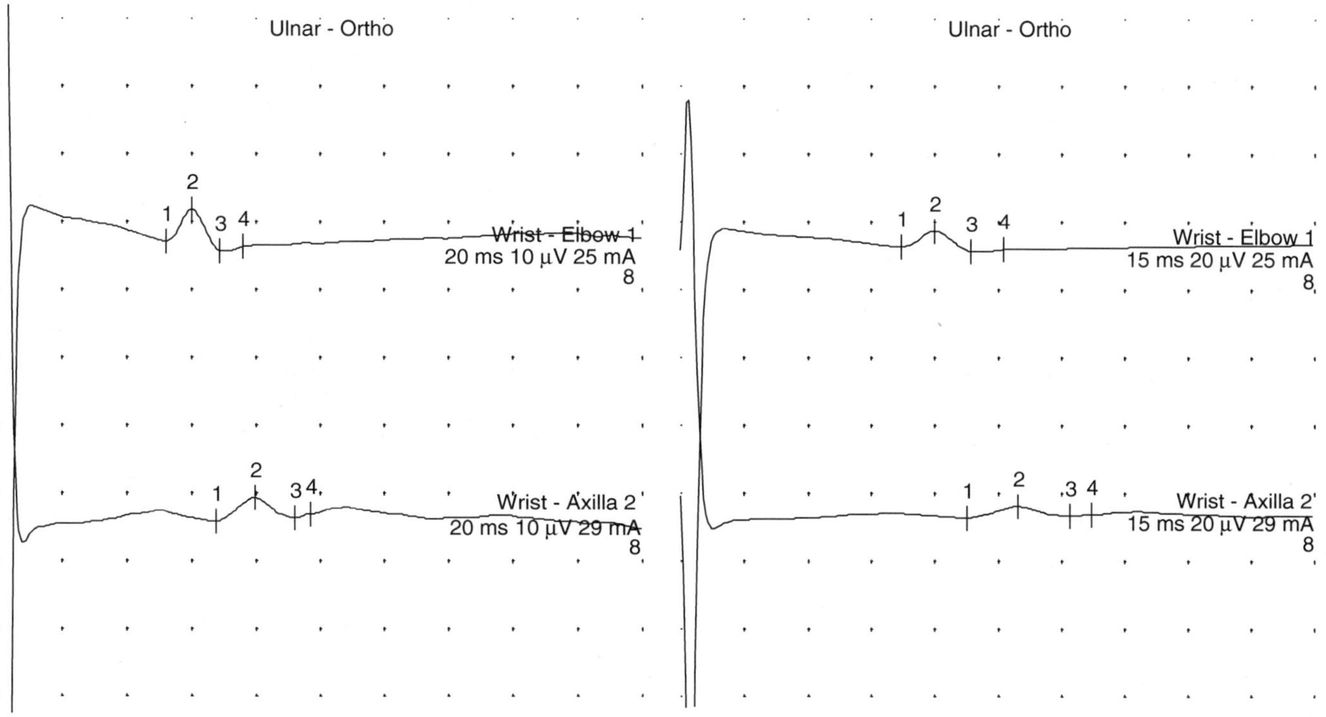

Sensitivity 10 µV/div, sweep speed 2 ms/div   Sensitivity 20 µV/div, sweep speed 1.5 ms/div

Onset latency (wrist – elbow): 3.40 ms; **Onset latency** (wrist – axilla): 6.90 ms; **Peak latency** (wrist – elbow): 6.0 ms **Peak latency** (wrist – axilla): 7.90 ms; **Onset to peak amplitude** (wrist – elbow): 6.6 µV; **Onset to peak amplitude** (wrist – axilla): 3.8 µV; **Peak to peak amplitude** (wrist – elbow): 6.3 µV; **Peak to peak amplitude** (wrist – axilla): 3.3 µV; **SNCV** (wrist – elbow, 19 cm): 55.9 m/s; **SNCV** (wrist – axilla, 30 cm): 43.5 m/s

Fig. 4 Orthodromic sensory nerve action potentials (*SNAPs*) recorded at the elbow (*upper trace*) and at the axilla (*lower trace*) in a patient (39 years old) with diabetes mellitus with clinical evidence of peripheral neuropathy, stimulation of the wrist

References

1. Mayer RF (1963) Nerve conduction studies in man. Neurology 13:1021–1030
2. Dawson GD (1956) The relative excitability and conduction velocity of sensory and motor nerve fibres in man. J Physiol 131:431–451

Digit IV, V – Wrist, Elbow; Digit IV, V – Wrist

S4

Surface Recording Technique, Orthodromic Study

Original Settings Sensitivity was 10–20 μV/division, and the machine used was a TECA model B. Sweep speed, low-frequency filter, high-frequency filter, duration of pulse, and rate of pulse were not specified.

Position This study was performed in the supine position, with the elbow and fingers slightly flexed.

Recording Following the orthodromic method (Figs. 1 and 2), signals have been register at the wrist (R1) and at the elbow (R2). At the wrist (R1), the active electrode (A) was placed proximally to the distal crease at the wrist, 14 cm proximal to the stimulating cathode. The reference electrode (R) was placed 2 cm proximally to the active electrode [1]. At the elbow (R), the active (A) electrode was placed proximally to the ulnar groove, with a 2–4 cm separation between the proximal cathode (−) and distal anode (+). The authors [1] used a pair of standard 0.6 cm diameter electroencephalograph electrodes mounted on a plastic block for recording sensory responses. Other routine electrodes (like clips) can be used successfully. In the article, ground (G) electrode was showed on the *thenar* muscle, but the authors usually placed the ground electrode over the mid-forearm, and on occasion it was placed between the stimulating cathode and the active electrode to diminish a shock artifact, on the palm of the hand.

Stimulation The ulnar nerve was stimulated 14 cm distally from the wrist to digit IV and digit V (S). Each digit was stimulated separately (Figs. 3 and 4). The authors used moistened pipe cleaner for stimulating the sensory fibers of the ulnar nerve. In digit IV and digit V, the active stimulating electrode (cathode) was applied at the base of the digit (proximal to the recording site); the anode (+) was positioned slightly distal to the distal interphalangeal joint (distal to the recording site). Supramaximal stimulation was used; stimulus in all determinations was at least 30 % greater than that elicited a maximum response.

Measurements Distal peak latency (ms) was measured from the beginning of the shock artifact to the peak of the negative deflection of the evoked sensory nerve action potential (SNAP). Measurement of length was made using a metal tape measure; the nerve length was measured with the arm in the position of stimulation. The finger–wrist segment was measured with the wrist as close as possible to a 180° position. The temperature was not controlled (studies were done at room temperature). The authors obtained normal values from 48 normal (Table 1) subjects (separate digit IV and digit V values were not reported).

Comment

For Melvin et al. [1] the ulnar fastest sensory fibers tended to be faster than the ulnar fastest motor fibers. In healthy subjects, orthodromic sensory responses were recorded in 98 % of trials; despite this high degree of success in obtaining normal values, the authors observed that greater difficulties were encountered in obtaining SNAPs in those patients with peripheral nerve disease, in which an ulnar nerve conduction velocity below 45 m/s irrespective of age was frequently observed.

G. Gentili, M. Di Napoli, *The Ulnar Nerve*, DOI 10.1007/978-3-319-43292-2_4

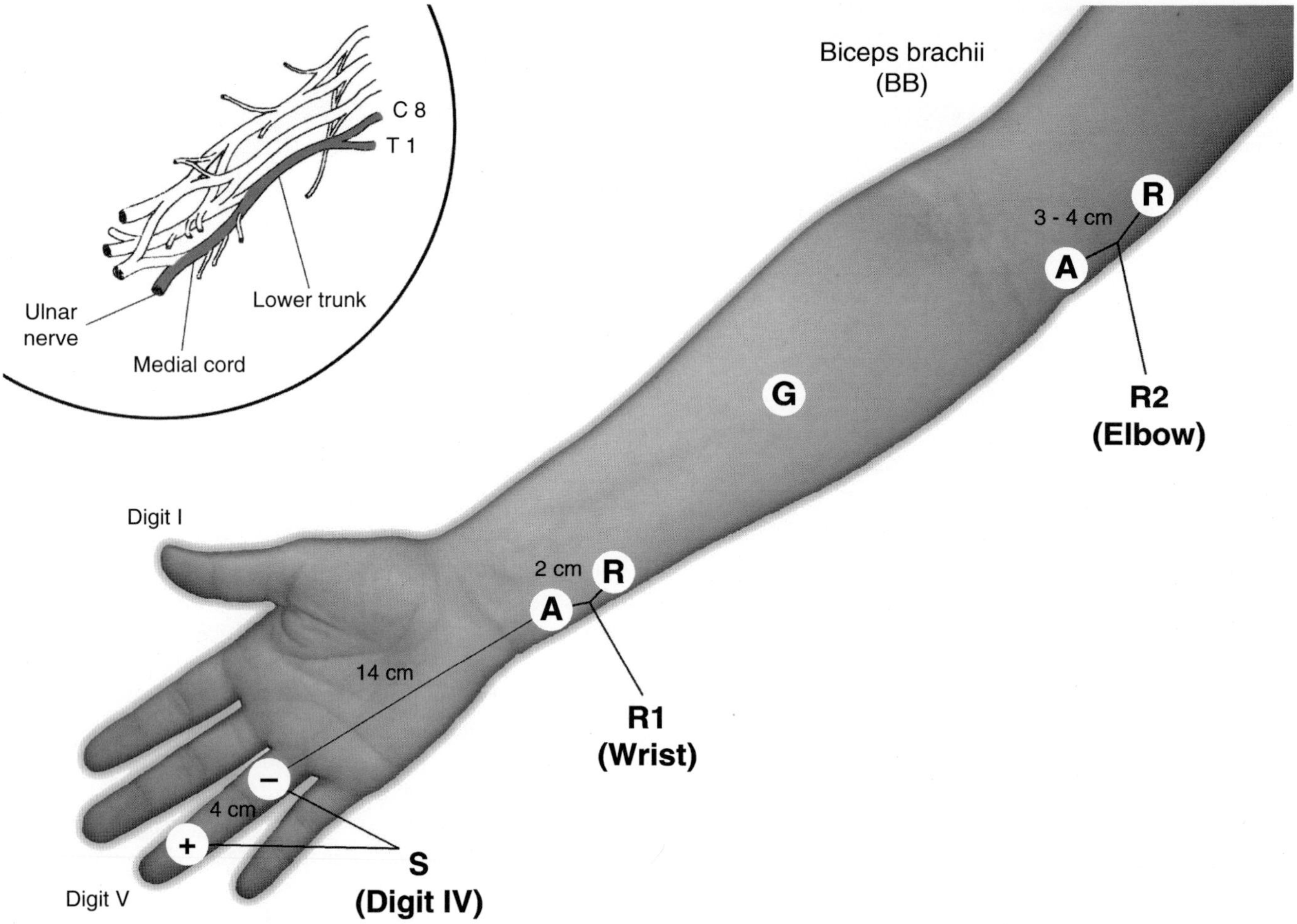

Typical waveform (digit IV – wrist, elbow):

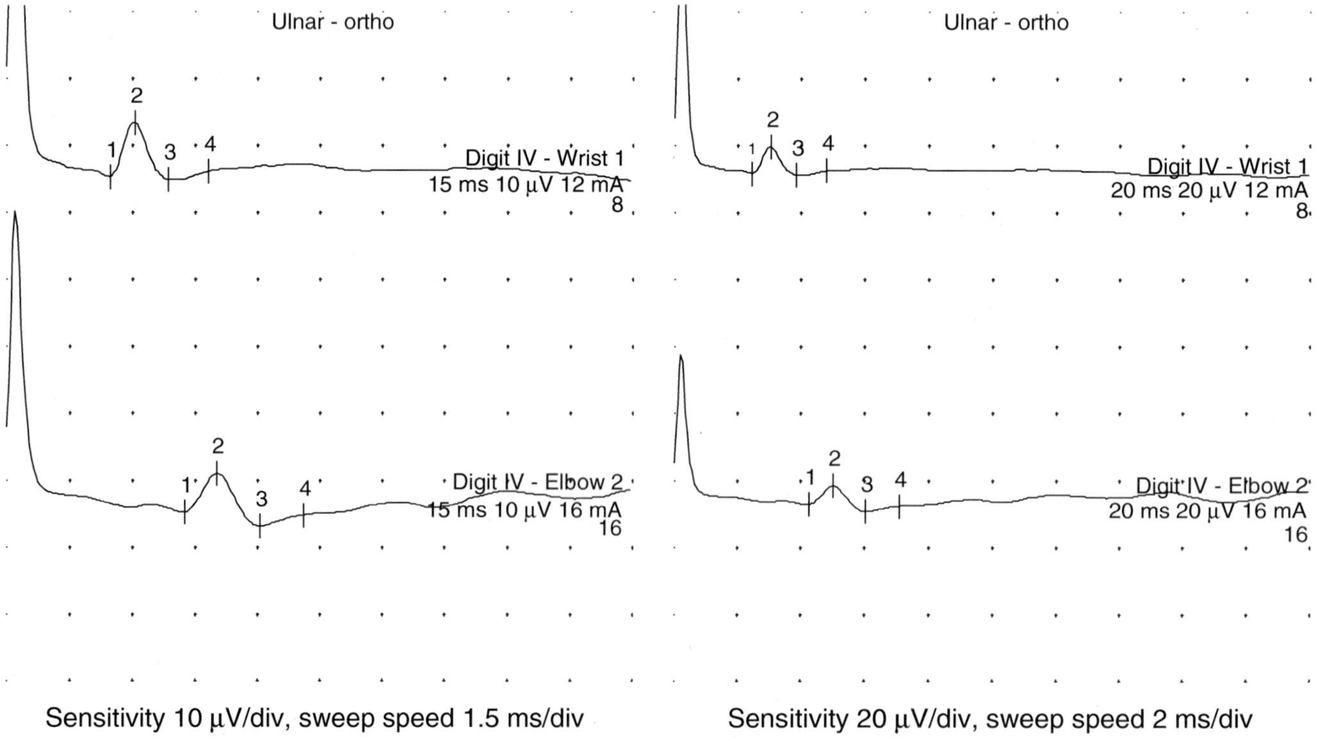

Fig. 1 Orthodromic sensory nerve action potentials (*SNAPs*) recorded at the wrist (*upper trace*) and at the elbow (*lower trace*), stimulation of digit IV

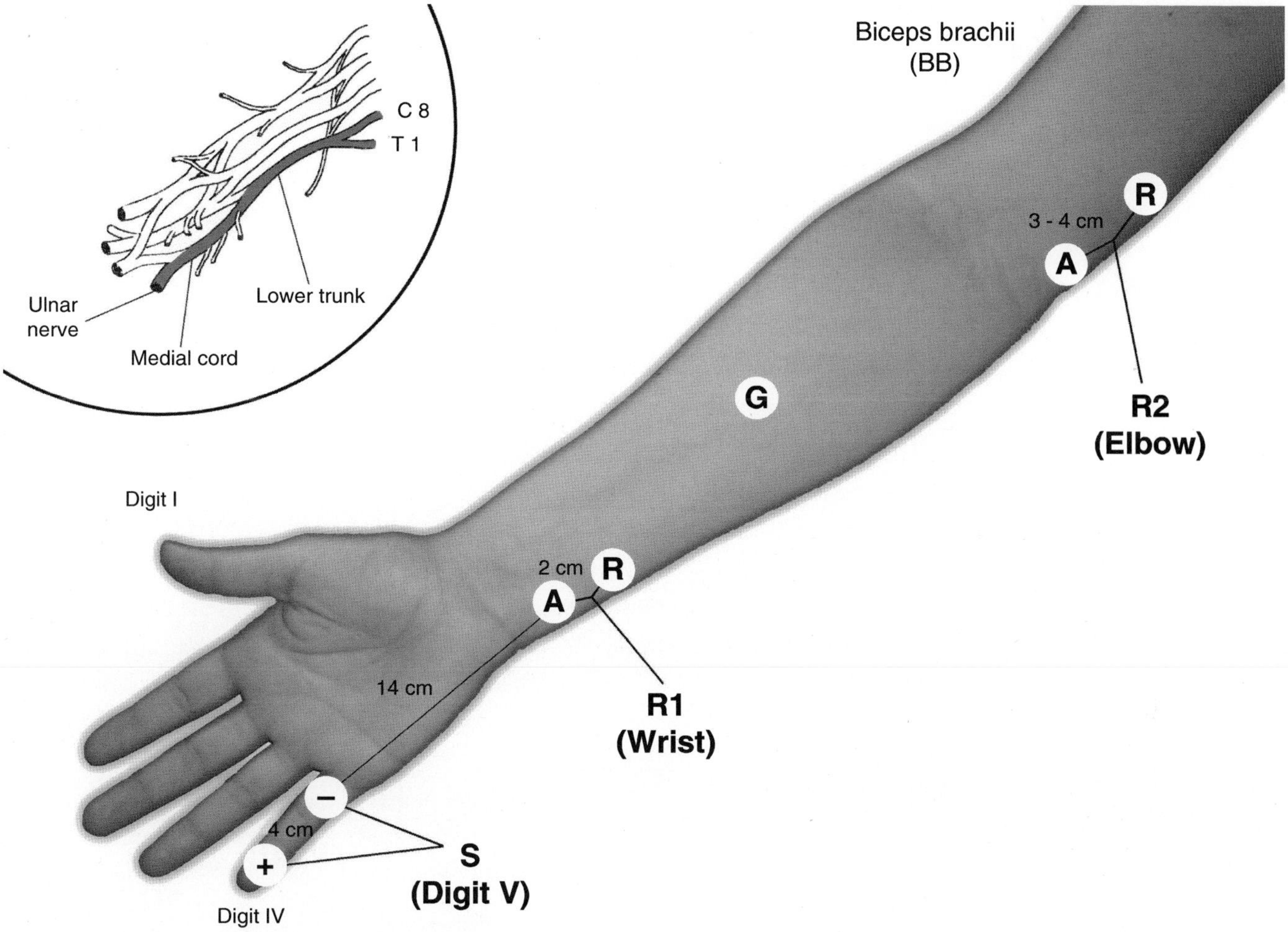

Typical waveform (digit V – wrist, elbow):

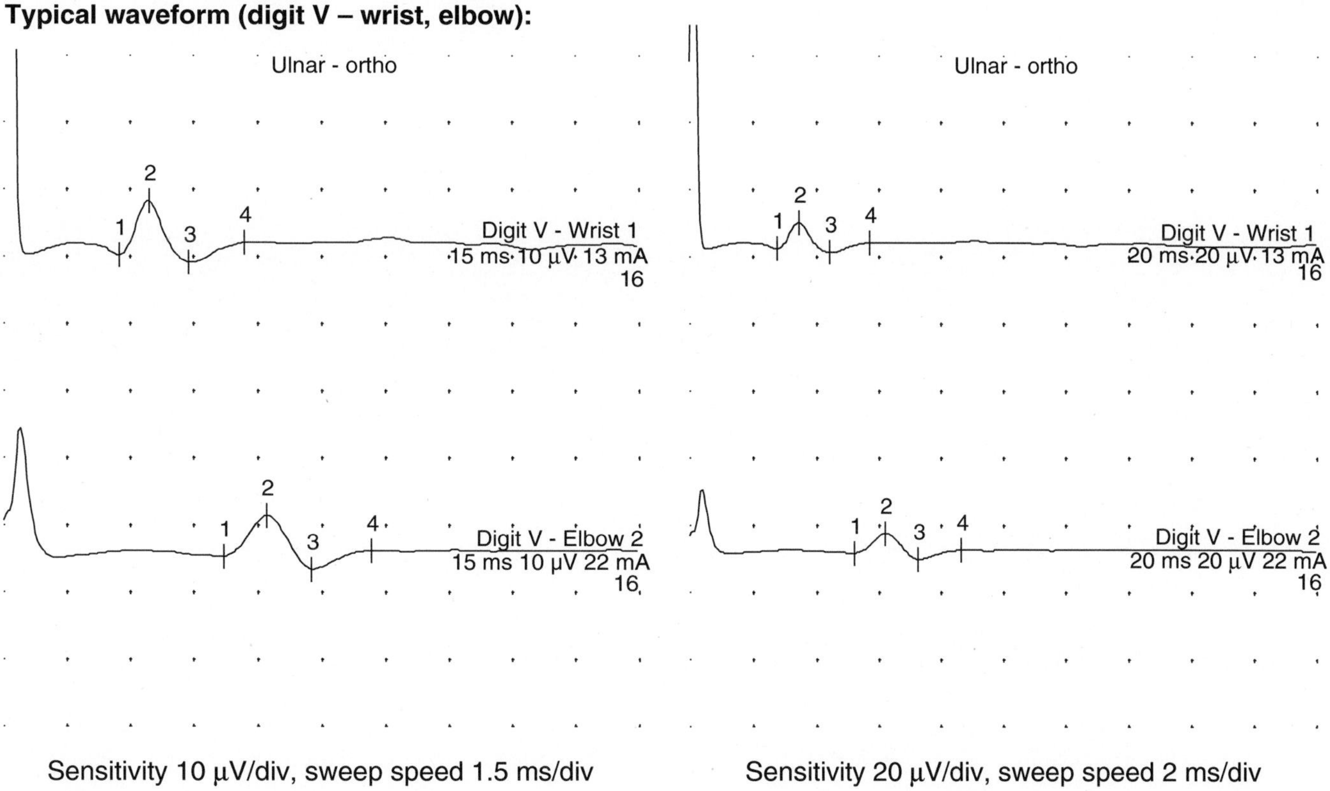

Fig. 2 Orthodromic sensory nerve action potentials (*SNAPs*) recorded at the wrist (*upper trace*) and at the elbow (*lower trace*), stimulation of digit V

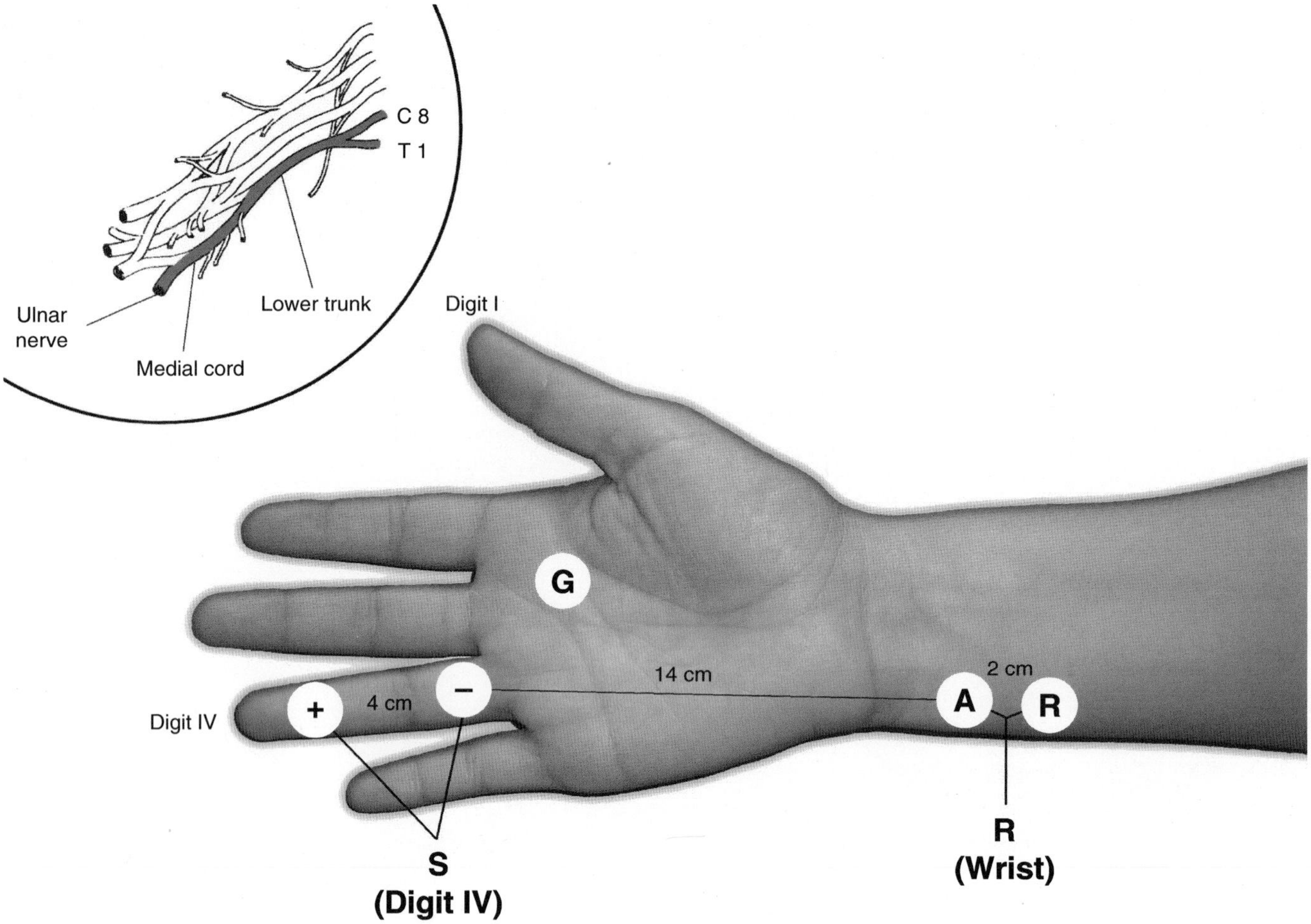

Typical waveform (digit IV – wrist):

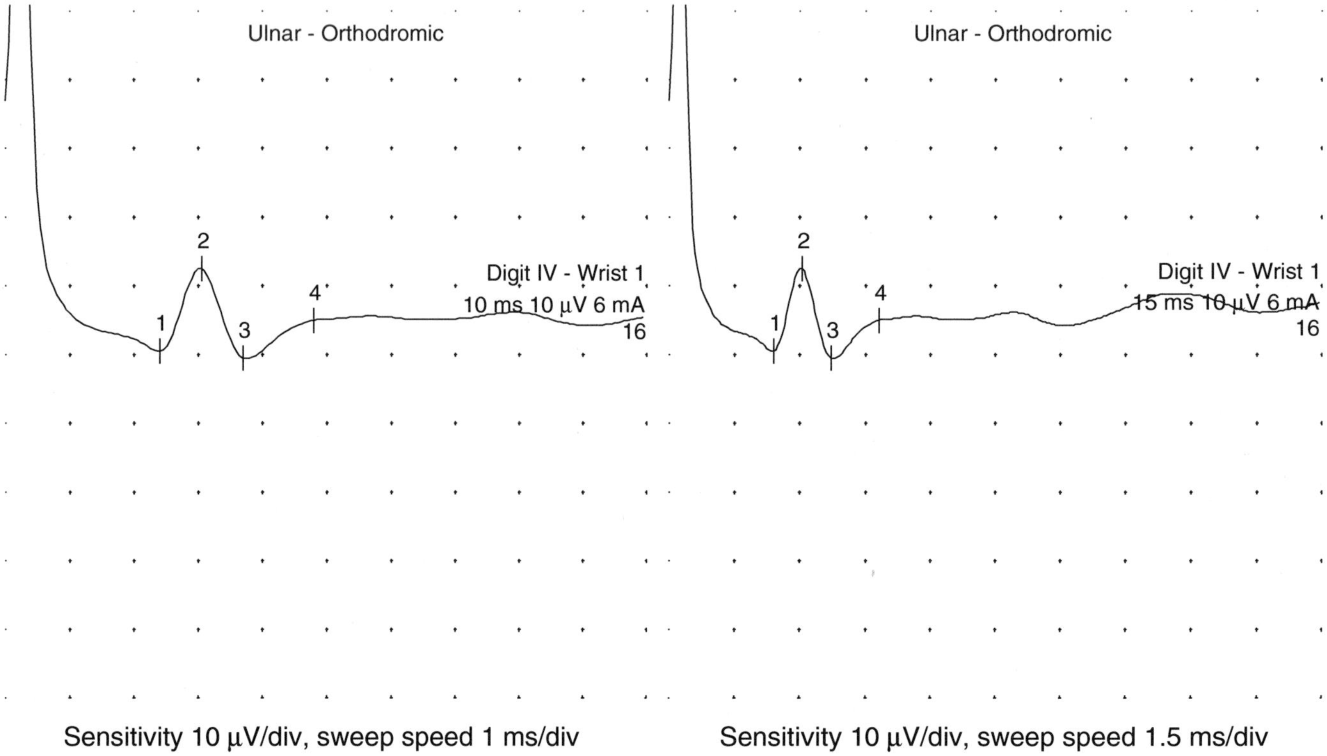

Fig. 3 Orthodromic sensory nerve action potential (*SNAP*) recorded at the wrist, stimulation of digit IV

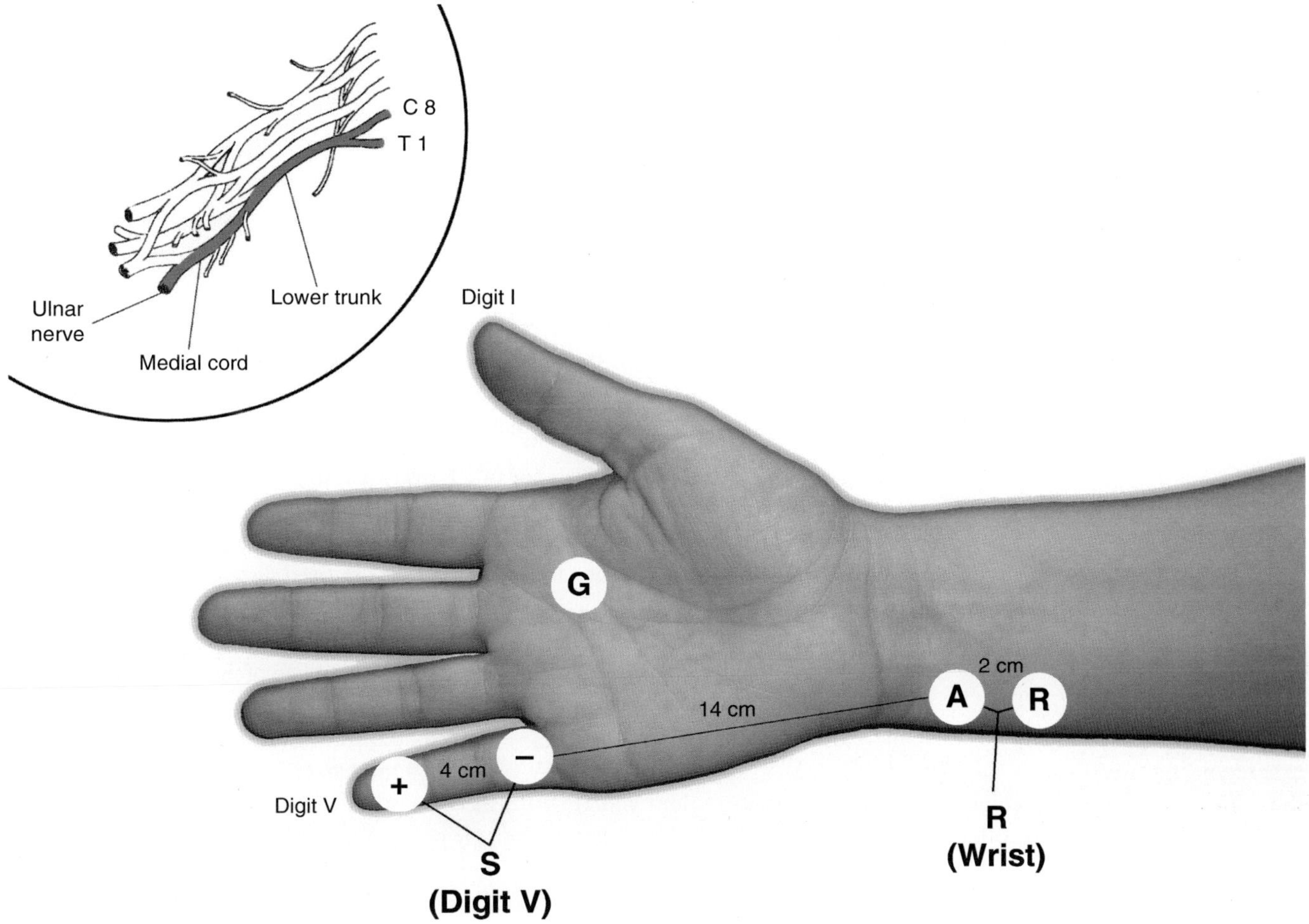

Typical waveform (digit V – wrist):

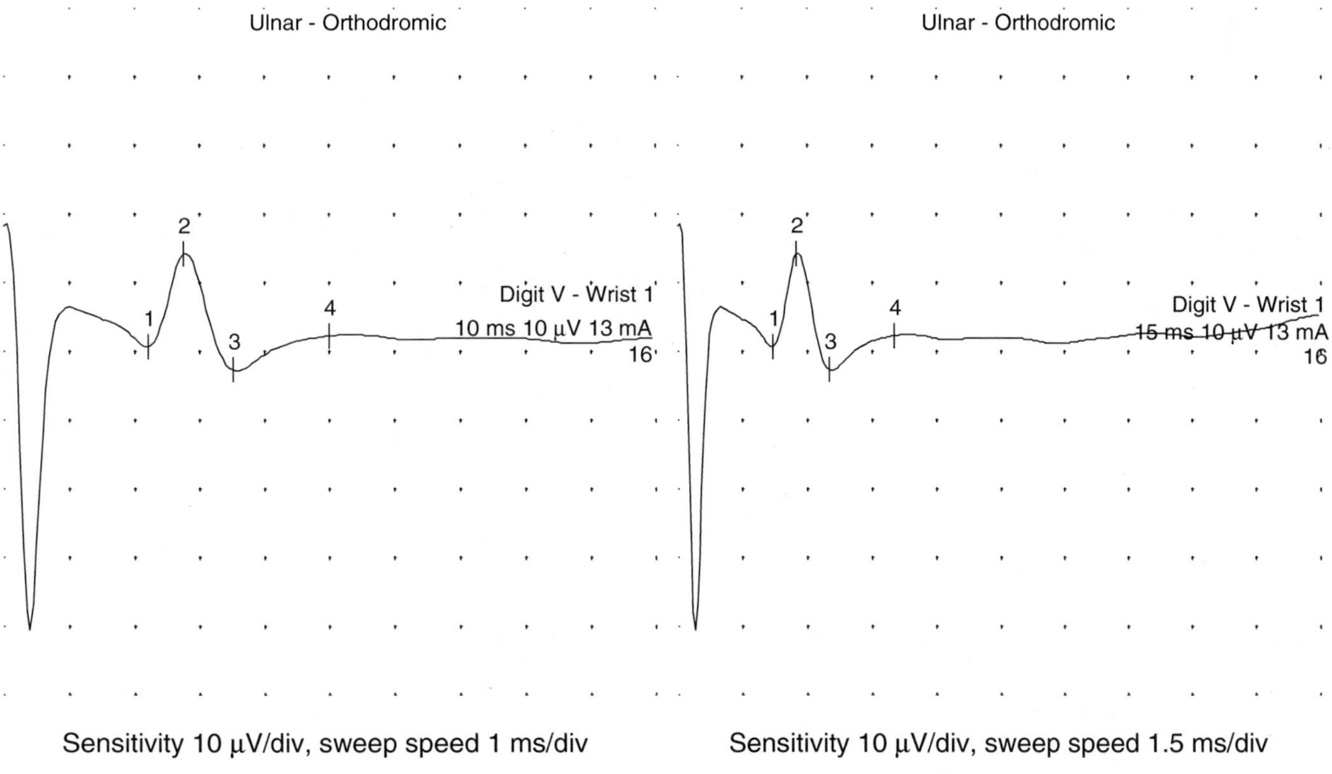

Fig. 4 Orthodromic sensory nerve action potential (*SNAP*) recorded at the wrist, stimulation of digit V

Table 1 Normal values [1]

	Mean ± SD
Digit–wrist, peak latency (ms)–20 subjects	3.0 ± 0.20
Digit–wrist, SNCV (m/s)–age range 16–24 (21 subjects)	57.0 ± 4.3
Digit–wrist, SNCV (m/s)–age range 25–39 (14 subjects)	56.3 ± 4.1
Digit–wrist, SNCV (m/s)–age range 40–69 (12 subjects)	56.8 ± 4.3
All subjects	56.7 ± 4.2

Table 2 Normal values [4]

	Mean ± SD	Range
Digit V–wrist, onset latency (ms)	1.9 ± 0.3	1.7–2.5
Digit V–wrist, amplitude (μV)	21 ± 4.3	8–45

The authors suggested to investigate both distal and proximal sensory conductions of the ulnar nerve; it would be helpful to be able to show differences in nerve conduction velocity between the proximal (wrist–elbow) and distal (digit–wrist) segments. Likewise, metabolic disorders such as diabetes may selectively involve sensory fibers (Fig. 6).

In the same year, Liberson et al. [2] compared two methods for determining sensory conduction velocity: (a) the "antidromic" technique by wrist and elbow stimulation and recording SNAPs over the skin of the digit V and (b) the "direct neurographic technique" (i.e., the orthodromic method) by distal stimulation (S) to digit V and proximal recording at the wrist (R1) and at the elbow (R2). The authors recorded SNAPs at the wrist by surface electrodes, but they found SNAPs recording at the elbow somewhat difficult because of their low amplitude, and needle electrodes or an averaging technique were required. They studied 20 upper extremities of 20 children from 5 to 16 years old, and a mean digit V–wrist SNCV of 63.4 ± 7.2 m/s was found.

In 1967, Johnson and Melvin [3] measured distal latency in 44 healthy control subjects (10–19 years, 1 subject; 20–29 years, 11 subjects; 30–39 years, 8 subjects; 40–49 years, 10 subjects; 50–59 years, 9 subjects; 60–69 years, 3 subjects; 70–79 years, 2 subjects). They found a mean orthodromic digit V–wrist latency of 2.6 ms (±0.4 SD). For the authors, a difference in hand size can account for considerable variation in sensory latency in healthy subjects, and in an attempt to reduce this variability, they adopted a 14 cm fixed distance between recording and active stimulating electrodes. They also observed, in clinical determinations of sensory latencies, other variables such as the temperature of the hand and age of the examined subject to be confounders in the accurate latency measurement.

Bhala [4] performed the orthodromic method (Table 2) in 61 healthy subjects (9 men and 52 women, age range 19–69 years). With the exception of one, all the healthy subjects were less than 50 years of age. The onset latency (ms) of the sensory response was measured from the stimulus artifact to the initial deflection of the SNAP. The amplitude of the SNAP was measured from peak to peak.

The author [4] used a TECA model TE-4 electromyograph, and ring electrodes were used for the digit V stimulations. Since the latency is dependent upon the distance between the points of stimulation and recording, the position of placement of the electrodes was standardized. The active recording electrode was placed at about 1–1.5 cm proximal to the pisiform bone, and the distance between the distal stimulating (cathode) and the proximal recording electrode was approximately 10.5 cm in the majority of the hands (Fig. 5).

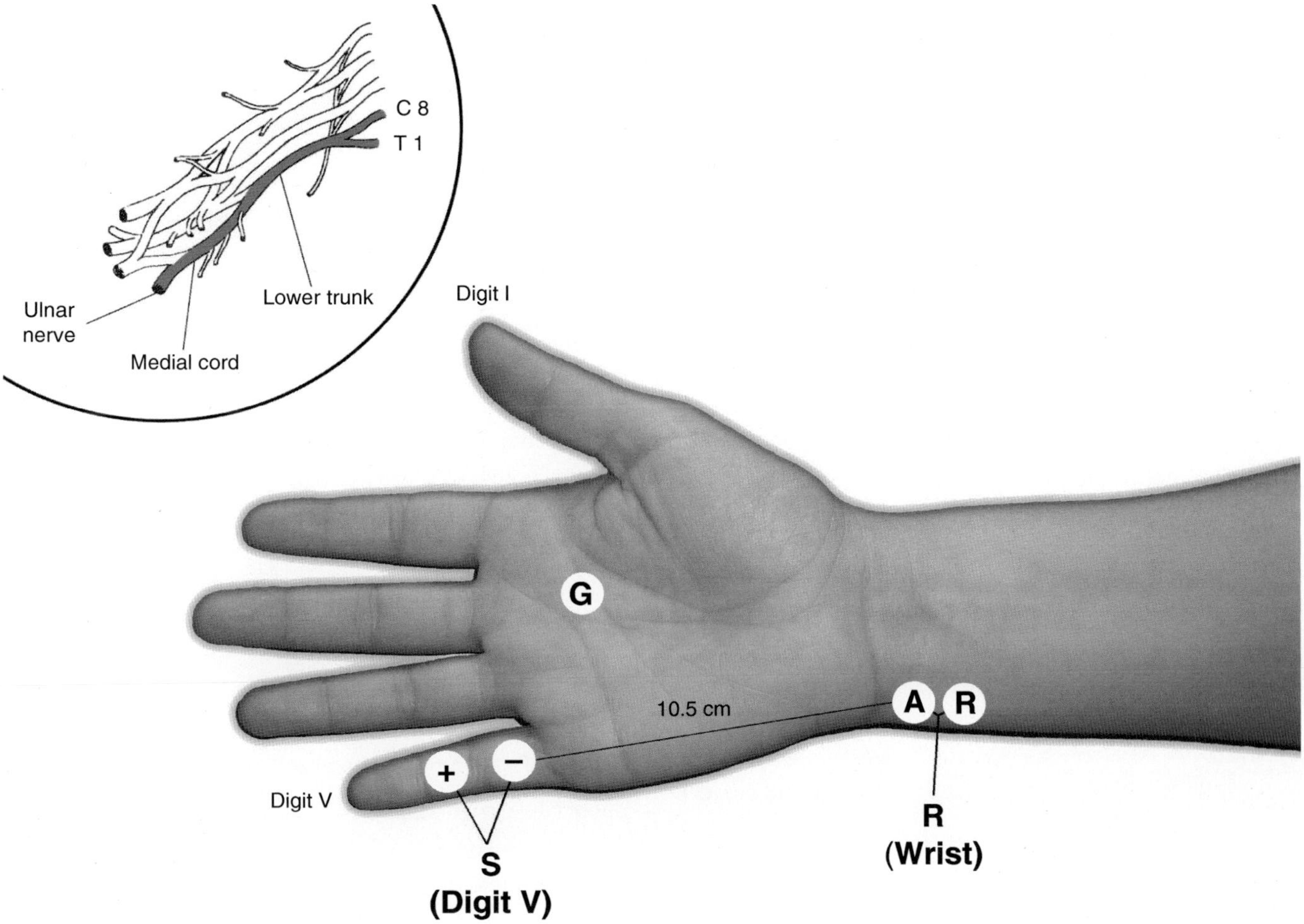

Fig. 5 Orthodromic method by Bhala [4]

Comment

For Bhala [4] the latency of the SNAP with antidromic and orthodromic technique was essentially the same. The amplitude, however, was considerably reduced (from approximately 30 to 55 %) when the orthodromic technique was used for recording. Also this technique was found to be more painful to electrical stimulation. The author also examined 78 patients with suspected ulnar nerve entrapment at the elbow (39 men and 39 women). The majority of the patients (75.6 %) were less than 50 years of age. Of the 40 patients who had motor nerve conduction velocity (MNCV) of less than 45 m/s in the across elbow segment of the ulnar nerve, 29 (72.5 %) showed absent or abnormal ulnar sensory response in the digit V. In the majority of the cases, the sensory response showed decreased amplitude (less than 8.0 μV) but normal latency (less than 2.5 ms) of the SNAP. Prolonged latency was not seen in any of the patients of his study. It is suggested that the amplitude of the ulnar sensory response should be routinely compared to that of the median sensory response in the same hand and to the ulnar response in the opposite hand whenever one suspects a compressive lesion of the ulnar nerve in the arm or hand. Pathological waveforms and values in a patient with a sensory polyneuropathy and ulnar neuropathy at the elbow (UNE) are reported (Figs. 6 and 7, respectively).

Pathological waveform (digit IV–wrist, elbow):

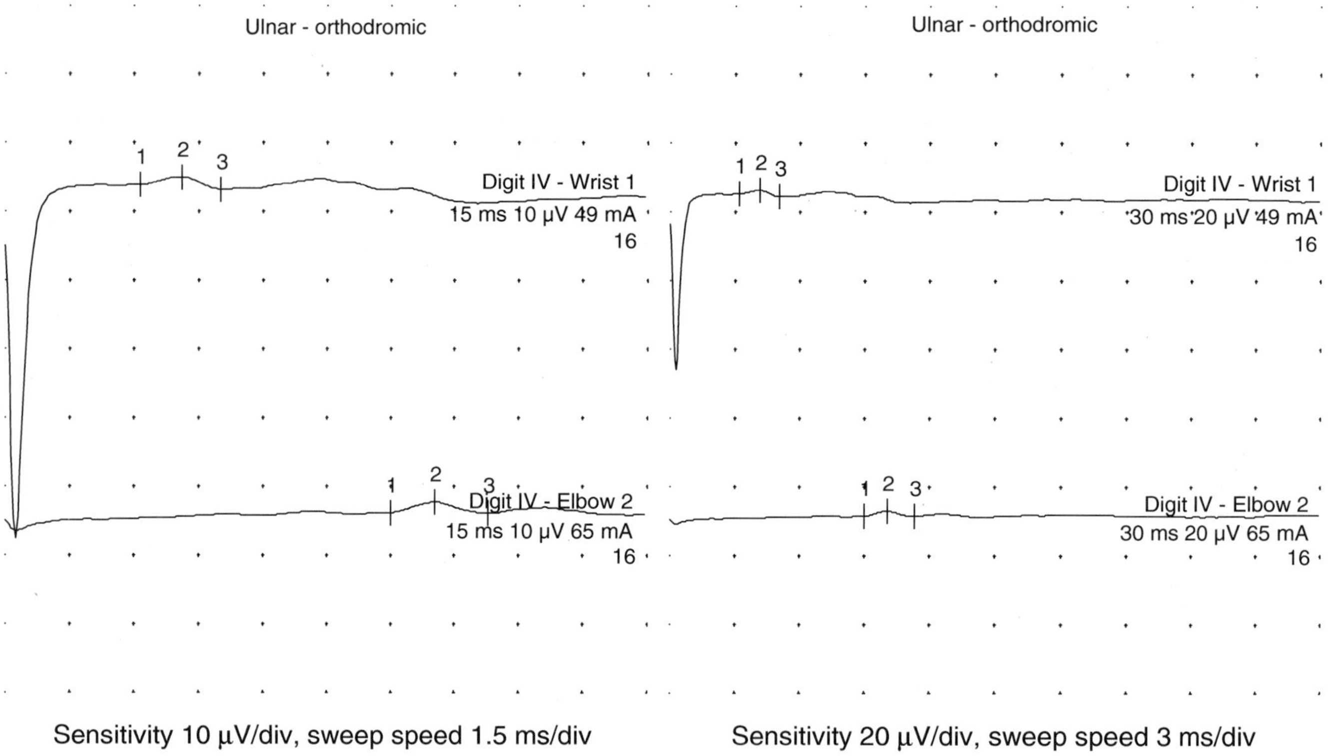

Onset latency (digit IV – wrist): 3.15 ms; **Onset latency** (digit IV – elbow): 9.0 ms; **Peak latency** (digit IV – wrist): 3.9 ms; **Peak latency** (digit IV – elbow): 9.85 ms; **Onset to peak amplitude** (digit IV – wrist): 1.2 μV; **Onset to peak amplitude** (digit IV – elbow): 1.6 μV; **Peak to peak amplitude** (digit IV – wrist): 1.8 μV; **Peak to peak amplitude** (digit IV – elbow): 1.7 μV; **SNCV** (digit IV – wrist, 14 cm distance): 44.4 m/s; **SNCV** (wrist – elbow, 22 cm distance): 37.6 m/s

Fig. 6 Orthodromic sensory nerve action potential (*SNAP*) recorded at the wrist (*upper trace*) and at the elbow (*lower trace*) in a sensory polyneuropathy patient, stimulation of digit IV

Pathological waveform (digit V – wrist):

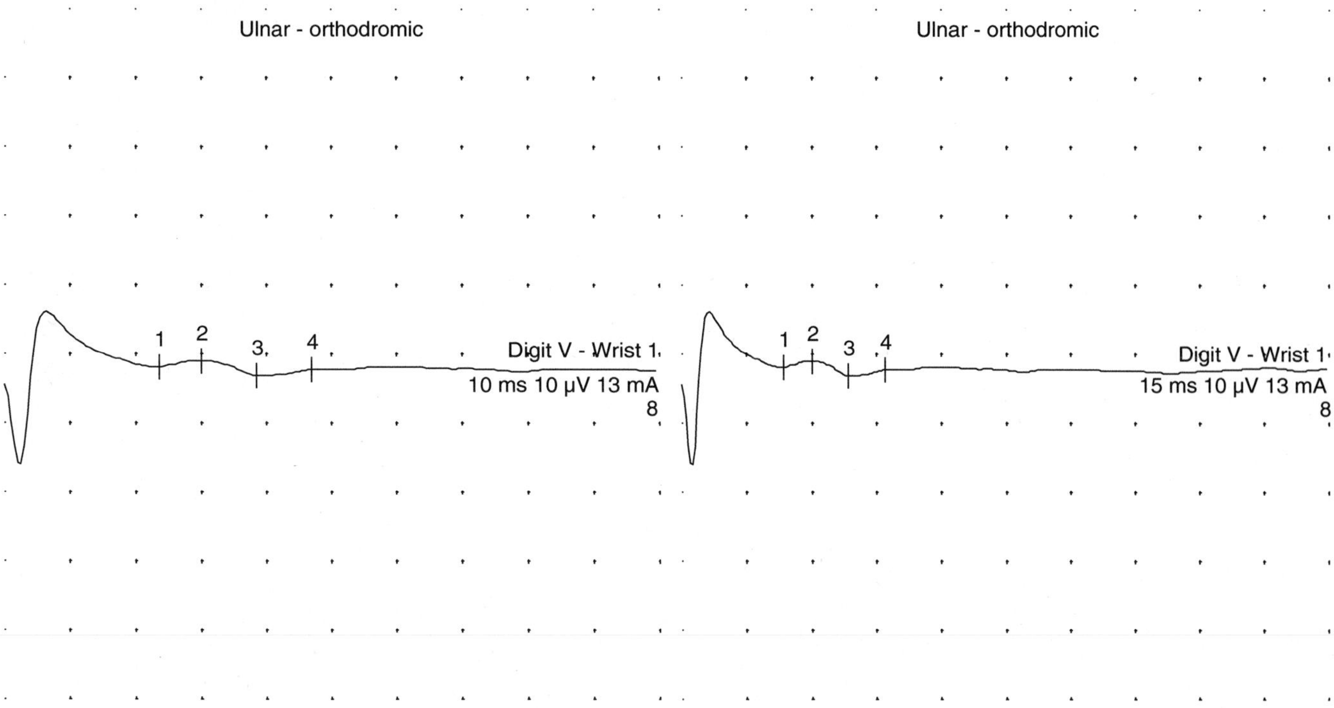

Sensitivity 10 µV/div, sweep speed 1 ms/div   Sensitivity 10 µV/div, sweep speed 1.5 ms/div

Onset latency (digit V – wrist): 2.35 ms; **Peak latency** (digit V – wrist): 3.0 ms; **Onset to peak amplitude** (digit V – wrist): 0.98 µV; **Peak to peak amplitude** (digit V – wrist): 2.2 µV; **SNCV** (digit V – wrist, 14 cm distance): 59.6 m/s;

Fig. 7 Orthodromic sensory nerve action potential (*SNAP*) recorded at the wrist (*upper trace*) in a patient with UNE, stimulation of digit V

References

1. Melvin JL, Harris DH, Johnson EW (1966) Sensory and motor conduction velocities in the ulnar and median nerves. Arch Phys Med Rehabil 47:511–519
2. Liberson WT, Gratzer M, Zalis A et al (1966) Comparison of conduction velocities of motor and sensory fibers determined by different methods. Arch Phys Med Rehabil 47:17–23
3. Johnson EW, Melvin JL (1967) Sensory conduction studies of median and ulnar nerves. Arch Phys Med Rehabil 48:25–30
4. Bhala RP (1976) Electrodiagnosis of ulnar nerve lesions at the elbow. Arch Phys Med Rehabil 57:206–212

Wrist, Elbow – Digit IV, V; Wrist – Digit IV, V

S5

Surface Recording Technique, Antidromic Study

Original Settings Sensitivity was 10–20 μV/division, and the machine used was a TECA model B. Sweep speed, low-frequency filter, high-frequency filter, duration of pulse, and rate of pulse were not specified.

Position This study was performed in the supine position, with the elbow and fingers slightly flexed.

Recording Following the antidromic method [1], signals were recorded 14 cm distally from the wrist to digit IV and digit V (Figs. 1 and 2). Each recording was made separately. The active electrode (A) was placed to the base of the digit, and the reference (R) was placed 4 cm distally to the active electrode, slightly distal to the distal interphalangeal joint. The ground (G) electrode was usually placed over the mid-forearm, and on occasion, it was placed between the stimulating cathode and the active electrode to diminish a shock artifact, on the palm of the hand (the figure shows the ground electrode placed on the palm). The authors used moistened pipe cleaner for recording from the distal sensory fibers of the ulnar nerve.

Stimulation For both digit IV and digit V recordings, the ulnar nerve was stimulated at the wrist (S1) and the elbow (S2). At the wrist (S1), the active stimulating electrode (cathode) was applied 14 cm proximal to the active electrode at the base of the digit, over the ulnar nerve at the wrist, near the tendon of the *flexor carpi ulnaris* (FCU) – ideally proximal to the distal wrist crease. The anode (+) was proximal. Distal stimulation (S1) determined distal evoked sensory nerve action potentials (SNAPs), while proximal stimulation at the elbow (S2), just above the ulnar groove, allowed the determination of the forearm mixed nerve conduction velocity. In the case of stimulation at the elbow (S2), the authors suggested 2–4 cm separation between the cathode (−) and anode (+), using a pair of standard 0.6 cm-diameter electroencephalograph electrodes mounted on a plastic block for stimulating the nerve. Sensory antidromic nerve conduction study to digit IV and digit V can also be performed by wrist stimulation alone (Figs. 3 and 4). Supramaximal stimulation was used; stimulus in all determinations was at least 30 % greater than that which elicited a maximum response.

Measurements Distal peak latency (ms) was measured from the beginning of the shock artifact to the peak of the negative deflection of the SNAP. Measurement of length was made using a metal tape measure; the nerve length was measured with the arm in the position of stimulation. The finger–wrist segment was measured with the wrist as close as possible to a 180° position. Sensory nerve conduction velocity (SNCV) was calculated in the conventional way and measured meter per second (m/s). The temperature was not controlled (studies were done at room temperature). The authors studied 48 healthy subjects (Table 1); separate digit IV and digit V values were not reported.

Comment

Melvin et al. [1] studied a group of 48 healthy subjects, and antidromic sensory studies were performed in 36 of them. They described no significant differences between antidromic and orthodromic conduction velocities, although the antidromic velocities were slower and required less time when done in conjunction with a motor latency and were usually somewhat easier to obtain. They did not specify sensory digit IV and digit V values. It can be useful if active recording electrode is placed slightly distal to the base of the digit in order to avoid volume-conducted muscle response from the activated adjacent hand intrinsic muscles innervated by the ulnar nerve. The authors did not perform this nerve conduction technique on patients, and pathological values were not reported.

G. Gentili, M. Di Napoli, *The Ulnar Nerve*, DOI 10.1007/978-3-319-43292-2_5

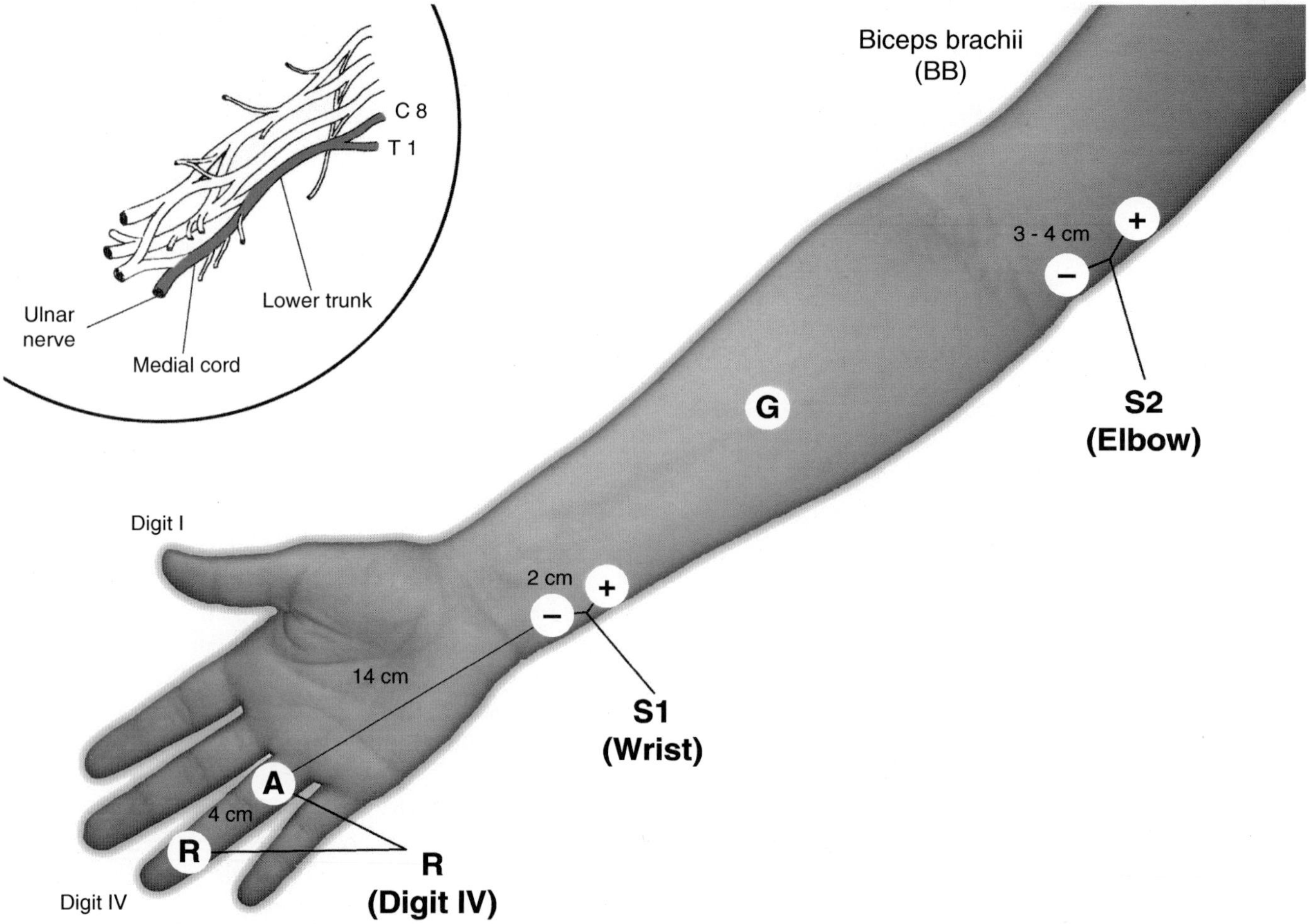

Typical waveform (wrist, elbow – digit IV):

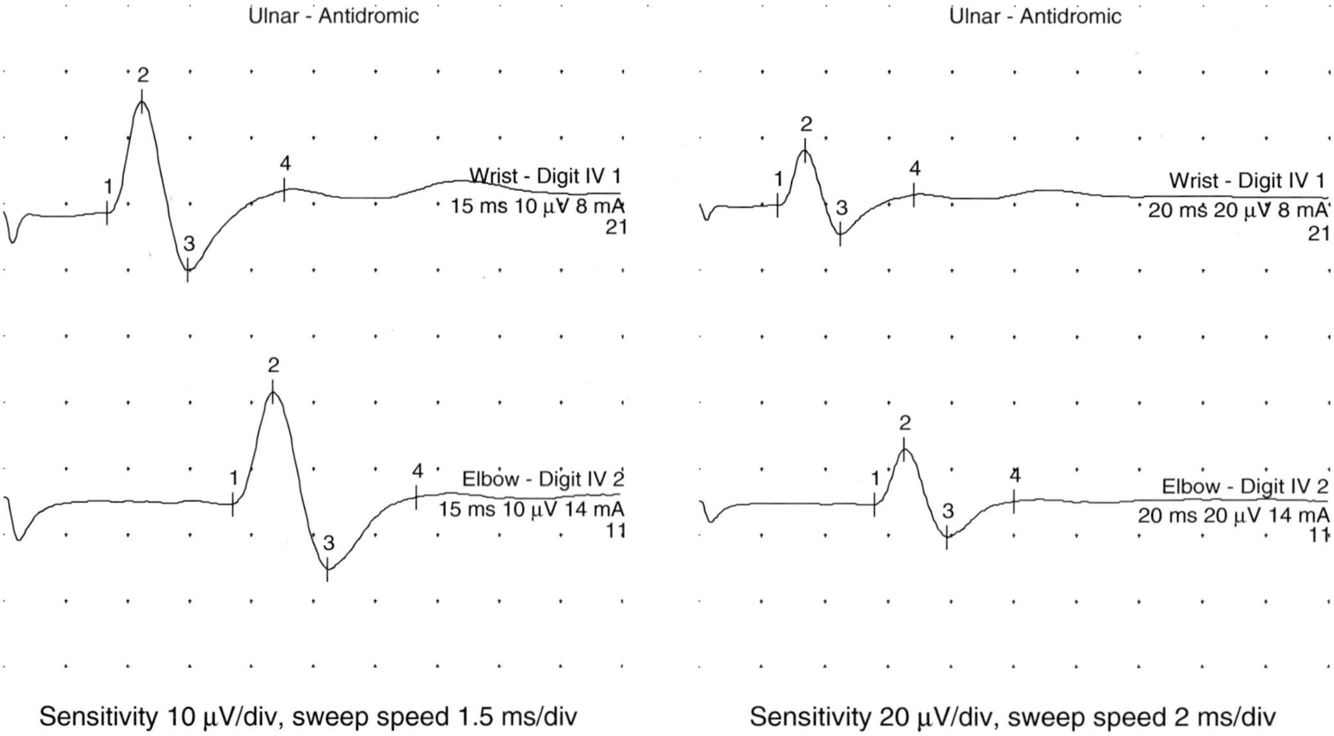

Fig. 1 Antidromic sensory nerve action potentials (*SNAPs*) recorded to digit IV, stimulation of the wrist (*upper trace*) and of the elbow (*lower trace*)

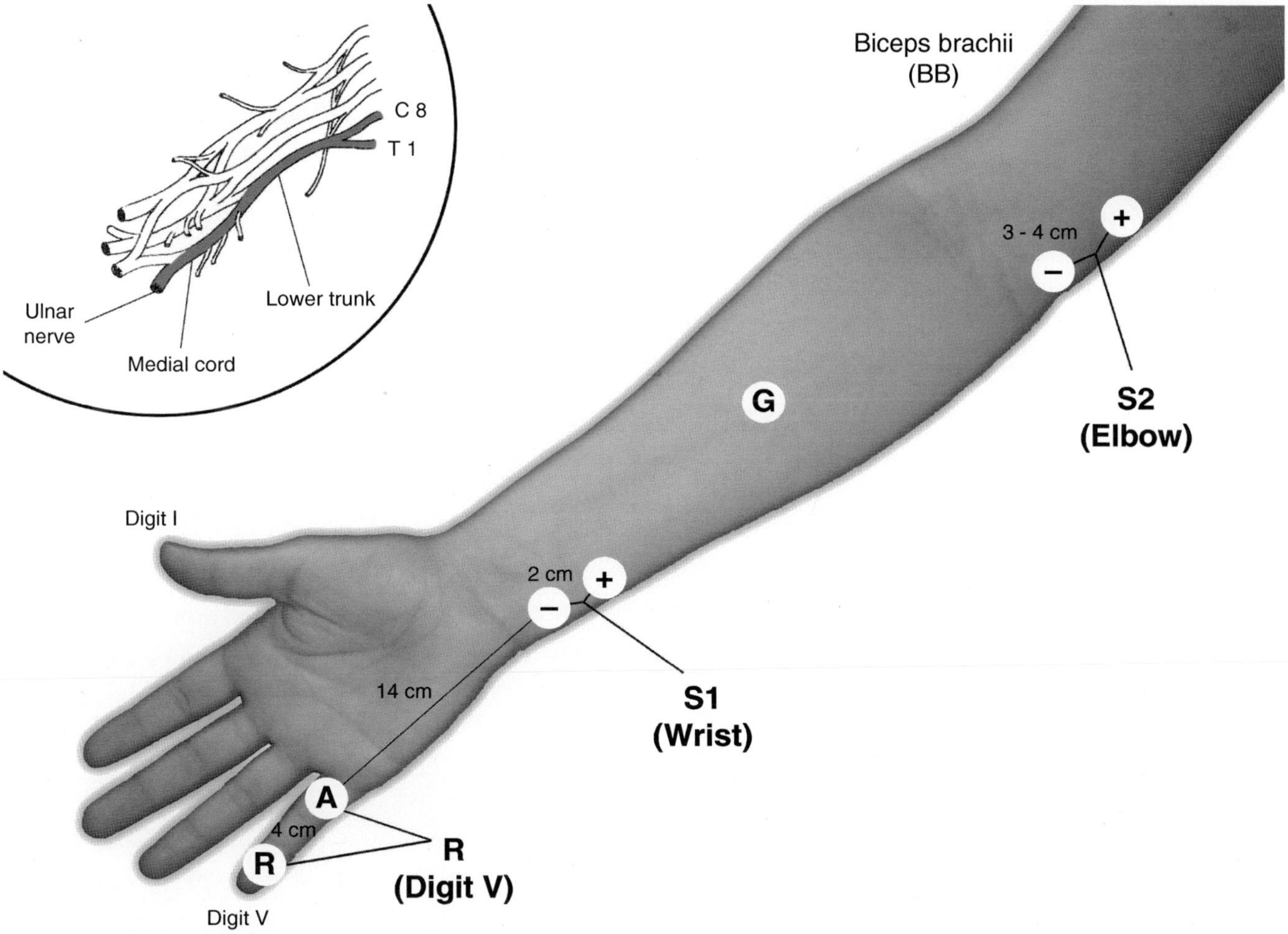

Typical waveform (wrist, elbow – digit V):

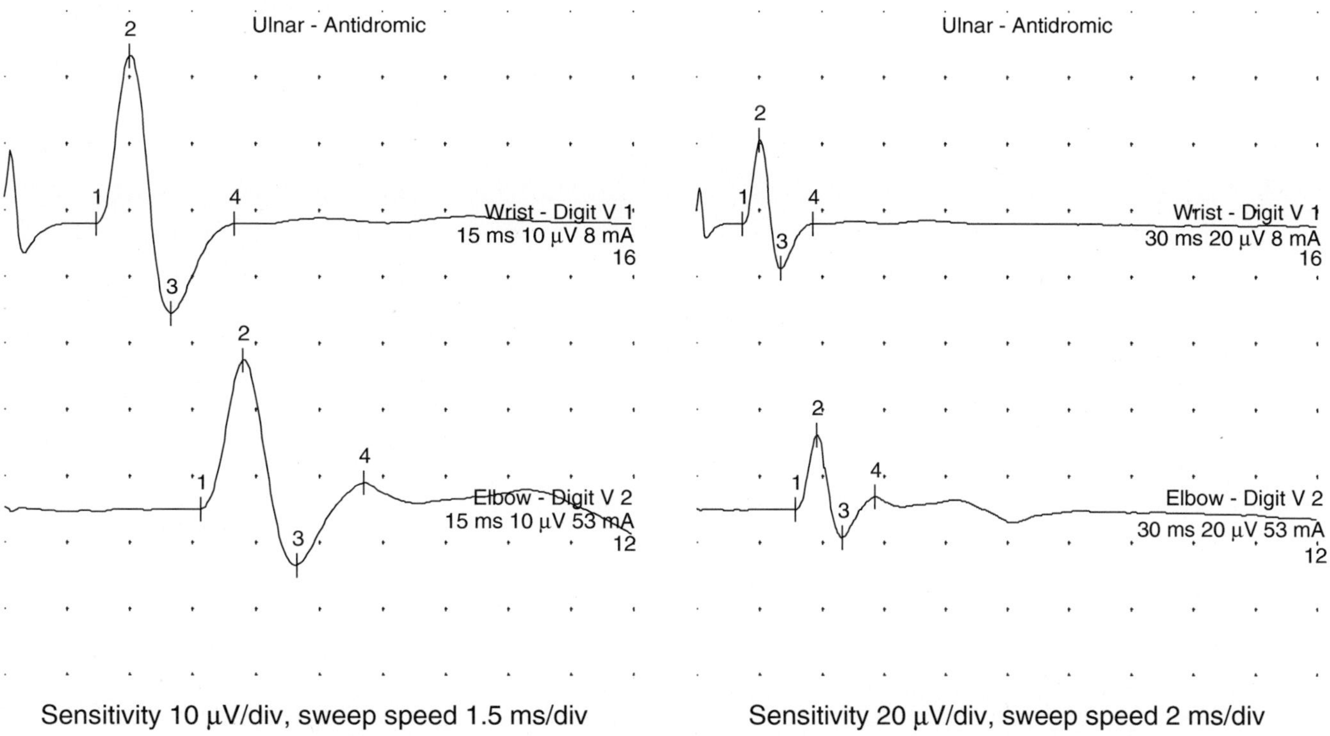

Fig. 2 Antidromic sensory nerve action potentials (*SNAPs*) recorded to digit V, stimulation of the wrist (*upper trace*) and of the elbow (*lower trace*)

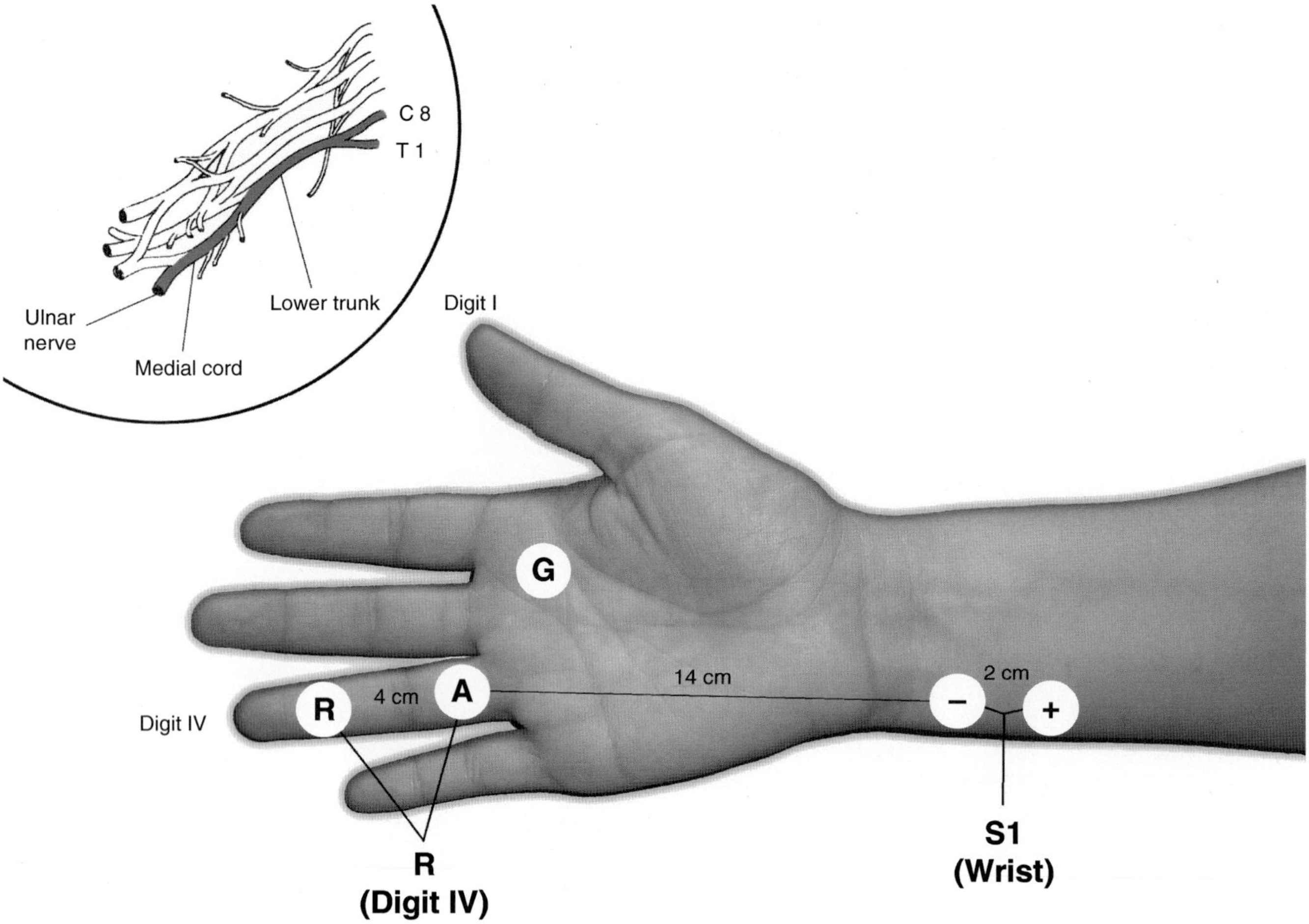

Typical waveform (wrist – digit IV):

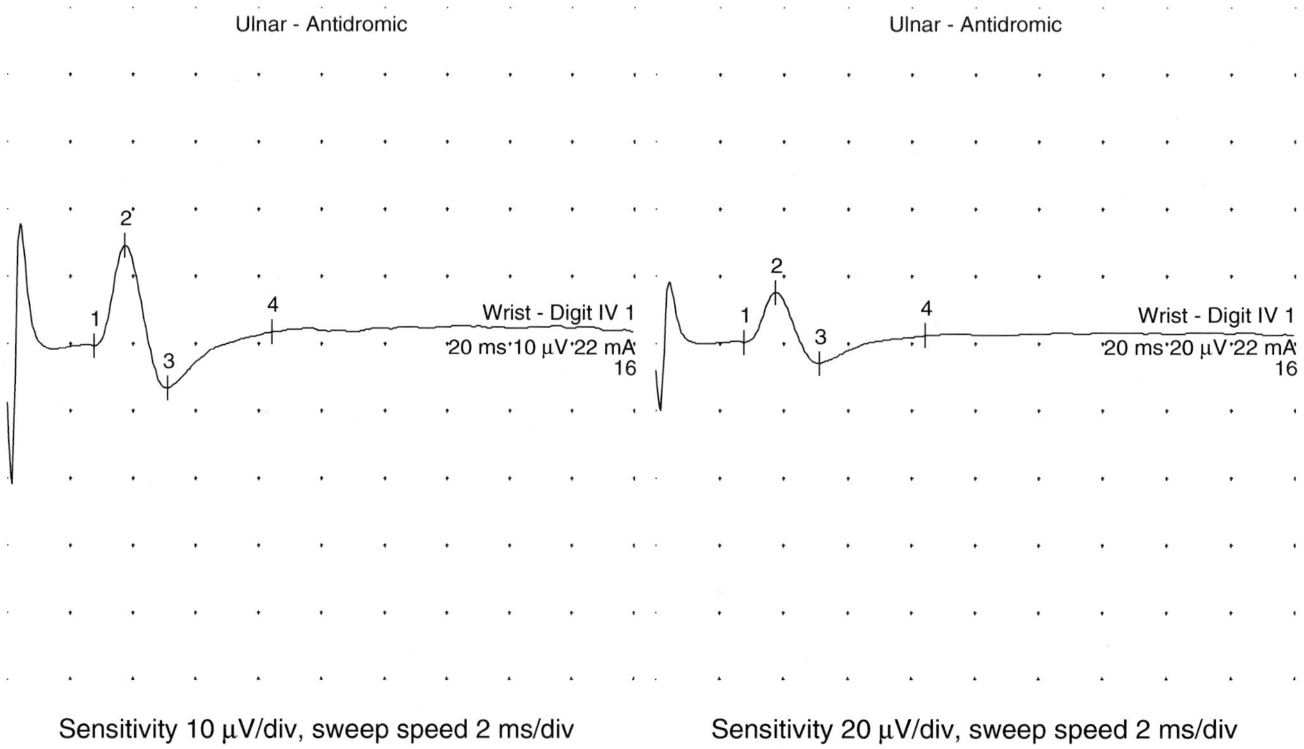

Fig. 3 Antidromic sensory nerve action potential (*SNAP*) recorded to digit IV, stimulation of the wrist

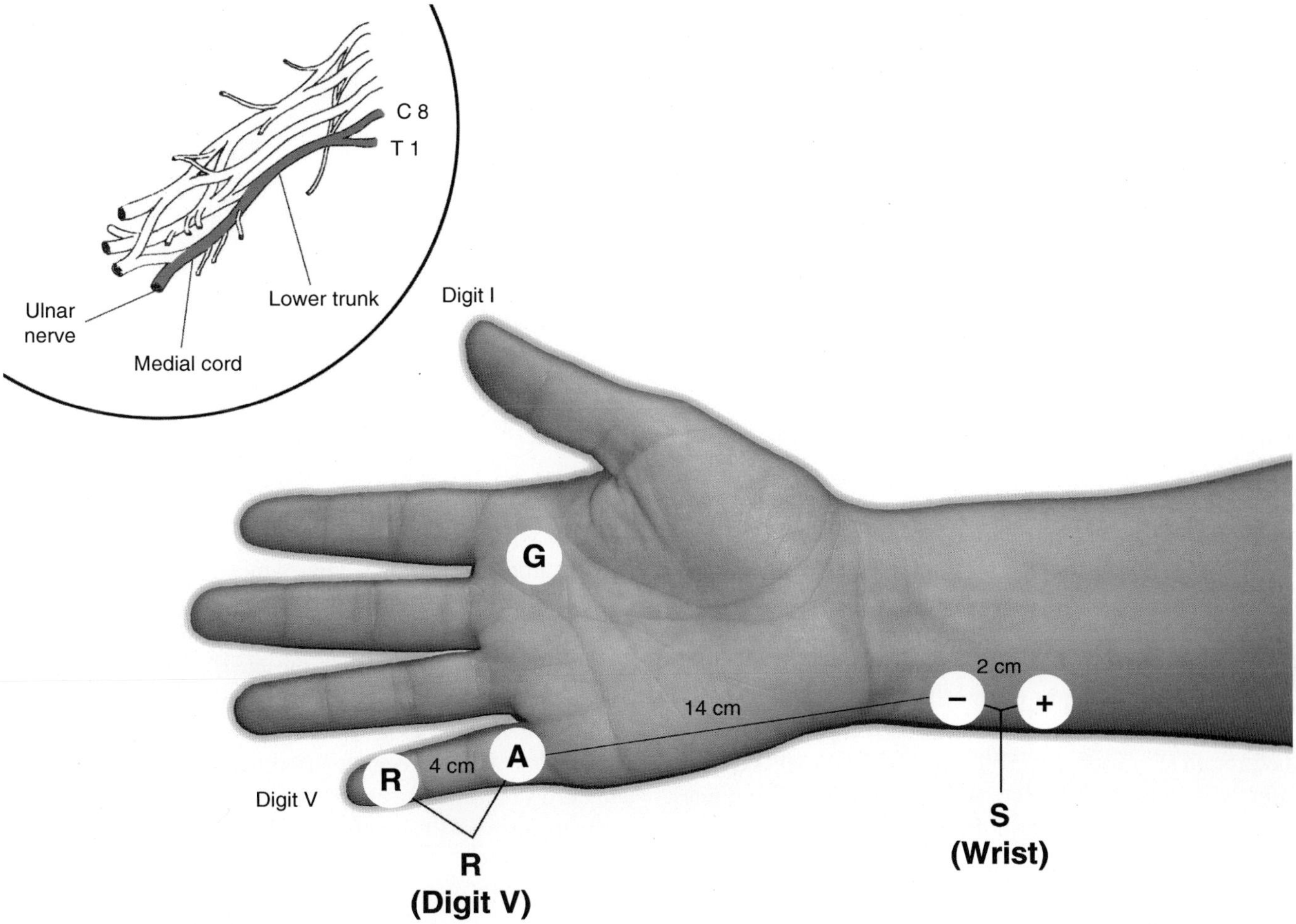

Typical waveform (wrist – digit V):

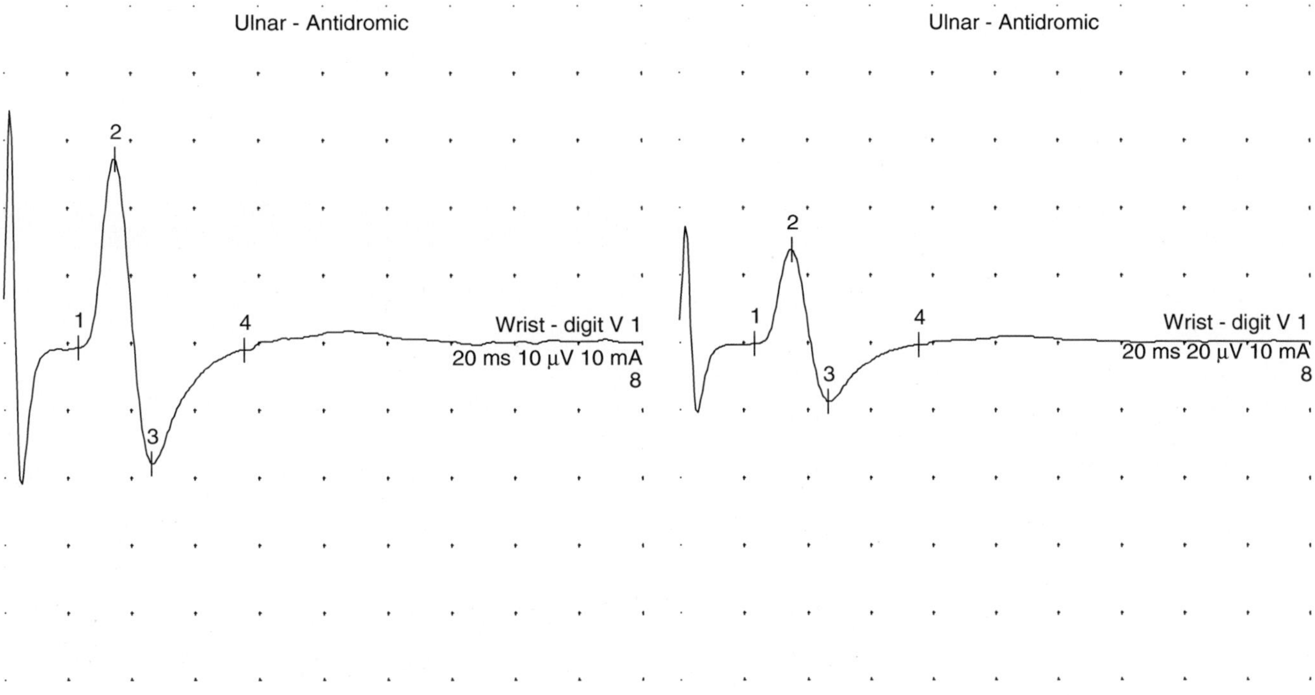

Sensitivity 10 µV/div, sweep speed 1.5 ms/div

Sensitivity 20 µV/div, sweep speed 2 ms/div

Fig. 4 Antidromic sensory nerve action potential (*SNAP*) recorded to digit V, stimulation of the wrist

Table 1 Normal values [1]

	Mean ± SD
Digit–wrist, peak latency (ms) – 20 subjects	3.2 ± 0.3
Digit–wrist, SNCV (m/s) – 36 subjects	54.9 ± 3.9

In the same year, Liberson et al. [2] compared two methods for determining sensory conduction velocity: (a) the "antidromic" technique by wrist and elbow stimulation and recording SNAPs over the skin of the digit V and (b) the "direct neurographic technique" (i.e., the orthodromic method) by distal stimulation (S) to the digit V and proximal recording at the wrist (R1) and at the elbow (R2). They studied 25 upper extremities of adults from 40 to 70 years old, and a mean wrist–digit V SNCV of 47.7 ± 8.9 m/s was found.

Mavor and Shiozawa [3] recorded antidromic ulnar sensory responses in 13 healthy (Table 2) subjects (age range 15–71 years) using surface ring electrodes for recording to the first interphalangeal joint of digit V. Reference electrode was placed 3 cm distally to the same finger. The authors used 30 % supramaximal stimulation with 0.2 ms duration, at the level of the wrist and the elbow. Skin temperature was measured using a thermistor at the elbow, wrist, palm, and finger before and after each test in all subjects. When necessary, the arm was immersed in 40 °C hot water for 10 minutes. Latency was measured from the onset of the stimulus to the initial negative deflection of the sensory response and subtraction calculated latencies between the elbow and wrist. The distances between cathode stimulation sites and active recording sites at the levels of the wrist and elbow were measured using a millimeter tape. They did not use a fixed distance between the recording and stimulation sites. Onset latency, negative peak latency, negative peak amplitude, peak-to-peak amplitude, and nerve conduction velocity (in the wrist–elbow and wrist–digit V segments) were calculated. Values for two subjects were eliminated because of substantial interference by muscle action potential after elbow stimulations.

Table 2 Normal values [3]

	Mean ± SD	Range
Wrist–digit V, SNCV – onset latency (m/s)	47 ± 3.3	44–54
Wrist–digit V, SNCV – peak latency (m/s)	40 ± 3.2	35–45
Elbow–wrist, SNCV – onset latency (m/s)	64 ± 5.4	56–76
Elbow–wrist, SNCV – peak latency (m/s)	62 ± 4.7	53–71
Wrist–digit V, negative peak amplitude (μV)	21 ± 10.9	5–44
Elbow–digit V, negative peak amplitude (μV)	10	2–15

Comment
Mavor and Shiozawa [3] observed a proximal–distal gradient in conduction velocity with higher values for the more proximal nerve segment; they also observed a significant decrease in the amplitude of the antidromically SNAPs with increasing age.

Bhala [4] performed the antidromic method (Table 3) in 61 healthy subjects (9 men and 52 women, age range 19–69 years). With the exception of one, all the healthy subjects were less than 50 years of age. The onset latency (ms) of the sensory response was measured from the stimulus artifact to the initial deflection of the SNAP. The amplitude of the SNAP was measured from peak to peak.

Table 3 Normal values [4]

	Mean ± SD	Range
Wrist–digit V, onset latency (ms)	1.9 ± 0.3	1.5–2.5
Wrist–digit V, amplitude (μV)	46 ± 8.7	25–80

The author [4] used a TECA model TE-4 electromyograph, and ring electrodes were used for recordings. Since the latency is dependent upon the distance between the points of stimulation and recording, the position of placement of the electrodes was standardized. The stimulation cathode (−) was placed at about 1–1.5 cm proximal to the pisiform bone, and the distance between the proximal stimulating (cathode) and the distal recording electrode was approximately 10.5 cm in the majority of the hands (Fig. 5).

Comment
For Bhala [4] the latency of the SNAP with antidromic and orthodromic technique was essentially the same. The amplitude, however, was considerably increased (from approximately 30–55 %) when the antidromic technique was used for recording. Also this technique was found to be less painful to electrical stimulation. The author also examined 78 patients with suspected ulnar nerve entrapment at the elbow (39 men and 39 women). The majority of the patients (75.6 %) were less than 50 years of age. Of the 40 patients who had motor nerve conduction velocity (MNCV) of less than 45 m/s in the across elbow segment of the ulnar nerve, 29 (72.5 %) showed absent or abnormal ulnar sensory response in the digit V. In the majority of the cases the sensory response showed decreased amplitude (less than 8.0 μV) but normal latency (less than 2.5 ms) of the SNAP. Prolonged latency was not seen in any of the patients of his study. It is suggested that the amplitude of the ulnar sensory response should be routinely compared to that of the median sensory response in the same hand and to the ulnar response in the opposite hand whenever one suspects a compressive lesion of the ulnar nerve in the arm or hand.

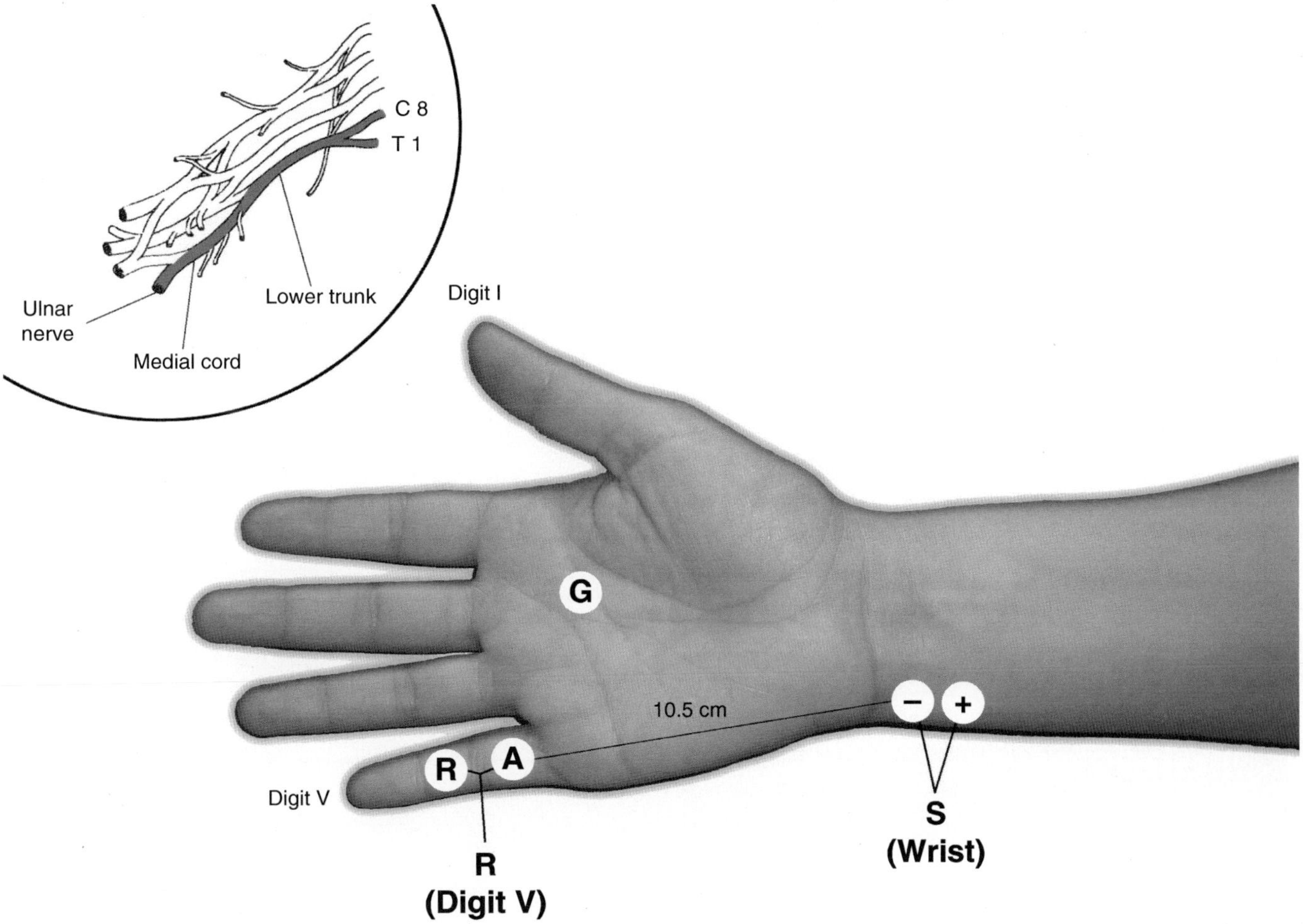

Fig. 5 Antidromic method by Bhala [4]

Table 4 Normal values [5]

	Mean ± SD
Wrist–digit V, onset latency (ms)	2.4 ± 0.2
Wrist–digit V, SNCV (m/s)	64.0 ± 6.9
Wrist–digit V, distal amplitude (μV)	3.4 ± 0.3

Hennessey et al. [5] provided normative electrodiagnostic data for the median and ulnar nerves based on several factors, such as temperature control and standardized technique. They performed bilateral sensory conduction studies (Table 4) on 44 healthy subjects (23 men and 21 women, 37 right-handed, age range 19–43 years) using a TECA Neurostar EMG instrument. Low-frequency filter was 20 Hz and high-frequency filter was 2 kHz; sweep speed was 2 ms/division. Skin temperature was monitored in all subjects by an EMG Associates Dermatemp DT-1000 infrared temperature scanner before each study, at the touch pad of the digit II, midpalm, and antecubital fossa (mean skin temperature was 32.4 °C, 33.2 °C, and 33.4 °C, respectively). Subjects with a skin temperature lower than 31 °C at any recording site had their extremities warmed with an electric heating pad. The authors [5] performed sensory antidromic nerve conduction studies using the method previously described by Melvin et al. [1]. The active recording ring electrodes were placed over the digit V (the reference was placed 4 cm distally to the active recording electrode). Ulnar nerve stimulation was performed distally at the wrist – posterior to the flexor carpi ulnaris tendon (14 cm proximally to the active recording electrode) and proximally below the elbow – just distal to the medial epicondyle with the elbow in 70° of flexion, as suggested by Checkles et al. [6]. They [5] placed the ground electrode on the dorsum of the hand. Sensory distal latency (SDL) and sensory nerve conduction velocity (SNCV) were reported; the author found significant amplitude differences between men (27.0 ± 7.8 μV) and women (52.9 ± 13.9 μV), likely caused by the influence of finger circumference rather than a direct influence of gender [7, 8].

Buschbacher [9] derived a normative database for the ulnar digital antidromic sensory conduction from a large and varied subject population. He studied 258 healthy subjects (Table 5); 57 % were women and 43 % were men, 76 % were white, 22 % were black, and 2 % were of other races (age range was 19–76 years; mean age was 44 ± 15 years for the men and 40 ± 14 years for the women). The mean height was 176 ± 7 cm for the men (height range 157–204 cm) and 163 ± 7 cm for the women (height range 138–178 cm). The mean weight was 182 ± 31 lb. for the men (weight range 112–279 lb.) and 154 ± 33 lb. for the women (weight range 103–290 lb.). Skin surface temperatures were measured over the dorsum of the hand and maintained above 32 °C. The limb was warmed if necessary using hot water bottles. All recordings were performed using a sensitivity of 20 μV/division (filter settings 20 Hz-2 kHz) and a sweep speed of 1 ms/division. Stimulation was 14 cm proximal to the active electrode. The 14 cm stimulation point was over the ulnar nerve at the wrist. Ulnar nerve recordings were made from digit V. The author measured the onset latency, peak latency, negative peak amplitude, peak-to-peak amplitude, area, rise time, and duration of the evoked sensory responses.

Table 5 Normal values [9]

	Mean ± SD	Mean ± 2SD	Range
Wrist–digit V, onset latency (ms)	2.6 ± 0.2	3.0	2.0–3.3
Wrist–digit V, peak latency (ms)	3.4 ± 0.3	4.1	2.7–4.8
Onset to peak amplitude (μV) – *wrist–digit V*	Mean ± SD	Mean ± 2SD	Range
Age range 19–49 (BMI < 24)	43 ± 17	14	
Age range 19–49 (BMI ≥ 24)	36 ± 16	11	
Age range 50–79 (BMI < 24)	27 ± 11	10	
Age range 50–79 (BMI ≥ 24)	20 ± 9	5	
All subjects	33 ± 17	6	1–85
Peak-to-peak amplitude (μV) – *wrist–digit V*	Mean ± SD	Mean ± 2SD	Range
Age range 19–49 (BMI < 24)	70 ± 36	13	
Age range 19–49 (BMI ≥ 24)	54 ± 31	8	
Age range 50–79 (BMI < 24)	37 ± 13	13	
Age range 50–79 (BMI ≥ 24)	27 ± 15	4	
All subjects	50 ± 32	4	1–179
Area (ηV/s) – *wrist–digit V*	Mean ± SD	Mean ± 2SD	Range
Age range 19–49 (BMI < 24)	53 ± 29	14	
Age range 19–49 (BMI ≥ 24)	38 ± 22	10	
Age range 50–79 (BMI < 24)	28 ± 11	12	
Age range 50–79 (BMI ≥ 24)	22 ± 14	6	
All subjects	37 ± 24	8	3–138
Rise time – all subjects	Mean ± SD	Mean ± 2SD	Range
Wrist–digit V (ms)	0.8 ± 0.2	1.2	0.5–2.1
Duration – all subjects	Mean ± SD	Mean ± 2SD	Range
Wrist–digit V (ms)	1.9 ± 0.4	2.7	1.0–3.5

Comment

In the past, the latency for sensory study usually has been measured to the negative peak of the response, but for Buschbacher [9], this was mainly due to technical factors, and with modern equipment and technique, the onset latency is almost always easily measured. Side-to-side differences are useful in diagnosing unilateral, though not systemic, disease (0.4 ± 2SD onset latency, 0.5 ± 2SD peak latency, 0.4 ± 2SD rise time, 0.7 ± 2SD duration, 53 %, 97th percentile onset-to-peak amplitude, 64 %, 97th percentile peak-to-peak amplitude). The author observed that amplitude values were traditionally measured from the negative to the positive peak (peak to peak) because of technical factors, and the onset-to-peak amplitude value may be preferable as an approximation of the number of axons conducting impulses. A decrease of more than approximately 53 % in onset-to-peak amplitude or 64 % in peak-to-peak amplitude from one side to the other should probably be considered pathologic. Pathological waveform and values in a patient with ulnar neuropathy at the wrist (UNW), in a patient with ulnar neuropathy at the elbow (UNE), and in a patient with a chronic inflammatory demyelinating polyneuropathy (CIDP) are reported (Figs. 6, 7, and 8, respectively).

Pathological waveform (wrist – digit V):

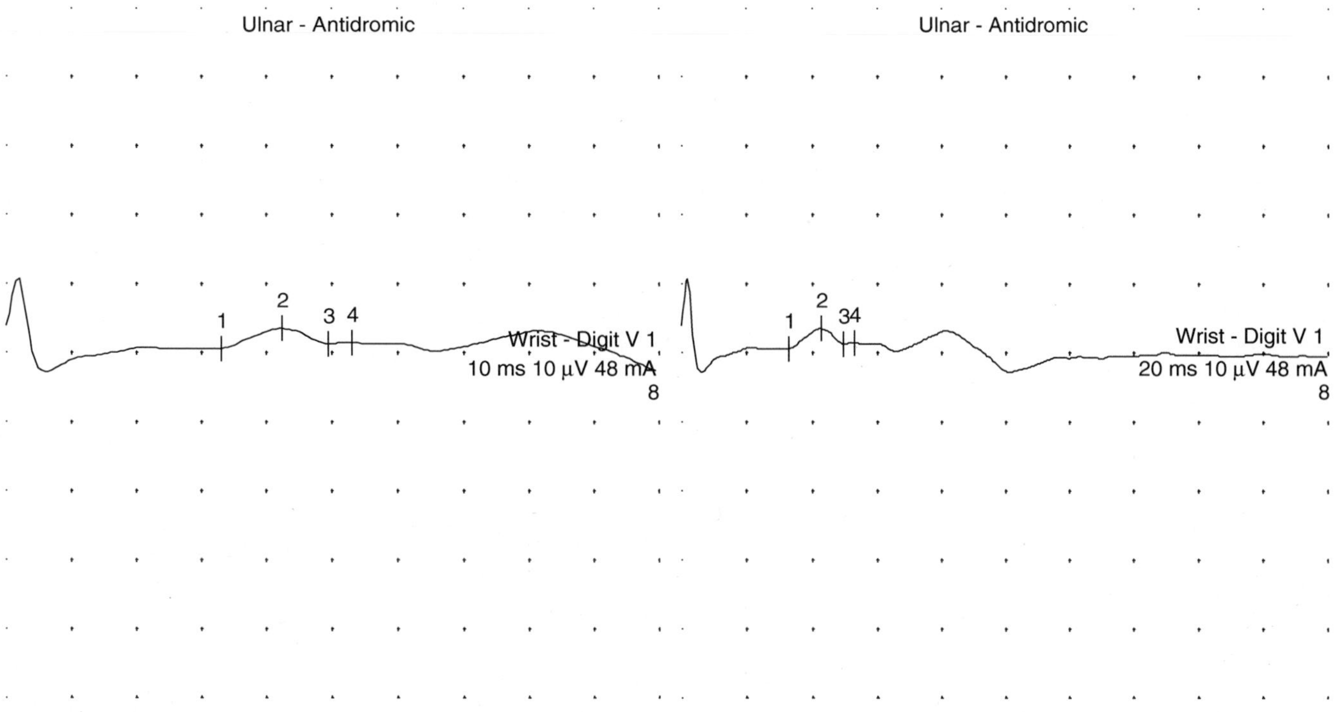

Sensitivity 10 µV/div, sweep speed 1 ms/div   Sensitivity 10 µV/div, sweep speed 2 ms/div

Onset latency: 3.30 ms; **Peak latency**: 4.20 ms; **Onset to peak amplitude**: 2.9 µV; **Peak to peak amplitude**: 2.2 µV; **SNCV** (wrist – digit V, 14 cm distance): 42.4 m/s

Fig. 6 Antidromic sensory nerve action potential (*SNAP*) recorded to digit V in a Guyon's canal syndrome, stimulation of the wrist

Pathological waveform (wrist – digit V):

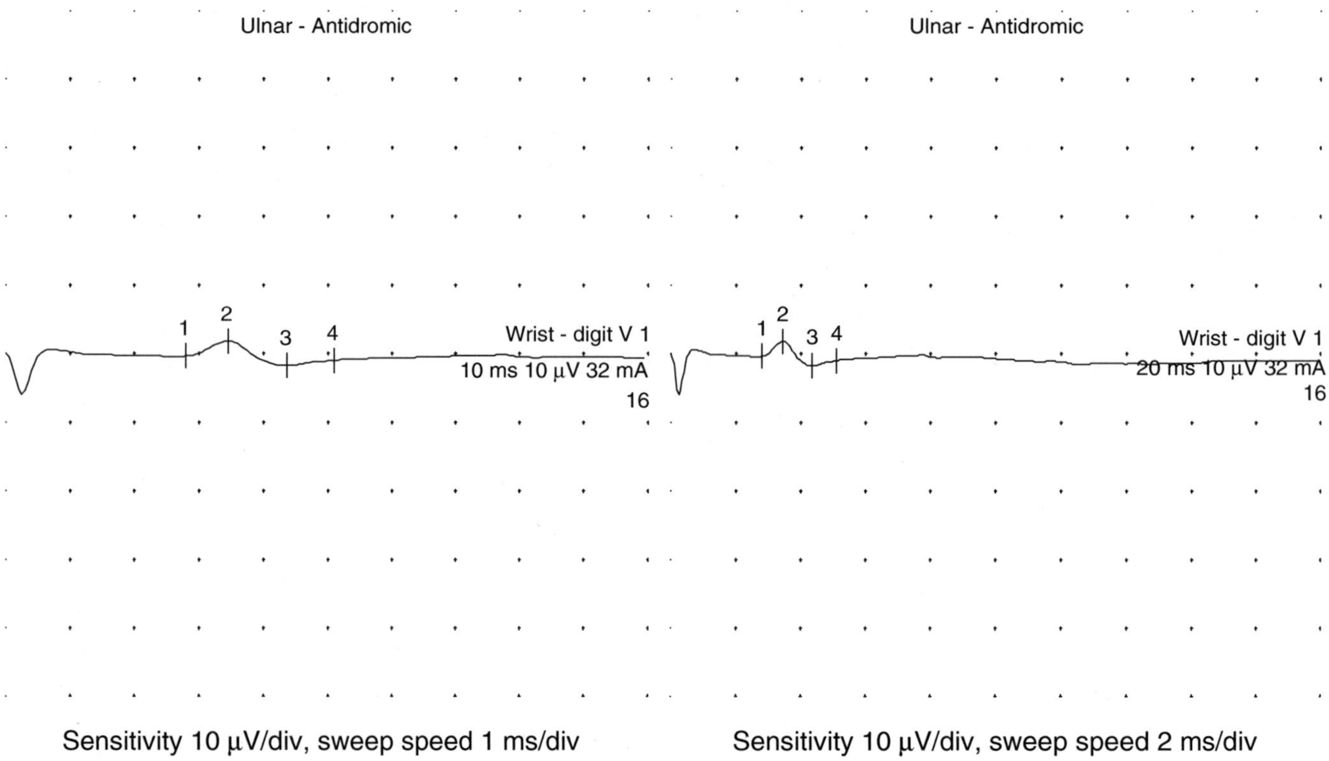

Onset latency: 2.80 ms; **Peak latency**: 4.30 ms; **Onset to peak amplitude**: 2.2 µV; **Peak to peak amplitude**: 3.7 µV; **SNCV** (wrist – digit V, 14 cm distance): 50.0m/s

Fig. 7 Antidromic sensory nerve action potential (*SNAP*) recorded to digit V in a patient with UNE, stimulation of the wrist

Pathological waveform (wrist – digit V):

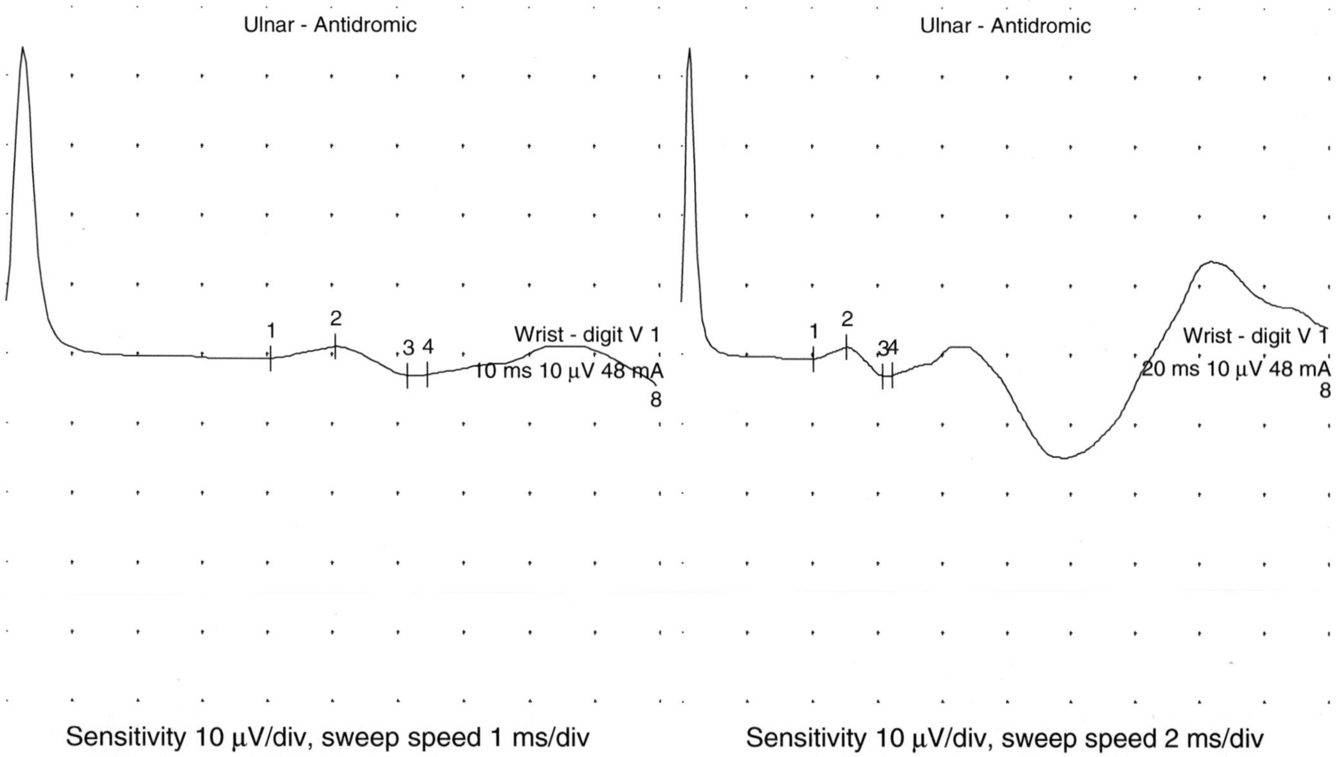

Onset latency: 4.05 ms; **Peak latency**: 5.15 ms; **Onset to peak amplitude**: 1.6 μV; **Peak to peak amplitude**: 4.1 μV; **SNCV** (wrist – digit V, 14 cm distance): 34.6 m/s

Fig. 8 Antidromic sensory nerve action potential (*SNAP*) recorded to digit V in a patient with CIDP associated with anti-MAG/SGPG (pathogenic IgM antibodies – Abs against myelin-associated glycoprotein/peripheral nerve glycolipid sulfated glucuronyl paragloboside), stimulation of the wrist

References

1. Melvin JL, Harris DH, Johnson EW (1966) Sensory and motor conduction velocities in the ulnar and median nerves. Arch Phys Med Rehabil 47:511–519
2. Liberson WT, Gratzer M, Zalis A et al (1966) Comparison of conduction velocities of motor and sensory fibers determined by different methods. Arch Phys Med Rehabil 47:17–23
3. Mavor H, Shiozawa R (1971) Antidromic digital and palmar nerve action potentials. Electroencephalogr Clin Neurophysiol 30:210–221
4. Bhala RP (1976) Electrodiagnosis of ulnar nerve lesions at the elbow. Arch Phys Med Rehabil 57:206–212
5. Hennessey WJ, Falco FJE, Braddom RL (1994) Median and ulnar nerve conduction studies: normative data for young adults. Arch Phys Med Rehabil 75:259–264
6. Checkles NS, Russakov AV, Piero DL (1971) Ulnar nerve conduction velocity: effect of elbow position on measurement. Arch Phys Med Rehabil 52:362–365
7. Stetson DS, Albers JW, Silverstein BA et al (1992) Effects of age, sex, and anthropometric factors on nerve conduction measures. Muscle Nerve 15:1095–1104
8. Bolton CF, Carter KM (1980) Human sensory nerve compound action potential amplitude: variation with sex and finger circumference. J Neurol Neurosurg Psychiatry 43:925–928
9. Buschbacher RM (1999) Ulnar 14-cm and 7-cm antidromic sensory studies to the fifth digit: reference values derived from a large population of normal subjects. Am J Phys Med Rehabil 78:S63–S68

Digit V – Wrist, Below Elbow, Above Elbow

S6

Needle Recording Technique, Orthodromic Study

Original Settings Sensitivity was 1–20 μV/division, sweep speed was 0.5–2 ms/division, and duration was 0.2 ms. Low-frequency filter, high-frequency filter, rate of pulse, and the machine used were not specified.

Position This study was performed with the subjects lying supine with the arm at 45° to the trunk, the elbow extended, and the forearm supinated.

Recording Following orthodromic method [1], the sensory nerve action potentials (SNAPs) were recorded at the wrist (R1), below the sulcus (i.e., below the elbow), about 5 cm distal to the medial epicondyle (R2), and above the sulcus (i.e., above the elbow), about 5 cm proximal to the medial epicondyle (R3). The authors also recorded sensory signals in the axilla – about 15 cm proximal to the above elbow stimulation site, but data were not reported. The author used stainless steel needles (0.6 mm in diameter) insulated except at the tip for recordings (the same needle electrodes used for motor study – see motor conduction studies). In all recordings, the active needle electrode (A) was placed close to the nerve adjusting its position until the lowest threshold for the compound muscle action potential (CMAP) was reached when the electrode was used to stimulate the nerve (<1 mA). The reference electrode (R) was placed subcutaneously at a transverse distance of at least 4 cm from the active needle electrode (Fig. 1). The ground (G) electrode position was not specified in the text; the figure shows the ground electrode positioned on the palm.

Stimulation Following the orthodromic method [1], the ulnar nerve was stimulated to digit V (S) using surface electrode; the active stimulating electrode (cathode) was applied proximally at the base of the digit; the anode (+) was positioned distally on the same digit. A current of not less than 60 mA was used, which is at least three times the current required to obtain a maximal response in a healthy subject.

Measurements Distal peak latency (ms) was measured from the onset of the stimulus to the peak of the positive deflection of the SNAP. The distances were measured between the stimulating cathode (−) and the active recording electrode (A) at each site. The sensory nerve conduction velocity (SNCV) in the fastest fibers was calculated for the segments digit V–wrist, wrist–below sulcus (i.e., wrist–below elbow), and below sulcus–above sulcus (i.e., below elbow–above elbow). In ten cases, the author calculated the SNCV in the above sulcus–axilla segment, but nerve conduction values were not reported. The SNCV was measured in meter per second (m/s). The amplitude (mV) was measured peak to peak, and SNAPs were evaluated according to duration and degree of synchronization. The temperature was monitored on the skin and near the nerve by a thermocouple, and it was maintained between 34 and 36°C; each limb was warmed before and throughout the investigation. Normal values (Table 1) were reported; the author [1] studied 36 healthy subjects (Group 1, 21 subjects, age range 18–65 years; Group 2, 15 subjects, age range 70–89 years). Pathological values were recorded from 47 cases, but pathological sample data were not reported.

G. Gentili, M. Di Napoli, *The Ulnar Nerve*, DOI 10.1007/978-3-319-43292-2_6

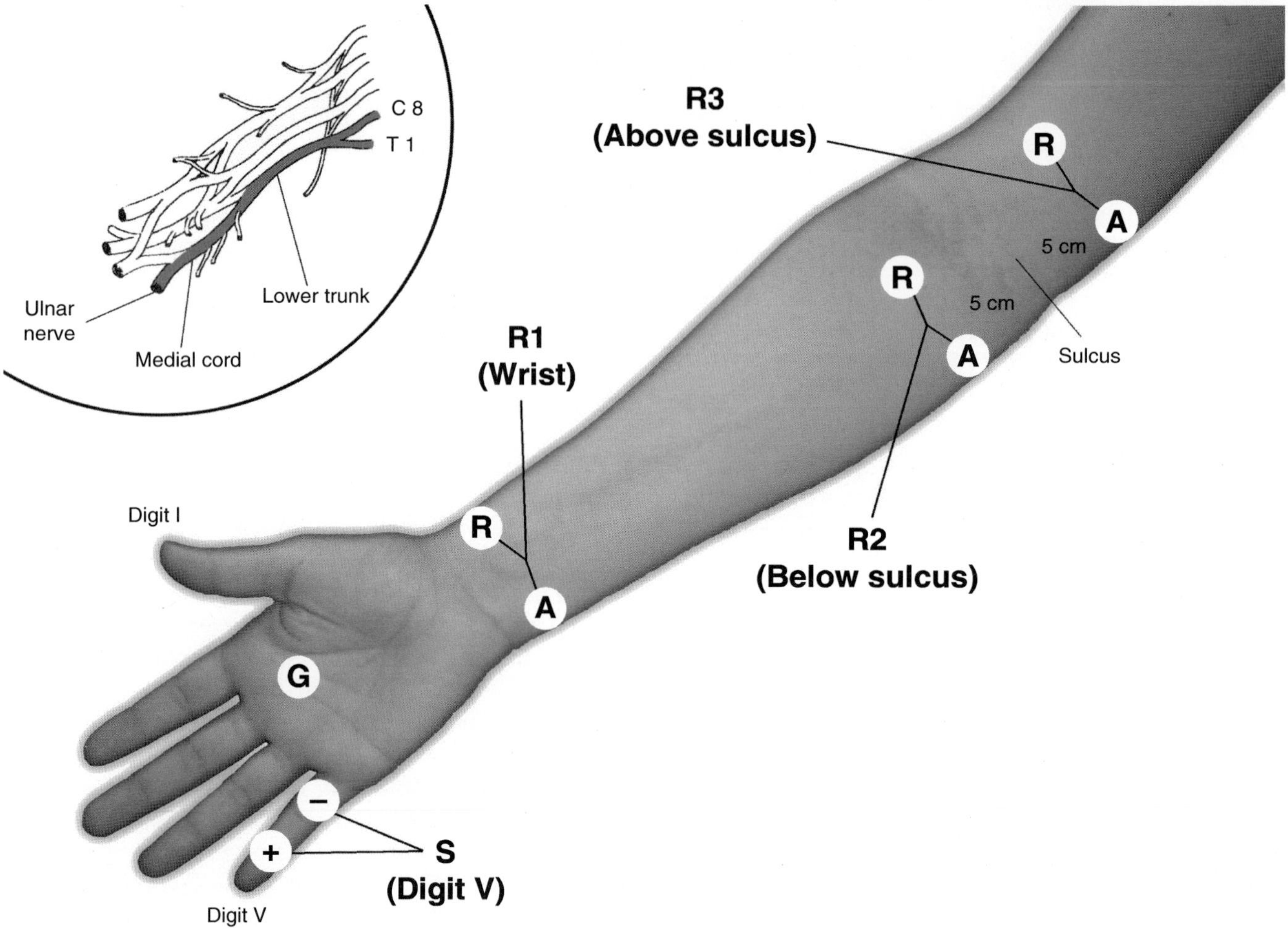

Typical waveform (digit V – wrist, below sulcus, above sulcus):

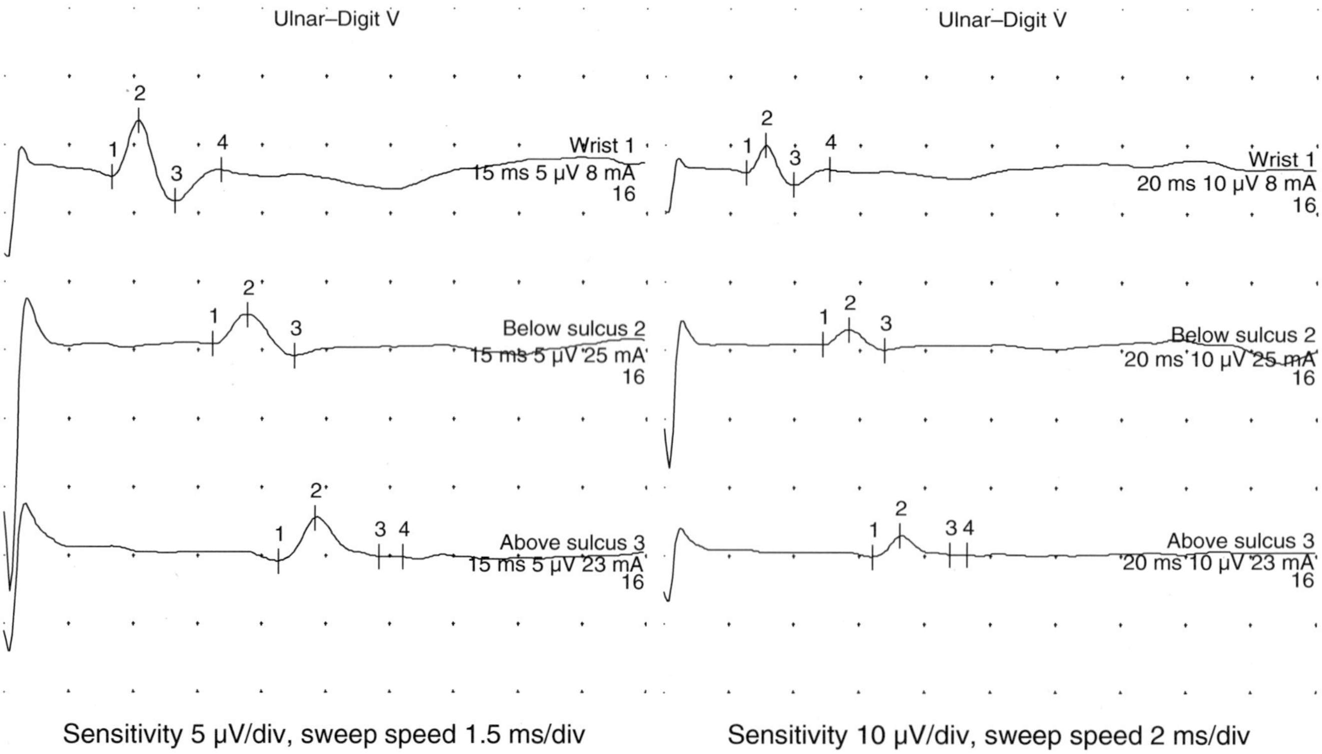

Fig. 1 Orthodromic sensory nerve action potentials (*SNAPs*) recorded at the wrist (*upper trace*), below the sulcus (*middle trace*), and above the sulcus (*lower trace*), stimulation of digit V

Table 1 Normal values [1]

Group 1	Mean ± SD	95 %
Digit V–wrist, SNCV (m/s)	55 ± 4.5	49
Wrist–below sulcus, SNCV (m/s)	71.0 ± 4.0	63
Wrist–above sulcus, SNCV (m/s)	66.0 ± 3.0	60
Below sulcus–above sulcus, SNCV (m/s)	58.0 ± 4.0	50
Digit V–wrist – amplitude (μV)	14.5 ± 7.0	8.5
Digit V–below sulcus – amplitude (μV)	6.5 ± 2.5	3
Digit V–above sulcus – amplitude (μV)	5.0 ± 2.0	2
Digit V–wrist – duration (ms)	2.7 ± 0.5	3.7
Digit V–below sulcus – duration (ms)	6.6 ± 1.4	9.4
Digit V–above sulcus – duration (ms)	9.0 ± 2.0	13.0
Group 2	Mean ± SD	95 %
Digit V–wrist, SNCV (m/s)	50.0 ± 4.0	42
Wrist–below sulcus, SNCV (m/s)	64.0 ± 2.0	60
Wrist–above sulcus, SNCV (m/s)	59.0 ± 2.0	55
Below sulcus–above sulcus, SNCV (m/s)	54.0 ± 5.0	44
Digit V–wrist – amplitude (μV)	6.0 ± 3.0	1.5
Digit V–below sulcus – amplitude (μV)	3.0 ± 2.0	0.5
Digit V–above sulcus – amplitude (μV)	3.0 ± 1.0	0.5

Comment

In 18–65 years old healthy subjects (Group 1), Payan [1] observed that the SNAP recorded above the sulcus was 35 % smaller (maximum 50 %) than that recorded below the sulcus, and none contained a late phase exceeding 20 % of the amplitude of the main component. In subjects over 70, since the degree of desynchronization varied widely, the author found the amplitude the only useful parameter. In all control subjects, he also observed a slower conduction velocity across the sulcus than in the forearm. SNAPs were recorded at the wrist in 45 of 50 cases; 39 SNAPs showed low amplitude, of which 29 were also desynchronized and prolonged. SNAPs with low amplitude value (≤1 μV) were associated with a severe decrease in conduction velocity across the sulcus (40 m/s or less). In 67 % of all cases, the author observed a conduction slowing from digit V to the wrist, though the mean velocity was only 13 % below the normal limit; slowing in the distal segment was frequently associated to the slowing in the forearm and across the sulcus. SNAPs were recorded below the sulcus in 41 of 50 cases; 38 SNAPs were smaller than normal, of which 33 were desynchronized but only 23 prolonged. Slowing of SNCV in the wrist–below sulcus segment was present in 83 %, and the mean was 20 % below normal, a higher incidence and degree than in the distal segment. SNAPs were recorded above the sulcus in 38 of 50 cases; 37 SNAPs were smaller than normal, of which 35 were desynchronized but only 23 prolonged. Slowing of SNCV in the below sulcus–above sulcus segment was present in 86 %, and it was more pronounced than in either forearm or distal segment, the mean being 33 % lower than normal.

Reference

1. Payan J (1969) Electrophysiological localization of ulnar nerve lesions. J Neurol Neurosurg Psychiatry 32:208–220

Digit II, V – Wrist

S7

Surface Recording Technique, Orthodromic Study

Original Settings Sensitivity was 15 μV/division, sweep speed was 2 ms/division, duration of pulse was 0.1–0.4 ms, and the machines used were a Tektronix type 122 preamplifier and a Tektronix type 564 B storage oscilloscope; low-frequency filter, high-frequency filter, and rate of pulse were not specified.

Position This study was performed in the supine position.

Recording Following the orthodromic method [1], signals were recorded using surface electrodes (10 mm disks or standard rings) from the median and ulnar nerves separately (Fig. 1). For the median nerve (R1) and ulnar nerve (R2), the active recording electrodes (A) were placed proximal to the distal crease at the wrist, on the median and ulnar trunks, respectively. No fixed distances were mentioned in the text. The cathodes were placed proximally. Great care was taken to prevent the spread of stimulus to the adjacent digits by covering them with a nonconducting adhesive plaster, as it was found that a spread of stimulus through contact of stimulating electrodes with the adjacent digits could significantly increase the size of the sensory potential. Ground (G) electrode position was not specified in the report; the figure shows the ground electrode placed on the palm.

Stimulation Stimulation was applied over the median and ulnar nerves to the base of the digit II (index finger) and digit V (little finger), respectively. The cathodes were placed distally on the proximal interphalangeal joint of digit II and digit V for median (S1) and ulnar (S2) stimulations, respectively. The authors used only supramaximal stimulation with a duration of 0.1–0.4 ms.

Measurements Onset latency (ms) was measured from the stimulus onset to the onset of the negative deflection of the evoked sensory nerve action potential (SNAP); peak latency (ms) was measured from the stimulus onset to the peak of the negative deflection of SNAP. The onset latency was used for the purpose of estimation of sensory nerve conduction velocity (SNCV). The amplitudes (μV) of the sensory potentials were measured from the negative peak to the positive peak of the SNAP (peak-to-peak amplitude). All examinations were performed in a room with a constant ambient temperature of 23–25°C. Median-to-ulnar amplitude ratio (Amp Rat) was determined by dividing the median digital sensory amplitude (digit II stimulation) by the ulnar digital sensory amplitude (digit V stimulation). Normal values (Table 1) were obtained from 50 healthy women (age range 30–60 years). The authors also studied 15 patients (Table 2) with a clinical diagnosis of carpal tunnel syndrome (CTS) – 7 bilateral, 22 affected hands (all women, age range 31–60 years).

Comment
For Loong and Seah [1], the study indicated that the measurement of the ratio of median to ulnar sensory action potential amplitude was a useful addition to the battery of tests available for the diagnosis of the carpal tunnel syndrome (CTS), especially in those patients who showed a normal latency and sensory conduction. For the authors, the estimation of the median–ulnar sensory action potential amplitude ratio had a greater sensitivity than that of the motor latency measurement or sensory conduction estimation alone. However, the method suffers from the disadvantage that this sensitivity is reduced when it is performed on patients with an associated distal (wrist) or proximal (elbow) ulnar nerve lesion in the same extremity.

Redmond and Rivner [2] performed orthodromic sensory conduction studies stimulating digit II (index finger) and digit V (little finger) and recording sensory nerve action potentials (SNAPs) from the median nerve and ulnar nerve at the wrist – at 13 cm and 11 cm fixed distance from the active recording electrode, respectively (Fig. 2). They tested the specificity of the various electrodiagnostic tests for CTS in a group of 50 healthy (Table 3) subjects (30 women and 20

G. Gentili, M. Di Napoli, *The Ulnar Nerve*, DOI 10.1007/978-3-319-43292-2_7

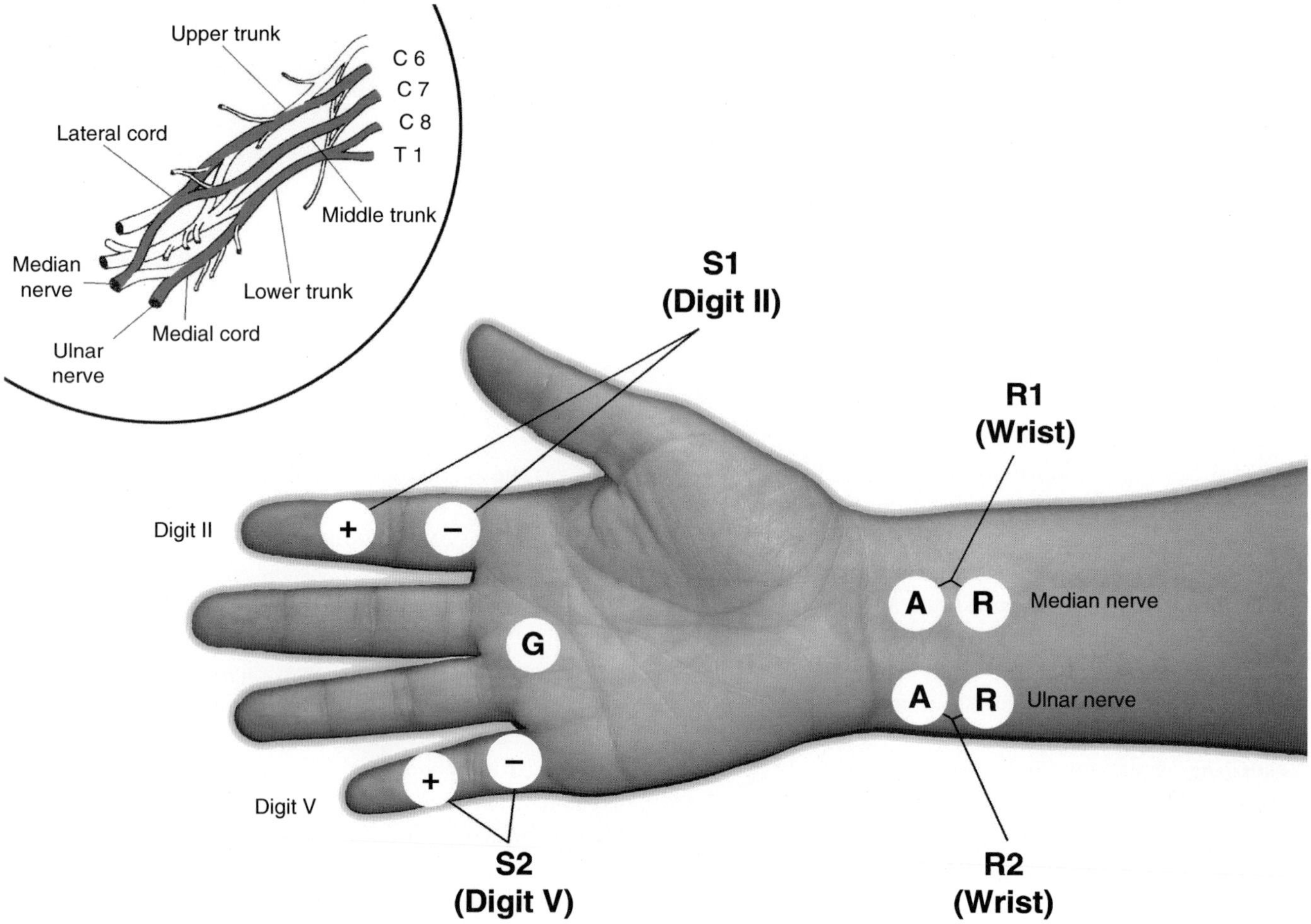

Typical waveform (digit II - wrist, digit V – wrist):

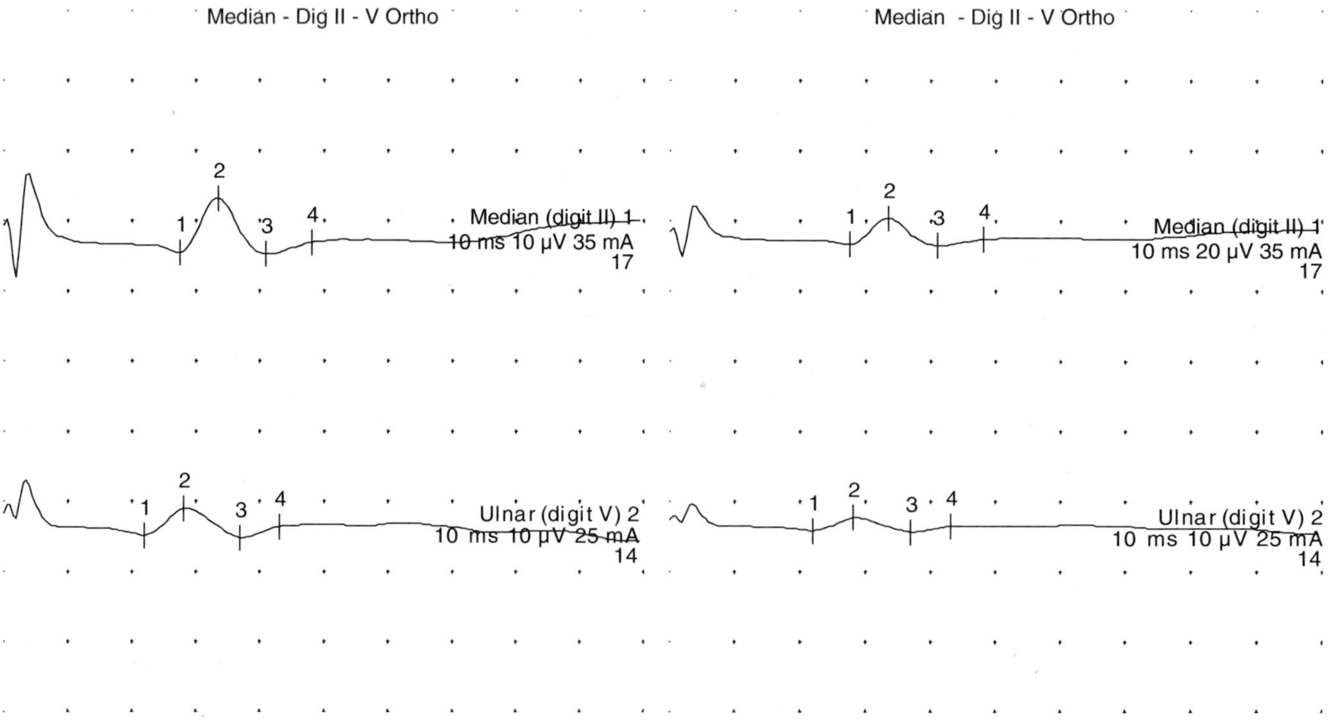

Sensitivity 10 µV/div, sweep speed 1 ms/div

Sensitivity 20 µV/div, sweep speed 1 ms/div

Fig. 1 Orthodromic sensory nerve action potentials (*SNAPs*) recorded at the wrist, stimulation of digit II (*upper trace*) and digit V (*lower trace*)

Table 1 Normal values [1]

Median nerve – onset latency (ms)		Range
Digit II–wrist		2.2–3.6
Median nerve – SNCV (m/s)	Mean ± SD	Range
Digit II–wrist	52.3 ± 4.1	44–59
Median nerve – peak-to-peak amplitude (μV)		Range
Digit II–wrist		9–48
Ulnar nerve – peak-to-peak amplitude (μV)		Range
Digit V–wrist		7.5–33
Median–ulnar – amplitude ratio	Limit of normal	Range
– All subjects –	≥1.1	1.1–2.4

men, 49 subjects were right-hand dominant, average age 34 ± 11 years, age range 18–62 years). All nerve conduction studies were performed with a TECA TD-20 electromyograph (filters were set at 20 Hz and 2 kHz), and the surface skin temperature was measured over the palm. When necessary, a heat lamp was used to achieve a temperature of at least 34°C.

Foresti et al. [3] studied 25 healthy (Table 4) subjects (average age 42 years, age range 18–69 years, male–female ratio 2.5:1) and 100 consecutive patients (Table 5) with suspected CTS (mean age 49 ± 11.9 years, age range 27–78 years, male–female ratio 3:1) using Loong and Seah's method, but they preferred the near-nerve recording technique. They performed orthodromic study of the median and ulnar nerves using ring electrodes for stimulation to digit II and digit V, respectively, and subdermal needle electrodes for recording SAPs at the wrist. No fixed distance was used between the stimulation site (digit II and digit V) and recording site (wrist). They performed bilateral studies using a five-channel MYSTRO-PLUS electromyograph. Hand temperature was monitored, and if it was less than 32°C, the limb was warmed. Onset latency and peak latency were measured, and sensory nerve conduction velocity was calculated from the onset latency.

Comment

Foresti et al. [3] in a sample of 200 hands from 100 patients with suspected CTS found 159 hands with a clinical suspicion of CTS, and 149 hands of these were found to have electrophysiological signs of CTS (ten hands were normal), and 61 patients had bilateral CTS. For median orthodromic sensory conduction (digit II–wrist segment), the authors found high values of sensibility (93.19 %, and 90.84 %) and specificity (>99 %, 90.30 %), on the base of an electrophysiological gold standard and using a clinical gold standard independent of the electrodiagnostic procedures, respectively. Some patients had no recording response to the orthodromic stimuli of the digit II (12 hands), and the difference between the ulnar nerve's parameters of the patients and control group was not significant, even if in 17 hands, in which CTS was associated with ulnar nerve entrapment at the wrist level (Guyon's canal syndrome). They suggested performing always bilateral studies performing median and ulnar orthodromic sensitive conduction test on both hands, symptomatic or not. Pathological waveform and values in a case of CTS and Guyon's canal syndrome are reported (Figs. 3 and 4).

Table 2 Pathological values [1]

Median nerve – onset latency (ms) *Digit II–wrist*	Right hand	Left hand
Patient (40 years)	2.9	2.5
Patient (42 years)	3.2	2.9
Patient (50 years)	2.4	2.5
Patient (59 years)	4.6	–
Patient (45 years)	2.9	–
Patient (37 years)	1.8	–
Patient (37 years)	2.3	–
Patient (60 years)	3.3	–
Patient (35 years)	3.2	2.8
Patient (37 years)	3.0	–
Patient (37 years)	–	–
Patient (31 years)	–	–
Patient (42 years)	2.8	3.4
Patient (45 years)	3.2	–
Patient (43 years)	3.5	3.3
Median nerve – peak latency (ms) *Digit II–wrist*	**Right hand**	**Left hand**
Patient (40 years)	4.0	3.1
Patient (42 years)	4.2	3.8
Patient (50 years)	3.0	3.4
Patient (59 years)	5.5	–
Patient (45 years)	3.7	–
Patient (37 years)	2.4	–
Patient (37 years)	2.8	–
Patient (60 years)	4.5	–
Patient (35 years)	4.4	3.2
Patient (37 years)	3.9	–
Patient (37 years)	–	–
Patient (31 years)	–	–
Patient (42 years)	3.3	4.1
Patient (45 years)	3.9	–
Patient (43 years)	4.4	4.2

(continued)

Table 2 (continued)

Median nerve – peak amplitude (μV) *Digit II–wrist*	Right hand	Left hand
Patient (40 years)	6.0	10.5
Patient (42 years)	8.0	18.0
Patient (50 years)	10.5	16.5
Patient (59 years)	3.8	–
Patient (45 years)	16.5	–
Patient (37 years)	13.5	–
Patient (37 years)	16.5	–
Patient (60 years)	4.0	–
Patient (35 years)	8.0	18.0
Patient (37 years)	15.0	–
Patient (37 years)	–	–
Patient (31 years)	–	–
Patient (42 years)	11.3	12.0
Patient (45 years)	16.5	–
Patient (43 years)	10.5	13.5

Median nerve – SNCV (m/s) *Digit II–wrist*	Right hand	Left hand
Patient (40 years)	38	46
Patient (42 years)	34	38
Patient (50 years)	48	42
Patient (59 years)	25	–
Patient (45 years)	38	–
Patient (37 years)	50	—
Patient (37 years)	48	–
Patient (60 years)	36	–
Patient (35 years)	36	45
Patient (37 years)	38	–
Patient (37 years)	–	–
Patient (31 years)	–	–
Patient (42 years)	42	35
Patient (45 years)	30	–
Patient (43 years)	33	36

Table 2 (continued)

Ulnar nerve – peak amplitude (μV) *Digit V–wrist*	Right hand	Left hand
Patient (40 years)	16.5	22.5
Patient (42 years)	22.5	19.5
Patient (50 years)	11.3	19.5
Patient (59 years)	19.0	–
Patient (45 years)	22.5	–
Patient (37 years)	18.0	–
Patient (37 years)	24.0	–
Patient (60 years)	10.0	–
Patient (35 years)	16.5	16.5
Patient (37 years)	16.5	–
Patient (37 years)	–	18.5
Patient (31 years)	–	15.0
Patient (42 years)	16.5	13.5
Patient (45 years)	18.6	18.0
Patient (43 years)	14.3	10.5

Median–ulnar nerves – amplitude ratio (μV)	Right hand	Left hand
Patient (40 years)	0.4	0.5
Patient (42 years)	0.4	0.9
Patient (50 years)	0.9	0.9
Patient (59 years)	0.2	–
Patient (45 years)	0.7	–
Patient (37 years)	0.8	–
Patient (37 years)	0.7	–
Patient (60 years)	0.4	–
Patient (35 years)	0.5	1.1
Patient (37 years)	0.9	–
Patient (37 years)	–	–
Patient (31 years)	–	–
Patient (42 years)	0.9	0.9
Patient (45 years)	0.9	–
Patient (43 years)	0.7	1.3

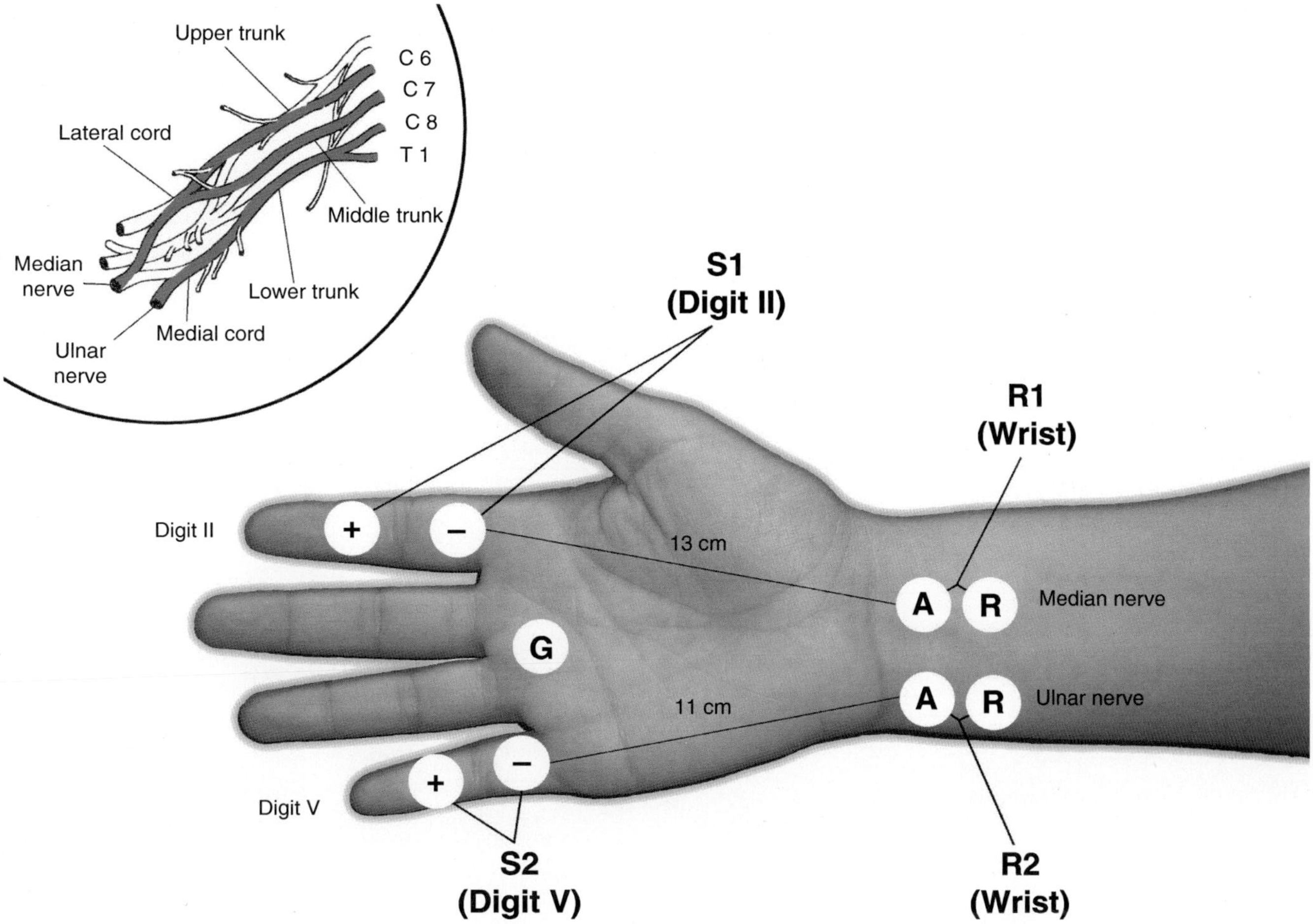

Fig. 2 Fixed recording distances: 13 cm (R1, median nerve) and 11 cm (R2, ulnar nerve)

Table 3 Normal values [2]

Median nerve – onset latency (ms) Right hand – *all subjects*	Distance (cm)	Mean ± 2SD	Range
Digit II–wrist	13	2.0 ± 0.14	2.3–3.3

Median nerve – onset latency (ms) Left hand – *all subjects*	Distance (cm)	Mean ± 2SD	Range
Digit II–wrist	13	2.0 ± 0.12	2.2–3.3

Ulnar nerve – onset latency (ms) Right hand – *all subjects*	Distance (cm)	Mean ± 2SD	Range
Digit V–wrist	11	2.5 ± 0.14	2.0–2.8

Ulnar nerve – onset latency (ms) Left hand – *all subjects*	Distance (cm)	Mean ± 2SD	Range
Digit V–wrist	11	2.5 ± 0.12	2.0–3.2

Median nerve – peak-to-peak amplitude (µV) Right hand – *all subjects*	Distance (cm)	Mean ± 2SD	Range
Digit II–wrist	13	28	12–60

Median nerve – peak-to-peak amplitude (µV) Left hand – *all subjects*	Distance (cm)	Mean ± 2SD	Range
Digit II–wrist	13	30	11–64

Ulnar nerve – peak-to-peak amplitude (µV) Right hand – *all subjects*	Distance (cm)	Mean ± 2SD	Range
Digit V–wrist	11	18	5–66

Ulnar nerve – peak-to-peak amplitude (µV) Left hand – *all subjects*	Distance (cm)	Mean ± 2SD	Range
Digit V–wrist	11	18	5–66

Median–ulnar nerves – amplitude ratio	Mean ± 2SD
Right hand – *all subjects*	1.8 ± 0.8

Median–ulnar nerves – amplitude ratio	Mean ± 2SD
Left hand – *all subjects*	1.9 ± 0.9

Table 4 Normal values [3]

Median nerve – *all subjects* Digit II–wrist	Mean	Range	Limit of normal
Onset latency (ms)	2.60 ± 0.30	2.14–3.40	3.50
Peak latency (ms)	3.20 ± 0.30	2.60–3.80	3.80
SNCV (m/s)	54.15 ± 3.50	48–59.30	47.15

Ulnar nerve – *all subjects* Digit V–wrist	Mean	Range	Limit of normal
Onset latency (ms)	2.10 ± 0.30	1.54–2.55	2.70
Peak latency (ms)	2.60 ± 0.30	2.00–3.10	3.10
SNCV (m/s)	56.06 ± 4.40	48–75.30	47.26

Table 5 Pathological values [3]

Median nerve – *all patients* Digit II–wrist	Mean	Range
Onset latency (ms)	3.62 ± 0.88	2.16–7.12
Peak latency (ms)	4.39 ± 1.34	2.54–10.80
SNCV (m/s)	39.30 ± 8.62	18.20–59.30

Ulnar nerve – *all patients* Digit V–wrist	Mean	Range
Onset latency (ms)	2.12 ± 0.26	1.54–3.04
Peak latency (ms)	2.64 ± 0.31	2.00–3.78
SNCV (m/s)	53.45 ± 5.75	36.70–75.20

Pathological waveform (digit II – wrist, digit V – wrist):

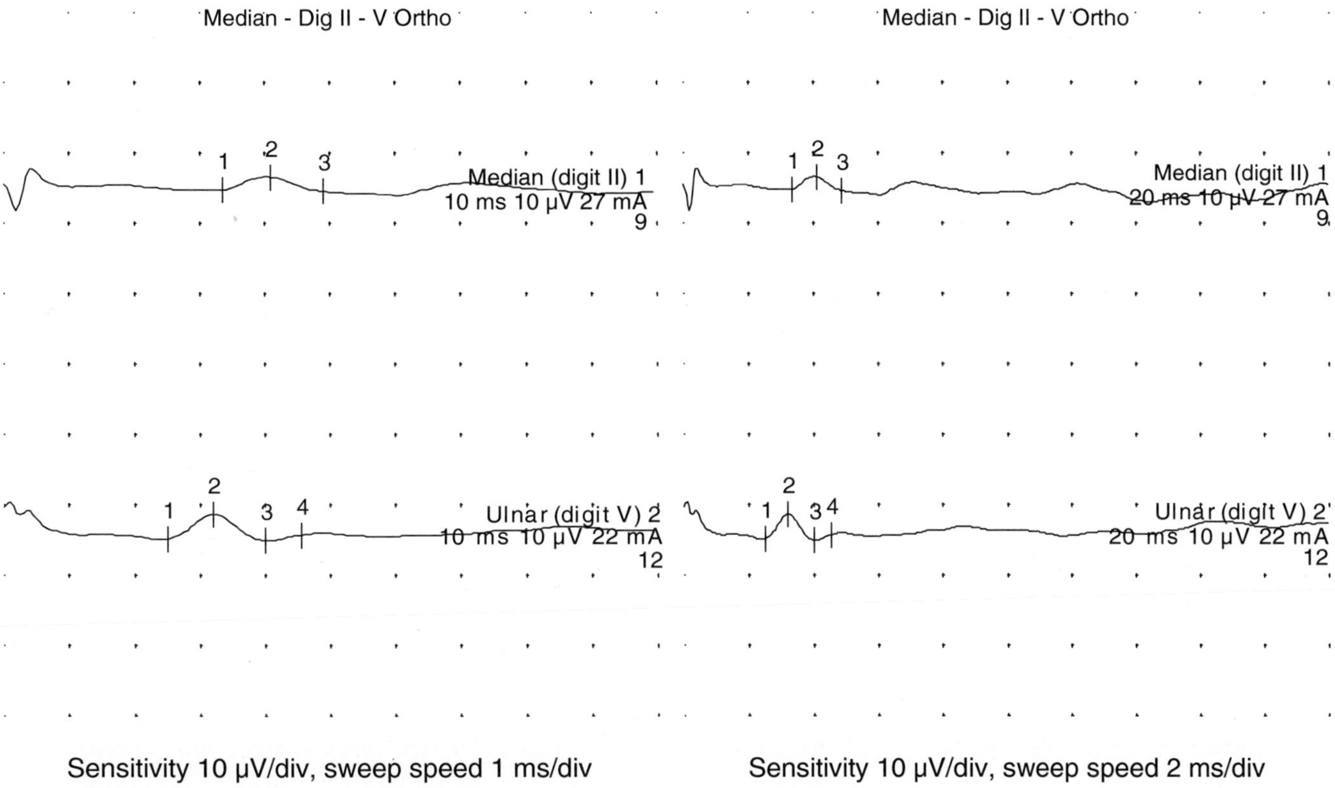

Onset latency (digit II – wrist): 3.35 ms; **Onset latency** (digit V – wrist): 2.5 ms; **Peak latency** (digit II – wrist): 4.1 ms; **Peak latency** (digit V – wrist): 3.25 ms; **Onset to peak amplitude** (digit II – wrist): 1.9 µV; **Onset to peak amplitude** (digit V – wrist): 3.6 µV; **Peak to peak amplitude** (digit II – wrist): 2.2 µV; **Peak to peak amplitude** (digit V – wrist): 3.9 µV; **SNCV** (digit II – wrist, 13 cm distance): 38.8 m/s; **SNCV** (digit V – wrist, 13 cm distance): 52.0 m/s

Fig. 3 Orthodromic sensory nerve action potentials (*SNAPs*) recorded at the wrist (surface electrodes) in mild CTS – grade 2 by Bland's CTS classification scale [4], stimulation of digit II (*upper trace*) and digit V (*lower trace*)

Pathological waveform (digit II – wrist, digit V – wrist):

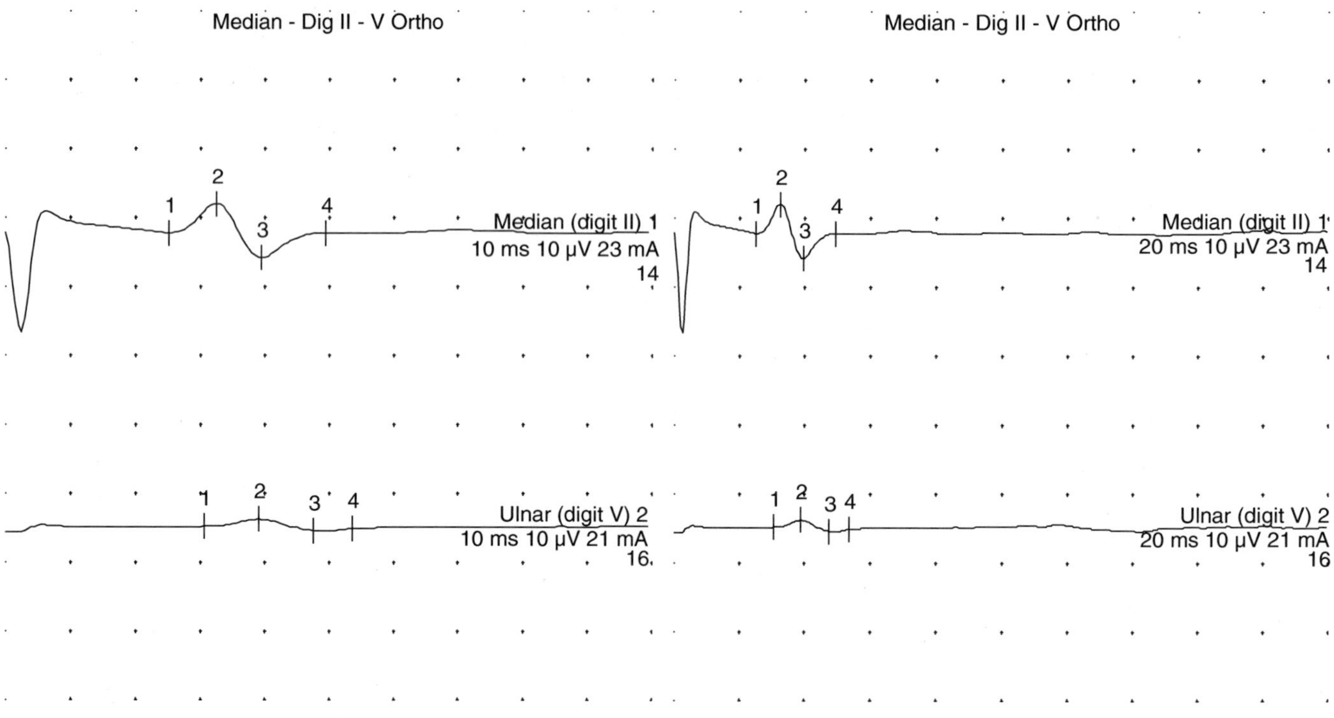

Sensitivity 10 μV/div, sweep speed 1 ms/div   Sensitivity 10 μV/div, sweep speed 2 ms/div

Onset latency (digit II – wrist): 2.5 ms; **Onset latency** (digit V – wrist): 3.05 ms; **Peak latency** (digit II – wrist): 3.35 ms; **Peak latency** (digit V – wrist): 3.9 ms; **Onset to peak amplitude** (digit II – wrist): 4.2 μV; **Onset to peak amplitude** (digit V – wrist): 1.1 μV; **Peak to peak amplitude** (digit II – wrist): 8.0 μV; **Peak to peak amplitude** (digit V – wrist): 1.8 μV; **SNCV** (digit II – wrist, 13 cm distance): 52.0 m/s; **SNCV** (digit V – wrist, 11 cm distance): 36.1 m/s

Fig. 4 Orthodromic sensory nerve action potentials (*SNAPs*) recorded at the wrist (surface electrodes) in Guyon's canal syndrome, stimulation of digit II (*upper trace*) and digit V (*lower trace*)

References

1. Loong SC, Seah CS (1971) Comparison of median and ulnar sensory nerve action potentials in the diagnosis of the carpal tunnel syndrome. J Neurol Neurosurg Psychiatry 34:750–754
2. Redmond MD, Rivner MH (1988) False positive electrodiagnostic tests in carpal tunnel syndrome. Muscle Nerve 11:511–517
3. Foresti C, Quadri S, Rasella M et al (1996) Carpal tunnel syndrome: which electrodiagnostic path should we follow? A prospective study of 100 consecutive patients. Electromyogr Clin Neurophysiol 36:377–384
4. Bland JDP (2000) A neurophysiological grading scale for carpal tunnel syndrome. Muscle Nerve 23:1280–1283

Wrist, Elbow – Digit V

Surface Recording Technique, Antidromic Study

S8

Original Settings Sensitivity was 10–20 μV/division, and the machine used was a TECA model B. Low-frequency filter, high-frequency filter, sweep speed, duration of pulse, and rate of pulse were not specified.

Position This study was performed in the supine position, with the elbow and fingers slightly flexed.

Recording According to a previous standardized technique [1], the author used the antidromic method [2]. Signals were recorded 14 cm distally from the wrist to a not specified finger, using moistened pipe cleaner. Probably he recorded sensory potentials from digit V (R), placing the active electrode (A) to the base of the digit and placing the reference (R) 4 cm proximally to the active electrode, slightly distal to the distal interphalangeal joint (Fig. 1). The ground (G) electrode was not specified in the paper; it was probably placed over the mid-forearm or on the palm of the hand (the figure shows the ground electrode placed on the palm).

Stimulation Following the method described by Melvin et al. [1], the ulnar nerve was stimulated at the wrist (S1) and at the elbow (S2). At the wrist (S1), the active stimulating electrode (cathode) was applied 14 cm proximal to the active electrode at the base of the digit, over the ulnar nerve at the wrist, near the tendon of the *flexor carpi ulnaris* (FCU) – ideally proximal to the distal wrist crease.

The anode (+) was proximal. Distal stimulation (S1) determined distal evoked sensory nerve action potentials (SNAPs), while proximal stimulation at the elbow (S2), just above the ulnar groove, allowed determination of forearm mixed nerve conduction velocity. In the case of stimulation at the elbow (S2), the author suggested 2–4 cm separation between the cathode (−) and anode (+), using a pair of standard 0.6 cm diameter electroencephalograph electrodes mounted on a plastic block for stimulating the nerve. Supramaximal stimulation was used; stimulus in all determinations was at least 30 % greater than that which elicited a maximum response.

Measurements The author measured the "terminal latency" and calculated "residual latency." The distal peak latency (ms) was called "terminal latency"; it measured from the beginning of the shock artifact to the peak of the negative deflection of the evoked sensory nerve action potential (SNAP). The sensory nerve conduction velocity (SNCV) was calculated from the nerve using stimulating and recording sites at above the elbow and at the wrist, respectively. From this elbow–wrist conduction velocity, an expected terminal latency from the wrist to finger was calculated and compared with the measured sensory terminal latency. The difference between these values was termed the "residual latency" for the sensory fibers. The temperature of the forearm was determined using a skin surface thermistor thermometer; all forearm temperatures equaled or exceeded 34 °C.

Expected distal latency (DL) of the SNAP recorded from digit V after stimulation at the wrist (14 cm segment) and residual latency (RL) for SNAP recorded from digit V after stimulation at the wrist (14 cm segment) were calculated as follows (distance between the cathode placed at the wrist and the active electrode placed at the base of digit V was 140 mm fixed):

G. Gentili, M. Di Napoli, *The Ulnar Nerve*, DOI 10.1007/978-3-319-43292-2_8

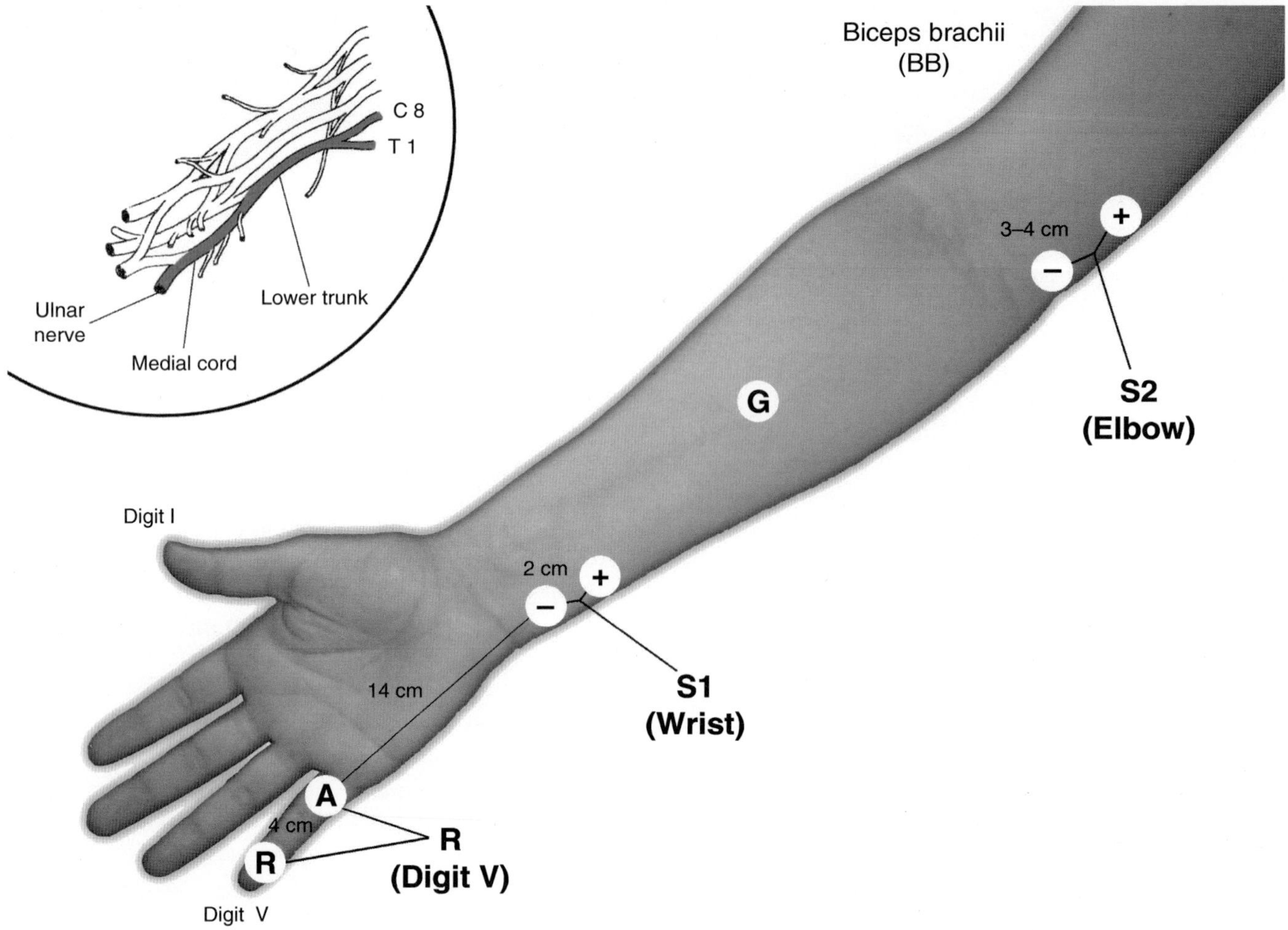

Typical waveform (wrist, elbow – digit V):

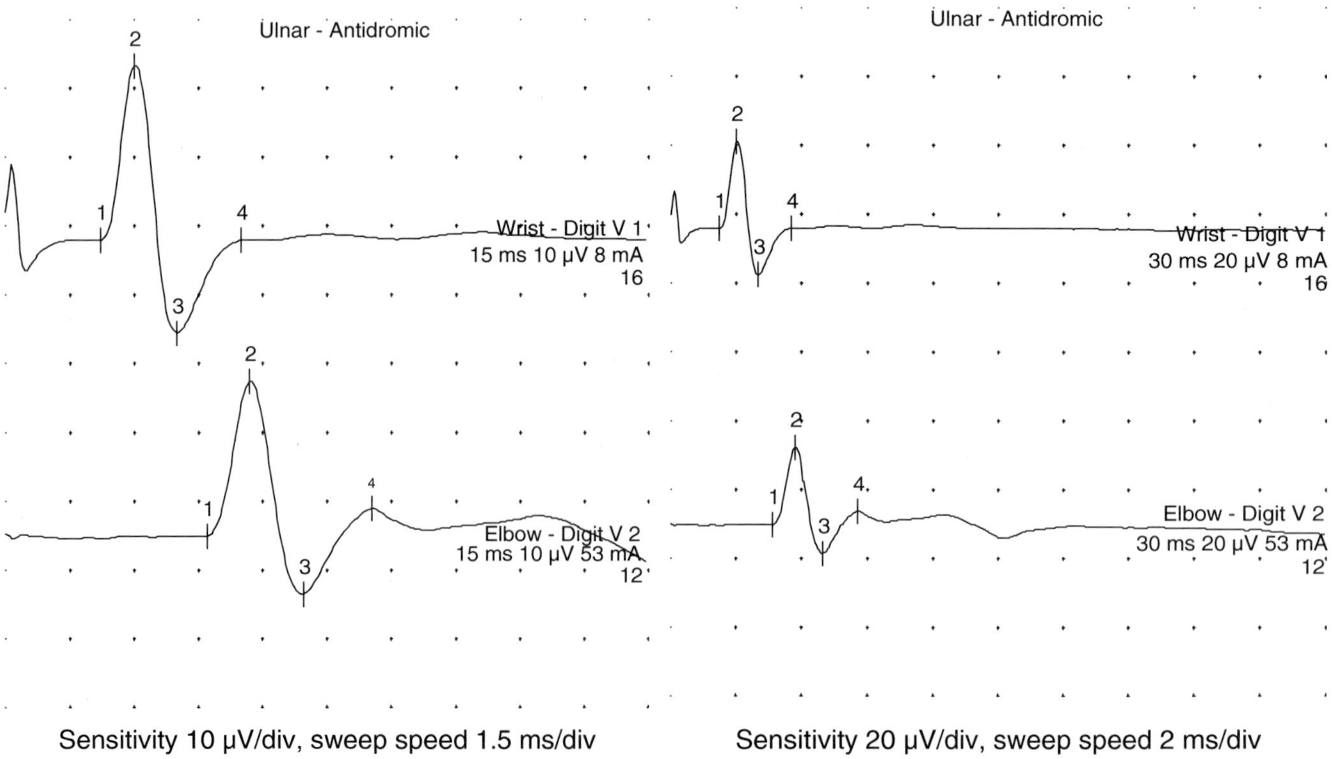

Fig. 1 Antidromic sensory nerve action potentials (*SNAPs*) recorded to digit V, stimulation of the wrist (*upper trace*) and of the elbow (*lower trace*)

Table 1 Normal values [1]

SNCV (m/s) – *elbow–wrist*	Mean ± SD	Range
10 subjects	59 ± 5	55–61
Terminal latency (ms) – *wrist–digit V*	Mean ± SD	Range
10 subjects	3.4 ± 0.3	3.2–3.6
Residual latency (ms) – *wrist–digit V*	Mean ± SD	Range
10 subjects	1.5 ± 0.5	0.9–2.1

Table 2 Pathological values [1]

SNCV (m/s) – *elbow–wrist*	Mean ± SD	Range
12 subjects	51 ± 18	38–64
Terminal latency (ms) – *wrist–digit V*	Mean ± SD	Range
12 subjects	4.6 ± 1.2	4.1–6.5
Residual latency (ms) – *wrist–digit V*	Mean ± SD	Range
12 subjects	3.2 ± 2.0	1.8–4.8

$$\text{Expected DL}(\text{ms}) = \frac{\text{distance between cathode (wrist) and active electrode (digit V), (mm)}}{\text{nerve conduction velocity (wrist–elbow), (m/s)}}$$

$$\text{Residual Latency, RL}(\text{ms}) = \text{observed DL}(\text{ms}) - \text{expected DL}(\text{ms})$$

For normal values, the author studied 20 patients (Table 1) with no history of alcoholism or diabetes (age range 21–71 years, mean age 45 years); pathological values were obtained from 20 patients (Table 2) with neuropathies selected completely at random from a much larger population of patients with neuropathies (age range 28–73 years, mean age 50 years).

Comment

For Kaplan [1], the values for the residual latency measurements in the control population of the sensory conduction were virtually identical with the residual latency measurements of the motor conduction in the same nerve and also from nerve to nerve. The author found that the residual latency measurement was as effective and as accurate as terminal latency measurement in determining the presence of a neuropathy distal to the wrist. He concluded describing the RL technique as an accurate way of measuring the state of the delicately tapering nerve endings distal to the wrist. Pathological waveform and values in Guyon's canal syndrome are reported (Fig. 2).

Pathological waveform (wrist, elbow – digit V):

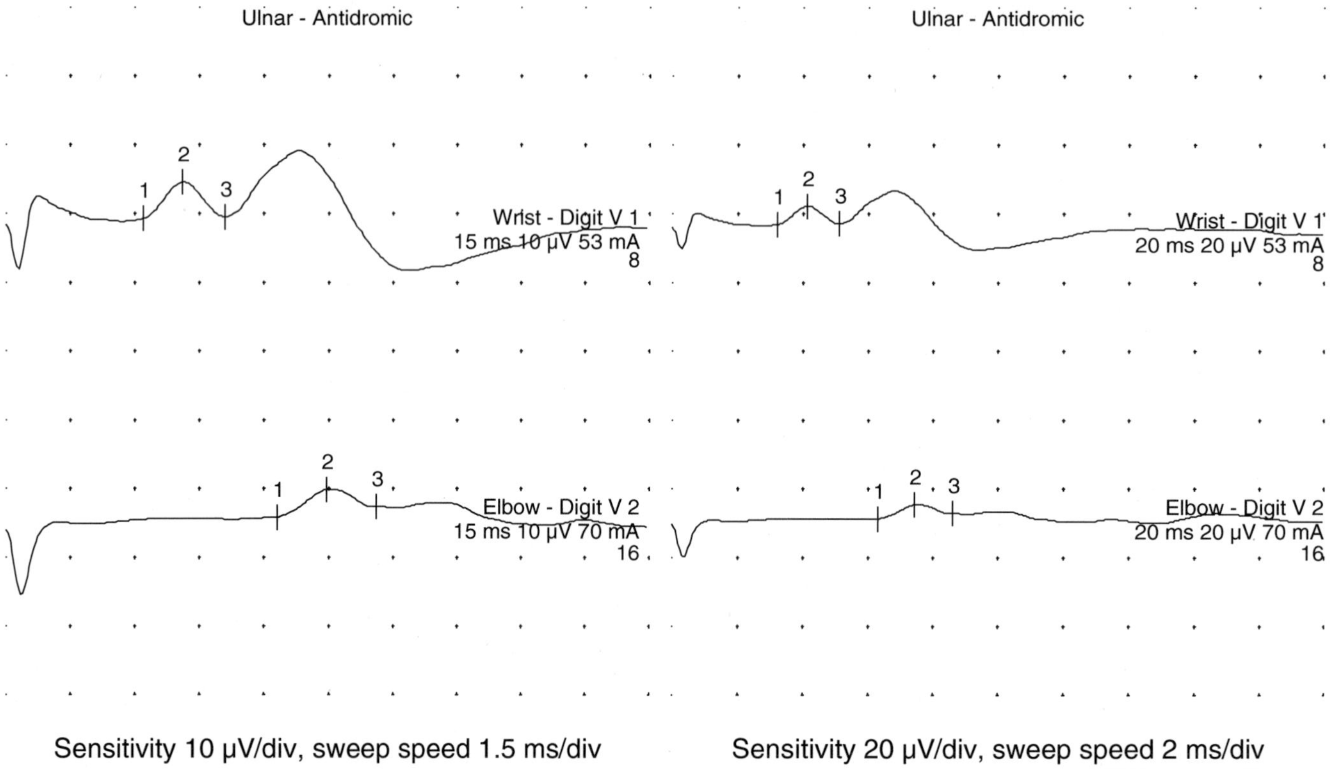

Onset latency (wrist – digit V): 3.2 ms; **Onset latency** (elbow – digit V): 6.3 ms; **Peak latency** (wrist – digit V): 4.15 ms; **Peak latency** (elbow – digit V): 7.4 ms; **Onset to peak amplitude** (wrist – digit V): 5.3 µV; **Onset to peak amplitude** (elbow – digit V): 4.0 µV; **Peak to peal amplitude** (wrist – digit V): 5.1 µV; **Peak to peak amplitude** (elbow – digit V): 2.5 µV; **SNCV** (wrist – digit V, 14 cm distance): 43.8 m/s; **SNCV** (wrist – elbow, 21 cm distance): 67.7 m/s

Fig. 2 Antidromic sensory nerve action potentials (*SNAP*) recorded to digit V wrist in a patient with Guyon's canal syndrome, stimulation of the wrist (*upper trace*) and of the elbow (*lower trace*)

References

1. Melvin JL, Harris DH, Johnson EW (1966) Sensory and motor conduction velocities in the ulnar and median nerves. Arch Phys Med Rehabil 47:511–519
2. Kaplan PE (1976) Sensory and motor residual latency measurements in healthy patients and patients with neuropathy – part 1. J Neurol Neurosurg Psychiatry 39:338–340

S9

Wrist – Digit II, V

Surface Recording Technique, Antidromic Study

Original Settings Sensitivity was 20 μV/division, sweep speed was 2 ms/division, duration of pulse was 0.1–0.4 ms, and the machine used was a TEKTRONIX electromyograph. Low-frequency filter, high-frequency filter, and rate of pulse were not specified.

Position This study was performed in the supine position.

Recording Following the antidromic method [1], signals were recorded using surface electrodes from the median and ulnar nerves separately (Fig. 1). For the median nerve (R1), the active recording electrode (A) was placed around the proximal interphalangeal joint of digit II; the reference electrode (R) was placed over the distal interphalangeal joint of digit II. For the ulnar nerve (R2), the active recording electrode (A) was placed around the proximal interphalangeal joint of digit V; the reference electrode (R) was placed over the distal interphalangeal joint of digit V. The authors chose the proximal interphalangeal joint rather than the base of the finger as the site for the proximal ring electrode to reduce or eliminate the volume-conducted motor response. Adjacent digits were kept from inadvertently contacting each other. The ground (G) electrode was placed on the dorsum of the hand, between the recording and the stimulating electrodes (the figure shows the ground electrode placed on the palm).

Stimulation Stimulations were applied over the median and ulnar nerves at the wrist. Median and ulnar nerves were stimulated separately (S1 for median nerve, S2 for ulnar nerve), both at 14 cm from the active recording electrode (A) placed around the proximal interphalangeal joint of digit II (R1) and digit V (R2). The cathodes were placed proximally. The author used only supramaximal stimulation with a duration of 0.1–0.4 ms, using the same sites of motor stimulations at the wrist (both median and ulnar nerves, S1 and S2, respectively) to compare median and ulnar sensory latencies in the same and opposite hands (see motor conduction studies). Felsenthal [1] chose the same stimulation points in an attempt to achieve some standardization.

Measurements Onset latency (ms) was measured from the stimulus onset to the initial deflection of the sensory nerve action potential (SNAP). Peak latency (ms) was measured from the stimulus onset to the peak of the negative deflection of the SNAP. All latencies were determined at the same session and at room temperature. Skin and room-temperature data was not given. All latencies of the median and ulnar nerves were compared in all subjects in the same hand and the opposite hand. Normal values (Table 1) were obtained from 100 normal hands of 50 healthy volunteers.

Comment

For Felsenthal [1] in patients with long hands, it may be necessary to increase the distance used for the median latency determinations, and for accuracy in constructing comparisons, a similar increase should be made when obtaining the ulnar latencies. By determining the conduction velocity of the nerve, a correction factor could be calculated for such a case that would allow comparison between the latency value of a long distal segment and the normal latency value reported for that nerve. If the distance between the stimulation site and the recording electrode affects the onset latency of SNAP directly, the distance between the active and the reference electrodes affects the peak latency. The peak latency decreases as the interelectrode distance decreases and increases with the increasing interelectrode distance; so the shape of a bipolar recorded potential depends on the distance between the electrodes.

G. Gentili, M. Di Napoli, *The Ulnar Nerve*, DOI 10.1007/978-3-319-43292-2_9

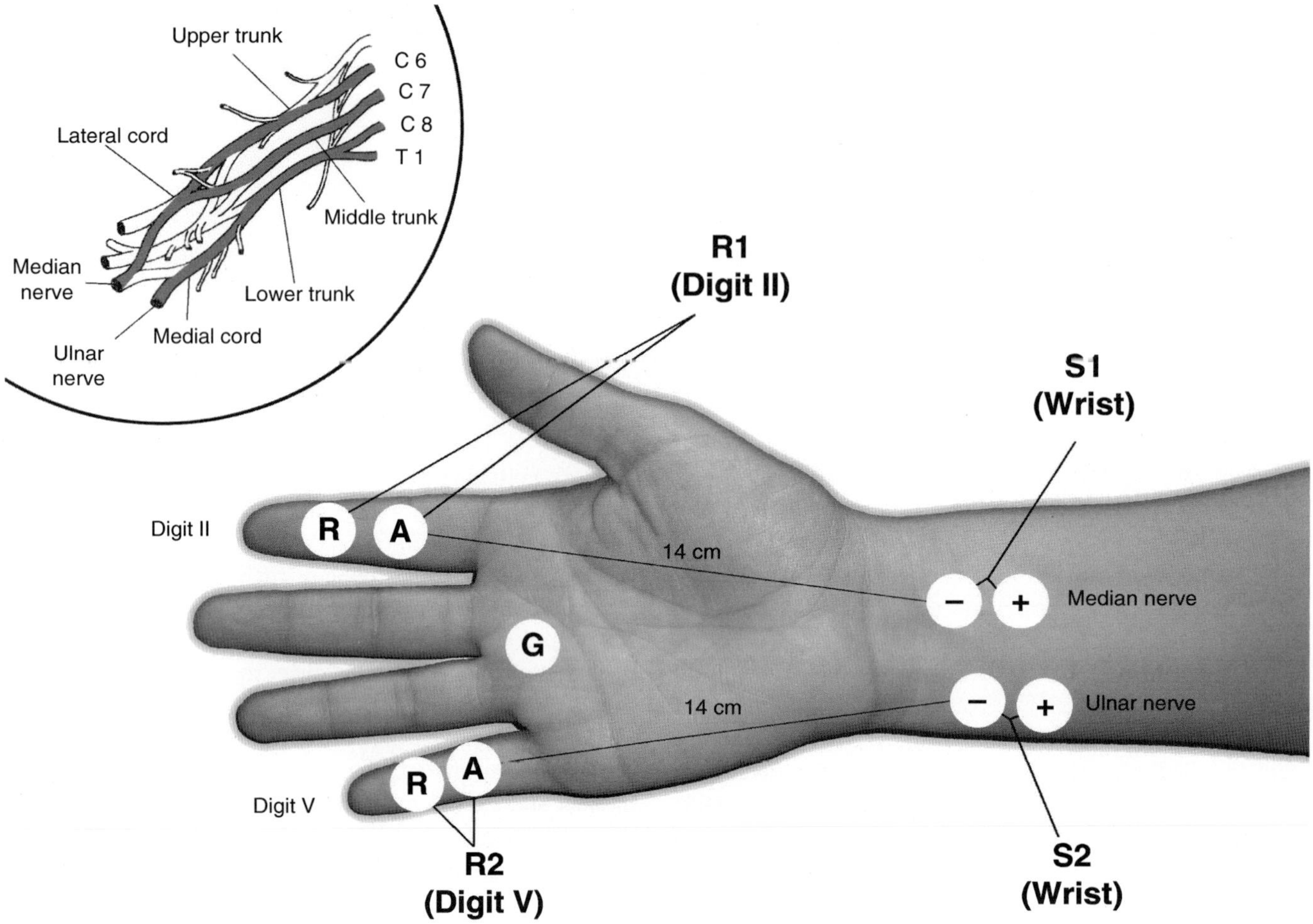

Typical waveform (wrist – digit II, wrist – digit V):

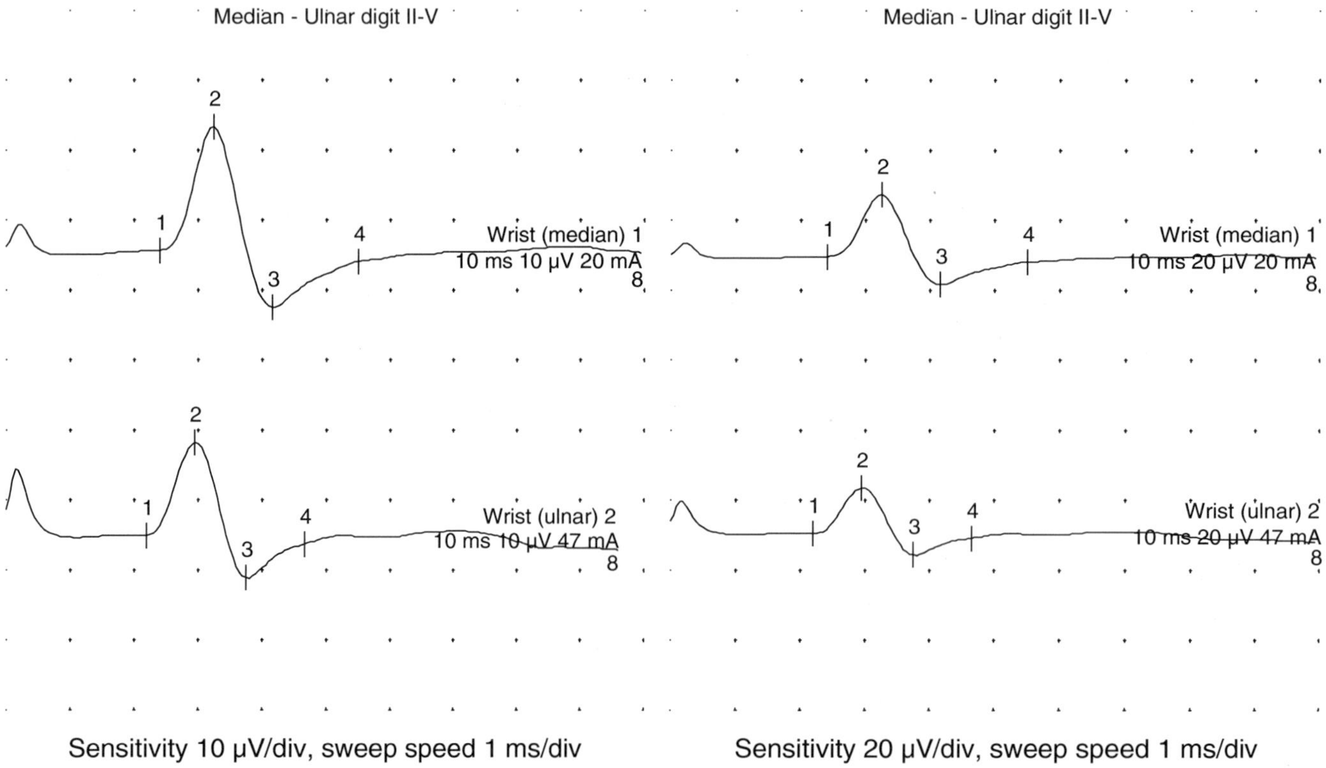

Fig. 1 Antidromic sensory nerve action potentials (*SNAPs*) recorded to digit II (*upper trace*) and to digit V (*lower trace*), stimulation of the median and ulnar nerves at the wrist

Table 1 Normal values [1]

Median nerve Onset latency (ms) – *wrist–digit II*	Mean ± SD	Mean ± 2SD	Range
All subjects	2.6 ± 0.23	3.1	2.1–3.3

Ulnar nerve Onset latency (ms) – *wrist–digit V*	Mean ± SD	Mean ± 2SD	Range
All subjects	2.6 ± 0.25	3.1	2.1–3.2

Median nerve Negative peak latency (ms) – *wrist–digit II*	Mean ± SD	Mean ± 2SD	Range
All subjects	3.3 ± 0.26	3.8	2.7–4.2

Ulnar nerve Negative peak latency (ms) – *wrist–digit V*	Mean ± SD	Mean ± 2SD	Range
All subjects	3.2 ± 0.31	3.8	2.6–4.1

Median nerve Onset latency difference (ms) – *wrist–digit II* *Opposite hands*	Mean ± SD	Mean ± 2SD	Range
All subjects	0.14 ± 0.12	0.38	0.0–0.4

Ulnar nerve Onset latency difference (ms) – *wrist–digit V* *Opposite hands*	Mean ± SD	Mean ± 2SD	Range
All subjects	0.14 ± 0.11	0.36	0.0–0.4

Median nerve Negative peak latency difference (ms) – *wrist–digit II* *Opposite hands*	Mean ± SD	Mean ± 2SD	Range
All subjects	0.15 ± 0.12	0.39	0.0–0.4

Ulnar nerve Negative peak latency difference (ms) – *wrist–digit V* *Opposite hands*	Mean ± SD	Mean ± 2SD	Range
All subjects	0.16 ± 0.12	0.40	0.0–0.5

Median – ulnar nerves Onset latency difference (ms) *Same hand*	Mean ± SD	Mean ± 2SD	Range
All subjects	0.17 ± 0.12	0.41	0.0–0.5

Median – ulnar nerves Peak latency difference (ms) *Same hand*	Mean ± SD	Mean ± 2SD	Range
All subjects	0.18 ± 0.13	0.44	0.0–0.5

Table 2 Normal values [2]

Median – ulnar nerves Amplitude ratio *(peak to peak)*	Mean ± SD	Range	Limit of normal (–2SD)
All subjects	1.254 ± 0.326	0.67–2.0	≥0.60

Median – ulnar nerves: amplitude ratio *(%)*	< 50	50–59	60–69	70–79	80–89	90–99	> 100
All subjects (100 hands)	–	–	1	2	4	4	89

In 1979, Felsenthal [2] performed the same technique [1] on 100 hands from 60 patients (Table 3) with the diagnosis of carpal tunnel syndrome (CTS), and he compared results with those previously obtained from 100 normal hands of 50 healthy volunteers (Tables 1 and 2). He measured peak-to-peak amplitude (μV) of the SNAP, from the negative peak to the positive peak of the SNAP. The author performed this method to assess the CTS (Table 3). In each hand of each

Table 3 Pathological values [2]

Median nerve – peak-to-peak amplitude (μV) *wrist–digit II*	Mean ± SD	Range
Group I – CTS with normal DSL	27.0 ± 15.5	2.4–70
Group II – peak DSL between 3.9 and 4.9 ms	22.7 ± 12.3	8–60
Group III – peak DSL ≥ 5.0 ms	15.2 ± 7.9	4–30
Group IV – absent median DSL	–	–

Ulnar nerve – peak-to-peak amplitude (μV) *wrist–digit V*	Mean ± SD	Range
Group I – CTS with normal DSL	32.9 ± 13.5	12–60
Group II – peak DSL between 3.9 and 4.9 ms	28.1 ± 15.4	10–76
Group III – peak DSL ≥ 5.0 ms	26.9 ± 12.2	10–50
Group IV – absent median DSL	19.4 ± 6.9	10–30

Median–ulnar nerves amplitude ratio (%)	< 50	50–59	60–69	70–79	80–89	90–99	> 100
Group I (17 hands)	3	2	1	3	1	–	7
Group II (31 hands)	4	5	4	4	4	–	10
Group III (14 hands)	5	4	–	–	–	–	5
Group IV (20 hands)	–	–	–	–	–	–	–

subject, the amplitude of the median SNAP was also expressed as a percentage of the amplitude of the ulnar SNAP. Skin and room-temperature data were not given. Median to ulnar amplitude ratio (Amp Rat) was determined by dividing the median digital sensory amplitude (stimulating at the wrist and recording to digit II) by the ulnar digital sensory amplitude (stimulating at the wrist and recording to digit V) as follows:

$$\text{Amplitude ratio}(\%) = \frac{\text{digital sensory amplitude to digit II, } (\mu\text{V})}{\text{digital sensory amplitude to digit V, } (\mu\text{V})}$$

Comment

For Felsenthal [2], with reference to the normal values, the amplitude of the median SNAP usually equaled or exceeded the amplitude of the ulnar response (89 of 100 normal hands). The author compared the antidromic technique with the orthodromic technique described by Loong and Seah [3]. Since 1965, the difference in amplitude has been observed using orthodromic or antidromic techniques: the amplitudes of the evoked potential are typically smaller using the former technique. The amplitude of the observed response decreases with increasing distance from the source of the action potential. If one electrode is closer to the nerve than the other, its contribution will be increased so that mainly upgoing or down-going potentials may be recorded from the same nerve merely due to lateral displacement of one electrode. With the antidromic technique, the ring electrodes about digit II and digit V are probably equidistant from the digital nerves, but with surface electrodes, it is difficult to assure that both electrodes are equidistant from the nerve. The interelectrode distance is another factor that affects the amplitude of the SNAPs. The increase in amplitude secondary to increased interelectrode distance helps explain why the amplitude of the ulnar SNAP in some volunteers equals or exceeds that of the median SNAP (11 of 100 normal hands). Other factors like a smaller median contribution to the sensory innervation of the digit II or an excessive skin impedance may reduce the amplitude of the SNAP.

Comment

The criteria used by Felsenthal [2] for the diagnosis of CTS included: distal sensory latency (DSL) to onset >3.1 ms (onset latency was measured from the stimulus onset to the onset of the negative deflection of SNAP, for both digit II and digit V recordings); DSL to peak >3.8 ms (peak latency was measured from the stimulus onset to the peak of the negative deflection of SNAP, for both digit II and digit V recordings); DSL difference >0.4 ms between the median and ulnar DSL measured either to onset or peak; distal motor latency (DML) >4.0 ms to onset; and DML difference >1.0 ms between the median and ulnar DML in the same hand. The patients were subdivided into four groups. Group I was composed of all patients with normal median sensory latencies and either greater than 0.4 ms difference between median and ulnar DSL or in two patients greater than 1.0 ms difference between the median and ulnar DML. Group II included all cases of CTS with DSL to the peak of between 3.9 and 4.9 ms. Group III was composed of all patients with a latency to peak greater than 4.9 ms. Group IV included all cases with absent median DSL.

Some authors [4, 5] performed this method in obese patients [4] and in polyneuropathic patients with or without association with CTS [5] founding the comparison between SNAPs obtained from antidromic stimulation of the median and ulnar nerves, a useful tool to make diagnosis of CTS with good sensitivity (72 %) and specificity (82 %) values. However, this method suffers from the disadvantage that the sensitivity in CTS diagnosis can be reduced when performed on patients with an associated ulnar nerve lesion on its course in the same extremity. Pathological waveform and values in a case of CTS and Guyon's canal syndrome are reported (Figs. 2 and 3).

Pathological waveform (wrist – digit II, wrist – digit V):

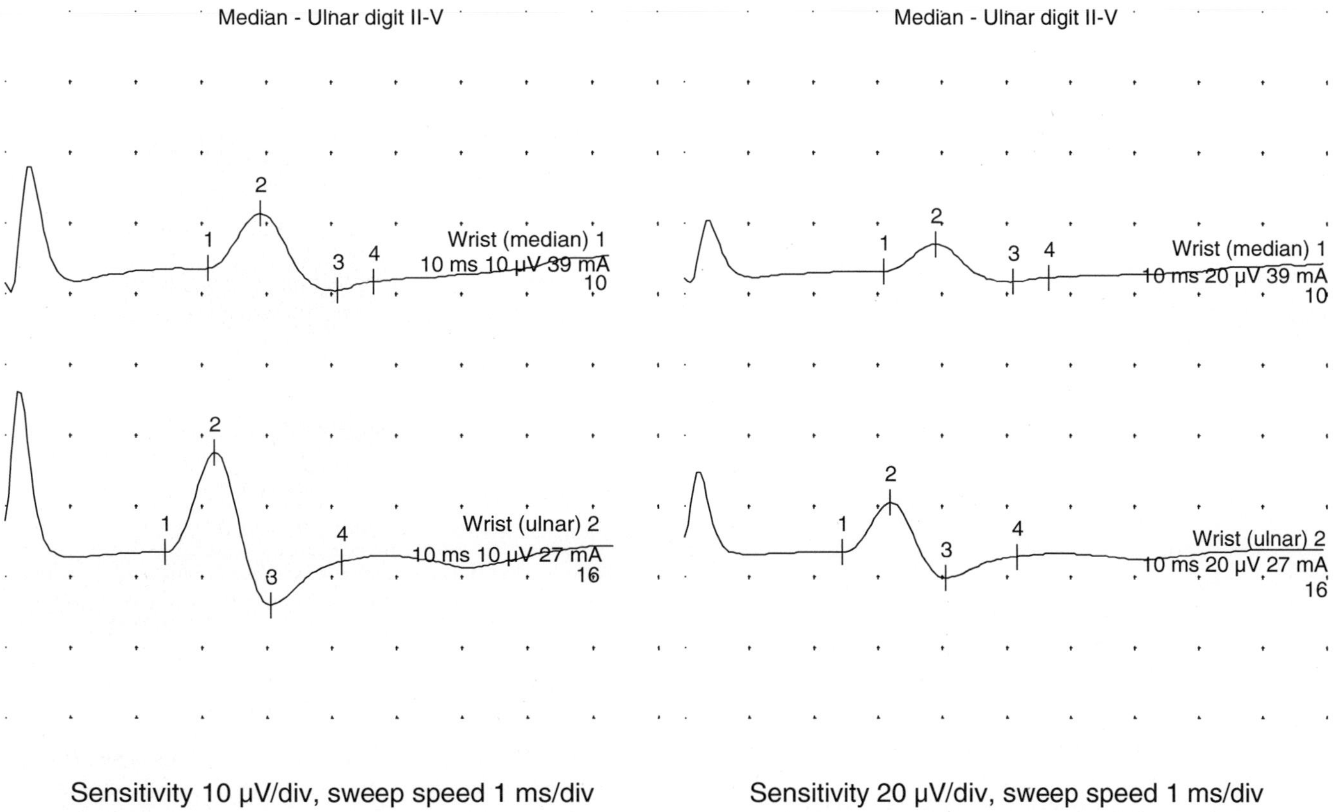

Onset latency (wrist – digit II): 3.10 ms; **Onset latency** (wrist – digit V): 2.45 ms; **Peak latency** (wrist – digit II): 3.90ms; **Peak latency** (wrist – digit V): 3.15 ms; **Onset to peak amplitude** (wrist – digit II): 7.7 µV; **Onset to peak amplitude** (wrist – digit V): 14.2 µV; **Peak to peak amplitude** (wrist – digit II): 11.1 µV; **Peak to peak amplitude** (wrist – digit V): 21.8 µV; **SNCV** (wrist – digit II, 14 cm distance): 45.2 m/s; **SNCV** (wrist – digit V, 14 cm distance): 57.1 m/s

Fig. 2 Antidromic sensory nerve action potentials (*SNAPs*) recorded to digit II (*upper trace*) and to digit V (*lower trace*) in very mild *CTS* – grade 1 by Bland's CTS classification scale [5], stimulation of the median and ulnar nerves at the wrist

Pathological waveform (wrist – digit II, wrist – digit V):

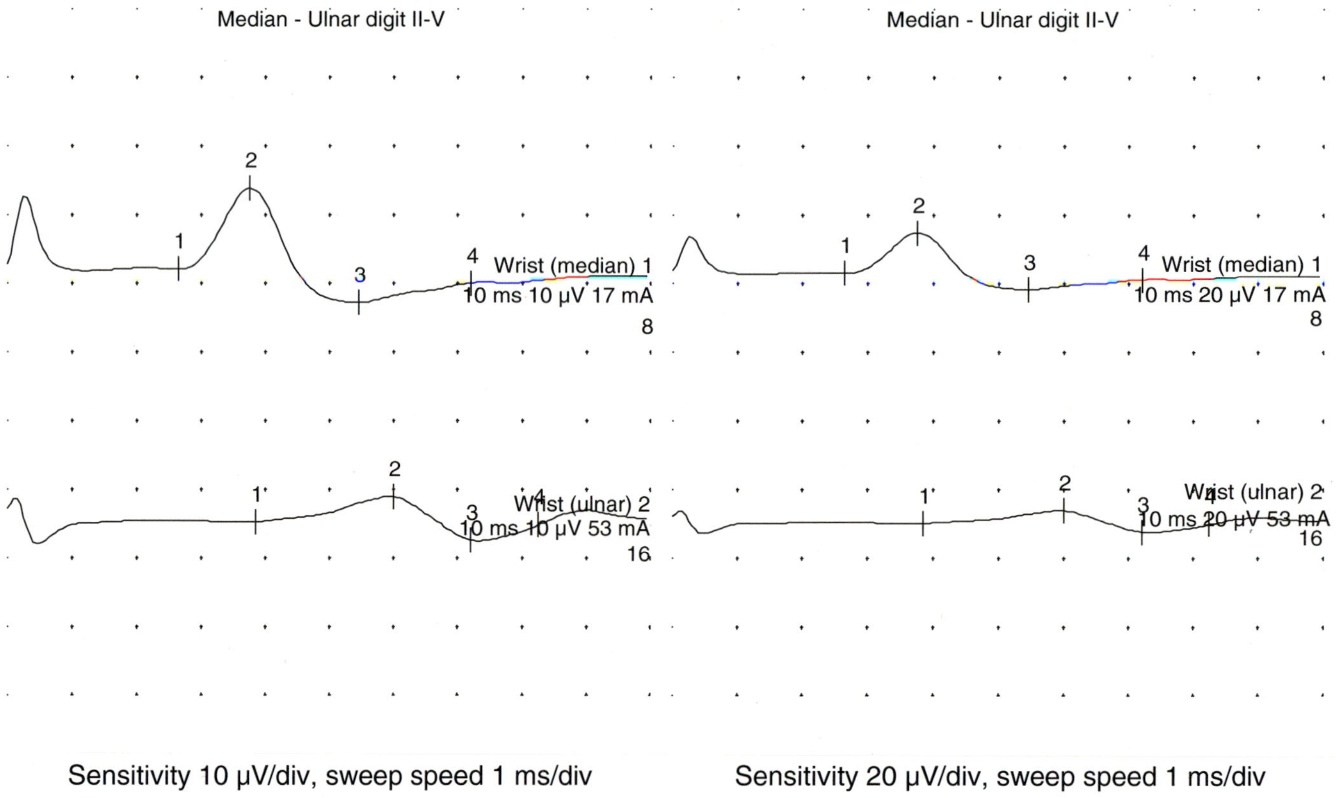

Onset latency (wrist – digit II): 2.65 ms; **Onset latency** (wrist – digit V): 3.85 ms; **Peak latency** (wrist – digit II): 3.75 ms; **Peak latency** (wrist – digit V): 6.0 ms; **Onset to peak amplitude** (wrist – digit II): 11.8 µV; **Onset to peak amplitude** (wrist – digit V): 3.7 µV; **Peak to peak amplitude** (wrist – digit II): 16.9 µV; **Peak to peak amplitude** (wrist – digit V): 6.5 µV; **SNCV** (wrist – digit II, 14 cm distance): 52.8 m/s; **SNCV** (wrist – digit V, 14 cm distance): 36.4 m/s

Fig. 3 Antidromic sensory nerve action potentials (*SNAPs*) recorded to digit II (*upper trace*) and to digit V (*lower trace*) in Guyon's canal syndrome, stimulation of the median and ulnar nerves at the wrist

References

1. Felsenthal G (1977) Median and ulnar distal motor and sensory latencies in the same normal subjects. Arch Phys Med Rehabil 58:297–302
2. Felsenthal G (1978) Comparison of evoked potentials in the same hand in normal subjects and in patients with carpal tunnel syndrome. Am J Phys Med 57:228–232
3. Loong SC, Seah CS (1971) Comparison of median and ulnar sensory nerve action potentials in the diagnosis of the carpal tunnel syndrome. J Neurol Neurosurg Psychiatry 34:750–754
4. Aygül R, Ulvi H, Kotan D et al (2009) Sensitivities of conventional and new electrophysiological techniques in carpal tunnel syndrome and their relationship to body mass index. J Brachial Plex Peripher Nerve Inj 4:12
5. Gazioglu S, Boz C, Altunayoglu Cakmak A (2011) Electrodiagnosis of carpal tunnel syndrome in patients with diabetic polyneuropathy. Clin Neurophysiol 122:1463–1469
6. Bland JDP (2000) A neurophysiological grading scale for carpal tunnel syndrome. Muscle Nerve 23:1280–128

Digit V – Palm, Wrist, Elbow; Wrist – Elbow, Axilla

S10

Surface Recording Technique, Orthodromic Study

Original Settings Duration of pulse was 0.1 ms and rate of pulse was 3/s. Sensitivity, sweep speed, low-frequency filter, high-frequency filter, and the machine used were not specified.

Position Test position was not specified in the text; we performed the study with the subject lying on a comfortable couch with the arm slightly abducted, the forearm in the supine position, and elbow fully extended.

Recording Following the orthodromic method [1], sensory nerve action potentials (SNAPs) were recorded on the ulnar nerve course (Fig. 1), using surface bipolar electrodes (2 cm fixed interelectrode distance) at the palm (R1), wrist (R2), and elbow (R3). The authors also recorded mixed nerve action potentials (MNAPs) at the elbow and axilla (Figs. 2 and 3), following Mayer's method [2] with some modifications. The distance between recording and stimulating electrodes was not fixed. The ground (G) electrode position was not specified in the report; the figures show the ground electrode placed on the palm (Fig. 1) and over the forearm (Fig. 2).

Stimulation For SNAPs recording, the ulnar nerve was stimulated distally to the digit V (S) with the fingers covered and separated each other with a cotton (Fig. 1). For MNAPs recording, the ulnar nerve was stimulated more proximal at the wrist (Fig. 2). A fixed distance, between stimulating and recording electrodes, was not used.

Measurements Maximum sensory nerve conduction velocity (SNCV) was measured in meter per second (m/s), and it was determined for digit–palm and wrist–elbow segments. Peak-to-peak amplitude (μV) was measured from the negative to the positive peak. The mixed nerve conduction velocity (MxCV) was measured in the cubital nerve, stimulating the ulnar nerve at the wrist and recording the MxEP at the elbow and axilla. The skin was prepared before exploration, and when necessary, the hand was warmed by immersion in hot water. The mean temperature of the skin over the nerves was 33 °C (palm and wrist) and 33.5 °C (elbow). Normal values were obtained from 47 ulnar nerves of 47 healthy subjects (25 men and 22 men, age range 21–77 years). Explorations were performed on the right hand of all subjects, divided into three groups: ≤ 30 years (Table 1), 30–49 years (Table 2), and ≥ 50 years (Table 3).

Comment

Cruz Martinez et al. [1] observed a faster SNCV in the proximal (palm–wrist) than in the distal (digit–palm) segment (15 % in the group aged 21–29 years and 11 % in the other age groups). The SNCV decreased as a linear function of age, but the decrement was more marked in the palm–wrist segment than in the digit–palm segment. The amplitude of the SNAP was higher at the palm than at the wrist, and the authors observed shorter duration at the distal than proximal recordings.

G. Gentili, M. Di Napoli, *The Ulnar Nerve*, DOI 10.1007/978-3-319-43292-2_10

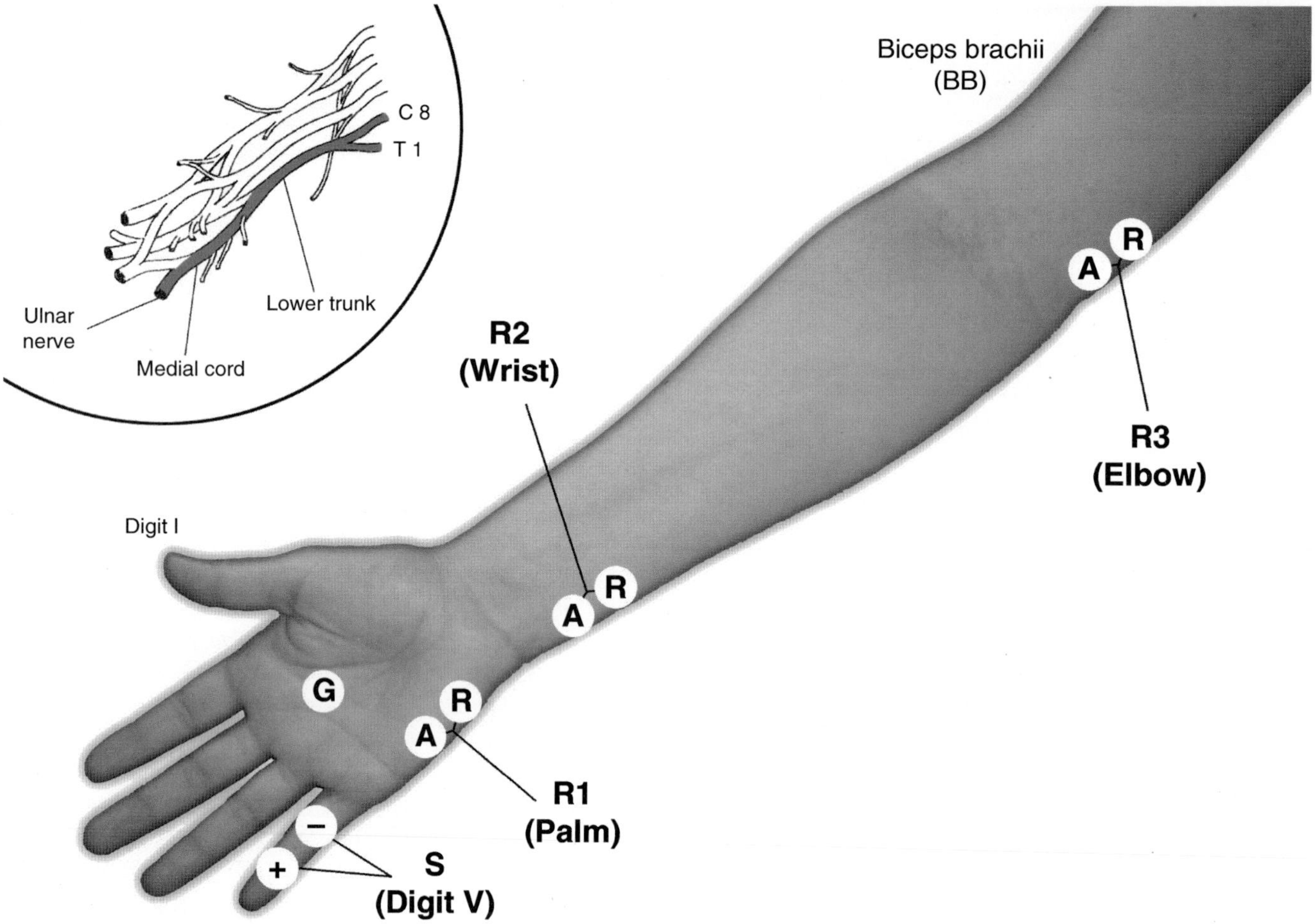

Typical waveform (digit V – palm, wrist, elbow):

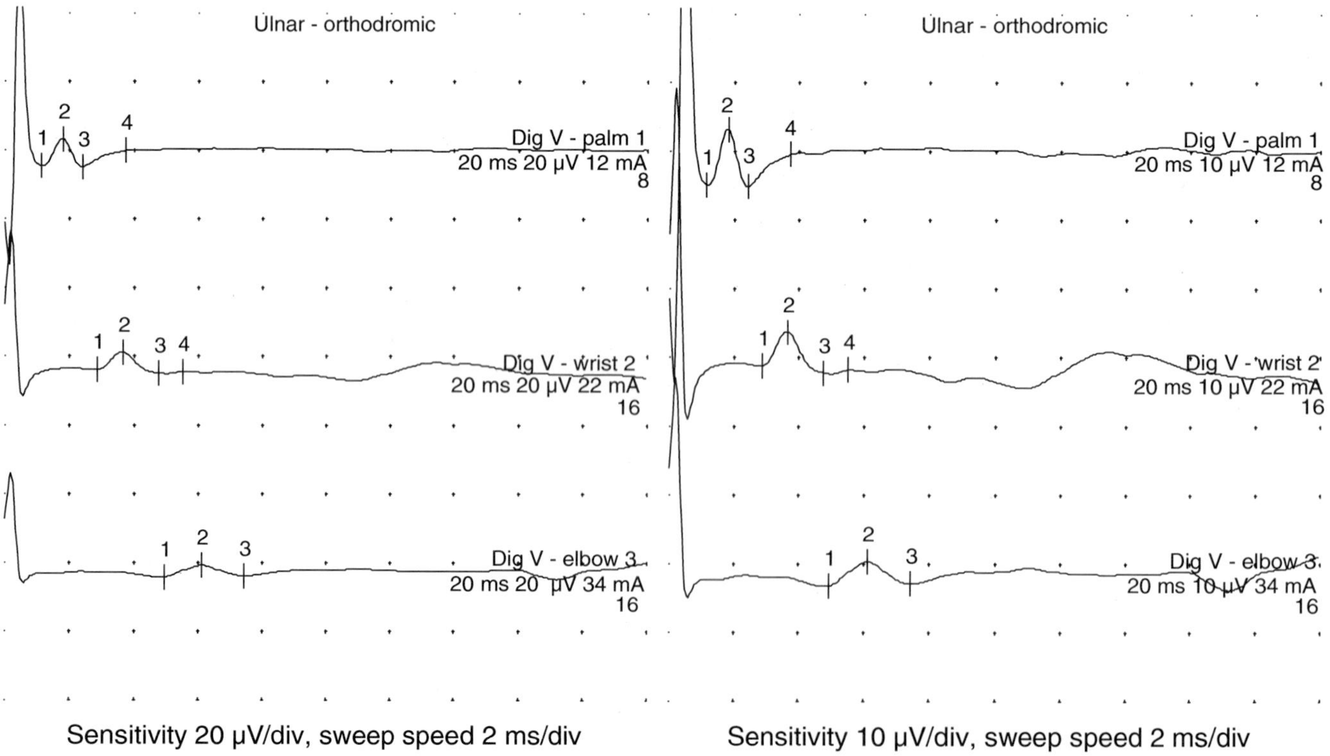

Fig. 1 Orthodromic sensory nerve action potentials (*SNAPs*) recorded on the palm (*upper trace*), at the wrist (*middle trace*), and at the elbow (*lower trace*), stimulation to digit V

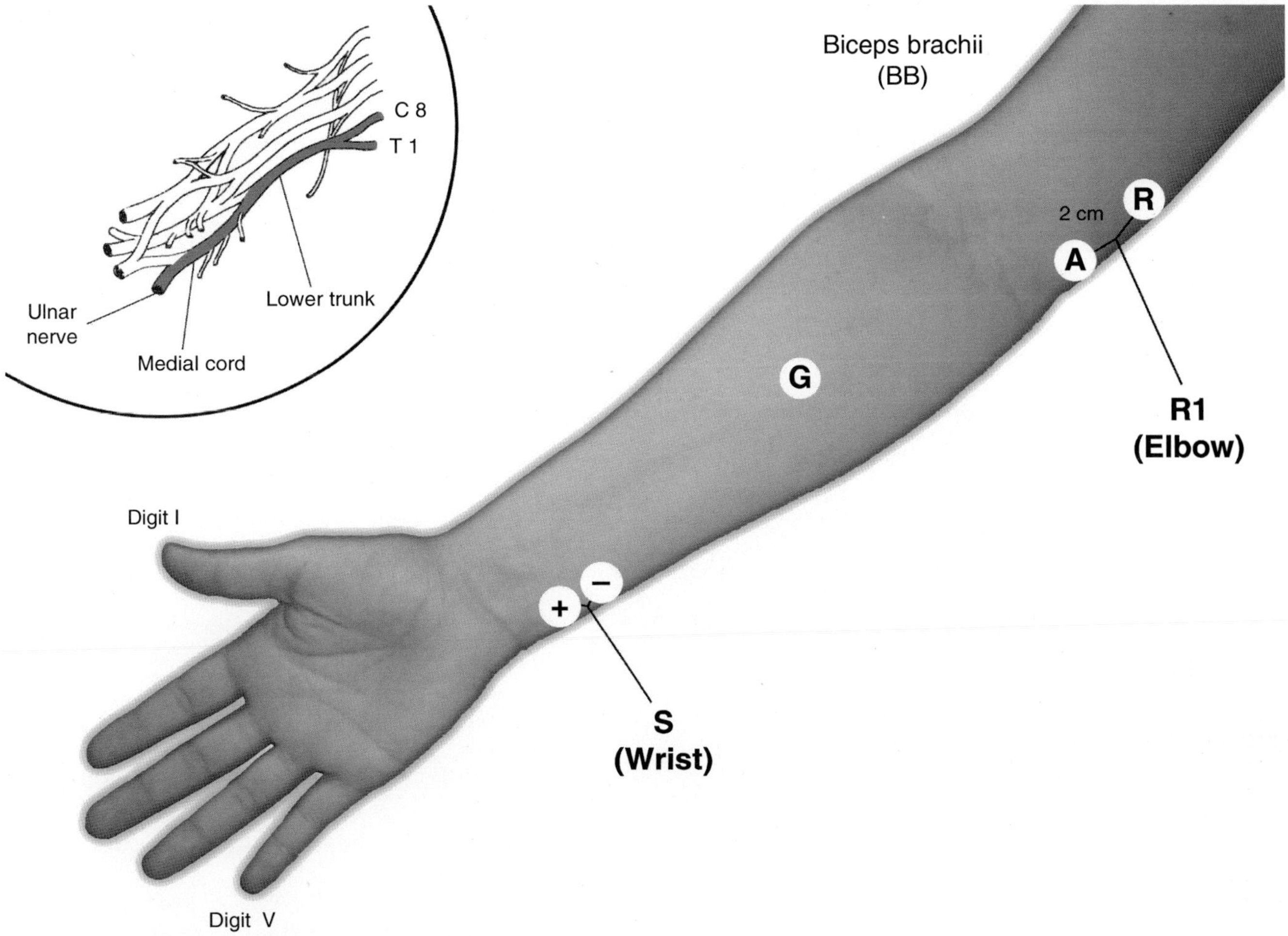

Typical waveform (wrist – elbow, axilla):

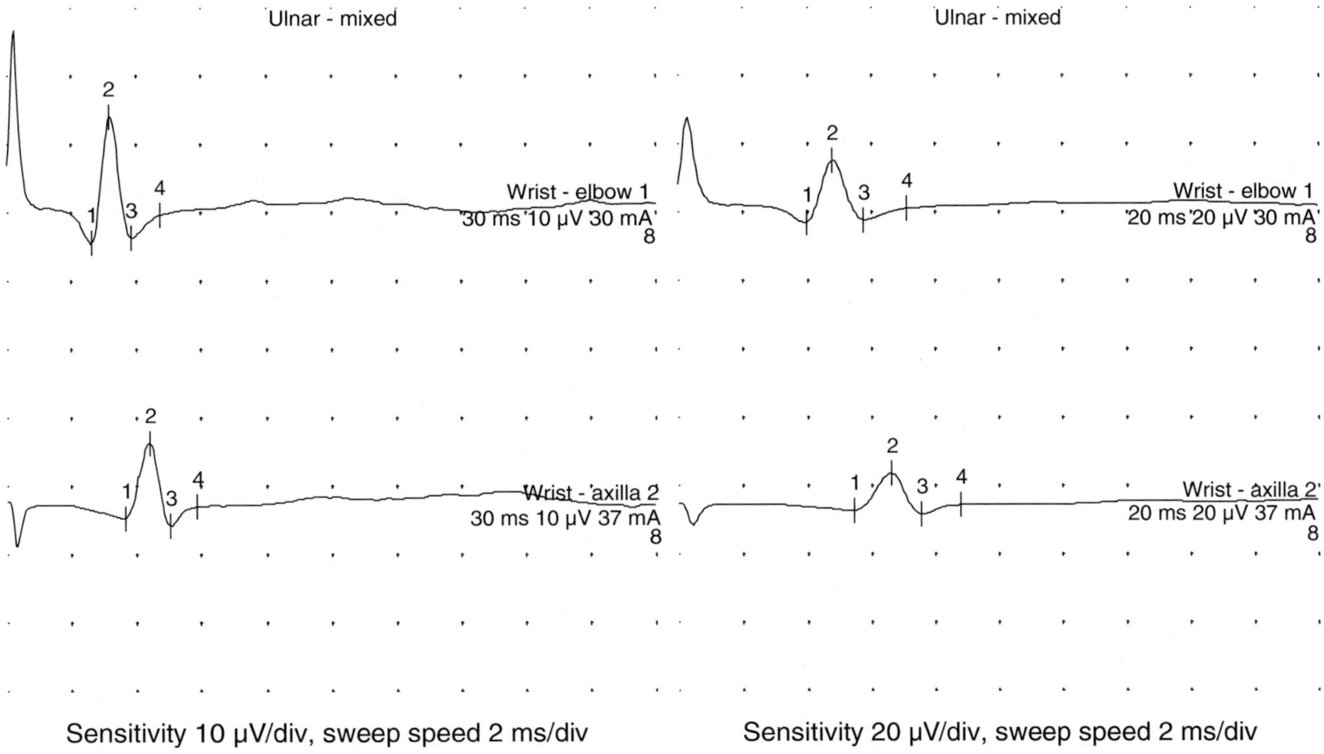

Fig. 2 Mixed nerve action potentials (*MNAPs*) recorded at the elbow (*upper trace*) and at the axilla (*lower trace*), stimulation of the wrist

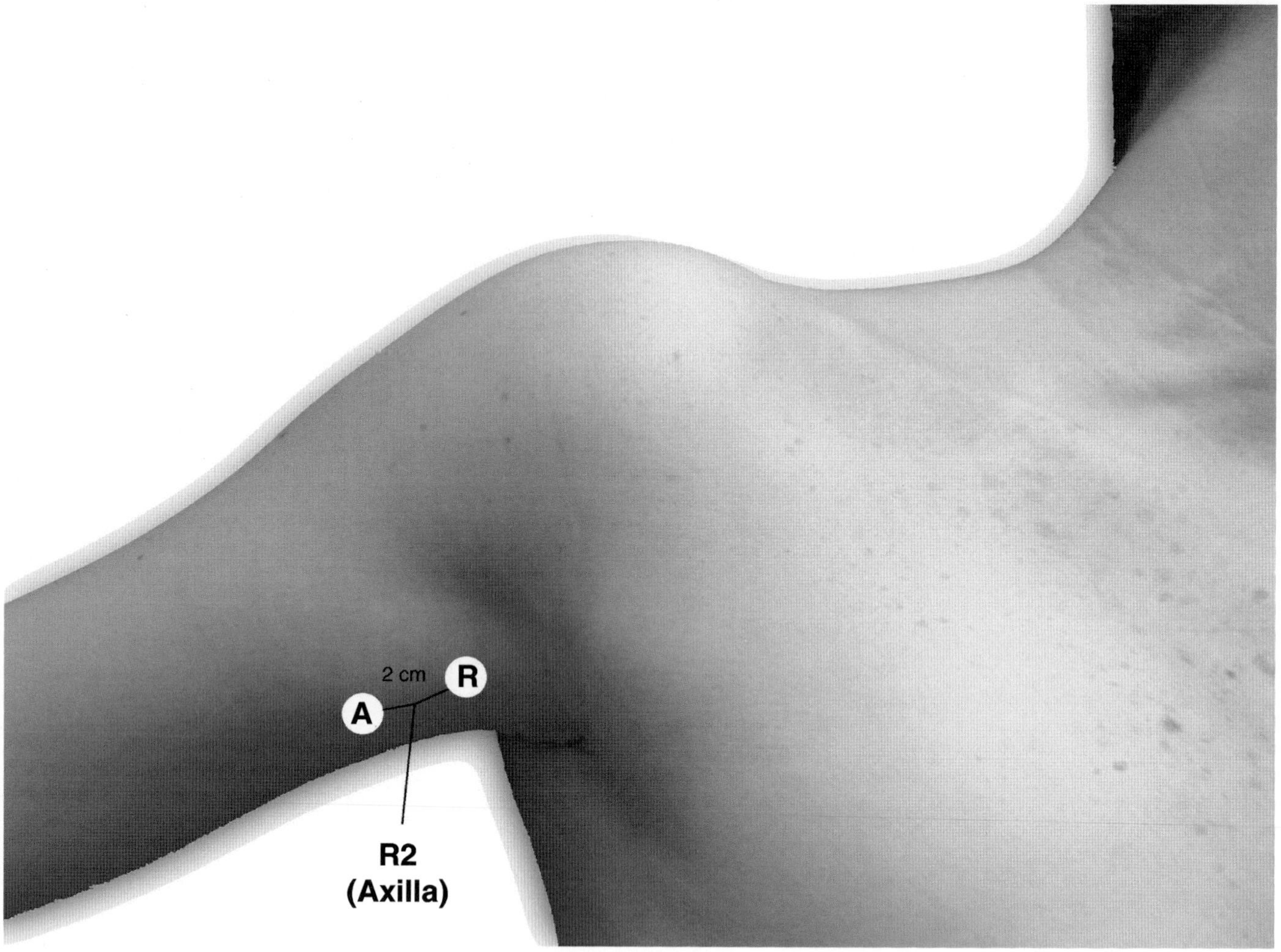

Fig. 3 Ulnar proximal recording (axilla), stimulation of the wrist

Table 1 Normal values [1], subjects ≤30 years

	Mean ± SD
Digit–palm, SNCV (m/s)	50.0 ± 4.20
Digit–palm, peak-to-peak amplitude (μV)	17.5 ± 6.15
Digit–palm, duration (m/s)	0.91 ± 0.10
Palm–wrist, SNCV (m/s)	57.6 ± 4.49
Digit–wrist, SNCV (m/s)	53.1 ± 2.36
Wrist–below elbow, SNCV (m/s)	70.5 ± 3.86
Wrist–above elbow, SNCV (m/s)	61.4 ± 4.31
Below elbow–above elbow, SNCV (m/s)	45.7 ± 6.39
Above elbow–axilla, SNCV (m/s)	69.3 ± 5.74
Wrist–below elbow, MxCV (m/s)	73.0 ± 5.21
Wrist–above elbow, MxCV (m/s)	63.2 ± 3.56
Below elbow–above elbow, MxCV (m/s)	46.0 ± 5.57
Above elbow–axilla, MxCV (m/s)	73.3 ± 7.16

Table 2 Normal values [1], subjects 31–49 years

	Mean ± SD
Digit–palm, SNCV (m/s)	50.6 ± 5.10
Digit–palm, peak-to-peak amplitude (μV)	17.4 ± 6.08
Digit–palm, duration (m/s)	0.90 ± 0.10
Palm–wrist, SNCV (m/s)	56.3 ± 4.96
Digit–wrist, SNCV (m/s)	53.8 ± 4.61
Wrist–below elbow, SNCV (m/s)	69.8 ± 4.52
Wrist–above elbow, SNCV (m/s)	60.3 ± 3.99
Below elbow–above elbow, SNCV (m/s)	41.7 ± 6.03
Above elbow–axilla, SNCV (m/s)	69.2 ± 6.11
Wrist–below elbow, MxCV (m/s)	70.8 ± 4.42
Wrist–above elbow, MxCV (m/s)	60.7 ± 3.49
Below elbow–above elbow, MxCV (m/s)	43.2 ± 6.74
Above elbow–axilla, MxCV (m/s)	70.9 ± 7.50

Table 3 Normal values [1], subjects ≥50 years

	Mean ± SD
Digit–palm, SNCV (m/s)	48.4 ± 5.57
Digit–palm, peak-to-peak amplitude (μV)	11.2 ± 4.98
Digit–palm, duration (m/s)	0.97 ± 0.11
Palm–wrist, SNCV (m/s)	54.1 ± 5.42
Digit–wrist, SNCV (m/s)	51.4 ± 3.80
Wrist–below elbow, SNCV (m/s)	64.9 ± 4.69
Wrist–above elbow, SNCV (m/s)	56.2 ± 6.16
Below elbow–above elbow, SNCV (m/s)	39.3 ± 5.69
Above elbow–axilla, SNCV (m/s)	67.6 ± 8.30
Wrist–below elbow, MxCV (m/s)	68.1 ± 6.47
Wrist–above elbow, MxCV (m/s)	58.9 ± 6.74
Below elbow–above elbow, MxCV (m/s)	39.6 ± 6.22
Above elbow–axilla, MxCV (m/s)	70.2 ± 7.25

The amplitude was 35 % greater in the palm, and the duration was 13 % larger in the wrist. The amplitude and duration of the SNAP were a linear function of the age, the former decreasing and the latter increasing. The SNCV and MxCV were faster in the wrist–below-elbow and above-elbow–axilla segments than in the wrist–above-elbow segment. The SNCV across the cubital tunnel averaged 17.3 m/s slower than in the wrist–above-elbow segment. The MxCV across the elbow was also slower than in other segments of the nerve. In their series, the authors [1] observed that an abnormal shape of the SNAP at the elbow and the decrement of the SNCV in the below-elbow segment predominated in older subjects. For the authors [1], both SNCV and MxCV calculated in the wrist–below-elbow, wrist–above-elbow, and below-elbow–above-elbow segments support the main information for diagnosis of ulnar nerve entrapment at the elbow, but their results suggest that age causes decrement of the SNCV, changes in the shape of the SNAP, or both in all peripheral nerve segments, but primarily in the sites where the nerves are habitually compressed. Pathological waveforms and values in a patient with diabetic polyneuropathy and associated ulnar neuropathy at the elbow (UNE) on the left side are reported (Figs. 4 and 5).

Pathological waveform (digit V – palm, wrist, elbow):

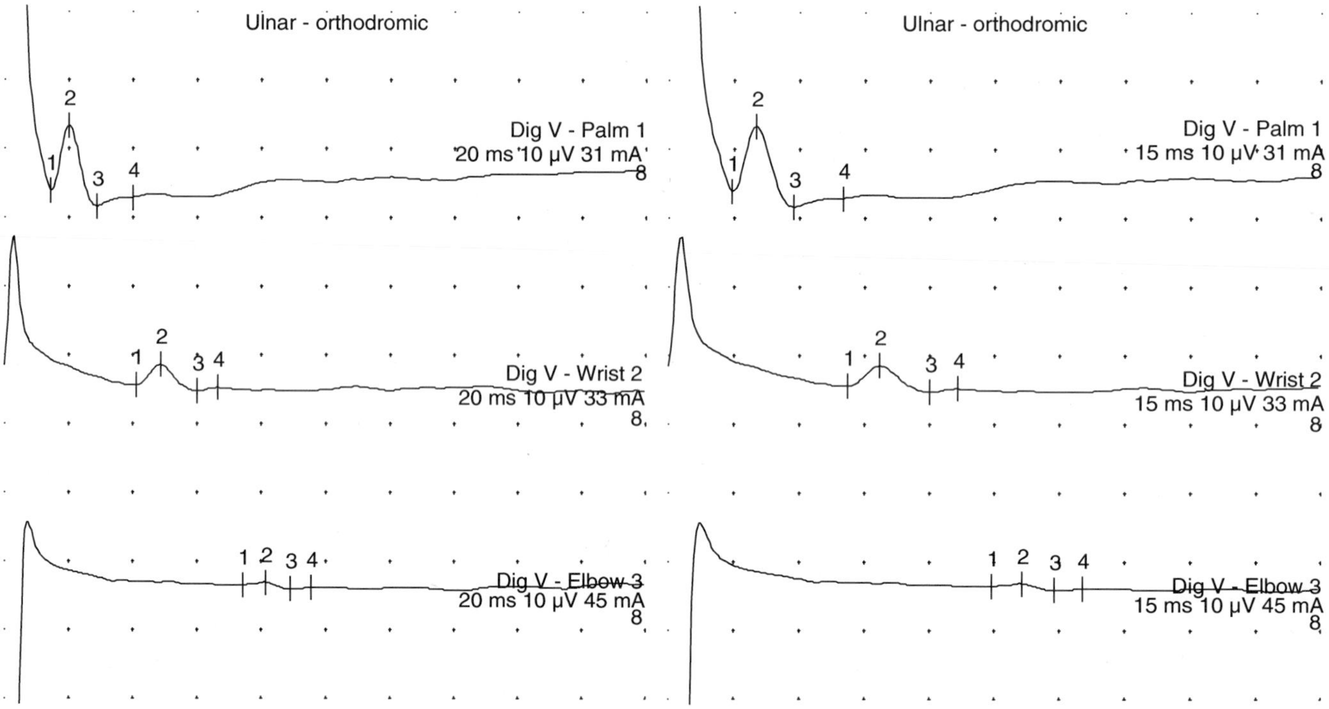

Sensitivity 10 µV/div, sweep speed 2 ms/div   Sensitivity 10 µV/div, sweep speed 1.5 ms/div

Onset latency (digit V – palm): 1.45 ms; **Onset latency** (digit V – wrist): 4.10 ms; **Onset latency** (digit V – elbow): 7.45 ms; **Peak latency** (digit V – palm): 2.0 ms; **Peak latency** (digit V – wrist): 4.90 ms; **Peak latency** (digit V – elbow): 8.10 ms; **Onset to peak amplitude** (digit V – palm): 9.6 µV; **Onset to peak amplitude** (digit V – wrist): 3.1 µV; **Onset to peak amplitude** (digit V – elbow): 0.37 µV; **Peak to peak amplitude** (digit V – palm): 11.9 µV; **Peak to peak amplitude** (digit V – wrist): 4.0 µV; **Peak to peak amplitude** (digit V – elbow): 1.0 µV; **SNCV** (digit V – palm, 7 cm distance): 48.3 m/s; **SNCV** (digit V – wrist, 18 cm distance): 43.9 m/s; **SNCV** (digit V – elbow, 32 cm distance): 43.0 m/s

Fig. 4 Orthodromic sensory nerve action potentials (*SNAPs*) recorded at the wrist (*upper trace*), below the elbow (*middle trace*), and above the elbow (*lower trace*) in a patient with diabetic polyneuropathy (*right side*), stimulation to digit V

Pathological waveform (digit V – palm, wrist, elbow):

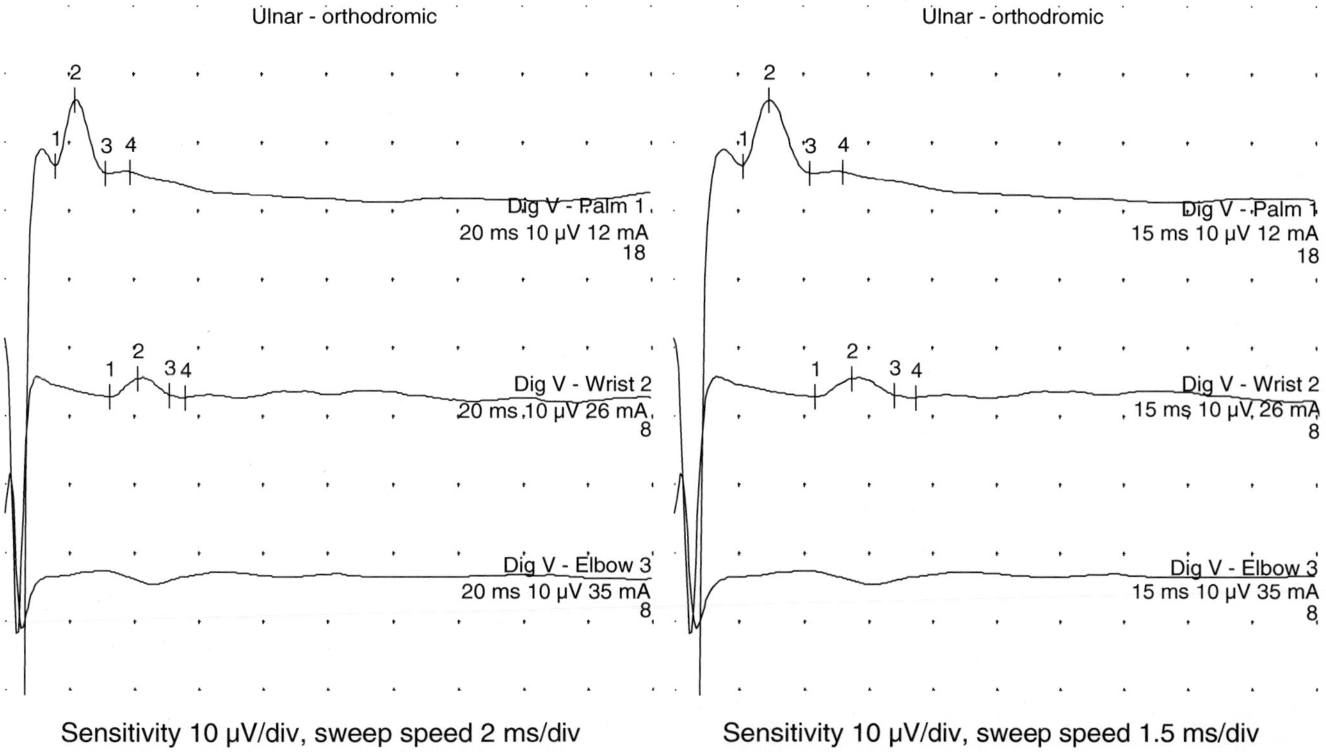

Onset latency (digit V – palm): 1.45 ms; **Onset latency** (digit V – wrist): 4.10 ms; **Onset latency** (digit V – elbow): 7.45 ms; **Peak latency** (digit V – palm): 2.0 ms; **Peak latency** (digit V – wrist): 4.90 ms; **Peak latency** (digit V – elbow): 8.10 ms; **Onset to peak amplitude** (digit V – palm): 9.6 μV; **Onset to peak amplitude** (digit V – wrist): 3.1 μV; **Onset to peak amplitude** (digit V – elbow): 0.37 μV; **Peak to peak amplitude** (digit V – palm): 11.9 μV; **Peak to peak amplitude** (digit V – wrist): 4.0 μV; **Peak to peak amplitude** (digit V – elbow): 1.0 μV; **SNCV** (digit V – palm, 7 cm distance): 48.3 m/s; **SNCV** (digit V – wrist, 18 cm distance): 43.9 m/s; **SNCV** (digit V – elbow, 32 cm distance): 43.0 m/s

Fig. 5 Orthodromic sensory nerve action potentials (*SNAPs*) recorded at the wrist (*upper trace*), below the elbow (*middle trace*), and above the elbow (*lower trace*) in a patient with diabetic polyneuropathy and associated UNE (*left side*), stimulation to digit V

References

1. Cruz Martinez A, Barrio M, Pérez Conde MC et al (1978) Electrophysiological aspects of sensory conduction velocity in healthy adults. 1. Conduction velocity from digit to palm, from palm to wrist, and across the elbow, as a function of age. J Neurol Neurosurg Psychiatry 41:1092–1096
2. Mayer RF (1963) Nerve conduction studies in man. Neurology 13:1021–1030

Digit I, II, III, V – Wrist

S11

Surface Recording Technique, Orthodromic Study

Original Settings Duration of pulse was 0.1 ms and rate of pulse was 3/s. Sensitivity, sweep speed, low-frequency filter, high-frequency filter, and the machine used were not specified.

Position Test position was not specified in the text; we performed the study with the subject lying on a comfortable couch with the arm slightly abducted, the forearm in the supine position, and elbow fully extended.

Recording Following the orthodromic method [1], sensory nerve action potentials (SNAPs) have been register at the wrist, on the median and ulnar nerves course (Figs. 1, 2, 3, and 4). The authors used surface bipolar electrodes (2 cm fixed interelectrode distance) for each recording. The ground (G) electrode position was not specified in the text; all figures show the ground electrode placed on the palm. The authors used a protocol previously described, with some modifications [2].

Stimulation The median nerve was stimulated distally to the digit I (Fig. 1), digit II (Fig. 2), and digit III (Fig. 3); the ulnar nerve was stimulated distally to the digit V (Fig. 4). In each digit stimulation, the other fingers were covered and separated each other with cotton. A fixed distance, between stimulating and recording electrodes, was not used.

Measurements Maximum sensory nerve conduction velocity (SNCV) was measured in meter per second (m/s), and it was determined for the digit–wrist segment. Peak-to-peak amplitude (μV) was measured from the negative to the positive peak. The duration (ms) of the SNAP was also measured for both median (digit I, II, and III stimulations) and ulnar (digit V stimulation) nerves. The amplitude ratio between the SNAP amplitudes of digits 2:1, 3:1, 5:2, and 5:3 was established in both hands, separately, in all subjects. The skin was prepared before exploration, and when necessary, the hand was warmed by immersion in hot water. The temperature of the skin over the nerves was 33 °C (palm and wrist) and 33.5 °C (elbow). Normal values were obtained from 47 ulnar nerves of 47 healthy subjects (25 men and 22 men, age range 21–77 years). Explorations were performed on the right hand of all subjects, divided into three groups: ≤ 30 years (Table 1), 30–49 years (Table 2), and ≥ 50 years (Table 3). There were no significant differences between the hands in the circumference of the fingers and wrist.

G. Gentili, M. Di Napoli, *The Ulnar Nerve*, DOI 10.1007/978-3-319-43292-2_11

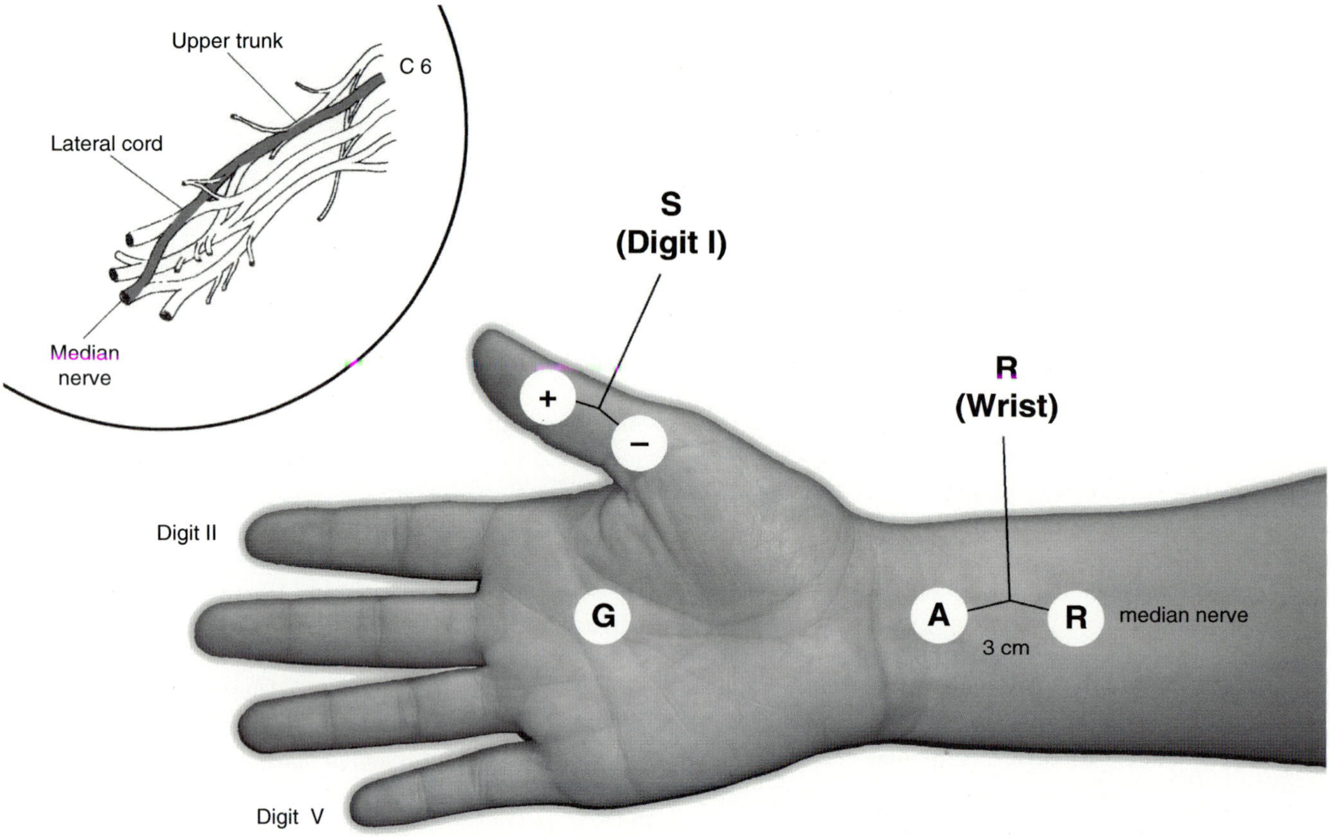

Typical waveform (digit I – wrist):

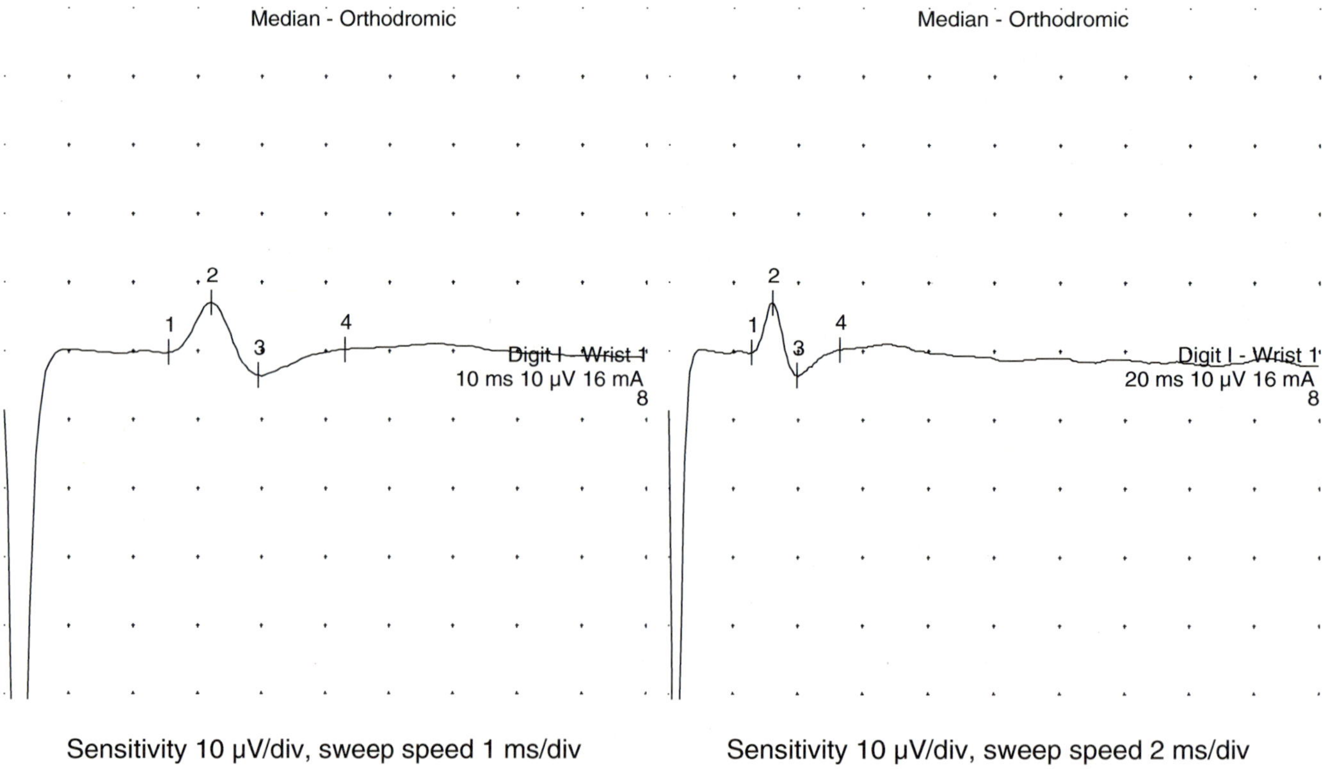

Fig. 1 Orthodromic sensory nerve action potential (*SNAPs*) recorded at the wrist, stimulation of digit I

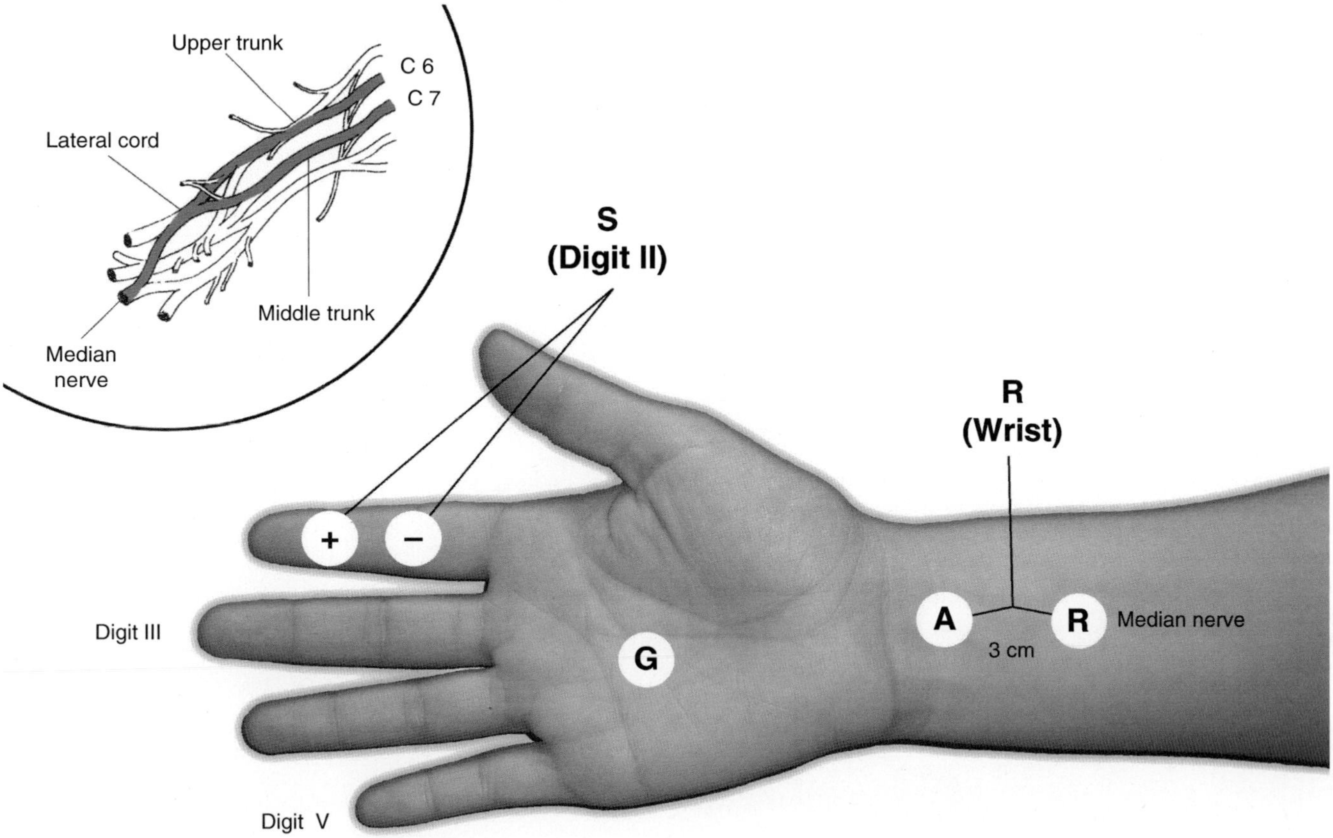

Typical waveform (digit II – wrist):

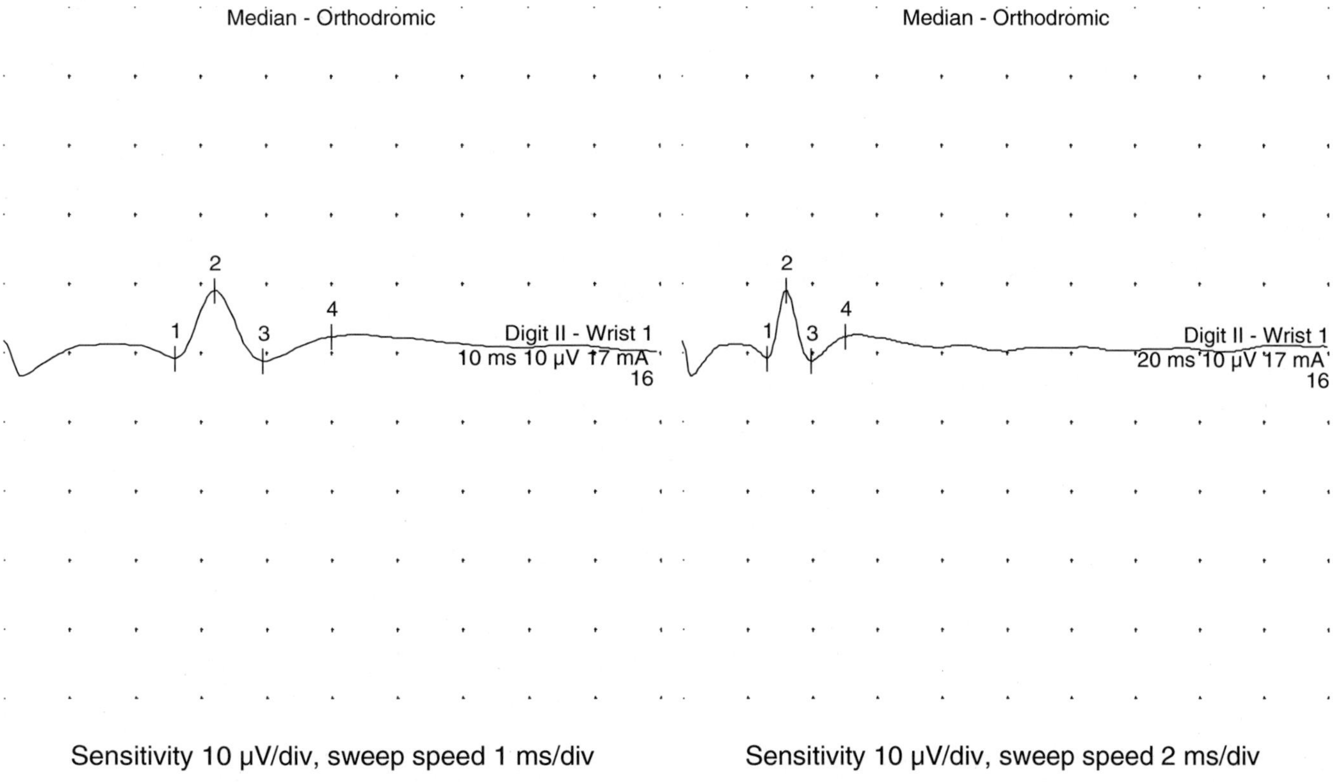

Fig. 2 Orthodromic sensory nerve action potential (*SNAPs*) recorded at the wrist, stimulation of digit II

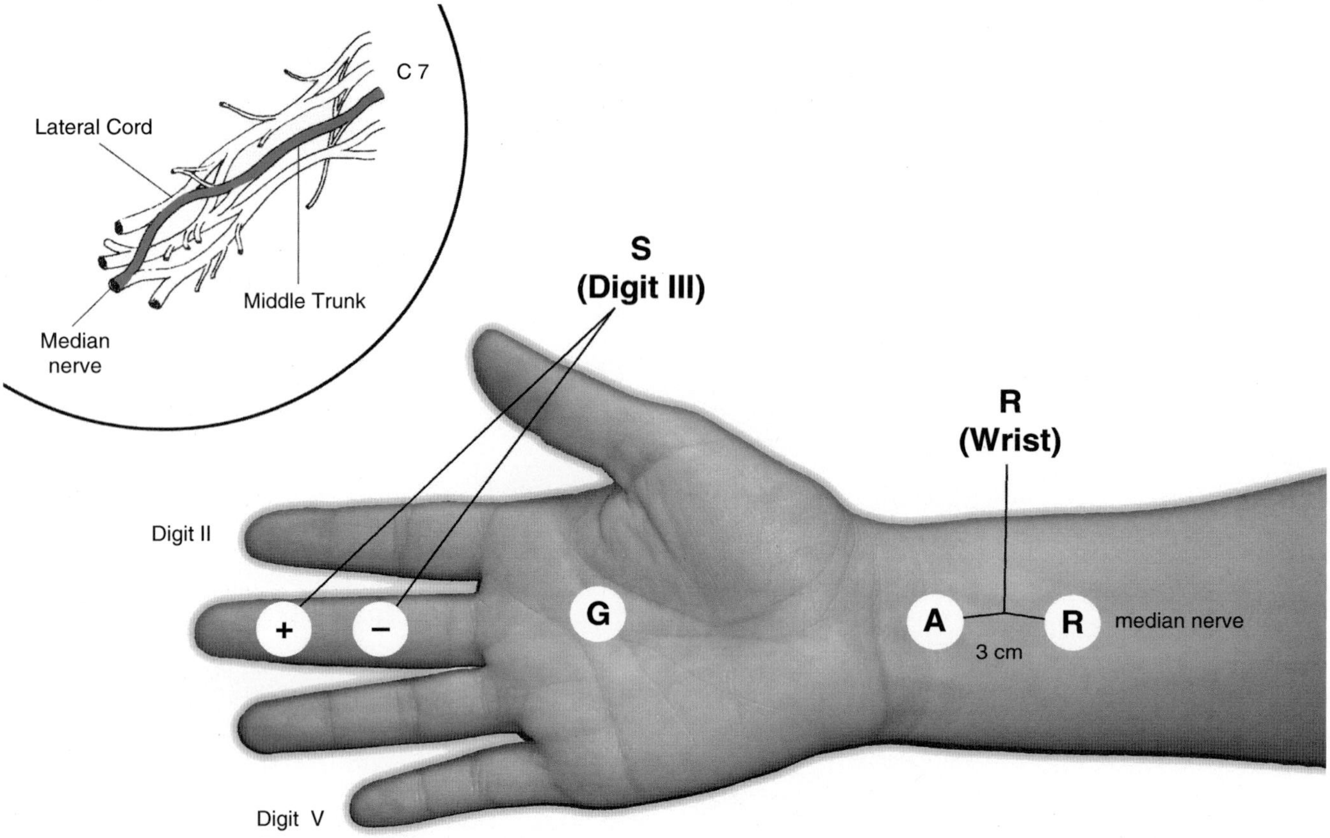

Typical waveform (digit III – wrist):

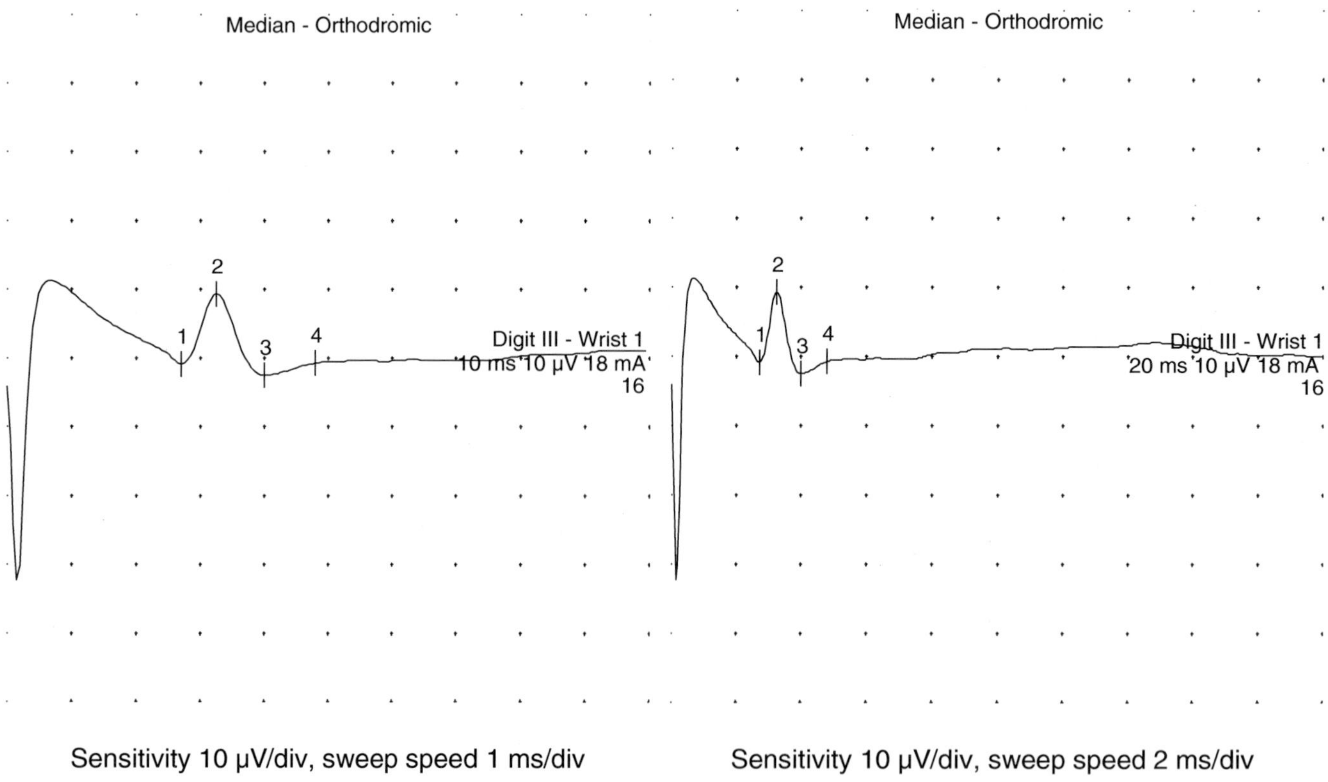

Fig. 3 Orthodromic sensory nerve action potential (*SNAPs*) recorded at the wrist, stimulation of digit III

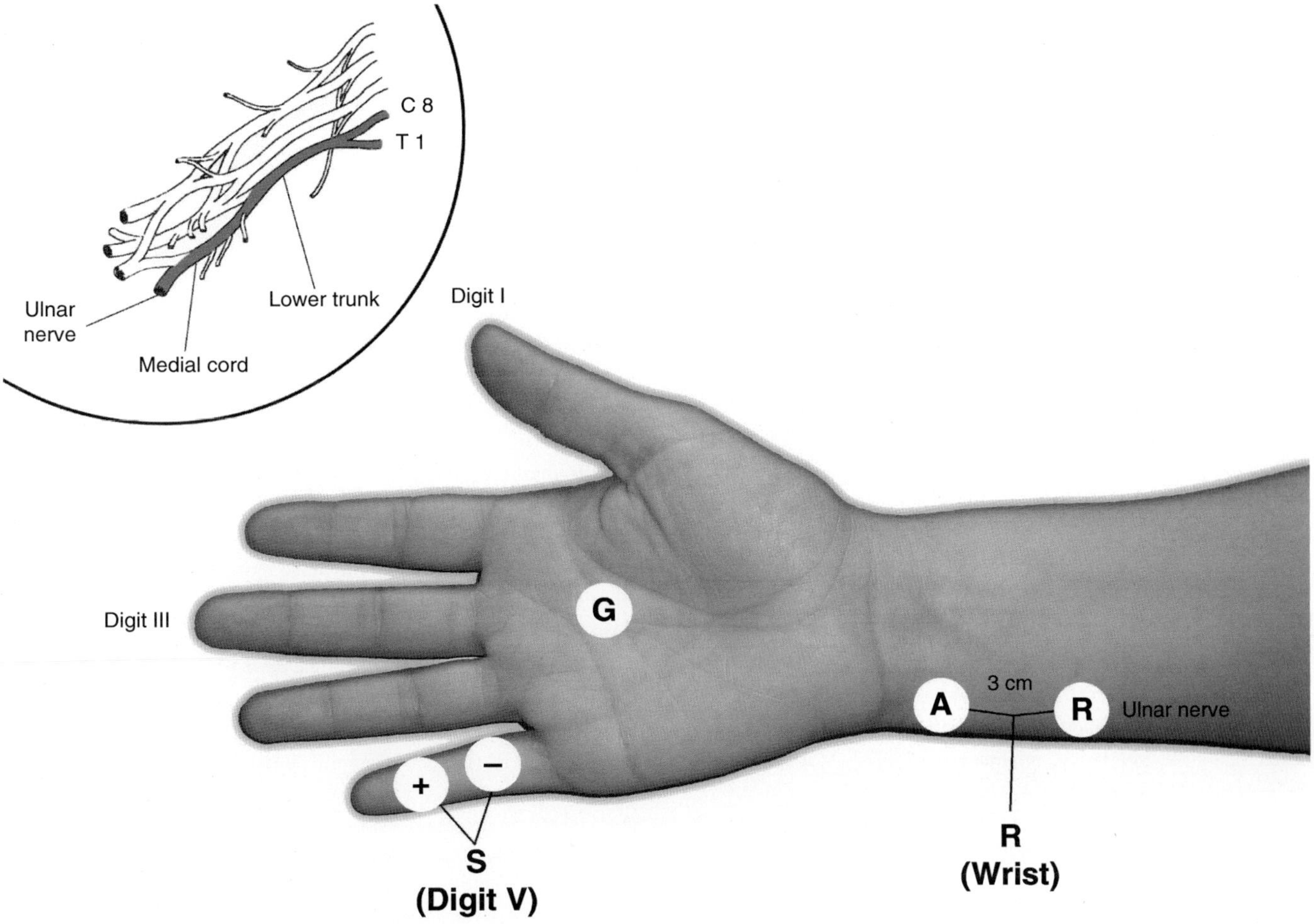

Typical waveform (digit V – wrist):

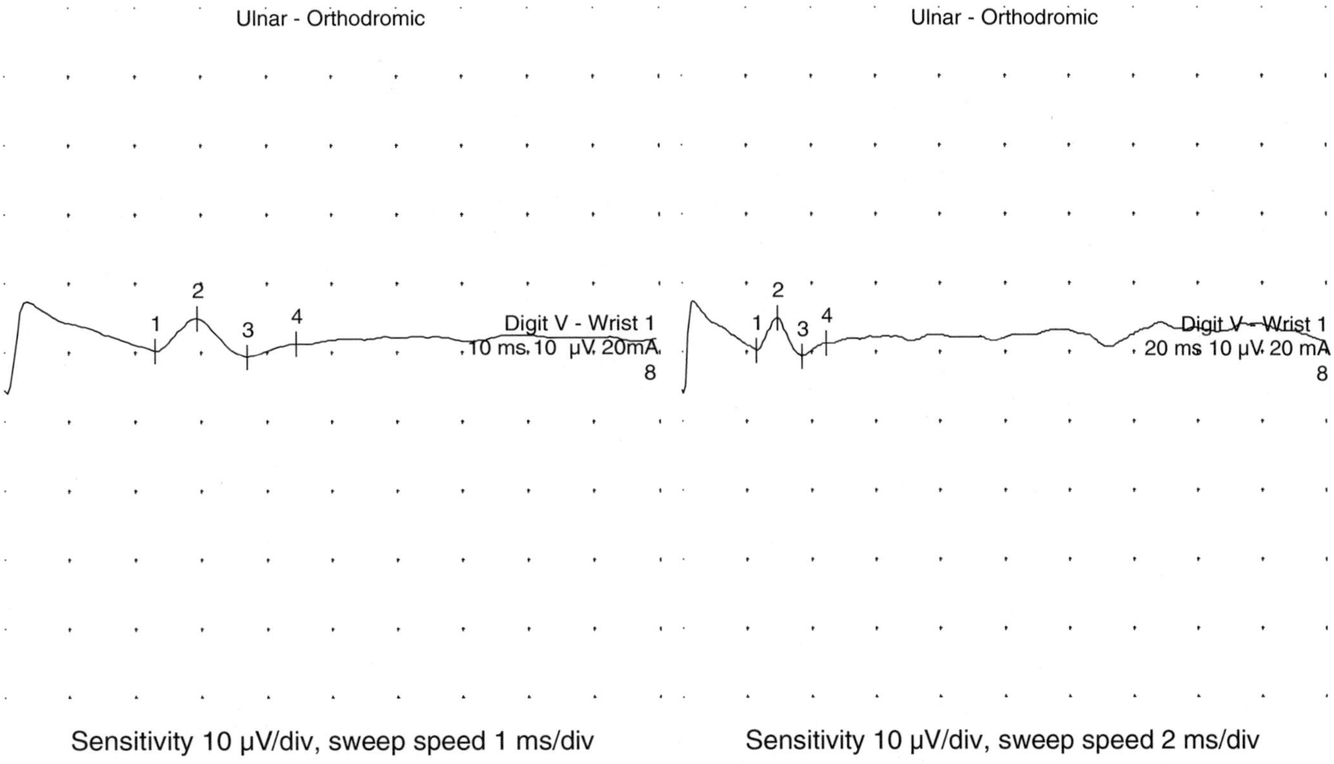

Fig. 4 Orthodromic sensory nerve action potential (*SNAPs*) recorded at the wrist, stimulation of digit V

Table 1 Normal values [1] – subjects ≤ 30 years

	Mean ± SD
Digit I–wrist, SNCV (m/s)	44.2 ± 3.28
Digit I–wrist, peak-to-peak amplitude (μV)	40.3 ± 9.30
Digit I–wrist, duration (m/s)	0.94 ± 0.10
Digit II–wrist, SNCV (m/s)	54.6 ± 4.10
Digit II–wrist, peak-to-peak amplitude (μV)	16.3 ± 4.40
Digit II–wrist, duration (m/s)	1.06 ± 0.07
Digit III–wrist, SNCV (m/s)	53.4 ± 3.0
Digit III–wrist, peak-to-peak amplitude (μV)	22.2 ± 7.30
Digit III–wrist, duration (m/s)	1.05 ± 0.10
Digit V–wrist, SNCV (m/s)	53.1 ± 2.36
Digit V–wrist, peak-to-peak amplitude (μV)	11.5 ± 2.68
Digit V–wrist, duration (m/s)	1.01 ± 0.08
Digit II–digit I (right hand), amplitude ratio	0.41 ± 0.08
Digit III–digit I (right hand), amplitude ratio	0.55 ± 0.17
Digit V–digit I (right hand), amplitude ratio	0.28 ± 0.09
Digit V–digit II (right hand), amplitude ratio	0.69 ± 0.18
Digit V–digit III (right hand), amplitude ratio	0.51 ± 0.16
Digit II–digit I (left hand), amplitude ratio	0.40 ± 0.09
Digit III–digit I (left hand), amplitude ratio	0.55 ± 0.12
Digit V–digit I (left hand), amplitude ratio	0.27 ± 0.08
Digit V–digit II (left hand), amplitude ratio	0.68 ± 0.21
Digit V–digit III (left hand), amplitude ratio	0.50 ± 0.13

Table 2 Normal values [1] – subjects 31–49 years

	Mean ± SD
Digit I–wrist, SNCV (m/s)	43.5 ± 3.24
Digit I–wrist, peak-to-peak amplitude (μV)	33.7 ± 11.20
Digit I–wrist, duration (m/s)	0.96 ± 0.09
Digit II–wrist, SNCV (m/s)	54.2 ± 3.66
Digit II–wrist, peak-to-peak amplitude (μV)	13.7 ± 3.40
Digit II–wrist, duration (m/s)	1.07 ± 0.08
Digit III–wrist, SNCV (m/s)	53.6 ± 2.97
Digit III–wrist, peak-to-peak amplitude (μV)	18.0 ± 5.70
Digit III–wrist, duration (m/s)	1.05 ± 0.09
Digit V–wrist, SNCV (m/s)	53.8 ± 4.61
Digit V–wrist, peak-to-peak amplitude (μV)	11.0 ± 2.58
Digit V–wrist, duration (m/s)	1.02 ± 0.07
Digit II–digit I (right hand), amplitude ratio	0.39 ± 0.08
Digit III–digit I (right hand), amplitude ratio	0.51 ± 0.09
Digit V–digit I (right hand), amplitude ratio	0.32 ± 0.08
Digit V–digit II (right hand), amplitude ratio	0.81 ± 0.13
Digit V–digit III (right hand), amplitude ratio	0.62 ± 0.13
Digit II–digit I (left hand), amplitude ratio	0.41 ± 0.06
Digit III–digit I (left hand), amplitude ratio	0.66 ± 0.12
Digit V–digit I (left hand), amplitude ratio	0.30 ± 0.06
Digit V–digit II (left hand), amplitude ratio	0.74 ± 0.17
Digit V–digit III (left hand), amplitude ratio	0.46 ± 0.15

Table 3 Normal values [1] – subjects ≥ 50 years

	Mean ± SD
Digit I–wrist, SNCV (m/s)	41.2 ± 3.63
Digit I–wrist, peak-to-peak amplitude (μV)	19.8 ± 4.30
Digit I–wrist, duration (m/s)	1.04 ± 0.10
Digit II–wrist, SNCV (m/s)	52.5 ± 3.90
Digit II–wrist, peak-to-peak amplitude (μV)	10.1 ± 1.70
Digit II–wrist, duration (m/s)	1.12 ± 0.12
Digit III–wrist, SNCV (m/s)	51.5 ± 2.99
Digit III–wrist, peak-to-peak amplitude (μV)	12.0 ± 2.50
Digit III–wrist, duration (m/s)	1.12 ± 0.12
Digit V–wrist, SNCV (m/s)	51.4 ± 3.80
Digit V–wrist, peak-to-peak amplitude (μV)	7.5 ± 2.42
Digit V–wrist, duration (m/s)	1.13 ± 0.13
Digit II–digit I (right hand), amplitude ratio	0.51 ± 0.08
Digit III–digit I (right hand), amplitude ratio	0.61 ± 0.16
Digit V–digit I (right hand), amplitude ratio	0.38 ± 0.08
Digit V–digit II (right hand), amplitude ratio	0.75 ± 0.16
Digit V–digit III (right hand), amplitude ratio	0.62 ± 0.19
Digit II–digit I (left hand), amplitude ratio	0.45 ± 0.10
Digit III–digit I (left hand), amplitude ratio	0.56 ± 0.12
Digit V–digit I (left hand), amplitude ratio	0.33 ± 0.05
Digit V–digit II (left hand), amplitude ratio	0.72 ± 0.15
Digit V–digit III (left hand), amplitude ratio	0.58 ± 0.13

Comment

Cruz Martinez et al. [1] observed that the ratio of amplitude changed according to the age. The ratios 2:1, 3:1, 5:1, 5:2, and 5:3 increased in elderly subjects. According to their results, the amplitude of SNAP of the digits innervated by the median nerve decreased more than in the digits innervated by the ulnar nerve, and this reduction was different for each finger. These findings suggest that age causes changes particularly in the points where the nerves are more frequently compressed, such as the carpal or the cubital tunnels. In the 44 right-handed subjects in their series, the authors [1] compared the amplitude of the SNAPs in both hands. The SNCV and duration of the SNAP were the same in each hand, whereas the amplitude of SNAP at the wrist was larger in the left hand than in the right one, especially for digit III, digit I, and digit II. The amplitude at the wrist was 30.4 (digit I), 28.1 (digit II), and 21.1 % (digit V) higher in the left hand than in the right. The circumference of the fingers and wrist, the skin temperature, and the distance between stimulating and recording points did not show significant differences between hands. For Cruz Martinez et al. [1], the analysis of the ratio between the amplitude of the SNAP that corresponds to each finger can be more exact than the study of the absolute values of the amplitude. This method permits a more precise and earlier topographic localization in radicular or brachial plexus lesions, and according to a segmental sensory innervations (digit I, C6; digit II, C6–C7; digit III, C7; digit V, C8–T1), the variations of the SNAP of each finger provide information about which spinal nerve is damaged.

References

1. Cruz Martinez A, Barrio M, Pérez Conde MC et al (1978) Electrophysiological aspects of sensory conduction velocity in healthy adults. 2. Ratio between the amplitude of sensory evoked potentials at the wrist on stimulating different fingers in both hands. J Neurol Neurosurg Psychiatry 41:1097–1101
2. Cruz Martinez A, Barrio M, Pérez Conde MC et al (1978) Electrophysiological aspects of sensory conduction velocity in healthy adults. 1. Conduction velocity from digit to palm, from palm to wrist, and across the elbow, as a function of age. J Neurol Neurosurg Psychiatry 41:1092–1096

Wrist, Palm – Digit II, V

Surface Recording Technique, Antidromic Study

S12

Original Settings Sensitivity was 20 μV/division, sweep speed was 2 ms/division, duration of pulse was 0.1–0.4 ms, and the machine used was a Tektronix electromyograph. Low-frequency filter, high-frequency filter, and rate of pulse were not specified.

Position This study was performed in the supine position.

Recording Following the antidromic method [1], signals were recorded using surface electrodes from the median and ulnar nerves separately. For the median nerve (R1), the active recording electrode (A) was placed around the proximal interphalangeal joint of digit II; the reference electrode (R) was placed over the distal interphalangeal joint of digit II (Fig. 1). For the ulnar nerve (R1), the active recording electrode (A) was placed around the proximal interphalangeal joint of digit V; the reference electrode (R) was placed over the distal interphalangeal joint of digit V (Fig. 2). Adjacent digits were kept from inadvertently contacting each other. The ground (G) was placed on the dorsum of the hand, between the recording and stimulating electrodes; the figure shows the ground electrode placed on the palm.

Stimulation Stimulation with the cathode distal was applied over the median and ulnar nerves at the wrist, at 7 and 14 cm fixed distance proximal to the proximal interphalangeal joint of digit II and digit V, respectively. Conduction time for the wrist–digit segment of the nerve was obtained by stimulating at 14 cm (wrist level), while the palm–digit conduction time was obtained by stimulating at 7 cm (midpalmar level). For the median and ulnar nerves, stimulation was applied at the wrist level (S1) and the midpalmar level (S2). For each nerve, the conduction time for the wrist–palm segment was obtained by subtracting the palm–digit conduction time from the wrist–digit conduction time. The authors used only supramaximal stimulation.

Measurements Onset latency (ms) was measured from the stimulus onset to the onset of the negative deflection of the evoked sensory nerve action potential (SNAP); peak latency (ms) was measured from the stimulus onset to the peak of the negative deflection of the SNAP. All latencies for each subject were determined at the same session and room temperature. The authors studied 30 normal (Table 1) subjects (age range 22–68 years) and 33 patients (Table 2) suspected clinically of having carpal tunnel syndrome (CTS) and normal distal sensory latency (sample data were not given).

Comment

Felsenthal and Spindler [1] for the wrist–digit, palm–digit, and wrist–palm segment sensory conduction measured the initial deflection (onset latency) and the peak of the initial negative deflection (peak latency) for both the median and ulnar nerves. The sensory conduction times of the wrist–digit, palm–digit, and wrist–palm segments of the median and ulnar nerves in opposite hands (side to side difference) were compared; the mean difference and SD values of paired median and ulnar segment conduction times (onset and peak latency) in the same hand were also calculated.

Comment

For Felsenthal and Spindler [1], with reference to the pathological values, the wrist–palm conduction time for the median nerve ranged from 1.7 to 2.1 ms or less to the peak of the initial negative deflection. In ten patients, the median and ulnar comparison for the wrist–digit conduction time ranged from 0.5 to 1.1 ms for the initial deflection from the baseline and 0.6–1.1 ms for the peak of the initial deflection and 1.8–2.3 ms for measurement of the peak deflection. In each of these patients, both comparisons of wrist–digit and the wrist–palm conduction times exceeded the normal standards of 0.5 for wrist–digit comparisons and 1.6 ms to initial deflection or 1.7 ms to the peak deflection for wrist–palm conduction times. For the authors, the results of these two electrodiagnostic methods paralleled each other, and neither appeared more sensitive than the other in establishing the diagnosis.

G. Gentili, M. Di Napoli, *The Ulnar Nerve*, DOI 10.1007/978-3-319-43292-2_12

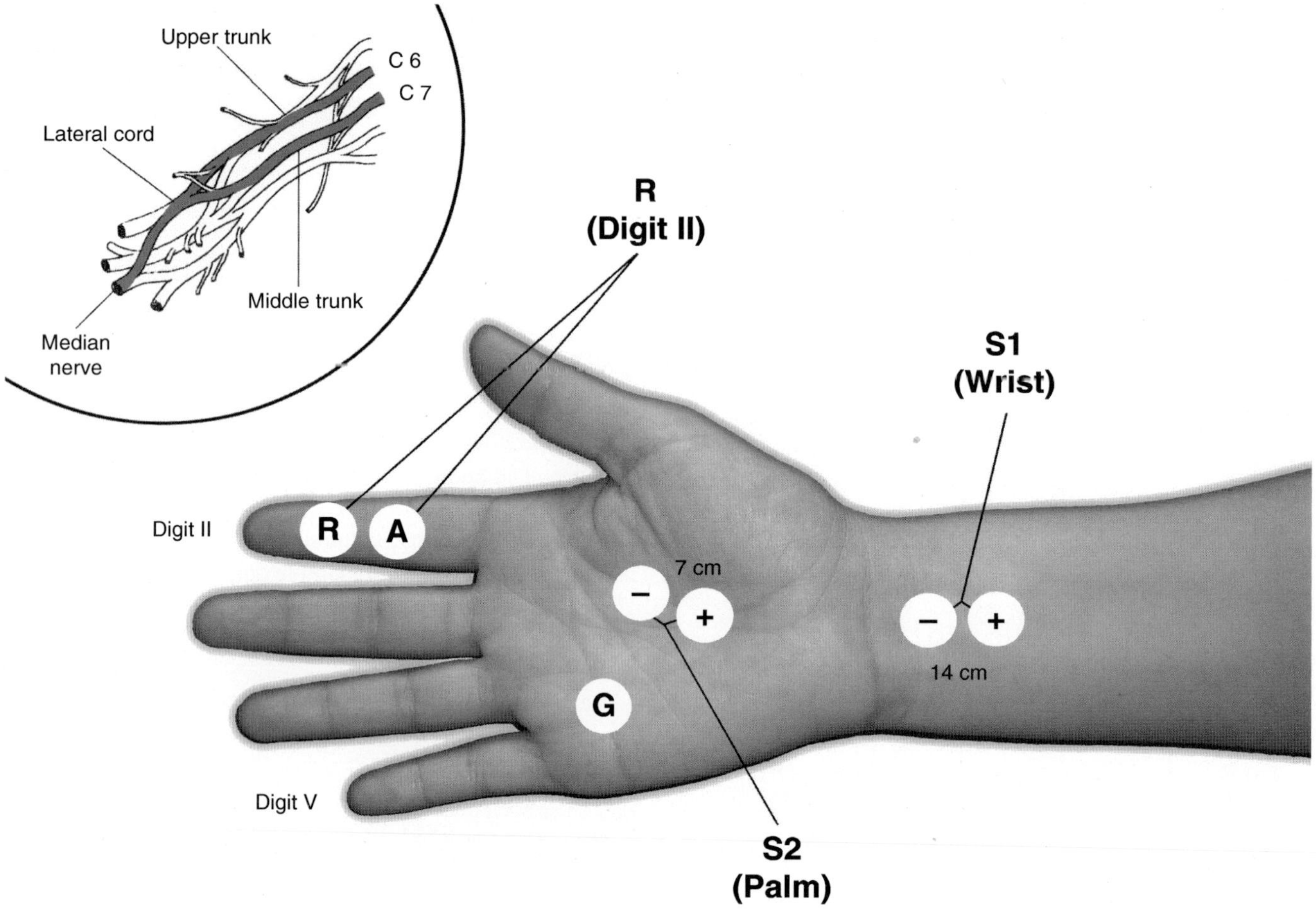

Typical waveform (wrist, palm – digit II):

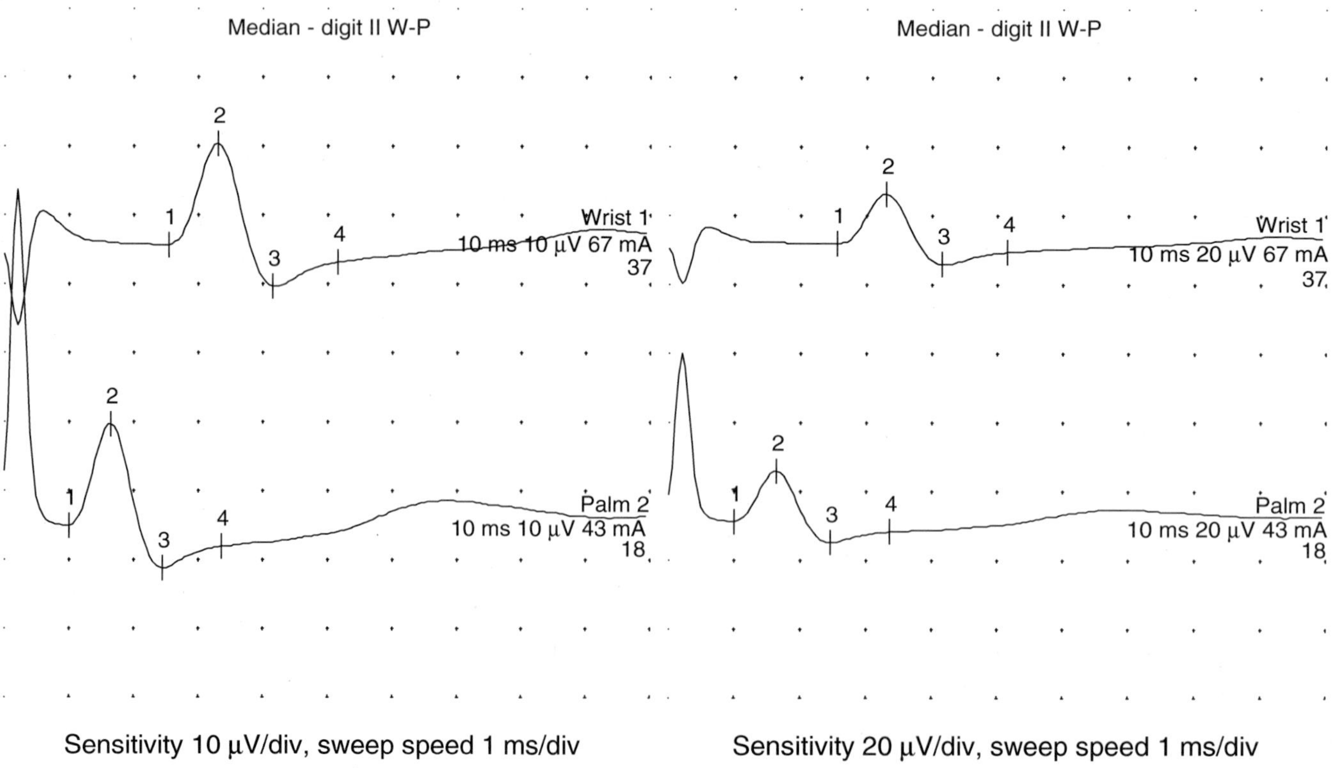

Fig. 1 Median nerve: antidromic sensory nerve action potentials (*SNAPs*) to digit II; stimulation of the wrist (*upper trace*) and on the palm (*lower trace*)

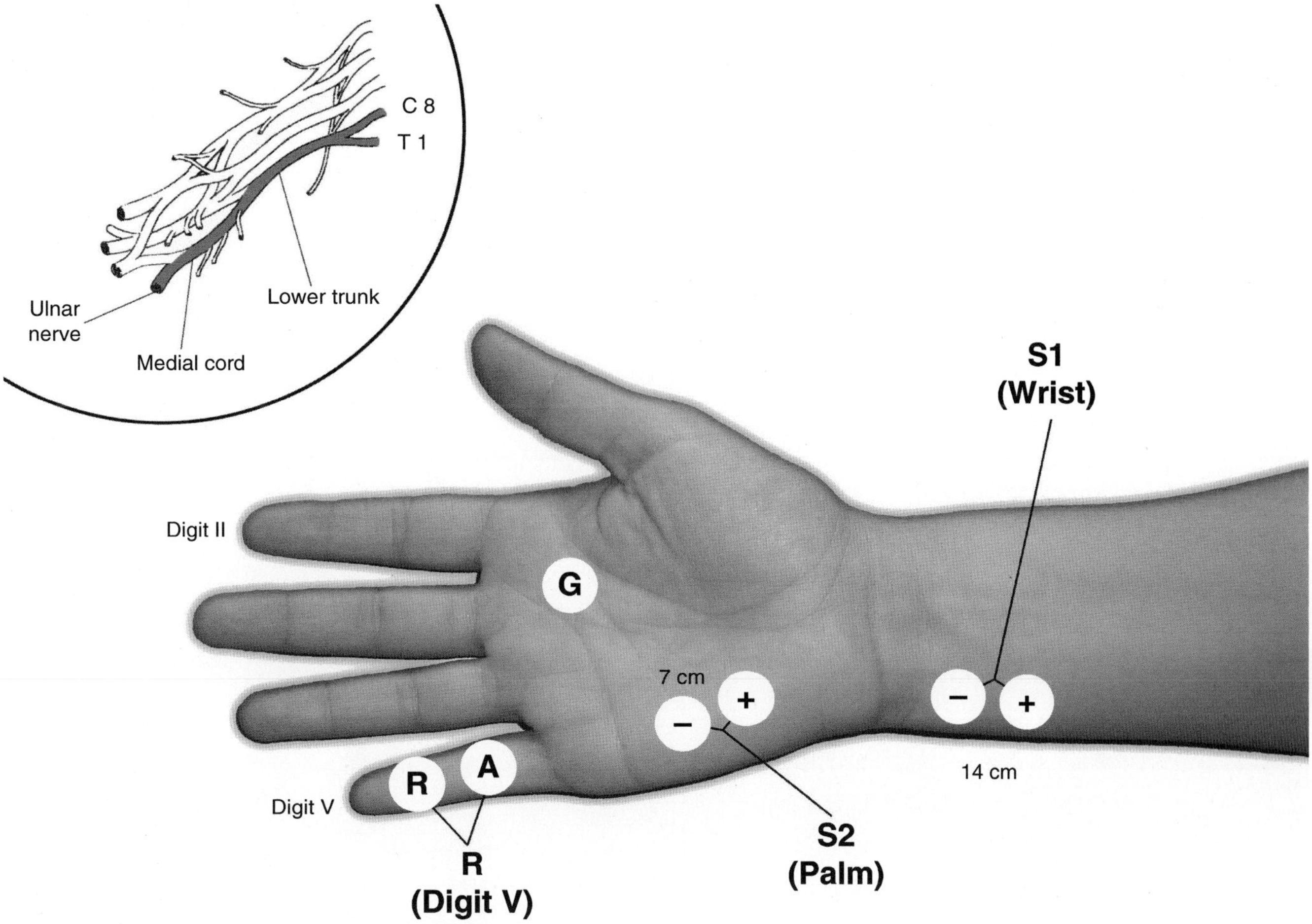

Typical waveform (wrist, palm – digit V):

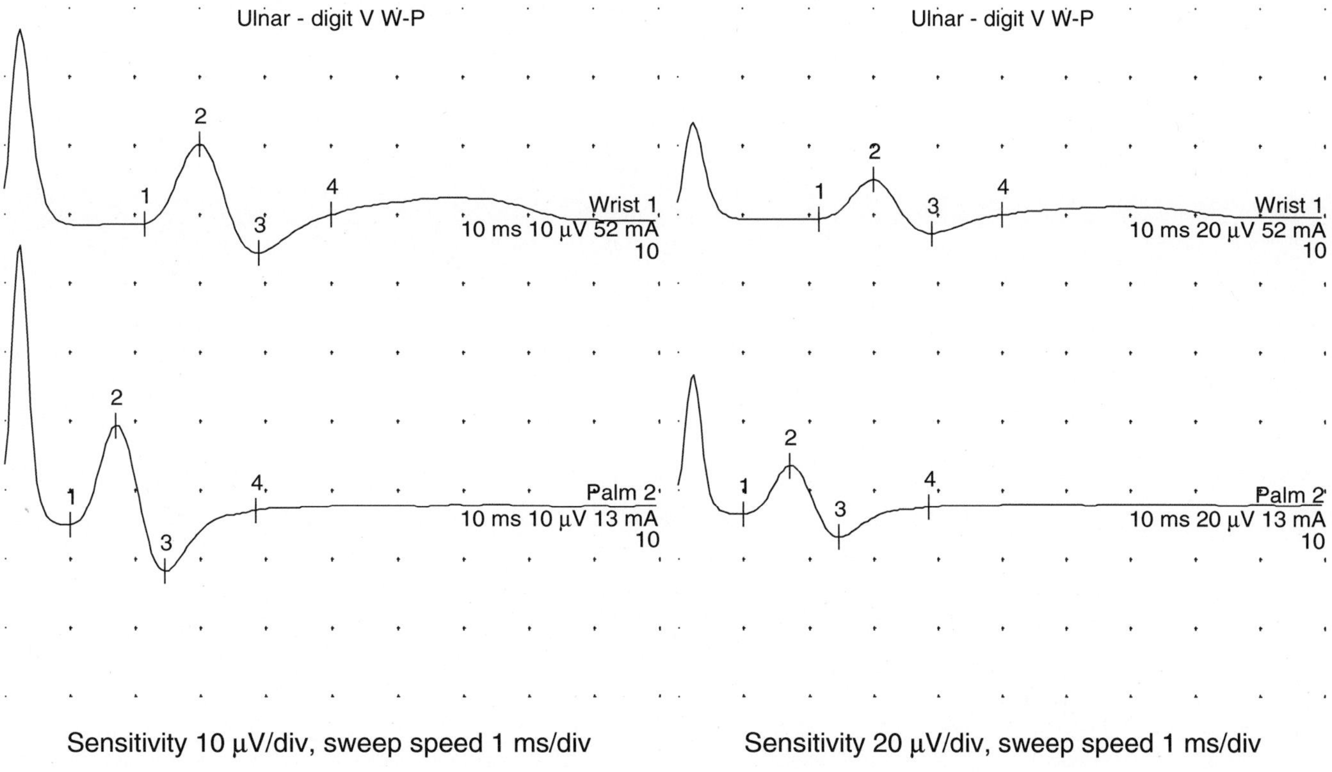

Fig. 2 Ulnar nerve: antidromic sensory nerve action potentials (*SNAPs*) to digit V; stimulation of the wrist (*upper trace*) and on the palm (*lower trace*)

Table 1 Normal values [1]

Median nerve – onset latency (ms)	Distance (cm)	Mean ± SD	Mean ± 2 SD	Range
Wrist–digit II	14	2.6 ± 0.20	3.0	2.1–3.0
Palm–digit II	7	1.3 ± 0.13	1.6	1.0–1.6
Wrist–palm	7	1.3 ± 0.13	1.6	1.0–1.7
Ulnar nerve – onset latency (ms)	Distance (cm)	Mean ± SD	Mean ± 2 SD	Range
Wrist–digit V	14	2.5 ± 0.19	2.9	2.1–2.9
Palm–digit V	7	1.3 ± 0.14	1.6	1.1–1.7
Wrist–palm	7	1.2 ± 0.11	1.4	1.0–1.4
Median nerve – peak latency (ms)	Distance (cm)	Mean ± SD	Mean ± 2 SD	Range
Wrist–digit II	14	3.2 ± 0.25	3.7	2.7–3.7
Palm–digit II	7	1.8 ± 0.19	2.2	1.5–2.2
Wrist–palm	7	1.4 ± 0.16	1.7	1.0–1.9
Ulnar nerve – peak latency (ms)	Distance (cm)	Mean ± SD	Mean ± 2 SD	Range
Wrist–digit V	14	3.1 ± 0.25	3.6	2.8–3.7
Palm–digit V	7	1.9 ± 0.19	2.3	1.6–2.4
Wrist–palm	7	1.2 ± 0.14	1.5	1.0–1.6
Median nerve – onset latency (ms) *Differences side to side*	Distance (cm)	Mean ± SD	Mean ± SD	Range
Wrist–digit II	14	0.17 ± 0.14	0.45	0.0–0.5
Palm–digit II	7	0.08 ± 0.07	0.22	0.0–0.4
Wrist–palm	7	0.13 ± 0.11	0.35	0.0–0.3
Ulnar nerve – onset latency (ms) *Differences side to side*	Distance (cm)	Mean ± SD	Mean ± SD	Range
Wrist–digit V	14	0.11 ± 0.08	0.27	0.0–0.4
Palm–digit V	7	0.04 ± 0.05	0.14	0.0–0.1
Wrist–palm	7	0.10 ± 0.08	0.26	0.0–0.3
Median nerve – peak latency *(ms)* *Differences side to side*	Distance (cm)	Mean ± SD	Mean ± 2 SD	Range
Wrist–digit II	14	0.19 ± 0.15	0.49	0.0–0.5
Palm–digit II	7	0.11 ± 0.11	0.33	0.0–0.5
Wrist–palm	7	0.13 ± 0.13	0.40	0.0–0.4
Ulnar nerve – peak latency (ms) *Differences side to side*	Distance (cm)	Mean ± SD	Mean ± 2 SD	Range
Wrist–digit V	14	0.12 ± 0.11	0.34	0.0–0.4
Palm–digit V	7	0.07 ± 0.05	0.17	0.0–0.2
Wrist–palm	7	0.12 ± 0.07	0.26	0.0–0.3
Median–ulnar nerves Onset latency (ms) *Differences in the same side*	Distance (cm)	Mean ± SD	Mean ± 2 SD	Range
Wrist–digit	14	0.14 ± 0.12	0.38	0.0–0.4
Palm–digit	7	0.09 ± 0.07	0.23	0.0–0.3
Wrist–palm	7	0.14 ± 0.12	0.38	0.0–0.5
Median–ulnar nerves Peak latency (ms) *Differences in the same side*	Distance (cm)	Mean ± SD	Mean ± 2 SD	Range
Wrist–digit	14	0.17 ± 0.15	0.46	0.0–0.5
Palm–digit	7	0.14 ± 0.09	0.32	0.0–0.3
Wrist–palm	7	0.18 ± 0.15	0.49	0.0–0.7
Median–ulnar nerves Palm latency difference (ms)	Mean ± SD	Range	Limit of normal (−2SD)	
Onset latency – all subjects	0.12 ± 0.13	0–0.5	≤0.38	
Peak latency – all subjects	0.15 ± 0.17	0–0.7	≤0.49	

Table 2 Pathological values [1]

Median nerve Wrist–digit II	Onset latency (ms)	Peak latency (ms)
Normal limit (≤)	3.1	3.8
Patient 1	2.8	3.5
Patient 2	3.1	3.8
Patient 3	3.0	3.6
Patient 4	3.0	3.5
Patient 5	3.0	3.9
Patient 6	3.1	3.7
Patient 7	3.1	3.8
Patient 8	3.0	3.6
Patient 9	3.1	3.7
Patient 10	2.9	3.4
Median nerve Wrist–palm	Onset latency (ms)	Peak latency (ms)
Normal limit (≤)	1.6	1.7
Patient 1	1.9	2.2
Patient 2	1.9	2.1
Patient 3	2.1	2.3
Patient 4	1.7	1.8
Patient 5	1.7	1.9
Patient 6	2.0	2.1
Patient 7	2.0	2.0
Patient 8	1.9	2.0
Patient 9	1.9	2.1
Patient 10	1.8	1.9
Median–ulnar nerves Wrist–digit latency difference	Onset latency (ms)	Peak latency (ms)
Normal limit (≤)	0.5	0.5
Patient 1	0.6	0.6
Patient 2	0.6	0.7
Patient 3	0.8	0.9
Patient 4	0.6	0.6
Patient 5	0.5	0.6
Patient 6	1.1	1.1
Patient 7	0.6	0.7
Patient 8	0.7	0.7
Patient 9	0.9	0.9
Patient 10	0.8	1.8

Andary et al. [2] in a prospective study assessed a technique of comparing median–ulnar sensory latency differences across the wrist to diagnose CTS and compared the relative sensitivities of other previously reported techniques (antidromic median–radial sensory latency difference, antidromic median–ulnar sensory difference, orthodromic median–ulnar mixed palm difference). They studied 33 hands (Table 3) in 17 asymptomatic control subjects (19 women and 14 men, average age 36 years) and 90 hands in 81 patients with electrodiagnosis of CTS (mean age 42 years). The authors performed conduction studies on Cadwell 5200 A, Cadwell Quantum 84, and Nicolet Viking II electromyography instruments (filter settings: 10–2 kHz). Skin temperature was measured during the exam; all temperatures over the proximal carpal tunnel were maintained above 30 °C (hands were warmed if necessary).

Andary et al. [2] used the same protocol by Felsenthal and Spindler [1] with a modified fixed distance for palm stimulation (Figs. 3 and 4) which was set on the palm at 6 cm from the active recording electrode to digit II (median nerve study) and digit V (ulnar nerve study). Standard antidromic median sensory latencies were recorded over digit II (wrist–digit II segment) after stimulation at 14 cm proximal to the distal wrist crease (called "Sensory Index"); median

Table 3 Pathological values [2]

Median Onset latency (ms)	Distance (cm)	Mean ± SD	Mean ± 2 SD	Limit of normal
Wrist–digit II	14	3.3 ± 0.22	3.74	≤3.9
Palm–digit II	6	1.72 ± 0.24	2.20	
Wrist–palm	8	1.59 ± 0.15	1.89	≤2.0

Ulnar Onset latency (ms)	Distance (cm)	Mean ± SD		Mean ± 2 SD
Wrist–digit V	14	3.44 ± 0.33		4.10
Palm–digit V	6	2.03 ± 0.27		2.57
Wrist–palm	8	1.41 ± 0.17		1.75

Median–ulnar Onset latency difference (ms)	Distance (cm)	Mean ± SD	Mean ± 2 SD	Limit of normal
Median–palm – ulnar–palm	8	0.18 ± 0.25	0.68	≤0.8

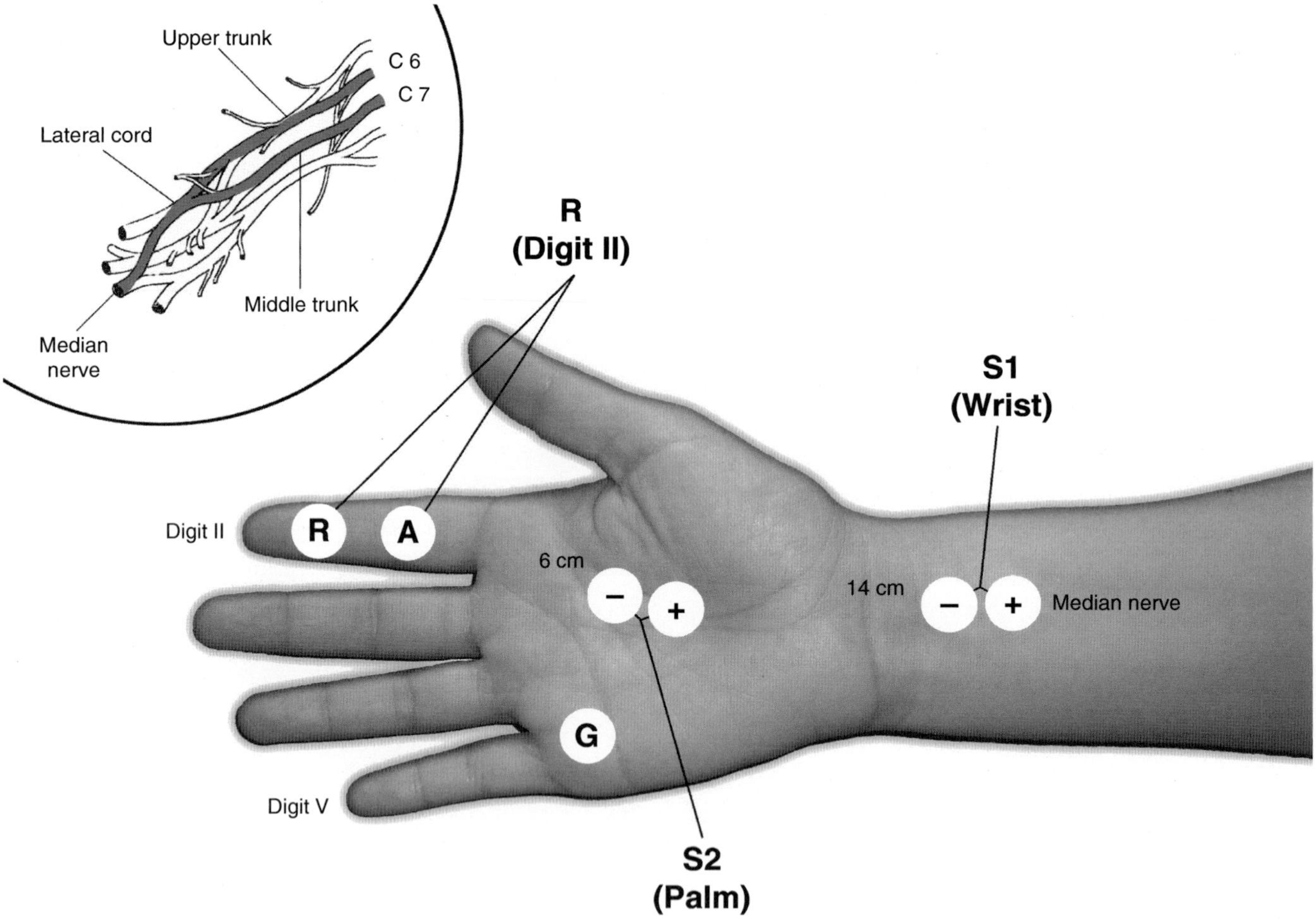

Fig. 3 Median nerve: stimulation of the wrist, S1 (14 cm), and on the palm, S2 (6 cm)

sensory palmar latency to digit II (palm–digit II segment) was calculated by recording the antidromic distal latency to digit II after stimulation on the palm, 6 cm proximal to the active recording electrode to digit II (called "SPL Index"). The same was made for the ulnar nerve conduction study (wrist–digit V and palm–digit V segments); the authors also calculated the median and ulnar sensory latencies across the wrist (median–palm and ulnar–palm segments, respectively) subtracting the "SPL Index" to the "Sensory Index" (called "Sensory Palm"):

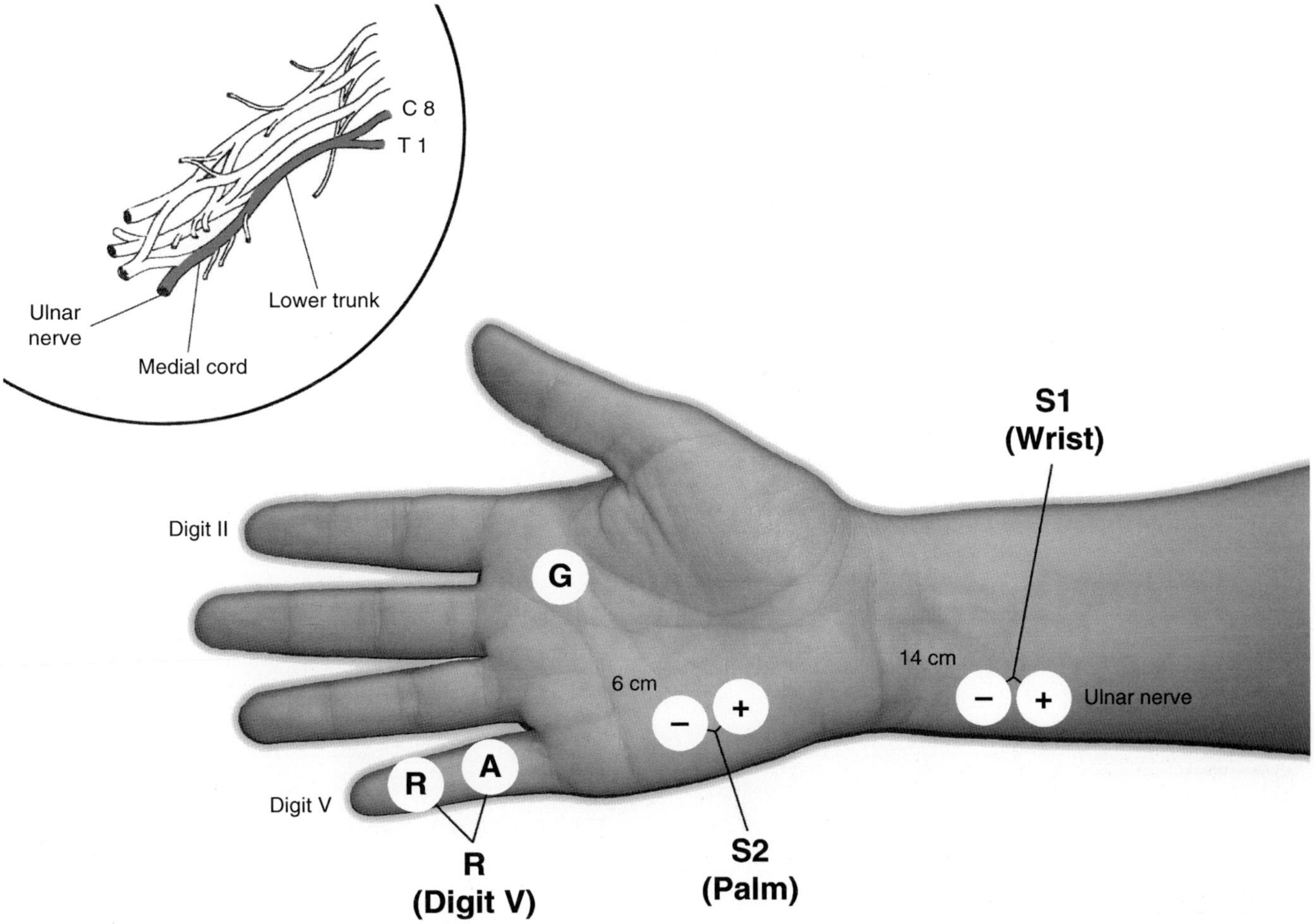

Fig. 4 Ulnar nerve: stimulation of the wrist, S1 (14 cm), and on the palm, S2 (6 cm)

$$\text{Median} - \text{palm}(\text{ms}) = (\text{palm} - \text{digit II}, \text{ms}) - (\text{wrist} - \text{digit II}, \text{ms}) \text{ for median nerve}$$

$$\text{Ulnar} - \text{palm}(\text{ms}) = (\text{palm} - \text{digit V}, \text{ms}) - (\text{wrist} - \text{digit V}, \text{ms}) \text{ for ulnar nerve}$$

The sensory palmar–median–ulnar difference was calculated by subtracting ulnar "Sensory Palm" to median "Sensory Palm" (called "Sensory Palm M-U"); it was the difference between the median and ulnar sensory 8 cm segments across the wrist. It was calculated as follows:

$$\text{Median} - \text{ulnar palm difference}(\text{ms}) = (\text{median} - \text{palm}, \text{ms}) - (\text{ulnar} - \text{palm}, \text{ms})$$

They also calculated the ratio of median sensory latency across the wrist to the total latency from the wrist to digit II (called "Wrist/Total"), as follows:

$$\text{Wrist / Total} = \frac{\text{wrist} - \text{palm } 8\,\text{cm }(\text{ms})}{\text{wrist} - \text{digit II } 14\,\text{cm }(\text{ms})}$$

Comment

For Andary et al. [2], care must be taken in case of coexistent ulnar lesion at the wrist such as Guyon's canal syndrome; in presence of ulnar neuropathy at the wrist in fact, a pathological increasing of the distal sensory latency (DSL) after wrist stimulation can be observed. The sensory palmar median–ulnar difference (Sensory Palm M-U) is the difference in sensory conduction between the median and ulnar sensory fibers across the wrist (8 cm segment). If a prolonged ulnar DSL is recorded, and normal median nerve sensory values are observed, they suggested to perform the Wrist/Total ratio for the digit V recordings to assess a pathologic ulnar involvement at the wrist, as follows:

$$\text{Wrist / Total} = \frac{\text{wrist} - \text{palm } 8\,\text{cm }(\text{ms})}{\text{wrist} - \text{digit V } 14\,\text{cm }(\text{ms})}$$

Buschbacher [3, 4] derived a normative database for the median and ulnar digital antidromic sensory conduction from a large and varied subject population. He studied 258 healthy subjects (Table 4) – 57 % were women and 43 % were men, 76 % were white, 22 % were black, and 2 % were of other races (age range was 19–76 years, mean age was 44 ± 15 years for the men and 40 ± 14 years for the women). The mean height was 176 ± 7 cm for the men (height range 157–204 cm) and 163 ± 7 cm for the women (height range 138–178 cm). The mean weight was 182 ± 31 lb. for men (weight range 112–279 lb.) and 154 ± 33 lb. for women (weight range 103–290 lb.). Skin surface temperatures were measured over the dorsum of the hand and maintained above 32 °C. The limb was warmed if necessary using hot water bottles. All recordings were performed using a sensitivity of 20 μV (filter settings 20 Hz-kHz) and a sweep speed of 1 ms/division. Stimulation was applied 7 cm and 14 cm proximal to the active electrode. The 14 cm stimulation point was over the median and ulnar nerves at the wrist; the 7 cm stimulation point was midway between the proximal site of stimulation and the active electrode. Recordings were made from digit II and digit III (median nerve) and from digit V (ulnar nerve). The author measured the onset latency, peak latency, negative peak amplitude, peak-to-peak amplitude, area, rise time, and duration of the SNAPs. Pathological waveforms and values in cases of CTS and Guyon's canal syndrome are reported (Figs. 5, 6, and 7).

Table 4 Normal values [3, 4]

Median nerve Onset latency (ms)	Distance (cm)	Mean ± SD	Mean ± 2 SD	Range
Wrist–digit II	14	2.6 ± 0.3	3.2	2.0–3.5
Palm–digit II	7	1.3 ± 0.2	1.7	1.0–1.8
Ulnar nerve Onset latency (ms)	**Distance (cm)**	**Mean ± SD**	**Mean ± 2 SD**	**Range**
Wrist–digit V	14	2.6 ± 0.2	3.0	2.0–3.3
Palm–digit V	7	1.4 ± 0.2	1.8	0.9–1.9
Median nerve Peak latency (ms)	**Distance (cm)**	**Mean ± SD**	**Mean ± 2 SD**	**Range**
Wrist–digit II	14	3.4 ± 0.3	4.1	2.7–4.7
Palm–digit II	7	1.9 ± 0.2	2.4	1.5–2.8
Ulnar nerve Peak latency (ms)	**Distance (cm)**	**Mean ± SD**	**Mean ± 2 SD**	**Range**
Wrist–digit V	14	3.4 ± 0.3	4.1	2.7–4.8
Palm–digit V	7	2.0 ± 0.2	2.5	1.6–2.8
Median nerve Onset-to-peak amplitude (μV)	**Distance (cm)**	**Mean ± SD**	**Mean ± 2 SD**	**Range**
Age range 19–49 (BMI < 24)	14 (wrist–digit II)	49 ± 19	17	
Age range 19–49 (BMI ≥ 24)	14 (wrist–digit II)	41 ± 19	11	
Age range 50–79 (BMI < 24)	14 (wrist–digit II)	28 ± 11	9	
Age range 50–79 (BMI ≥ 24)	14 (wrist–digit II)	24 ± 11	7	
All subjects	14 (wrist–digit II)	37 ± 19	8	6 – 92
Median nerve Onset-to-peak amplitude (μV)	**Distance (cm)**	**Mean ± SD**	**Mean ± 2 SD**	**Range**
Age range 19–49 (BMI < 24)	7 (palm–digit II)	53 ± 30	7	
Age range 19–49 (BMI ≥ 24)	7 (palm–digit II)	41 ± 22	6	
Age range 50–79 (BMI < 24)	7 (palm–digit II)	29 ± 13	9	
Age range 50–79 (BMI ≥ 24)	7 (palm–digit II)	24 ± 13	4	
All subjects	7 (palm–digit II)	38 ± 24	48	2–135
Median nerve Peak-to-peak amplitude (μV)	**Distance (cm)**	**Mean ± SD**	**Mean ± 2 SD**	**Range**
Age range 19–49 (BMI < 24)	14 (wrist–digit II)	75 ± 35	19	
Age range 19–49 (BMI ≥ 24)	14 (wrist–digit II)	62 ± 31	13	
Age range 50–79 (BMI < 24)	14 (wrist–digit II)	43 ± 16	15	

Table 4 (continued)

Median nerve Peak-to-peak amplitude (μV)	Distance (cm)	Mean ± SD	Mean ± 2 SD	Range
Age range 50–79 (BMI ≥ 24)	14 (wrist–digit II)	35 ± 16	8	
All subjects	14 (wrist–digit II)	56 ± 31	9	5–158

Median nerve Peak-to-peak amplitude (μV)	Distance (cm)	Mean ± SD	Mean ± 2 SD	Range
Age range 19–49 (BMI < 24)	7 (palm–digit II)	75 ± 39	15	
Age range 19–49 (BMI ≥ 24)	7 (palm–digit II)	58 ± 38	4	
Age range 50–79 (BMI < 24)	7 (palm–digit II)	46 ± 21	13	
Age range 50–79 (BMI ≥ 24)	7 (palm–digit II)	30 ± 18	5	
All subjects	7 (palm–digit II)	53 ± 36	4	2–190

Median nerve – area (ηV/s)	Distance (cm)	Mean ± SD	Mean ± 2 SD	Range
Age range 19–49 (BMI < 24)	14 (wrist–digit II)	56 ± 30	17	
Age range 19–49 (BMI ≥ 24)	14 (wrist–digit II)	45 ± 25	12	
Age range 50–79 (BMI < 24)	14 (wrist–digit II)	33 ± 14	11	
Age range 50–79 (BMI ≥ 24)	14 (wrist–digit II)	26 ± 12	9	
All subjects	14 (wrist–digit II)	41 ± 25	11	7–141

Median nerve – area (ηV/s)	Distance (cm)	Mean ± SD	Mean ± 2 SD	Range
Age range 19–49 (BMI < 24)	7 (palm–digit II)	55 ± 37	9	
Age range 19–49 (BMI ≥ 24)	7 (palm–digit II)	39 ± 25	7	
Age range 50–79 (BMI < 24)	7 (palm–digit II)	30 ± 12	12	
Age range 50–79 (BMI ≥ 24)	7 (palm–digit II)	21 ± 13	5	
All subjects	7 (palm–digit II)	37 ± 28	6	3–160

Median nerve – rise time (ms)	Distance (cm)	Mean ± SD	Mean ± 2 SD	Range
Wrist–digit II	14	0.8 ± 0.2	1.1	0.5–1.4
Palm–digit II	7	0.6 ± 0.1	0.8	0.2–1.1

Median nerve – duration (ms)	Distance (cm)	Mean ± SD	Mean ± 2 SD	Range
Wrist–digit II	14	2.0 ± 0.4	2.9	1.2–3.8
Palm–digit II	7	1.7 ± 0.4	2.5	0.8–3.0

Pathological waveform (wrist, palm – digit II):

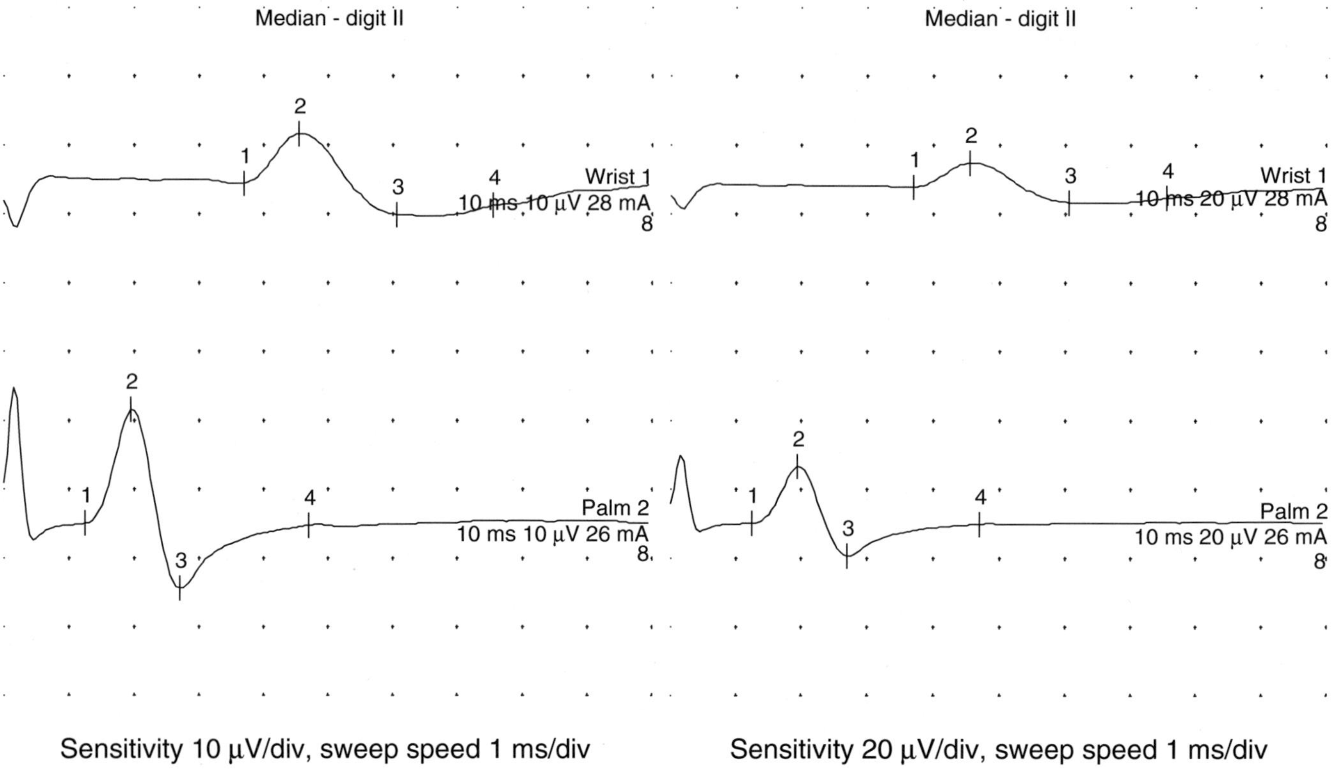

Onset latency (wrist – digit II): 3.70 ms; **Onset latency** (palm – digit II): 1.25 ms; **Peak latency** (wrist – digit II): 4.45 ms; **Peak latency** (palm – digit II): 1.95 ms; **Onset to peak amplitude** (wrist – digit II): 7.2 μV; **Onset to peak amplitude** (palm – digit II): 16.6 μV; **Peak to peak amplitude** (wrist – digit II): 11.9 μV; **Peak to peak amplitude** (palm – digit II): 26.2 μV; **SNCV** (wrist – digit II): 37.8 m/s; **SNCV** (palm – digit II): 56.0 m/s

Fig. 5 Antidromic sensory nerve action potentials (*SNAPs*) recorded to digit II after stimulation at the wrist (*upper trace*) and on the palm (*lower trace*) in mild CTS – grade 2 by Bland's CTS classification scale [5]

Pathological waveform (wrist, palm – digit II):

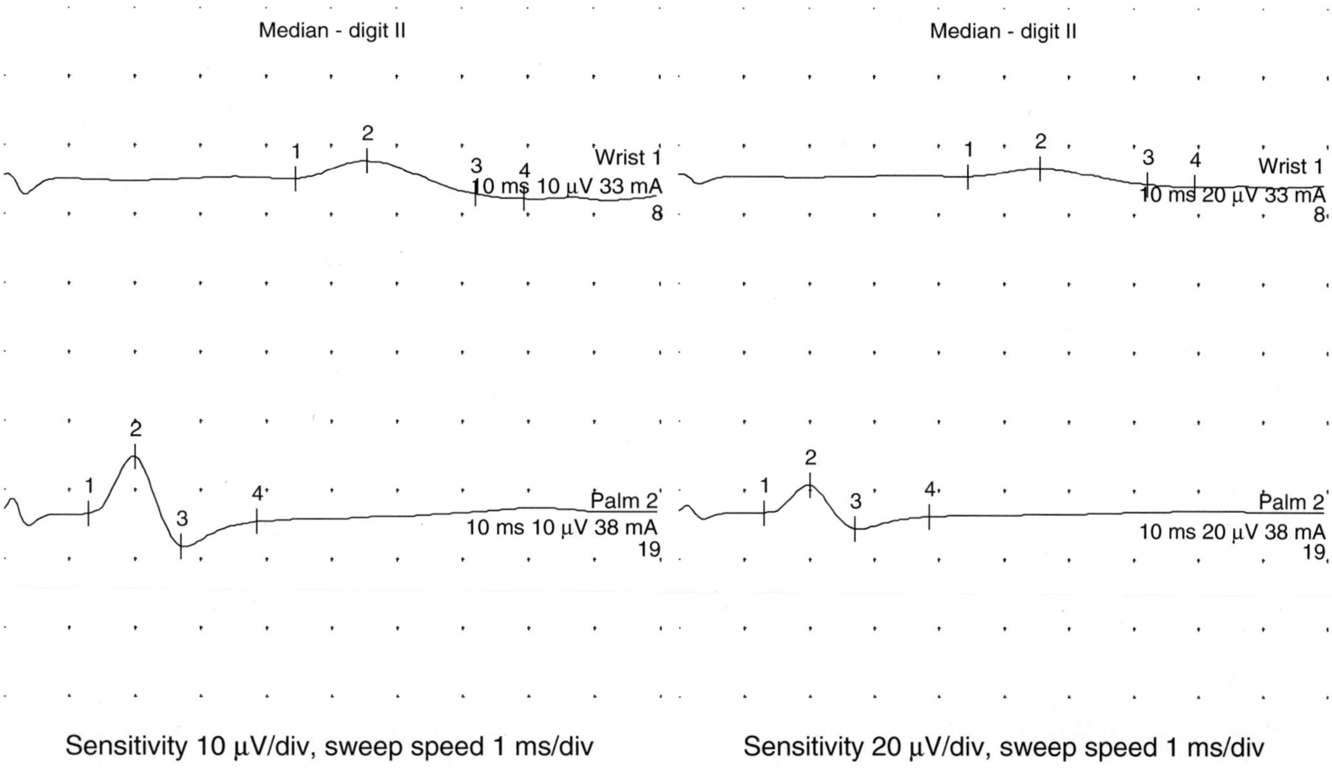

Onset latency (wrist – digit II): 4.45 ms; **Onset latency** (palm – digit II): 1.30 ms; **Peak latency** (wrist – digit II): 5.55 ms; **Peak latency** (palm – digit II): 2.00 ms; **Onset to peak amplitude** (wrist – digit II): 2.7 μV; **Onset to peak amplitude** (palm – digit II): 8.4 μV; **Peak to peak amplitude** (wrist – digit II): 4.9 μV; **Peak to peak amplitude** (palm – digit II): 13.0 μV; **SNCV** (wrist – digit II): 31.5 m/s; **SNCV** (palm – digit II): 53.8 m/s

Fig. 6 Antidromic sensory nerve action potentials (*SNAPs*) recorded to digit II after stimulation at the wrist (*upper trace*) and on the palm (*lower trace*) in moderately severe CTS – grade 3 by Bland's CTS classification scale [5]

Pathological waveform (wrist, palm – digit V):

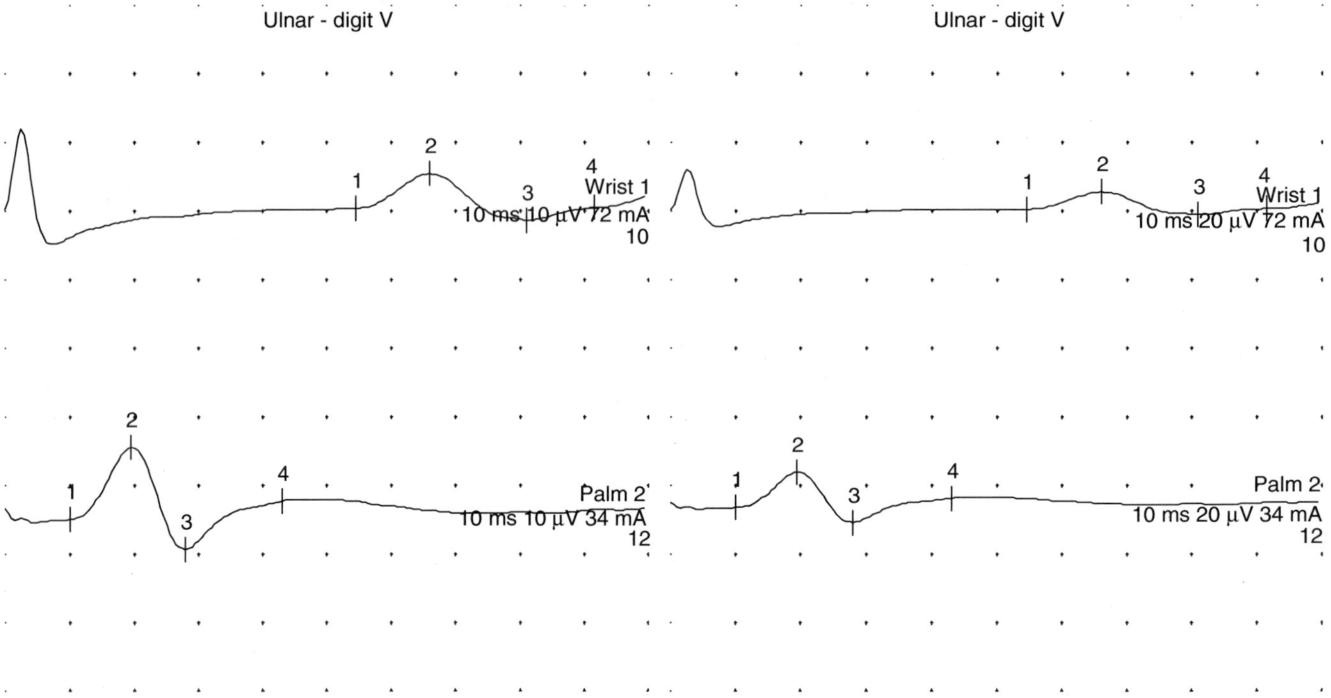

Sensitivity 10 μV/div, sweep speed 1 ms/div Sensitivity 20 μV/div, sweep speed 1 ms/div

Onset latency (wrist – digit V): 5.45 ms; **Onset latency** (palm – digit V): 1.05 ms; **Peak latency** (wrist – digit V): 6.75 ms; **Peak latency** (palm – digit V): 1.95 ms; **Onset to peak amplitude** (wrist – digit V): 5.0 μV; **Onset to peak amplitude** (palm – digit V): 10.6 μV; **Peak to peak amplitude** (wrist – digit V): 6.9 μV; **Peak to peak amplitude** (palm – digit V): 15.0 μV; **SNCV** (wrist – digit V, 14 cm distance): 25.7 m/s; **SNCV** (palm – digit V, 7 cm distance): 66.7 m/s

Fig. 7 Antidromic sensory nerve action potentials (*SNAPs*) recorded to digit V after stimulation at the wrist (*upper trace*) and on the palm (*lower trace*) in Guyon's canal syndrome

References

1. Felsenthal G, Spindler H (1979) Palmar conduction time of median and ulnar nerves of normal subjects and patients with carpal tunnel syndrome. Am J Phys Med Rehabil 58:131–138
2. Andary MT, Fankhauser MJ, Ritson JL et al (1996) Comparison of sensory mid-palm studies to other techniques in carpal tunnel syndrome. Electromyogr Clin Neurophysiol 36:279–285
3. Buschbacher RM (1999) Median 14-cm and 7-cm antidromic sensory studies to digits two and three. Am J Phys Med Rehabil 78:S53–62
4. Buschbacher RM (1999) Ulnar 14-cm and 7-cm antidromic sensory studies to the fifth digit: reference values derived from a large population of normal subjects. Am J Phys Med Rehabil 78:S63–68
5. Bland JDP (2000) A neurophysiological grading scale for carpal tunnel syndrome. Muscle Nerve 23:1280–1283

Wrist – Dorsal Hand

Surface Recording Technique, Antidromic Study

S13

Original Settings Sensitivity was 10 μV/division, and sweep speed was 2 ms/division. Low-frequency filter, high-frequency filter, duration of pulse, rate of pulse, and the machine used were not specified (author only reported that voltage, duration of the stimulus, and filter settings were those used for routine sensory studies).

Position This study was performed with the patient lying on a bed in the supine position, with the arm alongside the body and the hand pronated. When relaxation is hard to obtain in prone position, the hand can be supinated (palm up).

Recording Following the antidromic method [1], signals were recorded using surface electrodes from the dorsal branch of the ulnar nerve. The active recording electrode (A) was placed on the bottom of the "V" located between the fourth and fifth metacarpals, and the reference electrode (R) was placed distally at the base of the digit V (little finger) to avoid artifactual recording from the palmar branch. The ground (G) electrode was placed on the dorsum of the hand, between the recording and the stimulating electrodes (Fig. 1). The author suggested this method to stimulate and record from the dorsal branch of the ulnar nerve because it leaves the common trunk of the ulnar nerve about 5–8 cm proximal to the ulnar styloid process, under cover of the *flexor carpi ulnaris* (FCU) muscle. It crosses the medial side of the ulna at an acute angle and becomes dorsal 4–5 cm proximal to the styloid process (Fig. 2).

As it crosses at that point, it is easily accessible to stimulation between the tendon of the FCU and the ulna. The dorsal branch of the ulnar nerve supplies the sensory innervations of the ulnar part of the back of the hand and parts of the dorsum of the digit IV (ring finger) and digit V (little finger).

Stimulation Stimulation was applied to the dorsal sensory branch of the ulnar nerve at the wrist, at a distance of 8 cm from the active (A) recording electrode (Fig. 1). The stimulator was placed in the space between the ulna and the FCU muscle tendon.

Measurements Onset latency (ms) was measured from the stimulus onset to the initial deflection of the sensory nerve action potential (SNAP). Peak amplitude (μV) was measured from the stimulus onset to the peak of the negative deflection of the SNAP. Sensory nerve conduction velocity (SNCV) was calculated in only 16 healthy subjects, and it was measured in meter per second (m/s). Skin temperature was not monitored; all studies were performed at room temperature. Normal values (Table 1) were obtained from 50 hands of 30 healthy volunteers and asymptomatic patients (age range 10–66 years) with normal routine sensory and motor conduction studies.

Comment

For Jabre [1], the dorsal branch of the ulnar nerve may be involved in injury more frequently than is currently thought, because the ulnar branch can be easily traumatized in its course over the bony ulna when the forearm is put against a sharp edge, as in writing or a table or desk. He found in about 11 % of the healthy subjects a difference of 50 % or more in SNAP amplitude between the two arms, and the author explained these asymmetric amplitude responses as a possible direct sign of a dorsal sensory neuropathy. Therefore, patients with vague sensory complaints about the ulnar distribution of the dorsum of the hand have normal results on routine sensory and motor nerve conduction studies, but performing the method suggested by the author [1], a dorsal neuropathy can be diagnosed. The method described is a simple means for localizing a lesion proximal or distal to the takeoff of the dorsal sensory branch of the ulnar nerve and studying the ulnar nerve segment proximal to a wrist lesion when the routine sensory and motor amplitudes are depressed or absent.

Young and Kalantri [2] performed the same technique [1] on 54 hands of 27 healthy subjects (Table 2) and patients, who did not have signs or symptoms of ulnar neuropathy

G. Gentili, M. Di Napoli, *The Ulnar Nerve*, DOI 10.1007/978-3-319-43292-2_13

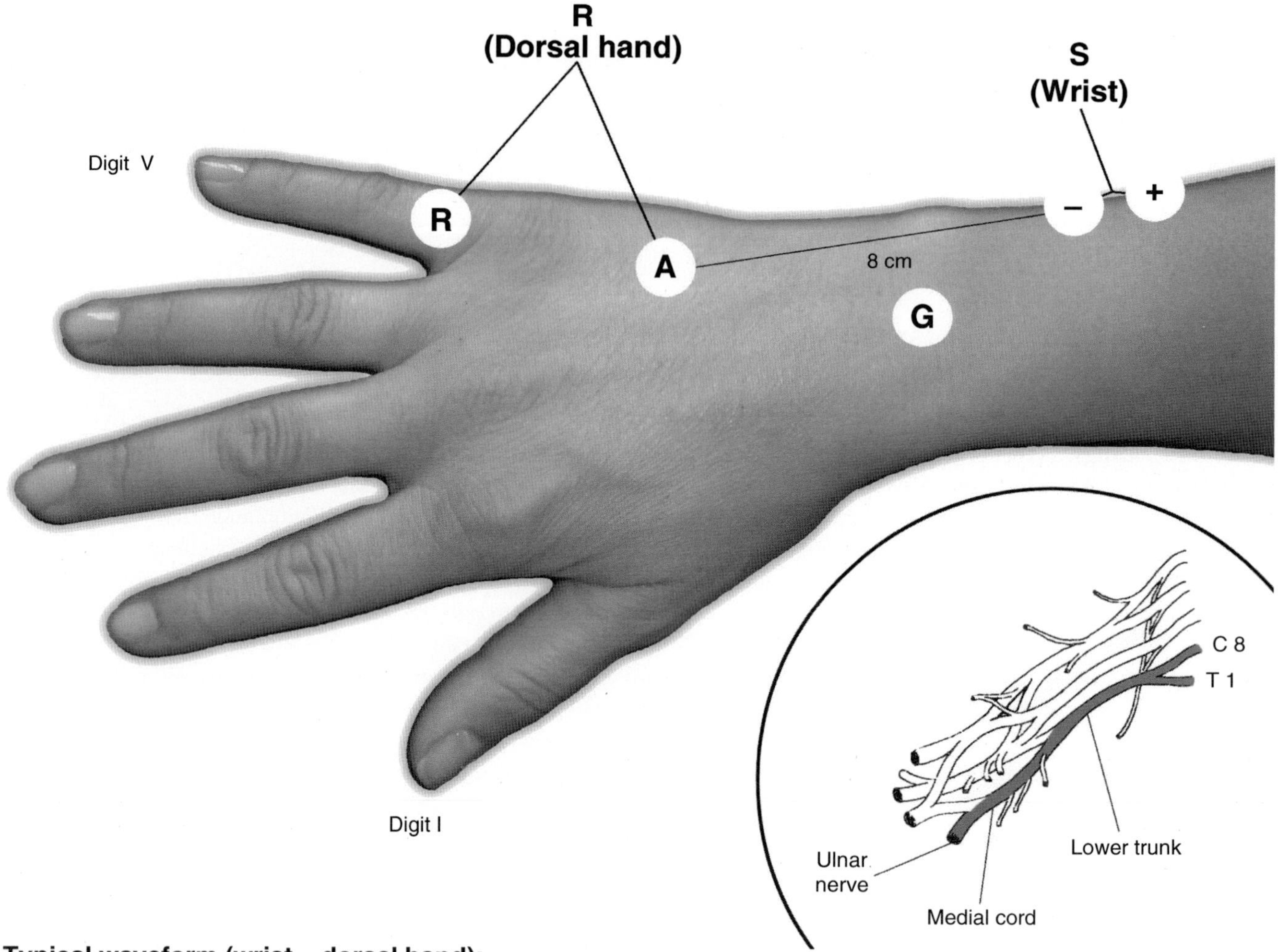

Typical waveform (wrist – dorsal hand):

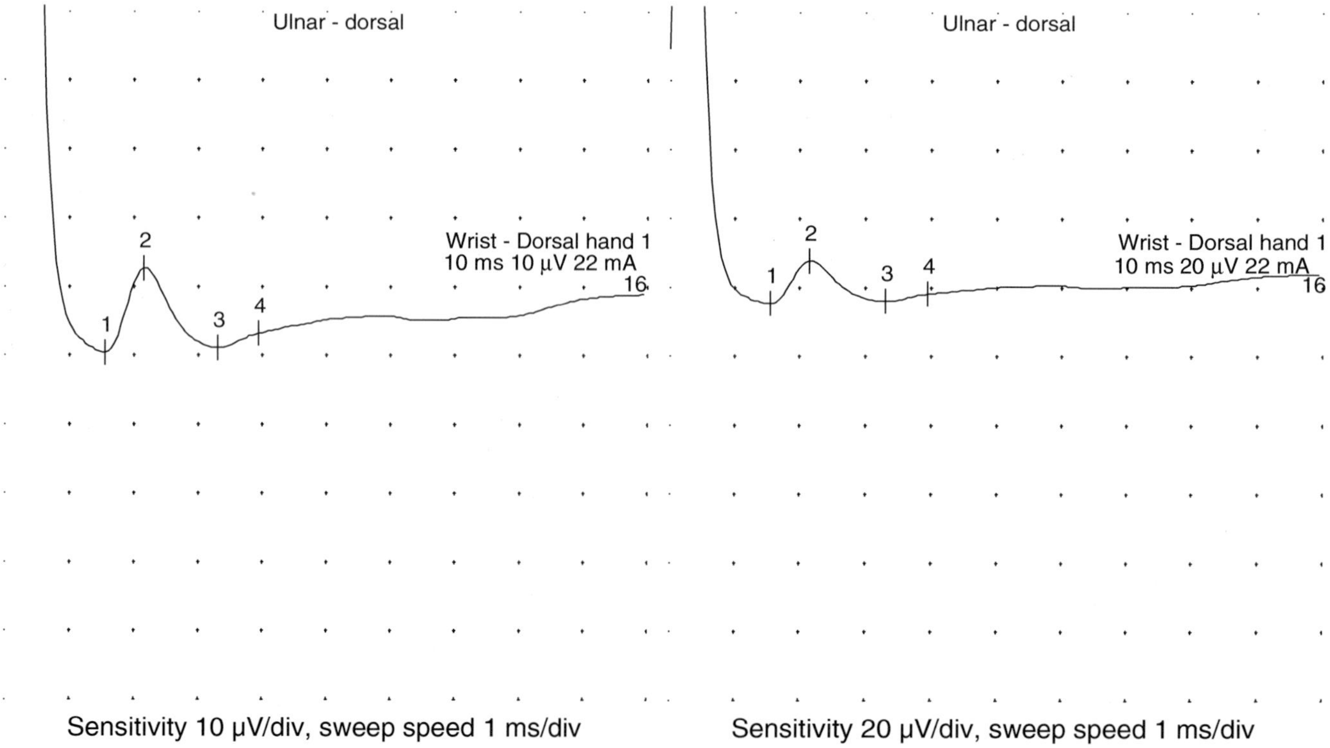

Fig. 1 Antidromic sensory nerve action potential (*SNAPs*) recorded on the dorsal hand, stimulation of the dorsal sensory branch of the ulnar nerve at the wrist (8 cm distance)

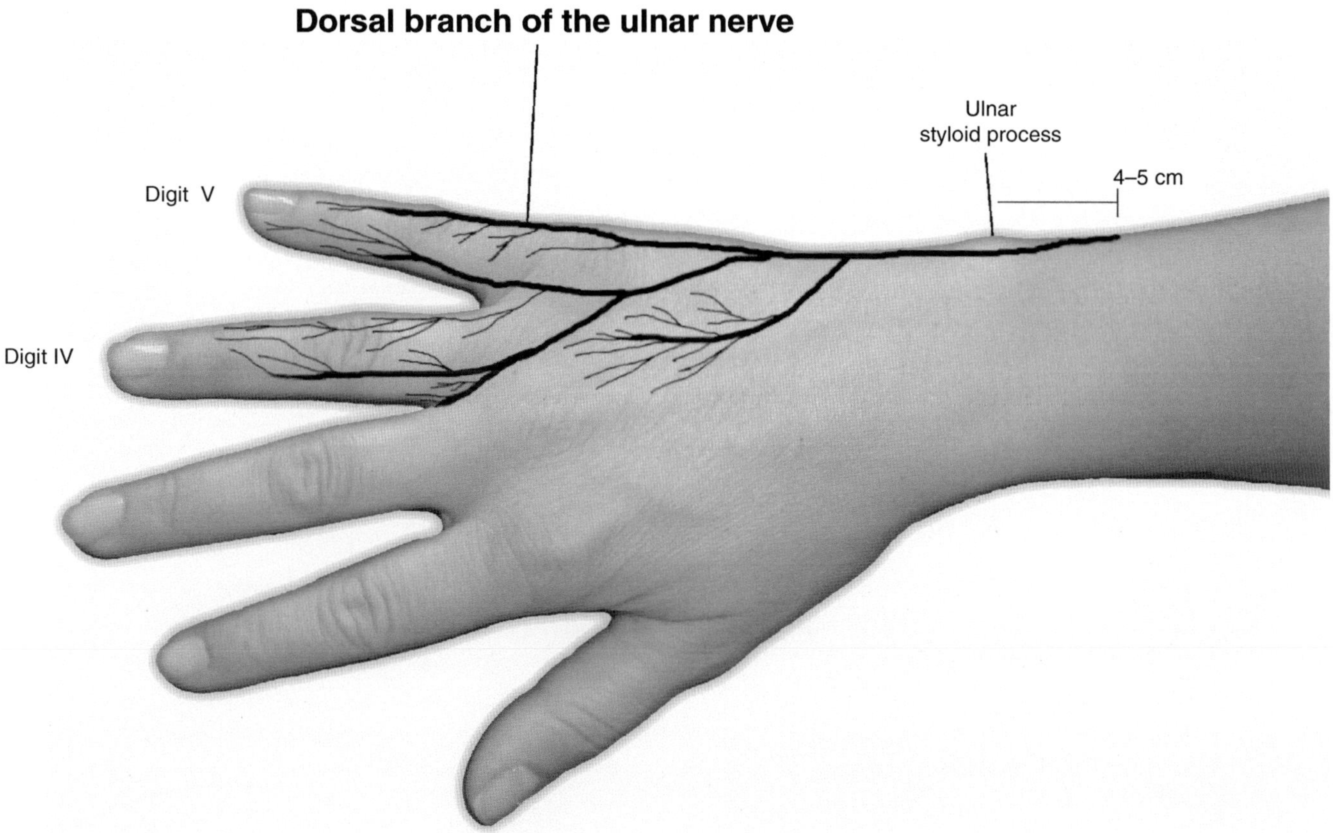

Fig. 2 Distribution of dorsal sensory branch of the ulnar nerve

Table 1 Normal values [1]

	Mean ± SD
Wrist–dorsal hand, onset latency – 8 cm (ms)	2.0 ± 0.3
Wrist–dorsal hand, peak amplitude – 8 cm (μV)	20.0 ± 6.0
Wrist–dorsal hand, SNCV – 8 cm (m/s)	60.0 ± 4.0

(17 men and 10 women, age range 22–71 years, mean age 40.7 ± 13 years). A bar-type bipolar electrode with 3 cm interelectrode distance was used to record the SNAPs, and it was placed between fourth and fifth metacarpals – 8–10 cm from the cathode stimulating electrode between the ulna and the FCU tendon (Fig. 3).

They used a Cadwell 5200A electromyographic machine (low-frequency filter was 10 Hz and high-frequency filter was 2 kHz); all SNAPs were recorded at 20 μV/division. An average was used only if the amplitude was lower than 5 μV. Peak latencies (ms) were measured to the peak of the SNAPs, and the SNAP amplitudes were measured from the baseline of the negative peak. The temperature of the palm of the hand was measured by a surface thermistor before each conduction study, and it ranged from 32 to 34.5 °C.

Comment

For Young and Kalantri [2], the use of the bar electrode has a disadvantage because of the 3 cm interelectrode distance than the optimal 4 cm, and this has been shown to reduce the amplitudes of the SNAPs. With the bar electrode, however, the technique is easily standardized and can be reproduced accurately, even in subjects with small hands in which may not be possible to maintain the 4 cm distance. According to a previous study [3] on temperature effects on the sensory nerve conduction, for which the conduction velocity increases almost linearly, by 1.6 m/s/°C for the sensory fibers of the ulnar nerve, as the surface temperature measured near the nerve increases from 26 to 33 °C, their study results highlighted the importance of keeping the limb under study at optimal temperatures to obtain faster responses and prevent erroneous results and misdiagnosis, since Dumitru [3] showed that. Pathological waveform and values in a patient with an ulnar lesion at the distal forearm are reported (Fig. 4).

Table 2 Normal values [2]

	Mean ± SD	Range
Wrist–dorsal hand, peak latency – 8 cm (ms)	1.84 ± 0.2	1.51–2.35
Wrist–dorsal hand, peak latency – 10 cm (ms)	2.09 ± 0.21	1.76–2.68
Wrist–dorsal hand, peak amplitude – 8 cm (μV)	26.5 ± 9.7	9.0–48.0
Wrist–dorsal hand, peak amplitude – 10 cm (μV)	23.5 ± 8.8	7.0–43.0

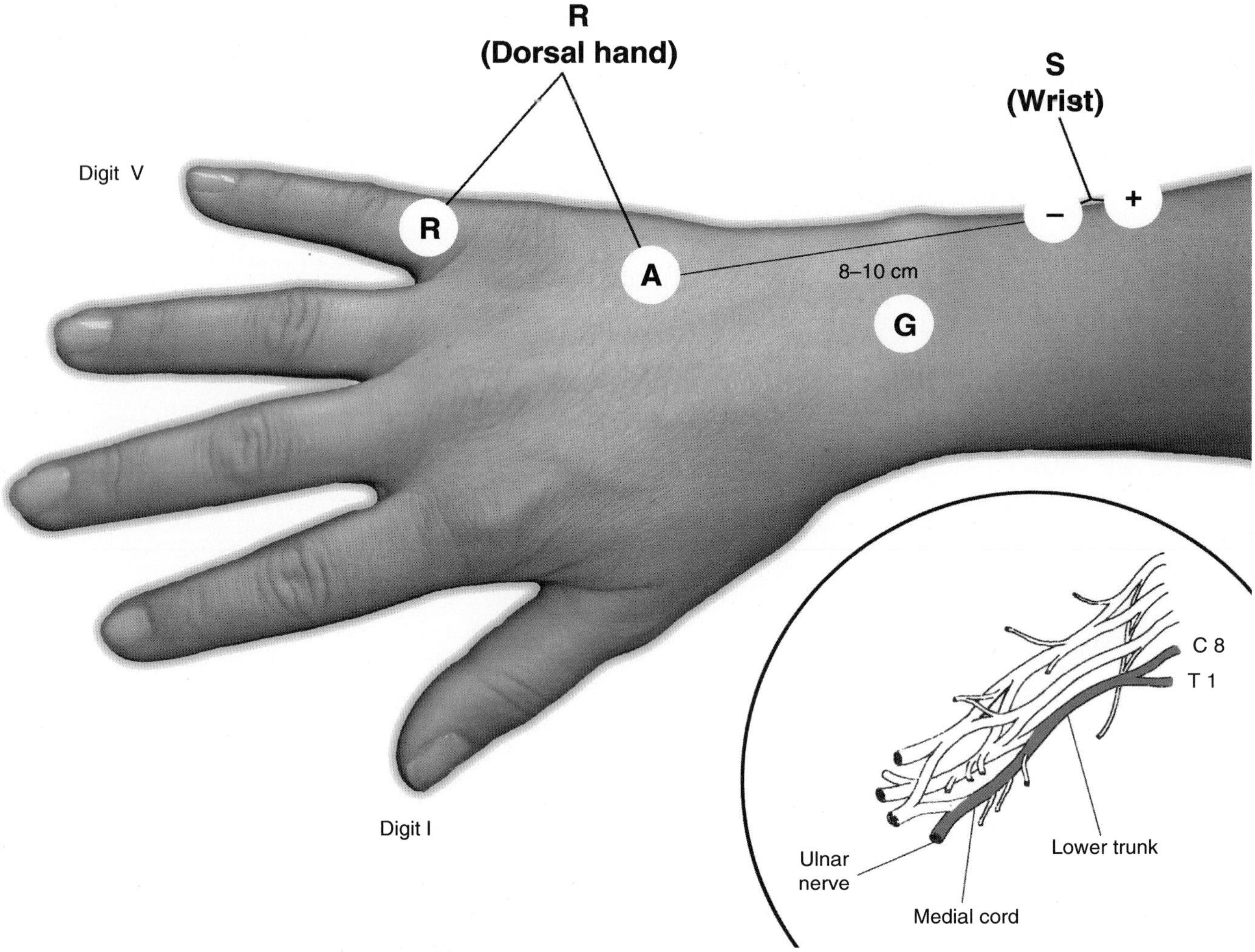

Fig. 3 Antidromic method by Young and Kalantri [2], stimulation of the dorsal sensory branch of the ulnar nerve at the wrist (8–10 cm distance)

Some authors [1, 4] proposed to distinguish UNE from UNW based on the fact that the dorsal ulnar cutaneous nerve joins the ulnar nerve proximal to the wrist. However, for Venkatesh et al. [5], the dorsal ulnar cutaneous response is not uncommonly spared in UNE. They studied 25 consecutive patients (13 men and 12 women, age range 27–77 years, average age 48 years) with clinical ulnar neuropathy and unequivocal electrophysiologic evidence of ulnar neuropathy at the elbow using Jabre's method. Onset-to-peak amplitudes of < 8 μV or < 50 % in amplitude when compared to the asymptomatic limb were considered abnormal. The dorsal ulnar cutaneous response was spared in 12 (55 %) patients with axonal loss, 5 of whom had reduced digital ulnar sensory responses; in 3 patients without axonal loss, the dorsal ulnar cutaneous response was normal. They postulate that these findings are likely due to the fascicular arrangement and sparing of dorsal ulnar cutaneous sensory nerve fibers at the elbow. Sparing of some fibers

Pathological waveform (wrist – dorsal hand):

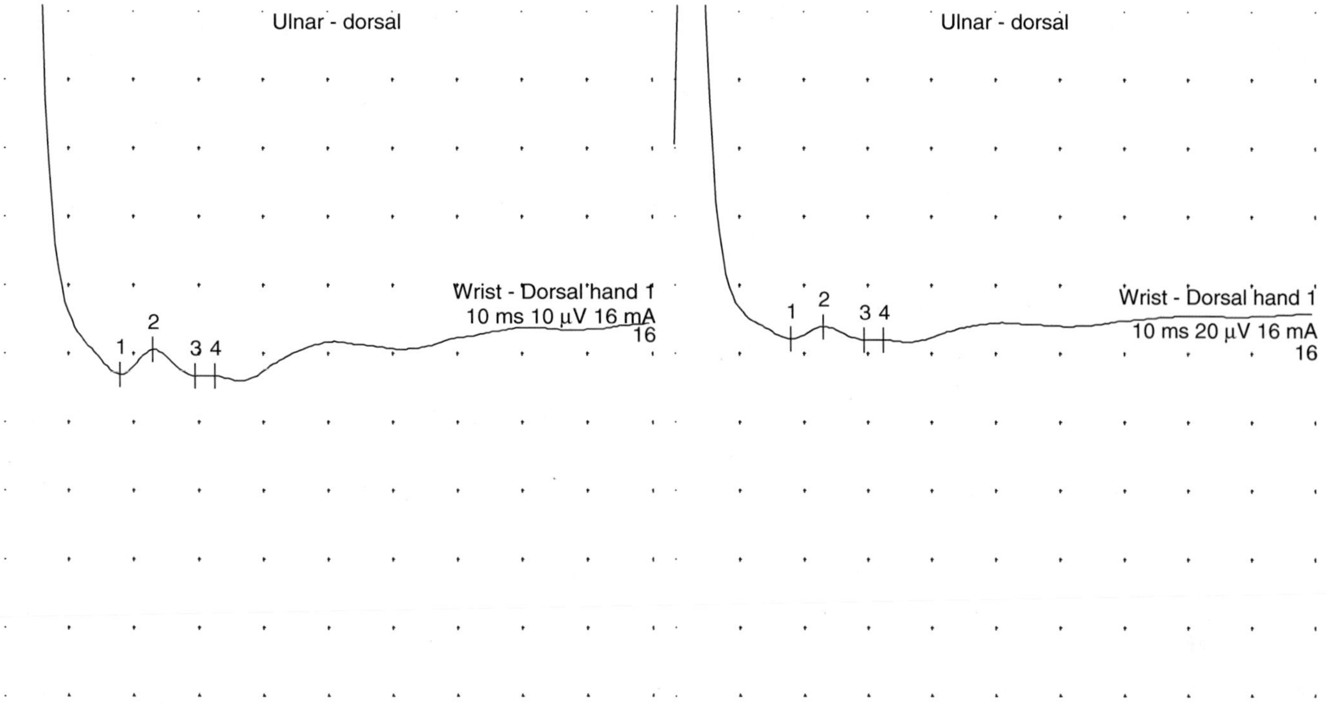

Sensitivity 10 µV/div, sweep speed 1 ms/div | Sensitivity 20 µV/div, sweep speed 1 ms/div

Onset latency (wrist – dorsal hand): 1.80 ms; **Peak latency** (wrist – dorsal hand): 2.30 ms; **Onset to peak amplitude** (wrist – dorsal hand): 3.7 µV; **Peak to peak amplitude** (wrist – dorsal hand): 3.9 µV; **SNCV** (wrist – dorsal hand, 8 cm distance): 44.4 m/s

Fig. 4 Antidromic sensory nerve action potential (*SNAP*) recorded on the dorsal hand in a patient with an injury of the ulnar nerve at the distal forearm, stimulation of the ulnar nerve at the wrist

with preferential fascicular involvement of others has been documented in other entrapment neuropathies [6, 7], such as in the carpal tunnel syndrome (CTS) or in the UNE. Jabley et al. [8] found that the fascicle forming the dorsal ulnar sensory branch commonly separates from the main ulnar trunk above the elbow (about 20 cm proximally to the medial epicondyle) and travels as a separate nerve next to the ulnar nerve in the forearm. They noted, "it was as though the dorsal cutaneous nerve and the ulnar nerve were two separate nerves travelling within a common epineural conduit while still retaining their autonomy." These anatomical findings explained the electrophysiologic sparing of the dorsal ulnar cutaneous sensory responses in some cases of UNE, even in those cases with evidence of axonal loss. The combination of a spared dorsal ulnar cutaneous sensory response and an abnormal digital ulnar sensory response to digit V cannot reliably be used alone to localize the site of the lesion to the wrist. The electrophysiologic measurement of the dorsal ulnar cutaneous nerve remains useful but only in those cases where it is abnormal, implying localization of the lesion to a more proximal site [5].

Cowdery et al. [9], using a Nicolet Viking IV and a Dantec Keypoint electromyography machines, studied 20 (Table 2) patients (15 men and 5 women, age range 22–77 years, mean age 54 years) with ulnar neuropathy at the wrist (UNW). They reported only amplitude data from 15 patients (mean 18 µV, range 14–30 µV), and dorsal ulnar cutaneous sensory response was normal in all 15 patients with UNW tested. They did not confirm systematically in their study the observation of a dorsal ulnar cutaneous response sparing in UNE [5], but they underlined that (1) the dorsal ulnar cutaneous sensory response was of normal amplitude in one of the two patients with UNW with incidental UNE; (2) even the antidromic SNAP from digit V was within the normal range in amplitude and latency in five patients with mild UNE (50 %) and two patients with severe UNE (20 %). Thus, for the authors [9], a normal digit V or dorsal ulnar cutaneous sensory response does not rule out UNE.

References

1. Jabre JF (1980) Ulnar nerve lesions at the wrist: a new technique for recording from the sensory dorsal branch of the ulnar nerve. Neurology 30:873–876
2. Young SHY, Kalantri A (2000) Dorsal ulnar cutaneous nerve conduction studies in an asymptomatic population. Arch Phys Med Rehabil 81:1171–1172
3. Dumitru D (1995) Electrodiagnostic medicine. Hanley & Belfus, Philadelphia
4. Kim DJ, Kalantri A, Guha S et al (1981) Dorsal cutaneous ulnar nerve conduction. Diagnostic aid in ulnar neuropathy. Arch Neurol 38:321–322
5. Venkatesh S, Kothari MJ, Preston DC (1995) The limitations of the dorsal ulnar cutaneous sensory response in patients with ulnar neuropathy at the elbow. Muscle Nerve 18:345–347
6. Logigian EL, Busis NA, Berger AR et al (1987) Lumbrical sparing in carpal tunnel syndrome: anatomic, physiologic and diagnostic implications. Neurology 37:1499–1505
7. Uncini A, Di Muzio A, Awad J et al (1993) Sensitivity of three median to ulnar comparative studies in diagnosis of mild ulnar carpal tunnel syndrome. Muscle Nerve 16:1366–1373
8. Jabley ME, Wallace WH, Heckler FR (1980) Internal topography of the major nerves of the forearm and hand. J Hand Surg Am 5:1–21
9. Cowdery SR, Preston DC, Herrmann DN et al (2002) Electrodiagnosis of ulnar neuropathy at the wrist. Neurology 59:420–427

Forearm – Dorsal Hand; Dorsal Hand – Forearm

S14

Surface Recording Technique, Antidromic and Orthodromic Study

Original Settings The machine used was a two-channel electromyograph TECA model TE-4. Sensitivity, sweep speed, low-frequency filter, high-frequency filter, duration of pulse, and rate of pulse were not specified (the author only reported that calibration and stimulation settings were identical to those used for routine sensory conduction studies).

Position This study was performed with the patient seated, with the arm supported on a level surface, the forearm fully pronated, the wrist in neutral position, and the fingers slightly flexed.

Recording Following the antidromic method [1], signals were recorded using a plastic-mounted, bipolar electrode (surface disks were 3 cm apart) placed along the fifth metacarpal bone, with the reference electrode level with the fifth metacarpophalangeal joint (Fig. 1). The authors used this montage to record from the dorsal branch of the ulnar nerve, which leaves the main trunk some centimeters proximal to the ulnar styloid and becomes cutaneous as it passes between the *flexor carpi ulnaris* (FCU) muscle tendon and the ulna bone. It supplies sensation to the dorsoulnar aspect of the wrist and hand and to the dorsal surface of the digit IV (ring finger) and digit V (little finger). If some difficulty is encountered in obtaining a sensory response, the authors suggested using an orthodromic technique rather than an antidromic technique (Fig. 2).

Stimulation Stimulation was applied at the forearm, approximately 8–10 cm above the ulnar styloid, with the stimulator firmly pressed between the FCU muscle tendon and the ulna bone. If some difficulty is encountered in obtaining a sensory response, the authors suggested moving slowly 2–3 cm above and below the stimulating point.

Measurements Peak latency (ms) was measured from the stimulus onset to the peak of the first negative deflection of the sensory nerve action potential (SNAP). The peak-to-peak amplitude (μV) was measured from the negative to the positive peak. Sensory nerve conduction velocity (SNCV) was calculated by dividing the interelectrode distance by the sensory latency, and it was measured in meter per second (m/s). All conduction studies were performed in a thermostatically controlled environment (22.8 °C). Normal values (Table 1) were obtained from 66 hands of 33 healthy subjects (21 women and 12 men, age range 21–71 years, mean age 32 years).

Comment

The method proposed by Kim et al. [1] differs from that reported by Jabre [2] in the following aspects: (1) the active recording electrode is placed on the fifth metacarpal rather than between the fourth and fifth metacarpals; (2) the site of stimulation is more proximal (11–12 cm from the active electrode) than Jabre's method (8 cm from the active electrode). The authors believed that dorsal cutaneous ulnar conduction should be useful in identifying distal ulnar lesions, especially those within or just proximal to the Guyon's canal; on the contrary, dorsal cutaneous conduction would be expected to be abnormal or absent in the presence of a lesion at, or above, the elbow. Pathological waveform and values in patients with ulnar neuropathy at the elbow (UNE) and ulnar neuropathy at the wrist (UNW) are reported (Figs. 3 and 4).

G. Gentili, M. Di Napoli, *The Ulnar Nerve*, DOI 10.1007/978-3-319-43292-2_14

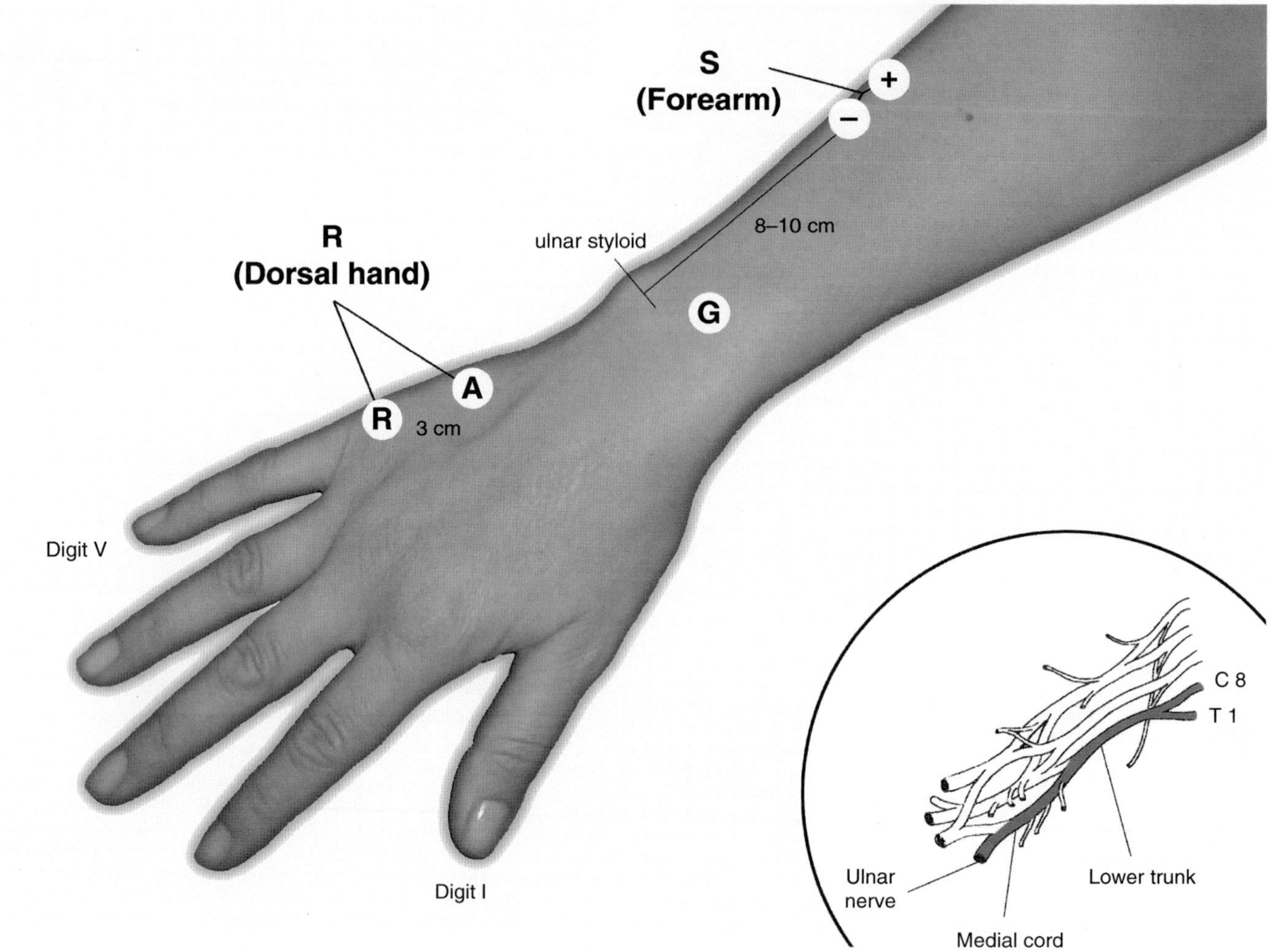

Typical waveform (forearm – dorsal hand):

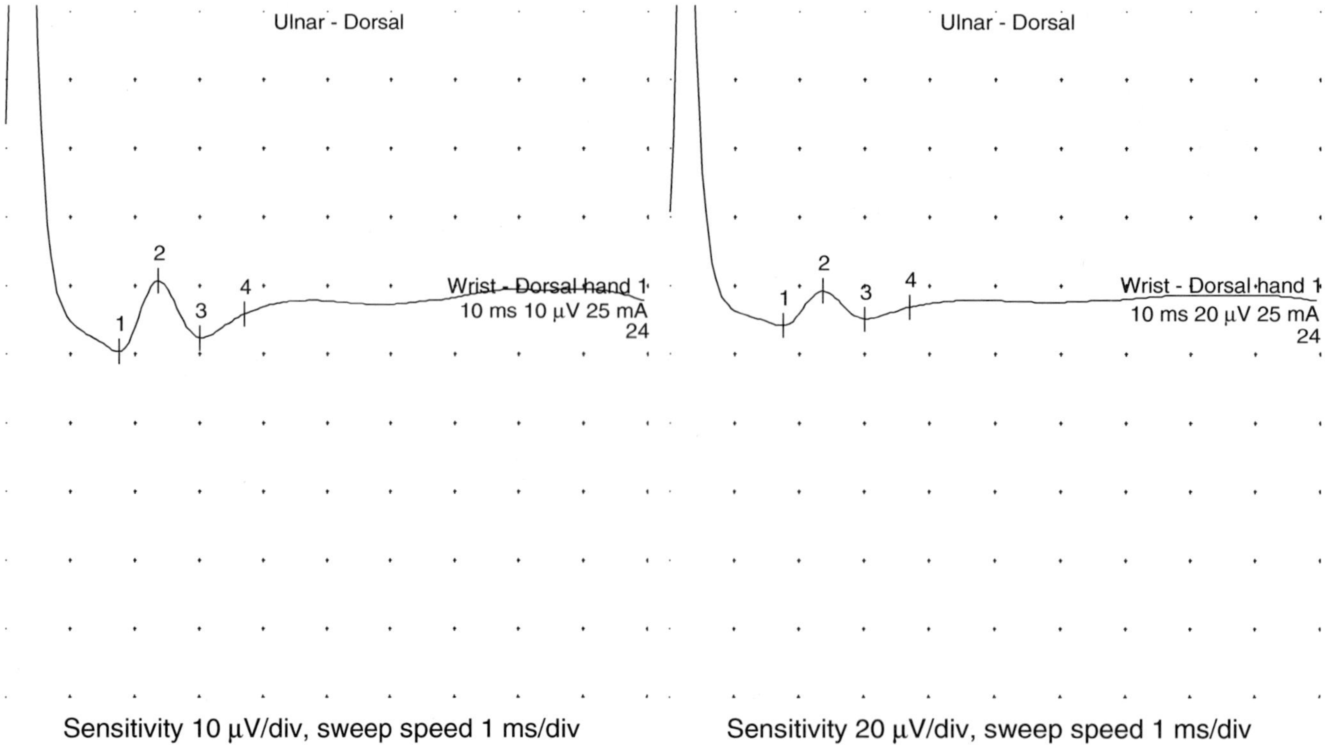

Fig. 1 Antidromic sensory nerve action potential (*SNAP*) recorded on the dorsal hand, stimulation of the ulnar nerve at the forearm

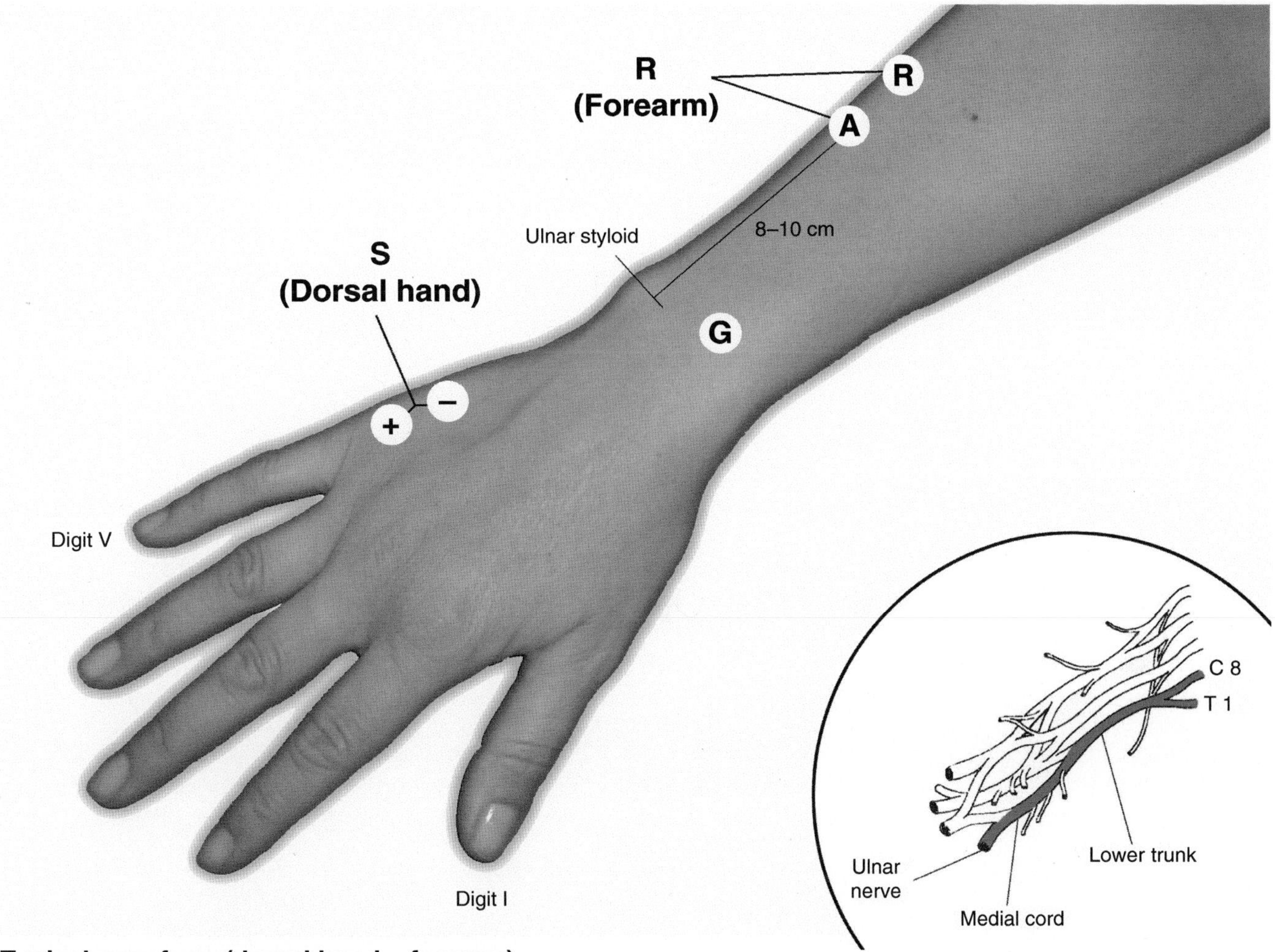

Typical waveform (dorsal hand – forearm):

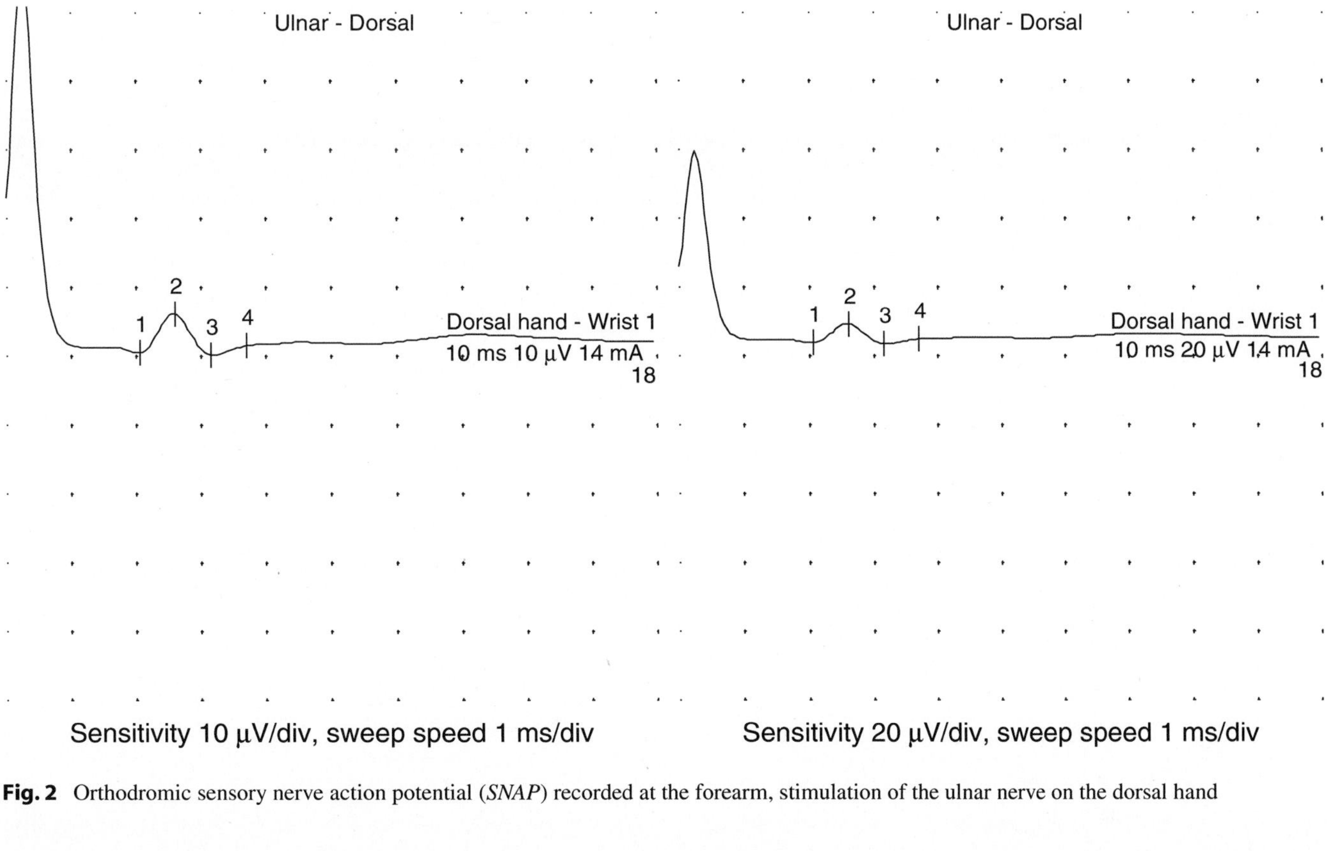

Fig. 2 Orthodromic sensory nerve action potential (*SNAP*) recorded at the forearm, stimulation of the ulnar nerve on the dorsal hand

Table 1 Normal values [1]

	Mean ± SD
Wrist–dorsal hand, peak latency (ms)	2.1 ± 0.3
Wrist–dorsal hand, peak-to-peak amplitude (μV)	24.2 ± 10.8
Wrist–dorsal hand, SNCV (m/s)	47.8 ± 3.8

Pathological waveform (forearm – dorsal hand):

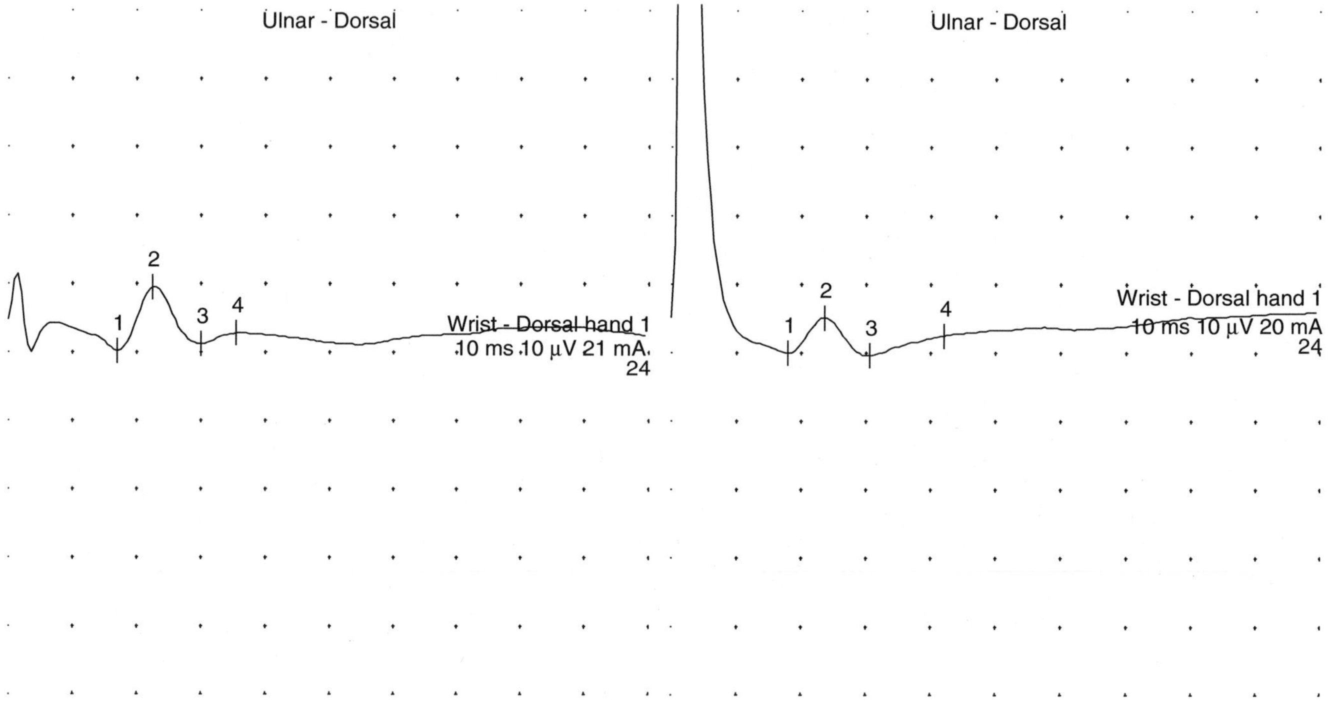

Onset latency (forearm – dorsal hand, affected side): 1.80 ms; **Onset latency** (forearm – dorsal hand, unaffected side): 1.70 ms; **Peak latency** (forearm – dorsal hand, affected side): 2.25 ms; **Peak latency** (forearm – dorsal hand, unaffected side): 2.20 ms; **Onset to peak amplitude** (forearm – dorsal hand, affected side): 5.1 μV; **Onset to peak amplitude** (forearm – dorsal hand, unaffected side): 9.7 μV; **Peak to peak amplitude** (forearm – dorsal hand, affected side): 5.4 μV; **Peak to peak amplitude** (forearm – dorsal hand, unaffected side): 8.9 μV; **SNCV** (forearm – dorsal hand, affected side -10 cm distance): 55.6 m/s; **SNCV** (forearm – dorsal hand, unaffected side - 10 cm distance): 58.8 m/s

Fig. 3 Antidromic sensory nerve action potential (*SNAP*) recorded on the dorsal hand in a patient with right UNE (affected side, *right trace*; unaffected side, *left trace*), stimulation of the ulnar nerve at the forearm

Pathological waveform (dorsal hand – forearm):

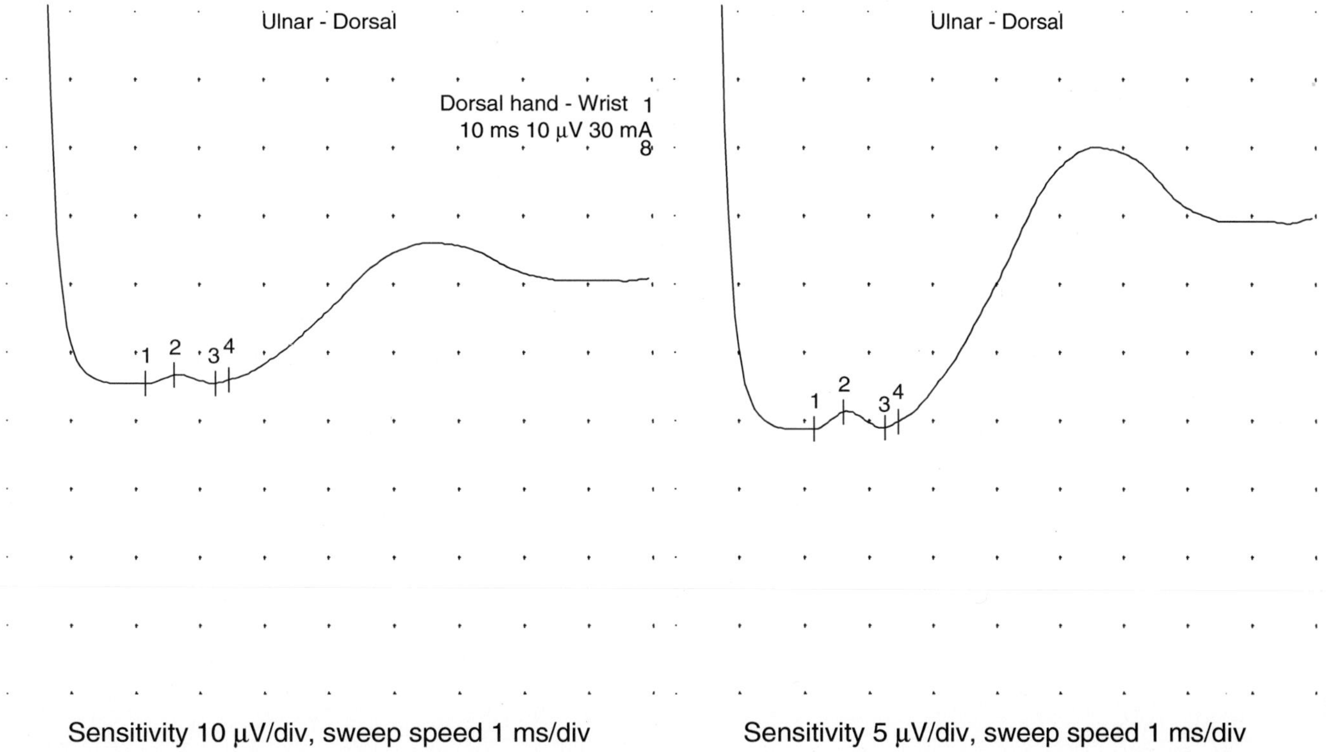

Onset latency (dorsal hand – forearm): 2.15 ms; **Peak latency** (dorsal hand – forearm): 2.65 ms; **Onset to peak amplitude** (dorsal hand – forearm): 1.2 µV; **Peak to peak amplitude** (dorsal hand – forearm): 1.1 µV; **SNCV** (dorsal hand – forearm, 11 cm distance): 51.2 m/s

Fig. 4 Orthodromic sensory nerve action potential (*SNAP*) recorded over the forearm in a patient with UNW, stimulation of the ulnar nerve on the dorsal hand

References

1. Kim DJ, Kalantri A, Guha S et al (1981) Dorsal cutaneous ulnar nerve conduction. Diagnostic aid in ulnar neuropathy. Arch Neurol 38:321–322
2. Jabre JF (1980) Ulnar nerve lesions at the wrist: new technique for recording from the sensory dorsal branch of the ulnar nerve. Neurology 30:873–876

Wrist, Below Elbow, Above Elbow, Upper Arm – Digit V

S15

Surface Recording Technique, Antidromic Study

Original Settings Low-frequency filter was 32 Hz, high-frequency filter was 3.2 kHz, and the machine used was a TECA electromyography (model TE-4). Sensitivity, sweep speed, duration, and rate of pulse were not specified.

Position This study was performed in the supine position, with the arm 75° abducted to the trunk, the elbow extended, and forearm supinated.

Recording Following the antidromic method [1], signals were recorded distally from the wrist to digit V (Fig. 1). Sensory nerve action potentials (SNAPs) were recorded with digital ring electrodes (TECA 6032). The authors placed ring electrodes around the digit V (little finger), but the exact position was not specified. We placed the active electrode (A) to the base of digit V and the reference (R) slightly distal to the distal interphalangeal joint of the same digit. The ground (G) electrode was not specified in the text; the figure shows the ground electrode placed on the palm.

Stimulation The ulnar nerve was stimulated at four sites: at the wrist (S1), below the elbow (S2), above the elbow (S3), and at the upper arm (S4). At wrist (S1), we placed the active stimulating electrode (cathode) 14 cm proximal to the active electrode at the base of the digit, over the ulnar nerve at the wrist, near the tendon of the *flexor carpi ulnaris* (FCU) – ideally proximal to the distal wrist crease. The anode (+) was proximal.

Proximal stimulations were applied below the elbow (S2, just below the ulnar groove), above the elbow (S3, just above the ulnar groove), and more proximally (S4, upper arm). The anode (+) was proximal at each stimulation.

Measurements Onset latency (ms) was measured from the stimulating artifact to the onset of the negative deflection of the SNAP. Sensory nerve conduction velocity (SNCV) of each segment (wrist–below elbow, below elbow–above elbow, above elbow–upper arm) was calculated dividing the distance between two different stimulating sites by difference of onset latencies, and it was measured in meter per second (m/s). Percentage SNCV (% SNCV) across the elbow (below-elbow–above-elbow segment) was calculated with SNCV of the forearm segment being taken as 100 %. Peak amplitude (μV) of the SNAP was measured from peak to peak. The skin temperature was controlled over the forearm and maintained at 34 °C or higher. Normal values (Table 1) of the SNCVs of all segments were measured in 25 healthy volunteers – Group 1 (32 hands of 15 men and 10 women, mean age 40.8 years, age range 20–66 years). The authors also studied lesions of 64 ulnar nerves (Table 2) in 44 patients (33 men and 14 women, 35 right and 29 left hands, mean age 41.6 years, age range 18–64 years) with clinical signs and symptoms suggestive of the ulnar neuropathy at the elbow (UNE). Patients were divided into two groups according to the motor nerve conduction velocity (MNCV): Group 2, abnormal SNCV with normal MNCV (12 hands), and Group 3, abnormal SNCV and MNCV (27 hands).

Comment

In 32 non-symptomatic hands of 25 healthy control subjects, Kimura et al. [1] observed that SNCV across the elbow was slightly slowed when compared to the SNCV of the forearm segment (below 10 % slowing), suggesting them that every non-symptomatic extremity may be exposed to great risk of the ulnar entrapment lesion at the elbow. The authors also observed a significantly slowed SNCV across the elbow (below elbow–above elbow) in 12 of 64 symptomatic hands when compared to SNCVs of the forearm (wrist–below elbow) and of the arm (above elbow–upper arm) segments, but normal MNCV in all segments examined. Their series showed 27 symptomatic hands with significant slowing of both SNCV and MNCV across the elbow, and in some patients, the SNAPs were

G. Gentili, M. Di Napoli, *The Ulnar Nerve*, DOI 10.1007/978-3-319-43292-2_15

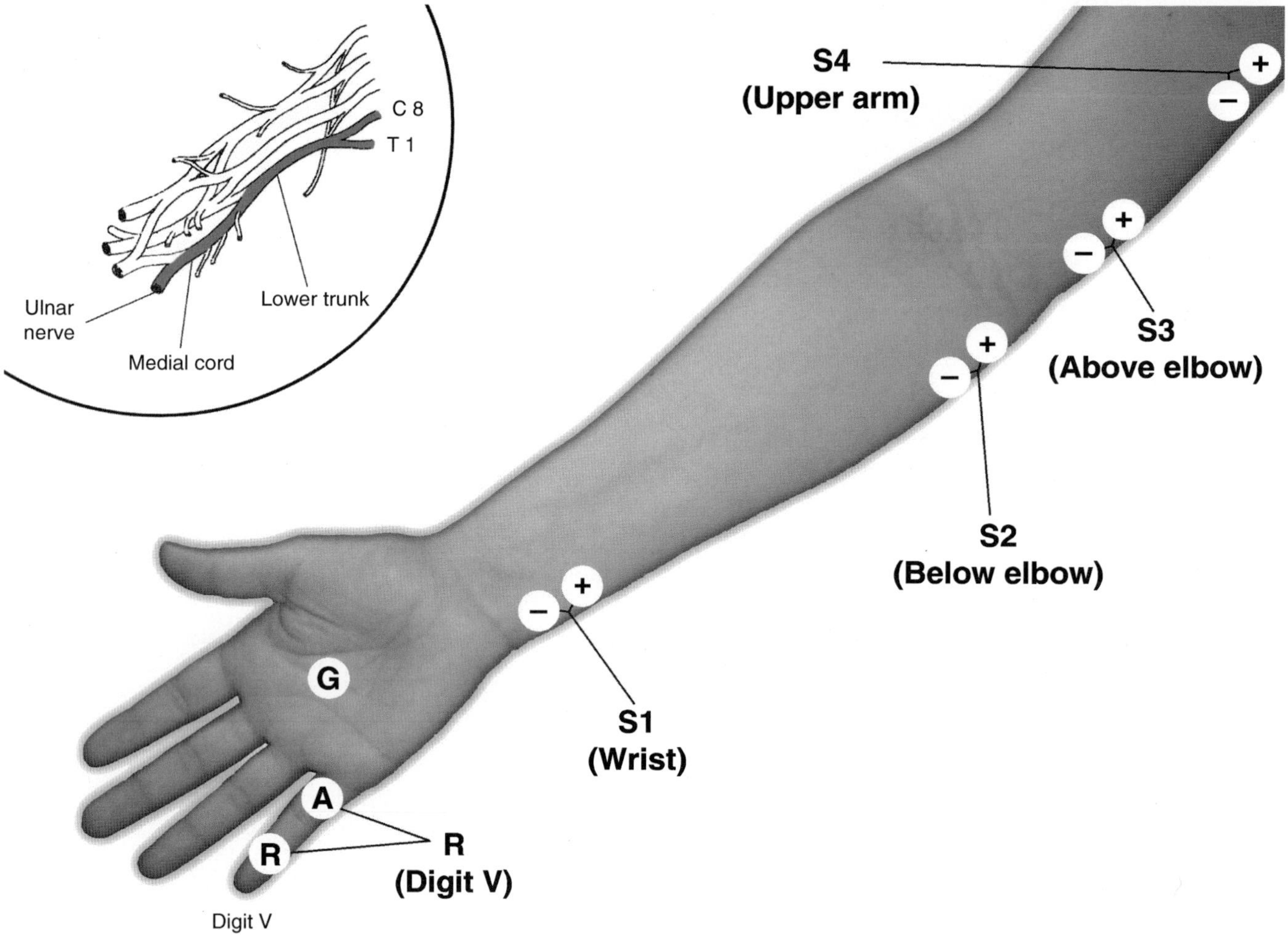

Typical waveform (wrist, below elbow, above elbow, upper arm – digit V):

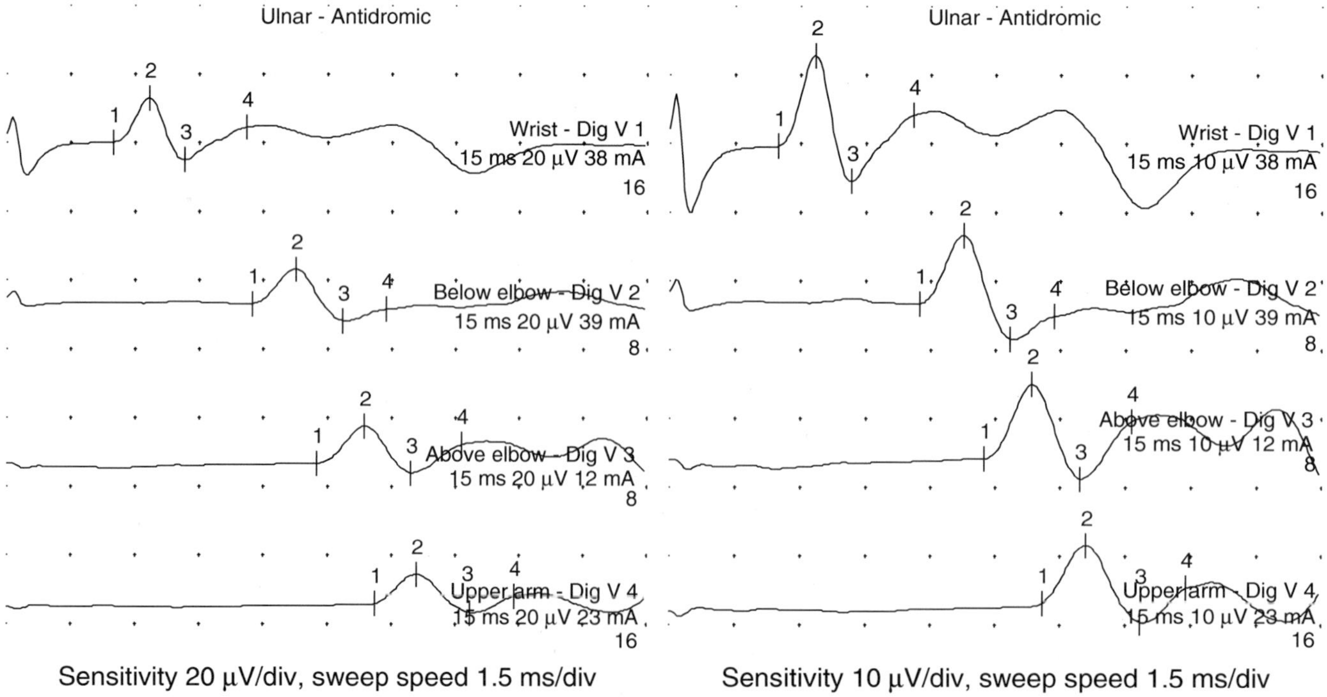

Fig. 1 Antidromic sensory nerve action potentials (*SNAPs*) recorded to digit V, stimulation of the wrist, below the elbow, above the elbow, and at upper arm (*top to bottom*)

Table 1 Normal values [1]

32 hands	Mean ± SD	% SNCV
Wrist–below elbow, SNCV (m/s)	62.0 ± 5.4	
Below elbow–above elbow, SNCV (m/s)	57.3 ± 5.5	91.7 ± 8.9
Above elbow–upper arm, SNCV (m/s)	60.1 ± 4.6	

Table 2 Pathological values [1]

Group 2–12 hands	*Mean ± SD*	*% SNCV*
Wrist–below elbow, SNCV (m/s)	60.4 ± 6.2	
Below elbow–above elbow, SNCV (m/s)	41.0 ± 7.6	67.3 ± 10.9
Above elbow–upper arm, SNCV (m/s)	64.0 ± 6.6	
Group 3–27 hands	*Mean ± SD*	*% SNCV*
Wrist–below elbow, SNCV (m/s)	61.8 ± 3.1	
Below elbow–above elbow, SNCV (m/s)	39.5 ± 3.5	63.5 ± 6.3
Above elbow–upper arm, SNCV (m/s)	63.6 ± 5.6	

absent. The presence of the purely sensory lesions that only sensory fibers show electrophysiological abnormalities and that motor fibers are not involved alone led the authors to the concept that sensory fibers are the first to be damaged in patients with UNE. For the authors [1], the early detection of abnormal conduction of the ulnar nerve at the cubital tunnel is regarded as essential, especially if protective measures could prevent nerve damage or surgical interventions can be performed while damage is still reversible. They concluded that the measurement of the SNCV across the elbow and representation of percentage slowing of SNCV across the elbow as SNCV of the forearm segment (100 %) may be a more sensitive and reliable parameter than other more established parameters. Pathological waveform and values in a patient with ulnar neuropathy at the elbow (UNE) are reported (Fig. 2).

Pathological waveform (wrist, below elbow, above elbow, upper arm – digit V):

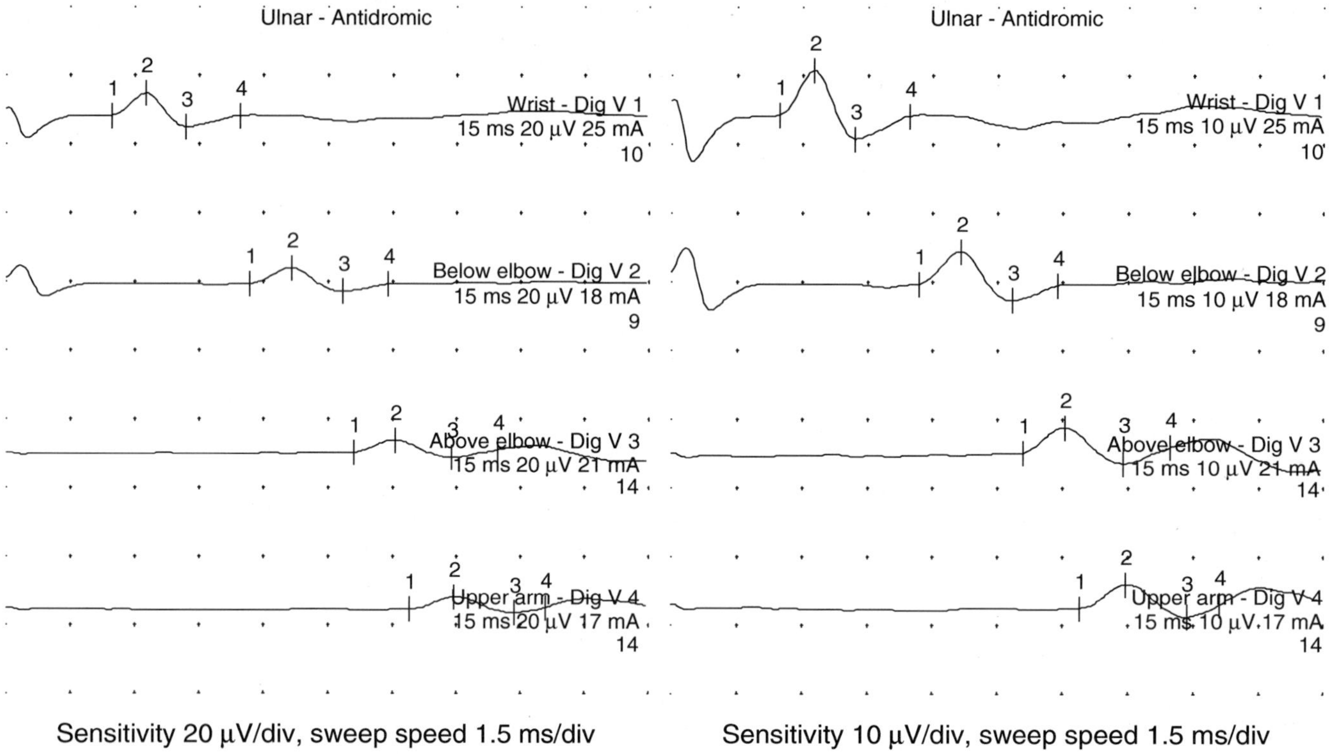

Onset latency (wrist – digit V): 2.45 ms; **Onset latency** (below elbow – digit V): 5.7 ms; **Onset latency** (above elbow – digit V) : 8.1 ms; **Onset latency** (upper arm – digit V): 9.4 ms; **Peak latency** (wrist – digit V): 3.2 ms; **Peak latency** (below elbow – digit V): 6.5 ms; **Peak latency** (above elbow – digit V): 9.05 ms; **Peak latency** (upper arm – digit V): 10.3 ms; **Onset to peak amplitude** (wrist – digit V): 6.5 μV; **Onset to peak amplitude** (below elbow – digit V): 4.8 μV; **Onset to peak amplitude** (above elbow – digit V): 3.7 μV; **Onset to peak amplitude** (upper arm – digit V): 3.5 μV; **Peak to peak amplitude** (wrist – digit V): 10.1 μV; **Peak to peak amplitude** (below elbow – digit V): 7.3 μV; **Peak to peak amplitude** (above elbow – digit V): 5.4 μV ; **Peak to peak amplitude** (upper arm – digit V): 5.0 μV; **SNCV** (wrist – below elbow, 19 cm distance): 58.5 m/s; **SNCV** (below elbow-above elbow, 10 cm distance): 41.7 m/s; **SNCV** (above elbow – upper arm, 9 cm distance): 69.2 m/s

Fig. 2 Antidromic sensory nerve action potentials (*SNAPs*) recorded to digit V in UNE, stimulation of the wrist, below the elbow, above the elbow, and at the upper arm (*top to bottom*)

Reference

1. Kimura I, Ayyar DR, Lippmann SM (1984) Early electrodiagnosis of the ulnar entrapment neuropathy at the elbow. Tohoku J Exp Med 142:165–172

Wrist, Below Elbow, Above Elbow, Axilla – Digit V

S16

Surface Recording Technique, Antidromic Study

Original Settings Sensitivity was 10–20 μV/division, low-frequency filter was 32 Hz, high-frequency filter was 3.2 kHz, and duration was 0.1–0.5 ms. Sweep speed, rate of pulse, and the machine used were not specified.

Position This study was performed with elbow flexed to 135° (full extension, 0°), with the forearm supinated.

Recording Following the antidromic method [1], signals were recorded distally from the wrist to digit V (Fig. 1). Sensory nerve action potentials (SNAPs) were recorded by finger clips. The active electrode (A) was placed midway along the proximal phalanx of the fifth finger, and the reference (R) was placed at the junction of the middle phalanx of the same finger. A lead band placed on the ventral surface of the palm was the ground (G) electrode.

Stimulation The ulnar nerve was stimulated percutaneously at four sites [1]: at the wrist, below the elbow, above the elbow, and in the axilla (Figs. 1, 2, and 3). At wrist (S1), the active stimulating electrode (cathode) was located 11 cm proximal to the active electrode. Proximal stimulations were applied below the elbow (S2, 4 cm distal to the medial epicondyle), above the elbow (S3, 10 cm proximal to the below-elbow site with the elbow extended), and more proximally in the axilla (S4, approximately 11 cm proximal to the above-elbow site). The anode (+) was proximal at each.

Stimulation The authors used supramaximal square-wave pulses of 100–300 V and 0.1–0.5 ms duration. All studies were done with the elbow flexed to 135° (full extension, 0°), with the forearm supinated (Fig. 2); the 10 cm below-elbow–above-elbow segment length was taken, in all subjects, with the elbow fully extended, the forearm supinated, and the arm abducted 30° (Fig. 1).

Measurements Onset latency (ms) was measured from the stimulating artifact to the onset of the negative deflection of the SNAP. Sensory nerve conduction velocity (SNCV) of each segment (wrist–below elbow, below elbow–above elbow, above elbow–axilla) was calculated dividing distance between two different stimulating sites by the difference of onset latencies, and it was measured in meter per second (m/s). Negative peak amplitude (μV) of the SNAP was measured from the baseline to the negative peak or initial positive peak if present. Surface distances were measured by a steel tape from cathodal stimulation sites. Electronic averaging was not used. Hand surface was measured, and if less than 33 °C, warming was performed by immersion in 40 °C water for 10 min. Normal values (Table 1) were obtained from 49 healthy volunteers without symptoms or signs of peripheral nerve disease or nerve injury (35 women and 14 men, age range 22–69 years, with ten subjects in each decade).

Kincaid et al. [1] reported values for the right side only because they observed no significant differences between right and left sides. SNAPs could be evoked by stimulation at all sites, including the axilla, in all but one subject (the SNAP was not clearly separated from the volume conducted motor response). The mean SNCV of the wrist–above-elbow segment was not significantly different from wrist–below-elbow segment, whereas the below-elbow–above-elbow segment was significantly faster in the same comparison. The authors [1] observed that the data produced for the calculated individual SNCV differences were not normally distributed and, in such circumstances, the SD was not an adequate measure of data distribution. For this reason, cumulative percentage tables and distribution polygons were constructed to evaluate segmental SNCV differences, and a value of 96 % of the normal subjects was used as cutoff. The segmental velocity differences using a 96 % cutoff were reported in Table 2 (B indicates wrist–below-elbow segment; E, wrist–above-elbow segment; EB, below-elbow–above-elbow segment; XE, above-elbow–axilla segment). Using this cutoff, the slowest below-elbow–

G. Gentili, M. Di Napoli, *The Ulnar Nerve*, DOI 10.1007/978-3-319-43292-2_16

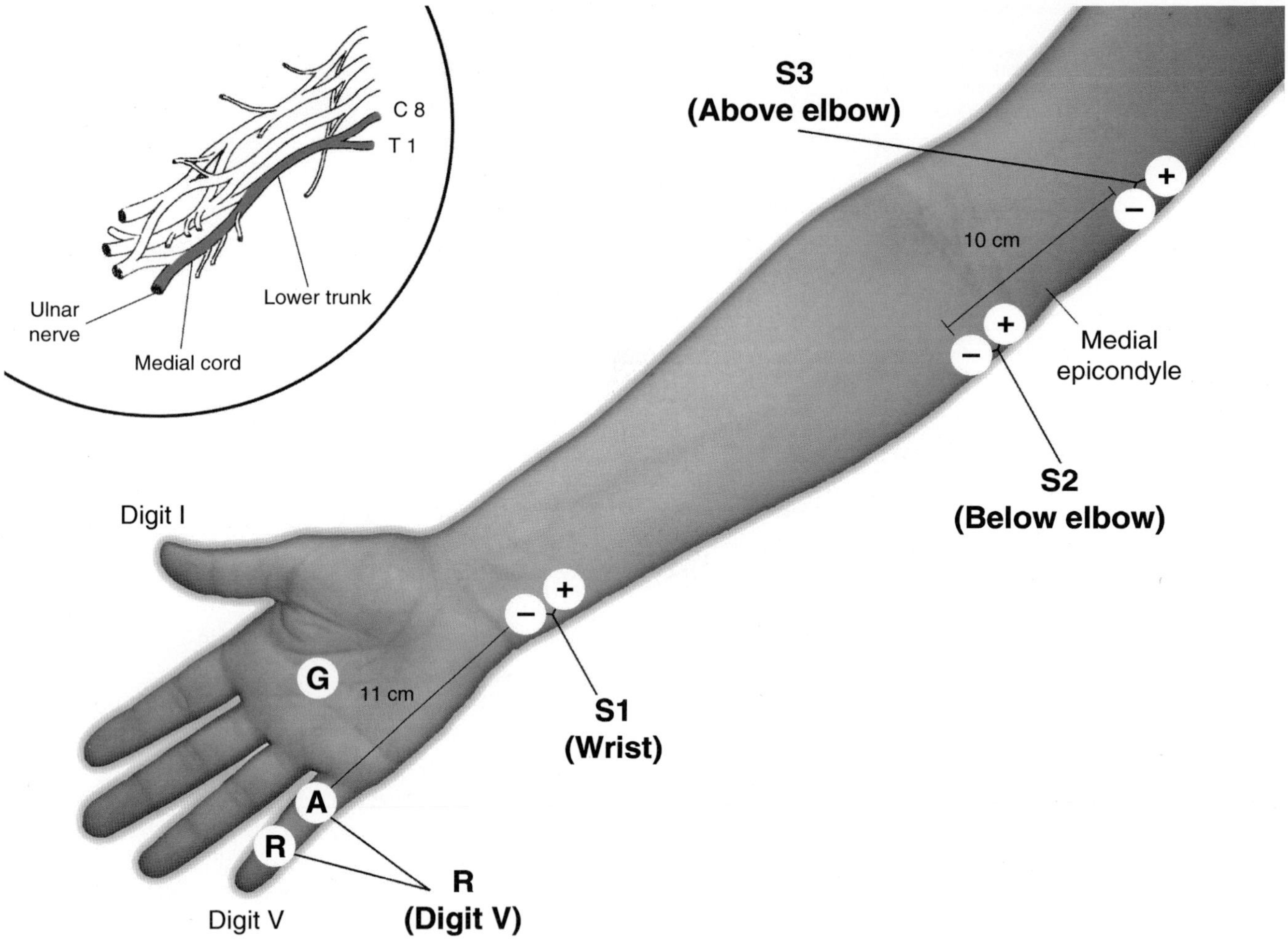

Typical waveform (wrist, below elbow, above elbow, axilla – digit V):

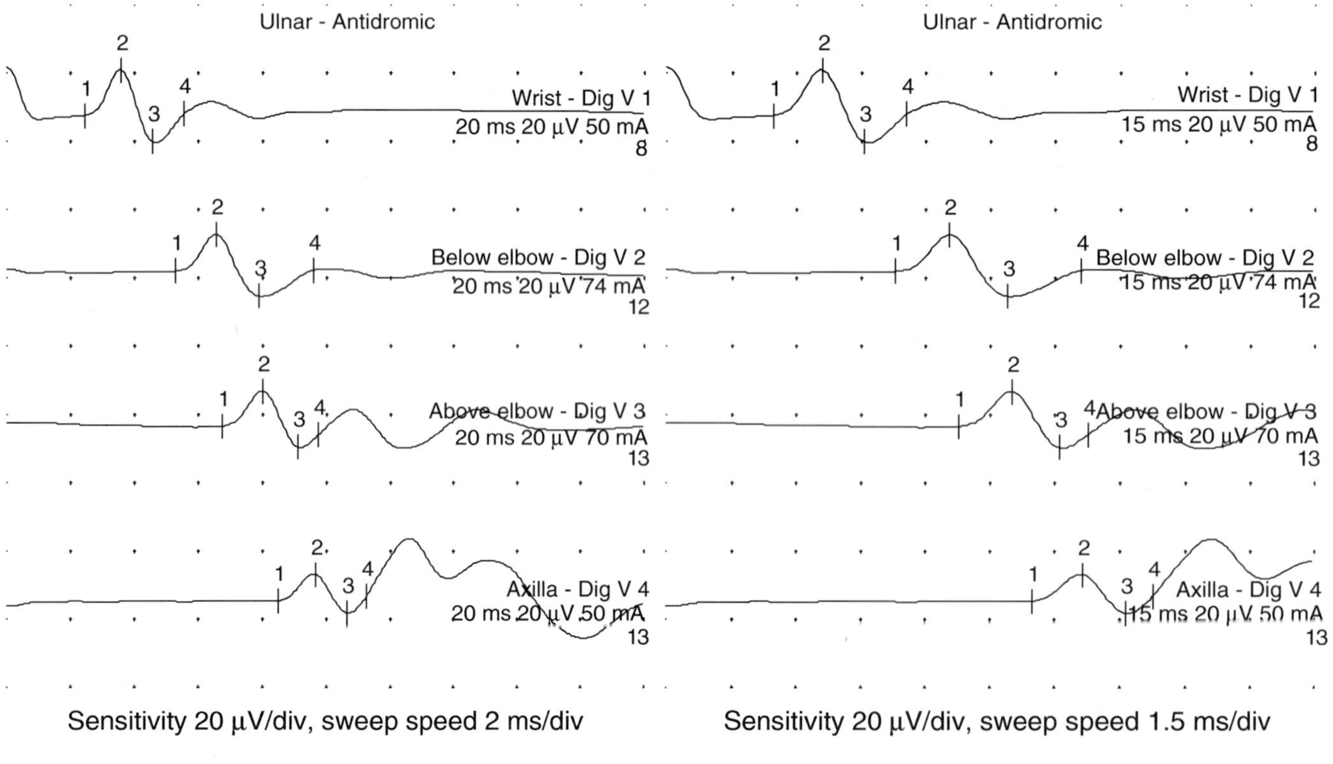

Fig. 1 Antidromic sensory nerve action potentials (*SNAPs*) recorded to digit V, stimulation of the wrist, below the elbow, above the elbow and in the axilla (*top to bottom*)

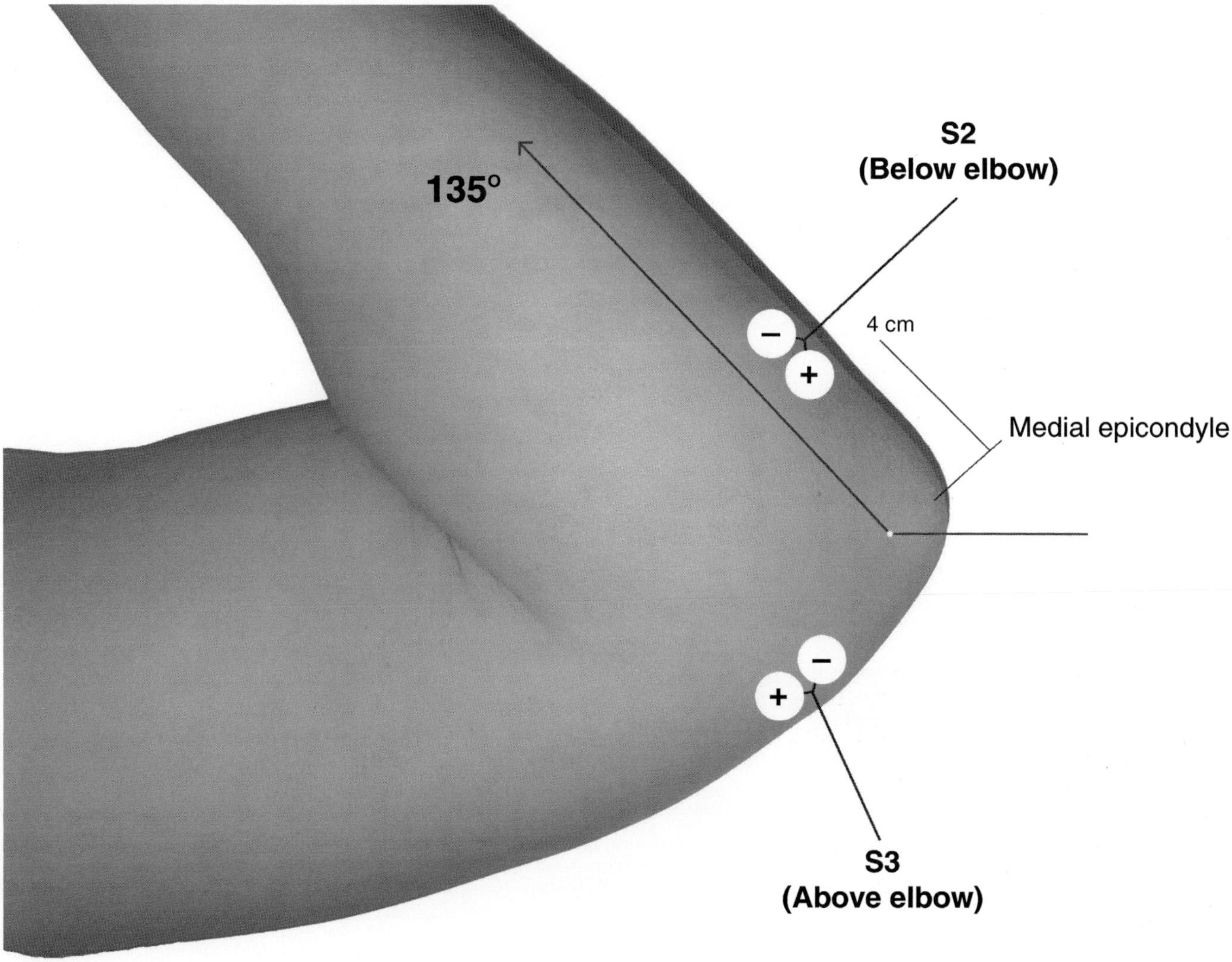

Fig. 2 Below- and above-elbow stimulation points (elbow flexed to 135°)

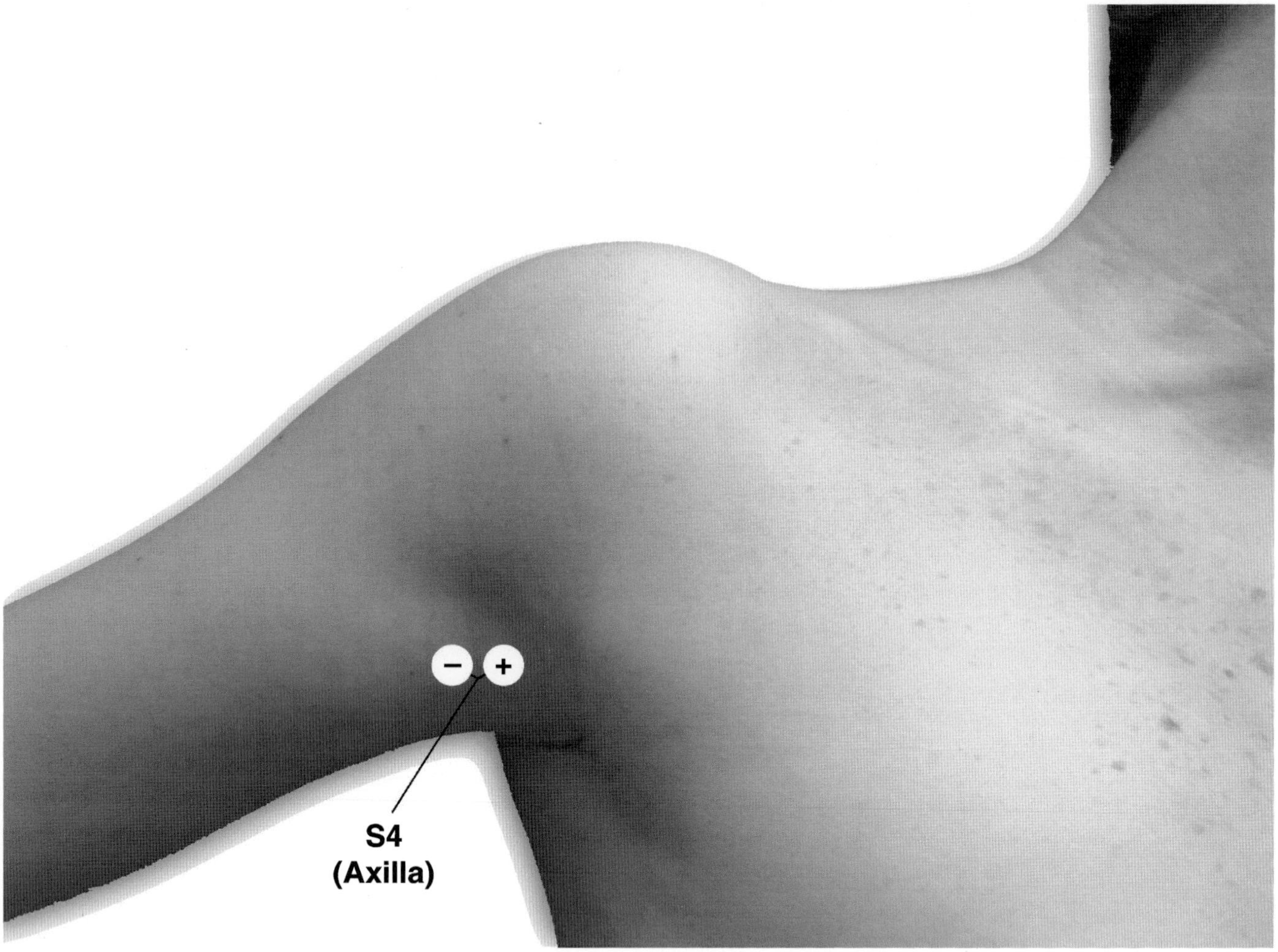

Fig. 3 Proximal stimulation point (S4 – axilla)

Table 1 Normal values [1]

	Mean ± SD	Range
Wrist–below elbow, SNCV (m/s)	64.6 ± 5.1	53.5–73.9
Wrist–above elbow, SNCV (m/s)	65.8 ± 4.9	55.1–73.5
Wrist–axilla, SNCV (m/s)	66.2 ± 5.0	53.3–73.9
Below elbow–above elbow, SNCV (m/s)	68.5 ± 7.5	50.8–88.3
Above elbow–axilla, SNCV (m/s)	67.9 ± 9.3	47.9–88.6

above-elbow SNCV was 51 m/s. The amplitude of the SNAP by stimulation at the wrist–above-elbow site was a mean of 21 % lower than the wrist–below-elbow SNAP, and the maximum difference was 43 %. The flexed elbow position allowed supramaximal stimulation (stimulus duration of greater than 0.1 ms was rarely required) at the elbow site at a much lower stimulus strength than in the elbow-extended position, despite the stimulus site being at approximately the same location. Another advantage of elbow flexion suggested by the authors [1] was that all portions of the ulnar nerve were directly accessible for locating stimulation points and for performing subsequent distance measurement. With the elbow extended, visualizing the below-elbow, above-elbow, and axillary stimulation points required the examiner to stand to view the arm's medial aspect.

Comment

For Kincaid et al. [1], short segments used for the nerve conduction studies have a higher potential for error in SNCV calculation producing erroneous values because of the proportionately higher experimental error inherent in measuring distance and latency. The SNCV determinations on nerve segments with a length of less than 10 cm or with conduction time less than 2.5 ms were shown to be susceptible to a high degree of error. For these reasons they chose to examine SNCV differences between both the shorter across elbow segment (mean length 12.9 cm) and the forearm (mean length 19.5 cm) and the longer wrist–above-elbow segment (mean length 32.5 cm), also compared with the forearm. Underestimation of the true length of the elbow nerve segment as assessed by surface measurement has been suggested as the explanation for such findings with the elbow in extension [2–4]. The degree of flexion necessary to eliminate this measurement error has been suggested to be 45° [2], 90° [3], and 110° [4], but Kincaid and colleagues [1] chose 135°, presuming that this would provide a maximum elongation of the nerve across the elbow. They concluded that performing ulnar nerve conduction studies with the elbow flexed to 135° and comparing conduction values for the wrist–above-elbow and wrist–forearm segment provides values with the least normal conduction values variation between these segments. For the authors, the application of such values to the evaluation of patients with suspected ulnar nerve lesion at the elbow may allow for increased reliability in the electrodiagnosis of these lesions. Pathological waveform and values in a patient with ulnar neuropathy at the elbow (UNE) are reported (Fig. 4).

Table 2 Normal values [1]

	Mean ± SD	Range	Limit
EB, segmental velocity differences (m/s)	1.3 ± 2.6	−3.7–7.7	−3.4–6.8
EB – B, segmental velocity differences (m/s)	3.9 ± 7.4	−9.0–25.6	−8.5–25.3
XE – EB, segmental velocity differences (m/s)	−0.7 ± 10.0	−26.9–21.7	−18.5–20.9

Pathological waveform (wrist, below elbow, above elbow, axilla – digit V):

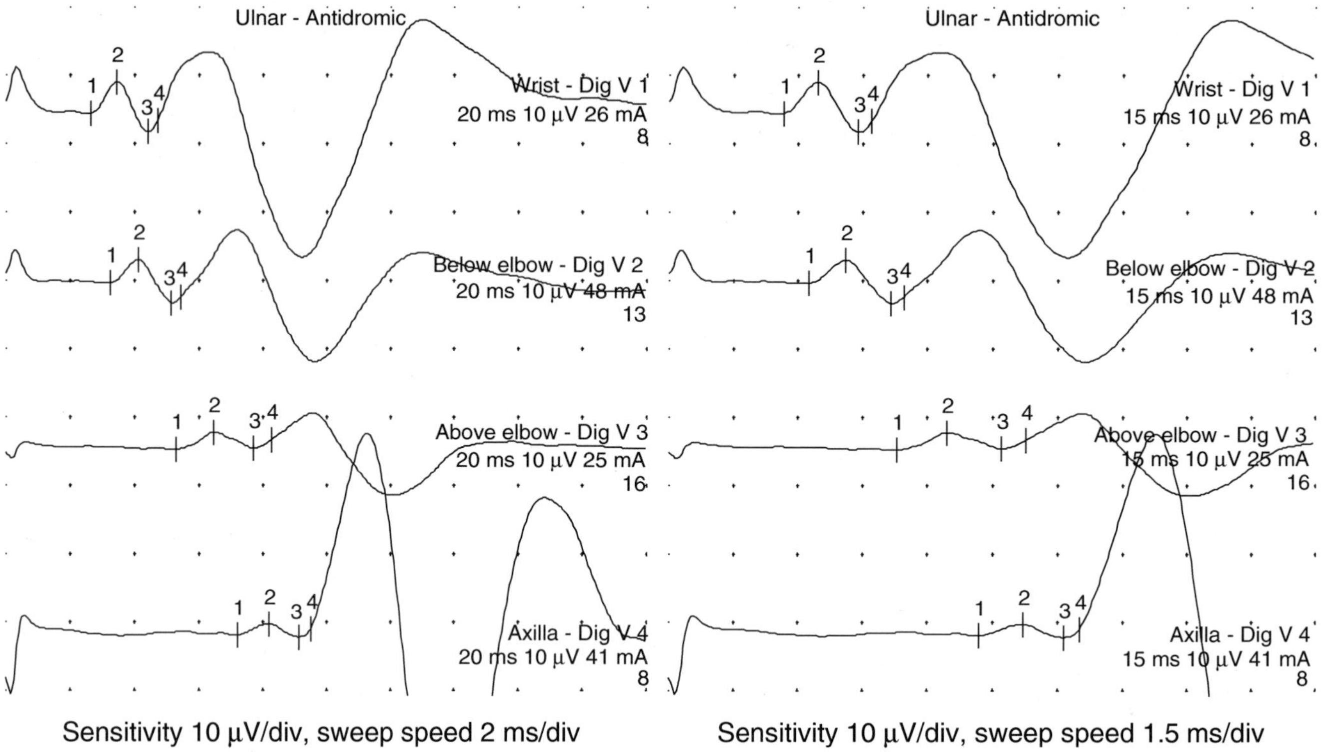

Onset latency (wrist – digit V): 2.65 ms; **Onset latency** (below elbow – digit V): 3.25 ms; **Onset latency** (above elbow – digit V): 5.3 ms; **Onset latency** (axilla – digit V): 7.2 ms; **Peak latency** (wrist – digit V): 3.4 ms; **Peak latency** (below elbow – digit V): 4.1 ms; **Peak latency** (above elbow – digit V): 6.45 ms; **Peak latency** (axilla – digit V): 8.2 ms; **Onset to peak amplitude** (wrist – digit V): 4.7 μV; **Onset to peak amplitude** (below elbow – digit V): 3.4 μV; **Onset to peak amplitude** (above elbow – digit V): 2.7 μV; **Onset to peak amplitude** (axilla – digit V): 1.7 μV; **Peak to peak amplitude** (wrist – digit V): 7.5 μV; **Peak to peak amplitude** (below elbow – digit V): 6.5 μV; **Peak to peak amplitude** (above elbow – digit V): 2.6 μV; **Peak to peak amplitude** (axilla – digit V): 1.9 μV; **SNCV** (wrist – digit V, 14 cm distance): 52.8 m/s; **SNCV** (wrist – below elbow, 18 cm distance): 55.4 m/s; **SNCV** (below elbow – above elbow, 8 cm distance): 39.0 m/s; **SNCV** (above elbow – axilla, 11 cm distance): 57.9 m/s

Fig. 4 Antidromic sensory nerve action potentials (*SNAPs*) recorded to digit V in UNE, stimulation of the wrist, below the elbow, above the elbow, and in the axilla (*top to bottom*)

References

1. Kincaid JC, Phillips LH, Daube JR (1986) The evaluation of suspected ulnar neuropathy at the elbow. Normal conduction study values. Arch Neurol 43:44–47
2. Harding C, Halar E (1983) Motor and sensory ulnar nerve conduction velocities: effect of elbow position. Arch Phys Med Rehabil 64:227–232
3. Nelson RM (1980) Effects of elbow position on motor conduction velocity of the ulnar nerve. Phys Ther 60:780–783
4. Checkles NS, Russakov AD, Piero DL (1971) Ulnar nerve conduction velocity: effect of elbow position on measurement. Arch Phys Med Rehabil 52:362–365

Wrist – Arm, Axilla; Below Elbow – Axilla; Digit V – Arm

S17

Surface Recording Technique, Orthodromic Study

Original Settings Sensitivity was 4–10 μV/division, low-frequency filter was 10 Hz, high-frequency filter was 2 kHz, sweep speed was 2 ms/division, duration of pulse was 0.2 ms, and the machine used was a TECA TD20 two-channel electrodiagnostic unit. The rate of pulse was not specified.

Position This study was performed in the supine position; the authors performed all tests with the elbow flexed 70°, but we obtained better results with the elbow fully extended in a neutral position (Figs. 1, 2, and 3).

Recording The authors [1] performed three different nerve conduction techniques. First, following the orthodromic method, mixed nerve action potentials (MNAPs) were recorded proximally at the arm (R1) and at the axilla (R2), 35 cm and 25 cm distal to the sternal notch (measured with the arm externally rotated and abducted to 90 °C), respectively (Fig. 1). The authors used stainless steel disks 1 cm in diameter applied with conductive gel, and the optimal location of the active recording (A) electrode was determined as the site at which submaximal stimulation produced the largest visible contraction in the distal ulnar innervated muscles. The reference (R) electrode was placed 6 cm from the active recording electrode, perpendicular to the course of the ulnar nerve. The authors used unipolar recordings to avoid variable summation and subtraction from bipolar recordings, in which both electrodes record a significant nerve potential. Following the orthodromic method [1], the authors suggested an additional ulnar MNAP study recording in the axilla, in order to evaluate the across-elbow (cubital tunnel) ulnar nerve conduction (Fig. 2). Usually 8–12 potentials (in some cases 16) were averaged to record the MNAPs proximally from the arm and axillary electrodes. With a third orthodromic method [1], sensory nerve action potentials (SNAPs) were then recorded at the arm (Fig. 3), using the same electrode placement for MNAPs recording (orthodromic method 1 and 2). The ground (G) electrode position was always placed between the stimulation and recording sites; the figures show ground electrode placed on the forearm.

Stimulation First, following the orthodromic method, the ulnar nerve was stimulated at the wrist (S), 14 cm proximal to the midproximal phalanx of digit V. Following the orthodromic method, the ulnar nerve was then stimulated 4 cm distal to the medial epicondyle. For both methods, the supramaximal nerve stimulation was performed (the stimulus intensity was gradually increased to elicit appropriate muscle contractions and until no appreciable increase in MNAP amplitude occurred, using 0.2 ms duration). Following the orthodromic method, for the sensory nerve response (SNAP) recordings, the ulnar nerve was stimulated distally to the digit V by ring electrodes, with the cathode placed at the midproximal phalanx and the anode at the distal phalanx.

Measurements The MNAP and SNAPs latencies (ms) were measured to the onset of the initial positive deflection, the intersection of the upward deflection with the baseline, and the return to the baseline of the terminal tail. The authors chose these points to represent the fastest (1), intermediate (2), and slowest (3) conduction velocities, respectively. For the earliest initial positive deflection (representing approaching depolarization in the fastest conducting fibers), the authors used the marker "1"; for the baseline crossing between the initial positive and the subsequent negative deflection (representing the intermediate conduction latency), they used the marker "2"; and for the return to the baseline after the negative deflection (representing the latency of the slowest conducting fibers), they used the marker "3." Conduction velocities were calculated as the quotient of the distance from the stimulation site to the active recording electrode, divided by the appropriate latency. MNAP and SNAP amplitudes were measured as the difference between the maximum positive and negative deflections (peak to peak), the amplitude ratio comparing the

G. Gentili, M. Di Napoli, *The Ulnar Nerve*, DOI 10.1007/978-3-319-43292-2_17

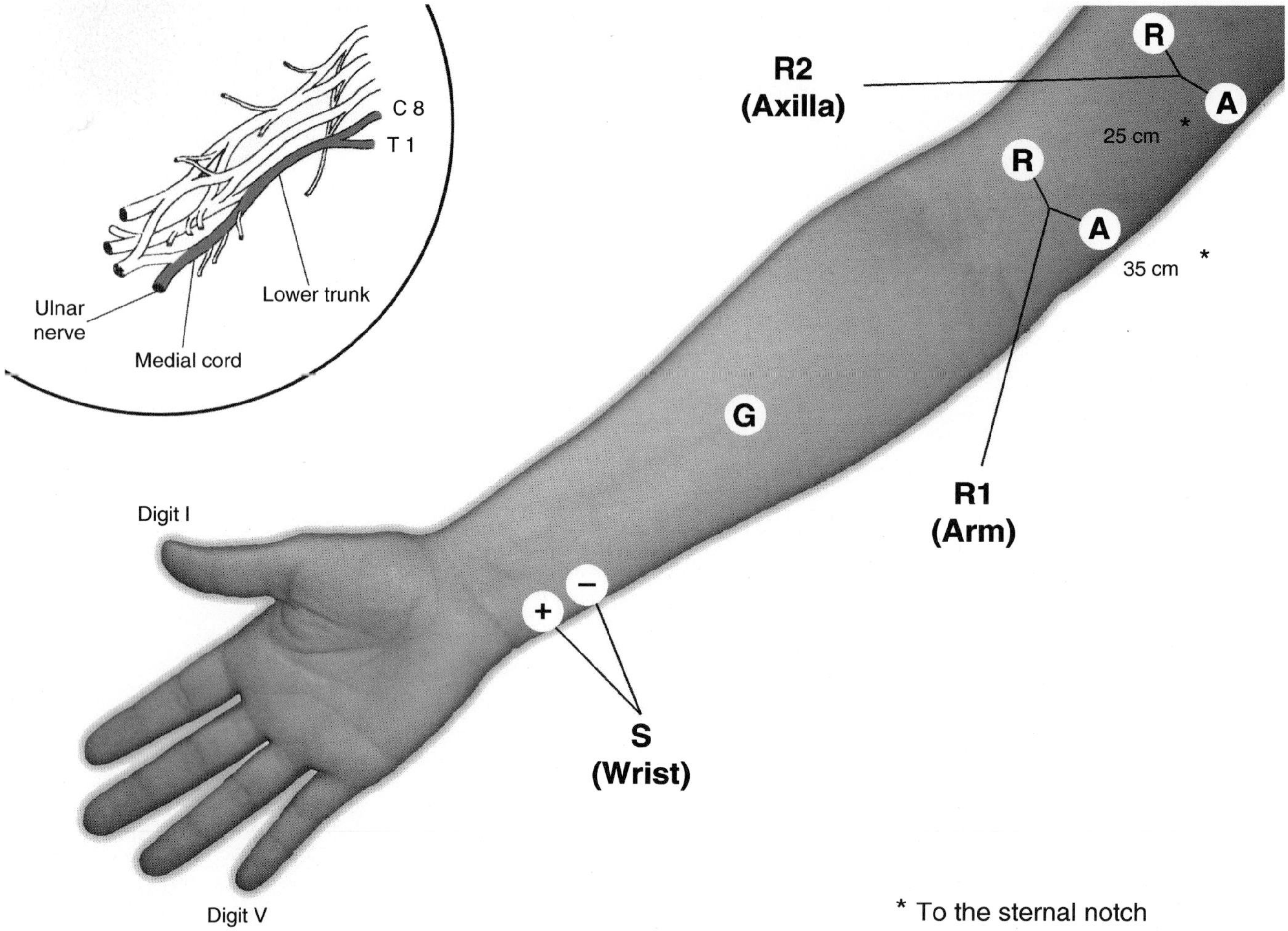

Typical waveform (wrist – arm, axilla):

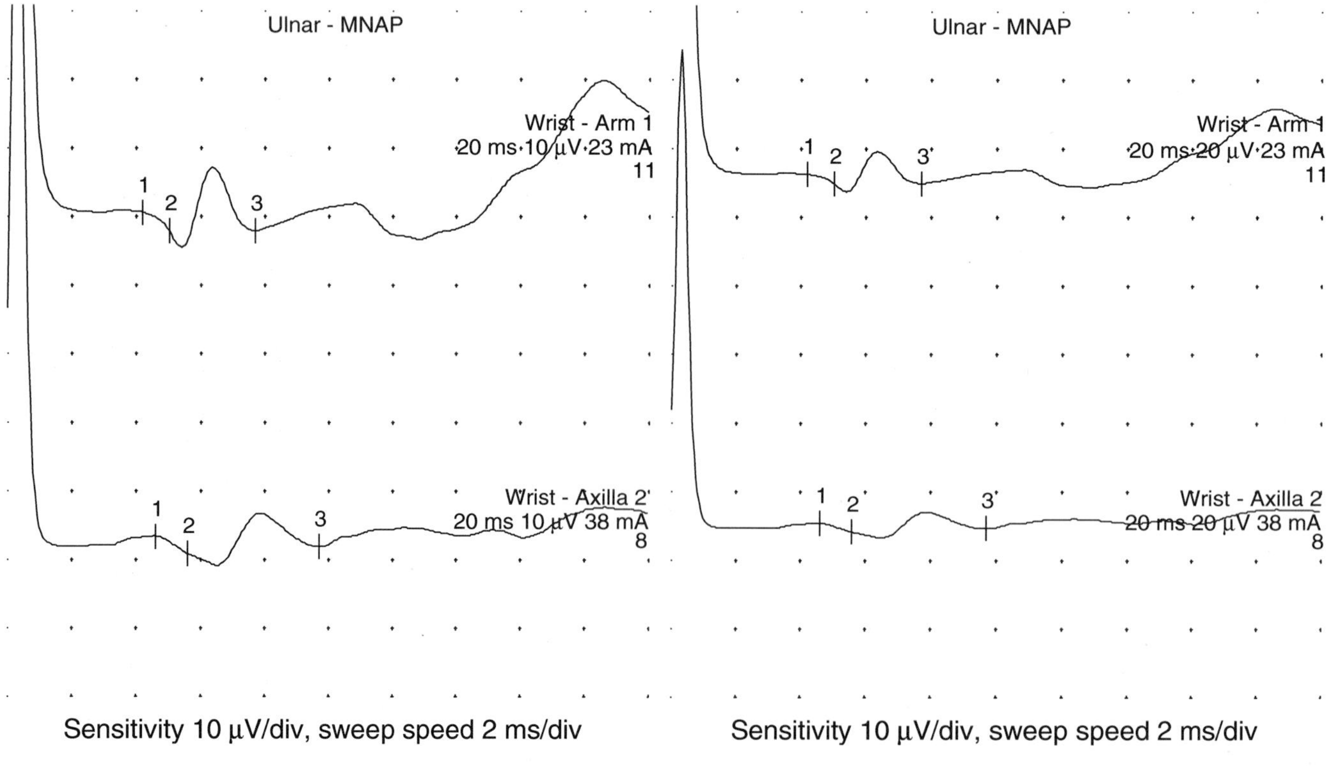

Fig. 1 Orthodromic mixed ulnar nerve responses (*MNAPs*) recorded at the arm (*upper trace*) and at the axilla (*lower trace*), stimulation of the wrist

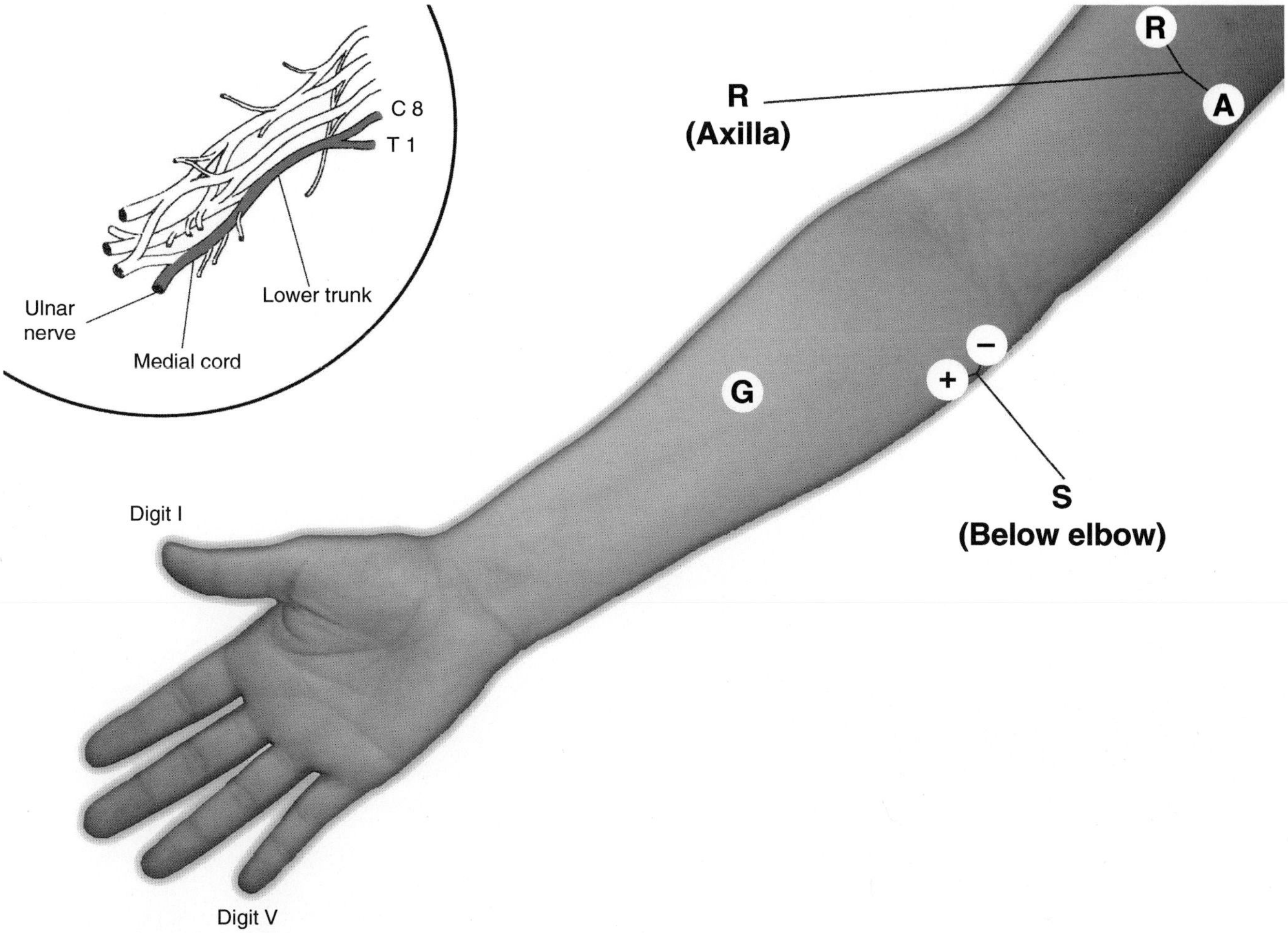

Typical waveform (below elbow–axilla):

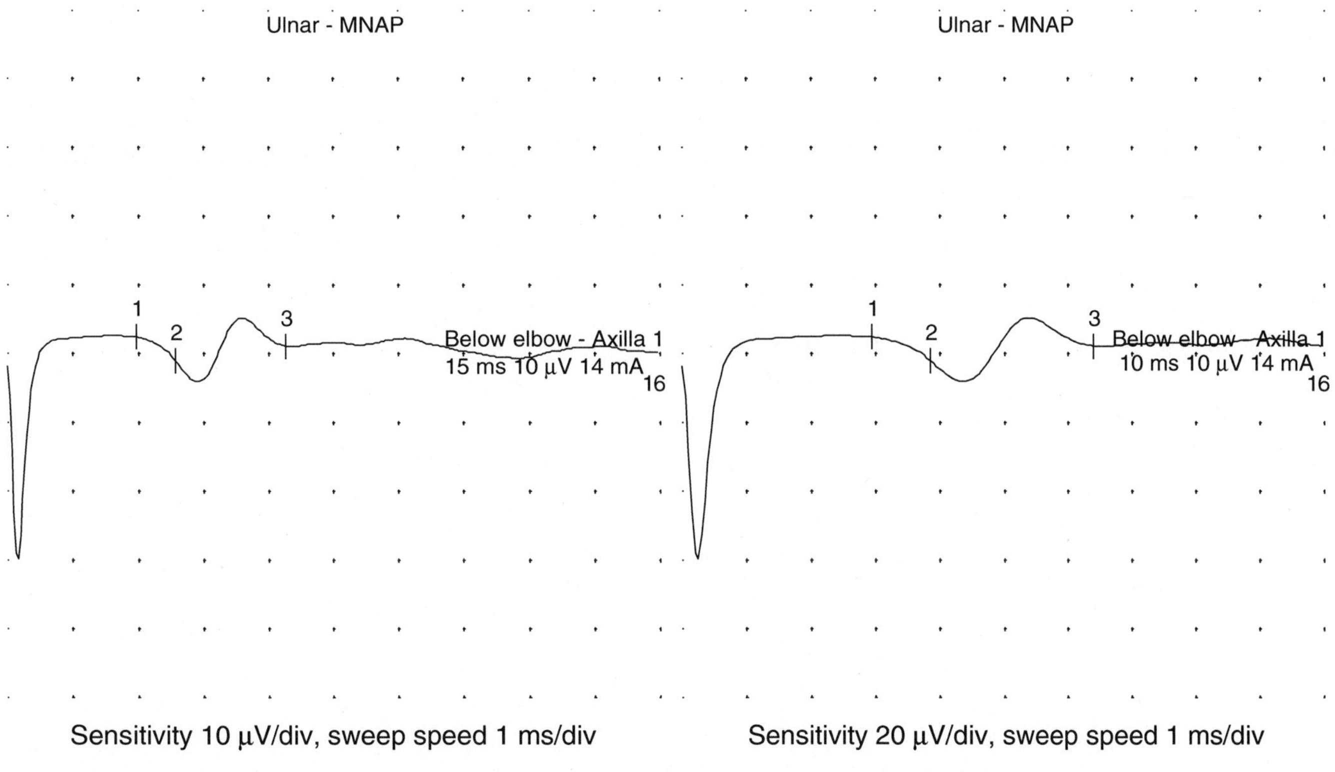

Fig. 2 Orthodromic mixed ulnar nerve response (*MNAP*) recorded at the axilla, stimulation below the elbow

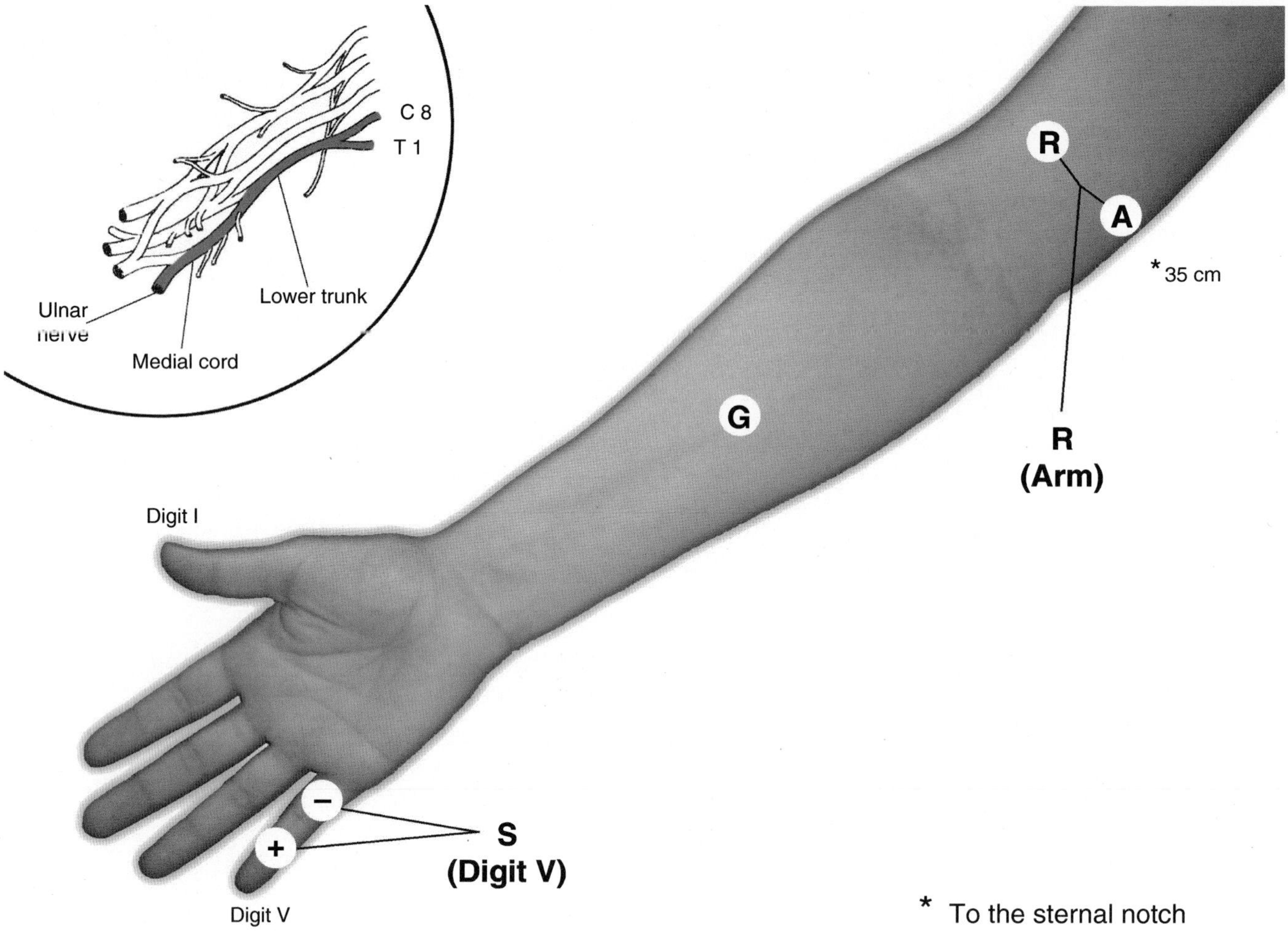

Typical waveform (digit V – arm):

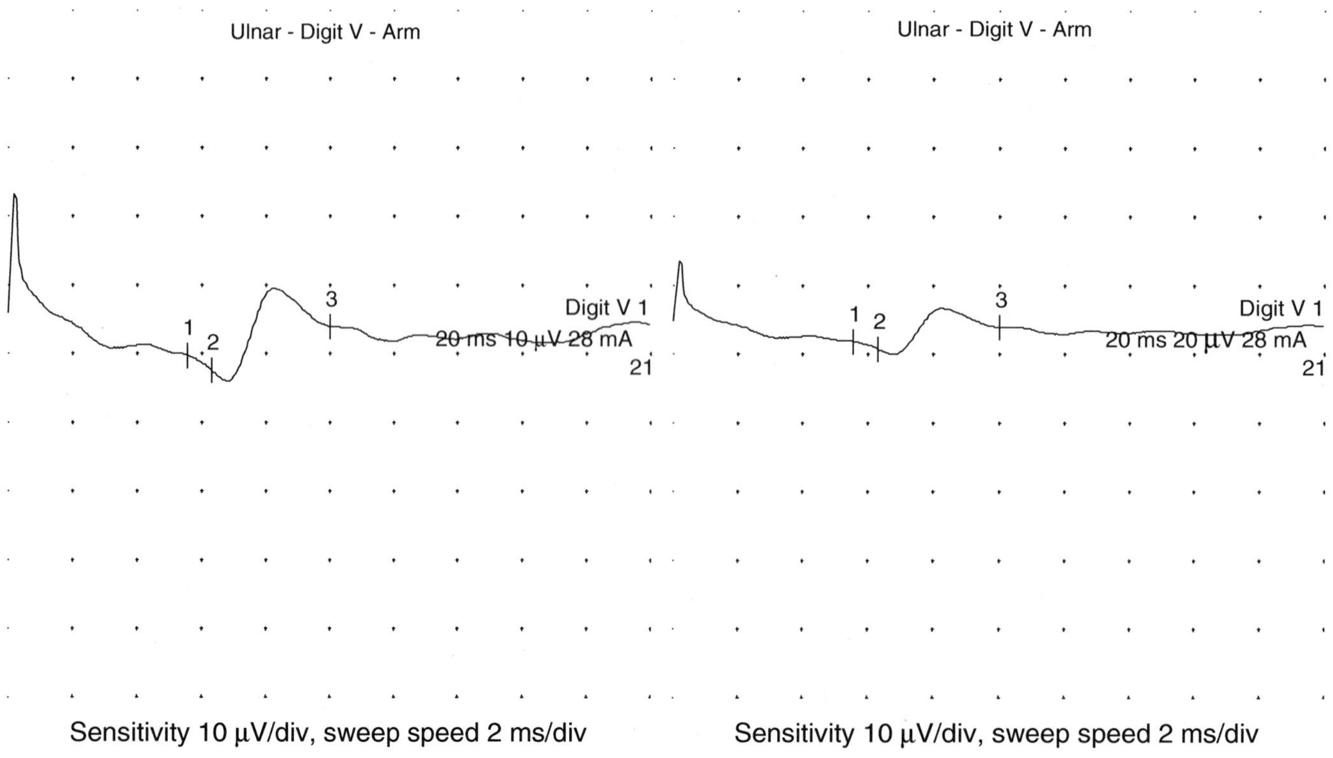

Fig. 3 Orthodromic sensory ulnar nerve response (*SNAP*) recorded at the arm, stimulation to digit V

Table 1 Normal values [1]

MNAP, mixed conduction velocity (m/s)	Mean + SD
Wrist–arm (fastest), right arm	79.3 + 8.7
Wrist–arm (intermediate), right arm	58.3 + 3.6
Wrist–arm (slowest), right arm	45.7 + 3.1
Wrist–arm (fastest), left arm	82.3 + 11.0
Wrist–arm (intermediate), left arm	58.7 + 4.8
Wrist–arm (slowest), left arm	43.7 + 4.4

MNAP, peak-to-peak amplitude (μV)	Mean + SD	Mean + 2SD	Mean − 2SD
Wrist–arm, right arm	25.9 + 12.9		
Wrist–arm, left arm	22.7 + 10.6		
Right/left RATIO	1.28 + 0.49	2.26	0.29

SNAP, sensory conduction velocity (m/s)	Mean + SD
Wrist–arm (fastest), right arm	63.8 + 6.5
Wrist–arm (intermediate), right arm	54.9 + 4.7
Wrist–arm (slowest), right arm	47.0 + 4.1
Wrist–arm (fastest), left arm	65.5 + 6.1
Wrist–arm (intermediate), left arm	54.7 + 4.6
Wrist–arm (slowest), left arm	45.7 + 4.0

SNAP, peak-to-peak amplitude (μV)	Mean + SD	Mean + 2SD	Mean − 2SD
Wrist–arm, right arm	3.0 + 1.5		
Wrist–arm, left arm	2.5 + 1.2		
SNAP/MNAP ratio, right arm	0.12 + 0.04	0.20	0.04
SNAP/MNAP ratio, left arm	0.12 + 0.05	0.22	0.02

arm, and the elbow MNAPs and SNAPs obtained from the same arm and those of the contralateral extremity. Surface cutaneous temperatures were recorded (18 subjects) over the neurovascular bundle in the axilla (mean temperature 34.8 ± 0.5 °C, range temperature 33.3–35.5 °C) and over the median nerve at the wrist (mean temperature 31.7 ± 1.7 °C, range temperature 28.0–33.9 °C). The authors studied 40 ulnar nerves (Table 1) from 20 control subjects (age range 23–62 years, mean age 30.4 ± 8.4 years).

Comment

For Chassin et al. [1], although the amplitudes demonstrated larger peak-to-peak values for the right as compared with the left arm, these differences were not statistically significant. The conduction study across the cubital tunnel revealed an ulnar nerve conduction velocity (NCV) from below-elbow stimulation to the axilla of 59.7 ± 3.8 m/s. The latency of this value was measured to the point of the intersection with the baseline, corresponding to the intermediate conduction velocity. The authors suggested that the NCV calculated for the intermediate conduction fibers group in their study was an estimate of modal conduction velocities in mixed peripheral nerves. They measured the orthodromic SNAP (digit V–wrist segment) and MNAP (wrist–arm and wrist–axilla segments) conduction velocities and amplitudes. They suggested using several electrophysiological criteria to identify the peripheral ulnar nerve axonal atrophy: (1) ipsilateral ulnar to median MNAP ratio, (2) ipsilateral to contralateral ulnar MNAP ratio, and (3) ipsilateral ulnar SNAP/MNAP ratio. In patients with entrapments of individual nerves, the authors found small amplitudes when they compared the ulnar nerve in the ipsilateral arm or with the contralateral arm.

Reference

1. Chassin SL, Little JW, DeLisa JA (1987) Compound nerve action potentials from the median and ulnar nerves. Arch Phys Med Rehabil 68:31–35

Forearm – Dorsal Hand

S18

Surface Recording Technique, Antidromic Study

Original Settings Sensitivity was 20–50 μV/division, sweep speed was 1 ms/division, low-frequency filter was 32 Hz, high-frequency filter was 3.2 kHz, duration of pulse was 0.1–0.2 ms, and the machine used was a TECA TE4 model. The rate of pulse was not specified.

Position This study was performed with the patient seated, with the arm supported on a level surface, the forearm fully pronated, the wrist in neutral position, and the fingers slightly flexed.

Recording Following the antidromic method [1], signals were recorded using a plastic-mounted, bipolar electrode (surface disks were 3 cm apart) placed along the fifth metacarpal bone, with the reference electrode level with the fifth metacarpophalangeal joint (Fig. 1). The authors used this montage to record from the dorsal branch of the ulnar nerve, which leaves the main trunk some centimeters proximal to the ulnar styloid and becomes cutaneous as it passes between the flexor carpi ulnaris (FCU) muscle tendon and the ulna bone. It supplies sensation to the dorsoulnar aspect of the wrist and hand and to the dorsal surface of the digit IV (ring finger) and digit V (little finger). The authors compared the conduction parameters of the dorsal cutaneous branch of the ulnar nerve with the superficial radial nerve, the sensory branch of the radial nerve supplying the dorsolateral hand; following the antidromic method, the sensory nerve action potentials (SNAPs) were recorded using plastic-mounted, bipolar electrode (surface disks were 3 cm apart) placed along the first metacarpal bone, with the reference electrode level over the first dorsal web space (Fig. 1).

Stimulation Stimulation was applied at the forearm – 14 cm proximally to the recording electrodes over the dorsum of the hand, on the ulnar, and radial nerves [1]. The anode was proximal in both stimulations (Fig. 1). The ground (G) electrode was placed between the stimulation and recording sites; the figures show ground electrode placed over the forearm.

Measurements Onset latency (ms) was measured from the stimulating artifact to the onset of the negative deflection of the SNAP. Peak latency (ms) was measured from the stimulus onset to the peak of the first negative deflection of the sensory nerve action potential (SNAP). The negative peak amplitude (μV) was measured from the onset of the SNAP to the peak of the negative deflection. The skin temperature of the dorsum of the hand was measured using an electronic thermometer with surface probe. The hand and forearm were warmed with an infrared lamp when necessary, and all studies were performed with skin temperatures between 31 and 36 °C. Normal values (Table 1) were obtained from 55 hands of 33 healthy subjects (age range 22–69 years, mean age 37 ± 13 years).

Comment

For Hoffman et al. [1] the sensory responses from the dorsal cutaneous branch of the ulnar nerve were obtained without significant difficulty in all cases. Muscle contraction artifact did occasionally interfere with visualizing the sensory response, and, in such cases, they suggested to stimulate the forearm fully supinated, rather than pronated, provided an easier method for obtaining a SNAP. The mean latencies to onset and peak were similar, and the mean sensory nerve conduction velocity (SNCV) was 63.6 m/s. The authors reported two cases to demonstrate the utility of paired nerve conduction studies of the dorsal cutaneous branch of the ulnar nerve and superficial radial nerve. In their first case, the paired sensory conduction study of the ulnar and radial nerves showed significant slowing of the ulnar sensory indicating that the injury was distal to the dorsal cutaneous branch

G. Gentili, M. Di Napoli, *The Ulnar Nerve*, DOI 10.1007/978-3-319-43292-2_18

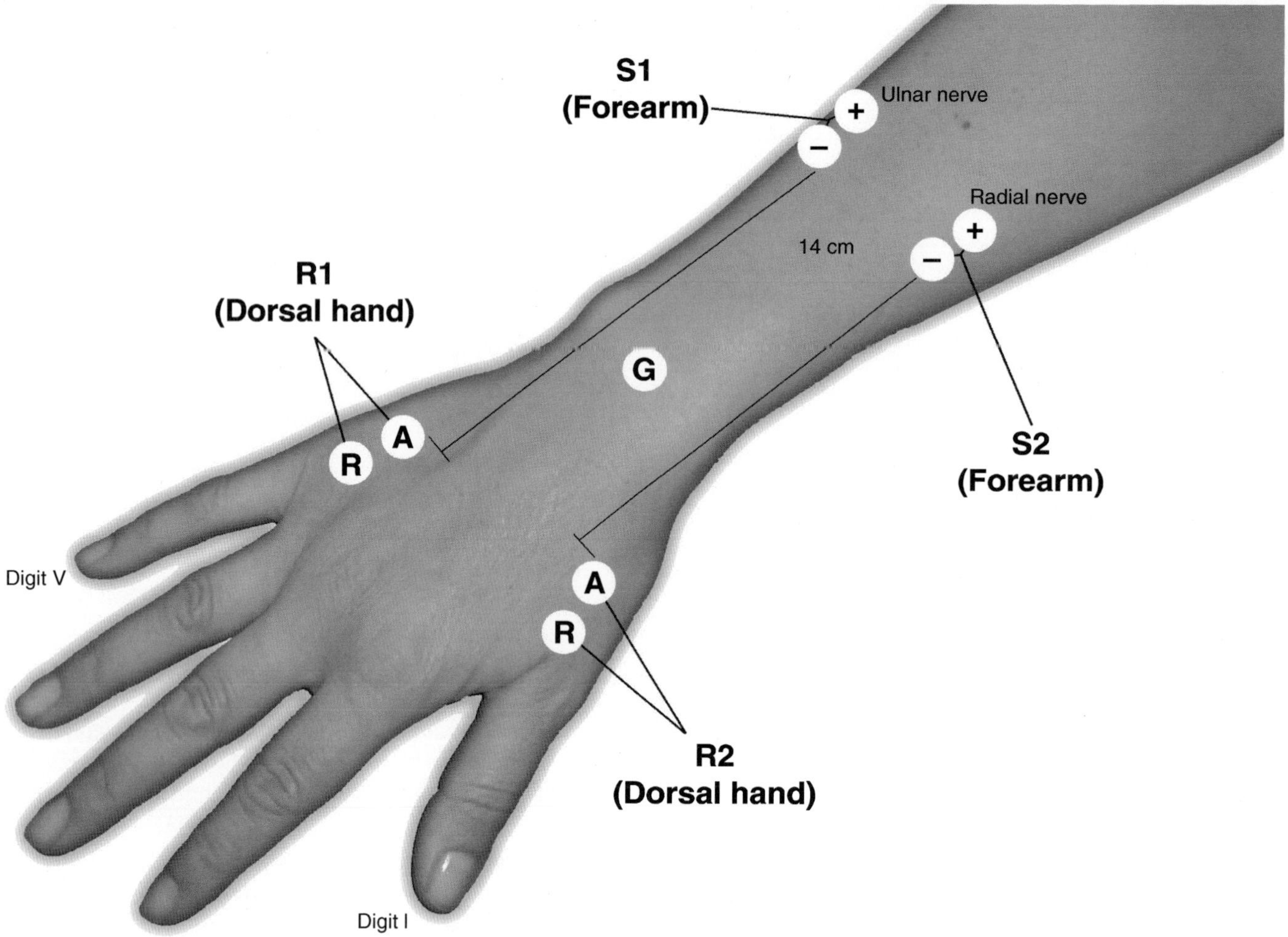

Typical waveform (forearm – dorsal hand):

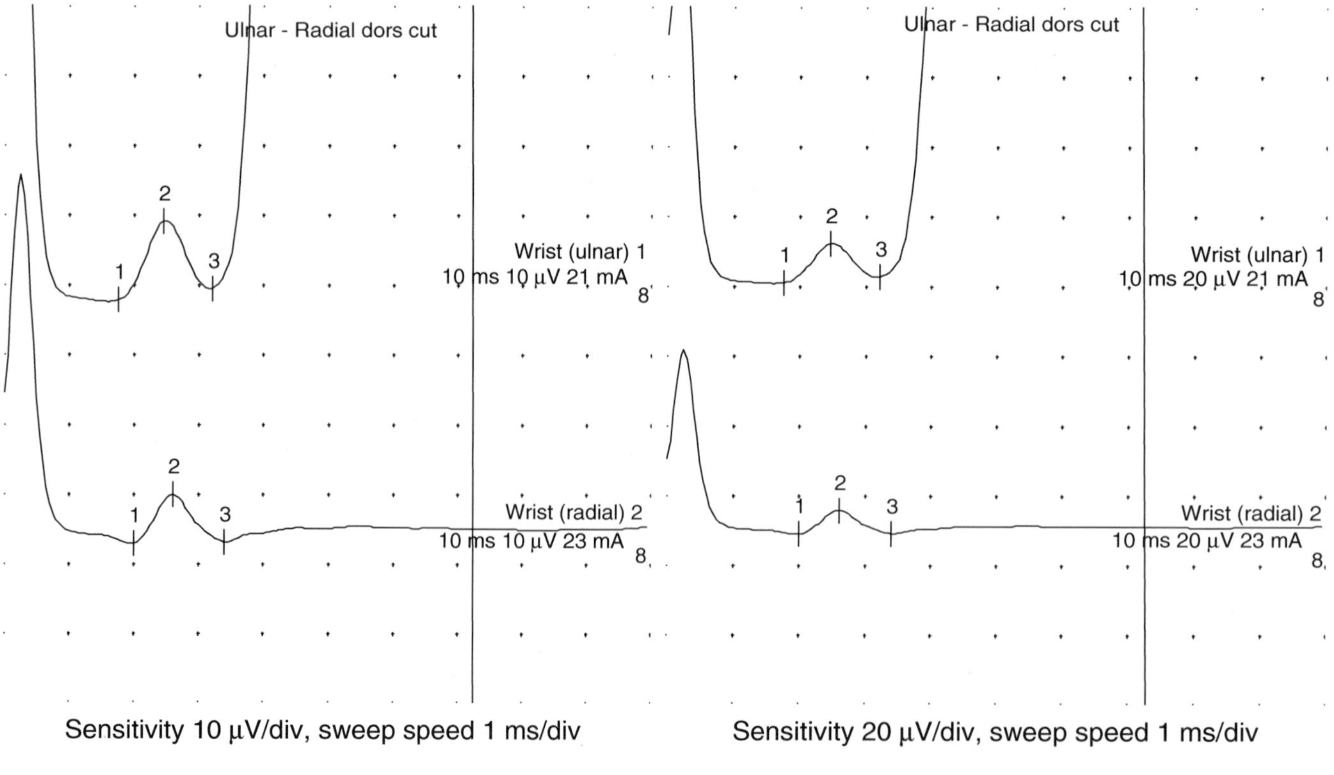

Fig. 1 Antidromic sensory nerve action potentials (*SNAPs*) recorded on the dorsal hand, stimulation of the ulnar and radial nerves at the forearm

Table 1 Normal values [1]

	Mean ± 2SD
Forearm–dorsal hand, onset latency – ulnar nerve (ms)	2.2 ± 0.3
Forearm–dorsal hand, peak latency – ulnar nerve (ms)	2.8 ± 0.5
Forearm–dorsal hand, negative peak amplitude – ulnar nerve (μV)	24 ± 17
Forearm–dorsal hand, onset latency – radial nerve (ms)	2.2 ± 0.3
Forearm–dorsal hand, peak latency – radial nerve (ms)	2.8 ± 0.3
Forearm–dorsal hand, negative peak amplitude – radial nerve (μV)	32 ± 18

of the ulnar nerve, at or near the wrist. In the second case, they described electrophysiological findings suggesting injury with complete conduction block of the right superficial radial nerve (no response could be obtained with stimulation of the right radial nerve) and axonal damage of the right dorsal cutaneous branch of the ulnar nerve (low amplitude of the SNAP – 5 μV). Their findings indicate that distal sensory latencies of the dorsal cutaneous branch of the ulnar nerve and superficial radial nerve are similar and that comparison of these nerves may be useful in distinguishing distal ulnar nerve entrapment syndromes when routine studies are equivocal. For the authors, the paired study of the dorsal branch of the ulnar and superficial radial nerves is an easy and reproducible study that requires a minimal amount of time and can be helpful in identifying isolated injuries to the dorsal cutaneous branch of the ulnar nerve and differentiating ulnar nerve lesions at the wrist from more proximal lesions. Pathological waveform and values in a patient with ulnar neuropathy at the wrist (UNW) are reported (Fig. 2).

Pathological waveform (forearm – dorsal hand):

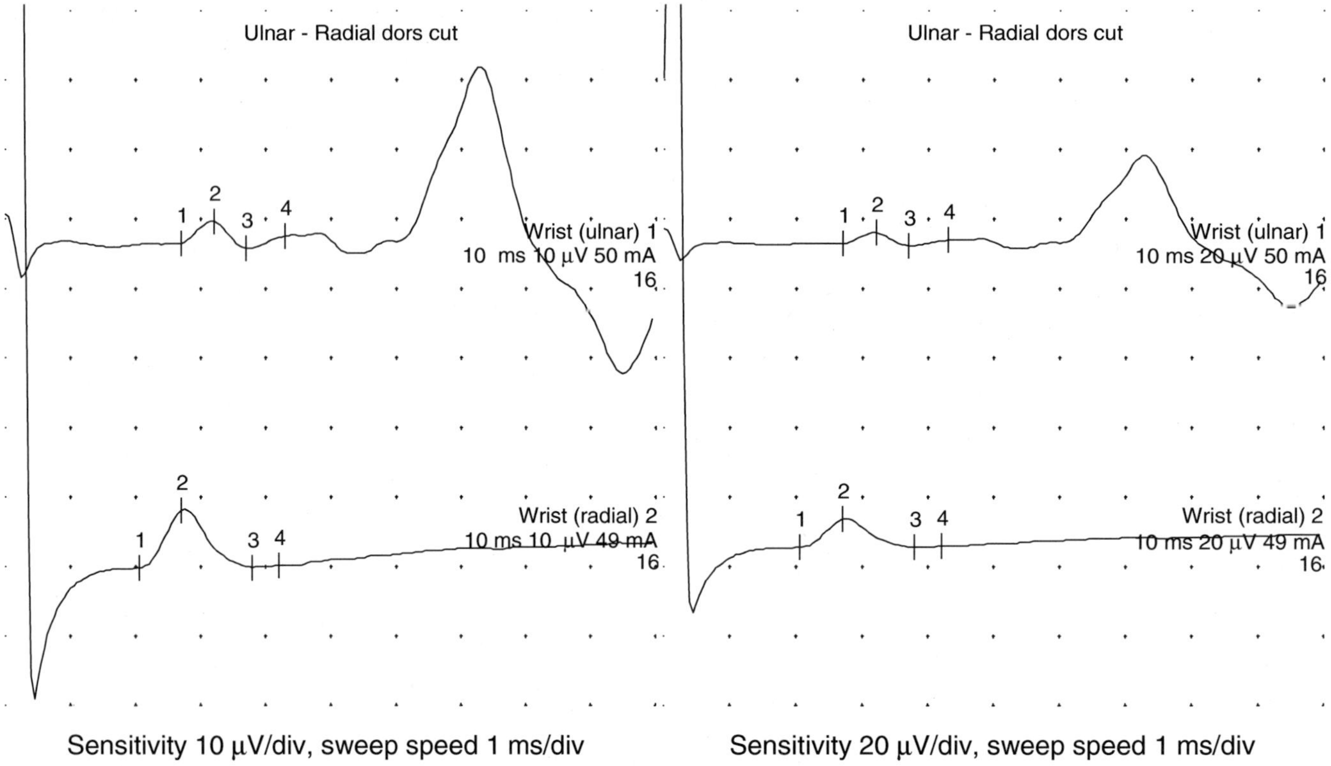

Onset latency (forearm – dorsal hand, ulnar nerve): 2.70 ms; **Onset latency** (forearm – dorsal hand, radial nerve): 2.05 ms; **Peak latency** (forearm – dorsal hand, ulnar nerve): 3.30 ms; **Peak latency** (forearm – dorsal hand, radial nerve): 2.75 ms; **Onset to peak amplitude** (forearm – dorsal hand, ulnar nerve): 3.1 μV; **Onset to peak amplitude** (forearm – dorsal hand, radial nerve): 8.4 μV; **Peak to peak amplitude** (forearm – dorsal hand, ulnar nerve): 4.0 μV; **Peak to peak amplitude** (forearm – dorsal hand, radial nerve): 8.3 μV; **SNCV** (forearm – dorsal hand, ulnar nerve -14 cm distance): 51.9 m/s; **SNCV** (forearm – dorsal hand, radial nerve - 14 cm distance): 68.3 m/s

Fig. 2 Antidromic sensory nerve action potentials (*SNAP*) recorded on the dorsal hand in a patient with UNW, stimulation of the ulnar (*upper trace*) and radial (*lower trace*) nerves

Reference

1. Hoffman MD, Mitz M, Luisi M et al (1988) Paired of the dorsal cutaneous ulnar and superficial radial sensory nerves. Arch Phys Med Rehabil 69:591–594

Wrist, Below Elbow, Above Elbow – Digit V

S19

Surface Recording Technique, Antidromic Study

Original Settings Sensitivity was 10 μV/division, sweep speed was 2 ms/division, and the machine used was a neurodiagnostic NDI electromyograph. Low-frequency filter, high-frequency filter, duration, and rate of pulse were not specified.

Position This study was performed with the arm abducted to the trunk and forearm supinated. The elbow was flexed to 90°.

Recording Following the antidromic method [1], signals were recorded 14 cm distally from the wrist stimulation point to digit V (Fig. 1). Sensory nerve action potentials (SNAPs) were recorded using pickup ring electrodes placed over the proximal and distal interphalangeal joints of digit V (little finger). The ground (G) electrode position was not specified in the text; the figure shows the ground electrode placed on the palm.

Stimulation The ulnar nerve was stimulated at three sites (Figs. 1 and 2): at the wrist (S1), 14 cm proximal to the proximal interphalangeal joint of digit V over the course of the ulnar nerve; below the elbow (S2), immediately distal to the ulnar groove; and above the elbow, 10 cm proximal to the below-elbow stimulation point over the course of the ulnar nerve (S3). The anode (+) was proximal in all stimulations.

Measurements Onset latency (ms) was measured from the stimulating artifact to the onset of the negative deflection of the SNAP. Peak latency (ms) was measured from the stimulating artifact to the peak of the negative deflection of the SNAP. If the response was triphasic, latency was measured from the peak of the initial positive deflection from the baseline. Sensory nerve conduction velocity (SNCV) of the forearm and cubital segments (wrist–below elbow and below elbow–above elbow) was calculated dividing the distance between two different stimulating sites by the difference of onset latencies, and it was measured in meter per second (m/s). Peak amplitude (μV) of the SNAPs was measured from the peak of the negative deflection to the peak of the aftercoming positive deflection. Percentage amplitude decrement (%) of the forearm and across the elbow (below-elbow–above-elbow segment) was calculated with an amplitude of the forearm segment being taken as 100 %. The skin temperature was controlled bilaterally using a thermistor taped over the wrist crease and averaged at 31.9 °C (range 30.0–33.8 °C). The authors selected a relatively short across elbow segment of 10 cm because a longer segment would include more normal nerve tending to average out potential abnormalities and leading to a false normal result, while a shorter segment would increase the chance of measurement error and the possibility of finding a falsely abnormal result. Normal values (Table 1) were obtained from 20 healthy subjects (14 women and 6 men, age range 18–46 years, mean age 33.8 years).

Comment

For Felsenthal et al. [1], in 20 healthy subjects, the mean + 2SD latency for the 10 cm segment of the ulnar nerve across the elbow was 1.8 ms (onset latency) and 1.9 ms (peak latency). The mean + 2 SD side-to-side comparison of the same nerve segment was 0.6 ms (onset latency) and 0.3 ms (peak latency). The mean + 2 SD amplitude decrement was 41 %. They set the angle of the elbow at 90° because of the clinical observation that when the elbow was maintained at a different angle of flexion, the actual angle used by various electromyographers was variable. They reviewed the literature and reported techniques with the elbow fully extended, 30° from full extension, 45° from full extension, 90° from full extension, and 135° of flexion. For the authors, more consistent results could be obtained placing the elbow at 90° of flexion,

G. Gentili, M. Di Napoli, *The Ulnar Nerve*, DOI 10.1007/978-3-319-43292-2_19

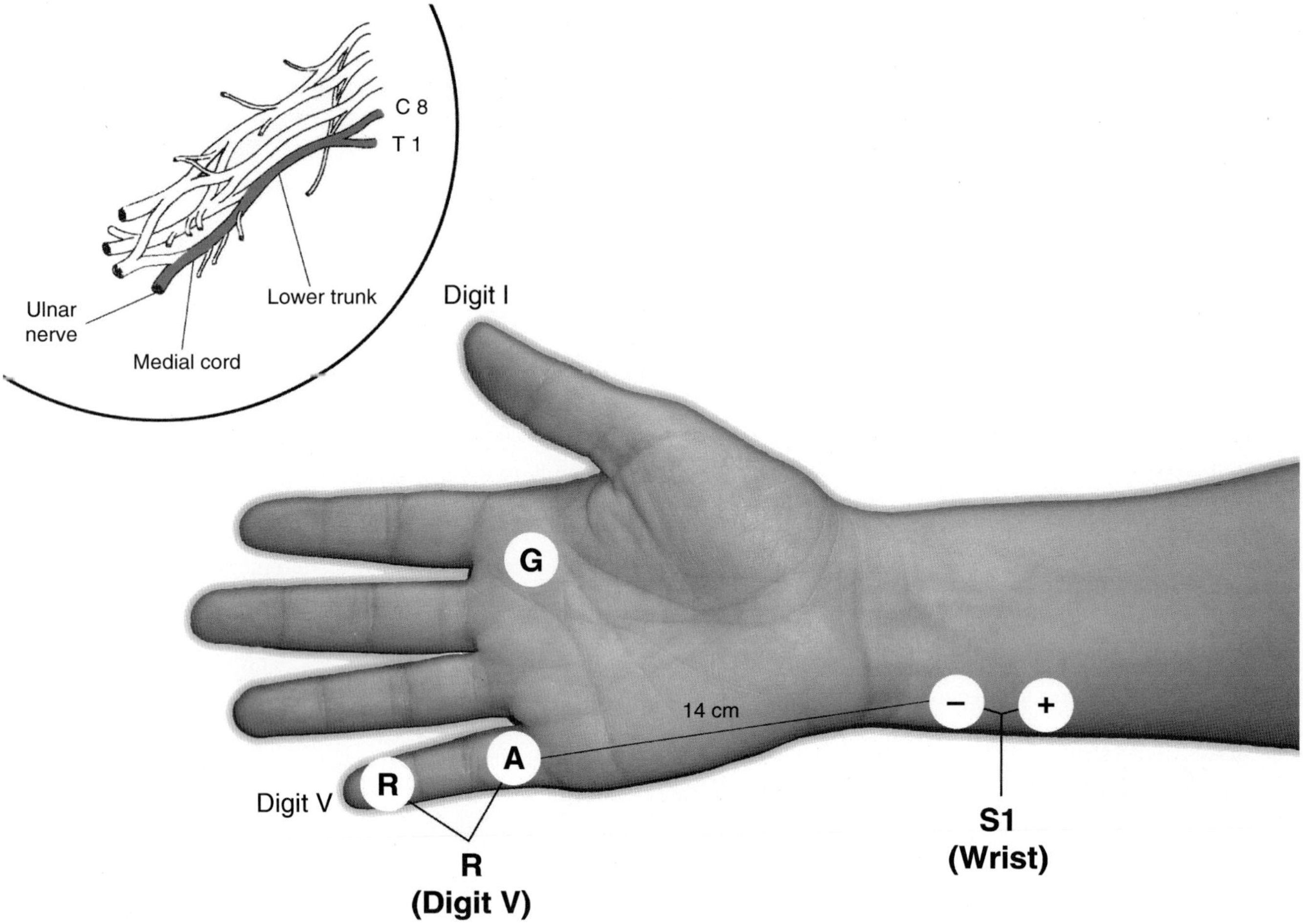

Typical waveform (wrist, below elbow, above elbow – digit V):

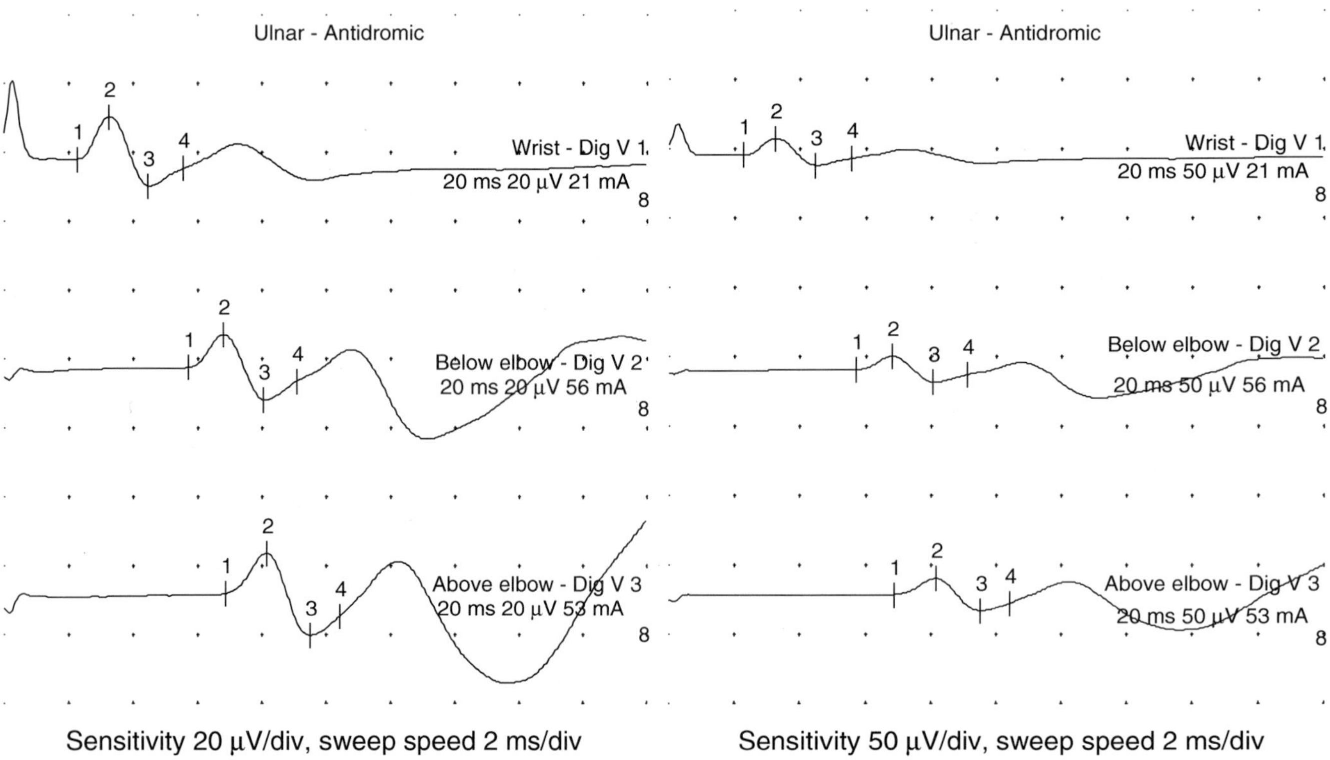

Fig. 1 Antidromic sensory nerve action potentials (*SNAPs*) recorded to digit V, stimulation of the wrist, below the elbow, and above the elbow (*top to bottom*)

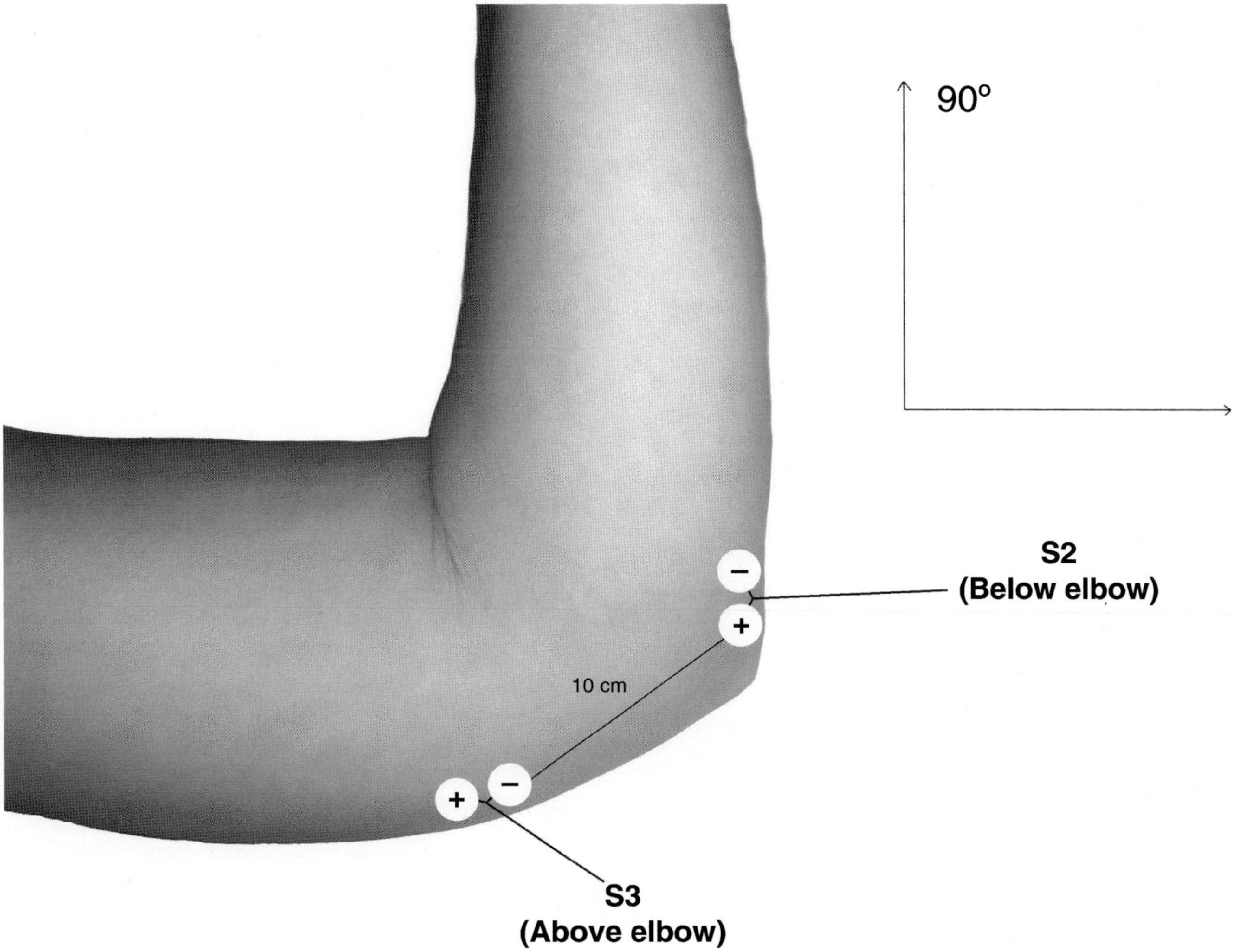

Fig. 2 Below- and above-elbow stimulation points (elbow flexed to 90°)

Table 1 Normal values [1]

	Mean ± SD	Range
Wrist–digit V, onset latency (ms)	2.6 ± 0.2	2.2–3.2
Wrist–digit V, peak latency (ms)	3.2 ± 0.5	2.6–4.0
Below elbow–digit V, onset latency (ms)	5.8 ± 0.2	4.9–7.1
Below elbow–digit V, peak latency (ms)	6.6 ± 0.5	5.6–7.7
Above elbow–digit V, onset latency (ms)	7.2 ± 0.5	6.2–8.8
Above elbow–digit V, peak latency (ms)	8.1 ± 0.5	7.2–9.5
Across elbow–digit V, onset latency (ms)	1.4 ± 0.2	1.0–2.2
Across elbow–digit V, peak latency (ms)	1.5 ± 0.2	1.2–1.9
Same nerve segment, onset latency side to side (ms)	0.2 ± 0.2	0–0.9
Same nerve segment, peak latency side to side (ms)	0.1 ± 0.1	0–0.4
Wrist–below elbow, SNCV to the onset (m/s)	67.5 ± 4.5	54.3–76.0
Wrist–below elbow, SNCV to the peak (m/s)	63.6 ± 4.0	55.9–70.4
Below elbow–above elbow, SNCV to the onset (m/s)	70.9 ± 10.4	45.4–100.0
Below elbow–above elbow, SNCV to the peak (m/s)	64.9 ± 6.8	52.6–83.3
Wrist–digit V, peak amplitude (μV)	65.1 ± 23.9	20–110
Below elbow–digit V, peak amplitude (μV)	31.1 ± 11.7	6–60
Above elbow–digit V, peak amplitude (μV)	25.4 ± 10.1	6–48
Below elbow–digit V, amplitude decrement (%)	51.7 ± 11.1	12–72
Below elbow–above elbow, amplitude decrement (%)	18.6 ± 11.2	0–50

because the sensory fibers which lie superficially within the nerve in the cubital tunnel would become more stretched than deeper-lying motor fibers when the elbow is flexed [2]. The authors also performed sensory and motor conduction studies on three patients with ulnar nerve compressive lesions across the elbow. They reported that two patients had purely sensory complaints; the third also had some evidence of weakness of the ulnar innervated muscles in the symptomatic arm. In all three cases, abnormalities of sensory conduction studies were observed than normal motor conduction findings. Most striking was the amplitude decrement of the SNAP across the elbow. For this reason, the authors suggested to perform sensory nerve conduction study across the elbow because the sensory parameters may show a greater degree of abnormality than the motor parameters when done early after the onset of symptoms. Pathological waveform and values in a patient with ulnar neuropathy at the elbow (UNE) are reported (Fig. 2).

Harding and Halar [2] bilaterally tested 20 healthy subjects (12 men and 8 women, mean age 32 years, age range 21–68 years) with no history or signs of ulnar nerve entrapment syndromes. All subjects were tested supine with the shoulder abducted to 45°, forearm supinated, and wrist in a neutral position. The study was performed to assess the effect of four different elbow positions on sensory nerve conduction velocity (SNCV), so the elbow was tested in extension (0° flexion) and flexion (45°, 90°, and 135°). It was maintained in position during testing by specially designed splints. Authors [2] followed the method of measuring and recording suggested by the American Academy of Orthopedic Surgeons [3]. Sensory nerve action potentials (SNAPs) were recorded by ring or pipe cleaner electrodes placed around the digit V (little finger), but the exact position was not specified. The ulnar nerve was stimulated at two sites: below the elbow (S1) and above the elbow (S2). The below-elbow stimulation point was 5 cm distal to the medial epicondyle of the humerus and was used for each of the four elbow positions (0°, 45°, 90°, and 135°). The above-elbow stimulation point was positioned midway between the olecranon and the medial epicondyle of the humerus. The anode (+) was proximal at all stimulations (duration of the pulse was 0.05–0.1 ms). With the elbow in extension (0° flexion), the authors [2] observed that the below-elbow–digit V (BE) and above-elbow–below-elbow (AE) segments were made equal in length. They performed SNCV determinations of two different AE segments in each flexed position (45°, 90°, and 135°), one longer than and one equal to the BE segment. The longer segment was, in fact, the same AE segment that was measured and studied in extension but which had become progressively longer with increasing elbow flexion. Thus, the longer AE segment was termed AE uncorrected (AE-u) and the one of equal length AE corrected (AE-c). For the AE-u segment, the skin marker for the proximal stimulation remained constant, while the skin distance of the segment became longer with elbow flexion. For the AE-c segment, the stimulation point had to be moved progressively distally with each position of the elbow flexion. Thus, the AE-c segment was always remeasured to be equal to the BE segment. Harding and Halar [1] calculated SNCV of each segment (BE–digit V, AE-c–digit V, and AE-u–digit V) dividing distance between stimulating cathode and active recording electrode by peak latency, and it was measured in meter per second (m/s). Skin distance was measured using a flexible steel tape. The skin temperature was controlled 8 cm proximal to the distal wrist crease (BE segment) and midway between the medial epicondyle of the humerus and the olecranon (AE segment). BE skin temperature measurement varied between 31.8 and 35 °C (mean 33.6 °C), and AE segment temperature ranged from 31 to 35.5 °C (mean 32.9 °C). The authors also performed motor conduction study to the *abductor digiti minimi* (ADM) muscle using same stimulating method (see motor conduction studies).

Comment

Harding and Halar [2] observed for the AE segments that the SNCV increased with elbow flexion at a progressively faster rate compared with the motor nerve conduction velocity (MNCV). For the authors this SNCV increment was due to a decrease in conduction time with increasing elbow flexion. Also, for the BE segment, the increase in SNCV between 0° and 45° elbow flexion was significant, and considerable SNCV variation between right and left arms was found at all positions of elbow flexion. The AE and BE SNCV showed the least variation at 0° elbow flexion position. At 90° and 135° elbow flexion, percentage variation was greater than at 45°. At 0° flexion using a 19 cm segment length, mean SNCVs were 61.2 ± 3.7 m/s and 62.4 ± 10.3 m/s for BE and AE segments, respectively. The authors tried to explain the significantly greater increase in AE SNCV with elbow flexion when compared with AE MNCV by that the sensory fibers, which lie superficially within the nerve in the cubital tunnel, would become more stretched than the deeper-lying motor fibers when the elbow was flexed. At each elbow flexion position, the SNCV percentage variation was much greater than for motor response at the AE stimulation site (Fig. 3).

Pathological waveform (wrist, below elbow, above elbow – digit V):

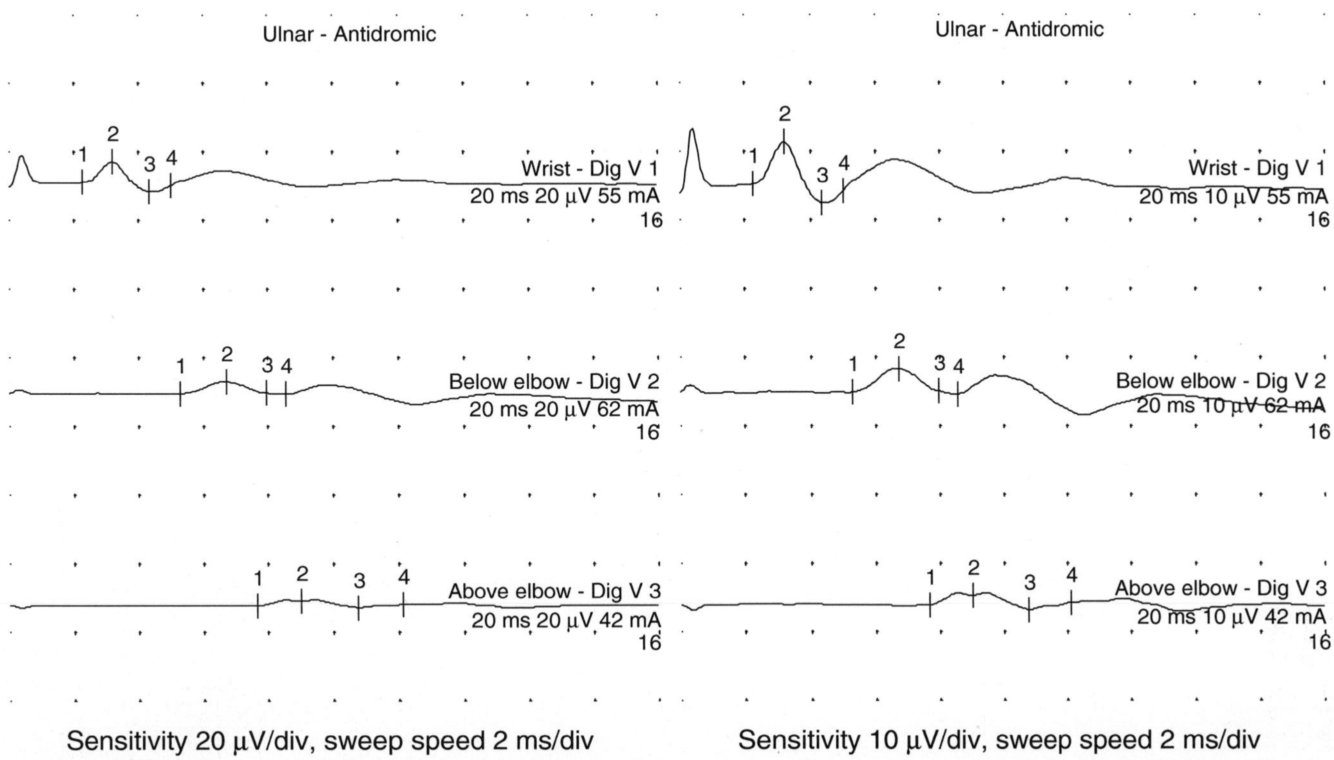

Onset latency (wrist – digit V): 2.25 ms; **Onset latency** (below elbow – digit V): 5.3 ms; **Onset latency** (above elbow – digit V): 7.65 ms; **Peak latency** (wrist – digit V): 3.35 ms; **Peak latency** (below elbow – digit V): 6.7 ms; **Peak latency** (above elbow – digit V): 9.0 ms; **Onset to peak amplitude** (wrist – digit V): 6.3 μV; **Onset to peak amplitude** (below elbow – digit V): 3.4 μV; **Onset to peak amplitude** (above elbow – digit V): 1.5 μV; **Peak to peak amplitude** (wrist – digit V): 9.0 μV; **Peak to peak amplitude** (below elbow – digit V): 3.2 μV; **Peak to peak amplitude** (above elbow – digit V): 2.3 μV; **SNCV**(wrist – digit, 14 cm distance): 62.2 m/s; **SNCV** (wrist – below elbow, 17 cm distance): 55.7 m/s; **SNCV** (below elbow – above elbow, 10 cm distance): 42.6 m/s

Fig. 3 Antidromic sensory nerve action potentials (*SNAPs*) recorded to digit V in UNE, stimulation of the wrist, below the elbow, and above the elbow (*top to bottom*)

References

1. Felsenthal G, Freed MJ, Kalafut R et al (1989) Across-elbow ulnar nerve sensory conduction technique. Arch Phys Med Rehabil 70:668–672
2. Harding C, Halar E (1983) Motor and sensory ulnar nerve conduction velocities: effect of elbow position. Arch Phys Med Rehabil 64:227–232
3. American Academy of Orthopedic Surgeons (AAOS) (1965) Joint motion: method of measuring and recording. American Academy of Orthopedic Surgeons (AAOS), Chicago

Palm – Distal Forearm, Elbow

Surface Recording Technique, Orthodromic Study

S20

Original Settings Sensitivity was 10–20 μV/division, sweep speed was 10 ms/division, low-frequency filter was 20 Hz, high-frequency filter was 2 kHz, and the machine used was a TECA Neurostar. Duration of pulse and rate of pulse were not specified.

Position This study was performed in the supine position, with the elbow and fingers slightly flexed.

Recording Following the orthodromic method (Fig. 1), recording electrodes (mounted on a plastic bar) were placed 10 cm proximally between the ulna and *flexor carpi ulnaris* (FCU) muscle tendon (R1) and over the ulnar groove at the elbow (R2). The authors [1] chose the orthodromic method because of the motor artifact observed with the antidromic recording technique. The ground (G) electrode position was not specified in the text; the figure shows the ground electrode placed on the palm.

Stimulation It was applied with a bar electrode over the *hypothenar* area to digit IV ray. In each study, stimulus intensity was adjusted to obtain the maximal sensory nerve action potential (SNAP) amplitude and minimize stimulus artifact. They performed an additional study in several subjects, using the collision technique [2], with simultaneous stimulation of the *hypothenar* area and digit V, in order to examine the possibility if value conduction to the superficial branch being responsible for SNAPs recorded with *hypothenar* stimulation. They observed two distinct SNAPs with little change in amplitude represented two distinct sensory branches of the ulnar nerve: the ulnar palmar cutaneous nerve (UPCN) and the superficial sensory branch of the ulnar nerve (SSUN).

Measurements Distal peak latency (ms) was measured from the beginning of the shock artifact to the peak of the negative deflection of the SNAP. Peak amplitude (μV) was measured from the baseline to the peak of the negative deflection of the SNAP. The authors recorded distal (forearm) and proximal (elbow) SNAPs, and sensory nerve conduction velocity (SNCV) for the wrist to elbow segment was also calculated. Skin and room temperatures were not reported. The authors performed normal values in 20 healthy (Table 1) volunteers (11 men and 9 women, mean age 35.2 ± 9.1 years, age range 22–58 years); some subjects had both upper extremities tested.

Comment

Stowell and Gnatz [1] reported a case in which injury to the superficial sensory branch of the ulnar nerve, with sparing of the UPCN, was demonstrated electrophysiologically. Because the UPCN traverses the wrist superficial to Guyon's canal, it may be a helpful adjunct in localizing lesions of the ulnar nerve in the forearm, wrist, and hand. For the authors, when present, the UPCN may provide another distal ulnar nerve branch crossing the wrist outside Guyon's canal that can be electrodiagnostically used in isolated cases of ulnar nerve pathology, but variability in *hypothenar* sensory supply may limit the usefulness of UPCN testing in routine nerve conduction studies. Pathological waveform and values in a case of injury to the UPCN at the forearm (Fig. 2) and in a patient (Fig. 3) with ulnar neuropathy at the elbow (UNE) are reported.

G. Gentili, M. Di Napoli, *The Ulnar Nerve*, DOI 10.1007/978-3-319-43292-2_20

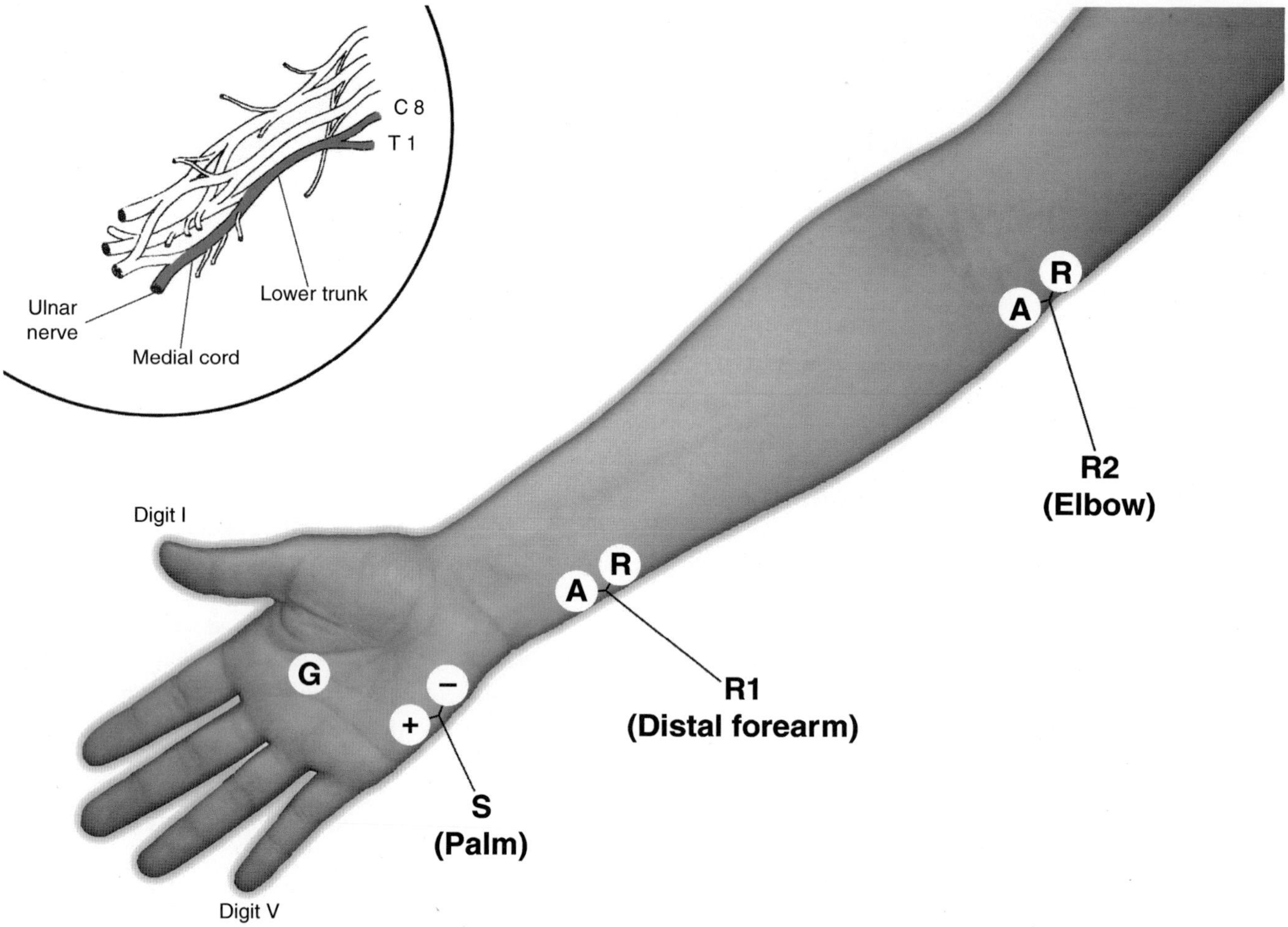

Typical waveform (palm – distal forearm, elbow):

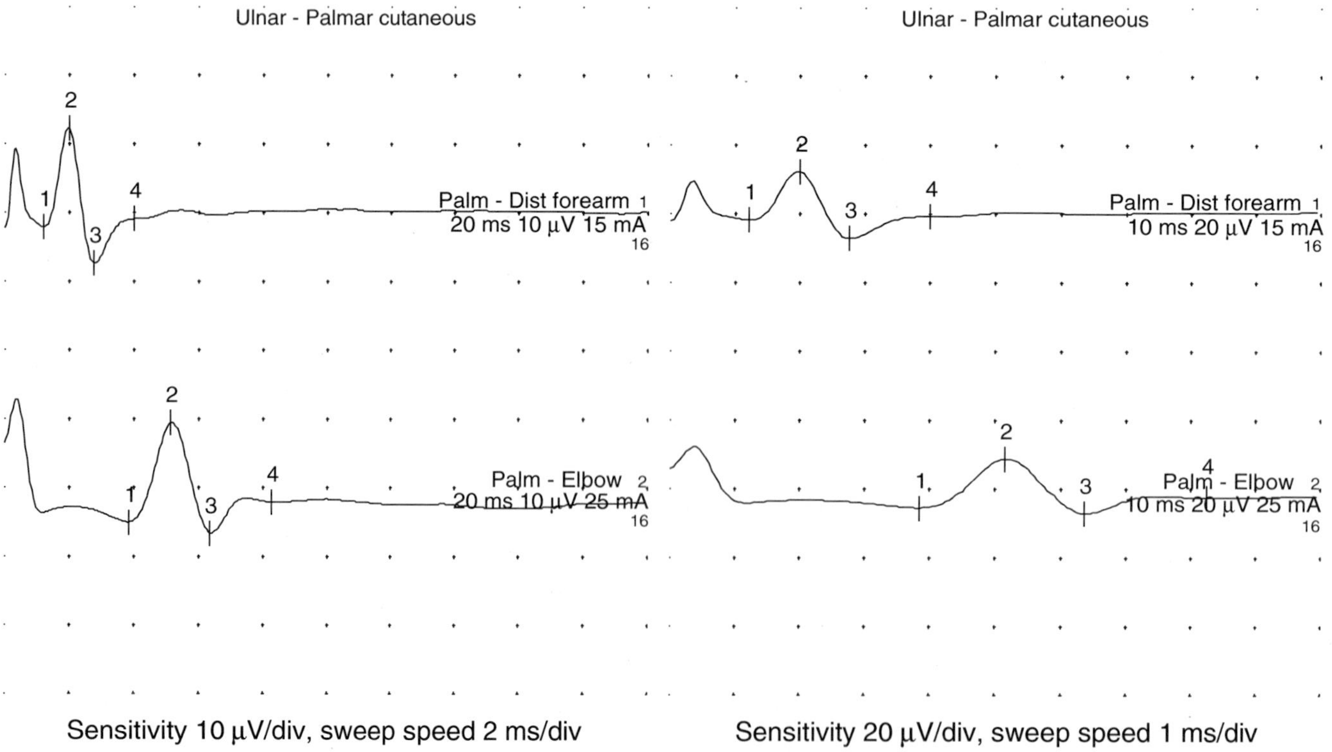

Fig. 1 Orthodromic sensory nerve action potentials (*SNAPs*) recorded at the distal forearm (*upper trace*) and at the elbow (*lower trace*); stimulation on the palm amplitude represented two distinct sensory branches of the ulnar nerve: the ulnar palmar cutaneous nerve (UPCN) and the superficial sensory branch of the ulnar nerve (SSUN)

Table 1 Normal values [1]

	Mean ± SD	Range	Limit of normal
Palm–distal forearm, distal latency (ms)	2.19 ± 0.17	1.88–2.48	<2.5
Palm–distal forearm, distal amplitude (μV)	12.7 ± 6.9	6.0–34.0	>6.0
Palm–elbow, proximal latency (ms)	5.45 ± 0.54	4.52–6.92	
Palm–elbow, proximal amplitude (μV)	14.5 ± 9.2	8.0–50.0	>8.0
Distal forearm–elbow, SNCV (m/s)	61.9 ± 4.9	54.4–71.0	>52.1

Pathological waveform (palm – distal forearm, elbow):

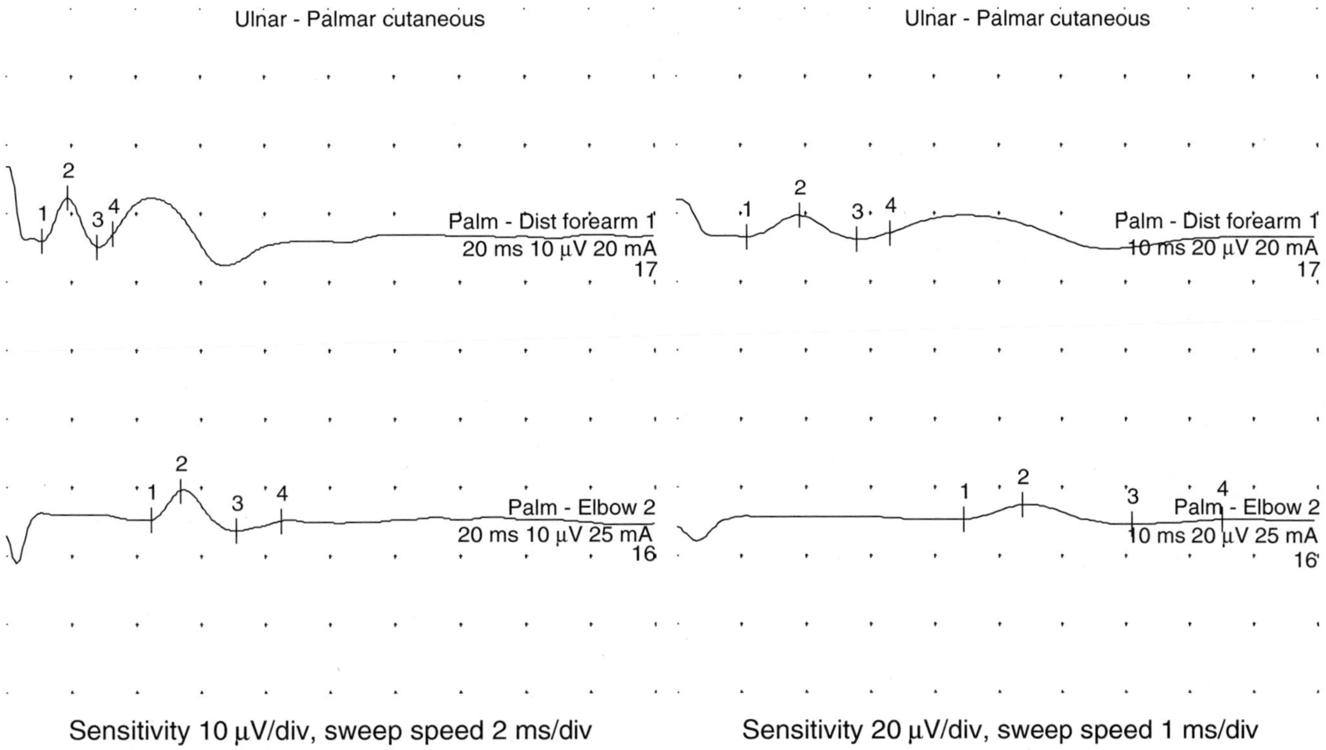

Onset latency (palm – distal forearm): 1.10 ms; **Onset latency** (palm – elbow): 4.45 ms; **Peak latency** (palm – distal forearm): 1.9 ms; **Peak latency** (palm – elbow): 5.35 ms; **Onset to peak amplitude** (palm – distal forearm): 6.5 μV; **Onset to peak amplitude** (palm – elbow): 4.3 μV; **Peak to peak amplitude** (palm – distal forearm): 7.3 μV; **Peak to peak amplitude** (palm – elbow): 6.1 μV; **SNCV** (palm – distal forearm, 7 cm distance): 63.6 m/s; **SNCV** (palm – elbow, 18 cm distance): 40.4 m/s

Fig. 2 Orthodromic sensory nerve action potentials (*SNAPs*) recorded at the distal forearm (*upper trace*) and at the elbow (*lower trace*) in a case of injury to the UPCN at the forearm, stimulation on the palm

Pathological waveform (palm – distal forearm, elbow):

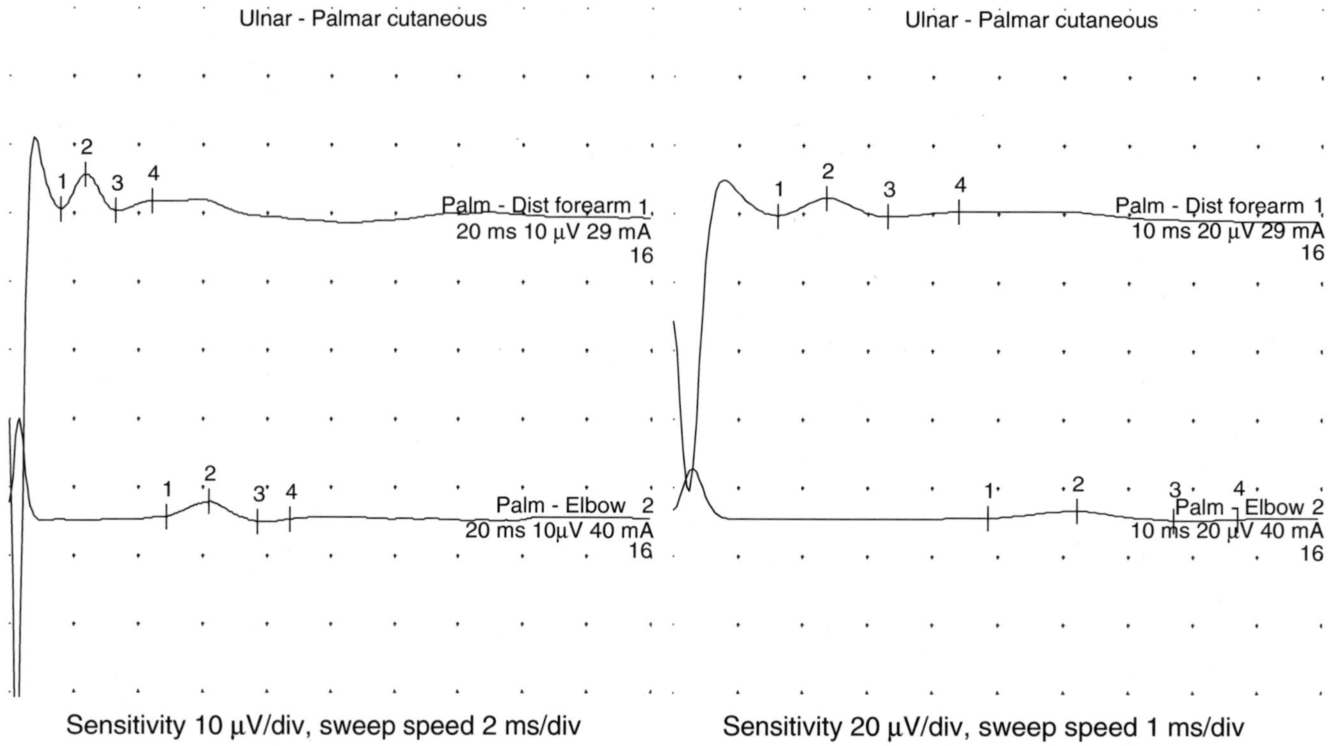

Onset latency (palm – distal forearm): 1.60 ms; **Onset latency** (palm – elbow): 4.85 ms; **Peak latency** (palm – distal forearm): 2.3 ms; **Peak latency** (palm – elbow): 6.25 ms; **Onset to peak amplitude** (palm – distal forearm): 5.1 µV; **Onset to peak amplitude** (palm – elbow): 2.2 µV; **Peak to peak amplitude** (palm – distal forearm): 5.4 µV; **Peak to peak amplitude** (palm – elbow): 3.0 µV; **SNCV** (palm – distal forearm, 8 cm distance): 50.0 m/s; **SNCV** (palm – elbow, 22 cm distance): 45.4 m/s

Fig. 3 Orthodromic sensory nerve action potentials (*SNAPs*) recorded at the distal forearm (*upper trace*) and at the elbow (*lower trace*) in a case of UNE, stimulation on the palm

References

1. Stowell ED, Gnatz SM (1992) Ulnar palmar cutaneous nerve and hypothenar sensory conduction studies. Arch Phys Med Rehabil 73:842–846
2. Kimura J (1976) Physiologic block of nerve impulses in studies of motor nerve conduction velocity. Neurology 26:680–682

Wrist, Below Elbow, Above Elbow – Digit V; Digit V – Wrist, Below Elbow, Above Elbow; Wrist – Below Elbow, Above Elbow

S21

Surface Recording Technique, Antidromic and Orthodromic Study

Original Settings The machine used was a DANTEC Counterpoint four-channel electromyograph. Sensitivity, sweep speed, low-frequency filter, high-frequency filter, duration, and rate of pulse were not specified.

Position This study was performed in the supine position with the elbow extended and arm fully supinated and abducted 45° at the shoulder.

Recording The authors [1] performed three several conduction studies: (1) following the antidromic method (Fig. 1), sensory nerve action potentials (SNAPs) were recorded with Velcro®/flannel ribbon electrodes (5 mm widths) placed around the proximal and middle phalanges of the digit V; (2) following the orthodromic method (Fig. 2), SNAPs were recorded with bipolar felt-tip (which had been previously applied for motor conduction study by the authors) placed at three sites: at the wrist (R1), below the elbow (R2), and above the elbow (R3); (3) mixed nerve potentials (Fig. 3) were recorded from the below-elbow (R1) and above-elbow (R2) sites with bipolar felt-tip electrodes applied previously as shown in Fig. 2. The ground (G) electrode was ever placed on the dorsum of the hand; the figures show ground electrode placed on the palm.

Stimulation (1) Following the antidromic method (Fig. 1), the ulnar nerve was stimulated at the wrist (S1), below the elbow (S2), and above the elbow (S3). The authors interchanged stimulating and recording electrodes, allowing recording from the digit V and utilizing the same three stimulation sites used in the motor conduction study. The anode (+) was proximal in all stimulations. All supramaximal SNAPs were averaged over 10–100 stimulus repetitions; (2) following the orthodromic method (Fig. 2), digit V was supramaximally stimulated using Velcro®/flannel ribbon electrodes (5 mm widths) placed around the proximal and middle phalanges using a 0.2 ms pulse duration. The anode (+) was proximal in all stimulations. The below-elbow stimulating site was set at least 3.5 cm distal to the medial epicondyle, and the above-elbow stimulating site was set at least 10 cm from the below-elbow site; (3) for mixed nerve potential recording, the authors used percutaneous supramaximal stimuli to the ulnar nerve at the wrist using a 0.2 ms pulse duration.

Measurements Onset latency (ms) was measured from the stimulating artifact to the onset of the negative deflection (peak) of the SNAP. If the response was triphasic, latency was measured from the peak of the initial positive deflection (peak) to the baseline. Sensory nerve conduction velocity (SNCV) of the forearm and cubital segments (wrist–below elbow and below elbow–above elbow) was calculated dividing the distance between two different stimulating sites by difference of onset latencies, and it was measured in meter per second (m/s). Peak amplitude (μV) of the SNAPs was measured from the baseline to the peak of the negative deflection. The temperature of the limb was monitored by a sensor placed on the palm, and it was maintained at 34–36 °C using a feedback-controlled infrared heating lamp. Normal values (Tables 1, 2, and 3) were performed in 27 healthy subjects and 9 patients with unrelated problems (40 ulnar nerves, 20 women and 16 men, age range 21–62 years, average age

G. Gentili, M. Di Napoli, *The Ulnar Nerve*, DOI 10.1007/978-3-319-43292-2_21

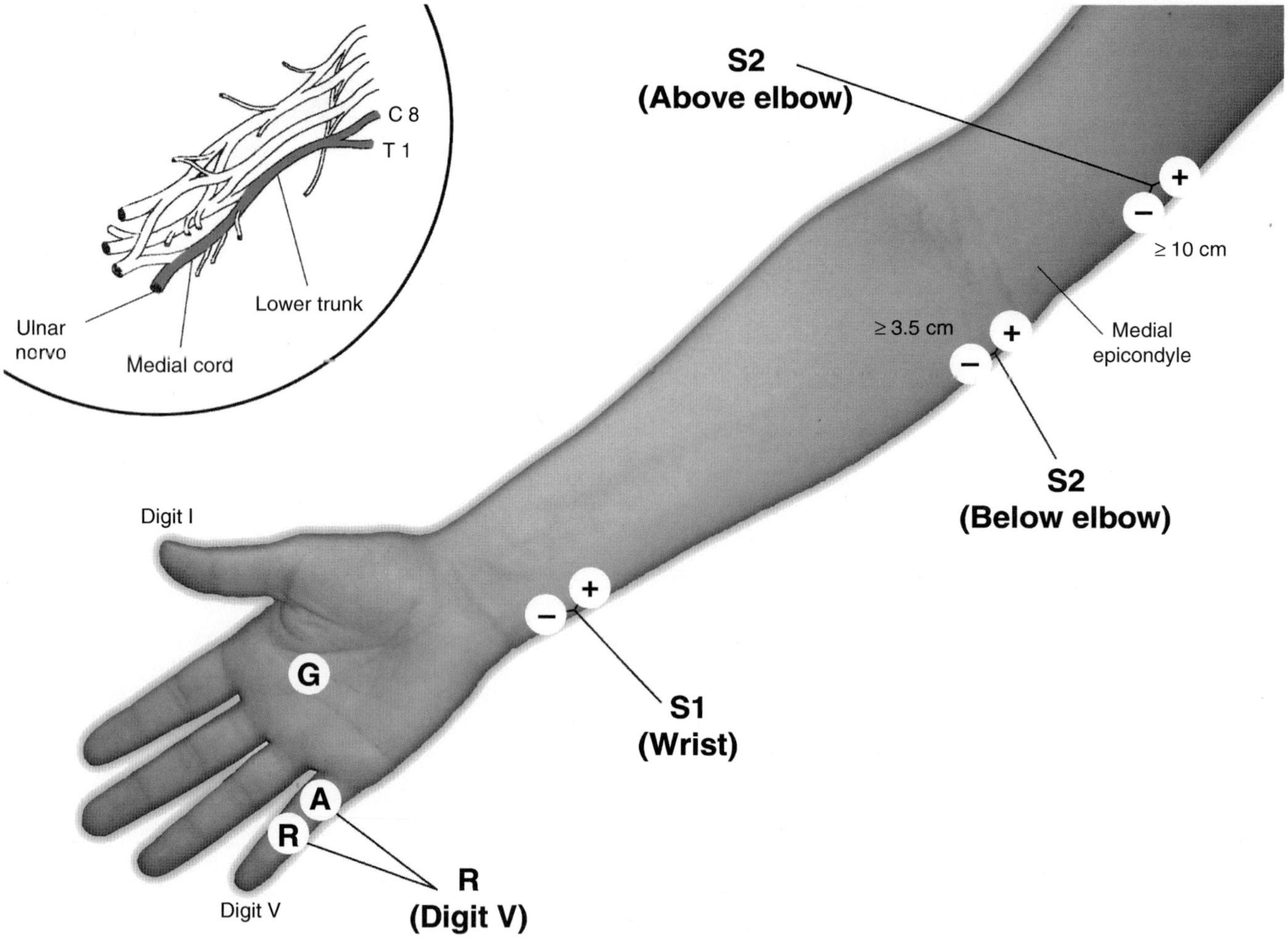

Typical waveform (wrist, below elbow, above elbow – digit V):

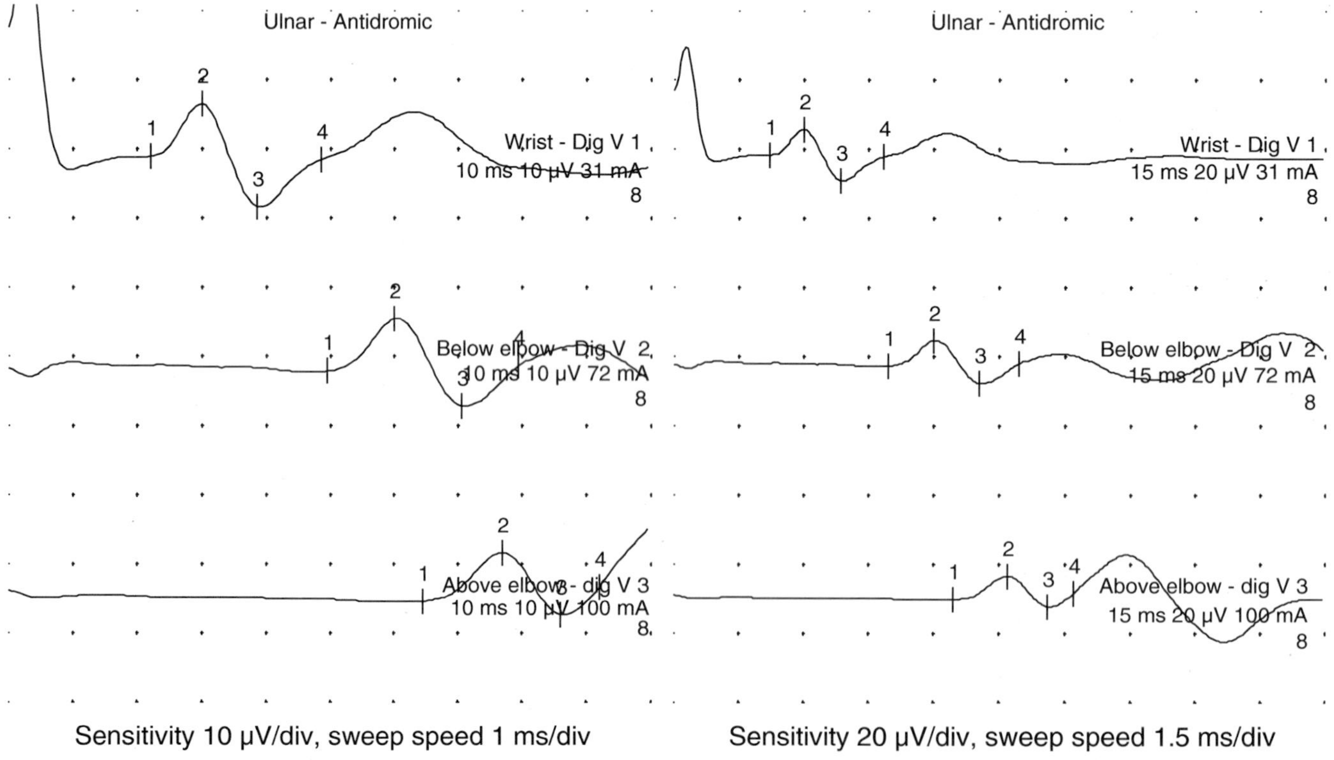

Fig. 1 Antidromic sensory nerve action potentials (*SNAPs*) recorded to digit V, stimulation of the wrist, below the elbow, and above the elbow (*top to bottom*)

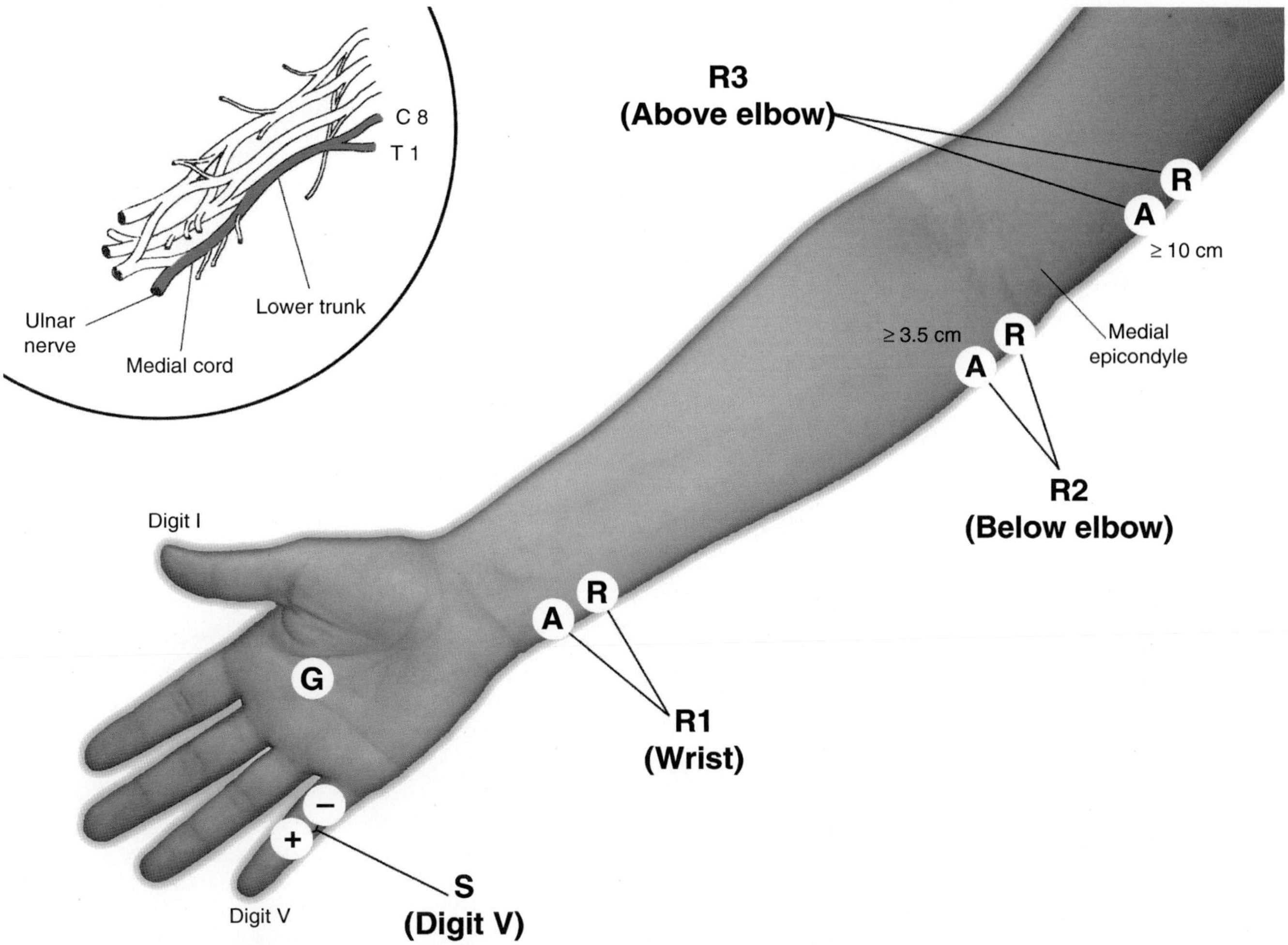

Typical waveform (digit V – wrist, below elbow, above elbow):

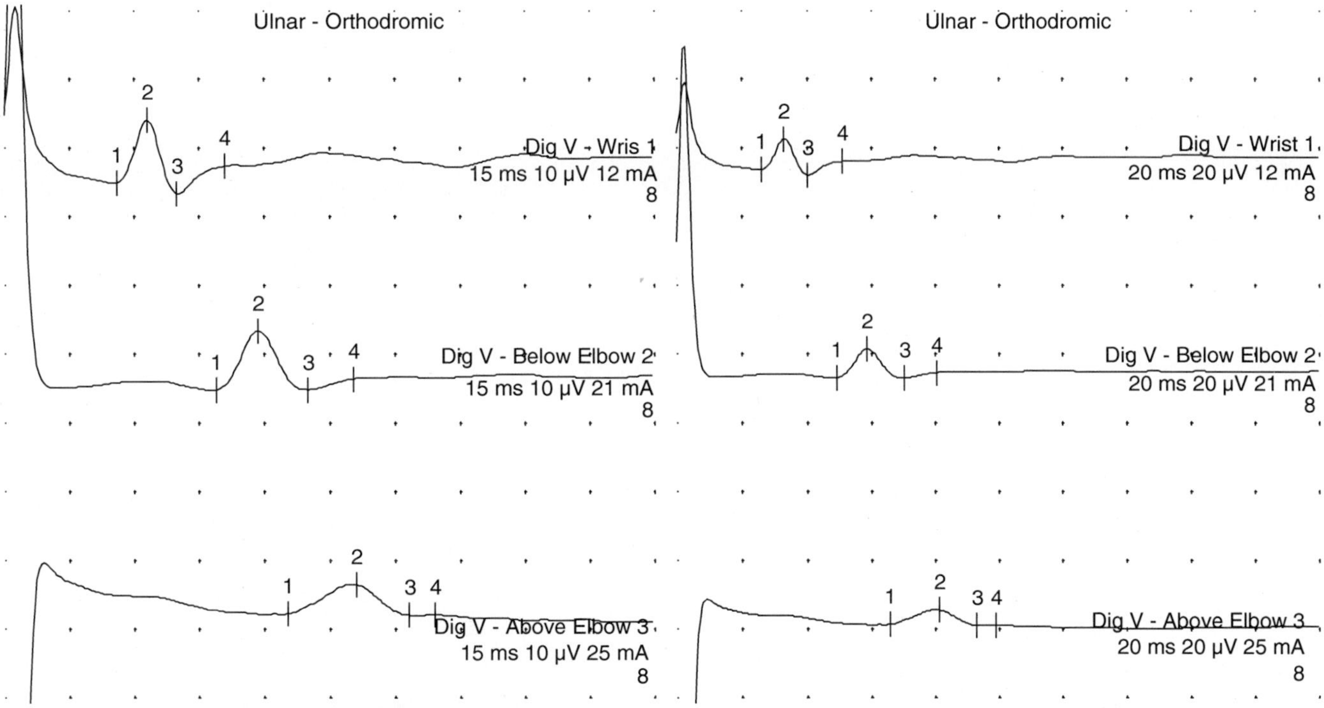

Sensitivity 10 µV/div, sweep speed 1.5ms/div   Sensitivity 20 µV/div, sweep speed 2 ms/div

Fig. 2 Orthodromic sensory nerve action potentials (*SNAPs*) recorded at the wrist (*upper trace*), below the elbow (*middle trace*), and above the elbow (*lower trace*), stimulation to digit V

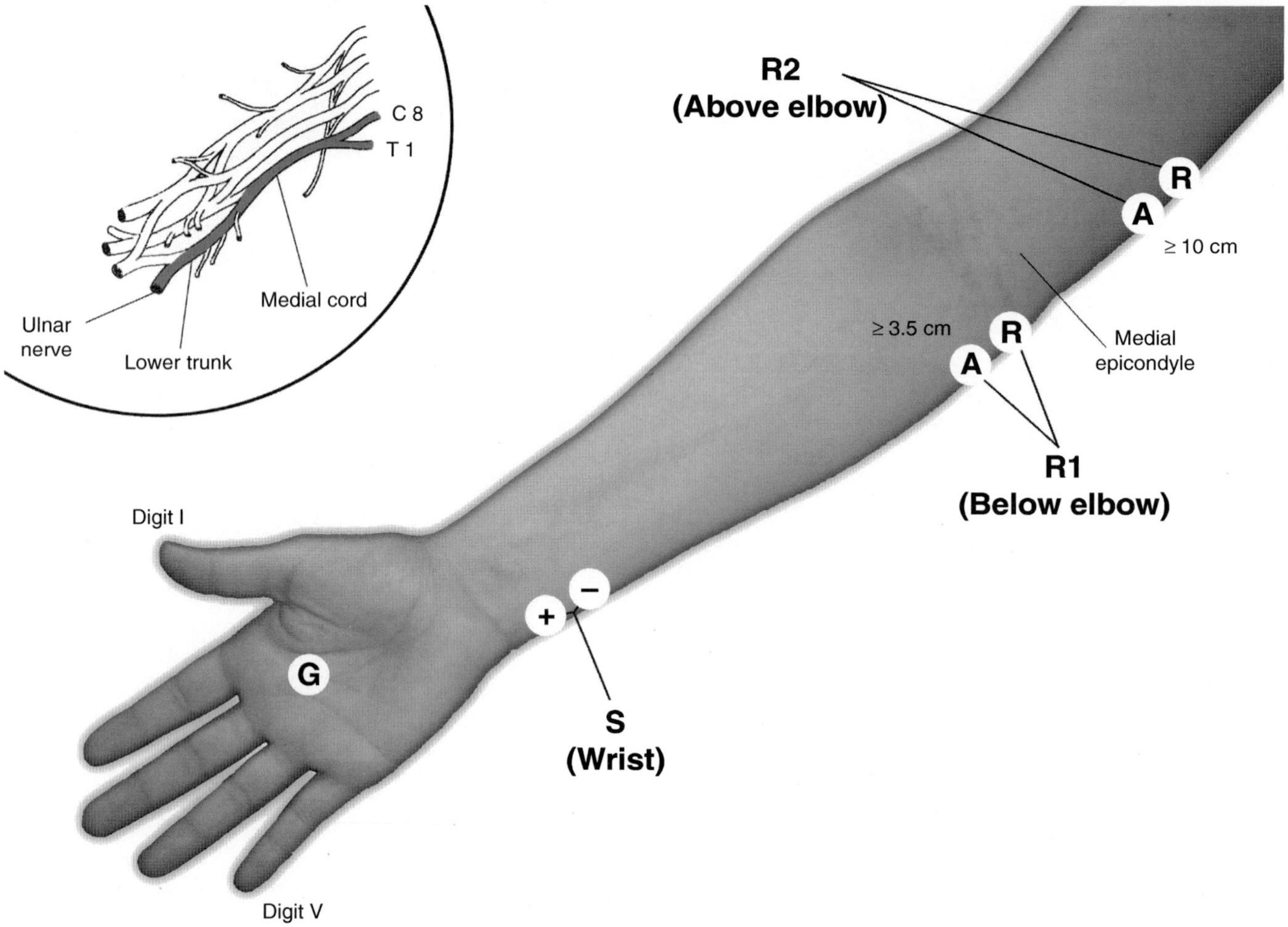

Typical waveform (wrist – below elbow, above elbow):

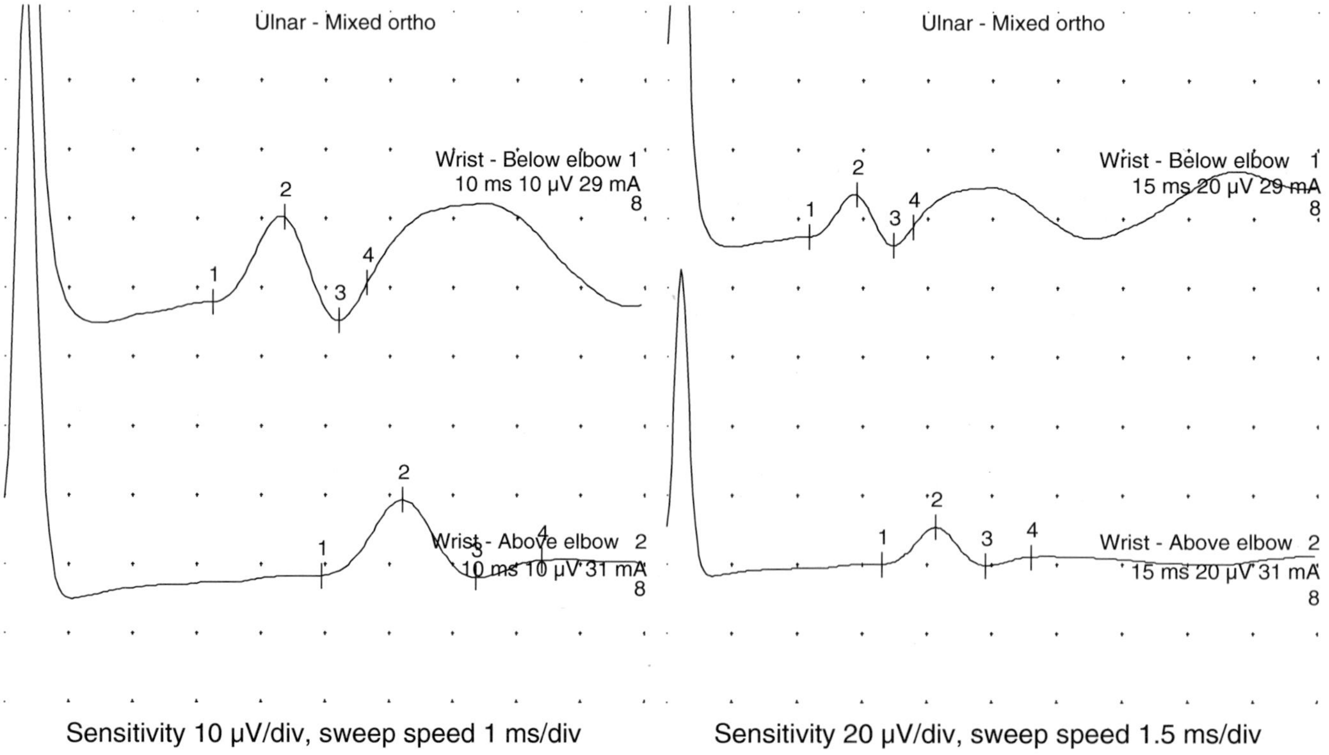

Fig. 3 Mixed nerve action potentials (*MNAPs*) recorded below the elbow (*upper trace*) and above the elbow (*lower trace*), stimulation of the wrist

35 years). Only below-elbow–wrist (BE–W segment) and above-elbow–below-elbow (AE–BE segment) conduction velocities were reported. The authors also performed all three methods on 43 patients (24 women and 19 men, age range 18–77 years, average age 42 years) referred to the EMG laboratory for evaluation of possible ulnar neuropathy at the elbow (UNE); patients were clinically divided into two groups, "definite UNE" (Tables 4, 5, and 6) and "possible UNE" (Tables 7, 8, and 9).

Table 1 Normal values [1] – antidromic method

	Mean ± SD	Range
Below elbow–wrist (BE–W), SNCV (m/s)	64.3 ± 4.1	53.4–73.6
Above elbow–below elbow (AE–BE), SNCV (m/s)	60.7 ± 4.9	50.0–71.0
AE–BE, SNCV difference (m/s)	3.6 ± 3.6	−1.5 to 9.8

Table 2 Normal values [1] – orthodromic method

	Mean ± SD	Range
Below elbow–wrist (BE–W), SNCV (m/s)	65.8 ± 4.1	57.5–74.0
Above elbow–below elbow (AE–BE), SNCV (m/s)	59.0 ± 5.2	47.8–69.0
AE–BE, SNCV difference (m/s)	6.8 ± 6.0	−3.8 to 13.2

Table 3 Normal values [1] – mixed method

	Mean ± SD	Range
Below elbow–wrist (BE–W), SNCV (m/s)	70.9 ± 4.2	63.9–81.4
Above elbow–below elbow (AE–BE), SNCV (m/s)	60.4 ± 6.3	44.0–75.0
AE–BE, SNCV difference (m/s)	10.6 ± 6.3	−10.5 to 21.5

Table 4 Pathological values, *definite UNE* [1] – antidromic method

	Mean ± SD	Range
Below elbow–wrist (BE–W), SNCV (m/s)	63.5 ± 3.6	57.8–70.3
Above elbow–below elbow (AE–BE), SNCV (m/s)	54.9 ± 8.9	39.0–70.0
AE–BE, SNCV difference (m/s)	8.6 ± 7.9	−8.0 to 18.8

Table 5 Pathological values, *definite UNE* [1] – orthodromic method

	Mean ± SD	Range
Below elbow–wrist (BE–W), SNCV (m/s)	65.9 ± 6.8	59.4–80.6
Above elbow–below elbow (AE–BE), SNCV (m/s)	48.9 ± 6.9	40.0–61.9
AE–BE, SNCV difference (m/s)	17.0 ± 10.2	3.7–38.0

Table 6 Pathological values, *definite UNE* [1] – mixed method

	Mean ± SD	Range
Below elbow–wrist (BE–W), SNCV (m/s)	71.0 ± 7.9	57.6–86.3
Above elbow–below elbow (AE–BE), SNCV (m/s)	52.8 ± 12.0	36.6–77.7
AE–BE, SNCV difference (m/s)	18.2 ± 11.7	−8.5 to 40.2

Table 7 Pathological values, *possible UNE* [1] – antidromic method

	Mean ± SD	Range
Below elbow–wrist (BE–W), SNCV (m/s)	66.7 ± 5.4	50.7–73.3
Above elbow–below elbow (AE–BE), SNCV (m/s)	61.6 ± 5.6	50.0–65.0
AE–BE, SNCV difference (m/s)	5.1 ± 7.0	−13.6 to 15.8

Table 8 Pathological values, *possible UNE* [1] – orthodromic method

	Mean ± SD	Range
Below elbow–wrist (BE–W), SNCV (m/s)	70.9 ± 6.4	55.4–78.2
Above elbow–below elbow (AE–BE), SNCV (m/s)	55.9 ± 4.4	50.0–65.0
AE–BE, SNCV difference (m/s)	15.0 ± 7.0	3.2–25.0

Table 9 Pathological values, *possible UNE* [1] – mixed method

	Mean ± SD	Range
Below elbow–wrist (BE–W), SNCV (m/s)	76.7 ± 8.7	58.8–92.8
Above elbow–below elbow (AE–BE), SNCV (m/s)	55.4 ± 5.2	47.6–64.2
AE–BE, SNCV difference (m/s)	21.4 ± 9.8	6.6–45.2

Comment

In each subject, the difference in conduction velocity (called by the authors "drop-off") across the elbow segment was calculated as the difference between AE and BE segmental SNCV; in 36 control subjects, Raynor et al. [1] observed a higher mean drop-off in SNCV across the elbow segment than one would expect, and they believed this drop in conduction velocity as a result of underestimating nerve length in view of slack and wrinkling in the nerve with the elbow in full extension [2,3].

Comment

In "definite UNE patients" group, Raynor et al. [1] frequently observed unobtainable, at one or more site, sensory and mixed potentials. They localized conduction abnormalities by motor conduction studies. In "possible UNE patients," the authors found the sensory and mixed conduction studies more sensitive of the motor conduction studies, which localized the ulnar compression at the elbow in 16 of 43 patients (37 %): the sensory conduction studies (both the orthodromic and antidromic methods) localized the lesion in additional 10 of the remaining 27 patients (37 %), and the mixed nerve studies identified another 8 patients of the 17 (47 %) who had not been previously localized. The overall sensitivity for "definite UNE patients" was 86 % and 68 % for sensory and motor conduction studies, respectively. In the "possible UNE patients" group, they found highest sensitivity value (67 %) for sensory and mixed conduction studies than motor studies (9 %). In their study, Raynor et al. [1] performed reliable antidromic and orthodromic sensory conduction studies, using surface electrodes for recording below and above the elbow in all healthy subjects and in all but one of the patients with mild clinical deficits. Mixed nerve potentials recorded by surface electrodes were obtained from below- and above-elbow sites easily in all healthy subjects and in all but three patients with severe UNE.

Kothari et al. [4] performed the mixed nerve conduction study in the flexed elbow position to test if that would potentially eliminate the problem in determining the onset latency for the below-elbow segment and marked variability in healthy subjects observed by Raynor and colleagues [1]. The authors also performed motor conduction studies (with the 90° flexed elbow position), recording from the *abductor digiti minimi* (ADM) and *first dorsal interosseous* (FDI) muscles, and antidromic sensory conduction study to digit V. All sensory and mixed conduction studies were performed using 10 mm silver disk recording electrodes. The temperature of the limb was maintained at 34 °C with a sensor placed on the palm. Normal data were gathered in 20 healthy volunteers (12 women and 8 men, age range 21–72 years, mean age 37 years), and pathological values were obtained from 21 patients with clinical symptoms and signs of ulnar neuropathy at the elbow (11 men and 10 women, age range 17–70 years, mean age 47 years). Ulnar nerve was stimulated at the wrist; mixed nerve potentials were recorded, following the method by Raynor et al. [1], from the below-elbow and above-elbow sites with the 90° flexed elbow position (Fig. 4). Supramaximal stimulation was used in all recordings, and responses were averaged at least five times.

Onset latency was measured from the stimulus artifact to the onset of the negative deflection of the mixed response, and segmental conduction velocities were calculated for the forearm (below elbow–wrist, BE–W) and cubital tunnel (below elbow–above elbow, BE–AE) segments. In 20 healthy subjects, the mean mixed conduction velocity for the BE–W segment was 73.9 m/s (range 64.5–82.6 m/s), and the mean mixed conduction velocity for the BE–AE segment was 64.6 m/s (range 57.1–80 m/s). The mean decrease in mixed conduction velocity between the BE–W and BE–AE segments was 9.3 ± 3.9 m/s; the upper limit of normal (mean ± 2SD) was 17.1 m/s. In the patient group, the authors [4] found a mixed nerve conduction velocity in the BE–W segment of 65.3 m/s, and velocity in the BE–AE segment was 45.8 m/s. The mean decrease in mixed conduction was 17.4 m/s (range −2.3 to 36.2 m/s). Of the 21 patients, 12 (57 %) showed abnormal mixed nerve conduction studies and conventional antidromic sensory study to digit V, and in only one patient (5 %), the mixed nerve study was abnormal with all other conduction studies completely normal. Conversely, they observed abnormal motor conduction from the ADQ muscle in 15 (71 %) patients and the FDI muscle in 17 (81 %) patients, with a mean conduction velocity for the BE–AE of 47.2 m/s and 40.9 m/s, respectively. The average decrease in motor conduction in the BE–AE segment was 14.1 m/s from the ADQ muscle, and it was 13.6 m/s from the FDI muscle. The authors compared their results with those by Raynor and colleagues [1], and they found a smaller decrease in conduction velocity with the elbow 90° flexed position than that identified with the arm straight. Also, they observed a low variability of conduction velocity in control subjects with a SD value of only 3.9 m/s and an upper limit of normal of 17.1 m/s. In light of their findings, for the authors, although the mixed nerve study performed in the elbow flexed position did produce more consistent results than in the elbow-straight position, the sensitivity of this test for the diagnosis of UNE is very low if compared with motor conduction studies. In all 21 patients, they performed a needle examination in the FDI, ADM, *flexor carpi ulnaris* (FCU), and the *flexor digiti profundis* of the digit IV and digit V (FDP) muscles. All 21 patients had evidence of chronic denervation (increased amplitude, duration, and a number of phases with a reduced recruitment pattern), and two patients showed sign of active denervation (positive sharp waves and fibrillations); a coexisting C8-T1 radiculopathy was excluded by sampling other non-ulnar innervated C8-T1 muscles. For Kothari et al. [4], despite their patient's sample seemed to have a chronic (severe?) compression of the ulnar nerve at the elbow (chronic denervation at the EMG studies was observed in all patients), the motor conduction study recording from the FDI muscle provided a slightly higher sensitivity than recording from the ADQ muscle, and they suggested to perform both studies to identify very mild cases of UNE. Pathological waveform and values in patients with ulnar neuropathy at the elbow (UNE) are reported (Figs. 5 and 6).

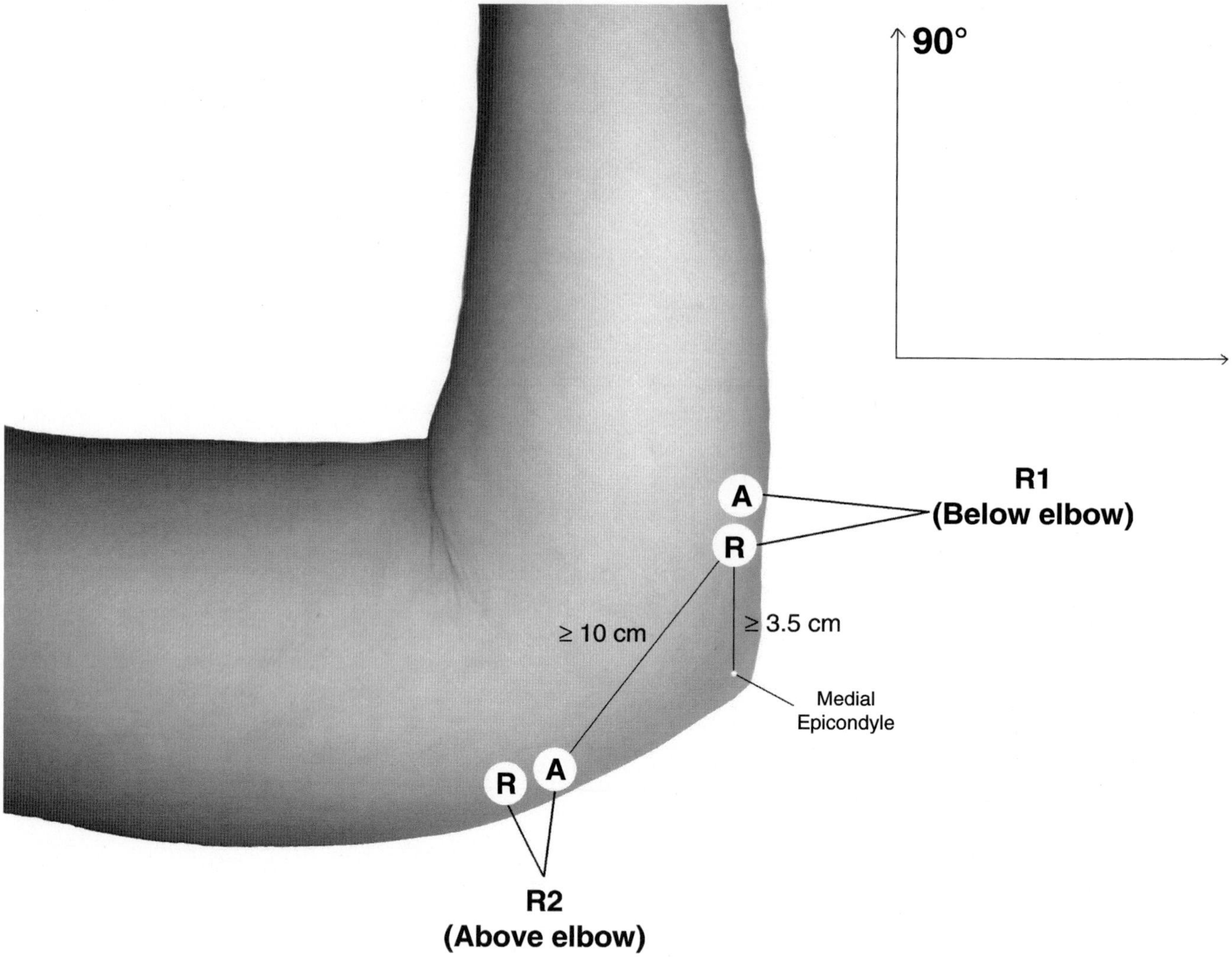

Fig. 4 Mixed ulnar nerve proximal recording sites – elbow flexed to 90° [4]

Pathological waveform (wrist – below elbow, above elbow):

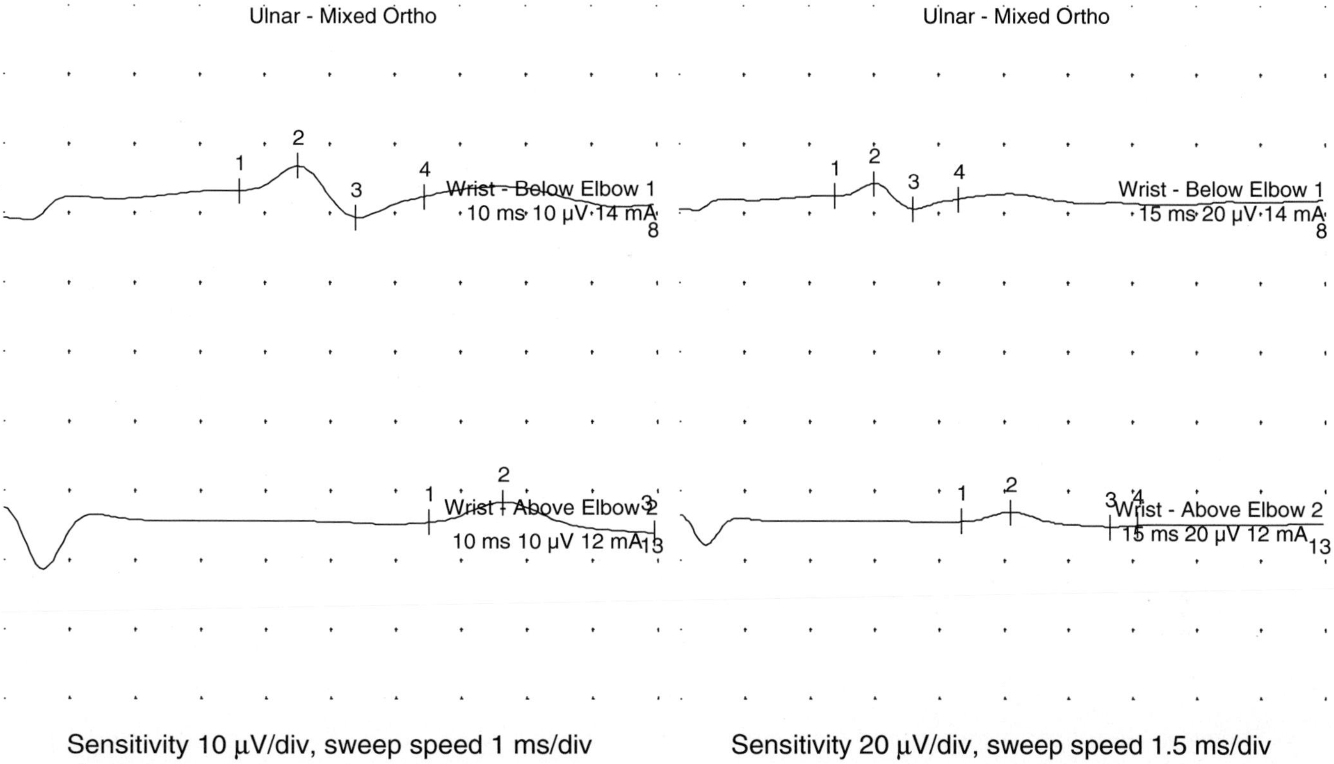

Onset latency (wrist – below elbow): 3.6 ms; **Onset latency** (wrist – above elbow): 6.5 ms; **Peak latency** (wrist – below elbow): 4.45 ms; **Peak latency** (wrist – above elbow): 7.7 ms; **Onset to peak amplitude** (wrist – below elbow): 3.6 μV; **Onset to peak amplitude** (wrist – above elbow): 3.1 μV; **Peak to peak amplitude** (wrist – below elbow): 7.6 μV; **Peak to peak amplitude** (wrist – above elbow): 4.6 μV; **SNCV** (wrist – below elbow, 19 cm distance): 52.8 m/s; **SNCV** (below elbow – above elbow, 10 cm distance): 34.5 m/s

Fig. 5 Mixed nerve action potentials (*MNAPs*) recorded below (*upper trace*) and above the elbow (*lower trace*) in a patient with UNE, stimulation of the wrist

Pathological waveform (digit V – wrist, below elbow, above elbow):

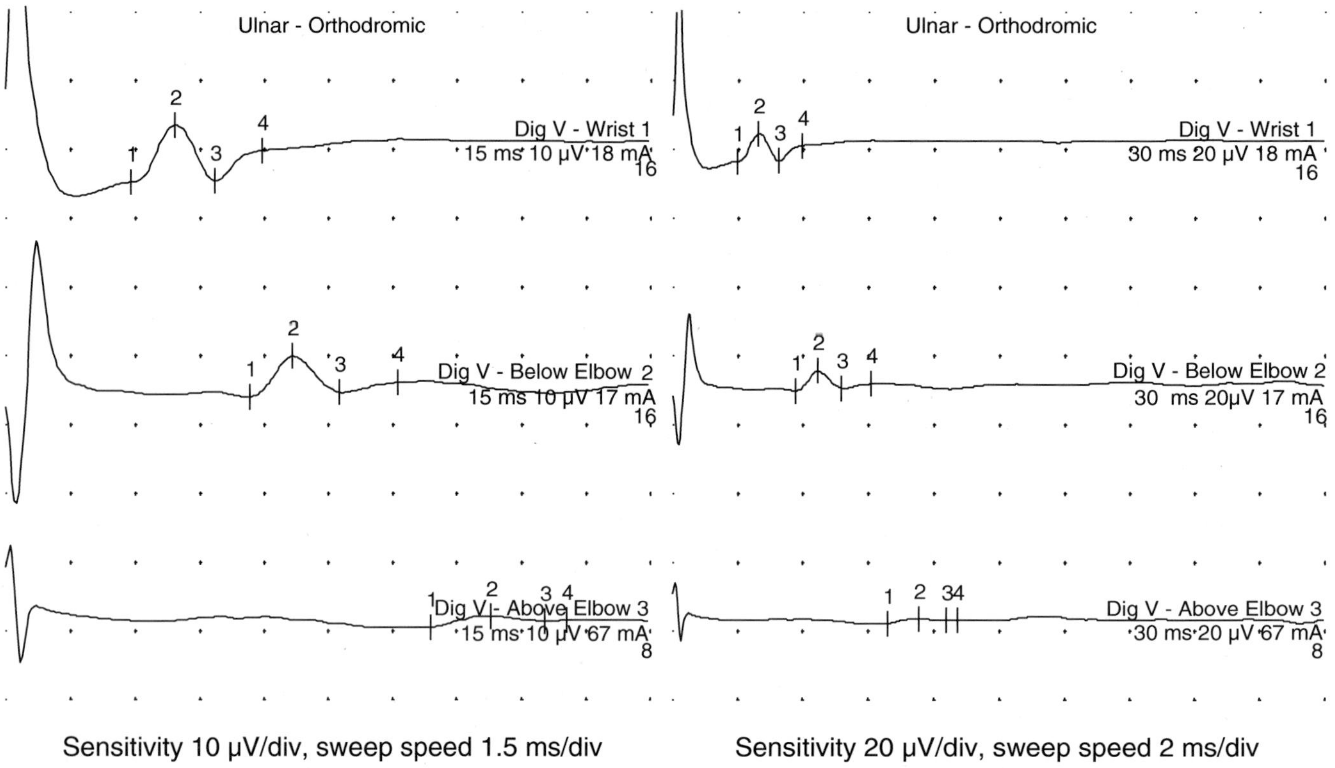

Onset latency (digit V – wrist): 2.90 ms; **Onset latency** (digit V – below elbow): 5.65 ms; **Onset latency** (digit V – above elbow): 9.90 ms; **Peak latency** (digit V – wrist): 3.90 ms; **Peak latency** (digit V – below elbow): 6.60 ms; **Peak latency** (digit V – above elbow): 11.10 ms; **Onset to peak amplitude** (digit V – wrist): 8.5 µV; **Onset to peak amplitude** (digit V – below elbow): 6.1 µV; **Onset to peak amplitude** (digit V – above elbow): 1.7 µV; **Peak to peak amplitude** (digit V – wrist): 8.3 µV; **Peak to peak amplitude** (digit V – below elbow): 5.4 µV; **Peak to peak amplitude** (digit V – above elbow): 0.88 µV; **SNCV** (digit V – wrist, 15 cm distance): 51.7 m/s; **SNCV** (wrist – below elbow, 17 cm distance): 61.8 m/s; **SNCV** (below elbow-above elbow, 10 cm distance): 23.5 m/s

Fig. 6 Orthodromic sensory nerve action potentials (*SNAPs*) recorded at the wrist (*upper trace*), below the elbow (*middle trace*), and above the elbow (*lower trace*) in a patient with UNE, stimulation to digit V

References

1. Raynor EM, Shefner JM, Preston DC et al (1994) Sensory and mixed nerve conduction studies in the evaluation of ulnar neuropathy at the elbow. Muscle Nerve 17:785–792
2. Checkles NS, Russakov AD, Piero DL (1971) Ulnar nerve conduction velocities: effect of elbow position on measurement. Arch Phys Med Rehabil 52:362–365
3. Cruz-Martinez A, Barrio M, Perez-Conde MC et al (1978) Electrophysiological aspects of sensory conduction velocity in healthy adults. J Neurol Neurosurg Psychiatry 41:1092–1096
4. Kothari MJ, Heistand M, Rutkove SB (1998) Three ulnar nerve conduction studies in patients with ulnar neuropathy at the elbow. Arch Phys Med Rehabil 79:87–89

Wrist – Above Elbow

S22

Surface Recording Technique, Orthodromic Study

Original Settings Sensitivity was 10 μV/division, sweep speed was 1 ms/division, and the machine used was an MS Mystro EMG machine; low-frequency filter, high-frequency filter, duration of pulse, and rate of pulse were not specified.

Position This study was performed with the arm maintained in a slightly abducted and internally rotated position, the elbow flexed (to an angle of 15° from horizontal), and the forearm pronated.

Recording Following the orthodromic method [1], the compound nerve action potentials (CNAPs) were recorded using surface recording electrodes above the elbow (Fig. 1), at the location 10 cm proximal to the medial epicondyle. The authors chose this recording site because this location covers the anatomical area of both the ulnar and median nerves, and CNAPs of both nerves could thus be recorded without moving the surface recording electrodes. Each recording was made separately. The ground (G) electrode position was not specified in the report; the figure shows the ground electrode placed over the forearm.

Stimulation The median and ulnar nerves were stimulated at the wrist (proximal skin fold) using a surface stimulator and placing it on the course of both nerves (Fig. 2). Each nerve was stimulated separately (S1 and S2, respectively).

Measurements Onset latency (ms) was measured from the onset of the stimulus artifact to the onset of the negative deflection of the CNAP, and the latencies of the two CNAPs were compared subtracting the latency observed for the median nerve from the value for the ulnar nerve and named this the "difference in the latency index" (DLat index). In all controls and patients, the limb temperature was measured at the elbow with an electronic thermometer model EPT 200 (Medilog, England), and the average temperature over the elbow was 30.5 °C (range 28–34 °C). The authors compared the DLat index with motor nerve conduction study to the *abductor digiti minimi* (ADM) muscle performed in the across-elbow segment (below elbow, 2 cm distal to the medial epicondyle; above elbow, 10 cm proximal to the medial epicondyle). The motor nerve conduction velocity (MNCV) was measured in meter per second (m/s), and it was from 52 to 88.9 m/s in the across-elbow segment. Merlevede et al. [1] recorded CNAPs at the elbow in 60 healthy (Table 1) subjects (38 women and 22 men, age range 13–61 years, mean age 33.6 years). To determine the value of their test, the authors obtained the DLat index in a group of 10 patients with ulnar neuropathy of varying clinical severity and in a heterogeneous group of 24 patients with other neurological histories (sample data of both groups were not reported).

Comment

Merlevede et al. [1] elicited the ulnar and median CNAPs by surface stimulation at the wrist and recording 10 cm proximal to the medial epicondyle. In 57 normal controls, they observed a mean DLat index value of 0.72 ± 0.30 ms, with a DLat index for the 95 % limit of the normal population set at 1.30 ms. They found all but three of their controls had "normal" values below the set limit of 1.3 ms (1.32, 1.36, and 1.38 ms), and in order to avoid false-positive values, the authors set the "normal limit" at 1.4 ms. They found the DLat index highly sensitive, because 9 of 10 patients with ulnar neuropathy at the elbow (UNE) had DLat index values of more than 1.4 ms or an absent ulnar CNAP, and in only one patient with UNE, they observed a DLat index of 1.37 ms, which was just at the upper limit of normal. They also found the DLat index very specific, because in all 20 nondiabetic patients, the DLat index was normal; increased DLat index values were observed in three diabetic patients without complaints of possible ulnar neuropathy.

G. Gentili, M. Di Napoli, *The Ulnar Nerve*, DOI 10.1007/978-3-319-43292-2_22

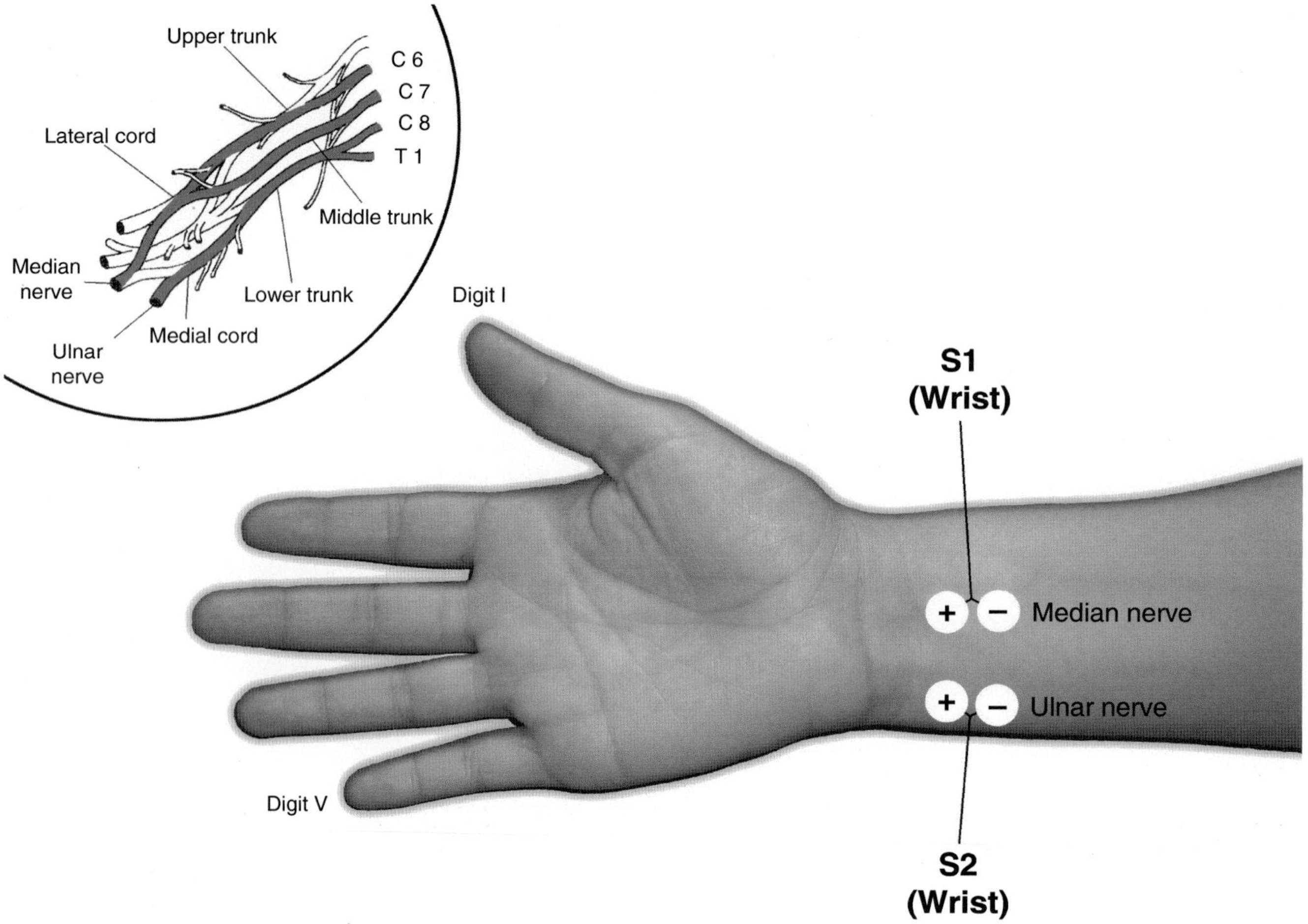

Typical waveform (wrist – above elbow):

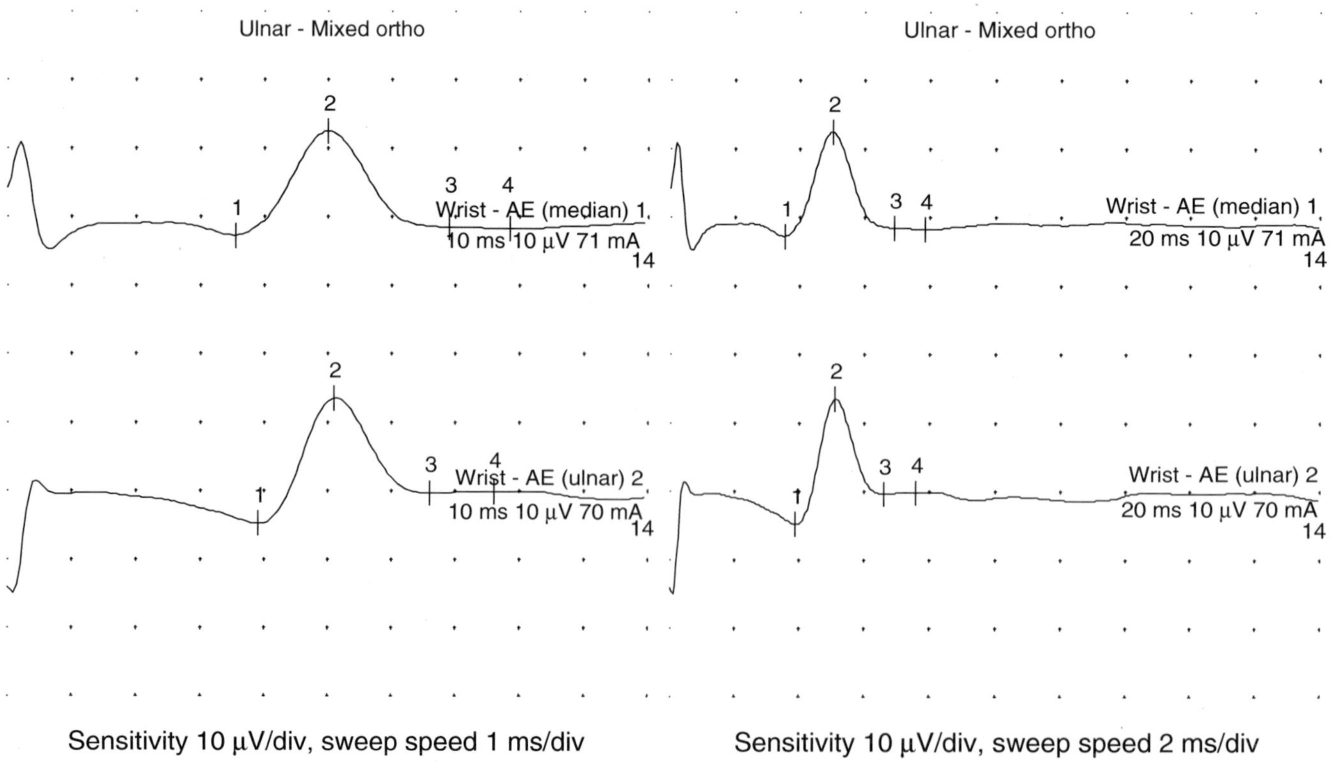

Fig. 1 Orthodromic compound nerve action potentials (*CNAPs*) recorded above the elbow from the median nerve (*upper trace*) and ulnar nerve (*lower trace*), stimulation of the wrist

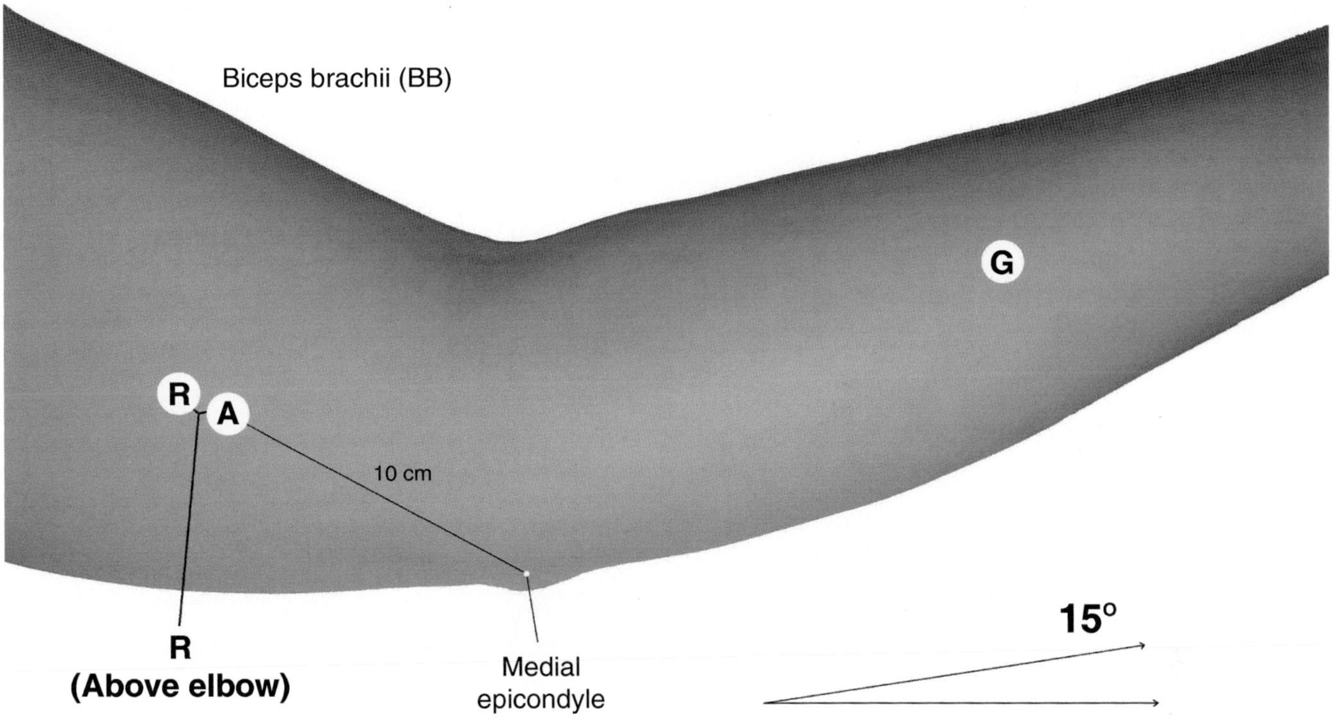

Fig. 2 Mixed ulnar nerve proximal recording site (elbow flexed to 15°)

Table 1 Normal values [1]

	Mean ± SD	Limit of normal
Wrist–elbow, DLat index – onset latency (ms)	0.72 ± 0.3	1.4

In order to avoid the difficulty in obtaining an accurate determination of the length of the ulnar nerve, Poza [2] performed a similar method by Merlevede et al. [1] with some modifications. He placed the active recording bar electrode (A) 5 cm proximal to the medial epicondyle (reference electrode proximal); the stimulation site for median and ulnar nerves was set proximal on the wrist crease, on the course of both nerves, respectively. Conventional MNCV study was also performed in all subjects performing stimulation at the wrist, below the elbow, and above the elbow using a variable distance for the across-elbow segment (greater than 10 cm) and recording from the *abductor digiti minimi* (ADM) muscle. All studies were performed with the elbow in moderate flexion (90°) and forearm in the supine position (Fig. 3).

The MNAPs were averaged from 6 to 32 acquisitions, and the skin was accurately prepared and cleaned by alcohol and abrasive cream to avoid the volume conduction effects due to the median and ulnar nerve stimulation. Mixed latency difference (Dmix) was calculated using onset latencies, and side-to-side differences were also calculated in all subjects. The author used a MEDELEC Sapphire 4ME and a MEDELEC Synergy electromyography machines (sensibility 10 μV, sweep speed 1 ms/division) to study 407 ulnar nerves (Tables 2, 3, and 4) from 407 healthy subjects (223 men and 184 women – age range 16–85 years and mean age 46.60 ± 13.66 years, men age range 17–79 years and women age range 16–85 years, men mean age 45.79 ± 13.86 and women mean age 47.58 ± 13.39) and 172 ulnar nerves (Tables 5, 6, and 7) from 172 patients with UNE clinically diagnosed (114 men and 58 women; age range 16–87 years and mean age 47.99 ± 14.27 years, men age range 18–87 years and women age range 16–85 years, men mean age 47.88 ± 14.58 and women mean age 48.21 ± 13.78).

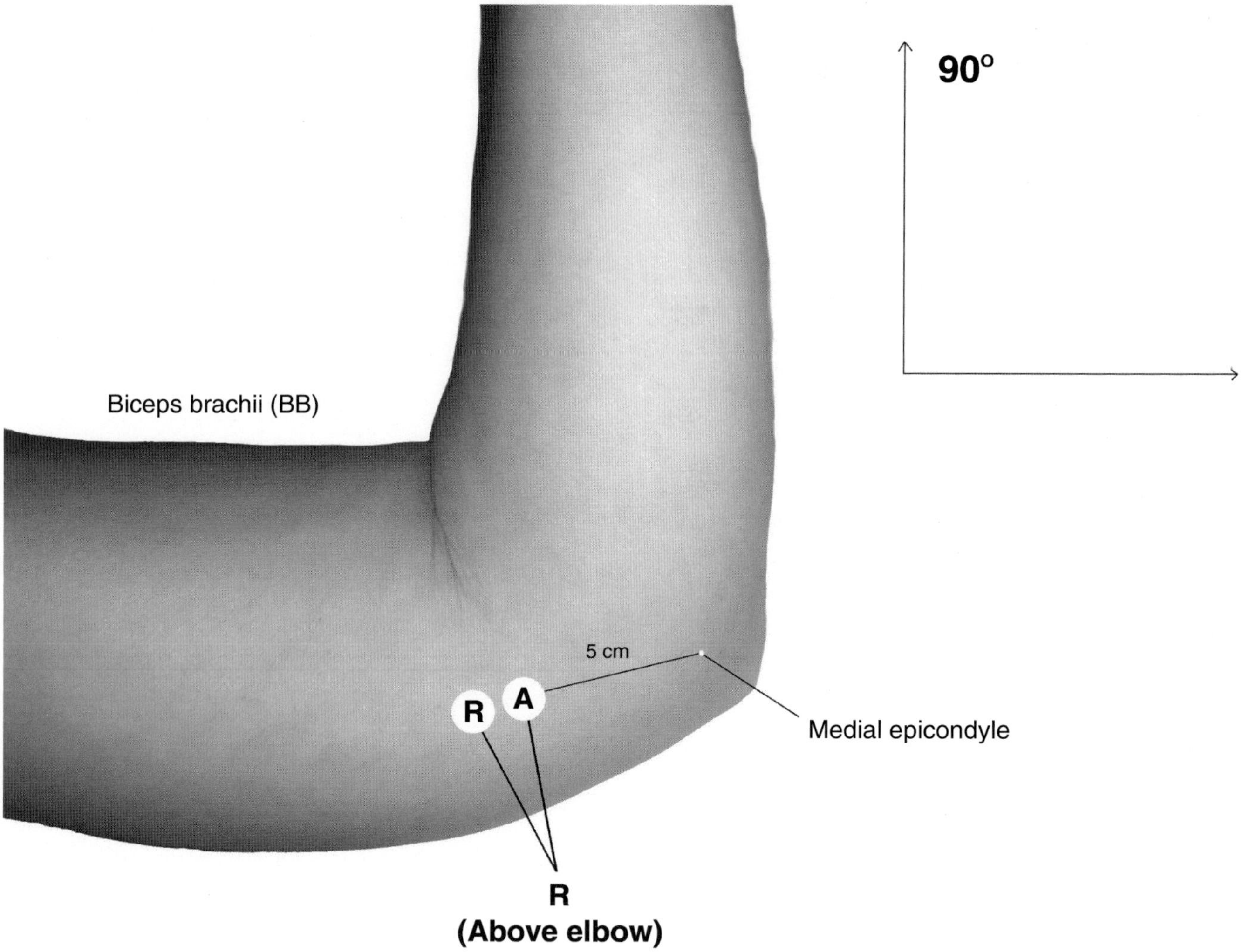

Fig. 3 Mixed ulnar nerve proximal recording site – elbow flexed to 90° [2]

Table 2 Normal values [2]

	Mean ± SD	Range
Below elbow–above elbow, MNCV (m/s) – all subjects	62.50 ± 6.37	50–80
Below elbow–above elbow, MNCV (m/s) – men	61.76 ± 6.68	50–80
Below elbow–above elbow, MNCV (m/s) – women	63.41 ± 5.87	50–77
Below elbow–above elbow, MNCV (m/s) – 16–40 years	63.88 ± 6.35	51–76
Below elbow–above elbow, MNCV (m/s) – 40–60 years	62.78 ± 6.03	50–80
Below elbow–above elbow, MNCV (m/s) – >60 years	58.90 ± 6.50	50–75

Table 3 Normal values [2]

	Mean ± SD	Range
Wrist–above elbow, Dmix (ms) – all subjects	0.64 ± 0.25	0.09–1.88
Wrist–above elbow, Dmix (ms) – men	0.67 ± 0.25	0.11–1.88
Wrist–above elbow, Dmix (ms) – women	0.60 ± 0.24	0.09–1.32
Wrist–above elbow, Dmix (ms) – 16–40 years	0.64 ± 0.24	0.10–1.32
Wrist–above elbow, Dmix (ms) – 40–60 years	0.62 ± 0.23	0.09–1.48
Wrist–above elbow, Dmix (ms) – >60 years	0.67 ± 0.31	0.10–1.88

Table 4 Normal values [2]

	Mean ± SD	Range
Wrist–above elbow, Dmix side to side (ms) – all subjects	0.15 ± 0.14	0–1.12
Wrist–above elbow, Dmix side to side (ms) – men	0.14 ± 0.16	0–1.12
Wrist–above elbow, Dmix side to side (ms) – women	0.15 ± 0.13	0–0.61
Wrist–above elbow, Dmix side to side (ms) – 16–40 years	0.13 ± 0.13	0–0.61
Wrist–above elbow, Dmix side to side (ms) – 40–60 years	0.14 ± 0.12	0–0.59
Wrist–above elbow, Dmix side to side (ms) – >60 years	0.20 ± 0.23	0–1.12

Table 5 Pathological values [2]

	Mean ± SD	Range
Below elbow–above elbow, MNCV (m/s) – all patients	47.13 ± 12.19	18–73
Below elbow–above elbow, MNCV (m/s) – men	45.97 ± 12.82	18–73
Below elbow–above elbow, MNCV (m/s) – women	49.50 ± 10.53	30–70
Below elbow–above elbow, MNCV (m/s) – 16–40 years	49.83 ± 12.44	19–70
Below elbow–above elbow, MNCV (m/s) – 40–60 years	48.30 ± 11.13	24–73
Below elbow–above elbow, MNCV (m/s) – >60 years	39.94 ± 12.47	18–68

Table 6 Pathological values [2]

	Mean ± SD	Range
Wrist–above elbow, Dmix (ms) – all patients	1.58 ± 0.70	0.54–4.68
Wrist–above elbow, Dmix (ms) – men	1.68 ± 0.74	0.54–4.68
Wrist–above elbow, Dmix (ms) – women	1.37 ± 0.57	0.60–3.24
Wrist–above elbow, Dmix (ms) – 16–40 years	1.48 ± 0.67	0.62–3.60
Wrist–above elbow, Dmix (ms) – 40–60 years	1.54 ± 0.73	0.54–4.68
Wrist–above elbow, Dmix (ms) – >60 years	1.82 ± 0.64	1.16–3.80

Table 7 Pathological values [2]

	Mean ± SD	Range
Wrist–above elbow, Dmix side to side (ms) – all patients	0.79 ± 0.68	0.05–4.0
Wrist–above elbow, Dmix side to side (ms) – men	0.87 ± 0.73	0.04–4.0
Wrist–above elbow, Dmix side to side (ms) – women	0.62 ± 0.54	0.05–2.48
Wrist–above elbow, Dmix side to side (ms) – 16–40 years	0.72 ± 0.49	0.03–2.41
Wrist–above elbow, Dmix side to side (ms) – 40–60 years	0.76 ± 0.75	0.05–4.0
Wrist–above elbow, Dmix side to side (ms) – >60 years	0.90 ± 0.76	0.19–3.40

Comment

Poza [2] found a 1.1 ms (95 % percentile) as the upper limit of the normal population for Dmix. The Dmix showed higher sensitivity than the conventional study of motor conduction through the elbow to diagnose UNE because, in 54 (31 %) symptomatic nerves, a normal MNCV in the across-elbow segment was observed (>50 m/s) and Dmix was higher than 1 ms. The author determined that a difference of the Dmix ≥1 ms had a sensitivity of 87 % and a specificity of 91 % for the diagnosis of UNE, but using the Dmix >1.4 ms proposed by Merlevede et al. [1], he observed a higher specificity (99 %) but a very low sensibility (48 %). Furthermore, a difference between both arms equal or greater than 0.3 ms had a sensitivity of 80 % and specificity of 91 %. If both conditions are present, the test is very specific. In conclusion, Poza [2] found the measurement of the difference in mixed latency (Dmix) between median and ulnar nerves from the wrist to above-elbow segment, a valuable tool for evaluating patients with suspected ulnar neuropathy at the elbow without motor involvement.

Heise and Toledo [3] stimulated the median and ulnar nerves at the wrist placing the surface recording electrodes 3 cm apart on the medial aspect of the arm and 10 cm above the medial epicondyle. The authors used a similar method by Merlevede et al. [1], but they preferred to examine each subject lying in a supine position with the shoulder slightly abducted, the elbow extended, and the forearm supinated (Fig. 4). A portable DANTEC Keypoint electromyography machine was used for all recordings, and the normal values (Table 8) for the MNAP difference (MLD) were calculated in 57 healthy volunteers (age range 20–77 years, average 39 years).

The MLD was also calculated in 100 patients retrospectively evaluated with persistent sensory complaints at the fourth and fifth fingers, with or without Tinel sign at the elbow. All patients were divided into two groups: (a) 45 patients with ulnar neuropathy at the elbow (UNE) confirmed by conventional MNCV study and (b) 55 patients who did not fit two or three required criteria for a positive MNCV test (MNCV in the across-elbow segment <50 m/s, MNCV difference >10 m/s between forearm and across-elbow segments, and drop of CMAP amplitude across the elbow at least 20 %). These ulnar MCV criteria for the diagnosis of UNE were based on the American Association of Electrodiagnostic Medicine's (AAEM) guideline [4]. The standard MNCV was performed stimulating the ulnar nerve at the wrist, below the elbow, and above the elbow (10 cm across-elbow segment) and recording from the *abductor digiti minimi* (ADM) muscle. All MNCV studies were performed with the shoulder abducted, the arm externally rotated, and the elbow flexed at 90°. Many of the B group patients were tested using motor short-segment incremental studies (also called "inching") across the elbow [5], including all patients with an abnormal mixed latency difference. The authors considered abnormal an increase of CMAP latency of at least 0.7 ms in a 2 cm step, and all abnormal inching studies showed a clear increased gap or abrupt change of amplitude comparing the continuous recording by visual inspection. For Heise and Toledo [3], the MLD in the control group ranged from 0 to 1.1 ms. The 98th percentile was also 1.1. ms, which was set as their upper normative value. They noted a significant correlation between the MLD value and the height of the examined subject, and the following equation was

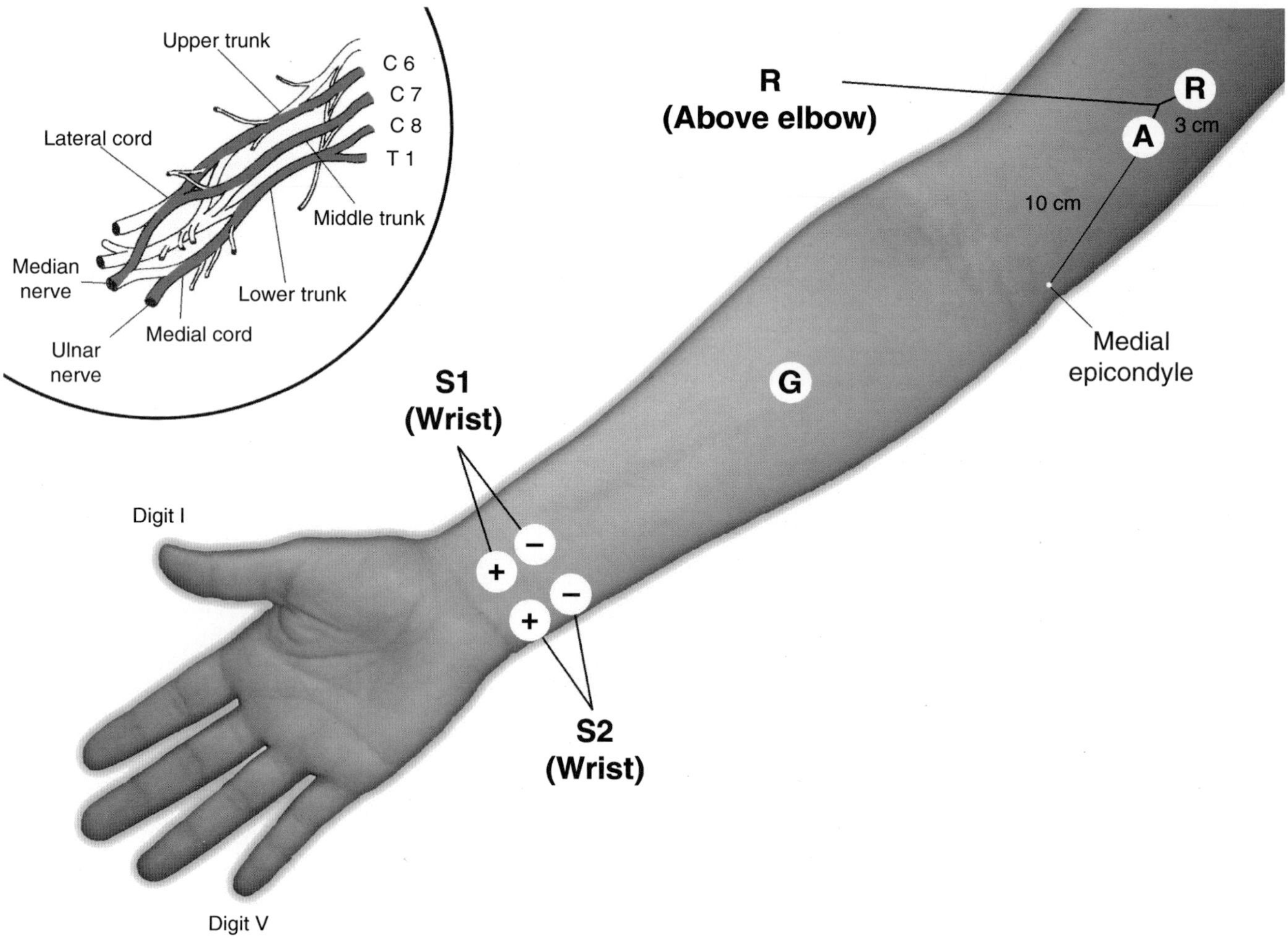

Fig. 4 Mixed ulnar nerve proximal recording site – elbow extended [3]

Table 8 Normal values [3]

	Range	Limit of normal
Wrist–elbow, MLD – peak latency (ms)	0 ± 1.1	1.1

developed for height correction: MLD – 0.015×H+2.67, where MLD is the mixed latency difference and H is height (in centimeters). The authors modified the original technique developed by Merlevede and colleagues [1]. They changed the limb position preferring the elbow extended and the forearm supinated (Fig. 4) because a mixed latency difference increasing was observed with the elbow flexed at 15° suggesting them that, contrary to the belief of some, the elbow position does influence the MLD. Because 15° can be difficult to judge subjectively, they placed the elbow in full extension and the forearm in the supine position made easier to stimulate the median and ulnar nerves than the prone position. Also, they used peak latencies for MLD calculation instead of the onset latencies because automatic markers can easily place the peak latencies in position even with low-amplitude CMAPs or noisy baselines. Care must be taken not to overstimulate the ulnar nerve because the spread of current from the ulnar nerve to the median nerve is a common pitfall of this technique. For the authors [3], the MLD test was not more sensitive than conventional MNCV of the ulnar nerve, but the MLD can assure the diagnosis in cases of mild MNCV of the ulnar nerve abnormalities by providing multiple internally consistent findings [6]. The amplitudes of the MNAPs were too variable in the healthy subjects to have any diagnostic utility, and even the MNAPs comparison between median and ulnar nerves in the same subject showed a significant variation. Like others [1], they also suggested to be careful to make the diagnosis of UNE based only on abnormal MLD, because it is not possible to localize the lesion by this method. Further, patients with Martin–Gruber anastomosis were excluded from the analysis because its presence is a potential threat for the MLD method in patients with possible UNE. In these patients in fact, the ulnar MNAP amplitude is reduced because of sensory axonal degeneration and temporal dispersion of the MNAP at the elbow. In this situation, the motor fibers traveling through the median nerve and Martin–Gruber anastomosis may produce a "double-peaked" ulnar MNAP, and, in severe UNE cases, only the first peak may be recorded producing a normal mixed latency difference. Pathological waveform and values in patient with UNE are reported (Fig. 5).

Pathological waveform (wrist – above elbow):

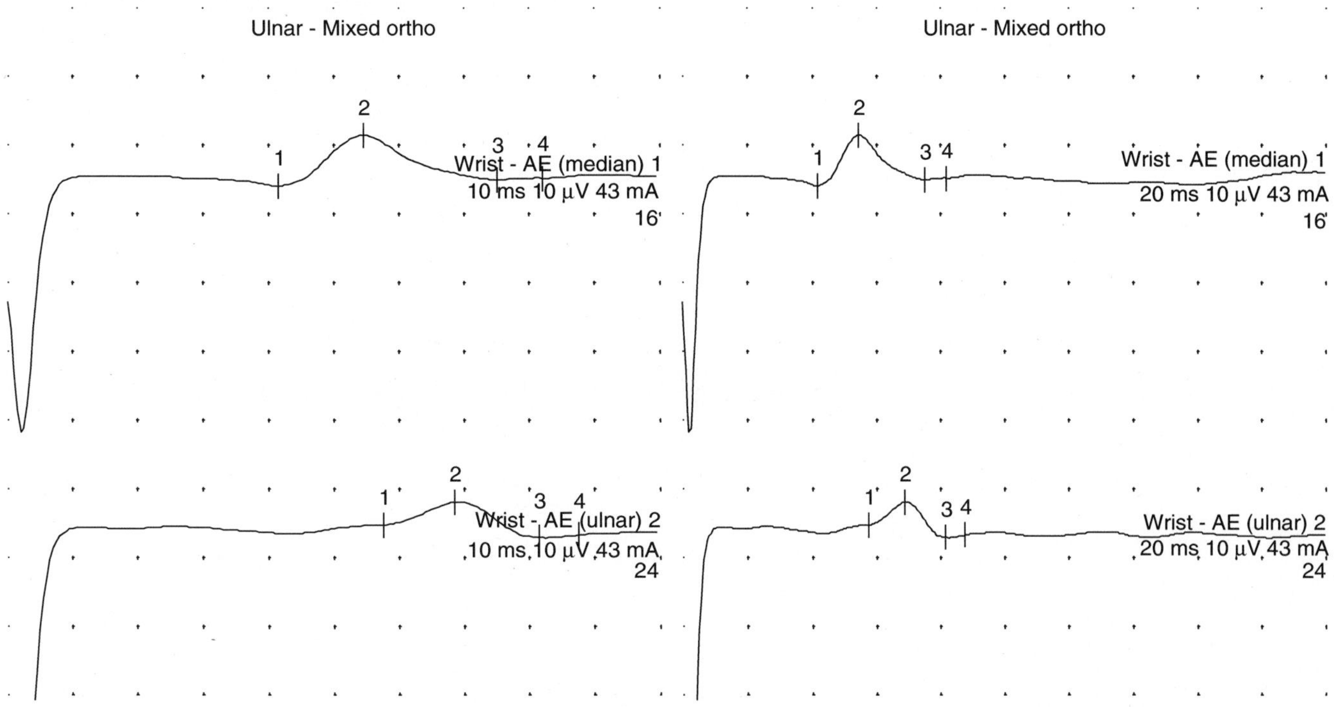

Onset latency (wrist – above elbow – median nerve): 4.15 ms; **Onset latency** (wrist – above elbow – ulnar nerve): 5.75 ms; **Peak latency** (wrist – above elbow – median nerve): 5.40 ms; **Peak latency** (wrist – above elbow – ulnar nerve): 6.85 ms; **Onset to peak amplitude** (wrist – above elbow – median nerve): 7.5 µV; **Onset to peak amplitude** (wrist – above elbow – ulnar nerve): 3.3 µV; **Peak to peak amplitude** (wrist – above elbow – median nerve): 6.4 µV; **Peak to peak amplitude** (wrist – above elbow – ulnar nerve): 5.3 µV; **NCV** (wrist-above elbow – median nerve, 28 cm): 67.5 m/s; **NCV** (wrist – above elbow – ulnar nerve, 28 cm): 48.7 m/s

Fig. 5 Orthodromic compound nerve action potentials (*CNAPs*) recorded above the elbow from the median nerve (*upper trace*) and ulnar nerve (*lower trace*) in a patient with UNE, stimulation of the wrist

References

1. Merlevede K, Theys P, Van Hees J (2000) Diagnosis of ulnar neuropathy: a new approach. Muscle Nerve 23:478–481
2. Poza JJ (2005) Comparación de la latencia mixta de muñeca a codo de nervios mediano y cubital como test diagnóstico de la neuropatia cubital en el codo. Neurologia 20:65–70
3. Heise CO, Toledo SM (2006) Mixed latency difference for diagnosis of ulnar neuropathy at the elbow. Arch Phys Med Rehabil 87:408–410
4. Kincaid JC (2003) Ulnar neuropathy at the elbow. In: AAEM Course H (p H29-36): Peripheral neuropathy. Joint Meeting of American Association of Electrodiagnostic Medicine and International Federation of Clinical Neurophysiology; 16–20 Sept, San Francisco
5. Azrieli Y, Weimer L, Lovelace R et al (2003) The utility of segmental nerve conduction studies in ulnar mononeuropathy at the elbow. Muscle Nerve 27:46–50
6. American Association of Electrodiagnostic Medicine (AAEM), American Academy of Physical Medicin and Rehabilitation (AAPR), American Academy of Neurology (AAN) (1999) Practice parameter for electrodiagnostic studies in ulnar neuropathy at the elbow: summary statement. Arch Phys Med Rehabil 80:357–359

Wrist – Elbow; Elbow – Axilla

S23

Surface Recording Technique, Orthodromic Study

Original Settings Sensitivity was 20 μV/division, sweep speed was 1 ms/division, low-frequency filter was 10 Hz, high-frequency filter was 2 kHz, duration of pulse was 0.5 ms, rate of pulse was 1/s, and the machine used was a DANTEC Counterpoint EMG machine.

Position Throughout the study, the elbow was kept straight.

Recording Following the orthodromic method [1], segmental mixed nerve conduction studies were performed at the wrist–elbow (ES) and elbow–axilla (AS) segments. For ES segment studies (Fig. 1), the mixed response was recorded at the elbow (at the level of the medial humeral epicondyle) using DANTEC electrodes held firmly with fixation straps. The distance between the active and reference electrodes was 23 mm. For AS segment studies (Figs. 2 and 3), the mixed response was recorded at the axilla (along the medial humeral border) using DANTEC electrodes fixed by straps. The axilla recording position was adjusted to obtain the highest supramaximal response. Twenty averages were made for all responses. The ground (G) electrode position was not specified in the report; the figures show the ground electrode placed on the palm (Fig. 1) and over the forearm (Fig. 2).

Stimulation For ES segment study, the stimulation was performed at the wrist (Fig. 1), while for AS study, the stimulation was performed at the elbow (Fig. 2). The authors [1] used a mean 15 cm distance between stimulation and recording points. Supramaximal responses were recorded with stimulation intensities below 20 mA. The cathode (−) was proximal, and the anode (+) was distal for both ES and AS studies.

Measurements The amplitude of the mixed responses was measured in μV, and ES to AS amplitude ratio was calculated as follows:

G. Gentili, M. Di Napoli, *The Ulnar Nerve*, DOI 10.1007/978-3-319-43292-2_23

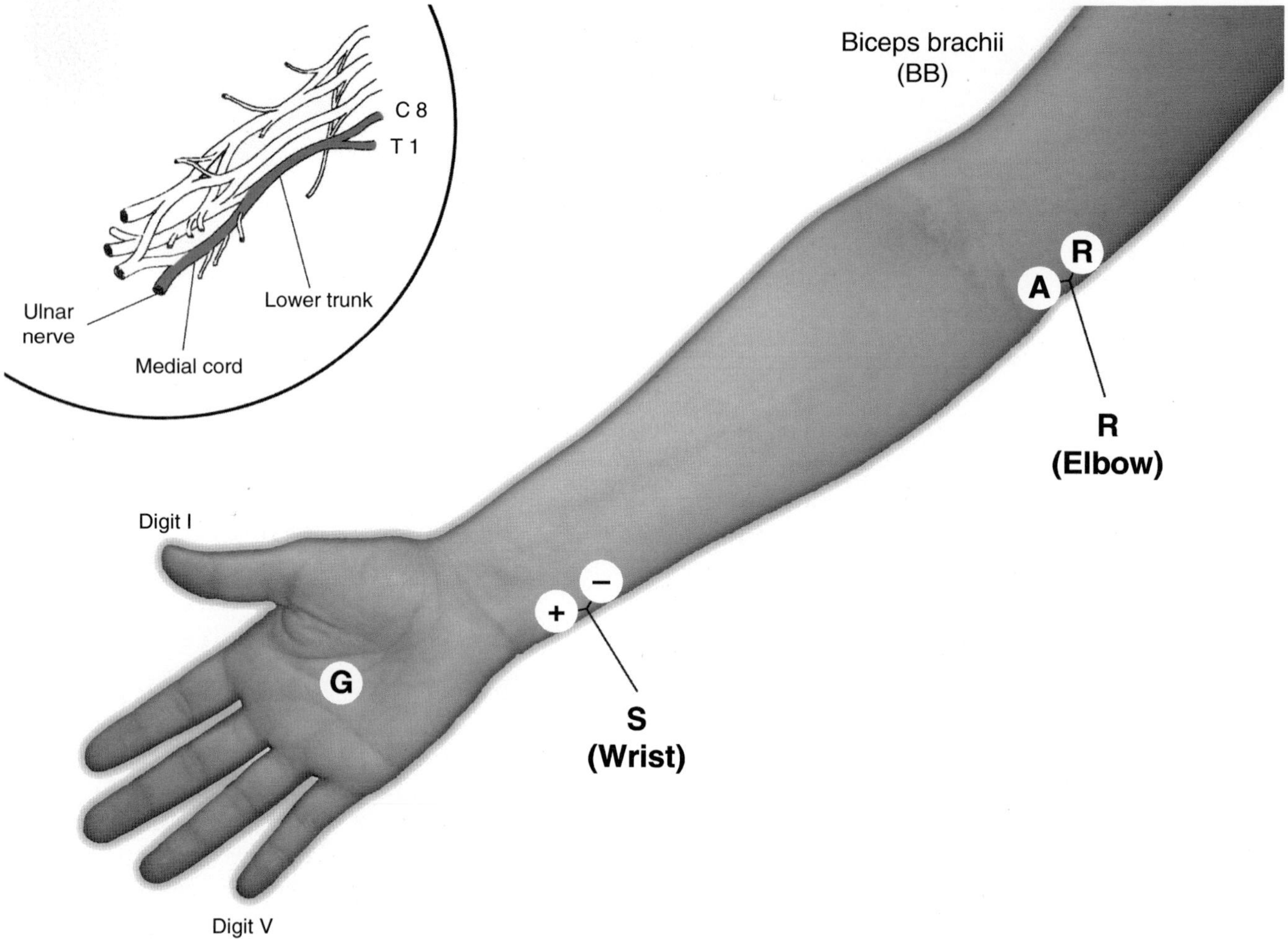

Typical waveform (wrist – elbow):

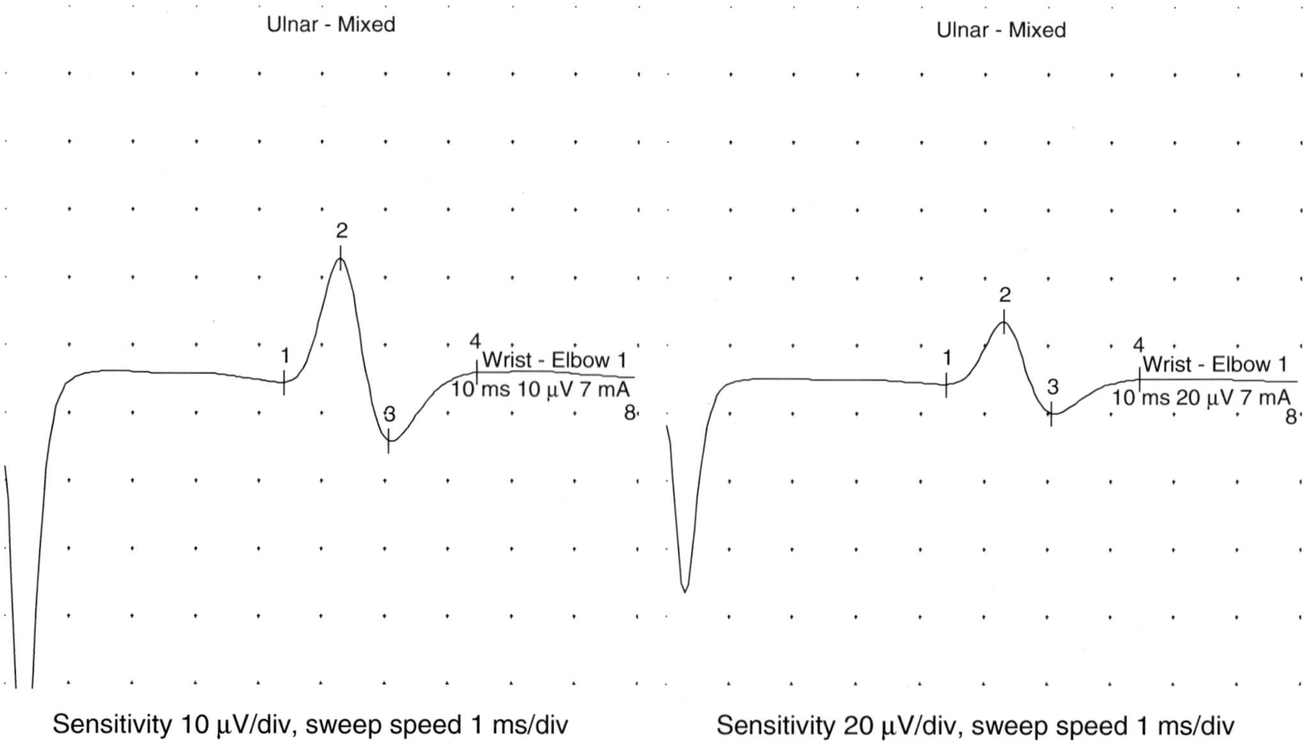

Fig. 1 Mixed nerve action potentials (*MNAPs*) recorded at the elbow, stimulation of the wrist

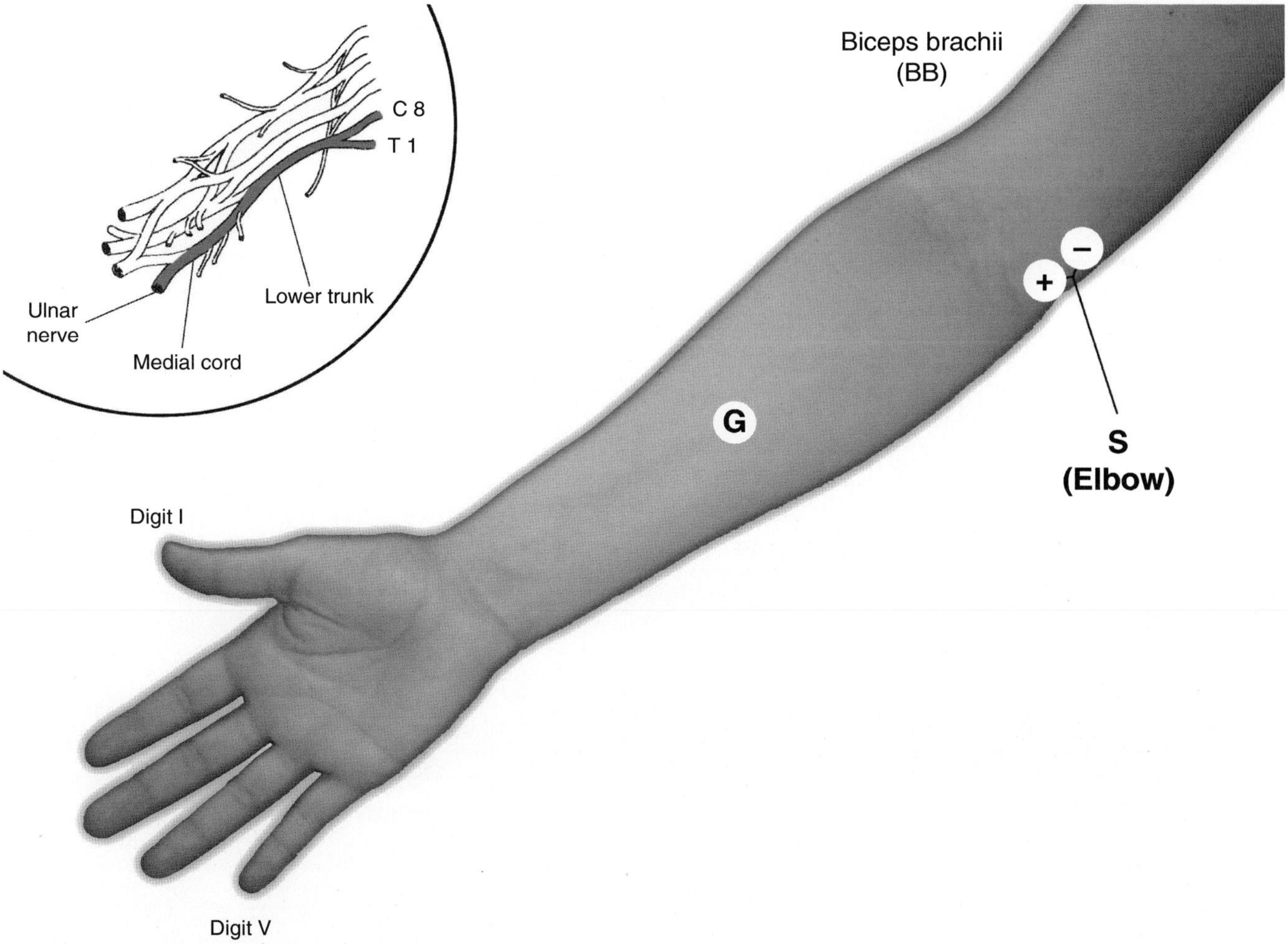

Typical waveform (elbow – axilla):

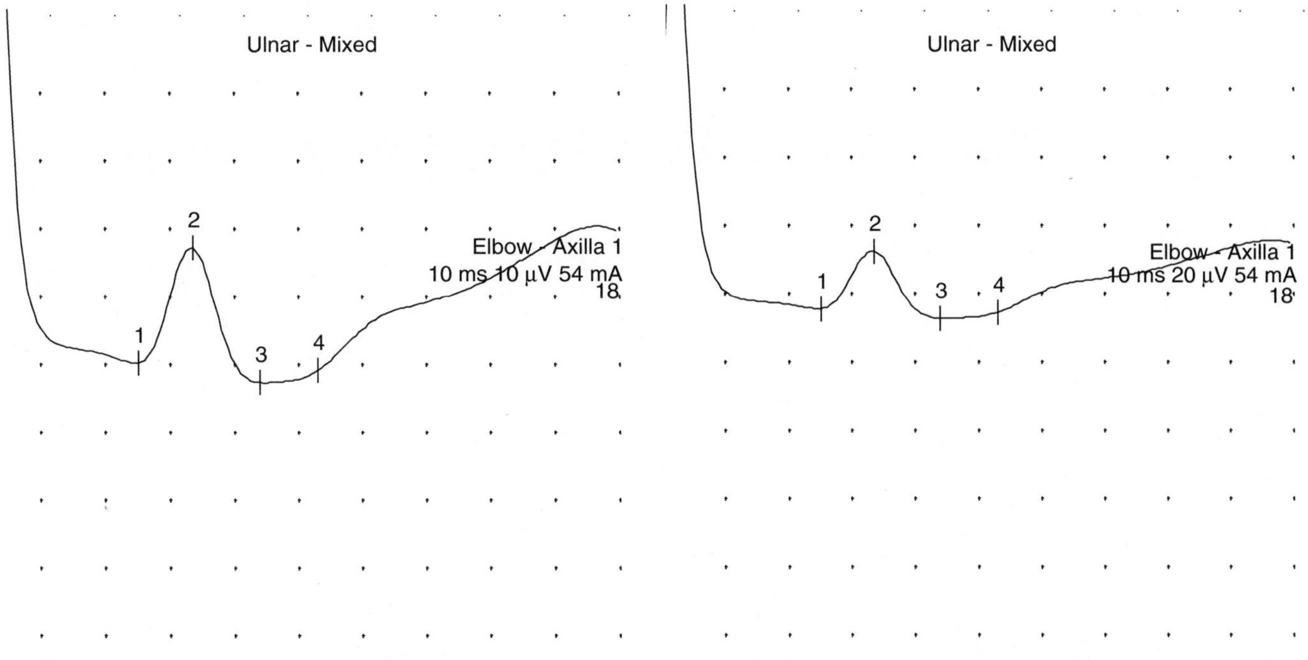

Sensitivity 10 µV/div, sweep speed 1 ms/div

Sensitivity 20 µV/div, sweep speed 1 ms/div

Fig. 2 Mixed nerve action potentials (*MNAPs*) recorded at the axilla, stimulation at the elbow

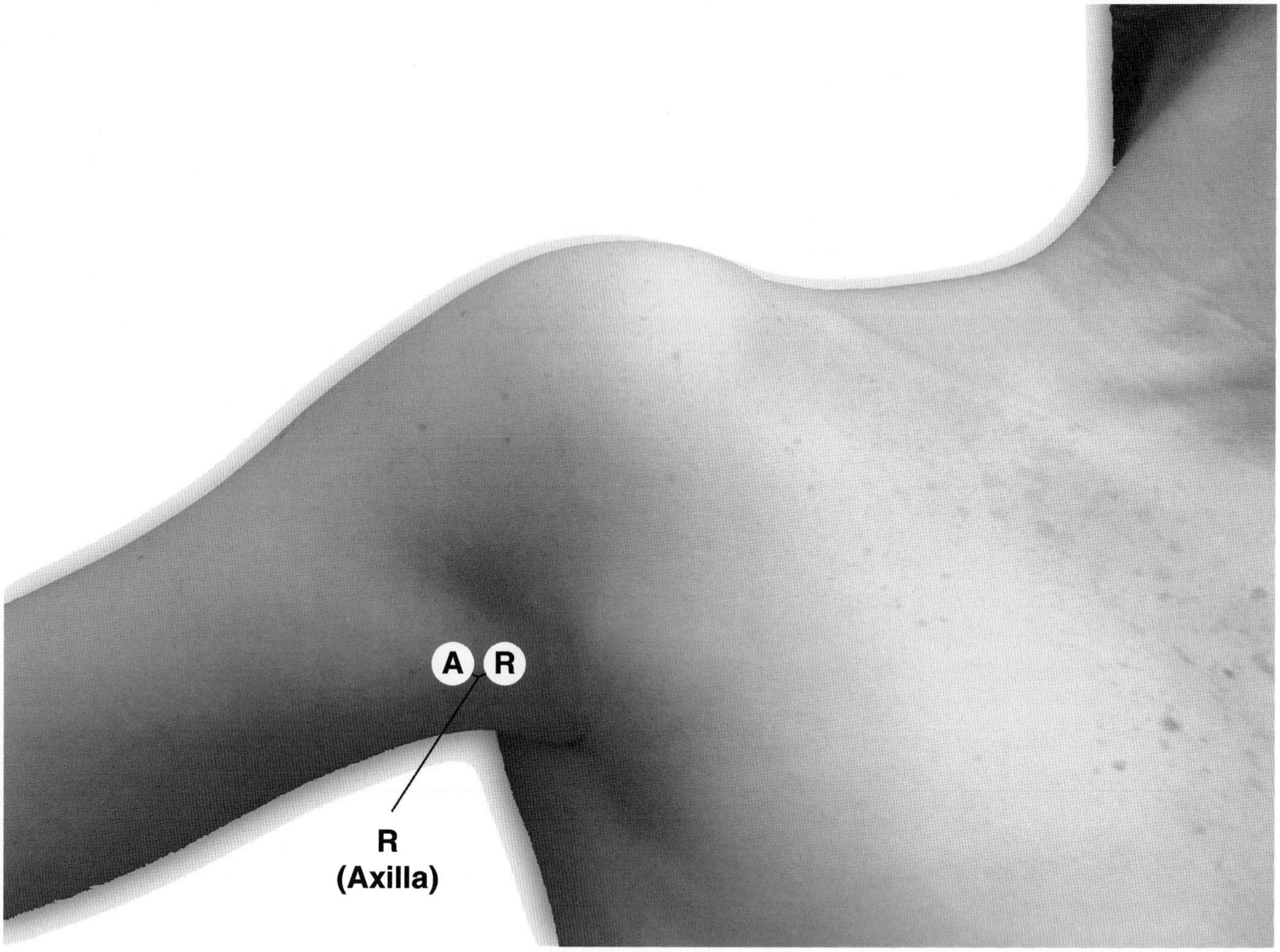

Fig. 3 Elbow–axilla (AS) segment study, ulnar proximal recording (axilla)

$$\text{ES / AS Amp RATIO} = \frac{\text{wrist} - \text{elbow (ES) mixed amplitude } (\mu V)}{\text{elbow} - \text{axilla (AS) mixed amplitude } (\mu V)}$$

Temperatures of the laboratory and skin were not reported. Lo et al. [1] recorded bilaterally MNAPs at the elbow and in the axilla in 20 control subjects (12 men and 8 women, age range 22–72 years). The mean value in the control subjects of ES/AS ratio was 1.06 ± 0.25 (SD), and the limit of normal was 0.56 ± 0.50 (2SD). The authors also studied a small sample of patients (Table 1) with symptoms and signs of ulnar sensory loss and unilateral wasting or weakness of ulnar-innervated muscles (11 patients, 7 men and 4 women, average age 45 years).

Table 1 Pathological values [1]

	Value
PATIENT 1 – wrist–elbow segment/elbow–axilla segment ratio (ES/AS ratio)	9.4/27
PATIENT 2 – wrist–elbow segment/elbow–axilla segment ratio (ES/AS ratio)	5.8/17
PATIENT 3 – wrist–elbow segment/elbow–axilla segment ratio (ES/AS ratio)	0/15
PATIENT 4 – wrist–elbow segment/elbow–axilla segment ratio (ES/AS ratio)	0/8.3
PATIENT 5 – wrist–elbow segment/elbow–axilla segment ratio (ES/AS ratio)	27/53
PATIENT 6 – wrist–elbow segment/elbow–axilla segment ratio (ES/AS ratio)	0/102
PATIENT 7 – wrist–elbow segment/elbow–axilla segment ratio (ES/AS ratio)	0/9.9
PATIENT 8 – wrist–elbow segment/elbow–axilla segment ratio (ES/AS ratio)	5/18
PATIENT 9 – wrist–elbow segment/elbow–axilla segment ratio (ES/AS ratio)	10/30
PATIENT 10 – wrist–elbow segment/elbow–axilla segment ratio (ES/AS ratio)	4.4/22.2
PATIENT 11 – wrist–elbow segment/elbow–axilla segment ratio (ES/AS ratio)	0/0

Comment

Lo et al. [1] based their study on the anatomic and physiologic process related to a severe conduction block at the elbow, resulting in marked secondary axonal degeneration. This kind of lesion results in a severe axon loss of sensory and motor fascicles below the elbow at the wrist–elbow (ES) segment, with relative sparing of the elbow–axilla (AS) segment. In all patients, they observed absent or small-amplitude responses at the ES and distinctly recordable AS potentials, consistent with elbow lesions. All patients had ratios less than 2SDs of the control mean (i.e., <0.56). Routine nerve conduction studies showed (1) unrecordable SNAP recording from the digit V (patients 3, 4, 6, 7, 9, 10, and 11), indicating ulnar sensory axonal neuropathy; (2) absent dorsal ulnar SNAPs (patients 3, 4, 5, and 7), suggesting a lesion of the ulnar nerve above the wrist; and (3) in patients 4 and 9, CMAPs from *abductor digiti minimi* (ADM) muscle which could not be elicited because of long-standing severe axon loss. The authors [1] advocated the use of the ES/AS ratio in addition to routine nerve conduction studies for selected cases, especially when conventional methods or needle electromyography may not be helpful in the diagnosis of an ulnar neuropathy at the elbow (UNE), or a rare high ulnar lesion above the elbow is suspected. Pathological waveform and values in a case (39-year-old male) of diabetes mellitus with peripheral neuropathy are reported (Fig. 4).

Pathological waveform (wrist – elbow, elbow – axilla):

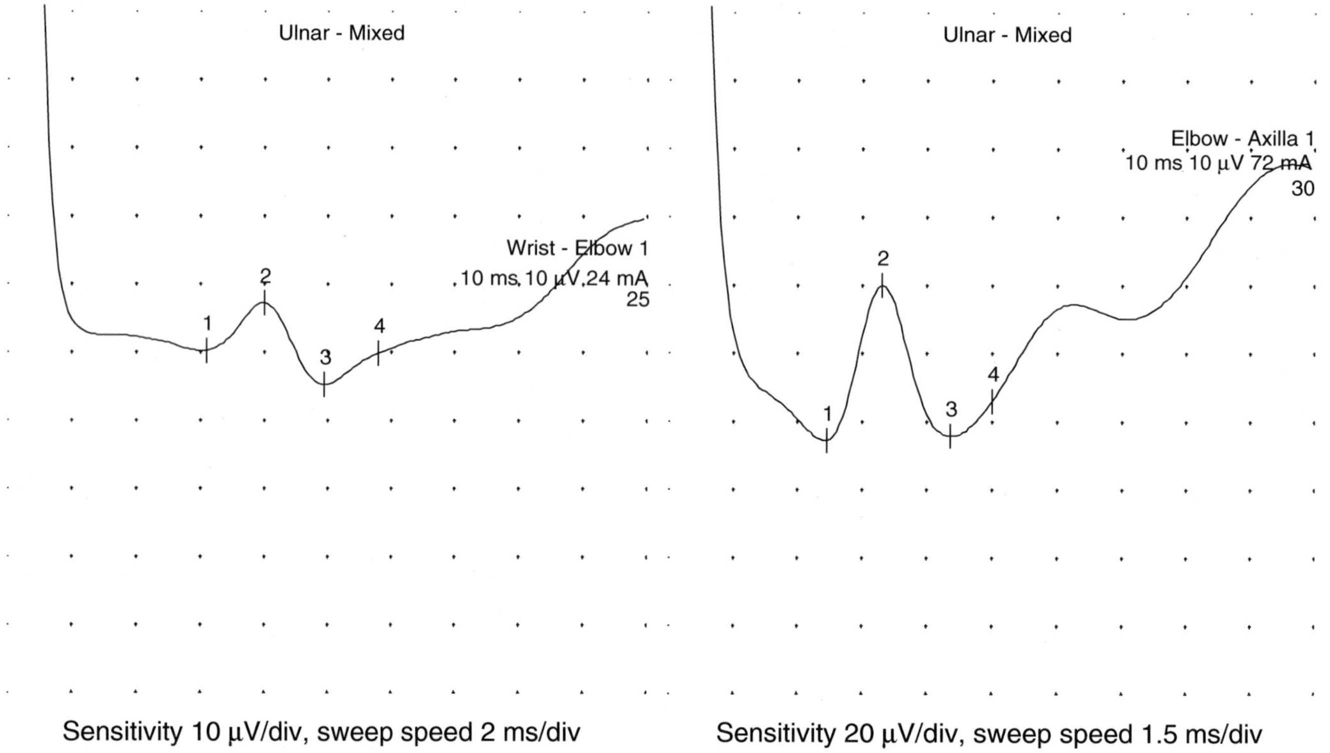

Onset latency (wrist – elbow): 3.10 ms; **Onset latency** (elbow – axilla): 2.45 ms; **Peak latency** (wrist – elbow): 3.30 ms **Peak latency** (elbow – axilla): 3.20 ms; **Onset to peak amplitude** (wrist – elbow): 6.9 µV; **Onset to peak amplitude** (elbow – axilla): 22.8 µV; **Peak to peak amplitude** (wrist – elbow): 11.9 µV; **Peak to peak amplitude** (elbow – axilla): 22.2 µV; **SNCV** (wrist – elbow, 15 cm): 48.4 m/s; **SNCV** (elbow – axilla, 15 cm): 61.2 m/s

Fig. 4 Mixed nerve action potentials (*MNAPs*) in a patient with UNE, recorded at the elbow (*left trace*), stimulation of the wrist and at the axilla (*right trace*), stimulation at the elbow

Reference

1. Lo YL, Dan YF, Lee MP et al (2001) Segmental mixed nerve conduction studies in ulnar neuropathy at the elbowman. J Clin Neurophysiol 18:456–459

Wrist, Below Elbow, Above Elbow – Digit V

S24

Surface Recording Technique, Antidromic Study

Original Settings Sensitivity, sweep speed, low-frequency filter, high-frequency filter, duration, rate of pulse, and the machine used were not specified.

Position The elbow position was kept at 90° flexion.

Recording Following the antidromic method [1], segmental ulnar sensory responses were recorded with a bar electrode placed on the volar aspect of the digit V (Fig. 1). The authors [1] used recording disks with 0.8 cm in diameter and a 3.5 cm interelectrode distance. The ground (G) electrode position was not specified in the text; the figure shows ground electrode placed on the palm.

Stimulation Following the antidromic method [1], the ulnar nerve was stimulated at three sites: at the wrist (S1), 14 cm proximal to the active recording electrode; below the elbow (S2), 2 cm below the medial epicondyle; and above the elbow (S3), 8 cm above the medial epicondyle (Figs. 1 and 2). The anode (+) was proximal in all stimulations.

Measurements Negative-peak amplitude (μV) was measured from the baseline to the negative peak of the sensory nerve action potentials (SNAPs). Peak latency (ms) was measured from the stimulus artifact to the peak of the SNAP. The authors [1] calculated the amplitude ratio between the SNAPs evocated by the below-elbow and wrist (BE–W) stimulations and by the above-elbow and below-elbow (AE–BE) stimulations.

The skin temperature was measured just proximal to the metacarpophalangeal joint of the palmar aspect of the digit V, and it was maintained at or above 32°C. Normal values were obtained from 71 asymptomatic (Tables 1, 2, 3, and 4) subjects (40 women and 31 men, mean age 46 years, age range 24–68 years), and abnormal findings were detected in 28 patients with sensory symptoms in the fourth and fifth digits and/or weakness in the hand intrinsic (mean age 47 years, age range 21–73 years). Twenty-two patients were diagnosed with ulnar neuropathy, and six patients were diagnosed with C8 radiculopathy. Ulnar neuropathy at the elbow (UNE) was diagnosed on the electrodiagnostic findings following the AAEM Quality-Assurance Committee criteria [2]:

1. Absolute motor conduction velocity (MNCV) across the elbow (AE–BE segments) < 50 m/s
2. Sensory nerve conduction velocity (SNCV) in the AE–BE segment more than 10 m/s slower than the below-elbow–wrist (BE–W) segment
3. Decrease of the negative amplitude of the compound muscle action potential (CMAP) from the BE to AE greater than 20 %
4. Significant waveform change in CMAP configuration at the AE site compared with the BE site
5. Amplitude ratio of the SNAP BE–W stimulation < 0.45 or in AE–BE stimulation < 0.65
6. Abnormal EMG spontaneous activities and/or polyphasic or large-amplitude motor unit potentials (MUP) with reduced recruitment pattern in the *flexor carpi ulnaris* (FCU), *abductor digiti minimi* (ADM), and *first dorsal interosseous* (FDI) muscles.

C8 radiculopathy was diagnosed by needle EMG when abnormal spontaneous activities were present in at least two muscles of different nerve innervations but same C8 myotome.

Lee et al. [3] recorded SNAPs at the digit V using ring electrodes with stimulation at 14 cm from the active recording electrode (on the proximal interphalangeal joint of the digit V), 3 cm below the elbow, and 10 cm above the elbow. They studied 100 healthy subjects (58 women and 42 men, mean age 34.7 ± 8.4 years, age range 23–64 years) using a

G. Gentili, M. Di Napoli, *The Ulnar Nerve*, DOI 10.1007/978-3-319-43292-2_24

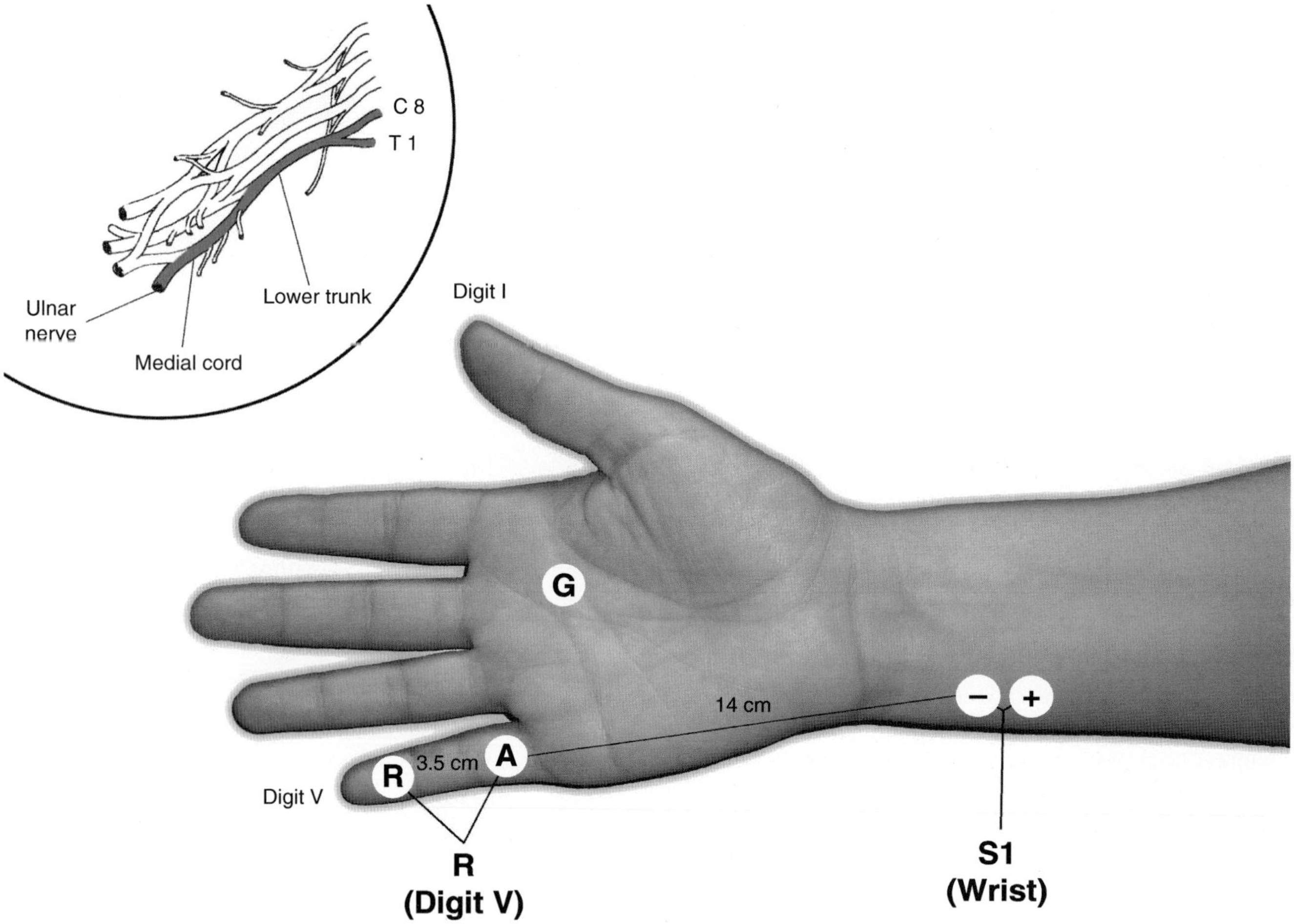

Typical waveform (wrist, below elbow, above elbow – digit V):

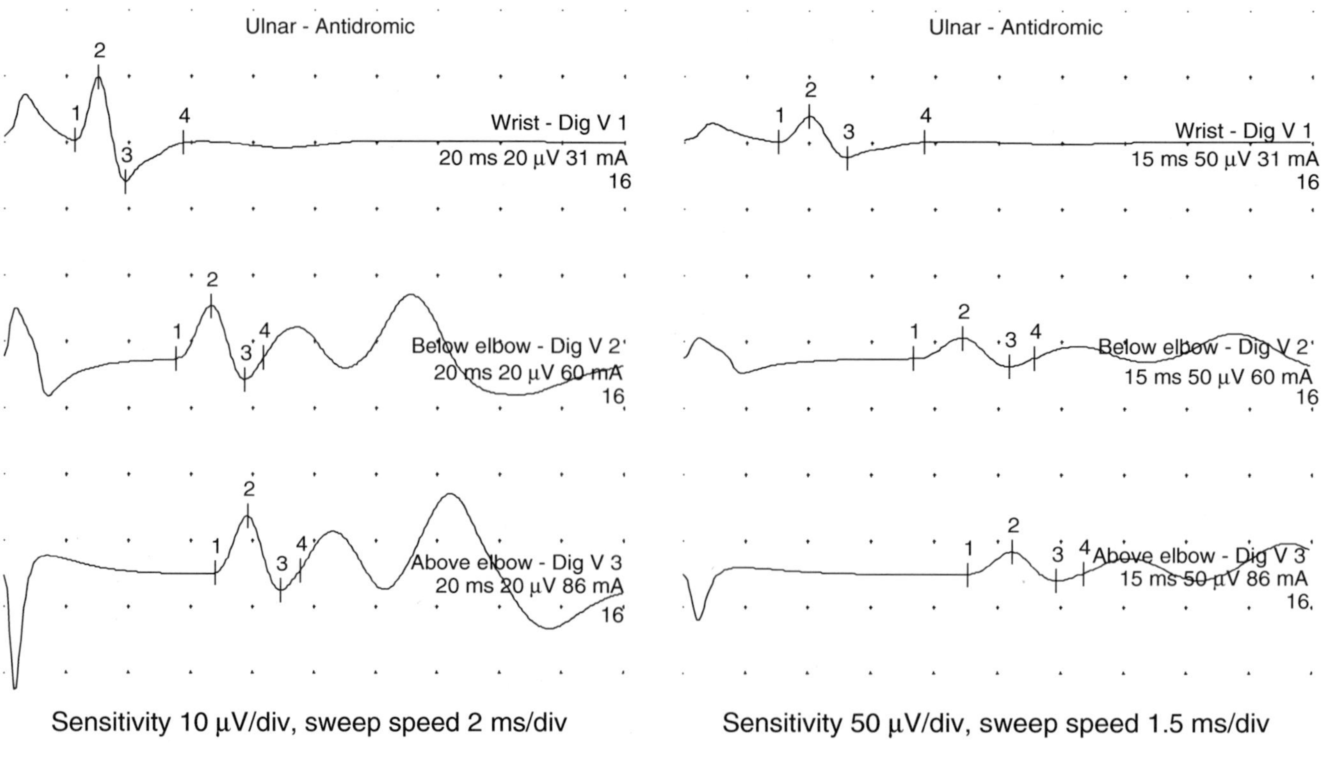

Fig. 1 Antidromic sensory nerve action potentials (*SNAPs*) recorded to digit V, stimulation of the wrist, below the elbow, and above the elbow (*top to bottom*)

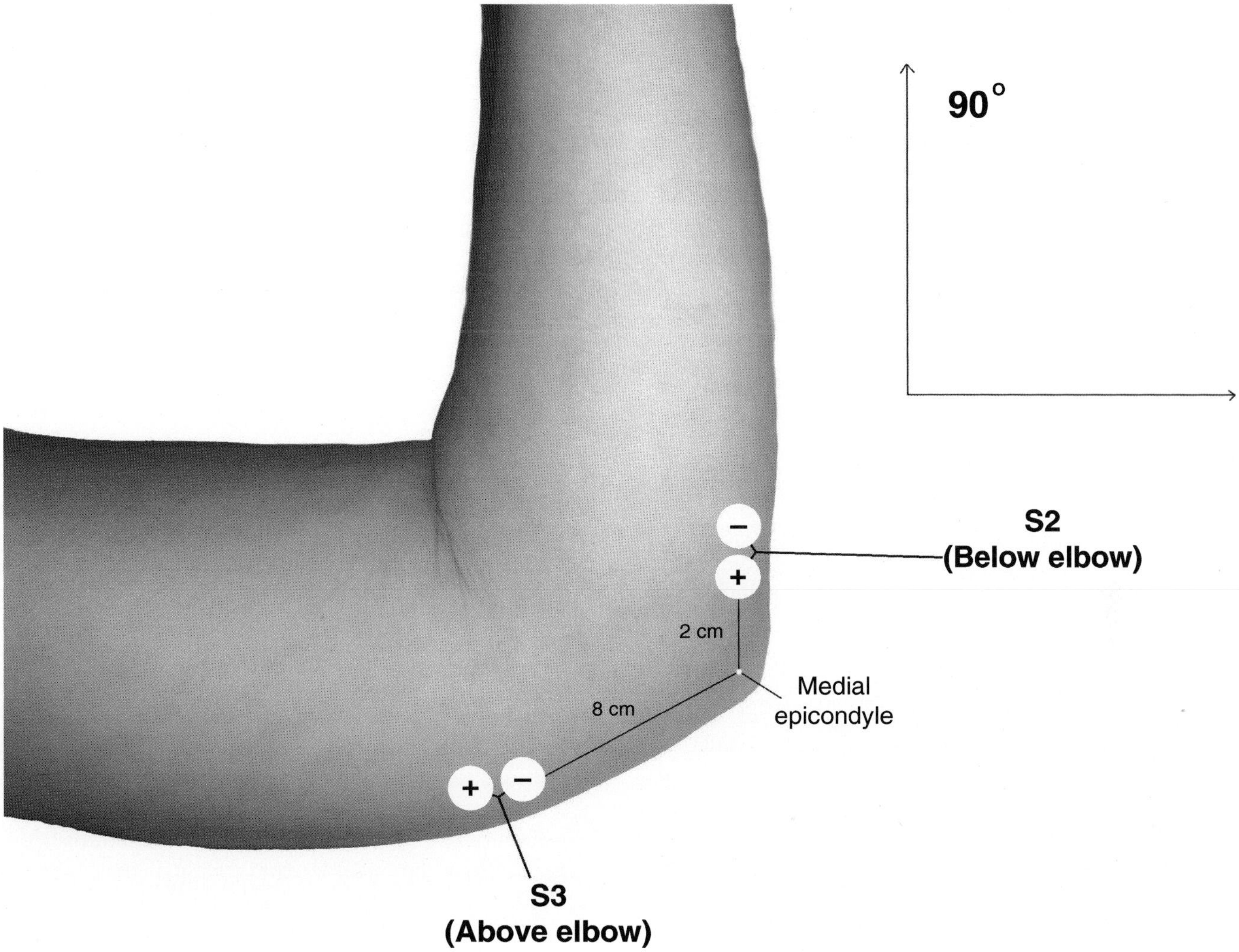

Fig. 2 Below- and above-elbow stimulation points (elbow flexed to 90°)

Table 1 Normal values [1]

	Mean ± SD
Wrist–digit V, SNAP latency (ms) – right arm	3.3 ± 0.3
Wrist–digit V, SNAP latency (ms) – left arm	3.3 ± 0.3
Wrist–digit V, SNAP latency (ms) – both arms	3.3 ± 0.3
Below elbow–digit V, SNAP latency (ms) – right arm	6.1 ± 0.6
Below elbow–digit V, SNAP latency (ms) – left arm	6.1 ± 0.8
Below elbow–digit V, SNAP latency (ms) – both arms	6.1 ± 0.7
Above elbow–digit V, SNAP latency (ms) – right arm	7.6 ± 0.7
Above elbow–digit V, SNAP latency (ms) – left arm	7.6 ± 0.7
Above elbow–digit V, SNAP latency (ms) – both arms	7.6 ± 0.7

Table 2 Normal values [1]

	Mean ± SD
Wrist–digit V, SNAP amplitude (μV) – right arm	38.6 ± 13.6
Wrist–digit V, SNAP amplitude (μV) – left arm	38.9 ± 13.7
Wrist–digit V, SNAP amplitude (μV) – both arms	38.7 ± 13.6
Below elbow–digit V, SNAP amplitude (μV) – right arm	23.7 ± 9.5
Below elbow–digit V, SNAP amplitude (μV) – left arm	24.2 ± 9.8
Below elbow–digit V, SNAP amplitude (μV) – both arms	23.9 ± 9.6
Above elbow–digit V, SNAP amplitude (μV) – right arm	19.1 ± 7.5
Above elbow–digit V, SNAP amplitude (μV) – left arm	19.7 ± 8.0
Above elbow–digit V, SNAP amplitude (μV) – both arms	19.4 ± 7.8

Table 3 Normal values [1]

	Mean ± SD	Limit
Below elbow–wrist, SNAP amplitude ratio – right arm	0.61 ± 0.08	
Below elbow–wrist, SNAP amplitude ratio – left arm	0.61 ± 0.08	
Below elbow–wrist, SNAP amplitude ratio – both arms	0.61 ± 0.08	0.45
Above elbow–below elbow, SNAP amplitude ratio – right arm	0.81 ± 0.08	
Above elbow–below elbow, SNAP amplitude ratio – left arm	0.81 ± 0.09	
Above elbow–below elbow, SNAP amplitude ratio – both arms	0.82 ± 0.08	0.65

Table 4 Normal values [1]

	Mean ± SD
Below elbow–wrist, SNAP SNCV (m/s) – right arm	67.3 ± 5.6
Below elbow–wrist, SNAP SNCV (m/s) – left arm	67.4 ± 5.3
Below elbow–wrist, SNAP SNCV (m/s) – both arms	67.4 ± 5.4
Above elbow–below elbow, SNAP SNCV (m/s) – right arm	71.2 ± 5.4
Above elbow–below elbow, SNAP SNCV (m/s) – left arm	71.1 ± 5.6
Above elbow–below elbow, SNAP SNCV (m/s) – both arms	71.2 ± 5.5

Comment The authors [1] categorized the electrophysiologic findings of UNE into four types:

Type 1: normal CMAP and SNAP by wrist stimulation and CMAP/SNAP amplitude reduction and/or decreased MNCV and SNCV across the elbow (five patients). Needle EMG revealed no abnormal spontaneous activity at rest and polyphasic or large-amplitude MUPs with reduced recruitment pattern.

Type 2: normal CMAP and low-amplitude SNAP by wrist stimulation and CMAP/SNAP amplitude reduction and/or decreased MNCV and SNCV across the elbow (one patient). Needle EMG revealed abnormal spontaneous activity at rest and polyphasic or large-amplitude MUPs with reduced recruitment pattern.

Type 3: low-amplitude CMAP and low-amplitude or absent SNAP by wrist stimulation and amplitude reduction of CMAP and/or decreased MNCV across the elbow (11 patients). Needle EMG revealed abnormal spontaneous activity at rest and polyphasic or large-amplitude MUPs with discrete recruitment pattern.

Type 4: normal CMAP and SNAP with wrist stimulation and only amplitude reduction of the SNAP across the elbow without change of CMAP. Needle EMG showed normal findings (five patients).

In these five patients, Kwon et al. [1] observed abnormal sensory nerve action potential amplitude ratio only: the amplitude ratios of BE–W (two patients) and AE–BE (three patients) were decreased in each lesion in the symptomatic side, whereas the amplitude ratios were normal in the asymptomatic side. Ultrasound examination in the patients of Type IV revealed ulnar nerve swelling at the retroepicondyle level (three patients – low AE–BE amplitude ratio) and at the mid- to proximal forearm (two patients – low-amplitude BE–W ratio). For the authors, the amplitude ratios of ulnar SNAPs among the wrist and elbow (BE–W and AE–BE ratios) may be useful in the diagnosis of ulnar neuropathy with only sensory symptoms and normal segmental motor conduction findings. Because the authors found abnormal SNAP amplitude ratios in only five patients – Type IV (22.7 %) of the 22 electrophysiologically diagnosed UNE patients – they suggested further investigation to confirm the hypothesis that a different pathophysiologic mechanism involves the sensory fibers of the ulnar nerve in the across-elbow segment.

TECA Synergy electromyograph for nerve conduction studies. Only SNAP peak latency with above-elbow stimulation was reported (mean 8.3 ± 0.7 ms, range 6.6–9.7 cm), and it was measured with a sensitivity of 5 μV/division. The skin temperature was maintained at 32°C or higher on the palm and at the elbow (retrocondylar groove) using hydrocolloid packs and a heating lamp in a temperature-controlled room. In all subjects, in addition to height (mean 167.5 ± 8.9 cm, range 147–189 cm), finger-to-arm length (mean 47.3 cm, range 42.5–54.0 cm) was measured from the proximal interphalangeal joint at the digit V to the above-elbow stimulation site with a 90° elbow flexion in each subject (Fig. 3). Elbow segment was measured from 3 cm distal to midpoint of medial epicondyle and olecranon to 10 cm proximal along the intermuscular septum toward midaxilla.

Their study reported the reference values of sensory nerve latencies at 10 cm above-elbow stimulation and a formula predicting the latencies based on finger-to-arm length, as follows:

$$\text{SNAP peak latency}(\text{ms}) = 0.46 + 1.6 \times (\text{finger} - \text{to} - \text{arm length, cm})$$

They proposed their test as a complementary parameter to self-assess the validity of segmental conduction velocity in subjects with suspected UNE, since the AAEM practice guidelines for the diagnosis of UNE are (1) MNCV across the elbow < 50 m/s, (2) differential slowing of more than 10 m/s across the elbow than the forearm segment, and (3) change of configuration and reduction in CMAP negative-peak amplitude (mV) of more than 20 % across the elbow [2]. Among these criteria, the MNCV slowing across the elbow showed the highest sensitivity than the other methods, such as the differential slowing between across-elbow (below elbow–above elbow) and forearm (wrist–below-elbow segment) segments [4], but some authors [5] found distance measurement errors more common in the across-elbow than in the forearm measurement because of the shorter distance

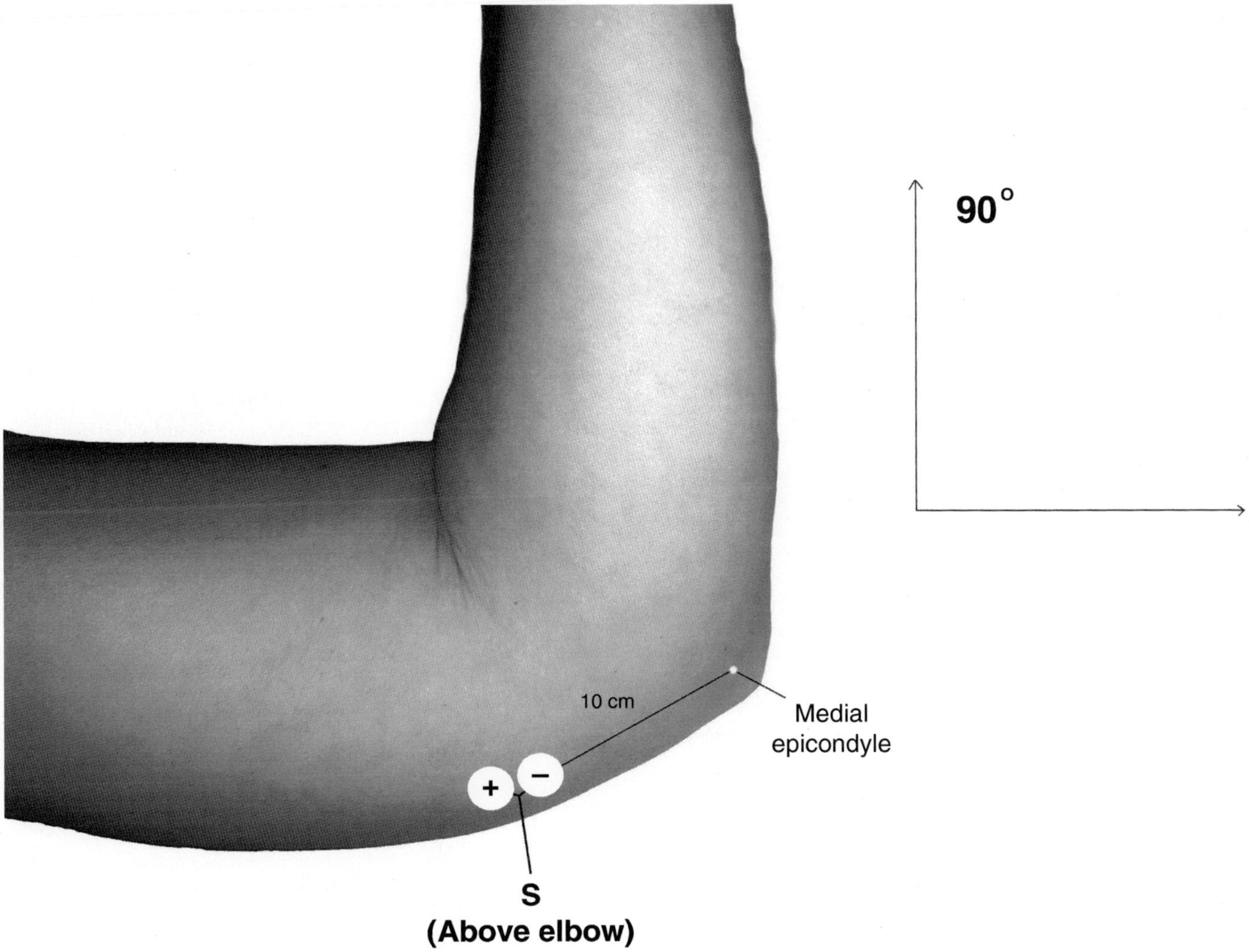

Fig. 3 Above-elbow stimulation point (elbow flexed to 90°)

and position of the elbow. The measurement errors alone can explain conduction velocity differences of up to 14 m/s between the forearm and the across-elbow segments. Lee et al. [3] provided the formula for predicted norms of SNAP latency at above-elbow stimulation based on the finger-to-arm length. In terms of cutoff latency of above-elbow stimulation, the upper limit of 95 % confidence interval in sensory latencies, approximately 1 ms longer than the predicted value, may be applied using a prediction formula based on the finger-to-arm length. Alternatively, applying the absolute cutoff value, 2SDs above the mean (9.7 ms), can be done. If the latency at 10 cm above-elbow stimulation is longer than 2SDs above the mean or 1 ms longer than the predicted value, then the slow SNCV obtained across the elbow is likely a valid value. However, the cutoff value of 2SDs above the mean was an empirical value, and they did not report the specificity or predictive value of this cutoff latency because of the small sample of subjects with suspected UNE (14 patients). For widespread use of the cutoff latencies, the authors [3] suggested to validate the cutoff values in a larger healthy population (the study did not include subjects older than 65 years) and a larger sample of patients with UNE. Pathological waveform and values in a patient with ulnar neuropathy at the elbow (UNE) are reported (Fig. 4).

Pathological waveform (wrist, below elbow, above elbow – digit V):

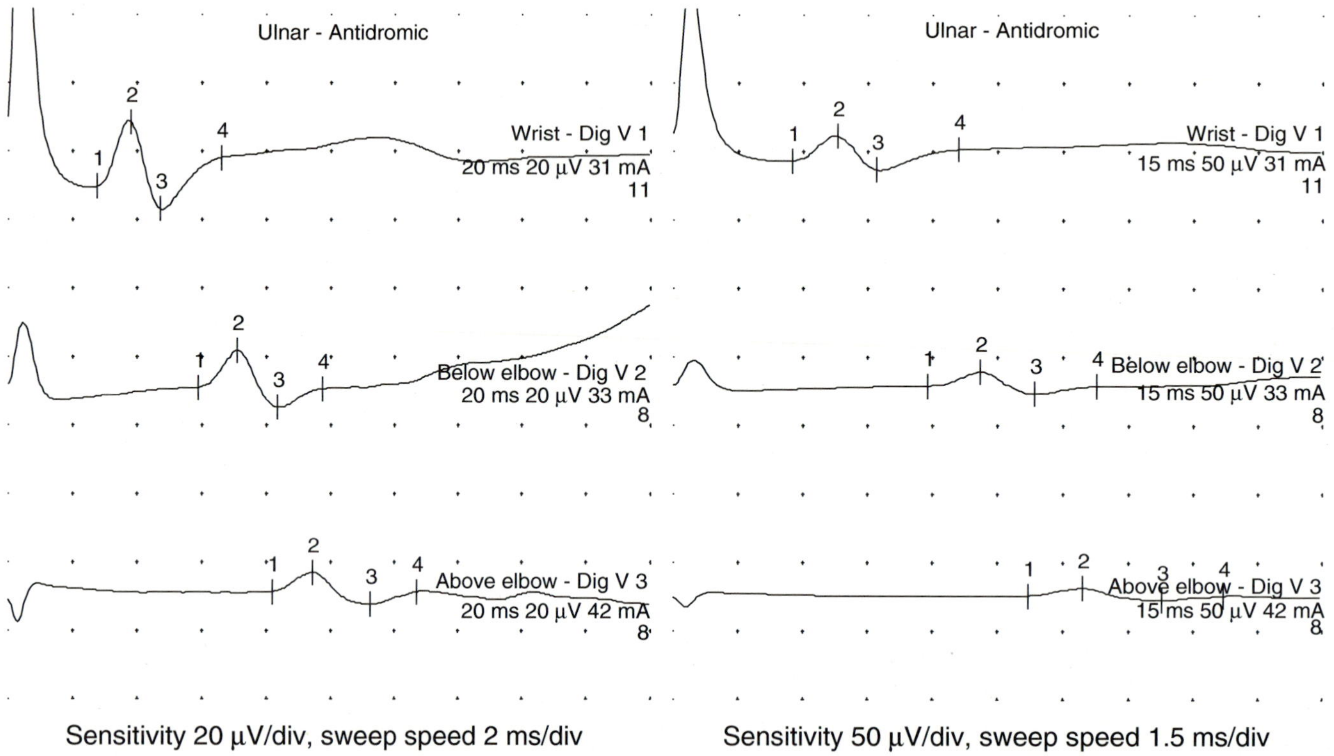

Onset latency (wrist – digit V): 2.75 ms; **Onset latency** (below elbow – digit V): 5.90 ms; **Onset latency** (above elbow – digit V): 8.20 ms; **Peak latency** (wrist – digit V): 4.45 ms; **Peak latency** (below elbow – digit V): 7.15 ms; **Peak latency** (above elbow – digit V): 9.20 ms; **Onset to peak amplitude** (wrist – digit V): 19.2 µV; **Onset to peak amplitude** (below elbow – digit V): 11.1 µV; **Onset to peak amplitude** (above elbow – digit V): 5.5 µV; **Peak to peak amplitude** (wrist – digit V): 25.8 µV; **Peak to peak amplitude** (below elbow – digit V): 16.9 µV; **Peak to peak amplitude** (above elbow – digit V): 9.6 µV; **BE/W amplitude ratio** (below elbow/wrist, 11.1/19.2): 0.58; **AE/BE amplitude** (above elbow/below elbow, 5.5/11.1): 0.49; **SNCV** (wrist – digit V, 18cm distance): 57.1 m/s; **SNCV** (below elbow – above elbow, 9 cm distance): 39.1 m/s

Fig. 4 Antidromic sensory nerve action potentials (*SNAPs*) recorded to digit V, stimulation of the wrist (*upper trace*), below the elbow (*middle trace*), and above the elbow (*lower trace*) in a patient with UNE

References

1. Kwon HK, Lee HJ, Hwang M et al (2008) Amplitude ratio of ulnar sensory nerve action potentials in segmental conduction study. Am J Phys Med Rehabil 87:642–646
2. AAEM Quality Assurance Committee (1999) Practice parameter for electrodiagnostic studies in ulnar neuropathy at the elbow: summary statement. Muscle Nerve 22:408–411
3. Lee SW, Kim DDJ, Pedersen AL et al (2011) Ulnar nerve latency at above elbow stimulation. A complementary tool for segmental conduction velocity across the elbow. Am J Phys Med Rehabil 90:302–307
4. Shakir A, Micklesen P, Robinson L (2004) Which motor nerve conduction study is best in ulnar neuropathy at the elbow? Muscle Nerve 29:585–590
5. Landau ME, Diaz MI, Barner KC et al (2002) Changes in nerve conduction velocity across the elbow due to experimental error. Muscle Nerve 26:838–840

Part II

Motor Conduction Studies

Wrist, Elbow, Axilla – Hand

M1

Surface Recording Technique, Study from *Abductor Digiti Minimi* (ADM) Muscle

Original Settings Duration of pulse was 0.1 ms; sensitivity, sweep speed, low-frequency filter, high-frequency filter, and the machine used were not specified.

Position The authors did not specified the arm position; we performed the study in the supine position, with the subject comfortably seated on a chair.

Recording Compound muscle action potentials (CMAPs) were recorded percutaneously from the *abductor digiti minimi* (ADM) muscle by surface electrodes (Fig. 1). The active (A) electrode was placed over the belly of the ADM muscle; the reference (R) was placed distally over the tendon of the ADM muscle [1]. Distances between the recording and stimulating electrodes were not fixed. The ground (G) electrode position was not specified in the text; the figure shows the ground electrode placed on the forearm.

Stimulation Electrical stimuli (supramaximal square wave shocks of 0.1 ms duration) were applied to the left arm over three separate points of the ulnar nerve [1]: at the wrist (S1), at the elbow (S2), and 5 cm below the axilla (S3). The stimulating cathode (-) was placed proximal to the recording site; the anode (+) was positioned distally (Figs. 1 and 2).

Measurements Distal latency (ms) was measured from the stimulus artifact to the onset of the CMAP, and it was measured to the nearest 0.1 ms. Maximum motor nerve conduction velocity (MNCV) of the wrist–elbow and elbow–axilla segments was calculated as the distance between two appropriate stimulating electrodes, divided by the difference in their CMAP latencies from each point. The maximum MNCV, so calculated, is actually a mean value for the fastest conducting fibers of the ulnar nerve trunk. The velocity estimate is in part dependent on the uniformity of the response of the neuromuscular system between the stimulating electrode (wrist) and the pickup electrode (hand). To test the uniformity of the neuromuscular system, residual latency (RL) was calculated as the difference between the actual latency following stimulation 5 cm below the axilla and the latency derived by assuming that the impulse was conducted to the pickup electrode at the average velocity of conduction in the axilla–wrist segment. The authors [1] obtained normal values (Table 1) from 175 healthy male subjects (ambulatory, employees, and staff members of the Baltimore City Hospitals). There were 25 subjects in each age decade from the third to the ninth. Limb and room temperatures were not controlled.

Comment

Norris et al. [1] observed that the MNCV decreased with increasing age in all segments of nerve measured, and the variability of the estimate of velocity increased with decreasing length of the segment measured. Furthermore, no significant difference was found between MNCV measured in the elbow–axilla segment and those measured in the wrist–elbow segment. Residual latency showed no change with age. Pathological waves and values in patients with ulnar neuropathy at the elbow (UNE) are reported (Figs. 3 and 4).

G. Gentili, M. Di Napoli, *The Ulnar Nerve*, DOI 10.1007/978-3-319-43292-2_25

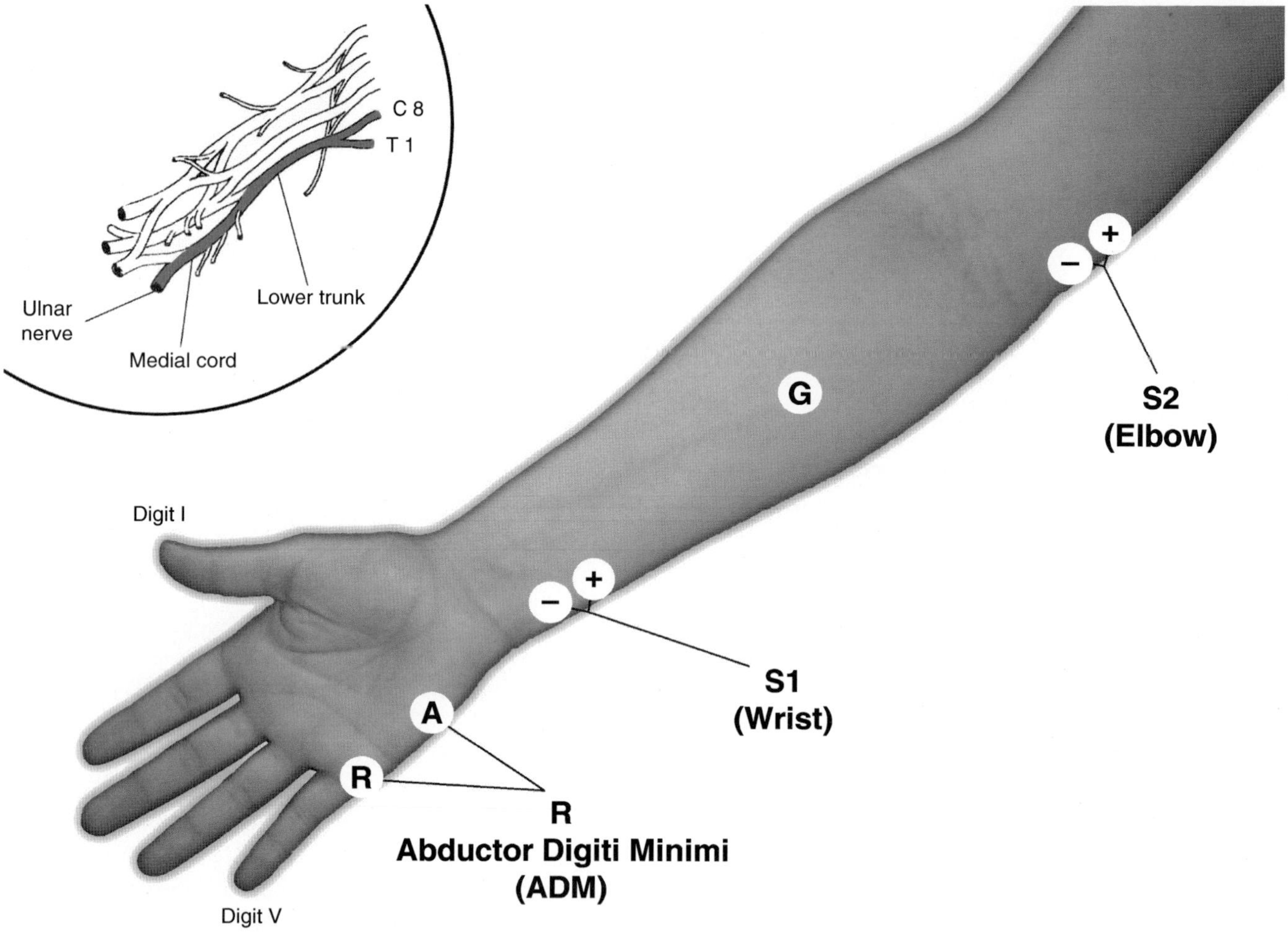

Typical waveform (wrist, elbow, axilla – ADM muscle):

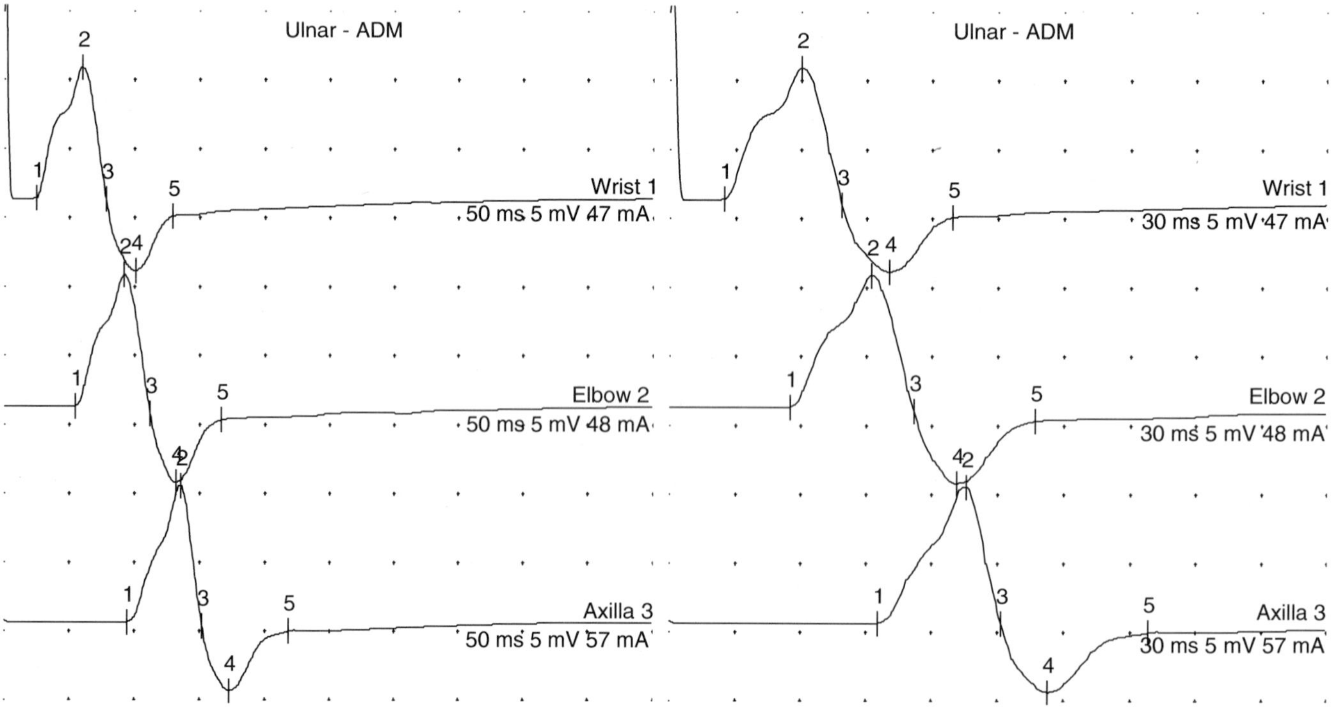

Sensitivity 5 mV/div, sweep speed 5 ms/div

Sensitivity 5 mV/div, sweep speed 3 ms/div

Fig. 1 Compound muscle action potentials (*CMAPs*) recorded at the hand from the ADM muscle, stimulation of the wrist (*upper trace*), elbow (*middle trace*), and axilla (*lower trace*)

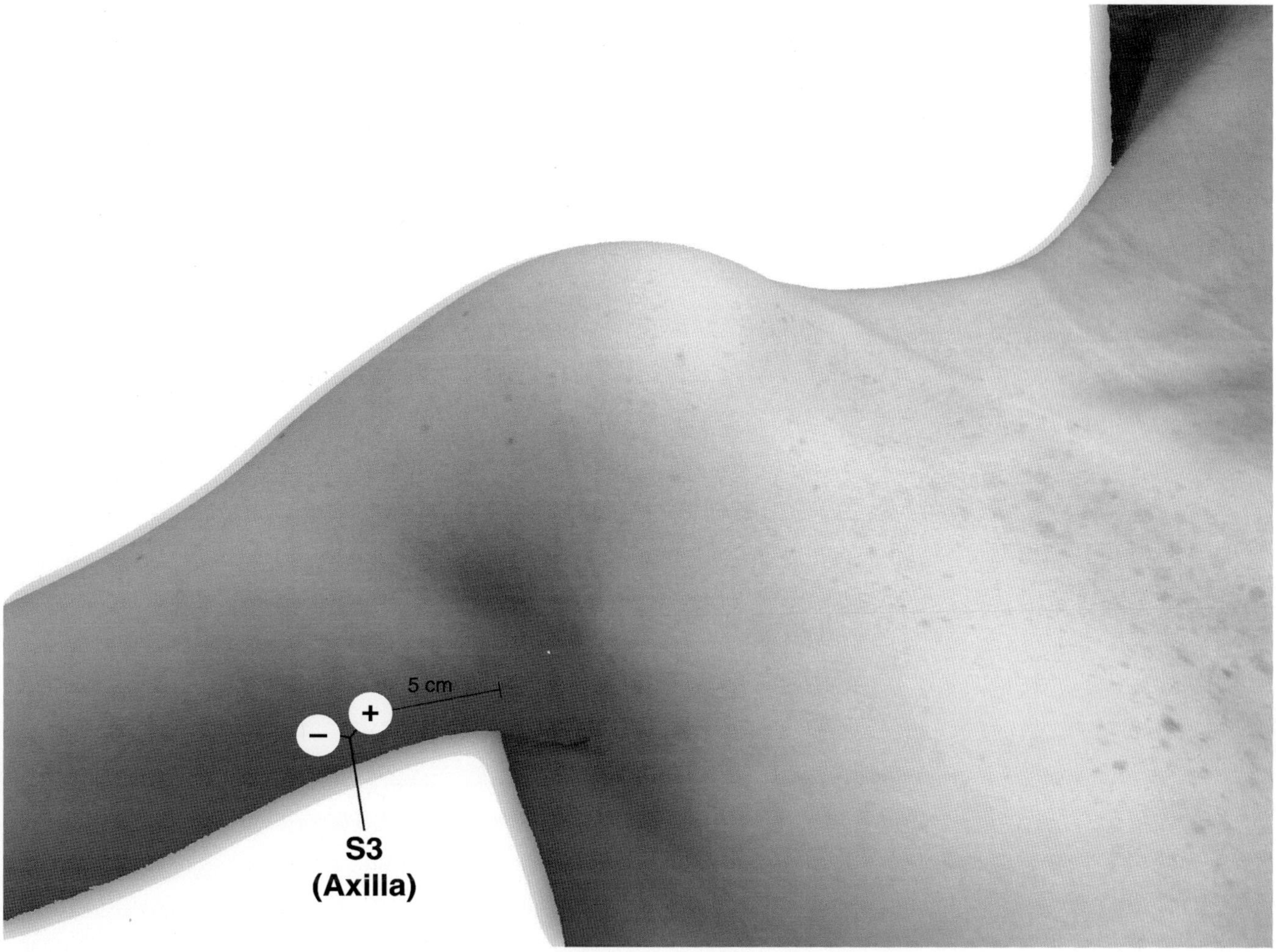

Fig. 2 Proximal stimulation point (S3 – axilla)

Table 1 Normal values (1)

	Mean ± SD
Wrist–elbow, MNCV (m/s) – 20–29 years	58.1 ± 2.8
Wrist–axilla, MNCV (m/s) – 20–29 years	57.7 ± 2.1
Elbow–axilla, MNCV (m/s) – 20–29 years	57.9 ± 3.0
Residual latency (RL) – 20–29 years	1.7 ± 0.1
Wrist–elbow, MNCV (m/s) – 30–39 years	59.9 ± 1.8
Wrist–axilla, MNCV (m/s) – 30–39 years	58.8 ± 1.4
Elbow–axilla, MNCV (m/s) – 30–39 years	57.7 ± 2.2
Residual latency (RL) – 30–39 years	1.9 ± 0.1
Wrist–elbow, MNCV (m/s) – 40–49 years	55.9 ± 2.6
Wrist–axilla, MNCV (m/s) – 40–49 years	55.6 ± 2.0
Elbow–axilla, MNCV (m/s) – 40–49 years	56.6 ± 3.6
Residual latency (RL) – 40–49 years	1.8 ± 0.1
Wrist–elbow, MNCV (m/s) – 50–59 years	54.2 ± 2.6
Wrist–axilla, MNCV (m/s) – 50–59 years	53.6 ± 1.7
Elbow–axilla, MNCV (m/s) – 50–59 years	53.2 ± 2.4
Residual latency (RL) – 50–59 years	1.9 ± 0.1
Wrist–elbow, MNCV (m/s) – 60–69 years	51.7 ± 2.3
Wrist–axilla, MNCV (m/s) – 60–69 years	52.0 ± 1.7
Elbow–axilla, MNCV (m/s) – 60–69 years	53.0 ± 2.8
Residual latency (RL) – 60–69 years	1.8 ± 0.1
Wrist–elbow, MNCV (m/s) – 70–79 years	53.0 ± 2.4
Wrist–axilla, MNCV (m/s) – 70–79 years	52.6 ± 2.1
Elbow–axilla, MNCV (m/s) – 70–79 years	52.7 ± 2.3
Residual latency (RL) – 70–79 years	1.7 ± 0.1
Wrist–elbow, MNCV (m/s) – 80–89 years	48.9 ± 2.9
Wrist–axilla, MNCV (m/s) – 80–89 years	48.5 ± 1.8
Elbow–axilla, MNCV (m/s) – 80–89 years	48.6 ± 1.9
Residual latency (RL) – 80–89 years	2.0 ± 0.1

Pathological waveform (wrist, elbow, axilla – ADM muscle):

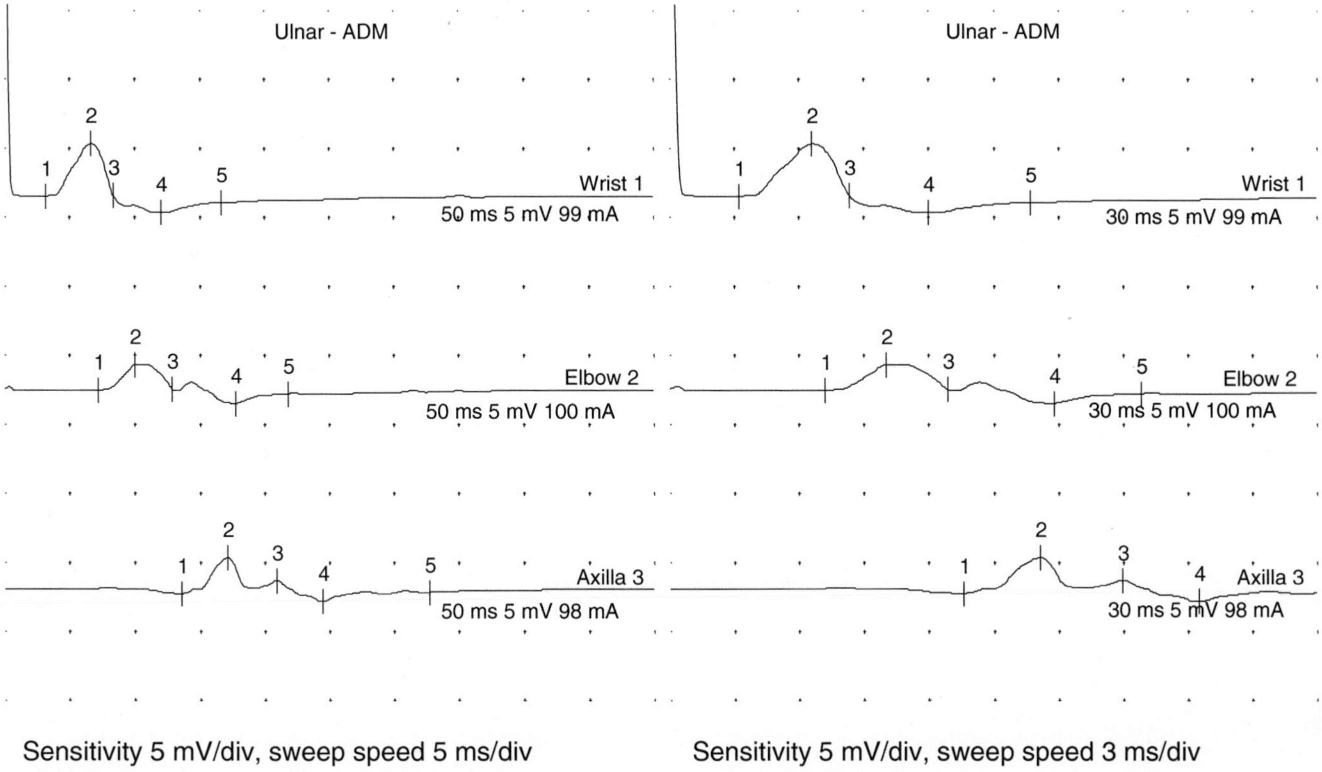

Onset latency (wrist– ADM): 3.20 ms; **Onset latency** (elbow – ADM): 7.20 ms; **Onset latency** (axilla – ADM): 13.55 ms; **Peak latency** (wrist – ADM): 6.90 ms; **Peak latency** (elbow – ADM): 10.15 ms; **Peak latency** (axilla – ADM): 16.75 ms; **Onset to peak amplitude** (wrist – ADM): 3.9 mV; **Onset to peak amplitude** (elbow – ADM): 1.9 mV; **Onset to peak amplitude** (axilla – ADM): 2.7 mV; **Peak to peak amplitude** (wrist – ADM): 5.1 mV; **Peak to peak amplitude** (elbow – ADM): 2.9 mV; **Peak to peak amplitude** (axilla – ADM): 3.3 mV; **MNCV** (wrist – elbow, 19 cm distance): 47.5 m/s; **MNCV** (elbow – axilla, 22 cm distance): 34.6 m/s

Fig. 3 Compound muscle action potentials (*CMAPs*) recorded from the ADM muscle using surface electrodes [1] in a patient with diabetic polyneuropathy and severe UNE, stimulation of the wrist (*upper trace*), elbow (*middle trace*), and axilla (*lower trace*)

Pathological waveform (wrist, elbow, axilla – ADM muscle):

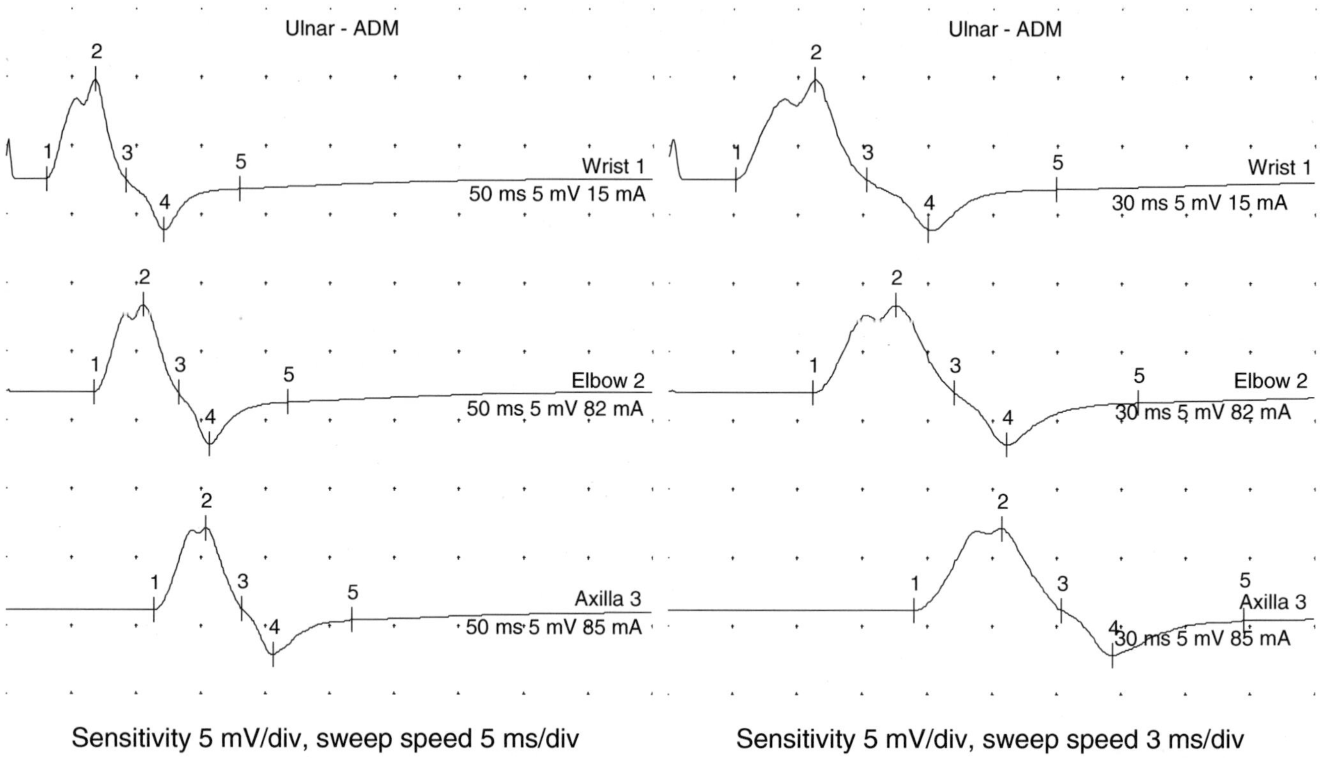

Onset latency (wrist – ADM): 3.10 ms; **Onset latency** (elbow – ADM): 6.70 ms; **Onset latency** (axilla – ADM): 11.40 ms; **Peak latency** (wrist – ADM): 6.85 ms; **Peak latency** (elbow – ADM): 10.45 ms; **Peak latency** (axilla – ADM): 15.25 ms; **Onset to peak amplitude** (wrist – ADM): 7.3 mV; **Onset to peak amplitude** (elbow – ADM): 6.4 mV; **Onset to peak amplitude** (axilla – ADM): 6.0 mV; **Peak to peak amplitude** (wrist – ADM): 11.0 mV; **Peak to peak amplitude** (elbow – ADM): 10.2 mV; **Peak to peak amplitude** (axilla – ADM): 9.3 mV; **MNCV** (wrist – elbow, 20 cm distance): 55.6 m/s; **MNCV** (elbow – axilla, 20 cm distance): 42.6 m/s

Fig. 4 Compound muscle action potentials (*CMAPs*) recorded from the ADM muscle using surface electrodes [1] in a patient with UNE, stimulation of the wrist (*upper trace*), elbow (*middle trace*), and axilla (*lower trace*)

Reference

1. Norris AH, Shock NW, Wagman IH (1953) Age changes in the maximum conduction velocity of the motor fibers of human ulnar nerves. J Appl Physiol 5:589–593

Elbow, Wrist – Hand

M2

Surface Recording Technique, Study from *Abductor Digiti Minimi* (ADM) Muscle

Original Settings Sensitivity was 5 mV/division; sweep speed, low-frequency filter, high-frequency filter, duration of pulse, and the machine used were not specified.

Position This study was performed in the supine position, with the subject lying on a comfortable couch with the arm slightly abducted and supported at three or four places by broad webbing straps hung from the beam.

Recording Motor responses were recorded from the *abductor digiti minimi* (ADM) muscle by surface electrodes (Fig. 1). The author used chloride-coated silver plates 16 by 24 mm, held on with adhesive tape and made contact through electrode paste. The active (A) electrode was placed on the motor point of the ADM muscle; the reference (R) was placed distally at the base of the digit V [1]. Distances between the recording and stimulating electrodes were not fixed. The ground (G) electrode was placed on the dorsum of the hand; the figure shows the ground electrode placed on the forearm.

Stimulation For measuring the motor conduction time from the elbow to the wrist, the author used surface electrodes (spherical pads 15 mm in diameter). The first stimulation was performed above the elbow (S1); the second was performed at the wrist (S2), both on the course of the ulnar nerve. The stimulating cathode (–) was placed proximal to the recording site; the anode (+) was positioned distally [1].

The author used a different intensity of stimulation in order to study the threshold of excitability of nerves fibers; supramaximal stimulation was mainly used. He showed that the motor fibers to the small muscles of the hand had a higher threshold to electrical stimulation than that of the more excitable sensory afferent fibers from the fingers.

Measurements
Distal latency (ms) was measured from the stimulus artifact to the initial deflection of the compound muscle action potential (CMAP). The author adjusted the recording electrode positions until the nerve action potentials had as near as possible the same shape whether stimulated at the elbow or at the wrist. The motor conduction times were taken as the difference in time, between the corresponding points on the action potentials recorded at the hand by stimulation at the elbow and at the wrist. No control of limb temperature was attempted. Normal values (Table 1) were obtained from ten ulnar nerves of healthy subjects.

Comment
Dawson [1] examined the relative electrical excitability and conduction velocity of sensory and motor nerve fibers in man. The author used a previously described method [2] with little modifications. He observed that there was no evidence of electrical nerve activity since the stimulus was below threshold for motor fibers. For the author, the motor conduction time derived from the first rise of the muscle action potentials was probably related to the fastest conducting motor fibers, and the fastest conducting motor fibers have a lower conduction velocity than the fastest conducting sensory fibers, a difference consistent with the higher threshold of the motor fibers to electrical stimulation.

G. Gentili, M. Di Napoli, *The Ulnar Nerve*, DOI 10.1007/978-3-319-43292-2_26

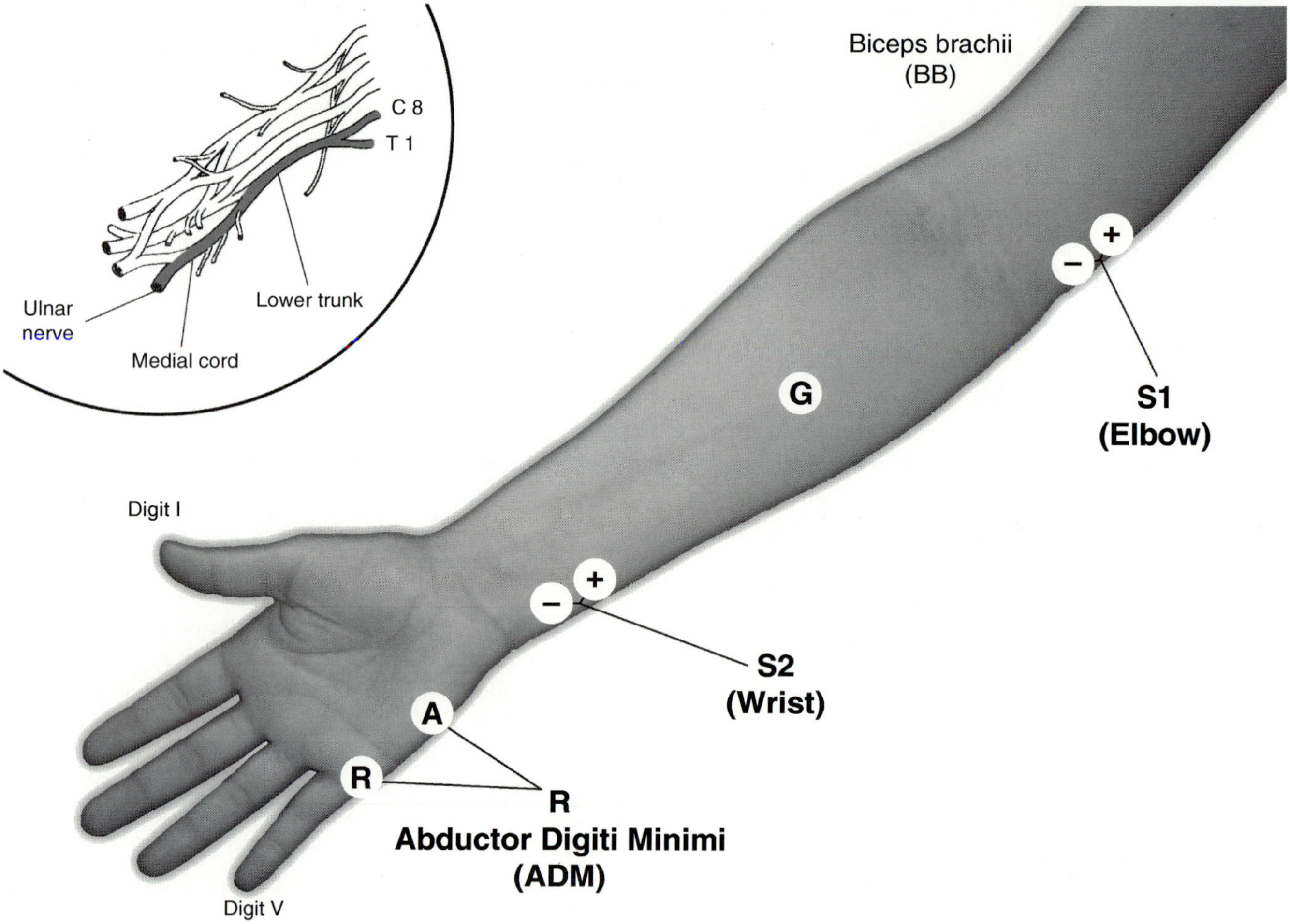

Typical waveform (elbow, wrist – ADM muscle):

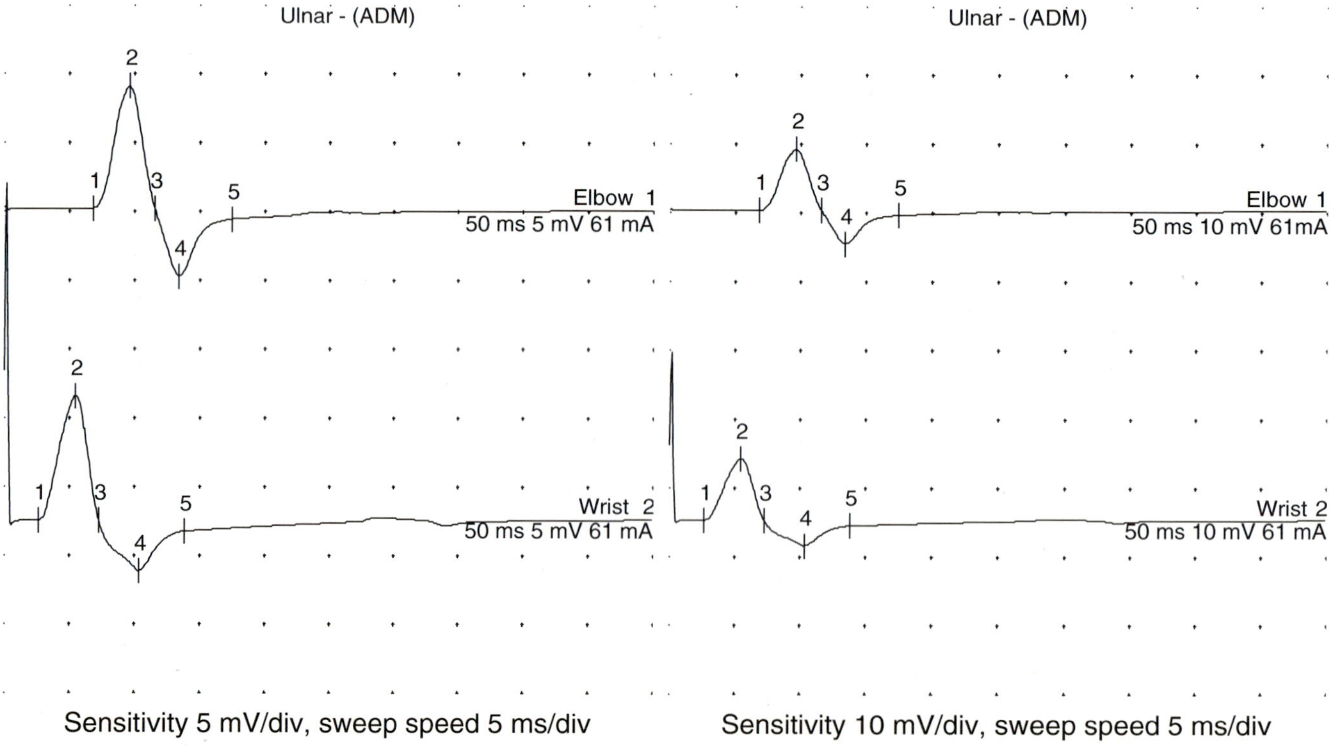

Fig. 1 Compound muscle action potentials (*CMAPs*) recorded at the hand from the ADM muscle, stimulation of the elbow (*upper trace*) and of the wrist (*lower trace*)

Table 1 Normal values [1]

	Mean
Wrist–elbow, motor conduction time (ms)	5.49

McQuillen and Gorin [3] stimulated five left ulnar nerves of five healthy (Table 2) subjects (age range 22–33 years, mean age 27 years). CMAPs were recorded using surface disk electrodes (DISA 13 K 60) placed over the belly and tendon of the ADM muscle. The authors [2] used a DISA 14 A 20 amplifier and motor potentials were displayed on a storage oscilloscope (TEKTRONIX RM564). Following the Dawson's method [1], maximal stimulation was applied at the elbow and wrist by bipolar, saline-moistened pad, stimulating electrode (DISA 13 K 62). All measurements were obtained in an air-conditioned laboratory, and the room temperature was maintained at a temperature of 23–24 °C. When stimulating at the wrist, the surface stimulating electrode at the elbow was converted to a surface recording electrode (see sensory conduction studies). The nerve conduction distance was measured from the center of the stimulating cathode (-) at the wrist to the center of the recording electrode (A) over the belly of the ADM muscle. Motor nerve conduction velocity (MNCV) was derived by dividing conduction distance by conduction time, and it was measured in meter per second (m/s). The motor conduction time was the difference between the proximal latency (from stimulus at the elbow to the response in the hand) and distal latency (from stimulus at the wrist to the response at the hand).

Comment

In order to assess the ulnar nerve function in the forearm, McQuillen and Gorin [3] expressed the motor conduction velocity as the velocity obtained by employing motor latencies to the onset of the CMAP and as that obtained from the latencies to the negative peak of the motor response. They suggested that the time course of the CMAP recorded by surface electrodes reflects the sequential arrival of nerve impulses in the end-plate region, and impulses in the fastest nerve fibers reach the muscle first and give rise to the onset of the grouped CMAPs. By the time the initial negative peak of the CMAP is reached, the modal arrival of nerve impulses has peaked, and more muscle fibers are beginning to repolarize than are continuing to depolarize. The authors observed that in the forearm segment of the ulnar nerve, the velocity obtained by the onset latency was faster than that obtained by the negative peak latency.

Table 2 Normal values [3]

	Mean ± SD	Range
Subject 1 (33 years): wrist–ADM, distal latency (ms)	3.1 ± 0.3	2.6–3.4
Subject 1 (33 years): wrist–elbow, MNCV – onset latency (m/s)	55.4 ± 2.9	49.5–58.3
Subject 1 (33 years): wrist–elbow, MNCV – peak latency (m/s)	55.3 ± 3.7	51.9–57.8
Subject 2 (25 years): wrist–ADM, distal latency (ms)	2.3 ± 0.2	2.0–2.6
Subject 2 (25 years): wrist–elbow, MNCV – onset latency (m/s)	56.6 ± 1.1	54.6–58.2
Subject 2 (25 years): wrist–elbow, MNCV – peak latency (m/s)	54.6 ± 2.2	51.9–57.8
Subject 3 (24 years): wrist–ADM, distal latency (ms)	2.4 ± 0.3	2.0–3.0
Subject 3 (24 years): wrist–elbow, MNCV – onset latency (m/s)	61.7 ± 2.9	58.4–63.6
Subject 3 (24 years): wrist–elbow, MNCV – peak latency (m/s)	57.4 ± 3.3	52.3–64.3
Subject 4 (33 years): wrist–ADM, distal latency (ms)	2.4 ± 0.2	2.1–2.7
Subject 4 (33 years): wrist–elbow, MNCV – onset latency (m/s)	57.9 ± 1.5	55.0–60.2
Subject 4 (33 years): wrist–elbow, MNCV – peak latency (m/s)	57.1 ± 0.8	55.8–58.1
Subject 5 (22 years): wrist–ADM, distal latency (ms)	2.6 ± 0.1	2.5–2.8
Subject 5 (22 years): wrist–elbow, MNCV – onset latency (m/s)	56.0 ± 1.6	53.6–58.0
Subject 5 (22 years): wrist–elbow, MNCV – peak latency (m/s)	49.9 ± 4.6	38.1–52.6
All subjects (27 years): wrist–ADM, distal latency (ms)	2.6 ± 0.4	2.0–3.4
All subjects (27 years): wrist–elbow, MNCV – onset latency (m/s)	57.5 ± 3.1	49.5–63.6
All subjects (27 years): wrist–elbow, MNCV – peak latency (m/s)	54.4 ± 4.1	38.1–64.3

References

1. Dawson GD (1956) The relative excitability and conduction velocity of sensory and motor nerve fibres in man. J Physiol 131:431–451
2. Dawson GD, Scott JW (1949) The recording of nerve action potentials through skin in man. J Neurol 12:259–267
3. McQuillen MP, Gorin FJ (1969) Serial ulnar nerve conduction velocity measurements in normal subjects. J Neurol Neurosurg Psychiatry 32:144–148

Wrist, Below Elbow, Above Elbow, Mid-Arm, Axilla – Hand

M3

Needle Recording Technique, Study from *Abductor Digiti Minimi* (ADM) and *First Dorsal Interosseous* (FDI) Muscles

Original Settings Sweep speed was 1–5 ms/division; sensitivity, low-frequency filter, high-frequency filter, duration of pulse, and the machine used were not specified.

Position This study was performed in the supine position.

Recording The recording electrode was a coaxial needle electrode inserted into the *abductor digiti minimi* (ADM) or *first dorsal interosseous* (FDI) muscles (Figs. 1 and 2), on the motor point [1]. The ground (G) electrode was positioned between the stimulating cathode and the recording needle electrode to diminish the stimulus artifact in the records; the figures show the ground electrode positioned on the distal wrist crease.

Needle Insertion For the ADM muscle, with the hand in a neutral position to full supination, the recording needle electrode was inserted at the midpoint between the fifth metacarpophalangeal joint and the pisiform bone. For the FDI muscle, with the hand in a neutral position to full pronation, the recording needle electrode was inserted at the midpoint between the proximal phalanx of the digit I (thumb) and the metacarpal bone.

Stimulation The ulnar nerve was stimulated at the wrist (S1), below the elbow (S2), and above the elbow (S3). In a case of a deep-branch neuritis of the right ulnar with sparing of *hypothenar* muscle, Simpson [1] performed stimulation at these three points, and compound muscle action potentials (CMAPs) were recorded from ADM and FDI muscles, separately. The author also studied a patient with chronic traumatic ulnar neuritis at right elbow, and additional proximal stimulations at the mid-arm level and at the axilla were performed (Fig. 3). The anode was proximal in all stimulations [1]. The distance between the wrist-stimulating electrode (–) and the recording needle in the ADM muscle was 8 cm in patient with ulnar deep-branch neuritis, and it was 6.5 cm in patient with chronic traumatic ulnar neuritis at the elbow. For stimulation, the author used an EEG electrode 1 cm in diameter (cathode) and a large metal plate (anode) placed over the upper arm. Stimuli were delivered through an isolating transformer (100–150 μs duration), one per second. Only supramaximal stimuli were used because, despite an increase in the intensity of the stimulus, no increase in the CMAP amplitude was observed. We use conventional bipolar stimulator, and the anode (+) was proximal in all stimulations (Figs. 1, 2, and 3).

Measurements Distal latency (ms) of the CMAP was measured from the stimulus artifact to the onset of the negative deflection of CMAP. The motor conduction time of the impulse in each segment was obtained by subtraction (wrist–below-elbow, below-elbow–above-elbow, and above-elbow–axilla segments). The length of nerve from the stimulating cathode in each situation to the recording needle electrode was estimated by surface measurement (the path to the ADM muscle was 4 cm, shorter by surface measurement than the path to the FDI muscle). Symmetrical points of stimulation and recording were compared in both arms. No control of temperature was attempted. Conduction values were obtained from a patient with right ulnar deep-branch neuritis (Tables 1 and 2) and from a patient with chronic traumatic ulnar neuritis at the right elbow (Tables 3 and 4).

Comment

In a patient with right ulnar deep-branch neuritis and in a patient with chronic traumatic ulnar neuritis at the right elbow, Simpson [1] found that slowed nerve conduction was confined in the damaged segment of the nerve. In the former he observed that the branch to the ADM muscle escaped the conduction block which was found in fibers to the FDI muscle. The application, using needle [1] and surface electrodes recording, in a case of ulnar neuropathy at the elbow (UNE) is here reported (Figs. 4, 5, and 6).

G. Gentili, M. Di Napoli, *The Ulnar Nerve*, DOI 10.1007/978-3-319-43292-2_27

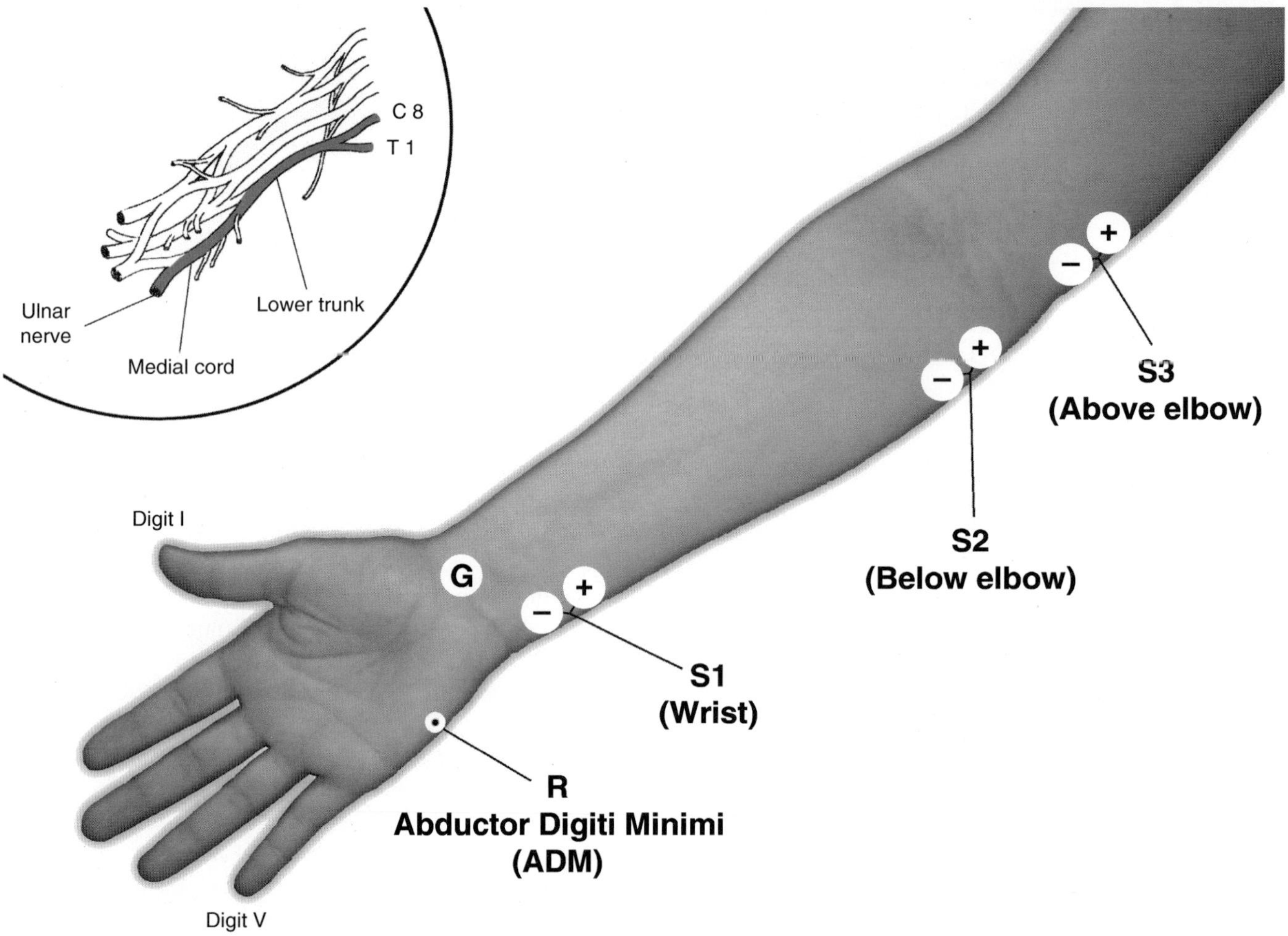

Typical waveform (wrist, below elbow, above elbow, mid-arm, axilla – ADM muscle):

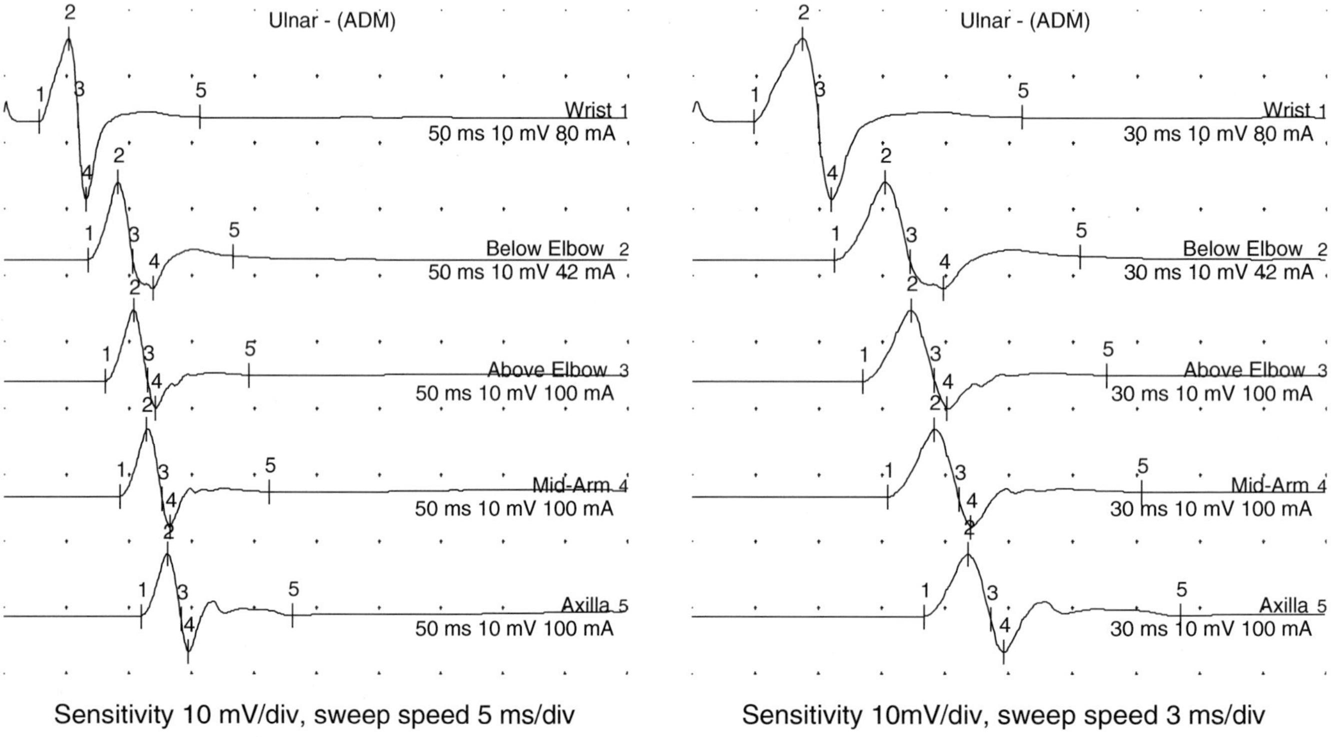

Fig. 1 Compound muscle action potentials (*CMAPs*) recorded at the hand from the ADM muscle, *top to bottom*: stimulation of the wrist, below the elbow, above the elbow, of the mid-arm and axilla

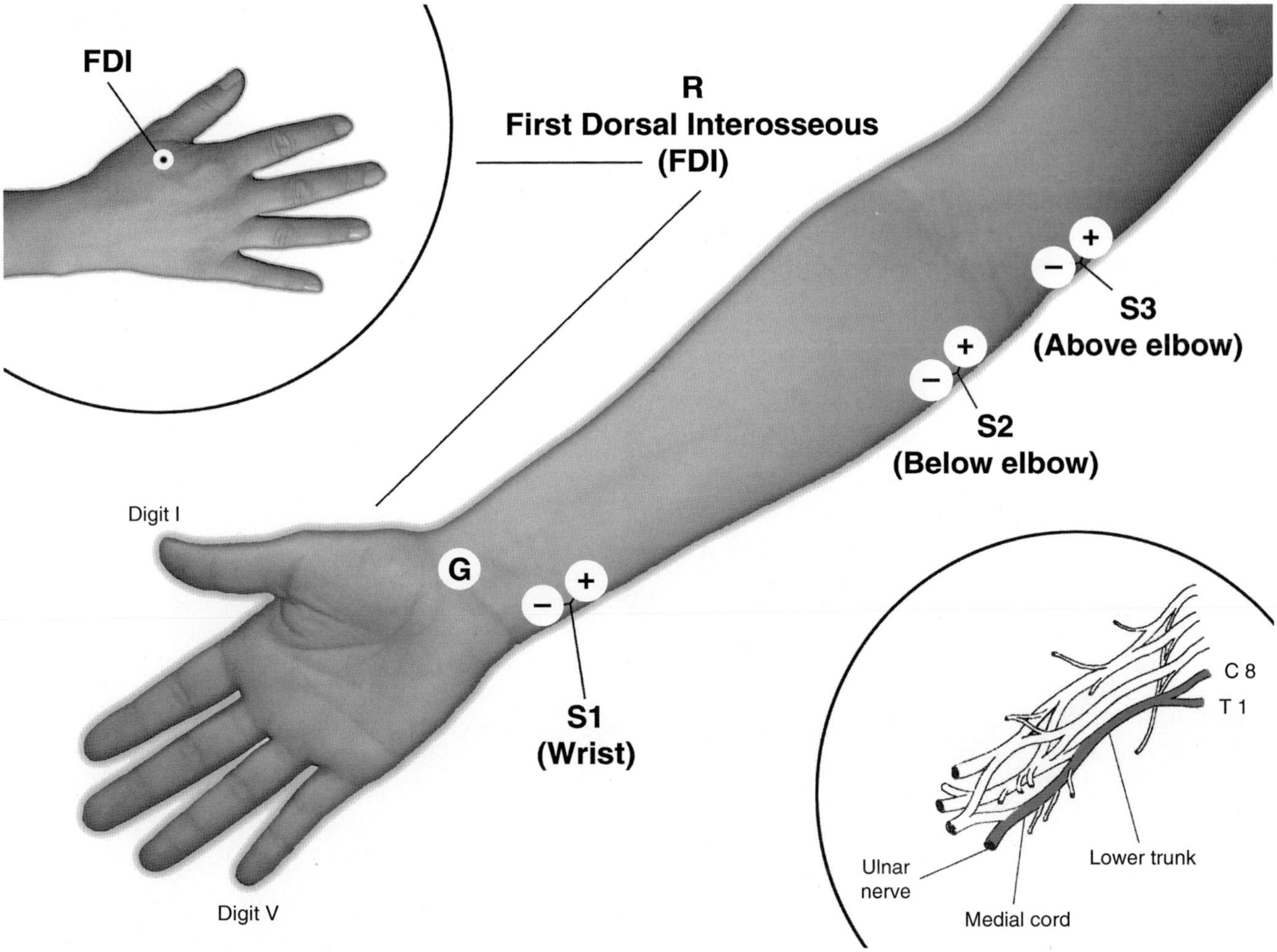

Typical waveform (wrist, below elbow, above elbow, mid-arm, axilla – FDI muscle):

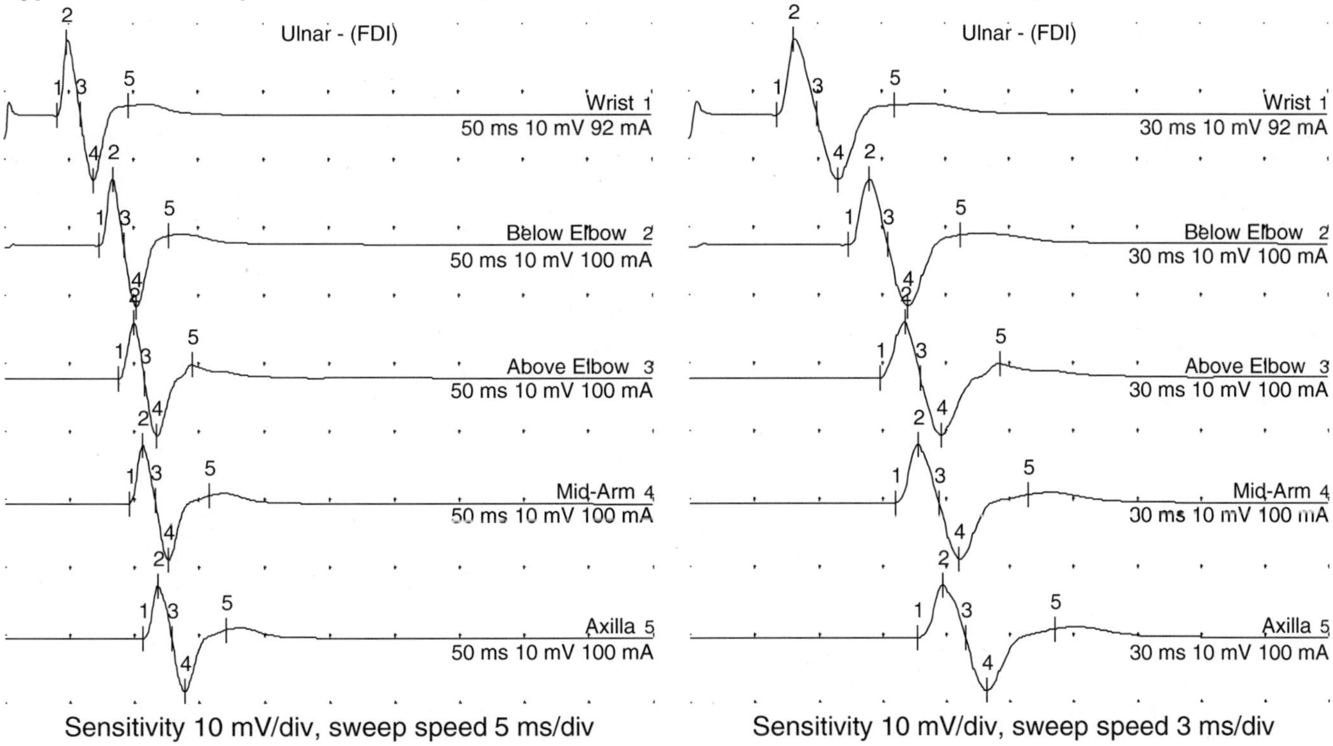

Fig. 2 Compound muscle action potentials (*CMAPs*) recorded at the hand from the FDI muscle, *top to bottom*: stimulation of the wrist, below the elbow, above the elbow, of the mid-arm and axilla

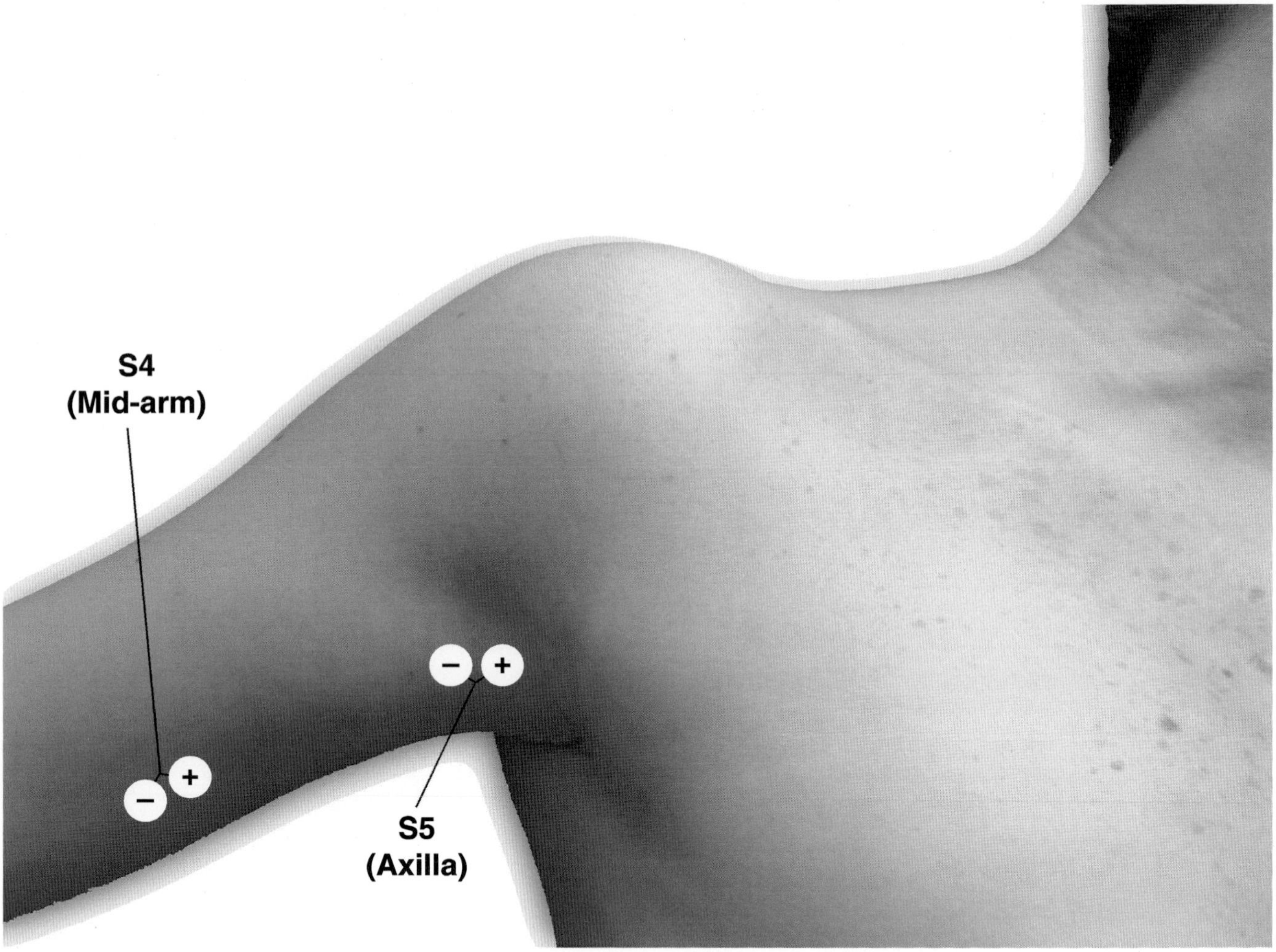

Fig. 3 Proximal stimulation points (*S4, mid-arm*; *S5, axilla*)

Table 1 Normal values [1]

Right hand	Mean
Wrist–ADM (8 cm), distal latency (ms)	3
Below elbow–ADM (17 cm), distal latency (ms)	4
Above elbow–ADM (34 cm), distal latency (ms)	8.5

Table 2 Pathological values [1]

Right hand	Mean
Wrist–FDI (12 cm), distal latency (ms)	12.23
Below elbow–FDI (21 cm), distal latency (ms)	15.26
Above elbow–FDI (38 cm), distal latency (ms)	19.30

Table 3 Normal values [1]

Left hand	Mean
Wrist–ADM (6.5 cm), distal latency (ms)	2.5
Below elbow–ADM (17.5 cm), distal latency (ms)	3.5
Above elbow–ADM (39 cm), distal latency (ms)	8.7
Axilla–ADM (46.5 cm), distal latency (ms)	10

Table 4 Pathological values [1]

Left hand	Mean
Wrist–ADM (6.5 cm), distal latency (ms)	2.5
Below elbow–ADM (17 cm), distal latency (ms)	5
Above elbow–ADM (37 cm),distal latency (ms)	17.5
Axilla–ADM (43.5 cm), distal latency (ms)	20.5

Pathological waveform (wrist, below elbow, above elbow, mid-arm, axilla – ADM muscle):

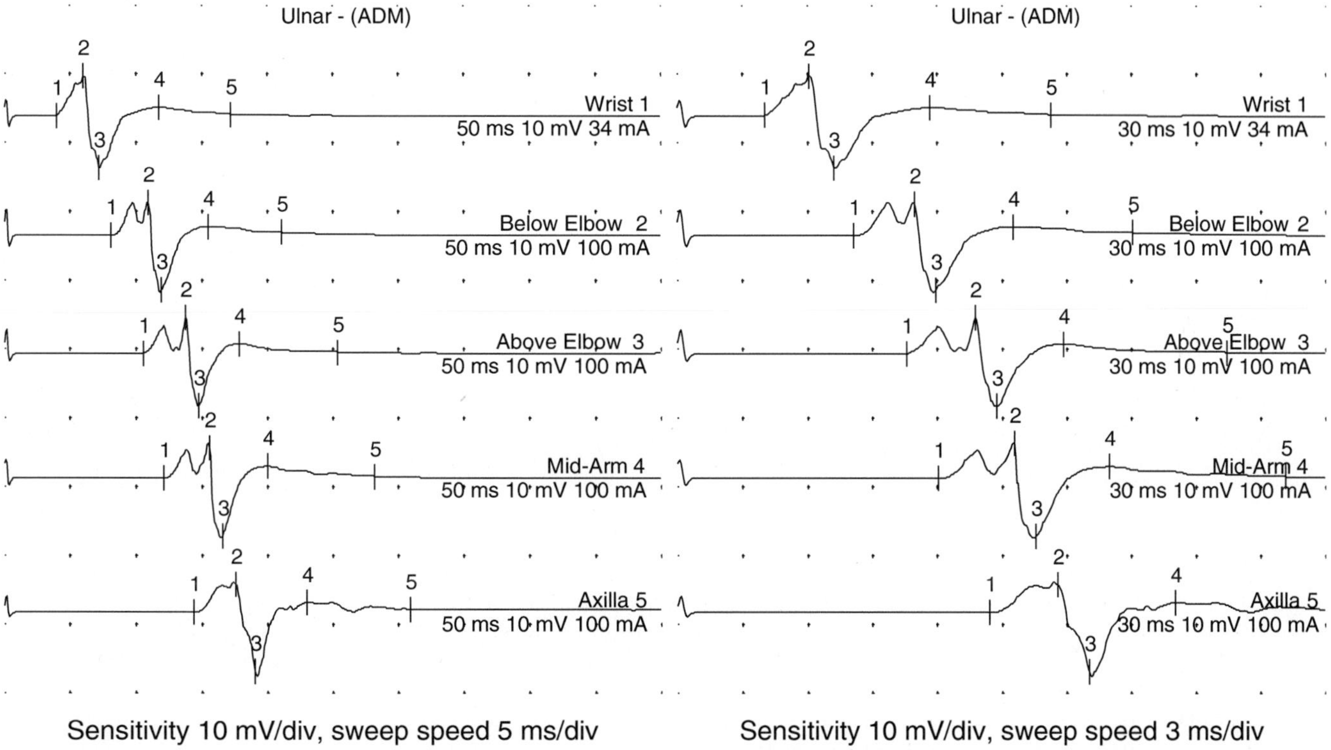

Onset latency (wrist – ADM): 4.0 ms; **Onset latency** (below elbow – ADM): 8.10 ms; **Onset latency** (above elbow – ADM): 10.55 ms; **Onset latency** (mid-arm – ADM): 12.10 ms; **Onset latency** (axilla – ADM): 14.40 ms; **Peak latency** (wrist – ADM): 5.95 ms; **Peak latency** (below elbow – ADM): 10.85 ms; **Peak latency** (above elbow – ADM): 13.60 ms; **Peak latency** (mid-arm – ADM): 15.70 ms; **Peak latency** (axilla – ADM): 17.65 ms; **Onset to peak amplitude** (wrist – ADM): 5.7 mV; **Onset to peak amplitude** (below elbow – ADM): 4.7 mV; **Onset to peak amplitude** (above elbow – ADM): 5.3 mV; **Onset to peak amplitude** (mid-arm – ADM): 4.5 mV; **Onset to peak amplitude** (axilla – ADM): 4.1 mV; **Peak to peak amplitude** (wrist – ADM): 4.6 mV; **Peak to peak amplitude** (below elbow – ADM): 3.3 mV; **Peak to peak amplitude** (above elbow – ADM): 3.8 mV; **Peak to peak amplitude** (mid-arm – ADM): 2.8 mV; **Peak to peak amplitude** (axilla – ADM): 2.5 mV; **MNCV** (wrist – below elbow, 21 cm distance): 51.2 m/s; **MNCV** (below elbow – above elbow, 8 cm distance): 32.7 m/s; **MNCV** (above elbow – mid-arm, 9 cm distance): 58.1 m/s; **MNCV** (mid-arm – axilla, 15 cm distance): 65.2 m/s

Fig. 4 Compound muscle action potentials (*CMAPs*) recorded from the ADM muscle using needle electrode [1] in UNE, *top to bottom*: stimulation of the wrist, below the elbow, above the elbow, of the mid-arm and axilla

Pathological waveform (wrist, below elbow, above elbow, mid-arm, axilla – ADM muscle):

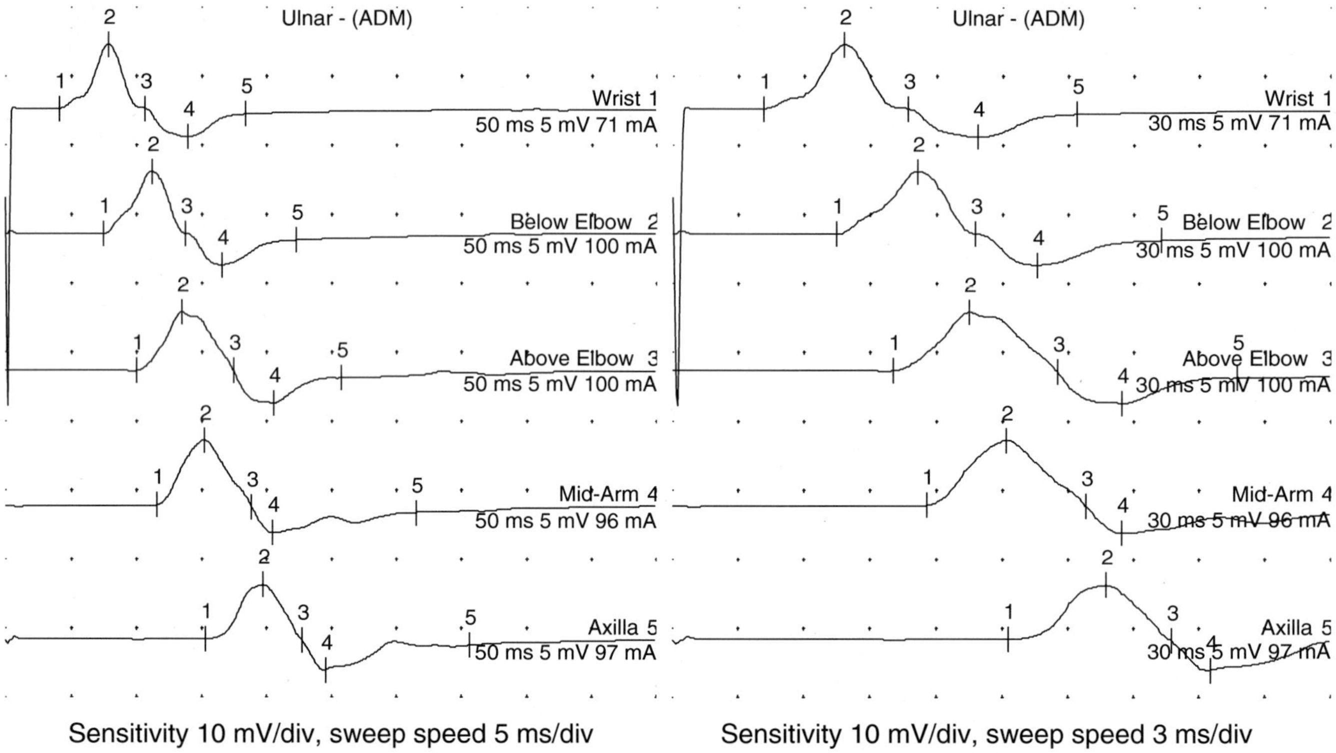

Onset latency (wrist – ADM): 4.1 ms; **Onset latency** (below elbow – ADM): 7.45 ms; **Onset latency** (above elbow – ADM): 10.0 ms; **Onset latency** (mid-arm – ADM): 11.55 ms; **Onset latency** (axilla – ADM): 15.25 ms; **Peak latency** (wrist – ADM): 7.65 ms; **Peak latency** (below elbow – ADM): 10.70 ms; **Peak latency** (above elbow – ADM): 13.40 ms; **Peak latency** (mid-arm – ADM): 15.10 ms; **Peak latency** (axilla ADM): 19.15 ms; **Onset to peak amplitude** (wrist – ADM): 4.6 mV; **Onset to peak amplitude** (below elbow – ADM): 4.5 mV; **Onset to peak amplitude** (above elbow – ADM): 4.3 mV; **Onset to peak amplitude** (mid-arm – ADM): 4.7 mV; **Onset to peak amplitude** (axilla – ADM): 3.9 mV; **Peak to peak amplitude** (wrist – ADM): 6.8 mV; **Peak to peak amplitude** (below elbow – ADM): 6.9 mV; **Peak to peak amplitude** (above elbow – ADM): 6.6 mV; **Peak to peak amplitude** (mid-arm – ADM): 6.8 mV; **Peak to peak amplitude** (axilla – ADM): 6.2 mV; **MNCV** (wrist – below elbow, 18 cm distance): 53.7 m/s; **MNCV** (below elbow – above elbow, 8 cm distance): 31.4 m/s; **MNCV** (above elbow – mid-arm, 9 cm distance): 58.1 m/s; **MNCV** (mid-arm – axilla, 18.5 cm distance): 50.0 m/s

Fig. 5 Compound muscle action potentials (*CMAPs*) recorded from the ADM muscle using surface electrodes in UNE, *top to bottom*: stimulation of the wrist, below the elbow, above the elbow, of the mid-arm and axilla

Pathological waveform (wrist, below elbow, above elbow, mid-arm, axilla – FDI muscle):

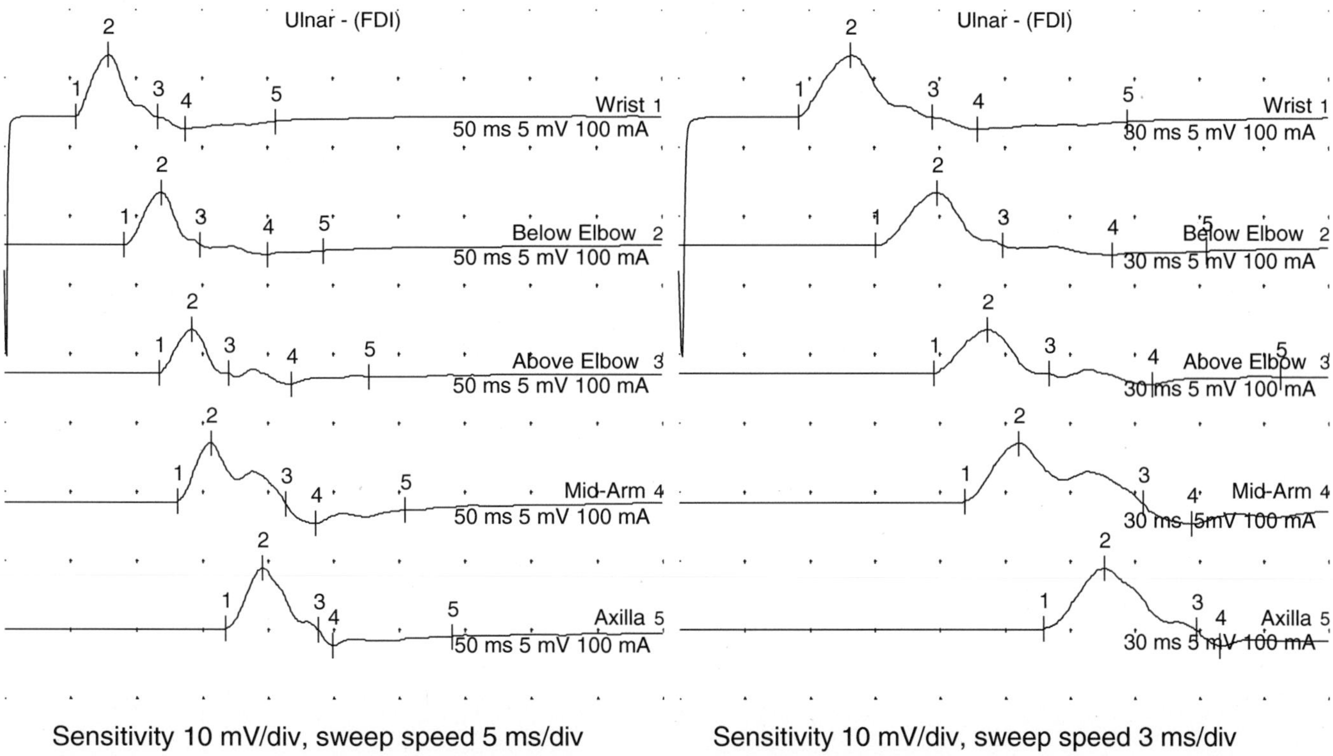

Onset latency (wrist–FDI): 5.50 ms; **Onset latency** (below elbow – FDI): 9.05 ms; **Onset latency** (above elbow –FDI): 11.70 ms; **Onset latency** (mid-arm–FDI): 13.10 ms; **Onset latency** (axilla–FDI): 16.70 ms;**Peak latency** (wrist –FDI): 7.70 ms; **Peak latency** (below elbow –FDI): 12.0 ms; **Peak latency** (above elbow–FDI): 13.95 ms; **Peak latency** (mid-arm–FDI): 15.40 ms; **Peak latency** (axilla–FDI): 19.45 ms; **Onset to peak amplitude** (wrist –FDI): 4.4 mV; **Onset to peak amplitude** (below elbow –FDI): 3.8 mV; **Onset to peak amplitude** (above elbow –FDI): 3.2 mV; **Onset to peak amplitude** (mid-arm–FDI): 4.2 mV; **Onset to peak amplitude** (axilla–FDI): 4.4 mV; **Peak to peak amplitude** (wrist –FDI): 5.4 mV; **Peak to peak amplitude** (below elbow –FDI): 4.6mV; **Peak to peak amplitude** (above elbow –FDI): 4.0 mV; **Peak to peak amplitude** (mid-arm–FDI): 5.9 mV; **Peak to peak amplitude** (axilla–FDI): 5.7 mV; **MNCV** (wrist – below elbow, 18 cm distance): 50.7 m/s; **MNCV** (below elbow –above elbow, 8 cm distance): 30.2 m/s; **MNCV** (above elbow –mid-arm, 8 cm distance): 57.1 m/s; **MNCV** (mid-arm–axilla, 18 cm distance): 50.0 m/s

Fig. 6 Compound muscle action potentials (*CMAPs*) recorded from the FDI muscle using surface electrodes in UNE, *top to bottom*: stimulation of the wrist, below the elbow, above the elbow, of the mid-arm and axilla

Reference

1. Simpson JA (1956) Electrical signs in the diagnosis of carpal tunnel and related syndromes. J Neurol Neurosurg Psychiatry 19:275–280

Wrist, Above Elbow – Hand

M4

Needle Recording Technique, Study from *Abductor Digiti Minimi* (ADM) and *First Dorsal Interosseous* (FDI) Muscles

Original Settings Sensitivity was 2 mV/division, sweep speed was 1–5 ms/division, duration of pulse was 70–100 μs, and the machine used was a double-channel electromyograph; low-frequency filter and high-frequency filter were not specified.

Position This study was performed in the supine position.

Recording The recording electrode was a coaxial needle electrode inserted into the *abductor digiti minimi* (APB) or first dorsal *interosseous* (FDI) muscles (Figs. 1 and 2). In each case the position of the needle tip in the muscle was adjusted until the earliest response to a given stimulus was found, and care was taken to ensure that the initial phase of the evoked potential was similar in appearance for both proximal and distal stimuli [1]. The ground (G) position was not mentioned in the text; it can be placed between the stimulating cathode and the recording needle electrode to diminish stimulus artifact in the records (the figures show ground positioned on the forearm).

Needle Insertion For the ADM muscle, with the hand in a neutral position to full supination, the recording needle electrode was inserted at the midpoint between the fifth metacarpophalangeal joint and the pisiform bone. For the FDI muscle, with the hand in a neutral position to full pronation, the recording needle electrode was inserted at the midpoint between the proximal phalanx of the digit I (thumb) and the metacarpal bone.

Stimulation The ulnar nerve was stimulated at the wrist (S1) and above the elbow (S2). The anode (+) was proximal in both stimulations. The distal stimulating cathode (S1) was placed over the ulnar nerve at a point 2–3 cm above the distal wrist crease. The proximal stimulation (S2) was delivered 3–5 cm above the elbow. The length of nerve between the two points stimulated was estimated by surface measurement and was usually between 25 and 30 cm. For stimulating cathodes (−), the authors used EEG electrodes 1 cm in diameter [1]. The anode (+) was a plate electrode 2.5 × 5 cm placed over the flexor surface of the forearm or over the deltoid insertion; these anode positions were chosen for convenience and did not affect nerve conduction time. We use conventional bipolar stimulator, and the anode was proximal in all stimulations (Figs. 1 and 2). In measurements of maximum conduction velocity, an isolating transformer was used for stimulation (on 77 subjects authors used a brief condenser discharge with a time constant of 70 μs; on 73 subjects they used a rectangular pulse of 100 μs duration).

Measurements Distal latency (ms) of the compound muscle action potential (CMAP) was measured from the stimulus artifact to the onset of the negative deflection of CMAP. The motor conduction time for the segment of nerve between the two stimulating cathodes (wrist and elbow) was then obtained by the subtraction of the shorter from the longer figure. The length of nerve from the stimulating cathode in each situation to the recording needle electrode was estimated by surface measurement. Motor nerve conduction velocity (MNCV) was calculated in the elbow–wrist (forearm) segment and it was measured in meter per second (m/s). No control of temperature was attempted; all tests were performed in a warm room with the subject lying on a couch and covered with blankets. In a few subjects with cold hands, the limbs to be tested were immersed in hot water for 5 or 10 min before the session. Normal values (Table 1) were obtained from 46 control subjects (ADM muscle) and from 49 control subjects (FDI muscle) formed by healthy subjects working in laboratory and patients presenting isolated peripheral nerve lesions (age range 18–65 years). All subjects were divided into five age groups: (1) ADM muscle, 18–25 years (3 subjects), 26–35 years (6 subjects), 36–45 years (15 subjects), 46–55 years (15 subjects), and 56–65 years (7 subjects), and (2) FDI muscle, 18–25 years (2 subjects), 26–35 years (5 subjects), 36–45 years (20 subjects), 46–55 years (14 subjects), and 56–65 years (8 subjects).

G. Gentili, M. Di Napoli, *The Ulnar Nerve*, DOI 10.1007/978-3-319-43292-2_28

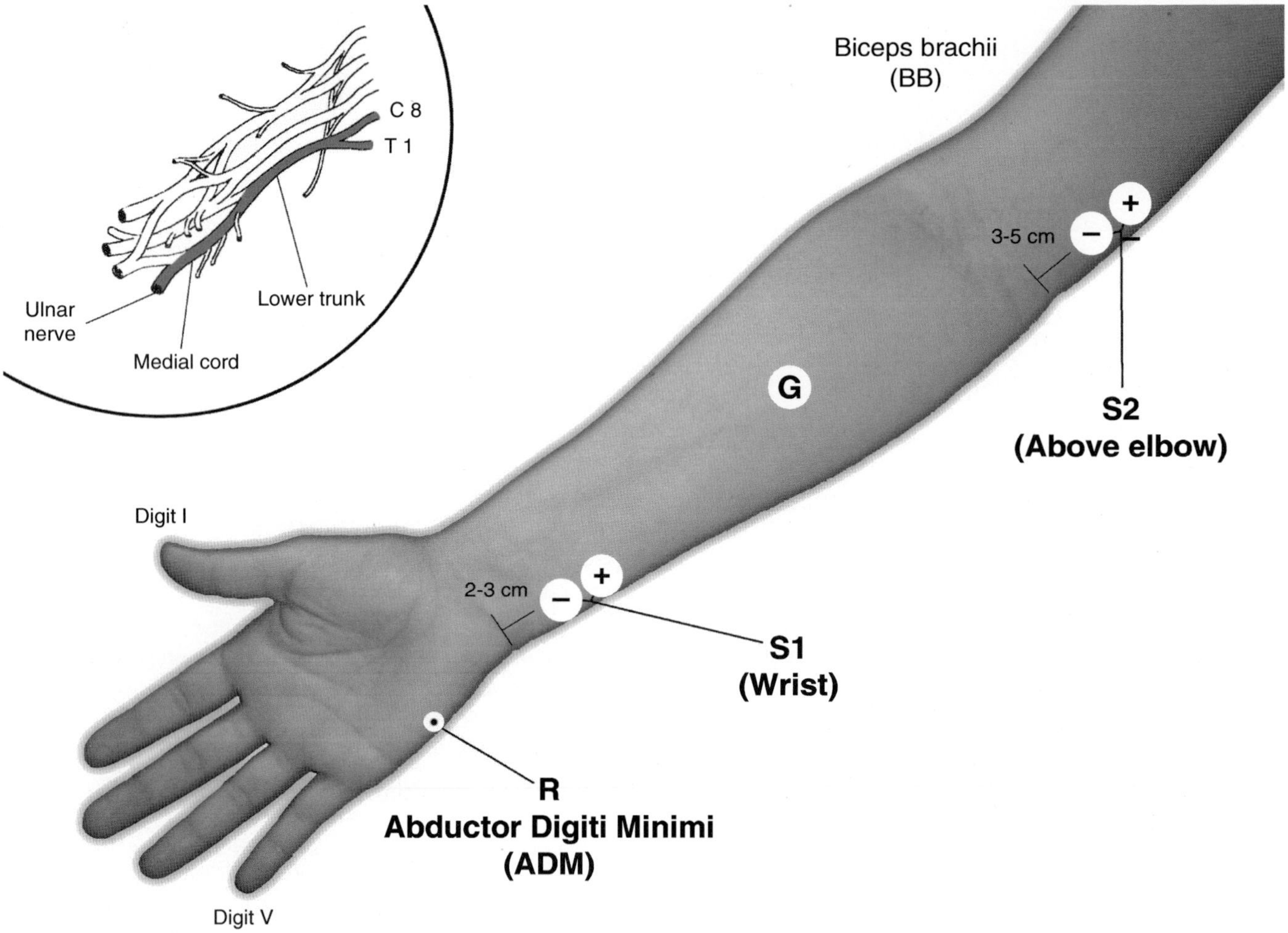

Typical waveform (wrist, above elbow – ADM muscle):

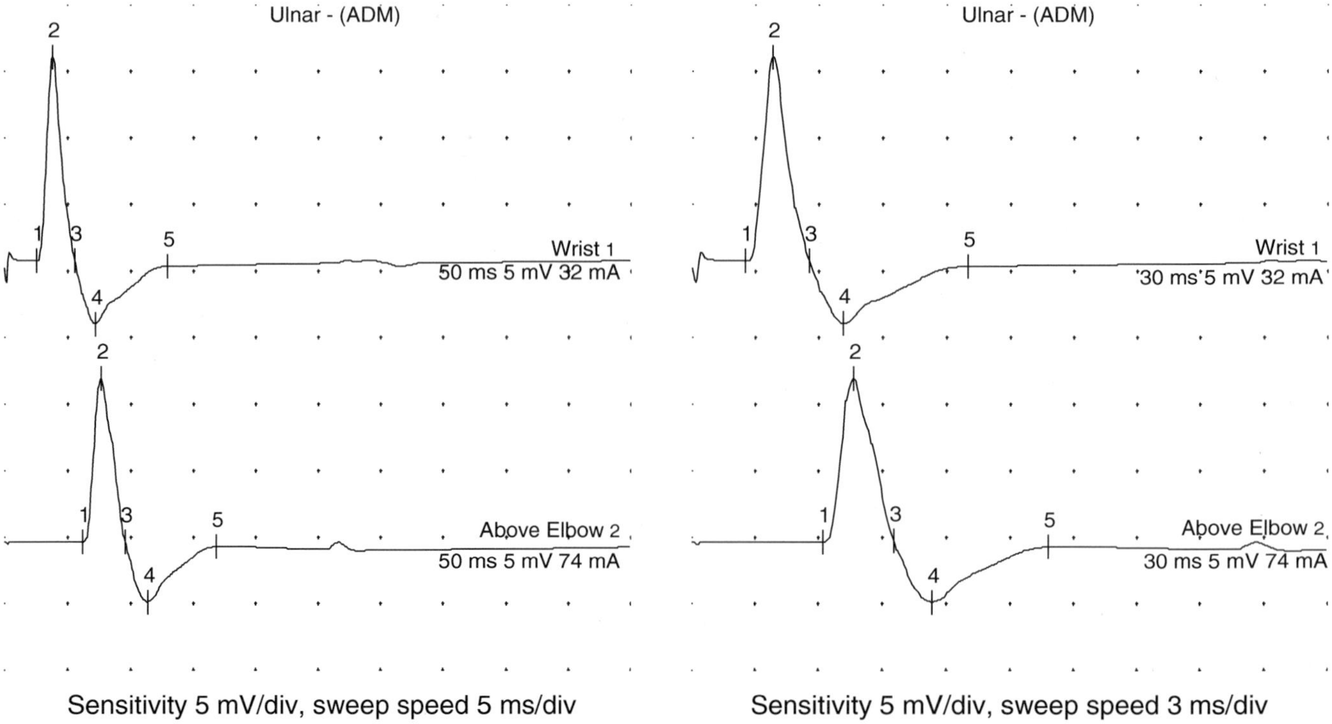

Fig. 1 Compound muscle action potentials (*CMAPs*) recorded at the hand from the ADM muscle, stimulation of the wrist (*upper trace*) and above the elbow (*lower trace*)

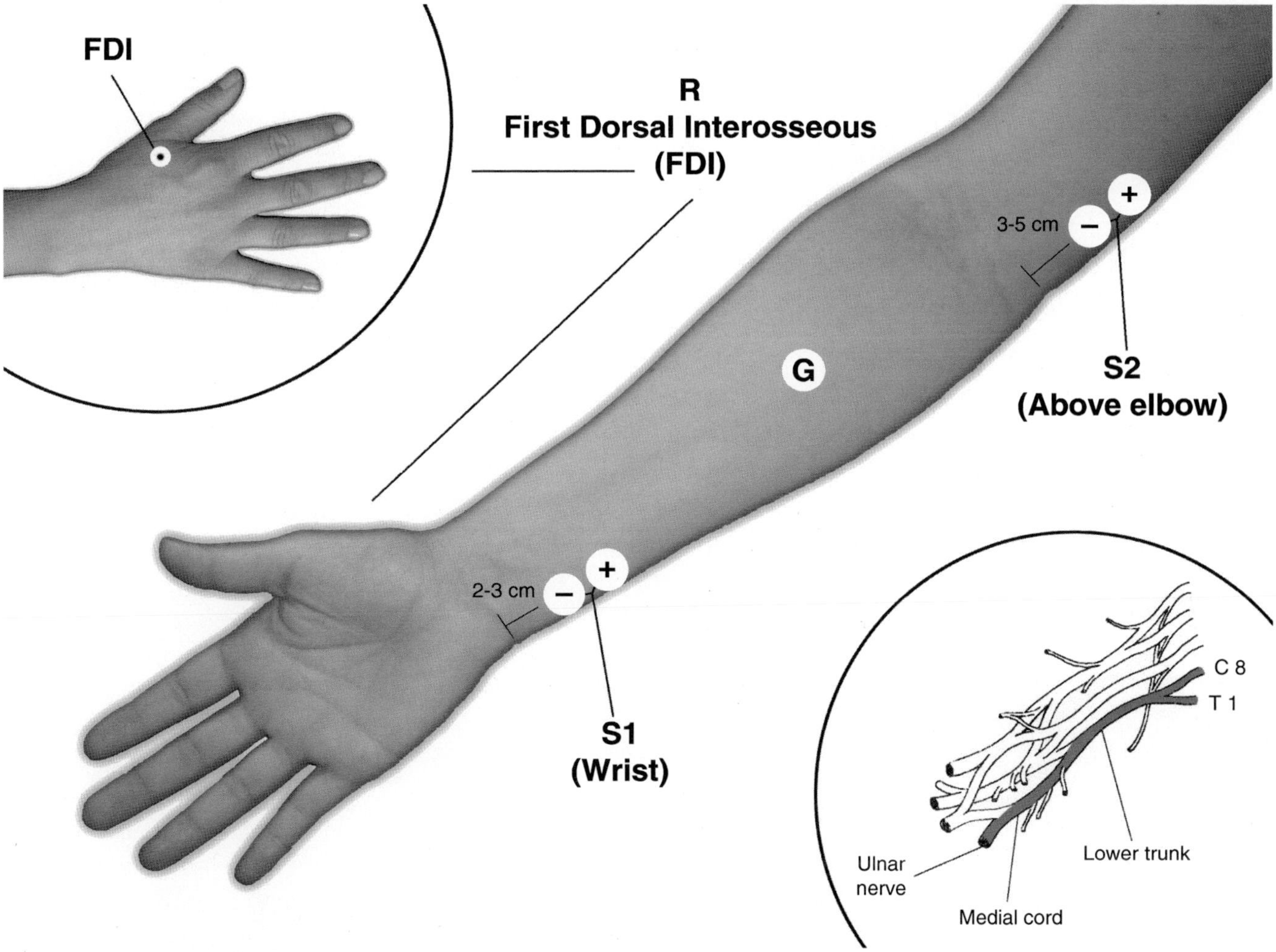

Typical waveform (wrist, above elbow – FDI muscle):

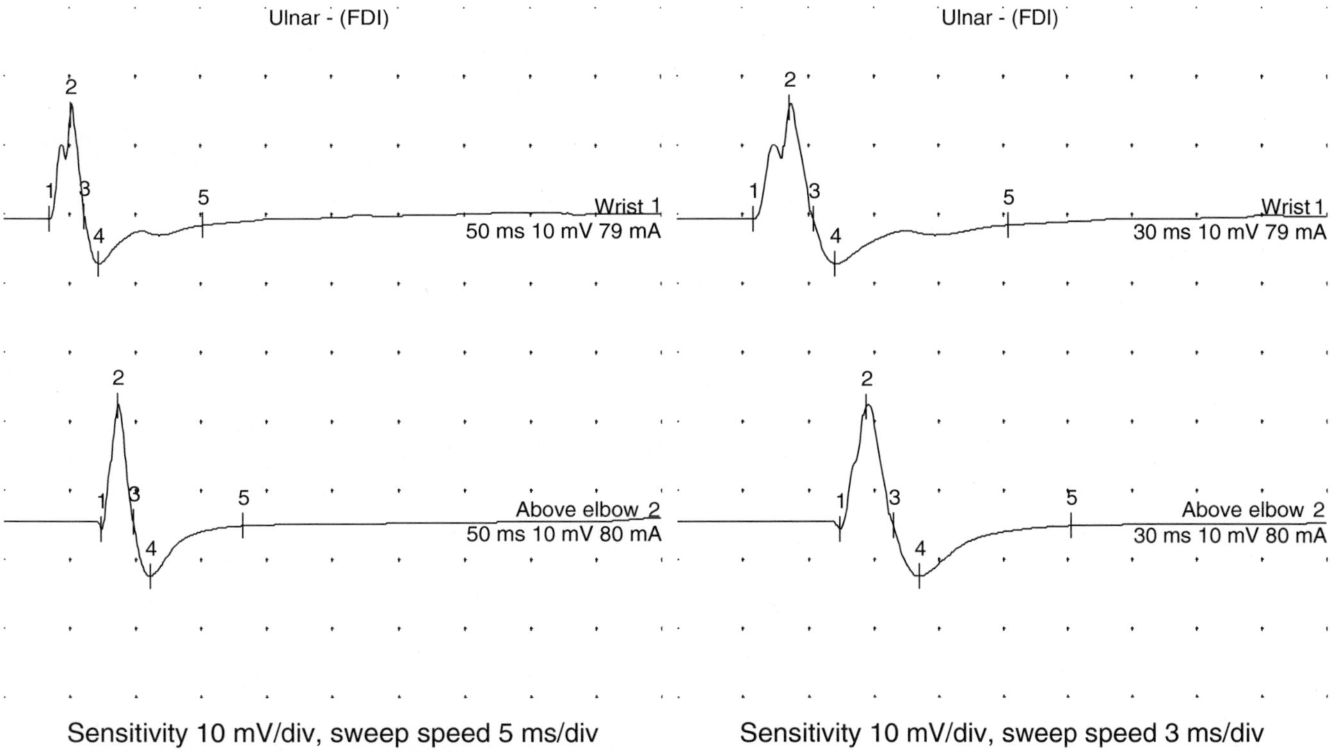

Fig. 2 Compound muscle action potentials (*CMAPs*) recorded at the hand from the FDI muscle, stimulation of the wrist (*upper trace*) and above the elbow (*lower trace*)

Comment

For Thomas et al. [1] the use of the coaxial needle electrode instead of surface electrodes showed clear advantages in the examination of patients with nerve lesions because of the restricted area of recording, and a single surviving unit in an affected muscle can be localized by exploration during voluntary contraction and then studied in isolation during nerve stimulation. He found coaxial needle electrode useful in patients with lesions of the deep branch of the ulnar nerve in the hand, because he observed that the delayed response of a few surviving units in the FDI muscle may be completely obscured by the response of unaffected ADM muscle using surface electrodes for recording. The application, in both ADM and FDI muscles recording, in a case of ulnar neuropathy at the elbow (UNE) is here reported (Figs. 3 and 4).

Table 1 Normal values [1]

	Mean ± SD	Range
Wrist–above elbow (ADM), MNCV (m/s)	56.2 ± 4.6	49.0–65.6
Wrist–above elbow (FDI), MNCV (m/s)	55.0 ± 4.9	46.2–66.2

Pathological waveform (wrist, above elbow – ADM muscle):

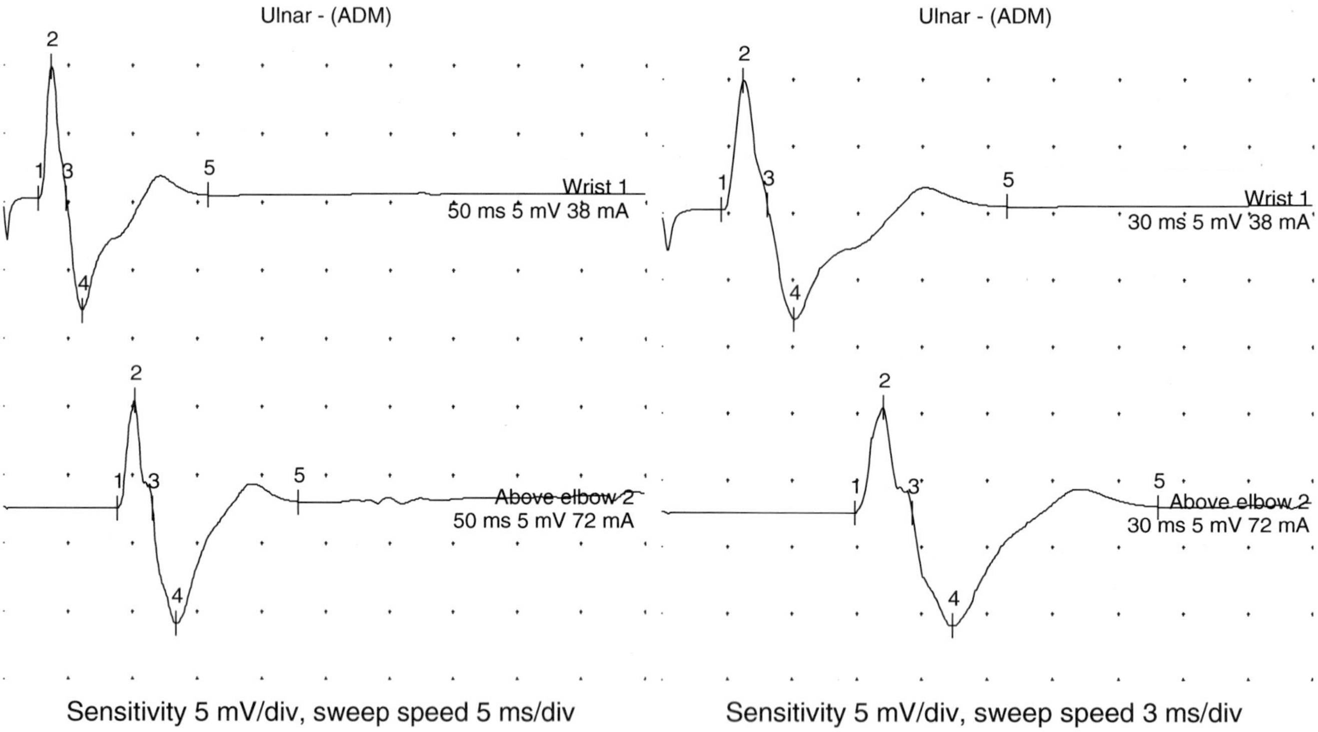

Onset latency (wrist – ADM): 2.65 ms; **Onset latency** (above elbow – ADM): 8.85 ms; **Peak latency** (wrist – ADM): 3.80 ms; **Peak latency** (above elbow – ADM): 11.10 ms; **Onset to peak amplitude** (wrist – ADM): 9.7mV; **Onset to peak amplitude** (above elbow – ADM): 7.9mV; **Peak to peak amplitude** (wrist – ADM): 18.0 mV; **Peak to peak amplitude** (above elbow – ADM): 16.4 mV; **MNCV** (wrist – above elbow, 26 cm): 41.9 m/s

Fig. 3 Compound muscle action potentials (*CMAPs*) recorded at the hand from the ADM muscle using needle recording method [1] in UNE, stimulation of the wrist (*upper trace*) and above the elbow (*lower trace*)

Pathological waveform (wrist, above elbow – FDI muscle):

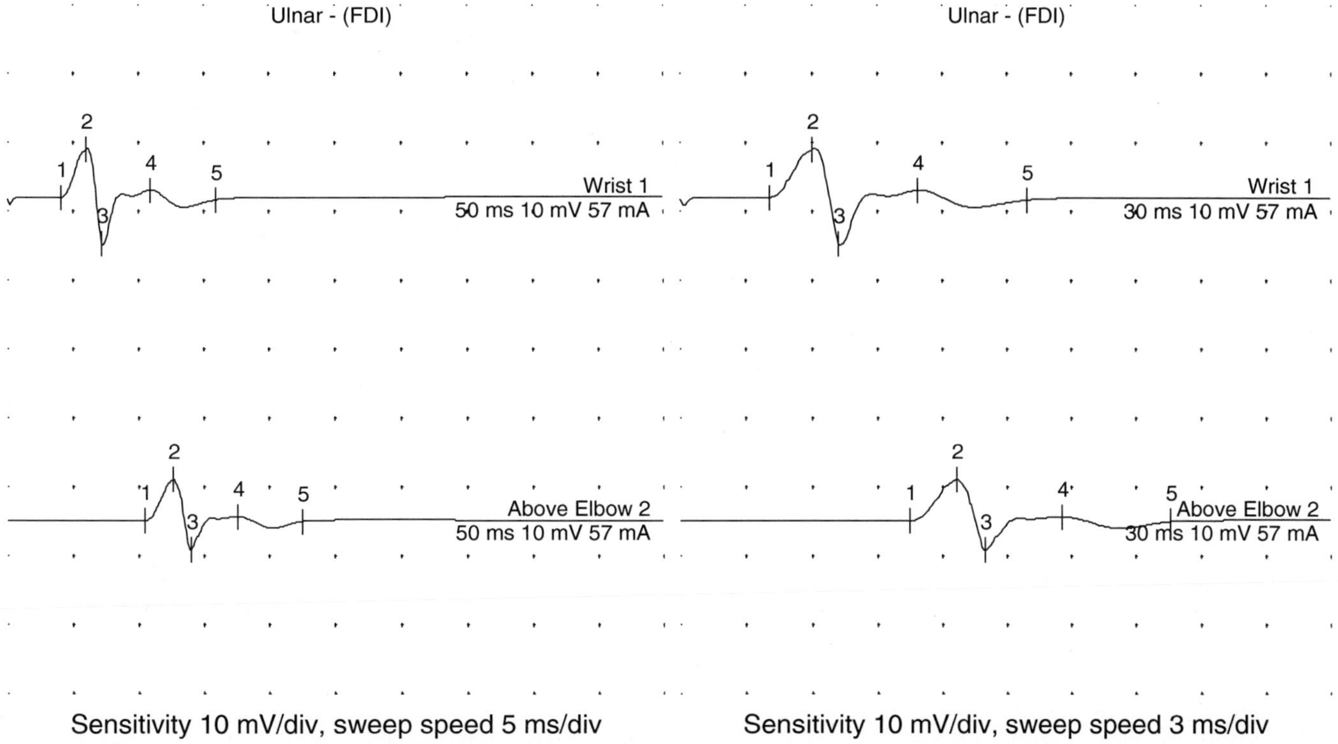

Onset latency (wrist – FDI): 4.05 ms; **Onset latency** (above elbow – FDI): 10.45 ms; **Peak latency** (wrist – FDI): 6.0 ms;**Peak latency** (above elbow – FDI): 12.25 ms; **Onset to peak amplitude** (wrist – FDI): 7.1 mV; **Onset to peak amplitude** (above elbow – FDI): 6.0 mV; **Peak to peak amplitude** (wrist – FDI): 6.0 mV; **Peak to peak amplitude** (above elbow – FDI): 5.5 mV; **MNCV** (wrist – above elbow, 26 cm): 40.6 m/s

Fig. 4 Compound muscle action potentials (CMAPs) recorded at the hand from the FDI muscle using needle recording method [1] in UNE, stimulation of the wrist (*upper trace*) and above the elbow (*lower trace*)

Reference

1. Thomas PK, Sears TA, Gilliatt RW (1959) The range of conduction velocity in normal motor nerve fibres to the small muscles of the hand and foot. J Neurol Neurosurg Psychiatry 22:175–181

Wrist, Above Elbow, Axilla – Hand

M5

Needle Recording Technique, Study from *Abductor Digiti Minimi* (ADM) and *First Dorsal Interosseous* (FDI) Muscles

Original Settings Sweep speed was 2 ms/division, sensitivity was 1 mV/division, and duration of pulse was 1.5 ms; low-frequency filter, high-frequency filter, and the machine used were not specified.

Position This study was performed in the supine position.

Recording The recording electrode was a coaxial needle electrode placed in the *abductor digiti minimi* (ADM) and *first dorsal interosseous* (FDI) muscles (Figs. 1 and 2). In each case the position of the needle tip in the muscle was adjusted to ensure that the initial deflection of the evoked compound muscle action potential (CMAP) was as sharp as possible, thereby aiding the measurement of the motor latency [1]. The ground (G) position was not mentioned in the text; it can be placed between the stimulating cathode and the recording needle electrode to diminish stimulus artifact in the records (the figures show ground positioned on the forearm).

Needle Insertion For the ADM muscle, with the hand in a neutral position to full supination, the recording needle electrode was inserted at the midpoint between the fifth metacarpophalangeal joint and the pisiform bone. For the FDI muscle, with the hand in a neutral position to full pronation, the recording needle electrode was inserted at the midpoint between the proximal phalanx of the digit I (thumb) and the metacarpal bone.

Stimulation The ulnar nerve was stimulated at the wrist (S1), above the elbow (S2), and at the axilla (S3). The authors placed the cathode (−) over the ulnar nerve trunk in all stimulation sites. They used a plate electrode (2.5 × 5 cm) as anode (+), placed over the flexor surface of the forearm (for S1 and S2 stimulations) or over the deltoid insertion (for S3 stimulation). For the authors these positions did not affect nerve conduction times. The distal stimulation was 2 cm above the wrist (S1); stimulation at the elbow was 3–6 cm above the medial epicondyle (S2). The distance between the wrist-stimulating electrode (−) and the recording needle in the ADM muscle varied from 5.5 to 8.5 cm in different patients (Figs. 1, 2, and 3). The distance between stimulation above the wrist (S1) and above the elbow (S2) varied from 23 to 30 cm, and the distance between stimulation above the elbow (S2) and at the axilla (S3) varied from 12 to 22 cm in different patients. Stimuli were delivered through an isolating transformer (150 μs duration). Only supramaximal stimuli were used because, despite an increase in the intensity of the stimulus, no increase in the CMAP amplitude was observed. We use conventional bipolar stimulator, and the anode was proximal in all stimulations.

Measurements Distal latency (ms) of the CMAP was measured from the stimulus artifact to the initial deflection of CMAP. The motor conduction time of the impulse in each segment was obtained by subtraction (axilla–elbow, elbow–wrist, and wrist–ADM segments). The length of nerve from the stimulating cathode in each situation to the recording needle electrode was estimated by surface measurement, but it was not possible to estimate accurately the length of the deep branch of the ulnar nerve to the FDI muscle. No control of temperature was attempted. Normal conduction values (Table 1) were obtained from 50 healthy nerves (sample data were not reported). The authors also studied latencies and conduction times in nine patients with ulnar nerve lesions in the hand comparing data from the ADM and FDI muscles. In some patients no electrical response to the nerve stimulation was obtained from the FDI muscle, and latency was measured to the third or fourth dorsal *interosseous* muscles. A reexamination in some cases was performed some months after the initial examination, and in seven patients with sparing of the ADM muscle, the CMAP latencies after wrist stimulation (S1) were normal (range 2.1–3.2 ms), while the latency of the CMAP from FDI muscle was markedly increased in nearly every case (range 16.0–29 ms).

G. Gentili, M. Di Napoli, *The Ulnar Nerve*, DOI 10.1007/978-3-319-43292-2_29

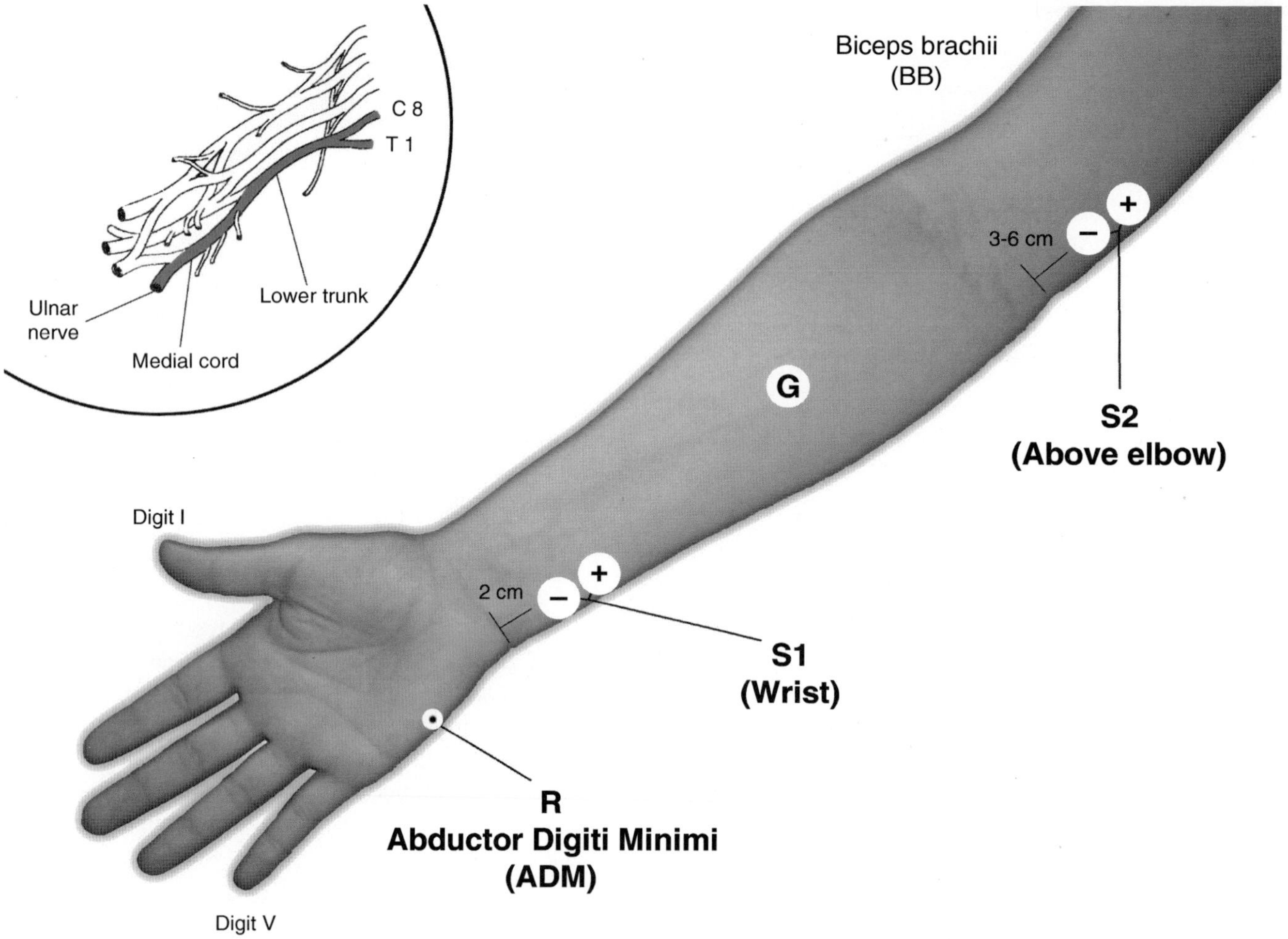

Typical waveform (wrist, above elbow, axilla – ADM muscle):

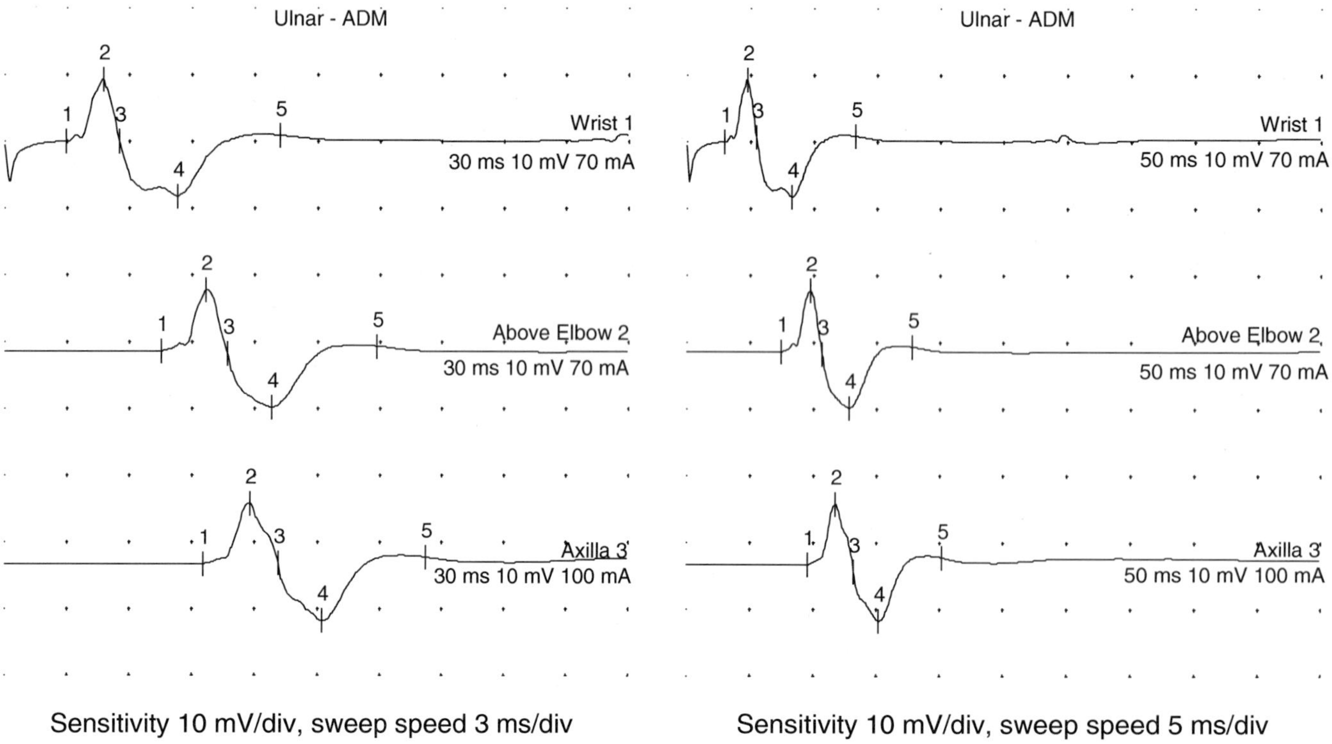

Fig. 1 Compound muscle action potentials (*CMAPs*) recorded at the hand from the ADM muscle, stimulation of the wrist (*upper trace*), above the elbow (*middle trace*), and axilla (*lower trace*)

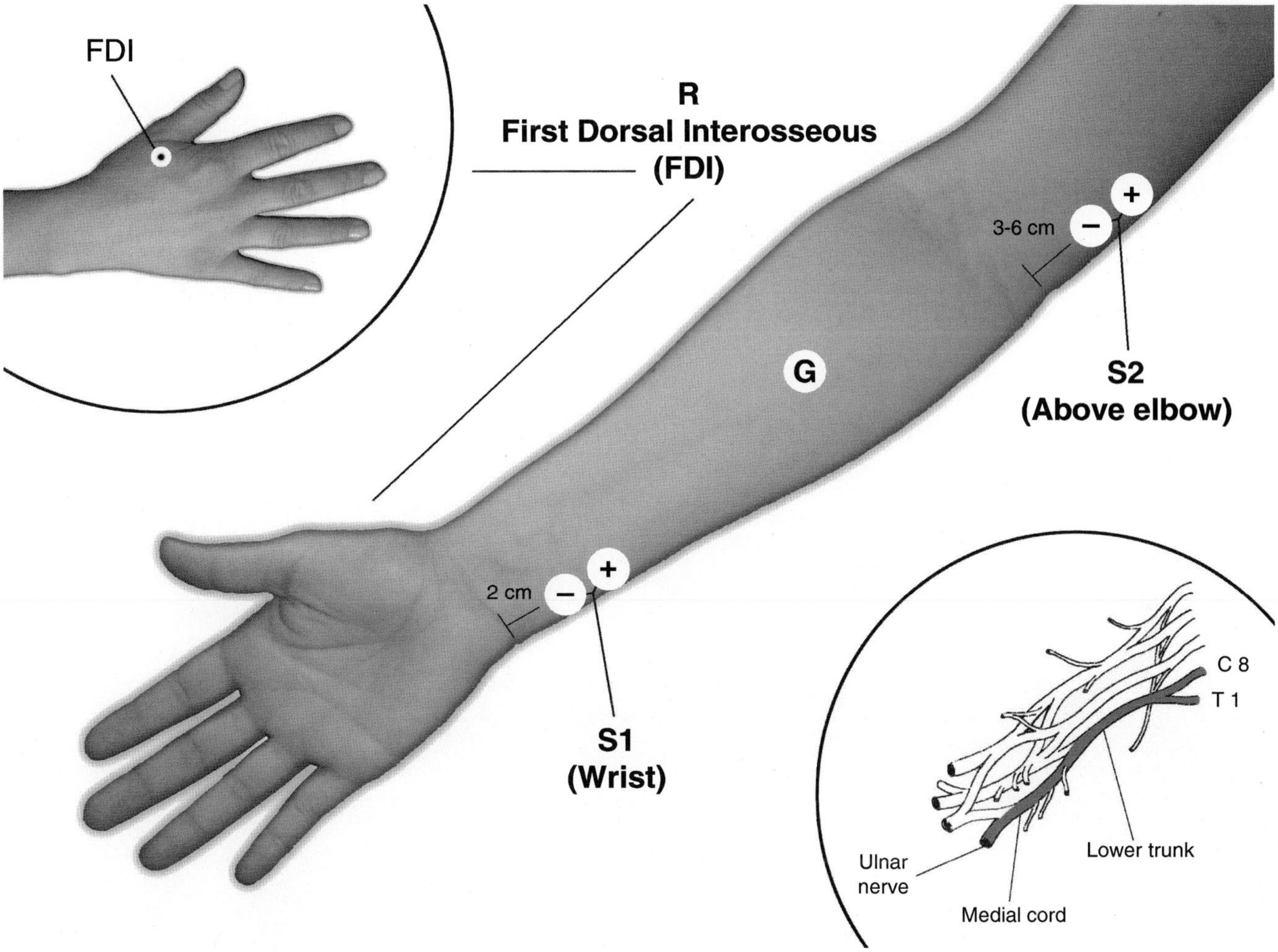

Typical waveform (wrist, above elbow, axilla – FDI muscle):

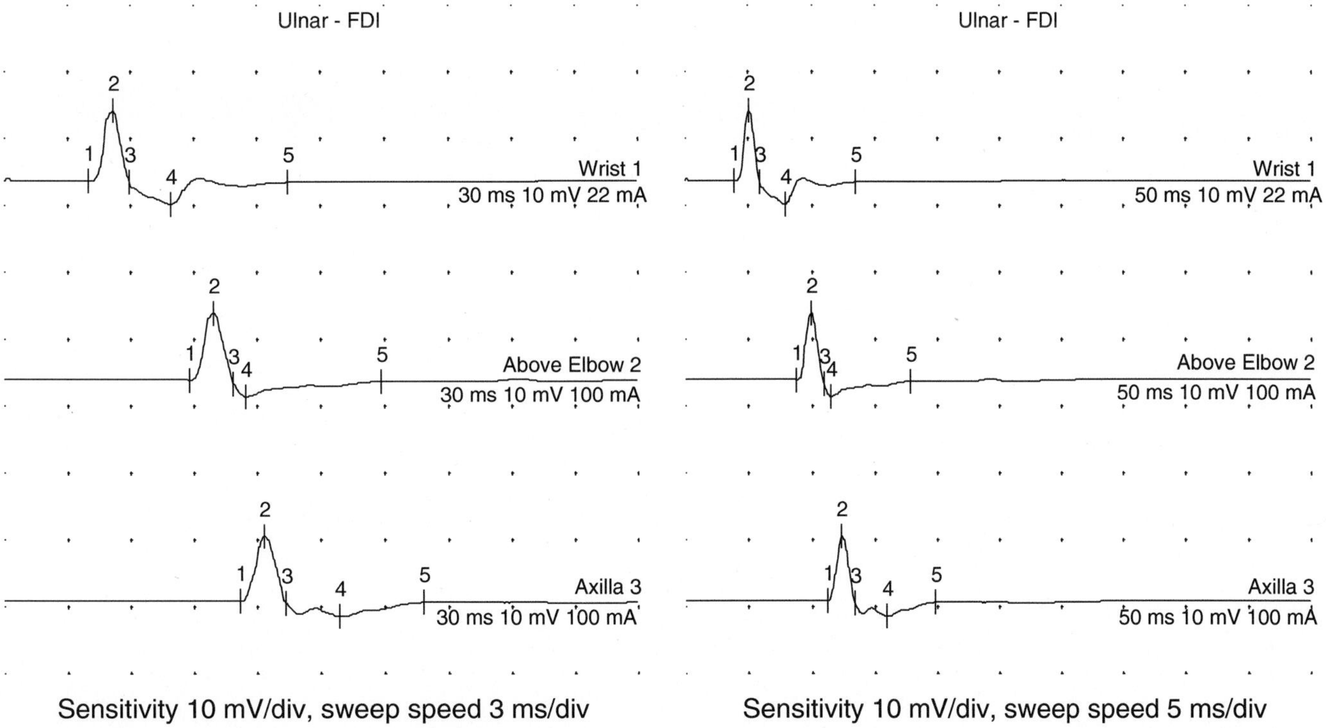

Fig. 2 Compound muscle action potentials (*CMAPs*) recorded at the hand from the FDI muscle, stimulation of the wrist (*upper trace*), above the elbow (*middle trace*), and axilla (*lower trace*)

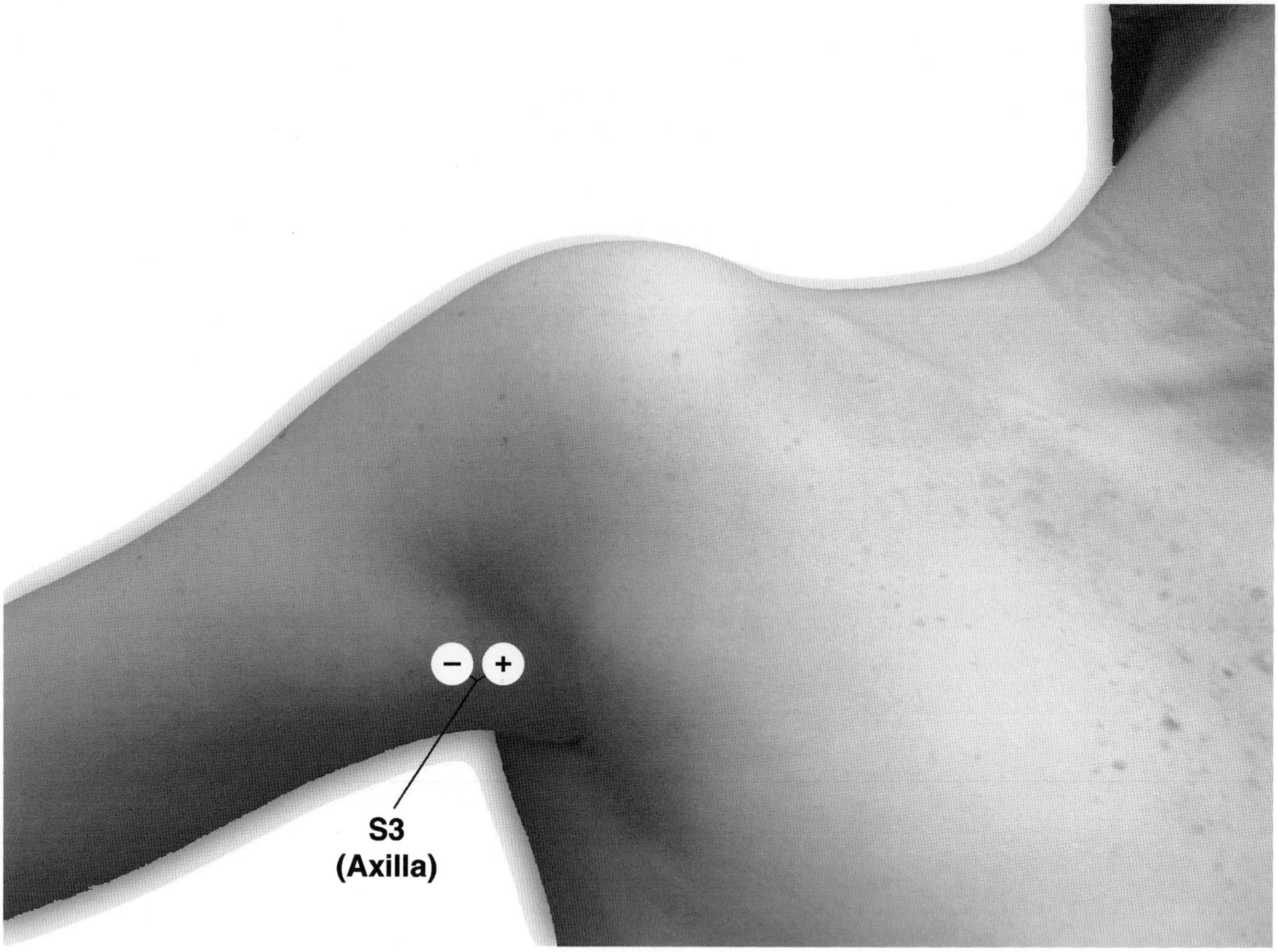

Fig. 3 Proximal stimulation point (*S3 – axilla*)

Table 1 Normal values [1]

	Mean ± SD	Range
Wrist–ADM (5.5–8 cm), distal latency (ms)	2.9 ± 0.39	2–3.7
Wrist–FDI, distal latency (ms)	3.8 ± 0.53	3.0–5.0

Comment

Ebeling et al. [1] found prolonged distal latencies recording from FDI muscle after wrist stimulation in seven patients, and they confirmed the findings previously described by Simpson [2], supporting the clinical diagnosis of compression of the deep branch of the ulnar nerve in its distal part. Moreover, the conduction times between the elbow and the wrist, and between the axilla and the elbow recording from the FDI muscle, showed slightly slower conduction above the wrist than fibers to the ADM muscle. For the authors, as possible explanation of the slowed nerve conduction proximal to the distal lesion at the hand observed in patients, some changes occur proximal to the site of damage in the nerve fibers involved in the lesion process. They examined electrically the ulnar nerve function, and the measurement of the motor nerve conduction time provided a quantitative estimate of the severity of nerve damage.

Gilliatt and Thomas [3] performed the nerve conduction method by Ebeling et al. [1] on 15 healthy subjects (Table 2), and results for nerve conduction values from ADM (12 controls) and FDI (3 controls) were presented together in view of the small total number. Control observations were made on healthy subjects working in the laboratory and on the unaffected nerves of patients presenting isolated peripheral nerve lesions (age range 18–62 years). A group of 14 patients (sample data were not reported) with clinical confirmed ulnar neuropathy at the elbow was also enrolled in the study. In 10 of these 14 patients, the ulnar nerve was explored surgically and pathological changes were observed; radiographic abnormalities of the elbow joint were described in 12 patients. In these 12 patients, the authors found a diffuse slowing of conduction in the forearm (wrist–above elbow) segment (less than 45 m/s), and in four of them, the nerve conduction velocity was less than 25 m/s. Gilliatt and Thomas [3] also compared distal latencies at the wrist with those of the control subjects, and they found statistically significant differences in seven patients (in six patients distal latencies were within the normal range but consistently above the mean value). The application, in both ADM and FDI muscles recording, in a case of ulnar neuropathy at the elbow (UNE) is here reported (Figs. 4 and 5).

Table 2 Normal values [3]

	Mean ± SD	Range
Elbow–wrist, forearm conduction velocity (m/s)	54.5 ± 5.5	47.5–64.8
Axilla–elbow, arm conduction velocity (m/s)	59.8 ± 6.1	50.0–73.9

Comment

Gilliatt and Thomas [3] observed a small but statistically significant difference in the motor nerve conduction velocity between the upper arm (axilla–above elbow) and the forearm (wrist–above elbow) segment, founding the former a slightly higher velocity than the latter. For the authors, this difference was due to cooling of the ulnar nerve in the elbow region where it occupies a superficial position. In all their patients, the most constant finding was a reduction in conduction velocity in the wrist–above-elbow segment, with more variable slowing in nerve conduction velocity distal to the wrist. They suggested that distal latency increasing was strictly related to the severity of the ulnar neuropathy at the elbow (UNE), but their sample was not large enough to establish this point with certainty.

Pathological waveform (wrist, above elbow, axilla – ADM muscle):

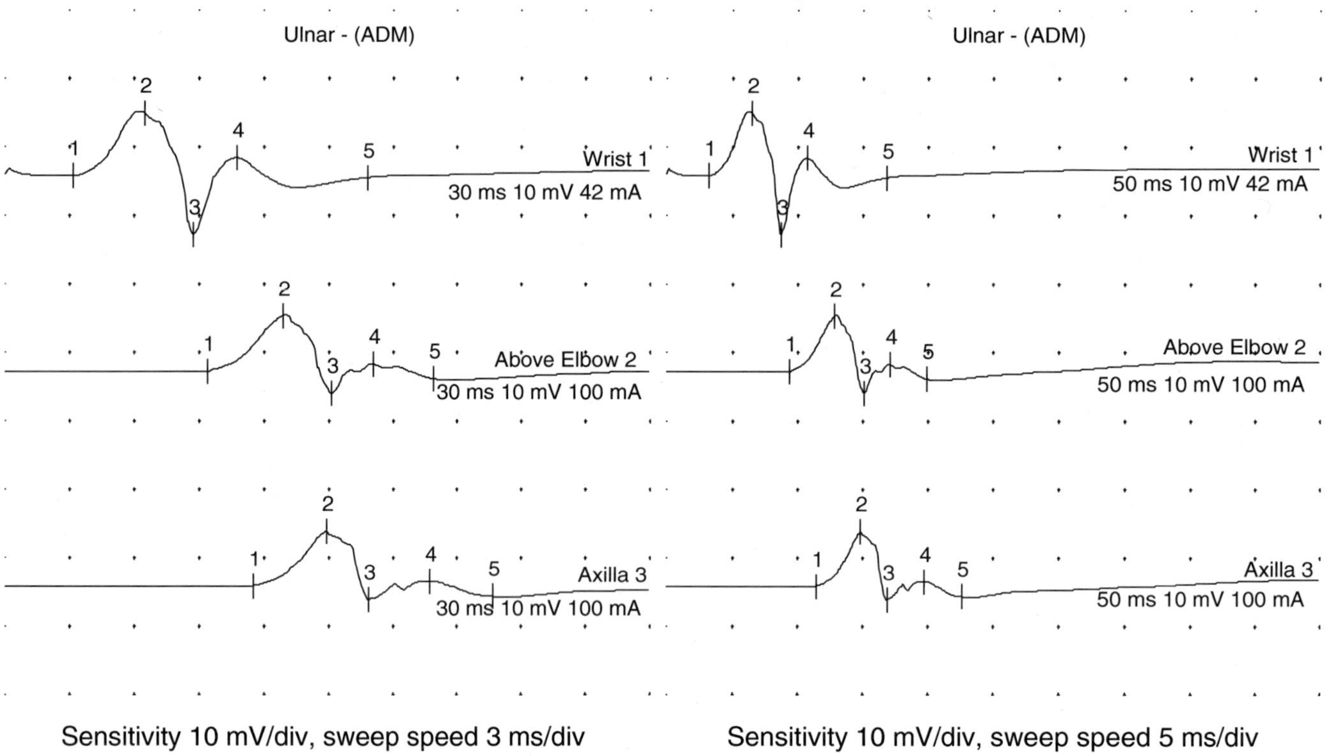

Onset latency (wrist – ADM): 3.25 ms; **Onset latency** (above elbow – ADM): 9.45 ms; **Onset latency** (axilla – ADM): 11.20 ms; **Peak LATENCY** (wrist – ADM): 8.90 ms; **Peak latency** (above elbow – ADM): 15.0 ms; **Peak latency** (axilla – ADM): 16.75 ms; **Onset to peak amplitude** (wrist – ADM): 9.2 mV; **Onset to peak amplitude** (above elbow – ADM): 8.0 mV; **Onset to peak amplitude** (axilla – ADM): 8.0 mV; **Peak to peak amplitude** (wrist – ADM): 6.5 mV; **Peak to peak amplitude** (above elbow – ADM): 7.1 mV; **Peak to peak amplitude** (axilla – ADM): 7.2 mV; **MNCV** (wrist – above elbow, 24 cm): 38.7 m/s; **MNCV** (axilla – above elbow, 12 cm): 68.6 m/s

Fig. 4 Compound muscle action potentials (*CMAPs*) recorded at the hand from the ADM muscle using needle recording method [1] in UNE, stimulation of the wrist (*upper trace*), above the elbow (*middle trace*), and axilla (*lower trace*)

Pathological waveform (wrist, above elbow, axilla – FDI muscle):

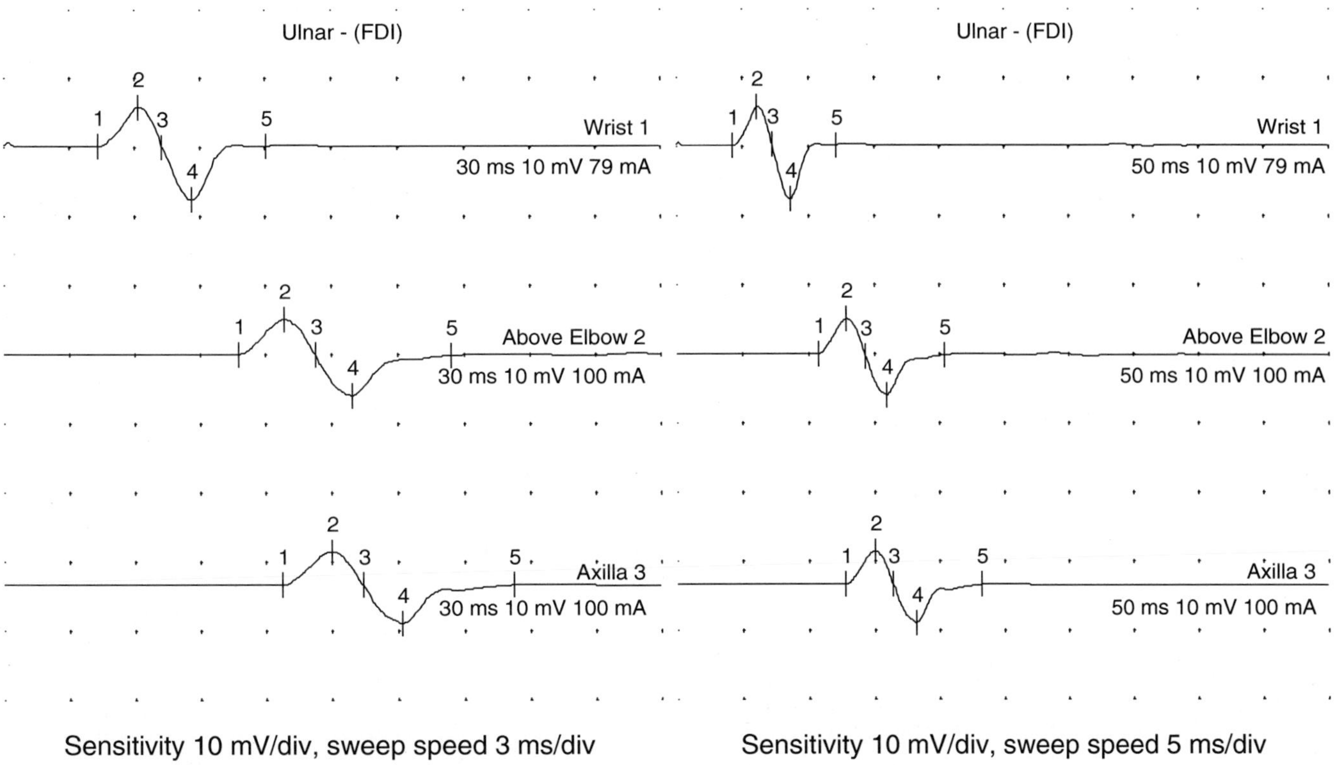

Onset latency (wrist – FDI): 4.30 ms; **Onset latency** (above elbow – FDI): 10.70 ms; **Onset latency** (axilla – FDI): 12.75 ms; **Peak latency** (wrist – FDI): 6.10 ms; **Peak latency** (above elbow – FDI): 12.95 ms; **Peak latency** (axilla – FDI): 15.0 ms; **Onset to peak amplitude** (wrist – FDI): 5.7 mV; **Onset to peak amplitude** (above elbow – FDI): 5.1 mV; **Peak to peak amplitude** (wrist – FDI): 13.6 mV; **Peak to peak amplitude** (above elbow – FDI): 11.1 mV; **Peak to peak amplitude** (axilla – FDI): 10.5 mV; **MNCV** (wrist – above elbow, 24 cm): 37.5 m/s; **MNCV** (axilla – above elbow, 12 cm): 58.5 m/s

Fig. 5 Compound muscle action potentials (*CMAPs*) recorded at the hand from the FDI muscle using needle recording method [1] in UNE, stimulation of the wrist (*upper trace*), above the elbow (*middle trace*), and axilla (*lower trace*)

References

1. Ebeling P, Gilliatt RW, Thomas PK (1960) A clinical and electrical study of ulnar nerve lesions in the hand. J Neurol Neurosurg Psychiatry 23:1–9
2. Simpson JA (1956) Electrical signs in the diagnosis of carpal tunnel and related syndromes. J Neurol Neurosurg Psychiatry 19:275–280
3. Gilliatt RW, Thomas PK (1960) Changes in nerve conduction with ulnar lesions at the elbow. J Neurol Neurosurg Psychiatry 23:312–320

Axilla, Above Elbow, Mid-Forearm, Wrist – Hand

M6

Surface Recording Technique, Study from *Abductor Digiti Minimi* (ADM) Muscle

Original Settings Sensitivity was 5 mV/division, duration of pulse was 0.2–0.5 ms; sweep speed, low-frequency filter, high-frequency filter, and the machine used were not specified.

Position Following the text, this study was performed in the supine position with the arm held in approximately 70° of flexion, both during stimulation and during measurements of distances.

Recording The active electrode (A) was placed over the belly (motor point) of the *abductor digiti minimi* (ADM) muscle (Fig. 1); the reference (R) was placed distally near its tendon of insertion [1]. The author used silver disk skin-recording electrodes, applied to the skin with conducting jelly and attached with collodion on gummed tape. Distances between the recording and stimulating electrodes were not fixed. The ground (G) electrode was placed on the forearm (Fig. 2).

Stimulation The ulnar nerve was stimulated at four sites: at the axilla (S1), above the elbow (S2), at the mid-forearm (S3), and at the wrist (S4). The cathode (−) stimulating electrode was positioned on the nerve trunk at each site of stimulation; a metal plate stimulating anode (+) was attached proximal to cathode in each stimulation (Figs. 1, 2, and 3). Single-pulse, square wave stimuli of from 0.2 to 0.5 ms duration were applied at half-second intervals. Effective stimulation voltages for just maximal responses were determined. These varied considerably for normal individuals, both for the same and different sites of stimulation, depending on variable skin resistances and the amount of adipose, tendon, or muscle tissue overlying the nerve at the point of stimulation.

No fixed distances were reported in the article. In all normal controls, the stimulation voltage was raised above that necessary to elicit a just maximal response (usually less than 100 V). In normal individuals, stimuli of greater duration than 0.5 ms frequently gave rise to considerable pain in the distribution of the stimulated nerve or local skin pain at the site of stimulation. Care must be taken in the spread of stimulation of concomitant activation of nearby nerves (in the upper arm, the median and ulnar nerve lies close to one another) resulting in simultaneous recording of contraction activity from nearby muscles. We use conventional bipolar stimulator, and the anode was proximal in all stimulations.

Measurements The authors measured the "terminal conduction time" and "nerve conduction velocity." They explained that by subtracting the conduction times obtained with stimulation of the same nerve at two different points, one obtains the time taken for conduction of the nerve action potential along the segment of the nerve between the two points, a true nerve conduction time. The distance between the two points measured along the skin is a fairly accurate estimate of the length of the underlying segment of nerve. By dividing the distance by time, the motor nerve conduction velocity (MNCV) along the nerve segment can be determined and measured in meter per second (m/s). When conduction time is specified as the time from the stimulus artifact to the initial deflection, positive or negative, of the muscle potential, this time and the corresponding calculated velocity refer to the fastest conducting alpha motor fibers of the nerve. This meaning was implicit in the term "nerve conduction velocity" when used in their paper. Velocities can be determined along the proximal and distal segments or along the whole nerve, for example, from the axilla to the wrist in the case of the ulnar nerve. The greater the distance and time, the more reproducible and accurate is the measurement. Because there was no practical way of making a direct clinical measurement of the true nerve conduction time along the terminal branches of a peripheral motor nerve, for example, those distal to the wrist, "terminal conduction times" (which include the motor endplate delay), rather

G. Gentili, M. Di Napoli, *The Ulnar Nerve*, DOI 10.1007/978-3-319-43292-2_30

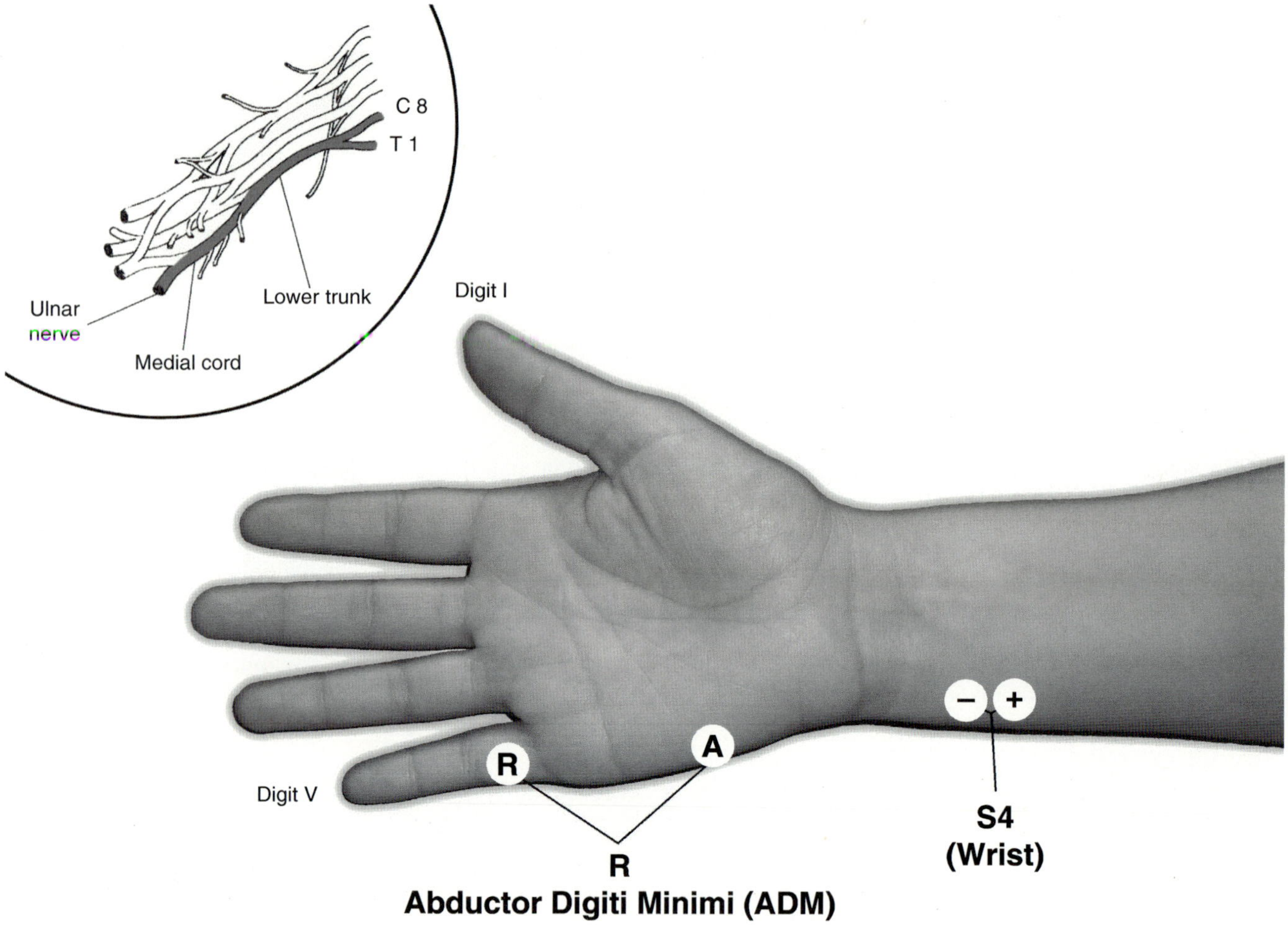

Typical waveform (axilla, above elbow, mid-forearm, wrist – ADM muscle):

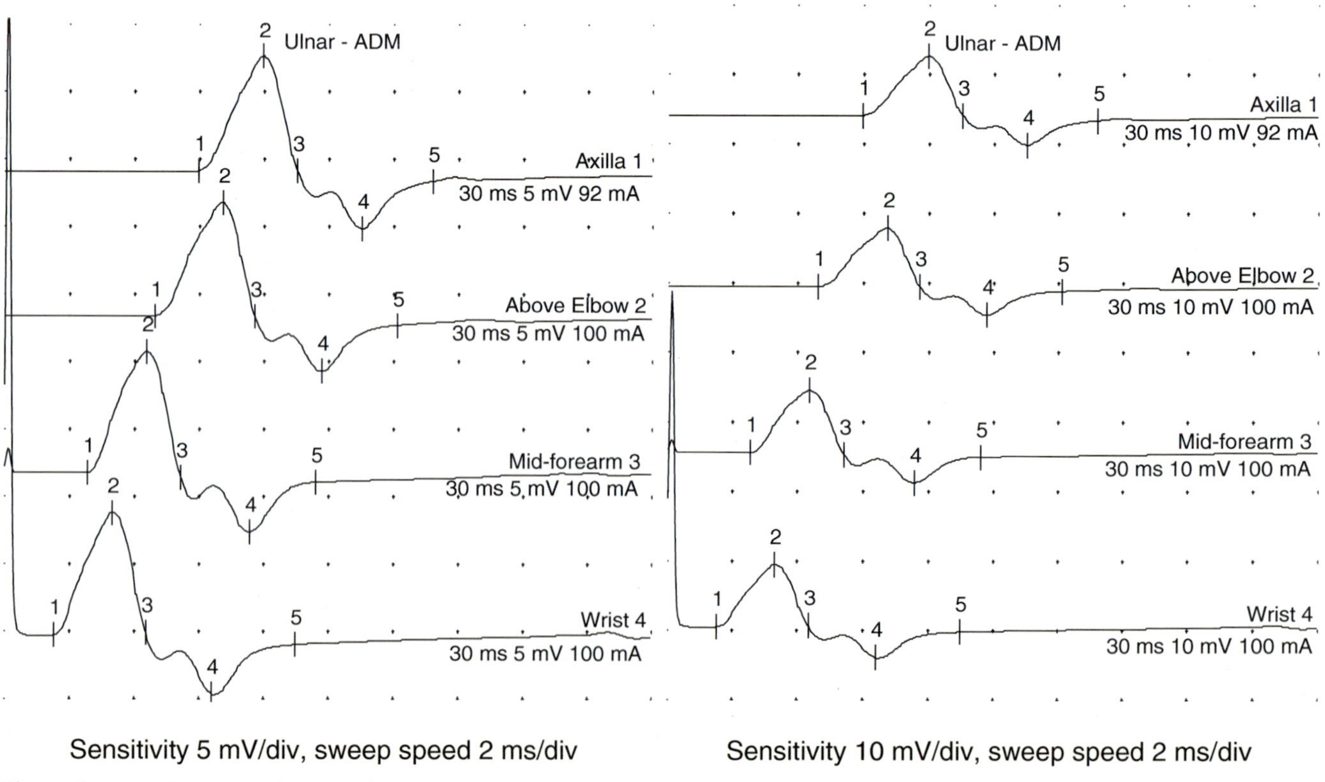

Fig. 1 Compound muscle action potentials (*CMAPs*) recorded at the hand from the ADM muscle, *top to bottom*: stimulation of the axilla, above the elbow, mid-forearm, and wrist

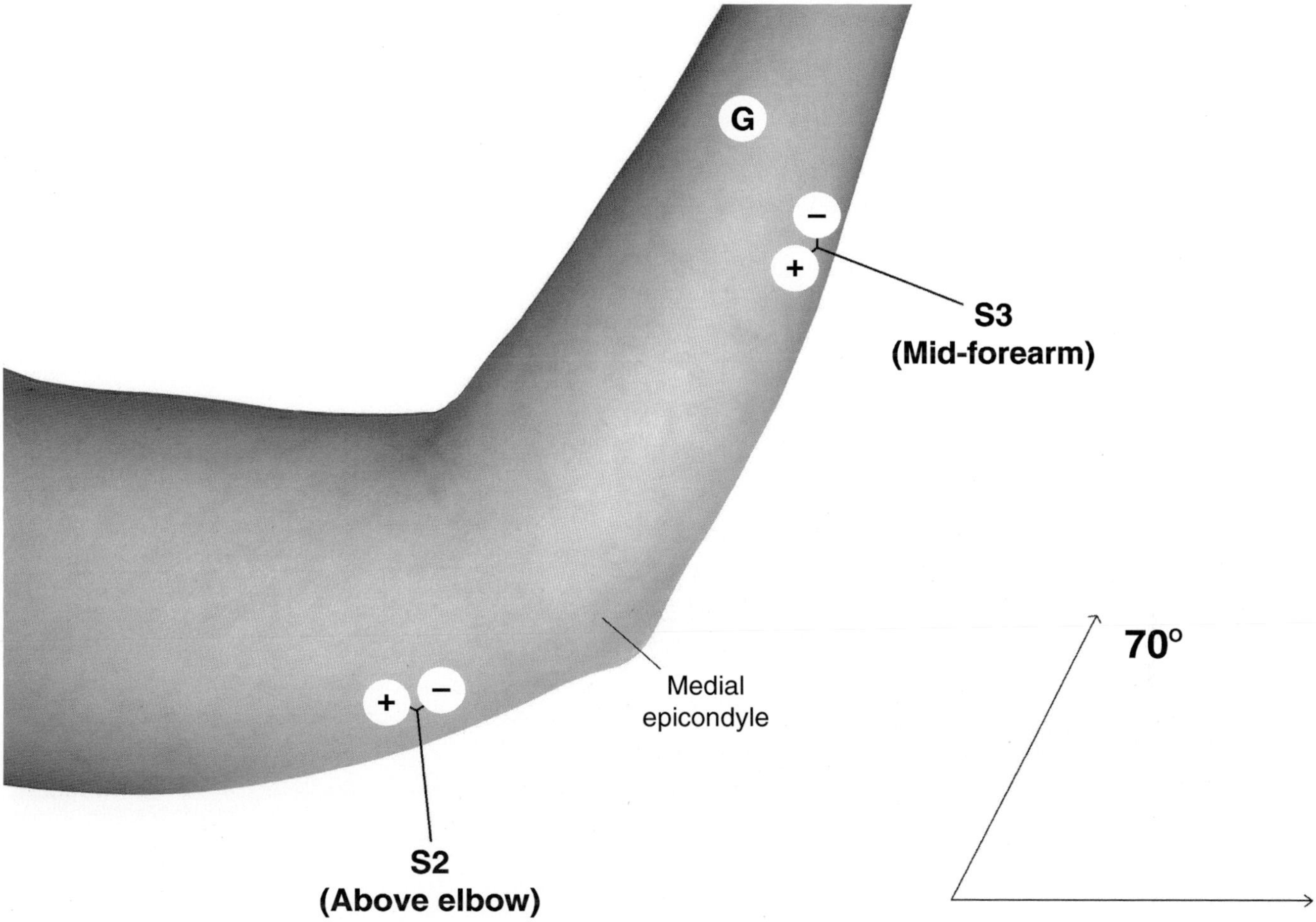

Fig. 2 Mid-forearm (*S3*) and above elbow (*S2*) stimulation point (elbow flexed to 70°)

than the conduction velocities, have been reported for distal portions of the ulnar nerve. Hodes and associates [2] calculated "residual latencies" for normal ulnar and median nerves. The term "residual latency" refers to the difference between a measured terminal conduction time and an estimated time for propagation of the nerve action potential from the wrist to the motor end plate, assuming the velocity of propagation to be the same in this terminal nerve segment as in a more proximal segment of the nerve. This estimated time is obtained by dividing the distance from the stimulation point at the wrist to the active muscle recording electrode by the nerve conduction velocity from the elbow to wrist. Distal latency (ms) of the compound muscle action potential (CMAP) was measured from the stimulus artifact to the onset of the negative deflection of CMAP. No control of temperature was attempted. Distances between the midpoints of each stimulation site on the skin were measured to the nearest millimeter. Normal values (Table 1) were obtained from nine ulnar nerves of seven normal subjects (residual latency was calculated from seven nerves of five subjects).

Comment

Mavor and Libman [1] found that average conduction velocity for the axilla to wrist segment of nine ulnar nerves in seven healthy subjects was 60.0 m/s (range 56.0–62.7 m/s). The average terminal conduction time on stimulation at the wrist was 2.8 ms (range 2.3–3.4 ms). Residual latencies were calculated for seven ulnar nerves in five individuals and averaged 1.5 ms (range 1.0–1.9 ms). The authors compared normal values obtained from seven normal controls with motor conduction values from a sample of patients affected by chronic peripheral neuropathy (three cases of hypertrophic interstitial neuropathy). They found a marked decrease of the conduction velocities of all nerves tested and increased thresholds to stimulation. Frequently, in this group of patients, the authors observed prolonged duration and fragmentation in CMAPs. In a small sample of patients (a total of four cases) with suspected cubital tunnel syndrome (CTS),

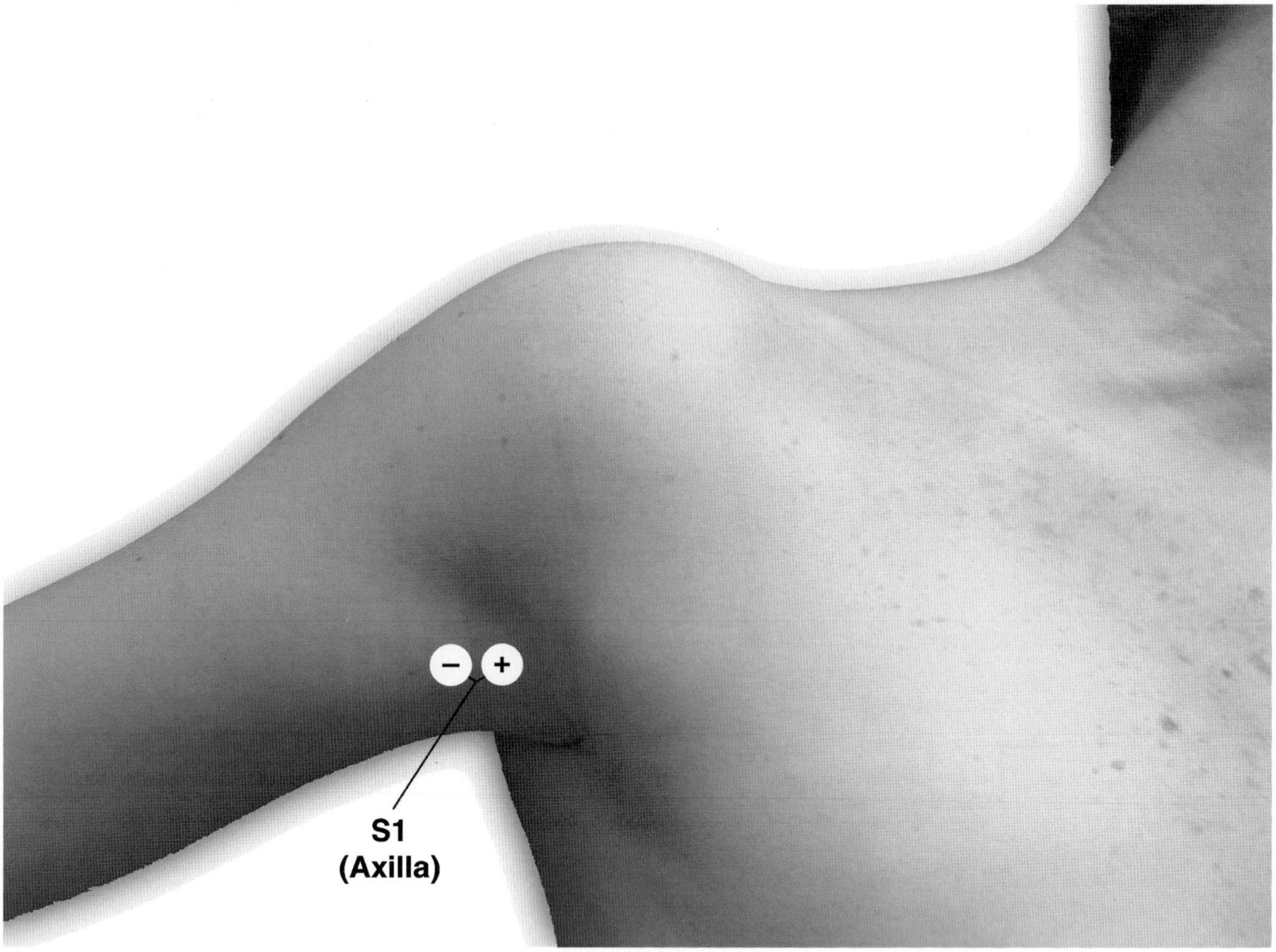

Fig. 3 Proximal stimulation point (*S1 – axilla*)

Table 1 Normal values [1]

	Average	Range
Axilla–wrist, MNCV (m/s)	60.0	56.0–62.7
Wrist–ADM, distal latency (ms)	2.8	2.3–3.4
Wrist–ADM, residual latency (ms)	1.5	1.0–1.9

the authors recorded decreased conduction velocities by 4–17 m/s along segments of the ulnar nerve on the affected side as compared with segments on the normal contralateral side, either from above the elbow to the wrist or to the mid-forearm. The application in a patient with a chronic inflammatory demyelinating polyneuropathy (CIDP) and in a case of ulnar neuropathy at the elbow (UNE) are here reported (Figs. 4 and 5, respectively).

Pathological waveform (axilla, above elbow, mid-forearm, wrist – ADM muscle):

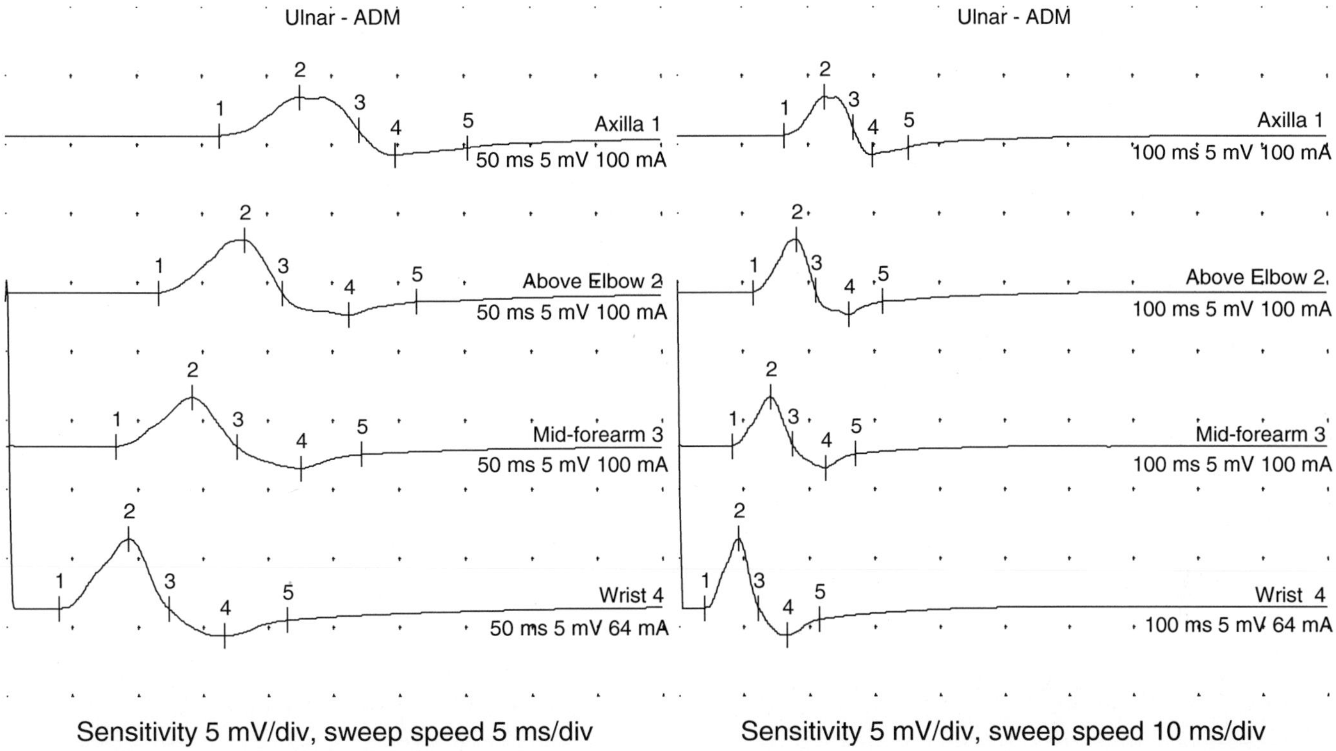

Onset latency (axilla – ADM): 16.35 ms; **Onset latency** (above elbow – ADM): 11.60 ms; **Onset latency** (mid-forearm – ADM): 8.35 ms; **Onset latency** (wrist – ADM): 3.95 ms; **Peak latency** (axilla – ADM): 22.50 ms; **Peak latency** (above elbow – ADM): 17.75 ms; **Peak latency** (mid-forearm – ADM): 14.05 ms; **Peak latency** (wrist – ADM): 9.10 ms; **Onset to peak amplitude** (axilla – ADM): 2.8 mV; **Onset to peak amplitude** (above elbow – ADM): 3.9 mV; **Onset to peak amplitude** (mid-forearm – ADM): 3.6 mV; **Onset to peak amplitude** (wrist – ADM): 5.2 mV; **Peak to peak amplitude** (axilla – ADM): 4.3 mV; **Peak to peak amplitude** (above elbow – ADM): 5.5 mV; **Peak to peak amplitude** (mid-forearm – ADM): 5.3 mV; **Peak to peak amplitude** (wrist – ADM): 7.1 mV; **MNCV** (axilla – above elbow, 14 cm): 29.5 m/s; **MNCV** (above elbow – mid-forearm, 10 cm): 30.8 m/s; **MNCV** (mid-forearm – wrist, 15 cm): 34.1 m/s

Fig. 4 Compound muscle action potentials (*CMAPs*) recorded at the hand from the ADM muscle using surface recording method [1] in a patient with CIDP associated with antiMAG/SGPG (pathogenic IgM antibodies – Abs against myelin-associated glycoprotein/peripheral nerve glycolipid sulfated glucuronyl paragloboside), *top to bottom*: stimulation of the axilla, above the elbow, mid-forearm, and wrist

Pathological waveform (axilla, above elbow, mid-forearm, wrist – ADM muscle):

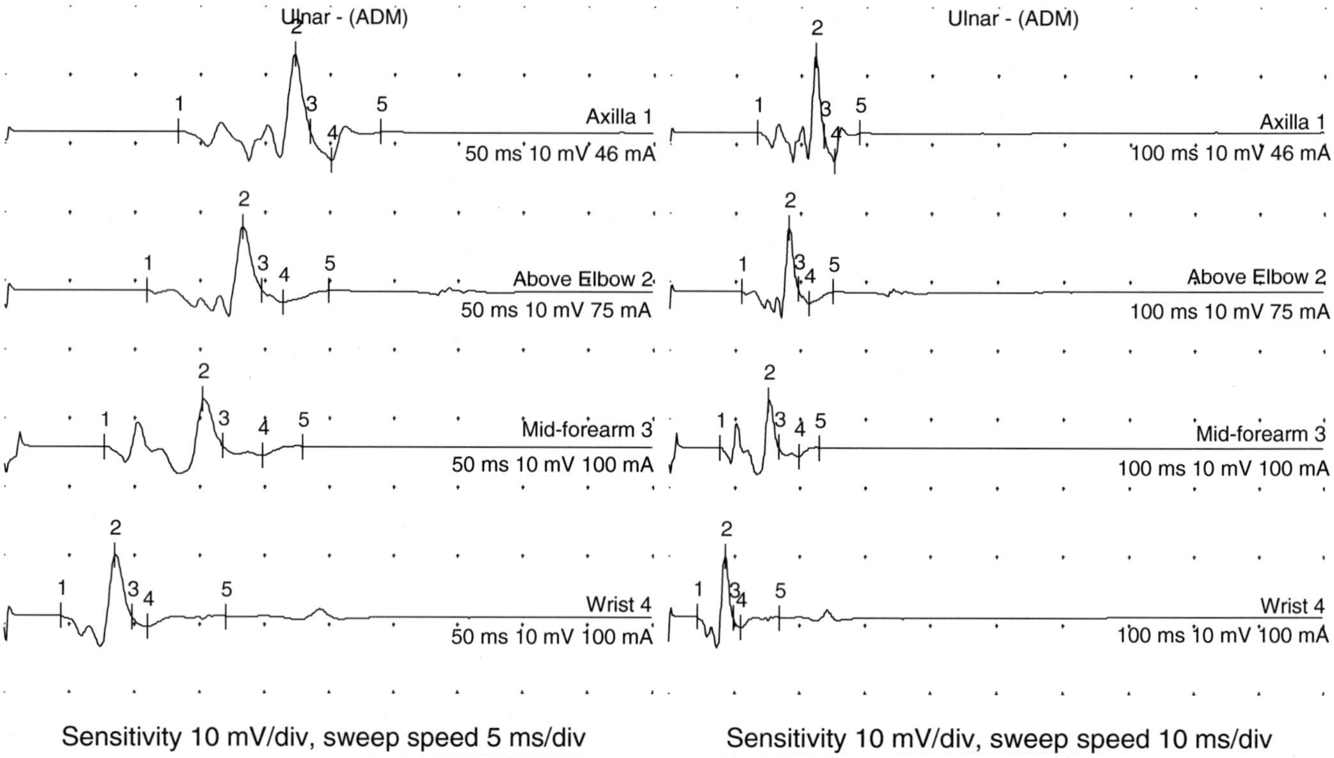

Onset latency (axilla – ADM): 13.25 ms; **Onset latency** (above elbow – ADM): 10.90 ms; **Onset latency** (mid-forearm – ADM): 7.60 ms; **Onset latency** (wrist – ADM): 5.0 ms; **Peak latency** (axilla – ADM): 22.35 ms; **Peak latency** (above elbow – ADM): 18.10 ms; **Peak latency** (mid-forearm – ADM): 15.05 ms; **Peak latency** (wrist – ADM): 8.20 ms; **Onset to peak amplitude** (axilla – ADM): 11.3 mV; **Onset to peak amplitude** (above elbow – ADM): 9.5 mV; **Onset to peak amplitude** (mid-forearm – ADM): 7.1 mV; **Onset to peak amplitude** (wrist – ADM): 9.1 mV; **Peak to peak amplitude** (axilla – ADM): 15.6 mV; **Peak to peak amplitude** (above elbow – ADM): 11.3 mV; **Peak to peak amplitude** (mid-forearm – ADM): 8.5 mV; **Peak to peak amplitude** (wrist – ADM): 10.6 mV; **MNCV** (axilla–above elbow, 14 cm): 68.1 m/s; **MNCV** (above elbow – mid-forearm, 12 cm): 36.4 m/s; **MNCV** (mid-forearm – wrist, 17 cm): 52.3 m/s

Fig. 5 Compound muscle action potentials (*CMAPs*) recorded at the hand from the ADM muscle using needle recording method in UNE (patient with failed submuscular transposition of the ulnar nerve), *top to bottom*: stimulation of the axilla, above the elbow, mid-forearm, and wrist

References

1. Mavor H, Libman I (1962) Motor nerve conduction velocity measurement as a diagnostic tool. Neurology 12:733–744
2. Hodes R, Larrabee MG, German W (1948) The human electromyogram in response to nerve stimulation and the conduction velocity of motor axons; studies on normal and on injured peripheral nerves. Arch Neurol Psychiatry 60:340–365

Axilla, Elbow, Wrist – Hand

M7

Surface Recording Technique, Study from *Abductor Digiti Minimi* (ADM) Muscle

Original Settings Duration was 0.1 ms; sensitivity, sweep speed, low-frequency filter, high-frequency filter, and the machine used were not specified.

Position This study was performed in the supine position with the arm and the elbow fully extended.

Recording The active electrode (A) was placed over the belly (motor point) of the *abductor digiti minimi* (ADM) muscle (Fig. 1); the reference (R) was placed distally near its tendon of insertion [1]. Distances between the recording and stimulating electrodes were not fixed. The ground (G) electrode position was not specified in the text; the figure shows the ground electrode placed on the forearm.

Stimulation The ulnar nerve was stimulated at three sites (Figs. 1 and 2): at the axilla (S1), at the elbow (S2), and at the wrist (S3). The cathode (–) stimulating electrode was distal; supramaximal stimuli of short duration (0.1 ms) and of a square wave pulse were delivered at each stimulation point. Intensity of stimulation ranged from 50 to 100 V, but in all cases supramaximal stimuli were applied by observing the maximal response and then increasing the voltage from 30 to 50 % or doubling the duration of stimulation [1].

Measurements Latency (ms) was determined from the stimulus artifact to the onset of the first negative deflection of the compound muscle action potential (CMAP). The distance (cm) between the stimulating and recording electrodes was measured on the skin to the nearest 0.1 cm. Arm temperature (°C) was checked with a needle thermistor inserted in the midportion of the upper arm and forearm along the course of the ulnar nerve. Temperature recordings were done on 14 extremities, and the mean value was 35.2 °C (upper arm) and 34.06 °C (forearm). The authors [1] calculated the motor nerve conduction velocity (MNCV) of each segment, as well as the total average conduction velocity, as the length of the segment measured, divided by the difference in time of conduction between any two successive stimulation points. The MNCV was measured in meter per second (m/s). The MNCV value so obtained represents a mean value for the fastest conducting motor fibers of the ulnar nerve. They chose to express the relation between velocities in the proximal (axilla–elbow) and distal (elbow–wrist) segments as a ratio of

$$\text{MNCV ratio} = \frac{\text{elbow} - \text{wrist MNCV, (m/s)}}{\text{axilla} - \text{elbow MNCV, (m/s)}}$$

Spiegel and Johnson [1] observed that only when the ratio was less than 1, the proximal MNCV was faster than the distal. They studied 38 ulnar nerves of 21 healthy subjects (age range 19–37 years, average age 26 years) founding a ratio <1 in only one-third of the sample, and a mean MNCV ratio of 1.05 ± 0.148. The mean proximal (axilla–elbow) segment was 60.3 m/s, and the mean distal (elbow–wrist) segment was 63.2 m/s.

Comment

Spiegel and Johnson [1] found a difference of 2.9 m/s between the mean proximal and distal MNCVs, with the velocity slightly faster in the distal than in the proximal segments. They observed some potential sources of error in clinical determinations of the MNCV which could have effects on the results obtained. The first was in counting the CMAP latency by two different investigators; it will be seen that the estimation of the latency to the nearest 0.2 ms provided accuracy to 1 part in 20, or 5 % error. Another source of error was inaccurate measurement of the distances on the skin between the sites of stimulation, particularly in the upper arm and the axilla, where the skin is loose and can be

G. Gentili, M. Di Napoli, *The Ulnar Nerve*, DOI 10.1007/978-3-319-43292-2_31

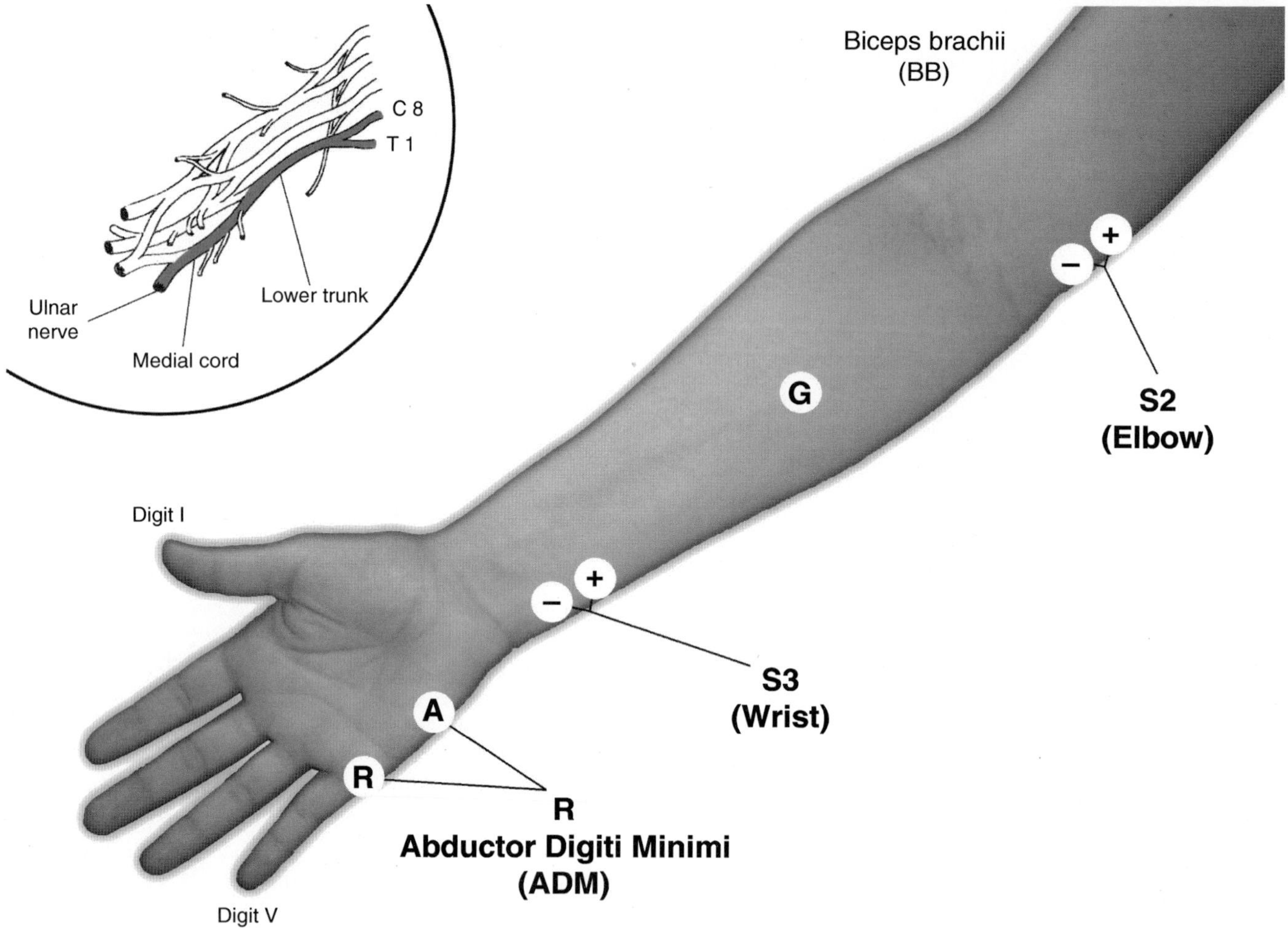

Typical waveform (axilla, elbow, wrist – ADM muscle):

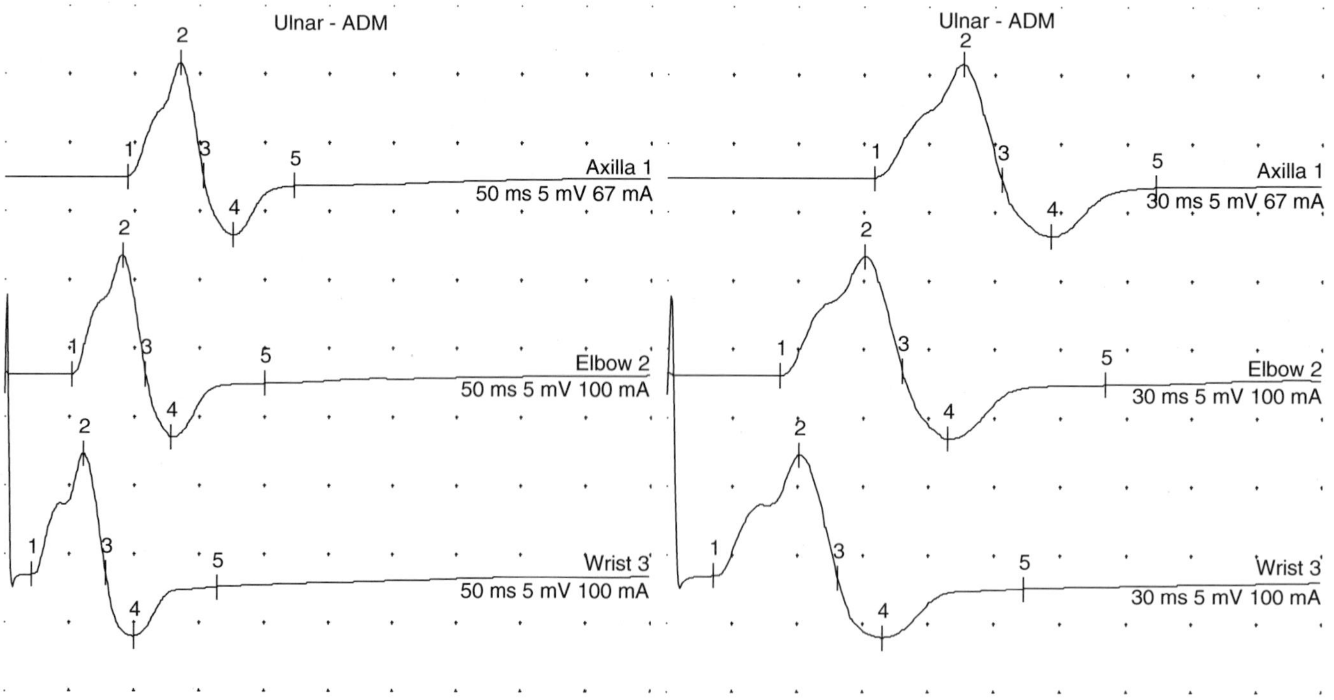

Sensitivity 5 mV/div, sweep speed 5 ms/div   Sensitivity 5 mV/div, sweep speed 3 ms/div

Fig. 1 Compound muscle action potentials (*CMAPs*) recorded at the hand from the ADM muscle, *top to bottom*: stimulation of the axilla, elbow, and wrist

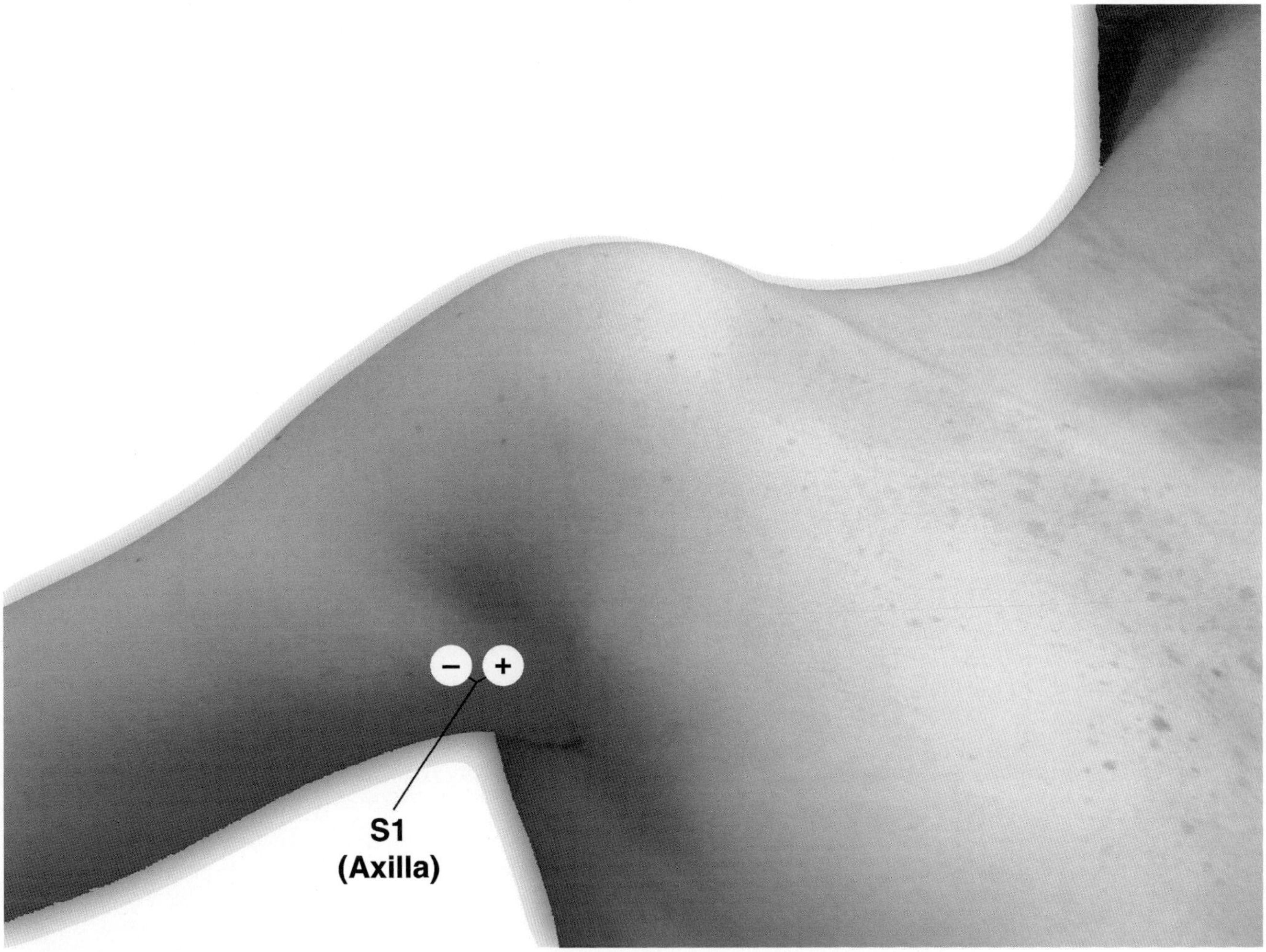

Fig. 2 Proximal stimulation point (S1 – axilla)

displaced easily by the pressure of the stimulating electrodes. An error of 1 cm in measurement of the distance could influence the final MNCV obtained by approximately 2–3 m/s. The authors [1] observed, in a 25 cm representative value for the length of the segment measured on the skin, accurate determination of the distance to the nearest 0.5 cm provided an accuracy of 1 part in 50, or 2 % error. Since this was a considerably smaller error than that obtained in determination of the CMAP latency (5 %), they found in the latency measurement a greater influence in precise determinations. Pathological waves and values in patients with ulnar neuropathy at the elbow (UNE) are here reported (Figs. 3 and 4).

Pathological waveform (axilla, elbow, wrist – ADM muscle):

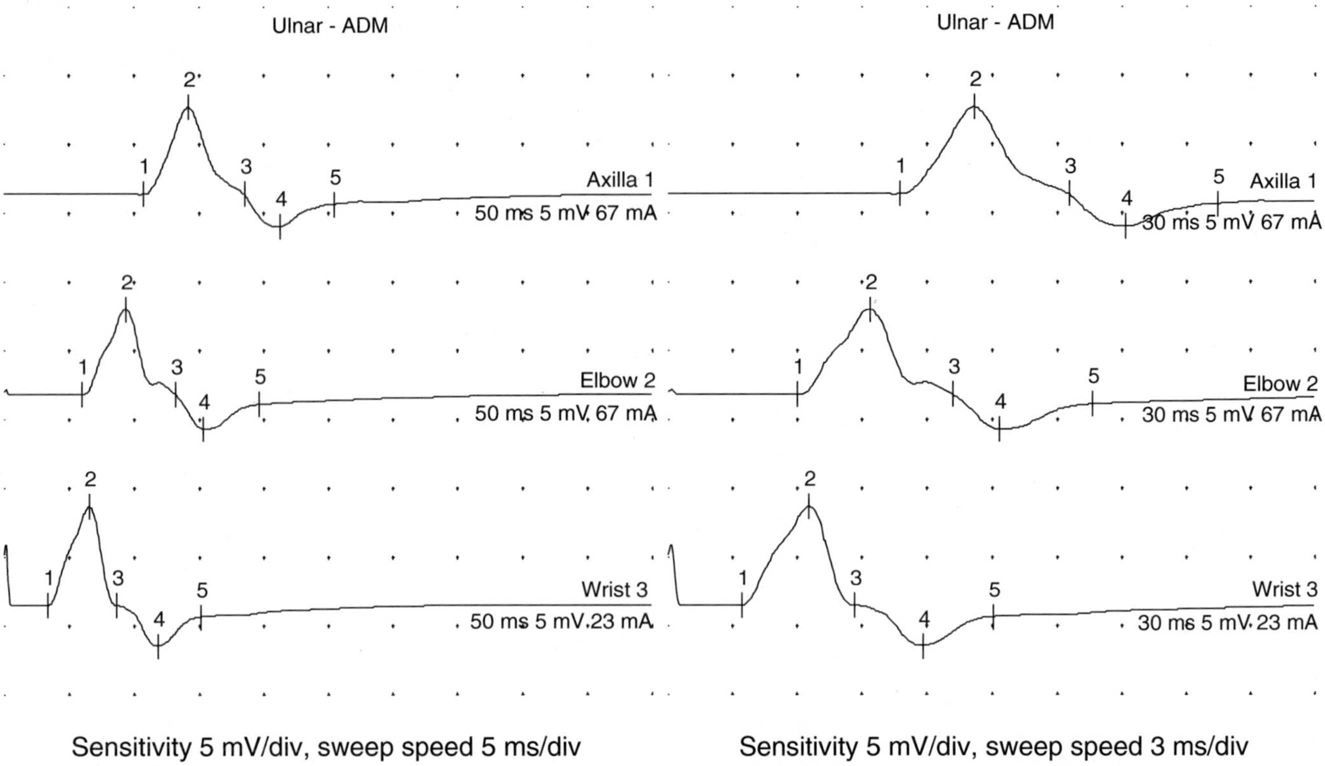

Onset latency (axilla – ADM): 10.75 ms; **Onset latency** (elbow – ADM): 6.0 ms; **Onset latency** (wrist – ADM): 3.40 ms; **Peak latency** (axilla – ADM): 13.80 ms; **Peak latency** (elbow – ADM): 9.65 ms; **Peak latency** (wrist – ADM): 6.45 ms; **Onset to peak amplitude** (axilla – ADM): 6.4 mV; **Onset to peak amplitude** (elbow – ADM): 6.2 mV; **Onset to peak amplitude** (wrist – ADM): 7.2 mV; **Peak to peak amplitude** (axilla – ADM): 8.8 mV; **Peak to peak amplitude** (elbow – ADM): 8.8 mV; **Peak to peak amplitude** (wrist – ADM): 10.2 mV; **MNCV** (axilla – elbow, 20 cm): 42.1 m/s; **MNCV** (elbow – wrist, 16 cm): 61.5 m/s

Fig. 3 Compound muscle action potentials (*CMAPs*) recorded at the hand using surface recording method [1] from the ADM muscle in a patient with UNE, *top to bottom*: stimulation of the axilla, elbow, and wrist

Pathological waveform (axilla, elbow, wrist – ADM muscle):

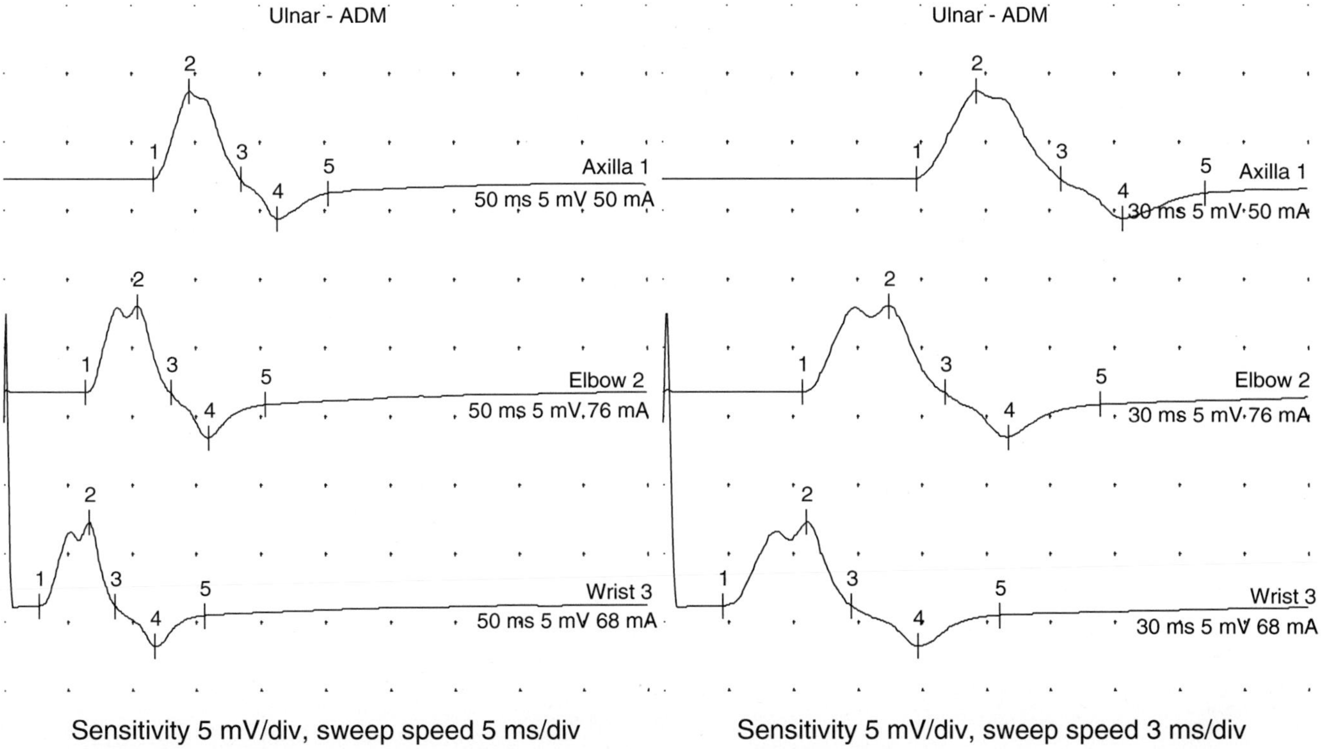

Onset latency (axilla – ADM): 11.75 ms; **Onset latency** (elbow – ADM): 6.40 ms; **Onset latency** (wrist – ADM): 2.70 ms; **Peak latency** (axilla – ADM): 14.85 ms; **Peak latency** (elbow – ADM): 10.55 ms; **Peak latency** (wrist – ADM): 6.55 ms; **Onset to peak amplitude** (axilla – ADM): 6.4 mV; **Onset to peak amplitude** (elbow – ADM): 6.3 mV; **Onset to peak amplitude** (wrist – ADM): 6.2 mV; **Peak to peak amplitude** (axilla – ADM): 9.4 mV; **Peak to peak amplitude** (elbow – ADM): 9.6 mV; **Peak to peak amplitude** (wrist – ADM): 9.2 mV; **MNCV** (axilla – elbow, 20 cm): 37.4 m/s; **MNCV** (elbow – wrist, 20 cm): 54.1 m/s

Fig. 4 Compound muscle action potentials (*CMAPs*) recorded at the hand using surface recording method [1] from the ADM muscle in a patient with UNE, *top to bottom*: stimulation of the axilla, elbow, and wrist

Reference

1. Spiegel MH, Johnson EW (1962) Conduction velocity in the proximal and distal segments of the motor fibers of the ulnar nerve of human beings. Arch Phys Med Rehabil 43:57–61

Wrist, Lower Forearm, Upper Forearm, Elbow, Arm – Hand

M8

Surface Recording Technique, Study from *Abductor Digiti Minimi* (ADM) Muscle

Original Settings The machine used was a TECA model B; sensitivity, sweep speed, low-frequency filter, high-frequency filter, and duration of pulse were not specified.

Position Following the text, this study was performed in the supine position with the arm in full extension. The forearm was supinated in the first part of the study and then pronated.

Recording The author [1] used silver disk surface electrodes to record the compound muscle action potentials (CMAPs). The active electrode (A) was placed over the belly (motor point) of the *abductor digiti minimi* (ADM) muscle; the reference (R) was placed distally near its tendon of insertion (Fig. 1). The ground (G) electrode was not specified in the text; the figure shows ground electrode placed on the palm.

Stimulation The ulnar nerve was stimulated at five sites: at the wrist (S1), lower forearm (S2), upper forearm (S3), just above the elbow (S4), and high in the upper arm (S5). The anode was proximal in all stimulations (Figs. 1 and 2). Distances between the recording and stimulating electrodes ranged between 7 and 12 cm, and attempts were made to keep the stimuli points equidistant and yet maintain close proximity to the superficial course of the nerve [1].

Measurements The purposes of the study were: (1) to determine whether or not a velocity decrement could be demonstrated in the normal ulnar nerve, and the answer to this question was important for the author [1], because in certain peripheral neuropathies, the peripheral fibers may be affected first and be reflected in the velocity change in distal segments or terminal fibers first, rather than affect the average velocity over a large segment of nerve trunk [2–4]; (2) to investigate how a velocity gradient could affect the residual latency. The velocity gradient may have clinical importance because the residual latency may be altered and thus have diagnostic value, particularly with reference to physiological or pathological nerve changes distal to the most distal stimulus point [5, 6]; (3) to determine if alterations in nerve velocity gradient may reflect or suggest localized pathology along the course of a superficial nerve; (4) to investigate the practical limitations of the accuracy in determining the velocity gradient and better understand all the factors involved. Motor nerve conduction velocity (MNCV) was computed by the formula "velocity = distance/time," and it was measured in meter per second (m/s). The distance was measured from each stimulus point to the pickup electrode over the muscle. Normal values (Tables 1 and 2) were obtained from 20 healthy subjects (age range 20–54 years, mean age 39.1 years). These were volunteers from secretaries, technicians, physicians, nurses, and hospital maintenance workers, all of whom had no signs or symptoms of any known neurological disease. Only the left ulnar was used because the author [1] observed no significant difference in the velocities between normal right and left sides. All subjects were tested on two occasions, several weeks apart. In the first phase of testing (Table 1), Schubert [1] used multiple points for stimulation (wrist, lower forearm, upper forearm, elbow, arm) ranging from 7 to 12 cm apart, and the CMAPs were recorded from the ADM muscle. The MNCV of four segments (wrist–lower forearm, lower forearm–upper forearm, upper forearm–elbow, elbow–arm) was calculated in each subject. A few weeks later, in the second phase of testing (Table 2), all subjects were retested but this time were used in stimulation at only three points along the ulnar nerve course (wrist, elbow, arm). The MNCV of two segments (wrist–elbow, elbow–arm) was calculated in each subject. In order to maintain the highest possible accuracy in the distance measurements during the second phase of testing, three ruler measurements were made at each shock point and pickup electrode point, and these three measurements were averaged and the average measurement was used in the calculation. The temperature of the laboratory was checked and maintained between 22 and 26°C.

G. Gentili, M. Di Napoli, *The Ulnar Nerve*, DOI 10.1007/978-3-319-43292-2_32

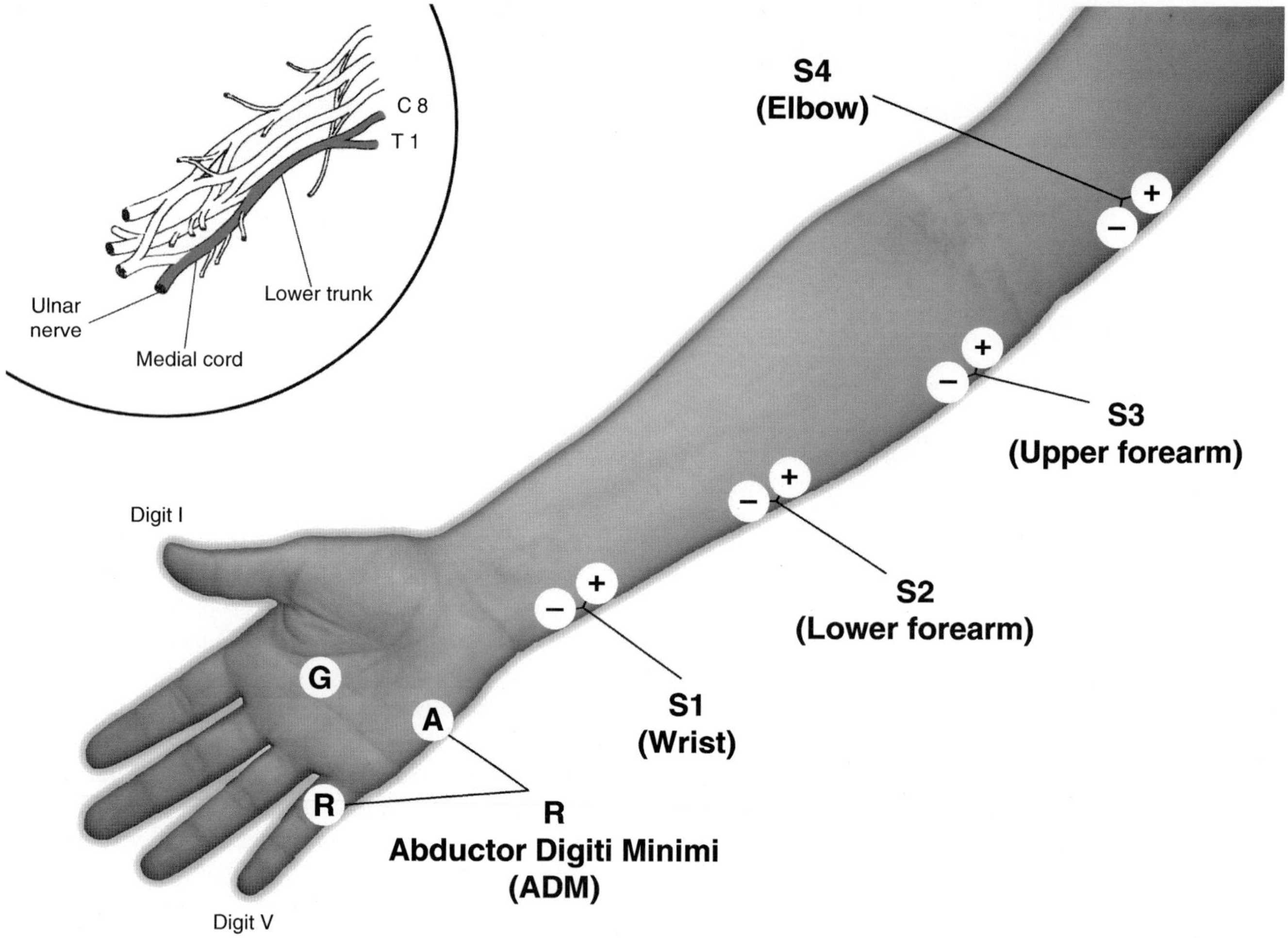

Typical waveform (wrist, lower forearm, upper forearm, elbow, arm – ADM muscle):

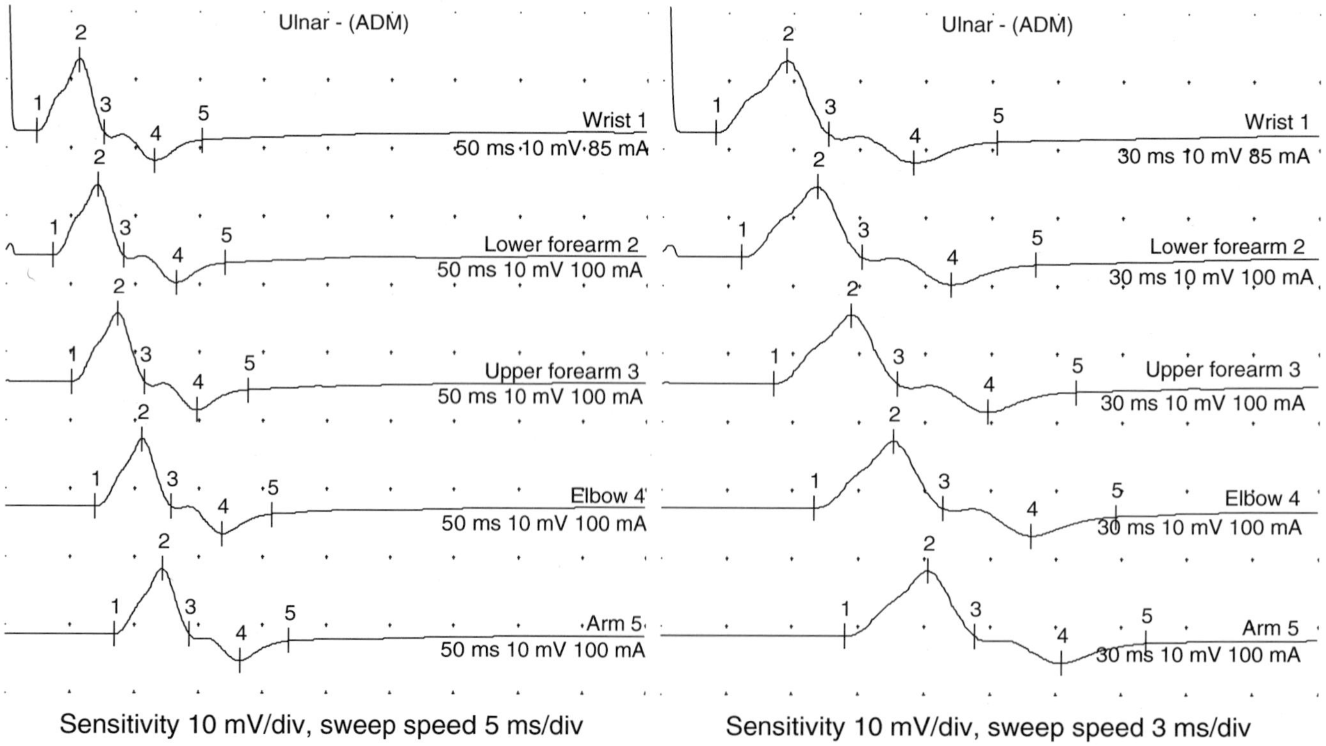

Fig. 1 Compound muscle action potentials (*CMAPs*) recorded at the hand from the ADM muscle, *top to bottom*: stimulation of the wrist, lower forearm, upper forearm, elbow, and of the arm

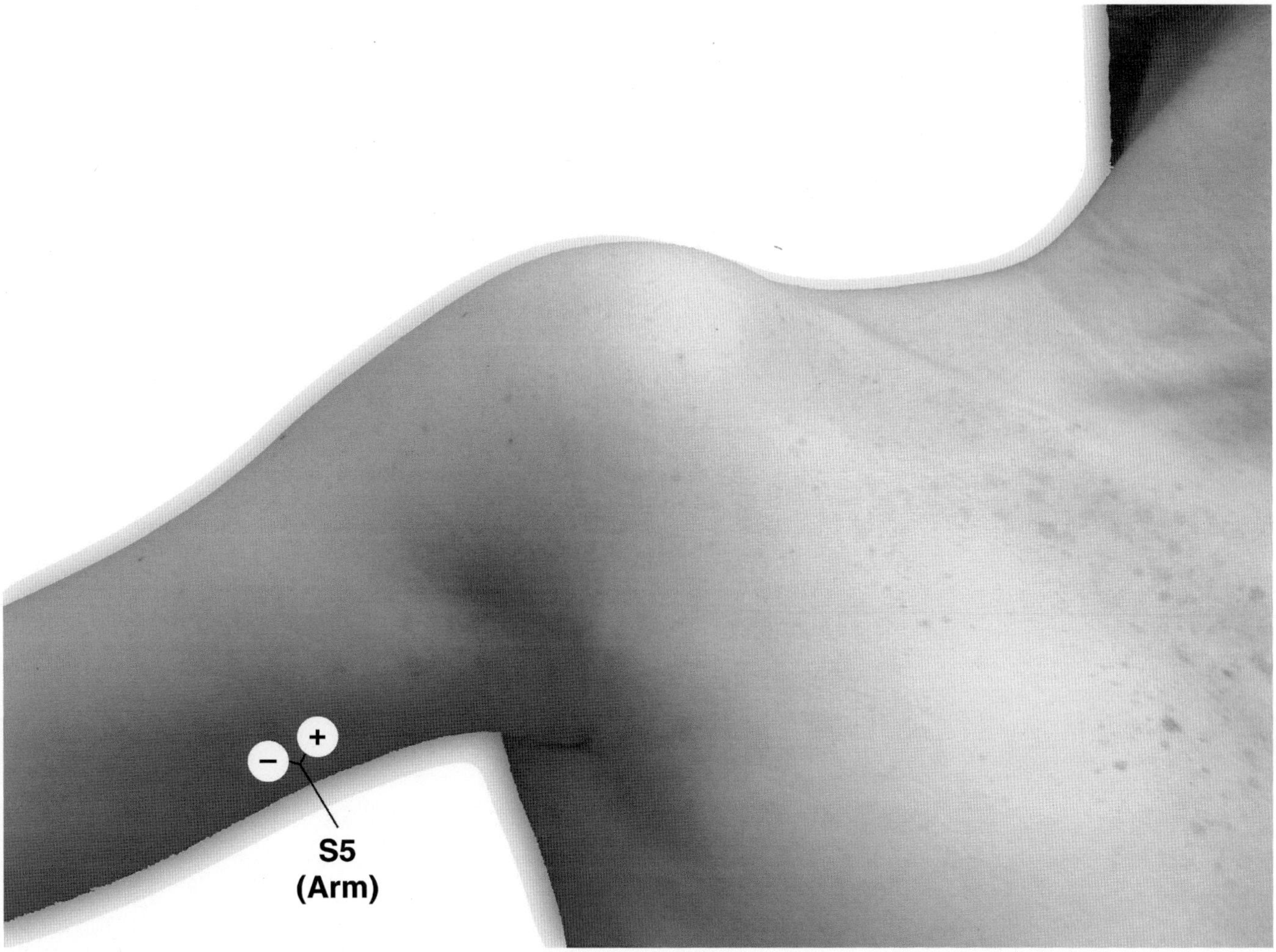

Fig. 2 Proximal stimulation point (*S5 – arm*)

Table 1 Normal values [1]

	Mean ± SD	Range
Wrist–lower forearm, MNCV (m/s)	60.44 ± 7.42	51.2–75.0
Wrist–lower forearm, distance (cm)	7.36 ± 5.65	4.1–9.75
Lower forearm–upper forearm, MNCV (m/s)	73.79 ± 16.84	51.2–110.7
Lower forearm–upper forearm, distance (cm)	7.55 ± 4.45	6.25–10.7
Upper forearm–elbow, MNCV (m/s)	54.63 ± 9.64	35.0–76.3
Upper forearm–elbow, distance (cm)	10.31 ± 6.7	7.8–14.5
Elbow–arm, MNCV (m/s)	53.93 ± 8.08	38.5–68.1
Elbow–arm, distance (cm)	10.06 ± 6.0	7.0–13.0

Table 2 Normal values [1]

	Mean ± SD	Range
Wrist–elbow, MNCV (m/s)	61.99 ± 4.25	53.8–69.7
Wrist–elbow, distance (cm)	25.78 ± 10.1	23.0–33.1
Elbow–arm, MNCV (m/s)	62.51 ± 5.87	53.2–71.5
Elbow–arm, distance (cm)	15.69 ± 9.1	10.0–19.1

Comment

For Schubert [1], due to softness of the tissue and elasticity of the skin, a ruler measurement at the surface may vary 2 cm in any direction, and the measurement variation is also enhanced further by pressure and posture changes appearing during the application of the stimulating electrode. If velocity calculations are made with interstimulus points 10 cm distant, for the author [1] the error in the velocity calculation may be 40%, when the skin measurement varied 2 cm in opposite directions at both stimuli points, and when calculation is based on 24 cm interstimulus distance, an error of 16% may be expected with a 4 cm variation in measurement. The author observed that a large error was introduced in calculating the velocity when the stimuli were applied at points where the nerve trunk was not superficial, such as lower and upper forearm stimulation sites. The velocity calculations are accurate when the interstimulus distances are over 10 or even 15 cm, when the stimuli are applied at a superficial course of the nerve and when careful measurements are made. Residual latency [6] can be defined as the time difference between the measured latency (observed) and the calculated latency (expected), as follows:

$$\text{Residual latency, RL}(\text{ms}) = \text{observed latency}(\text{ms}) - \text{expected latency}(\text{ms})$$

$$\text{Expected latency}(\text{ms}) = \frac{\text{distance between stimulating cathode and active electrode (ADM muscle), (mm)}}{\text{nerve conduction velocity of the tested segment, (m/s)}}$$

For Schubert [1], since the residual latency represents a measured period of time in excess of a calculated period of time, it is subject to several variables. These include (a) accuracy of measurement and calculations; (b) nerve velocity; (c) slowing of impulses in terminal fibers; (d) synaptic delay, including depolarizations at the end plate; and (e) conduction time of the action potential through muscle tissue and skin to pickup electrodes. For the author, there is a direct effect of the velocity gradient on the residual latency, and if the conduction velocity decreases peripherally in the nerve segments between the stimulus points, the residual latency will decrease, and if the conduction time is delayed (increased) distal to the most distal stimulus point, the residual latency will be increased with the same average velocity.

References

1. Schubert HA (1964) Conduction velocities along course of ulnar nerve. J Appl Physiol 19:423–426
2. Magladery JW, McDougal DB (1950) Electrophysiological studies of nerve and reflex activities in normal man; identification of certain reflexes in electromyogram and conduction velocity of peripheral nerve fibers. Bull Johns Hopkins Hosp 86:265–290
3. Magladery JW, McDougal DB, Stoll J (1950) Electrophysiological studies of nerve and reflex activities in normal man; effects of peripheral ischemia. Bull Johns Hopkins Hosp 86:291–312
4. Magladery JW, McDougal DB, Stoll J (1950) Electrophysiological studies of nerve and reflex activities in normal man; post-ischemic state. Bull Johns Hopkins Hosp 86:313–340
5. Hodes R, Larrabee MG, German WU (1948) Human electromyogram in response to nerve stimulation and conduction velocity of motor axons: studies on normal and on injured peripheral nerves. Arch Neurol Psychiatry 60:340–365
6. Low MD, Basmajian JV, Lyons GM (1962) Conduction velocity and residual latency in the human ulnar nerve and the effects on them of ethyl alcohol. Am J Med Sci 244:720–730

Axilla, Above Elbow, Wrist – Hand

M9

Surface Recording Technique, Study from *Abductor Digiti Minimi* (ADM) Muscle

Original Settings Sensitivity was 5 mV/division, duration of pulse was 0.1–0.5 ms, and sweep speed was 10 ms/division; low-frequency filter, high-frequency filter, and the machine used were not specified.

Position Following the text, this study was performed in the supine position with the arm in the extended position.

Recording The active electrode (A) was placed over the belly (motor point) of the *abductor digiti minimi* (ADM) muscle (Fig. 1). The author used for *hypothenar* muscle the term *abductor digiti quinti*; they used silver disk surface electrodes, and on occasion recordings were made using monopolar concentric needles in the muscle [1]. Needle electrodes were especially used when distal muscles were atrophic. The figure shows the active (A) electrode placed on the motor point of the ADM muscle and the reference (R) placed distally near its tendon of insertion. Distances between the recording and stimulating electrodes were not fixed. The ground (G) electrode position was not specified in the text; the figure shows ground placed on the forearm.

Stimulation The ulnar nerve was stimulated at three sites: at the axilla (S1), above the elbow (S2), and at the wrist (S3). A supramaximal electric stimulus from 0.1 to 0.5 ms in duration was delivered at each site of stimulation. No fixed distance was reported in the article. The cathode (−) and anode (+) stimulating electrodes were not mentioned in the text; we used conventional bipolar stimulator, and the anode was proximal in all stimulations (Figs. 1 and 2).

Measurements Distal latency (ms) of the compound muscle action potential (CMAP) was measured from the onset of the stimulus artifact to the junction of the negative deflection of CMAP and the baseline. By dividing the distance by time, the motor nerve conduction velocity (MNCV) along the nerve segment can be determined and measured in meter per second (m/s). The amplitude (mV) of the CMAP was also observed and measured. The temperature of the laboratory was monitored and maintained between 26 and 30°C. Skin temperature of the extremities ranged between 33 and 36°C (surface measurements). Normal conduction values (Table 1) were obtained from 64 healthy subjects (age range 10–80 years) divided into three groups: Group 1 (30 cases), age range 10–35 years; Group 2 (16 cases), age range 36–50 years; and Group 3 (18 cases), age range 51–80 years. Mayer [1] also recorded CMAPs in 41 patients (Table 2) with diabetes mellitus without clinical evidence of peripheral neuropathy divided into three groups: Group 1 (14 cases), age range 10–35 years; Group 2 (19 cases), age range 36–50 years; and Group 3 (8 cases), age range 51–80 years. 64 patients (Table 3) with diabetes mellitus with clinical evidence of peripheral neuropathy were also studied and divided into three groups: Group 1 (9 cases), age range 10–35 years; Group 2 (22 cases), age range 36–50 years; and Group 3 (33 cases), age range 51–80 years.

Comment

The values of normal conduction velocity in the largest motor fibers to the small muscles of the hand, as the *abductor digiti minimi* (ADM) muscle, determined by Mayer [1] were similar to those reported by Thomas et al. [2]. The author measured the intramuscular temperature in the arm (34 °C) and in the forearm (37°C) with a needle thermistor; when the extremity was warmed, he observed that the velocity increased uniformly and the proximal segment remained faster. After the age of 50 and mostly in elderly patients, Mayer [1] frequently observed a decreased conduction velocity in the elbow segment, and he suggested that repeated trauma and secondary ischemia related to a chronic compression may be a factor in slowing conduction distally in the aged.

G. Gentili, M. Di Napoli, *The Ulnar Nerve*, DOI 10.1007/978-3-319-43292-2_33

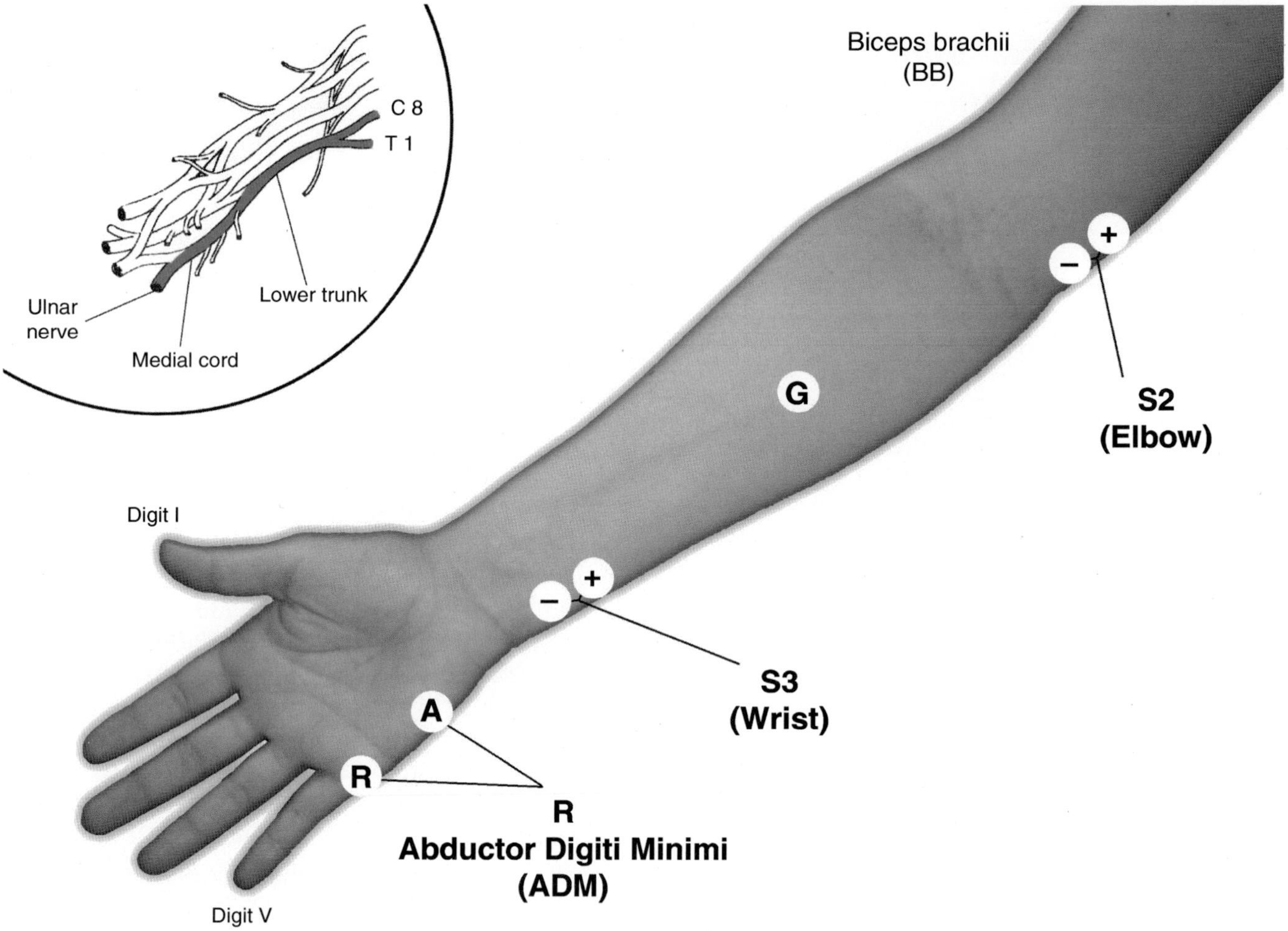

Typical waveform (axilla, above elbow, wrist – ADM muscle):

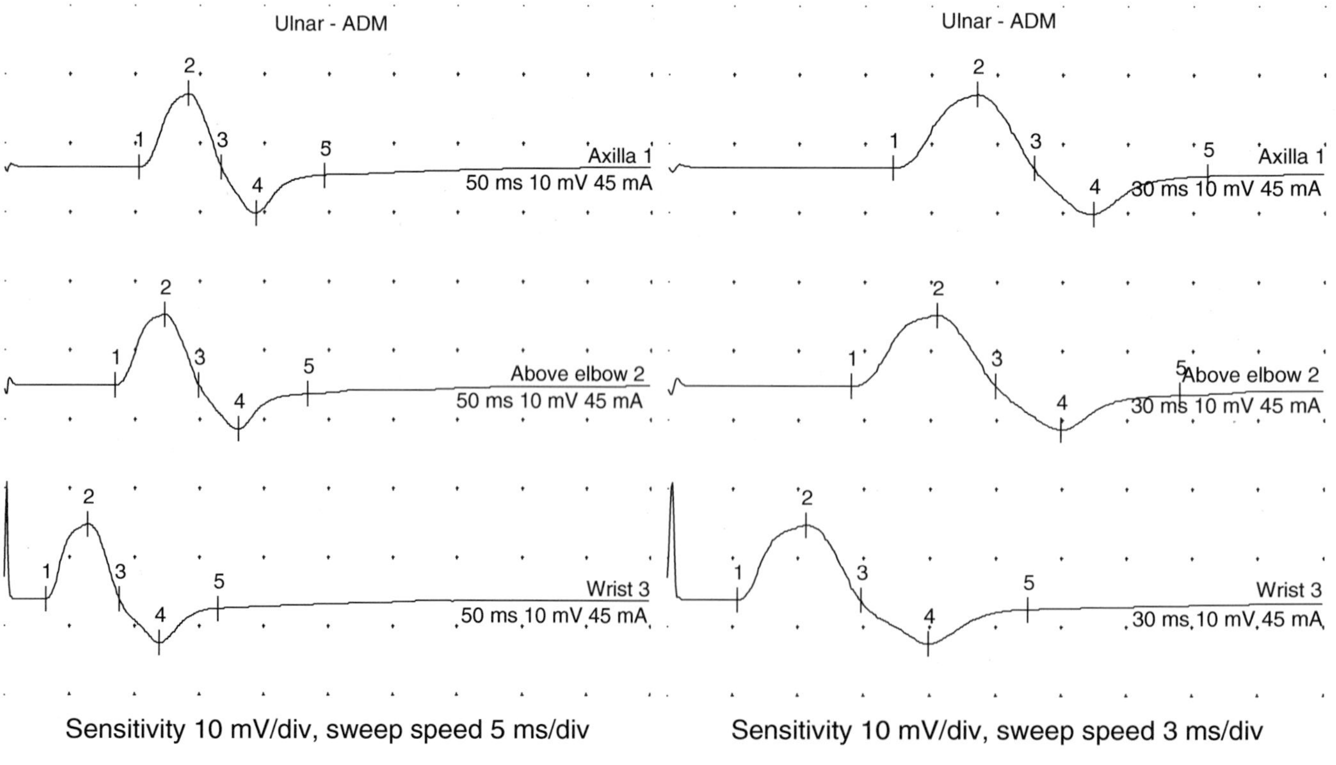

Fig. 1 Compound muscle action potentials (*CMAPs*) recorded at the hand from the ADM muscle, stimulation of the axilla (*upper trace*), above the elbow (*middle trace*), and of the wrist (*lower trace*)

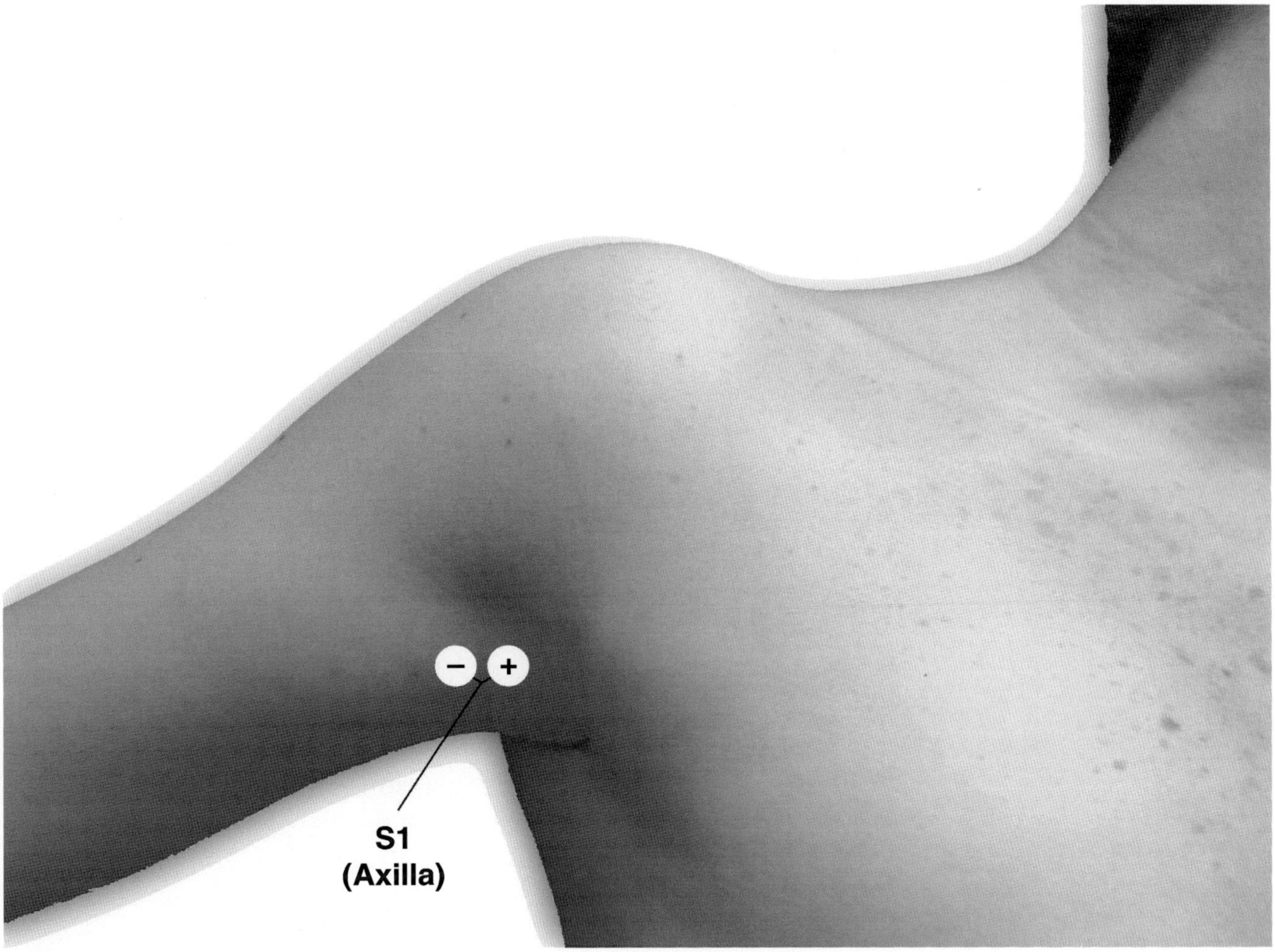

Fig. 2 Proximal stimulation point (S1 – axilla)

Table 1 Normal values [1]

Group 1	Mean ± SD
Wrist–ADM, distal latency (ms)	2.7 ± 0.3
Wrist–elbow, MNCV (m/s)	58.9 ± 2.2
Elbow–axilla, MNCV (m/s)	64.4 ± 2.6
Group 2	Mean ± SD
Wrist–ADM, distal latency (ms)	2.7 ± 0.3
Wrist–elbow, MNCV (m/s)	57.8 ± 2.1
Elbow–axilla, MNCV (m/s)	63.3 ± 2.0
Group 3	Mean ± SD
Wrist–ADM, distal latency (ms)	3.0 ± 0.35
Wrist–elbow, MNCV (m/s)	53.3 ± 3.2
Elbow–axilla, MNCV (m/s)	59.9 ± 0.7

Table 2 Pathological values [1]

Group 1	Mean ± SD
Wrist–ADM, distal latency (ms)	2.9 ± 0.3
Wrist–elbow, MNCV (m/s)	51.5 ± 4.9
Elbow–axilla, MNCV (m/s)	60.6 ± 3.5
Group 2	Mean ± SD
Wrist–ADM, distal latency (ms)	3.1 ± 0.35
Wrist–elbow, MNCV (m/s)	54.3 ± 3.8
Elbow–axilla, MNCV (m/s)	59.8 ± 4.1
Group 3	Mean ± SD
Wrist–ADM, distal latency (ms)	3.3 ± 0.7
Wrist–elbow, MNCV (m/s)	51.2 ± 4.3
Elbow–axilla, MNCV (m/s)	58.3 ± 3.3

Table 3 Pathological values [1]

Group 1	Mean ± SD
Wrist–ADM, distal latency (ms)	2.9 ± 0.6
Wrist–elbow, MNCV (m/s)	49.2 ± 8.2
Elbow–axilla, MNCV (m/s)	57.8 ± 1.9
Group 2	Mean ± SD
Wrist–ADM, distal latency (ms)	3.5 ± 0.4
Wrist–elbow, MNCV (m/s)	46.0 ± 5.7
Elbow–axilla, MNCV (m/s)	55.9 ± 6.2
Group 3	Mean ± SD
Wrist–ADM, distal latency (ms)	3.2 ± 0.6
Wrist–elbow, MNCV (m/s)	47.5 ± 4.3
Elbow–axilla, MNCV (m/s)	55.3 ± 8.4

In 1964, Trojaborg [3] stimulated the ulnar nerve at three points along its course (Figs. 3 and 4): at the level of the wrist (6–8 cm above the leading-off electrode), elbow (19–25 cm above the wrist), and axilla (17–25 cm above the elbow). He measured the conduction velocities of the motor fibers of the ulnar nerve using both surface and concentric needle electrodes for recording (12 ulnar nerves), and also he tested surface and needle electrodes for stimulation; the author concluded there was no significant difference in conduction velocity when recording or stimulating with surface or needle electrodes (conduction velocities in the elbow–wrist segment was 53.5 ± 7.1 m/s for surface electrodes and 54.7 ± 6.0 m/s for needle electrodes, respectively). The author [3] reported only median nerve data; conduction velocities (for both methods) in the axilla–elbow segment (22 ulnar nerves) was 63.4 ± 5.3 m/s, in the elbow–wrist segment (30 ulnar nerves) was 56.4 ± 4.8 m/s, and in the axilla–wrist segment (22 nerves) was 59.4 ± 4.1 m/s. The difference in conduction velocity between right and left side was 6.0 ± 5.0 m/s in the axilla–elbow segment (11 ulnar nerves), 2.8 ± 2.2 m/s in the elbow–wrist segment (15 ulnar nerves), and 4.4 ± 2.9 m/s in the axilla–wrist segment (11 ulnar nerves).

For normal values (Table 4), the author [3] studied 47 control subjects (age range of 36 subjects was 20–50 years, 8 subjects were over 50 years, 3 subjects were below 20 years).

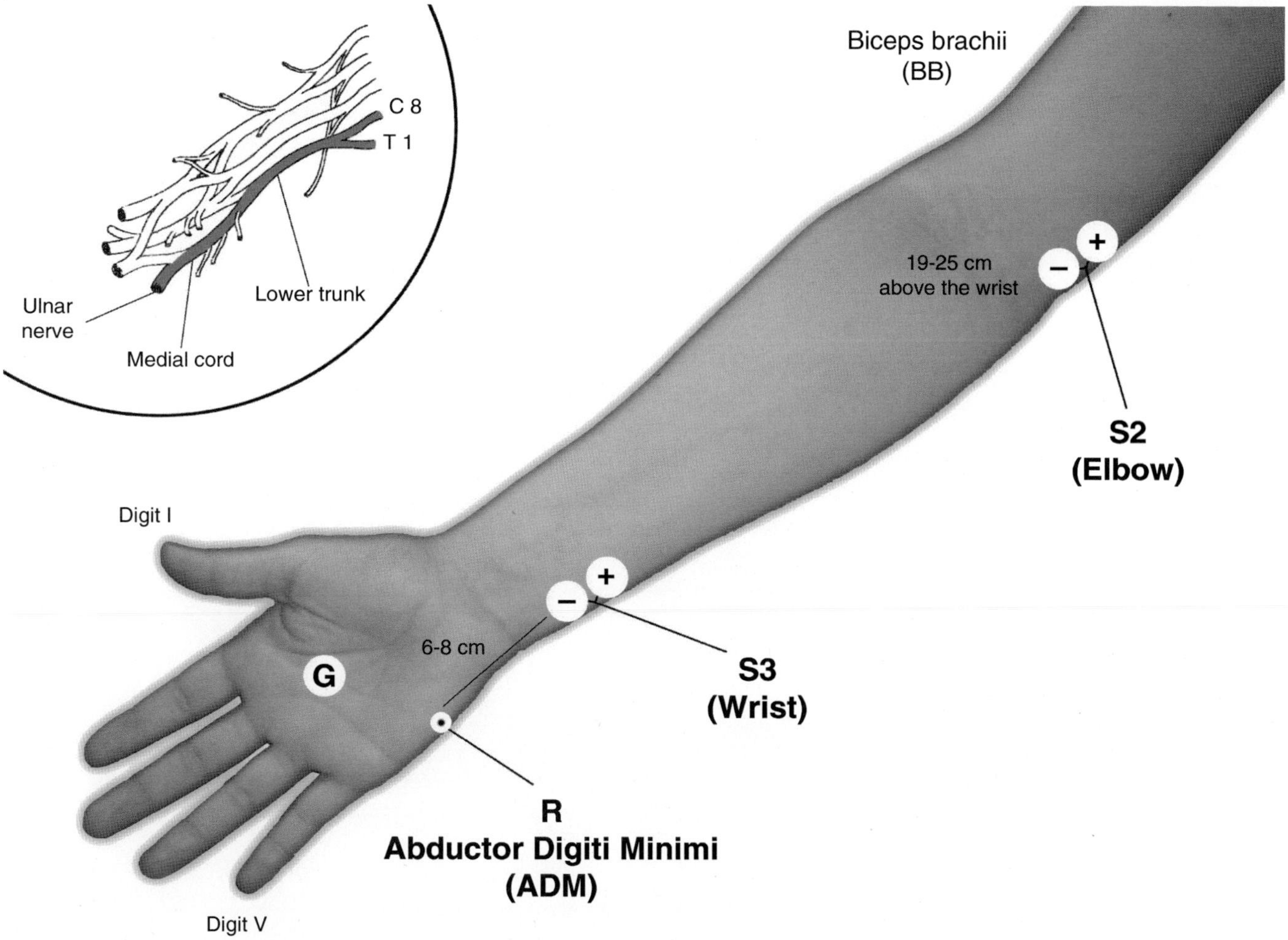

Typical waveform (axilla, above elbow, wrist – ADM muscle):

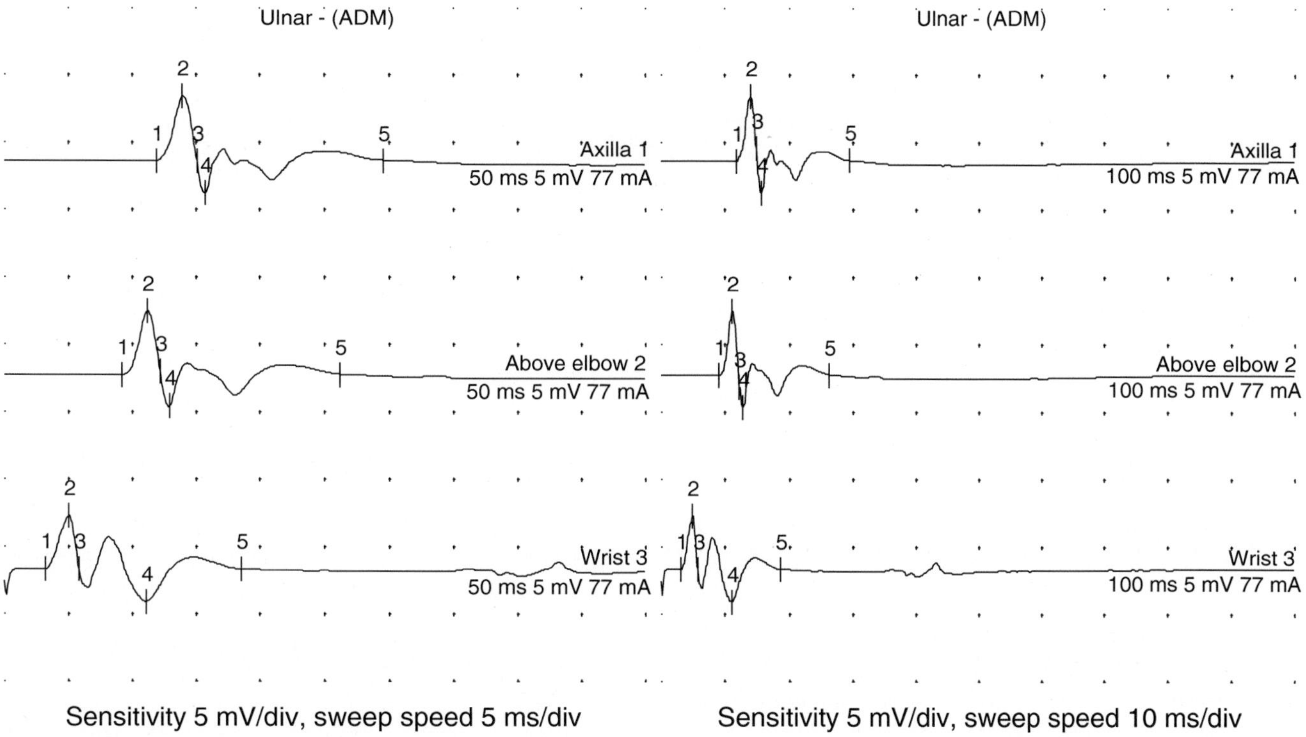

Fig. 3 Compound muscle action potentials (*CMAPs*) recorded at the hand from the ADM muscle by needle electrode, stimulation of the axilla (*upper trace*), above the elbow (*middle trace*), and of the wrist (*lower trace*)

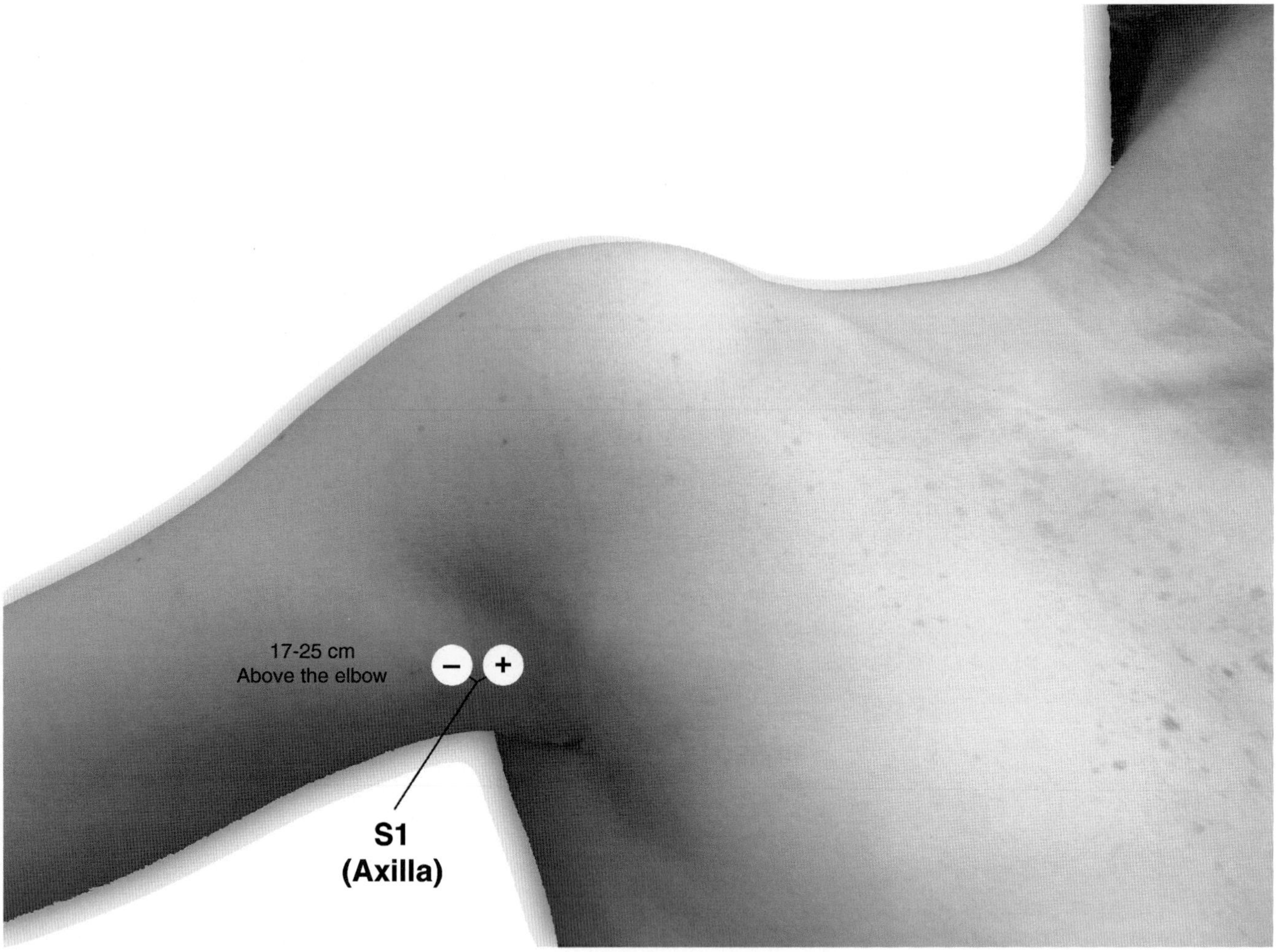

Fig. 4 Proximal stimulation point (S1 – axilla) by Trojaborg [3]

Table 4 Normal values [3]

	Mean value	SD
Axilla–elbow, MNCV (m/s) – 22 ulnar nerves	63.4 ± 1.1	5.3
Elbow–wrist, MNCV (m/s) – 30 ulnar nerves	56.4 ± 0.9	4.8
Axilla–wrist, MNCV (m/s) – 22 ulnar nerves	59.4 ± 0.9	4.1
	Mean value	SD
Negative-peak amplitude (mV) – 22 proximal electrode recordings	14.1 ± 0.9	4.4
Negative-peak amplitude (mV) – 30 middle electrode recordings	14.4 ± 0.9	4.9
Negative-peak amplitude (mV) – 22 distal electrode recordings	16.8 ± 0.8	4.7

Comment

For Trojaborg [3], the conduction velocity was significantly slower (11 %) in the distal than in the proximal portion of the nerve; for the author, the difference in conduction velocities between the proximal and distal segments was due to differences in temperature and the decrease in fiber diameter from the proximal to the distal part of the nerve. The difference in conduction velocity between the right and left extremities in the same subject and the variation from person to person averaged 5–10 %. The application, using surface and needle recordings, in a case of ulnar neuropathy at the elbow (UNE) is here reported (Figs. 5 and 6).

Pathological waveform (axilla, above elbow, wrist – ADM muscle):

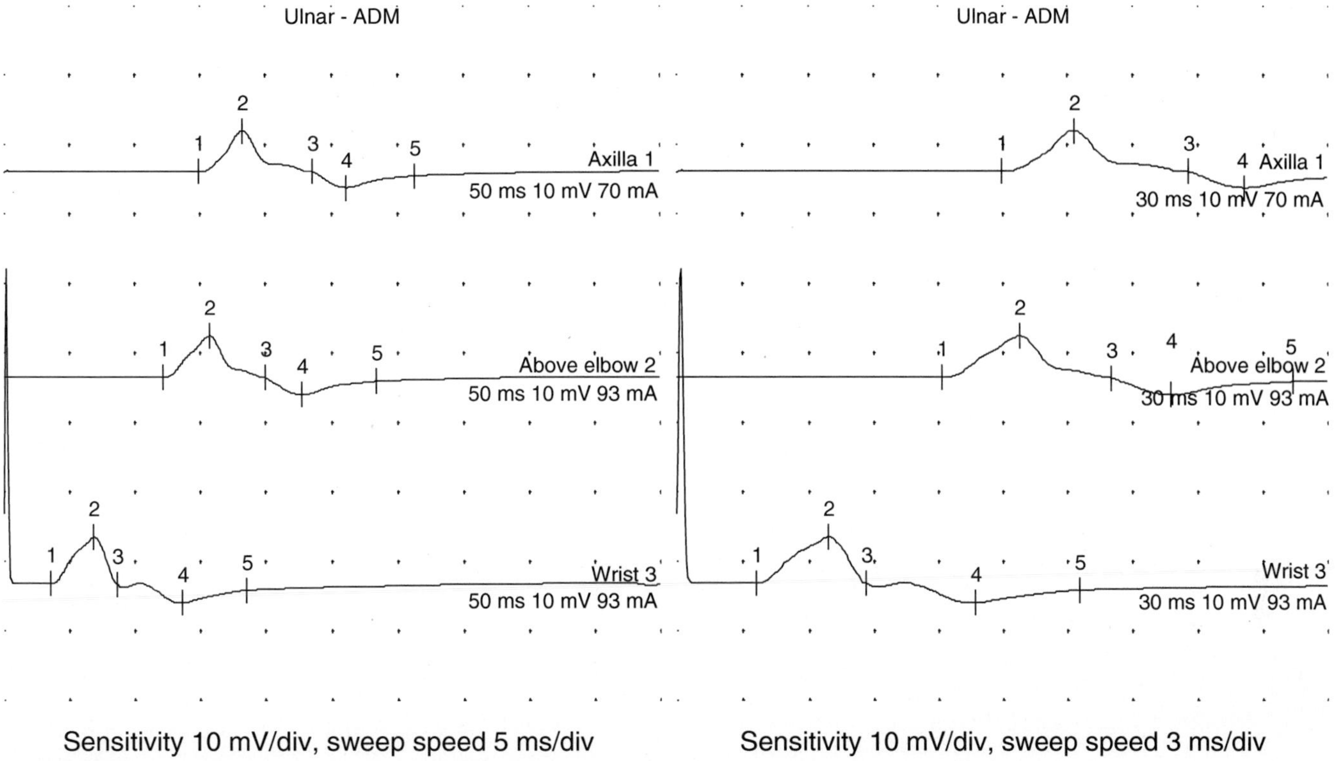

Onset latency (axilla – ADM): 14.95 ms; **Onset latency** (above elbow – ADM): 12.15 ms; **Onset latency** (wrist – ADM): 3.55 ms; **Peak latency** (axilla – ADM): 18.10 ms; **Peak latency** (above elbow – ADM): 15.40 ms; **Peak latency** (wrist – ADM): 7.10 ms; **Onset to peak amplitude** (axilla – ADM): 6.1 mV; **Onset to peak amplitude** (above elbow – ADM): 6.1 mV; **Onset to peak amplitude** (wrist – ADM): 6.7 mV; **Peak to peak amplitude** (axilla – ADM): 8.4 mV; **Peak to peak amplitude** (above elbow – ADM): 8.6 mV; **Peak to peak amplitude** (wrist – ADM): 9.7 mV; **MNCV** (wrist – above elbow, 29 cm): 33.7 m/s; **MNCV** (axilla – above elbow, 15 cm): 53.6 m/s

Fig. 5 Compound muscle action potentials (*CMAPs*) recorded at the hand from the ADM muscle using surface recording method [1] in UNE, stimulation of the axilla (*upper trace*), above the elbow (*middle trace*), and of the wrist (*lower trace*)

Pathological waveform (axilla, above elbow, wrist – ADM muscle):

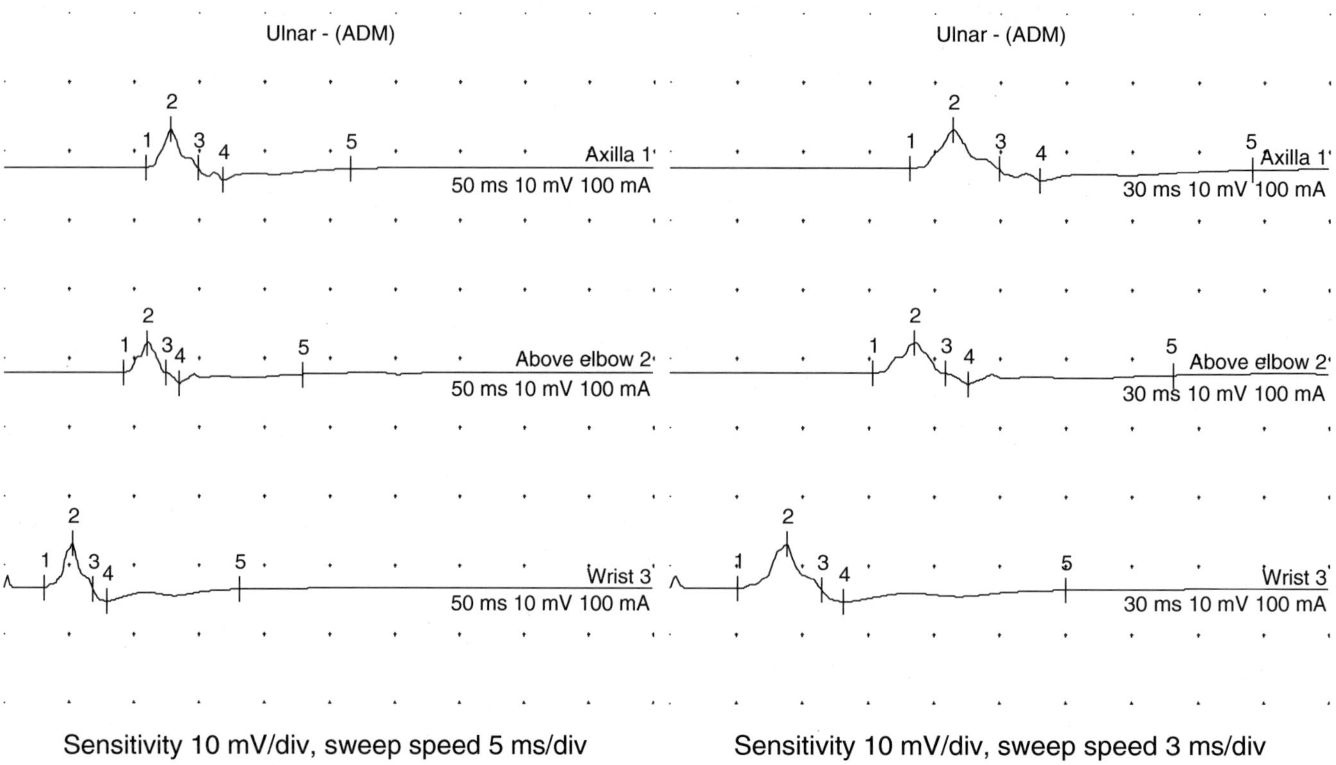

Onset latency (axilla – ADM): 10.90 ms; **Onset latency** (above elbow – ADM): 9.15 ms; **Onset latency** (wrist – ADM): 3.05 ms; **Peak latency** (axilla – ADM): 12.95 ms; **Peak latency** (above elbow – ADM): 10.80 ms; **Peak latency** (wrist – ADM): 4.75 ms; **Onset to peak amplitude** (axilla – ADM): 5.7 mV; **Onset to peak amplitude** (above elbow – ADM): 4.5 mV; **Onset to peak amplitude** (wrist – ADM): 6.5 mV; **Peak to peak amplitude** (axilla – ADM): 7.5 mV; **Peak to peak amplitude** (above elbow – ADM): 6.1 mV; **Peak to peak amplitude** (wrist – ADM): 8.5 mV; **MNCV** (wrist – above elbow, 25 cm): 41.0 m/s; **MNCV** (axilla – above elbow, 12 cm): 68.6 m/s

Fig. 6 Compound muscle action potentials (*CMAPs*) recorded at the hand from the ADM muscle using needle recording method [3] in UNE, stimulation of the axilla (*upper trace*), above the elbow (*middle trace*), and of the wrist (*lower trace*)

References

1. Mayer RF (1963) Nerve conduction studies in man. Neurology 13:1021–1030
2. Thomas PK, Sears RA, Gilliatt RW (1959) The range of conduction velocity in normal motor nerve fibers for the small muscles of the hand and foot. J Neurol Neurosurg Psychiatry 22:175–181
3. Trojaborg W (1964) Motor nerve conduction velocities in normal subjects with particular reference to the conduction in proximal and distal segments of median and ulnar nerve. Electroenceph Clin Neurophysiol 17:314–321

Wrist, Elbow – Hand

M10

Surface Recording Technique, Study from *Abductor Digiti Minimi* (ADM) Muscle

Original Settings The machine used was a TECA model B; sensitivity, low-frequency filter, high-frequency filter, sweep speed, and duration of pulse were not specified.

Position This study was performed in the supine position.

Recording The active electrode (A) was placed over the belly (motor point) of the *abductor digiti minimi* (ADM) muscle (Fig. 1). The author used for *hypothenar* muscle the term *abductor digiti quinti*; they used silver disk surface electrodes [1]. Distances between the recording and stimulating electrodes were not fixed. The ground (G) electrode was 2 × 1.5 cm lead plate which was usually placed over the mid-forearm (Fig. 1); Fig. 2 shows the ground electrode positioned on the palm.

Stimulation The ulnar nerve was stimulated at two sites: at the wrist (S1) and at the elbow (S2). The cathode (–) and anode (+) stimulating electrodes were initially positioned just lateral to the tendon of the *flexor carpi ulnaris* (FCU) muscle (S1) and then were positioned in the more proximal portion of the ulnar groove (S2). We used conventional bipolar stimulator, and the anode was proximal in all stimulations (Figs. 1 and 2). No fixed distance was reported in the article. A supramaximal electric stimulus was used in all determinations (it was at least 30 % greater than that which elicited a maximum motor response).

Measurements Distal latency (ms) of the compound muscle action potential (CMAP) was measured from the stimulus onset to the onset of the negative deflection of CMAP. Motor nerve conduction velocity (MNCV) was calculated in the elbow–wrist (forearm) segment and measured in meter per second (m/s). No control of temperature was attempted. Normal values (Table 1) were obtained from 47 healthy subjects. Pathological values were not reported and the clinical application of the technique was not described.

Hennessey et al. [2] provided normative electrodiagnostic data for the median and ulnar nerves based on several factors, such as temperature control and standardized technique. They performed bilateral motor conduction studies (Table 2) on 44 paid healthy subjects (23 men and 21 women, 37 right-handed, age range 19–43 years) using a TECA Neurostar EMG instrument. Low-frequency filter was set at 20 Hz and high-frequency filter was set at 2 kHz; sweep speed was set at 2 ms/division. Skin temperature was monitored in all subjects by an EMG Associates Dermatemp DT-1000 infrared temperature scanner before each study at the touch pad of the digit II, midpalm, and antecubital fossa (mean surface skin temperature was 32.4°C, 33.2°C, and 33.4°C, respectively). Subjects with a skin temperature of less than 31°C at any recording site had their extremities warmed with an electric heating pad. The authors [2] performed motor nerve conduction studies by the method previously described by Melvin et al. [1]. The active and recording electrode was placed over the motor point of the *abductor digiti minimi* (ADM) muscle (one-half the distance between the metacarpophalangeal joint and the distal wrist crease). The reference electrode was placed 4 cm distally over the digit V (little finger). Ulnar nerve stimulation was performed distally at the wrist, posterior to the *flexor carpi ulnaris* tendon (8 cm proximal to the active recording electrode), and proximally below the elbow, just distal to the medial epicondyle with the elbow in 70° of flexion, as suggested by Checkles et al. [3]. The distance was measured along the ulnar surface of the forearm. The authors [2] placed the ground electrode on the dorsum of the hand; the figure shows ground electrode placed on the palm (Fig. 2). Motor distal latency (MDL), motor nerve conduction velocity (MNCV), and distal amplitude (DAMP) were reported. The application in a case of Charcot Marie-Tooth (PMP22-CMT1A) is here reported (Fig. 3).

G. Gentili, M. Di Napoli, *The Ulnar Nerve*, DOI 10.1007/978-3-319-43292-2_34

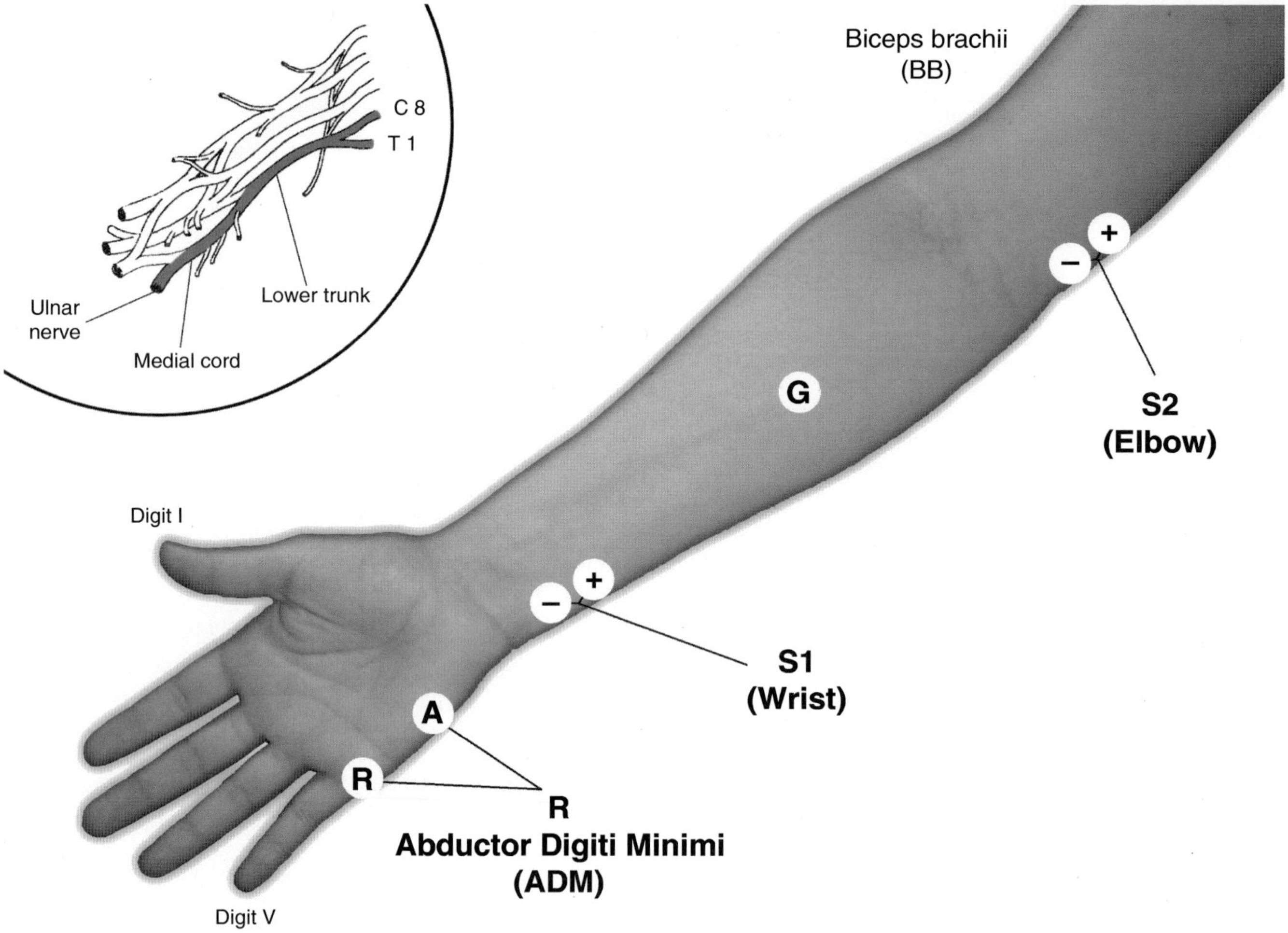

Typical waveform (wrist, elbow – ADM muscle):

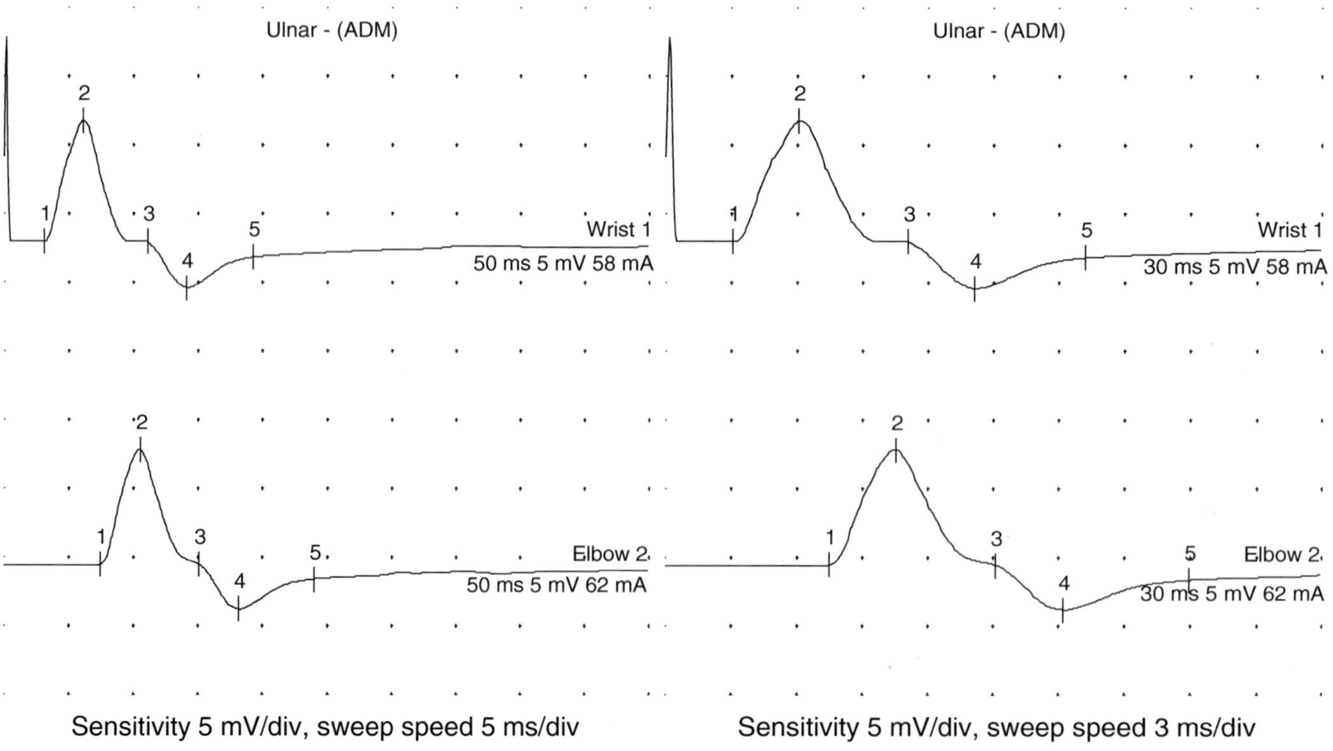

Fig. 1 Compound muscle action potentials (*CMAPs*) recorded at the hand from the ADM muscle, stimulation of the wrist (*upper trace*) and of the elbow (*lower trace*)

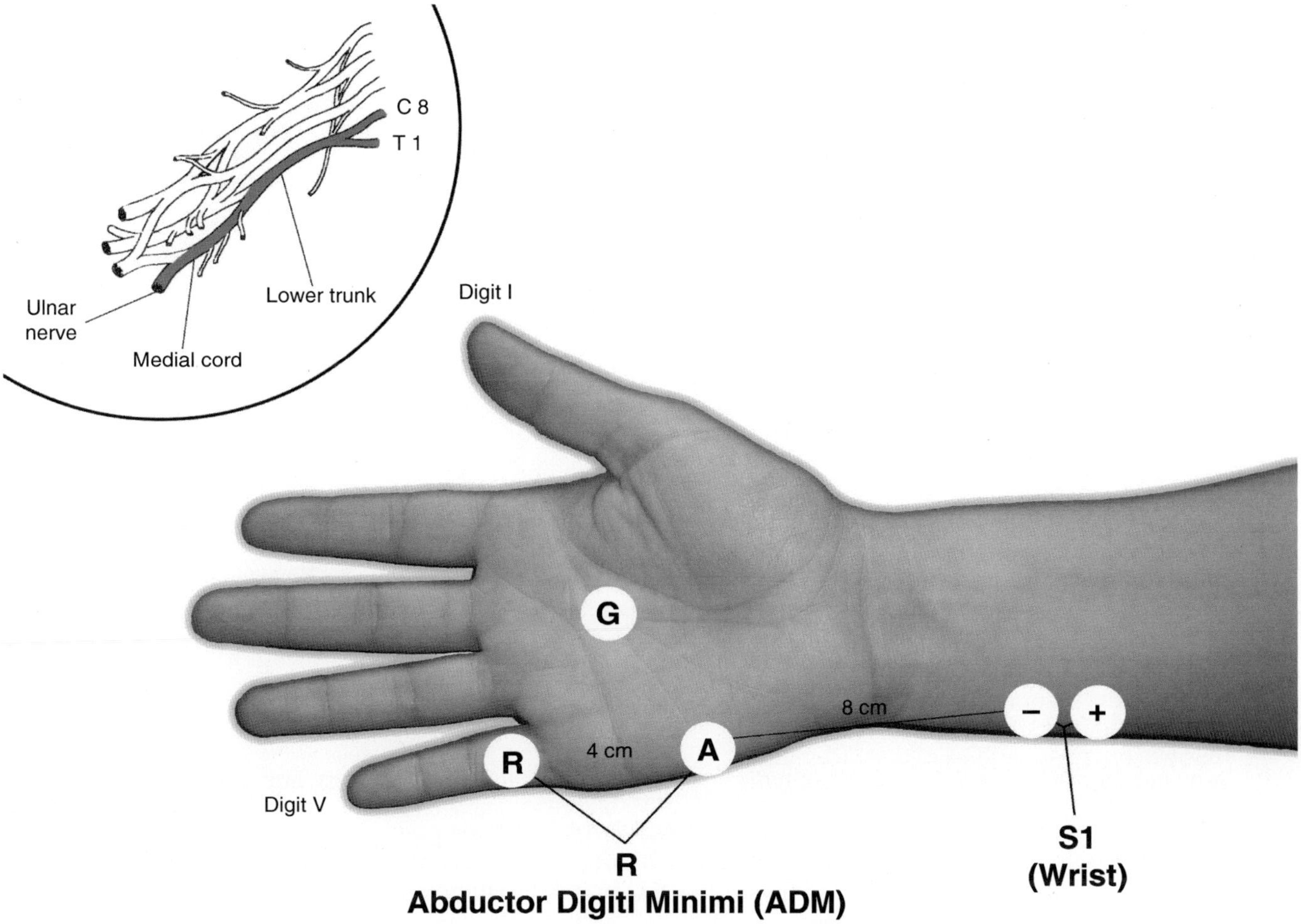

Fig. 2 Distal stimulation point (S1 – wrist), 8 cm distance from the active recording electrode [2]

Table 1 Normal values [1]

	Mean ± SD
Elbow–wrist, MNCV (m/s)	57.0 ± 4.7

Table 2 Normal values [2]

	Mean ± SD
Wrist–ADM, MDL (ms)	2.6 ± 0.3
Wrist–ADM, MNCV (m/s)	63.0 ± 4.8
Wrist–ADM, DAMP (mV)	12.6 ± 2.3

Pathological waveform (wrist, elbow – ADM muscle):

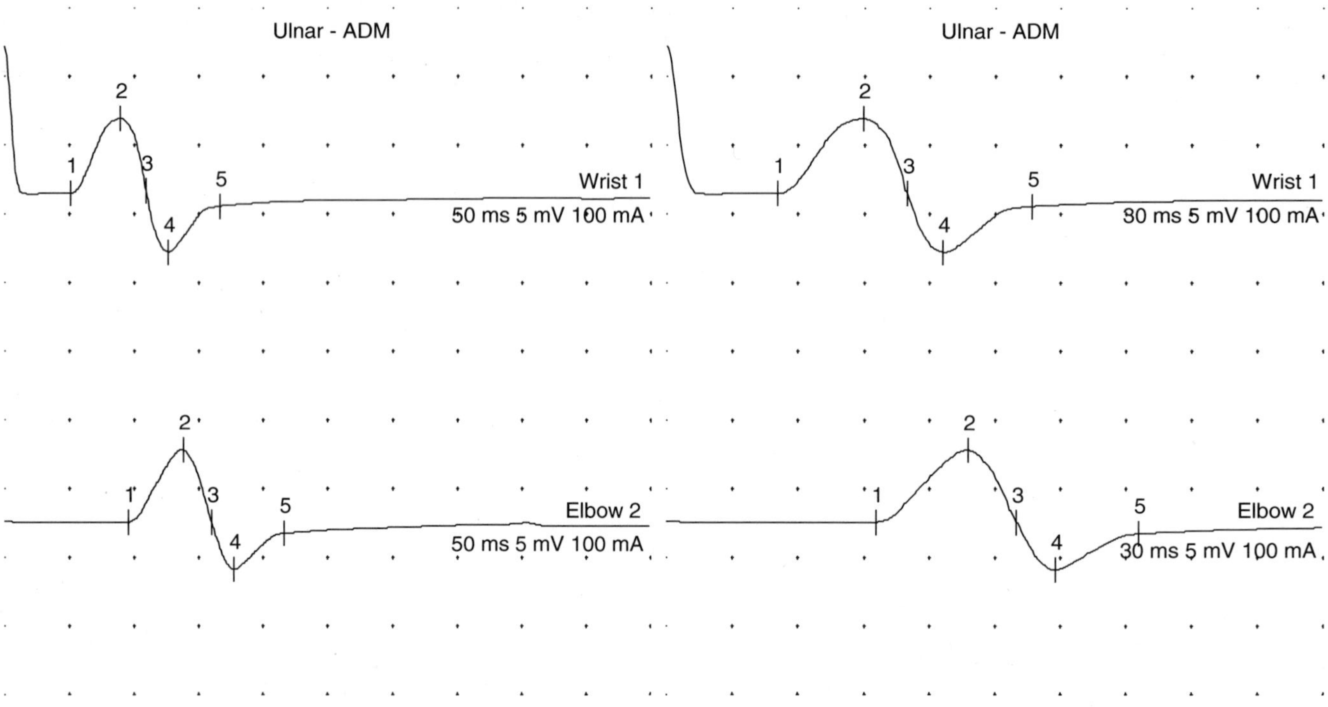

Sensitivity 5 mV/div, sweep speed 5 ms/div

Sensitivity 5 mV/div, sweep speed 3 ms/div

Onset latency (wrist – ADM): 5.05 ms; **Onset latency** (elbow – ADM): 9.55 ms; **Peak latency** (wrist – ADM): 8.95 ms; **Peak latency** (elbow – ADM): 13.75 ms; **Onset to peak amplitude** (wrist – ADM): 5.5 mV; **Onset to peak amplitude** (elbow – ADM): 5.2 mV; **Peak to peak amplitude** (wrist – ADM): 9.8 mV; **Peak to peak amplitude** (elbow – ADM): 8.8 mV; **MNCV** (elbow – wrist, 17 cm): 37.8 m/s

Fig. 3 Compound muscle action potentials (*CMAPs*) recorded at the hand from the ADM muscle in a PMP22-CMT1A patient, stimulation of the wrist (*upper trace*) and of the elbow (*lower trace*)

References

1. Melvin JL, Harris DH, Johnson EW (1966) Sensory and motor conduction velocities in the ulnar and median nerves. Arch Phys Med Rehabil 47:511–519
2. Hennessey WJ, Falco FJE, Braddom RL (1994) Median and ulnar nerve conduction studies: normative data for young adults. Arch Phys Med Rehabil 75:259–264
3. Checkles NS, Russakov AD, Piero DL (1971) Ulnar nerve conduction velocity – effect of elbow position on measurement. Arch Phys Med Rehabil 52:362–365

Wrist, Below Elbow, Above Elbow, Axilla – Hand

M11

Surface Recording Technique, Study from *Abductor Digiti Minimi* (ADM) and *Adductor Pollicis* (AP) Muscle

Original Settings Duration was 0.1–1 ms; sensitivity, sweep speed, low-frequency filter, high-frequency filter, duration of pulse, and the machine used were not specified.

Position Position was not specified in the text, because the author [1] used the same method previously described in detail in a thesis by Tascon-Alonzo [2]. We performed this study with the forearm in the supine position and the elbow in full extension.

Recording Compound muscle action potentials (CMAPs) were recorded simultaneously (Fig. 1), but separately on different channels of a triple-beam oscilloscope from both the *hypothenar* and *thenar* muscles [1]. The first recording was from the *abductor digiti minimi* (ADM) muscle, with the active electrode (A) placed over the belly (motor point) of the ADM muscle (*hypothenar* eminence) and the reference (R) placed distally near its tendon of insertion; the second recording was from the *adductor pollicis* (AP) muscle, with the active electrode (A) placed over the belly (motor point) of the AP muscle (*thenar* eminence). The ground (G) electrode was not specified in the text; the figure shows ground electrode placed over the forearm.

Stimulation The ulnar nerve was stimulated at four points [1]: at the wrist (S1), below elbow (S2), above elbow (S3), and in the axilla (S4). The anode was proximal in all stimulations (Figs. 1 and 2). The points at the elbow were approximately 2 cm distal to the olecranon (distal to elbow, DE) and 2 cm proximal to the olecranon (proximal to elbow, PE). Fixed distances were not reported in the paper.

Measurements The conduction time (ms) was the time interval between the beginning of the stimulus artifact and the beginning of the first deflection, positive or negative, of the CMAP (i.e., the distal latency). The difference in latency (ms) between the responses from any two points equaled the conduction time between these two points (wrist–below elbow, below elbow–above elbow, above elbow–axilla). The conduction distance (cm) was measured between the position of the center of the stimulating cathode (-) on the skin and the position of the center of the recording electrode over the belly of the muscle. The motor nerve conduction velocity (m/s) between any two points on the ulnar nerve was calculated by dividing the conduction distance by the conduction time between these two parts. In the segment of the nerve distal to the wrist, the author [1] recorded only conduction time because of errors in calculation of conduction velocity. The conduction time in the deep branch of the ulnar nerve was recorded by the author as the conduction time from the wrist to the AP muscle less the conduction time from the wrist to the ADM muscle. This difference (AP–ADM latency difference) was usually of the order of 0.5 ms. Differences greater than 1 ms were considered abnormal. The author reported two pathological cases referred for electroneuromyography: case (1) a 25-year-old man, a seaman, with a numb sensation along the ulnar border of the left hand and of the digit IV (ring finger) and the digit V (little finger) of the left hand, which was associated with paresthesia in the ulnar nerve distribution as well as with some weakness in the hand. Nerve conduction study showed a marked decline in CMAP amplitude following stimulation at the axilla and also a prolonged conduction time between the axilla and above-elbow stimulation points (Table 1). These findings were suggestive for a partial conduction block in the middle of the upper arm. The patient was referred for surgery, and at operation a fibrous band was found constricting the nerve where it passed from the anterior to posterior compartment through the medial intermuscular septum. The surgeon divided this band, and the patient 4 months following the operation was again

G. Gentili, M. Di Napoli, *The Ulnar Nerve*, DOI 10.1007/978-3-319-43292-2_35

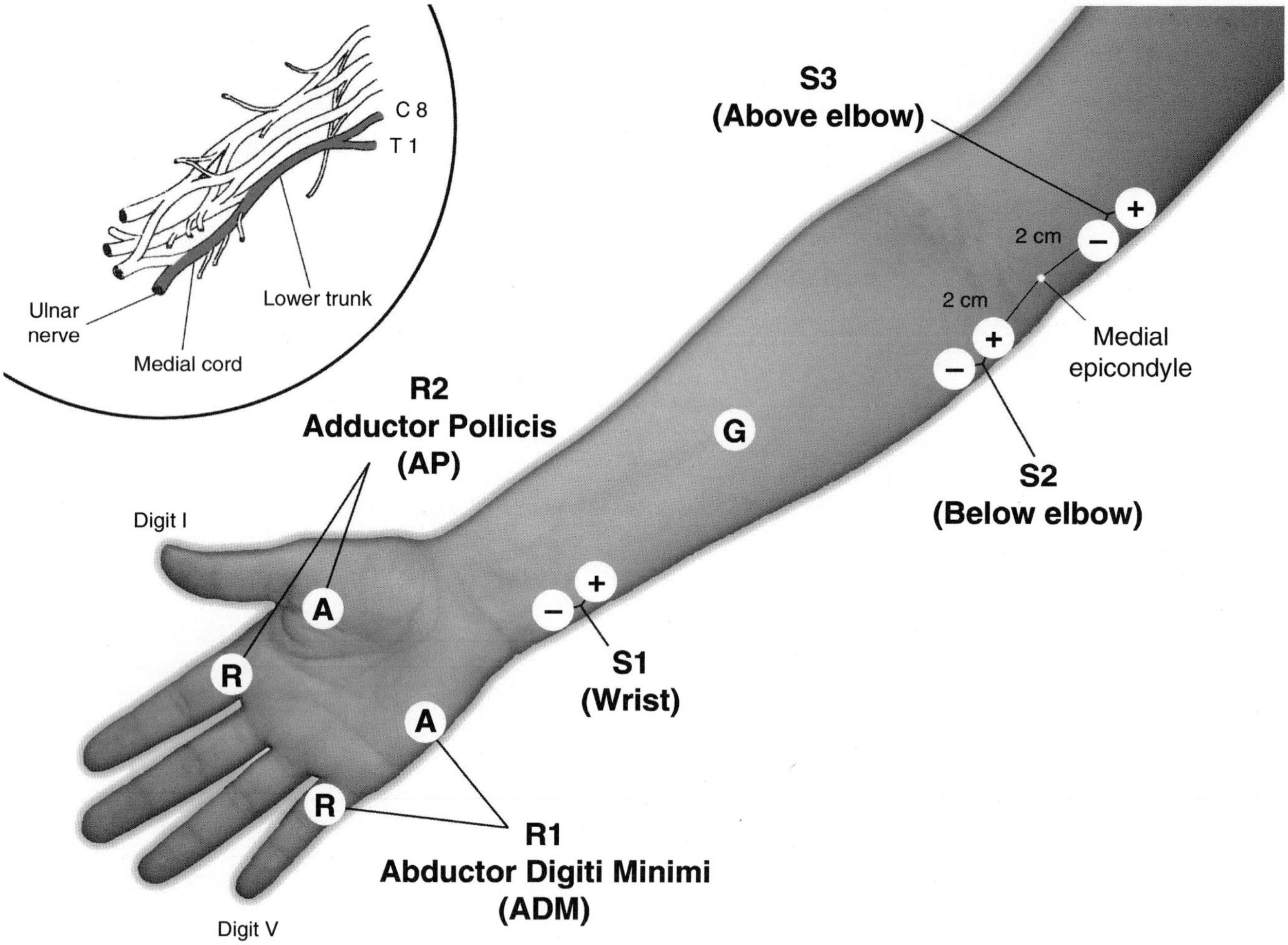

Typical waveform (wrist, below elbow, above elbow, axilla – ADM, AP muscles):

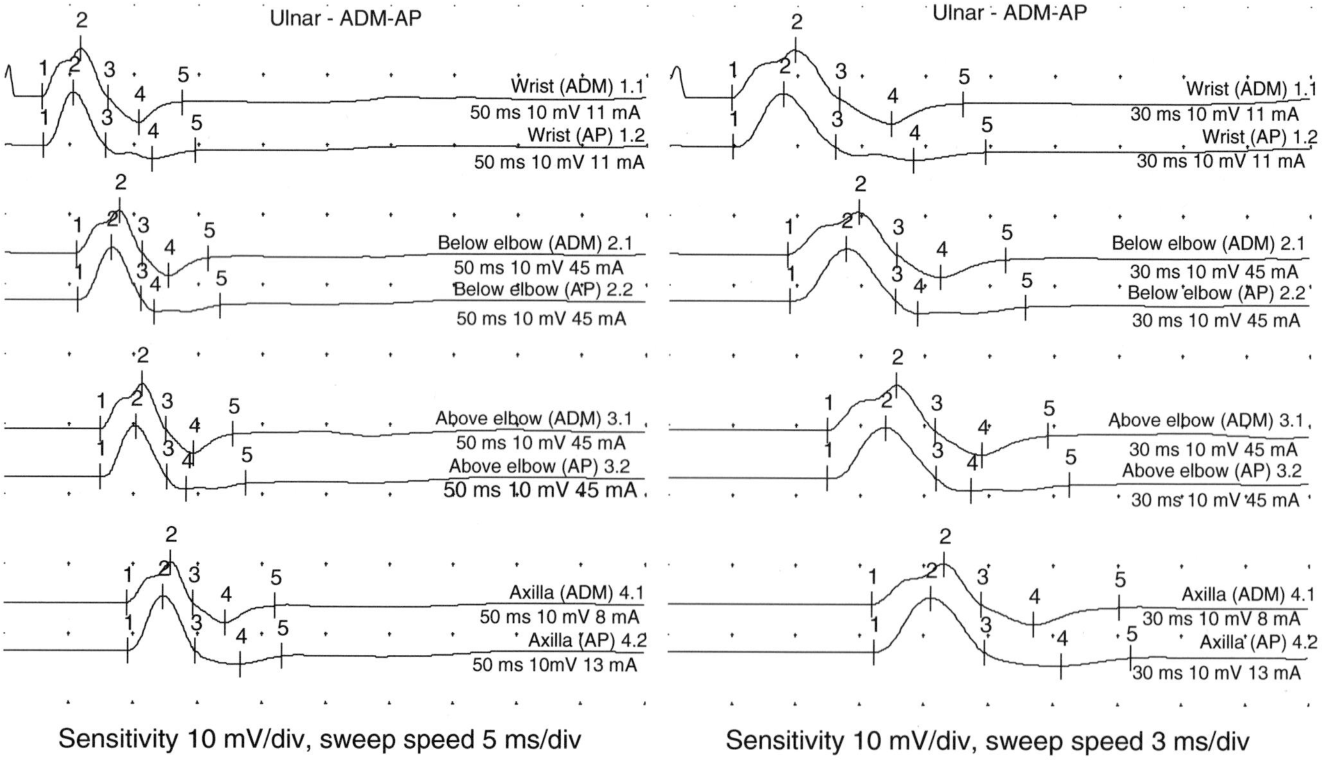

Fig. 1 Compound muscle action potentials (*CMAPs*) recorded at the hand from the ADM and AP muscles, *top to bottom*: stimulation of the wrist, below the elbow, above the elbow, and the axilla

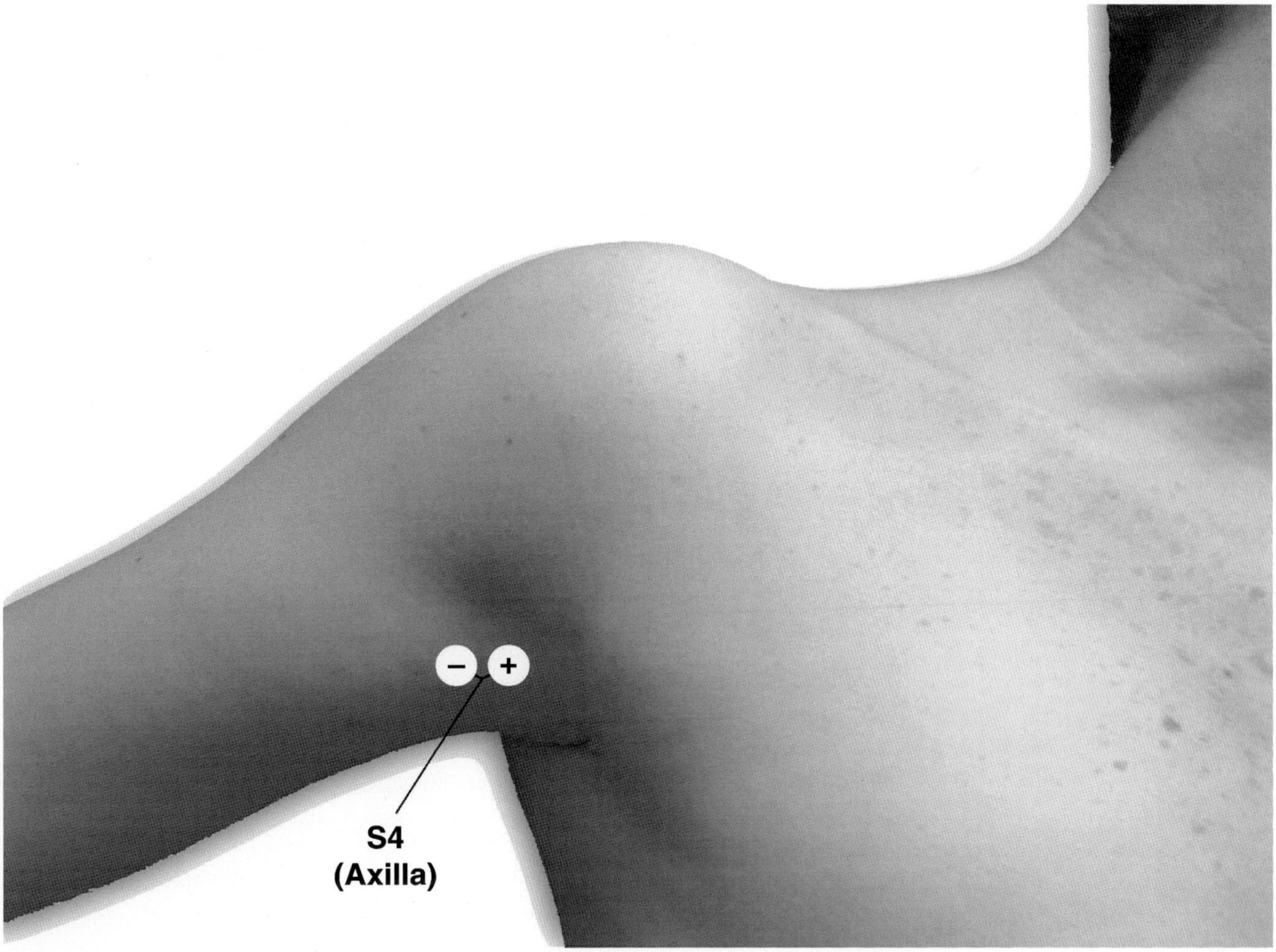

Fig. 2 Proximal stimulation point (S4 – axilla)

Table 1 Case 1 values [1]

	Mean
Wrist–below elbow, MNCV (m/s) – right arm	58.0
Wrist–above elbow, MNCV (m/s) – right arm	52.8
Above elbow–below elbow, MNCV (m/s) – right arm	38.3
Above elbow–axilla, MNCV (m/s) – right arm	52.4
Wrist–below elbow, MNCV (m/s) – left arm (preop)	60.0
Wrist–above elbow, MNCV (m/s) – left arm (preop)	52.4
Above elbow–below elbow, MNCV (m/s) – left arm (preop)	29.1
Above elbow–axilla, MNCV (m/s) – left arm (preop)	29.4
Wrist–below elbow, MNCV (m/s) – left arm (post-op)	60.0
Wrist–above elbow, MNCV (m/s) – left arm (post-op)	59.6
Above elbow–below elbow, MNCV (m/s) – left arm (post-op)	47.8
Above elbow–axilla, MNCV (m/s) – left arm (post-op)	52.6

seen for nerve conduction studies. These showed a completely restored amplitude of CMAP and a markedly improved conduction time between the axilla and elbow and across the level of the block: (Case 2) A 37-year-old woman, a schoolteacher, had a 6-month history of pain on the ulnar side of the right palm and weakness of the little finger. Nerve conduction study on the right ulnar nerve revealed that the conduction time from the wrist to AP muscle was greatly prolonged as compared with that from the wrist to ADM muscle (normal). These findings suggested the presence of a lesion of the deep branch of the ulnar nerve in the hand, and the patient was referred for surgery. At operation a ganglion compressing the deep branch of the ulnar nerve against the deep fascia of the palm was successfully removed. Two months postoperatively, nerve conduction studies were repeated. These showed marked improvement in conduction time from the wrist to AP muscle, and the patient remarked to feel a great improvement in most of her finger movements and the overall function of her hand.

Comment

For Carpendale [1], the ulnar nerve is liable to compression at three levels in the arm. These are, in sequence from the axilla to the hand, as follows: (1) in the middle of the upper arm where it passes from the anterior to posterior compartment through the medial intermuscular septum; (2) at the elbow where it passes through the ulnar groove, posterior to the medial epicondyle, and also in the ulnar tunnel where it passes between the two heads of *flexor carpi ulnaris* (FCU) muscle; (3) in the hand as it divides over the pisohamate ligament, into a superficial cutaneous branch and a deep muscular branch. Compression at these sites is difficult to localize by clinical signs alone, and standard methods for measuring conduction velocity in the ulnar nerve consist of stimulation at the level of the elbow and wrist and recording CMAPs from the ADM muscle. However, a method such as this will fail to reveal lesions involving the ulnar nerve in the upper arm or in the deep branch in the hand. In both pathological cases, the author [1] found normal conduction values, and routine methods failed to localize the lesion (in the upper arm – case 1, in the hand – case 2). For instance, in case 1, the conduction velocity across the elbow was only 29.1 m/s, compared to 38.3 m/s in the unaffected right ulnar. This finding might have been considered abnormal and have been the basis for a false diagnosis of a tardy ulnar palsy at the level of the left elbow. In the second patient, no localization of the lesion would have been possible by the standard technique because no recording would have been made from those muscles supplied by the deep branch of the ulnar nerve more distal than ADM muscle. Pathological waves and values in patients with ulnar neuropathy at the elbow (UNE) and lesion of the deep branch of the ulnar nerve with associated UNE (diabetic patient) are here reported (Figs. 3 and 4).

Pathological waveform (wrist, below elbow, above elbow, axilla – ADM, AP muscles):

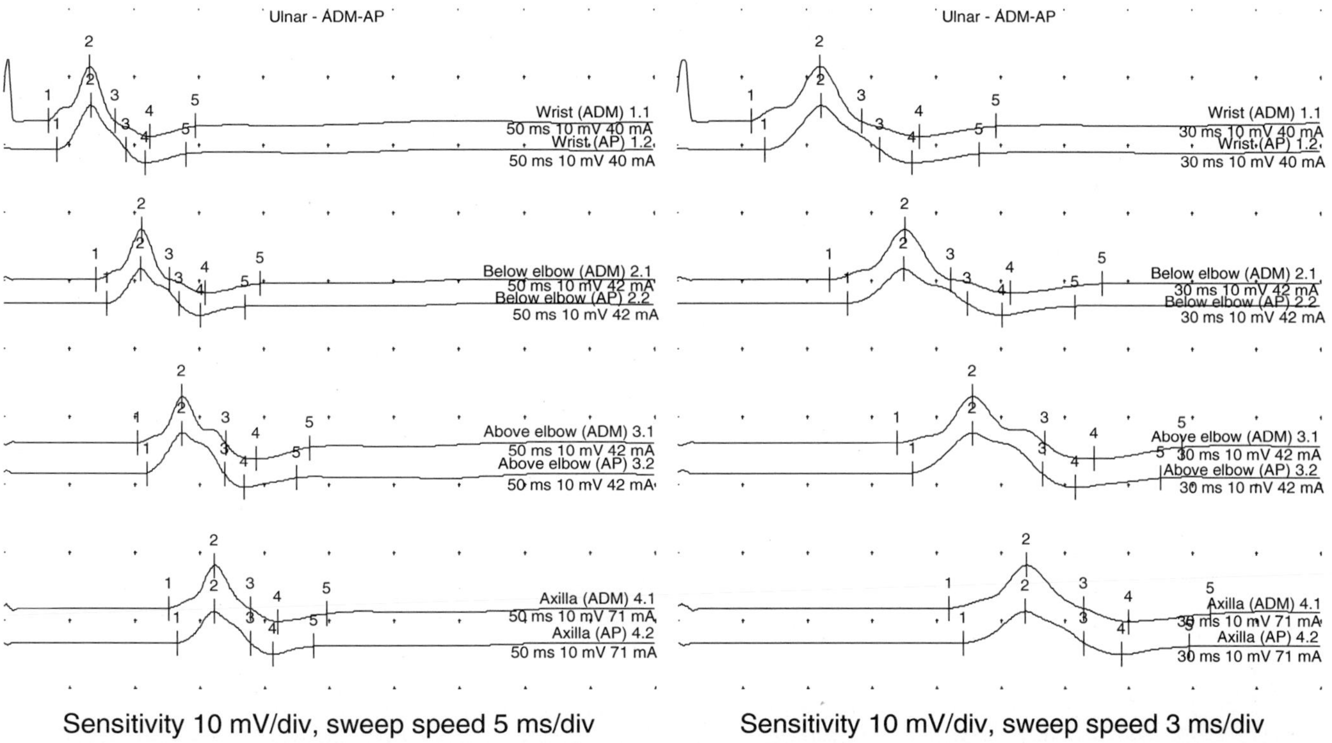

Onset latency (wrist – ADM): 3.45 ms; **Onset latency** (wrist – AP): 4.05 ms; **Onset latency** (below elbow – ADM): 7.05 ms; **Onset latency** (below elbow – AP): 7.85 ms; **Onset latency** (above elbow – ADM): 10.15 ms; **Onset latency** (above elbow – AP): 10.90 ms; **Onset latency** (axilla – ADM): 12.55 ms; **Onset latency** (axilla – AP): 13.20 ms; **Peak latency** (wrist – ADM): 6.0 ms; **Peak latency** (wrist – AP): 5.75 ms; **Peak latency** (below elbow – ADM): 8.90 ms; **Peak latency** (below elbow – AP): 8.55 ms; **Peak latency** (above elbow – ADM): 10.70 ms; **Peak latency** (above elbow – AP): 10.15 ms;**Peak latency** (axilla – ADM): 12.35 ms; **Peak latency** (axilla – AP): 12.10 ms; **Onset to peak amplitude** (wrist – ADM): 8.0 ms; **Onset to peak amplitude** (wrist – AP): 6.5 ms; **Onset to peak amplitude** (below elbow – ADM): 7.4 ms; **Onset to peak amplitude** (below elbow – AP): 5.1 ms; **Onset to peak amplitude** (above elbow – ADM): 6.9 ms; **Onset to peak amplitude** (above elbow – AP): 6.0 ms; **Onset to peak amplitude** (axilla – ADM): 6.4 ms; **Onset to peak amplitude** (axilla – AP): 4.7 ms; **Peak to peak amplitude** (wrist – ADM): 10.4 ms; **Peak to peak amplitude** (wrist – AP): 8.6 ms; **Peak to peak amplitude** (below elbow – ADM): 9.5 ms; **Peak to peak amplitude** (below elbow – AP): 7.0 ms; **Peak to peak amplitude** (above elbow – ADM): 9.3 ms; **Peak to peak amplitude** (above elbow – AP): 8.2 ms; **Peak to peak amplitude** (axilla – ADM): 8.5 ms; **Peak to peak amplitude** (axilla – AP): 6.5 ms; **MNCV** (ADM recording; wrist – below elbow, 19 cm): 52.8 m/s; **MNCV** (AP recording; wrist – below elbow, 18 cm): 50.0 m/s; **MNCV** (ADM recording; below elbow – above elbow, 10 cm): 32.3 m/s; **MNCV** (AP recording; below elbow – above elbow, 29 cm): 32.8 m/s; **MNCV** (ADM recording; above elbow – axilla, 13 cm): 54.2 m/s; **MNCV** (AP recording; above elbow – axilla, 13 cm): 56.5 m/s

Fig. 3 Compound muscle action potentials (*CMAPs*) recorded at the hand from the ADM and AP muscles in a patient with UNE, *top to bottom*: stimulation of the wrist, below the elbow, above the elbow, and the axilla

Pathological waveform (wrist, below elbow, above elbow, axilla – ADM, AP muscles):

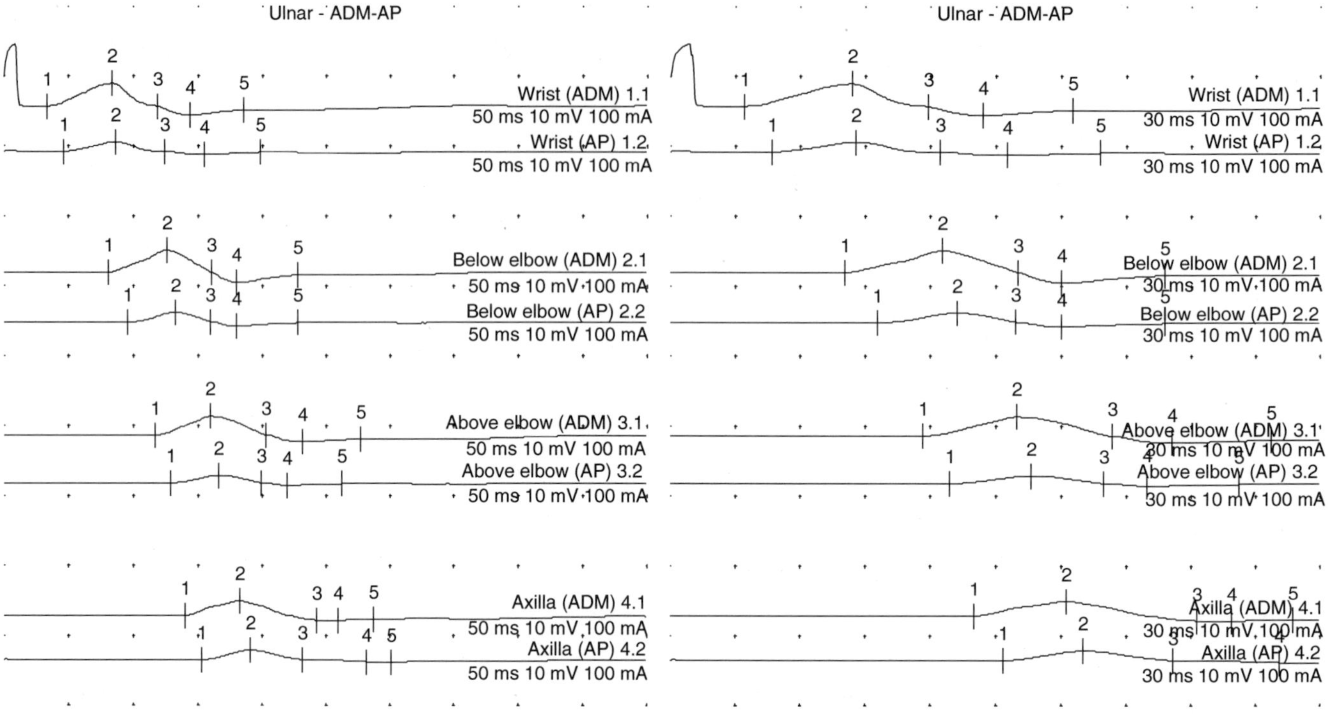

Sensitivity 10 mV/div, sweep speed 5 ms/div | Sensitivity 10 mV/div, sweep speed 3 ms/div

Onset latency (wrist – ADM): 3.35 ms; **Onset latency** (wrist – AP): 4.65 ms; **Onset latency** (below elbow – ADM): 8.05 ms; **Onset latency** (below elbow – AP): 9.55 ms; **Onset latency** (above elbow – ADM): 11.65 ms; **Onset latency** (above elbow – AP): 12.90 ms; **Onset latency** (axilla – ADM): 14.0 ms; **Onset latency** (axilla – AP): 15.3 ms; **Peak latency** (wrist – ADM): 7.1 ms; **Peak latency** (wrist – AP): 7.1 ms; **Peak latency** (below elbow – ADM): 12.40 ms; **Peak latency** (below elbow – AP): 12.45 ms; **Peak latency** (above elbow – ADM): 13.45 ms; **Peak latency** (above elbow – AP): 13.50 ms; **Peak latency** (axilla – ADM): 15.90 ms; **Peak latency** (axilla – AP): 15.80 ms; **Onset to peak amplitude** (wrist – ADM): 3.4 ms; **Onset to peak amplitude** (wrist – AP): 1.5 ms; **Onset to peak amplitude** (below elbow – ADM): 3.1 ms; **Onset to peak amplitude** (below elbow – AP): 1.4 ms; **Onset to peak amplitude** (above elbow – ADM): 2.8 ms; **Onset to peak amplitude** (above elbow – AP): 1.2 ms; **Onset to peak amplitude** (axilla – ADM): 2.1 ms; **Onset to peak amplitude** (axilla – AP): 1.4 ms; **Peak to peak amplitude** (wrist – ADM): 4.7 ms; **Peak to peak amplitude** (wrist – AP): 2.1 ms; **Peak to peak amplitude** (below elbow – ADM): 4.8 ms; **Peak to peak amplitude** (below elbow – AP): 2.2 ms; **Peak to peak amplitude** (above elbow – ADM): 3.8 ms; **Peak to peak amplitude** (above elbow – AP): 1.6 ms; **Peak to peak amplitude** (axilla – ADM): 3.0 ms; **Peak to peak amplitude** (axilla – AP): 1.8 ms; **MNCV** (ADM recording; wrist – below elbow, 22 cm): 46.8 m/s; **MNCV** (AP recording; wrist – below elbow, 22 cm): 44.9 m/s; **MNCV** (ADM recording; below elbow – above elbow, 10 cm): 27.8 m/s; **MNCV** (AP recording; below elbow – above elbow, 10 cm): 29.9 m/s; **MNCV** (ADM recording; above elbow – axilla, 14 cm): 59.6 m/s; **MNCV** (AP recording; above elbow – axilla, 14 cm): 58.3 m/s

Fig. 4 Compound muscle action potentials (*CMAPs*) recorded at the hand from the ADM and AP muscles in a patient with lesion of deep branch of the ulnar nerve associated with UNE (diabetic patient), *top to bottom*: stimulation of the wrist, below the elbow, above the elbow, and the axilla

References

1. Carpendale MT (1966) The localization of ulnar nerve compression in the hand and arm: an improved method of electroneuromyography. Arch Phys Med Rehabil 47:325–330
2. Tascon-Alonzo M (1963) Conduction time in the terminal portion of the motor nerve fibres of the ulnar, median, and peroneal nerves in healthy subjects and patients with neuropathy. Thesis. Graduate School, University of Minnesota

Wrist, Below Elbow, Elbow, Above Elbow, Erb's Point – Hand

M12

Needle Recording Technique, Study from *Abductor Digiti Minimi* (ADM) Muscle

Original Settings Duration of pulse was varied (0.1 ms at the wrist, 0.2–0.5 ms below the elbow, 0.1–0.2 ms at the elbow, 0.2 ms above the elbow, 0.2–0.2 ms at the Erb's point), rate of pulse was 1/s, and the machine used was a TECA TE 2-7; sensitivity, sweep speed, low-frequency filter, and high-frequency filter were not specified.

Position This study was performed with the subjects lying on a bed in the supine position, with the arm placed on a padded arm board, and a 2.5 cm wide strip of transparent tape was used over the palm of the hand and mid-forearm to attach the extremity to the board.

Recording The active needle electrode was placed over the motor point [1] of the *abductor digiti minimi* (ADM) muscle (Fig. 1). The ground (G) electrode was placed on the dorsum of the hand; the figure shows the ground electrode positioned on the palm.

Needle Insertion For the ADM muscle, with the hand in a neutral position to full supination, the recording needle electrode was inserted at the midpoint between the fifth metacarpophalangeal joint and the pisiform bone.

Stimulation The ulnar nerve was stimulated at five several points (Figs. 1 and 2): at the proximal wrist crease (S1), at a point 5 cm distal below the elbow (S2), at the elbow (S3), at a point 10 cm proximal above the elbow (S4), and at the Erb's point (S5), the angle formed by the clavicle and the posterolateral fibers of the sternocleidomastoid muscle [1]. The anode was proximal. Stimulus durations were 0.1 ms at the wrist, 0.2–0.5 ms below the elbow, 0.1–0.2 ms at the elbow, 0.2 ms above the elbow, and 0.2–05 at the Erb's point. For most sites, the stimuli were applied at a rate of one per second while the response was evaluated. Such repetitive stimulation at Erb's point was uncomfortable due to the vigorous muscular contractions in the extremity; stimuli at this site were controlled manually at about one each 5 s. This was well tolerated by the subjects.

Measurements Distal latency (ms) of the compound muscle action potential (CMAP) was measured from the stimulus onset to the onset of the negative deflection of the CMAP. The distances between the sites of stimulation were measured on the skin using a steel tape measure, except for measurements between Erb's point and the elbow and between Erb's point and the site 10 cm above the elbow. These segments were measured by an obstetric caliper. Velocities were calculated for the Erb's point–above-elbow, Erb's point–elbow, elbow–wrist, and below-elbow–wrist segments. For each segment, the difference in latencies of response to stimulation was divided into the distance between the two sites of stimulation, providing the motor nerve conduction velocity (MNCV) in meter per second (m/s). No temperature recording was made, the room temperature was relatively constant for all examinations, and subjects were covered except for the extremity being tested. Normal values (Table 1) were reported; the author [1] studied 50 ulnar nerves from 50 healthy volunteer subjects (28 right and 22 left, 29 women and 21 men, mean age 35.7 ± 10.5 years, age range 18–56 years). In order to illustrate the clinical application of the technique described in his article, Jebsen included some case reports: a patient affected by ulnar neuropathy at the elbow (UNE) and a Guillain–Barré syndrome case.

G. Gentili, M. Di Napoli, *The Ulnar Nerve*, DOI 10.1007/978-3-319-43292-2_36

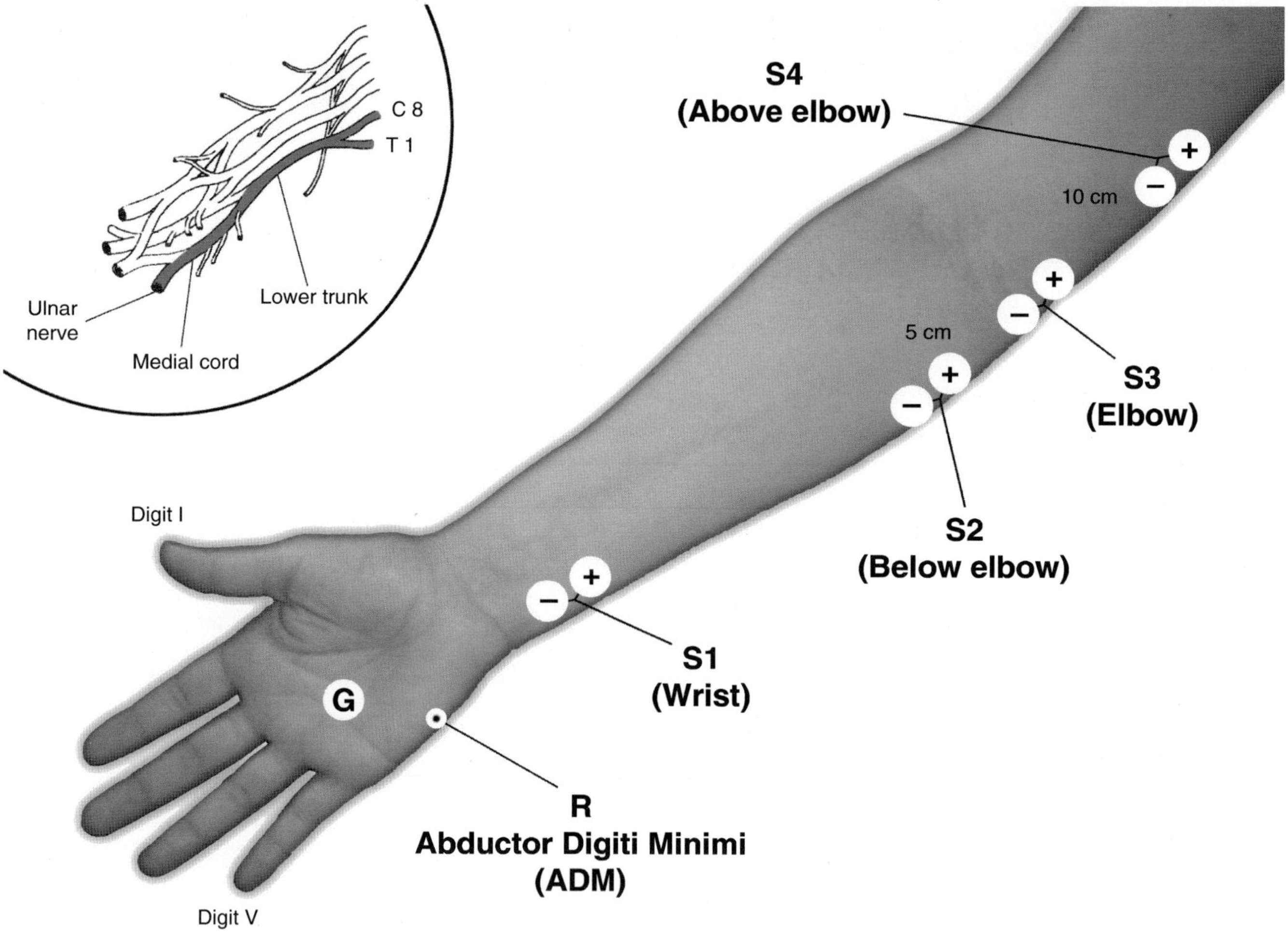

Typical waveform (wrist, below elbow, elbow, above elbow, Erb's point – ADM muscle):

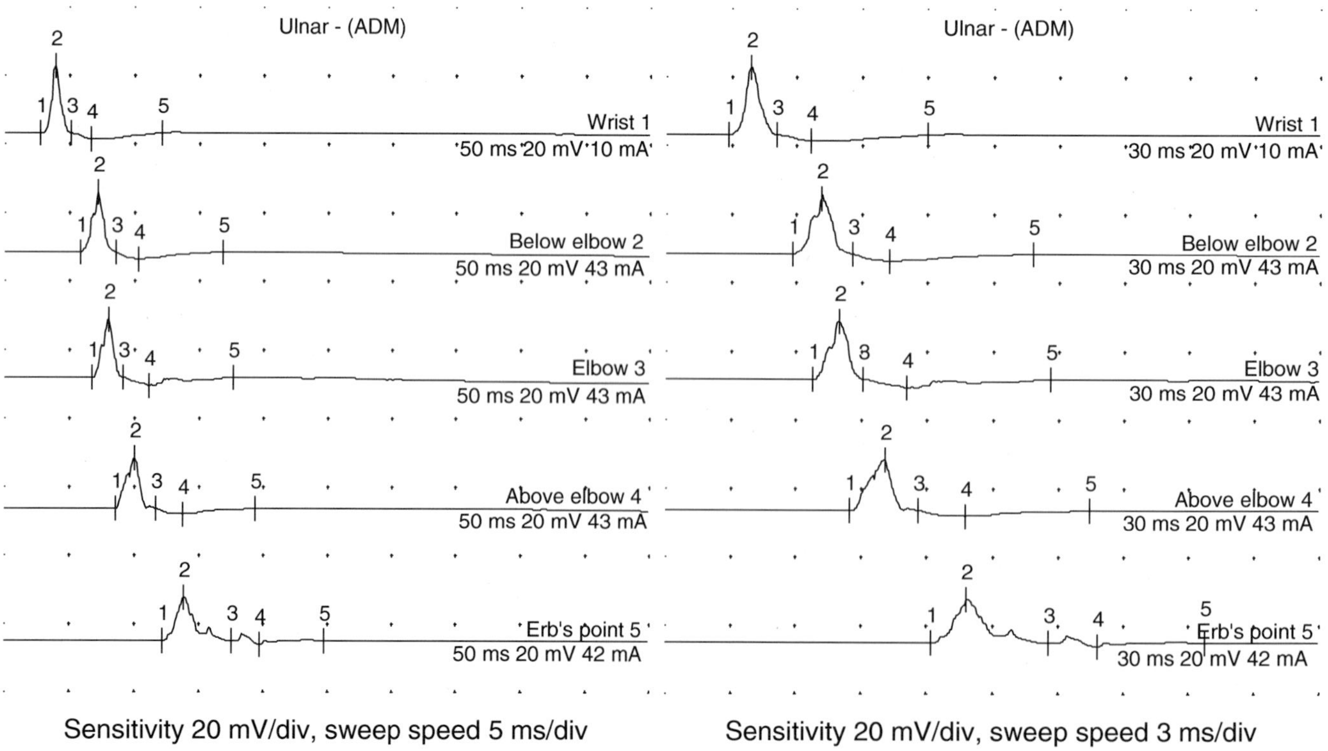

Fig. 1 Compound muscle action potentials (*CMAPs*) recorded at the hand from the ADM muscle; *top to bottom*: stimulation of the wrist, below the elbow, at the elbow, above the elbow, and of the Erb's point

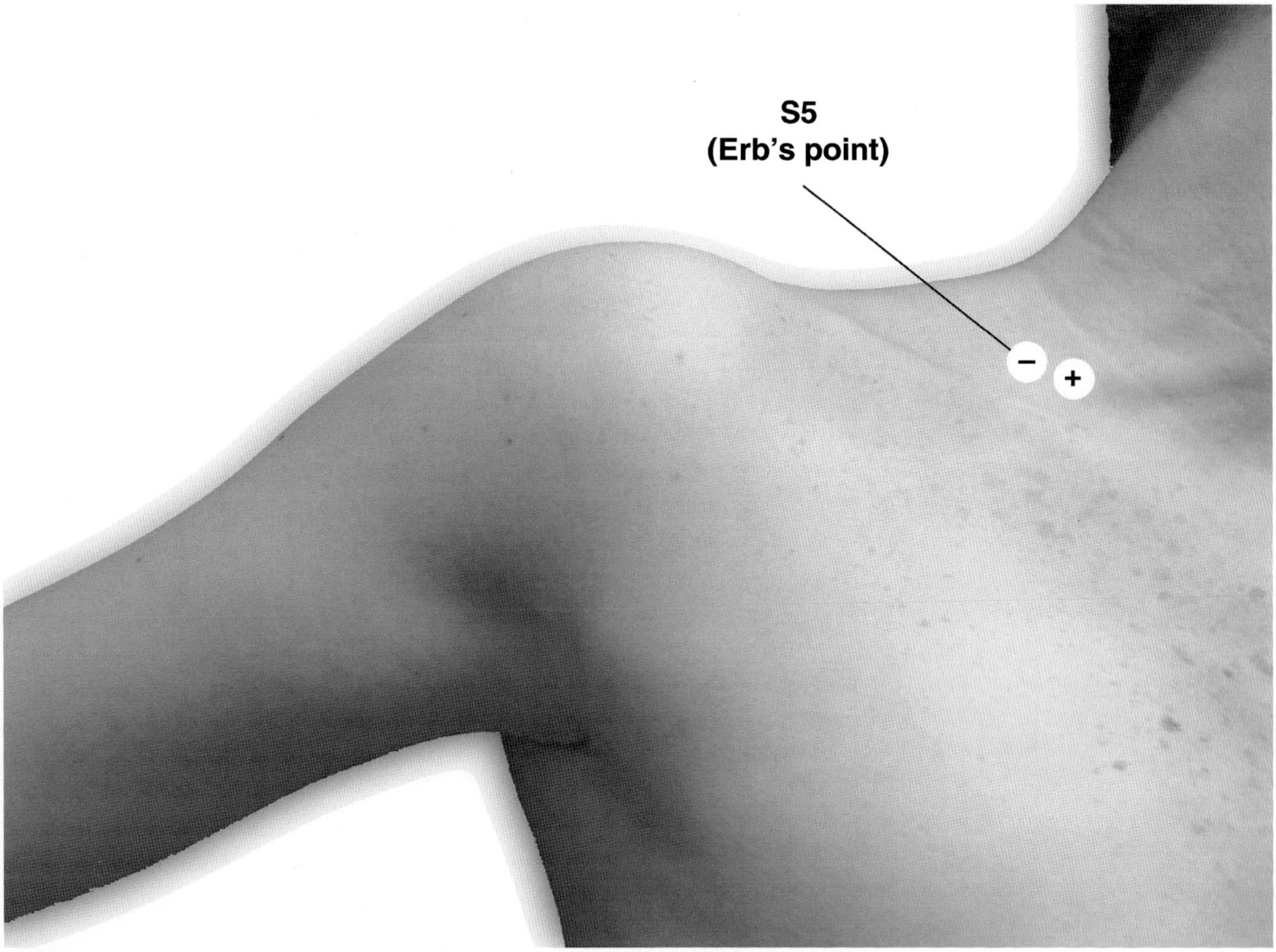

Fig. 2 Proximal stimulation point (S5 – Erb's point)

Table 1 Normal values [1]

	Mean ± SD	Mean ± 2SD	Range
Below elbow–wrist, MNCV (m/s)	59.4 ± 5.3	48.8	48–71
Elbow–wrist, MNCV (m/s)	57.9 ± 4.3	49.3	50–67
Erb's point–elbow, MNCV (m/s)	58.4 ± 4.1	50.2	51–69
Erb's point–above elbow, MNCV (m/s)	61.3 ± 5.4	50.5	52–78

Comment

Jebsen [1] collected data only from a normal control sample; three case reports were included just to illustrate the clinical application of the technique described in the paper. In the controls, the lower limit of the motor conduction velocity in the proximal segment of the ulnar nerve was approximately 50 m/s (Erb's point–elbow and Erb's point–above-elbow segments) which was similar to the MNCV in the distal segment (49 m/s). The author chose a 15 cm length for the above-elbow–below-elbow segment to minimize measurement errors and because in this way the most pathological lesions which affect the ulnar nerve would be included. He considered, for the ulnar nerve, a slow conduction velocity in the above-elbow–below-elbow segment when it was below 41.1 m/s (mean minus twice the SD). In the Erb's point–elbow and Erb's point–above-elbow segments instead, the ulnar nerve showed virtually identical motor conduction velocities, and the author suggested to compare the Erb's point–elbow with the elbow–wrist segment than the Erb's point–above elbow since it was technically simpler. He suggested using as routine points of stimulation for such comparisons the Erb's point, the elbow, and the wrist; the use of Erb's point as the proximal point of stimulation allows the inclusion of a more proximal segment than does stimulation in the axilla, thus accentuating any proximal–distal gradient. It may be useful in the diagnosis of local proximal neuropathy and in generalized neuropathies in which a proximal–distal conduction velocity gradient may exist. For Jebsen [1], a comparison of the mean ulnar above-elbow–below-elbow conduction velocity with the mean below-elbow–wrist conduction velocity would be ever done with the thought that such comparison might provide useful additional information. He often found in patients with ulnar lesion in the above-elbow–below-elbow segment an increased threshold to stimulation at the site of the lesion, and so the author strongly recommended to extend ulnar nerve evaluation to the proximal nerve segment performing stimulation at all five sites: wrist, below elbow, elbow, above elbow, and Erb's point. Pathological waveform and values in a case of ulnar neuropathy at the elbow (UNE) are reported (Fig. 3).

Pathological waveform (wrist, below elbow, elbow, above elbow, Erb's point – ADM muscle):

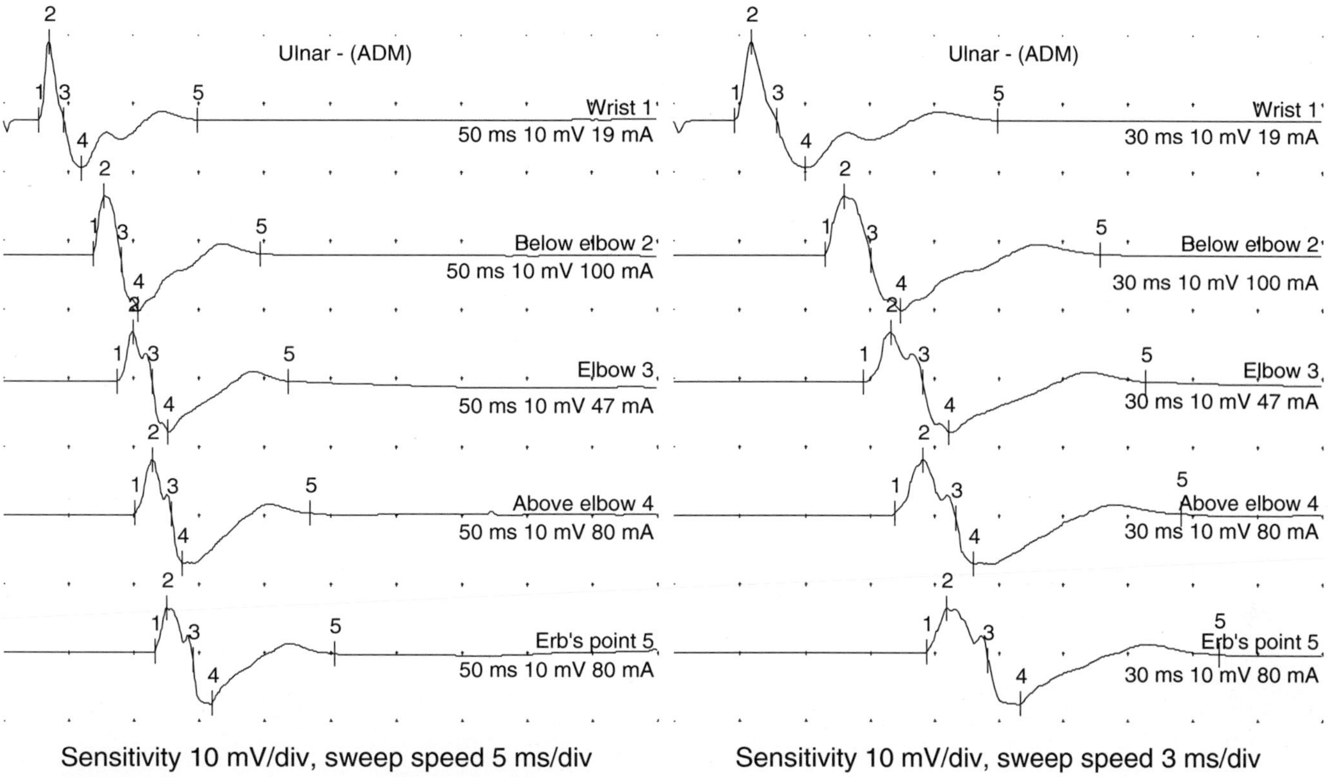

Onset latency (wrist – ADM): 2.75 ms; **Onset latency** (below elbow – ADM): 6.90 ms; **Onset latency** (elbow – ADM): 8.65 ms; **Onset latency** (above elbow – ADM): 10.10 ms; **Onset latency** (Erb's point – ADM): 11.60 ms; **Peak latency** (wrist – ADM): 3.45 ms; **Peak latency** (below elbow – ADM): 7.50 ms; **Peak latency** (elbow – ADM): 10.05 ms; **Peak latency** (above elbow – ADM): 11.15 ms; **Peak latency** (Erb's point – ADM): 12.30 ms; **Onset to peak amplitude** (wrist – ADM): 11.3 mV; **Onset to peak amplitude** (below elbow – ADM): 8.4 mV; **Onset to peak amplitude** (elbow – ADM): 7.2 mV; **Onset to peak amplitude** (above elbow – ADM): 7.8 mV; **Onset to peak amplitude** (Erb's point – ADM): 6.3 mV; **Peak to peak amplitude** (wrist – ADM): 18.5 mV; **Peak to peak amplitude** (below elbow – ADM): 16.8 mV; **Peak to peak amplitude** (elbow – ADM): 14.6 mV; **Peak to peak amplitude** (above elbow – ADM): 15.3 mV; **Peak to peak amplitude** (Erb's point – ADM): 14.2 mV; **MNCV** (wrist – below elbow, 22 cm distance): 53.0 m/s; **MNCV** (below elbow – elbow, 5 cm distance): 28.6 m/s; **MNCV** (elbow – above elbow, 10 cm distance): 69.0 m/s; **MNCV** (above elbow – Erb's point, 11 cm): 73.3 m/s

Fig. 3 Compound muscle action potentials (*CMAPs*) recorded at the hand from the ADM muscle using needle recording method [1] in UNE; *top to bottom*: stimulation of the wrist, below the elbow, at the elbow, above the elbow, and of the Erb's point

Reference

1. Jebsen RH (1967) Motor conduction velocities in the median and ulnar nerves. Arch Phys Med Rehabil 48:185–194

Wrist – Hand

Surface Recording Technique, Study from *Abductor Digiti Minimi* (ADM) and *First Dorsal Interosseous* (FDI) Muscles

M13

Original Settings Sweep speed was 10 ms/division, duration of pulse was 0.05–2 ms, rate of pulse was 1/s, and the machine used was a TECA electromyograph; sensitivity, low-frequency filter, and high-frequency filter were not specified.

Position Hand position was not specified in the text; we perform the test with the hand in the supine position.

Recording For the *abductor digiti minimi* (ADM) muscle (R1), the active electrode was placed over the motor point of the ADM muscle (Fig. 1); the reference (R) was placed slightly distal to the tendon insertion of the same muscle [1]. For the *first dorsal interosseous* (FDI) muscle recording (R2), the active electrode (A) was placed over the belly of the FDI muscle (Fig. 1), between the first and second metacarpals (Fig. 1); the reference (R) was placed slightly distal to the second metacarpophalangeal joint (over the tendon of the FDI muscle). R2 recording was made on the dorsum of the hand. The skin over the *hypothenar* area and the FDI muscle was abraded lightly with sandpaper in order to reduce the electrical resistance. Electrode jelly was used as a contact medium between the skin and the electrode, and adhesive material was used to keep the electrodes in place. The compound muscle action potentials (CMAPs) from ADM and FDI muscles were recorded separately (the authors used a Tektronix storage oscilloscope) using BECKMAN-type surface electrodes with low cable noise and low impedance for recording. The ground (G) electrode was placed on the dorsum of the hand; the figure shows the ground electrode placed on the palm.

Stimulation The ulnar nerve was stimulated percutaneously at the wrist using a stimulus of supramaximal intensity with cathode placed distally [1]. The stimulator delivered a rectangular pulse of 0.05–2.0 ms at 1 s intervals. The shock artifact and the time sweep of the oscilloscope were synchronized.

Measurements The latency (ms) of response (CMAP) for the fastest conducting fibers was measured from the beginning of the stimulus artifact to the beginning of the first deflection from the baseline. Because of considerable error involved in the calculation of the conduction velocity in this segment of the ulnar nerve due to the delay at the neuromuscular junction, conduction along the nerve terminals and the muscle fibers, and also due to inaccurate surface measurement of the distance between the points of stimulation and recordings, only CMAP latency was recorded. Normal values (Table 1) were obtained from 31 healthy volunteers (26 men and 5 women, age range 26–65 years), 27 subjects were studied bilaterally (22 right handed, 5 left handed) and 4 unilaterally for a total of 58 samples.

Comment

For Bhala and Goodgold [1], the range of the latency for the ADM and FDI muscles showed a great majority of their cases with a conduction time from 2.0 to 3.4 ms with a peak between 2.5 and 2.9 ms for the ADM muscle and from 3.0 to 4.4 ms with a peak between 4.0 and 4.4 ms for the FDI muscle. In 14 subjects (51.9 %) out of 27 examined bilaterally, the conduction time in the deep branch of the ulnar nerve was found less in the dominant hand than in the nondominant hand. In eight subjects (29.6 %), the conduction time was found less in the nondominant hand, and in five subjects (18.5 %), the values were the same in both hands though only three of these subjects used both hands equally well with almost equal dexterity. No relation was seen between sex and nerve conduction time. For the authors [1], these findings suggested that the motor conduction in the deep palmar branch of the ulnar nerve is faster in the dominant than in the nondominant hand. They found the method a useful

G. Gentili, M. Di Napoli, *The Ulnar Nerve*, DOI 10.1007/978-3-319-43292-2_37

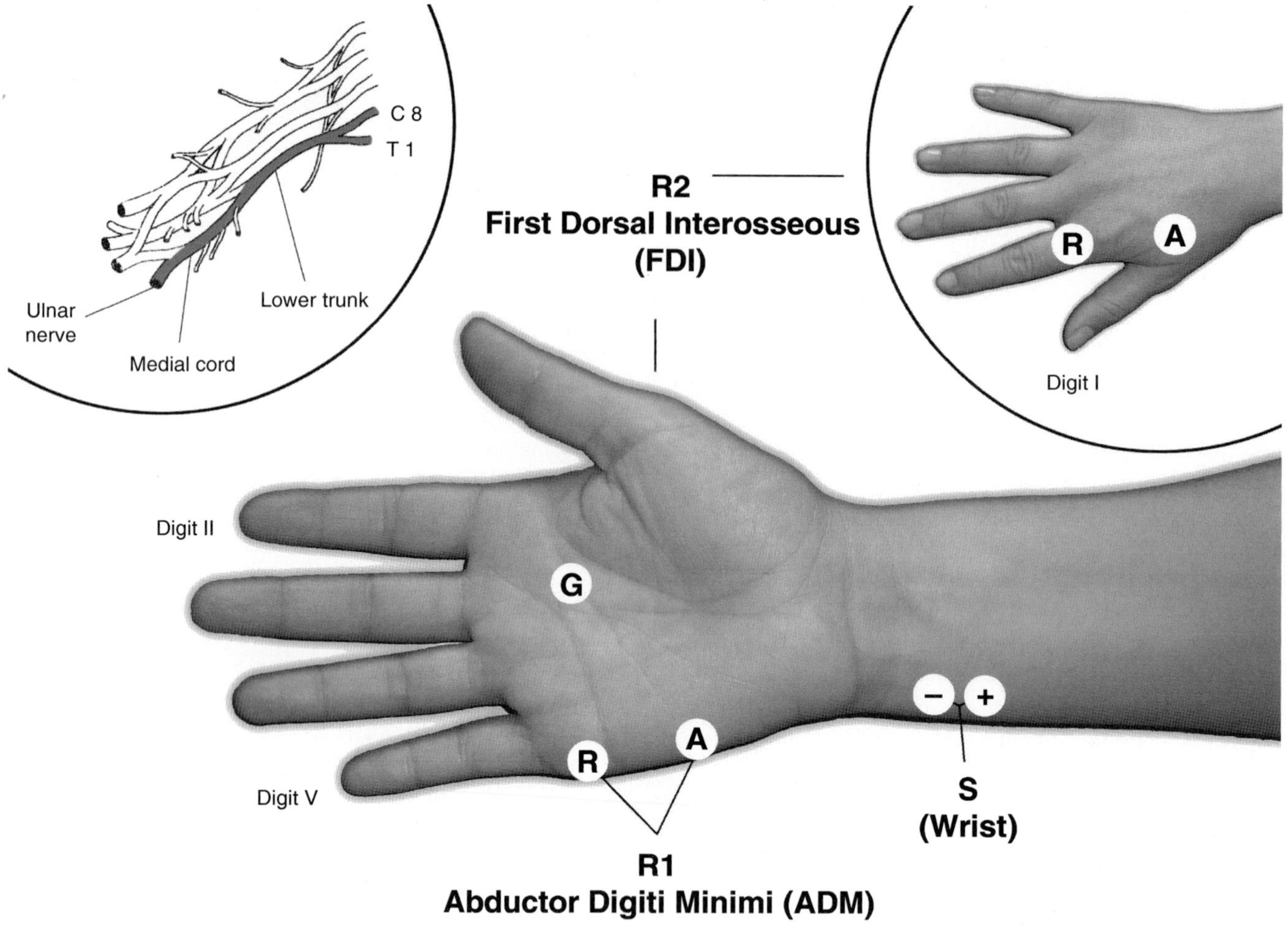

Typical waveform (wrist – ADM muscle, wrist – FDI muscle):

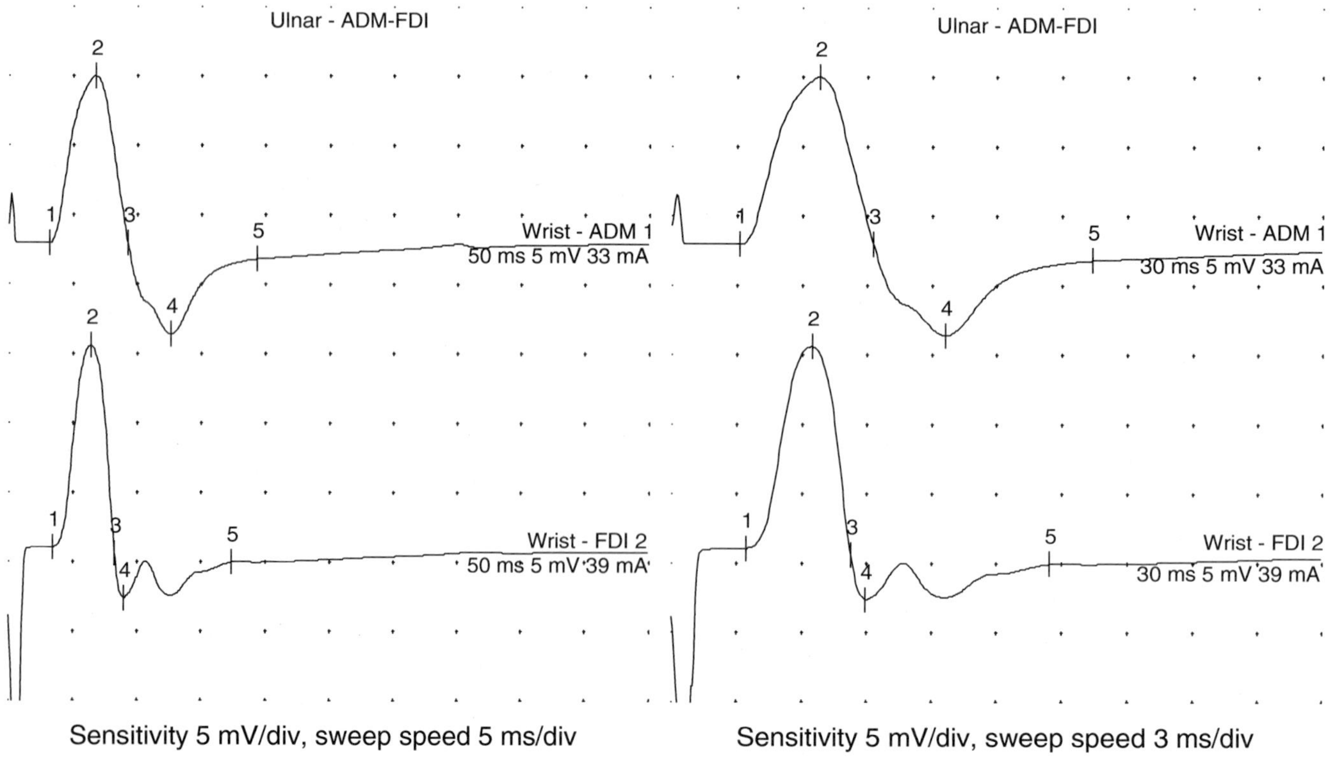

Fig. 1 Compound muscle action potentials (*CMAPs*) recorded at the hand from the ADM (*upper trace*) and FDI (*lower trace*) muscles, stimulation of the ulnar nerve at the wrist

Table 1 Normal values [1]

	Mean ± SD	Range
Wrist–ADM, distal latency (ms)	2.7 ± 0.49	1.9–4.0
Wrist–FDI, distal latency (ms)	4.0 ± 0.65	2.8–5.1
Wrist–ADM, wrist–FDI, distal latency difference (ms)	1.2 ± 0.48	0.4–2.4

tool in the evaluation of segmental neuropathies of the deep branch of the ulnar nerve secondary to chronic repeated trauma, ganglionic compression, fracture of the hook of the hamate, or other localized lesions in the hand.

In their study, Witmer et al. [2] examined a standardized technique assessing distal ulnar nerve conduction aimed to provide a more sensitive evaluation of ulnar nerve lesions at Guyon's canal. They compared the normal distal latency (DL) to the ADM and FDI muscles and response amplitudes (Table 2) of 50 healthy subjects (Group 1, age range 25–75 years) and 18 patients (Group 2) referred for electrodiagnostic evaluation of symptoms other than those characteristic of ulnar nerve lesions at the wrist (i.e., other neuropathies, including ulnar nerve compression at the elbow) (Table 3) to those obtained in 4 patients (Group 3) identified by chart review with ulnar lesions localized by history and/or surgery near the wrist but proximal to Guyon's canal (Table 4) and 6 patients (Group 4) identified upon referral or by chart review as having deep ulnar motor branch lesions verified by clinical symptoms and surgical observation (Table 5). All hands tested were maintained at a midpalm temperature > 30 °C using a YSI Precision 4000A skin thermometer. A DANTEC Counterpoint K2 was used to record CMAP responses simultaneously, from the ADM and FDI muscles by using two open channels. The ulnar nerve was stimulated using supramaximal stimulation at the wrist – 8 cm proximal to the active recording electrode over the ADM muscle. The authors used anatomical landmarks for active electrode placement. For the ADM muscle, the active recording electrode was placed at the midpoint of a line drawn from the most medial aspect of the fifth metacarpophalangeal (MCP) joint to the pisiform bone, with the reference placed at the medial fifth MCP joint. For the FDI muscle, the active electrode was placed at the midpoint of a line drawn between the most lateral point of the second MCP joint and the head of the first metacarpal at the carpal–metacarpal joint; the reference electrode was placed at the medial aspect of the interphalangeal joint of the digit I (Fig. 2).

Table 2 Normal values [2]: Group 1

	Mean ± SD	Range
Wrist–ADM, distal latency (ms) – right side	2.96 ± 0.28	2.4–3.6
Wrist–ADM, distal latency (ms) – left side	2.97 ± 0.25	2.4–3.5
Wrist–ADM, distal latency (ms) – variance side to side	0.01 ± 0.19	0–0.5
Wrist–FDI, distal latency (ms) – right side	3.91 ± 0.41	3.1–4.7
Wrist–FDI, distal latency (ms) – left side	3.87 ± 0.36	3.1–4.7
Wrist–FDI, distal latency (ms) – variance side to side	0.04 ± 0.22	0–0.5
Wrist–ADM, wrist–FDI, distal latency difference (ms) – right side	0.95 ± 0.25	0.5–1.4
Wrist–ADM, wrist–FDI, distal latency difference (ms) – left side	0.91 ± 0.25	0.4–1.3
Wrist–ADM, wrist–FDI, distal latency difference (ms) – side to side	0.04 ± 0.17	0–0.4
Wrist–ADM, amplitude (mV) – right side	10.6 ± 2.3	5.1–14.6
Wrist–ADM, amplitude (mV) – left side	10.1 ± 2.1	5.1–14.4
Wrist–ADM, amplitude (mV) – variance side to side	0.52 ± 1.64	−2.8–4.1
Wrist–FDI, amplitude (mV) – right side	14.9 ± 3.9	8.6–26.5
Wrist–FDI, amplitude (mV) – left side	14.6 ± 3.4	7.1–23
Wrist–FDI, amplitude (mV) – variance side to side	0.32 ± 3.36	−8.5–8.4
Wrist–ADM, wrist–FDI, amplitude difference (mV) – right side	4.3 ± 4.3	−4.2–16.4
Wrist–ADM, wrist–FDI, amplitude difference (mV) – left side	4.5 ± 3.7	−3.1–10.5

Table 3 Normal values [2]: Group 2

	Mean ± SD	Range
Wrist–ADM, distal latency (ms) – right side	3.26 ± 0.44	2.5–3.9
Wrist–ADM, distal latency (ms) – left side	3.17 ± 0.39	2.4–3.8
Wrist–ADM, distal latency (ms) – variance side to side	0.09 ± 0.24	0–0.8
Wrist–FDI, distal latency (ms) – right side	4.18 ± 0.58	2.8–5.3
Wrist–FDI, distal latency (ms) – left side	4.18 ± 0.51	3.1–5.2
Wrist–FDI, distal latency (ms) – variance side to side	–0.09 ± 0.24	0–0.8
Wrist–ADM, wrist–FDI, distal latency difference (ms) – right side	0.92 ± 0.32	0.3–1.5
Wrist–ADM, wrist–FDI, distal latency difference (ms) – left side	1.02 ± 0.23	0.7–1.4
Wrist–ADM, wrist–FDI, Distal latency difference (ms) – side to side	–0.09 ± 0.23	0–0.4
Wrist–ADM, amplitude (mV) – right side	9 ± 1.7	6.2–12
Wrist–ADM, amplitude (mV) – left side	8.3 ± 2.4	5.2–13.2
Wrist–ADM, amplitude (mV) – variance side to side	0.68 ± 1.46	–2.2–3
Wrist–FDI, amplitude (mV) – right side	12.3 ± 2.6	8.4–18.9
Wrist–FDI, amplitude (mV) – left side	11.2 ± 2.3	4.3–13.8
Wrist–FDI, amplitude (mV) – variance side to side	0.96 ± 2.71	–3.8–5.8
Wrist–ADM, wrist–FDI, amplitude difference (mV) – right side	3.1 ± 2.6	–0.6–8.5
Wrist–ADM, wrist–FDI, amplitude difference (mV) – left side	2.8 ± 2.7	–1.6–7.1

Table 4 Pathological values [2]: Group 3

	Mean ± SD	Range
Wrist–ADM, distal latency (ms) – symptomatic side	4.32 ± 0.79	3.5–5.3
Wrist–ADM, distal latency (ms) – asymptomatic side	2.97 ± 0.40	2.5–3.2
Wrist–ADM, distal latency (ms) – variance side to side	1.13 ± 0.76	0–0.3
Wrist–FDI, distal latency (ms) – symptomatic side	5.66 ± 1.22	4.5–7.6
Wrist–FDI, distal latency (ms) – asymptomatic side	4.43 ± 0.55	3.8–4.8
Wrist–FDI, distal latency (ms) – variance side to side	1.2 ± 1.51	0–0.6
Wrist–ADM, wrist–FDI, distal latency diff (ms) – symptomatic side	1.34 ± 0.8	0.6–2.6
Wrist–ADM, wrist–FDI, distal latency diff (ms) – asymptomatic side	1.47 ± 0.15	1.3–1.6
Wrist–ADM, wrist–FDI, distal latency difference (ms) – sym to asym	0.07 ± 0.91	–0.6–1.1
Wrist–ADM, amplitude (mV) – symptomatic side	2.2 ± 1.4	0.6–3.7
Wrist–ADM, amplitude (mV) – asymptomatic side	9.1 ± 1	7.9–9.8
Wrist–ADM, amplitude (mV) – variance side to side	6.64 ± 0.77	5.8–7.3
Wrist–FDI, amplitude (mV) – symptomatic side	1.3 ± 0.9	0.2–2.3
Wrist–FDI, amplitude (mV) – asymptomatic side	10.9 ± 3.6	7.9–14.9
Wrist–FDI, amplitude (mV) – variance sym to asym	9.45 ± 4.63	5.7–14.6
Wrist–ADM, wrist–FDI, amplitude diff (mV) – symptomatic side	–0.9 ± 1.4	–2.2–1.2
Wrist–ADM, wrist–FDI, amplitude diff (mV) – asymptomatic side	1.9 ± 2.8	0–5.1

Table 5 Pathological values [2]: Group 4

	Mean ± SD	Range
Wrist–ADM, distal latency (ms) – symptomatic side	3.26 ± 0.58	2.6–4.3
Wrist–ADM, distal latency (ms) – asymptomatic side	3.37 ± 0.12	3.3–3.5
Wrist–ADM, distal latency (ms) – variance side to side	0.17 ± 0.15	0–0.3
Wrist–FDI, distal latency (ms) – symptomatic side	5.42 ± 1.92	4.1–10.2
Wrist–FDI, distal latency (ms) – asymptomatic side	4.43 ± 0.31	4.1–4.7
Wrist–FDI, distal latency (ms) – variance sym to asym	0.4 ± 0.2	0–0.6
Wrist–ADM, wrist–FDI, distal latency diff (ms) – symptomatic side	2.17 ± 1.49	1.5–6.1
Wrist–ADM, wrist–FDI, distal latency diff (ms) – asymptomatic side	1.07 ± 0.31	0.8–1.4
Wrist–ADM, wrist–FDI, distal latency difference (ms) – sym to asym	0.57 ± 0.32	0.2–0.8
Wrist–ADM, amplitude (mV) – symptomatic side	9.4 ± 3.1	5.3–13.9
Wrist–ADM, amplitude (mV) – asymptomatic side	8.4 ± 1.7	7.3–10.4
Wrist–ADM, amplitude (mV) – variance sym to asym	0.87 ± 1.56	−2.3–0.8
Wrist–FDI, amplitude (mV) – symptomatic side	10.5 ± 6.5	3.5–19.9
Wrist–FDI, amplitude (mV) – asymptomatic side	10.6 ± 3.7	7.7–14.7
Wrist–FDI, amplitude (mV) – variance sym to asym	−1.5 ± 3.66	−5.7–1
Wrist–ADM, wrist–FDI, amplitude diff (mV) – symptomatic side	1.1 ± 4.6	−5.5–8
Wrist–ADM, wrist–FDI, amplitude diff (mV) – asymptomatic side	2.1 ± 5	−2.7–7.4

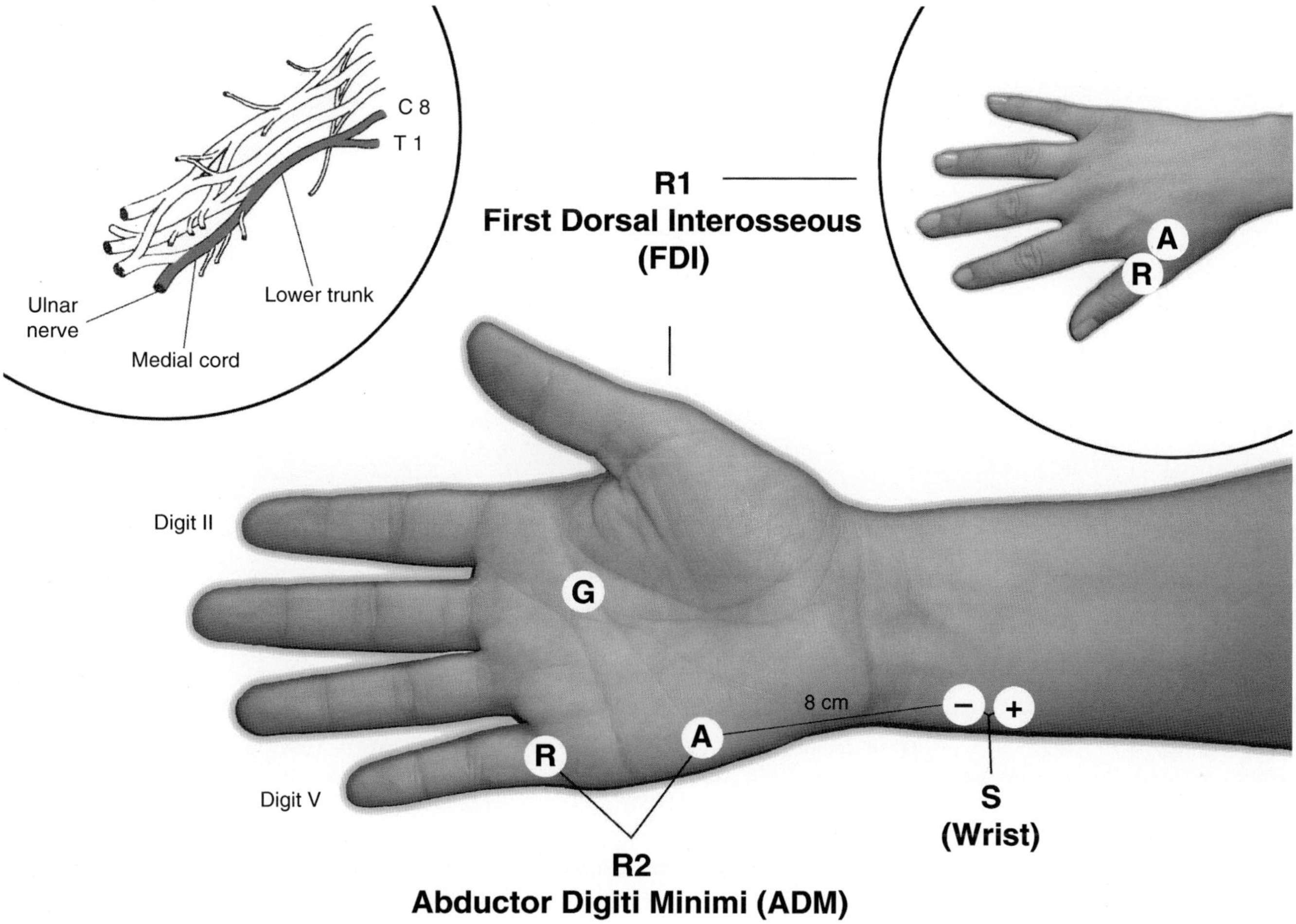

Fig. 2 ADM–FDI muscle CMAP comparison by Witmer et al. [2]

Comment

Witmer et al. [2] adjusted the active and reference electrodes in eliminating the frequent positive deflection prior to the main negative phase observed when recording the CMAP from the FDI muscle following the Olney and Wilbourn's method [3]. Their study showed that initial negative takeoff may be attained consistently, with preservation and often enhancement of CMAP amplitudes, and the side-to-side absolute latencies to both ADM and FDI muscles were reduced significantly. These values varied only to a maximum difference of 0.5 ms, up to 50 % less than that previously reported [3] for the ADM muscle (1 ms) and nearly 70 % less for the FDI muscle (1.3 ms). From their findings, Witmer et al. [1] proposed a new maximum acceptable side-to-side distal latency difference of 0.5 ms for both ADM and FDI muscles. An alternative diagnostic measure of the deep ulnar motor branch lesions is the ipsilateral FDI–ADM distal latency difference. Olney and Wilbourn [3] considered normal a latency difference at values up to 2 ms, but Witmer et al. [1] suggested to use a lower latency value (≥1.45 ms). Pathological waves and values in a diabetic patient with a lesion of the deep branch of the ulnar nerve and C8-T1 radiculopathy (Fig. 3) and a patient with a lesion of the deep branch of the ulnar nerve due to a cutting knife (Fig. 4) are here reported.

Pathological waveform (wrist – ADM muscle, wrist – FDI muscle):

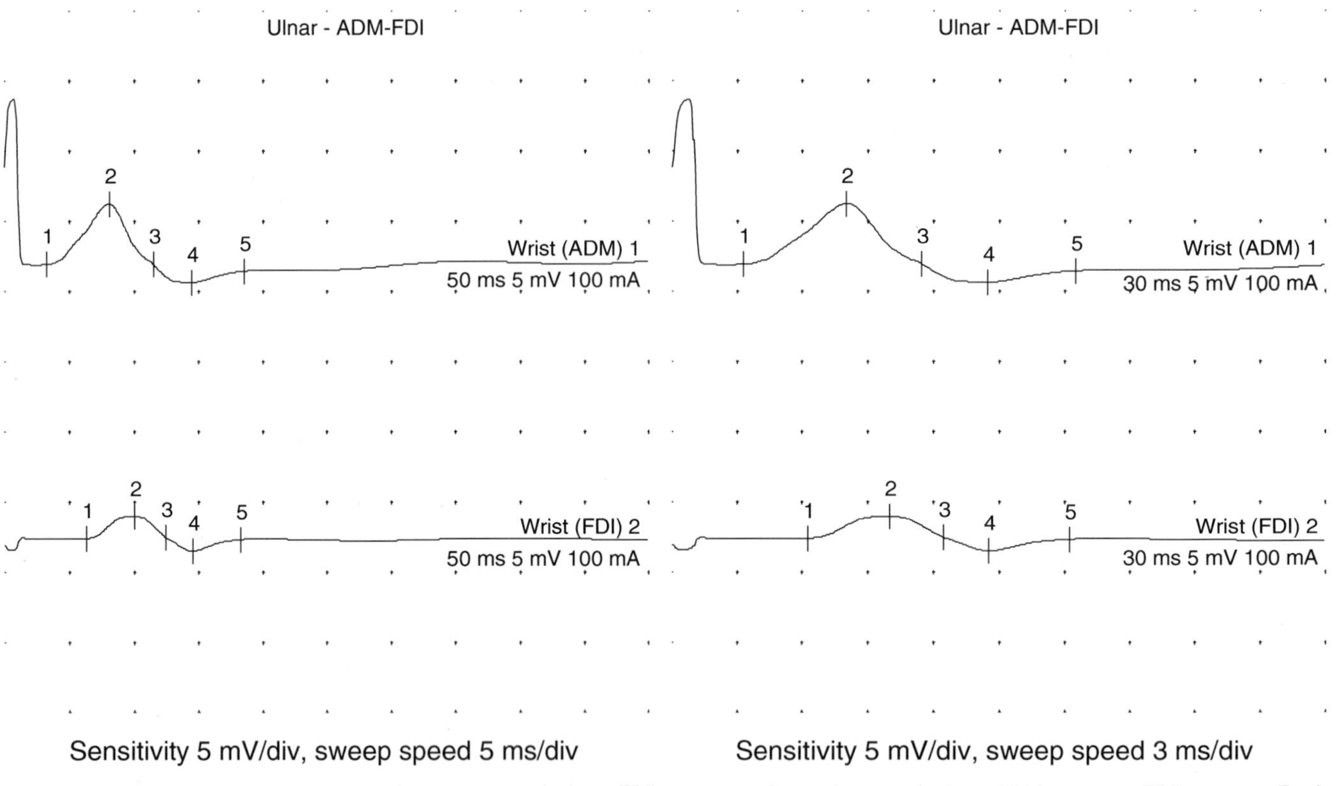

Onset latency (wrist – ADM): 3.25 ms; **Onset latency** (wrist – FDI): 6.25 ms; **Onset latency** (wrist – ADM – wrist – FDI): 3.0 ms; **Peak latency** (wrist – ADM): 7.75 ms; **Peak latency** (wrist – FDI): 10.20 ms; **Onset to peak amplitude** (wrist – ADM): 4.4 mV; **Onset to peak amplitude** (wrist – FDI): 1.6 mV; **Peak to peak amplitude** (wrist – ADM): 5.7 mV; **Peak to peak amplitude** (wrist – FDI): 2.6 mV

Fig. 3 Compound muscle action potentials (*CMAPs*) recorded at the hand from the ADM (*upper trace*) and FDI (*lower trace*) muscles in a patient with a lesion of the deep branch of the ulnar nerve and C8-T1 radiculopathy (diabetic patient), stimulation of the ulnar nerve at the wrist

Pathological waveform (wrist – ADM muscle, wrist – FDI muscle):

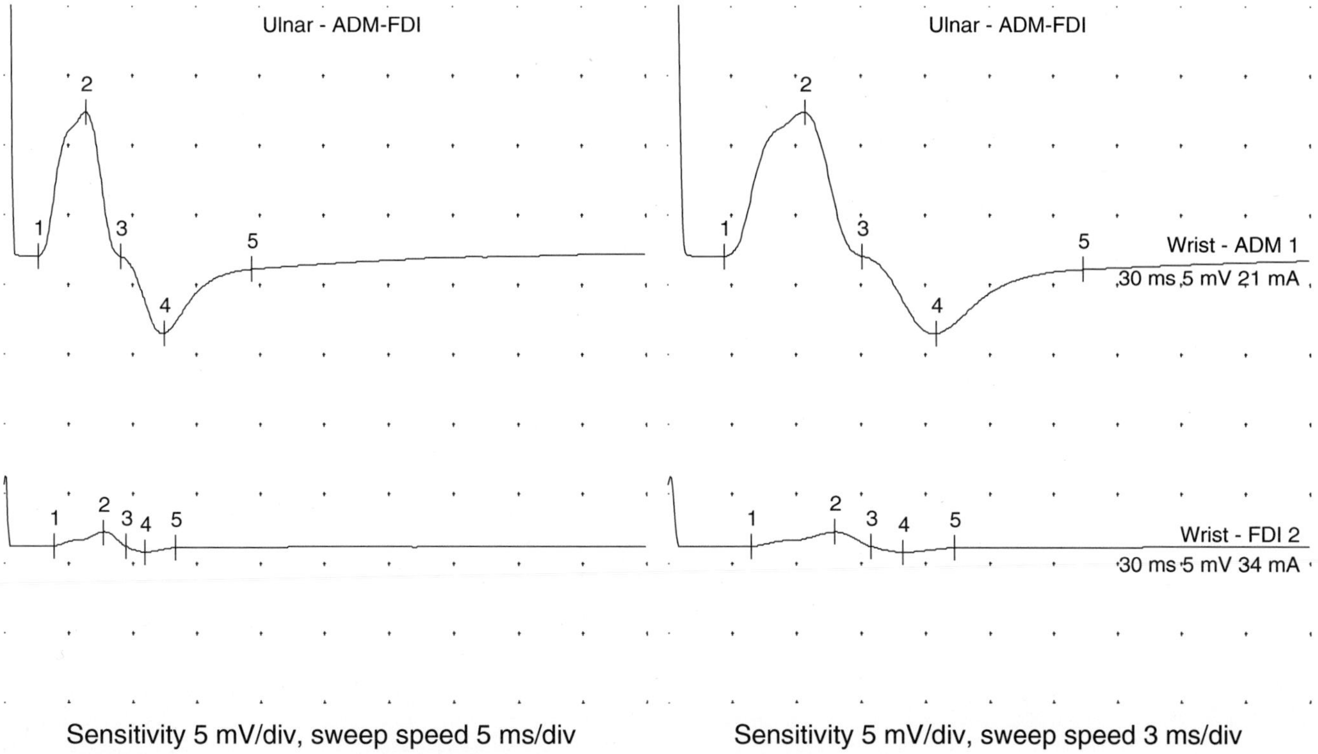

Onset latency (wrist – ADM): 2.65 ms; **Onset latency** (wrist – FDI): 3.95 ms; **Onset latency difference** (wrist – ADM – wrist – FDI): 1.3 ms; **Peak latency** (wrist – ADM): 6.55 ms; **Peak latency** (wrist – FDI): 7.70 ms; **Onset to peak amplitude** (wrist – ADM): 10.4 mV; **Onset to peak amplitude** (wrist – FDI): 1.0 mV; **Peak to peak amplitude** (wrist – ADM): 16.0 mV; **Peak to peak amplitude** (wrist – FDI): 1.5 mV

Fig. 4 Compound muscle action potentials (*CMAPs*) recorded at the hand from the ADM (*upper trace*) and FDI (*lower trace*) muscles in a patient with a lesion of the deep branch of the ulnar nerve (cutting knife), stimulation of the ulnar nerve at the wrist

References

1. Bhala RP, Goodgold J (1968) Motor conduction in the deep palmar branch of the ulnar nerve. Arch Phys Med Rehabil 49:460–466
2. Witmer B, Di Benedetto M, Gregory Kang C (2002) An improved approach to the evaluation of the deep motor branch of the ulnar nerve. Electromyogr Clin Neurophysiol 42:485–493
3. Olney RK, Wilbourn AJ (1985) Ulnar nerve conduction study of the first dorsal interosseous muscle. Arch Phys Med Rehabil 66:16–18

Axilla, Above Sulcus, Below Sulcus, Wrist – Hand; Above Sulcus – Forearm

M14

Needle Recording Technique, Study from *Abductor Digiti Minimi* (ADM), *First Dorsal Interosseous* (FDI), and *Flexor Carpi Ulnaris* (FCU) Muscles

Original Settings Duration of pulse was 0.2 ms, and the machine used was a DISA type 14 A 30; sensitivity, sweep speed, low-frequency filter, high-frequency filter, and rate of pulse were not specified.

Position This study was performed with the subjects lying on a bed in the supine position, with the arm at 45° to the trunk, the elbow extended, and the forearm supinated.

Recording The active needle electrode (0.45 mm in diameter) was placed over the motor point [1] of the ulnar *hypothenar* (UHT) muscle–*abductor digiti minimi* (ADM) muscle (Fig. 1) and *first dorsal interosseous* (FDI) muscle (Fig. 2). The ground (G) electrode position was not specified in the text; the figure shows the ground electrode positioned on the palm.

Needle Insertion For the ADM muscle, with the hand in a neutral position to full supination, the recording needle electrode was inserted at the midpoint between the fifth metacarpophalangeal joint and the pisiform bone. For the FDI muscle, with the hand in a neutral position to full pronation, the recording needle electrode was inserted at the midpoint between the proximal phalanx of the digit I (thumb) and the metacarpal bone.

Stimulation The ulnar nerve was stimulated at the axilla (S1), about 5 cm distal to the medial epicondyle ("below sulcus" (S2)), about 5 cm proximal to the medial epicondyle ("above sulcus" (S3)), and about 15 cm proximal to it (S4). The author used for stimulation the same needle electrodes used for the sensory orthodromic study (see sensory conduction studies). Stimulus duration was 0.2 ms; the stimulus was increased to at least 10 times the motor threshold (<1 mA), which was checked before the stimulation to ensure that the needles were still close to the nerve [1]. The "above-sulcus" stimulation site was also used for ulnar nerve conduction study to the *flexor carpi ulnaris* (FCU) muscle (Fig. 4). We use with similar results conventional bipolar stimulator in all sites of stimulation. The anode was proximal in all stimulations (Figs. 1, 2, 3, and 4).

Measurements Distal latency (ms) of the compound muscle action potential (CMAP) was measured from the stimulus onset to the onset of the initial deflection of the CMAP. Distal latency (ms) was also corrected to 7 cm for the wrist–UHT segment. The distances were measured between the stimulating cathode (-) and the active recording electrode (A) at each site, and motor nerve conduction velocity (MNCV) in the fastest fibers was calculated for the axilla–above-sulcus, above-sulcus–below-sulcus, and below-sulcus–wrist segments. The MNCV was measured in meter per second (m/s). The amplitude (mV) was measured peak to peak, and CMAPs were evaluated according to duration and degree of synchronization. The temperature was monitored on the skin and near the nerve by a thermocouple, and it was maintained between 34 and 36°C; each limb was warmed before and throughout the investigation. Normal values (Table 1) were reported; the author [1] studied 36 healthy subjects (Group 1, 21 subjects (age range 18–65 years); Group 2, 15 subjects (age range 70–89 years)). Pathological values were recorded from 47 cases, but pathological sample data were not reported. Pathological waveforms and values in a case of ulnar neuropathy at the elbow (UNE) are reported (Figs. 5, 6, 7, and 8).

G. Gentili, M. Di Napoli, *The Ulnar Nerve*, DOI 10.1007/978-3-319-43292-2_38

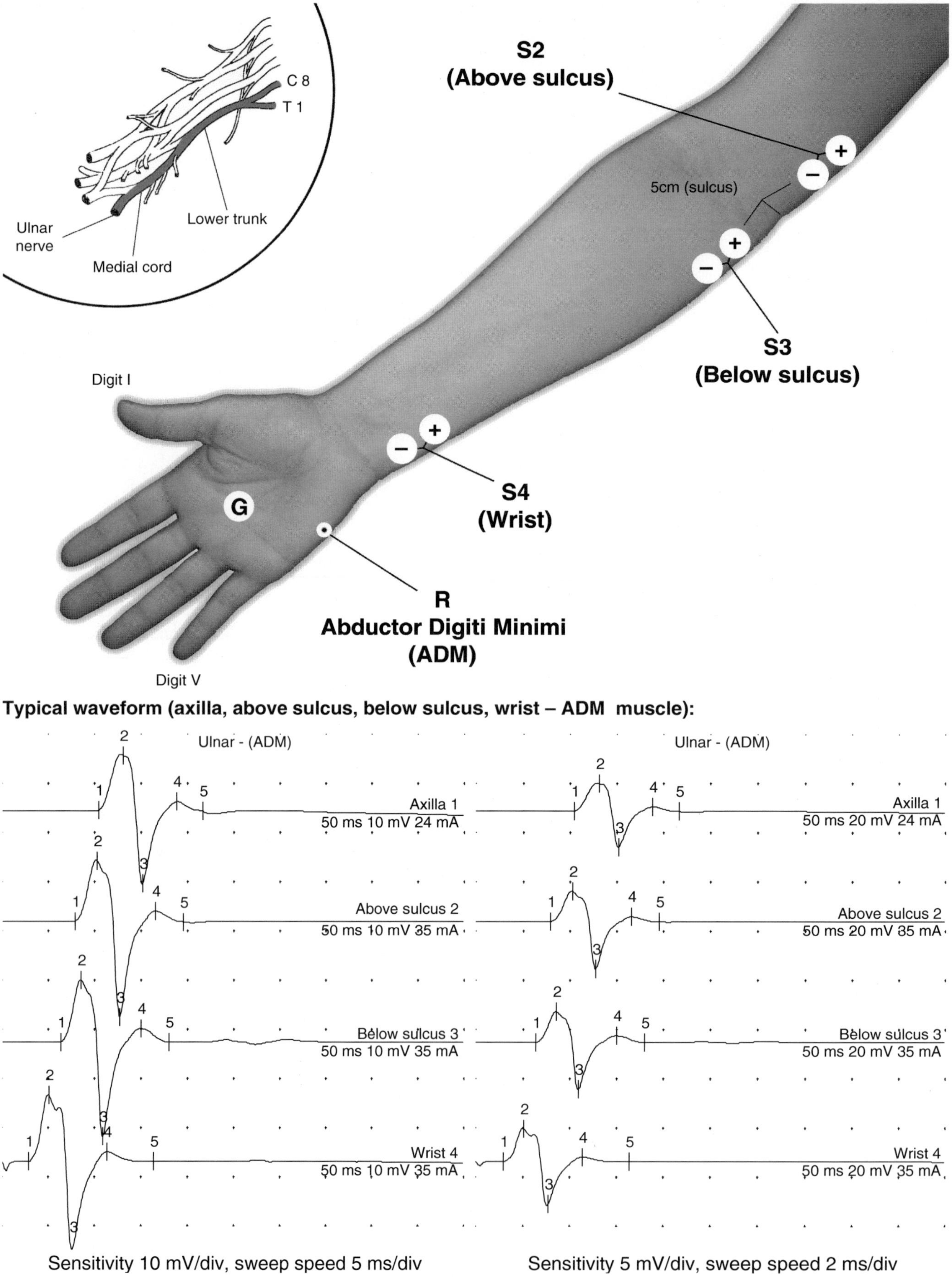

Fig. 1 Compound muscle action potentials (*CMAPs*) recorded at the hand from the ADM muscle, *top to bottom*: stimulation of the axilla, above sulcus, below sulcus, and of the wrist

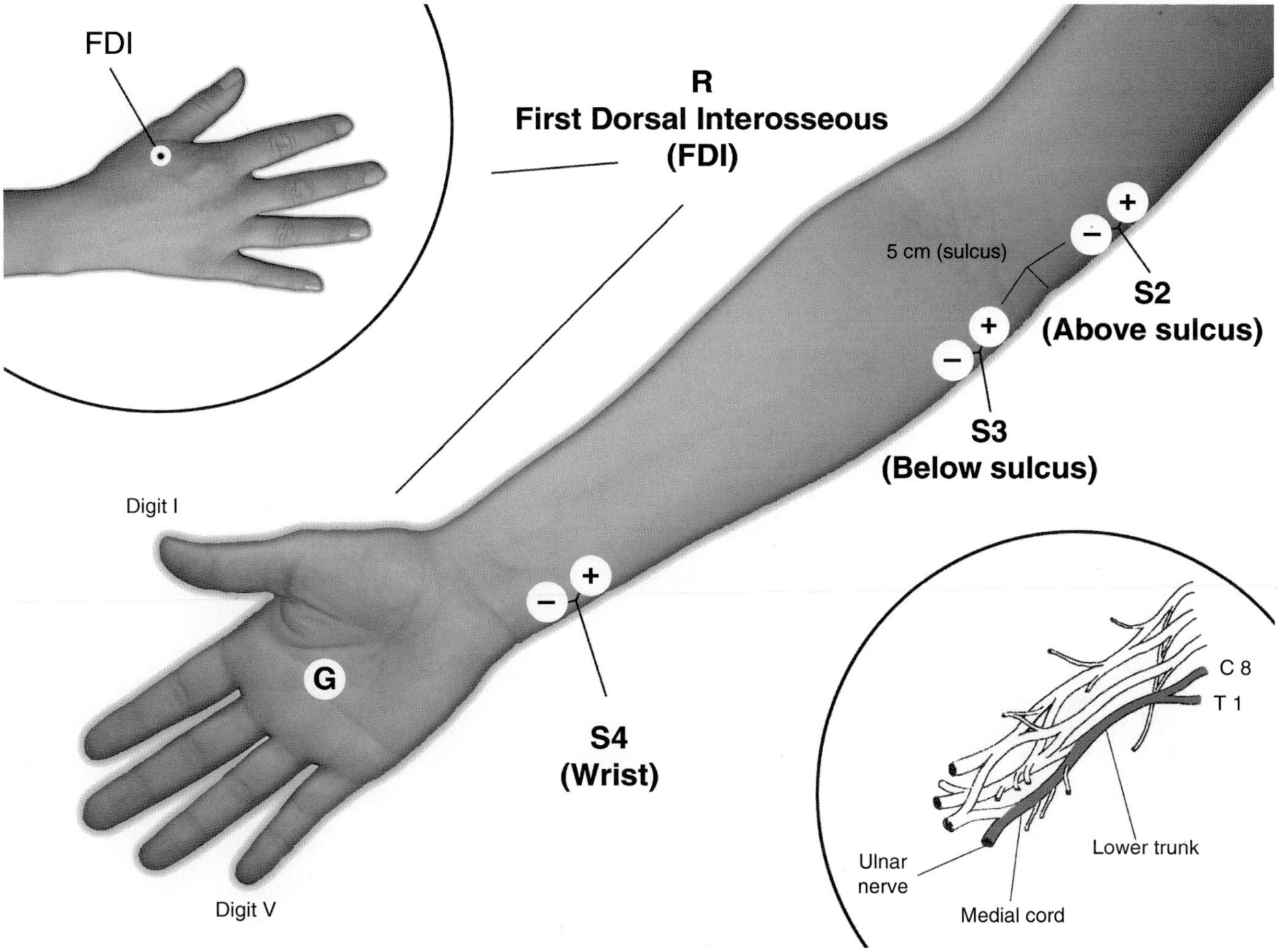

Typical waveform (axilla, above sulcus, below sulcus, wrist – FDI muscle):

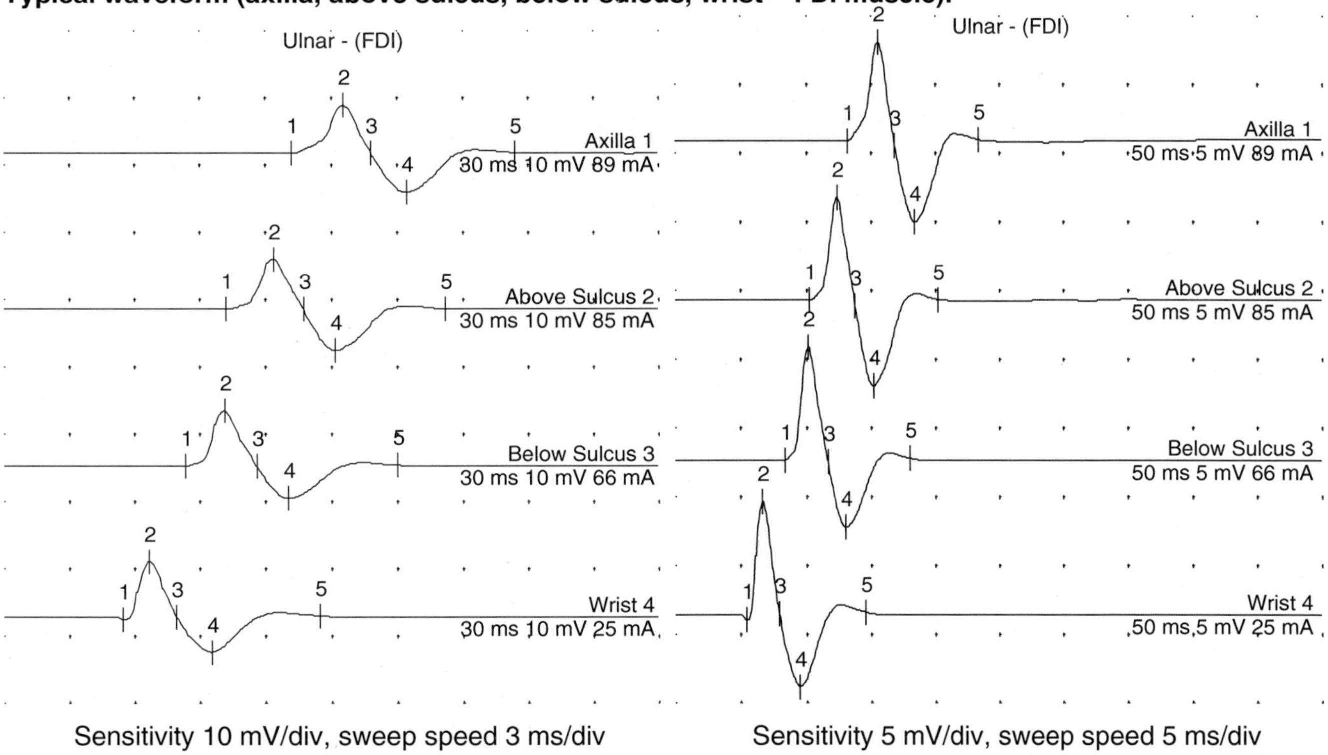

Fig. 2 Compound muscle action potentials (*CMAPs*) recorded at the hand from the FDI muscle, *top to bottom*: stimulation of the axilla, above sulcus, below sulcus, and of the wrist

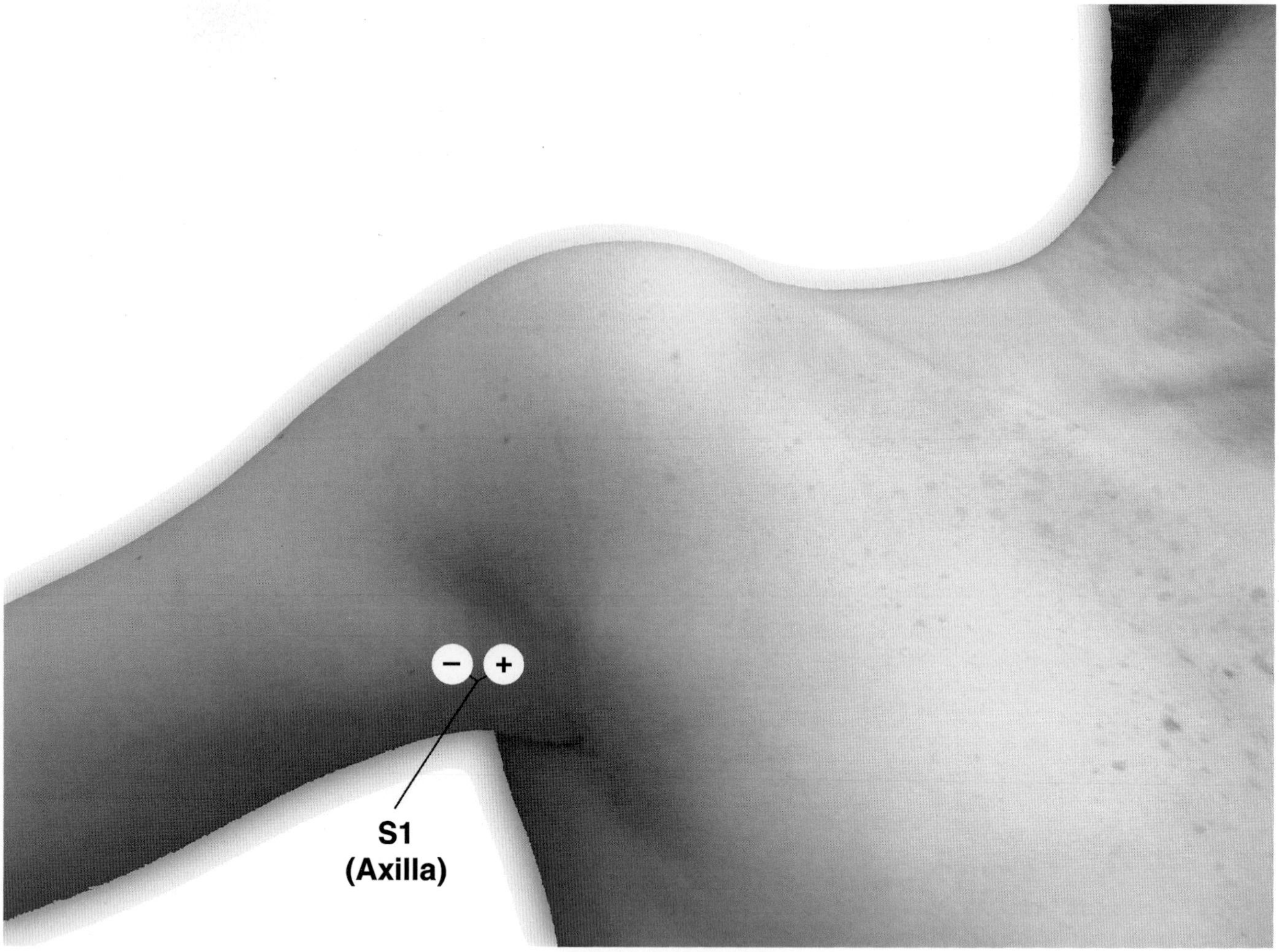

Fig. 3 Proximal stimulation point (S1 – Erb's point)

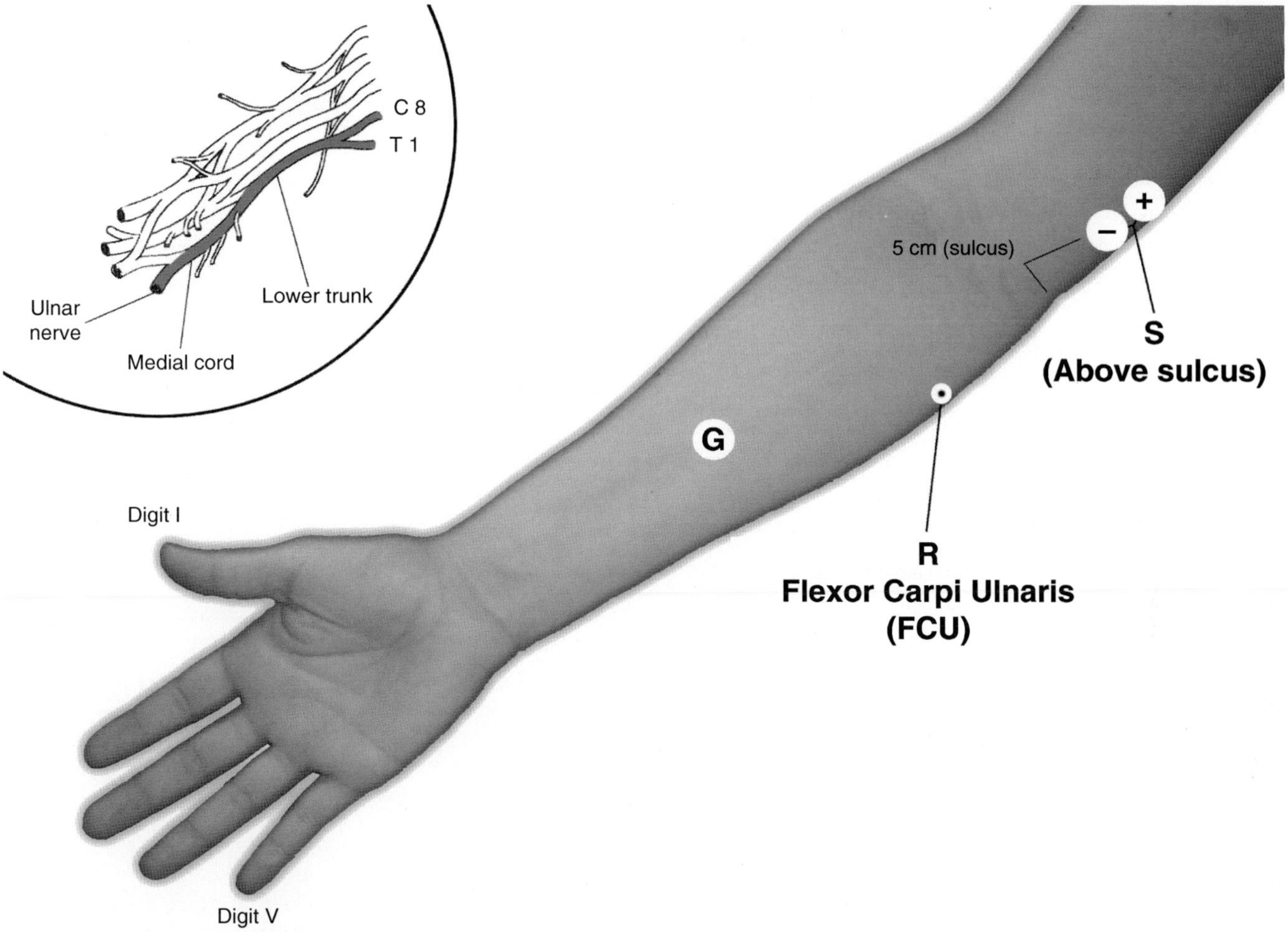

Typical waveform (above sulcus – FCU muscle):

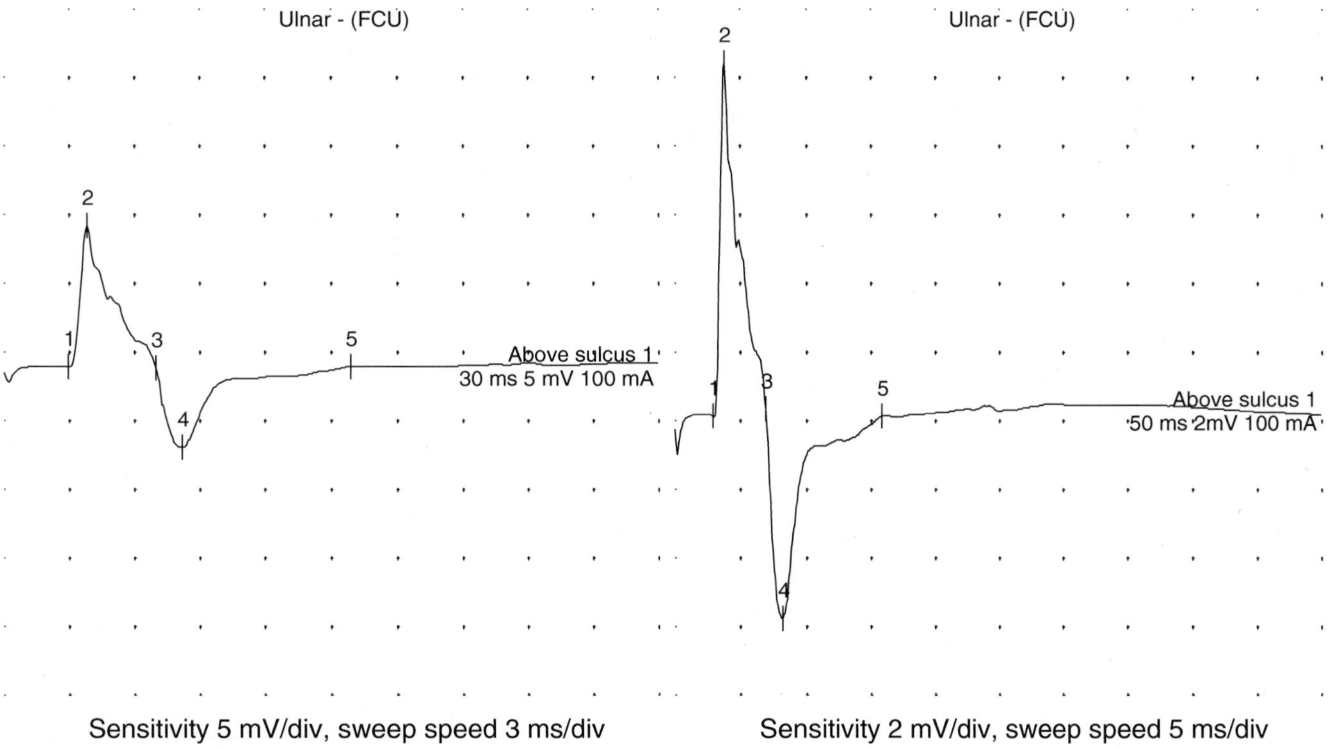

Fig. 4 Compound muscle action potential (*CMAP*) recorded at the hand from the FCU muscle, stimulation above the sulcus

Table 1 Normal values [1]

Group 1	Mean ± SD	Range
Above sulcus–wrist, MNCV (m/s)	63.0 ± 3.5	
Below sulcus–wrist, MNCV (m/s)	69.0 ± 5.5	
Above sulcus–below sulcus, MNCV (m/s)	52.0 ± 4.0	
Wrist–ADM (7 cm distance) – distal motor latency (ms)	2.4 ± 0.3	
Wrist–ADM (no fixed distance) – distal motor latency (ms)	2.3 ± 0.3	
Wrist–FDI (no fixed distance) – distal motor latency (ms)	3.2 ± 0.3	
Wrist–ADM (no fixed distance) – amplitude (mV)	21.0 ± 7.0	
Wrist–FDI (no fixed distance) – amplitude (mV)	26 ± 8.0	
Above sulcus–FCU (9–15 cm distance) – proximal motor latency (ms)	3.1 ± 0.3	2.5–3.8
Wrist–FCU (9–15 cm distance) – amplitude (mV)	18.0 ± 7.0	
Group 2	**Mean ± SD**	**Range**
Above sulcus–wrist, MNCV (m/s)	56.0 ± 3.0	
Below sulcus–wrist, MNCV (m/s)	61.0 ± 5.0	
Above sulcus–below sulcus, MNCV (m/s)	50.0 ± 2.0	
Wrist–ADM (7 cm distance) – distal motor latency (ms)	2.7 ± 0.3	
Wrist–ADM (no fixed distance) – distal motor latency (ms)	2.8 ± 0.3	
Wrist–FDI (no fixed distance) – distal motor latency (ms)	3.4 ± 0.3	
Wrist–ADM (no fixed distance) – amplitude (mV)	16.0 ± 7.0	
Wrist–FDI (no fixed distance) – amplitude (mV)	17.0 ± 10.0	6–30
Above sulcus–FCU (9–15 cm distance) – proximal motor latency (ms)	3.5 ± 0.4	3.0–4.1
Wrist–FCU (9–15 cm distance) – amplitude (mV)	13.0 ± 4.0	

Comment

Payan [1] recorded normal CMAPs from both ADM and FDI muscles in 35 cases when stimulating the wrist, and he observed a reduced CMAP amplitude in 12 cases, all but three of which were associated (85 %) with severe slowing of conduction across the sulcus (the author reported 30 m/s or fewer values), and in three cases, the CMAPs were absent. The author found a prolonged distal latency to both ADM and FDI muscles in 40 %, and all but one showed slowing in the forearm. For the author [1], when motor fibers to the hand muscles such as ADM or FDI muscles were blocked, a clear CMAP could be evoked in the FCU muscle; he found it of greater value than might be expected from the infrequent clinical involvement of this muscle. CMAP from FCU muscle was reduced in amplitude in 6 cases, desynchronized in 5 cases, and delayed in 21 cases.

Pathological waveform (axilla, above sulcus, below sulcus, wrist – ADM muscle):

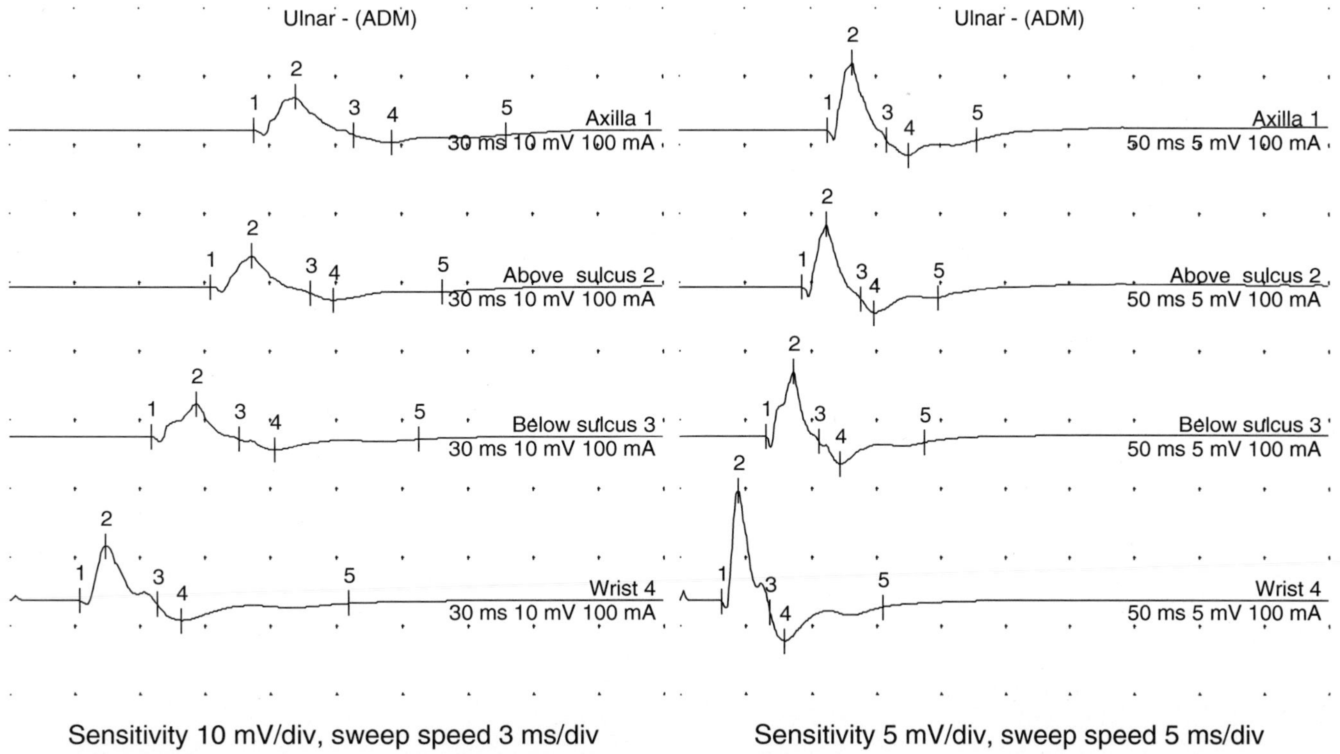

Onset latency (axilla – ADM): 11.25 ms; **Onset latency** (above sulcus – ADM): 9.25 ms; **Onset latency** (below sulcus – ADM): 6.50 ms; **Onset latency** (wrist – ADM): 3.20 ms; **Peak latency** (axilla – ADM): 13.40 ms; **Peak latency** (above sulcus – ADM): 11.30 ms; **Peak latency** (below sulcus – ADM): 8.85 ms; **Peak latency** (wrist – ADM): 4.40 ms; **Onset to peak amplitude** (axilla – ADM): 5.0 mV; **Onset to peak amplitude** (above sulcus – ADM): 4.6 mV; **Onset to peak amplitude** (below sulcus – ADM): 4.9 mV; **Onset to peak amplitude** (wrist – ADM): 8.2 mV; **Peak to peak amplitude** (axilla – ADM): 6.8 mV; **Peak to peak amplitude** (above sulcus – ADM): 6.6 mV; **Peak to peak amplitude** (below sulcus – ADM): 6.9 mV; **Peak to peak amplitude** (wrist – ADM): 11.0 mV; **MNCV** (axilla– above sulcus, 12 cm distance): 60.0 m/s; **MNCV** (above sulcus – below sulcus, 10 cm distance): 36.4 m/s; **MNCV** (below sulcus – wrist, 20 cm distance): 60.6 m/s

Fig. 5 Compound muscle action potentials (*CMAPs*) recorded at the hand from the ADM muscle using needle recording method [1] in UNE; *top to bottom*: stimulation of the axilla, above sulcus, below sulcus, and of the wrist

Pathological waveform (axilla, above sulcus, below sulcus, wrist – FDI muscle):

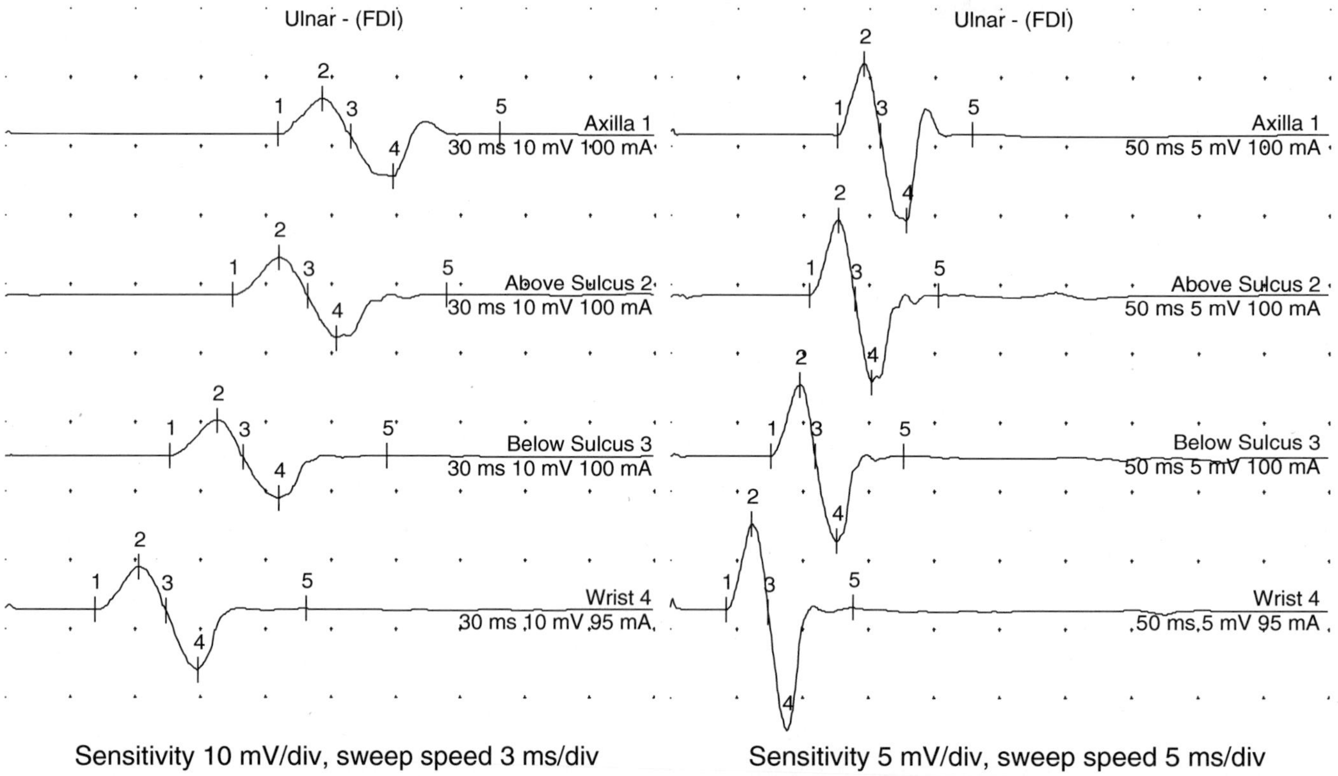

Onset latency (axilla – FDI): 12.55 ms; **Onset latency** (above sulcus – FDI): 10.45 ms; **Onset latency** (below sulcus – FDI): 7.55 ms; **Onset latency** (wrist – FDI): 4.15 ms; **Peak latency** (axilla – FDI): 14.85 ms; **Peak latency** (above sulcus – FDI): 12.50 ms; **Peak latency** (below sulcus – FDI): 10.05 ms; **Peak latency** (wrist – FDI): 6.10 ms; **Onset to peak amplitude** (axilla – FDI): 5.3 mV; **Onset to peak amplitude** (above sulcus – FDI): 5.5 mV; **Onset to peak amplitude** (below sulcus – FDI): 5.2 mV; **Onset to peak amplitude** (wrist – FDI): 6.3 mV; **Peak to peak amplitude** (axilla – FDI): 11.5 mV; **Peak to peak amplitude** (above sulcus – FDI): 11.8 mV; **Peak to peak amplitude** (below sulcus – FDI): 11.4 mV; **Peak to peak amplitude** (wrist – FDI): 15.1 mV; **MNCV** (axilla – above sulcus, 12 cm distance): 57.1 m/s; **MNCV** (above sulcus – below sulcus, 10 cm distance): 34.5 m/s; **MNCV** (below sulcus – wrist, 20 cm distance): 58.8 m/s

Fig. 6 Compound muscle action potentials (*CMAPs*) recorded at the hand from the FDI muscle using needle recording method [1] in UNE; *top to bottom*: stimulation of the axilla, above sulcus, below sulcus, and of the wrist

Pathological waveform (axilla, above sulcus, below sulcus, wrist – ADM muscle):

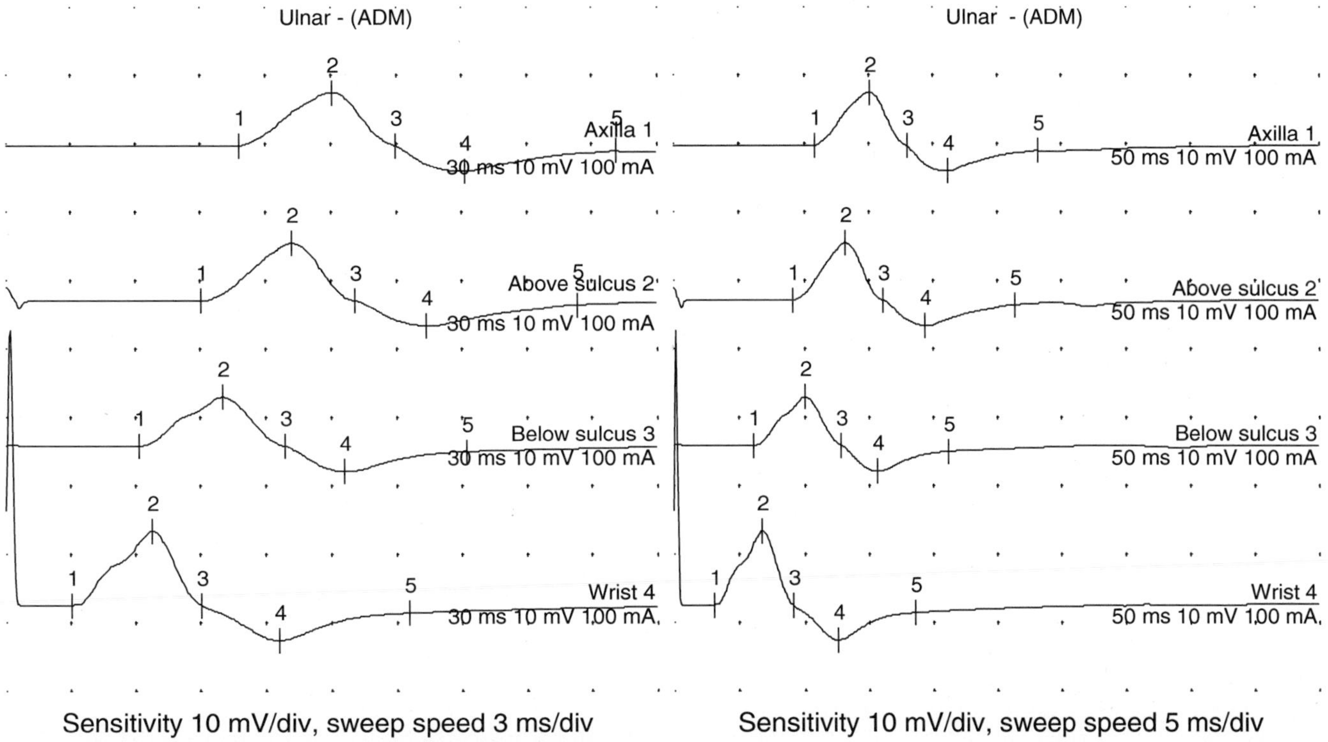

Onset latency (axilla – ADM): 10.80 ms; **Onset latency** (above sulcus – ADM): 9.05 ms; **Onset latency** (below sulcus – ADM): 6.15 ms; **Onset latency** (wrist – ADM): 3.05 ms; **Peak latency** (axilla – ADM): 15.05 ms; **Peak latency** (above sulcus – ADM): 13.40 ms; **Peak latency** (below sulcus – ADM): 10.0 ms; **Peak latency** (wrist – ADM): 6.90 ms; **Onset to peak amplitude** (axilla – ADM): 7.8 mV; **Onset to peak amplitude** (above sulcus – ADM): 8.4 mV; **Onset to peak amplitude** (below sulcus – ADM): 7.1 mV; **Onset to peak amplitude** (wrist – ADM): 11.0 mV; **Peak to peak amplitude** (axilla – ADM): 11.7 mV; **Peak to peak amplitude** (above sulcus – ADM): 12.3 mV; **Peak to peak amplitude** (below sulcus – ADM): 11.0 mV; **Peak to peak amplitude** (wrist – ADM): 16.1 mV; **MNCV** (axilla – above sulcus, 12 cm distance): 68.6 m/s; **MNCV** (above sulcus – below sulcus, 10 cm distance): 34.5 m/s; **MNCV** (below sulcus – wrist, 20 cm distance):64.5 m/s

Fig. 7 Compound muscle action potentials (*CMAPs*) recorded at the hand from the ADM muscle using surface recording method in UNE; *top to bottom*: stimulation of the axilla, above sulcus, below sulcus, and of the wrist

Pathological waveform (above sulcus – FCU muscle):

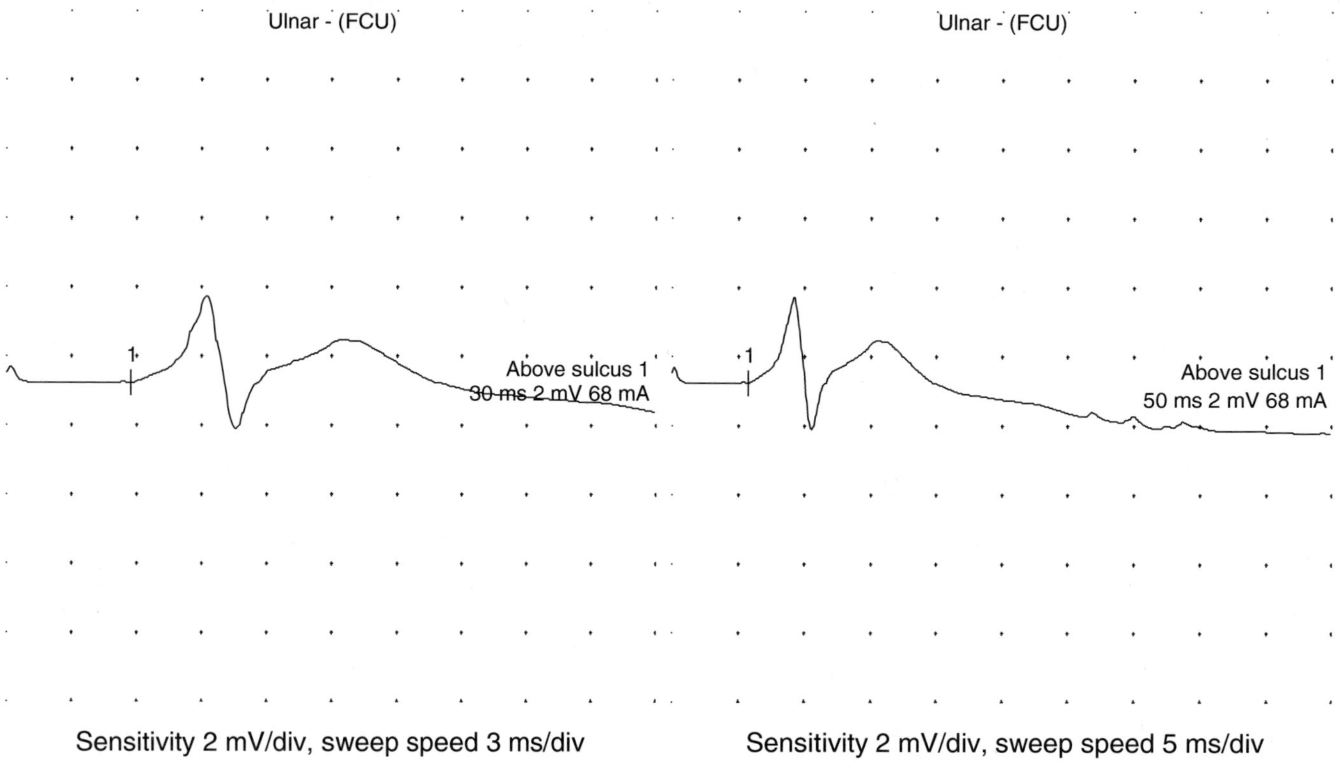

Onset latency (above sulcus – FCU): 5.70 ms; **Peak latency** (above sulcus – FCU): 9.15 ms; **Onset to peak amplitude** (above sulcus – FCU): 1.5 mV; **Peak to peak amplitude** (above sulcus – FCU): 4.0 mV

Fig. 8 Compound muscle action potential (*CMAP*) recorded at the forearm from the FCU muscle using needle recording method [1] in UNE, stimulation above sulcus

Reference

1. Payan J (1969) Electrophysiological localization of ulnar nerve lesions. J Neurol Neurosurg Psychiat 32:208–220

Wrist, Elbow, Arm – Hand

M15

Surface Recording Technique, Study from *Abductor Digiti Minimi* (ADM) Muscle

Original Settings Duration of pulse was 0.2 ms, and the machine used was a DISA type 14 A 30; sensitivity, sweep speed, low-frequency filter, high-frequency filter, and rate of pulse were not specified.

Position This study was performed with the subjects lying on a bed in the supine position; the authors performed the study on ulnar nerves in healthy subjects with the elbow extended (180°) and flexed (70°), in order to compare both methods (Figs. 1 and 2, respectively).

Recording The authors [1] performed ulnar nerve conduction study recording from the motor point of the *abductor digiti minimi* (ADM) muscle, using 0.5 cm diameter silver disk surface electrodes. The active electrode was placed on the *hypothenar* eminence, on a point midway between the distal wrist crease and the crease at the base of the digit V (little finger), and at the junction of dorsal and palmar skin; the reference electrode was placed medially at the base of the digit V (Fig. 1). A 2 cm diameter lead ground (G) electrode was secured with adhesive tape over the thenar eminence.

Stimulation The ulnar nerve was stimulated at three points along its course [1]: at the wrist (S1), 8 cm proximal to the active recording electrode placed on the belly of the ADM muscle; at the elbow (S2), just distal to the ulnar groove; and at the arm (S3), at the junction of the middle and distal third of the arm (Figs. 1 and 2).

Measurements The latency (ms) of the compound muscle action potential (CMAP) was measured from the onset of the stimulus artifact to the onset of the first deflection from the isoelectric line; it was determined to the nearest 0.1 ms from photographic records made of the oscilloscopic trace which was interrupted at 1 ms intervals by the blanking voltage at the *Z* axis from a tuning fork oscillator. Each ms equaled ½ cm. The average distance between the wrist and elbow stimulation points was 11.5 cm with the elbow flexed at 70° and 8.1 cm with the elbow extended to 180°. Surface skin temperature was not reported. By dividing the distance by time, the motor nerve conduction velocity (MNCV) along the nerve segment can be determined and measured in meter per second (m/s), and it was calculated for wrist–elbow (forearm) and elbow–arm (across-elbow) segments. The amplitude (mV) of the CMAP was measured from the onset of the negative deflection to the peak of the negative deflection (negative-peak amplitude). The duration (ms) of the CMAP was measured from the onset of the negative deflection to the point of its return to the isoelectric line. Skin and room temperatures were not reported. Normal conduction values from the ADM muscle (Tables 1 and 2) were obtained from 31 ulnar nerves of 18 healthy subjects (age range 20–58 years, mean age 28 years). In all subjects, the authors performed the motor conduction study with both the elbow extended (180°) (Table 1) and flexed (70°) (Table 2).

Hennessey et al. [2] performed an 8 cm distal stimulation of the ulnar nerve at the wrist in 44 paid healthy subjects (age range 19–43 years) using a TECA Neurostar EMG instrument. Low-frequency filter was set at 20 Hz and high-frequency filter was set at 2 kHz; sweep speed was set at 2 ms/division. They developed normative data (Table 3) based on several factors, such as temperature control, standardized technique, age, height, and finger circumference. An EMG Associates Dermatemp DT-1000 infrared temperature scanner with digital readout was used to record surface skin temperatures. The authors recorded skin temperatures before each study at the touch pad of the digit I, midpalm, and antecubital fossa; subjects with a skin temperature <31 °C at any recording site had their extremities warmed until all three sites were ≥31 °C. No recording site was warmer than 35 °C, and the heating time was approximately 10 min. All nerve conduction studies were performed bilaterally in each subject, with the subject sitting comfortably in the upright position. The active recording electrode was placed over the motor point of the ADM muscle, which was located one half the distance between the metacarpophalangeal joint and the distal wrist crease. The reference electrode was placed 4 cm distally over the digit V (little finger). The

G. Gentili, M. Di Napoli, *The Ulnar Nerve*, DOI 10.1007/978-3-319-43292-2_39

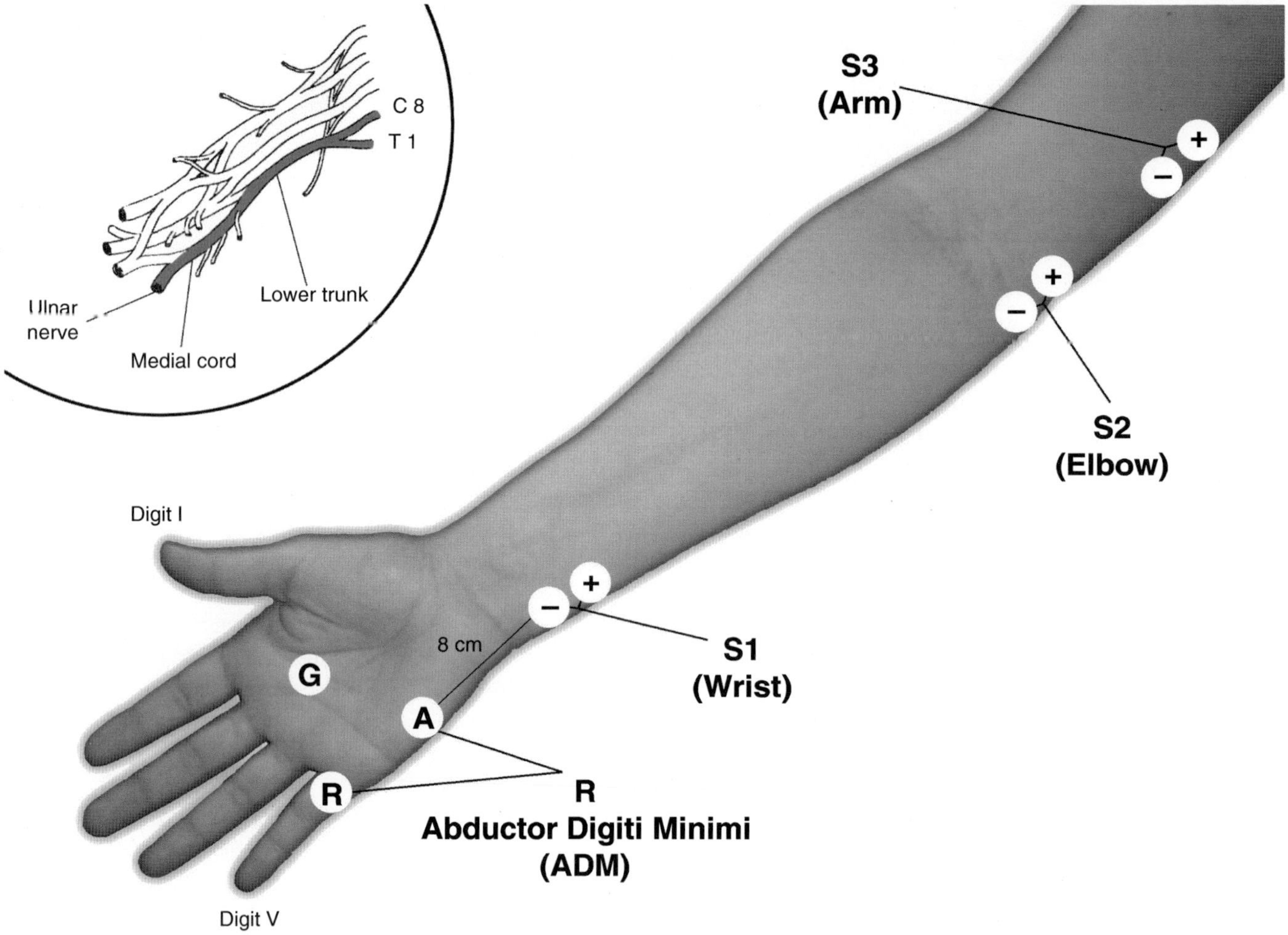

Typical waveform (wrist, elbow, arm – ADM muscle):

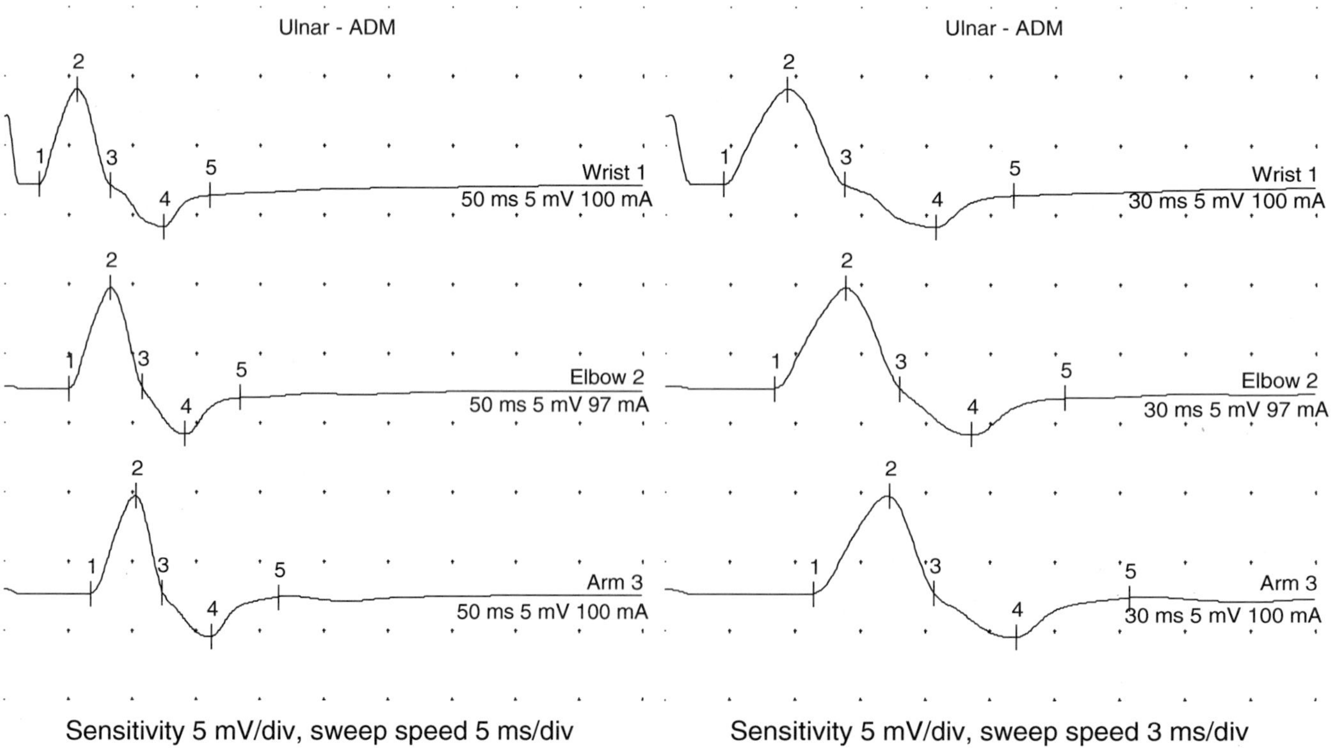

Fig. 1 Compound muscle action potentials (*CMAPs*) recorded at the hand from the ADM muscle, stimulation of the wrist (*upper trace*), the elbow (*middle trace*), and the arm (*lower trace*)

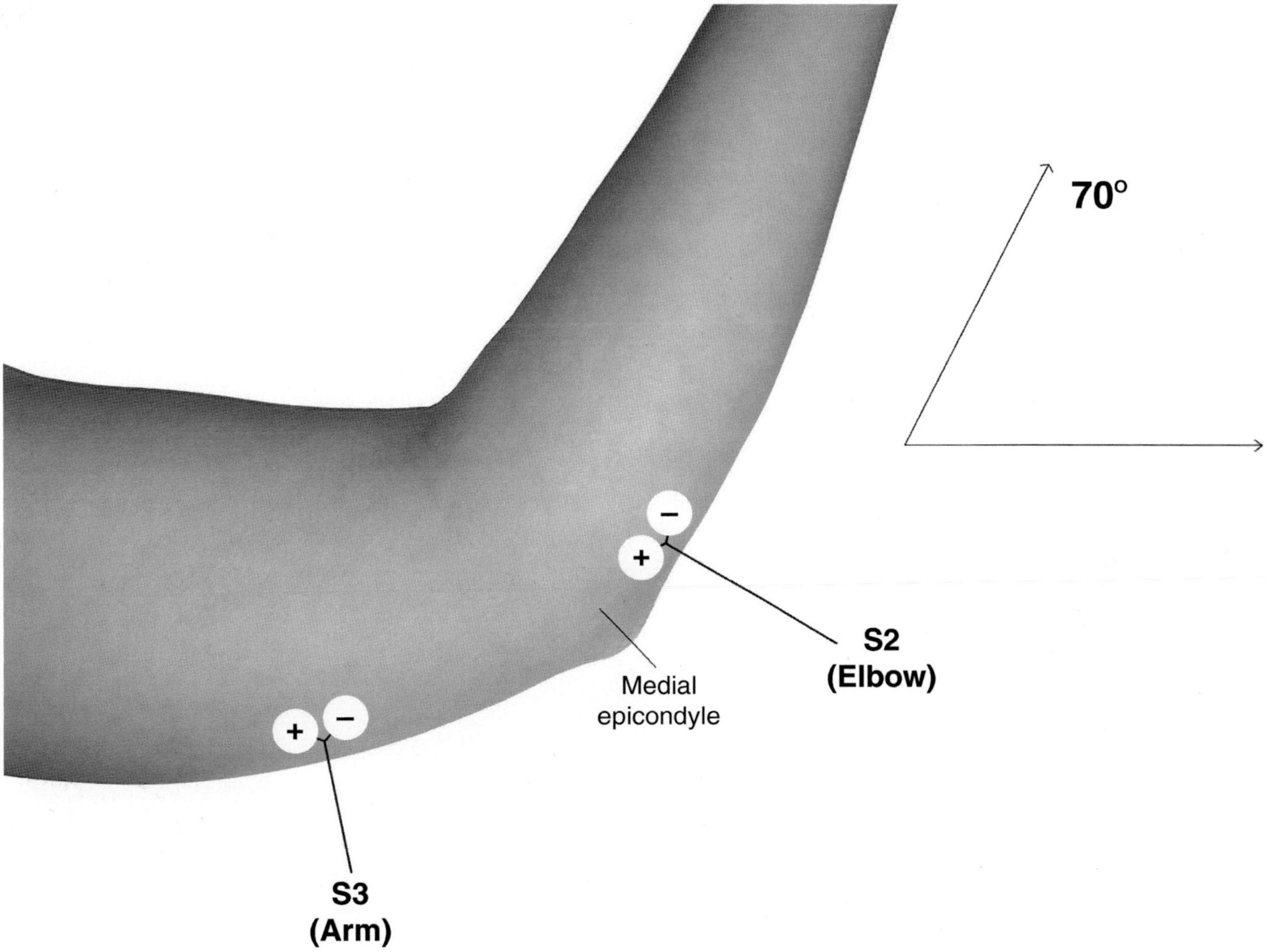

Fig. 2 Arm and elbow stimulation points (elbow flexed to 70°)

Table 1 Normal values [1] – elbow extended (180°)

	Mean ± SD	Range
Wrist–elbow, forearm MNCV (m/s)	62.5 ± 4.50	52–72
Elbow–arm, across-elbow MNCV (m/s)	49.9 ± 7.90	34–66
Wrist–ADM, negative-peak amplitude (mV)	6.08 ± 1.64	2.8–9.36
Elbow–ADM, negative-peak amplitude (mV)	5.25 ± 1.92	1.41–9.09
Arm–ADM, negative-peak amplitude (mV)	5.43 ± 1.50	2.43–8.43
Wrist–ADM, duration (ms)	4.20 ± 0.48	3.24–5.16
Elbow–ADM, duration (ms)	4.20 ± 0.50	3.20–5.20
Arm–ADM, duration (ms)	4.25 ± 0.49	3.27–5.23

Table 2 Normal values [1] – elbow flexed (70°)

	Mean ± SD	Range
Wrist–elbow, forearm MNCV (m/s)	61.8 ± 5.0	53–73
Elbow–arm, across-elbow MNCV (m/s)	62.7 ± 5.50	52–74
Wrist–ADM, negative-peak amplitude (mV)	6.14 ± 1.90	2.34–9.94
Elbow–ADM, negative-peak amplitude (mV)	5.60 ± 1.98	1.64–9.56
Arm–ADM, negative-peak amplitude (mV)	5.77 ± 1.79	2.19–9.35
Wrist–ADM, duration (ms)	4.15 ± 0.48	3.19–5.11
Elbow–ADM, duration (ms)	4.15 ± 0.50	3.15–5.15
Arm–ADM, duration (ms)	4.15 ± 0.48	3.19–5.11

distal ulnar motor nerve stimulation was performed 8 cm proximal to the active electrode, and it was delivered medial and posterior to the flexor carpi ulnaris (FCU) muscle tendon (Fig. 3). This distance was measured along the ulnar surface of the forearm. Proximal ulnar nerve stimulation was performed distal to the medial epicondyle with the elbow in 70° flexion, as suggested by Checkles et al. [1]. The authors [2] placed the ground electrode on the dorsum of the hand; the figure shows the ground electrode placed on the palm (Fig. 3). Motor distal latency (MDL), motor nerve conduction velocity (MNCV), and distal amplitude (DAMP) were reported.

Buschbacher [3] derived a normative database for the routine ulnar motor conduction from a large and varied subject population. He studied 248 healthy subjects (Table 4), 55 % were women, and 45 % were men, 75 % were white, 22 % were black, and 3 % were of other races (mean age was 44.7 ± 14.7 years for the men and 40.8 ± 14.8 years for the women). The mean height was 176 ± 7 cm for the men and 163 ± 7 cm for the women. The mean weight was 182 ± 30 lb for the men and 154 ± 33 lb for the women. Skin surface temperatures were measured over the dorsum of the hand and maintained above 32 °C. The limb was warmed if necessary with a hairdryer (this was required in only six subjects). All recordings were performed using a sensitivity of 5 mV/division (filter settings 2–3 Hz–10 kHz) and a sweep speed of 2 ms/division. The routine ulnar motor nerve conduction study was performed applying the active electrode on the ulnar surface of the *hypothenar* eminence – *abductor digiti minimi* (ADM) muscle, halfway between the pisiform bone and the fifth metacarpophalangeal joint. The reference electrode was placed slightly distal to this joint. The ground electrode was placed on the dorsum of the hand. Distal stimulation was at the wrist, 8 cm proximal to the active electrode, on a line measured to slightly radial to the tendon of the *flexor carpi ulnaris* (FCU) muscle (Fig. 3). Elbow stimulations were performed with the cathode approximately 4 cm distal to the medial epicondyle (below elbow) and approximately 10 cm proximal to the below-elbow stimulation site (above elbow), measured in a curve behind the medial epicondyle to a point slightly volar to the triceps muscle. The author also investigated the most proximal portion of the ulnar nerve performing axillary stimulation at least 10 cm proximal to the above-elbow stimulation site. Buschbacher [3] performed the motor nerve conduction study using distal stimulation (8 cm) as suggested by Checkles et al. [1], but the arm was positioned in a 45° abducted position and was externally rotated, with the elbow flexed to 90° and the forearm in the neutral position (thumb pointing to ear). The anode was proximal in all stimulations (Figs. 3 and 4). The author measured the distal motor latency, negative-peak amplitude, duration of the negative peak, and area under the negative peak.

Comment

Checkles et al. [1] observed no effect on the duration and negative-peak amplitude of the CMAP varying elbow position; the only difference in results between the two elbow positions was conduction velocity. The velocity of the across-elbow segment was faster and similar to that of the other segments when the elbow was flexed, presumably because the skin measurement was different from the actual length of the nerve segment when the elbow was extended [1]. For the authors, in order to accommodate for the change in length across the elbow in flexion, there must be some slack for the nerve in full extension. This would mean that measurement of the distances on the skin across the elbow with the elbow extended (180°) would be less than the true length of the ulnar nerve segment across the elbow. For Checkles et al. [1], the study clearly showed that false slowing across the ulnar groove may be reported if the ulnar nerve conduction study is done with the elbow extended (180°), and precision would be enhanced if the ulnar nerve conduction study is carried out with the elbow flexed (70°).

Table 3 Reference values [2]

Normal values	Mean ± SD
Wrist–ADM, MDL (ms)	2.6 ± 0.3
Wrist–ADM, MNCV (m/s)	63.0 ± 4.8
Wrist–ADM, DAMP (mV)	12.6 ± 2.3

Lee et al. [4] recorded CMAPs at the *abductor digiti minimi* (ADM) muscle using disposable electrodes (VIASYS Healthcare, Madison, WI) with stimulation at the wrist (8 cm from the active recording electrode), 3 cm below the elbow and 10 cm above the elbow. They studied 100 healthy subjects (58 women and 42 men, mean age 34.7 ± 8.4 years, age range 23–64 years) using a TECA Synergy electromyograph for nerve conduction studies. Only CMAP latency with above-elbow stimulation was reported (mean 8.0 ± 0.7 ms, range 6.6–9.7 cm), and it was measured with a sensitivity of 5 mV/division. The mean MNCV of the forearm (wrist–below-elbow segment) was 65.2 ± 6.6 m/s (range 52–90 m/s), and the mean MNCV across the elbow was 62.8 ± 9.4 m/s (34.5–84 m/s). The temperature was maintained at 32 °C or higher on the palm and at the elbow (retrocondylar groove) using hydrocolloid packs and a heating lamp in a temperature-controlled room. In all subjects, in addition to height (mean 167.5 ± 8.9 cm, range 147–189 cm), finger-to-arm length (mean 47.3 cm, range 42.5–54.0 cm) was measured from the proximal interphalangeal joint at the digit V to the above-elbow stimulation site with a 90° elbow flexion in each

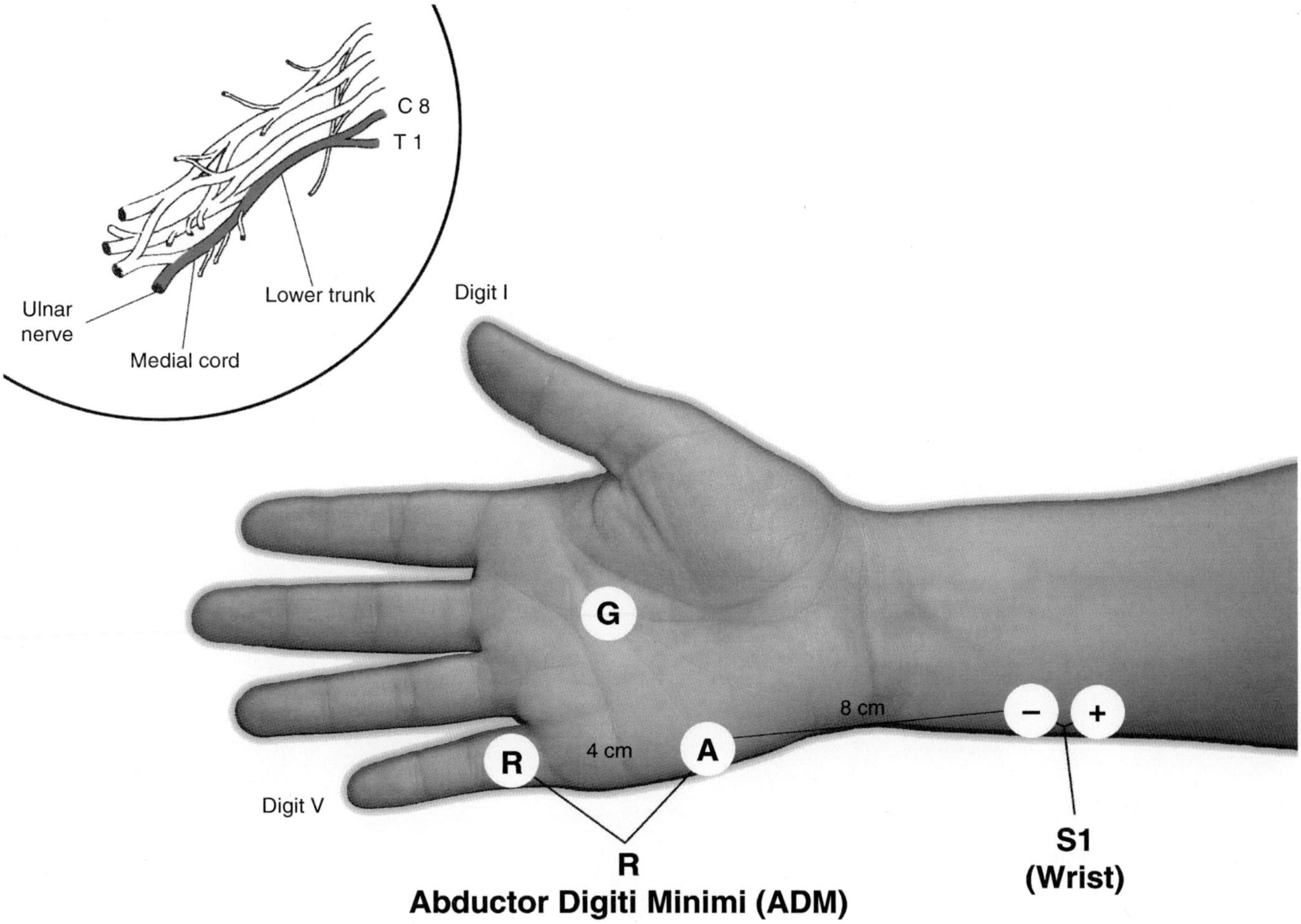

Fig. 3 Distal stimulation point (S1 – wrist), 8 cm distance from the active recording electrode [2]

> **Comment**
> For Hennessey et al. [2], the electrodiagnostician is better equipped to make appropriate decisions when using normative nerve conduction data that has been gathered with standardized techniques, controlled temperatures, and an adequate number of carefully screened healthy subjects. For the authors, the control of temperature and the use of standard distance techniques for distal latency determinations are essentials to develop normative data, and only normal values provided with a reliable method should be used to assess pathology.

Table 4 Normal values [3]

	Mean ± SD	Mean ± 2SD
Wrist–ADM, distal latency (ms)	3.0 ± 0.3	3.6
Wrist–ADM, duration (ms)	6.0 ± 0.9	7.8
Wrist–ADM, amplitude (mV)	11.6 ± 2.1	7.4
Wrist–ADM, area (μVs)	35.9 ± 7.1	21.7

subject (Fig. 5). Elbow segment was measured from 3 cm distal to the midpoint of medial epicondyle and olecranon to 10 cm proximal along the intermuscular septum toward the midaxilla.

Their study reported the reference values of sensory nerve latencies at 10 cm above-elbow stimulation and a formula predicting the latencies based on finger-to-arm length, as follows:

$$\text{CMAP onset latency}(\text{ms}) = -0.13 + 1.7 \times (\text{finger-to-arm length, cm})$$

They proposed their test as a complementary parameter to self-assess the validity of segmental conduction velocity in subjects with suspected UNE, since the AAEM practice guidelines for the diagnosis of UNE are (1) MNCV across the elbow <50 m/s, (2) differential slowing of more than 10 m/s across the elbow than the forearm segment, and (3) change of configuration and reduction in CMAP negative-peak amplitude (mV) of more than 20 % across the elbow [5]. Among these criteria, the MNCV slowing across the

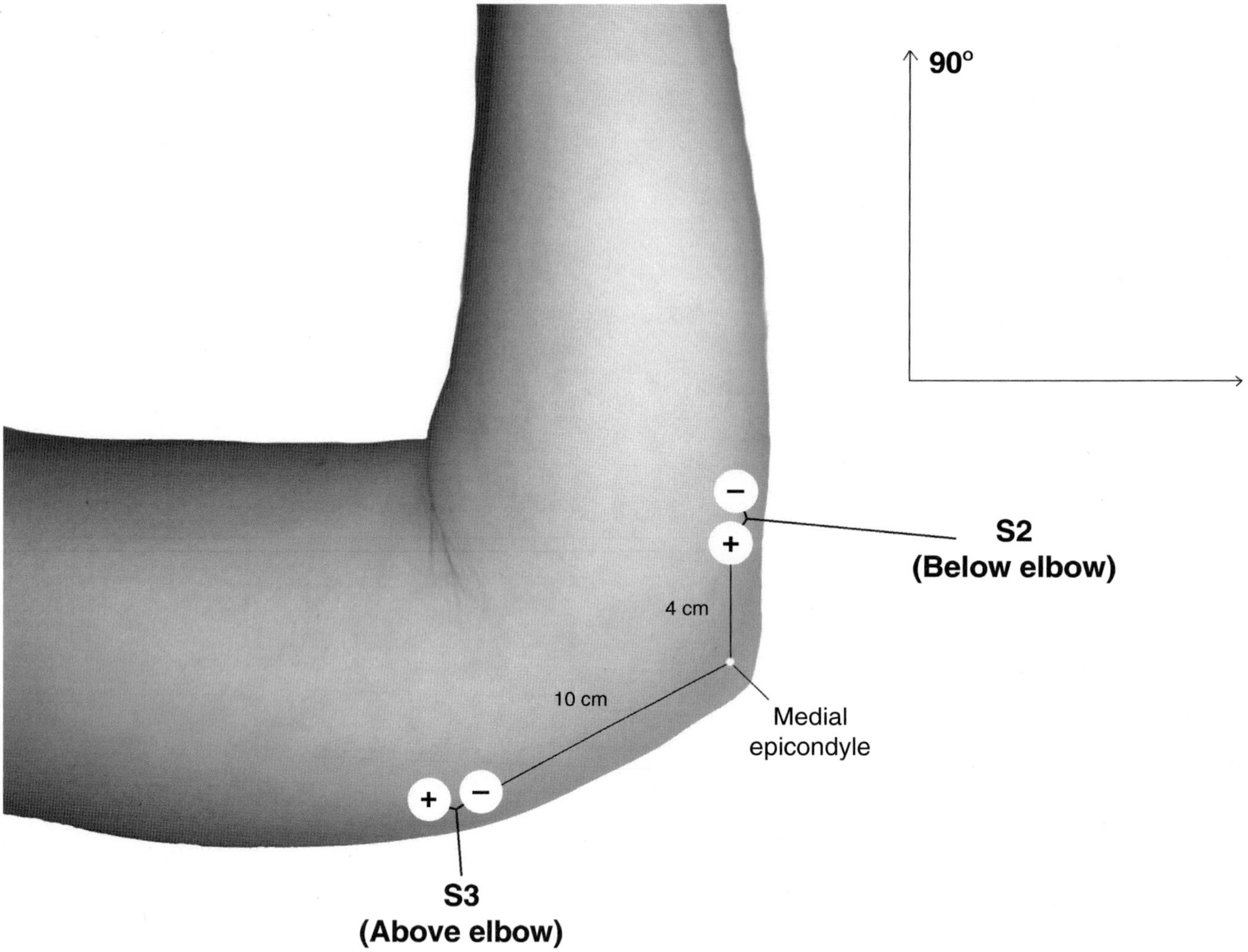

Fig. 4 Proximal stimulation points below (S2) and above (S3) the medial epicondyle [3]

Comment

The results obtained in Buschbacher's study [3] agree in general with those derived previously [1, 2]. Slight differences are most likely the result of different subject selection methods as well as methodological differences. In the clinical setting, he suggests to use the exact technique described in obtaining the reference values because a clinician must try to follow the exact protocol used in establishing normal ranges to best use them clinically. The authors [1] chose a method based on its ease of application and general acceptance among electromyographers. Side-to-side distal latency differences are useful in diagnosing some peripheral ulnar neuropathies (0.6 ms ± 2SD), and side-to-side distal CMAP amplitude differences are useful in diagnosing some peripheral ulnar neuropathies and radiculopathies (3.6 mV ± 2SD). A mean amplitude of 11.6 mV was reported in the study, and the upper limit of the normal drop in amplitude from below- to above-elbow stimulation was 2.2 mV ± 2SD; the upper limit of normal percentage drop (at the 97th percentile) was 16 %. The area is not routinely measured in clinical practice, although with modern electrodiagnostic machines, it is readily calculated. The author [3] reported a mean 35.9 μVs ± 2SD, and the acceptable drop in the area from below- to above-elbow stimulation was 6.7 μVs ± 2SD; the upper limit of normal percentage drop across the elbow (at the 97th percentile) was 16 %. The duration was 6 ms ± 2SD, and the upper limit of normal increase from below-elbow to above-elbow stimulation was 1.1 ms ± 2SD. Buschbacher [3] measured the ulnar nerve conduction velocity (NCV) with the elbow flexed (90°), because full extension allows the nerve to be slack and gives a false distance measurement, which results in the NCV measurements being slower than in the flexed position. The 90° elbow flexed position is also easy to use clinically. The upper limit of normal drop in NCV in the forearm segment to across the elbow segment was 18 m/s ± 2SD (side to side was 43 m/s ± 2SD), and the upper limit of normal percentage change was 23 %.

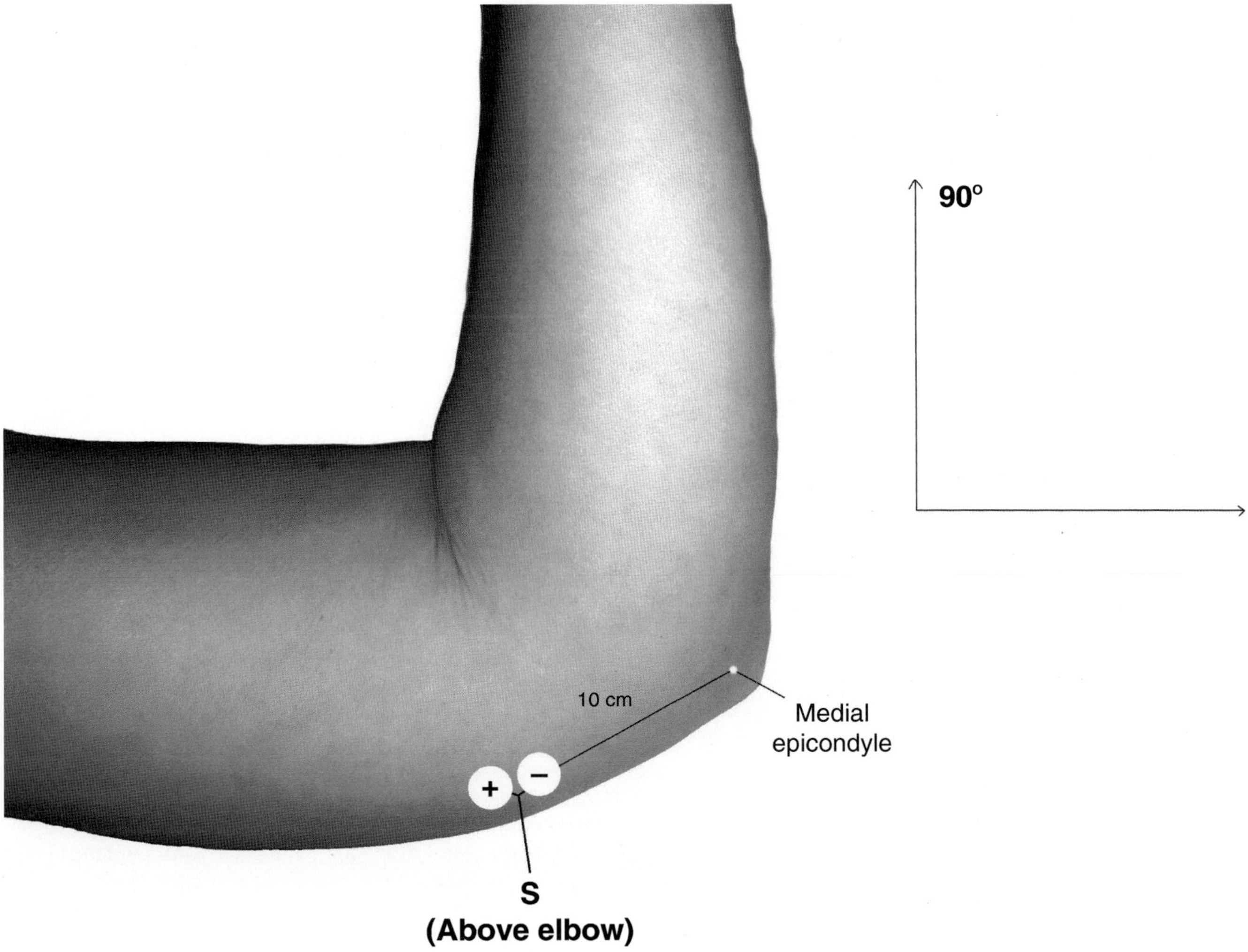

Fig. 5 Above-elbow stimulation point (elbow flexed to 90°)

elbow showed the highest sensitivity than the other methods, such as the differential slowing between across-elbow (below-elbow–above-elbow) and forearm (wrist–below-elbow segment) segments [6], but some authors [7] found distance measurement errors more common in the across-elbow than in the forearm measurement because of the shorter distance and position of the elbow. The measurement errors alone can explain conduction velocity differences of up to 14 m/s between the forearm and the across-elbow segments. Lee et al. [4] provided the formula for predicted norms of CMAP latency at above-elbow stimulation based on the finger-to-arm length. In terms of cutoff latency of above-elbow stimulation, the upper limit of 95 % confidence interval in motor latencies, approximately 1 ms longer than the predicted value, may be applied using a prediction formula based on the finger-to-arm length. Alternatively, applying the absolute cutoff value, 2SDs above the mean (9.4 ms) can be applied. If the latency at 10 cm above-elbow stimulation is longer than 2SDs above the mean or 1 ms longer than the predicted value, then the slow MNCV obtained across the elbow is likely a valid value. However, the cutoff value of 2SDs above the mean was an empirical value, and they did not report the specificity or predictive value of this cutoff latency because of the small sample of subjects with suspected UNE (14 patients); for widespread use of the cutoff latencies, the authors [4] suggested to validate the cutoff values in a larger healthy population (the study did not include subjects older than 65 years) and a larger sample of patients with UNE. Pathological waveform and values in a case of a diabetic patient with ulnar neuropathy at the elbow (UNE) are reported (Fig. 6).

Pathological waveform (wrist, elbow, arm – ADM muscle):

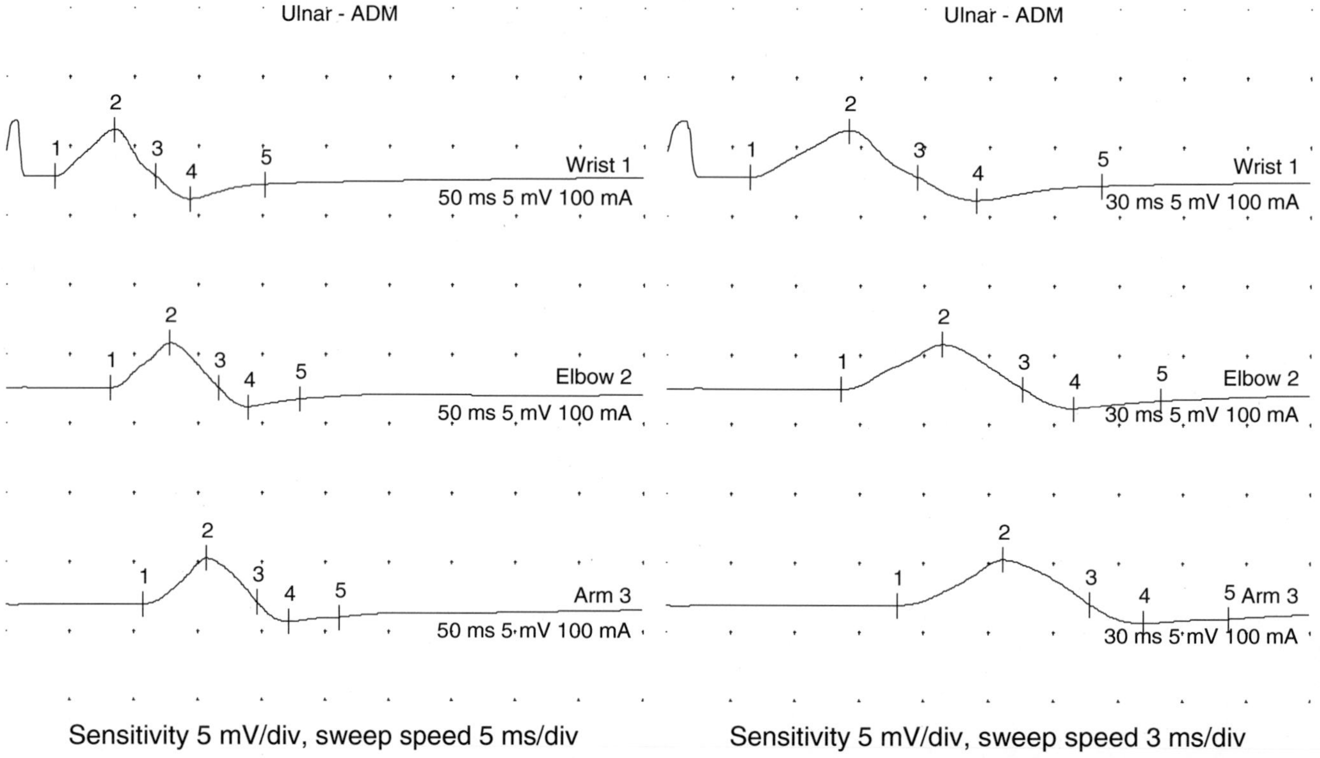

Onset latency (wrist – ADM): 3.80 ms; **Onset latency** (elbow – ADM): 8.10 ms; **Onset latency** (arm – ADM): 10.75 ms; **Peak latency** (wrist – ADM): 8.70 ms; **Peak latency** (elbow – ADM): 13.0 ms; **Peak latency** (arm – ADM): 15.25 ms; **Onset to peak amplitude** (wrist – ADM): 3.4 mV; **Onset to peak amplitude** (elbow – ADM): 3.2 mV; **Onset to peak amplitude** (arm – ADM): 3.3 mV; **Peak to peak amplitude** (wrist – ADM): 5.1 mV; **Peak to peak amplitude** (elbow – ADM): 4.7 mV; **Peak to peak amplitude** (arm – ADM): 4.6 mV; **MNCV** (wrist – elbow, 20 cm): 46.5 m/s; **MNCV** (elbow – arm, 9 cm): 34.0 m/s

Fig. 6 Compound muscle action potentials (*CMAPs*) recorded at the hand from the ADM muscle in diabetic patient with UNE, stimulation of the wrist (*upper trace*), the elbow (*middle trace*), and the arm (*lower trace*)

References

1. Checkles NS, Russakov AD, Piero DL (1971) Ulnar nerve conduction velocity – effect of elbow position on measurement. Arch Phys Med Rehabil 52:362–365
2. Hennessey WJ, Falco FJ, Braddom RL (1994) Median and ulnar nerve conduction studies: normative data for young adults. Arch Phys Med Rehabil 75:259–264
3. Buschbacher RM (1999) Ulnar nerve motor conduction to the abductor digiti minimi. Am J Phys Med Rehabil 78:S9–S14
4. Lee SW, Kim DDJ, Pedersen AL et al (2011) Ulnar nerve latency at above elbow stimulation. A complementary tool for segmental conduction velocity across the elbow. Am J Phys Med Rehabil 90:302–307
5. AAEM Quality Assurance Committee (1999) Practice parameter for electrodiagnostic studies in ulnar neuropathy at the elbow: summary statement. Muscle Nerve 22:408–411
6. Shakir A, Micklesen P, Robinson L (2004) Which motor nerve conduction study is best in ulnar neuropathy at the elbow? Muscle Nerve 29:585–590
7. Landau ME, Diaz MI, Barner KC et al (2002) Changes in nerve conduction velocity across the elbow due to experimental error. Muscle Nerve 26:838–840

Wrist – Hand, Elbow – Hand, Axilla – Hand

M16

Surface Recording Technique, Study from *Abductor Digiti Minimi* (ADM) Muscle

Original Settings Sensitivity was 2–5 mV/division for M-response and 500 μV/division for F-wave, low-frequency filter was 2–3 Hz, high-frequency filter was 10 kHz, sweep speed was 5 ms/division, and duration of pulse was 0.2–0.5 ms; the machine used was not specified.

Position This study was performed with the subject lying on a bed with the arm and hand immobilized in a supine position.

Recording The active electrode (A) was placed over the belly (motor point) of the *abductor digiti minimi* (ADM) muscle (Figs. 1 and 2), by surface electrodes [1]. The active (A) electrode was placed on the motor point of the ADM muscle; the reference (R) was placed distally near its tendon of insertion. Distances between the recording and stimulating electrodes were not fixed. The ground (G) electrode position was not mentioned in the text; the figure shows the ground electrode placed on the forearm.

Stimulation Using surface electrodes, the ulnar nerve was stimulated by the supramaximal electric shocks at the wrist, elbow, and axilla; same stimulation points were used for conventional motor conduction studies and F-wave studies [1]. For the wrist stimulation (Fig. 1), the cathode (−) and anode (+) stimulating electrodes were positioned just lateral to the tendon of the *flexor carpi ulnaris* (FCU) muscle.

For elbow stimulation (Fig. 2), the cathode (−) and anode (+) stimulating electrodes were positioned in the most proximal portion of the ulnar groove. The author performed also a more proximal stimulation, at the axilla level (Fig. 3). In all stimulations, the anode was proximal and the cathode was distal for conventional motor conduction studies; the cathode was proximal and the anode was distal for F-wave studies. No fixed distances were reported in the article.

Measurements Distal latency (ms) of the compound muscle action potential (CMAP) was measured from the stimulus artifact to the onset of the negative deflection of CMAP. In all subjects, the distance between the stimulating points was measured along the course of the nerve, and motor nerve conduction velocity (MNCV) was calculated. The approximate length of the nerve from the stimulus points to the spinal cord was obtained by measuring the distance from the stimulus points to the C7 spinal process. Surface measurements were taken along the course of the nerve from the wrist to the axilla with the subject in an upright position and the arm abducted to 90°.

The hand was supinated for the measurement of the median nerve so that stimulus points were easily accessible to the examiner standing behind the subject. Stimulus points at the axilla were projected posteriorly across the arm in the plane perpendicular to the axis of the arm, and distances between the projected points and the C7 spinous process were determined by surface measurements. Skin temperature was not controlled; all tests were performed in a warm room. The F-wave velocity of the central pathway was thus calculated as follows:

$$\text{F}-\text{wave velocity}\left(\text{m/s}\right)=\frac{\left(\begin{array}{l}\text{Distance from}\\ \text{stimulus}\\ \text{point}-\text{C7 spine}\end{array}\right)\times 2\ (\text{mm})}{\left(\text{F latency}-\text{M latency}\right)-1\ (\text{ms})}$$

The F-wave latency (ms) was measured from the stimulus artifact to the beginning of the evoked potential; it was constant in some but in others varied by a few milliseconds from one stimulus to the next. When the F-wave varied in latency, several trials were given at each stimulus point to determine the response with the shortest latency. Assuming that impulses carrying the F-wave traveled at the same velocity as motor impulses over the same nerve segment, the latency difference between the F-wave and the M-response must correspond to the time interval required for the passage of the impulse to and from the spinal cord plus a delay estimated to be 1.0 ms for the recurrent discharge to occur at the motoneurons [2]. Motor and late responses were obtained from 33 healthy (Table 1) volunteers (age range 22–58 years, mean age 35 years) and 8 patients (Table 2) with Charcot–Marie–Tooth (CMT) disease (age range 13–67 years).

G. Gentili, M. Di Napoli, *The Ulnar Nerve*, DOI 10.1007/978-3-319-43292-2_40

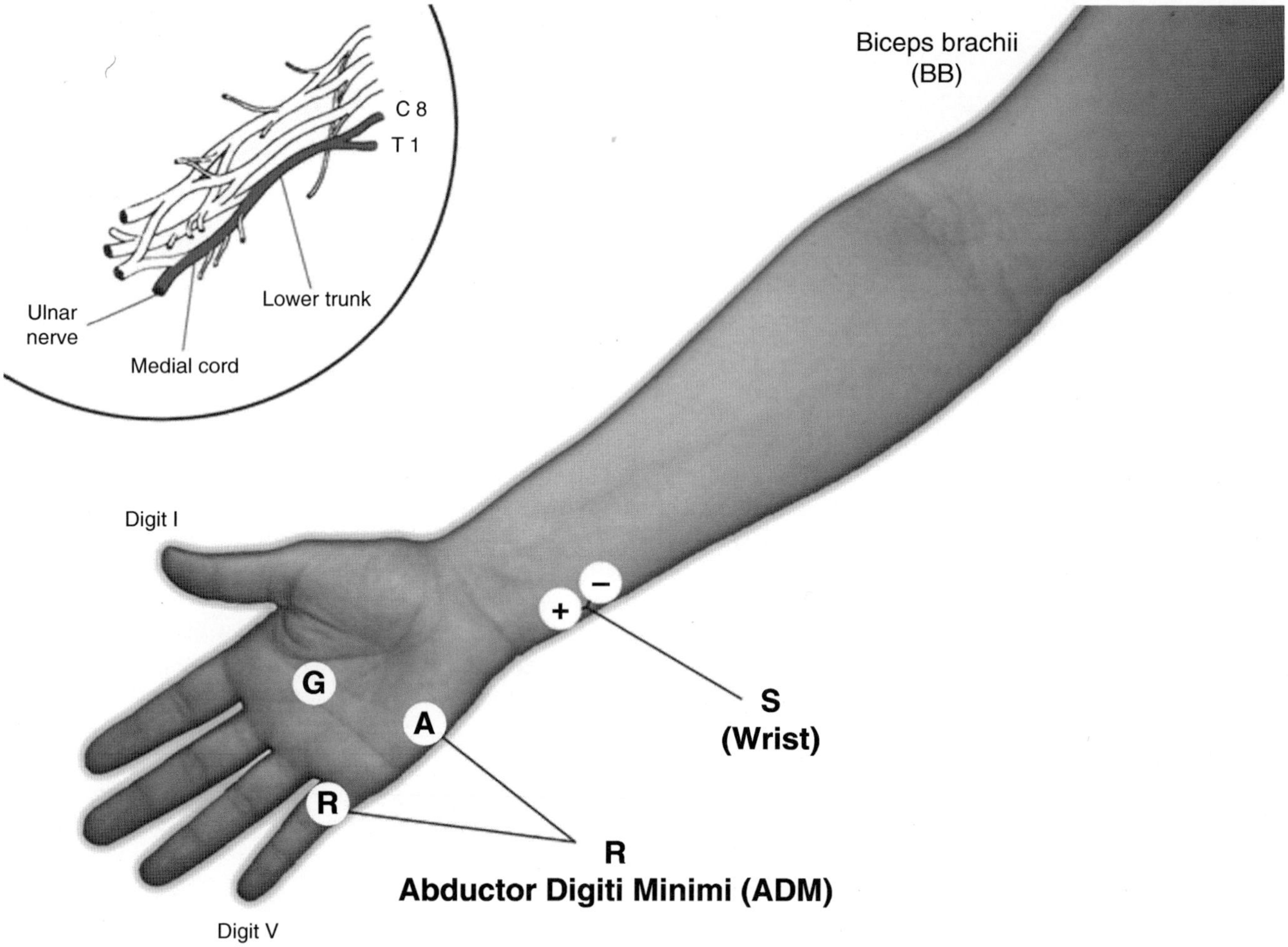

Typical waveform (wrist – ADM muscle):

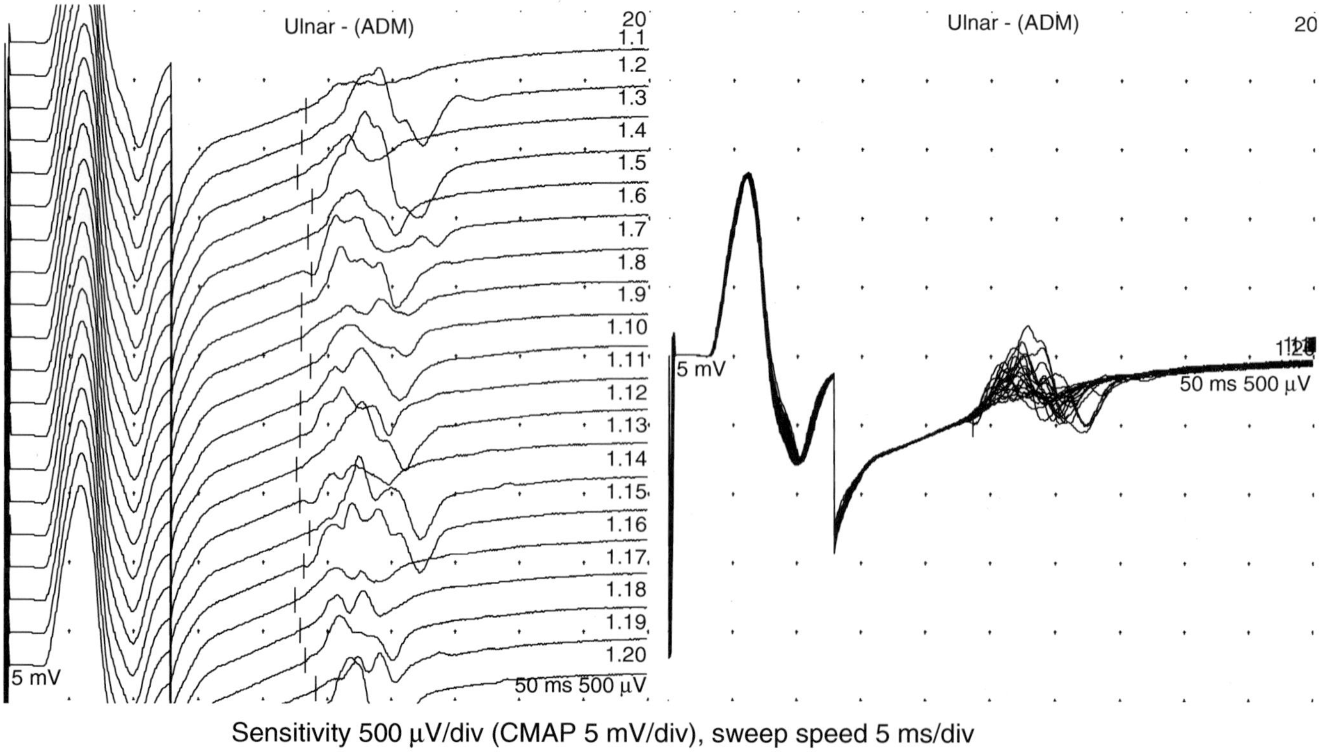

Fig. 1 F-waves recorded at the hand from the ADM muscle, stimulation of the wrist; raster mode, 20 stimuli (*left*); superimposed mode, 20 stimuli (*right*)

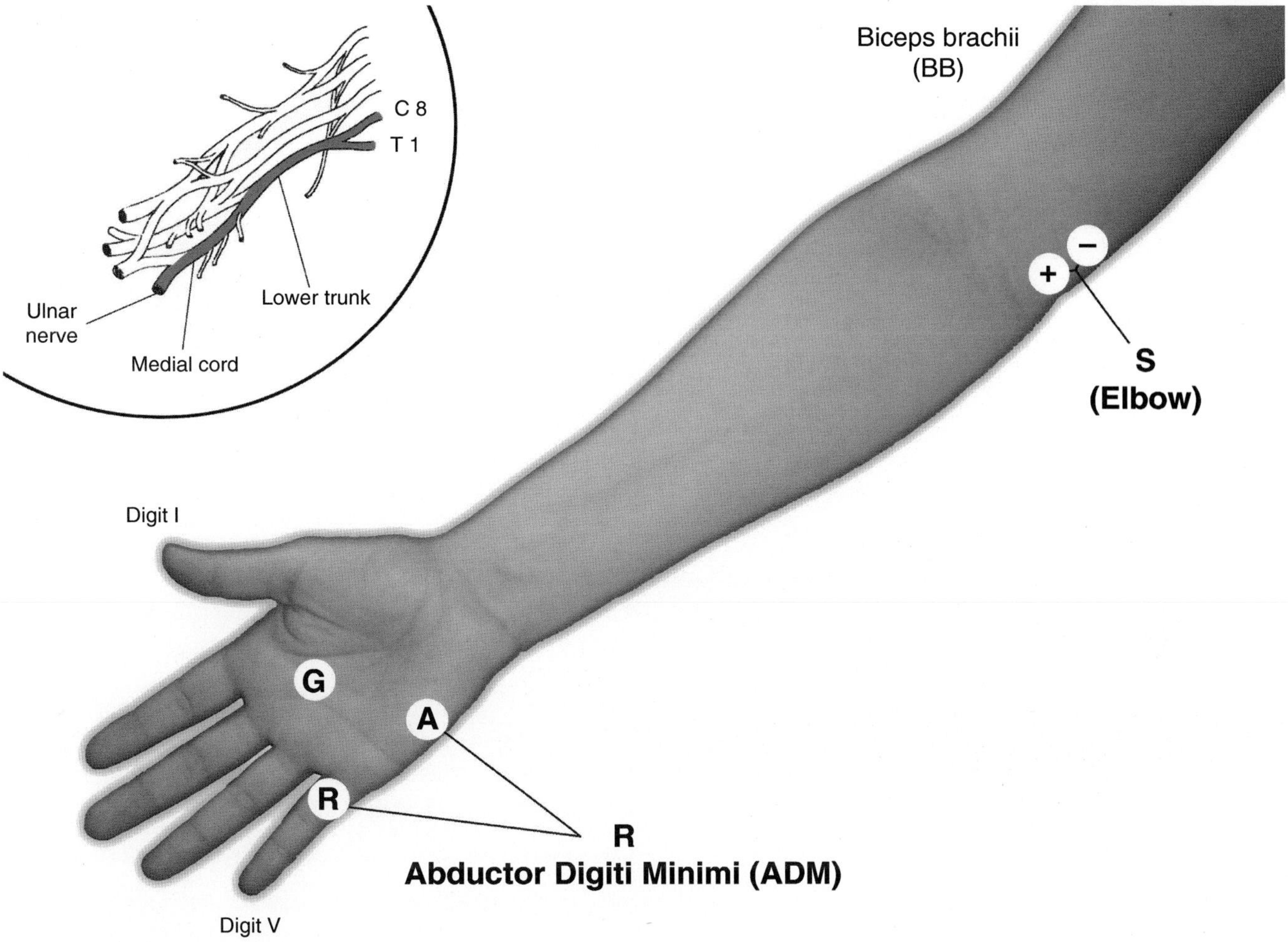

Typical waveform (elbow – ADM muscle):

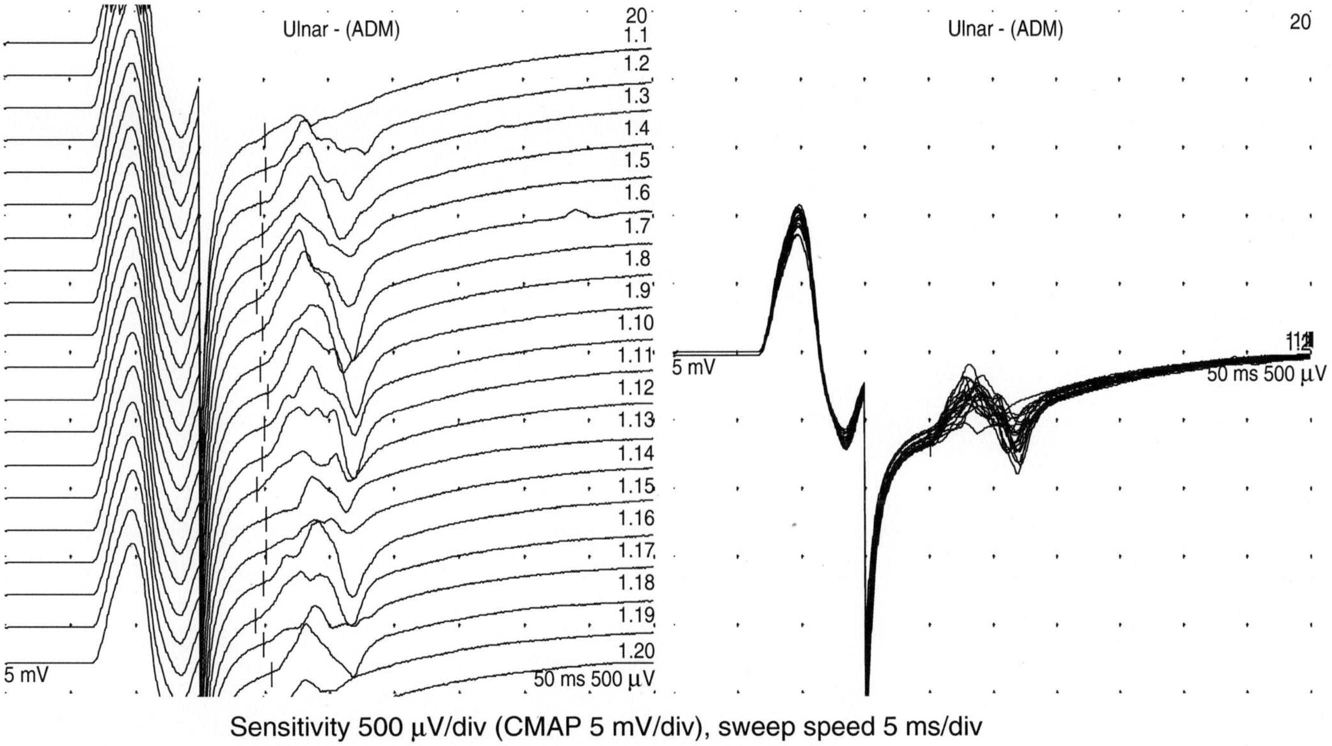

Fig. 2 F-waves recorded at the hand from the ADM muscle, stimulation of the elbow; raster mode, 20 stimuli (*left*); superimposed mode, 20 stimuli (*right*)

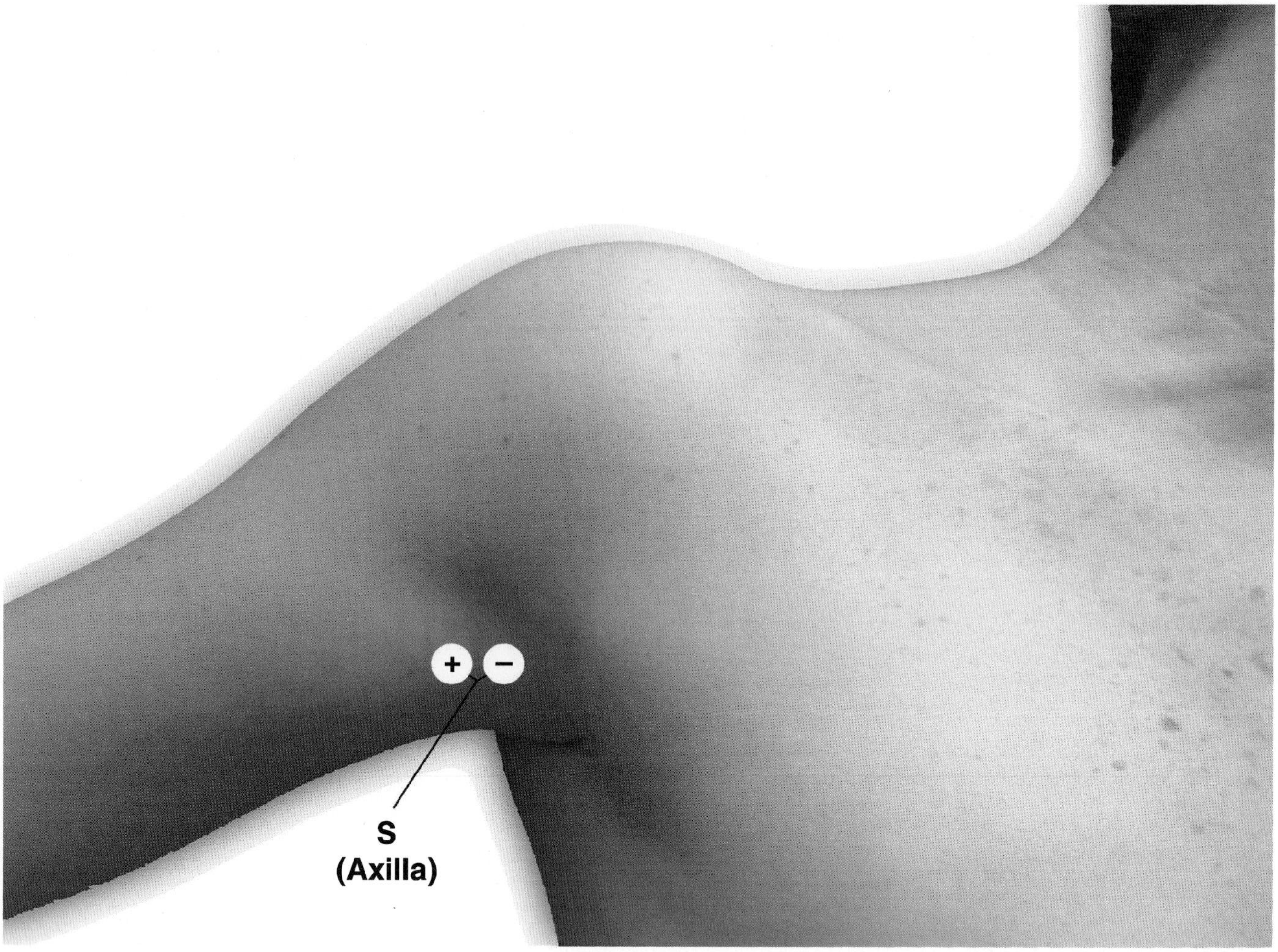

Fig. 3 Proximal stimulation (S3 – axilla)

Table 1 Normal values [1]

	Mean ± SD
Wrist–ADM, M-wave latency (ms)	2.9 ± 0.5
Elbow–ADM, M-wave latency (ms)	7.1 ± 1.2
Axilla–ADM, M-wave latency (ms)	11.2 ± 1.0
Elbow–wrist, M-wave latency difference (ms)	4.3 ± 0.7
Axilla–elbow, M-wave latency difference (ms)	4.0 ± 0.7
Elbow–wrist, MNCV (m/s)	55.9 ± 5.1
Axilla–elbow, MNCV (m/s)	61.3 ± 6.8
Wrist–ADM, F-wave latency (ms)	30.5 ± 3.0
Elbow–MT, F-wave latency (ms)	26.0 ± 1.8
Axilla–ADM, F-wave latency (ms)	21.9 ± 1.9
Elbow–wrist, F-wave latency difference (ms)	4.1 ± 0.8
Axilla–elbow, F-wave latency difference (ms)	4.1 ± 0.9
Wrist–spinal cord, F-wave velocity (m/s)	56.7 ± 2.9
Elbow–spinal cord, F-wave velocity (m/s)	59.4 ± 4.7
Axilla–spinal cord, F-wave velocity (m/s)	63.1 ± 5.9

Table 2 Pathological values [1]

	Mean ± SD
Wrist–ADM, M-wave latency (ms)	7.1 ± 3.0
Elbow–ADM, M-wave latency (ms)	20.6 ± 8.6
Axilla–ADM, M-wave latency (ms)	32.6 ± 12.9
Elbow–wrist, M-wave latency difference (ms)	13.5 ± 6.4
Axilla–elbow, M-wave latency difference (ms)	12.5 ± 4.8
Elbow–wrist, MNCV (m/s)	20.5 ± 11.1
Axilla–elbow, MNCV (m/s)	22.3 ± 11.6
Wrist–ADM, F-wave latency (ms)	85.5 ± 38.1
Elbow–ADM, F-wave latency (ms)	72.4 ± 31.9
Axilla–ADM F-wave latency (ms)	59.6 ± 27.4
Elbow–wrist, F-wave latency difference (ms)	13.2 ± 6.5
Axilla–elbow, F-wave latency difference (ms)	12.8 ± 6.0
Wrist–spinal cord, F-wave velocity (m/s)	21.9 ± 11.6
Elbow–spinal cord, F-wave velocity (m/s)	23.5 ± 12.5
Axilla–spinal cord, F-wave velocity (m/s)	28.3 ± 15.5

Comment

For Kimura [1], with supramaximal stimulation at the wrist, the F-wave was widely separate from the simultaneously recorded M-response. After proximal stimulation at the elbow, the F-wave occurred closer to the M-response, but these two responses were usually still distinctly apart (Fig. 2). When the nerve was stimulated at the axilla, however, the latency difference between the M-response and F-wave was normally so small that the latter appeared before the completion of the former, making an accurate latency measurement impossible. The F-wave elicited by axillary stimulation was recorded readily by the author using a "collision technique." With this technique, the orthodromic impulse from the axilla and the antidromic impulse from the wrist could be extinguished by collision (block of the M-response elicited by axillary stimulation by a second supramaximal stimulation at the wrist), leaving the M-response from the wrist and the F-wave from the axilla intact [3]. In most cases of the CMT disease included in the study, greatly decreased distal MNCVs were associated with a comparable reduction in F-wave velocities in the central segment. For the author, the study of the F-wave velocity could be useful in evaluating the central segment of the median and ulnar nerves. The possible diagnostic value was suggested by Kimura [1] as a supplement to the ordinary motor nerve conduction studies in an analysis of various neuropathies.

In order to determine which nerve conduction study (NCS) assessing the distal segments of the upper extremity motor nerves was more useful to differentiate anti-MAG/SGPG (pathogenic IgM antibodies-Abs against myelin-associated glycoprotein/peripheral nerve glycolipid sulfated glucuronyl paragloboside), Lupu et al. [4] performed several conduction techniques in patients with neuropathy due to IgM Abs against MAG and SGPG (MAG/SGPG-N). Electrophysiologically, conduction blocks characterize the chronic inflammatory demyelinating polyradiculoneuropathy (CIDP), whereas in MAG/SGPG-N, conduction blocks are rarely observed. MAG-SGPG-N and hereditary sensory-motor neuropathy Type 1 (HMSN1) have similar demyelinating features on nerve conduction studies, and two subtypes (HMSN1A, and HMSN1B) were described. HMSN1A is the most common subtype of HSMN1, and a mutation in the gene encoding for peripheral myelin protein 22 (PMP-22) was observed. HMSN1B is related to genetic abnormalities in the myelin protein zero. Median and ulnar nerve motor conduction studies were performed recording at 6 cm fixed distance from the recording electrode placed on the *abductor pollicis brevis* (APB) and the *abductor digiti minimi* (ADM) muscles, respectively. All nerve conduction studies were performed at a skin temperature of 34 °C. Terminal latency index (TLI), residual latency (RL), and F-wave latency ratio were calculated. Following a modified Kimura's method [1], the authors [4] recorded F-wave latency from the median *thenar* and ulnar *hypothenar* muscles after distal and proximal stimulations (S1 and S2 only) at the wrist and the elbow (3 cm above the medial epicondyle). Minimal F-wave latency was determined from at least 10 stimulations, and F-wave latency was not adjusted for the height because this information was missing and because of lack of normative data on F-wave latency at the elbow. They measured distal motor latency (DML) and proximal motor latency (PML), stimulating at the wrist and at the elbow, respectively. They calculated F-wave latency ratio at the wrist (which compares spinal cord to the wrist with wrist to *hypothenar* latency) and at the elbow (which compares spinal cord to elbow with elbow to *hypothenar* latency) using previously published formulas [5]. Following the method by Attarian et al. [6], the authors [4] also calculated a modified F-wave latency ratio (MFR), which compares spinal cord to elbow with wrist to *hypothenar* latency, as follows:

$$\text{F}-\text{wave wrist ratio}\left(\text{F}_{\text{wr}}\right)=\frac{\left(\begin{array}{c}\text{F}-\text{wave at the}\\ \text{wrist} - \text{DML}-1\end{array}\right)}{2\ \text{DML}}$$

$$\text{F}-\text{wave elbow ratio}\left(\text{F}_{\text{el}}\right)=\frac{\left(\begin{array}{c}\text{F}-\text{wave at the}\\ \text{elbow} - \text{PML}-1\end{array}\right)}{2\ \text{PML}}$$

$$\text{MFR}=\frac{\left(\text{F}-\text{wave at the wrist} - 2\ \text{PML} + \text{DML}-1\right)}{2\ \text{DML}}$$

The authors studied 12 healthy (Table 3) volunteers (58 % men and 42 % women, mean age 32.6 ± 11.3 years), 21 patients (Table 4) with MAG/SGPG-N (57 % men and 43 % women, mean age 64.6 ± 9.4 years), 26 patients (Table 5) with HMSN1 (58 % men and 42 % women, mean age 38.1 ± 13.7 years), and 20 patients (Table 6) with HMSN2 (70 % men and 30 % women, mean age 44.7 ± 14.5 years).

Table 3 Normal values [4]

	Mean ± SD
Wrist–ADM, F-wave latency (ms)	27 ± 2
Elbow–ADM, F-wave latency (ms)	21 ± 1
F_{el} (elbow–ADM)	0.8 ± 0.1
MFR	2.8 ± 0.3

Table 4 Pathological values [4]

MAG/SGPG-N, 21 patients	Mean ± SD
Wrist–ADM, F-wave latency (ms)	54 ± 17
Elbow–ADM, F-wave latency (ms)	46 ± 15
Fel (elbow–ADM)	0.9 ± 0.2
MFR	2.5 ± 0.5

Table 5 Pathological values [4]

HMSN1, 26 patients	Mean ± SD
Wrist–ADM, F-wave latency (ms)	71 ± 14
Elbow–ADM, F-wave latency (ms)	56 ± 11
Fel (elbow–ADM)	0.8 ± 0.2
MFR	2.5 ± 0.6

Table 6 Pathological values [4]

HMSN2, 20 patients	Mean ± SD
Wrist–ADM, F-wave latency (ms)	37 ± 9
Elbow–ADM, F-wave latency (ms)	31 ± 8
Fel (elbow–ADM)	0.9 ± 0.2
MFR	3.0 ± 0.5

Comment

For Lupu et al. [4], F-waves were markedly prolonged in MAG/SGPG-N (21 patients) and HMSN1 (26 patients). F-wave latency ratio (elbow–median *hypothenar* muscle) and MFR were both abnormal in the same groups. Their findings were consistent with those by Attarian et al. [6], who evaluated indexes calculated from standard electrophysiological data in differentiating CIDP. They studied 19 CIDP patients, 25 anti-myelin-associated glycoprotein/sulfated glucuronyl paragloboside (MAG/SGPG) antibody patients, 13 Charcot–Marie–Tooth disease Type 1A (CMT1A) patients, and 22 controls. They used MFR to compare the spinal cord–elbow segment (proximal segment) latency with that of the wrist–*hypothenar* muscle segment (distal segment). Compared with controls, MFR was increased in 16 CIDP patients, and it was normal in all patients affected by CMT1A. They suggested to use MFR index in differentiating CIDP because they found a good value of sensitivity (84 %) and specificity (89 %).

More recently, Rajabally and Varanasi [7] investigated the practical electrodiagnostic value of F-wave studies in CIDP. They studied 29 patients (Table 7) with axonal neuropathy (AN) as controls (18 men and 11 women, mean age 61.0 ± 16.6 years) and 18 patients (Table 8) affected by typical CIDP (12 men and 6 women, mean age 61.4 ± 12.8 years) using a MEDELEC Synergy mobile two-channel electromyographic device (Oxford Instruments Medical); the upper arm temperature was maintained at or above 32 °C. The authors performed conventional motor conduction study (including axilla and Erb's point as sites of stimulation in all cases), and F-waves were elicited antidromically from the wrist site using a supramaximal stimulation (about 150 % of that required to elicit maximum amplitude CMAP). For each F-wave study, a total of 10 stimulations were performed at a frequency of 1 Hz; F-wave latency (ms) was measured to the onset of the initial negative or positive deflection from baseline. Only late responses with a minimal amplitude of 20 μV at a gain of 200 μV and sweep speed of 10 ms were included in the study, and differentiation from A-waves was made by excluding all responses of exactly similar latency, amplitude, and morphology. The authors analyzed the minimum F-wave latency, maximum F-wave latency, mean F-wave latency, chronodispersion (max F latency–min F latency), persistence (percentage of recordable F-waves), and impersistence (<100 % recordable F-waves).

Table 7 Normal values [7]

	Mean ± SD
Wrist–ADM, F-wave minimum latency (ms)	28.8 ± 3.2
Wrist–ADM, F-wave mean latency (ms)	29.9 ± 3.3
Wrist–ADM, F-wave maximum latency (ms)	31.1 ± 3.4
Wrist–ADM, F-wave chronodispersion (ms)	2.3 ± 1.0
Wrist–ADM, F-wave impersistence (ms)	23/29

Table 8 Pathological values [7]

	Mean ± SD
Wrist–ADM, F-wave minimum latency (ms)	38.5 ± 9.6
Wrist–ADM, F-wave mean latency (ms)	41.3 ± 10.5
Wrist–ADM, F-wave maximum latency (ms)	44.0 ± 11.1
Wrist–ADM, F-wave chronodispersion (ms)	5.4 ± 2.9
Wrist–ADM, F-wave impersistence (ms)	10/18

Comment

For Rajabally and Varanasi [7] minimum ulnar F-wave latencies, mean F-wave latencies, and maximum F-wave latencies were significantly more delayed in CIDP patients than in AN controls. Chronodispersion was significantly greater in CIDP patients than in AN controls, and ROC curve analysis showed a cutoff of 3.8 ms for ulnar nerve, which offered a sensitivity of 77.8 % for CIDP with a specificity of 93.1 % versus AN. For the authors, the F-wave chronodispersion, which was significantly greater in upper limb CIDP nerves than AN nerves, was a significantly more useful F-wave parameter to utilize in the upper arms in the distinguishing CIDP from AN. Clinical application in a Charcot–Marie–Tooth (PMP22–CMT1A) disease, in a case of Guillain–Barré syndrome (GBS), and in a patient with a chronic inflammatory demyelinating polyneuropathy (CIDP) are reported (Figs. 4, 5, and 6, respectively).

Pathological waveform (wrist – ADM muscle):

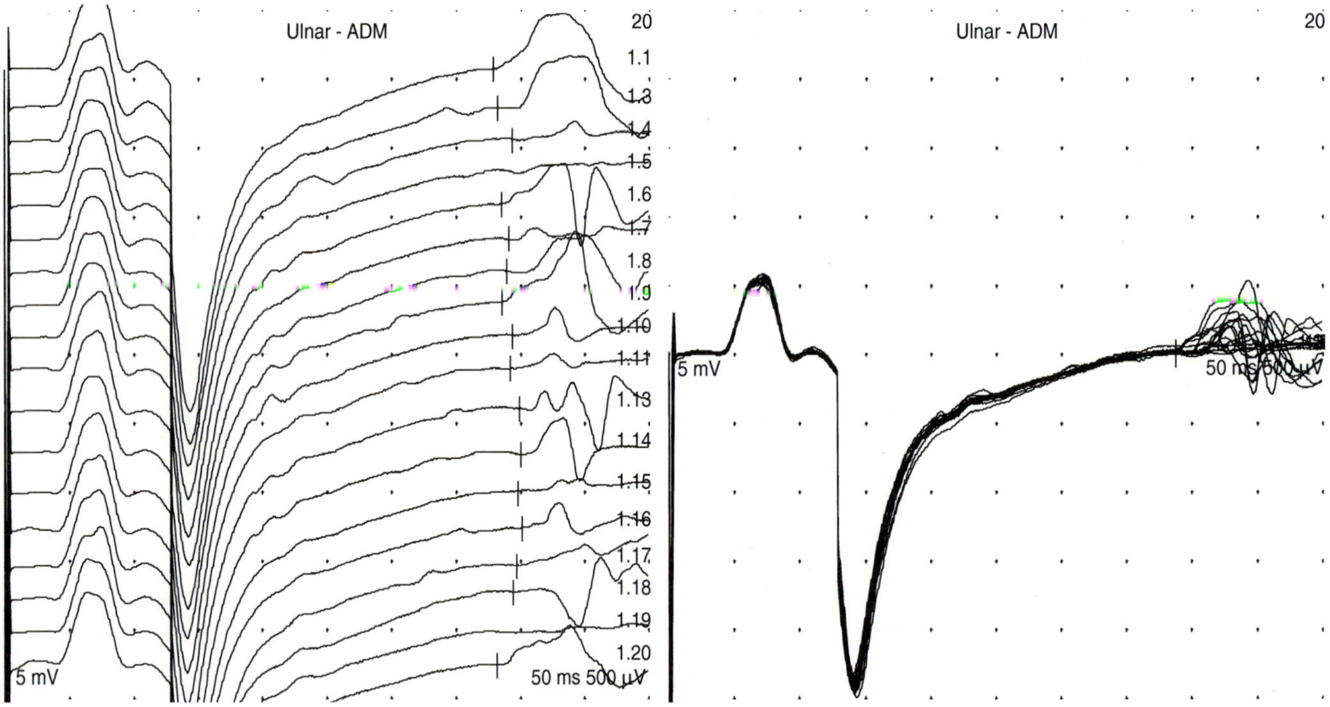

Sensitivity 500 µV/div (CMAP 5 mV/div), sweep speed 5 ms/div

Fig. 4 Prolonged F-waves recorded at the hand from the ADM muscle in a PMP22–CMT1A patient (mean F-wave latency 38.75 ms), stimulation of the wrist. Raster mode, 20 stimuli (*left*); superimposed mode, 20 stimuli (*right*)

Pathological waveform (wrist – ADM muscle):

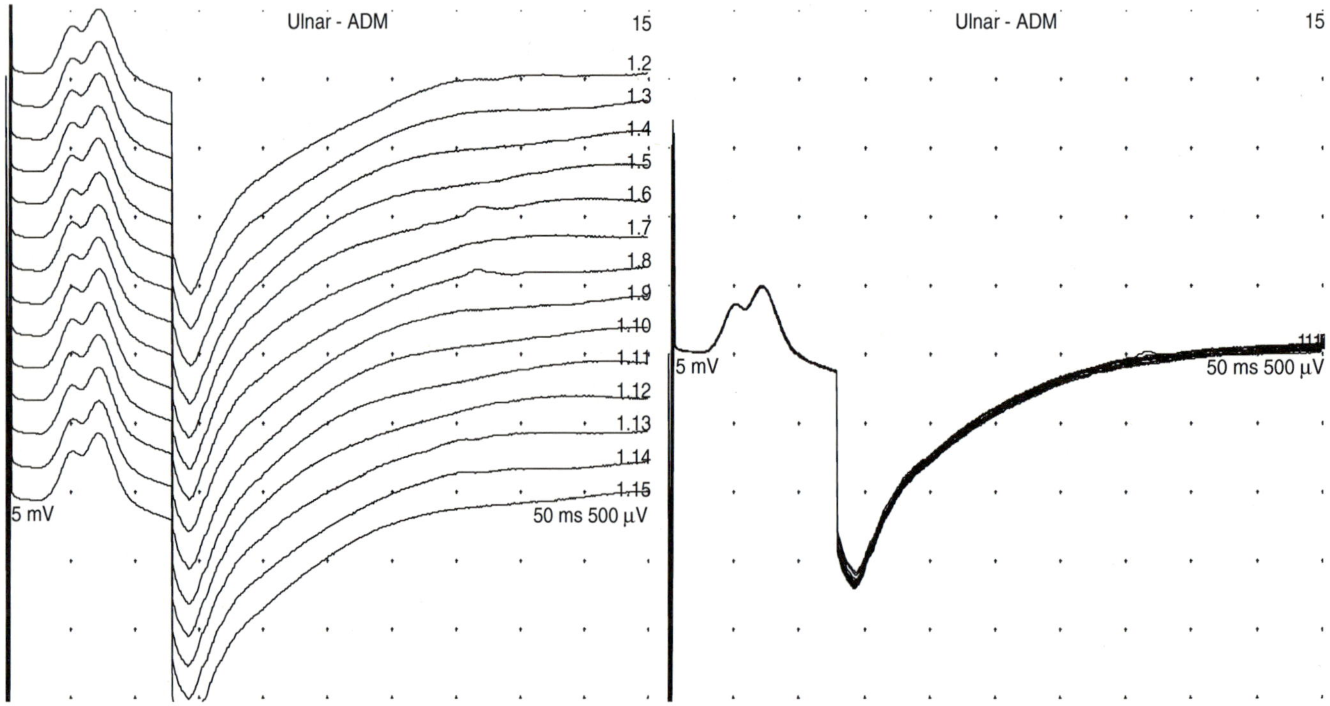

Sensitivity 500 µV/div (CMAP 5 mV/div), sweep speed 5 ms/div

Fig. 5 Prolonged F-waves recorded at the hand from the ADM muscle in a GBS patient (absent F-waves), stimulation of the wrist. Raster mode, 15 stimuli (*left*); superimposed mode, 15 stimuli (*right*)

Pathological waveform (wrist – ADM muscle):

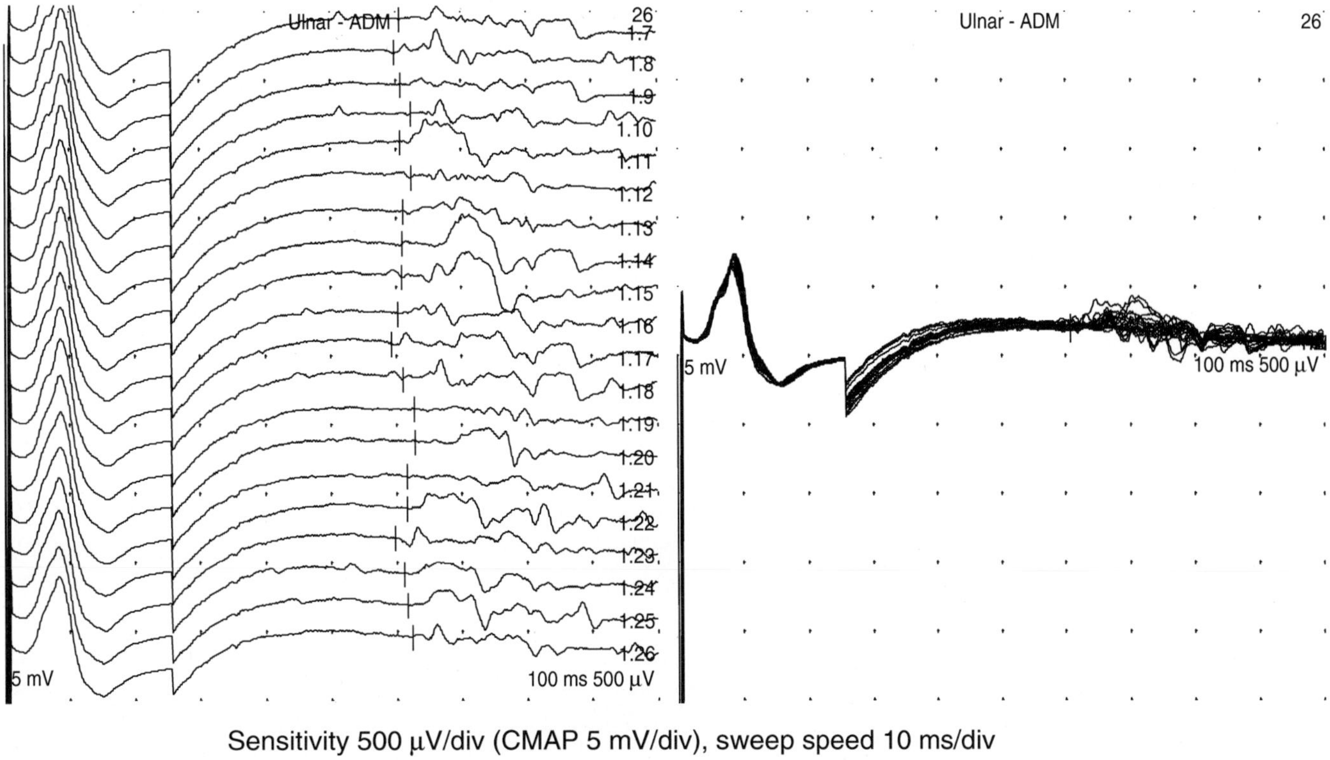

Fig. 6 Prolonged F-waves recorded at the hand from the ADM muscle in a patient with CIDP associated with anti-MAG/SGPG (pathogenic IgM antibodies – Abs against myelin-associated glycoprotein/peripheral nerve glycolipid sulfated glucuronyl paragloboside) – mean F-wave latency 58.36 ms, stimulation of the wrist. Raster mode, 20 stimuli (*left*); superimposed mode, 20 stimuli (*right*)

References

1. Kimura J (1974) F-Wave velocity in the central segment of the median and ulnar nerves: a study in normal subjects and in patients with Charcot-Marie-Tooth disease. Neurology 24:539–546
2. Renshaw B (1941) Influence of discharge of motoneurons upon excitation of neighboring motoneurons. J Neurophysiol 4: 167–183
3. Kimura J (1976) Collision technique Physiologic block of nerve impulses in studies of motor nerve conduction block. Neurology 26:680–682
4. Lupu VD, Mora CA, Dambrosia J et al (2007) Terminal latency index in neuropathy with antibodies against myelin-associated glycoproteins. Muscle Nerve 35:196–202
5. Kimura J (2001) Long and short of nerve conduction measures: reproducibility for sequential assessments. J Neurol Neurosurg Psychiatry 71:427–430
6. Attarian S, Azulay JP, Boucraut J et al (2001) Terminal latency index and modified F ratio in distinction of chronic demyelinating neuropathies. Clin Neurophysiol 112:457–463
7. Rajabally YA, Varanasi S (2013) Practical electrodiagnostic value of F-wave studies in chronic inflammatory demyelinating polyneuropathy. Clin Neurophysiol 124:171–175

Axilla, Above Elbow, Below Elbow, Wrist – Hand

M17

Surface Recording Technique, Study from *Abductor Digiti Minimi* (ADM) Muscle

Original Settings Duration of pulse was 0.1–0.5 ms; sensitivity, sweep speed, low-frequency filter, high-frequency filter, rate of pulse, and the machine used were not specified.

Position This study was performed with the subjects lying on a bed in the supine position, with the arm supinated at 45° to the trunk and the elbow comfortably extended.

Recording The active electrode (A) was placed over the belly (motor point) of the *abductor digiti minimi* (ADM) muscle (Fig. 1), by surface gold disk 5 mm diameter electrodes [1]. The active (A) electrode was placed over the belly of the ADM muscle; the reference (R) was placed distally over its tendon of insertion. Distances between the recording and stimulating electrodes were not fixed. The ground (G) electrode position was at the wrist, and the author used a copper ring electrode placed just distal to the S4 point of stimulation.

Stimulation The ulnar nerve was stimulated at four standard points along its length (Figs. 1 and 2): in the axilla (S1), above the elbow (S2), below the elbow (S3), and at the wrist (S4). A percutaneous, supramaximal stimulus of between 0.1 and 0.5 ms in duration and up to 150 V was delivered via an isolation unit to a bipolar stimulator; the cathode and anode were 5 mm in diameter, with a separation of 15 mm. Care was taken to assure that the distance between S2 and S3 was always at least 10 cm [1].

Measurements Distal latency (ms) and proximal latency (ms) of the compound muscle action potential (CMAP) were measured from the onset of the stimulus artifact to the onset of the initial deflection of the CMAP (i.e., wrist–ADM muscle and above elbow–ADM muscle, for distal and proximal latencies, respectively). The distance between the stimulating cathode and active recording electrode was measured and divided by the interval between the onset of the stimulus artifact and initial deflection of the evoked response. The motor nerve conduction velocity (MNCV) along three separate segments of the ulnar nerve was also calculated: from the axilla to above elbow (S1–S2), across the elbow (S2–S3), and from below the elbow to the wrist (S3–S4). Temperature data were not reported. Normal values (Table 1) were obtained from 48 ulnar nerves of healthy subjects (mean age 43.7 years, age range 19–78 years). Pathological values were recorded from 100 ulnar nerves of patients divided into two groups (Group 1 (severe ulnar lesions) and Group 2 (mild ulnar lesions)): Group 1–34 ulnar nerves from patients (Table 2) showing clear-cut clinical evidence of an ulnar nerve compression at the elbow (mean age 52.5 years, age range 21–77 years). All patients showed atrophy and significant weakness of the *first dorsal interosseous* (FDI) and *abductor digiti minimi* (ADM) muscles on the involved side and sensory deficit in an appropriate distribution; Group 2–56 ulnar nerves from patients (Table 3) are referred because of subjective complaints of numbness and tingling limited to the digits IV and V on the involved side and because of a presumptive diagnosis of ulnar palsy (mean age 44.1 years, age range 17–74 years).

G. Gentili, M. Di Napoli, *The Ulnar Nerve*, DOI 10.1007/978-3-319-43292-2_41

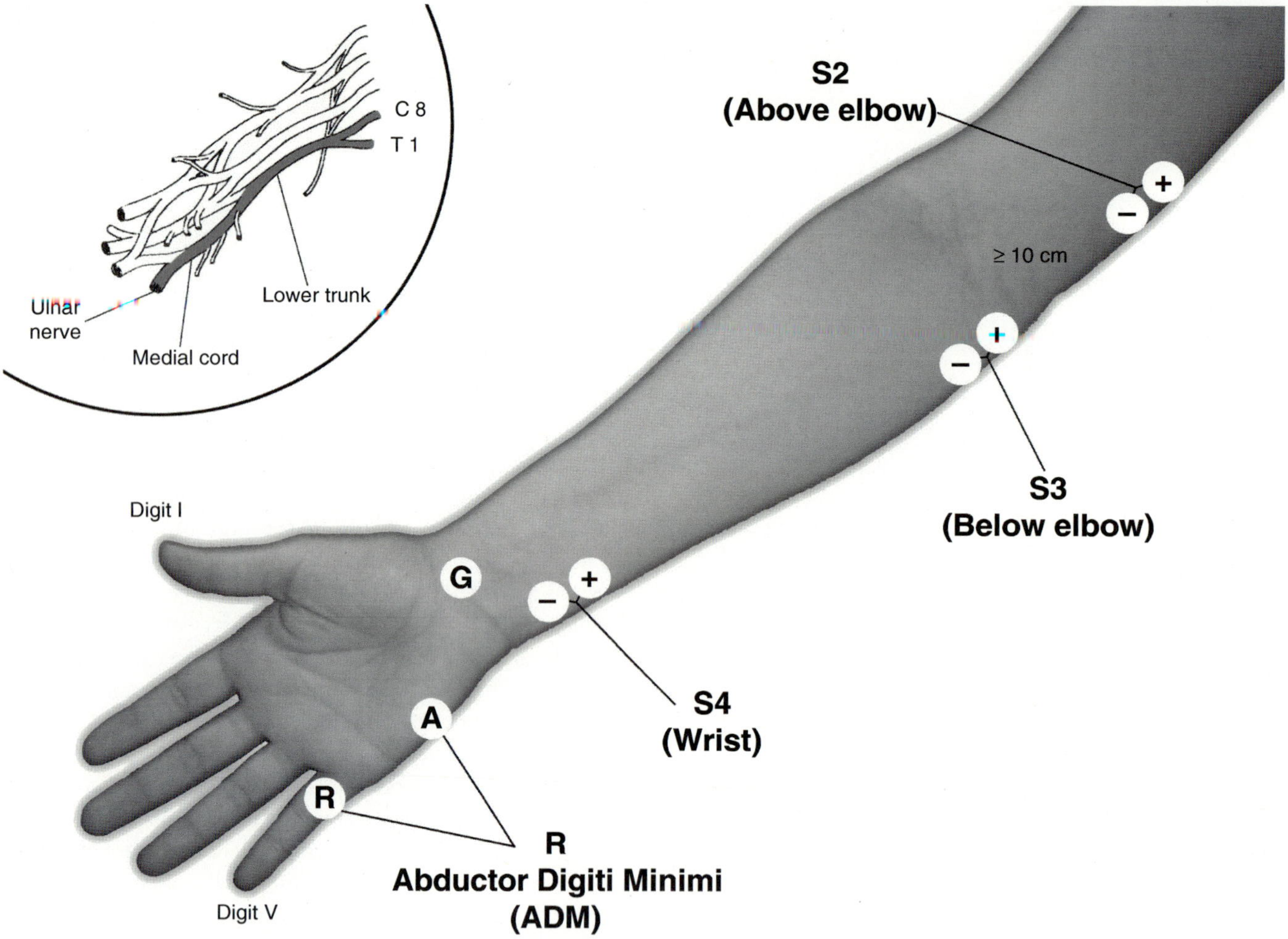

Typical waveform (axilla, above elbow, below elbow, wrist – ADM muscle):

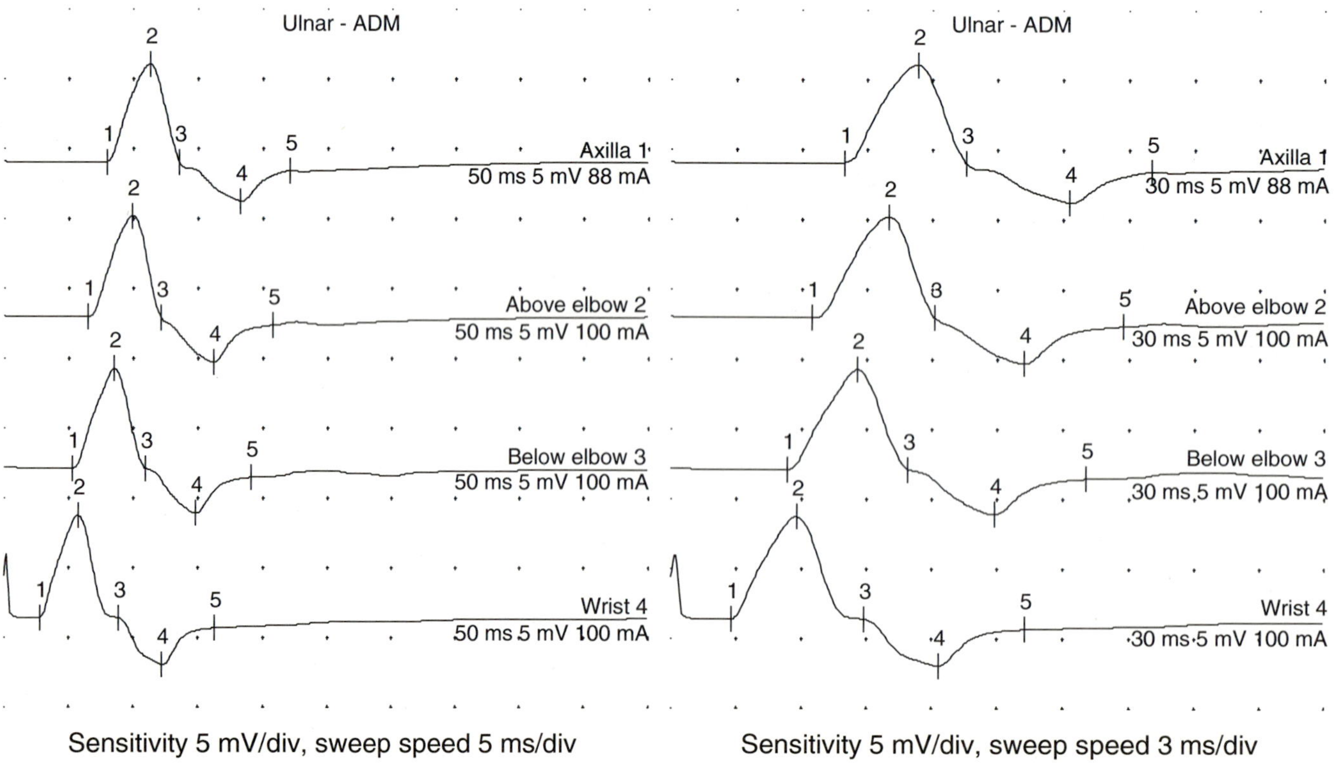

Fig. 1 Compound muscle action potentials (*CMAPs*) recorded at the hand from the ADM muscle; *top to bottom*: stimulation of the axilla, above elbow, below elbow, and of the wrist

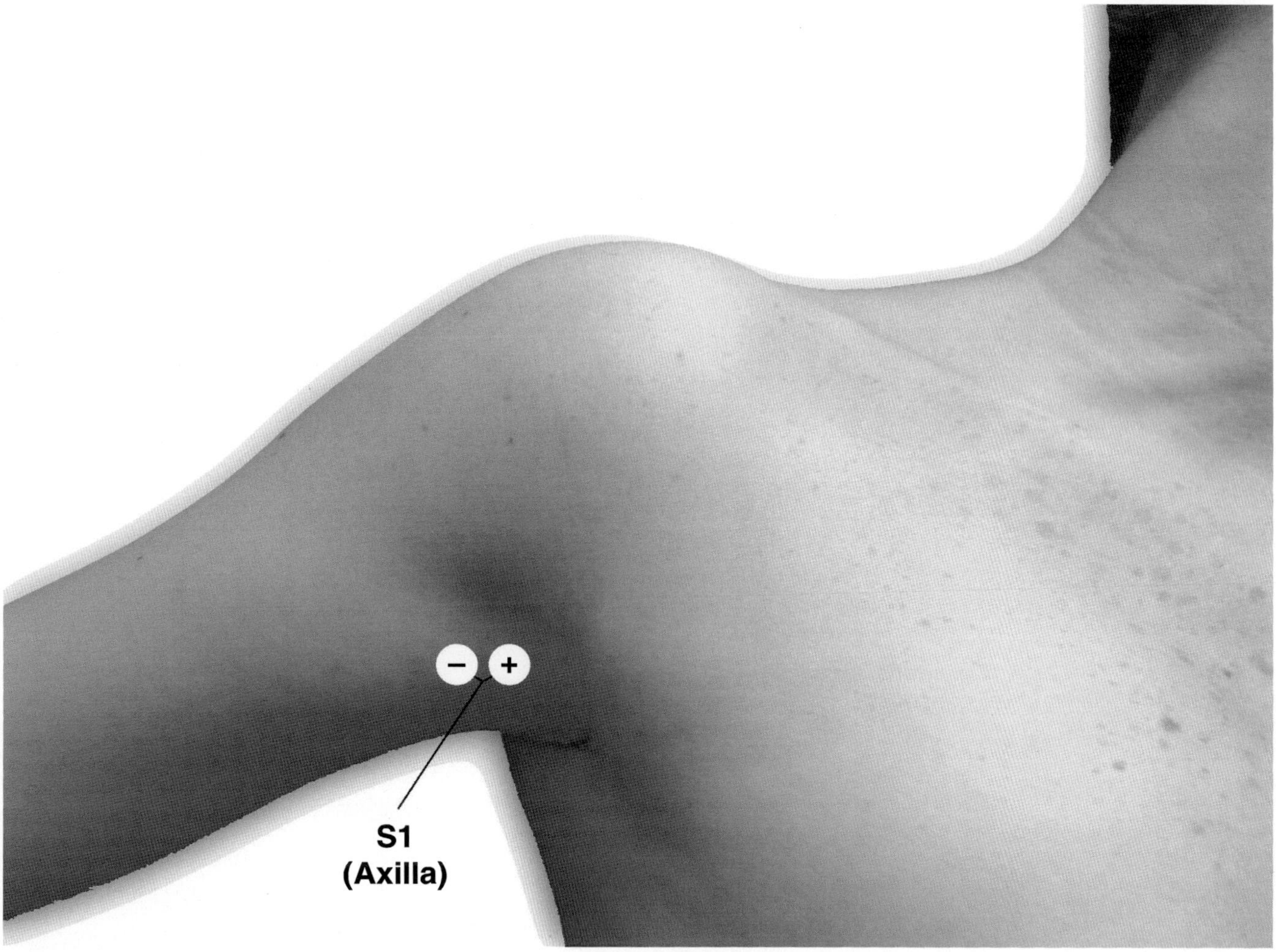

Fig. 2 Proximal stimulation point in the axilla (S1)

Table 1 Normal values [1]

	Mean ± SD	Range
Wrist–ADM – distal motor latency (ms)	3.0 ± 0.4	2.2–3.8
Above elbow–ADM – proximal motor latency (ms)	7.9 ± 0.85	6.6–10.0
Axilla–above elbow, MNCV (m/s)	59.2 ± 8.2	40.5–70.5
Above elbow–below elbow, MNCV (m/s)	53.2 ± 7.8	29.2–67.0
Below elbow–wrist, MNCV (m/s)	54.6 ± 6.7	41.5–68.5

Table 2 Pathological values [1]

Group 1	Mean ± SD	Range
Wrist–ADM – distal motor latency (ms)	3.9 ± 0.7	2.4–5.6
Above elbow–ADM – proximal motor latency (ms)	11.3 ± 2.0	7.9–16.5
Axilla–above elbow, MNCV (m/s)	52.8 ± 9.7	31.3–70.0
Above elbow–below elbow, MNCV (m/s)	34.8 ± 8.8	18.6–55.5
Below elbow–wrist, MNCV (m/s)	48.2 ± 9.4	25.2–62.3

Table 3 Pathological values [1]

Group 2	Mean ± SD	Range
Wrist–ADM – distal motor latency (ms)	3.3 ± 0.5	2.7–4.9
Above elbow–ADM – proximal motor latency (ms)	9.1 ± 1.2	6.5–12.4
Axilla–above elbow, MNCV (m/s)	52.7 ± 8.5	36.0–73.0
Above elbow–below elbow, MNCV (m/s)	47.7 ± 10.1	28.8–66.8
Below elbow–wrist, MNCV (m/s)	52.9 ± 6.4	40.0–67.5

Comment

Eisen [1], in control subjects, observed a small but significant slowing of the mean MNCV across the elbow and the forearm in relation to the upper arm; these findings were consistent with others [2–5], and the difference was probably due to a combination of lower temperatures and a decrease in nerve fiber diameters in the forearm as compared with the upper arm [5]. The author [1] also recorded CMAPs from ADM muscle in patients with severe ulnar nerve lesions (Group 1–34 nerves) and mild ulnar nerve lesions (Group 2–56 nerves). In the Group 1, 18 patients (53 %) demonstrated a fall in conduction velocity of 10 m/s or more across the elbow region in relation to values above and below the elbow. In the Group 2, 11 patients (19.7 %) demonstrated a fall in the MNCV of 10 m/s or more across the elbow in relation to the velocities above and below the elbow segment. The differences in the mean conduction velocities above and across the elbow of this group, in relation to the mean velocities of corresponding segments from control nerves, although small, were significant. No significant difference was demonstrated between the two groups in the MNCV from below the elbow to the wrist. Pathological waveforms and values in a case of ulnar neuropathy at the elbow (UNE) are reported (Fig. 3).

Pathological waveform (axilla, above elbow, below elbow, wrist – ADM muscle):

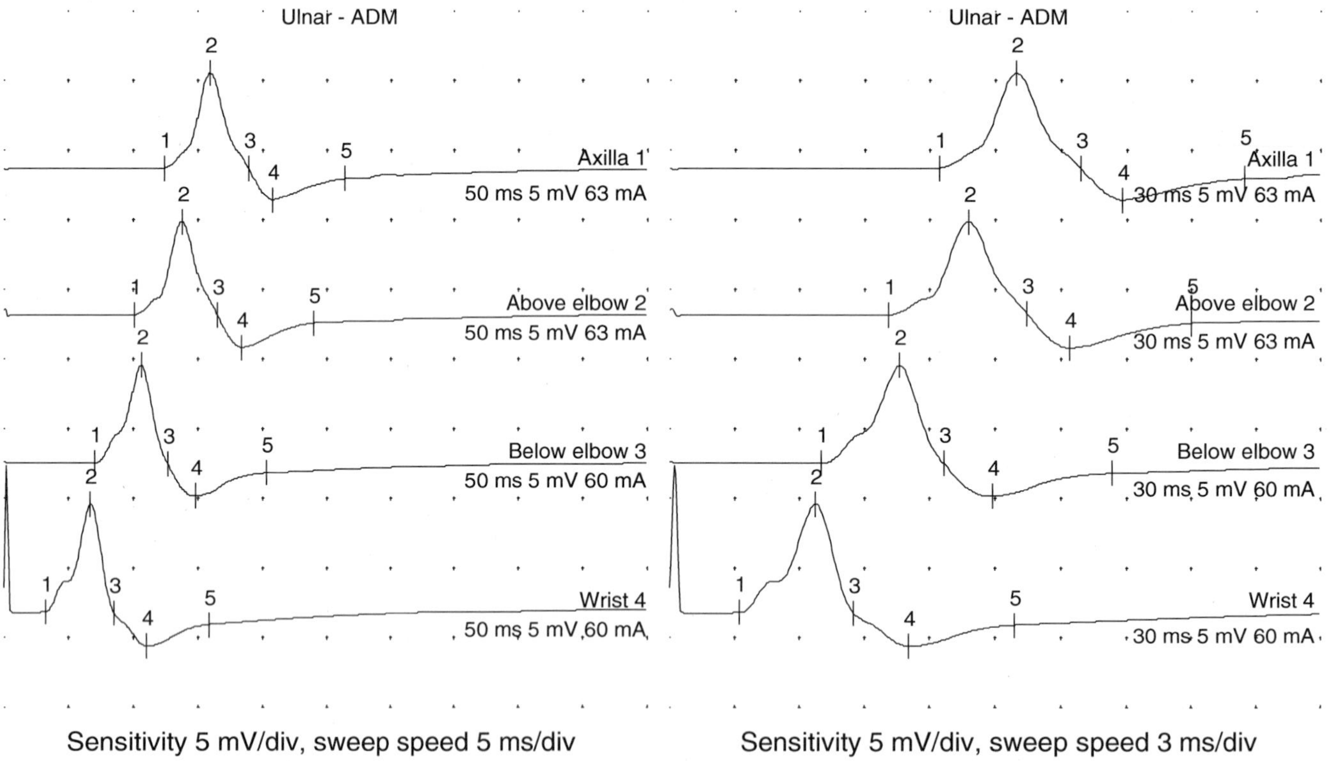

Onset latency (axilla – ADM): 12.40 ms; **Onset latency** (above elbow – ADM): 10.10 ms; **Onset latency** (below elbow – ADM): 7.0 ms; **Onset latency** (wrist – ADM): 3.20 ms; **Peak latency** (axilla – ADM): 15.90 ms; **Peak latency** (above elbow – ADM): 13.80 ms; **Peak latency** (below elbow – ADM): 10.60 ms; **Peak latency** (wrist – ADM): 6.65 ms; **Onset to peak amplitude** (axilla – ADM): 6.8 mV; **Onset to peak amplitude** (above elbow – ADM): 6.7 mV; **Onset to peak amplitude** (below elbow – ADM): 7.0 mV; **Onset to peak amplitude** (wrist – ADM): 7.8 mV; **Peak to peak amplitude** (axilla – ADM): 9.2 mV; **Peak to peak amplitude** (above elbow – ADM): 9.1 mV; **Peak to peak amplitude** (below elbow – ADM): 9.5 mV; **Peak to peak amplitude** (wrist – ADM): 10.3 mV; **MNCV** (axilla – above elbow, 12 cm): 52.2 m/s; **MNCV** (above elbow – below elbow, 10 cm): 32.3 m/s; **MNCV** (below elbow – wrist, 21 cm): 55.3 m/s

Fig. 3 Compound muscle action potentials (*CMAPs*) recorded at the hand from the ADM muscle in UNE; *top to bottom*: stimulation of the axilla, above elbow, below elbow, and of the wrist

References

1. Eisen A (1974) Early diagnosis of ulnar nerve palsy. Neurology 24:256–262
2. Payan J (1969) Electrophysiological localization of ulnar nerve lesions. J Neurol Neurosurg Psychiat 32:208–220
3. Gilliatt RW, Thomas PK (1960) Changes in nerve conduction with ulnar lesions at the elbow. J Neurol Neurosurg Psychiatry 23:312–320
4. Thompson EM, Coté J (1970) Conduction velocities in ulnar motor nerve using two separate muscles as indicators. South Med J 63:700–703
5. Trojaborg W (1964) Motor nerve conduction velocities in normal subjects with particular reference to the conduction in proximal and distal segments of median and ulnar nerve. Electroencephalogr Clin Neurophysiol 17:314–321

Wrist, Elbow, Above Elbow, Erb's Point – Hand

M18

Surface Recording Technique, Study from *Abductor Digiti Minimi* (ADM) Muscle

Original Settings Duration of pulse was 0.05–0.2 ms, and the machines used were a TECA TE-3 and TE-4 electromyograph; sensitivity, sweep speed, low-frequency filter, and high-frequency filter were not specified.

Position Following the text, this study was performed with the arm placed in 10° of abduction at the shoulder with the elbow extended and the arm supinated.

Recording Using silver disk surface electrodes, the active electrode (A) was placed over the belly (motor point) of the *abductor digiti minimi* (ADM) muscle (Fig. 1); the reference electrode was placed distally to its insertion tendon [1]. The ground (G) electrode was placed on the dorsum of the hand; the figure shows the ground electrode placed on the forearm.

Stimulation The ulnar nerve was stimulated at four sites: at the wrist (S1), at the elbow (S2), 10 cm above the elbow (S3), and at the Erb's point (S4) – which is at the angle formed by the clavicle and the posterolateral fibers of the sternocleidomastoid muscle (Fig. 2). Supramaximal stimulations were delivered at all sites using a 100–300 V potential.

Measurements Distal latency (ms) of the compound muscle action potential (CMAP) was measured from the onset of the stimulus artifact to the junction of the negative deflection of CMAP and the baseline. By dividing the distance by time, the motor nerve conduction velocity (MNCV) along the nerve segment can be determined and measured in meter per second (m/s); it was calculated for above-elbow–Erb's point and wrist–elbow segments. The amplitude (mV) of the CMAP was not reported. The temperature of the laboratory was monitored, and it was kept constant for all recording sessions. Surface skin temperature was measured in seven subjects before beginning the study using a telethermometer, and a range of 0–1°C between the two arms was reported. Normal conduction values (Table 1) were obtained from 60 ulnar nerves of 30 healthy subjects (18 men and 12 women, age range 20–44 years, mean age 28 years). The author performed the nerve conduction method in healthy subjects only, and he compared two measuring techniques to calculate the nerve conduction velocity across the thoracic outlet using a sharpened obstetric caliper as previously described [2] and steel tape measurements.

G. Gentili, M. Di Napoli, *The Ulnar Nerve*, DOI 10.1007/978-3-319-43292-2_42

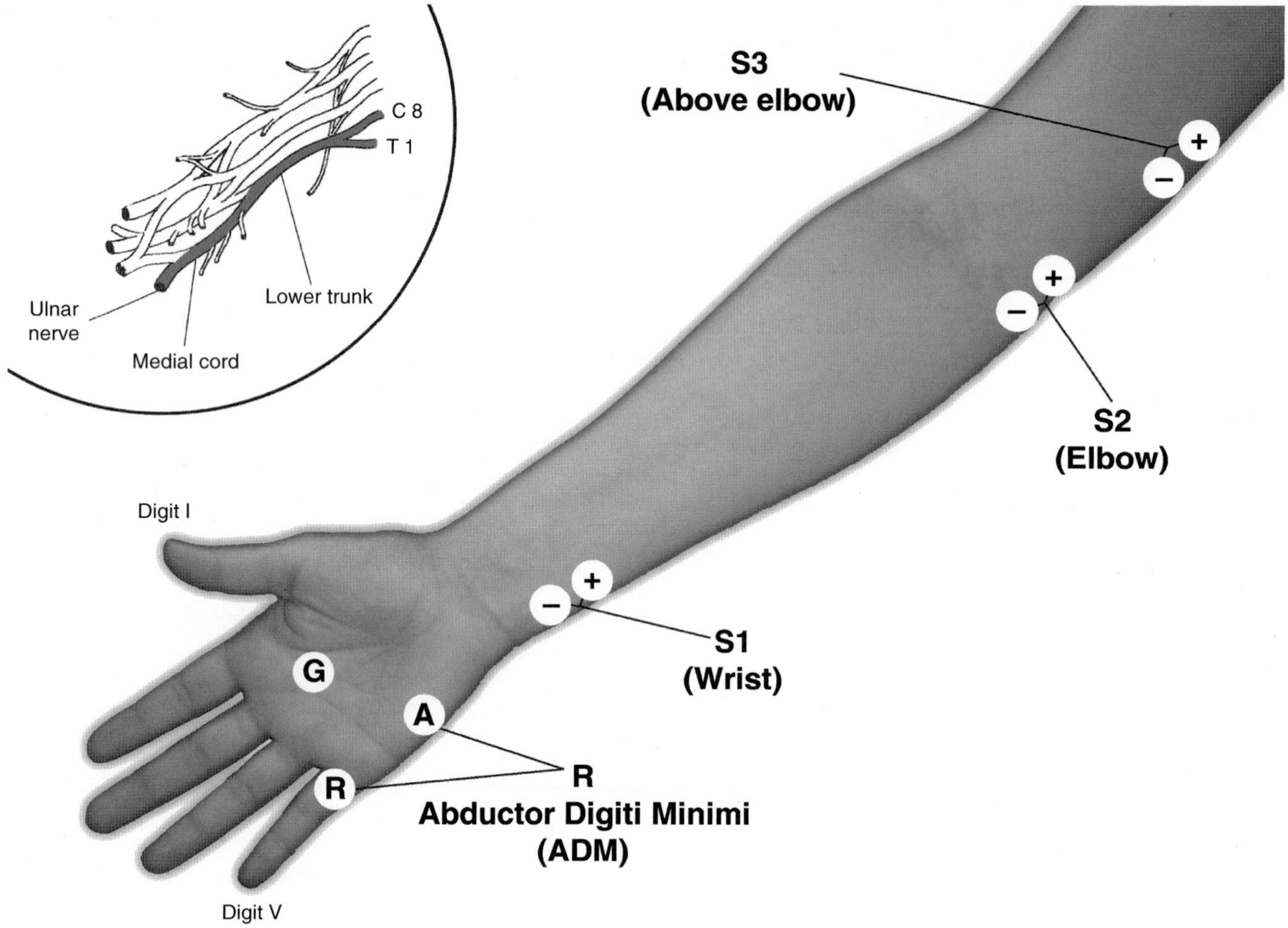

Typical waveform (wrist, elbow, above elbow, Erb's point – ADM muscle):

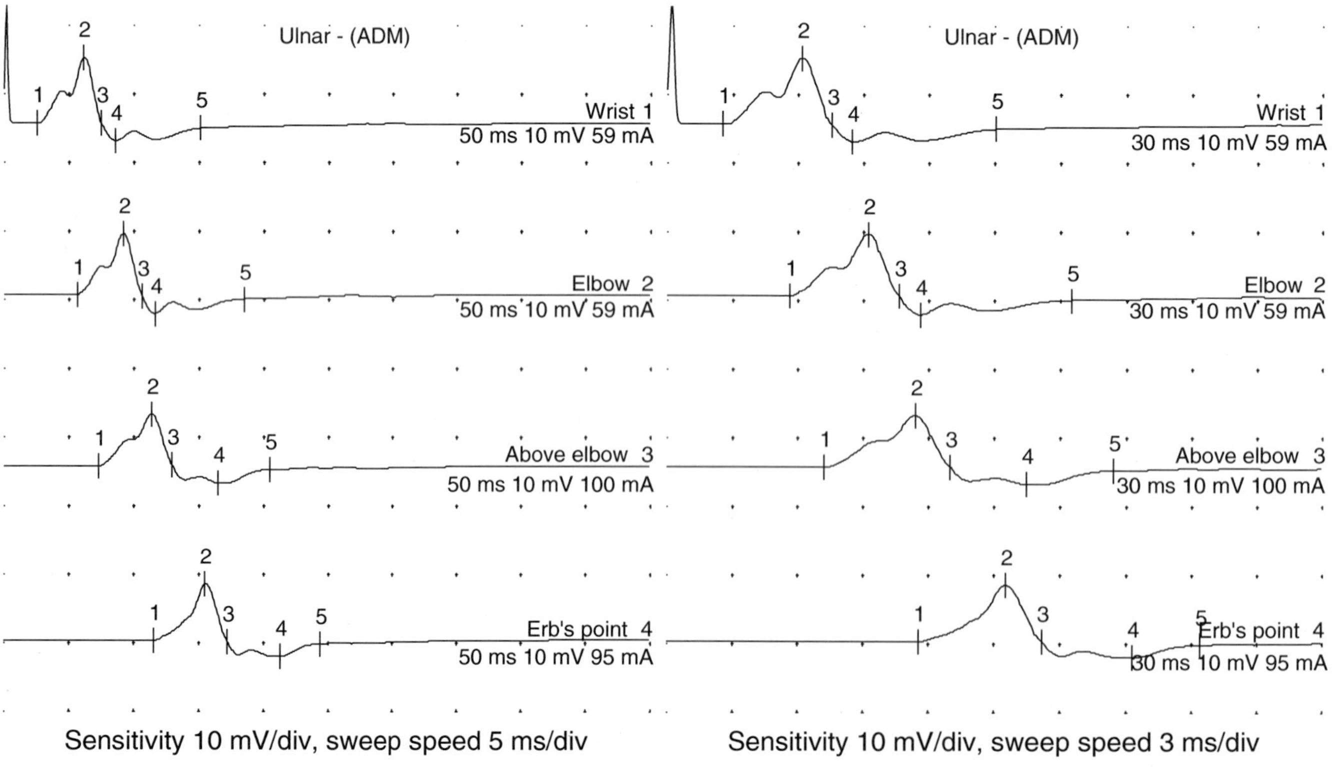

Fig. 1 Compound muscle action potentials (*CMAPs*) recorded at the hand from the ADM muscle; *top to bottom*: stimulation of the wrist, elbow, above elbow, and of the Erb's point

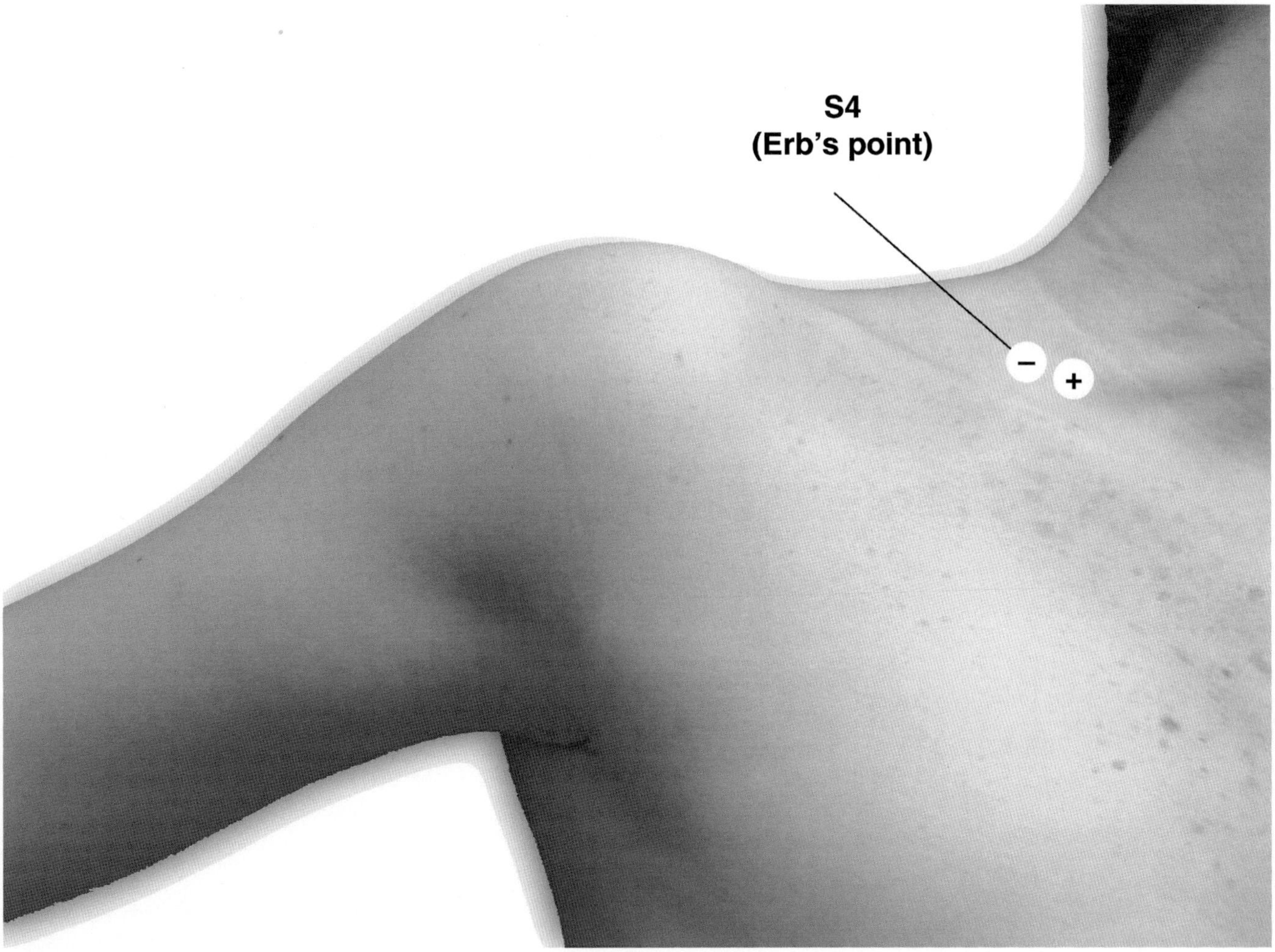

Fig. 2 Proximal stimulation point (S4 – Erb's point)

Table 1 Normal values [1]

	Mean ± SD	Range
Wrist–ADM, distal latency (ms)	2.45 ± 0.32	1.9–3.2
Wrist–elbow, MNCV (m/s)	58.1 ± 3.7	50–65.8
Above elbow–Erb's point, MNCV (m/s) – tape measure	70.2 ± 5.0	58–82.4
Above elbow–Erb's point, MNCV (m/s) – caliper measure	58.9 ± 4.2	50–67.7

Comment

London [1] found values for the wrist distal motor latency and motor conduction velocity in the wrist–elbow segment in agree with previous studies [3]. The mean distance for the above-elbow–Erb's point segment was 23.3 cm (obstetric caliper) and 27.8 cm (tape) showing a 4.5 cm difference. The mean difference between the right and left (side-to-side difference) was 3.46 ± 2.66 m/s (range 0.3–10 m/s) using the caliper and 3.65 ± 2.65 m/s (range 0–11.9 m/s) with the steel tape, and it was 2.16 ± 1.47 m/s (range 0.1–4.7 m/s). For the author, since 90 % of patients with a lower brachial plexus compression showed symptoms and signs in the ulnar nerve distribution, demonstration of delayed ulnar nerve conduction velocity across the thoracic outlet provides objective means of a neural compression. He found the rigid obstetric caliper easy to use and very suitable to measure the distance in this area of the body. Jebsen [2] demonstrated in cadaver studies that the mean in situ length of proximal ulnar nerve fibers was found to be longer than the caliper measurement by 5.1 ± 2.7 % (i.e., the mean caliper distance of 23.3 cm reported in the present study would be only 1.18 cm shorter than the true length). For London [1], comparing the conduction velocity in the thoracic outlet in both sides is a useful neurophysiological tool, and the opposite limb may be used as control; if both values are within the established normal range but differ by 9.0 m/s (mean ± 2SD), then the slower nerve might require careful serial study. Clinical application in a case of ulnar neuropathy at the elbow (UNE), in a patient affected by diabetes mellitus (DM) with diabetic polyneuropathy (DPN) and associated UNE, and in a patient with hand trauma are reported (Figs. 3, 4, and 5, respectively).

Pathological waveform (wrist, elbow, above elbow, Erb's point – ADM muscle):

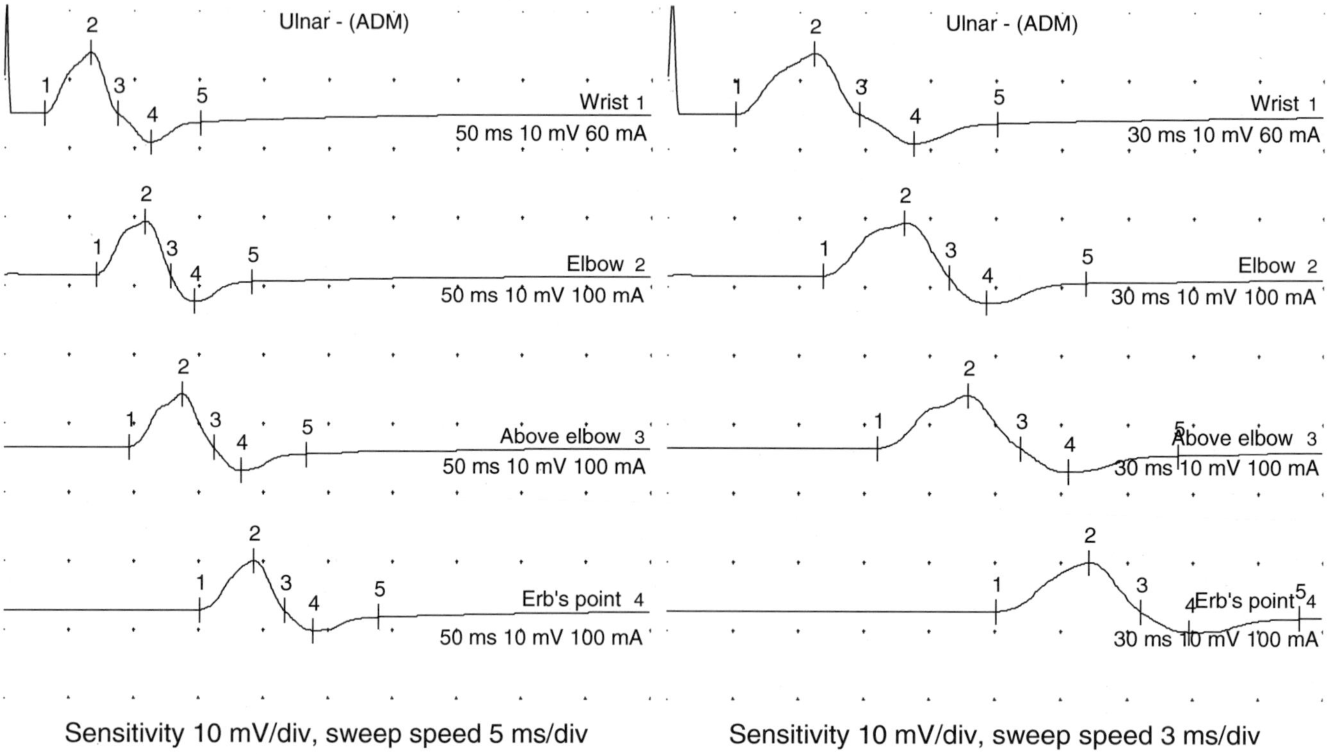

Onset latency (wrist – ADM): 2.95 ms; **Onset latency** (elbow – ADM): 7.05 ms; **Onset latency** (above elbow – ADM): 9.60 ms; **Onset atency** (Erb's point – ADM): 15.05 ms; **Peak latency** (wrist – ADM): 7.10 ms; **Peak latency** (elbow – ADM): 10.65 ms; **Peak latency** (above elbow – ADM): 13.60 ms; **Peak latency** (Erb's point – ADM): 19.30 ms; **Onset to peak amplitude** (wrist – ADM): 8.9 mV; **Onset to peak amplitude** (elbow – ADM): 7.8 mV; **Onset to peak amplitude** (above elbow – ADM): 7.8 mV; **Onset to peak amplitude** (Erb's point – ADM): 7.2 mV; **Peak to peak amplitude** (wrist – ADM): 13.2 mV; **Peak to peak amplitude** (elbow – ADM): 11.8 mV; **Peak to peak Amplitude** (above elbow – ADM): 11.4 mV; **Peak to peak amplitude** (Erb's point – ADM): 10.3 mV; **MNCV** (wrist – elbow, 21 cm): 52.5 m/s; **MNCV**(elbow – above elbow, 10 cm): 39.2 m/s; **MNCV** (Erb's point – above elbow, 28.5 cm): 52.3 m/s

Fig. 3 Compound muscle action potentials (*CMAPs*) recorded at the hand from the ADM muscle using surface recording method [1] in UNE; *top to bottom*: stimulation of the wrist, elbow, above elbow, and of the Erb's point

Pathological waveform (wrist, elbow, above elbow, Erb's point – ADM muscle):

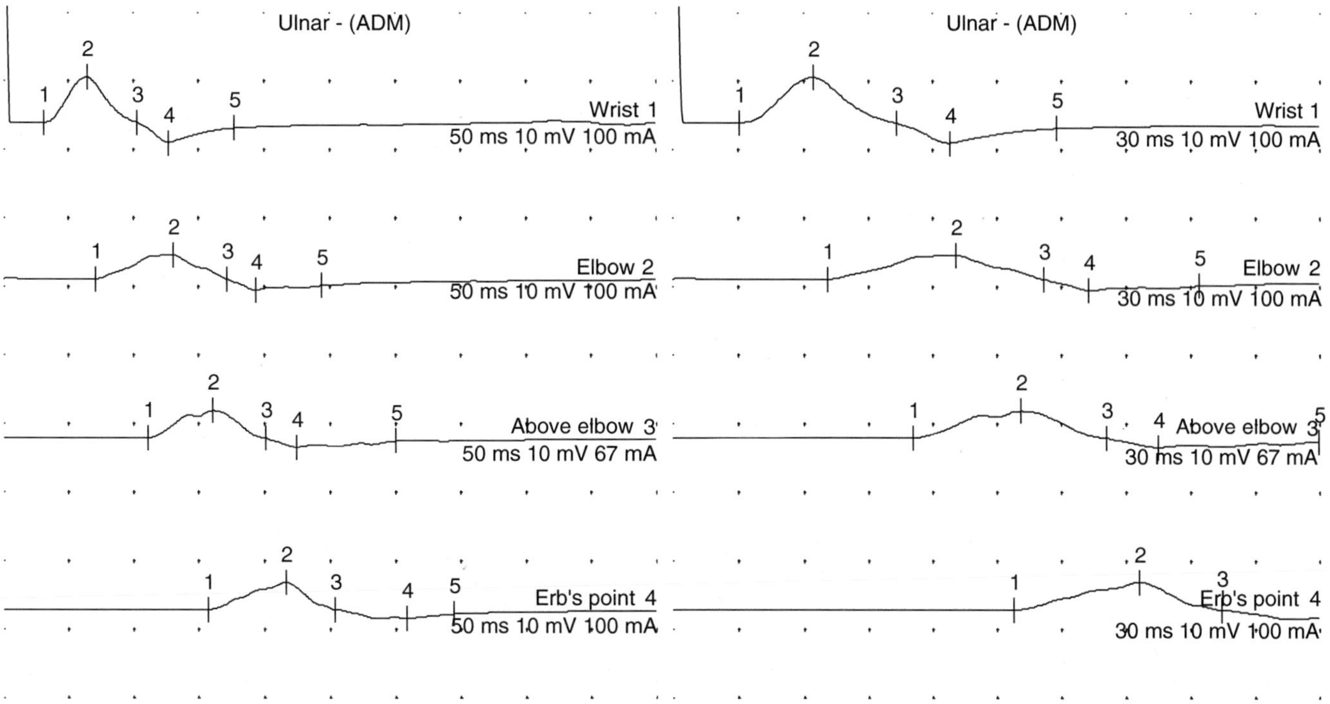

Sensitivity 10 mV/div, sweep speed 5 ms/div   Sensitivity 10 mV/div, sweep speed 3 ms/div

Onset latency (wrist – ADM): 3.0 ms; **Onset latency** (elbow – ADM): 7.05 ms; **Onset latency** (above elbow – ADM): 11.05 ms; **Onset latency** (Erb's point – ADM): 15.75 ms; **Peak latency** (wrist – ADM): 6.20 ms; **Peak latency** (elbow – ADM): 12.95 ms; **Peak latency** (above elbow – ADM): 16.10 ms; **Peak latency** (Erb's point – ADM): 15.65 ms; **Onset to peak amplitude** (wrist – ADM): 6.8 mV; **Onset to peak amplitude** (elbow – ADM): 3.6 mV; **Onset to peak amplitude** (above elbow – ADM): 4.1 mV; **Onset to peak amplitude** (Erb's point – ADM): 4.1 mV; **Peak to peak amplitude** (wrist – ADM): 9.7 mV; **Peak to peak amplitude** (elbow – ADM): 5.4 mV; **Peak to peak amplitude** (above elbow – ADM): 5.5 mV; **Peak to peak amplitude** (Erb's point – ADM): 5.6 mV; **MNCV** (wrist – elbow, 22 cm): 55.0 m/s; **MNCV** (elbow – above elbow, 10 cm): 25.0 m/s; **MNCV** (Erb's point – above elbow, 28 cm): 59.6 m/s

Fig. 4 Compound muscle action potentials (*CMAPs*) recorded at the hand from the ADM muscle using surface recording method [1] in UNE; *top to bottom*: stimulation of the wrist, elbow, above elbow, and of the Erb's point

Pathological waveform (wrist, elbow, above elbow, Erb's point – ADM muscle):

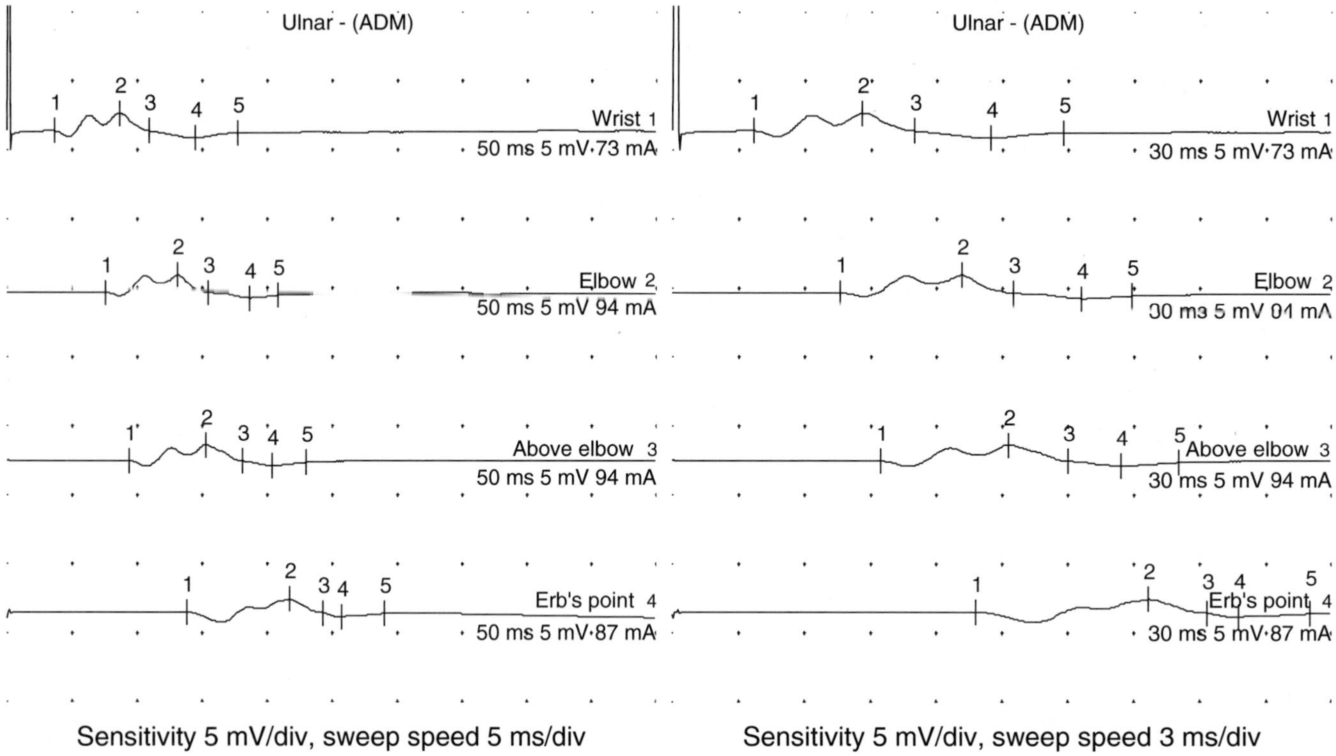

Onset latency (wrist – ADM): 3.65 ms; **Onset latency** (elbow – ADM): 7.55 ms; **Onset latency** (above elbow – ADM): 9.40 ms; **Onset latency** (Erb's point – ADM): 13.80 ms; **Peak latency** (wrist – ADM): 8.85 ms; **Peak latency** (elbow – ADM): 13.25 ms; **Peak latency** (above elbow – ADM): 15.20 ms; **Peak latency** (Erb's point – ADM): 21.35 ms; **Onset to peak amplitude** (wrist – ADM): 1.4 mV; **Onset to peak amplitude** (elbow – ADM): 1.4 mV; **Onset to peak amplitude** (above elbow – ADM): 1.2 mV; **Onset to peak amplitude** (Erb's point – ADM): 1.0 mV; **Peak to peak amplitude** (wrist – ADM): 1.9 mV; **Peak to peak amplitude** (elbow – ADM): 1.8 mV; **Peak to peak amplitude** (above elbow – ADM): 1.6 mV; **Peak to peak amplitude** (Erb's point – ADM): 1.3 mV; **MNCV** (wrist – elbow, 19 cm): 48.7 m/s; **MNCV** (elbow – above elbow, 10 cm): 54.1 m/s; **MNCV** (Erb's point – above elbow, 28 cm): 63.6 m/s

Fig. 5 Compound muscle action potentials (*CMAPs*) recorded at the hand from the ADM muscle using surface recording method [1] in hand trauma; *top to bottom*: stimulation of the wrist, elbow, above elbow, and of the Erb's point

References

1. London GW (1975) Normal ulnar nerve conduction velocity across the thoracic outlet: comparison of two measuring techniques. J Neurol Neurosurg Psychiatry 38:756–760
2. Jebsen RH (1967) Motor conduction velocities in the median and ulnar nerves. Arch Phys Med Rehabil 48:185–194
3. McQuillen MP, Gorin FJ (1969) Serial ulnar nerve conduction velocity measurements in normal subjects. J Neurol Neurosurg Psychiatry 32:144–148

Wrist, Elbow – Hand

M19

Surface Recording Technique, Study from *Abductor Digiti Minimi* (ADM) Muscle

Original Settings Sensitivity, low-frequency filter, high-frequency filter, sweep speed, duration of pulse, and the machine used were not specified.

Position This study was performed in the supine position, with the elbow and fingers slightly flexed.

Recording Kaplan [1], according to a previous standardized technique [2], placed the active electrode (A) over the belly (motor point) of the *abductor digiti minimi* (ADM) muscle (Fig. 1); the reference (R) was placed slightly distal to the first metacarpophalangeal joint, on the proximal phalanx of the digit V (little finger). The ground (G) electrode position was not specified in the paper; it was probably placed over the mid-forearm or on the palm of the hand; the figure shows the ground electrode placed on the palm.

Stimulation Following the method described by Melvin et al. [2], the ulnar nerve was stimulated at the wrist (S1) and at the elbow (S2). The cathode (−) and anode (+) stimulating electrodes were initially positioned just lateral to the tendon of the *flexor carpi ulnaris* (FCU) muscle (S1) and then were positioned in the more proximal portion of the ulnar groove (S2). We used conventional bipolar stimulator, and the anode was proximal in all stimulations (Fig. 1). No fixed distance, between stimulating and recording electrodes, was reported in the article.

A supramaximal electric stimulus was used in all determinations (it was at least 30 % greater than that which elicited a maximum motor response). Distal stimulation (S1) determined distal compound muscle action potential (CMAP), while proximal stimulation at the elbow (S2) allowed determination of the forearm motor nerve conduction velocity (MNCV). In the case of stimulation at the elbow (S2), the authors suggested 2–4 cm separation between the cathode (−) and anode (+), using a pair of standard 0.6 cm diameter electroencephalograph electrodes mounted on a plastic block for stimulating the nerve.

Measurements Distal peak latency (ms) was measured from the beginning of the shock artifact to the peak of the negative deflection of the CMAP, and it was called "terminal latency" (TL). The MNCV was calculated from the nerve using stimulating and recording sites at above the elbow and at the wrist, respectively. From this elbow–wrist conduction velocity, an expected terminal latency from the wrist to the ADM muscle was calculated and compared with the measured motor terminal latency. The difference between these values was termed the "residual latency" (RL) for the motor fibers. The temperature of the forearm was determined using a skin surface thermometer; all forearm temperatures equaled or exceeded 34 °C. Expected TL of the CMAP recorded from the ADM muscle after stimulation at the wrist (8 cm segment) and RL for CMAP recorded from the ADM muscle after stimulation at the wrist (8 cm segment) were calculated as follows (distance between cathode placed at the wrist and the active electrode placed on the OP muscle was 80 mm fixed):

$$\text{Expected DL}(\text{ms}) = \frac{\text{Distance between cathode (wrist) and active electrode (ADM), (mm)}}{\text{Nerve conduction velocity (elbow} - \text{wrist), (m/s)}}$$

$$\text{Residual Latency, RL}(\text{ms}) = \text{observed TL}(\text{ms}) - \text{expected TL}(\text{ms})$$

They measured TL and calculated RL in 20 patients (Table 1) with no history of alcoholism or diabetes as controls (age range 21–71 years, mean age 45 years), and 20 patients (Table 2) with neuropathies are selected completely at random from a much larger population of patients with neuropathies (age range 28–73 years, mean age 50 years).

G. Gentili, M. Di Napoli, *The Ulnar Nerve*, DOI 10.1007/978-3-319-43292-2_43

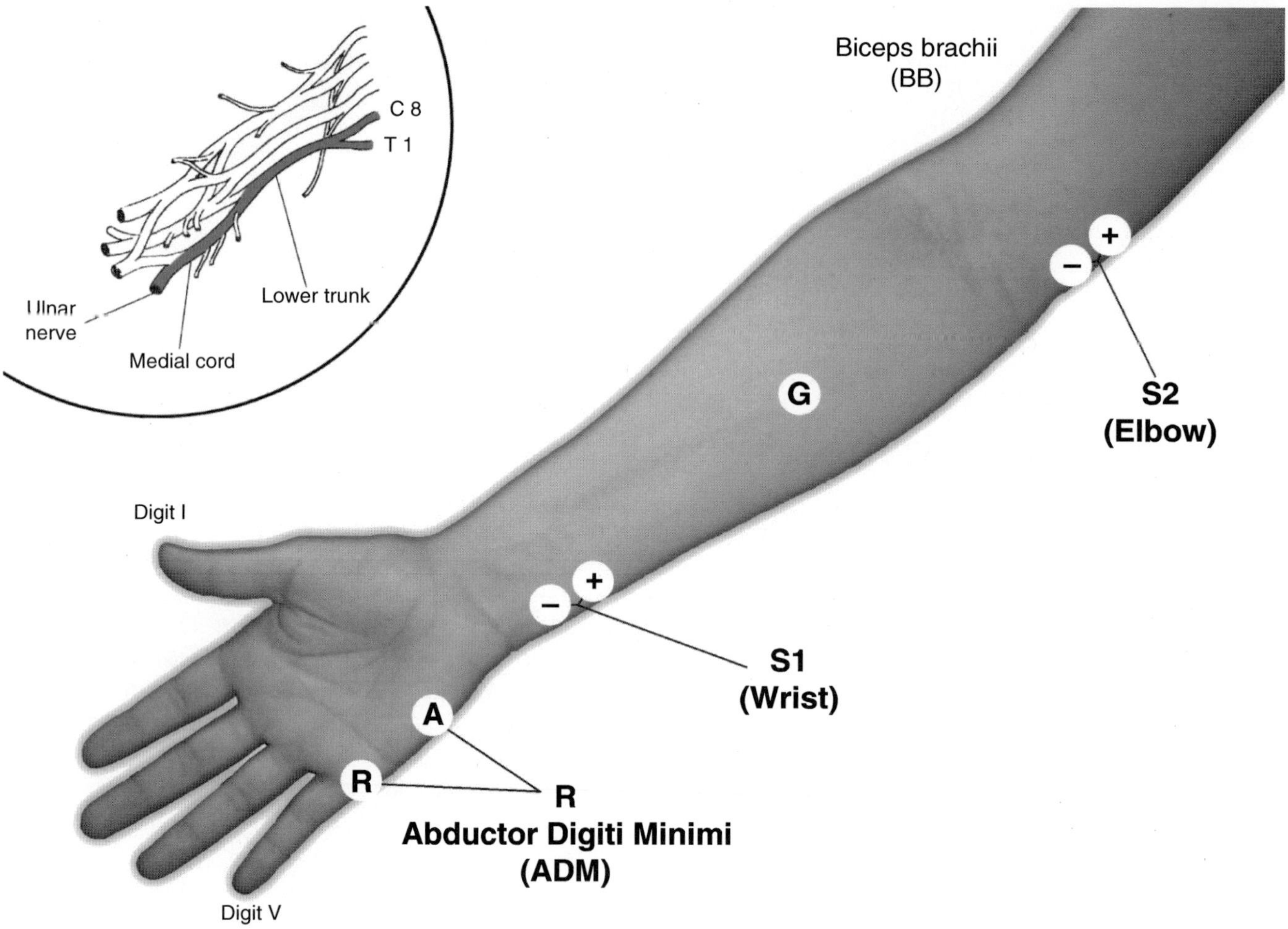

Typical waveform (wrist, elbow – ADM muscle):

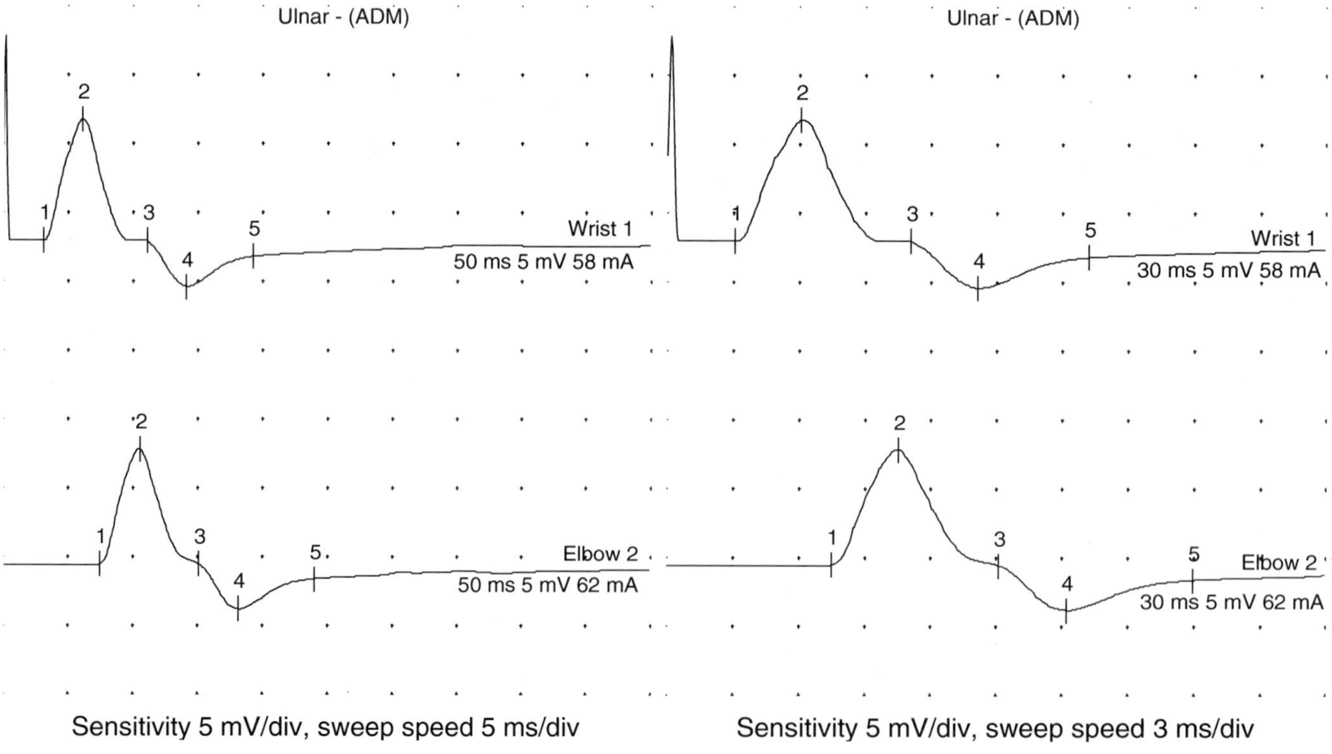

Fig. 1 Compound muscle action potentials (*CMAPs*) recorded at the hand from the ADM muscle; stimulation of the wrist (*upper trace*) and of the elbow (*lower trace*)

Table 1 Normal values [1]

	Mean ± SD	Range
Wrist–ADM, distal latency (ms)	3.3 ± 0.6	3.0–3.7
Elbow–wrist, residual latency (ms)	1.7 ± 0.5	1.0–2.5
Elbow–wrist, MNCV (m/s)	60 ± 6	54–65

Table 2 Pathological values [1]

	Mean ± SD	Range
Wrist–ADM, distal latency (ms)	4.4 ± 1.1	4.0–6.5
Elbow–wrist, residual latency(ms)	3.5 ± 1.7	1.5–4.2
Elbow–wrist, MNCV (m/s)	51 ± 19	39–61

Comment

For Kaplan [1], the values for the residual latency measurements in the motor conduction observed in the control population were virtually identical with the residual latency measurements of the sensory conduction in the same nerve and also from nerve to nerve. The result of the study indicated that the residual latency measurements were as effective and as accurate as the terminal latency measurements in determining the presence of a neuropathy distal to the wrist. For the author [1], the residual latency techniques seemed to be an accurate way of measuring the state of the delicately tapering nerve endings distal to the wrist. Pathological waveforms and values in a patient with Charcot–Marie–Tooth (CMT1A) disease are reported (Fig. 2).

Pathological waveform (wrist, elbow – ADM muscle):

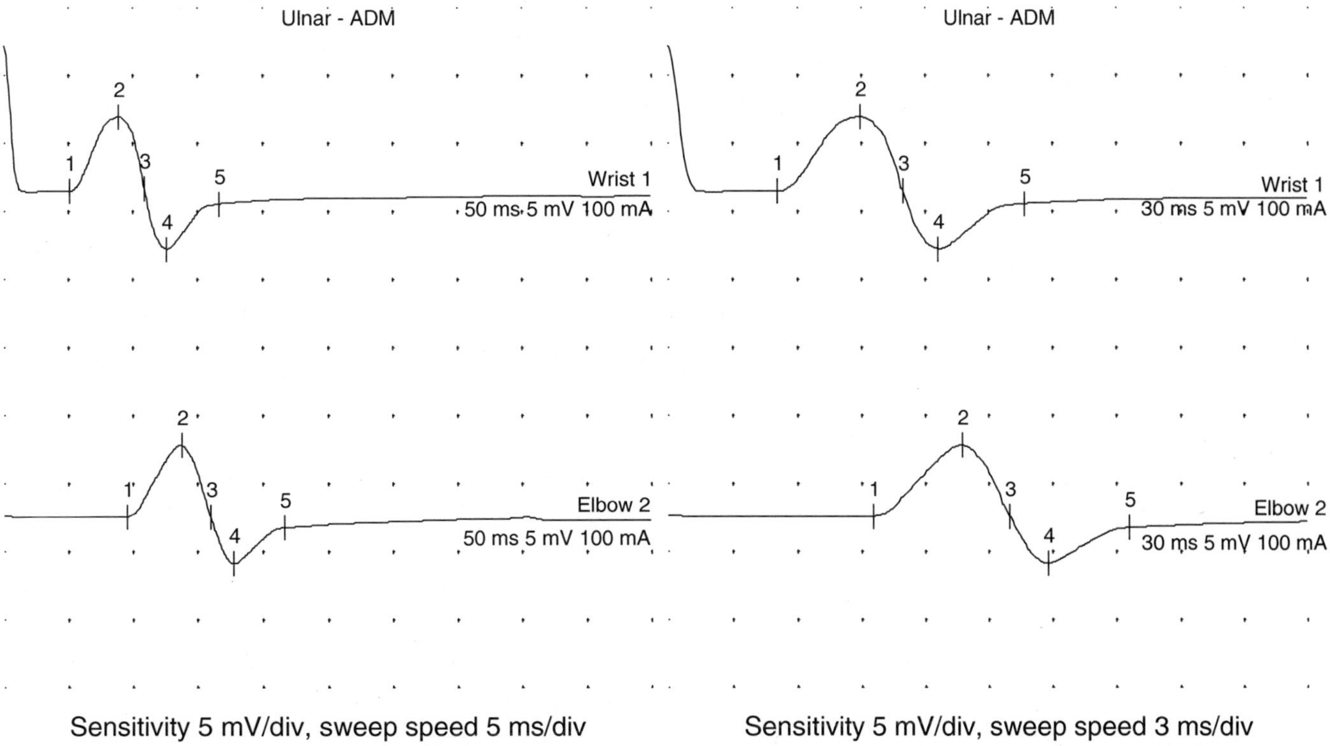

Onset latency (wrist – ADM): 5.05 ms; **Onset latency** (elbow – ADM): 9.55 ms; **Peak latency** (wrist – ADM): 8.95 ms; **Peak latency** (elbow – ADM): 13.70 ms; **Onset to peak amplitude** (wrist – ADM): 5.5 μV; **Onset to peak amplitude** (elbow – ADM): 5.2 μV; **Peak to peak amplitude** (wrist – ADM): 9.8 μV; **Peak to peak amplitude** (elbow – ADM): 8.8 μV; **MNCV** (elbow –wrist): 37.8 m/s

Fig. 2 Compound muscle action potentials (*CMAPs*) recorded at the hand from the ADM muscle in a patient with CMT1A; stimulation of the wrist (*upper trace*) and of the elbow (*lower trace*)

References

1. Kaplan PE (1976) Sensory and motor residual latency measurements in healthy patients and patients with neuropathy – Part 1. J Neurol Neurosurg Psychiatry 39:338–340
2. Melvin JL, Harris DH, Johnson EW (1966) Sensory and motor conduction velocities in the ulnar and median nerves. Arch Phys Med Rehabil 47:511–519

Axilla, Above Elbow, Below Elbow, Wrist – Hand

M20

Needle Recording Technique, Study from *Abductor Digiti Minimi* (ADM) Muscle

Original Settings The machine used was a TECA model TE-4 electromyograph; sweep speed, sensitivity, low-frequency filter, high-frequency filter, and duration of pulse were not specified.

Position This study was performed with the subject in the supine position with the shoulder abducted at 90° and the elbow flexed at 35°.

Recording The recording electrode was a coaxial needle electrode inserted in the midportion of the *abductor digiti minimi* (ADM) muscle (Fig. 1), on the motor point [1]. The ground (G) electrode position was not specified in the text; the figure shows the ground electrode placed on the palm.

Needle Insertion For the ADM muscle, with the hand in a neutral position to full supination, the recording needle electrode was inserted at the midpoint between the fifth metacarpophalangeal joint and the pisiform bone.

Stimulation Conduction in the forearm, across the elbow and upper arm segments of the ulnar nerve, was determined by stimulating the ulnar nerve at four standard points as follows (Figs. 1, 2, and 3): in the axilla (S1), above the elbow (S2) (5–7 cm proximal to the medial epicondyle), below the elbow (S3) (5–7 cm distal to the medial epicondyle, as distally as possible), and at the wrist (S4) – as distal as possible. The anode was proximal in all stimulations. To minimize errors in conduction velocity, the distance between two successive points of stimulation along the nerve, especially across the elbow, was kept more than 10 cm if possible. Supramaximal stimulus at a rate of one per second was used for ulnar nerve stimulation (0.01–0.2 ms duration).

Measurements Distal latency (ms) of the compound muscle action potential (CMAP) was measured from the onset of the stimulus artifact to the onset initial deflection of CMAP. Therefore, the conduction velocity in the most rapidly conducting fibers was measured. By dividing the distance by time, the motor nerve conduction velocity (MNCV) along the nerve segment can be determined and measured in meter per second (m/s); it was calculated for the upper arm (axilla–above elbow), across-elbow (above elbow–below elbow), and forearm (below elbow–wrist) segments. Distances were measured with a metal tape. The amplitude and temporal dispersion of the CMAP were not measured. All recordings were done at room temperature (skin and temperature data were not reported). Normal conduction values (Table 1) were obtained from 61 healthy subjects (9 men and 52 women, age range 19–69 years). With the exception of one, all the healthy subjects were less than 50 years of age.

G. Gentili, M. Di Napoli, *The Ulnar Nerve*, DOI 10.1007/978-3-319-43292-2_44

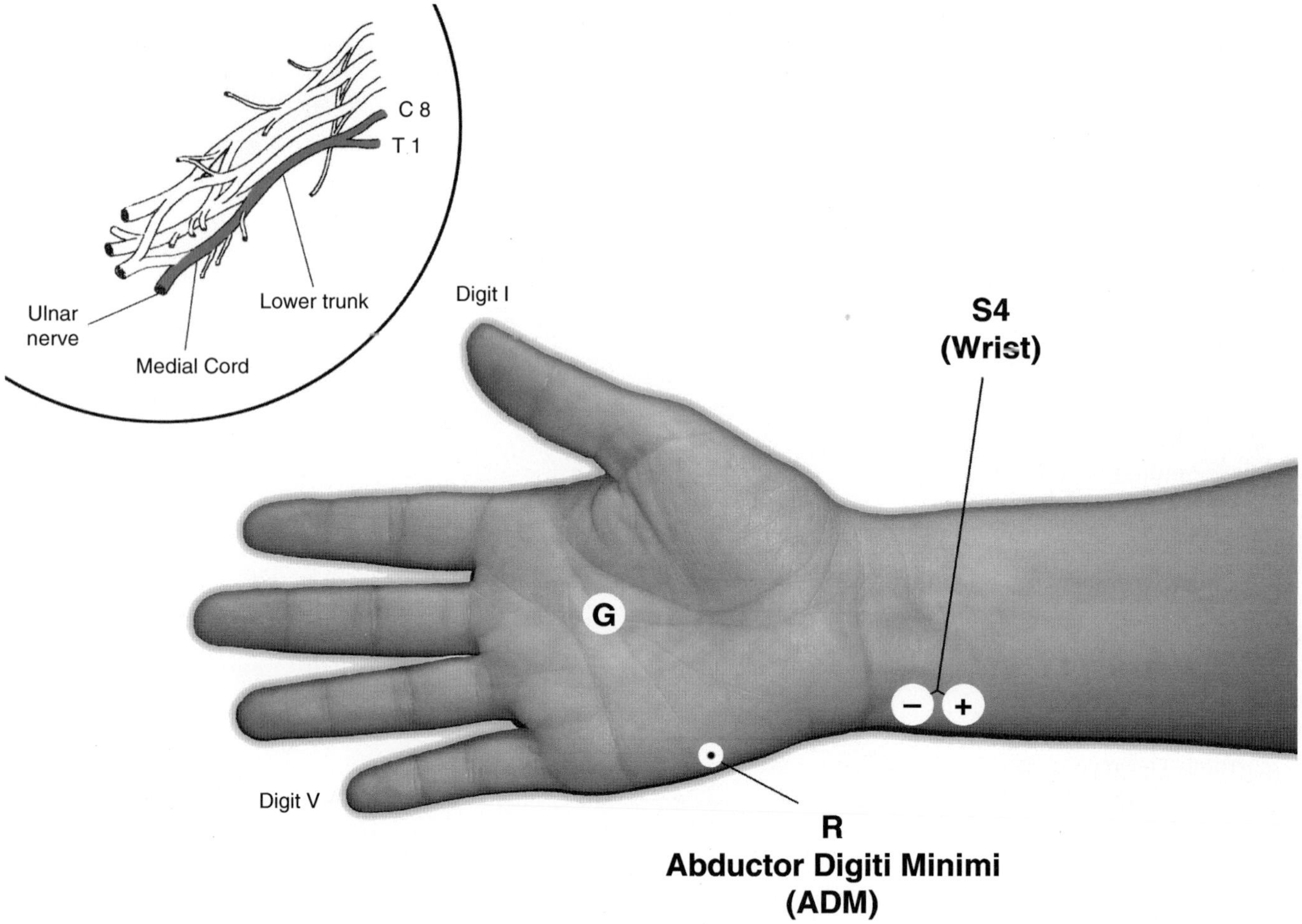

Typical waveform (axilla, above elbow, below elbow, wrist– ADM muscle):

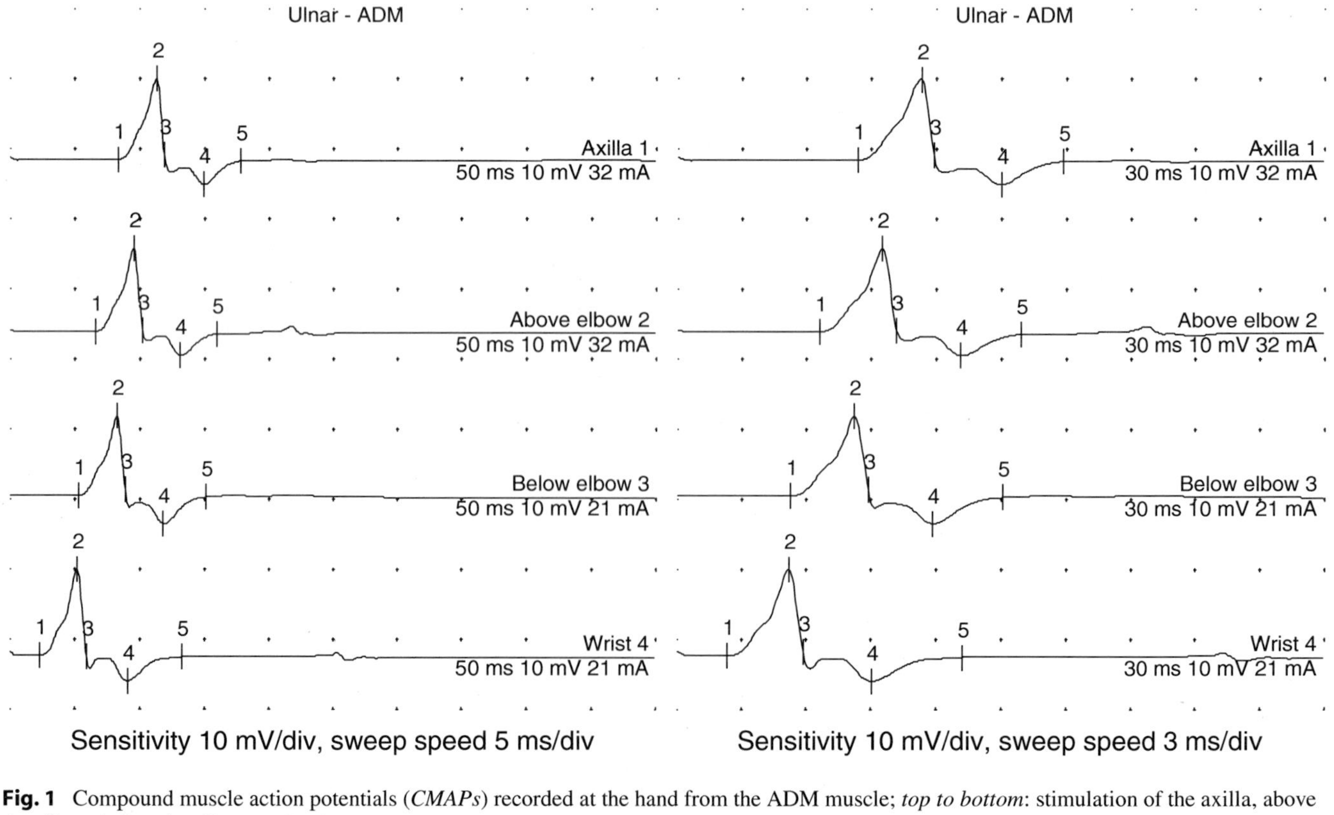

Fig. 1 Compound muscle action potentials (*CMAPs*) recorded at the hand from the ADM muscle; *top to bottom*: stimulation of the axilla, above the elbow, below the elbow, and wrist

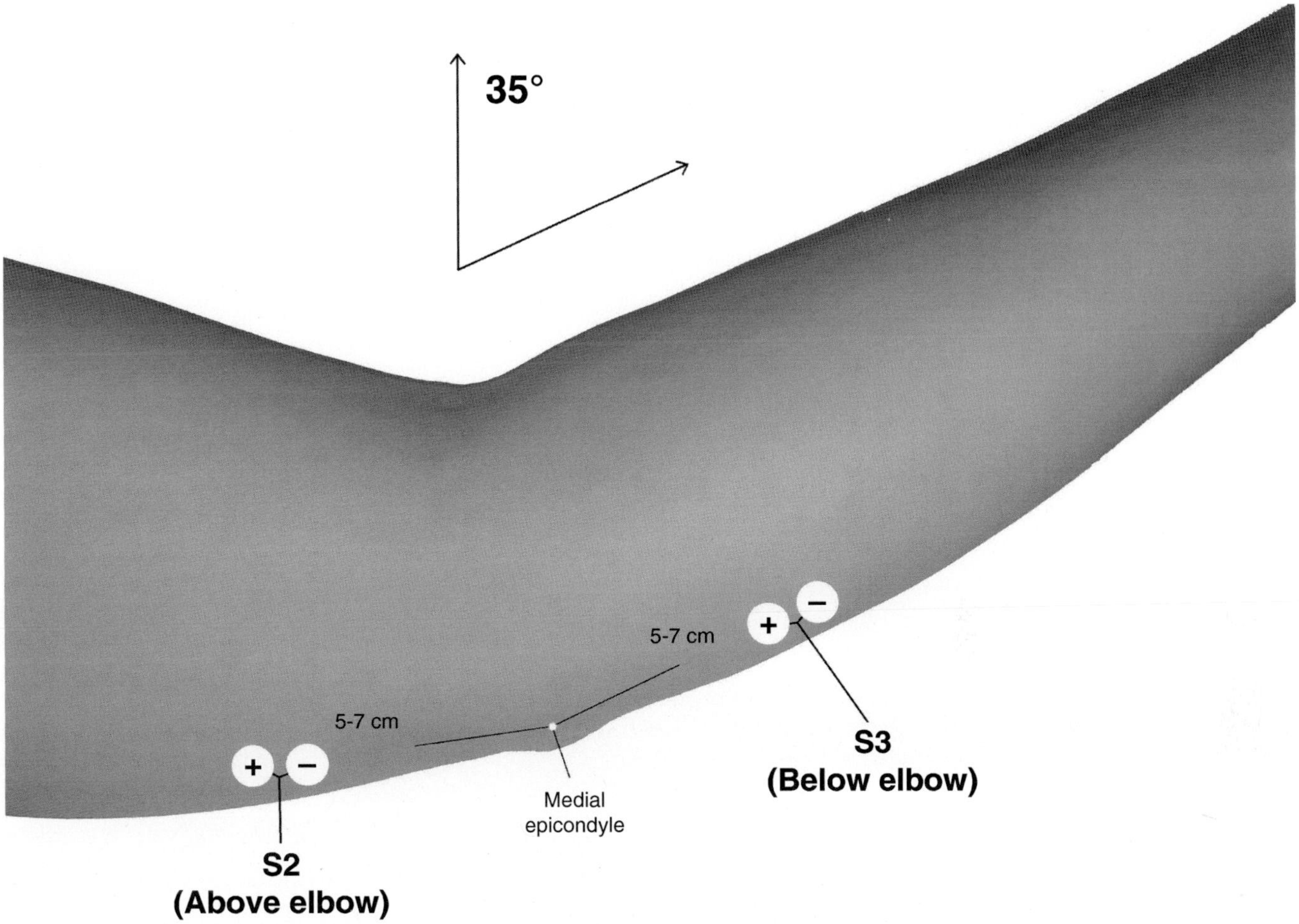

Fig. 2 Above- and below-elbow stimulation points (S2, above elbow; S3, below elbow)

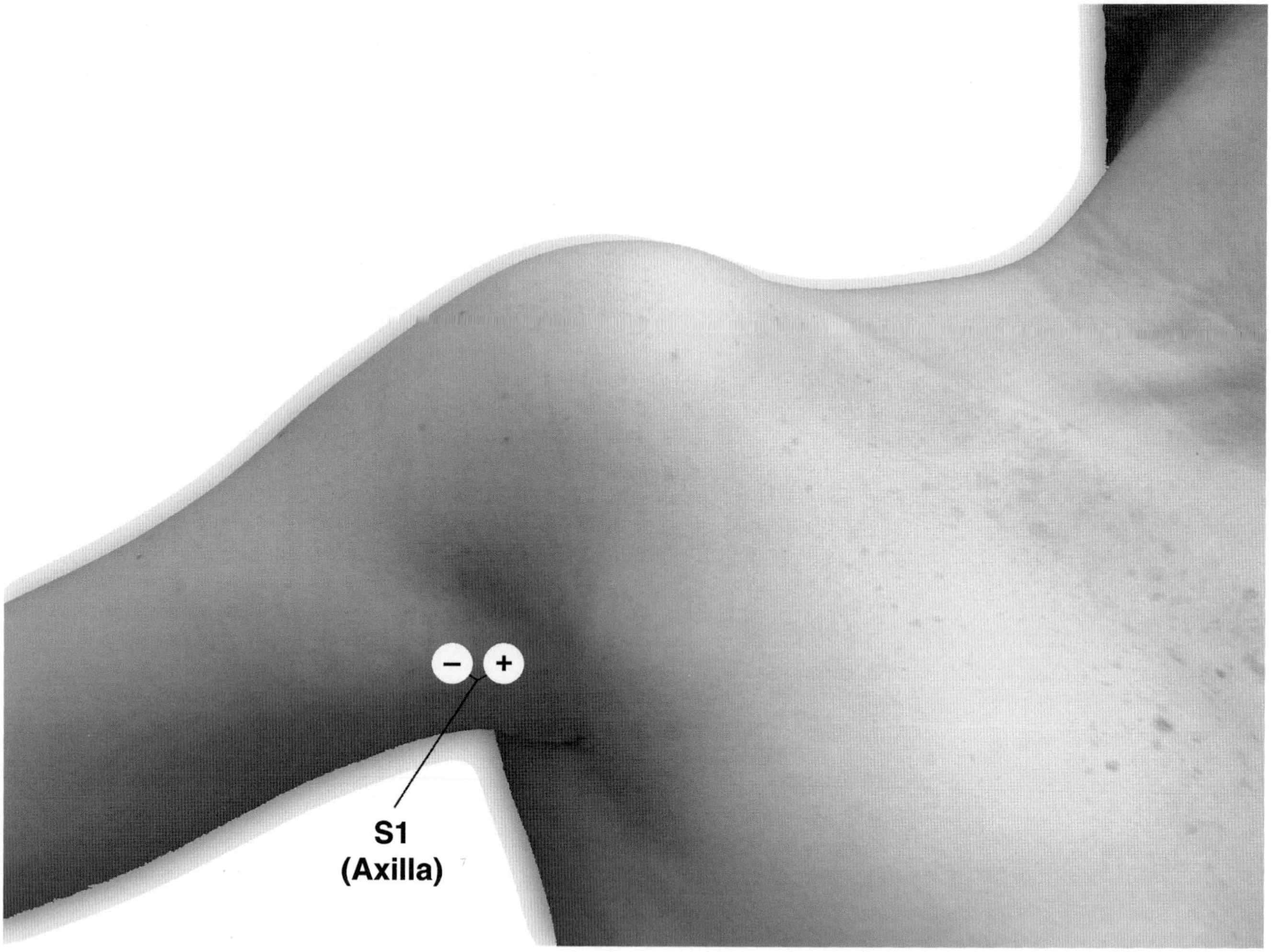

Fig. 3 Proximal stimulation (S1 – axilla)

Table 1 Normal values [1]

	Mean ± SD	Range
Axilla–above elbow, MNCV (m/s)	53.6 ± 5.9	46–76.6
Above elbow–below elbow, MNCV (m/s)	49.6 ± 3.4	45–60
Below elbow–wrist, MNCV (m/s)	51.7 ± 4.9	45–66

Comment

For Bhala [1], the determinations of ulnar nerve conduction velocity must be done with the elbow in 35° flexion because the true length of the ulnar nerve across the elbow segment is more accurately measured with the elbow flexed, and false slowing across the ulnar groove may be reported with determinations made with the elbow in full extension (the skin measurement is different from the actual length of the nerve segment when the elbow is extended). The author examined 78 patients with suspected ulnar nerve entrapment at the elbow (39 men and 39 women). The majority of the patients (75.6 %) were less than 50 years of age. Pathological waveforms and values in a diabetic patient with ulnar neuropathy at the elbow (UNE) are reported (Fig. 4).

Pathological waveform (axilla, above elbow, below elbow, wrist – ADM muscle):

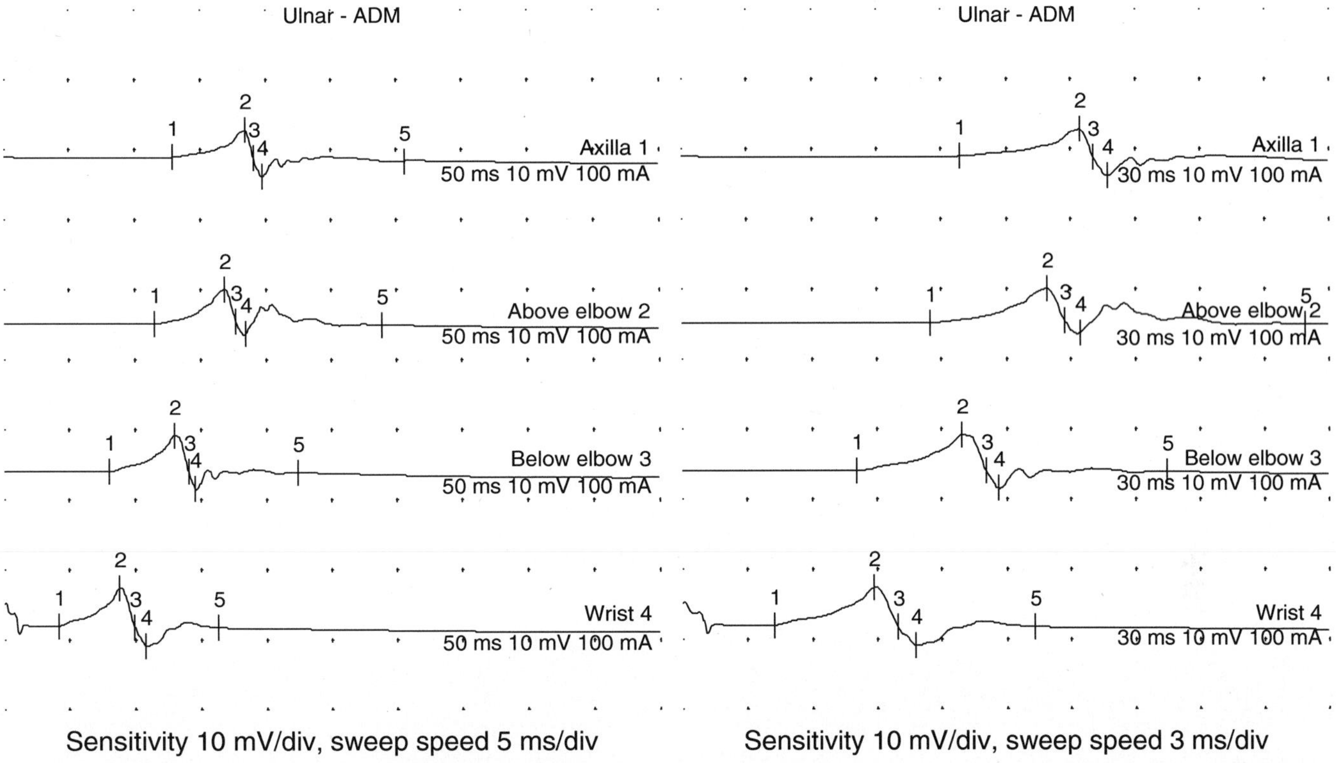

Onset latency (axilla – ADM): 13.30 ms; **Onset latency** (above elbow – ADM): 11.40 ms; **Onset latency** (below elbow – ADM): 8.05 ms; **Onset latency** (wrist – ADM): 4.10 ms; **Peak latency** (axilla – ADM): 18.30 ms; **Peak latency** (above elbow – ADM): 16.85 ms; **Peak latency** (below elbow – ADM): 12.95 ms; **Peak latency** (wrist – ADM): 8.95 ms; **Onset to peak amplitude** (axilla – ADM): 3.8 mV; **Onset to peak amplitude** (above elbow – ADM): 5.0 mV; **Onset to peak amplitude** (below elbow – ADM): 5.2 mV; **Onset to peak amplitude** (wrist – ADM): 5.5 mV; **Peak to peak amplitude** (axilla – ADM): 6.8 mV; **Peak to peak amplitude** (above elbow – ADM): 6.4 mV; **Peak to peak amplitude** (below elbow – ADM): 7.8 mV; **Peak to peak amplitude** (wrist – ADM): 8.5 mV; **MNCV** (axilla – above elbow, 11 cm): 57.9 m/s; **MNCV** (above elbow – below elbow, 10 cm): 29.9 m/s; **MNCV** (below elbow – wrist, 18 cm): 45.6 m/s

Fig. 4 Compound muscle action potentials (*CMAPs*) recorded at the hand from the ADM muscle in a diabetic patient with UNE; *top to bottom*: stimulation of the axilla, above elbow, below elbow, and of the wrist

Reference

1. Bhala RP (1976) Electrodiagnosis of ulnar nerve lesions at the elbow. Arch Phys Med Rehabil 57:206–212

Wrist – Hand; Elbow – Hand

M21

Surface Recording Technique, Study from *Abductor Digiti Minimi* (ADM) Muscle

Original Settings Sensitivity, low-frequency filter, high-frequency filter, sweep speed, duration of pulse, and the machine used were not specified.

Position This study was performed with the subject lying on a bed with the arm and hand immobilized in a supinated position.

Recording The active electrode (A) was placed over the belly (motor point) of the *abductor digiti minimi* (ADM) muscle (Figs. 1 and 2); the reference (R) was placed distally near its tendon of insertion [1]. The ground (G) electrode position was not mentioned in the text; the figure shows ground positioned on the forearm.

Stimulation Using surface electrodes, the ulnar nerve was stimulated by electric shocks at the wrist and at the elbow; same stimulation points were used for conventional motor conduction studies and F-wave studies [1]. For F-wave studies, the authors used a supramaximal stimulus intensity (increased by 20 %). The first stimulation was delivered at the wrist; the stimulating electrodes were positioned just lateral to the tendon of the *flexor carpi ulnaris* (FCU) muscle (Fig. 1).

The second stimulation was delivered at the elbow (Fig. 2), in the most proximal portion of the ulnar groove. In all stimulations the anode was proximal, and the cathode was distal for conventional motor conduction studies; the cathode was proximal, and the anode was distal for F-wave studies. No fixed distances were reported in the article.

Measurements Distal latency (ms) of the compound muscle action potential (CMAP) was measured from the stimulus artifact to the onset of the negative deflection of CMAP. Distances between the stimulating points were measured along the course of the nerve; distal motor nerve conduction time in this study was the difference between the proximal and distal motor nerve latencies, which, when divided by the distance between the two stimulating points, gives the motor nerve conduction velocity of the elbow to wrist segment of the arm. The latency of the F-wave response (ms) was taken as the interval between the stimulus artifact and the first deflection of the evoked response. This was found to vary slightly in given individuals so that the shortest of five responses (F_{min}) was chosen. In the majority of cases, no difficulty was encountered in eliciting an F-wave; however, when necessary the response could be augmented by a slight voluntary contraction of the particular muscle from which it was being recorded. It was assumed that the central delay of the F-wave response was 1 ms. Skin temperature was not controlled; all studies were carried out in an air-conditioned room, in which the temperature was maintained at between 23 and 25 °C. F-wave conduction velocity was computed from the following formula:

$$\text{F}-\text{wave velocity}(\text{m/s}) = \frac{\begin{pmatrix}\text{Distance from site}\\ \text{of stimulation} - \text{C7 spine}\end{pmatrix} \times 10\ (\text{cm})}{(\text{F conduction time})\ (\text{ms})}$$

$$\text{F conduction time}(\text{ms}) = \frac{(\text{F latency} - \text{M latency}) - 1\ (\text{ms})}{2}$$

The authors [1] studied 60 ulnar nerves from 60 normal (Table 1) subjects (age range 11–74 years, mean age 41.5 years) and 18 (Table 2) patients (age range 26–65 years, mean age 51.7 years) with cubital tunnel syndrome (ulnar neuropathy at the elbow–UNE).

G. Gentili, M. Di Napoli, *The Ulnar Nerve*, DOI 10.1007/978-3-319-43292-2_45

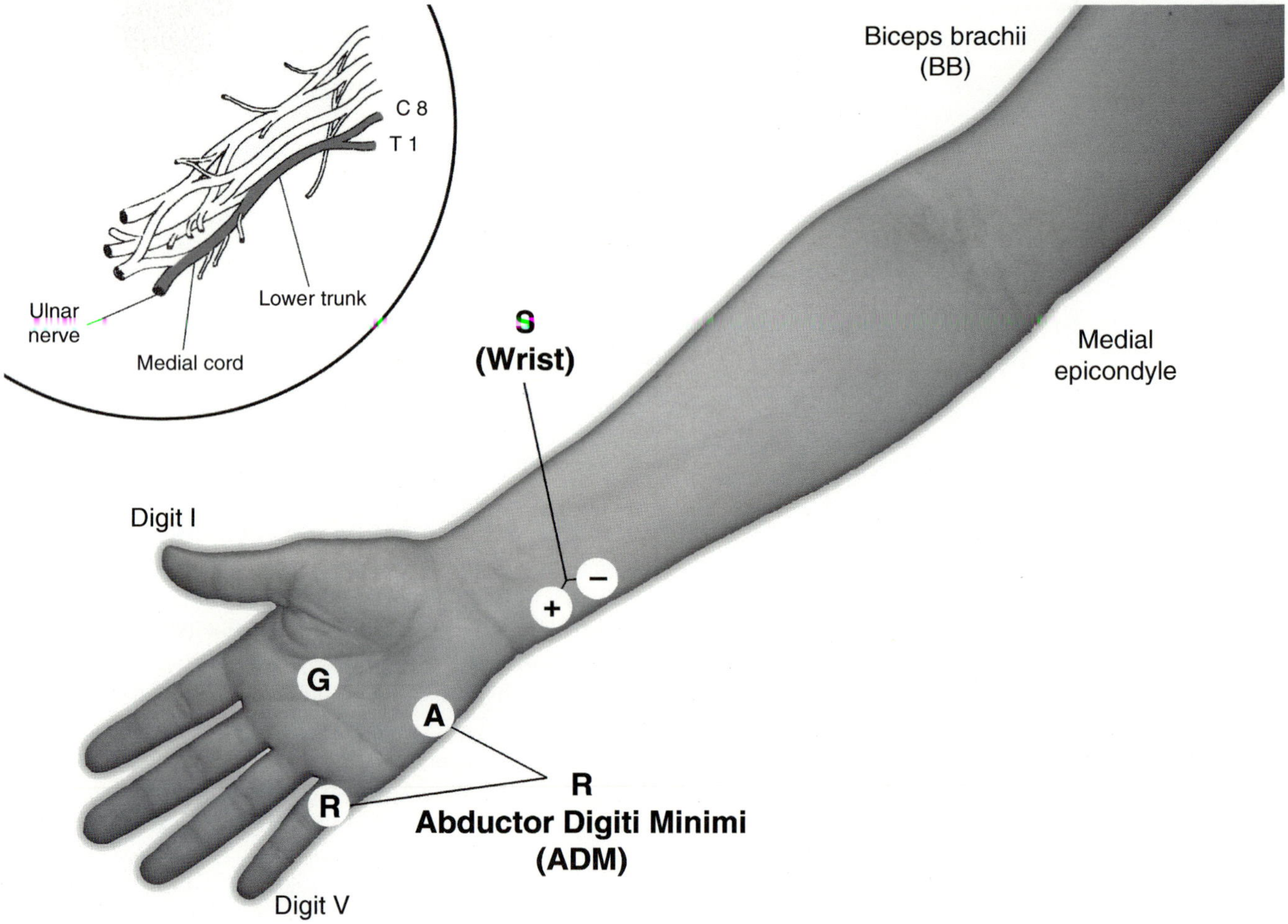

Typical waveform (wrist – ADM muscle):

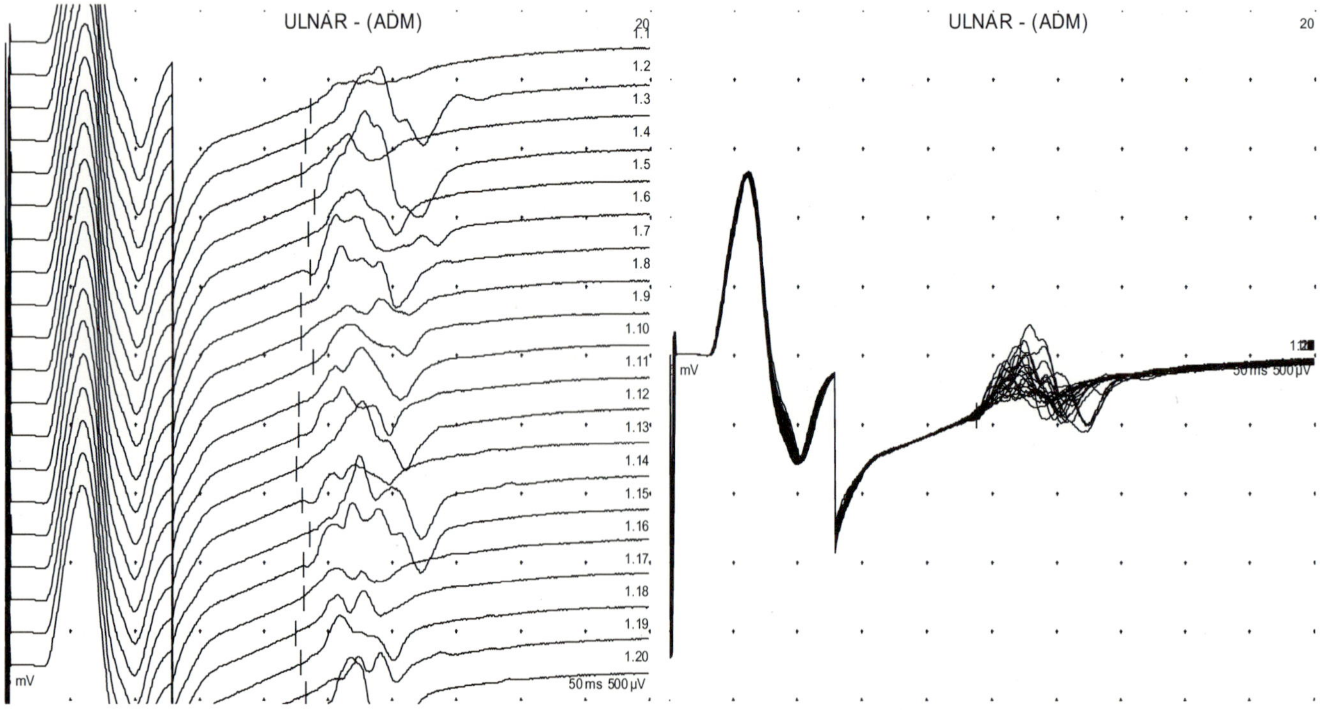

Sensitivity 500 μV/div (CMAP 5 mV/div), sweep speed 5 ms/div

Fig. 1 F-waves recorded at the hand from the ADM muscle, stimulation of the wrist; raster mode, 20 stimuli (*left*); superimposed mode, 20 stimuli (*right*)

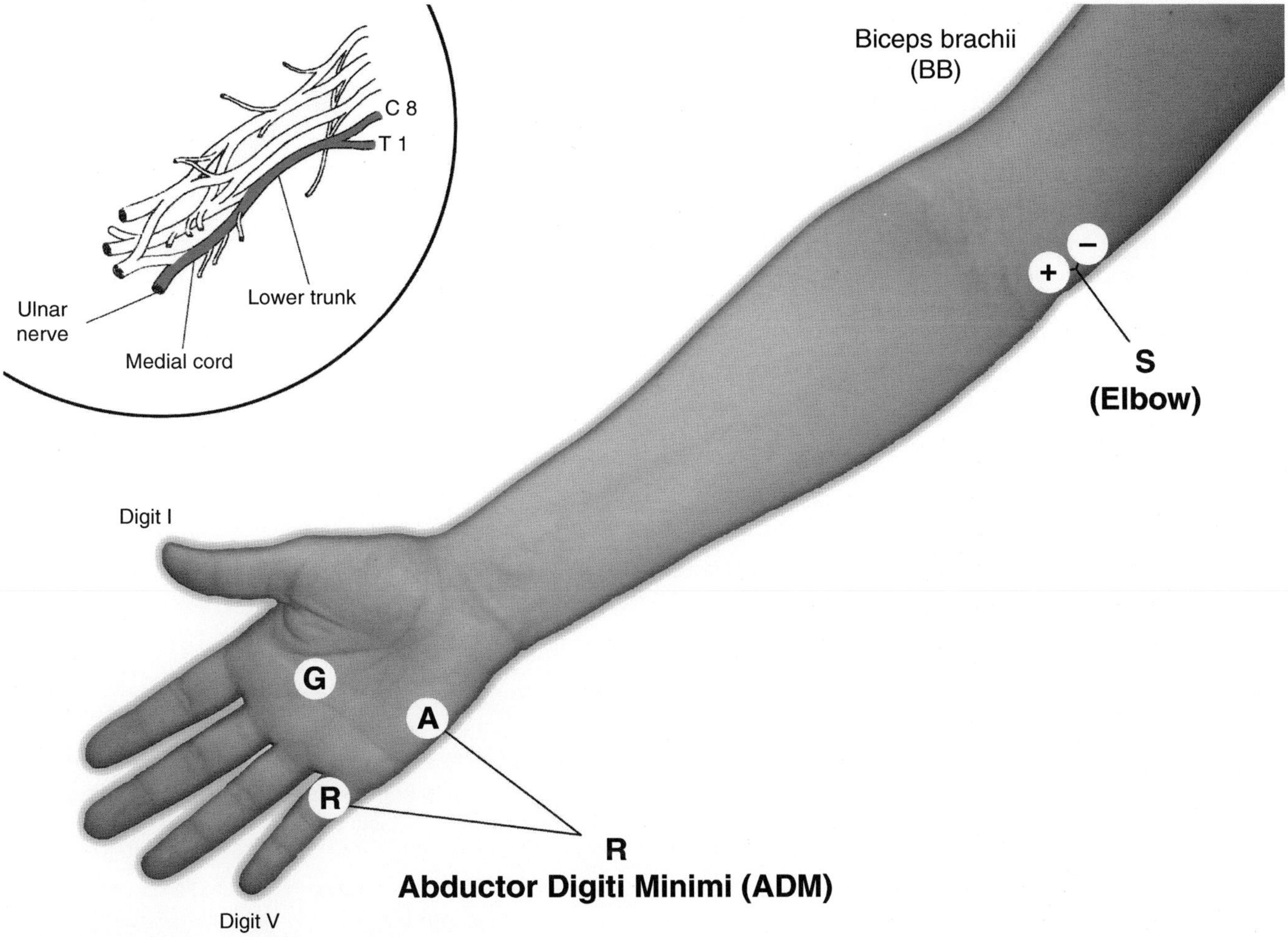

Typical waveform (elbow – ADM muscle):

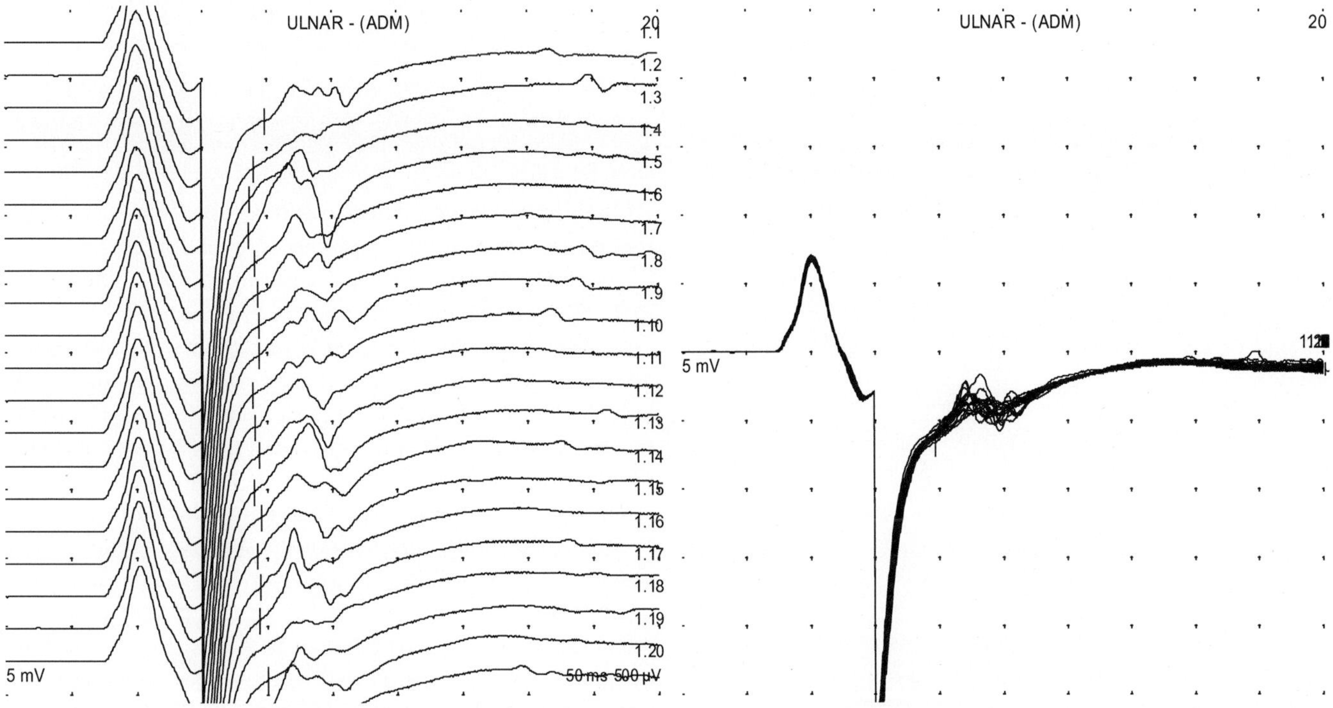

Sensitivity 500 µV/div (CMAP 5 mV/div), sweep speed 5 ms/div

Fig. 2 F-waves recorded at the hand from the ADM muscle, stimulation of the elbow; raster mode, 20 stimuli (*left*); superimposed mode, 20 stimuli (*right*)

Table 1 Normal values [1]

	Mean ± SD
Wrist–ADM, M-wave latency (ms)	2.73 ± 0.52
Elbow–ADM, M-wave latency (ms)	7.34 ± 0.63
Elbow–wrist, M-wave latency difference (ms)	57.2 ± 4.5
Wrist–ADM, F-wave latency (ms)	27.0 ± 2.0
Elbow–ADM, F-wave latency (ms)	23.0 ± 1.6
Wrist–spinal cord, F-wave velocity (m/s)	61.7 ± 3.5
Elbow–spinal cord, F-wave velocity (m/s)	62.3 ± 5.0
Elbow–spinal cord, conduction time (ms)	7.34 ± 0.69

Table 2 Pathological values [1]

	Mean ± SD
Wrist–ADM, M-wave latency (ms)	3.72 ± 0.71
Elbow–ADM, M-wave latency (ms)	10.26 ± 1.82
Elbow–wrist, M-wave latency difference (ms)	42.2 ± 8.9
Wrist–ADM, F-wave latency (ms)	31.7 ± 4.3
Elbow–ADM, F-wave latency (ms)	26.4 ± 2.7
Wrist–spinal cord, F-wave velocity (m/s)	55.2 ± 6.6
Elbow–spinal cord, F-wave velocity (m/s)	59.25 ± 10.15
Elbow–spinal cord, conduction time (ms)	7.59 ± 0.70

Comment

For Eisen et al. [1] the F-response can be evoked easily by stimulation of the ulnar nerve. An absent F-response, or a slowed conduction between the spinal cord and elbow in the presence of a normal conduction between the elbow and wrist, is indicative of a proximal lesion. Comparison of the F-response and the M-response latencies at a given stimulation point along the nerve is a sensitive indicator of whether the lesion is proximal, distal, or at both sites. For the authors [1], a conduction time between the cord and elbow greater than 9.0 ms can be considered abnormal; they found a mean F-wave latency obtained by stimulation at the elbow, as well as the respective mean F conduction velocity, abnormal when compared with the controls (66.7 % of patients had a significantly prolonged proximal motor latency, greater than 8.92 ms, and a slowed motor conduction velocity between the elbow and wrist, lower than 46 m/s). The mean F-wave latency obtained by stimulation at the wrist was prolonged in 33.3 % of patients. The F conduction time and velocity between the spinal cord and elbow was abnormal in 11 % of patients, compared with 52.9 % who had abnormally slowed motor conduction velocities between the elbow and wrist. F-waves could not be obtained by proximal (elbow) stimulation in two patients with cubital tunnel syndrome, and, using F-wave studies, the authors suggested that the absence of an F-response is also probably indicative of an additional pathology at a more proximal site. Eisen et al. [1] reported that F-waves were obtained easily by distal stimulation at the wrist (Fig. 1), and F-wave study by distal stimulation is routinely performed during an EMG session in order to investigate integrity and functionality of upper brachial structures such as trunks, cords, and roots.

In order to create a normative database for median nerve F-wave responses for a large subject population, Buschbacher [2] studied 193 asymptomatic subjects (56 % women, 44 % men, 80 white race, 19 black race, other races 1, age range 19–79 years, women mean age 42 ± 15 years, men mean age 48 ± 14 years, women mean height 163 ± 7 cm, men mean height 175 ± 8 cm, women mean weight 181 ± 31 lb, men mean weight 155 ± 33 lb) without risk factors for neuropathy (Tables 3 and 4) by stimulation at the wrist – 8 cm fixed distance between stimulation and recording sites (Fig. 3).

Buschbacher [2] applied ten consecutive supramaximal stimuli (25 % over maximal) to the wrist of each arm at a rate of approximately 1/s; the shortest of ten stimuli (F_{min}), the mean latency (F_{mean}), the range of latencies (F_{range}), and limit of normal were measured. All studies were performed at a temperature maintained at 32 °C or more.

Table 3 Normal values [2]

F-wave latency (F_{min}) (ms), age range 19–49 years			
Height in cm (in)	Mean ± SD	Range	Limit of normal
<160 (5' 3")	23.5 ± 1.3	20.3–26.3	26.1
≥160 and <180 (5' 3"–5' 11")	26.2 ± 2.0	22.4–31.1	30.2
≥180 (5' 11")	29.2 ± 1.8	26.2–32.9	32.8
F-wave latency (F_{min}) (ms), age range 50–79 years			
Height in cm (in)	Mean ± SD	Range	Limit of normal
<160 (5' 3")	25.0 ± 1.9	22.9–28.4	28.8
≥160 and <180 (5' 3"–5' 11")	28.1 ± 1.4	26.3–30.8	30.9
≥180 (5' 11")	30.4 ± 1.7	28.3–32.4	33.8
F-wave latency (F_{min}) (ms), age range 19–79 years			
	Mean ± SD	Range	Limit of normal
All subjects	26.5 ± 2.5	20.3–32.9	31.5

Table 4 Normal values [2]

F-wave latency (F_{mean}) (ms), age range 19–49 years			
Height in cm (in)	Mean ± SD	Range	Limit of normal
<160 (5' 3")	24.3 ± 0.6	22.8–25.3	25.5
≥160 and <180 (5' 3"–5' 11")	27.2 ± 2.0	23.9–31.7	31.2
≥180 (5' 11")	30.3 ± 1.9	27.1–33.3	34.1
F-wave latency (F_{mean}) (ms), age range 50–79 years			
Height in cm (in)	Mean ± SD	Range	Limit of normal
<160 (5' 3")	25.6 ± 1.9	21.9–28.7	29.4
≥160 and <180 (5' 3"–5' 11")	28.8 ± 1.6	26.2–33.0	32.0
≥180 (5' 11")	31.3 ± 1.9	28.9–34.5	35.1
F-wave latency (F_{mean}) (ms), age range 19–79 years			
	Mean ± SD	Range	Limit of normal
All subjects	27.7 ± 2.5	21.9–34.5	32.7

Comment

Buschbacher [2], for the purpose of his study, combined the categories for men and women because the gender effect was small. Side-to-side differences, which are particularly sensitive at detecting unilateral disease states, for the F_{min} and F_{mean} were 2.4 ms and 2.1 ms, respectively.

Nobrega et al. [3] performed F-wave studies on 100 healthy volunteers – right upper limb (57 men and 43 women, age range 16–53 years, mean age 32 ± 9 years, height range 170.5 ± 9.9 cm, upper limb length 74.2 ± 5.2 cm). All studies were conducted in a warm room; skin temperature was maintained at 32 °C. The authors used a NIHON-KOHDEN MEB 9102 with filter setting at 20 Hz–3 kHz. F-wave studies consisted of applying 32 supramaximal stimulations at 1 Hz to the ulnar nerve at the wrist and recording F-waves from the ADM muscle by disk electrodes placed on the belly of the ADM muscle (active electrode) and its tendon insertion (reference electrode). Several parameters were studied, such as persistence (i.e., the number of responses elicited by 32 supramaximal stimuli), peak-to-peak amplitude, duration (from the takeoff to the return to the baseline), minimum latency, mean latency, maximum latency, and chronodispersion (i.e., the latency difference between the minimum and maximum F-waves). The authors measured the upper limb length with a tape from the stimulus point (wrist) to the C7 spinous process via the axilla and midclavicular point, which approximates the nerve length under consideration [4]. The F-wave conduction velocity (FWCV) was measured in meter per second (m/s), and it was calculated using Kimura's formula [4]:

$$FWCV\,(m/s) = \frac{LL \times 2\ (cm)}{F - M - 1\ (ms)}$$

where LL represents the estimate upper limb length measured following method by Kimura [4], as described above, and F and M represent the latency of the F-wave and

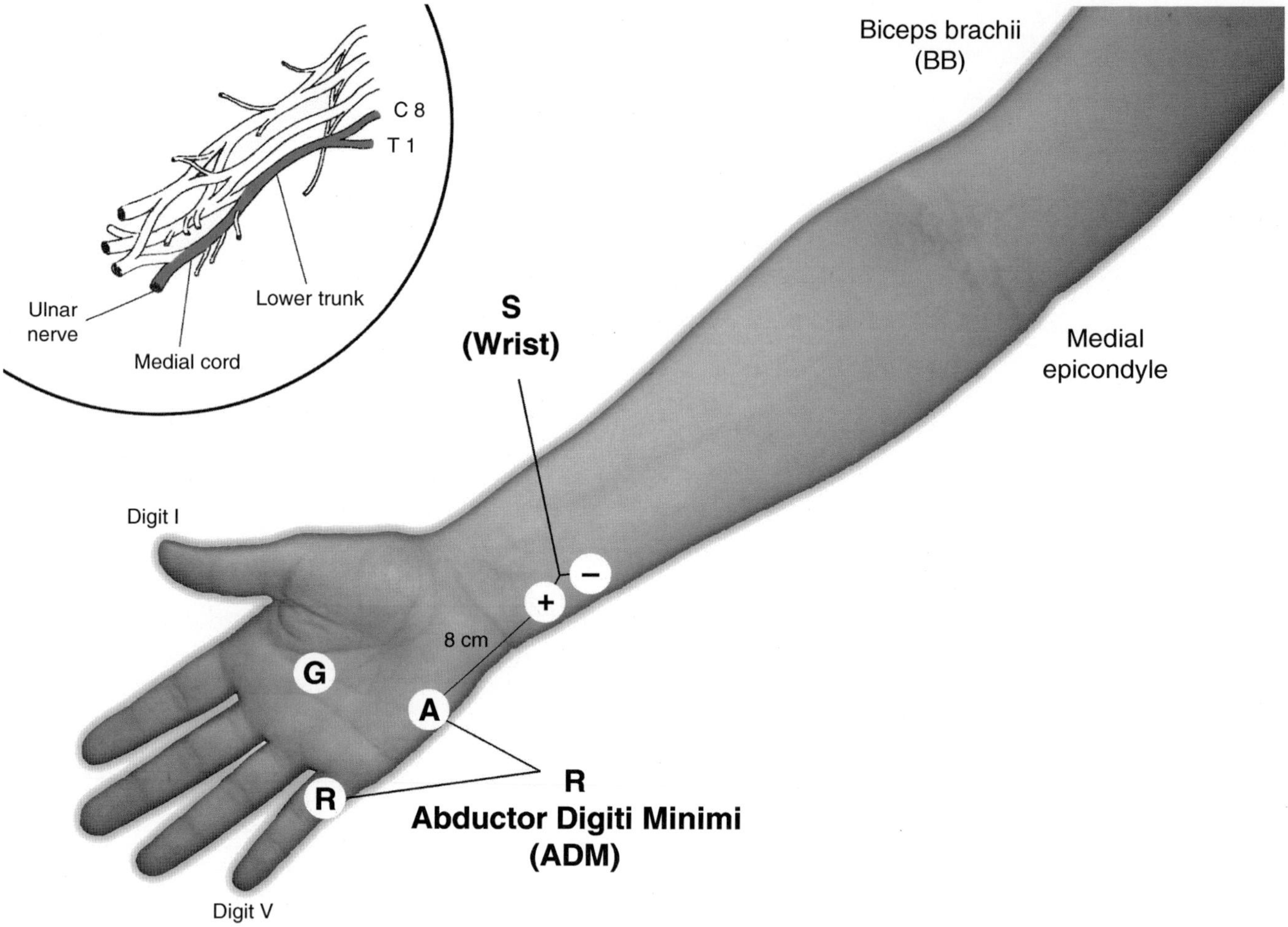

Fig. 3 F-wave study [2] recording from the ADM muscle, stimulation of the wrist

M-response, respectively. The authors also devised a new formula for F-wave height (FWH) index and F-wave limb (FWL) index as follows:

$$FWH = \frac{H\ (cm)}{F\ (ms)}$$

$$FWL = \frac{LL\ (cm)}{F\ (ms)}$$

where H represents the height measured using a standard wall scale. Various aspects of M-wave and F-wave, individually recorded responses with 32 stimuli, were reported (Tables 5, 6, and 7).

Table 5 Normal values [3]

	Mean ± SD
Wrist–ADM, M-wave latency (ms)	2.3 ± 0.3
Wrist–ADM, M-wave amplitude (mV)	15.2 ± 3.2
Elbow–wrist, M-wave conduction velocity (m/s)	60.8 ± 4.0
Wrist–ADM, F-wave minimum latency (ms)	26.5 ± 2.1
Wrist–ADM, F-wave mean latency (ms)	28.1 ± 2.2
Wrist–ADM, F-wave maximum latency (ms)	30.4 ± 2.3
Wrist–ADM, minimum FWCV (m/s)	55.0 ± 2.7
Wrist–ADM, mean FWCV (m/s)	60.0 ± 2.3
Wrist–ADM, maximum FWCV (m/s)	64.0 ± 3.0
Wrist–ADM, F-wave chronodispersion (ms)	3.9 ± 0.9
Wrist–ADM, F-wave persistence (%)	83 ± 19.0
Wrist–ADM, F-wave amplitude (μV)	347 ± 152
Wrist–ADM, F-wave duration (ms)	8.6 ± 2.9
Wrist–ADM, F-wave amplitude (μV)	347 ± 152

Table 6 Normal values [3]

FWH index	Mean ± SD	Limit of normal
Wrist–ADM, F-wave minimum latency	6.45 ± 0.30	5.85
Wrist–ADM, F-wave mean latency	6.07 ± 0.28	5.51
Wrist–ADM, F-wave maximum latency	5.63 ± 0.27	5.09

Table 7 Normal values [3]

FWL index	Mean ± SD	Limit of normal
Wrist–ADM, F-wave minimum latency	2.80 ± 0.12	2.56
Wrist–ADM, F-wave mean latency	2.64 ± 0.12	2.40
Wrist–ADM, F-wave maximum latency	2.45 ± 0.11	2.23

Comment

For Nobrega at al. [3], the minimum, mean, and maximum F-wave latencies showed a strong correlation to the height and to the limb length, and the mean latency remained the same whether calculated with the first 5 stimuli (28.2 ± 2.2 ms), first 10 stimuli (28.1 ± 2.2 ms), or with the total of the 32 stimuli (28.2 ± 2.2 ms). The authors also observed, with regards the indices, that the FWH was independent of the height, and the FWL showed a significant correlation for the F-wave maximum latency. The correlation of F-wave latencies to the limb length and height found by the authors suggested the necessity to adjust to this variable in the clinical evaluation because the use of F-wave latencies without such correction may lead to an erroneous conclusion. In assessing the minimum F-wave latency, for example, a value considered below an upper limit of normal for a 180 cm tall subject would clearly indicate an abnormal delay for a 160 cm short subject. With regards to the latency of the recorded F-waves, they [3] preferred the mean latency value to the extreme values of minimum (which showed normal values if at least one fast conducting fiber survives unaltered) or maximum (which may be prolonged if only one slow fiber is altered despite the integrity of the remaining axons) latencies. Electronic averaging also showed a number of problems for clinical use, such as absent responses because of dispersion and phase cancelation or reduced amplitude and delayed latency. Uncommonly measured but potentially useful aspects of F-wave include, in addition, mean latency, maximal latency, chronodispersion, and mean duration. Pathological waveform and values in a case of ulnar neuropathy at the elbow (UNE) (Figs. 4 and 5) and in a patient affected by chronic inflammatory demyelinating polyradiculoneuropathy (CIDP) associated with anti-MAG/SGPG (pathogenic IgM antibodies – Abs against myelin-associated glycoprotein/peripheral nerve glycolipid sulfated glucuronyl paragloboside) (Fig. 6) are reported.

Pathological waveform (wrist – ADM muscle):

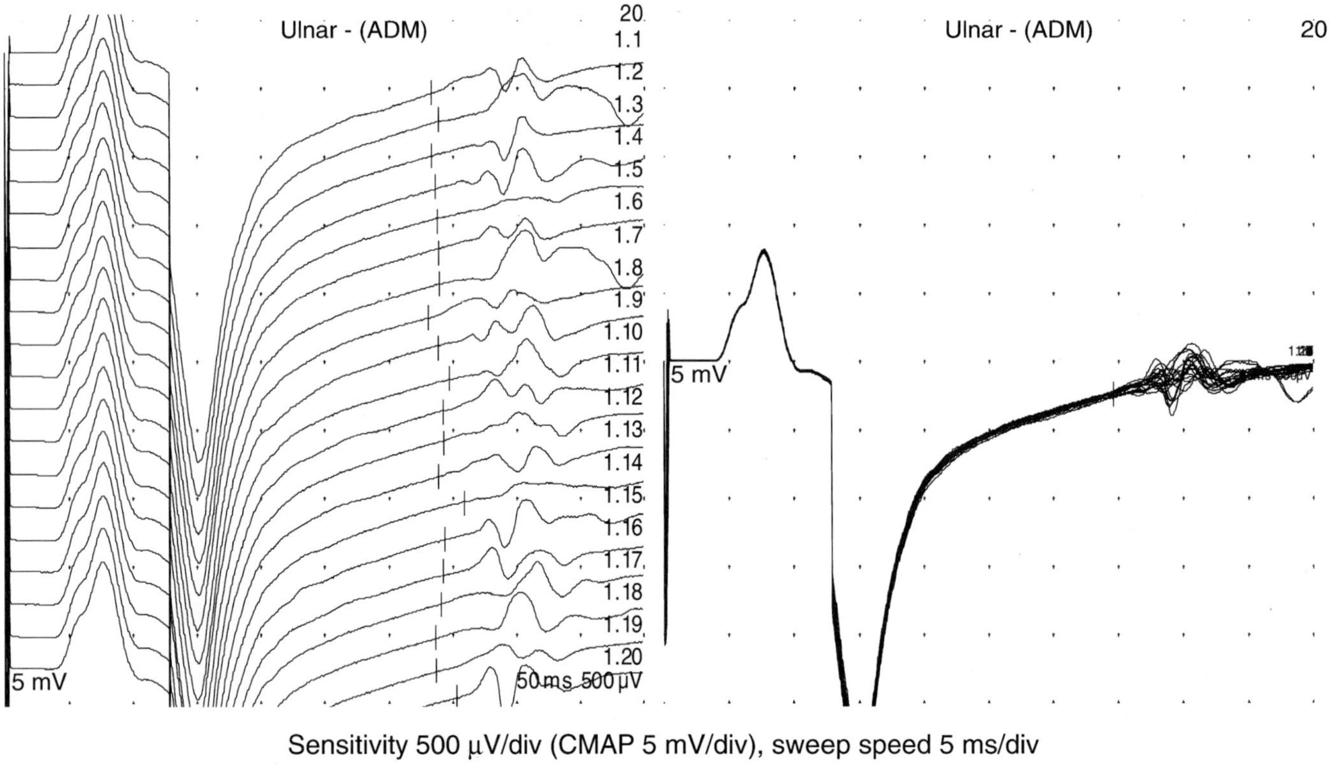

Fig. 4 F-waves recorded at the hand from the ADM muscle, stimulation of the wrist in UNE, mean F-wave latency 34.65 ms; raster mode, 20 stimuli (*left*); superimposed mode, 20 stimuli (*right*)

Pathological waveform (elbow – ADM muscle):

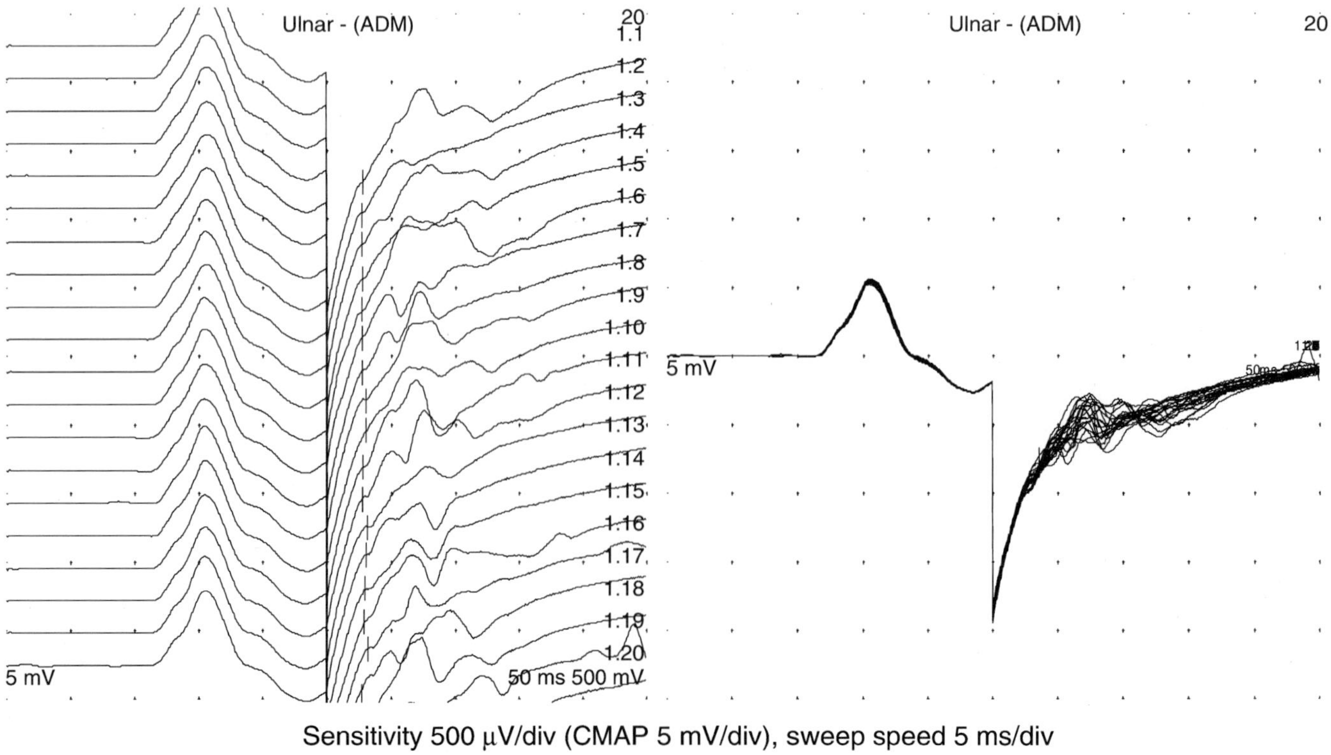

Fig. 5 F-waves recorded at the hand from the ADM muscle, stimulation at the elbow in UNE, mean F-wave latency 28.70 ms; raster mode, 20 stimuli (*left*); superimposed mode, 20 stimuli (*right*)

Pathological waveform (wrist – ADM muscle):

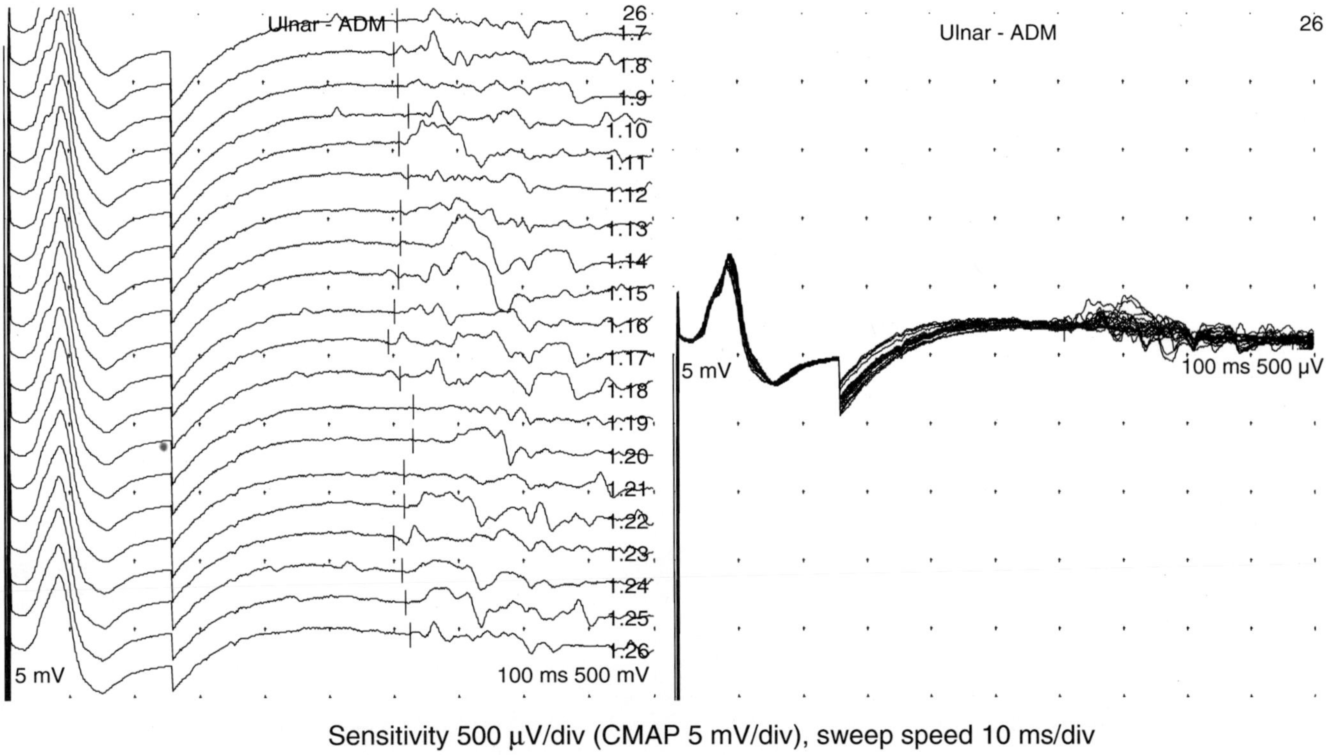

Fig. 6 F-waves recorded at the hand from the ADM muscle, stimulation of the wrist in a patient with CIDP, mean F-wave latency 58.36 ms; raster mode, 20 stimuli (*left*); superimposed mode, 20 stimuli (*right*)

References

1. Eisen A, Schomer D, Melmed C (1977) The application of F-wave measurements in the differentiation of proximal and distal upper limb entrapments. Neurology 27:662–668
2. Buschbacher RM (1999) Ulnar nerve f-wave latencies recorded from the abductor digiti minimi. Am J Phys Med Rehabil 78:S38–S42
3. Nobrega JA, Pinheiro DS, Manzano GM et al (2004) Various aspects of F-wave values in a healthy population. Clin Neurophysiol 115:2336–2342
4. Kimura J (1974) F-Wave velocity in the central segment of the median and ulnar nerves: a study in normal subjects and in patients with Charcot-Marie-Tooth disease. Neurology 24:539–546

Wrist – Hand

M22

Surface Recording Technique, Study from *Abductor Pollicis Brevis* (APB) and *Abductor Digiti Minimi* (ADM) Muscles

Original Settings Sensitivity was 1 mV/division, sweep speed was 2 ms/division, duration of pulse was 0.2 ms, and the machine used was a TEKTRONIX electromyograph; low-frequency filter and high-frequency filter were not specified.

Position This study was performed in the supine position.

Recording Following the median nerve stimulation (S1), the active electrode (A) was placed over the belly of the *abductor pollicis brevis* (APB) muscle (Fig. 1), halfway between the midpoint of the distal wrist crease and the midpoint of the first metacarpophalangeal joint [1]. The reference (R) was placed slightly distal to the first metacarpophalangeal joint (over the tendon of APB muscle). Following the ulnar nerve stimulation (S2), the active electrode was placed over the motor point of the *abductor digiti minimi* (ADM) muscle (Fig. 1). The reference (R) was placed slightly distal to the fifth metacarpophalangeal joint (over the tendon of the ADM muscle). The active electrode location was adjusted over the *thenar* eminence and the *hypothenar* eminence to obtain the greatest compound muscle action potential (CMAP) amplitude for both the median (S1) and ulnar (S2) nerve stimulations at the wrist. The author used a stainless steel disk electrode for recording.

The ground (G) electrode was placed on the dorsum of the hand, between the recording and stimulating electrodes; the figure shows the ground electrode placed on the palm.

Stimulation The median and ulnar CMAPs were evoked with a bar-stimulating electrode with the 1 cm diameter cathode (-) 2 cm distal to the 1 cm diameter anode (+). The stimulations were at the wrist, on the median nerve (S1), and on the ulnar nerve (S2), approximately 1 cm proximal to the proximal wrist crease [1]. Care was taken to ensure that the median and ulnar wrist stimulations were performed at the same distance proximal to the proximal wrist crease. The author used the same sites of motor stimulations at the wrist (both median and ulnar nerves, S1 and S2, respectively) to compare median and ulnar sensory latencies in the same and opposite hands (see sensory conduction studies). Felsenthal [1] chose the same stimulation points in an attempt to achieve some standardization. During a CMAP acquisition, the stimulation intensity was increased incrementally with careful observation of the waveform morphology to achieve supramaximal stimulation while minimizing the likelihood of stimulating unintended nerve segments.

Measurements Distal latency (ms) was measured from the stimulus onset to the onset of the initial deflection of the CMAP. Distal latency was measured for both the median and ulnar nerves, and differences between paired distal motor latencies were measured in the same hand and in the opposite hands in all subjects. All latencies were determined at the same session and at room temperature. Skin and room temperatures were not given. The author performed APB and ADM CMAP recording in 100 hands (Table 1) from 50 volunteers (age range 16–61 years, mean age 33.4 years).

Comment

Felsenthal [1] developed this method for carpal tunnel syndrome (CTS) diagnosis (see "The median nerve: motor conduction techniques"), and he showed the clinical application in two different cases of CTS. He suggested to compare the nerve pairs in each hand because in both CTS cases, side-to-side comparison were nondiagnostic showing values within the range of the mean plus 2 SD (0.61 ms). Subsequently, many authors [2–4] have performed Felsenthal's method [1] and compared it to other several nerve conduction techniques in order to measure its sensibility and specificity.

G. Gentili, M. Di Napoli, *The Ulnar Nerve*, DOI 10.1007/978-3-319-43292-2_46

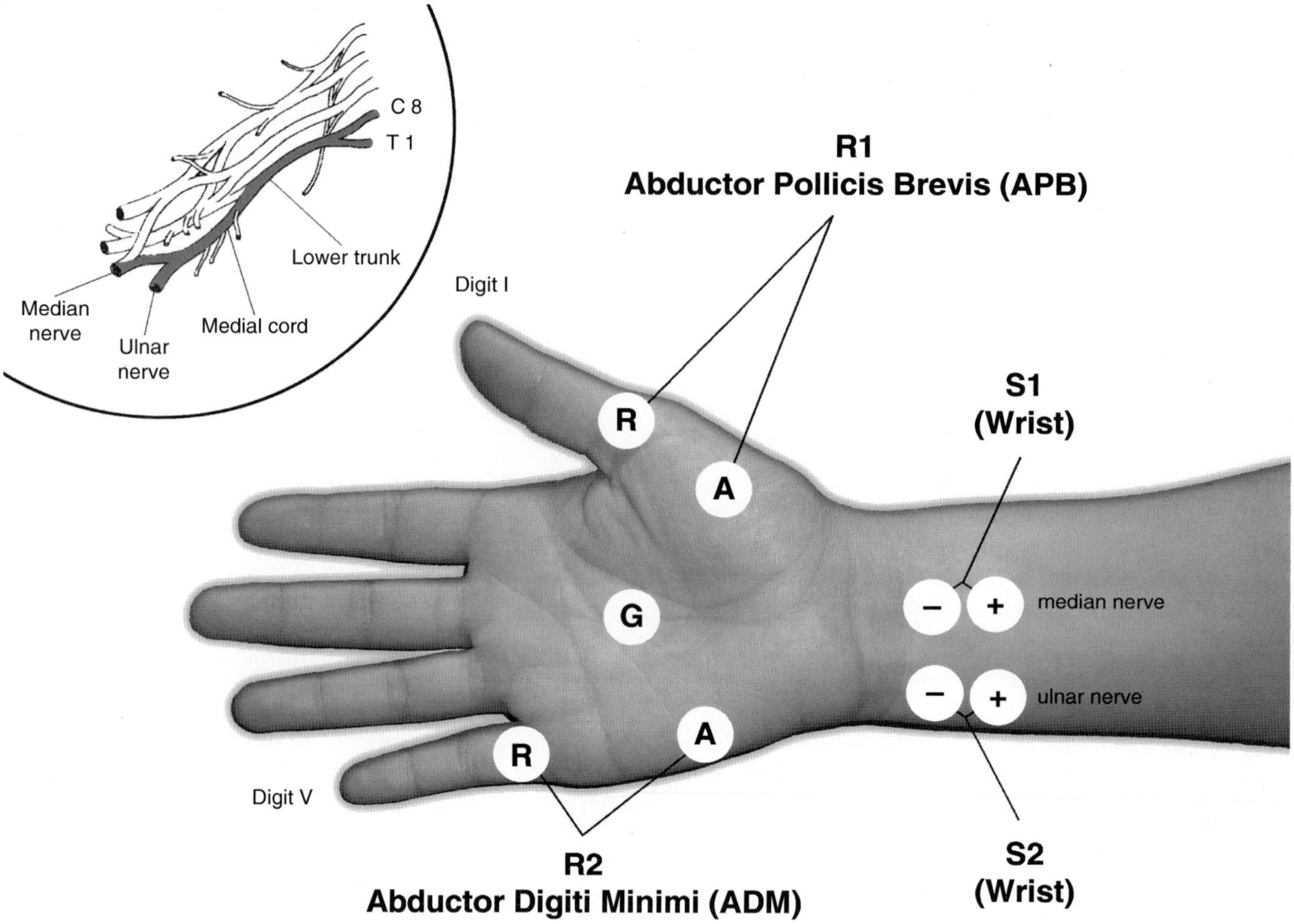

Typical waveform (wrist – APB muscle, wrist – ADM muscle):

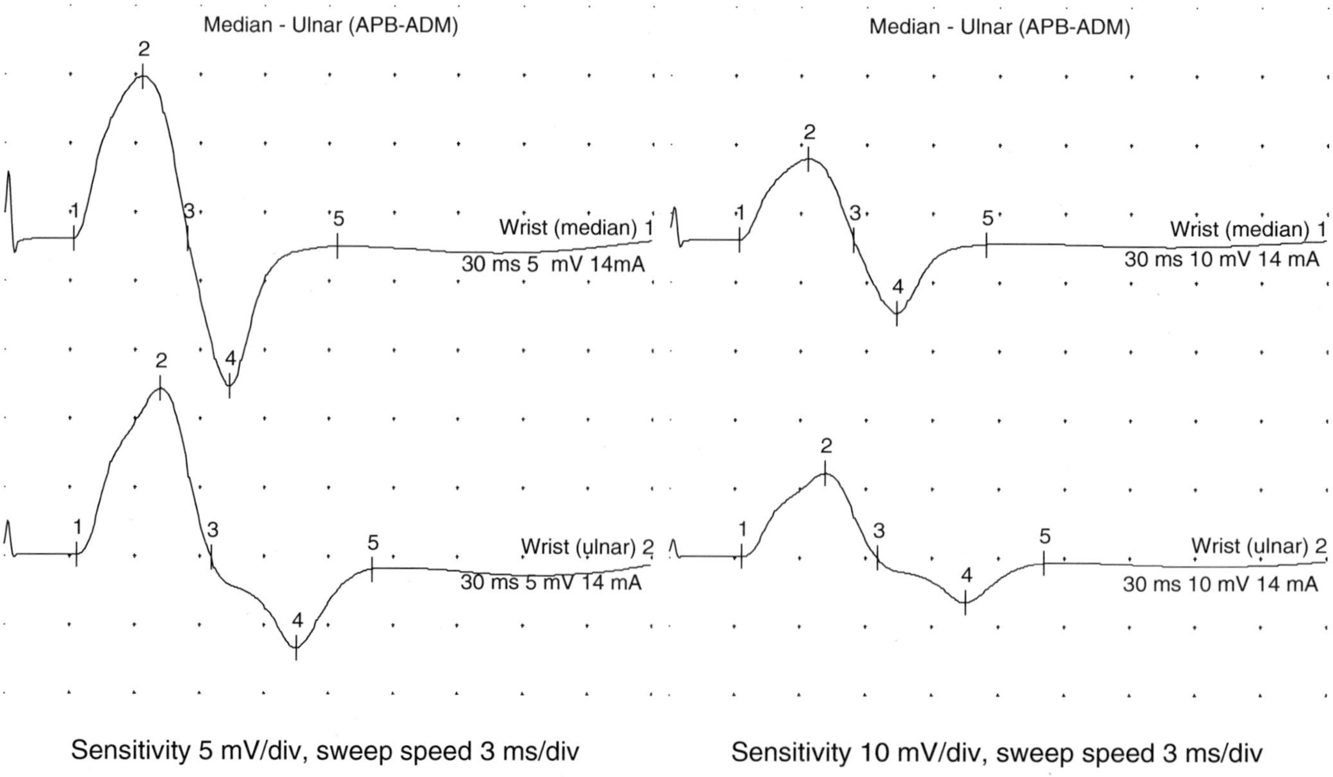

Fig. 1 Compound muscle action potentials (*CMAPs*) recorded at the hand from the APB and ADM muscles, stimulation of the wrist: median nerve (*upper trace*) and ulnar nerve (*lower trace*)

Sander et al. [2] performed the method described by Felsenthal [1], the median–*thenar* to ulnar–*hypothenar* latency difference (THLD) on 34 normal hands (Table 2) from 34 asymptomatic volunteers (15 right hands and 19 left hands, right-hand dominant in 33 controls – the handness of one control was not recorded, age range 26–71 years, mean age 41 years) and 79 hands from 50 patients (Table 3) with clinically and electrophysiologically diagnosed CTS. The patients were 29 bilateral CTS, 16 CTS on the right side, 5 CTS on the left side; the right hand dominant was observed in 38 CTS patients, and the left hand dominant was described in only 5 CTS patients – the handedness of seven patients was not recorded (age range 26–71 years, mean age 41 years). They found the test highly sensitive (85–88 %) if compared to the standard conduction techniques (i.e., median nerve palm–wrist latency).

Table 1 Normal values [1]

Median nerve	Mean ± SD	Mean ± 2SD	Range
Wrist–APB, distal latency (ms)	3.1 ± 0.41	3.9	2.2–4.2

Ulnar nerve	Mean ± SD	Mean ± 2SD	Range
Wrist–ADM, distal latency (ms)	2.9 ± 0.36	3.6	2.3–3.7

Median nerve	Mean ± SD	Mean ± 2SD	Range
Wrist–APB, distal latency difference (ms) – side to side	0.25 ± 0.18	0.61	0.0–0.9

Ulnar nerve	Mean ± SD	Mean ± 2SD	Range
Wrist–ADM, distal latency difference (ms) – side to side	0.22 ± 0.18	0.58	0.0–0.8

Median and ulnar nerve	Mean ± SD	Mean ± 2SD	Range
APB–ADM distal latency difference (ms) – side to side	0.40 ± 0.28	0.96	0.0–1.2

Table 2 Normal values [2]

	Mean ± SD	Range	Limit of normal	Abnormal cutoff
Wrist–APB, distal latency (ms)	2.92 ± 0.28	2.4–3.5	3.6	3.7
Wrist–THLD, distal latency (ms)	0.44 ± 0.30	-0.4–1.0	1.1	1.2

Table 3 Pathological values [2]

	Mean ± SD	Range	Sensitivity (%)
Wrist–APB, distal latency (ms) – all 79 hands	4.31 ± 0.94	2.9–6.6	72
Wrist–APB, distal latency (ms) – one hand	4.38 ± 0.93	2.9–6.6	72
Wrist–THLD, distal latency (ms) – all 79 hands	2.02 ± 0.93	0.0–4.4	85
Wrist–THLD, distal latency (ms) – one hand	2.12 ± 0.88	0.7–4.4	88

Comment

Sander et al. [2] measured the median–*thenar* to ulnar–*hypothenar* latency difference (THLD), and the ulnar–*hypothenar* latency was subtracted from the median–*thenar* distal latency to obtain the THLD. The latency of the median–*thenar* CMAP (ms) was measured from the stimulus onset to the onset of the negative deflection of the CMAP. When measuring the median–*thenar* and the ulnar–*hypothenar* CMAP onset latency, a small premotor potential with an initial negative deflection was occasionally present prior to the onset of the CMAP. This potential is likely generated by intramuscular nerve fibers. The CMAP onset latency in these circumstances was therefore measured at the onset of the subsequent CMAP negativity. The amplitude of the CMAP was measured from the baseline to the peak of negative deflection. Skin temperature on the hand was measured and maintained at or above 32 °C. The mean THLD of 0.44 ms indicates that, on average, the ulnar–*hypothenar* latency was shorter than the median–*thenar* latency. Using this cutoff, there were no false positives in the control group. Calculations involving CTS patients were performed using two methods: (1) all affected hands were included; (2) only one hand per patient was included (only the right hand of patients affected bilaterally). The authors compared their THLD technique with the median palm (the wrist mixed nerve action potential latency, MPWL). In accordance with Jackson and Clifford [6], they considered MPLW abnormal if the response was absent; the latency exceeded 1.7 ms or the latency exceeded the ipsilateral ulnar–palm–wrist latency by more than 0.3 ms. Among the 12 failures, the sensitivity of the mean THLD was 85 % in two hands (0.7 ms and 0.8 ms) and 88 % in one hand (0 ms, 0.6 ms, 0.9 ms, and 1.1 ms). The THLD, using author's methodology, is a highly sensitive (85–88 %) indicator of a median neuropathy at the wrist in patients with a clinical diagnosis of CTS; THLD test had a 100 % specificity of an abnormal value (no false positives) among the normal controls. The authors chose a CTS population with an electrophysiological evidence of a median neuropathy at the wrist to evaluate the usefulness of the described techniques in a set of patients with a documented median nerve dysfunction. The diagnostic sensitivity values are not directly comparable to values in some other reports that determined sensitivities in a population with purely clinically defined CTS.

During a 1-year period, Chang et al. [3] performed several sensory and motor conduction techniques to compare the sensitivities in the diagnosis of CTS. All studies were performed using a Nicolet Viking IV or Dantec Keypoint 4 electromyograph, and skin temperature at the hand was maintained at or above 32 °C. The mean THLD was calculated in 100 control (Table 4) subjects (64 women 36 men, age range 22–65 years, mean age 47.4 years).

Lee et al. [4] described a sensitive new method, the median terminal latency (MTL) ratio, and compared it to that of standard conduction techniques for diagnosing CTS, like the motor distal latency difference between the median–thenar and ulnar–*hypothenar* using Felsenthal's method [1]. They studied 100 volunteers (Table 5) and 153 patients with clinically suspected CTS, founding a low diagnostic sensitivity (63.9 %).

Gazioglu and colleagues [5] performed standard nerve conduction studies and segmental and comparative median nerve conduction tests in 86 hands (Table 6) from 43 healthy individuals (Group 4–32 women and 11 men, 42 right hands and 1 left hand, body mass index – BMI 27.5 ± 3.4, mean age 53.7 ± 5.5 years). They used a Nihon Kohden 9100 electromyograph at a room temperature of 25 °C and a palmar temperature maintained at approximately 32 °C. The authors compared sensitivity and specificity of some different tests in the diagnosis of CTS in diabetic polyneuropathy patients studying a total of 263 hands (Table 7) from 139 patients divided into three groups: Group 1 (CTS), 72 patients with CTS (140 hands from 72 patients, 59 women and 13 men, 71 right hands and 1 left hand, body mass index – BMI 29.0 ± 3.3, mean age 53.9 ± 3.5 years); Group 2 (DMPNP CTS-), 32 patients with polyneuropathy without associated CTS (61 hands from 32 patients, 19 women and 13 men, 30 right hands and 2 left hands, body mass index – BMI 29.1 ± 5.4, mean age 55.6 ± 9.5 years); and Group 3 (DMPNP CTS+), 35 patients with polyneuropathy and associated CTS (62 hands from 35 patients, 26 women and 9 men, 26 right hands and 9 left hands, body mass index – BMI 31.4 ± 7.1, mean age 55.5 ± 7.3 years). The sensitivities and specificities of the tests were used to compare the diagnosis of CTS in patients with diabetic polyneuropathy. A greater diagnostic accuracy can be observed in tests with higher sensitivity and specificity. Following the median–*thenar* and ulnar–*hypothenar* CMAP latency comparison [1], measurement of latency of the median–*thenar* and ulnar–*hypothenar* CMAPs was performed using the onset of first deflection (initial negative deflection for both the median–*thenar* and ulnar–*hypothenar* CMAPs), and the THLD was calculated by subtracting the ulnar–*hypothenar* distal latency from the median–*thenar* distal latency. They found normal THLD values (0.8 ± 0.2 ms) consistent with those reported by others [3, 4].

Table 4 Normal values [3]

	Mean ± SD	Limit of normal
Wrist–THLD, distal latency (ms)	0.82 ± 0.29	<1.55

Table 5 Normal values [4]

	Mean ± SD	Limit of normal (±2 SD)
Wrist–THLD, distal latency (ms)	0.76 ± 0.32	<1.40

Table 6 Normal values [5]

	Mean ± SD	Cut off
Wrist–THLD, distal latency (ms)	0.8 ± 0.2	≤1.3

Table 7 Pathological values [5]

	Mean ± SD
Wrist–THLD, distal latency (ms) – Group 1 (CTS)	2.1 ± 0.7
Wrist–THLD, distal latency (ms) – Group 2 (DMPNP CTS-)	1.0 ± 0.4
Wrist–THLD, distal latency (ms) – Group 3 (DMPNP CTS+)	1.8 ± 0.7

Comment

Gazioglu and colleagues [5] compared sensitivity and specificity of some different tests in the diagnosis of CTS in diabetic polyneuropathy patients in a group of patients with CTS, Group 1; on a group of diabetic patients with polyneuropathy without associated CTS, Group 2 (DMPNP CTS-); and on a group of diabetic patients with a polyneuropathy with associated CTS, Group 3 (DMPNP CTS+). The sensitivities and specificities of the tests were compared in the diagnosis of CTS in patients with diabetic polyneuropathy. The THLD was higher in Group 1 (CTS) than in Group 3 (polyneuropathy with CTS) and in Group 2 (polyneuropathy without CTS) than controls. These differences were significant between Group 3 and Group 2 and between the Group 1 and Group 3. Authors found, using a 1.3 ms cut off, a sensitivity of 74 % and a specificity of 77 %.

Although this method was developed by Felsenthal [1] for CTS diagnosis (Fig. 2), we found it very useful in Guyon's canal syndrome evaluation since that a delayed distal motor latency (DML) or a reduced amplitude for ulnar CMAP from the *abductor digiti minimi* (ADM) muscle can be easily observed in case of an ulnar nerve pathology at the wrist (Fig. 3). The method has significant limitations when both conditions are superimposed or in case of other pathological affections (Fig. 4).

Pathological waveform (wrist – APB muscle, wrist – ADM muscle):

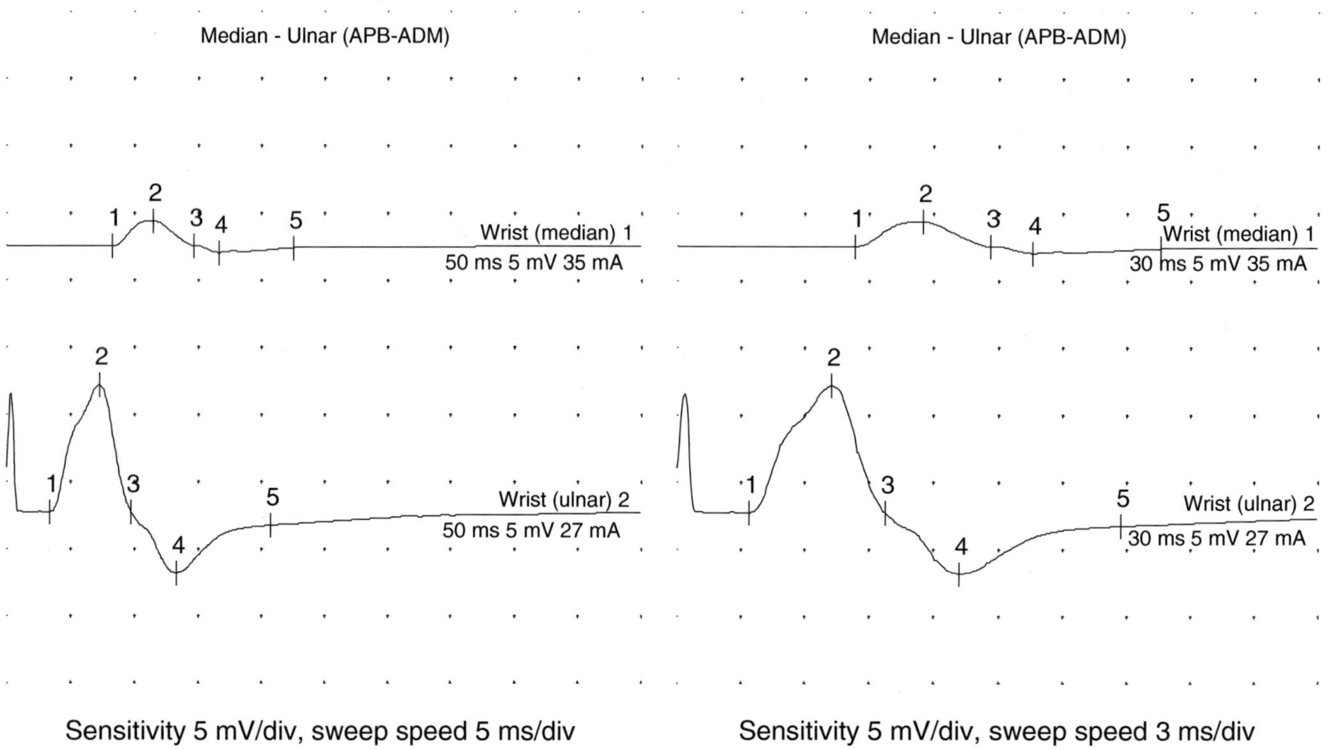

Onset latency (median – APB): 8.30 ms; **Onset latency** (ulnar – ADM): 3.35 ms; **Peak latency** (median – APB): 11.75 ms; **Onset latency difference** (median APB – ulnar ADM): 4.95 ms; **Peak latency** (ulnar – ADM): 7.25 ms; **Onset to peak amplitude** (median – APB): 1.9 mV; **Onset to peak amplitude** (ulnar – ADM): 9.5 mV; **Peak to peak amplitude** (median – APB): 2.4 mV; **Peak to peak amplitude** (ulnar – ADM): 14.1 mV

Fig. 2 Compound muscle action potentials (*CMAPs*) recorded at the hand from the APB and ADM muscles in very severe CTS, grade 5 by Bland's CTS classification scale [7], stimulation of the wrist: median nerve (*upper trace*) and ulnar nerve (*lower trace*)

Pathological waveform (wrist – APB muscle, wrist – ADM muscle):

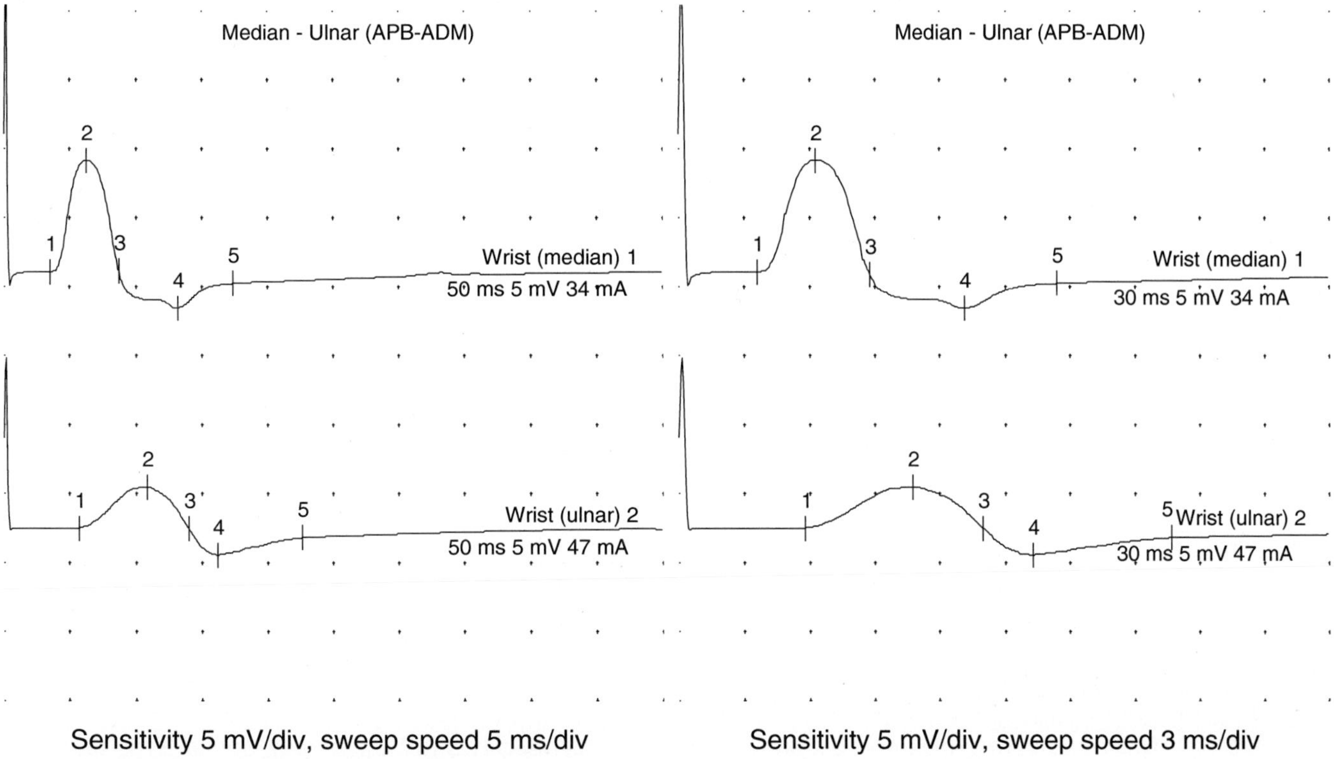

Onset latency (median – APB): 3.55 ms; **Onset latency** (ulnar – ADM): 5.75 ms; **Onset latency difference** (ulnar ADM – median APB): 2.20 ms; **Peak latency** (median – APB): 6.20 ms; **Peak latency** (ulnar – ADM): 10.60 ms; **Onset to peak amplitude** (median – APB): 8.2 mV; **Onset to peak amplitude** (ulnar – ADM): 3.0 mV; **Peak to peak amplitude** (median – APB): 10.8 mV; **Peak to peak amplitude** (ulnar – ADM): 5.0 mV

Fig. 3 Compound muscle action potentials (*CMAPs*) recorded at the hand from the APB and ADM muscles in a patient affected by Guyon's canal syndrome, stimulation of the wrist: median nerve (*upper trace*) and ulnar nerve (*lower trace*)

Pathological waveform (wrist – APB muscle, wrist – ADM muscle):

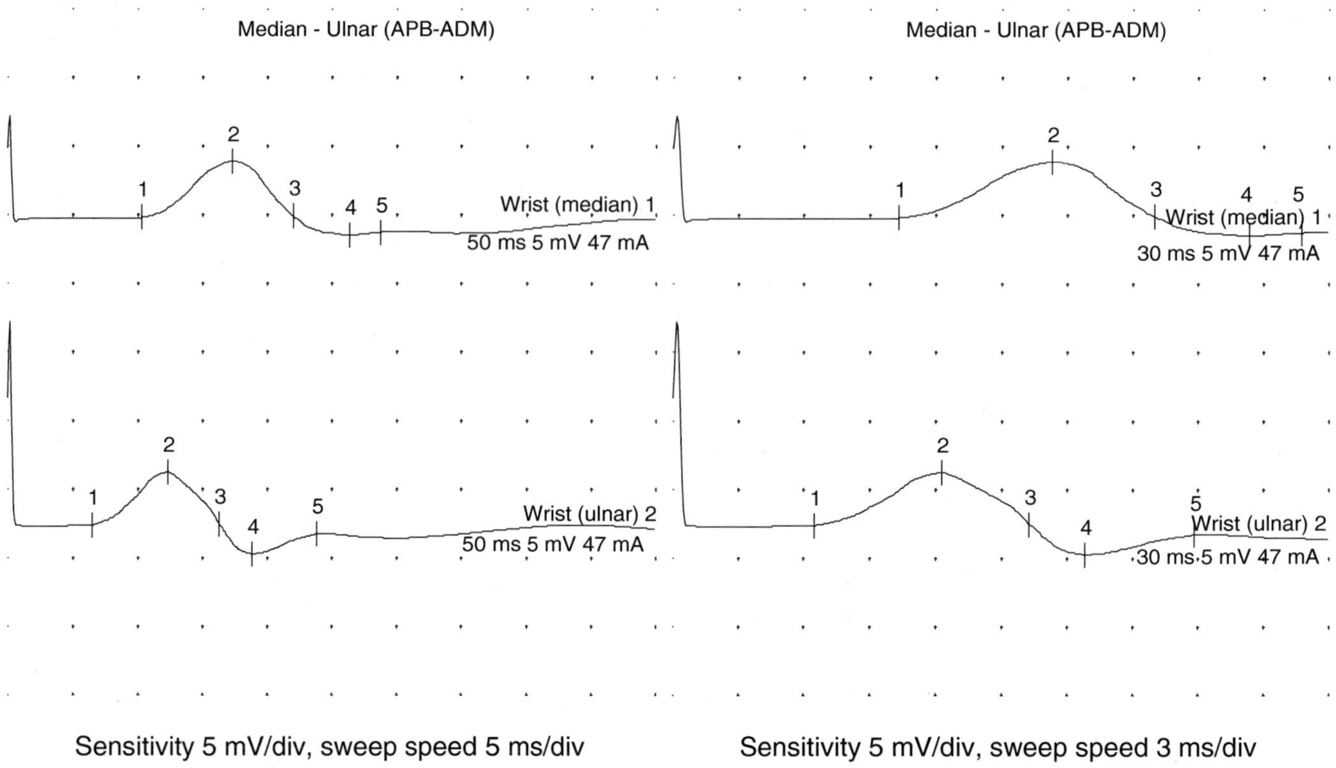

Onset latency (median – APB): 10.30 ms; **Onset latency** (ulnar – ADM): 6.45 ms; **Onset latency difference** (median APB – ulnar ADM): 3.55 ms; **Peak latency** (median – APB): 17.20 ms; **Peak latency** (ulnar – ADM): 12.10 ms; **Onset to peak amplitude** (median – APB): 4.1 mV; **Onset to peak amplitude** (ulnar – ADM): 3.8 mV; **Peak to peak amplitude** (median – APB): 5.4 mV; **Peak to peak amplitude** (ulnar – ADM): 6.0 mV

Fig. 4 Compound muscle action potentials (*CMAPs*) recorded at the hand from the APB and ADM muscles in very severe CTS, grade 5 by Bland's CTS classification scale [7] and Guillain–Barré syndrome (2 weeks of onset of symptoms), stimulation of the wrist: median nerve (*upper trace*) and ulnar nerve (*lower trace*)

References

1. Felsenthal G (1977) Median and ulnar distal motor and sensory latencies in the same normal subjects. Arch Phys Med Rehabil 58:297–302
2. Sander HW, Quinto C, Saadeh PB et al (1999) Sensitive median-ulnar motor comparative techniques in carpal tunnel syndrome. Muscle Nerve 22:88–98
3. Chang MH, Wei SJ, Chiang HL et al (2002) Comparison of motor conduction techniques in the diagnosis of carpal tunnel syndrome. Neurology 58:1603–1607
4. Lee KY, Lee YJ, Koh SH (2009) Usefulness of the median terminal latency ratio in the diagnosis of carpal tunnel syndrome. Clin Neurophysiol 120:765–769
5. Gazioglu S, Boz C, Altunayoglu Cakmak A (2011) Electrodiagnosis of carpal tunnel syndrome in patients with diabetic polyneuropathy. Clin Neurophysiol 122:1463–1469
6. Jackson DA, Clifford JC (1989) Electrodiagnosis of mild carpal tunnel syndrome. Arch Phys Med Rehabil 70:199–204
7. Bland JDP (2000) A neurophysiological grading scale for carpal tunnel syndrome. Muscle Nerve 23:1280–1283

Elbow, Wrist, Palm – Hand

M23

Surface Recording Technique, Study from *Abductor Digiti Minimi* (ADM) Muscle

Original Settings Sensitivity, low-frequency filter, high-frequency filter, sweep speed, duration of pulse, rate of pulse, and the machine used were not specified.

Position This study was performed in the supine position, with the elbow and fingers slightly flexed.

Recording In according to a previous standardized technique (2), the authors placed the active electrode (A) over the belly (motor point) of the *abductor digiti minimi* (ADM) muscle (Figs. 1 and 2); the reference electrode (R) was placed slightly distal to the fifth metacarpophalangeal joint (over the tendon of the ADM muscle). The ground (G) electrode position was not specified in the paper; the figure shows the ground electrode placed on the forearm.

Stimulation The ulnar nerve was stimulated at the elbow (S1), at the wrist (S2), and on the palm (S3). The first stimulation (S1) was delivered at the elbow, in the most proximal portion of the ulnar groove. The anode was proximal. The second stimulation (S2) was delivered at the wrist, just lateral to the tendon of the *flexor carpi ulnaris* (FCU) muscle. The anode was proximal. The authors used an 8 cm fixed distance between the cathode at the wrist and the active recording electrode on the ADM muscle. The third stimulation (S3) was delivered on the palm, on the midpoint of the distance from the recording electrode site to the wrist stimulation site, so the length of the short segment was 4 cm. This point, because of a 4 cm fixed distance from the recording electrode on the ADM muscle, it was just near the pisiform bone slightly distal to the distal wrist crease (Fig. 2).

Distal stimulation (S2) at the wrist determined distal compound muscle action potential (CMAP), while proximal stimulation at the elbow (S1), in the most proximal portion of the ulnar groove, allowed the determination of the forearm motor nerve conduction velocity. The length of the short segment (4 cm) was divided by the forearm motor nerve conduction velocity to determine the expected short-segment distal latency. After percutaneous stimulation of the nerve at the palmar site, the observed distal latency was obtained. The difference between the observed and expected distal latencies was the short-segment residual. The residual latency index (RLI) was determined by dividing the short-segment residual latency by the residual latency for the entire terminal nerve segment. In the case of stimulation at the elbow (S2), the authors suggested a 2–4 cm separation between the cathode (-) and the anode (+), using a pair of standard 0.6 cm diameter electroencephalograph electrodes mounted on a plastic block for stimulating the nerve. Supramaximal stimulation was used; a stimulus in all determinations was at least 30 percent greater than that which elicited a maximum response.

Measurements The authors measured distal peak latency (ms) from the beginning of the shock artifact to the peak of the negative deflection of CMAP, and they called this latency value with the term "terminal latency" (TL). The motor nerve conduction velocity (MNCV) was calculated from the nerve using stimulus at above the elbow and at the wrist, the recording sites placed to the ADM muscle. From this elbow–wrist conduction velocity, an expected terminal latency from the wrist to the ADM muscle was calculated and compared with the measured motor terminal latency. The difference between these values was termed the "residual latency" (RL) for the motor fibers. The temperature of the forearm was determined using a skin surface thermistor thermometer; all forearm temperatures equaled or exceeded 34 °C. Expected TL of the CMAP recorded from the ADM muscle after stimulation at the wrist (8 cm segment) and RL for the CMAP recorded from the ADM muscle after stimulation at the wrist (8 cm segment) were calculated as follows (the distance between the cathode placed at the wrist and the active electrode placed on the ADM muscle was 80 mm fixed):

G. Gentili, M. Di Napoli, *The Ulnar Nerve*, DOI 10.1007/978-3-319-43292-2_47

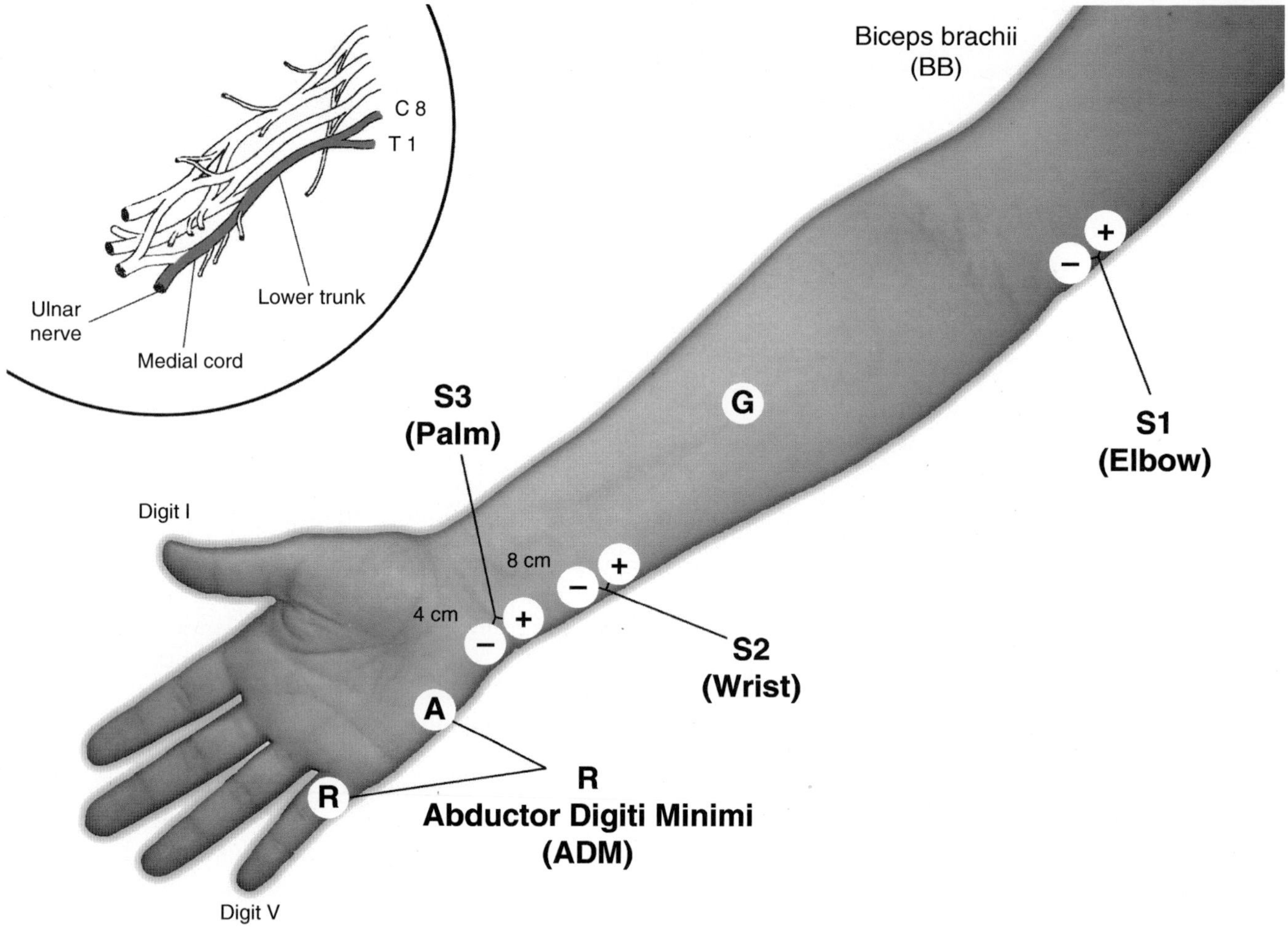

Typical waveform (elbow, wrist, palm – ADM muscle):

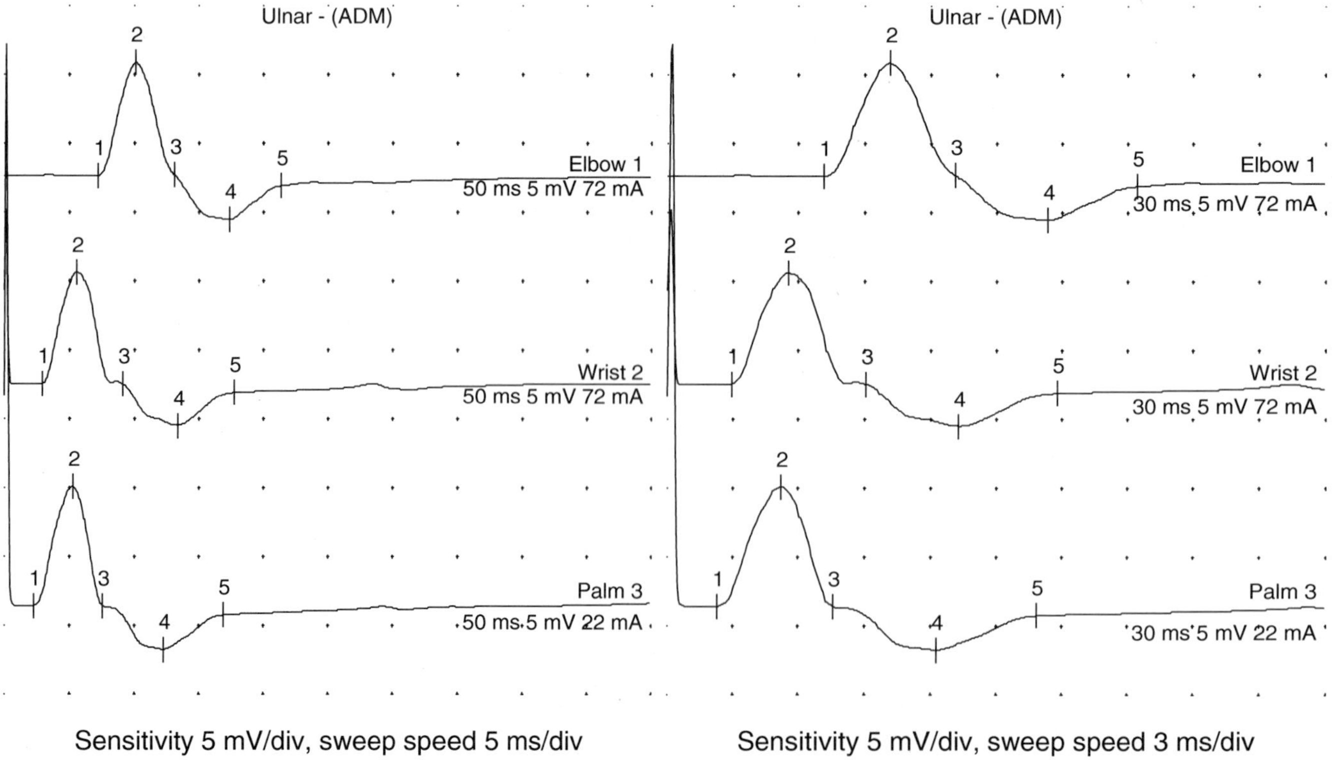

Fig. 1 Compound muscle action potentials (*CMAPs*) recorded from the ADM muscle, stimulation of the elbow (*upper trace*), wrist (*middle trace*), and on the palm (*lower trace*)

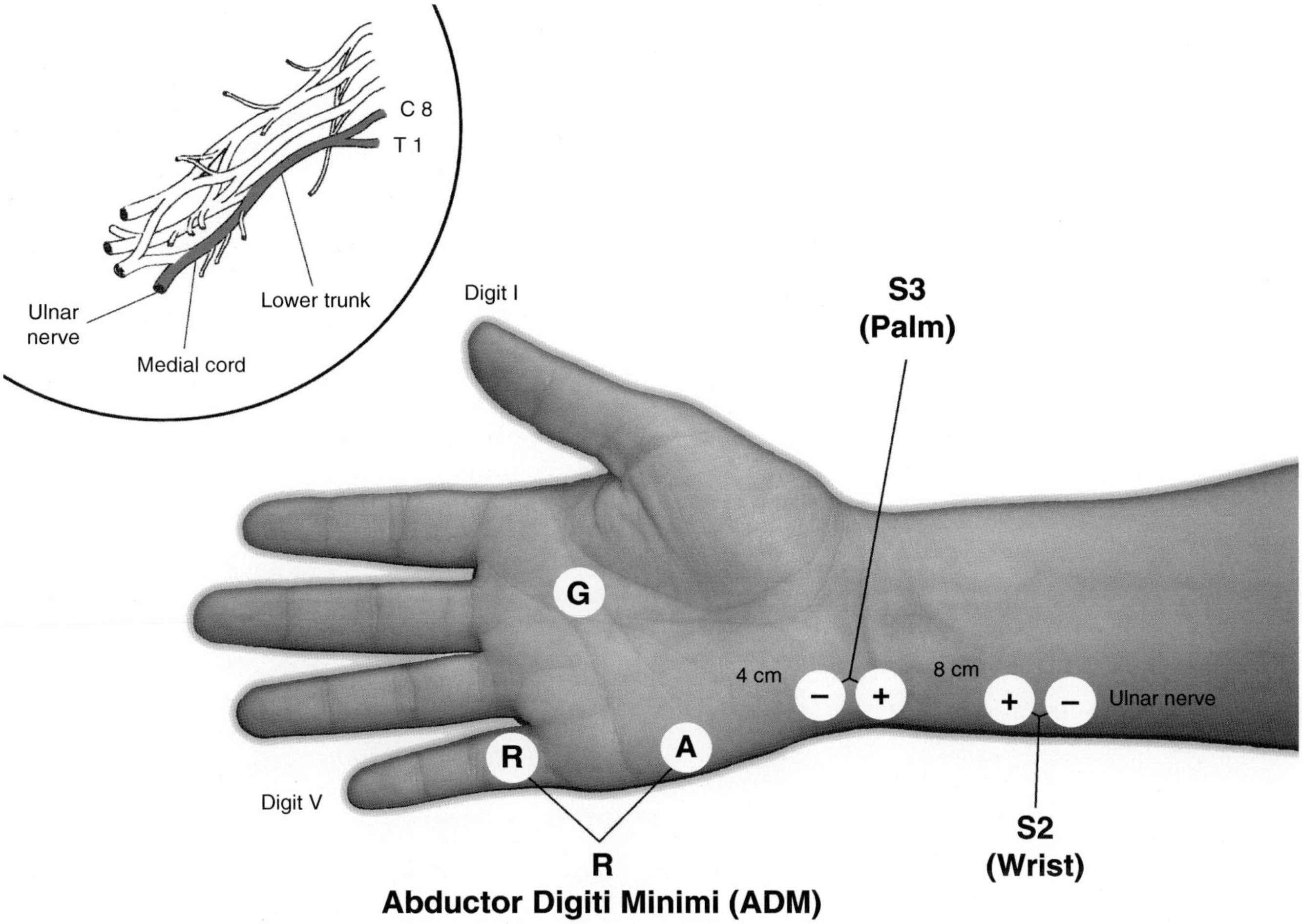

Fig. 2 Distal stimulations (S2, wrist; S3, palm), recording from the ADM muscle

$$\text{Expected TL}\,(\text{ms}) = \frac{\text{distance between cathode (wrist) and active electrode (ADM), (mm)}}{\text{nerve conduction velocity (elbow} - \text{wrist), (m/s)}}$$

$$\text{Residual Latency, RL (ms)} = \text{observed TL (ms)} - \text{expected TL (ms)}$$

Expected short-segment terminal latency (SSTL) of the CMAP recorded from the ADM muscle after stimulation on the palm (4 cm segment) and the short-segment residual latency (SSRL) for CMAP from the ADM muscle after stimulation on the palm (4 cm segment) were calculated as follows (the distance between the cathode placed on the palm and the active electrode placed on the ADM was 40 mm fixed):

$$\text{Expected SSTL (ms)} = \frac{\text{distance between cathode (palm) and active electrode (ADM), (mm)}}{\text{nerve conduction velocity (wrist} - \text{ADM), (m/s)}}$$

$$\text{Short Segment Residual Latency, SSRL (ms)} = \text{observed SSTL (ms)} - \text{expected SSTL (ms)}$$

$$\text{Residual Latency Index, RLI} = \frac{\text{short segment residual latency (SSRL), (ms)}}{\text{residual latency (RL), (ms)}}$$

The authors calculated SSRL and RLI in 20 healthy volunteers (Table 1) with no history of alcoholism or diabetes (age range 15–74 years, mean age 45 years) and in 5 patients (Table 2) with ulnar neuropathy at or distal to the wrist (age range 26–66 years, mean age 49 years). In order to quantitatively compare neuropathic and control populations, RLI values were converted to numbers used in the calculation of product–moment correlation coefficients. Coefficient values ≥ 0.48 indicated a statistically significant direct relationship. These higher correlation coefficient values were noted only in the control subjects, and because of these relationships would be consistent with a homogeneous population, the RLI values for the control population were thus standard or normative values.

Comment

Kaplan and Sahgal [1] compared the RLI values for the neuropathic population with the standard values. In 5 patients with a Guyon's canal syndrome, the RLI increased from 0.61 to 0.69. The short-segment residual latency values for the patients with distal ulnar neuropathy were greater than controls (to 2.0 ms). The authors observed occasional slight increases in insertional activity in the *first dorsal interosseous* (FDI) muscle at EMG examination. As a motor nerve conduction velocity in the ulnar nerve approaches the wrist, it slows perceptively. Residual latencies measure the time latency produced by that slowing (8 cm distance, from the wrist to the ADM muscle), and by adding a palmar stimulation site, the contribution of the short segment of the nerve distal to the wrist to the residual latency may be studied. Therefore, the RLI is a measure of the slowing of the motor nerve conduction velocity produced at the distal short segment (4 cm distance, from the palm to the ADM muscle). In those patients with neuropathy of the ulnar nerve distal to the wrist (UNW), the majority of the slowing of the motor nerve conduction velocity was observed in the distal short segment (4 cm distance, from palm to wrist). As a result, the amount contributed to the residual latency by the distal short segment increased relative to that of the proximal short segment. Therefore, the RLI increased (Fig. 3).

Table 1 Normal values [1]

	Mean ± SD	Range
Wrist–ADM, residual latency (ms) – 15 subjects	1.4 ± 0.8	1.0–1.9
Palm–ADM, short segment residual latency (ms)–15 subjects	0.9 ± 0.2	0.8–1.2
Residual latency index (ms) – 15 subjects	0.63 ± 0.5	0.60–0.65
Residual latency index (ms) – standard values	0.61	0.56–0.65

Table 2 Pathological values [1]

	Mean ± SD	Range
Wrist–ADM, residual latency (ms) – 5 subjects	3.0 ± 0.8	2.7–3.3
Palm–ADM, short-segment residual latency (ms) – 5 subjects	2.0 ± 0.4	1.8–2.2
Residual latency index (ms) – 5 subjects	0.68 ± 0.9	0.66–0.69
Residual latency index (ms) – standard values	0.69	0.65–0.75

Pathological waveform (elbow, wrist, palm – ADM muscle):

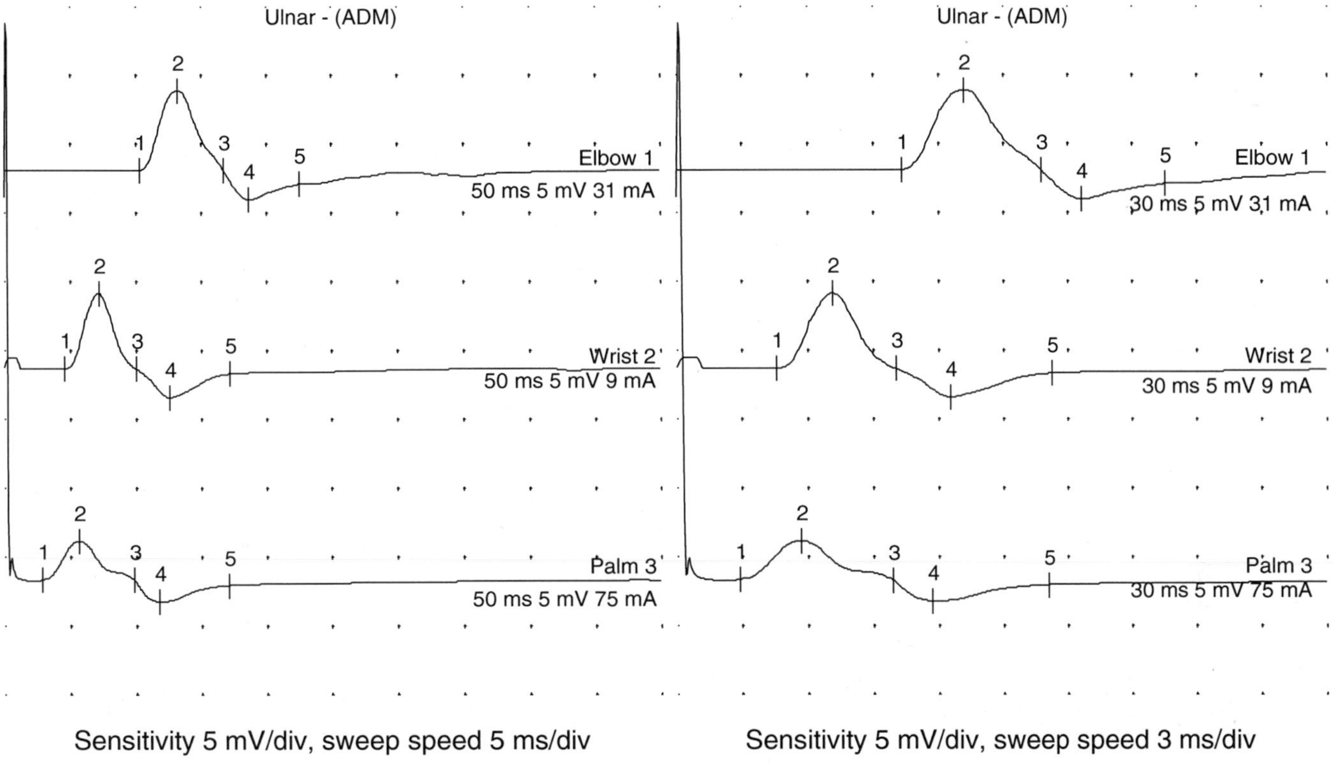

Onset latency (elbow – ADM muscle): 10.30 ms; **Onset latency** (wrist – ADM muscle): 4.55 ms; **Onset latency** (palm – ADM muscle): 2.85 ms; **Peak latency** (elbow – ADM muscle): 13.10 ms; **Peak latency** (wrist – ADM muscle): 7.05 ms; **Peak latency** (palm – ADM muscle): 5.85 ms; **Onset to peak amplitude** (elbow – ADM muscle): 5.8 mV; **Onset to peak amplitude** (wrist – ADM muscle): 5.5 mV; **Onset to peak amplitude** (palm – ADM muscle): 2.9 mV; **Peak to peak amplitude** (elbow – ADM muscle): 8.0 mV; **Peak to peak amplitude** (wrist – ADM muscle): 7.7 mV; **Peak to peak amplitude** (palm – ADM muscle): 4.5 mV; **MNCV** (elbow – wrist, 28 cm distance): 48.7 m/s

Residual latency (wrist – ADM muscle, 8 cm distance) calculation:

MNCV (elbow – wrist, 28 cm distance): 48.7 m/s; **Expected latency** (wrist – ADM muscle, 8 cm distance): 1.64 ms; **Observed latency** (wrist – ADM muscle, 8 cm distance): 4.55 ms; **Residual latency** (wrist – ADM muscle, 8 cm distance): 2.91 ms

Residual latency (palm – ADM muscle, 4 cm distance) calculation:

MNCV (elbow – wrist, 28 cm distance): 48.7 m/s; **Expected latency** (palm – ADM muscle, 4 cm distance): 0.82 ms; **Observed latency** (palm – ADM muscle, 4 cm distance): 2.85 ms; **Short segment residual latency** (palm – ADM muscle, 4 cm distance): 2.03 ms; **Residual latency index** (2.03/2.91): 0.70 ms

Fig. 3 Compound muscle action potentials (*CMAPs*) recorded from the ADM muscle in a patient with Guyon's canal syndrome, stimulation of the elbow (*upper trace*), wrist (*middle trace*), and on the palm (*lower trace*)

Reference

1. Kaplan PE, Sahgal V (1978) Residual latency: new applications of an old technique. Arch Phys Med Rehabil 59:24–27

Wrist, Elbow, Axilla, Erb's Point – Hand

M24

Surface Recording Technique, Study from *Abductor Digiti Minimi* (ADM) Muscle

Original Settings Sensitivity was 1 mV/division, sweep speed was 1 ms/division, and duration of pulse was 0.1 ms for stimulation of the wrist and of the elbow and 0.2–0.5 ms for stimulation of the axilla and of the Erb's point. The machine used was a two-channel TECA model TE2-7. Low-frequency filter and high-frequency filter were not specified.

Position This study was performed in the supine position. The subject was lying supine with his arm in 60° abduction at the shoulder, elbow extended, and the forearm in the supine position. This position placed the brachial plexus with the peripheral median nerve as close to a straight line as possible and to make the measurement of the Erb's point–axilla segment most reliable. This position also provides an ample approach to the axilla for the proper placement of the stimulator on individual nerves.

Recording Surface recording electrodes (pickup electrocardiograph electrodes, 6 mm diameter) for muscle potentials were placed over the motor point of the *abductor digiti minimi* (ADM) muscle (Fig. 1) and the tendon just distal to the fifth metacarpophalangeal joint. The ground (G) electrode was placed on the dorsum of the hand [1]; the figure shows ground electrode placed on the palm.

Stimulation The ulnar nerve [1] was stimulated using the surface electrodes at four points along its course (Figs. 1 and 2): at the wrist (S1), at the elbow (S2), at the axilla (S3), and at the Erb's point (S4).

The first stimulation (S1) was delivered at the wrist (at a constant distance of 5.5 cm from the corresponding muscle pickup electrode), just lateral to the tendon of the *flexor carpi ulnaris* (FCU) muscle. The second stimulation (S2) was delivered at the elbow, in the most proximal portion of the ulnar groove. The third stimulation (S3) was delivered at the axilla, and the fourth stimulation (S4) was delivered at the Erb's point (Fig. 2), lateral to the clavicular insertion of the lateral head of the sternocleidomastoid muscle (just above the clavicle).

Therefore, four motor latencies were obtained for each nerve tested. At all of the above sites of stimulation, the location of the stimulating cathode was carefully adjusted to obtain a perfectly matching configuration of the evoked action potential for each stimulation. In all stimulations the anode was proximal. The voltage required to activate a maximal response was generally supramaximal; the square pulses of current used for stimulation were 0.1 ms duration at the wrist (S1) and at the elbow (S2). At the axilla (S3) and at the Erb's point (S4), a longer pulse duration of 0.2 ms or 0.5 ms was needed (Fig. 2). The latter was the more frequently used.

Measurements Distal latency (ms) was measured from the stimulus artifact to the onset of the initial negative response (upward deflection). Motor nerve conduction velocity (MNCV) was measured in m/s and calculated in a conventional way for each of the three proximal tested segments (Erb's point–axilla, axilla–elbow, and elbow–wrist), dividing the distance by the latency difference between the two evoked potentials. The distances between cathodes were measured with a tape measure, except for the distance between the Erb's point and the axilla where a caliper was used. Skin temperature was not controlled; the room temperature was between 25 and 26.7 °C. The subject was resting for at least 30 min on a plinth before the start of the session and was covered with a blanket to the mid-chest. The authors studied 22 ulnar nerves from 12 healthy (Table 1) subjects (8 women and 4 men, age range 26–55 years, mean age 37.1 years).

G. Gentili, M. Di Napoli, *The Ulnar Nerve*, DOI 10.1007/978-3-319-43292-2_48

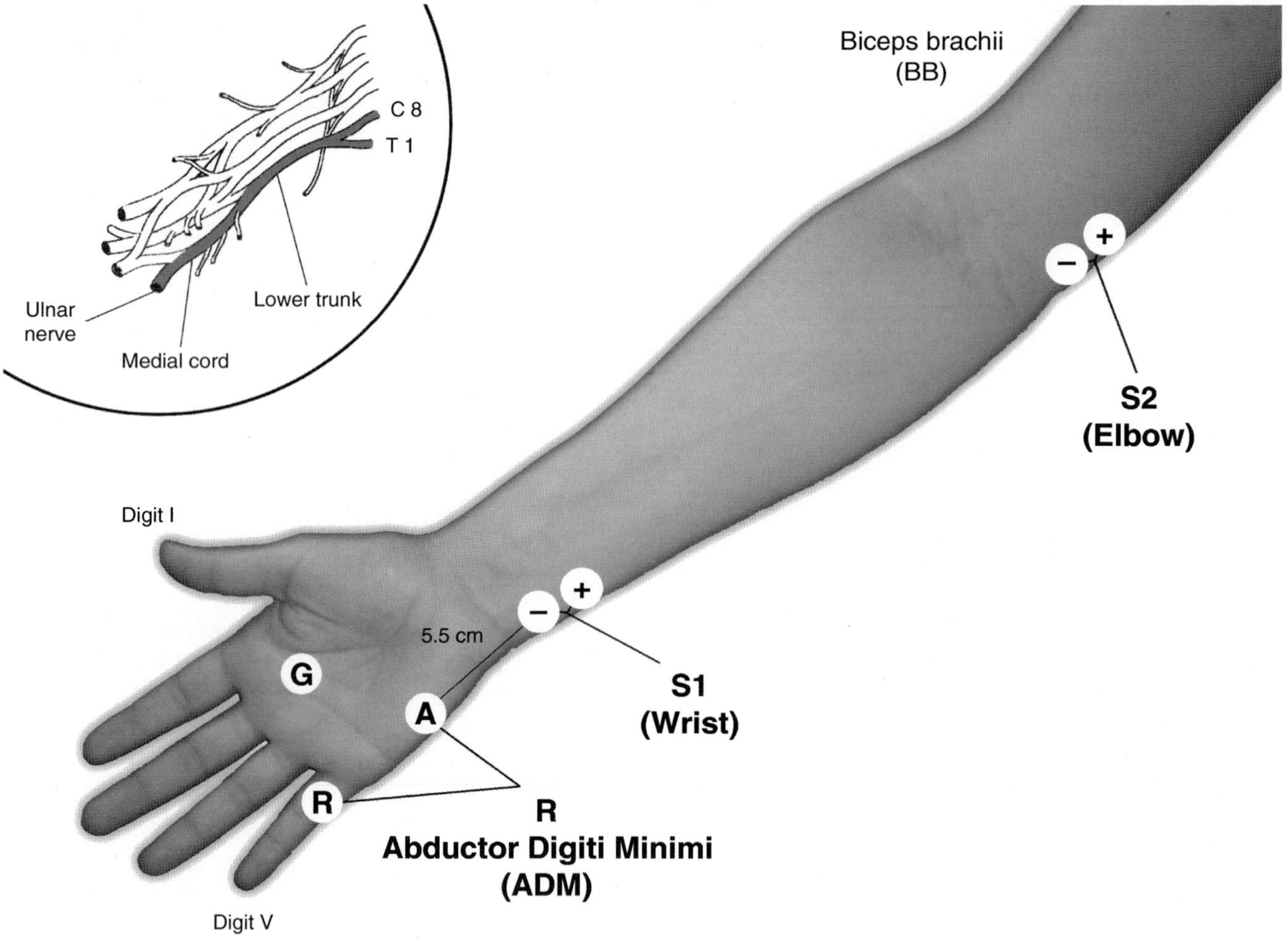

Typical waveform (wrist, elbow, axilla, Erb's point – ADM muscle):

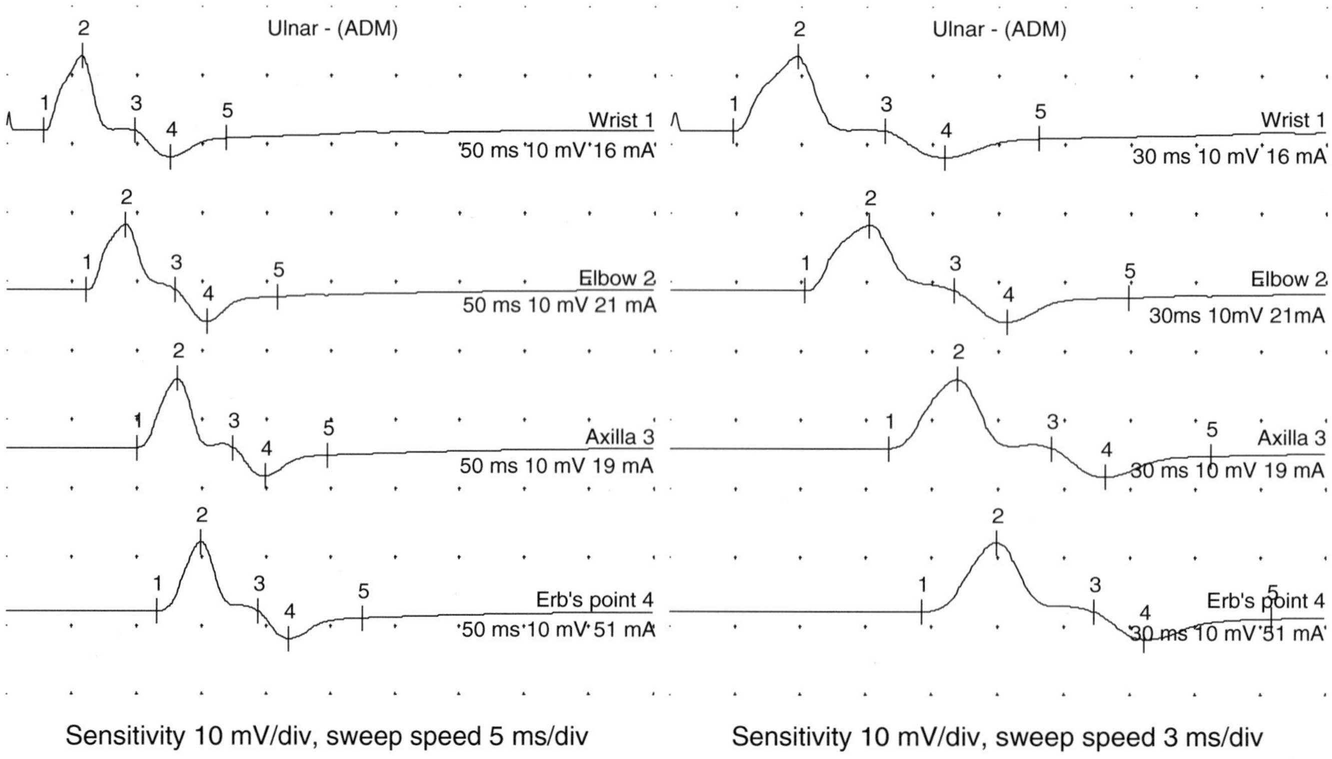

Fig. 1 Compound muscle action potentials (*CMAPs*) recorded at the hand from the ADM muscle, *top to bottom*: stimulation of the wrist, elbow, axilla, and Erb's point

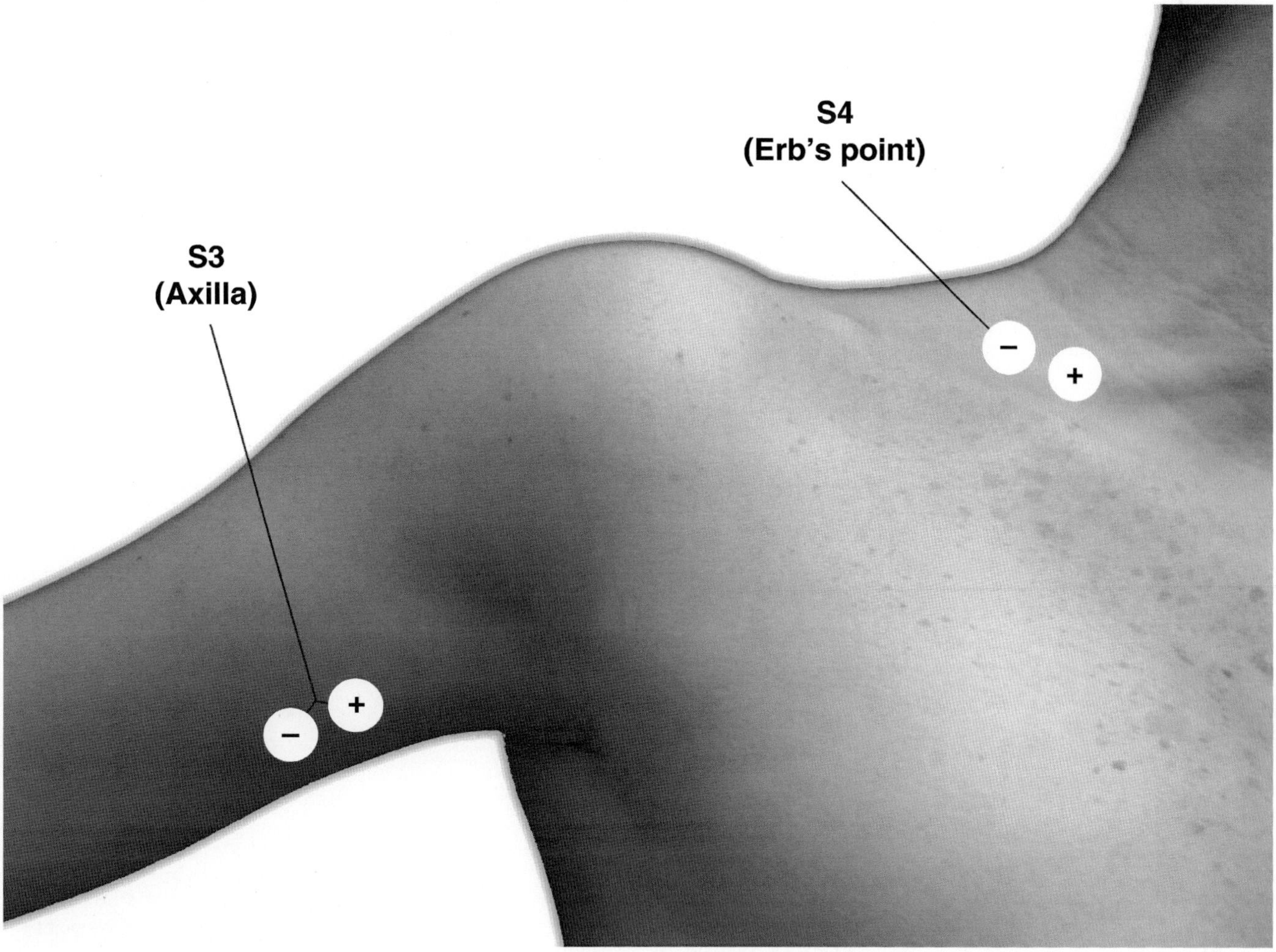

Fig. 2 Ulnar proximal stimulations (S3 on the axilla, S4 on the Erb's point), recording from the ADM muscle

Table 1 Normal values [1]

	Mean ± SD	Range	Mean ± 2SD
Wrist–ADM, distal latency (ms)	3.5 ± 0.36	2.8–4.2	
Elbow–wrist, MNCV (m/s)	48.9 ± 2.8	45.2–55.3	
Axilla–elbow, MNCV (m/s)	51.2 ± 4.2	44.9–60.6	
Erb's point–axilla, MNCV (m/s)	63.0 ± 5.5	55.0–73.2	52.0
Erb's point–axilla, conduction time (m/s)	3.1 ± 0.4	2.6–3.9	

	Mean ± SD
Elbow–wrist, MNCV difference side to side (m/s)	1.7 ± 2.1
Axilla–elbow, MNCV difference side to side (m/s)	4.9 ± 2.7
Erb's point–axilla, MNCV difference side to side (m/s)	4.0 ± 4.3

Comment

Ginzburg et al. [1] measured and compared the proximal MNCV values (Erb's point–axilla segment) with those distal (axilla–elbow and elbow–wrist segments). For the ulnar Erb's point–axilla segment, the upper limit of the normal difference between the right and left MNCV values was 12 m/s. These additional measurements also enabled to determine that a tested nerve was intact in its entire length and must be an essential part of the nerve conduction study in the case of neuropathy (Figs. 3 and 4).

Pathological waveform (wrist, elbow, axilla, Erb's point – ADM muscle):

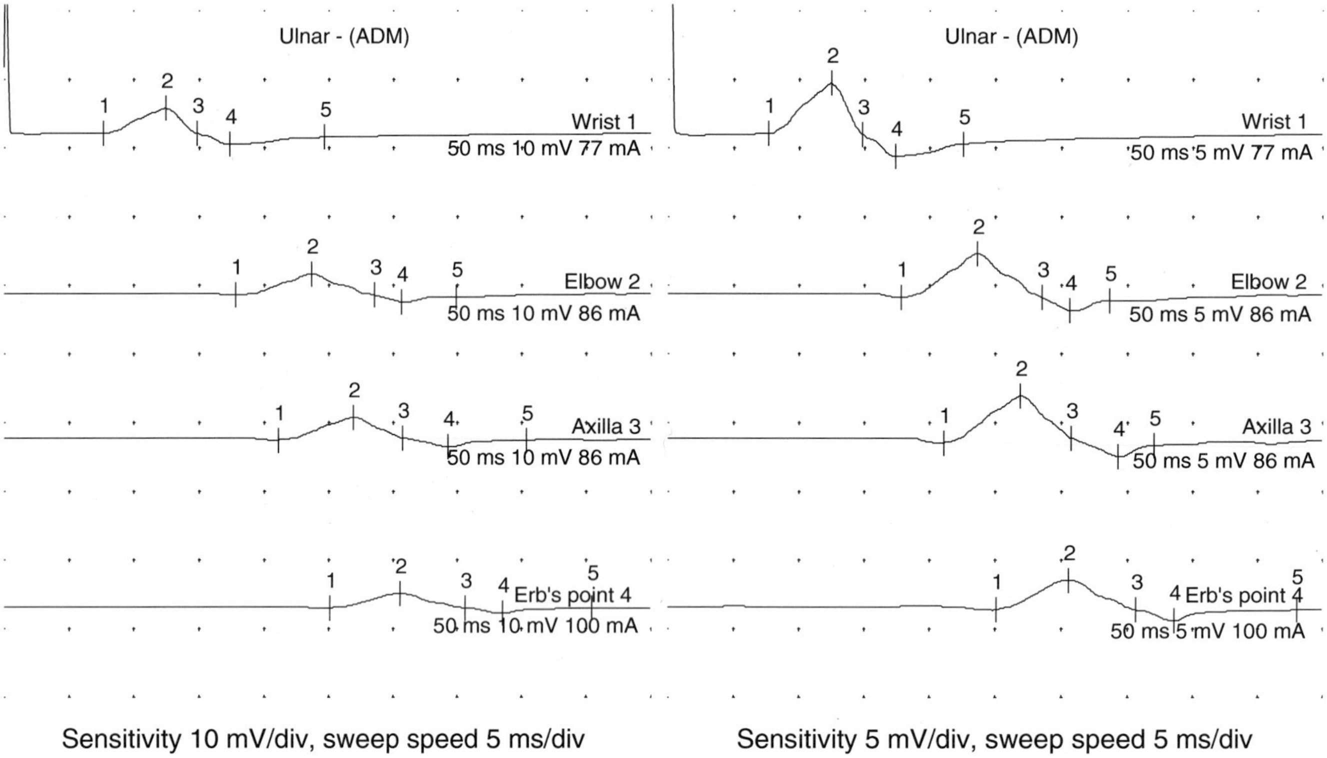

Onset latency (wrist – ADM): 7.65 ms; **Onset latency** (elbow – ADM): 17.75 ms; **Onset latency** (axilla – ADM): 21.05 ms; **Onset latency** (Erb's point – ADM): 25.10 ms; **Peak latency** (wrist – ADM): 12.40 ms; **Peak latency** (elbow – ADM): 23.10 ms; **Peak latency** (axilla – ADM): 27.05 ms; **Peak latency** (Erb's point – ADM): 30.15 ms; **Onset to peak amplitude** (wrist – ADM): 3.7 mV; **Onset to peak amplitude** (elbow – ADM): 3.3 mV; **Onset to peak amplitude** (axilla – ADM): 3.4 mV; **Onset to peak amplitude** (Erb's point – ADM): 2.2 mV; **Peak to peak amplitude** (wrist – ADM): 5.4mV; **Peak to peak amplitude** (elbow – ADM): 4.2 mV; **Peak to peak amplitude** (axilla – ADM): 4.4 mV; **Peak to peak amplitude** (Erb's point – ADM): 3.0 mV; **MNCV** (wrist – elbow): 27.7 m/s; **MNCV** (axilla – elbow): 39.4 m/s; **MNCV** (Erb's point – elbow): 39.5 m/s

Fig. 3 Compound muscle action potentials (*CMAPs*) recorded at the hand from the ADM muscle in a patient affected by chronic inflammatory demyelinating polyradiculoneuropathy (CIDP) and monoclonal gammopathy, *top to bottom*: stimulation of the wrist, elbow, axilla, and Erb's point

Pathological waveform (wrist, elbow, axilla, Erb's point – ADM muscle):

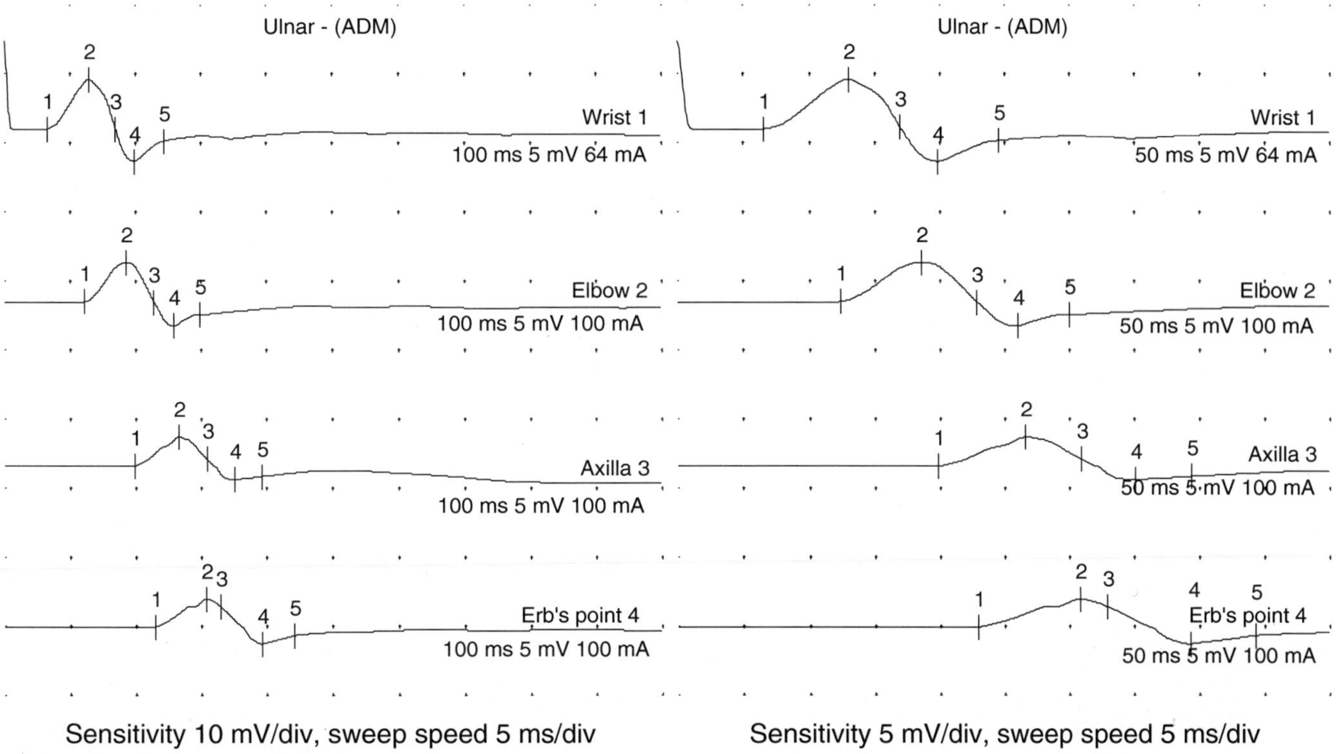

Onset latency (wrist – ADM): 6.50 ms; **Onset latency** (elbow – ADM): 12.35 ms; **Onset latency** (axilla – ADM): 19.80 ms; **Onset latency** (Erb's point – ADM): 22.90 ms; **Peak latency** (wrist – ADM): 12.60 ms; **Peak latency** (elbow – ADM): 18.15 ms; **Peak latency** (axilla – ADM): 26.95 ms; **Peak latency** (Erb's point – ADM): 30.20 ms; **Onset to peak amplitude** (wrist – ADM): 3.6 mV; **Onset to peak amplitude** (elbow – ADM): 2.8 mV; **Onset to peak amplitude** (axilla – ADM): 2.1 mV; **Onset to peak amplitude** (Erb's point – ADM): 2.0 mV; **Peak to peak amplitude** (wrist – ADM): 6.0 mV; **Peak to peak amplitude** (elbow – ADM): 4.7 mV; **Peak to peak amplitude** (axilla – ADM): 3.1 mV; **Peak to peak amplitude** (Erb's point – ADM): 3.3 mV; **MNCV** (wrist – elbow): 37.6 m/s; **MNCV** (axilla – elbow): 20.1 m/s; **MNCV** (Erb's point – elbow): 45.2 m/s

Fig. 4 Compound muscle action potentials (*CMAPs*) recorded at the hand from the ADM muscle in a case of Guillain–Barré syndrome (about 2 weeks of onset of symptoms), *top to bottom*: stimulation of the wrist, elbow, axilla, and Erb's point

Reference

1. Ginzburg M, Lee M, Ginzburg J et al (1978) Median and ulnar nerve conduction determinations in the Erb's point-axilla segment in normal subjects. J Neurol Neurosurg Psychiatry 41:444–448

Above Sulcus, Below Sulcus, Wrist – Hand

M25

Needle Recording Technique, Study from *Abductor Digiti Minimi* (ADM) Muscle

Original Settings Duration of pulse was 0.5 ms, and the machine used was a DISA 3-channel electromyograph; sensitivity, sweep speed, low-frequency filter, high-frequency filter, and rate of pulse were not specified.

Position This study was performed with the subjects lying on a bed in the supine position, with the elbow fully extended.

Recording The active needle electrode was placed over the motor point [1] of the *abductor digiti minimi* (ADM) muscle (Fig. 1). The ground (G) electrode position was not specified in the text; the figure shows the ground electrode positioned over the forearm.

Needle Insertion For the ADM muscle, with the hand in a neutral position to full supination, the recording needle electrode was inserted at the midpoint between the fifth metacarpophalangeal joint and the pisiform bone.

Stimulation The ulnar nerve was superficially stimulated at three points on its course: above the sulcus (i.e., above the elbow), below the sulcus (*i.e.*, below the elbow), and at the wrist. Supramaximal stimulation was used (0.5 ms of duration) in all subjects, but the EMG pattern was verified to be normal and fibrillation potentials at rest, therefore, not accepted.

Measurements Distal latency (ms) of the compound muscle action potential (CMAP) was measured from the stimulus onset to the onset of the negative deflection of the CMAP. The following distances between the sites of stimulation were measured in cm: above sulcus–wrist and below sulcus–wrist, with an accuracy of not more than 0.5 cm. The author [1] calculated the cubital ratio, which was defined as:

$$\text{Cubital ratio} = \frac{\text{above sulcus} - \text{wrist MNCV, (m/s)}}{\text{below sulcus} - \text{wrist MNCV, (m/s)}} \times 100$$

G. Gentili, M. Di Napoli, *The Ulnar Nerve*, DOI 10.1007/978-3-319-43292-2_49

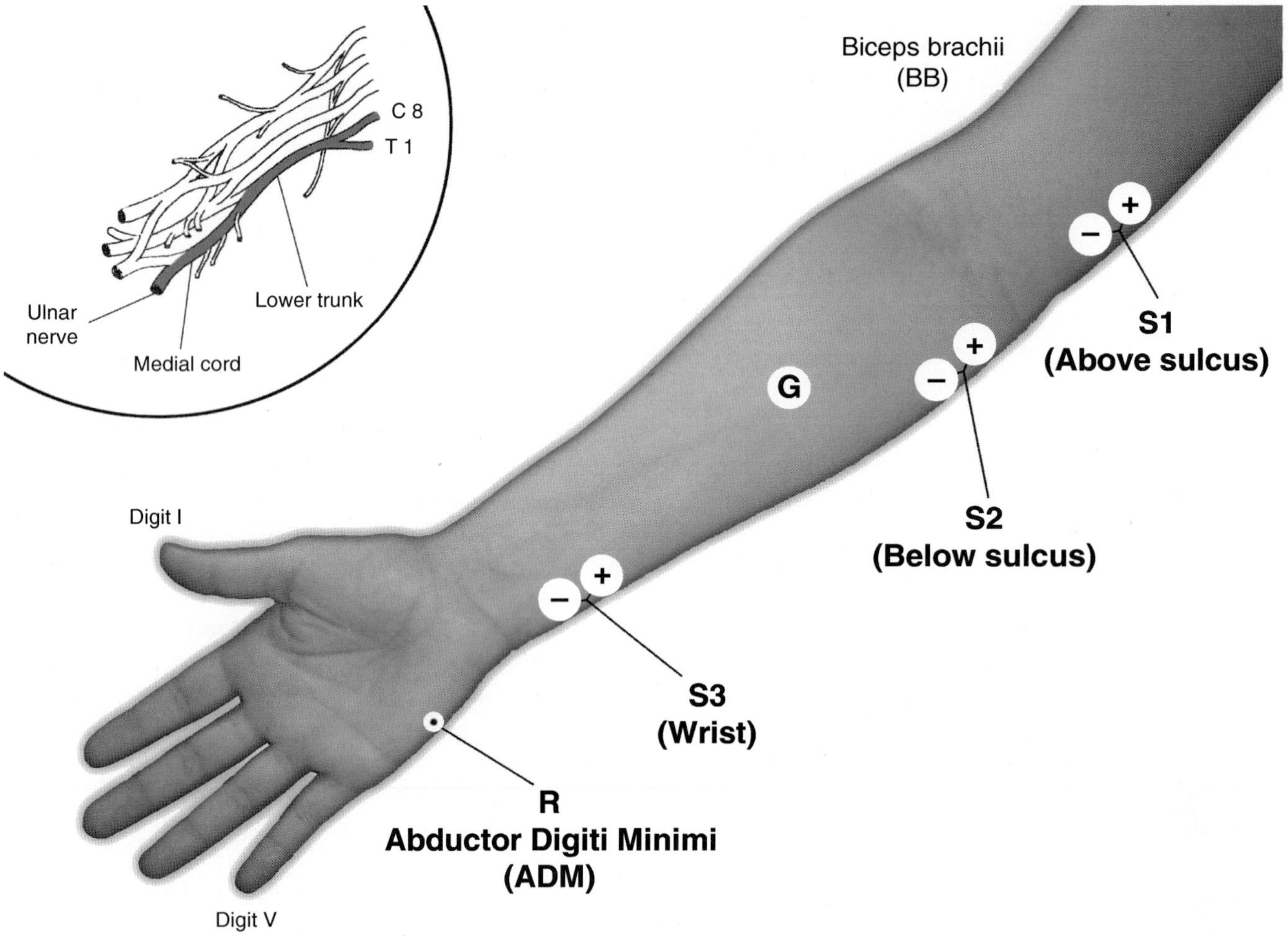

Typical waveform (above sulcus, below sulcus, wrist – ADM muscle):

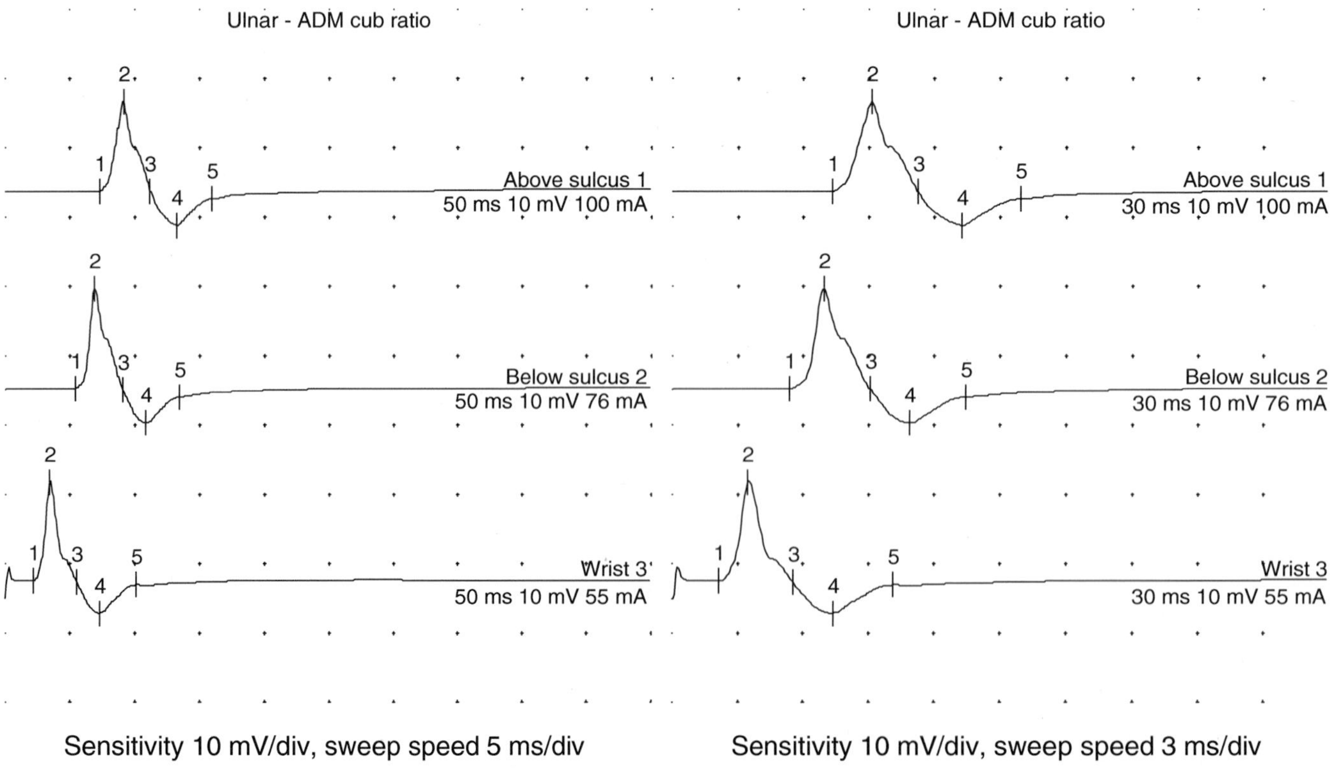

Fig. 1 Compound muscle action potentials (*CMAPs*) recorded at the hand from the ADM muscle, *top to bottom*: stimulation above the sulcus, below the sulcus, and at the wrist

Conduction velocity studies were performed in 76 healthy (Table 1) subjects (age range 14–72 years) divided into two groups: below 50 years (55 subjects) and above 50 years (21 subjects). The cubital ratio was calculated in each instance, and the mean value is given for each group (<50 years, > 50 years).

Table 1 Normal values [1]

	Mean ± SD	Mean	Range
Wrist–ADM, latency (ms) – <50 years	2.5 ± 0.4		
Below sulcus–wrist, MNCV (m/s) – <50 years	63.7 ± 5.6		
Above sulcus–wrist, MNCV (m/s) – <50 years	59.5 ± 5.2		
Above sulcus–below sulcus, MNCV (m/s) – <50 years	52.0 ± 9.1		
Above sulcus–below sulcus, MNCV ratio (%) – <50 years		93.4	
Wrist–ADM, latency (ms) – >50 years	2.9 ± 0.5		
Below sulcus–wrist, MNCV (m/s) – >50 years	58.3 ± 7.3		
Above sulcus–wrist, MNCV (m/s) – >50 years	54.0 ± 7.0		
Above sulcus–below sulcus, MNCV (m/s) – >50 years	44.4 ± 10.3		
Above sulcus–below sulcus, MNCV ratio (%) – >50 years		92.6	
Above sulcus–below sulcus, MNCV ratio (%) – all subjects		93	82–97

Comment

For Krogness [1], the cubital ratio values calculated below 82 % mean that a lesion exists in the ulnar nerve of the entrapment type (Fig. 2). The cubital ratio method was found by the author independent of age having the advantage that the MNCV determined in the cubital tunnel of an individual is related to the normal conduction velocity of the forearm of the same subject and not to a material of normal velocities with a very broad range.

Pathological waveform (above sulcus, below sulcus, wrist – ADM muscle):

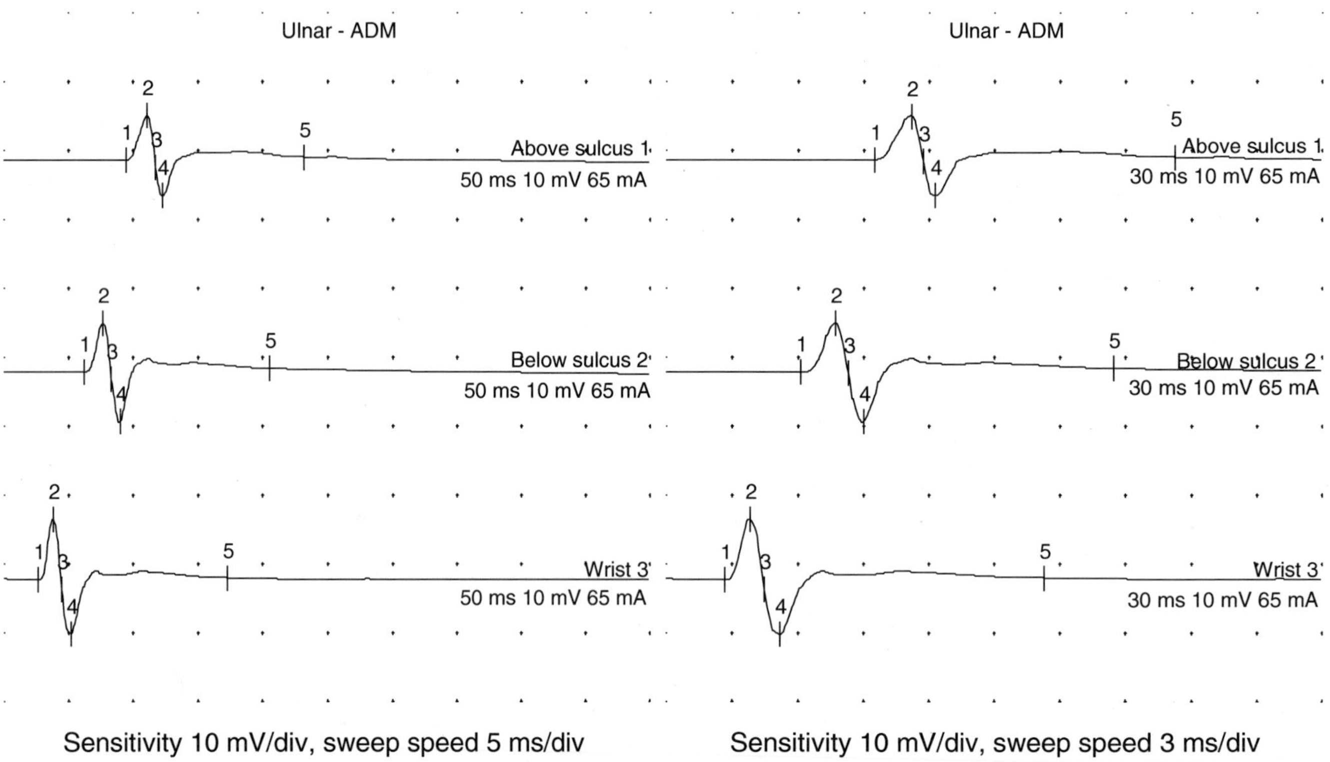

Onset latency (above sulcus – ADM): 9.45 ms; **Onset latency** (below sulcus – ADM): 6.15 ms; **Onset latency** (wrist – ADM): 2.65 ms; **Peak latency** (above sulcus – ADM): 10.95 ms; **Peak latency** (below sulcus – ADM): 7.65 ms; **Peak latency** (wrist – ADM): 4.10 ms; **Onset to peak amplitude** (above sulcus – ADM): 6.5 mV; **Onset to peak amplitude** (below sulcus – ADM): 7.2 mV; **Onset to peak amplitude** (wrist – ADM): 8.8 mV; **Peak to peak amplitude** (above sulcus – ADM): 11.7 mV; **Peak to peak amplitude** (below sulcus – ADM): 14.5 mV; **Peak to peak amplitude** (wrist – ADM): 16.8 mV; **MNCV** (above sulcus – below sulcus, 10 cm distance): 30.3 m/s; **MNCV** (above sulcus – wrist, 31 cm distance): 45.6 m/s; **MNCV** (below sulcus – wrist, 21 cm distance): 60.0 m/s

Fig. 2 Compound muscle action potentials (*CMAPs*) recorded at the hand from the ADM muscle using needle recording method [1] in UNE, *top to bottom*: stimulation above the sulcus, below the sulcus, and at the wrist

$$\text{Cubital ratio calculation} = \text{if} \quad \frac{\text{above sulcus} - \text{wrist MNCV, } (\text{m/s})}{\text{below sulcus} - \text{wrist MNCV, } (\text{m/s})} \times 100$$

$$\text{then} \quad \frac{45.6 \text{ m/s}}{60.0 \text{ m/s}} \times 100 = 0.76 \times 100 = 76$$

Reference

1. Krogness K (1978) Cubital ratio method: physiological slowing of MCV in sulcus of ulnar nerve. Electromyogr Clin Neurophysiol 18:213–216

Axilla, Above Elbow, Below Elbow, Wrist – Hand; Above Elbow to Below Elbow – Hand

M26

Surface Recording Technique, Study from *Abductor Digiti Minimi* (ADM) Muscle

Original Settings Sensitivity, sweep speed, low-frequency filter, high-frequency filter, duration of pulse, and the machine used were not specified.

Position Following the text, the electrophysiological study was performed with the patient lying supine and the arm at 45° to the trunk, the elbow extended, and the forearm supinated.

Recording Using surface electrodes, the active electrode (A) was placed over the *abductor digiti minimi* (ADM) muscle (Fig. 1), and the reference electrode position was not specified in the text; the figure shows reference electrode placed distally at the base of the digit V. The ground (G) electrode position was not specified in the text; the figure shows ground placed on the forearm.

Stimulation For conventional motor conduction study, the author performed ulnar nerve stimulations at four several sites: at the axilla (S1), above the elbow (S2), below the elbow (S3), and at the wrist (S4). A standard distance of 10 cm for the across-elbow (above-elbow–below-elbow) segment and 12 cm from the axilla to above the elbow was used (Figs. 1 and 3). For short-segment incremental stimulation (SSIS) technique (Fig. 2), the author moved the stimulator along the ulnar nerve in several 2 cm steps to detect the site at which an abrupt change occurred in the amplitude of the compound muscle action potential (CMAP) (Fig. 3).

Measurements For the purpose of his study, the author noted the amplitude of the CMAP and the motor nerve conduction velocity (MNCV). By dividing the distance by time, the MNCV along the nerve segment can be determined and measured in meter per second (m/s), and it was calculated for above-elbow–below-elbow (across-elbow) segment. A local demyelinative lesion was diagnosed when the following three criteria were present: (1) an abrupt amplitude change at the cubital tunnel, such that with stimulation just proximal to the tunnel, the CMAP amplitude was reduced by more than 40 % compared with stimulation just distal to the tunnel; (2) slowing of MNCV to less than 35 m/s in the across-elbow segment, which included the cubital tunnel; and (3) a change in form of the CMAP toward a desynchronized response with stimulation above the elbow compared with the wrist. The temperature of the skin and laboratory was not reported. Pathological conduction values (Table 1) were obtained from 15 ulnar nerves of 9 patients with weakness of ulnar-innervated hand muscles as well as sensory loss in the ulnar distribution (6 men and 3 women, age range 31–65 years, mean age 49 years).

Comment

For Miller [1], the precise localization to the cubital tunnel was possible in 12 nerves with an amplitude reduction exceeding 40 % in 9 nerves and 30 % in another 3. Precise localization of the conduction block depended on an abrupt CMAP amplitude change at the cubital tunnel using the SSIS technique, such that the CMAP amplitude was reduced when stimulation was compared just proximal and just distal to the cubital tunnel. The author reported that a change in the CMAP waveform (i.e., a desynchronized response with proximal stimulation) was present in most cases. Using the SSIS technique, abnormal nerve conduction was localized to the cubital tunnel, between 1.5 and 4 cm distal to the medial epicondyle, in each patient; the same procedure was then repeated by the author at surgical decompression in seven patients and confirmed the results obtained in the EMG laboratory. Pathological waveform and values in a patient with ulnar neuropathy at the elbow (UNE) are reported (Figs. 4 and 5).

G. Gentili, M. Di Napoli, *The Ulnar Nerve*, DOI 10.1007/978-3-319-43292-2_50

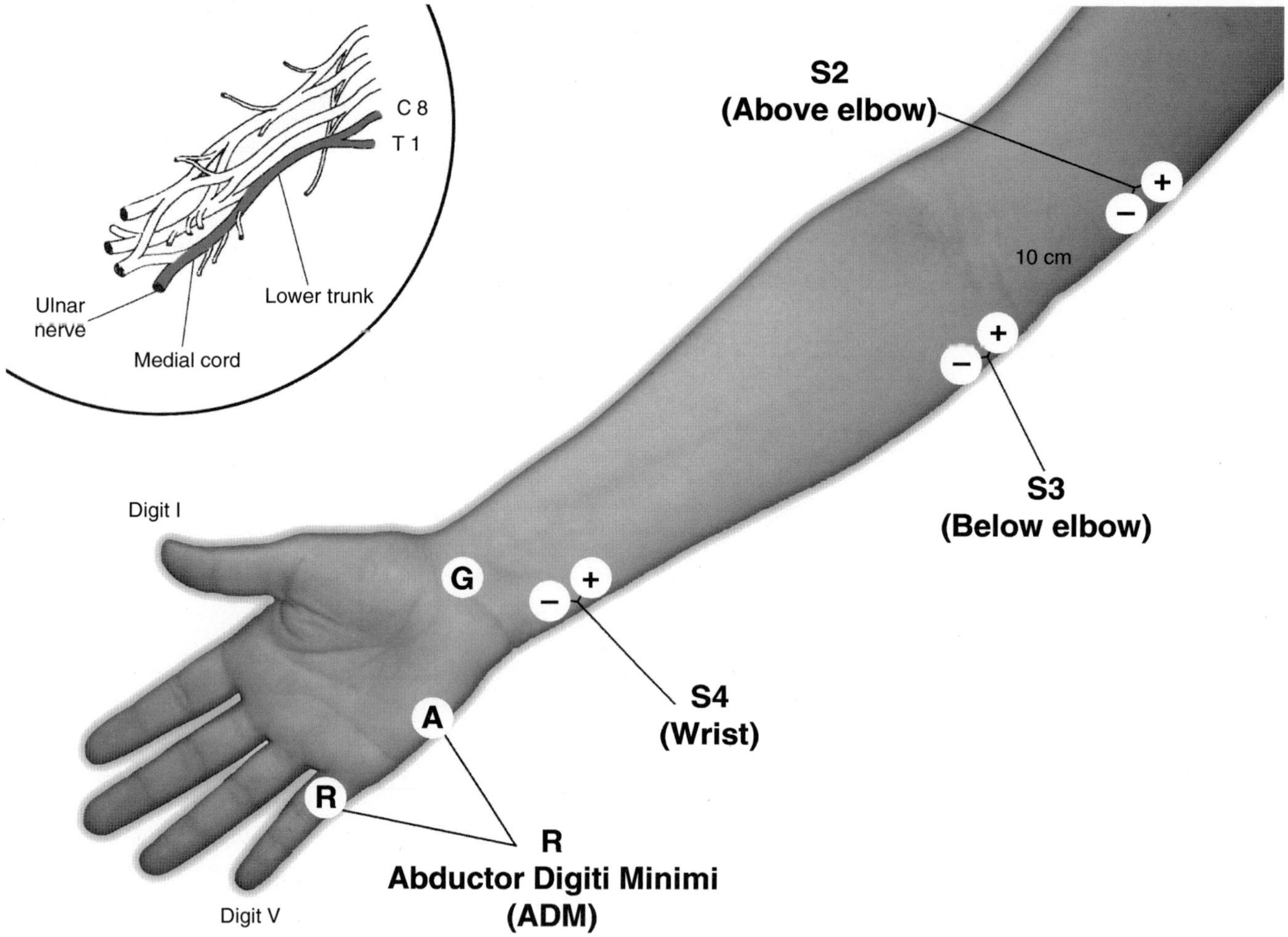

Typical waveform (axilla, above elbow, below elbow, wrist – ADM muscle):

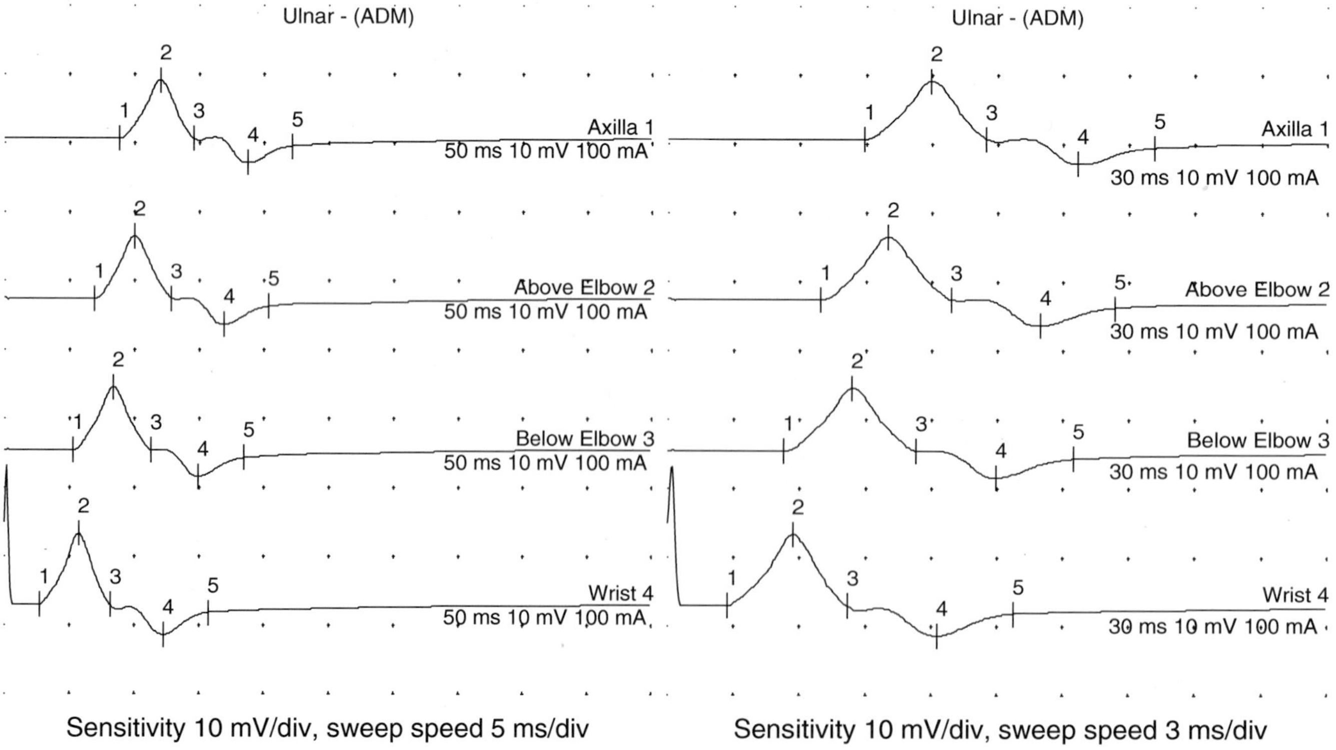

Fig. 1 Compound muscle action potentials (*CMAPs*) recorded at the hand from the ADM muscle, *top to bottom*: stimulation of the axilla, above elbow, below elbow, and of the wrist

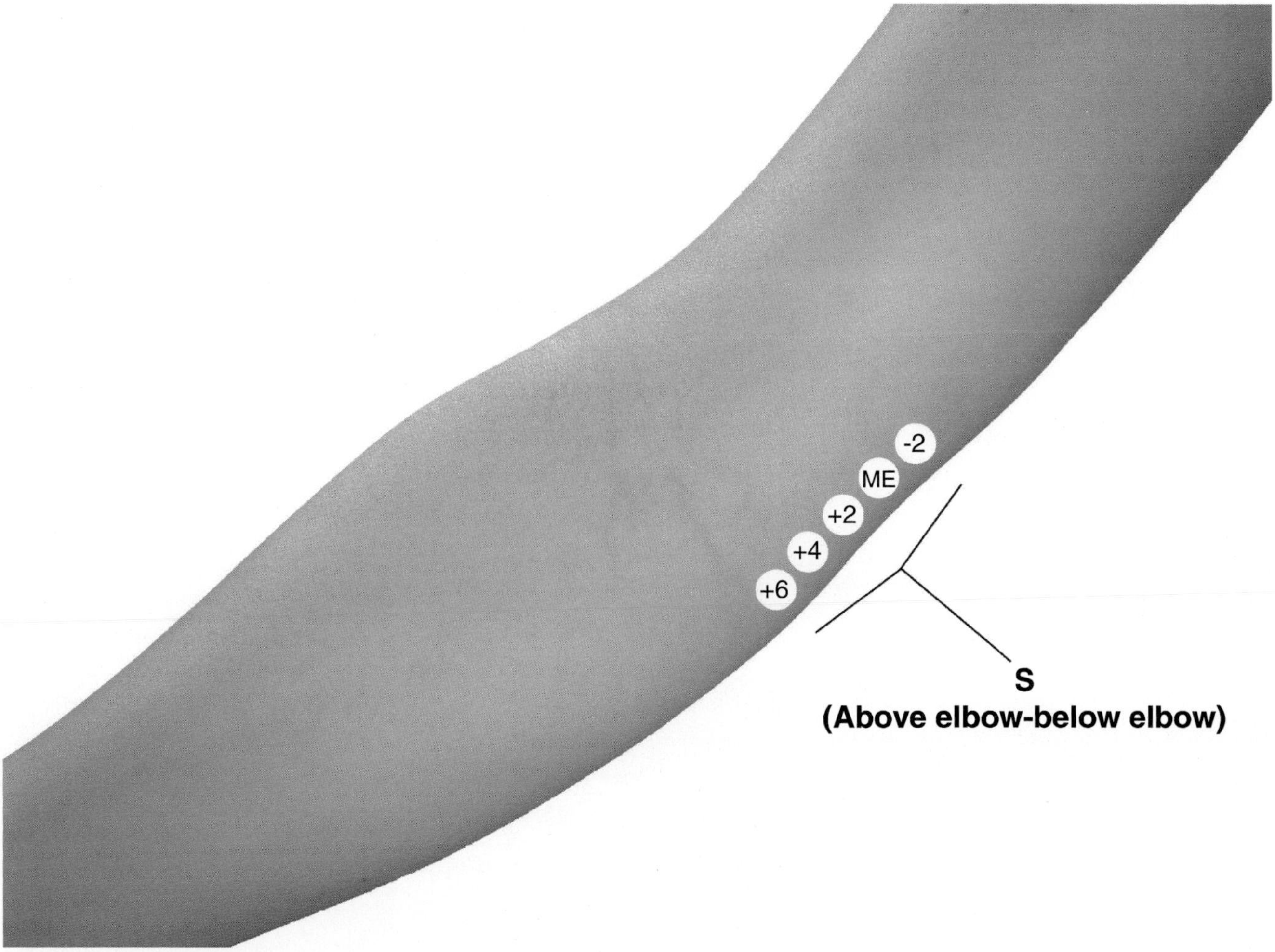

Typical waveform (above elbow to below elbow – ADM muscle):

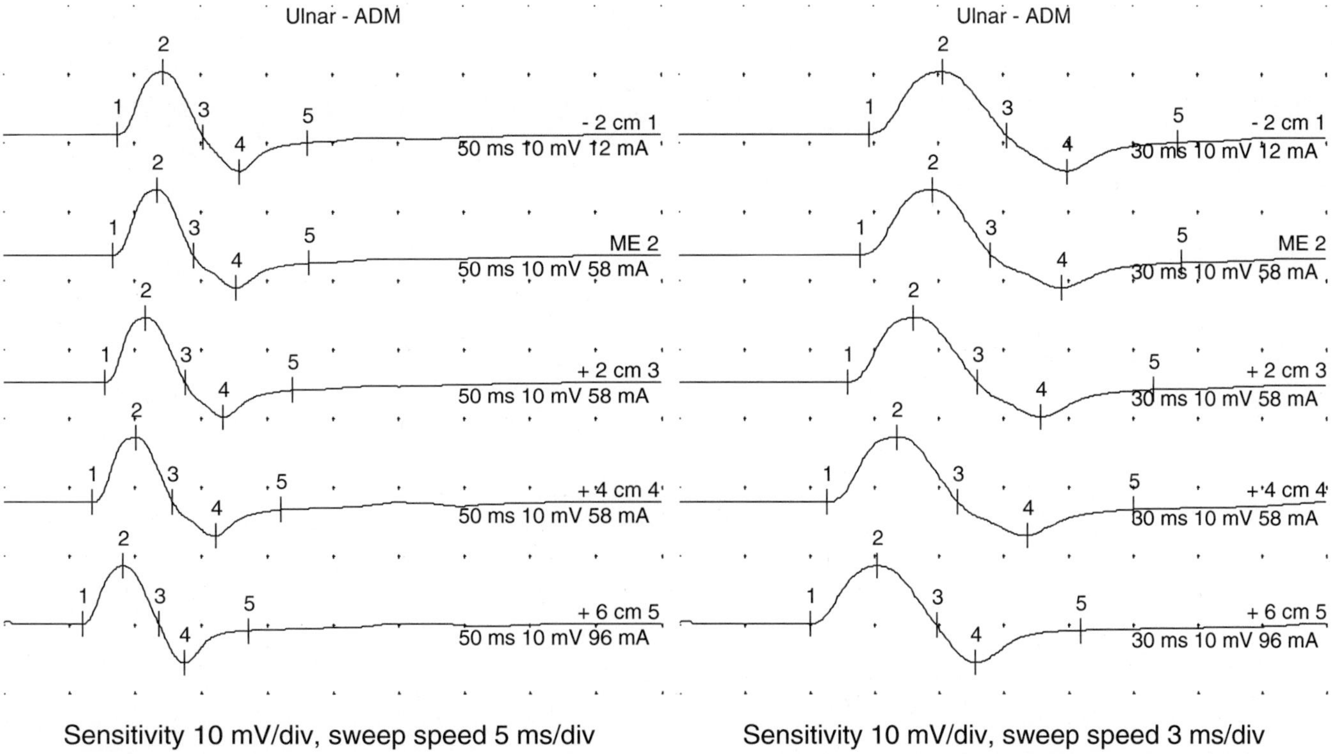

Fig. 2 Compound muscle action potentials (*CMAPs*) recorded at the hand from the ADM muscle, serial stimulation (inching) from the above-elbow to the below-elbow sites

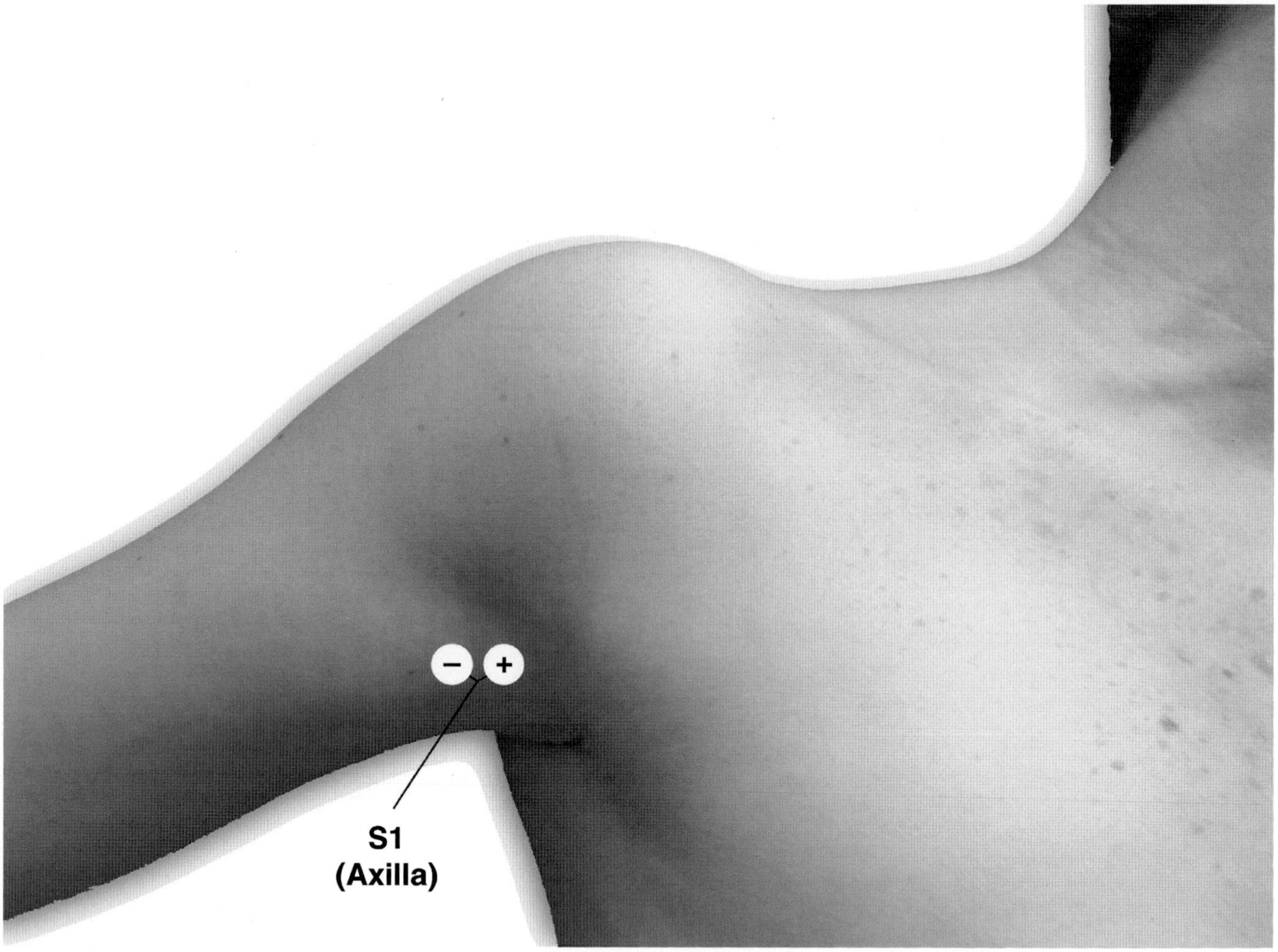

Fig. 3 Proximal stimulation point in the axilla (S1)

Table 1 Pathological values [1]

CMAP amplitude above/below (mV)	Amplitude reduction (%)	MNCV (m/s)
6.0/1.0	83	30
2.5/0.5	80	23
4.5/1.0	78	23
12.0/5.0	58	19
4.0/2.0	50	19
4.0/2.0	50	29
3.4/1.8	47	32
0.7/0.4	43	19
1.4/0.8	43	32
4.2/2.9	31	26
4.5/3.0	33	31
5.0/3.0	40	47
0.8/0.8	0	26
6.0/5.0	17	40
12.0/10.0	17	22

Pathological waveform (axilla, above elbow, below elbow, wrist – ADM muscle):

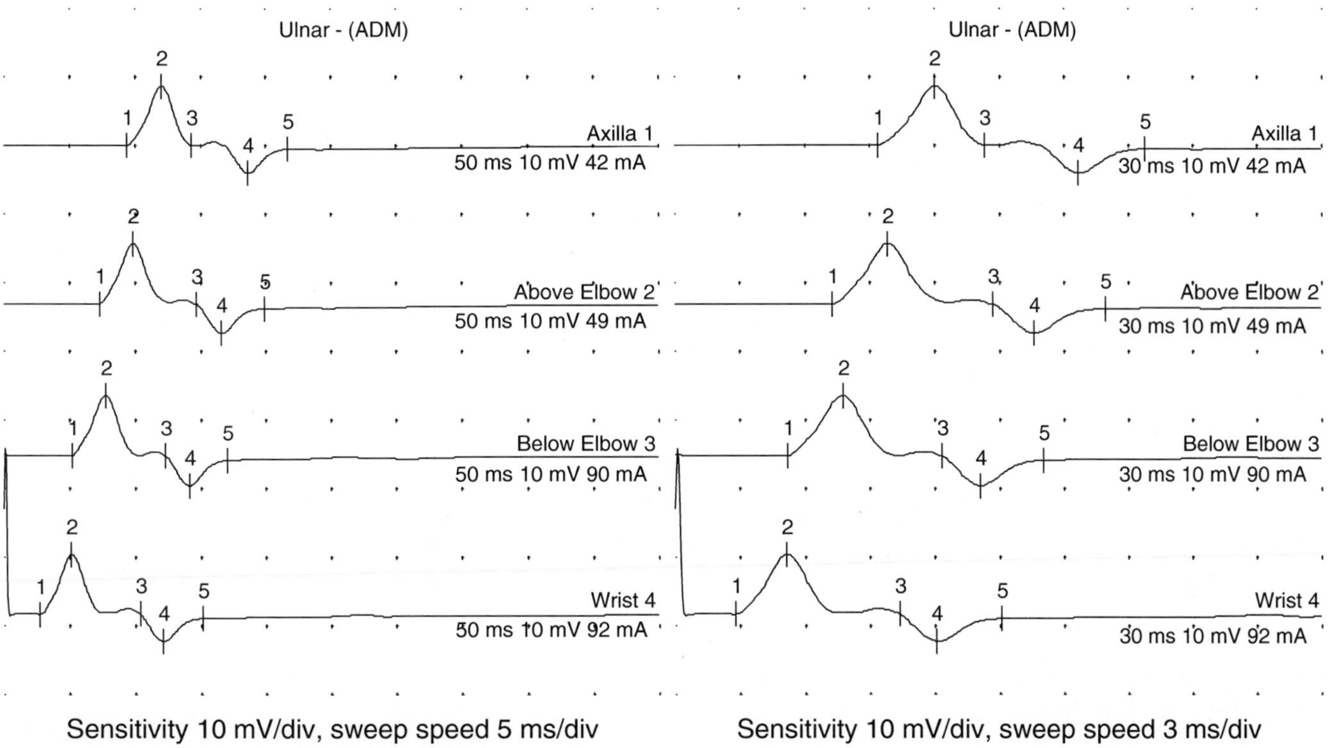

Onset latency (axilla – ADM): 9.15 ms; **Onset latency** (above elbow – ADM): 7.35 ms; **Onset latency** (below elbow – ADM): 5.05 ms; **Onset latency** (wrist – ADM): 2.75 ms; **Peak latency** (axilla – ADM): 12.0 ms; **Peak latency** (above elbow – ADM): 10.05 ms; **Peak latency** (below elbow – ADM): 7.70 ms; **Peak latency** (wrist – ADM): 5.65 ms; **Onset to peak amplitude** (axilla – ADM): 8.7 mV; **Onset to peak amplitude** (above elbow – ADM): 8.8 mV; **Onset to peak amplitude** (below elbow – ADM): 8.9 mV; **Onset to peak amplitude** (wrist – ADM): 8.9 mV; **Peak to peak amplitude** (axilla – ADM): 12.9 mV; **Peak to peak amplitude** (above elbow – ADM): 13.3 mV; **Peak to peak amplitude** (below elbow – ADM): 13.3 mV; **Peak to peak amplitude** (wrist – ADM): 12.8 mV; **MNCV** (axilla – above elbow, 12 cm): 63.6 m/s; **MNCV** (elbow – above elbow, 10 cm): 43.5 m/s; **MNCV** (Erb's point – above elbow, 15 cm): 62.5 m/s

Fig. 4 Compound muscle action potentials (*CMAPs*) recorded at the hand from the ADM muscle using surface recording method [1] in UNE, *top to bottom*: stimulation of the axilla, above elbow, below elbow, and of the wrist

Pathological waveform (above elbow to below elbow – ADM muscle):

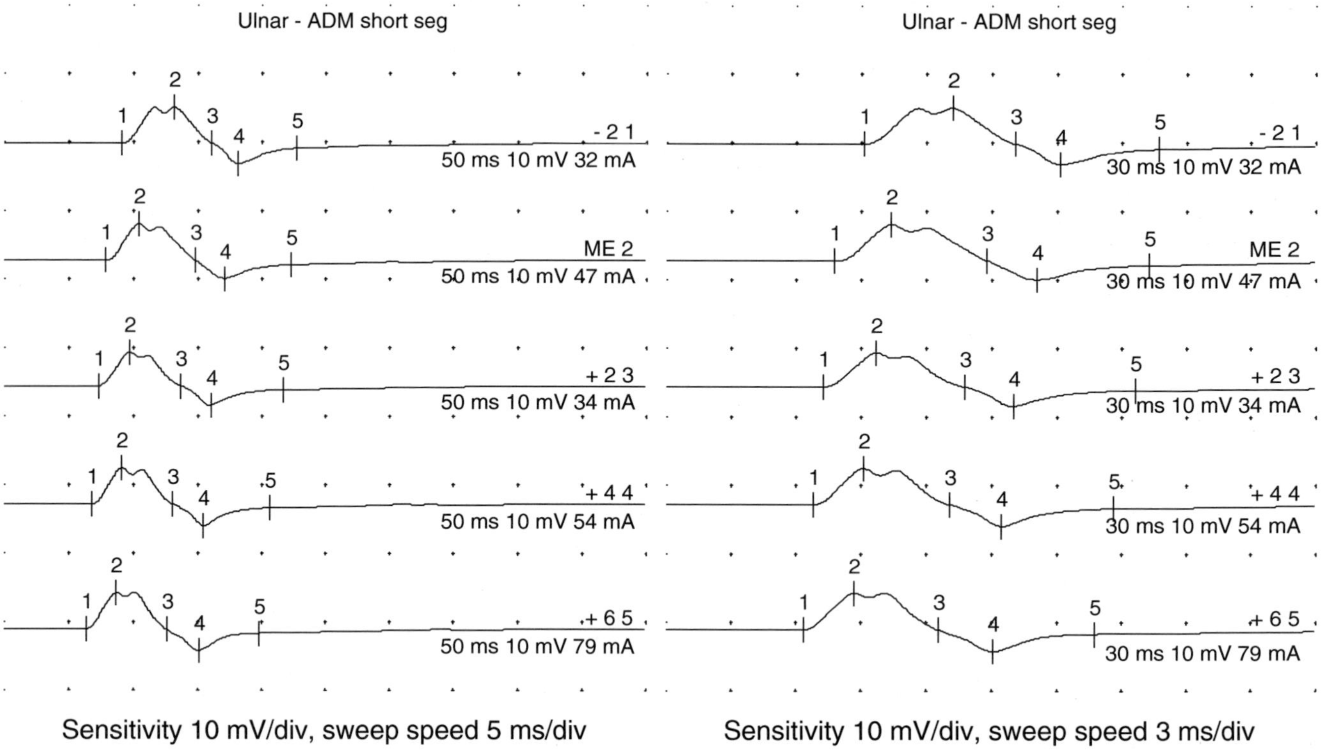

Onset latency (above elbow – ADM, -2): 9.10 ms; **Onset latency** (ME – ADM, level 0 – ME): 7.80 ms; **Onset latency** (below elbow – ADM, +2): 7.25 ms; **Onset latency** (below elbow – ADM, +4): 6.80 ms; **Onset latency** (below elbow – ADM, +6): 6.35 ms; **Peak latency** (above elbow – ADM, -2): 13.30 ms; **Peak latency** (ME – ADM, level 0): 10.25 ms; **Peak latency** (below elbow – ADM, +2): 9.35 ms; **Peak latency** (below elbow – ADM, +4): 9.10 ms; **Peak latency** (below elbow – ADM, +6): 8.85 ms; **Onset to peak amplitude** (above elbow – ADM, -2): 5.3 mV; **Onset to peak amplitude** (ME – ADM, level 0 – ME): 5.3 mV; **Onset to peak amplitude** (below elbow – ADM, +2): 5.0 mV; **Onset to peak amplitude** (below elbow – ADM, +4): 5.3 mV; **Onset to peak amplitude** (below elbow – ADM, +6): 5.4 mV

Fig. 5 Compound muscle action potentials (*CMAPs*) recorded at the hand from the ADM muscle using surface recording method [1] in a patient with UNE and diabetic polyneuropathy, serial stimulation (inching) from the above-elbow to the below-elbow sites (increased delay between –2 and ME points, no abrupt CMAP amplitude change)

Reference

1. Miller RG (1979) The cubital tunnel syndrome: diagnosis and precise localization. Ann Neurol 6:56–59

Wrist, Below Elbow, Above Elbow, Upper Arm – Hand; Upper Arm, Above Elbow – Forearm

M27

Needle Recording Technique, Study from *Abductor Digiti Minimi* (ADM) and *Flexor Carpi Ulnaris* (FCU) Muscles

Original Settings Duration of pulse was 0.1 ms; sensitivity, sweep speed, low-frequency filter, high-frequency filter, rate of pulse, and the machine used were not specified.

Position This study was performed with the subjects lying on a bed in the supine position, with the arm at 45° to the trunk, the elbow extended, and the forearm supinated.

Recording The active needle electrode (0.6 mm in diameter) was placed over the motor point [1] of the ulnar *hypothenar* (UHT) muscle–*abductor digiti minimi* (ADM) muscle (Fig. 1) and *flexor carpi ulnaris* (FCU) muscle (Fig. 2). The ground (G) electrode position was not specified in the text; the figure shows the ground electrode positioned over the forearm.

Needle Insertion For the ADM muscle, with the hand in a neutral position to full supination, the recording needle electrode was inserted at the midpoint between the fifth metacarpophalangeal joint and the pisiform bone. For the FCU muscle, with the forearm in the supine position, the recording needle electrode was inserted at about 10 cm distally from the medial epicondyle.

Stimulation Supramaximal stimulation of the ulnar nerve was effected by means of surface electrodes placed at the wrist, 2 cm above the distal wrist crease; below the elbow, 3–5 cm distal to the medial epicondyle; above the elbow, 2 cm above the medial epicondyle; and upper the arm, 12–18 cm above the medial epicondyle. The anode was proximal in all stimulations (Figs. 1, 2, and 3). For FCU muscle recordings, only the upper arm and above-elbow stimulations were used. The stimulus was a rectangular pulse of 0.1 ms duration.

Measurements Distal latency (ms) of the compound muscle action potential (CMAP) was measured from the stimulus onset to the onset of the initial deflection of the CMA. Motor nerve conduction velocity (MNCV) was calculated for the below-elbow–above-elbow and above-elbow–upper arm segments. The MNCV was measured in meter per second (m/s). The skin and room temperatures were not reported. The authors studied 44 ulnar nerves from 23 healthy subjects (age range 18–68 years) and 68 ulnar nerves from 64 patients with lesions at the elbow (40 right and 28 left, age range 18–84 years). Normal and pathological values were not reported.

Comment

Benecke and Conrad [1] observed that the latency increased with increasing distance between stimulating and recording electrodes. They conducted all measurements of the latency with the recording electrode positioned 10 cm below the medial epicondyle, because at this point the FCU muscle could most easily be identified and the standard deviation of the values obtained was low (3.12±0.3 ms). Using this recording technique, there was no significant relationship between forearm length and latency to the FCU muscle. A pathologically prolonged distal latency to FCU muscle (>4.0 ms) was found in 82.4% of the patients. For the authors [1], the evaluation of the distal latency to the FCU muscle is not only subject to fewer sources of error but is also technically much easier to perform. In the localization of a clinically suspected ulnar nerve lesion at the elbow, they suggest that the electrophysiological investigation should commence with an examination of the distal latency to the FCU muscle. Pathological waveforms and values in a case of ulnar neuropathy at the elbow (UNE) are reported (Fig. 4).

G. Gentili, M. Di Napoli, *The Ulnar Nerve*, DOI 10.1007/978-3-319-43292-2_51

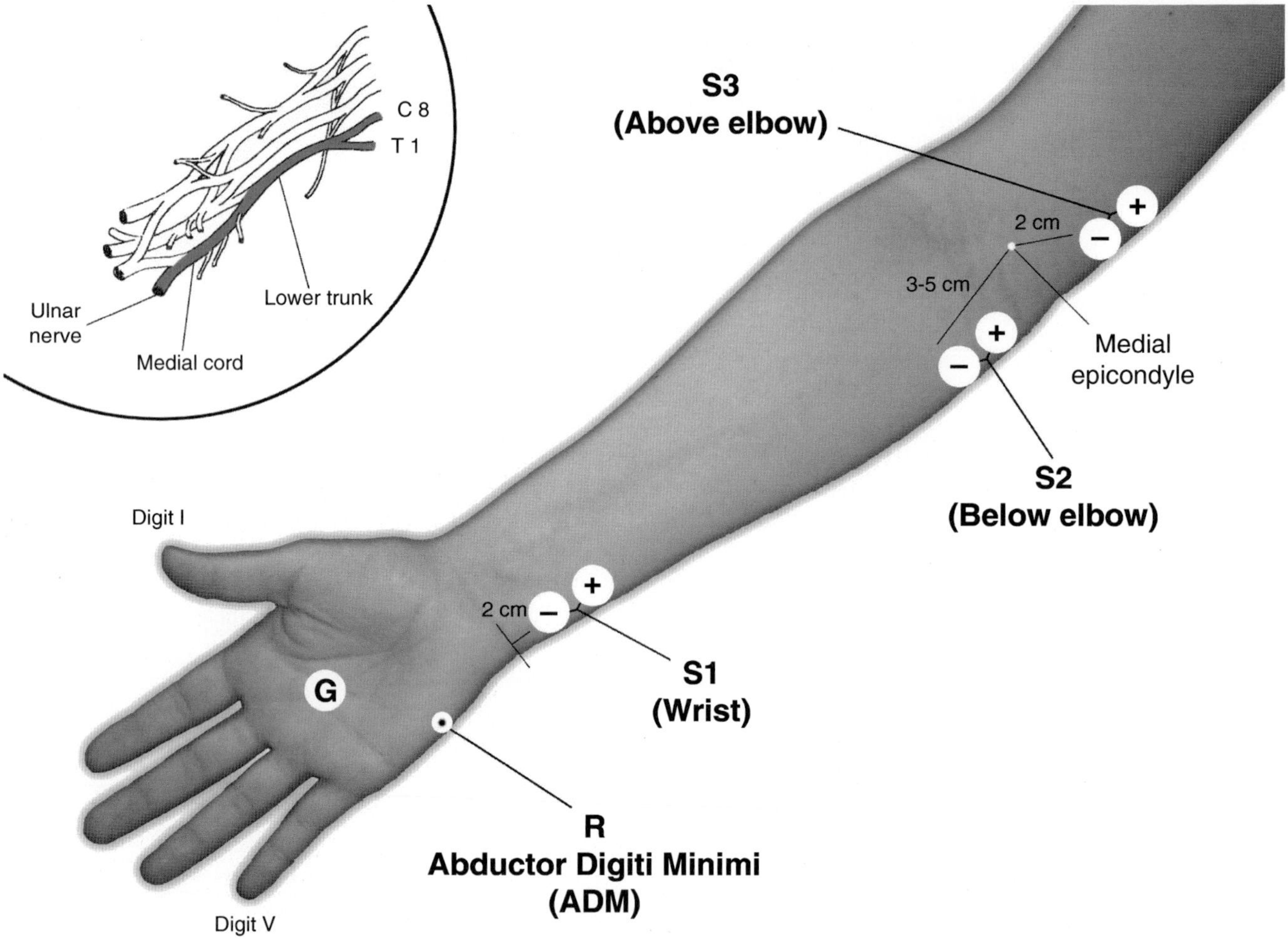

Typical waveform (wrist, below elbow, above elbow, upper arm – ADM muscle):

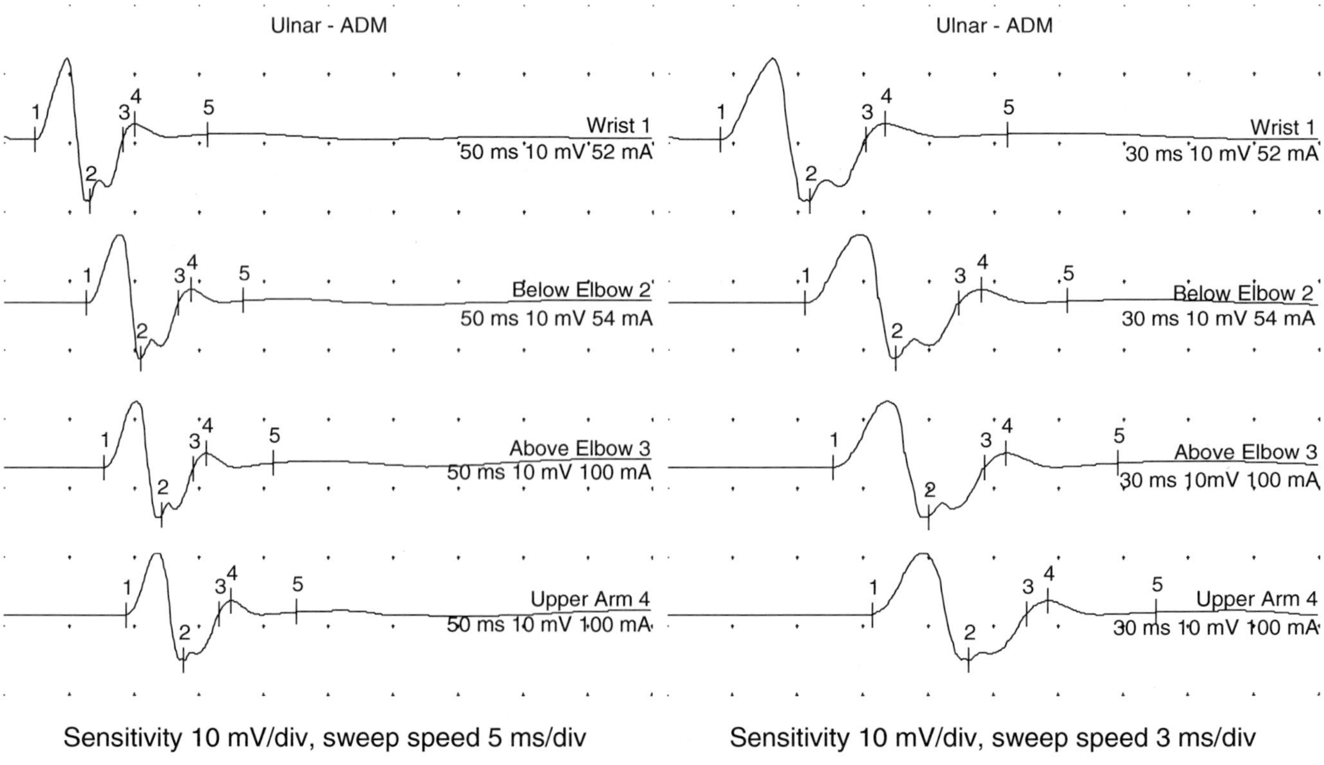

Fig. 1 Compound muscle action potentials (*CMAPs*) recorded at the hand from the ADM muscle, *top to bottom*: stimulation of the wrist, below elbow, above elbow, and upper arm

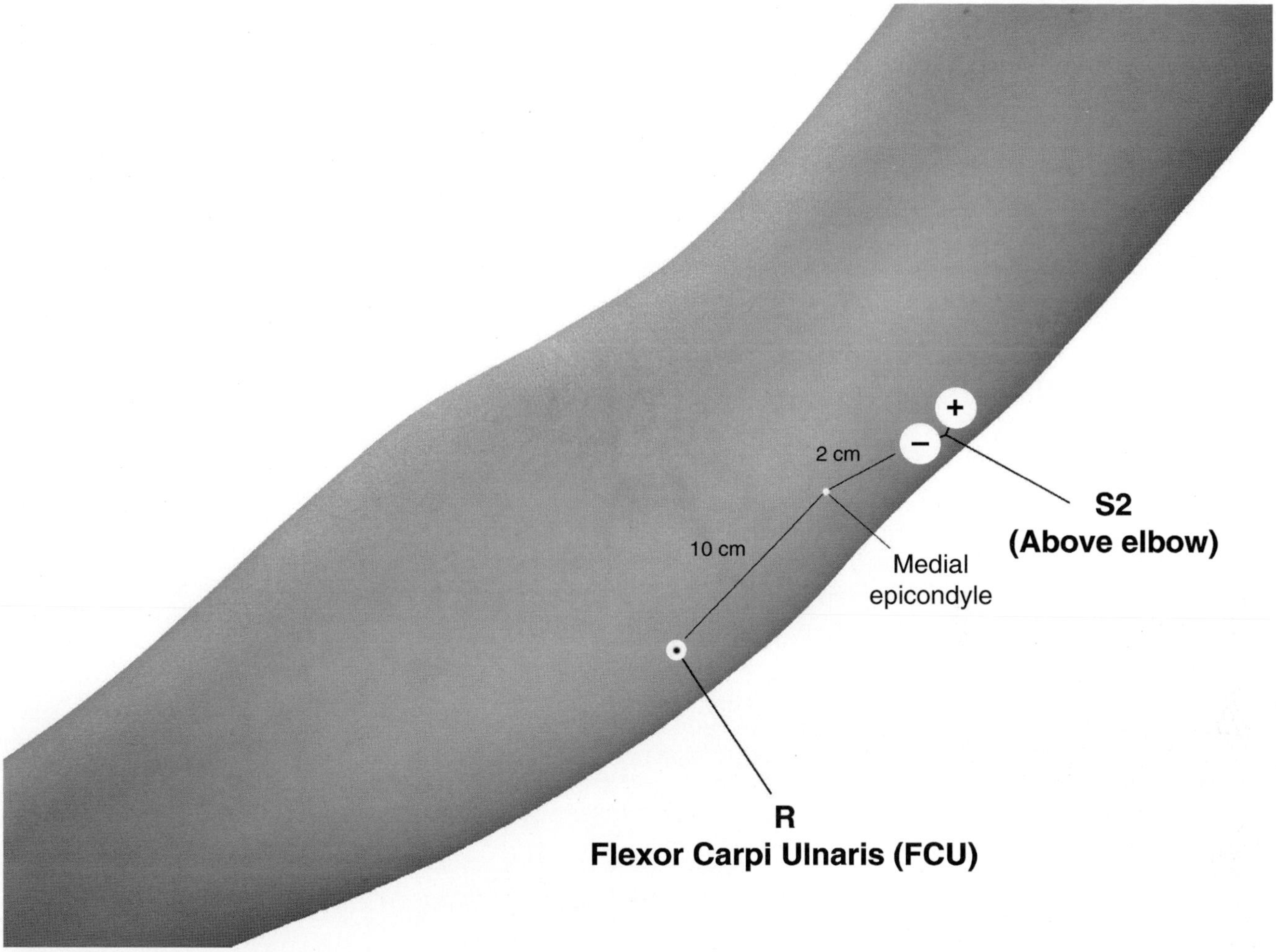

Typical waveform (upper arm, above elbow – FCU muscle):

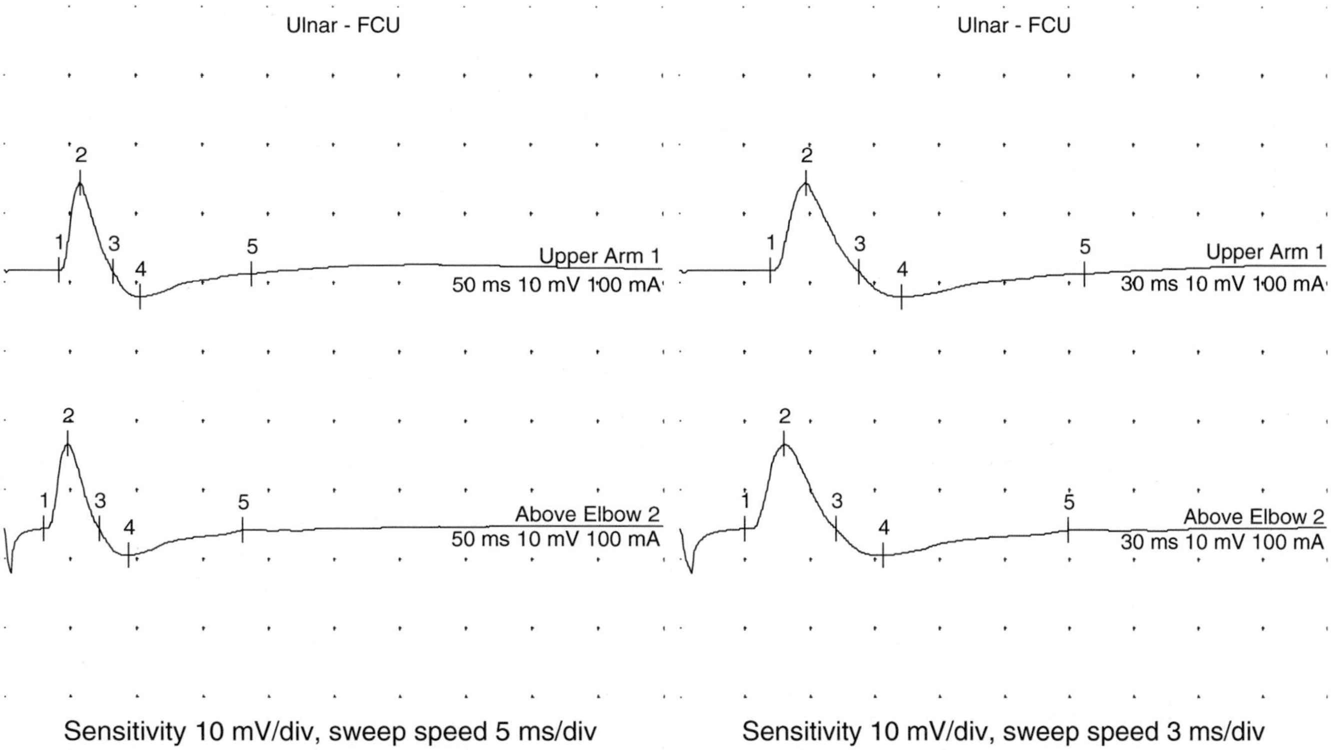

Fig. 2 Compound muscle action potentials (*CMAPs*) recorded at the hand from the FCU muscle, stimulation at the upper arm (*upper trace*), and above elbow (*lower trace*)

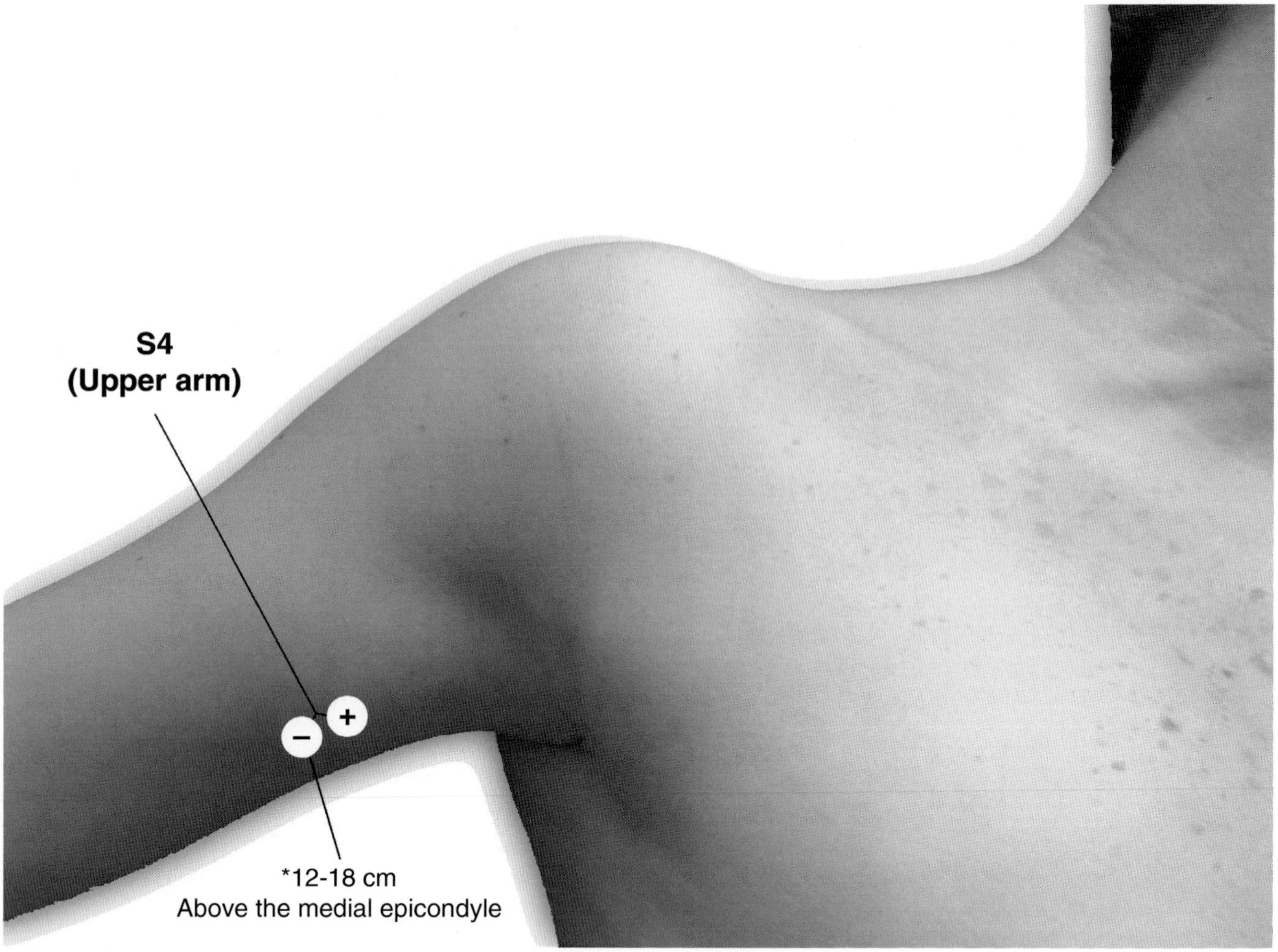

Fig. 3 Proximal stimulation point in the upper arm (S4 for the ADM study and S1 for the FCU study)

Pathological waveform (wrist, below elbow, above elbow, upper arm – ADM muscle):

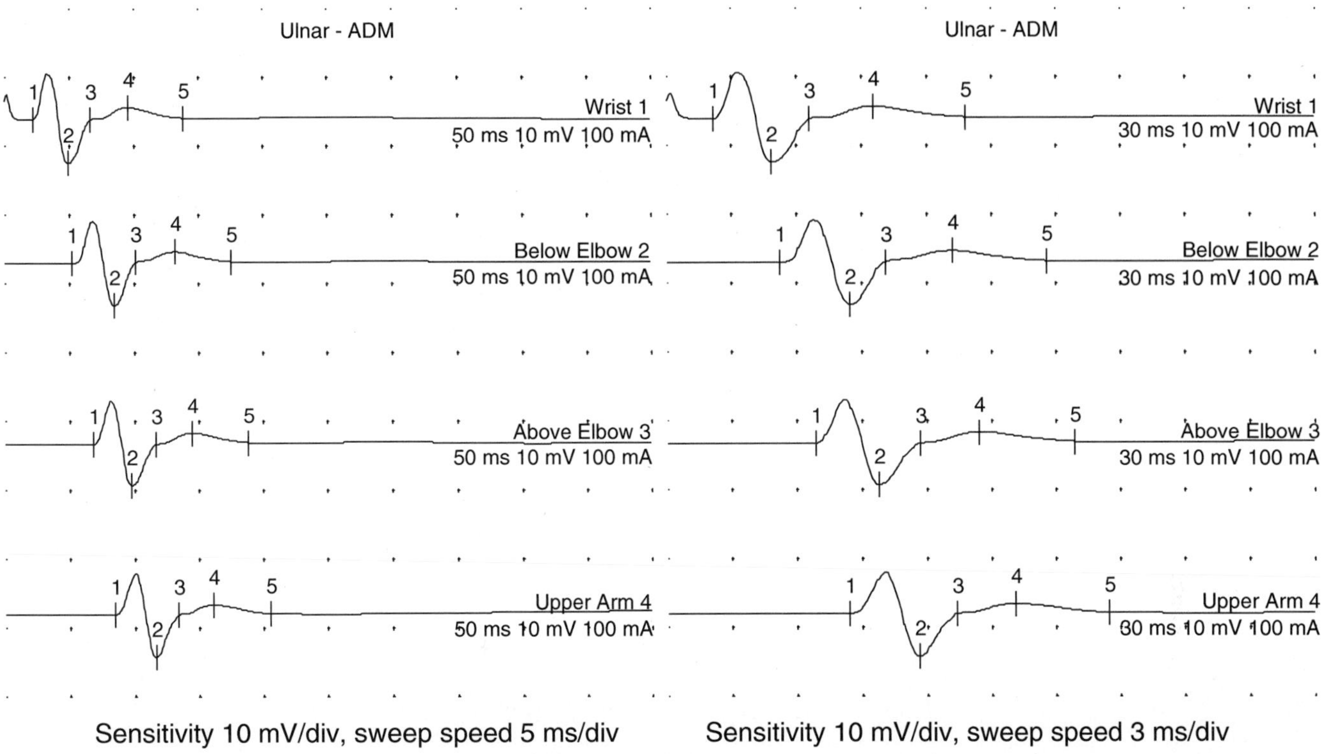

Onset latency (wrist – ADM): 2.15 ms; **Onset latency** (below elbow – ADM): 5.20 ms; **Onset latency** (above elbow – ADM): 6.85 ms; **Onset latency** (upper arm – ADM): 8.45 ms; **MNCV** (wrist – below elbow, 19 cm distance): 62.3 m/s; **MNCV** (below elbow – above elbow, 6 cm distance): 36.4 m/s; **MNCV** (above elbow – upper arm, 10 cm distance): 62.5 m/s

Fig. 4 Compound muscle action potentials (*CMAPs*) recorded at the hand from the ADM muscle in UNE, *top to bottom*: stimulation of the wrist, below elbow, above elbow, and upper arm

Reference

1. Benecke R, Conrad B (1980) Value of electrophysiological examination of flexor carpi ulnaris muscle in diagnosis of ulnar nerve lesions at elbow. J Neurol 223:207–217

Wrist – Hand

M28

Surface Recording Technique, Study from *First Dorsal Interosseous* (FDI) and *Abductor Digiti Minimi* (ADM) Muscles

Original Settings Sensitivity, sweep speed, low-frequency filter, high-frequency filter, duration of pulse, rate of pulse, and the machine used were not specified.

Position Arm and hand position were not specified in the text; we perform the test with the forearm and hand in the supine position.

Recording For the *first dorsal interosseous* (FDI) muscle recording (R1), the active electrode (A) was placed over the belly of the FDI muscle (Fig. 1), and the reference (R) was placed slightly distal to the second metacarpophalangeal joint (over the tendon of the FDI muscle). For the *abductor digiti minimi* (ADM) muscle (R2), the active electrode was placed over the motor point of the ADM muscle (Fig. 1), and the reference (R) was placed over the fifth metacarpophalangeal joint [1]. Prior to data collection and with reference electrode over the second metacarpophalangeal joint, variable locations for the active recording electrode over the FDI were studied by the authors, to determine the point which provided maximum amplitude. Localized where submaximal percutaneous stimulation over the FDI produced maximal visible twitch, the motor point was at the proximal edge of the muscle, near the apposition of the first and second carpometacarpal joints. Although this motor point placement of the active recording electrode usually resulted in initial negative deflection for the compound muscle action potential (CMAP), the amplitude of the CMAP was frequently increased one to several millivolts (mV) by moving the electrode over the center of the fleshy belly of the FDI muscle. The authors [1] observed that a larger CMAP from FDI muscle was frequently associated with a small initial positive deflection. In the same manner, the active recording electrode over the ADM muscle was placed at the motor point located by submaximal percutaneous stimulation. The FDI and ADM muscle recordings were performed simultaneously. The ground (G) electrode position was not specified in the text; the figure shows the ground electrode placed on the palm (Fig. 1).

Stimulation Supramaximal stimulation was performed at the wrist [1], on the course of the ulnar nerve (S), with the cathode at the proximal wrist crease (Fig. 1).

Measurements Distal motor latency (ms) was measured from the baseline to the onset of first negative deflection of the CMAP, for both ADM and FDI muscle recordings. This deflection was always negative for ADM muscle recordings but usually small and briefly positive for the FDI muscle recordings. Amplitude (mV) was measured from the baseline to the negative peak. Normal values (Tables 1, 2, 3, and 4) were obtained on 373 ulnar nerves of 188 healthy subjects (14 subjects < 20 years, 48 subjects 20–29 years, 103 subjects 30–39 years, 82 subjects 40–49 years, 84 50–59 years, 34 60–69 years, and 8 > 70 years). Subjects were evaluated in two laboratories, and 185 subjects were studied bilaterally. Subjects were free from signs and electrophysiologic evidence of polyneuropathy and ulnar neuropathy.

G. Gentili, M. Di Napoli, *The Ulnar Nerve*, DOI 10.1007/978-3-319-43292-2_52

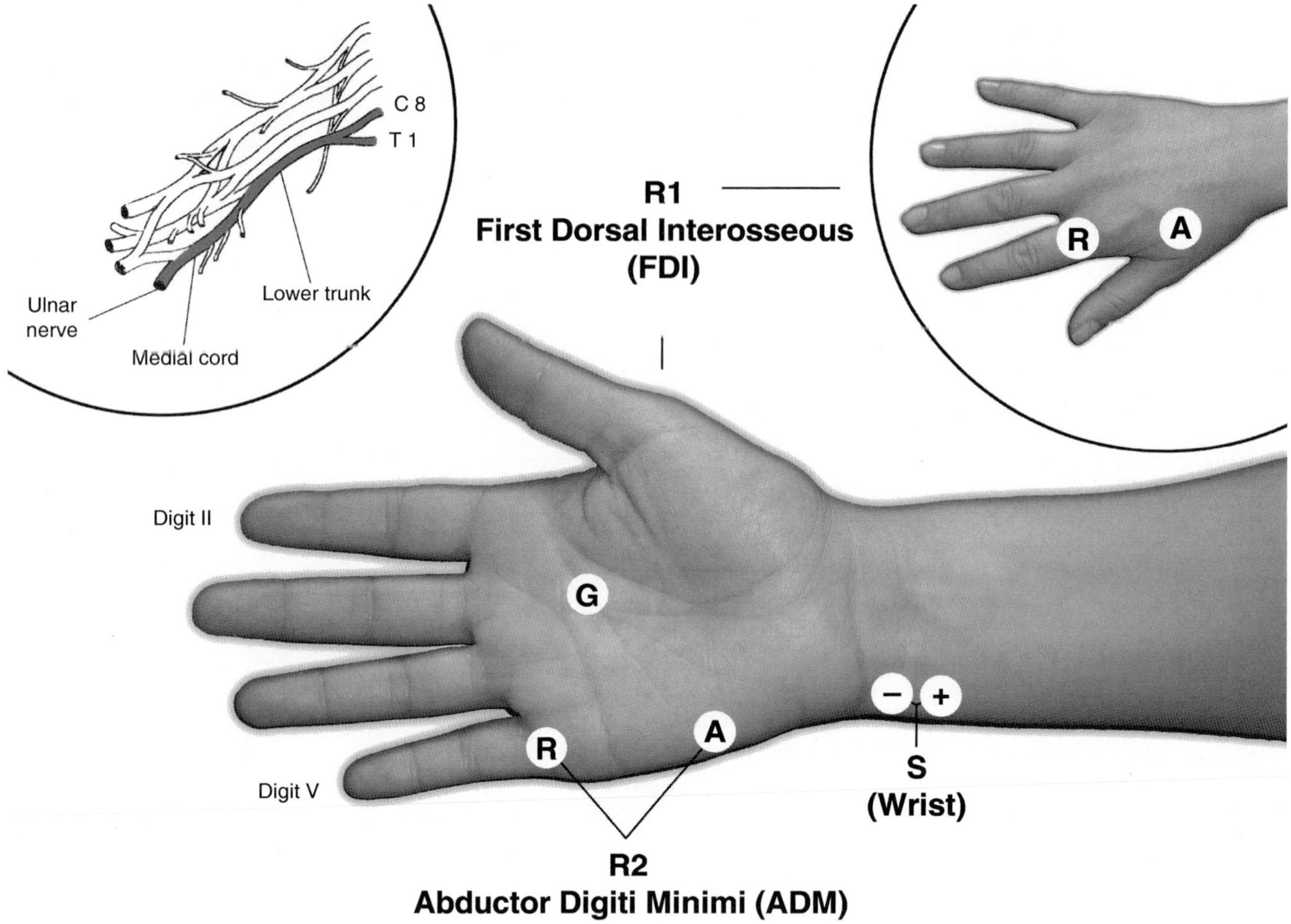

Typical waveform (wrist – FDI muscle, wrist – ADM muscle):

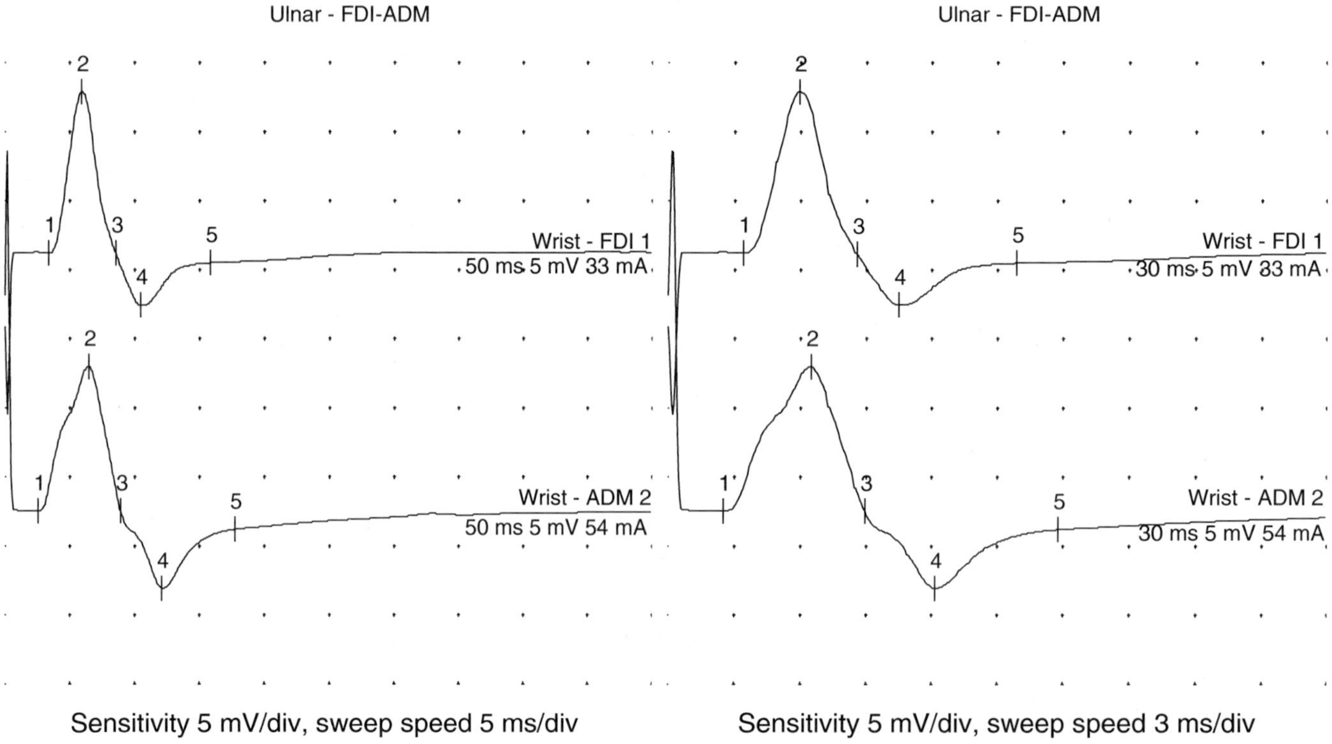

Fig. 1 Compound muscle action potentials (*CMAPs*) recorded at the hand from the FDI (*upper trace*) and ADM (*lower trace*) muscles, stimulation of the ulnar nerve at the wrist

Table 1 Normal values [1] – ADM muscle recordings

	Mean	Range
Wrist–ADM, distal latency (ms) – subjects < 20 years	2.5	2.2–2.9
Wrist–ADM, distal latency (ms) – subjects 20–29 years	2.5	2.0–3.0
Wrist–ADM, distal latency (ms) – subjects 30–39 years	2.4	1.8–3.2
Wrist–ADM, distal latency (ms) – subjects 40–49 years	2.5	2.0–3.0
Wrist–ADM, distal latency (ms) – subjects 50–59 years	2.6	2.0–3.4
Wrist–ADM, distal latency (ms) – subjects 60–69 years	2.7	2.2–3.1
Wrist–ADM, distal latency (ms) – subjects > 70 years	2.7	2.3–3.1

Table 2 Normal values [1] – FDI muscle recordings

	Mean	Range
Wrist–FDI, distal latency (ms) – subjects < 20 years	3.3	2.7–4.2
Wrist–FDI, distal latency (ms) – subjects 20–29 years	3.4	2.6–4.1
Wrist–FDI, distal latency (ms) – subjects 30–39 years	3.3	2.5–4.4
Wrist–FDI, distal latency (ms) – subjects 40–49 years	3.2	2.3–4.2
Wrist–FDI, distal latency (ms) – subjects 50–59 years	3.4	2.6–4.4
Wrist–FDI, distal latency (ms) – subjects 60–69 years	3.6	3.0–4.5
Wrist–FDI, distal latency (ms) – subjects > 70 years	3.6	3.0–4.2

Table 3 Normal values [1] – ADM muscle recordings

	Mean	Range
Wrist–ADM, amplitude (mV) – subjects < 20 years	13	11–16
Wrist–ADM, amplitude (mV) – subjects 20–29 years	12	5–20
Wrist–ADM, amplitude (mV) – subjects 30–39 years	12	6–21
Wrist–ADM, amplitude (mV) – subjects 40–49 years	12	6–19
Wrist–ADM, amplitude (mV) – subjects 50–59 years	11	7–17
Wrist–ADM, amplitude (mV) – subjects 60–69 years	12	6–15
Wrist–ADM, amplitude (mV) – subjects > 70 years	10	8–13

Table 4 Normal values [1] – FDI muscle recordings

	Mean	Range
Wrist–FDI, amplitude (mV) – subjects < 20 years	15	8–23
Wrist–FDI, amplitude (mV) – subjects 20–29 years	14	8–22
Wrist–FDI, amplitude (mV) – subjects 30–39 years	15	6–24
Wrist–FDI, amplitude (mV) – subjects 40–49 years	13	6–22
Wrist–FDI, amplitude (mV) – subjects 50–59 years	13	6–20
Wrist–FDI, amplitude (mV) – subjects 60–69 years	12	7–20
Wrist–FDI, amplitude (mV) – subjects > 70 years	12	8–15

Comment

For Olney and Wilbourn [1], the limit of normal for side-to-side latency difference was 0.2 ms (range 0–1.0 ms) for ADM muscle recording and 0.2 ms (range 0–1.3 ms) for the FDI muscle recording, and the limit of normal for ipsilateral latency difference between the FDI and ADM muscle recording was 0.9 ms (range 0.2–2.0 ms). Advancing age was associated with increasing distal motor latency for each muscle in the range of 0.05 ms per decade, but side-to-side latency difference between the FDI and ADM muscles did not significantly change with advancing age. Advancing age was associated with amplitude decreases of 0.5 mV per decade. Greater variability was found with amplitude measurements. They commonly observed greater FDI CMAP amplitude than that of ADM by an average of 3 mV (up to a maximum difference of 11 mV), but the ADM CMAP amplitude was in 20 % of the nerves studied greater than CMAP obtained from the FDI muscle. The authors [1] believed that the most precise and sensitive approach for utilizing latencies to detect distal compression is to compare the distal motor latency for each muscle with its contralateral counterpart and to contrast the ipsilateral difference in latency. They observed that side-to-side latencies do not normally differ by more than 1.0 ms for ADM muscle or 1.3 ms for FDI muscle, and the distal motor latency for the FDI does not normally exceed the ADM latency by more than 2.0 ms. For the authors, if motor conduction study to the FDI muscle is more routinely performed, ulnar neuropathy at the wrist (UNW) should become more frequently recognized, and the potential for surgical remediation should become better achieved.

For Olney and Hanson [2], with stimulation at the proximal wrist crease (typically 5.5.–6.5 cm proximal to the ADM active recording electrode), the distal motor latency does not normally exceed 3.4 ms (ADM muscle recording) and 4.5 ms (FDI muscle recording). The distal motor latency to the FDI muscle does not normally exceed the latency to the ipsilateral ADM muscle by more than 2 ms or the latency to the contralateral FDI muscle by more than 1.3 ms. The authors found that the amplitude of the CMAP was normally at least 5 mV over the ADM muscle and 6 mV over the FDI muscle. They described five different patterns of UNW based on the site of the lesion: (1) the lesion of the deep branch after it has given off the hypothenar branch produces weakness in the interosseous and lumbrical but not the ADM muscle; sensation is normal; (2) the lesion is at or just proximal to the hypothenar branch; severe interosseous and lumbrical weakness develops in association with at least mild ADM weakness; (3) the lesion is near or just proximal to the major bifurcation, and it causes weakness of all ulnar intrinsic hand muscles and impaired sensation in the superficial ulnar distribution; (4) the lesion is limited to the superficial branch, and it produces impaired sensation in the superficial, but not the dorsal cutaneous, ulnar distribution; (5) the lesion is due to an anomalous innervations patterns or from quite focal injuries, such as with sharp penetrating objects. In such cases, some authors observed the sensory branch involved in association with a pattern of weakness limited to the ADM or the interosseous muscles [3]. The recognition of the UNW pattern is obviously important for proper therapy. The more routine inclusion of the motor nerve conduction study of the FDI muscle and sensory nerve conduction study of the dorsal cutaneous branch were suggested to be included in the routine electrodiagnostic studies (EDX) session for UNW diagnosis [2]. The dorsal ulnar cutaneous sensory response, as a matter of fact, is not affected by a lesion at or near the canal of Guyon but is likely to be abnormal with an ulnar nerve lesion that occurs more than 6–8 cm proximal to the ulnar styloid, and ulnar motor nerve conduction study recording from the FDI muscle as well as the ADM muscle is more sensitive than other methods in detecting a distal ulnar lesion. Furthermore, for both the sensory and motor conduction studies of the ulnar nerve, a comparison with similar studies on the asymptomatic side may reveal significant asymmetries, even when latency and amplitude are within normal laboratory limits otherwise [1].

Kim [4] performed ulnar nerve conduction studies in 214 hands of 107 healthy subjects (54 women and 53 men, mean age 42.6 ± 11.2 years), and ulnar motor conduction study was performed separately with both ADM and FDI muscles in healthy subjects and patients with UNW. Among the various techniques developed to diagnose UNW, the ulnar nerve conduction study on the FDI muscle is very useful, and some authors [5] suggested that a conduction study to the FDI muscle should be done in all patients with suspected UNW. However, the shortcoming of FDI muscle recording is the initial positive deflection of CMAP, which poses a problem in measuring distal motor onset latency. Wertsch [6] demonstrated that the initial positive deflection was strongly influenced by the position of the reference recording electrode, and the reference electrode placement on the digit I (thumb) could avoid an initial positive deflection that may be seen when the reference is placed on the digit II (Fig. 2), following Olney and Wilbourn's method [1]. It is also demonstrated [7] that the CMAP waveform shape and onset latency could be affected by the different reference electrode position, and some researchers [7–9] used the base of the digit I as a reference electrode for FDI muscle recording (Fig. 3). The study proposed by Kim [4] was designed to derive normative values of ulnar nerve conduction study with the active recording electrode placed over the belly of the FDI muscle and the reference electrode on the proximal phalanx of the digit I. The author [4] performed both conventional ADM and FDI muscles recordings with the elbow placed in 90° flexion, and the ulnar nerve was supramaximally stimulated at the wrist (S1), 8 cm proximal to the active recording electrode for the ADM; below the elbow (S2), 3 cm distal to the medial epicondyle; and above the elbow (S3), 7 cm proximal to the medial epicondyle (Figs. 2, 3, and 4). Distal onset latency (ms), negative peak amplitude (mV), ipsilateral FDI-ADM muscle distal latency difference (ms), side-to-side latency (ms), and amplitude (mV) differences of the ulnar CMAP to the ADM and FDI muscles were calculated in healthy subjects (Tables 5, 6, 7, and 8) using a NICOLET Viking IV electrodiagnostic system (low-frequency and high-frequency filters were 2 Hz and 10 kHz, respectively).

Comment

Kim [4] in his study reported normative values for several parameters, such as the distal motor latency to the ADM muscle (mean ± 2SD, 3.5 ms; maximal value, 3.8 ms), the distal motor latency to the FDI muscle (mean ± 2SD, 4.2 ms; maximal value, 4.4 ms), the ipsilateral FDI-ADM latency difference (mean ± 2SD, 1.1 ms; maximal value, 1.4 ms), the side-to-side latency difference to the ADM muscle (mean ± 2SD, 0.4 ms; maximal value, 0.5 ms), and the side-to-side latency difference to the FDI muscle (mean ± 2SD, 0.4 ms; maximal value, 0.6 ms). There was no significant difference in the distal motor latency to the FDI and ADM muscles between Kim's method [4] and the previously developed technique by Olney and Wilbourn [1]. However, his study revealed that the ipsilateral latency difference between the FDI and ADM muscles, side-to-side latency difference to the ADM muscle, and side-to-side latency difference to the FDI were shorter than those of Olney and Wilbourn's study. The author postulated that the shorter side-to-side latency to the FDI and ADM muscles in his study might be caused by the differences of stimulation site and the different position of reference recording electrode. Olney and Wilbourn [1] used the proximal wrist crease as stimulation site, which may cause differences in stimulation distance between both sides, but the ulnar stimulation at 8 cm proximal to the active recording electrode placed over the belly of the FDI and ADM muscles allowed a more uniform stimulation distance. Also, Kim found the change of the reference electrode placement to the digit I from the digit II helpful to eliminate the measurement error by the initial positive deflection. The FDI conduction study with the reference electrode on the digit I has been used in the diagnosis of UNW, and because the incidence of UNW is rare, using the maximal values than mean + SD values may be a safer method that can decrease false positives.

Recently, Buschbacher et al. [10] tried to optimize the reference electrode position for the study of the ulnar nerve when recording from the FDI in order to create a normative database for this technique. They evaluated three previously published reference electrode positions using a MEDTRONIC Keypoint EMG machine with filters set at 2 Hz and 10 kHz, sweep speed of 5 ms/division, and sensitivity of 5 mV/division. First, a standard 8 cm motor conduction technique to the ADM muscle was performed [11–14], with the active recording electrode placed over the midpoint of the ADM and the reference placed just distal to the fifth metacarpophalangeal joint. Supramaximal stimulation (0.1 ms duration) was 8 cm proximal to the active electrode, just radial to the tendon of the *flexor carpi ulnaris* (FCU) muscle (Fig. 5). The stimulation site established by the standard 8 cm test to the ADM was used for all recordings from the FDI.

For recording from the FDI, the authors placed the active electrode on the dorsum of the first web space in the center of the triangle formed by the first carpometacarpal joint and the first and second metacarpal joints; the reference electrode was placed in several positions. It was placed on the digit II (Fig. 2), following Olney and Wilbourn's method [1]; on the proximal phalanx of the digit I (Fig. 6), following previously reported methods [7–9]; and on the metacarpophalangeal joint of the digit V (Fig. 7). The ground electrode was placed on the dorsum of the hand.

The authors compared surface recording techniques with needle recording from the FDI muscle because needle electrode is accurate for recording latency [15]. All recordings were done by a standard supramaximal stimulation with latency measured to the first deflection from the baseline. The amplitude was measured from baseline to peak. To determine which surface recorded latency most closely matched the needle recording latency, they determined for each subject which technique generated the latency most similar to the needle recording latency.

For the purpose of the study, Buschbacher et al. [10] developed normative data (Table 9) from 100 subjects (57 women and 43 men, age mean 32.2 ± 8 years, age range 19–65 years, height mean 167.3 ± 9.1 cm, height range 145–186 cm) reporting upper limit of normal (97th percentile) for ADM and FDI latency and lower limit of normal (3rd percentile) for FDI amplitude.

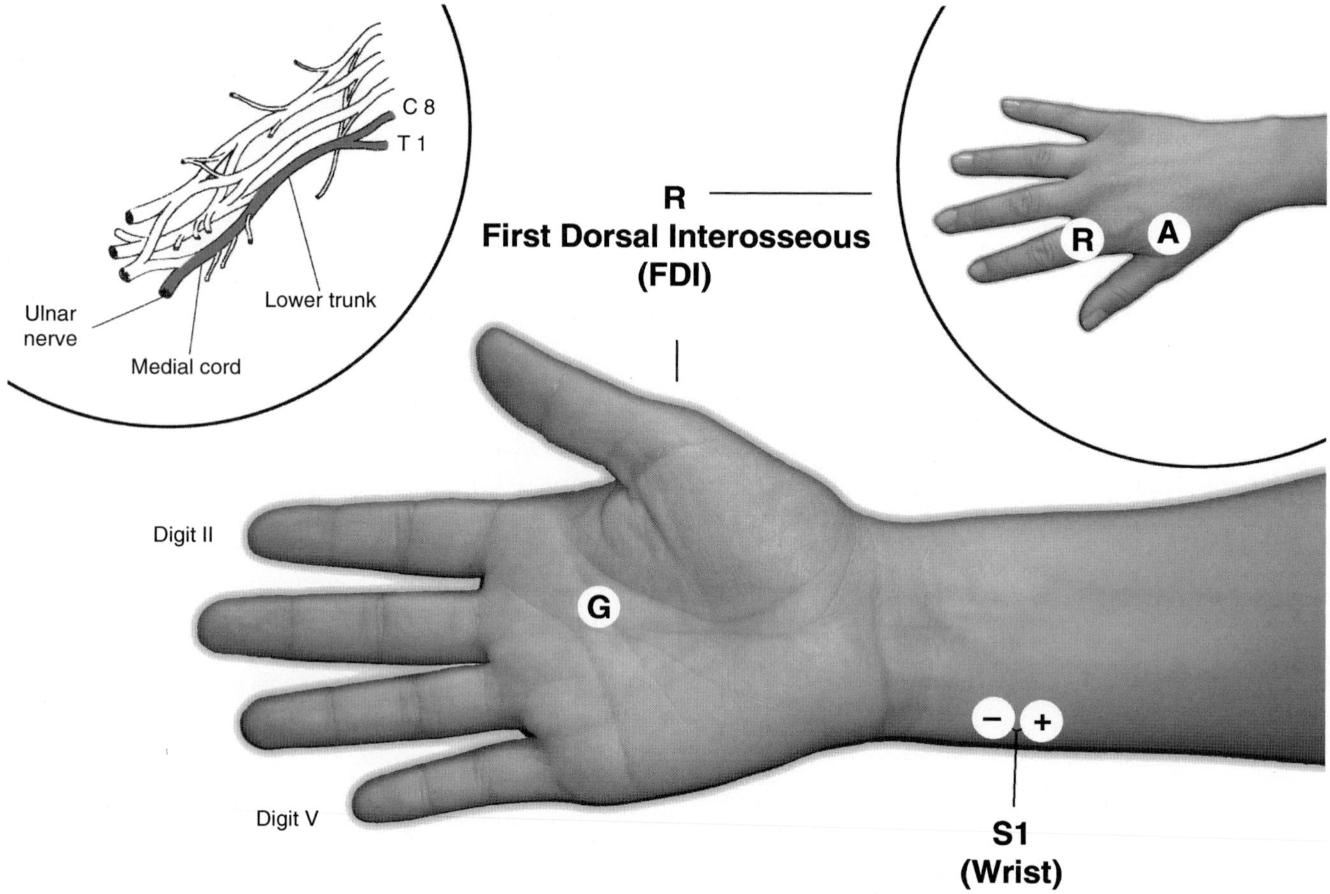

Typical waveform (wrist – FDI muscle):

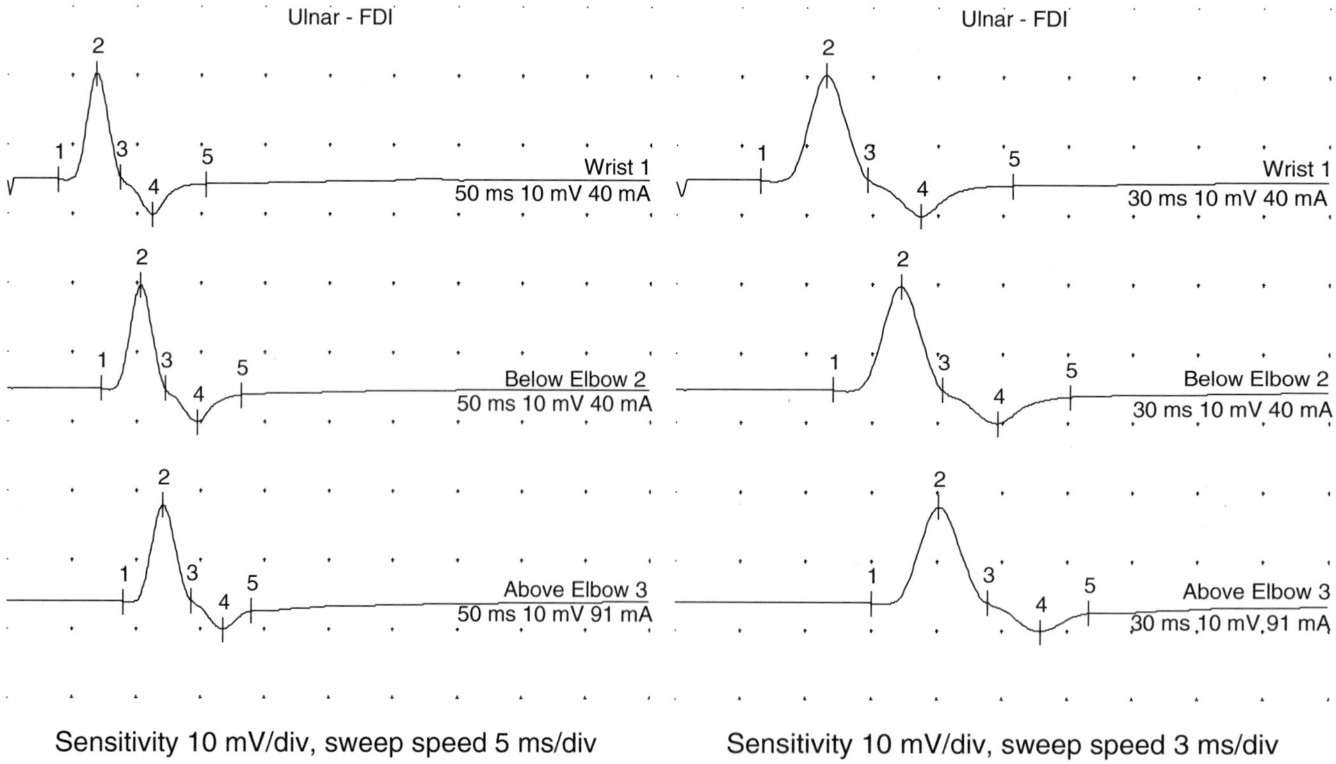

Fig. 2 Compound muscle action potentials (*CMAPs*) recorded at the hand from the FDI muscle, stimulation of the ulnar nerve at the wrist (*upper trace*), below elbow (*middle trace*), and above elbow (*lower trace*)

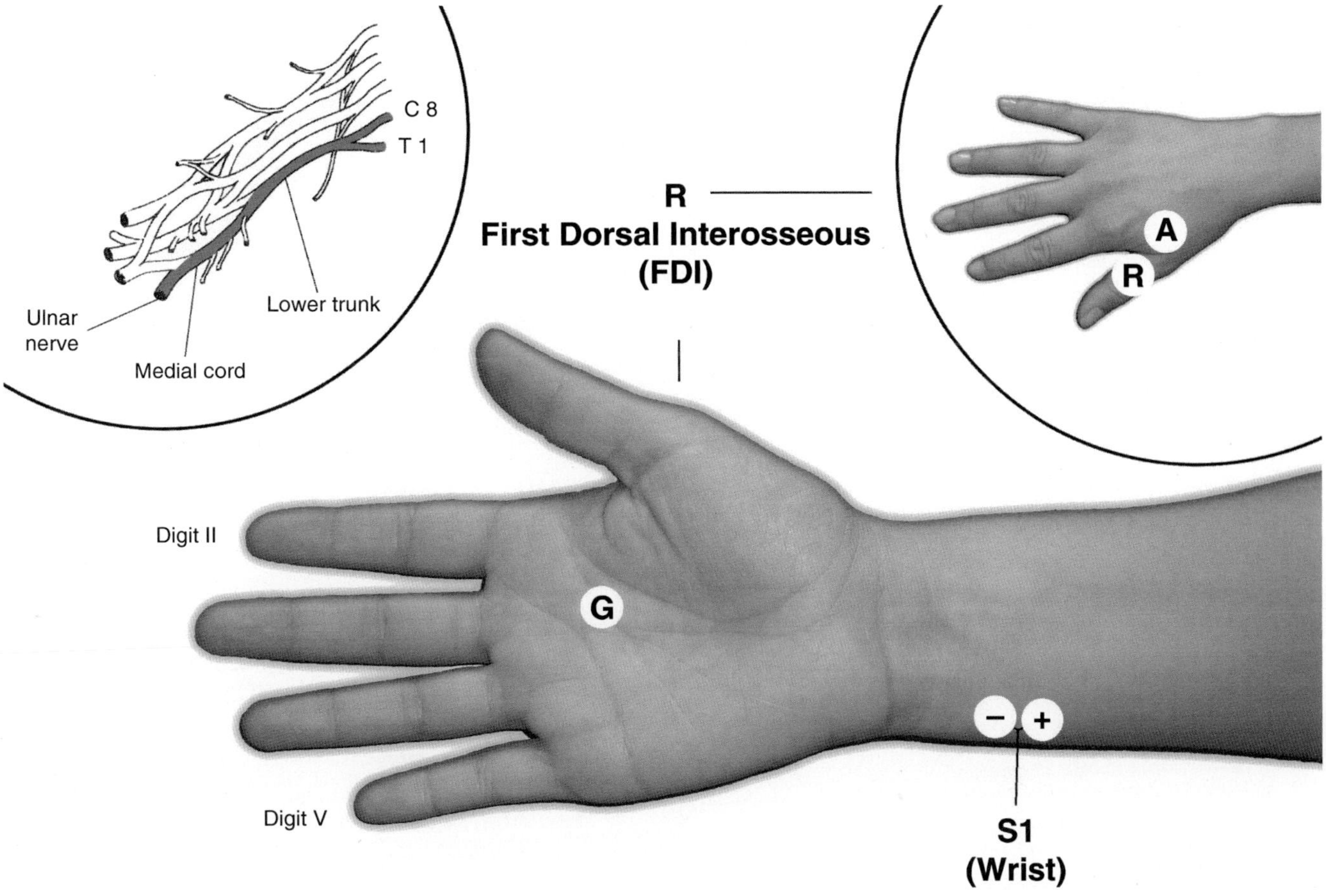

Typical waveform (wrist – FDI muscle):

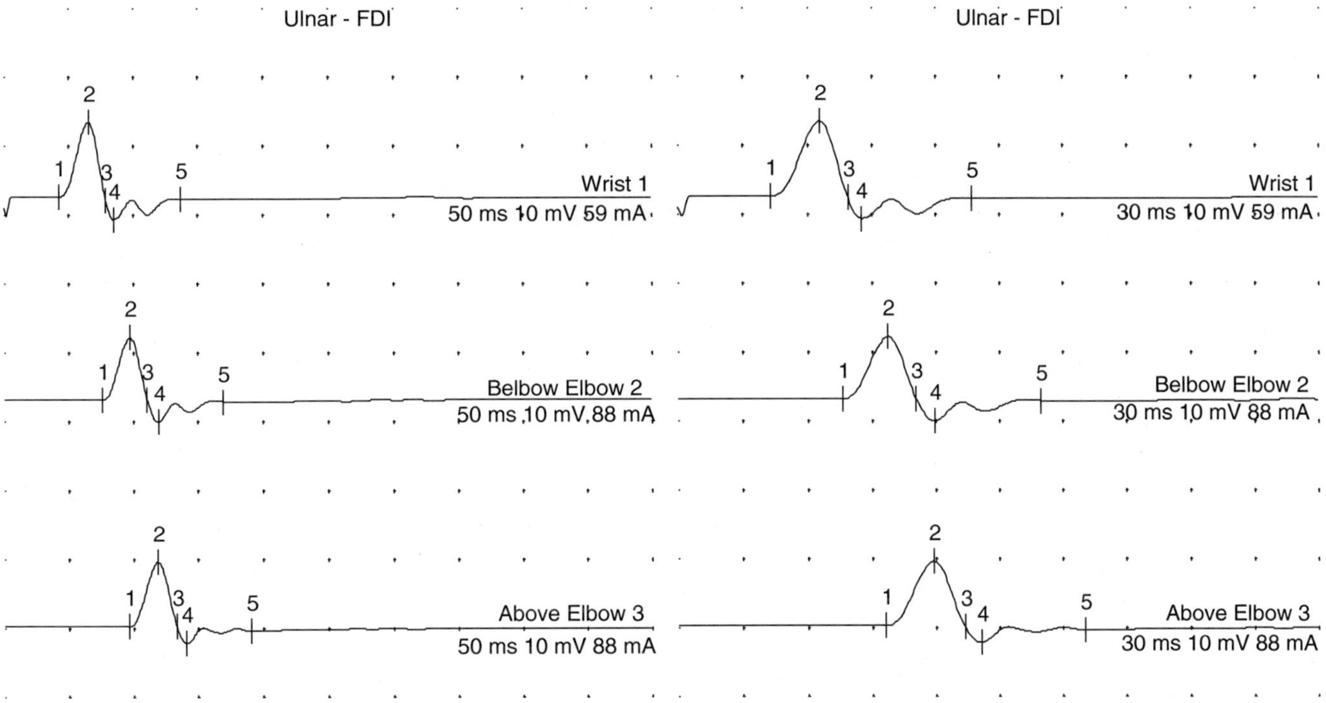

Sensitivity 10 mV/div, sweep speed 5 ms/div

Sensitivity 10 mV/div, sweep speed 3 ms/div

Fig. 3 Compound muscle action potentials (*CMAPs*) recorded at the hand from the FDI muscle, stimulation of the ulnar nerve at the wrist (*upper trace*), below elbow (*middle trace*), and above elbow (*lower trace*)

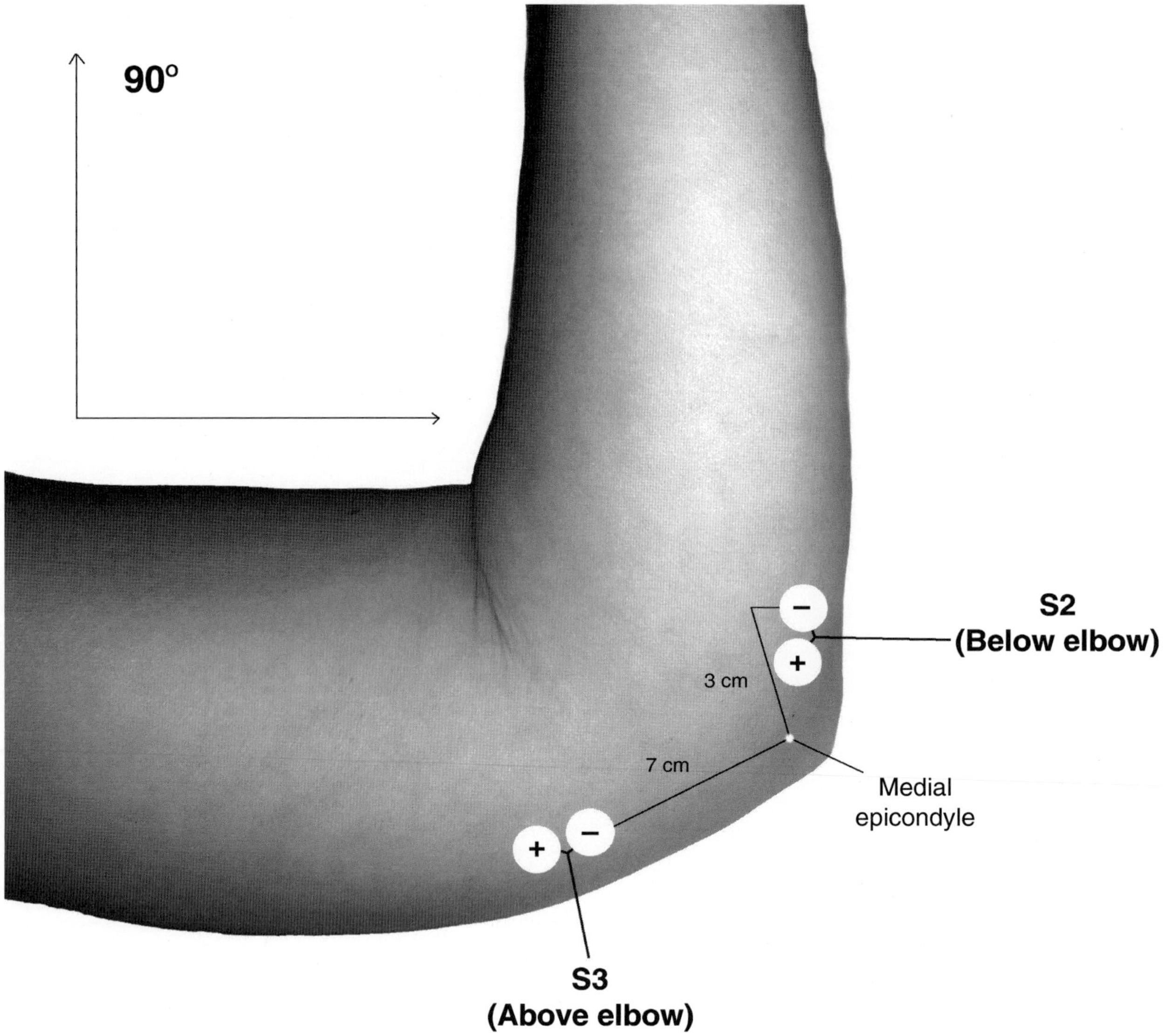

Fig. 4 Proximal stimulation points (S2, below elbow; S3, above elbow) with the elbow flexed to 90°

Table 5 Normal values [4] – ADM muscle recordings

	Mean	Range
Wrist–ADM, distal latency (ms) – subjects 20–29 years (30 hands)	2.8	2.4–3.4
Wrist–ADM, distal latency (ms) – subjects 30–39 years (66 hands)	2.8	2.3–3.5
Wrist–ADM, distal latency (ms) – subjects 40–49 years (58 hands)	2.9	2.4–3.6
Wrist–ADM, distal latency (ms) – subjects 50–59 years (40 hands)	2.9	2.5–3.8
Wrist–ADM, distal latency (ms) – subjects > 60 years (20 hands)	3.1	2.7–3.8
Wrist–ADM, distal latency (ms) – all subjects (214 hands)	2.9	2.3–3.8

Table 6 Normal values [4] – FDI muscle recordings

	Mean	Range
Wrist–FDI, distal latency (ms) – subjects 20–29 years (30 hands)	3.4	3.1–4.2
Wrist–FDI, distal latency (ms) – subjects 30–39 years (66 hands)	3.5	3.0–4.0
Wrist–FDI, distal latency (ms) – subjects 40–49 years (58 hands)	3.7	2.9–4.4
Wrist–FDI, distal latency (ms) – subjects 50–59 years (40 hands)	3.7	3.2–4.2
Wrist–FDI, distal latency (ms) – subjects > 60 years (20 hands)	3.7	3.3–4.1
Wrist–FDI, distal latency (ms) – all subjects (214 hands)	3.6	2.9–4.4

Table 7 Normal values [4] – ADM muscle recordings

	Mean	Range
Wrist–ADM, negative peak amplitude (mV) – subjects 20–29 years (30 hands)	12.9	8.8–16.4
Wrist–ADM, negative peak amplitude (mV) – subjects 30–39 years (66 hands)	12.4	8.3–18.1
Wrist–ADM, negative peak amplitude (mV) – subjects 40–49 years (58 hands)	11.9	7.0–15.4
Wrist–ADM, negative peak amplitude (mV) – subjects 50–59 years (40 hands)	11.6	7.5–17.5
Wrist–ADM, negative peak amplitude (mV) – subjects > 60 years (20 hands)	10.6	8.8–13.6
Wrist–ADM, negative peak amplitude (mV) – all subjects (214 hands)	11.9	7.0–18.1

Table 8 Normal values [4] – FDI muscle recordings

	Mean	Range
Wrist–FDI, negative peak amplitude (mV) – subjects 20–29 years (30 hands)	14.9	10.6–22.3
Wrist–FDI, negative peak amplitude (mV) – subjects 30–39 years (66 hands)	14.2	8.7–22.4
Wrist–FDI, negative peak amplitude (mV) – subjects 40–49 years (58 hands)	14.6	10.0–23.5
Wrist–FDI, negative peak amplitude (mV) – subjects 50–59 years (40 hands)	12.9	8.4–20.9
Wrist–FDI, negative peak amplitude (mV) – subjects > 60 years (20 hands)	12.5	10.2–16.8
Wrist–FDI, negative peak amplitude (mV) – all subjects (214 hands)	14.1	8.4–23.5

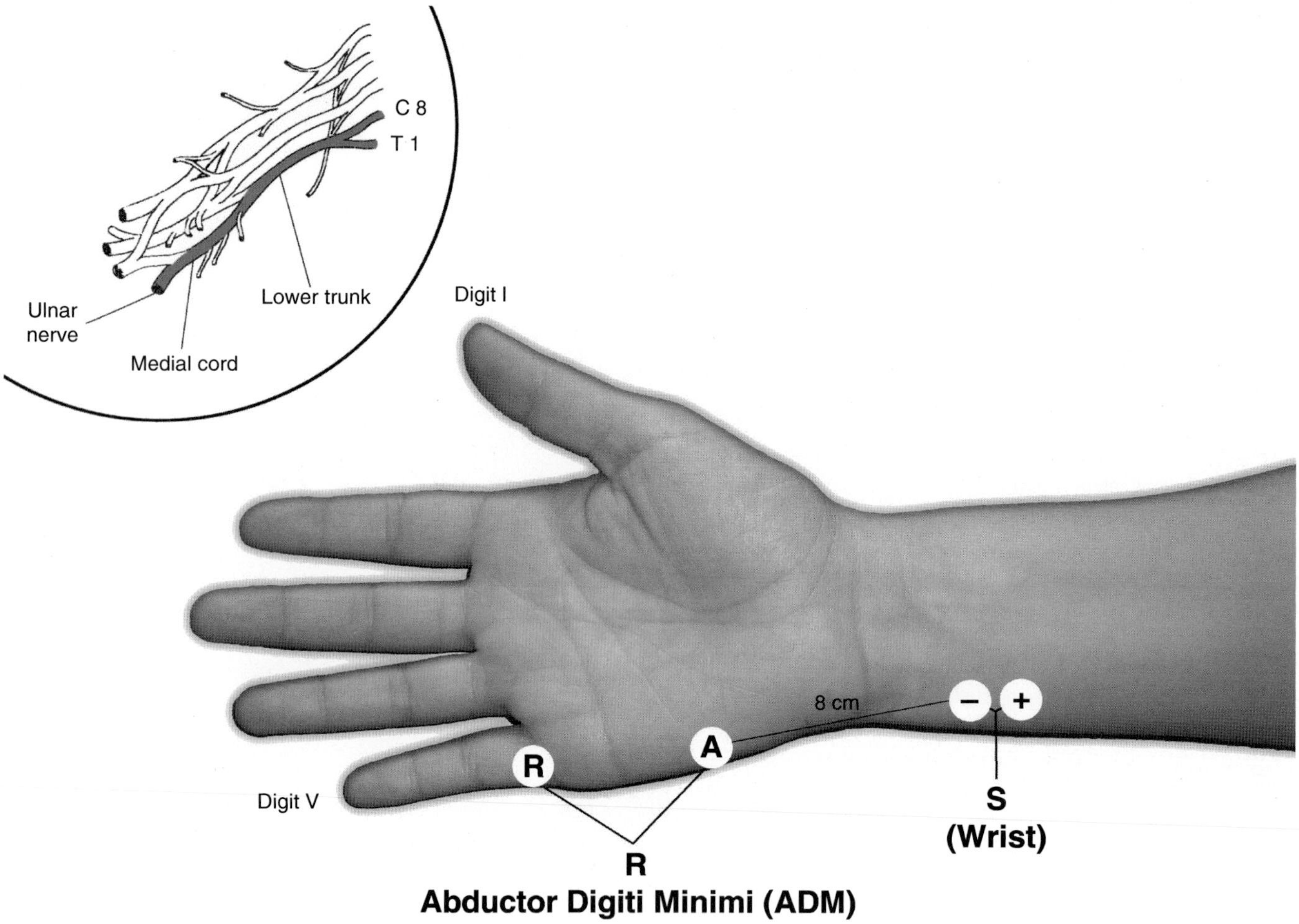

Fig. 5 Standard 8 cm motor conduction technique to the ADM muscle

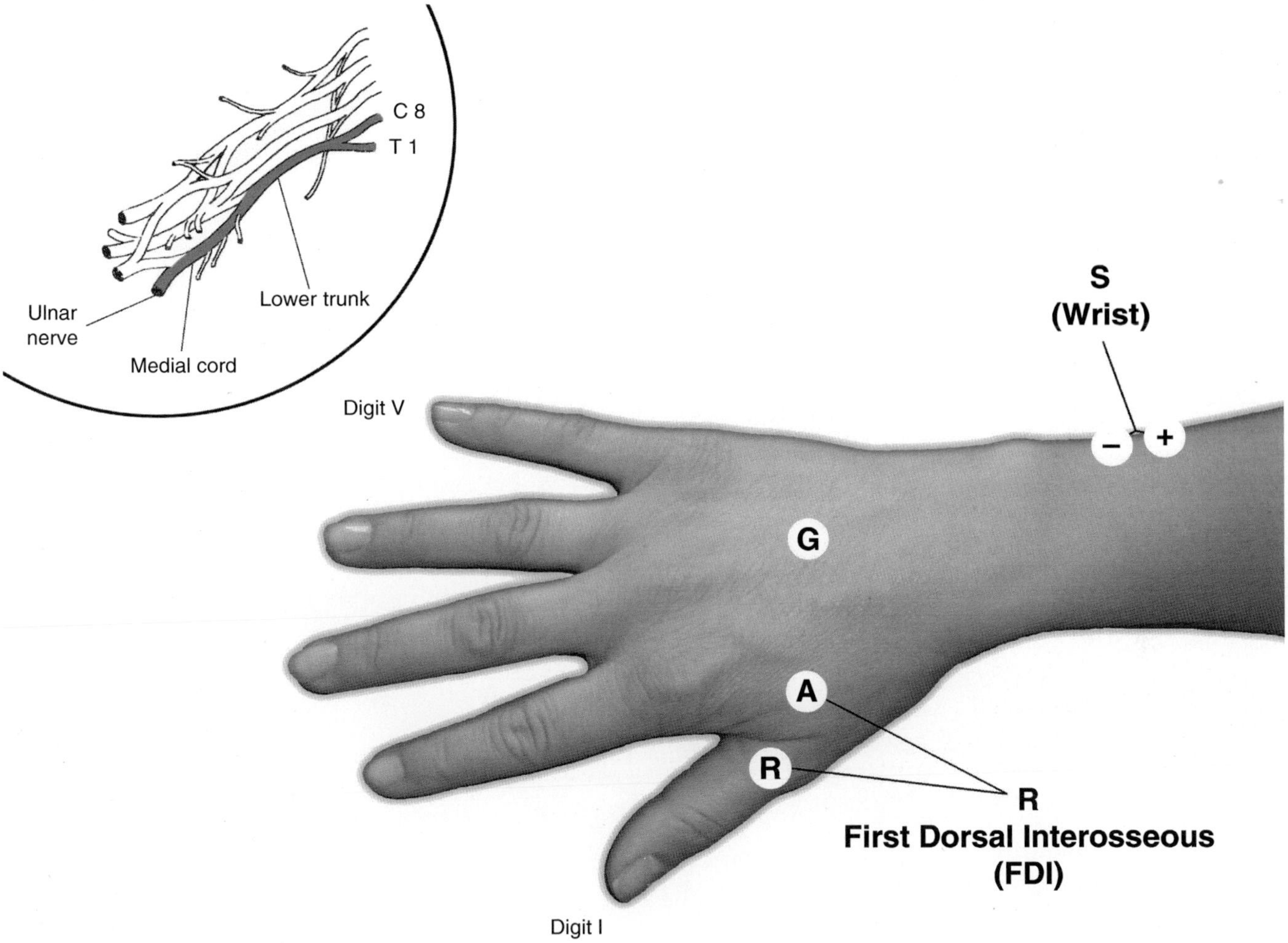

Fig. 6 Motor conduction to the FDI, reference on the proximal phalanx of the digit I

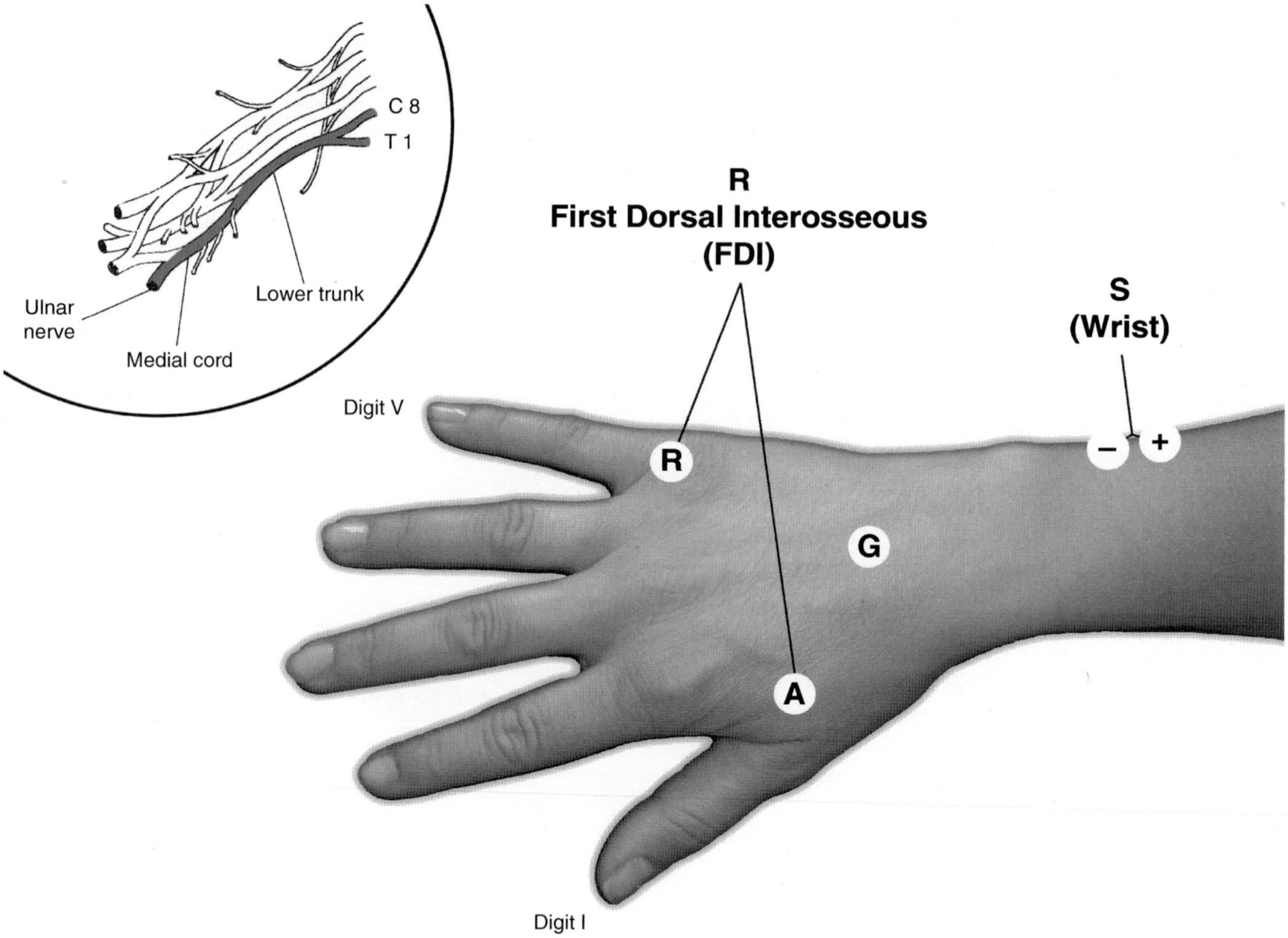

Fig. 7 Motor conduction to the FDI, reference on the metacarpophalangeal joint of the digit V

Table 9 Normal values [10]

	Mean ± SD	Range	Limit of normal
Wrist–ADM, latency (ms)	2.8 ± 0.3	2.1–3.7	3.4
Wrist–FDI, latency (ms)	3.4 ± 0.3	2.8–4.5	4.0
Wrist–ADM – wrist–FDI, latency difference (ms)	0.7 ± 0.3	0.2–1.4	1.3
Wrist–FDI, amplitude (mV) – 57 women	16.8 ± 4.0	8.8–26.1	9.0
Wrist–FDI, amplitude (mV) – 43 men	15.1 ± 4.4	9.2–25.4	9.3
Wrist–FDI, amplitude (mV) – all subjects	15.8 ± 4.2	8.8–26.1	9.2

Comment

Buschbacher et al. [10] compared the several surface recording methods with needle recording from the FDI muscle, and a strong correlation between the surface recording and the needle recording would make it likely that the surface recording was actually from the FDI. The authors found that the technique using the reference electrode just distal to the first metacarpophalangeal joint and the active electrode over the mid-web space is a good approximation of the "gold standard" of needle recording. Pathological waveform and values in a patient with ulnar neuropathy at the wrist (UNW) are reported (Fig. 8).

Pathological waveform (wrist – FDI muscle, wrist – ADM muscle):

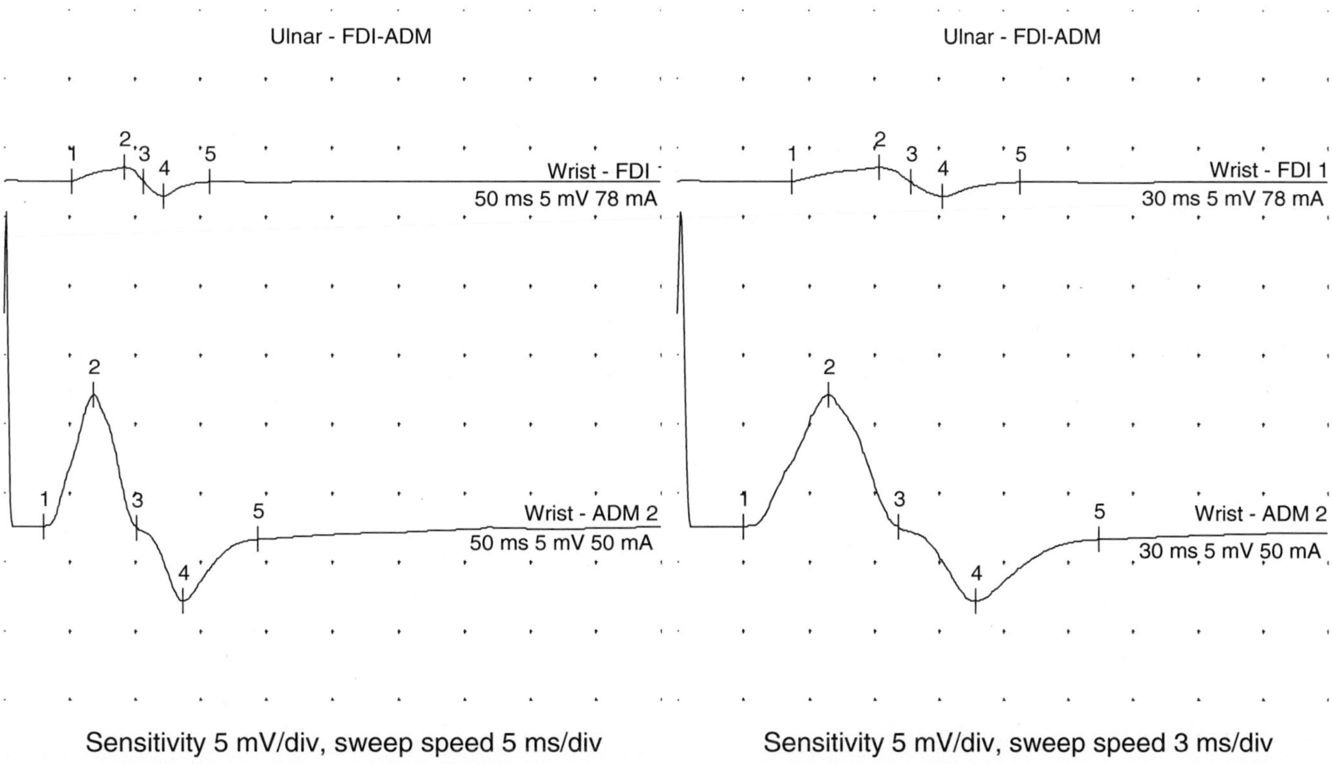

Onset latency (wrist – FDI): 5.25 ms; **Onset latency** (wrist – ADM): 3.05 ms; **Onset latency difference** (wrist – FDI – wrist – ADM): 2.20 ms; **Peak latency** (wrist – FDI): 9.10 ms; **Peak latency** (wrist – ADM): 6.10 ms; **Onset to peak amplitude** (wrist – FDI): 1.0 mV; **Onset to peak amplitude** (wrist – ADM): 9.6 mV; **Peak to peak amplitude** (wrist – FDI): 2.2 mV; **Peak to peak amplitude** (wrist – ADM): 15.0 mV

Fig. 8 Compound muscle action potentials (*CMAPs*) recorded at the hand from the FDI (*upper trace*) and ADM (*lower trace*) muscles in a patient with UNW, using Olney and Wilbourn's method [1] – stimulation of the ulnar nerve at the wrist

References

1. Olney RK, Wilbourn AJ (1985) Ulnar nerve conduction study of the first dorsal interosseous muscle. Arch Phys Med Rehabil 66:16–18
2. Olney RK, Hanson M (1988) AAEE case report #15: ulnar neuropathy at or distal to the wrist. Muscle Nerve 11:828–832
3. Olney RK, Wilbourn AJ, Miller RG (1983) Ulnar neuropathy at or distal to the wrist. Neurology 33(suppl 2):185
4. Kim DH (2011) Ulnar nerve conduction study of the first dorsal interosseous muscle in korean subjects. Ann Rehabil Med 35:658–663
5. Preston DC, Shapiro BE (2005) Electromyography and neuromuscular disorders: clinical-electrophysiologic correlations, 2nd edn. Elsevier, Philadelphia, pp 323–326
6. Wertsch JJ, Park TA, Lomas JN et al (1990) Effect of reference electrode position on deep ulnar nerve conduction studies. In: Abstracts of poster and platform presentations: 37th annual meeting, American Association of Electrodiagnostic Medicine. Chicago, 7–8 September . Muscle Nerve 13:862
7. Phongsamart G, Wertsch JJ, Ferdjallah M et al (2002) Effect of reference electrode position on the compound muscle action (CMAP) onset latency. Muscle Nerve 25:816–821
8. Cowdery SR, Preston DC, Herrmann DN et al (2002) Electrodiagnosis of ulnar neuropathy at the wrist. Conduction block versus traditional tests. Neurology 59:420–427
9. Park YK, Kwon HK, Lee HJ et al (2005) Abductor digiti minimi and first dorsal interosseous recordings for the localization of ulnar neuropathy at the elbow. J Korean Acad Rehab Med 29:598–601
10. Buschbacher RM, Bayindir O, Malec J et al (2015) Ulnar motor study to first dorsal interosseous: best reference electrode position and normative data. Muscle Nerve 52:231–233
11. Checkles NS, Russakov AD, Piero DL (1971) Ulnar nerve conduction velocity – effect of elbow position on measurement. Arch Phys Med Rehabil 52:362–365
12. Hennessey WJ, Falco FJ, Braddom RL (1994) Median and ulnar nerve conduction studies: normative data for young adults. Arch Phys Med Rehabil 75:259–264
13. Buschbacher RM (1999) Ulnar nerve motor conduction to the abductor digiti minimi. Am J Phys Med Rehabil 78:S9–S14
14. Lee SW, Kim DDJ, Pedersen AL et al (2011) Ulnar nerve latency at above elbow stimulation. A complementary tool for segmental conduction velocity across the elbow. Am J Phys Med Rehabil 90:302–307
15. Campbell WW (2014) Essentials of electrodiagnostic medicine, 2nd edn. Demos, New York

Wrist, Elbow, Above Elbow – Hand; Elbow, Above Elbow – Forearm

M29

Surface Recording Technique, Study from *Abductor Digiti Minimi* (ADM), *Flexor Carpi Ulnaris* (FCU), and *Flexor Digitorum Profundus* (FDP) Muscles

Original Settings Sensitivity was 1 mV/division (for ADM and FCU muscles recording) and 500 μV/division (for FDP muscle recording), sweep speed was 1 ms/division, and the machine used was a NEURODIAGNOSTIC NDI electromyograph; duration of pulse, low-frequency filter, and high-frequency filter were not specified.

Position This study was performed with the elbow flexed to 70° and the forearm abducted to 110° and supinated.

Recording Using silver disk surface electrodes, the authors [1] performed ulnar nerve conduction study recording from the motor point of the *abductor digiti minimi* (ADM) muscle, with the reference electrode placed slightly distal to the digit V. Conventional motor conduction study to the ADM muscle was performed with the elbow flexed to 70°; the figure shows elbow in full extended position (Fig. 1). Felsenthal et al. [1] performed two several studies recording from muscles innervated by the ulnar nerve at the forearm. They recorded from the *flexor carpi ulnaris* (FCU) muscle placing the active electrode (A) over the motor point of the muscle, two fingerbreadths volar to the ulna at the junction of the proximal and middle thirds of the forearm. The reference electrode was placed over the shaft of the ulna at the same distance. The motor conduction study to the FCU muscle was performed with the elbow flexed to 70° (Fig. 2).
The distance between the active recording electrode over the FCU muscle and the distal below elbow stimulation point varied with an individual's forearm length, and the recording active electrode was occasionally adjusted to obtain a clear initial negative deflection. They recorded from the *flexor digitorum profundus* (FDP) muscle placing a TECA bar mounted surface electrodes just volar to the shaft of the ulna over the FDP muscle, with the active recording electrode 10 cm distal to the distal elbow (DE) stimulation point, and the reference electrode placed farther distally. The motor conduction study to the FDP muscle was performed with the elbow flexed to 70° (Fig. 3). The authors chose this method because the use of the bar facilitated applying pressure and obtaining a clear initial negative deflection (for FDP muscle recordings, they use more gain than for ADM or FCU muscle recordings), obviating the need for separate disk electrodes. The compound muscle action potential (CMAP) waveform observed has an initial, sharp negative deflection corresponding to the onset of the CMAP, and it can be followed by highly variable muscle depolarizations from other portions of the ulnar-innervated FDP muscle. The ground (G) electrode was placed on the dorsum of the hand; the figure shows ground placed on the forearm.

Stimulation In ADM muscle recordings, the ulnar nerve was stimulated at 3 sites: at the wrist (S1), at the elbow (S2), and above the elbow (S3). In both FCU and FDP muscle recordings, stimulation was applied at the most proximal sites only. The distal elbow (DE) stimulation point was at the distal aspect of the ulnar groove (elbow stimulation point), and the proximal elbow (PE) stimulation point was 10 cm proximal to the DE point, along the course of the ulnar nerve, on the posteromedial aspect of the arm (above-elbow stimulation point). The DE and PE points were marked carefully and used in all studies (Figs. 1, 2, and 3).

Measurements Latency (ms) of the CMAP was measured from the onset of the stimulus artifact to the onset of the negative deflection of the CMAP for all three techniques; for FDP muscle recordings, the authors measured CMAP latency to both the onset and peak of the negative deflection of the CMAP. The amplitude (mV) of the CMAP was measured from the peak of the negative deflection to the peak of the positive deflection (peak-to-peak amplitude). By dividing the distance by time, the motor nerve conduction velocity (MNCV) along the nerve segment can be determined and measured in meter per second (m/s), and it was calculated for above-elbow–Erb's point and wrist–elbow segments. Surface

G. Gentili, M. Di Napoli, *The Ulnar Nerve*, DOI 10.1007/978-3-319-43292-2_53

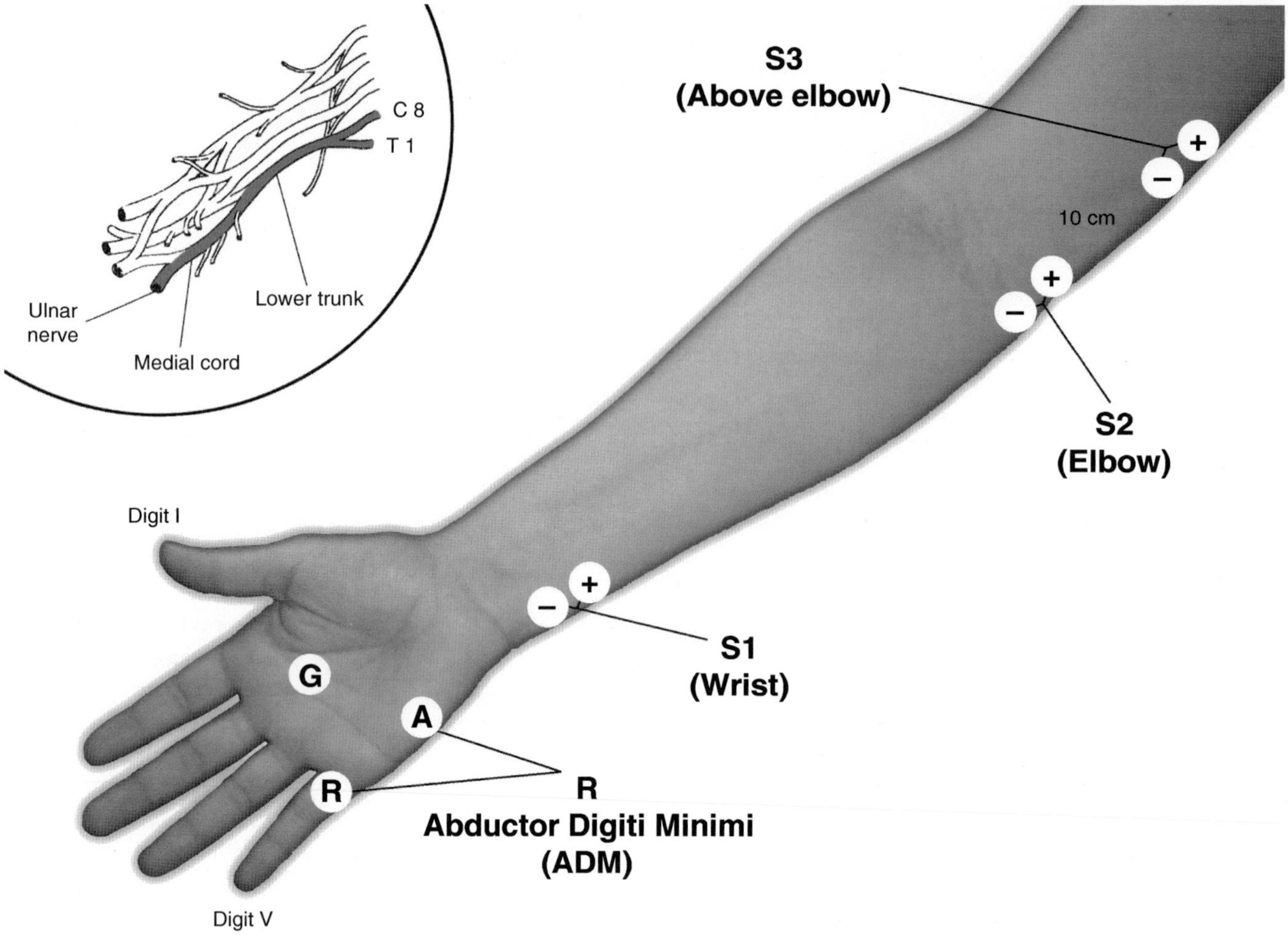

Typical waveform (wrist, elbow, above elbow – ADM muscle):

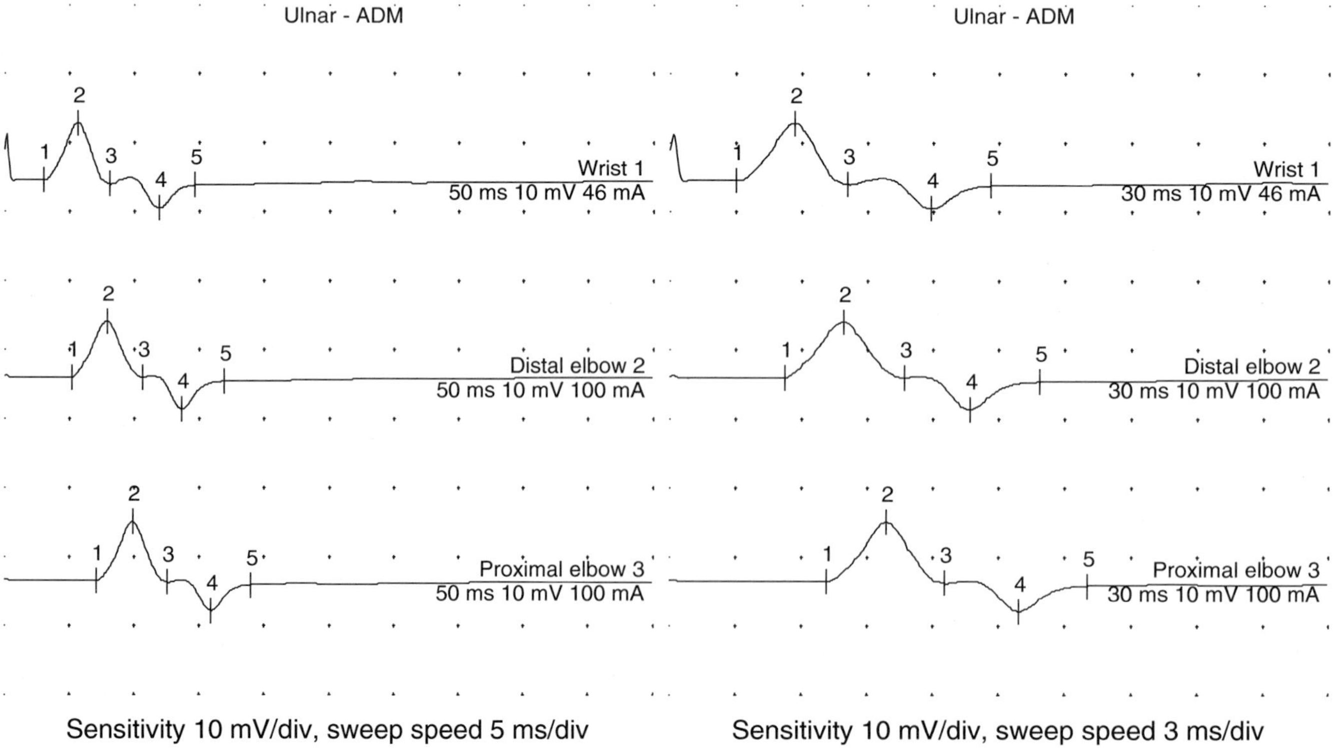

Fig. 1 Compound muscle action potentials (*CMAPs*) recorded at the hand from the ADM muscle, stimulation of the wrist (*upper trace*), elbow (*middle trace*), and above elbow (*lower trace*)

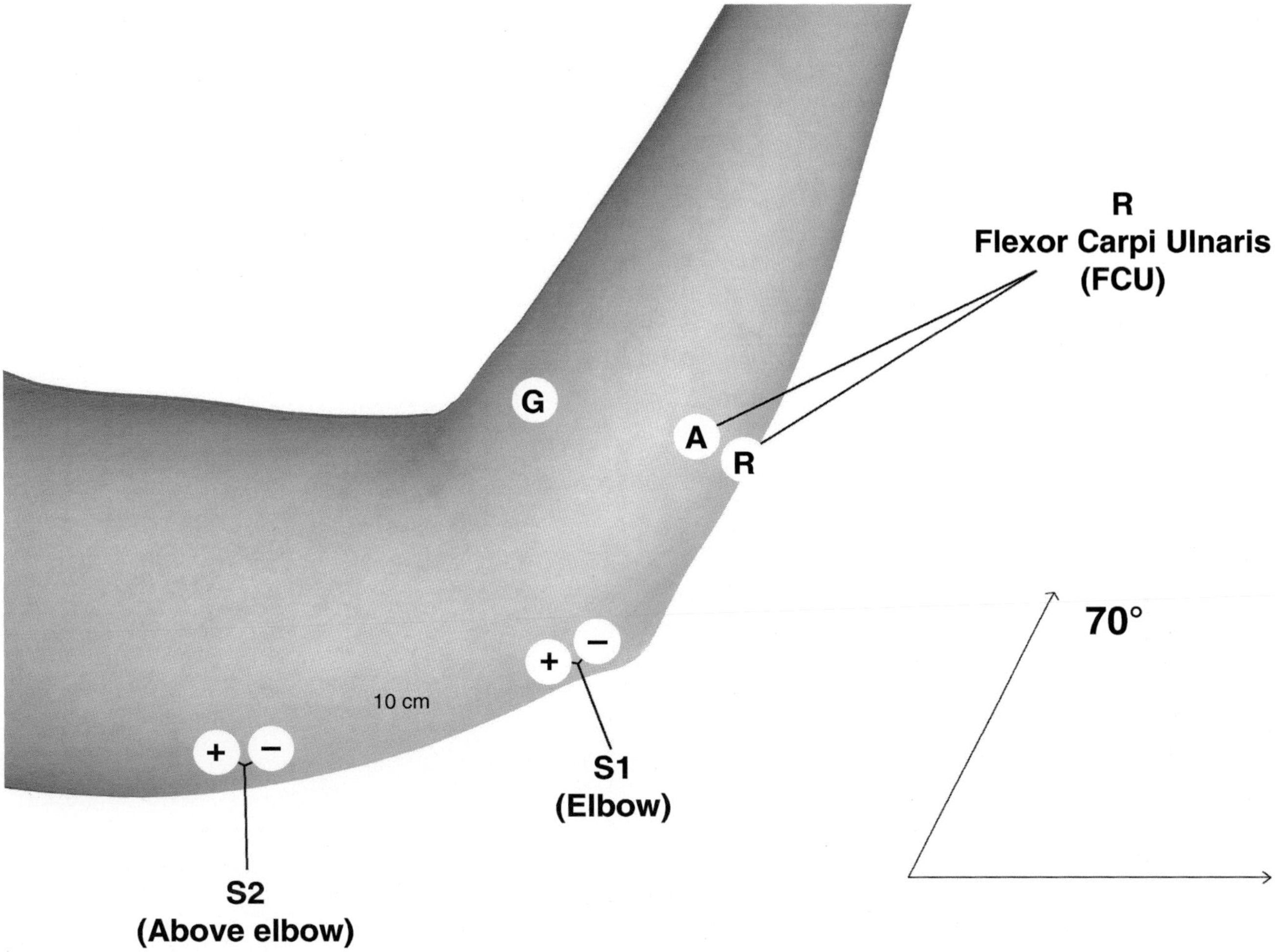

Typical waveform (elbow, above elbow – FCU muscle):

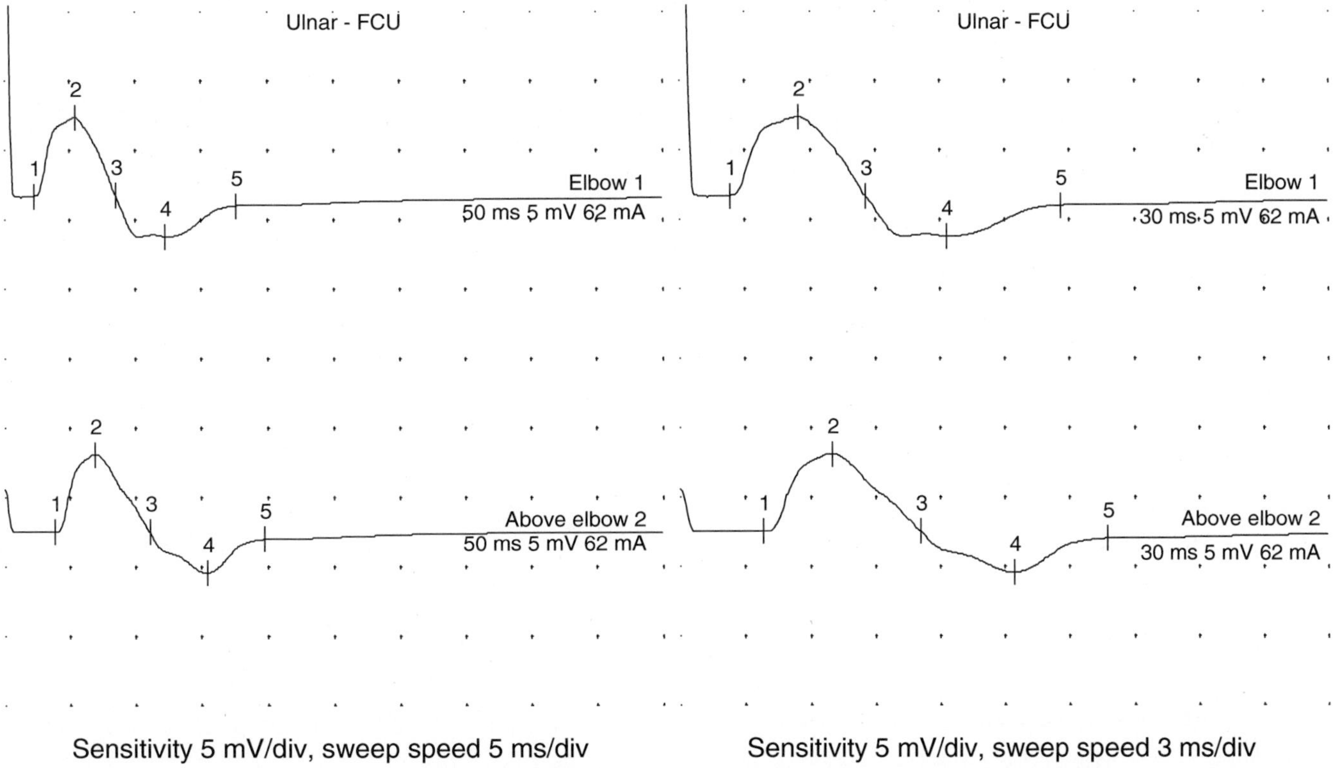

Fig. 2 Compound muscle action potentials (*CMAPs*) recorded at the hand from the FCU muscle, stimulation at the elbow (*upper trace*) and above elbow (*lower trace*)

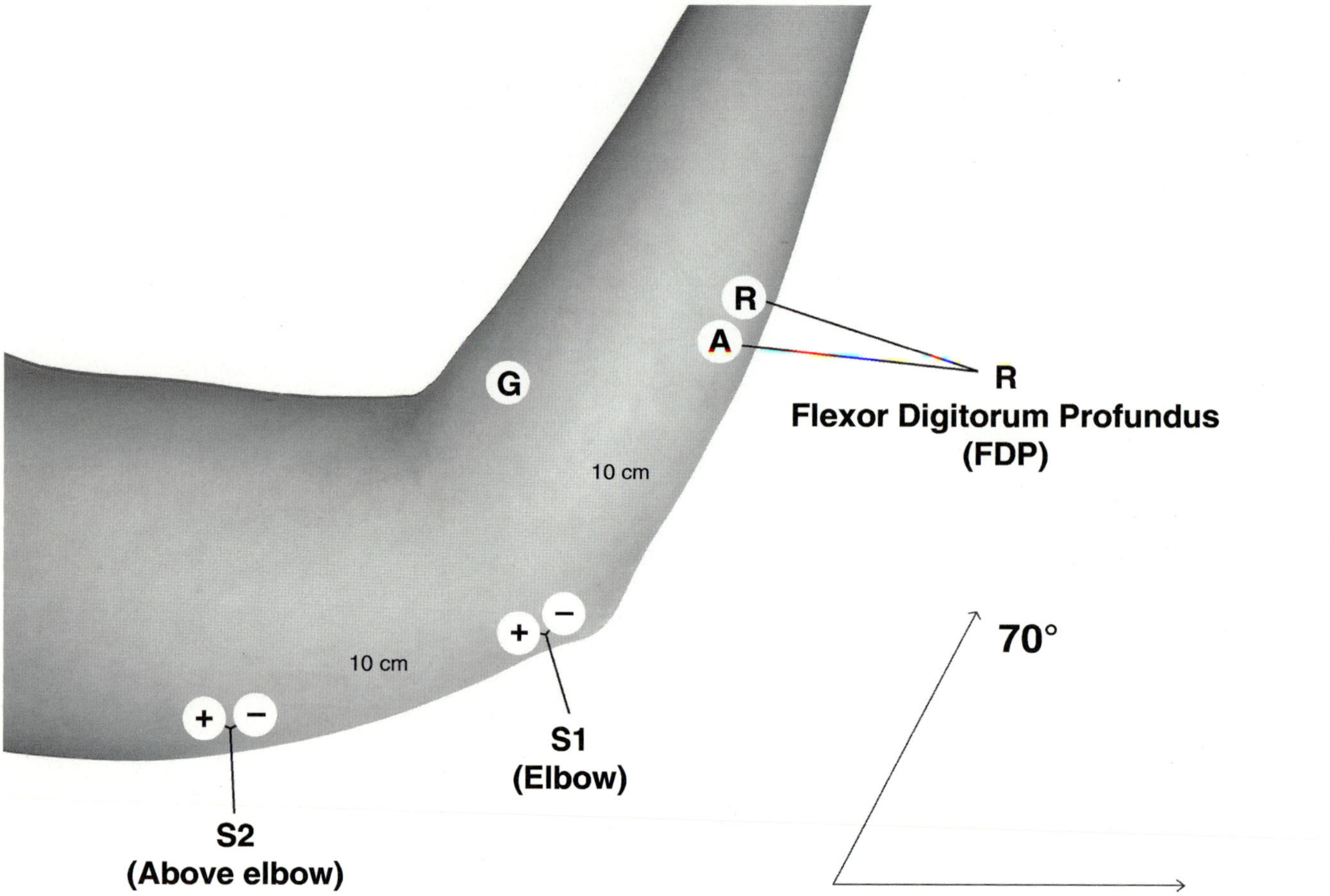

Typical waveform (elbow, above elbow – FDP muscle):

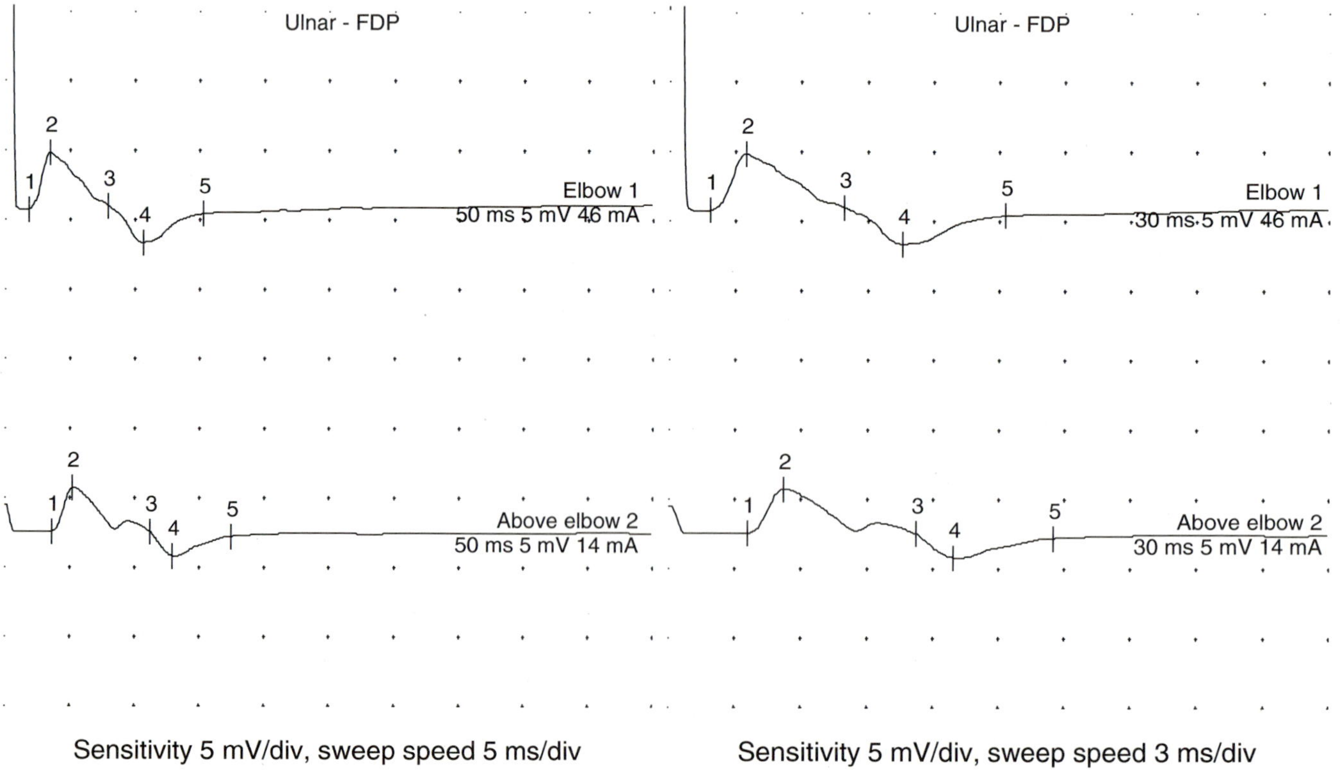

Fig. 3 Compound muscle action potentials (*CMAPs*) recorded at the hand from the FDP muscle, stimulation at the elbow (*upper trace*), and above elbow (*lower trace*)

skin temperature was measured using a thermistor taped over the wrist crease (average 32.5 °C, range 30–34 °C). Normal conduction values from the ADM muscle (Table 1), FCU muscle (Table 2), and FDP muscle (Table 3) were obtained from 48 ulnar nerves of 24 healthy subjects (19 women and 5 men, age range 19–52 years, mean age 29 years). The authors performed the three conduction methods in three cases of ulnar neuropathy at the elbow (UNE), and, in these three cases, all techniques detected abnormalities equally well.

Comment

The three techniques described by Felsenthal et al. [1] were performed easily and quickly and provided reliable and reproducible information. The distal elbow–proximal elbow segment (10 cm) had a conduction velocity of between 50 and 53 m/s using any method, with a conduction time across the elbow of 1.9–2.0 m/s. The side-to-side comparison of the ulnar motor conduction in the across-elbow segment revealed a difference of 8.1–9.5 m/s, and, for the authors, the small range of these normal values attested to the reliability and reproducibility of all three methods. During their study, they observed 17/48 ulnar nerve segments to have a larger CMAP recorded from the FCU muscle with stimulation at the PE stimulation point than the DE stimulation point, and this indicated that innervations to the FCU may arise proximal to the distal end of the olecranon sulcus. Thus, they suggested to the electromyographer to take care in making a judgment of amplitude abnormality unless a greater than normal amplitude decrement is observed in the distal elbow–proximal elbow segment. The FDP technique appeared to be the simplest, and probably the quickest, to perform of the three methods described. The two proximal forearm recording techniques from FCU and FDP muscles described by Felsenthal et al. [1] can improve the diagnosis and localization of the ulnar nerve lesions in the cubital tunnel and ulnar groove, especially when it is not possible to use the distal recording site over the ADM muscle. Pathological waveform and values in a case of ulnar neuropathy at the elbow (UNE) are reported (Figs. 4 and 5).

Table 1 Normal values [1]

	Mean ± SD	Range
Wrist–ADM, onset latency (ms)	2.7 ± 0.3	2.2–3.5
Elbow–ADM, onset latency (ms)	6.5 ± 0.75	5.4–8.7
Above elbow–ADM, onset latency (ms)	8.2 ± 0.8	7–10.5
Wrist–elbow, forearm MNCV (m/s)	62.0 ± 3.5	55–68
Elbow–above-elbow, elbow segment MNCV (m/s)	60.0 ± 5	50–71
Elbow–ADM, peak-to-peak amplitude (mV)	16.9 ± 3.3	10.5–27
Above elbow–ADM, peak-to-peak amplitude (mV)	16.6 ± 3.4	9.5–27

Table 2 Normal values [1]

	Mean ± SD	Range
Elbow–FCU, onset latency (ms)	2.6 ± 0.3	2–3.5
Above elbow–FCU, onset latency (ms)	4.2 ± 0.3	3.4–5
Elbow–above-elbow, elbow segment onset latency (ms)	1.6 ± 0.13	1.4–1.9
Elbow–above-elbow, elbow segment MNCV (m/s)	63.0 ± 4.7	52–71
Elbow–FCU, peak-to-peak amplitude (mV)	5.2 ± 1.8	2–9
Above elbow–FCU, peak-to-peak amplitude (mV)	5.2 ± 1.8	2.1–9.5

Table 3 Normal values [1]

	Mean ± SD	Range
Elbow–FDP, onset latency (ms)	2.4 ± 0.2	2.1–3.0
Above elbow–FDP, onset latency (ms)	4.0 ± 0.27	3.6–5
Elbow–FDP, peak latency (ms)	3.4 ± 0.2	3.2–4.1
Above elbow–FDP, peak latency (ms)	5.1 ± 0.28	4.7–6.1
Elbow–above-elbow, elbow segment onset latency (ms)	1.6 ± 0.14	1.3–2
Elbow–above-elbow, elbow segment MNCV (m/s)	63.0 ± 5.5	50–71
Elbow–FDP, peak-to-peak amplitude (mV)	2.0 ± 0.3	0.75–3
Above elbow–FDP, peak-to-peak amplitude (mV)	2.0 ± 0.7	0.75–3.1

Pathological waveform (elbow, above elbow – FCU muscle):

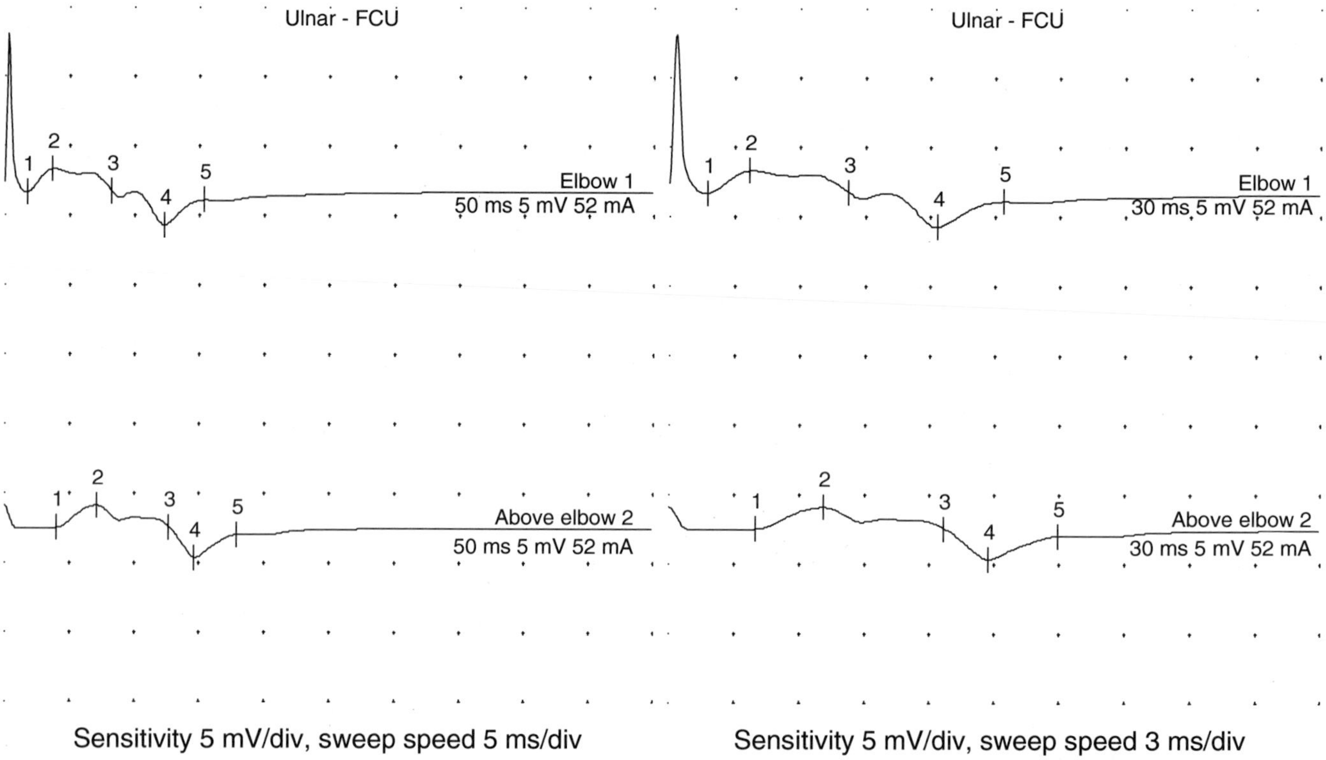

Onset latency (elbow – FCU): 1.75 ms; **Onset latency** (above elbow – FCU): 3.95 ms; **Peak latency** (elbow – FCU): 3.80 ms; **Peak latency** (above elbow – FCU): 7.15 ms; **Onset to peak amplitude** (elbow – FCU): 1.7 mV; **Onset to peak amplitude** (above elbow – FCU): 1.6 mV; **Peak to peak amplitude** (elbow – FCU): 4.1 mV; **Peak to peak amplitude** (above elbow – FCU): 3.8 mV; **MNCV** (above elbow – FCU, 10 cm): 45.5 m/s

Fig. 4 Compound muscle action potentials (*CMAPs*) recorded at the hand from the FCU muscle in UNE, stimulation of the elbow (*upper trace*) and above elbow (*lower trace*)

Pathological waveform (elbow, above elbow – FDP muscle):

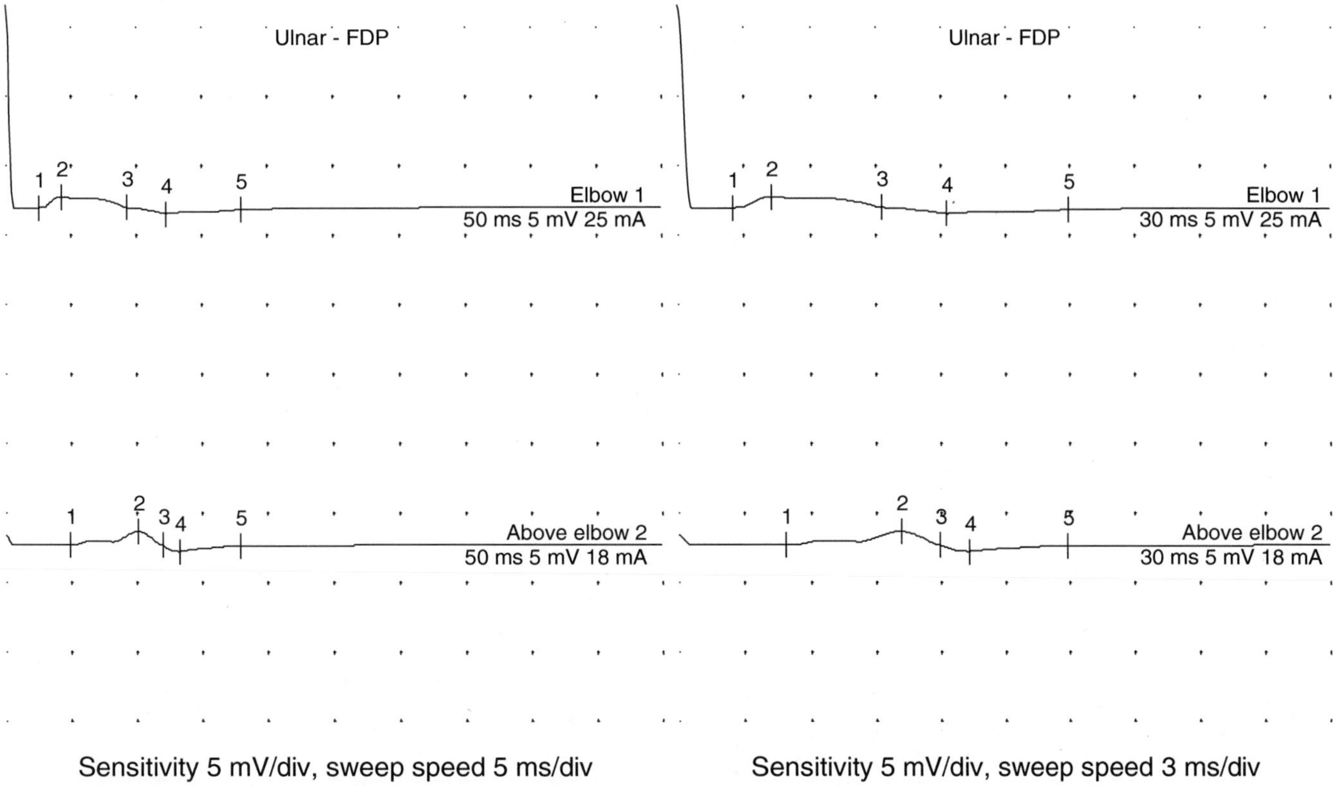

Onset latency (elbow – FDP): 2.5 ms; **Onset latency** (above elbow – FDP): 4.9 ms; **Peak latency** (elbow – FDP): 4.30 ms; **Peak latency** (above elbow – FDP): 10.1 ms; **Onset to peak amplitude** (elbow – FDP): 0.9 mV; **Onset to peak amplitude** (above elbow – FDP): 1.1 mV; **Peak to peak amplitude** (elbow – FDP): 1.3 mV; **Peak to peak amplitude** (above elbow – FDP): 1.5 mV; **MNCV** (above elbow – FDP, 10 cm): 41.7 m/s

Fig. 5 Compound muscle action potentials (*CMAPs*) recorded at the hand from the FDP muscle in UNE, stimulation of the elbow (*upper trace*) and above elbow (*lower trace*)

Reference

1. Felsenthal G, Brockman PS, Mondell DL et al (1986) Proximal forearm ulnar nerve conduction techniques. Arch Phys Med Rehabil 67:440–444

Wrist, Below Elbow, Above Elbow, Axilla – Hand

M30

Surface Recording Technique, Study from *Abductor Digiti Minimi* (ADM) Muscle

Original Settings Sensitivity was 5 mV/division, low-frequency filter was 1.6 Hz, high-frequency filter was 16 kHz, and duration was 0.1–0.5 ms. Sweep speed, rate of pulse, and the machine used were not specified.

Position This study was performed with the arm in two different positions: (1) the elbow fully extended (0°), the forearm supinated, and the arm abducted 30° (Fig. 1); (2) the elbow flexed to 135°, with the forearm supinated (Fig. 2).

Recording Using surface electrodes of 6 mm diameter, the authors [1] performed ulnar nerve conduction study recording from the *abductor digiti minimi* (ADM) muscle. The active (A) recording electrode was placed laterally over the *hypothenar* eminence midway between the pisiform bone and the metacarpophalangeal junction; the reference (R) electrode was placed on the proximal phalanx of the digit V (Fig. 1). The ground (G) electrode was fixed on the dorsum of the hand; the figure shows ground electrode placed on the palm.

Stimulation The ulnar nerve was stimulated percutaneously at four sites [1]: at the wrist, below the elbow, above the elbow, and in the axilla. At the wrist (S1), the active stimulating electrode (cathode) was located 6.5 cm proximal to the active electrode (Fig. 1). Proximal stimulations were applied below the elbow (S2 – 4 cm distal to the medial epicondyle), above the elbow (S3 – 10 cm proximal to the below-elbow site with the elbow extended), and more proximally in the axilla (S4 – approximately 11 cm proximal to the above elbow site). The anode (+) was proximal at each stimulation. Motor conduction studies were performed with the arm in two different positions: (1) the elbow fully extended (0°), the forearm supinated, and the arm abducted 30° (Fig. 1); (2) the elbow flexed to 135°, with the forearm supinated (Fig. 2). Stimulation in the axilla was performed in the same manner for both elbow fully extended and flexed positions (Fig. 3). The authors used supramaximal square-wave pulses of 100–300 V and 0.1–0.5 ms duration.

Measurements Onset latency (ms) was measured from the stimulating artifact to the onset of the initial negative deflection of the compound muscle action potential (CMAP). Motor nerve conduction velocity (MNCV) of each segment (wrist–below elbow, below elbow–above elbow, above elbow–axilla) was calculated dividing the distance between two different stimulating sites by difference of onset latencies, and it was measured in meter per second (m/s). Negative peak amplitude (μV) of the CMAP was measured from the baseline to the negative peak. Surface distances were measured by a steel tape from cathodal stimulation sites. Hand surface was measured, and if less than 33 °C, warming was performed by immersion in 40 °C water for 10 min. Normal values were obtained from 49 healthy volunteers without symptoms or signs of peripheral nerve disease or nerve injury (35 women and 14 men, age range 22–69 years, with ten subjects in each decade). All subjects were tested with the elbow fully extended (Table 1) and with the elbow flexed to 135° (Table 2).

G. Gentili, M. Di Napoli, *The Ulnar Nerve*, DOI 10.1007/978-3-319-43292-2_54

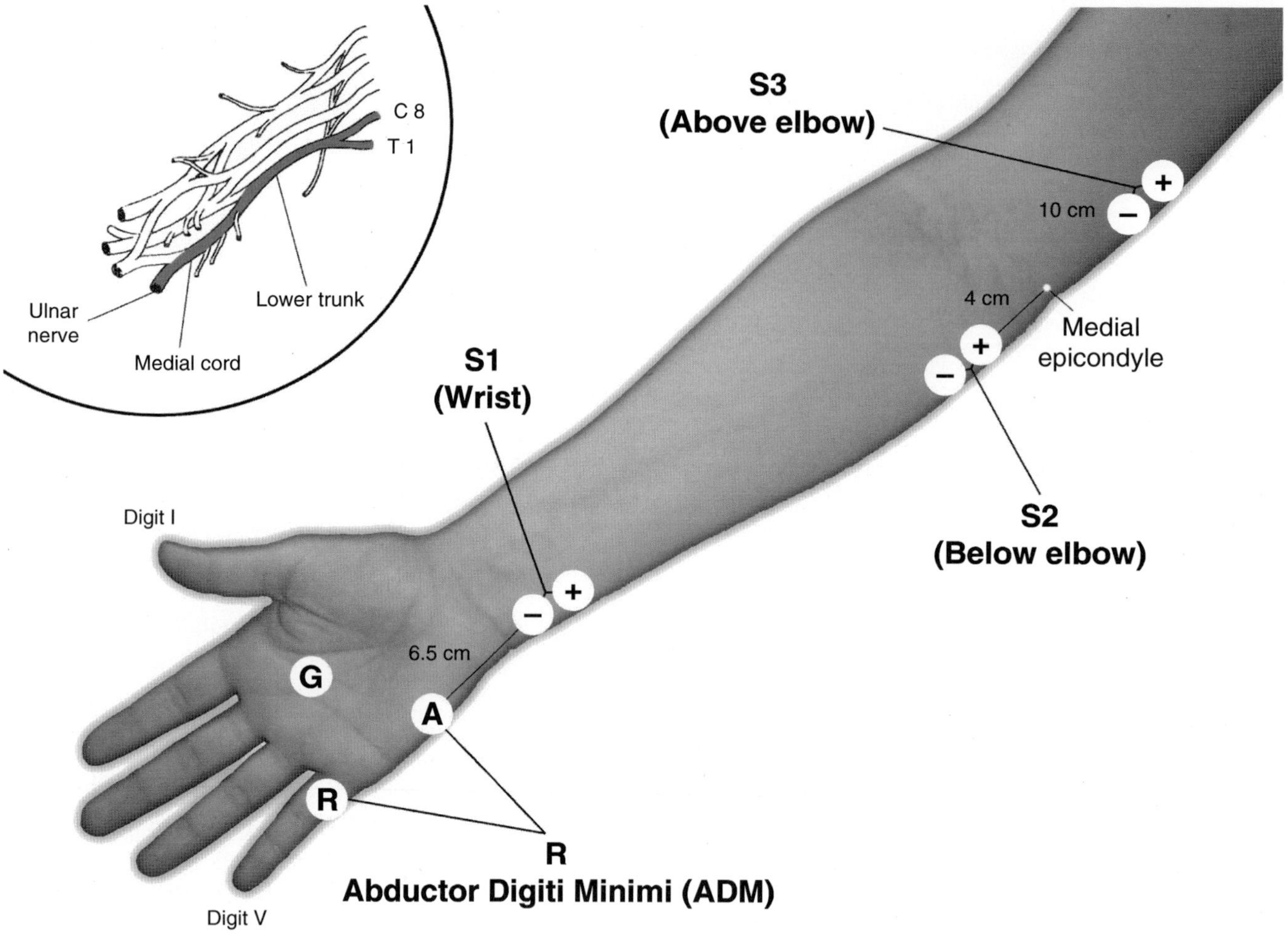

Typical waveform (wrist, below elbow, above elbow, axilla – ADM muscle):

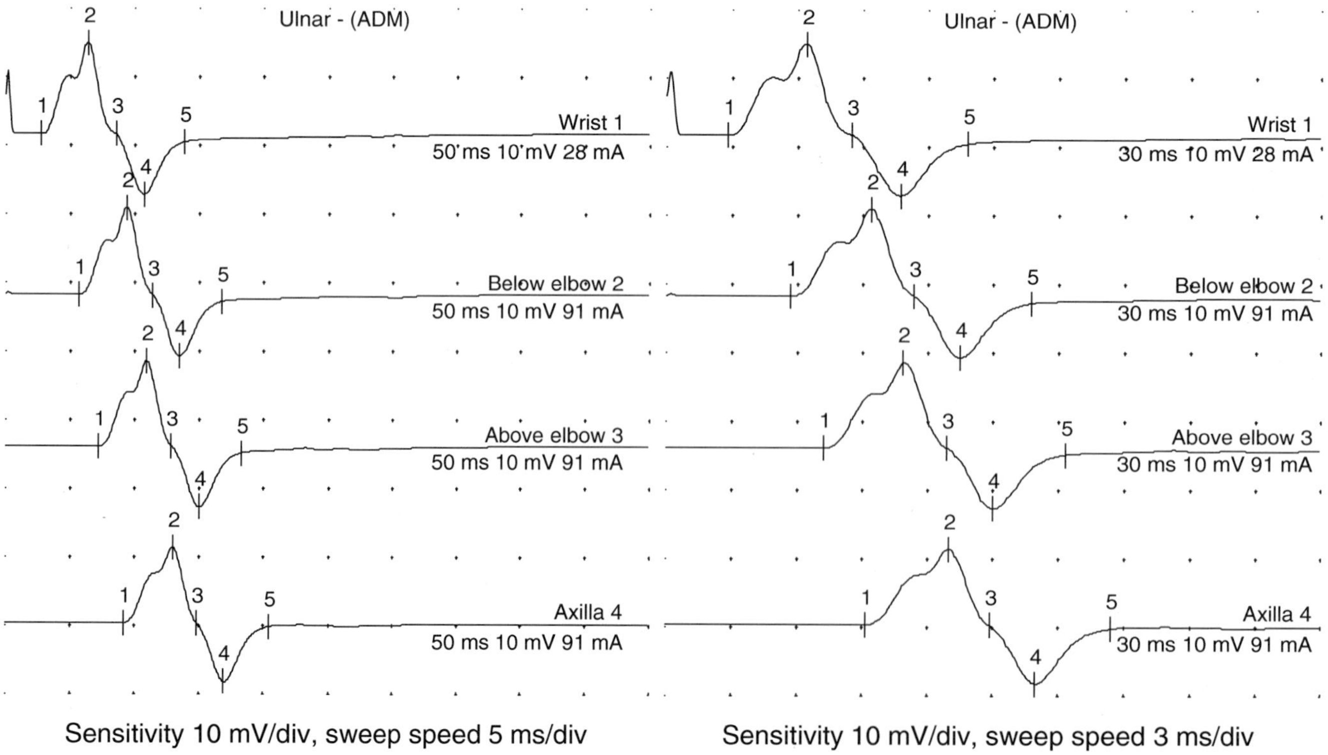

Fig. 1 Compound muscle action potentials (*CMAPs*) recorded at the hand from the ADM muscle, *top to bottom*: stimulation of the wrist, below elbow, above elbow and in the axilla

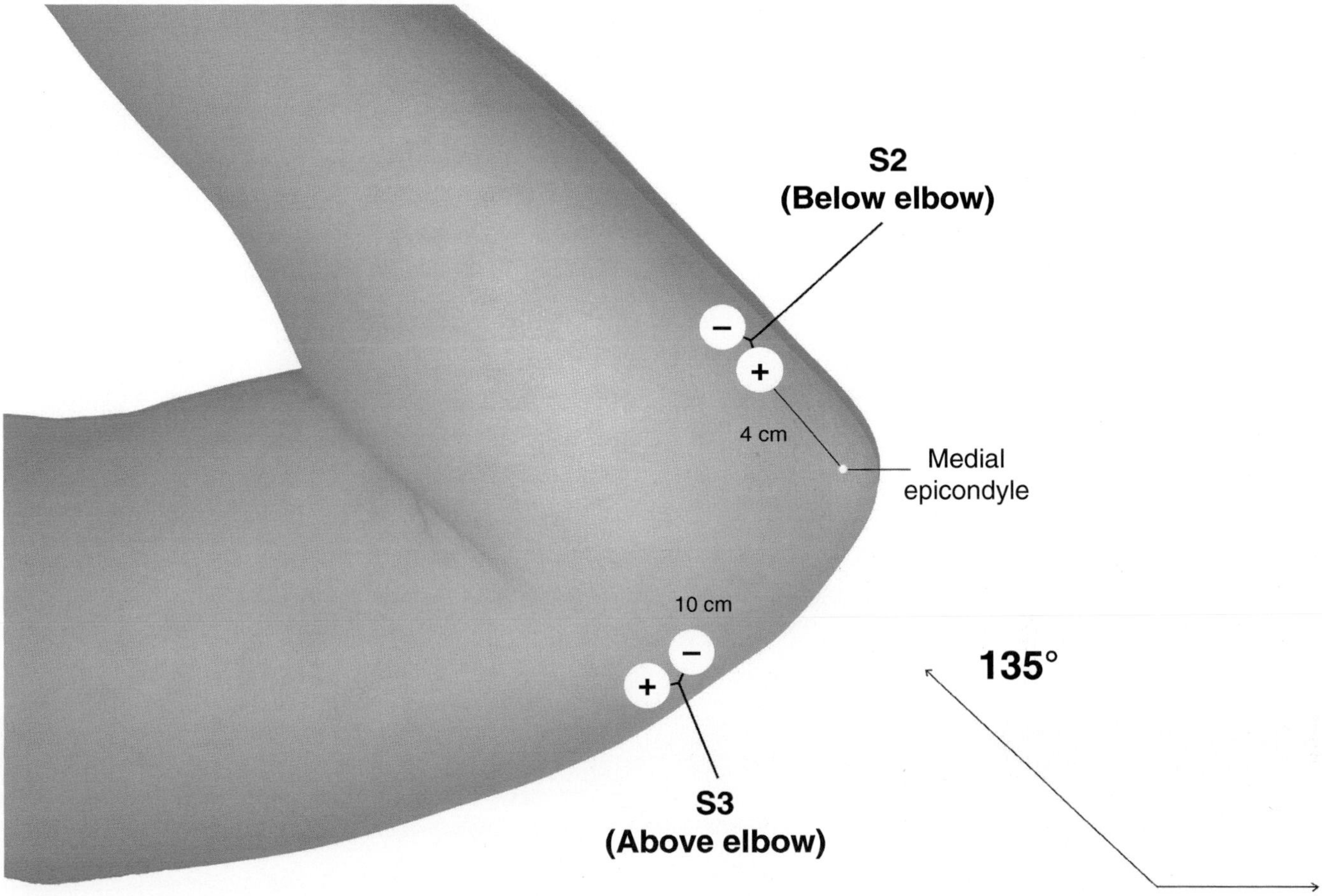

Fig. 2 Proximal stimulation points (S2, below elbow; S3, above elbow) with the elbow flexed to 135°

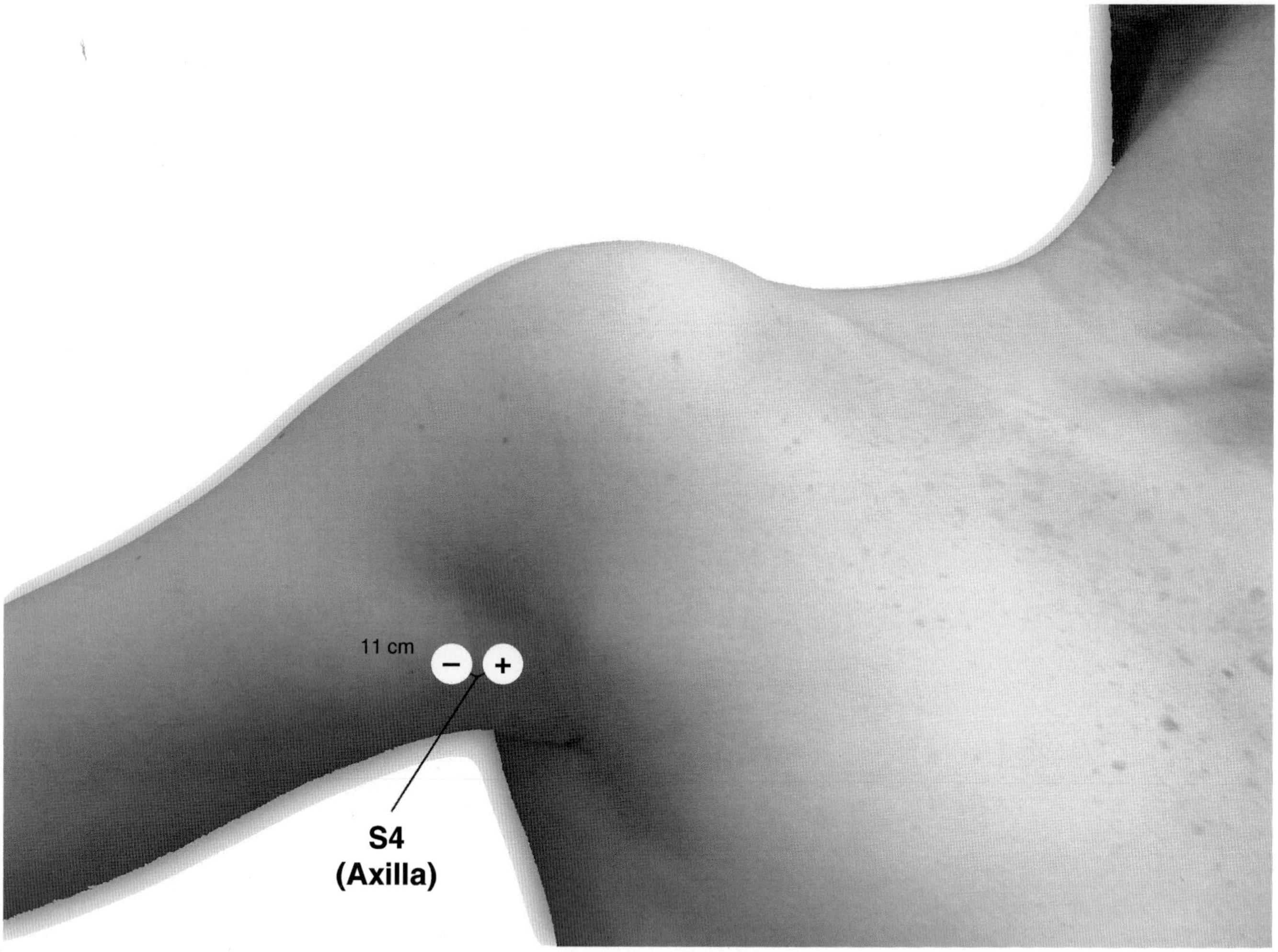

Fig. 3 Proximal stimulation point (S4 – axilla)

Table 1 Normal values [1] – *elbow fully extended (0°)*

	Mean ± SD	Range
Wrist–below elbow, MNCV (m/s)	65.7 ± 6.7	53.0–84.8
Wrist–above elbow, MNCV (m/s)	59.0 ± 4.9	47.4–71.5
Wrist–axilla, MNCV (m/s)	59.1 ± 4.2	50.5–69.7
Below elbow–above elbow, MNCV (m/s)	50.3 ± 5.9	37.9–63.3
Above elbow–axilla, MNCV (m/s)	60.9 ± 7.0	49.5–82.5

Table 2 Normal values [1] – *elbow flexed to 135°*

	Mean ± SD	Range
Wrist–below elbow, MNCV (m/s)	63.3 ± 5.2	49.5–76.7
Wrist–above elbow, MNCV (m/s)	63.0 ± 4.7	53.0–75.9
Wrist–axilla, MNCV (m/s)	62.6 ± 4.1	54.4–73.6
Below elbow–above elbow, MNCV (m/s)	62.8 ± 7.1	49.0–84.4
Above elbow–axilla, MNCV (m/s)	61.9 ± 6.0	50.0–77.8

Comment

Kincaid et al. [1] reported values for the right side only because they observed no significant differences between right and left sides. With the elbow extended, the wrist–above-elbow and below-elbow–above elbow segments showed mean motor velocities that were significantly slower than wrist–below-elbow segment. The authors [1] observed that the data produced by the calculated individual MNCV differences were not normally distributed, and, in such circumstances, the SD was not an adequate measure of data distribution. For this reason, cumulative percentage tables and distribution polygons were constructed to evaluate segmental MNCV differences, and a value of 96 % of the normal subjects was used as cutoff. The segmental velocity differences using a 96 % cutoff were reported in Table 3, elbow fully extended, and Table 4, elbow flexed to 135° (B indicates wrist–below-elbow segment; E, wrist–above-elbow segment; EB, below-elbow–above-elbow segment; XE, above-elbow–axilla segment). Using this cutoff, the slowest below-elbow–above-elbow MNCV was 38 m/s. The mean value of the above-elbow–axilla segment was significantly faster than below-elbow–above-elbow segment. Individual above-elbow–axilla conduction velocity within the 96 % limit was from 15.8 m/s slower to 18.1 m/s faster than the below-elbow–above-elbow segment. The amplitude of the CMAP recorded from the wrist–above-elbow and wrist–below-elbow sites with the elbow extended differed by a mean of 4 % (wrist–below elbow being lower). The maximum difference was 14 %. With the elbow flexed, the mean difference was 3.5 %, and the maximum was 10.5 %. The flexed elbow position allowed supramaximal stimulation (stimulus duration of greater than 0.1 ms was rarely required) at the elbow site at a much lower stimulus strength than in the elbow extended position, despite the stimulus site being at approximately the same location. Another advantage of elbow flexion suggested by the authors [1] was that all portions of the ulnar nerve were directly accessible for locating stimulation points and for performing subsequent distance measurement. With the elbow extended, visualizing the below elbow, above elbow, and axillary stimulation points required the examiner to stand to view the arm's medial aspect.

Table 3 Normal values [1] – *elbow fully extended*

	Mean ± SD	Range	Limit
EB, segmental velocity differences (m/s)	−6.7 ± 4.8	−23.2–0	−14.5–0.02
EB–B, segmental velocity differences (m/s)	−15.0 ± 8.7	−30.3–0.1	−29.8–0.2
XE–EB, segmental velocity differences (m/s)	10.2 ± 8.8	−7.4–33.7	−6.7–27.8

Table 4 Normal values [1] – *elbow flexed to 135°*

	Mean ± SD	Range	Limit
EB, segmental velocity differences (m/s)	−0.3 ± 2.9	−5.1–9.9	−4.7–7.6
EB–B, segmental velocity differences (m/s)	−0.4 ± 7.4	−12.3–25.8	−11.4–22.0
XE–EB, segmental velocity differences (m/s)	−1.0 ± 10.0	−18.2–23.4	−15.8–18.1

Comment

For Kincaid et al. [1], short segments used for nerve conduction studies have a higher potential for error in MNCV calculation producing erroneous values because of the proportionately higher experimental error inherent in measuring distance and latency. The MNCV determinations on nerve segments with a length of less than 10 cm or with conduction time less than 2.5 ms were shown to be susceptible to a high degree of error. For these reasons, they chose to examine MNCV differences between both the shorter across-elbow segment (mean length 12.9 cm), and the forearm (mean length 19.5 cm), and the longer wrist–above-elbow segment (mean length 32.5 cm), also compared with the forearm. Underestimation of the true length of the elbow nerve segment as assessed by surface measurement has been suggested as the explanation for such findings with the elbow in extension [2–4]. The degree of flexion necessary to eliminate this measurement error has been suggested to be 45° [2], 90° [3], and 110° [4], but Kincaid and colleagues [1] chose 135°, presuming that this would provide a maximum elongation of the nerve across the elbow. They concluded that performing ulnar nerve conduction studies with the elbow flexed to 135° and comparing conduction values for the wrist–above-elbow and wrist–forearm segment provide values with the least normal conduction values variation between these segments. For the authors, the application of such values to the evaluation of patients with suspected ulnar nerve lesion at the elbow may allow for increased reliability in the electrodiagnosis of these lesions. Pathological waveform and values in a patient with a chronic inflammatory demyelinating polyneuropathy (CIDP) are here reported (Fig. 4).

Pathological waveform (wrist, below elbow, above elbow, axilla – ADM muscle):

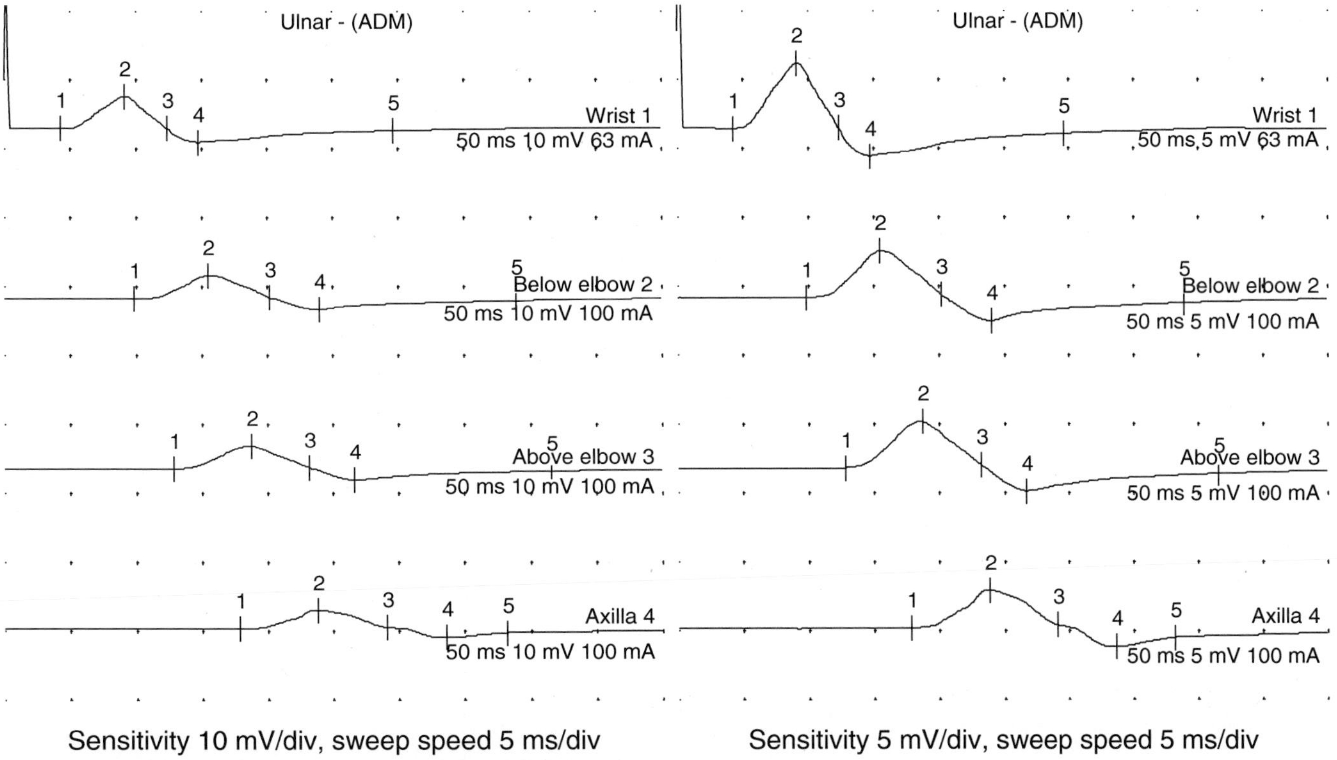

Onset latency (wrist – ADM): 4.30 ms; **Onset latency** (below elbow – ADM): 9.85 ms; **Onset latency** (above elbow – ADM): 12.85 ms; **Onset latency** (axilla – ADM): 17.80 ms; **Peak latency** (wrist – ADM): 9.75 ms; **Peak latency** (below elbow – ADM): 15.15 ms; **Peak latency** (above elbow – ADM): 18.05 ms; **Peak latency** (axilla – ADM): 24.60 ms; **Onset to peak amplitude** (wrist – ADM): 4.8 mV; **Onset to peak amplitude** (below elbow – ADM): 3.4 mV; **Onset to peak amplitude** (above elbow – ADM): 3.4 mV; **Onset to peak amplitude** (axilla – ADM): 2.8 mV; **Peak to peak amplitude** (wrist – ADM): 6.8 mV; **Peak to peak amplitude** (below elbow – ADM): 5.2 mV; **Peak to peak amplitude** (above elbow – ADM): 5.1 mV; **Peak to peak amplitude** (axilla – ADM): 4.2 mV; **MNCV** (wrist – below elbow, 17 cm distance): 30.6 m/s; **MNCV** (below elbow – above elbow, 9 cm distance): 30.0 m/s; **MNCV** (above elbow – axilla, 13 cm distance): 26.3 m/s

Fig. 4 Compound muscle action potentials (*CMAPs*) recorded at the hand from the ADM muscle using surface recording method (Kincaid) [1] in a patient with CIDP associated with anti-MAG/SGPG (pathogenic IgM antibodies – Abs against myelin-associated glycoprotein/peripheral nerve glycolipid sulfated glucuronyl paragloboside), *top to bottom*: stimulation of the wrist, below elbow, above elbow, and in the axilla

References

1. Kincaid JC, Phillips LH, Daube JR (1986) The evaluation of suspected ulnar neuropathy at the elbow. Normal conduction study values. Arch Neurol 43:44–47
2. Harding C, Halar E (1983) Motor and sensory ulnar nerve conduction velocities: effect of elbow position. Arch Phys Med Rehabil 64:227–232
3. Nelson RM (1980) Effects of elbow position on motor conduction velocity of the ulnar nerve. Phys Ther 60:780–783
4. Checkles NS, Russakov AD, Piero DL (1971) Ulnar nerve conduction velocity: effect of elbow position on measurement. Arch Phys Med Rehabil 52:362–365

Wrist – Hand

M31

Surface Recording Technique, Study from *Abductor Pollicis Brevis* (APB) and *First Dorsal Interosseous* (FDI) Muscles

Original Settings Duration of pulse was 0.2 ms and the machine used was a DISA 2000 M; sensitivity, sweep speed, low-frequency filter, and high-frequency filter were not specified.

Position This study was performed in the supine position.

Recording Following the median nerve stimulation (S1), the active electrode (A) was placed over the belly of the *abductor pollicis brevis* (APB) muscle (Fig. 1), halfway between the midpoint of the distal wrist crease and the midpoint of the first metacarpophalangeal joint [1]. The reference (R) was placed slightly distal to the first metacarpophalangeal joint (over the tendon of the APB muscle). Following the ulnar nerve stimulation (S2), the active electrode was placed over the motor point of the *first dorsal interosseous* (FDI) muscle (R2). The reference (R) was placed slightly distal to the second metacarpophalangeal joint (over the tendon of the FDI muscle). The active electrode location was adjusted over the muscular eminence to obtain the greatest compound muscle action potential (CMAP) amplitude for both the median (S1) and ulnar (S2) nerve stimulations at the wrist. The author used surface electrodes for recording (bar recording electrode fixed with a Velcro® strap). The distal motor latency (DML) of APB and FDI were recorded at the same time. The ground (G) electrode position was not specified in the text; the figure shows ground electrode placed on the palm.

Stimulation The median and ulnar CMAPs were evoked with a bar-stimulating electrode. The stimulations were at the wrist, on the median nerve (S1), and on the ulnar nerve (S2), proximal to the wrist crease. The author stimulated both the median and ulnar nerves at the wrist (S1 and S2, respectively) to compare the median and ulnar sensory latencies in the same and opposite hands [1]. The author used percutaneous supramaximal rectangular stimulation. In order to measure the forearm velocity both for the median and ulnar nerves, the proximal stimulation for the median nerve was applied at the internal bicipital groove and for the ulnar nerve above the median epicondylar groove. Only subjects with normal forearm velocity (over 50 m/s) were included in the study.

Measurements Distal latency (ms) was measured from stimulus onset to the onset of initial negative deflection of the CMAP. Onset latency was measured for both the median and ulnar nerves, and differences between the paired distal motor latencies were measured in the same hand and in opposite hands in all subjects. Skin temperature was tested before each investigation, but temperature data was not given. The author studied 102 normal median nerves (APB muscle) and 200 normal ulnar nerves (FDI muscle) from 124 healthy subjects (Table 1) composed of 26 volunteers providing 52 data and 98 patients referred to the laboratory either for spasmophilia or for atypical paresthesia or upper limb pain without any clinical feature of ulnar or median nerve entrapment lesions providing 148 data (83 women and 41 men, age range 22–84 years, mean age 53.73 ± 13.5 years).

G. Gentili, M. Di Napoli, *The Ulnar Nerve*, DOI 10.1007/978-3-319-43292-2_55

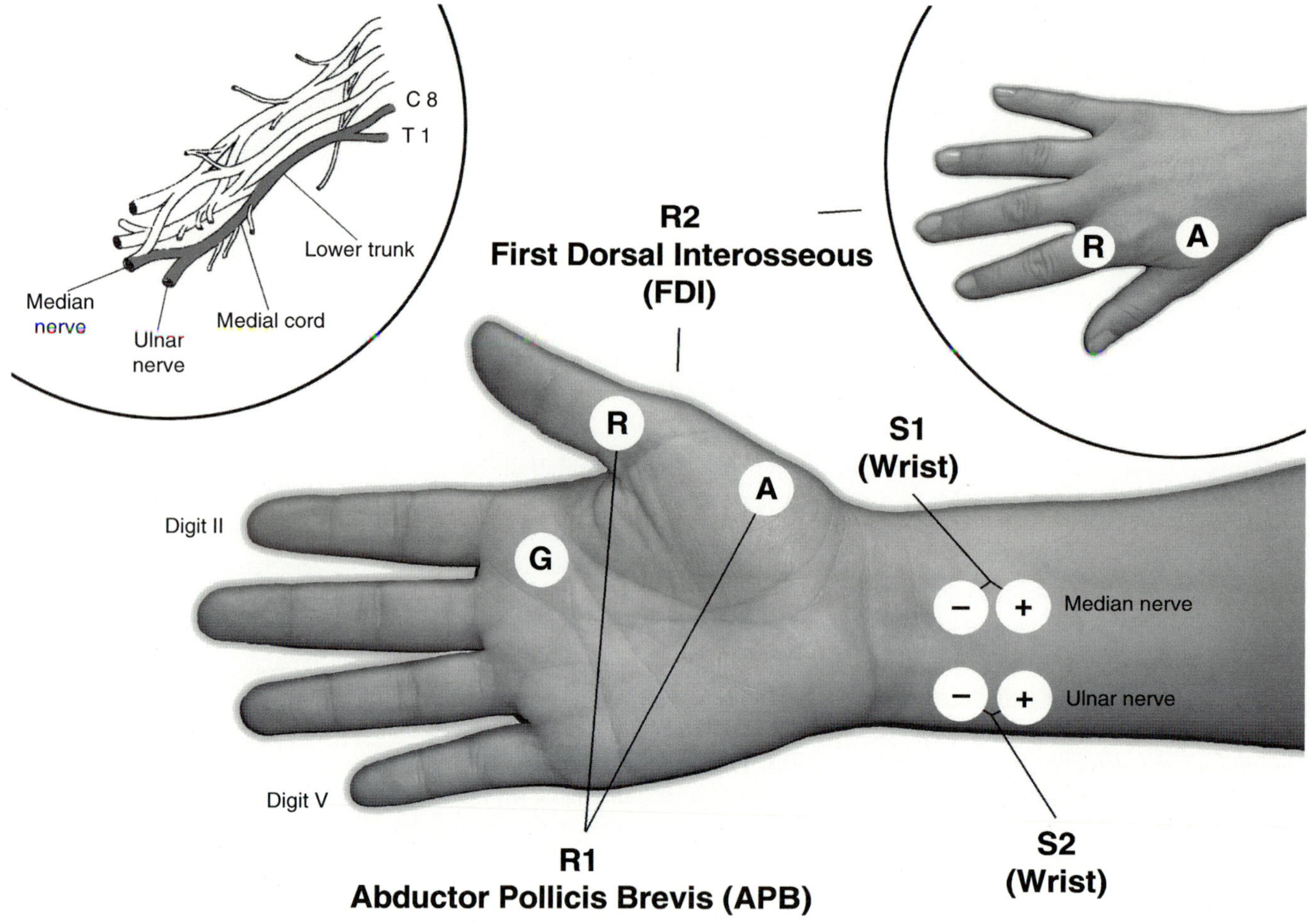

Typical waveform (wrist – APB muscle, wrist – FDI muscle):

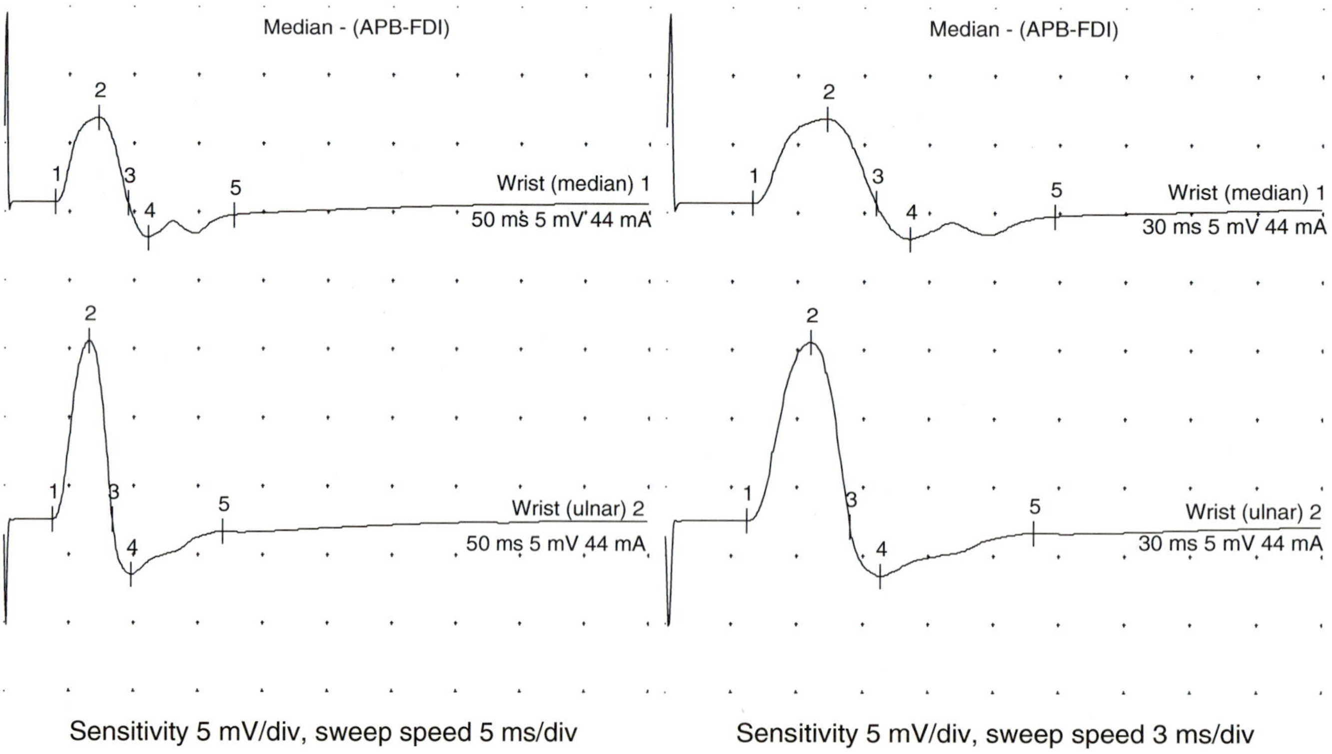

Fig. 1 Compound muscle action potentials (*CMAPs*) recorded at the hand from the APB and FDI muscles, stimulation of the wrist: median nerve (*upper trace*) and ulnar nerve (*lower trace*)

Table 1 Normal values [1]

	Mean ± SD
Wrist–APB, distal latency (ms) – 102 nerves	3.398 ± 0.29
Wrist–FDI, distal latency (ms) – 200 nerves	3.195 ± 0.30
Wrist–FDI, distal latency (ms) – 52 nerves (control group)	3.17 ± 0.28
Wrist–FDI, distal latency (ms) – 148 nerves (patients group)	3.20 ± 0.31
APB–FDI, distal latency difference – opposite hand (ms)–76 nerves	0.206 ± 0.15
APB–FDI, distal latency difference – same hand (ms) – 102 nerves	0.22 ± 0.32

Comment

Seror [1] found the same results in control subjects and in normal patients, so it seemed convenient to study them as one homogeneous group. The DML values to APB were greater than those to FDI in more than 80 % of all cases. The mean differential value was 0.22 ms; it was calculated from 102 wrist data. Mean value between the DML APB value (102 data) and DML FDI value (200 data) was rather similar and equal to 0.204. The study of bilateral data showed only a slight difference between the right and left arm, the mean value being 0.22 ms. The author found a difference of 0.5 ms or more in 5 % of the cases only. For ulnar nerve, the comparison between DML to FDI muscle and motor nerve conduction velocity (MNCV) showed a logical good correlation: MNCV 50-54 m/s, DML 3.287 (ms); MNCV 55–59 m/s, DML 3.232 (ms); MNCV 60–64 m/s, DML 3.157 (ms); and MNCV > 65 m/s, DML 3.105 (ms). For the author, comparing the motor latency between the median and ulnar nerves at the hand is of great interest in the study of pathological entrapment neuropathies of the median and ulnar nerves at the wrist (Figs. 2 and 3). Because in most of the cases, as only one nerve is entrapped, the other one can be taken as normal reference data, so the author asserted that (1) DML to FDI values are pathological when they are higher than 3.8 ms and higher than DML to APB with a difference of 0.4 ms or more; (2) DML to APB values are pathological when they are higher than 4 ms and higher than DML to FDI with a difference of 0.8 ms or more. Seror [1] emphasized that if a significant difference between DML is present, or both DML to APB and DML to FDI are elevated without a significant difference and a global decrease of MNCV, the other arm has to be examined (the difference between both arms is seldom more than 0.5 ms). Only then examiner may conclude that a single or double-entrapment neuropathy of the deep branch of the ulnar nerve and the median nerve at the wrist are present.

Pathological waveform (wrist – APB muscle, wrist – FDI muscle):

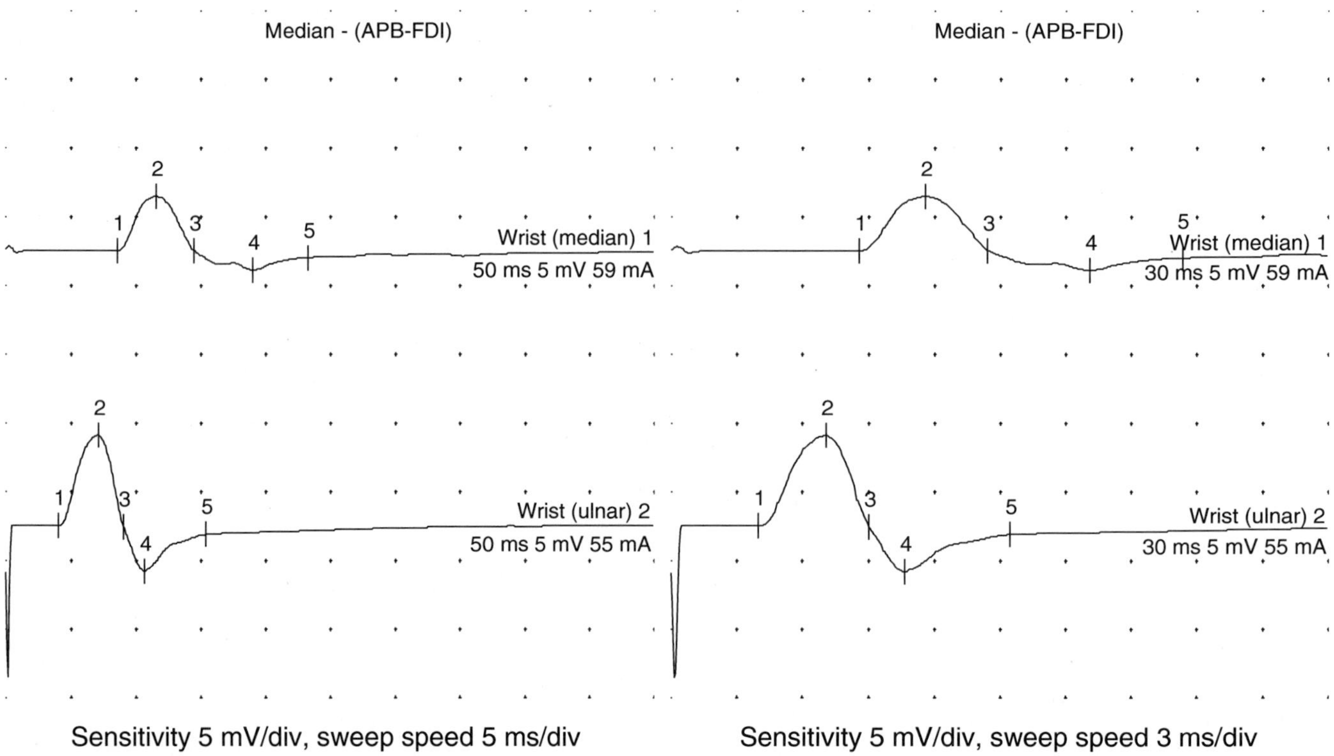

Onset latency (median – APB): 8.55 ms; **Onset latency** (ulnar – FDI): 3.85 ms; **Onset latency difference** (median APB – ulnar FDI): 4.60 ms; **Peak latency** (median – APB): 11.80 ms; **Peak latency** (ulnar – FDI): 7.20 ms; **Onset to peak amplitude** (median – APB): 3.9 mV; **Onset to peak amplitude** (ulnar – FDI): 6.6 mV; **Peak to peak amplitude** (median – APB): 5.4 mV; **Peak to peak amplitude** (ulnar – FDI): 9.9 mV

Fig. 2 Compound muscle action potentials (*CMAPs*) recorded at the hand from the APB and FDI muscles, in very severe CTS – grade 5 by Bland's CTS classification scale [2], stimulation of the wrist: median nerve (*upper trace*) and ulnar nerve (*lower trace*)

Pathological waveform (wrist – APB muscle, wrist – FDI muscle):

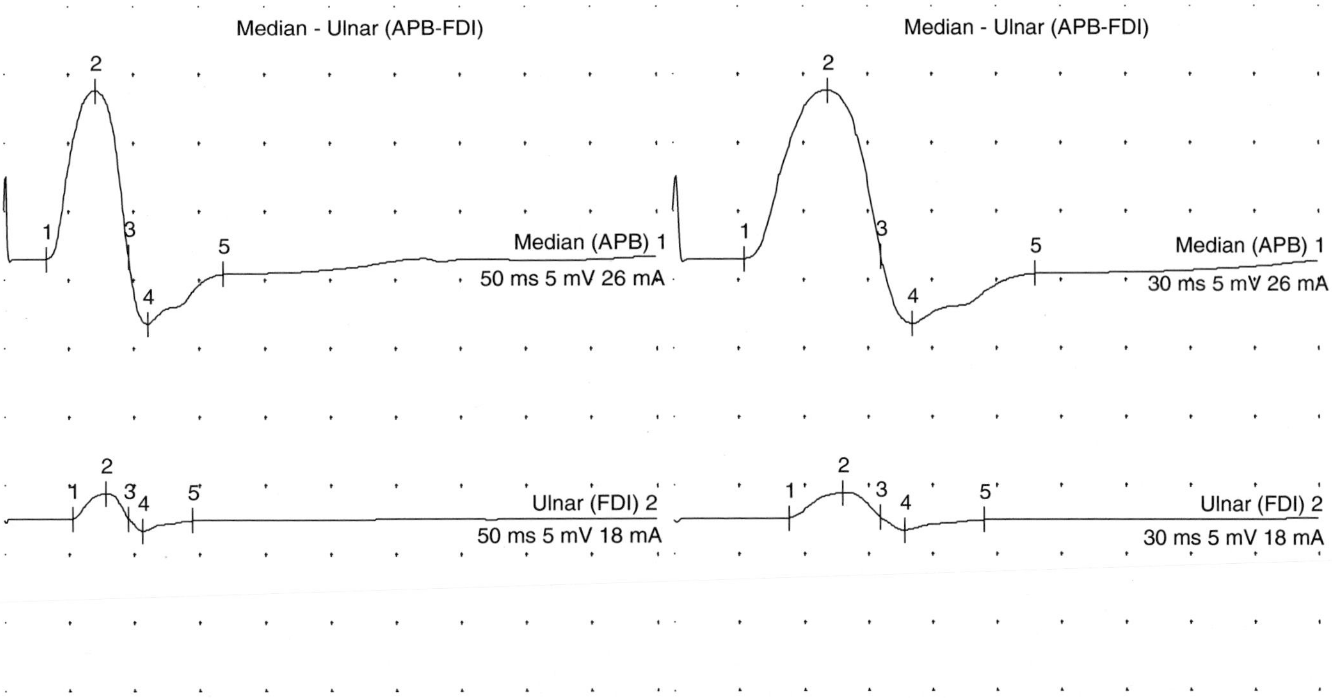

Sensitivity 5 mV/div, sweep speed 5 ms/div

Sensitivity 5 mV/div, sweep speed 3 ms/div

Onset latency (median – APB): 3.25 ms; **Onset latency** (ulnar – FDI): 5.25 ms; **Onset latency difference** (ulnar FDI – median APB): 2.0 ms; **Peak latency** (median – APB): 7.30 ms; **Peak latency** (ulnar – FDI): 7.65 ms; **Onset to peak amplitude** (median – APB): 12.4 mV; **Onset to peak amplitude** (ulnar – FDI): 1.9 mV; **Peak to peak amplitude** (median – APB): 17.1 mV; **Peak to peak amplitude** (ulnar – FDI): 2.8 mV

Fig. 3 Compound muscle action potentials (*CMAPs*) recorded at the hand from the APB and FDI muscles in a patient with Guyon's canal syndrome stimulation of the wrist: median nerve (*upper trace*) and ulnar nerve (*lower trace*)

References

1. Seror P (1988) Comparison of the distal motor latency of the first dorsal interosseous with abductor pollicis brevis. Report of 200 cases. Electromyogr Clin Neurophysiol 28:341–345
2. Bland JDP (2000) A neurophysiological grading scale for carpal tunnel syndrome. Muscle Nerve 23:1280–1283

Wrist – Hand

M32

Needle Recording Technique, Study from *Abductor Pollicis Brevis* (APB) and *Abductor Digiti Minimi* (ADM) Muscles

Original Settings Sensitivity, low-frequency filter, high-frequency filter, sweep speed, duration of pulse, and the machine used were not specified.

Position This study was performed in the supine position.

Recording Following the median nerve stimulation (S1), the recording needle electrode was inserted in the motor point of the *abductor pollicis brevis* (APB) muscle (Fig. 1), halfway between the midpoint of the distal wrist crease and the midpoint of the first metacarpophalangeal joint [1]. Following the ulnar nerve stimulation (S2), the recording needle electrode was inserted in the motor point of the *abductor digiti minimi* (ADM) muscle (R2). The ground (G) electrode position was not specified in the text; the figure shows the ground electrode placed on the palm.

Needle Insertion For the APB muscle, with the hand in a neutral position to full supination, the recording needle electrode was inserted on the lateral aspect of the base of the proximal phalanx of the thumb, using a medial-directed approach just anterior to the bone at the midpoint of the first metacarpal bone. For the ADM muscle, with the hand in a neutral position to full supination, the recording needle electrode was inserted at the midpoint between the fifth metacarpophalangeal joint and the pisiform bone.

Stimulation The median and ulnar nerves were stimulated with a surface stimulator, and the stimulation cathode (–) was placed at the proximal wrist both for the median (S1) and ulnar (S2) nerves stimulations. The authors did not use a constant distance between the stimulating cathode and the recording needle electrode [1].

Measurements Distal latency (ms) was measured from the stimulus onset to the onset of the initial deflection of the compound muscle action potential (CMAP). Onset latency was measured for both the median and ulnar nerves, and differences between paired distal motor latencies were measured in the same hand. As the authors did not use a constant distance between the stimulating cathode and the recording needle electrode, the range of distal motor latency was relatively large and the upper limits (mean ± 2SD) long. They considered a large number of subjects in each group, and therefore the distribution of the distance conducted across was similar in each group. The room temperature was kept not lower than 23 °C; skin temperature data was not reported (the authors specified that sufficient time was allowed for the patients to warm up). Normal nerve conduction values were obtained from 102 subjects (Table 1), selected on the basis of a completely normal motor conduction of the median and ulnar nerves in the forearm (66.7 % women and 33.3 % men, age range 22–86 years, mean age 47.5 years); for pathological values, the authors studied 116 patients (Table 2) suffering from median nerve entrapment at the wrist, as proven by a prolonged DML or distal sensory latency of the median nerve (67.2 % women and 32.8 % men, age range 20–82 years, mean age 49.2 years).

G. Gentili, M. Di Napoli, *The Ulnar Nerve*, DOI 10.1007/978-3-319-43292-2_56

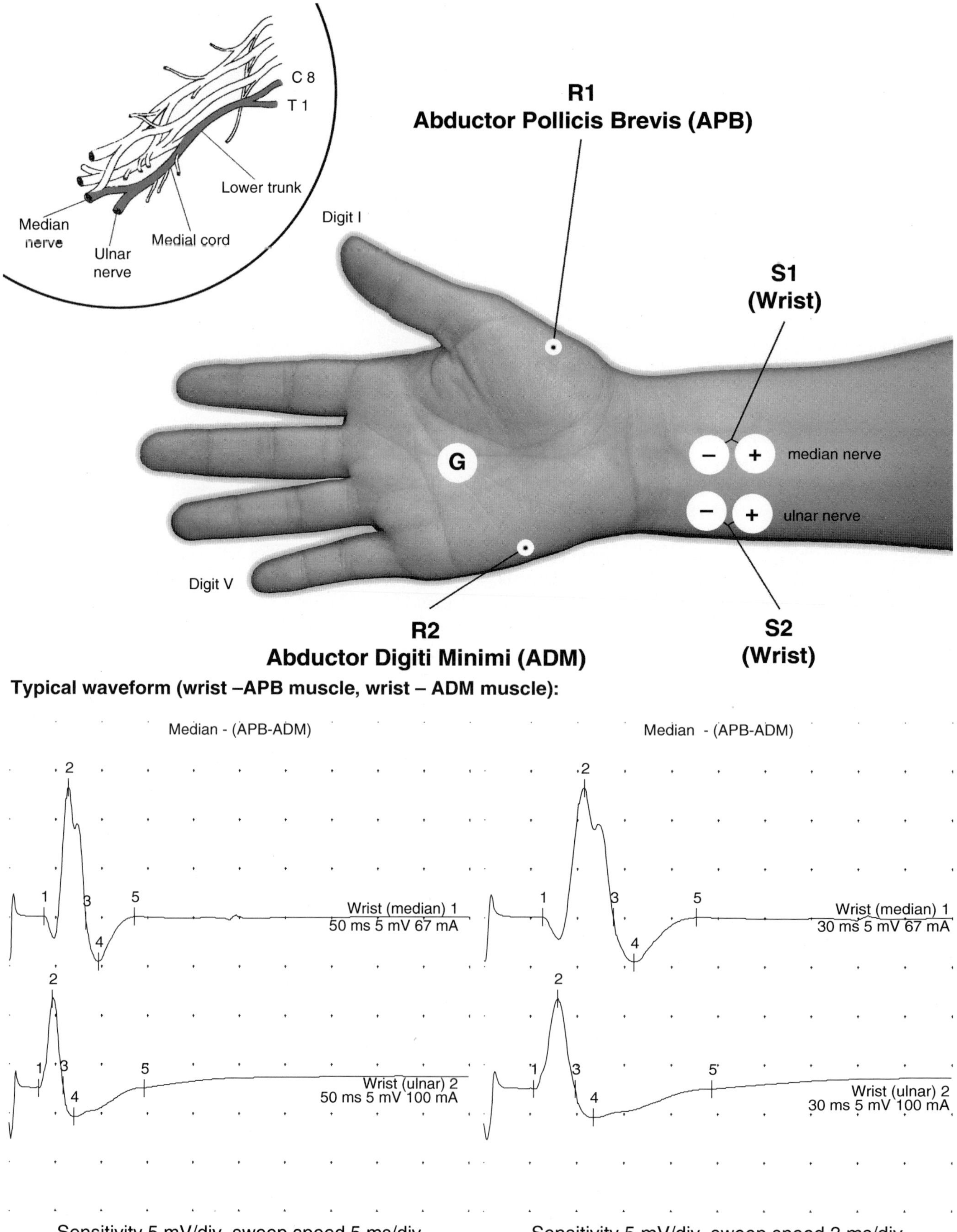

Fig. 1 Compound muscle action potentials (*CMAPs*) recorded at the hand from the APB and ADM muscles, stimulation of the wrist: median nerve (*upper trace*) and ulnar nerve (*lower trace*)

Table 1 Normal values [1]

	Mean ± SD	Mean ± 2SD	Range
Wrist–APB, distal latency (ms)	3.4 ± 0.49	4.5	2.4–4.4
Wrist–ADM, distal latency (ms)	2.5 ± 0.43	3.4	1.6–3.6
APB–ADM, distal latency difference – same hand (ms)	0.9 ± 0.5	1.9	−0.6–2.2

Table 2 Pathological values [1]

	Mean ± SD	Mean ± 2SD	Range
Wrist–APB, distal latency (ms)	5.0 ± 0.84	6.7	3.2–8.8
Wrist–ADM, distal latency (ms)	2.9 ± 0.56	4.0	1.9–5.0
APB–ADM, distal latency difference – same hand (ms)	2.1 ± 1.0	4.1	0.2–6.8

Comment

Mortier et al. [1] demonstrated a significant prolongation of the ulnar nerve DML in patients with carpal tunnel syndrome (CTS) compared to the DML in a control group. They found no relationship between the degree of reduction of the median nerve motor conduction velocity and that of the ulnar nerve, as it was demonstrated by the absence of the correlation between the two distal motor latencies in the patients group. Their result gave reason to the authors to doubt some methods, based on comparing the median and ulnar nerve motor and sensory latencies, used to diagnose a CTS. They suggested use of more reliable methods, based on abnormalities in only the median nerve sensory conduction and amplitude [2–5] or abnormalities in the median nerve distal motor latency (Fig. 2) when corrected for differences in conduction in the forearm [6]. A case of ulnar compressive neuropathy at the wrist – Guyon's canal syndrome is here reported (Fig. 3).

Pathological waveform (wrist – APB muscle, wrist – ADM muscle):

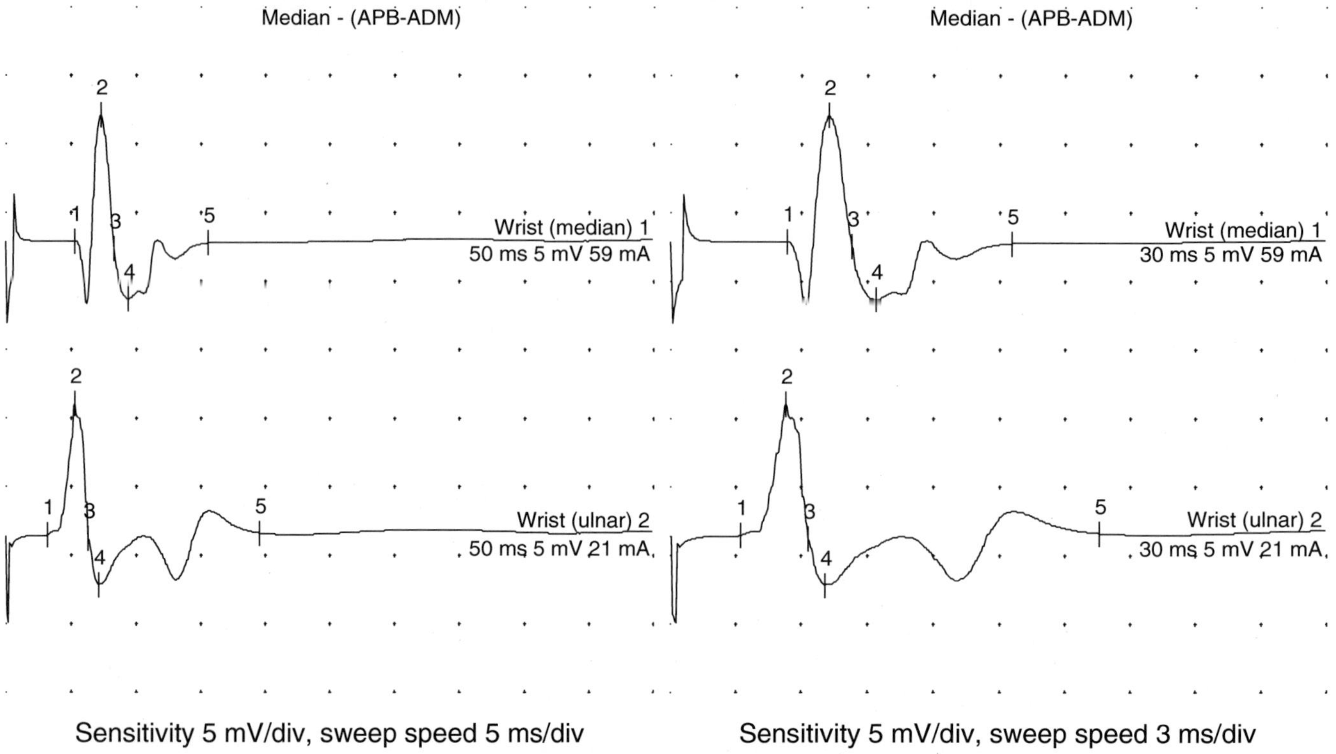

Onset latency (median – APB): 5.30 ms; **Onset latency** (ulnar – ADM): 3.20 ms; **Onset latency difference** (median APB – ulnar ADM): 2.10 ms; **Peak latency** (median – APB): 7.20 ms; **Peak latency** (ulnar – ADM): 5.60 ms; **Onset to peak amplitude** (median – APB): 9.2 mV; **Onset to peak amplitude** (ulnar – ADM): 9.6 mV; **Peak to peak amplitude** (median – APB): 13.5 mV; **Peak to peak amplitude** (ulnar –ADM): 13.3 mV

Fig. 2 Compound muscle action potentials (*CMAPs*) recorded at the hand from the APB and ADM muscles in moderately severe CTS – grade 3 by Bland's CTS classification scale [7], stimulation of the wrist: median nerve (*upper trace*) and ulnar nerve (*lower trace*)

Pathological waveform (wrist – APB muscle, wrist – ADM muscle):

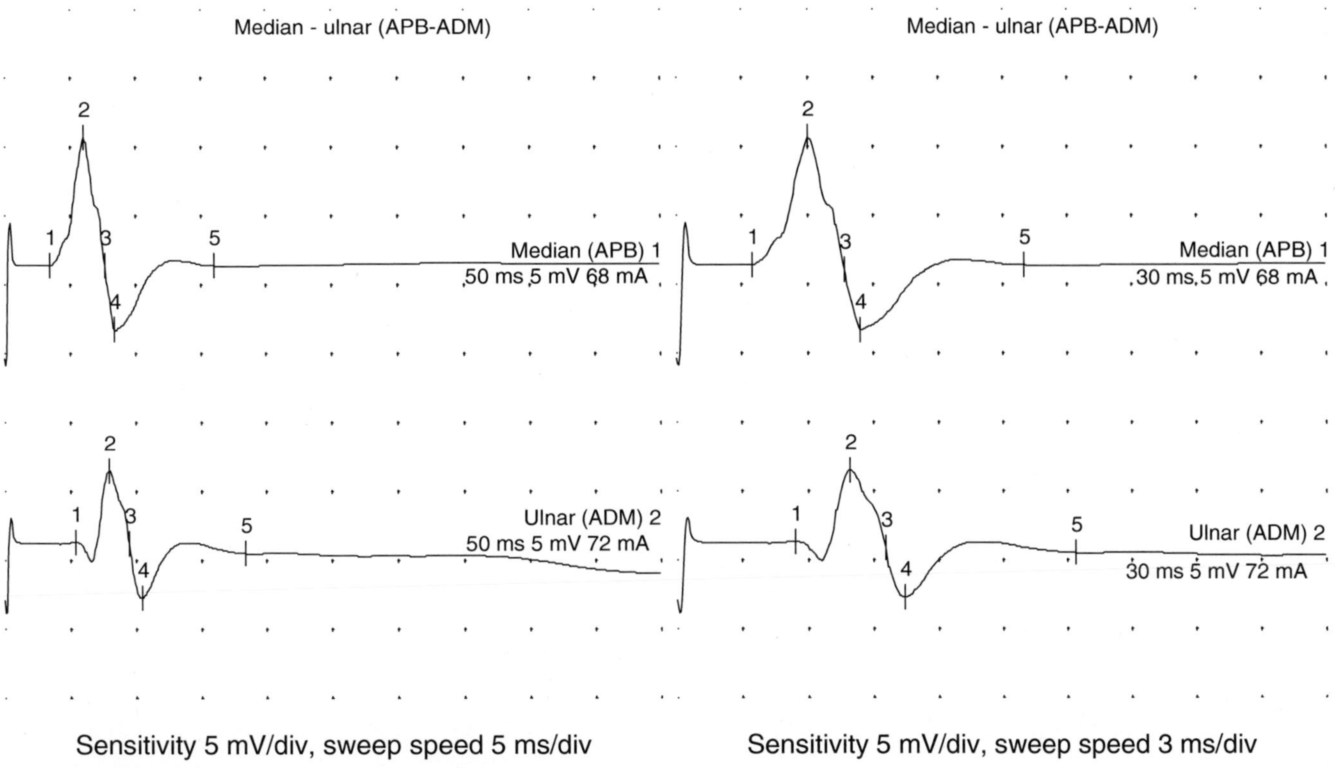

Onset latency (median – APB): 3.45 ms; **Onset latency** (ulnar – ADM): 5.35 ms; **Onset latency difference** (ulnar ADM – median APB): 1.90 ms; **Peak latency** (median – APB): 5.95 ms; **Peak latency** (ulnar – ADM): 7.70 ms; **Onset to peak amplitude** (median – APB): 9.3 mV; **Onset to peak amplitude** (ulnar – ADM): 5.2 mV; **Peak to peak amplitude** (median – APB): 14.1 mV; **Peak to peak amplitude** (ulnar – ADM): 9.3 mV

Fig. 3 Compound muscle action potentials (*CMAPs*) recorded at the hand from the APB and ADM muscles in a patient with Guyon's canal syndrome, stimulation of the wrist: median nerve (*upper trace*) and ulnar nerve (*lower trace*)

References

1. Mortier G, Deckers K, Dijs H et al (1988) Comparison of the distal motor latency of the ulnar nerve in carpal tunnel syndrome with a control group. Electromyogr Clin Neurophysiol 28:75–77
2. Buchthal F, Rosenfalck A, Trojaborg W (1974) Electrophysiological findings in entrapment of the median nerve at wrist and elbow. J Neurol Neurosurg Psychiatry 37:340–360
3. Kimura I, Ayyar DR (1985) The carpal tunnel syndrome: electrophysiological aspects of 639 symptomatic extremities. Electromyogr Clin Neurophysiol 25:151–164
4. Loong SC, Seah CS (1971) Comparison of median and ulnar sensory nerve action potentials in the diagnosis of the carpal tunnel syndrome. J Neurol Neurosurg Psychiatry 34:750–754
5. Monga TN, Shanks GL, Poole BJ (1985) Sensory palmar stimulation in the diagnosis of carpal tunnel syndrome. Arch Phys Med Rehabil 66:598–600
6. Kraft GH, Halvorson GA (1983) Median nerve residual latency: normal value and use in diagnosis of carpal tunnel syndrome. Arch Phys Med Rehabil 64:221–226
7. Bland JDP (2000) A neurophysiological grading scale for carpal tunnel syndrome. Muscle Nerve 23:1280–1283

Below Elbow to Above Elbow – Hand

M33

Surface Recording Technique, Study from *Abductor Digiti Minimi* (ADM) Muscle

Original Settings Sweep speed was 2 ms/division, duration of pulse was 0.1 ms, rate of pulse was 1/s, and the machines used were TECA models Mystro and TE 42; the sensitivity, low-frequency filter, and high-frequency filter were not specified.

Position The nerve conduction study was done with subjects in the supine position with the shoulder abducted 40° and externally rotated as needed. The elbow was flexed to 90° and the forearm supinated using a polypropylene splint with adjustable Velcro® fasteners to maintain this elbow position.

Recording Using surface electrodes, compound muscle action potentials (CMAPs) were recorded from the *abductor digiti minimi* (ADM) muscle [1]; the active electrode (A) was placed over the belly of the muscle, and the reference electrode was placed distally at the base of the digit V (Fig. 1). The ground (G) electrode position was not specified in the text; the figure shows ground placed on the palm.

Stimulation For the short-segment incremental study (SSIS), a point "P" was located over the course of the ulnar nerve on a line passing through the prominent point of the medial epicondyle perpendicular to the medial border of the ulna [1]. The distal points (D2 and D4) were 2 and 4 cm distal to the point "P," respectively. The proximal points (P2, P4, and P6) were 2, 4, and 6 cm proximal to the point "P," respectively (Fig. 2).

These stimulation points were situated 2 cm apart along the ulnar nerve. The distance between the points was meticulously measured with metal tape closely applied to the skin surface, and the points were marked on the skin. The authors tried to perform the more distal stimulation "D6" – 6 cm distal to the point "P" (Fig. 3), but the stimulation at this point was not done in all the control subjects and patients; therefore, this measurement was deleted. Stimulus duration was increased, if needed, especially at points D2 and D4, to obtain a supramaximal CMAP. They also performed conventional motor conduction study to the ADM muscle stimulating the ulnar nerve at the wrist – 8 cm proximal to the recording electrode, below the elbow, and above the elbow and at least 10 cm proximal to the point P6 – and in the axilla. The ulnar motor conduction study, along the 10 cm segment across the elbow with the elbow flexed to 90°, was obtained in the conventional way [2].

Measurements Motor nerve conduction velocity (MNCV) of each segment (wrist–below elbow, below elbow–above elbow, above elbow–axilla) was calculated dividing the distance between two different stimulating sites by the difference of onset latencies, and it was measured in meter per second (m/s). Negative peak amplitude (μV) of the compound muscle action potential (CMAP) was measured from the baseline to the negative peak. The differences in the amplitudes and conduction times between two successive responses obtained were compared and standard data established. All of the electrophysiologic recordings were made at a constant room temperature (27 °C). Normal values (Table 1) were obtained from 25 ulnar nerves (13 left and 12 right) of 20 healthy subjects (14 men, age range 27–58 years, mean age 35 years, and 6 women, age range 27–52 years, mean age 42 years).

G. Gentili, M. Di Napoli, *The Ulnar Nerve*, DOI 10.1007/978-3-319-43292-2_57

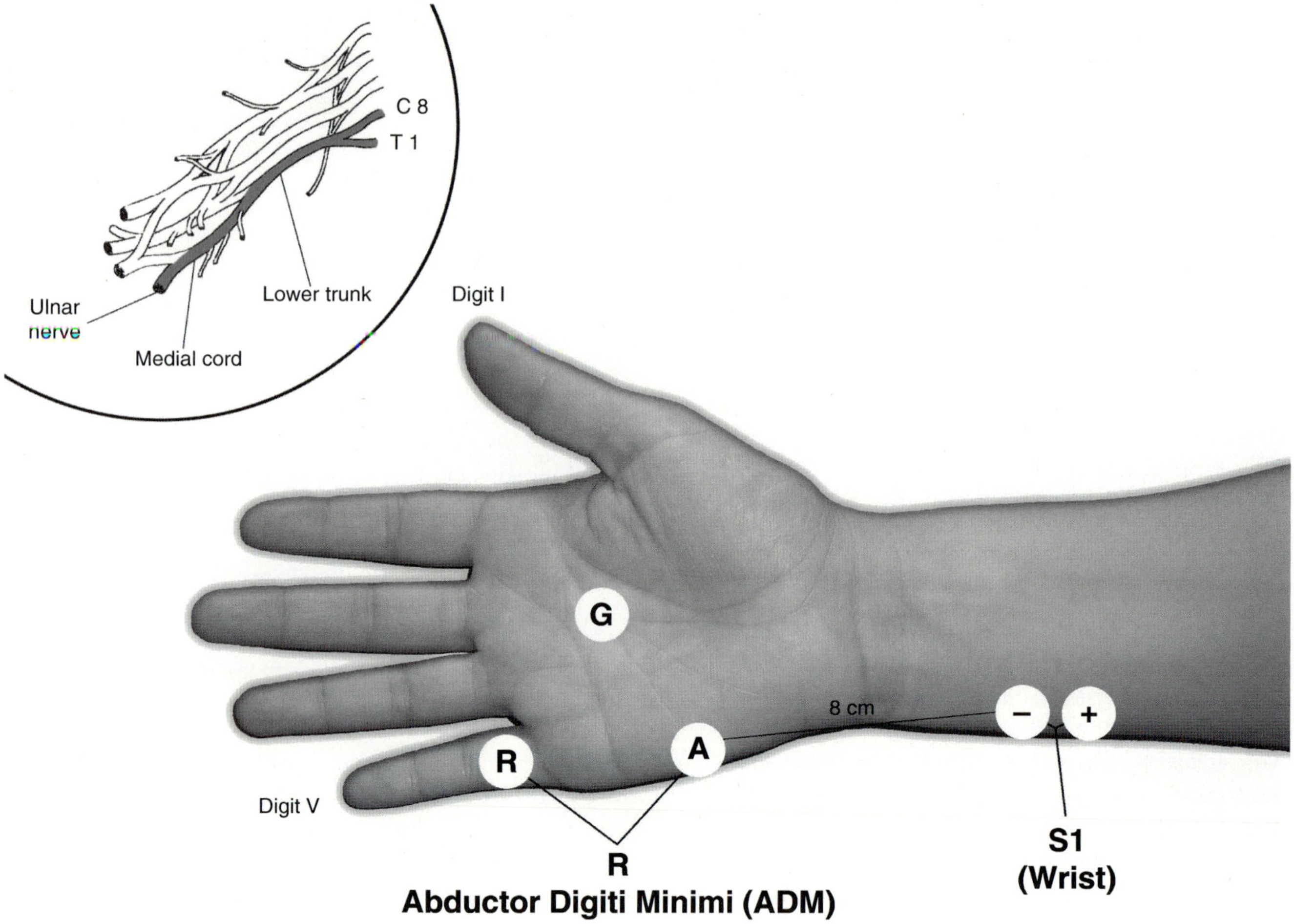

Fig. 1 Distal stimulation point (S1 – wrist), 8 cm distance from the active recording electrode

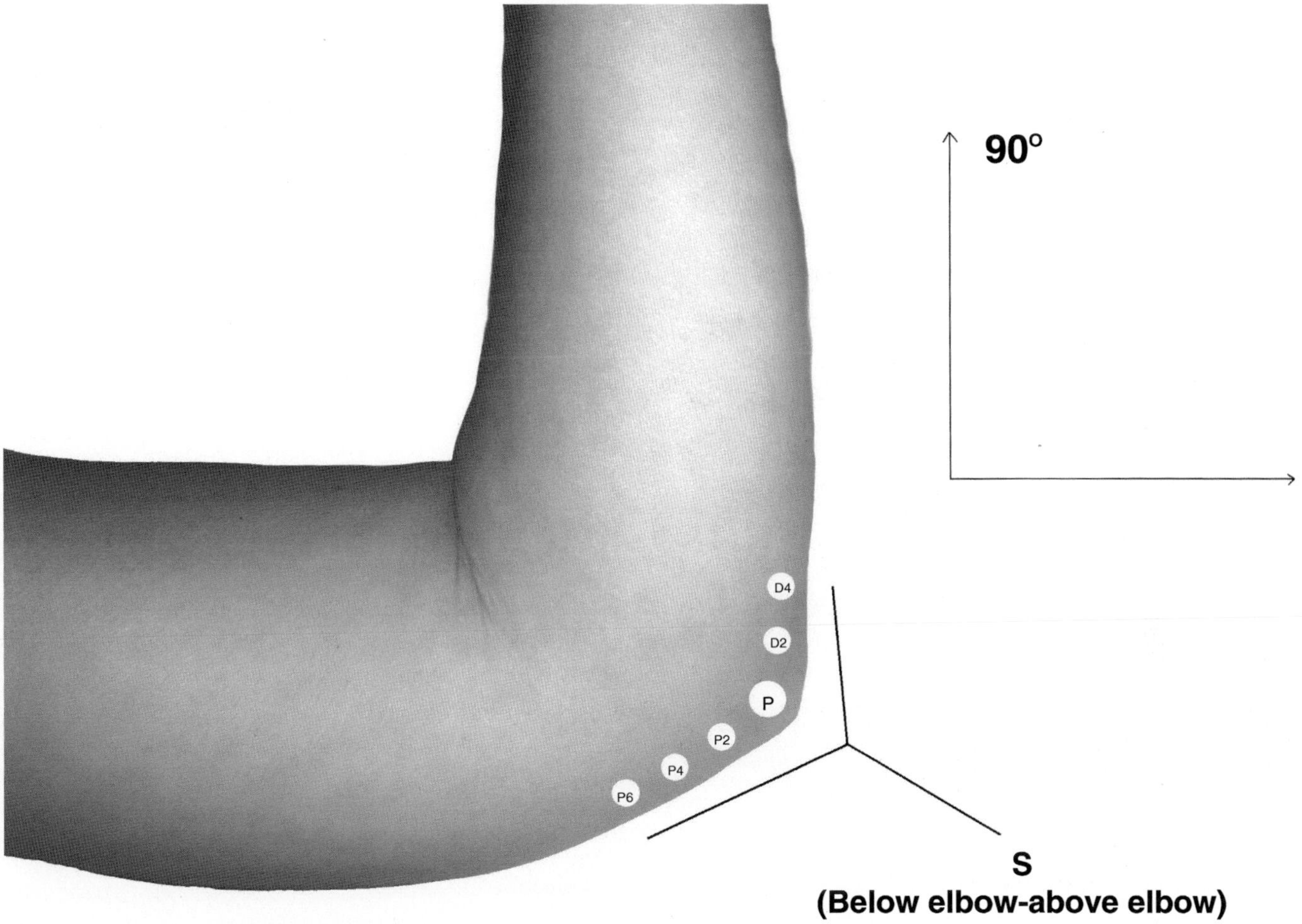

Typical waveform (below elbow to above elbow – ADM muscle):

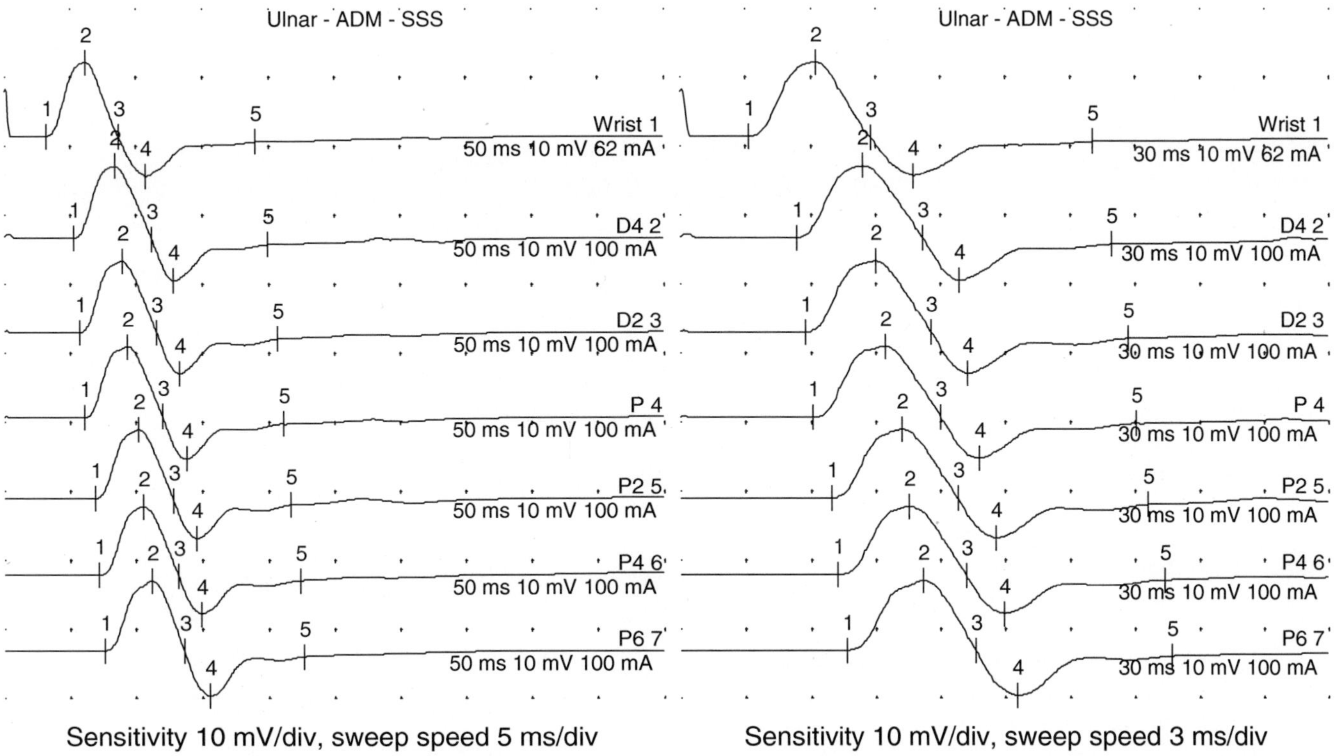

Fig. 2 Compound muscle action potentials (*CMAPs*) recorded at the hand from the ADM muscle, stimulation of the wrist and *short-segment incremental stimulation* (*SSIS*), from the below-elbow (D4, D2) to the above-elbow (P2, P4, P6) sites

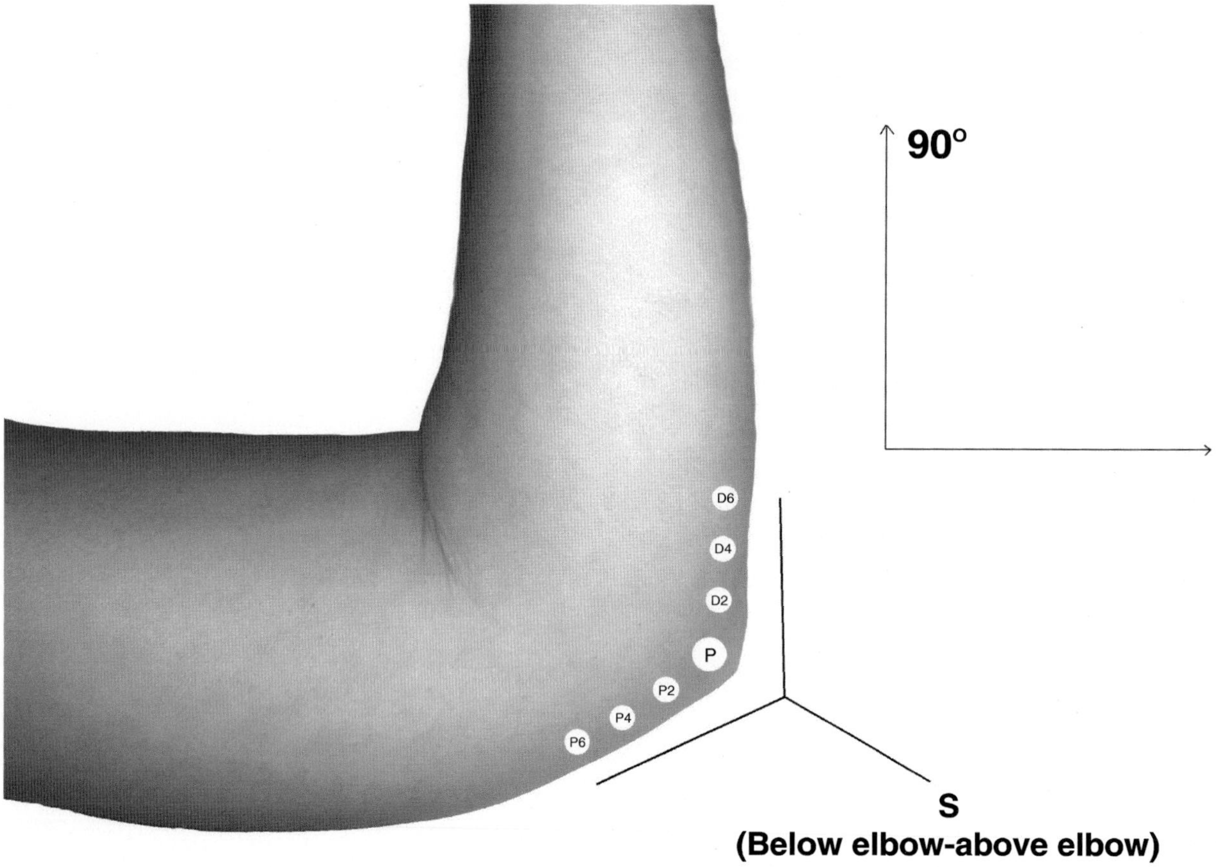

Typical waveform (below elbow to above elbow – ADM muscle):

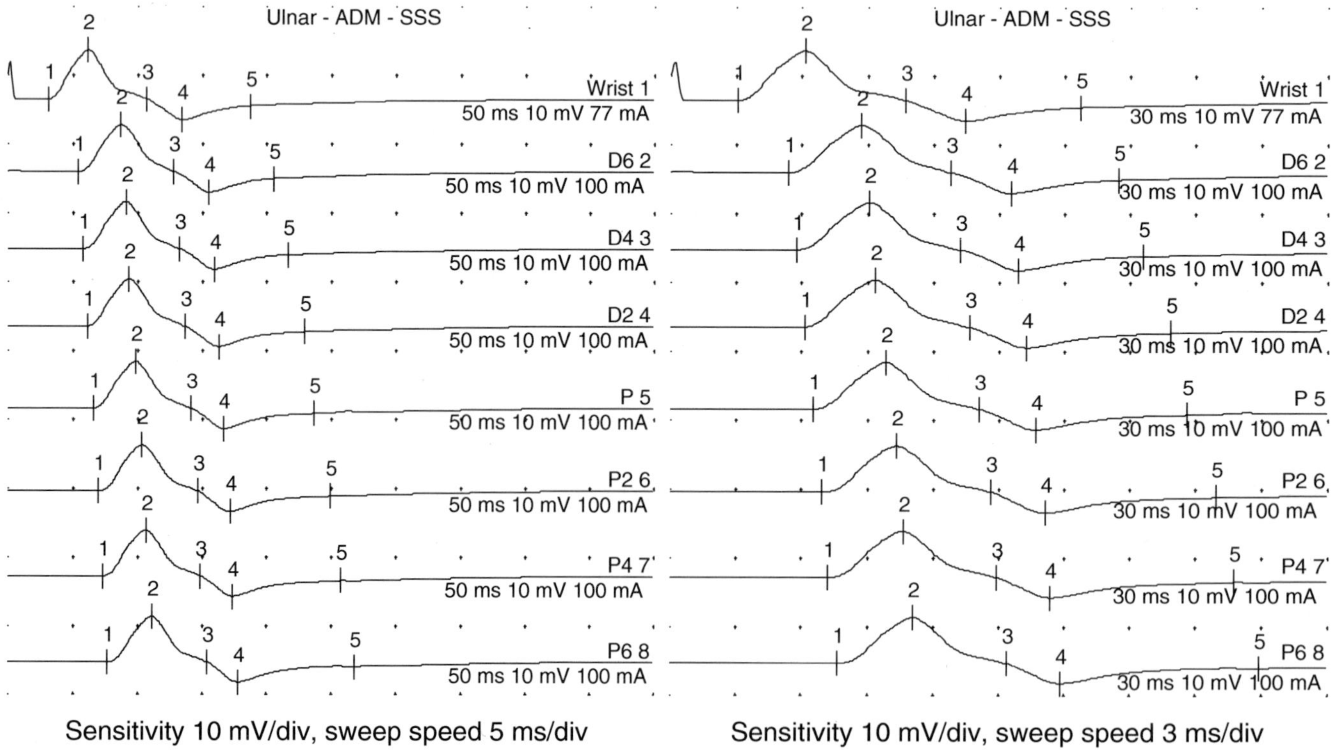

Fig. 3 Compound muscle action potentials (*CMAPs*) recorded at the hand from the ADM muscle, stimulation of the wrist and *short-segment incremental stimulation* (*SSIS*), from the below-elbow (D6, D4, D2) to the above-elbow (P2, P4, P6) sites

Table 1 Normal values [1]

	Mean ± SD	Range
Wrist–below elbow, MNCV (m/s) – right side	60.6 ± 5.2	53–68
Wrist–below elbow, MNCV (m/s) – left side	60.2 ± 7.7	53–65
Below elbow–above elbow, MNCV (m/s) – right side	58.2 ± 3.2	49–74
Below elbow–above elbow, MNCV (m/s) – left side	59.3 ± 7.1	50–76
Above elbow–axilla, MNCV (m/s) – right side	59.4 ± 4.3	50–65
Above elbow–axilla, MNCV (m/s) – left side	61.7 ± 6.0	53–71
D4-D2 segment, SSIS latency (ms) – right side	0.37 ± 0.08	0.24–0.60
D4-D2 segment, SSIS latency (ms) – left side	0.36 ± 0.07	0.23–0.48
D2-P segment, SSIS latency (ms) – right side	0.43 ± 0.10	0.29–0.48
D2-P segment, SSIS latency (ms) – left side	0.44 ± 0.08	0.3–0.60
P-P2 segment, SSIS latency (ms) – right side	0.38 ± 0.08	0.24–0.54
P-P2 segment, SSIS latency (ms) – left side	0.42 ± 0.09	0.32–0.58
P2-P4 segment, SSIS latency (ms) – right side	0.40 ± 0.08	0.24–0.48
P2-P4 segment, SSIS latency (ms) – left side	0.39 ± 0.10	0.24–0.60
P4-P6 segment, SSIS latency (ms) – right side	0.34 ± 0.07	0.23–0.48
P4-P6 segment, SSIS latency (ms) – left side	0.35 ± 0.05	0.24–0.44

Comment

Kanakamedala et al. [1], studied 13 patients (age range 29–66 years, mean age 50.6 years), referred for the evaluation of suspected ulnar neuropathy at the elbow, by conventional motor conduction study and SSIS. In all patients, they found a lower MNCV than in the controls, as were the below-elbow and above-elbow values. The conduction time was considerably prolonged in certain short segment (the maximum delay was 3.71 ms). It was mainly prolonged (greater than mean +2 SD) in the D2-P2 short segment alone in 3 patients, in the D2-P segment alone in two patient, and from P to P2 segment alone in 3 patients; a prolonged conduction time was observed in a patient alone in the D4-P, D2-P4, D2-P, P2-P4, and D4-D2 segments. For only one patient, the conduction time in all short segments was not prolonged. In summary, two patients showed a mild reduction in CMAP amplitude but markedly prolonged conduction times, and nine patients showed an abnormal delay in conduction in the short segments as well as considerable reduction in the amplitude of the CMAPs. One patient showed a reduction in amplitude only, and one patient had a delay in conduction time only. Analysis of their results indicates that in two patients the conduction slowing occurred within the cubital tunnel (D4-D2 segment), in association with a marked decline in the CMAP amplitude in these points. Overall, of the 13 patients studied, the CMAP was abnormal and conduction delayed within the cubital tunnel in three patients, abnormalities were more marked at the medial epicondyle in nine patients, and abnormalities were observed at both sides in only one patient. Some authors [3] found the SSIS technique a useful tool for the precise localization of conduction abnormalities around the elbow, and the conventional method of measuring the motor conduction across the entire segment may be inadequate. The SSIS permits the detection of the site at which an abrupt decline in the amplitude of the CMAPs or prolongation of conduction time between short segments occurs; both may indicate conduction block or conduction slowing due to focal demyelination. It is important to note that a conduction abnormality in a peripheral nerve may be found at, just proximal to, or just distal to the site of entrapment, and their [1] observations were similar to those of Brown et al. [4], who found localized conduction abnormalities just proximal to, at the level of, or within a segment 1–2 cm distal to the medial epicondyle.

Brown and Yates [5] suggested that an important technical limitation of the SSIS is the uncertainty of the exact point of excitation of the nerve, particularly at the cubital tunnel, where the nerve is deeper relative to the stimulating electrode. The ulnar nerve, just beyond the entrance to the cubital tunnel, penetrates between the two heads of the *flexor carpi ulnaris* (FCU) muscle to lie beneath that muscle and on the surface of the flexor digitorum profundus (FDP) muscle. For the authors, the result in both instances was unavoidable uncertainty, especially with percutaneous stimulation techniques, about the exact level at which the nerve was excited in relation to the overlying cathode. For this reason, attempts to stimulate with percutaneous electrodes more than 2 cm distal to the entrance of the cubital tunnel were abandoned. Despite this difficulty, localization of an abrupt reduction in the amplitude of CMAP and abnormal conduction delay between the evoked motor responses provides sufficient evidence to indicate the site of predominant entrapment. They included 17 ulnar neuropathies in their investigation, and the major conduction abnormalities were concentrated just proximal or distal to the medial epicondyle, those distal to the epicondyle probably originating at the cubital tunnel. Following the short-segment incremental study (SSIS) method, they stimulated the ulnar nerve at 2 cm intervals proximal (−) and distal (+) to the tip of the medial epicondyle (0) to include the retroepicondylar and trans-cubital tunnel segments (Fig. 4).

Azrieli et al. [6], following the Kanakamedala's method [1], performed short-segment incremental studies (SSISs) with a NICOLET Viking IV EMG system, using self-adhesive disk recording electrodes. A self-adhesive rectangular electrode was used as ground electrode. All the SSIS studies were performed with the patient supine and the arm abducted to 90° and externally rotated at the shoulder and with the elbow flexed at 90°; skin temperature was controlled over the volar forearm, and it was maintained at 32–34 °C. The midpoint of the ulnar groove was identified by a line drawn from the medial epicondyle to the olecranon. Markers were placed along the course of the ulnar nerve at 2 cm intervals to distances of 6 cm above and 6 cm below the midpoint, and the nerve was stimulated at each marker (Fig. 5). The authors [6] chose 2 cm short-segment distance because it is large enough to reliably incorporate damaged ulnar nerve segments as small as 5–10 mm, but small enough to avoid reduced sensitivity due to the incorporation of significant lengths of normally conducting nerve into latency measures and conduction velocity calculations [5]. The SSIS must be designed to minimize initial depolarization of the nerve distal to the point immediately underneath the cathode, also called "the virtual cathode effect"; for this reason, the authors used longer segments (2 cm instead of 1 cm) between stimulation points and stimulating electrodes with reasonable small surface areas (prong electrodes), while avoiding excessive stimulating current and excessive electrode gel (to prevent gel cross bridging), to circumvent the effect described by Brown and Yates [5].

Azrieli et al. [6] studied 25 arms in healthy volunteers, 21 symptomatic arms (Table 2) from 17 patients with clinical symptoms suggesting UNE (11 arms from 9 men and 10 arms from 7 women, average age 41 years – symptomatic group), and 25 asymptomatic arms (Table 3) in 15 subjects (14 arms from 8 men and 11 arms from 7 women, average age 36 years – asymptomatic group).

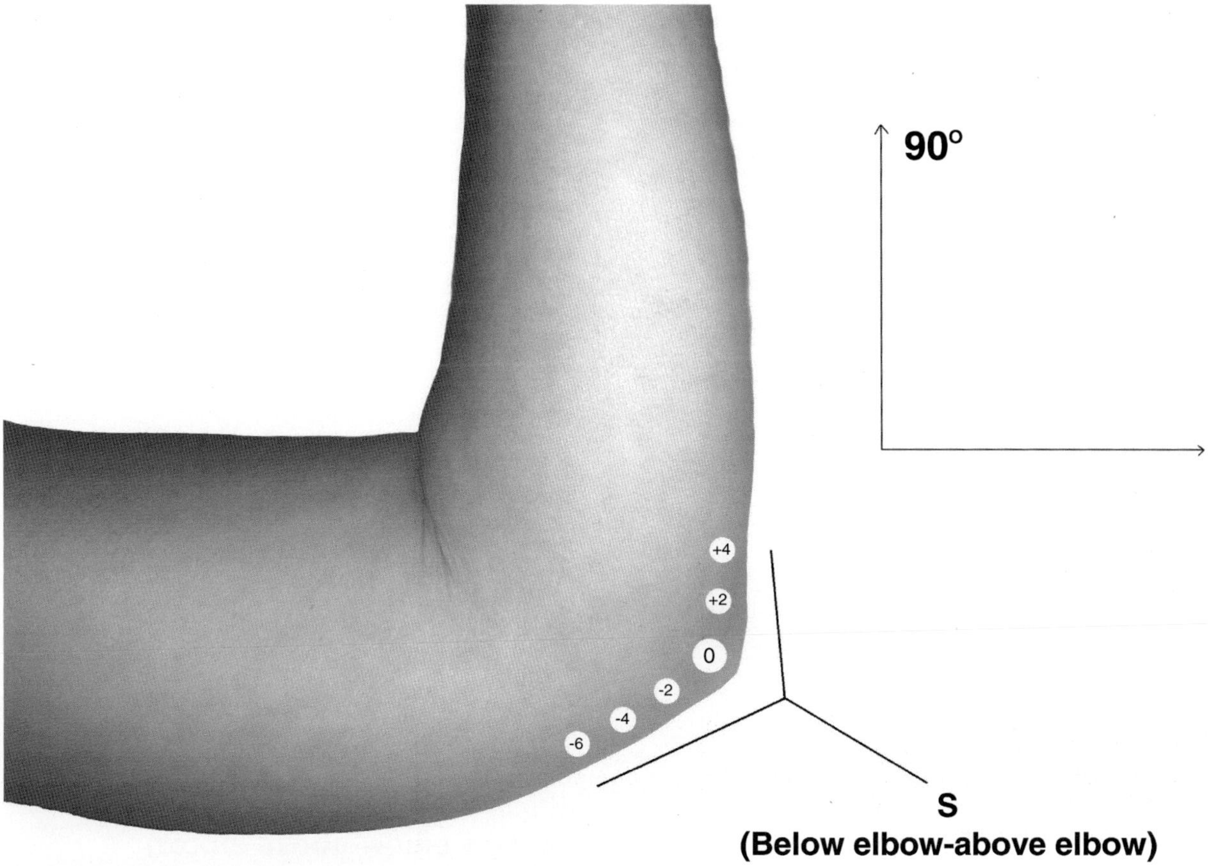

Typical waveform (below elbow to above elbow – ADM muscle):

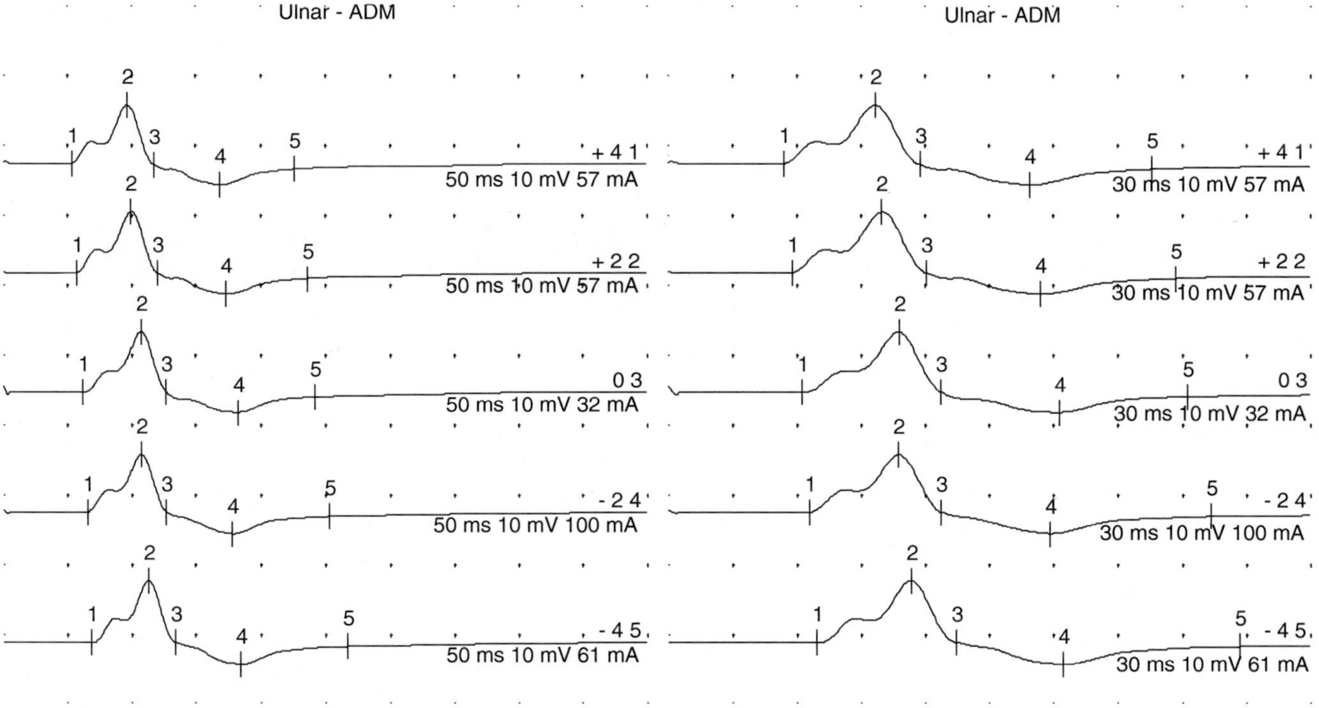

Sensitivity 10 mV/div, sweep speed 5 ms/div   Sensitivity 10 mV/div, sweep speed 3 ms/div

Fig. 4 Compound muscle action potentials (*CMAPs*) recorded at the hand from the ADM muscle, *short-segment incremental stimulation* (*SSIS*) from the below-elbow (+4, +2) to the above-elbow (−2, −4, −6) sites

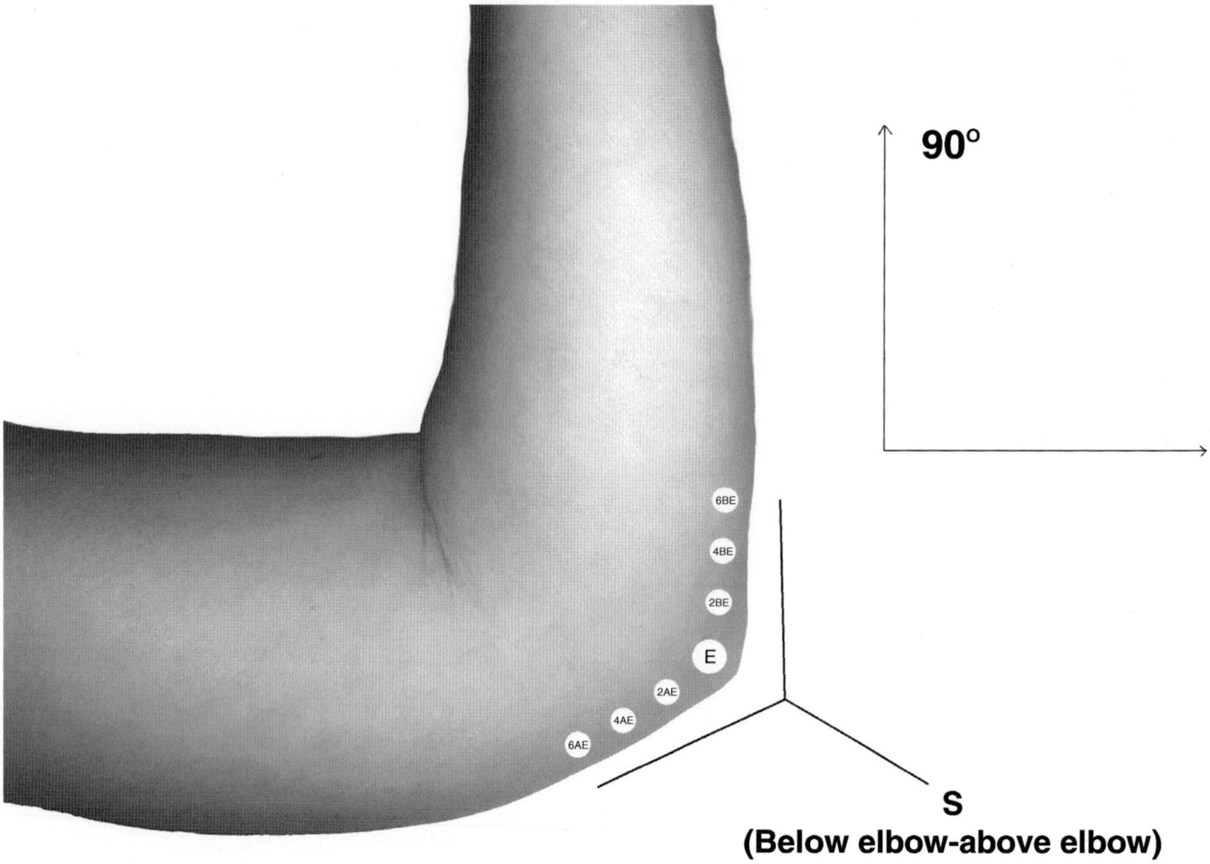

Typical waveform (below elbow to above elbow – ADM muscle):

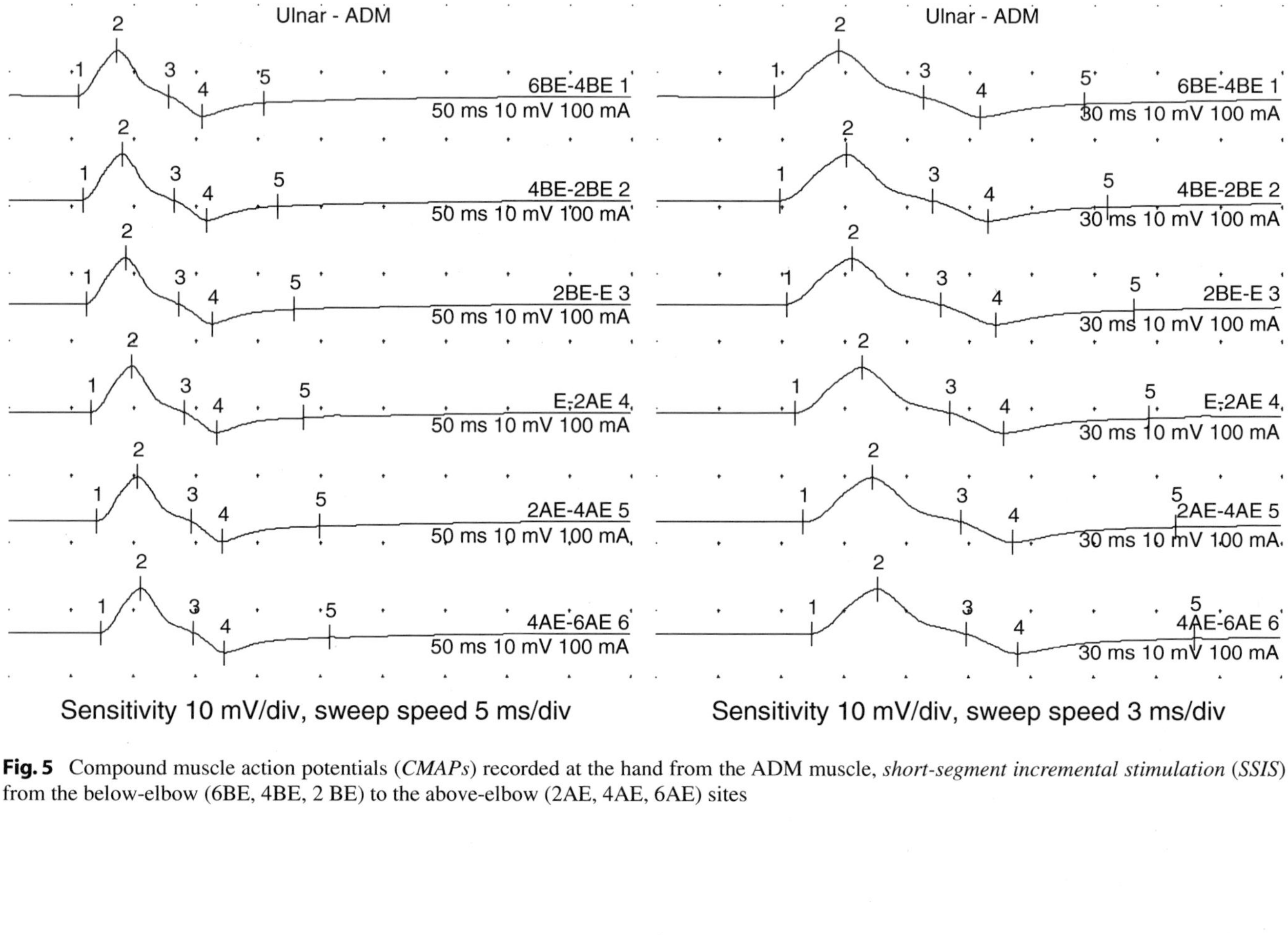

Fig. 5 Compound muscle action potentials (*CMAPs*) recorded at the hand from the ADM muscle, *short-segment incremental stimulation* (*SSIS*) from the below-elbow (6BE, 4BE, 2 BE) to the above-elbow (2AE, 4AE, 6AE) sites

Table 2 Pathological and normal values – symptomatic arms [6]

	Mean ± SD	Minimum	Maximum	Normal limit
6BE-4BE, latency change (ms)	0.29 ± 0.12	0.10	0.50	≤0.40
4BE-2BE, latency change (ms)	0.33 ± 0.09	0.20	0.50	≤0.40
2BE-E, latency change (ms)	0.44 ± 0.14	0.20	0.80	≤0.40
E-2AE, latency change (ms)	0.52 ± 0.25	0.20	1.30	≤0.50
2AE-4AE, latency change (ms)	0.45 ± 0.22	0.20	1.0	≤0.50
4AE-6AE, latency change (ms)	0.28 ± 0.06	0.20	0.40	≤0.40
6BE-4BE, amplitude change (mV)	0.30 ± 0.53	−0.50	1.40	
4BE-2BE, amplitude change (mV)	0.09 ± 0.49	−0.50	1.40	
2BE-E, amplitude change (mV)	−0.28 ± 0.40	−1.20	0.60	
E-2AE, amplitude change (mV)	−0.29 ± 1.24	−3.90	0.90	
2AE-4AE, amplitude change (mV)	−0.03 ± 0.44	−1.30	1	
4AE-6AE, amplitude change (mV)	−0.18 ± 0.47	−1.40	0.50	
6BE, absolute amplitude (mV)	9.09 ± 2.26	4.20	14.30	
4BE, absolute amplitude (mV)	9.46 ± 2.28	4.20	14.70	
2BE, absolute amplitude (mV)	9.55 ± 2.27	4.10	14.60	
E, absolute amplitude (mV)	9.27 ± 2.37	3.40	14.40	
2AE, absolute amplitude (mV)	8.98 ± 2.35	3.50	14.0	
4AE, absolute amplitude (mV)	8.95 ± 2.28	3.30	13.60	
6AE, absolute amplitude (mV)	8.77 ± 2.36	3.80	13.80	

Table 3 Pathological and normal values – asymptomatic arms [6]

	Mean ± SD	Minimum	Maximum	Normal limit
6BE-4BE, latency change (ms)	0.24 ± 0.08	0.10	0.40	≤0.40
4BE-2BE, latency change (ms)	0.27 ± 0.08	0.10	0.40	≤0.40
2BE-E, latency change (ms)	0.30 ± 0.07	0.20	0.40	≤0.40
E-2AE, latency change (ms)	0.36 ± 0.06	0.30	0.50	≤0.50
2AE-4AE, latency change (ms)	0.34 ± 0.06	0.20	0.40	≤0.50
4AE-6AE, latency change (ms)	0.29 ± 0.07	0.20	0.40	≤0.40
6BE-4BE, amplitude change (mV)	0.23 ± 0.76	−1.20	2.30	
4BE-2BE, amplitude change (mV)	0.16 ± 0.54	−0.50	1.90	
2BE-E, amplitude change (mV)	−0.07 ± 0.34	−1.20	0.50	
E-2AE, amplitude change (mV)	−0.01 ± 0.29	−0.60	0.80	
2AE-4AE, amplitude change (mV)	−0.02 ± 0.20	−0.40	0.70	
4AE-6AE, amplitude change (mV)	−0.10 ± 0.60	−1.70	1.40	
6BE, absolute amplitude (mV)	9.80 ± 2.02	6.30	15.30	
4BE, absolute amplitude (mV)	10.01 ± 2.22	6.20	15.80	
2BE, absolute amplitude (mV)	10.18 ± 2.28	6.30	15.80	
E, absolute amplitude (mV)	10.11 ± 2.18	6.30	16.0	
2AE, absolute amplitude (mV)	10.10 ± 2.17	6.30	15.90	
4AE, absolute amplitude (mV)	10.08 ± 2.16	6.30	15.80	
6AE, absolute amplitude (mV)	9.97 ± 2.32	6.50	15.70	

Comment

Azrieli et al. [6] found, for SSIS technique, high values of sensitivity (81 %) and specificity (100 %); among 21 asymptomatic arms, 17 were found abnormal, but despite its high sensitivity, SSIS technique was normal in every asymptomatic arm. Latency changes over short segments were the most useful measure in the mild to moderately symptomatic arms, as no conduction block or substantial amplitude changes were noted over the same segments. No substantial mean amplitude changes were noted in asymptomatic limbs across SSIS stimulation sites around the elbow, except at the most distal site, where slight reductions in mean amplitude were noted, most likely due to greater nerve depth at that location. All mean amplitude changes across short segments in asymptomatic limbs were insignificant (averaging ≤ 0.1 mV), while mild changes were observed in symptomatic limbs (averaging from 0.2 to 0.3 mV in each segment from 2 cm below to 6 cm above the elbow). The authors [6], however, in symptomatic limbs demonstrated slightly lower-absolute mean amplitudes at each segment below the elbow when compared to corresponding segments in asymptomatic limbs (about 0.5 mV less) and moderately lower-absolute mean amplitudes for each segment above the elbow (about 1.1 mV less in control than affected arms). During SSIS, they observed that changes in latency were more sensitive than changes in CMAP amplitude or configuration in detecting focal mononeuropathy at the elbow. For the authors, this is presumably due to the relatively low incidence of measurable conduction block at this location, as demonstrated during intraoperative evaluation [7]. SSISs were the sole abnormality in 10 of 21 symptomatic arms, and the segment 2 cm proximal (2AE) and the segment 2 cm distal (2BE) to the midpoint of the ulnar groove were most frequently abnormal by SSIS (10 and 11 limbs, respectively); for this reason, Azrieli et al. [6] found SSIS a useful tool for the sensitive detection of ulnar mononeuropathy at the elbow, particularly when routine studies are unrevealing and clinical suspicion remains high.

Visser et al. [8], following the Kanakamedala's method [1], performed all the short-segment incremental studies (SSISs) with the elbow flexed at 90°; skin temperature was measured at the wrist, and the arm was warmed to obtain a temperature above 32 °C, if necessary. The midpoint of the ulnar groove (ME) was identified by a line drawn from the medial epicondyle to the olecranon. From this level, markers were placed along the course of the ulnar nerve at 2 cm intervals from 4 cm proximal to 4 cm distal to the level of the medial epicondyle (ME). Stimulation was performed with the cathode at each marker, and focal supramaximal stimulation was applied without spread by adequate pressure to the stimulating electrodes. In order to obtain upper limits of normal for latency change and CMAP amplitude change over each of all 2 cm segments across the elbow, Visser et al. [8] studied 53 randomly selected asymptomatic subjects recording from the ADM muscle; *first dorsal interosseous* (FDI) muscle recordings were performed in 24 controls, following belly-tendon montage (Figs. 6 and 7). There were 25 men and 28 women (age range 17–72 years, age mean 42 years), and the reference values obtained by recording the ADM muscle did not differ from those obtained by recording the FDI muscle.

Normative data were determined using mean + 1.64 SD (95th percentile) and mean – 1.64 SD (5th percentile); latency change (ms) and amplitude change (%) for both ADM and FDI recordings were also reported (Table 4).

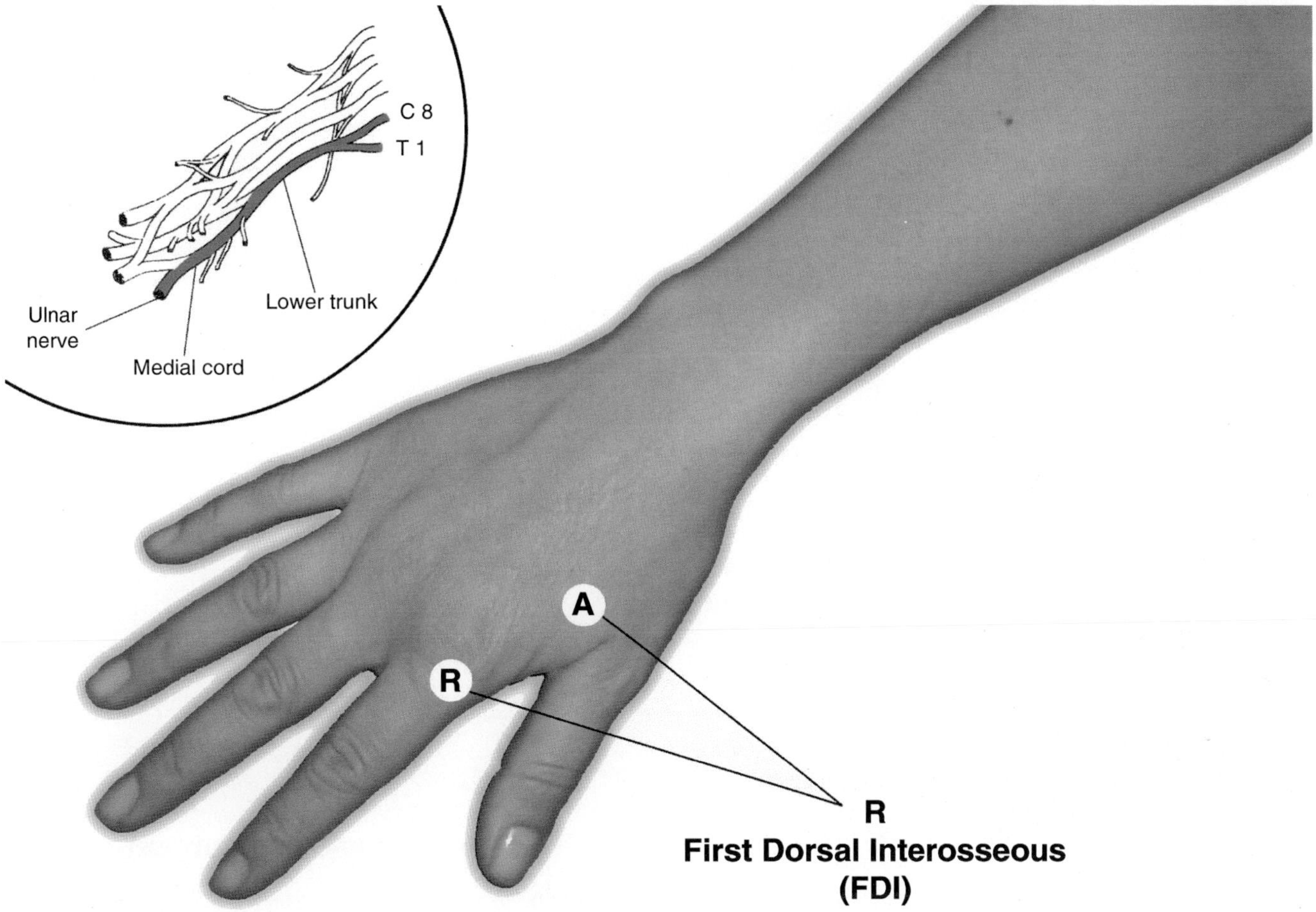

Fig. 6 *First dorsal interosseous* (*FDI*) recording (conventional belly-tendon montage)

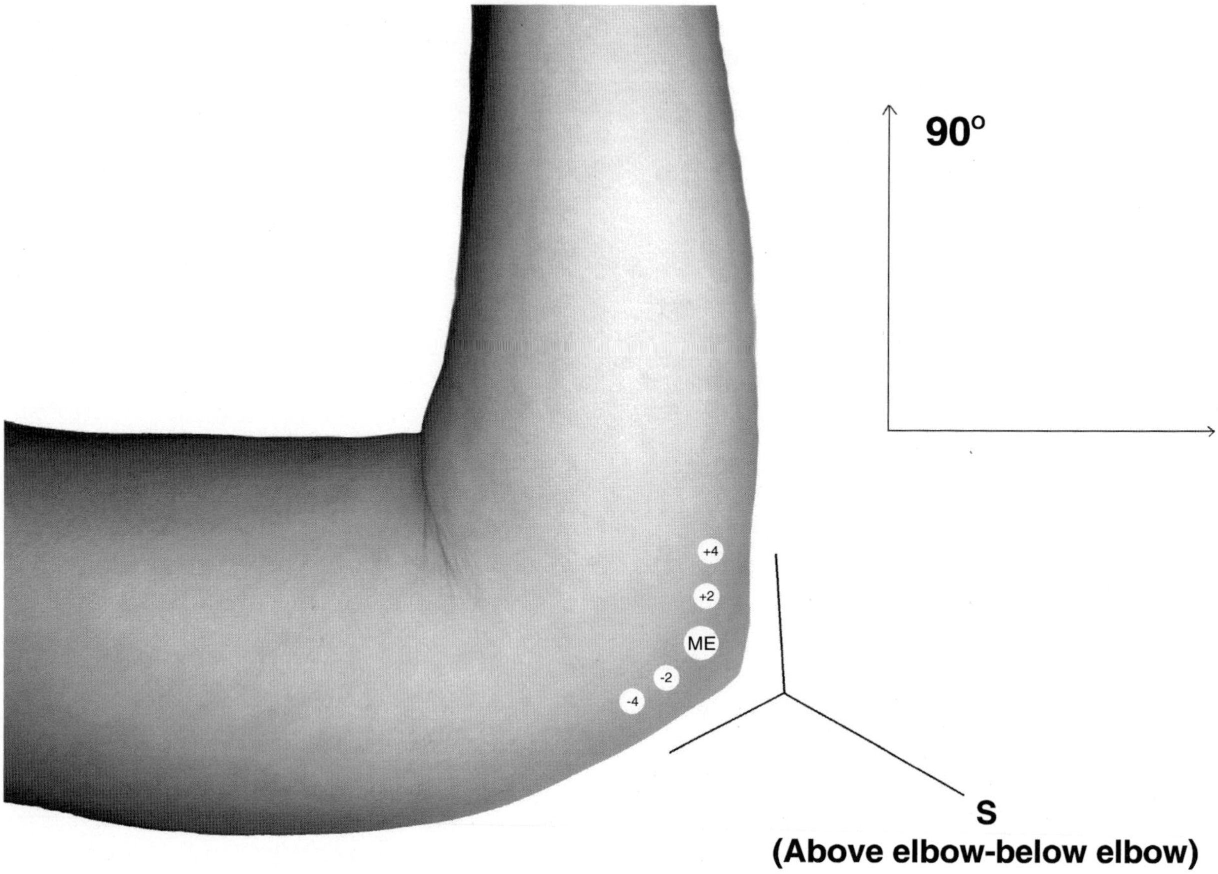

Typical waveform (above elbow to below elbow – FDI muscle):

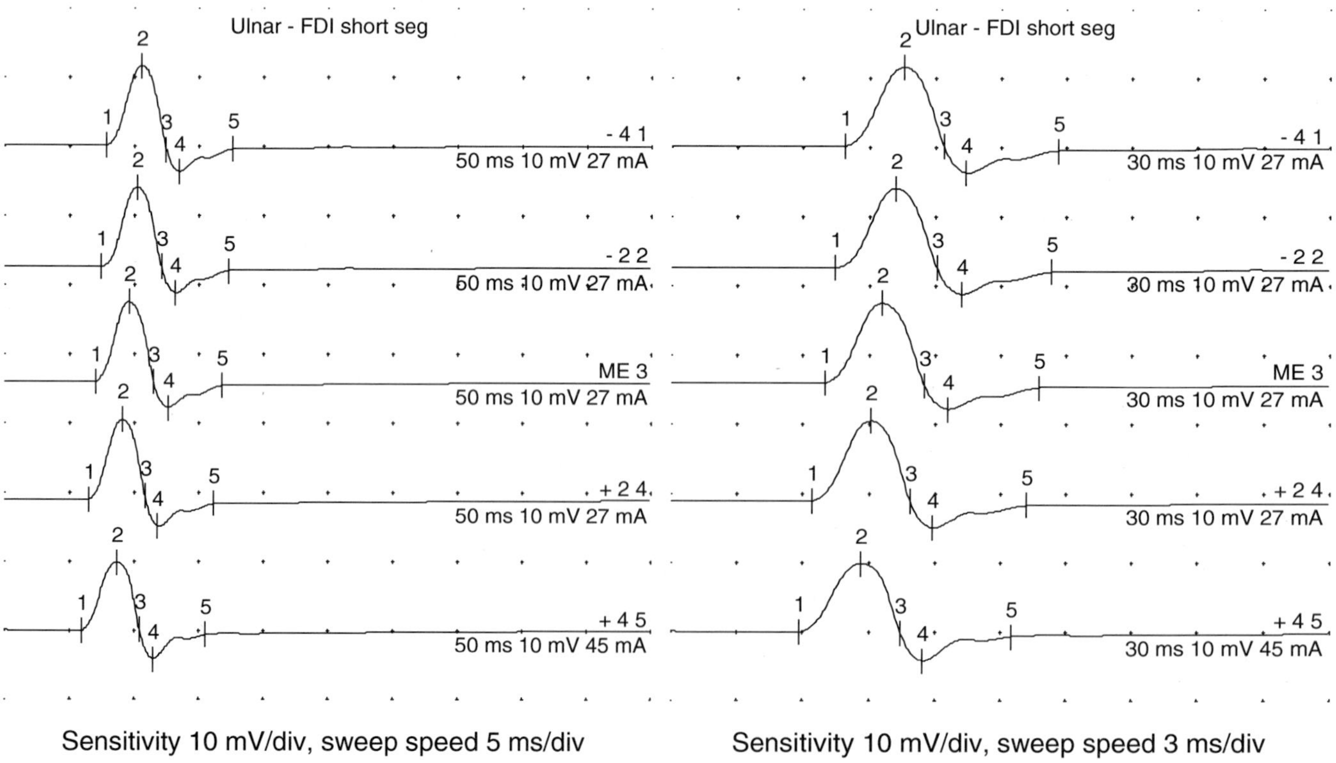

Fig. 7 Compound muscle action potentials (*CMAPs*) recorded at the hand from the FDI muscle, *short-segment incremental stimulation* (*SSIS*) from the above-elbow (–4, –2) to the below-elbow (+2, +4) sites

Table 4 Normal values [8]

	Mean ± SD	Limit of normal
4 cm to 2 cm above elbow, ADM latency change (ms)	0.36 ± 0.17	0.63
4 cm to 2 cm above elbow, ADM amplitude change (%)	4 ± 5.4	13
4 cm to 2 cm above elbow, FDI latency change (ms)	0.31 ± 0.16	0.57
4 cm to 2 cm above elbow, FDI amplitude change (%)	3.5 ± 4.6	12
2 cm above elbow to midpoint, ADM latency change (ms)	0.46 ± 0.23	0.84
2 cm above elbow to midpoint, ADM Amplitude change (%)	3.4 ± 5.3	12
2 cm above elbow to midpoint, FDI latency change (ms)	0.53 ± 0.21	0.87
2 cm above elbow to midpoint, FDI amplitude change (%)	2.5 ± 2.6	8
Midpoint to 2 cm below elbow, ADM latency change (ms)	0.45 ± 0.18	0.74
Midpoint to 2 cm below elbow, ADM amplitude change (%)	3.0 ± 4.3	10
Midpoint to 2 cm below elbow, FDI latency change (ms)	0.47 ± 0.13	0.68
Midpoint to 2 cm below elbow, FDI Amplitude change (%)	3.0 ± 3.6	9
2 cm to 4 cm below elbow, ADM latency change (ms)	0.22 ± 0.13	0.43
2 cm to 4 cm below elbow, ADM amplitude change (%)	5.4 ± 9.2	20
2 cm to 4 cm below elbow, FDI latency change (ms)	0.22 ± 0.2	0.55
2 cm to 4 cm below elbow, FDI amplitude change (%)	4.9 ± 4.5	12

Comment

In 35 arms, focal conduction block with an increased latency change was found by Visser et al. [8] in 34 arms; in two cases, it was not possible to make an accurate interpretation of SSIS because of the presence of increased latency change at two different segments (2–4 cm above midpoint of the elbow and also 4–2 cm below the elbow). Motor nerve conduction velocity (MNCV) slowing during routine nerve conduction studies (NCS) was the most common localizing finding, occurring in 66 arms (90 %), and SSIS had additional localizing value in 28 arms of the 37 patients (76 %) who had MNCV slowing alone or nonlocalizing electrodiagnostic findings. Visser et al. [8] found that SSIS can be performed accurately in almost all patients with ulnar neuropathy at the elbow (UNE), and in the patients who had conduction block on routine NCS, SSIS helped to localize the lesion exactly. In 32 of the 60 arms with abnormal SSIS (53 %), the lesion was localized proximal to the medial epicondyle; in 16 (27 %) the lesion was found at the level of the medial epicondyle, and in 12 (20 %), it was localized distal to the medial epicondyle. Eleven arms had abnormal SSIS over two segments of 2 cm and two arms over three segments. For the authors [8], when analyzing the results of SSIS, it is important to determine the upper limits of normal conduction times and amplitude changes over each 2 cm segment separately, because latency change and amplitude reduction differ over the various 2 cm segments across the elbow. They observed the highest values for latency changes in the segment between the midpoints of the medial epicondyle to 2 cm above or below the elbow. Therefore, the upper limits of normal should be determined according to the specific 2 cm segments studied across the elbow, as previously performed by Azrieli et al. [6]. They conclude that SSIS have additional diagnostic value in UNE, especially when the measurement of latency change is performed. SSISs are useful to locate the lesion, which is often located above the medial epicondyle, and they suggested to perform SSIS in every patient with UNE and not just those with conduction block on routine NCS.

To determine reference values for SSIS, Omejec and Podnar [9] recruited prospectively a control group of adult subjects; following previously proposed standards for normative data [10], normative values were derived from disease-free individuals (39 subjects, 54 % men and 46 % women, age mean 47 ± 14 years, age range 23–81 years, height mean 173 ± 8 cm, height range 158–188 cm, weight mean 77 ± 15 kg, weight range 54–105 kg, BMI mean 25 ± 4, BMI range 19–33), and reference values were derived from symptom-free individuals (49 subjects, 59 % men and 41 % women, age mean 47 ± 14 years, age range 23–81 years, height mean 174 ± 8 cm, height range 158–189 cm, weight mean 77 ± 15 kg, weight range 49–105 kg, BMI mean 25 ± 4, BMI range 19–33). To evaluate reference and normative values, a group of patients with clinically definite UNE was also recruited (67 patients, 51 % men and 49 % women, age mean 47 ± 15 years, age range 15–89 years, height mean 172 ± 8 cm, height range 155–192 cm, weight mean 79 ± 15 kg, weight range 45–118 kg, BMI mean 26±4, BMI range 17–37), and the authors used a single-channel NICOLET Synergy EMG system. Conventional NCS and SSIS were performed with the subject supine and the elbow flexed 90°. Skin temperature was checked at the wrist and at the elbow, and it was maintained above 32 °C. They identified the midpoint of the ulnar groove by drawing a line between the medial epicondyle and the olecranon (ME), and then several markers were placed 2 cm distal (D2), 4 cm distal (D4), 2 cm proximal (P2), 4 cm proximal (P4), and 6 cm proximal (P6) to the ME along the course of the ulnar nerve (Fig. 2), following Kanakamedala's method [1]. For normally distributed variables, such as the amplitude change for ADM and FDI muscles, the authors set the upper reference limit (URL) at mean + 1.65 SD; for non-normally distributed variables, such as the latency change for ADM and FDI, they set the URL at the 95th percentile. Reference limits for SSIS are shown in Tables 5, 6, 7, and 8. Latency change was measured in millisecond (ms), and nerve conduction velocity (NCV) was measured in meter per second (m/s).

Table 5 Normal values [9] – reference limits

	Mean
D4-D2 segment – ADM muscle, latency change (ms)	0.45
D4-D2 segment – ADM muscle, NCV (m/s)	44
D4-D2 segment – ADM muscle, amplitude drop (%)	6
D2-ME segment – ADM muscle, latency change (ms)	0.55
D2-ME segment – ADM muscle, NCV (m/s)	36
D2-ME segment – ADM muscle, amplitude drop (%)	10
ME-P2 segment – ADM muscle, latency change (ms)	0.80
ME-P2 segment – ADM muscle, NCV (m/s)	25
ME-P2 segment – ADM muscle, amplitude drop (%)	12
P2-P4 segment – ADM muscle, latency change (ms)	0.65
P2-P4 segment – ADM muscle, NCV (m/s)	31
P2-P4 segment – ADM muscle, amplitude drop (%)	13
P4-P6 segment – ADM muscle, latency change (ms)	0.45
P4-P6 segment – ADM muscle, NCV (m/s)	44
P4-P6 segment – ADM muscle, amplitude drop (%)	14
D4-P6 segment (10 cm) – ADM muscle, latency change (ms)	2.6
D4-P6 segment (10 cm) – ADM muscle, NCV (m/s)	38
D4-P6 segment (10 cm) – ADM muscle, amplitude drop (%)	14

Table 6 Normal values [9] – reference limits

	Mean
D4-D2 segment – FDI muscle, latency change (ms)	0.55
D4-D2 segment – FDI muscle, NCV (m/s)	36
D4-D2 segment – FDI muscle, amplitude drop (%)	12
D2-ME segment – FDI muscle, latency change (ms)	0.65
D2-ME segment – FDI muscle, NCV (m/s)	31
D2-ME segment – FDI muscle, amplitude drop (%)	15
ME-P2 segment – FDI muscle, latency change (ms)	0.80
ME-P2 segment – FDI muscle, NCV (m/s)	25
ME-P2 segment – FDI muscle, amplitude drop (%)	21
P2-P4 segment – FDI muscle, latency change (ms)	0.70
P2-P4 segment – FDI muscle, NCV (m/s)	29
P2-P4 segment – FDI muscle, amplitude drop (%)	22
P4-P6 segment – FDI muscle, latency change (ms)	0.50
P4-P6 segment – FDI muscle, NCV (m/s)	40
P4-P6 segment – FDI muscle, amplitude drop (%)	22
D4-P6 segment (10 cm) – FDI muscle, latency change (ms)	2.7
D4-P6 segment (10 cm) – FDI muscle, NCV (m/s)	36
D4-P6 segment (10 cm) – FDI muscle, amplitude drop (%)	22

Table 7 Normal values [9] – normative limits

	Mean
D4-D2 segment – ADM muscle, latency change (ms)	0.50
D4-D2 segment – ADM muscle, NCV (m/s)	40
D4-D2 segment – ADM muscle, amplitude drop (%)	6
D2-ME segment – ADM muscle, latency change (ms)	0.55
D2-ME segment – ADM muscle, NCV (m/s)	36
D2-ME segment – ADM muscle, amplitude drop (%)	8
ME-P2 segment – ADM muscle, latency change (ms)	0.65
ME-P2 segment – ADM muscle, NCV (m/s)	31
ME-P2 segment – ADM muscle, amplitude drop (%)	10
P2-P4 segment – ADM muscle, latency change (ms)	0.60
P2-P4 segment – ADM muscle, NCV (m/s)	33
P2-P4 segment – ADM muscle, amplitude drop (%)	11
P4-P6 segment – ADM muscle, latency change (ms)	0.50
P4-P6 segment – ADM muscle, NCV (m/s)	40
P4-P6 segment – ADM muscle, amplitude drop (%)	12
D4-P6 segment (10 cm) – ADM muscle, latency change (ms)	2.4
D4-P6 segment (10 cm) – ADM muscle, NCV (m/s)	42
D4-P6 segment (10 cm) – ADM muscle, amplitude drop (%)	12

Table 8 Normal values [9] – normative limits

	Mean
D4-D2 segment – FDI muscle, latency change (ms)	0.60
D4-D2 segment – FDI muscle, NCV (m/s)	33
D4-D2 segment – FDI muscle, amplitude drop (%)	9
D2-ME segment – FDI muscle, latency change (ms)	0.65
D2-ME segment – FDI muscle, NCV (m/s)	31
D2-ME segment – FDI muscle, amplitude drop (%)	11
ME-P2 segment – FDI muscle, latency change (ms)	0.65
ME-P2 segment – FDI muscle, NCV (m/s)	31
ME-P2 segment – FDI muscle, amplitude drop (%)	11
P2-P4 segment – FDI muscle, latency change (ms)	0.65
P2-P4 segment – FDI muscle, NCV (m/s)	31
P2-P4 segment – FDI muscle, amplitude drop (%)	12
P4-P6 segment – FDI muscle, latency change (ms)	0.50
P4-P6 segment – FDI muscle, NCV (m/s)	40
P4-P6 segment – FDI muscle, amplitude drop (%)	12
D4-P6 segment (10 cm) – FDI muscle, latency change (ms)	2.5
D4-P6 segment (10 cm) – FDI muscle, NCV (m/s)	40
D4-P6 segment (10 cm) – FDI muscle, amplitude drop (%)	12

Comment

Omejec and Podnar [9] calculated specificities of SSIS in control subjects and sensitivities in UNE patients, using both the reference and normative values for individual parameters and for arms. They found the largest URL for latency change (0.80 ms for ADM and FDI muscles) in the ME-P2 segment, and small amplitude changes between stimulation points across the elbow were observed. By using normative values instead of reference values, sensitivities for SSIS latency change and amplitude change increased from 88 to 95 %. They confirmed the ulnar lesion in the affected short segment in 80 % of UNE patients using reference limits and in 90 % using normative limits, showing a significant increase in the sensitivity of SSIS in the diagnosis and localization of UNE using normative limits. In the majority of these studies, the pathologic segment was ME-P2. Pathological waveform and values in patients with UNE are reported (Figs. 8, 9, 10, and 11).

Pathological waveform (below elbow to above elbow – ADM muscle):

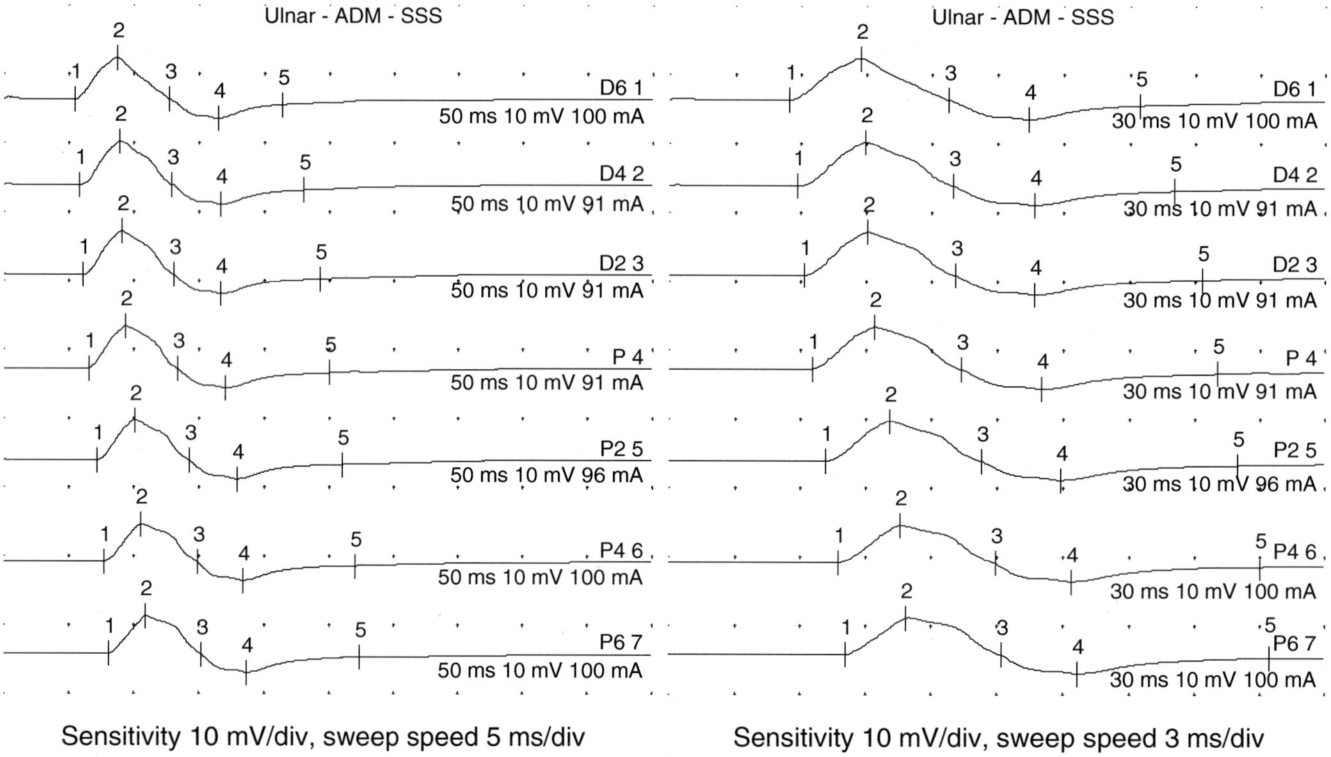

Onset latency (below elbow – ADM, D6): 5.45 ms; **Onset latency** (below elbow – ADM, D4): 5.80 ms; **Onset latency** (below elbow – ADM, D2): 6.15 ms; **Onset latency** (elbow – ADM, P): 6.50 ms; **Onset latency** (above elbow – ADM, P2): 7.15 ms; **Onset latency** (above elbow – ADM, P4): 7.70 ms; **Onset latency** (above elbow – ADM, P6): 8.05 ms; **Peak latency** (below elbow – ADM, D6): 8.90 ms; **Peak latency** (below elbow – ADM, D4): 9.0 ms; **Peak latency** (below elbow – ADM, D2): 9.15 ms; **Peak latency** (elbow – ADM, P): 9.30 ms; **Peak latency** (above elbow – ADM, P2): 10.00 ms; **Peak latency** (above elbow – ADM, P4): 10.50 ms; **Peak latency** (above elbow – ADM, P6): 10.65 ms; **Onset to peak amplitude** (below elbow – ADM, D6): 6.0 mV; **Onset to peak amplitude** (below elbow – ADM, D4): 6.4 mV; **Onset to peak amplitude** (below elbow – ADM, D2): 6.3 mV; **Onset to peak amplitude** (elbow – ADM, P): 6.0 mV; **Onset to peak amplitude** (above elbow – ADM, P2): 5.8 mV; **Onset to peak amplitude** (above elbow – ADM, P4): 5.5 mV; **Onset to peak amplitude** (above elbow – ADM, P6): 5.4 mV; **Peak to peak amplitude** (below elbow – ADM, D6): 8.9 mV; **Peak to peak amplitude** (below elbow – ADM, D4): 9.3 mV; **Peak to peak amplitude** (below elbow – ADM, D2): 9.2 mV; **Peak to peak amplitude** (elbow – ADM, P): 9.1 mV; **Peak to peak amplitude** (above elbow – ADM, P2): 8.7 mV; **Peak to peak amplitude** (above elbow – ADM, P4): 8.4 mV; **Peak to peak amplitude** (above elbow – ADM, P6): 8.3 mV

Fig. 8 Compound muscle action potentials (*CMAPs*) recorded at the hand from the ADM muscle in UNE, prolonged delay in P-P2 (0.65 ms) and P2-P4 (0.55 ms), *short-segment incremental stimulation* (SSIS) by Kanakamedala et al. [1], from the below-elbow (D4, D2) to the above-elbow (P2, P4, P6) sites

Pathological waveform (below elbow to above elbow – ADM muscle):

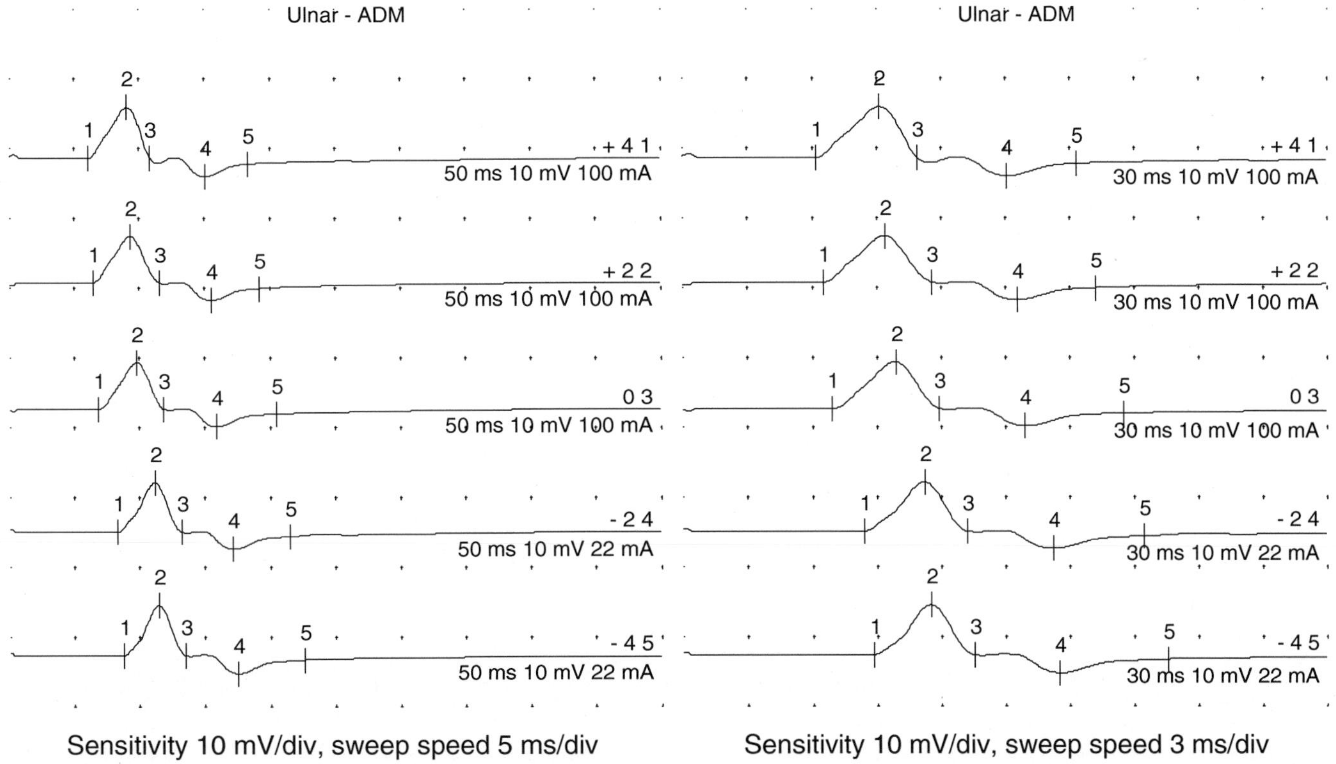

Onset latency (below elbow – ADM, +4): 6.15 ms; **Onset latency** (below elbow – ADM, +2): 6.50 ms; **Onset latency** (elbow – ADM, 0): 6.85 ms; **Onset latency** (above elbow – ADM, -2): 8.35 ms; **Onset latency** (above elbow – ADM, -4): 8.80 ms; **Peak latency** (below elbow – ADM, +4): 9.05 ms; **Peak latency** (below elbow – ADM, +2): 9.35 ms; **Peak latency** (elbow – ADM, 0): 9.75 ms; **Peak latency** (above elbow – ADM, -2): 11.10 ms; **Peak latency** (above elbow – ADM, -4): 11.85 ms; **Onset to peak amplitude** (below elbow – ADM, +4): 7.3 mV; **Onset to peak amplitude** (below elbow – ADM, +2): 6.7 mV; **Onset to peak amplitude** (elbow – ADM, 0): 6.8 mV; **Onset to peak amplitude** (above elbow – ADM, -2): 7.0 mV; **Onset to peak amplitude** (above elbow – ADM, -4): 7.2 mV; **Peak to peak amplitude** (below elbow – ADM, +4): 10.1 mV; **Peak to peak amplitude** (below elbow – ADM, +2): 9.4 mV; **Peak to peak amplitude** (elbow – ADM, 0): 9.4 mV; **Peak to peak amplitude** (above elbow – ADM, -2): 9.7 mV; **Peak to peak amplitude** (above elbow – ADM, -4): 9.9 mV

Fig. 9 Compound muscle action potentials (*CMAPs*) recorded at the hand from the ADM muscle in UNE, prolonged delay between "0" and "-2" stimulation sites (1.50 ms), *short-segment incremental stimulation* (SSIS) by Brown and Yates [5] from the below-elbow (+4, +2) to the above-elbow (−2, −4) sites

Pathological waveform (below elbow to above elbow – ADM muscle):

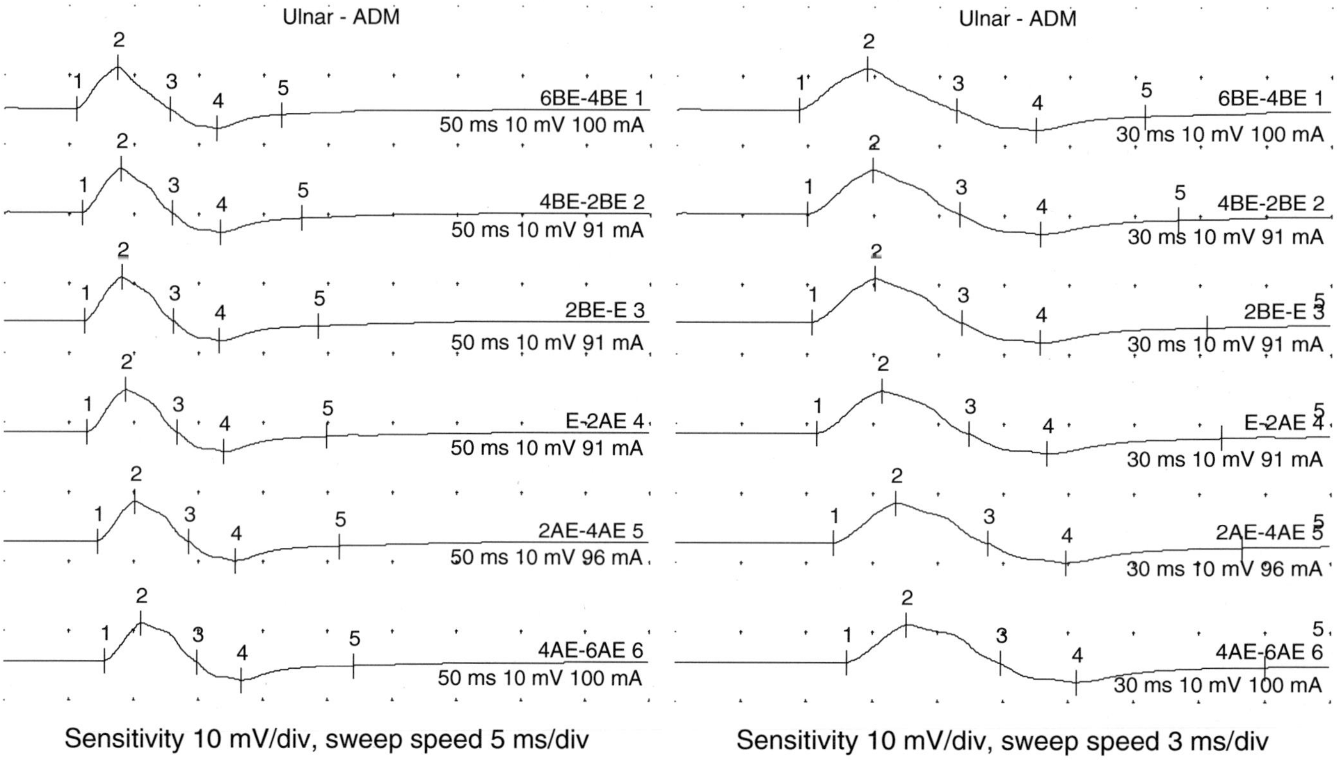

Onset latency (below elbow – ADM, 6BE): 5.60 ms; **Onset latency** (below elbow – ADM, 4BE): 6.00 ms; **Onset latency** (below elbow – ADM, 2BE): 6.30 ms; **Onset latency** (elbow – ADM, E): 6.80 ms; **Onset latency** (above elbow – ADM, 2AE): 7.40 ms; **Onset latency** (above elbow – ADM, 4AE): 8.00 ms; **Peak latency** (below elbow – ADM, 6BE): 8.80 ms; **Peak latency** (below elbow – ADM, 4BE): 9.00 ms; **Peak latency** (below elbow – ADM, 2BE): 9.20 ms; **Peak latency** (elbow – ADM, E): 9.40 ms; **Peak latency** (above elbow – ADM, 2AE): 9.70 ms; **Peak latency** (above elbow – ADM, 4AE): 10.10 ms; **Onset to peak amplitude** (below elbow – ADM, 6BE): 5.9 mV; **Onset to peak amplitude** (below elbow – ADM, 4BE): 6.3 mV; **Onset to peak amplitude** (below elbow – ADM, 2BE): 6.2 mV; **Onset to peak amplitude** (elbow – ADM, E): 6.1 mV; **Onset to peak amplitude** (above elbow – ADM, 2AE): 5.8 mV; **Onset to peak amplitude** (above elbow – ADM, 4AE): 5.4 mV; **Peak to peak amplitude** (below elbow – ADM, 6BE): 8.9 mV; **Peak to peak amplitude** (below elbow – ADM, 4BE): 9.3 mV; **Peak to peak amplitude** (below elbow – ADM, 2BE): 9.2 mV; **Peak to peak amplitude** (elbow – ADM, E): 9.1 mV; **Peak to peak amplitude** (above elbow – ADM, 2AE): 8.7 mV; **Peak to peak amplitude** (above elbow – ADM, 4AE): 8.4 mV

Fig. 10 Compound muscle action potentials (*CMAPs*) recorded at the hand from the ADM muscle in UNE, prolonged delay in the above-elbow segments (E-2AE 0.5 ms, 2AE-4AE 0.8 ms, 4AE-6AE 0.6 ms), *short-segment incremental stimulation* (SSIS) by Azrieli et al. [6] from the below-elbow (6BE, 4BE, 2 BE) to the above-elbow (2AE, 4AE, 6AE) sites

Pathological waveform (above elbow to below elbow – FDI muscle):

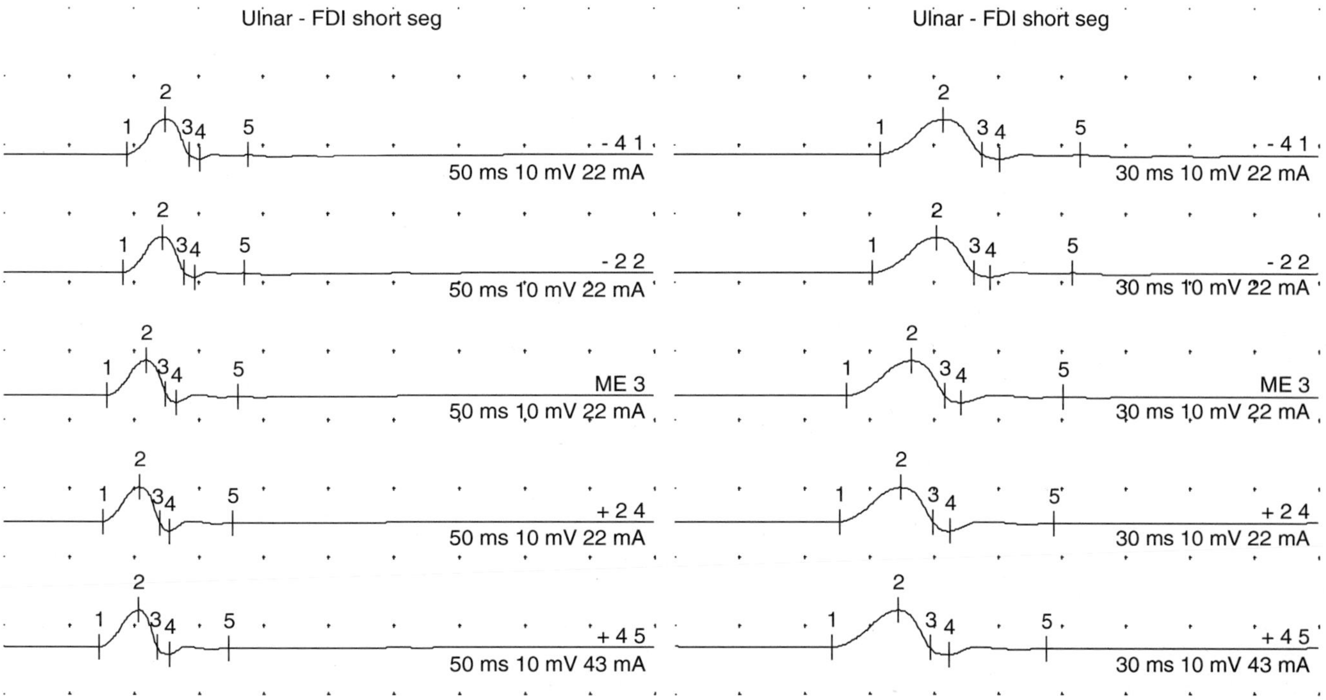

Sensitivity 10 mV/div, sweep speed 5 ms/div

Sensitivity 10 mV/div, sweep speed 3 ms/div

Onset latency (above elbow – ADM, -4): 9.50 ms; **Onset latency** (above elbow – ADM, -2): 9.15 ms; **Onset latency** (elbow – ADM, 0): 7.95 ms; **Onset latency** (below elbow – ADM,+2): 7.60 ms; **Onset latency** (below elbow – ADM, +4): 7.25 ms; **Peak latency** (above elbow – ADM, -4): 12.35 ms; **Peak latency** (above elbow – ADM, -2): 12.05 ms; **Peak latency** (elbow – ADM, 0): 10.75 ms; **Peak latency** (below elbow – ADM, +2): 10.40 ms; **Peak latency** (below elbow – ADM, +4): 10.15 ms; **Onset to peak amplitude** (above elbow – ADM, -4): 5.1 mV; **Onset to peak amplitude** (above elbow – ADM, -2): 5.2 mV; **Onset to peak amplitude** (elbow – ADM, 0): 5.1 mV; **Onset to peak amplitude** (below elbow – ADM, +2): 5.0 mV; **Onset to peak amplitude** (below elbow – ADM, +4): 5.5 mV; **Peak to peak aampltude** (above elbow – ADM, -4): 8.5 mV; **Peak to peak amplitude** (above elbow – ADM, -2): 8.7 mV; **Peak to peak amplitude** (elbow – ADM, 0): 8.4 mV; **Peak to peak amplitude** (below elbow – ADM, +2): 8.4 mV; **Peak topeak amplitude** (below elbow – ADM, +4): 8.3 mV

Fig. 11 Compound muscle action potentials (*CMAPs*) recorded at the hand from the FDI muscle in UNE, prolonged delay between "-2" and "ME" stimulation sites (1.20 ms), *short-segment incremental stimulation* (SSIS) by Visser et al. [8] from the above-elbow (–4, –2) to the below-elbow (+2, +4) sites

References

1. Kanakamedala RV, Simons DG, Porter RW et al (1988) Ulnar nerve entrapment at the elbow localized by short segment stimulation. Arch Phys Med Rehabil 69:959–963
2. Checkles NS, Russakov AD, Piero DL (1971) Ulnar nerve conduction velocity – effect of elbow position on measurement. Arch Phys Med Rehabil 52:362–365
3. Miller RG (1979) The cubital tunnel syndrome: diagnosis and precise localization. Ann Neurol 6:56–59
4. Brown WF, Ferguson GG, Jones MW et al (1976) Location of conduction abnormalities in human entrapment neuropathies. Can J Neurol Sci 3:111–122
5. Brown WF, Yates SK (1982) Percutaneous localization of conduction abnormalities in human entrapment neuropathies. Can J Neurol Sci 9:391–400
6. Azrieli Y, Weimer L, Lovelace R et al (2003) The utility of segmental nerve conduction studies in ulnar mononeuropathy at the elbow. Muscle Nerve 27:46–50
7. Jabre JF, Wilbourn AJ (1979) The EMG findings in 100 consecutive ulnar neuropathies. Acta Neurol Scand 60:91
8. Visser LH, Beekman R, Franssen H (2005) Short-segment nerve conduction studies in ulnar neuropathy at the elbow. Muscle Nerve 31:331–338
9. Omejec G, Podnar S (2015) Normative values for short-segment nerve conduction studies and ultrasonography of the ulnar nerve at the elbow. Muscle Nerve 51:370–377
10. Dorfman LJ, Robinson LR (1997) AAEM minimonograph #47. Normative data in electrodiagnostic medicine. Muscle Nerve 20:4–14

M34 Below Elbow to Above Elbow – Hand

Surface Recording Technique, Study from *Abductor Digiti Minimi* (ADM) Muscle

Original Settings Sensitivity, sweep speed, low-frequency filter, high-frequency filter, duration of pulse, and the machine used were not specified.

Position Following the text, conventional ulnar motor conduction study to the *abductor digiti minimi* (ADM) muscle was done palm up with the elbow flexed 70°–90° from the horizontal.

Recording Using surface electrodes, compound muscle action potentials (CMAPs) were recorded from the ADM muscle [1]; the active electrode (A) was placed over the belly of the muscle, and the reference electrode was placed distally at the base of the digit V (Fig. 1). The ground (G) electrode position was not specified in the text; the figure shows ground placed on the forearm.

Stimulation Conventional motor conduction study to the ADM muscle [1] was performed stimulating the ulnar nerve at the wrist, just below the medial epicondyle (ME) and just above the ME. The short-segment incremental study (SSIS) was performed mapping out the ulnar nerve's course with subthreshold stimuli by moving the stimulator perpendicular to the nerve's course until the maximal CMAP amplitude for a given subthreshold intensity was obtained. The point was marked with a dot. The process was then repeated at intervals from above to below the ME, after which the points were joined to outline the exact course of the ulnar nerve (Fig. 2). The authors [1] drawn a perpendicular line from the tip of the ME to the olecranon bisecting the nerve course, and they designed this point as "zero level." One-centimeter intervals were then measured proximal (-) and distal (-) to the "zero level." Stimulation was performed at each consecutive 1 cm segment from the distal to the proximal points using supramaximal stimulation (Fig. 2). For the authors [1], care must be taken to avoid excessively supramaximal stimuli because of the danger of distal migration of the effective cathode (-) producing erroneous latency determinations over these very short distances.

Measurements Latency (ms) was taken to the nearest 0.05 ms, and normal latency change ranges from 0.05 to 0.25 ms between consecutive segments. Latency change over 1 cm segment more than 0.40 ms was considered abnormal (mean + 2SD) for either surface or intraoperative study. The authors [1] performed routine nerve conduction studies, inching (moving the stimulator stepwise around the elbow in search of discrete points of change in CMAP amplitude or configuration), and SSIS (moving the stimulator at consecutive 1 cm segment from the proximal to the distal points) in 33 patients before the surgery (32 men and 1 woman, age range 35–76 years) referred to manage the ulnar neuropathy at the elbow (UNE). The temperatures of the skin and laboratory were not reported. They also performed intraoperative electroneurography using direct epineural SSIS previously described [2]. The recording was done with a TECA TE-42 electromyograph using surface recording electrodes. However, it proved more efficient to record with subcutaneous platinum needle electrodes. Grounding was accomplished with a saline-soaked cottonoid and a spring-mounted ground placed on the dorsum of the wrist. Stimuli were delivered by a constant current stimulus isolation unit using a stimulus duration usually of 0.05 ms (rarely 0.1 ms) and a stimulus intensity of 10–30 mA (supramaximal) for the nerve study. At surgery, the ulnar nerve was explored through a curvilinear incision passing just anterior to the ME. The position of the humeroulnar aponeurotic arcade (HUA) was noted and measured. The HUA was sharply divided, after which the belly of the *flexor carpi ulnaris* (FCU) muscle overlying the ulnar nerve was opened bluntly down to the exit point distally. The deep fascia/aponeurosis underlying the FCU muscle was then opened, and the ulnar nerve freed for another 1–2 cm. If SSIS conduction abnormalities were found about the HUA, no further dissection was done, and the procedure was terminated as an HUA release. If SSIS abnormalities were found in the retro epicondylar groove (RTC), further mobilization was carried out and anterior subcutaneous transposition performed.

G. Gentili, M. Di Napoli, *The Ulnar Nerve*, DOI 10.1007/978-3-319-43292-2_58

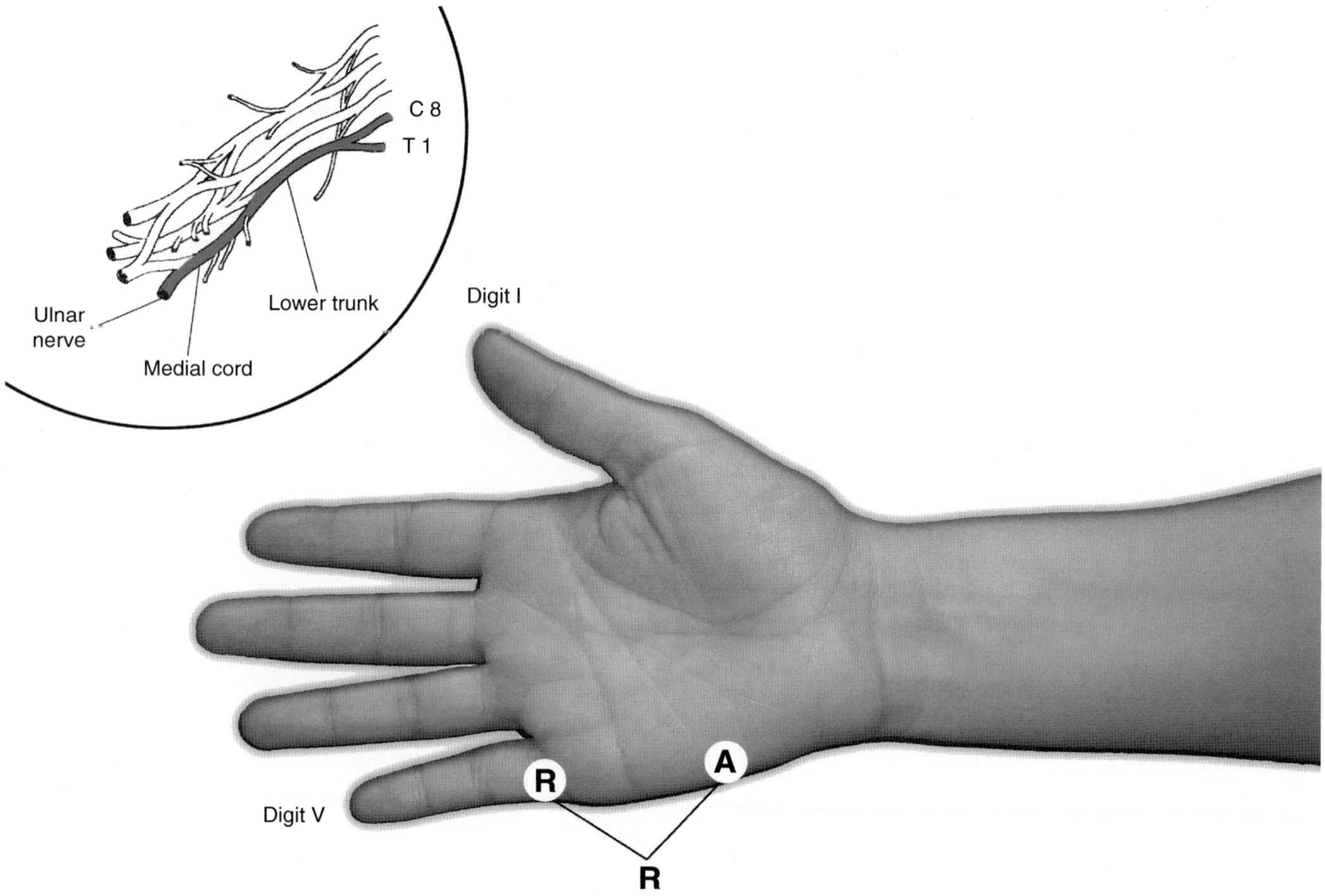

Abductor Digiti Minimi (ADM)

Typical waveform (below elbow to above elbow – ADM muscle):

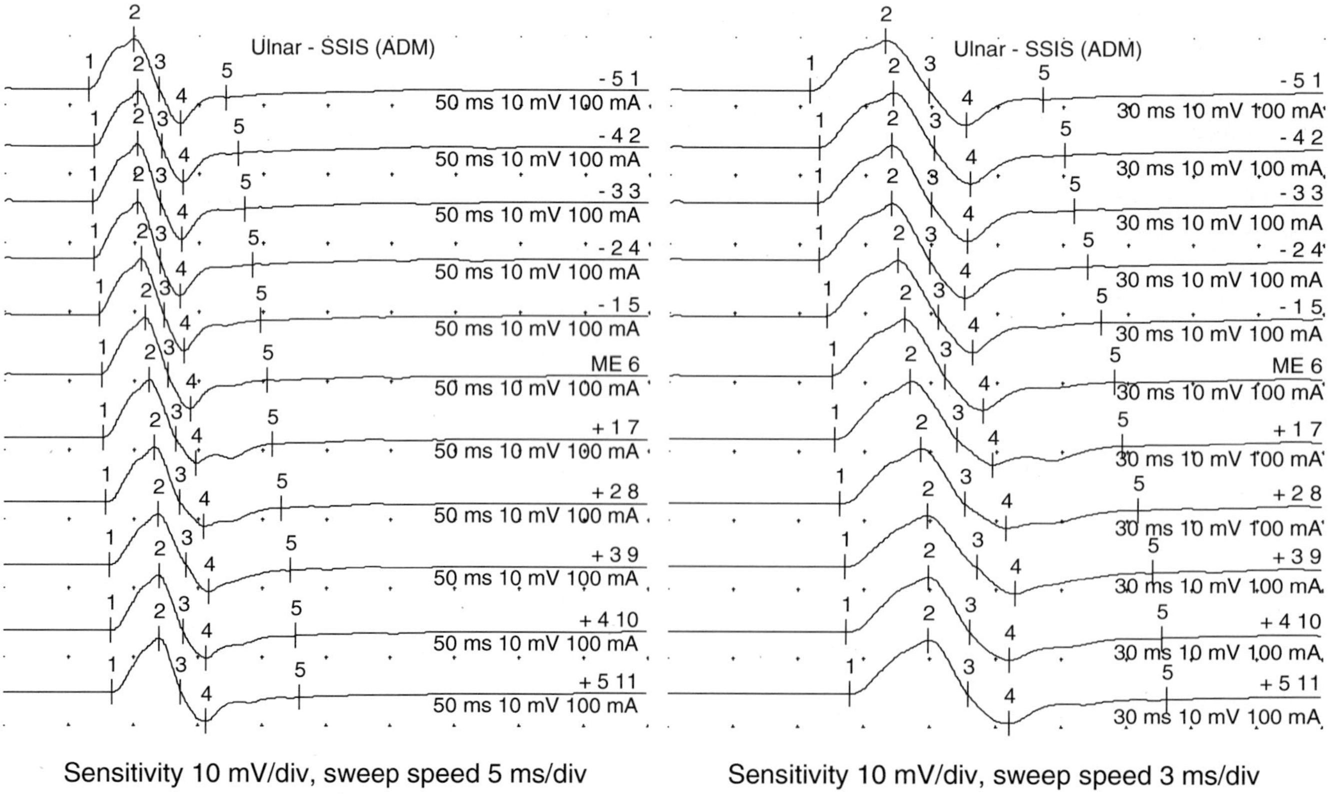

Fig. 1 Compound muscle action potentials (*CMAPs*) recorded at the hand from the ADM muscle, *short-segment incremental stimulation (SSIS)* from the below-elbow to the above-elbow sites

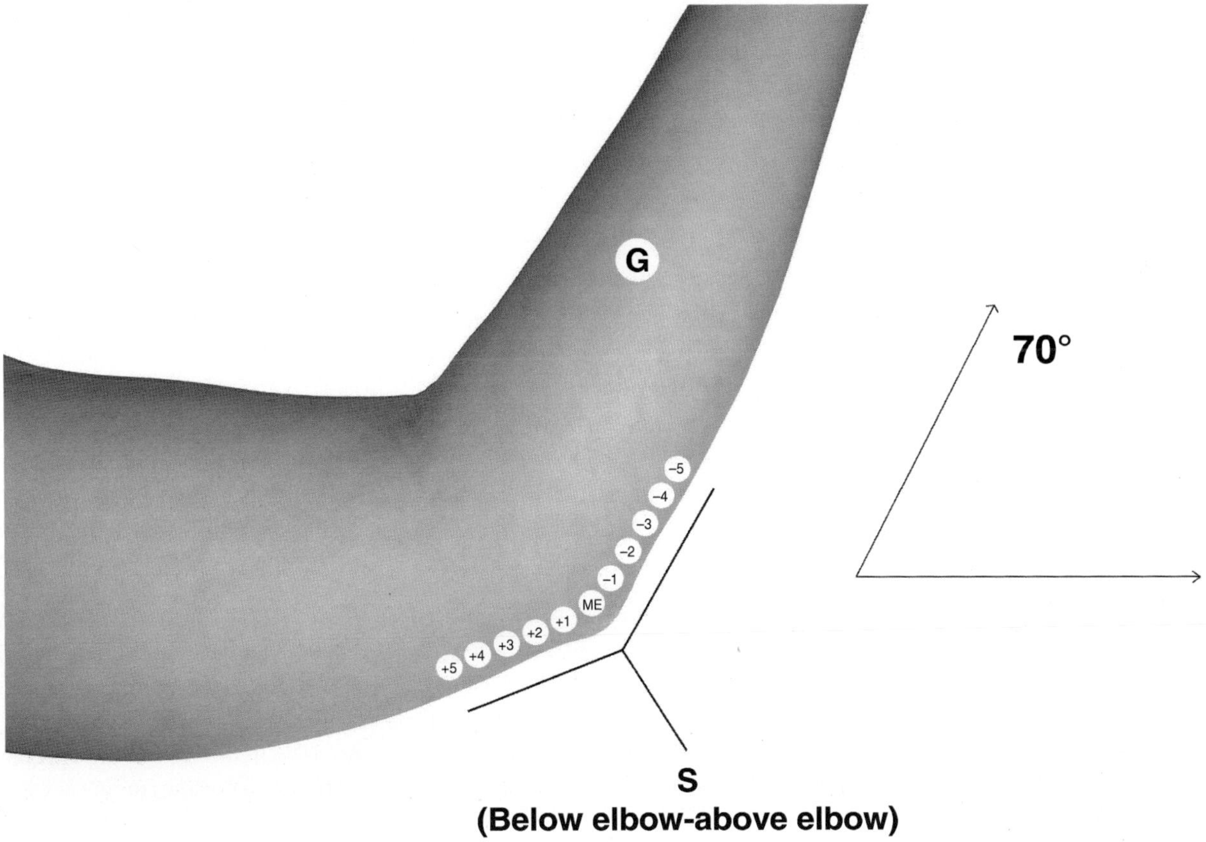

Fig. 2 *Short-segment incremental stimulation* (*SSIS*) from the below-elbow to the above-elbow sites with the elbow flexed to 70°

Comment

For Campbell et al. [1], the final diagnoses in the 33 primary ulnar nerve explorations were 14 pure RTC, three dual compression (RTC and HUA), 11 HUA compression (three cases at zero level), three compression at the FCU muscle exit, one compression due to epitrochlea anconeus muscle, and one focal injury in the upper arm. In these patients, 17 transpositions, 10 HUA releases, and six local releases were performed. In three patients with HUA compression at zero level, the HUA laid immediately behind the ME and were managed with the simple release (though the conduction abnormalities were retro-epicondylar). The three patients with compression at the FCU muscle exit were managed with local release. Local decompressions in a patient with compression of the ulnar nerve in the groove by an accessory epitrochlea anconeus muscle and in a patient with sustained focal trauma in the upper arm were performed. Conventional inching was done before surgery, but authors supplemented it with SSIS. Of the 16 ulnar nerves studied by inching, clear localizing changes occurred in only three (19 %). Of the 26 ulnar nerves studied by SSIS, authors localized abnormalities in 21 patients (81 %). Of the ten ulnar nerves studied by both techniques, inching was able to localize abnormalities in only one patient, whereas SSIS was able to localize abnormalities in all. The SSIS technique showed clear limitations because, in some instances, SSIS localized erroneously: one case seemed to be an RTC entrapment before surgery but proved at operation to be HUA entrapment, and one nerve seemed to have a focal latency change 5 cm proximal to the ME but had ordinary RTC compression at surgery. Conventional electrodiagnostic methods can localize an ulnar neuropathy to the general region of the elbow at varying levels of reliability, but separating the subgroups of UNE is much more difficult. For the authors [1], the inching technique is useful only when conduction block or differential slowing is present, but if the predominant pathophysiology is focally slowing, conventional surface inching which evaluates only amplitude and configuration changes will fail to detect the lesion. Because of this, they resorted to intraoperative electroneurography in an attempt at better localization [1, 2], and they performed epineural SSIS intraoperatively, measuring latency changes over consecutive 1 cm segments. By surface study, if abnormal latency changes center about the ME, they suspected RTC neuropathy. If the maximal change occurred 1–3 cm distal to the ME, HUA compression was most likely. Abnormal latency changes occurred greater than 3 cm distal to the ME were most likely to indicate compression at the exit from the FCU muscle [3]. This latter entrapment is important to recognize because (1) the treatment is local decompression; (2) surgeons rarely carry routine ulnar nerve exploration so far distally, and the syndrome is easily missed; (3) the usual ulnar nerve surgical procedures, HUA release and anterior transposition, do not help; and (4) it may be confused with HUA compression [3]. In conclusion, in order to separate the various UNE subgroups, the SSIS technique can accurately distinguish HUA from RTC compression 80 % of the time; correlation of electrodiagnostic findings with surgical observations is still required, and in questionable cases intraoperative epineural SSIS may be required for optimum management.

Herrmann et al. [4], following Campbell's method [1], performed SSIS to localize the site of conduction block on 13 patients (age range 18–83 years) with UNE and motor conduction block (≥50% decrease in ulnar CMAP amplitude between stimulation at the below-elbow and above-elbow sites) confirmed by standard electrodiagnostic evaluation. Ulnar nerve motor conduction study was performed with the elbow in 90° of flexion, and SSIS was performed on the affected ulnar nerve across the elbow using 1 cm segments, with the ME designed as station zero (Fig. 3).

The site of conduction block (on SSIS) was defined by the most distal 1 cm segment, in which a sudden drop in CMAP amplitude and negative area of 20 % occurred. The authors [4] considered abnormal a latency shift of ≥0.5 ms over a 1–cm segment. Of the 13 patients, SSIS localized the ulnar motor conduction block to above the epicondyle in 8 (62 %), at the epicondyle in 3 (23 %), and below the epicondyle (1–3 cm) in 2 (15 %). A prolonged segmental latency was found at the site of conduction block in 85 % of patients, and the conduction block evolved over a single segment (1 cm) in 9 patients, two segments (2 cm) in three cases, and three segments (3 cm) in 1 case. The conduction block occurred at ≥2 cm above the medial epicondyle in 54 % of their patients. Their observations [4] contrast with those reported [1, 5] for chronic UNE (in which mainly occurred a segmental motor slowing without conduction block), where the lesion is typically localized to either the retroepicondylar groove (at the level of the epicondyle) or the humeroulnar arcade (0–2.5 cm below the epicondyle). In conclusion, they found short-segment incremental

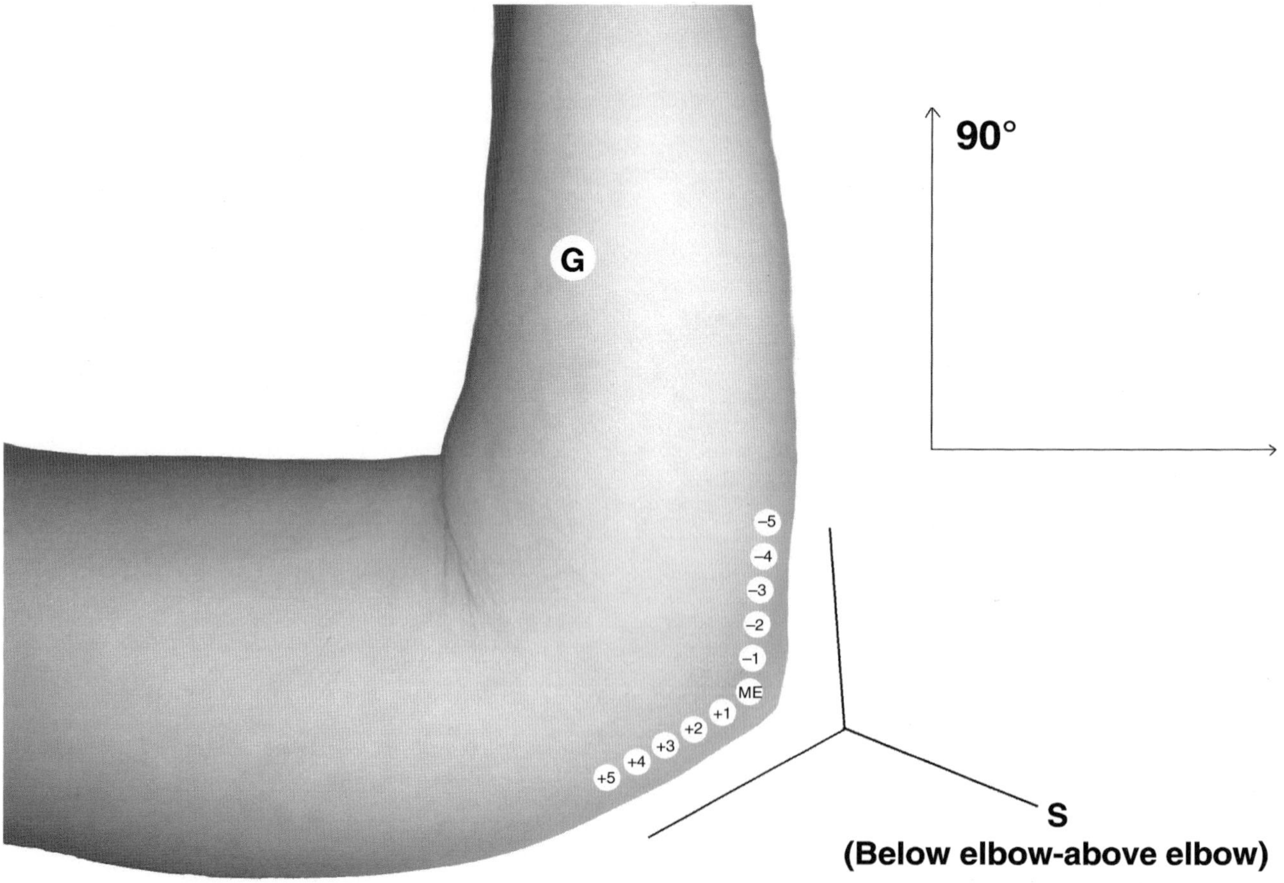

Fig. 3 *Short-segment incremental stimulation* (*SSIS*) from the below-elbow to the above-elbow sites with the elbow flexed to 90°

Pathological waveform (above elbow to below elbow – ADM muscle):

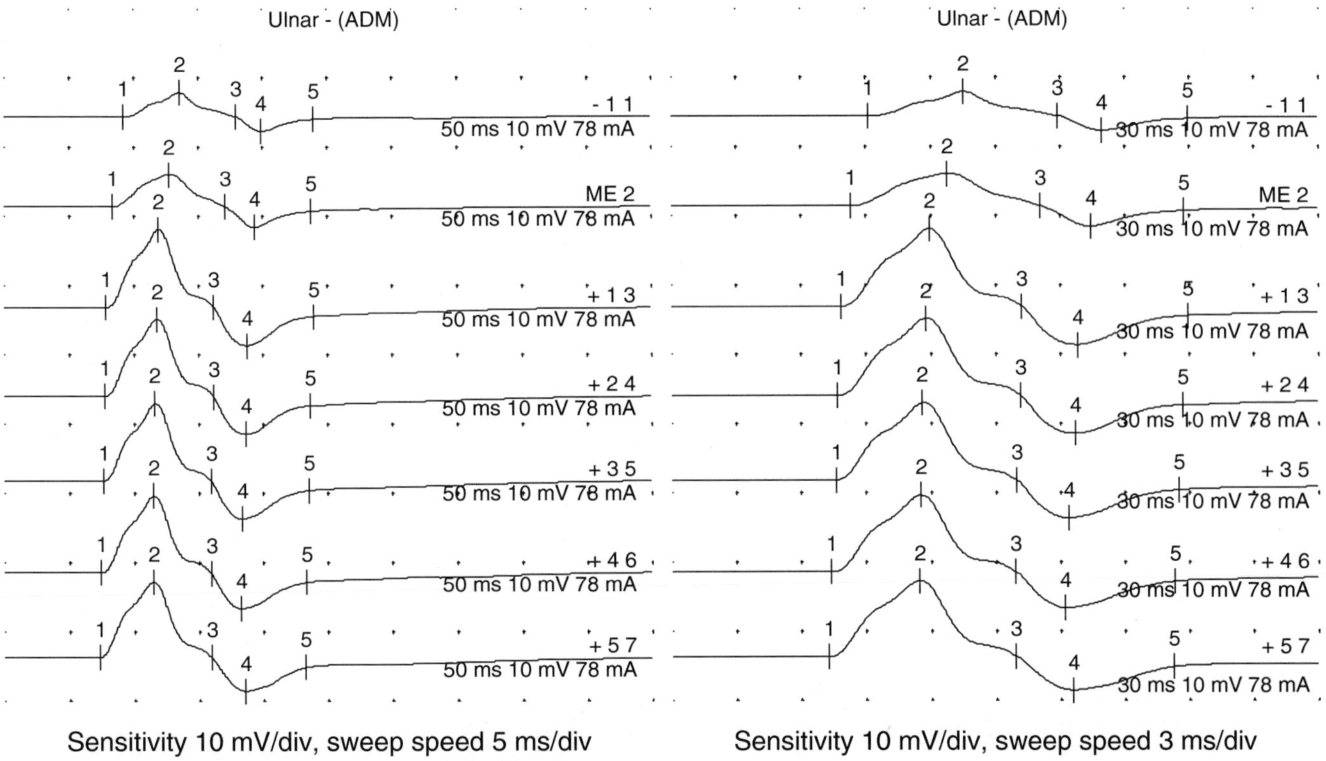

Onset latency (Above elbow – ADM, -1): 9.95 ms; **Onset latency** (ME – ADM, level 0 - ME): 8.95 ms; **Onset latency** (below elbow – ADM, +1): 7.85 ms; **Onset latency** (below elbow – ADM, +2): 7.70 ms; **Onset latency** (below elbow – ADM, +3): 7.60 ms; **Onset latency** (below elbow – ADM, +4): 7.40 ms; **Onset latency** (below elbow – ADM, +5): 7.25 ms; **Peak latency** (above elbow – ADM, -1): 13.40 ms; **Peak latency** (ME – ADM, level 0): 13.05 ms; **Peak latency** (below elbow – ADM, +1): 11.95 ms; **Peak latency** (below elbow – ADM, +2): 11.80 ms; **Peak latency** (below elbow – ADM, +3): 11.70 ms; **Peak latency** (below elbow – ADM, +4): 11.60 ms; **Peak latency** (below elbow – ADM, +5): 11.50 ms; **Onset to peak amplitude** (above elbow – ADM, -1): 3.30 mV; **Onset to peak amplitude** (ME – ADM, level 0 - ME): 4.3 mV; **Onset to peak amplitude** (below elbow – ADM, +1): 11.3 mV; **Onset to peak amplitude** (below elbow – ADM, +2): 11.10 mV; **Onset to peak amplitude** (below elbow – ADM, +3): 11.20 mV; **Onset to peak amplitude** (below elbow – ADM, +4): 11.10 mV; **Onset to peak amplitude** (below elbow – ADM, +5): 11.0 mV; **Peak to peak amplitude** (above elbow – ADM, -1): 3.30 mV; **Peak to peak amplitude** (ME – ADM, level 0 - ME): 4.3 mV; **Peak to peak amplitude** (below elbow – ADM, +1): 11.3 mV; **Peak to peak amplitude** (below elbow – ADM, +2): 11.10 mV; **Peak to peak amplitude** (below elbow – ADM, +3): 11.20 mV; **Peak to peak amplitude** (below elbow – ADM, +4): 11.10 mV; **Peak to peak amplitude** (below elbow – ADM, +5): 11.10 mV

Fig. 4 Compound muscle action potentials (*CMAPs*) recorded at the hand from the ADM muscle in UNE (RTC neuropathy) – prolonged delay (0.85 ms) between -1 and ME sites and *short-segment incremental stimulation* (*SSIS*) from the above-elbow to the below-elbow sites

(SSIS) study a useful tool to identify conduction block above the elbow in such cases. Pathological waveform and values in a patient with ulnar neuropathy at the elbow (UNE) are reported (Fig. 4).

References

1. Campbell WW, Pridgeon RM, Sahni SK (1992) Short segment incremental studies in the evaluation of ulnar neuropathy at the elbow. Muscle Nerve 15:1050–1054
2. Campbell WW, Sahni SK, Pridgeon RM et al (1988) Intraoperative electroneurography: management of ulnar neuropathy at the elbow. Muscle Nerve 11:75–81
3. Campbell WW, Sahni SK, Pridgeon RM (1988) Entrapment of the ulnar at its point of exit from the flexor carpi ulnaris. Muscle Nerve 11:467–470
4. Herrmann DN, Preston DC, McIntosh K et al (2001) Localization of ulnar neuropathy with conduction block across the elbow. Muscle Nerve 24:698–700
5. Campbell WW, Pridgeon RM, Riaz G et al (1991) Variations in anatomy of the ulnar nerve at the cubital tunnel: pitfalls in the diagnosis of ulnar neuropathy at the elbow. Muscle Nerve 14:733–738

Below Elbow, Above Elbow – Forearm; Wrist, Below Elbow, Above Elbow – Hand

M35

Surface Recording Technique, Study from *Flexor Carpi Ulnaris* (FCU), *Abductor Digiti Minimi* (ADM), and *First Dorsal Interosseous* (FDI) Muscles

Original Settings Low-frequency filter was 20 Hz, high-frequency filter was 2 kHz, duration of pulse was 0.2 ms, rate of pulse was 1/s, and the machine used was a DISA 1500 EMG system; sensitivity and sweep speed were not specified.

Position The nerve conduction study was done with the elbow flexed to 90°.

Recording Using surface recording electrodes, Uchida and Sugioka [1] performed two several studies, studying the ulnar nerve conduction across the elbow recording proximally recording from the *flexor carpi ulnaris* (FCU) muscle placing the active electrode (A) 12 cm distal to the medial epicondyle, over the motor point of the muscle (Fig. 1), and distally from the motor point of the *abductor digiti minimi* (ADM) and *first dorsal interosseous* (FDI) muscles (Figs. 2 and 3, respectively). The elbow was held at 90° of flexion during the examination because this position has been recommended by many authors [2, 3]. The ground (G) electrode position was not specified in the text; the figures show ground electrode placed on the forearm (Fig. 1) and on the palm (Figs. 2 and 3).

Stimulation In FCU muscle recordings, stimulation was applied at the proximal sites: below the elbow (S1), 4 cm distal to the medial epicondyle, and above the elbow (S3), 2 cm proximal to the medial epicondyle of the humerus (Fig. 1). To evoke a CMAP from the FCU muscle, in order to assess the presence of a conduction block in the across-elbow segment, only the proximal stimulation can be performed. In ADM and FDI muscles recordings, the ulnar nerve was stimulated at 3 sites: at the wrist (S1), 7 cm proximal to the recording electrode over the belly of the ADM muscle; below the elbow (S2), 4 cm distal to the medial epicondyle; and above the elbow (S3), 6 cm proximal to the medial epicondyle of the humerus (Figs. 2, 3, and 4).

Measurements Latency (ms) of the CMAP was measured from the onset of the stimulus artifact to the onset of the negative deflection of the CMAP for all techniques. The amplitude (mV) of the compound muscle action potential (CMAP) was measured from the negative to the positive peak of the CMAP (peak-to-peak amplitude). By dividing the distance by time, the motor nerve conduction velocity (MNCV) along the nerve segment can be determined and measured in meter per second (m/s), and it was calculated for the wrist–below-elbow (forearm), and the below-elbow–above-elbow (across-elbow) segments. Surface skin and room temperatures were not reported in the paper. The authors [1], between 1985 and 1992, studied 69 upper extremities of 60 patients (46 men, 55 arms, and 14 women, 14 arms) with clinically and electrophysiologically diagnosed ulnar neuropathy at the elbow (UNE). The patients were divided into two groups: Group 1, consisted of those with tardy ulnar nerve palsy and associated cubitus valgus resulting from pseudoarthrosis of the lateral humeral condyle (20 arms of 20 patients, age range 24–62 years, average age 42.4 years), and Group 2 (57 arms of 48 patients, age range 17–74 years, average age 50.1 years). Pathological data for both groups of patients are described in Table 1 (Group 1) and Table 2 (Group 2).

G. Gentili, M. Di Napoli, *The Ulnar Nerve*, DOI 10.1007/978-3-319-43292-2_59

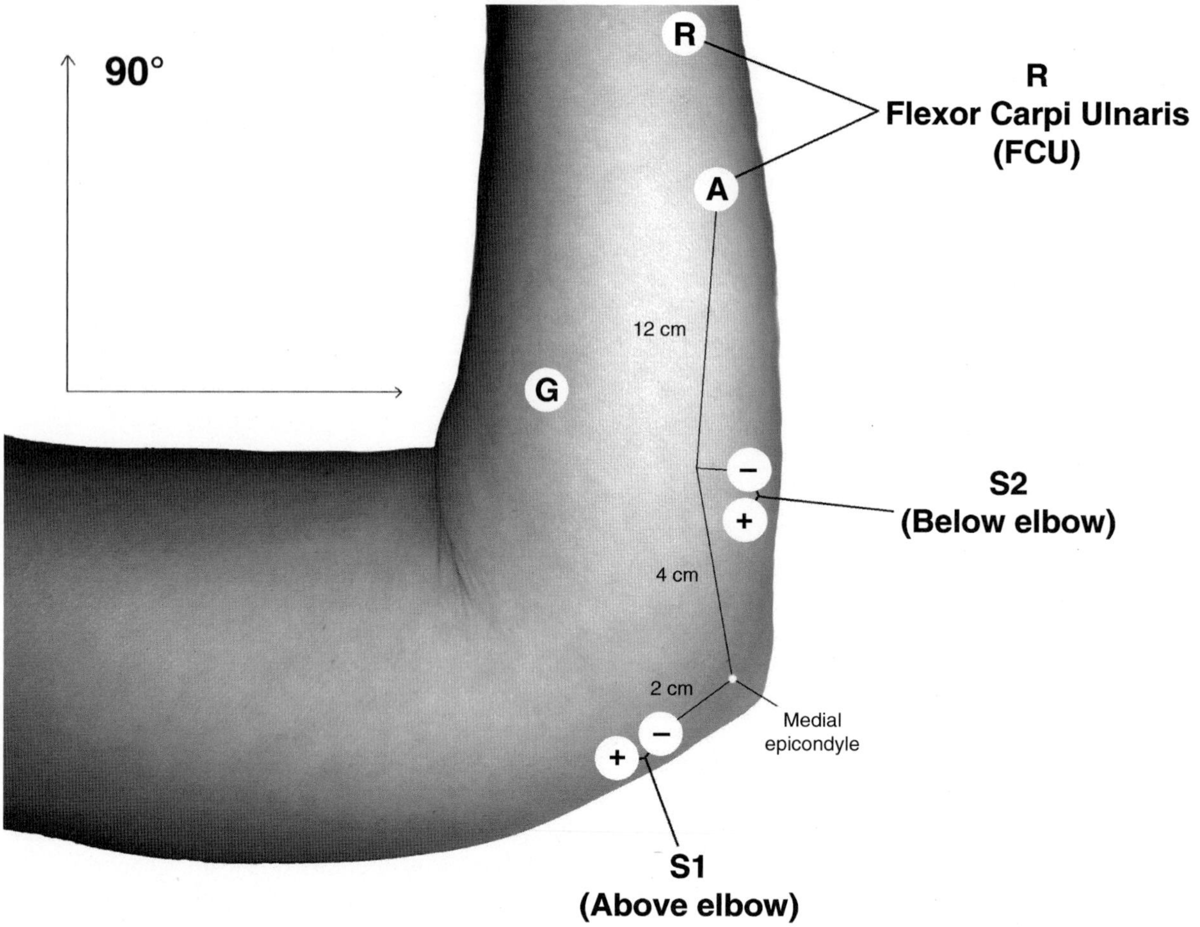

Typical waveform (below elbow, above elbow – FCU muscle):

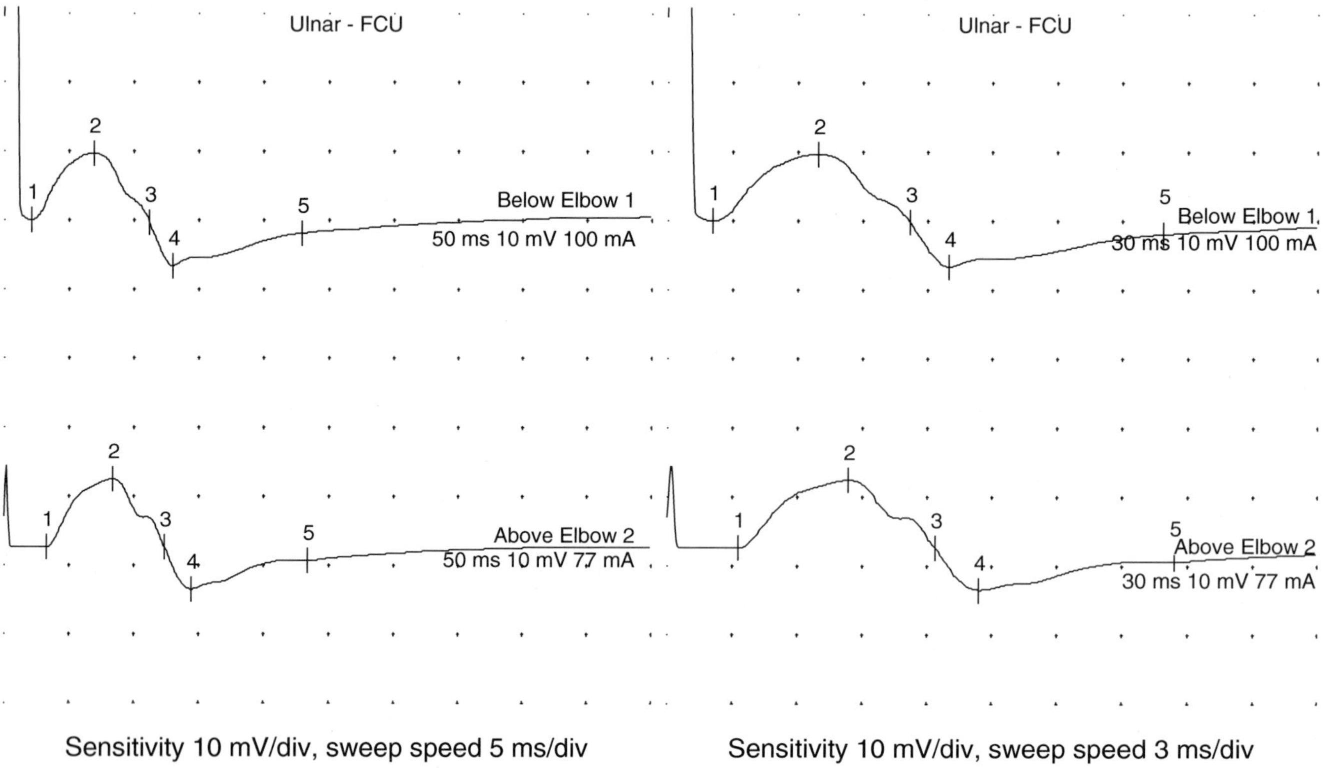

Fig. 1 Compound muscle action potentials (*CMAPs*) recorded at the hand from the FCU muscle, stimulation below the elbow (*upper trace*) and above the elbow (*lower trace*)

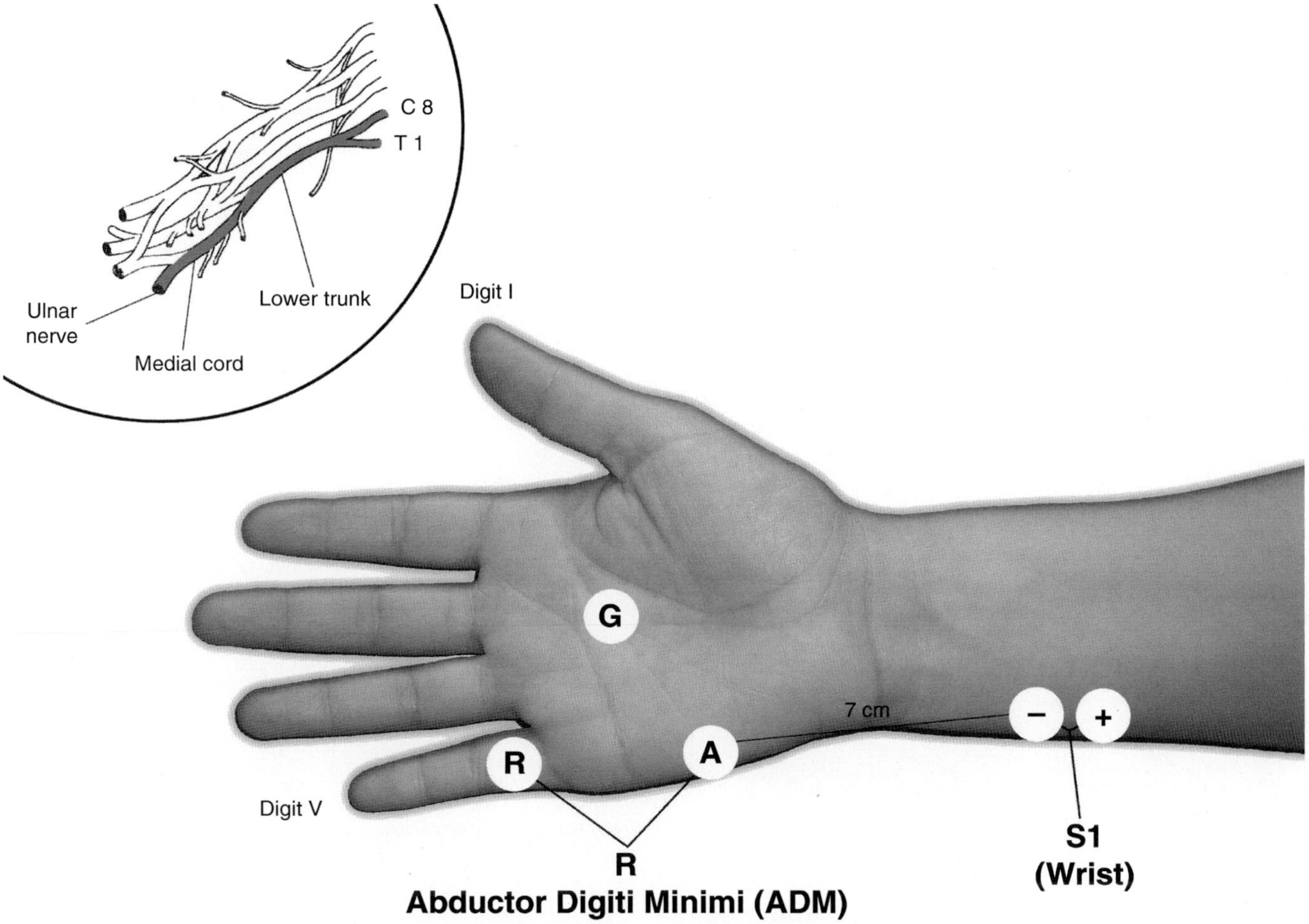

Typical waveform (wrist, below elbow, above elbow – ADM muscle):

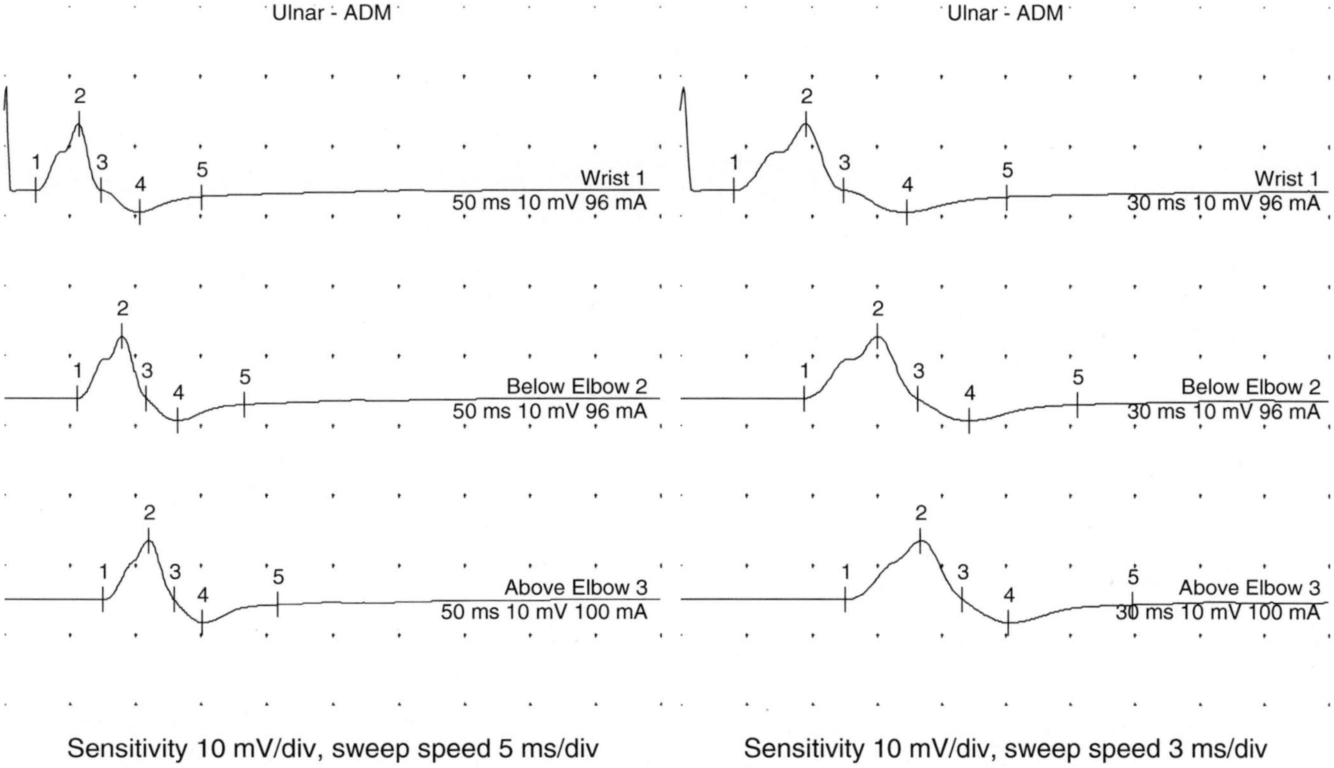

Fig. 2 Compound muscle action potentials (*CMAPs*) recorded at the hand from the ADM muscle, stimulation of the wrist (*upper trace*), below the elbow (*middle trace*), and above the elbow (*lower trace*)

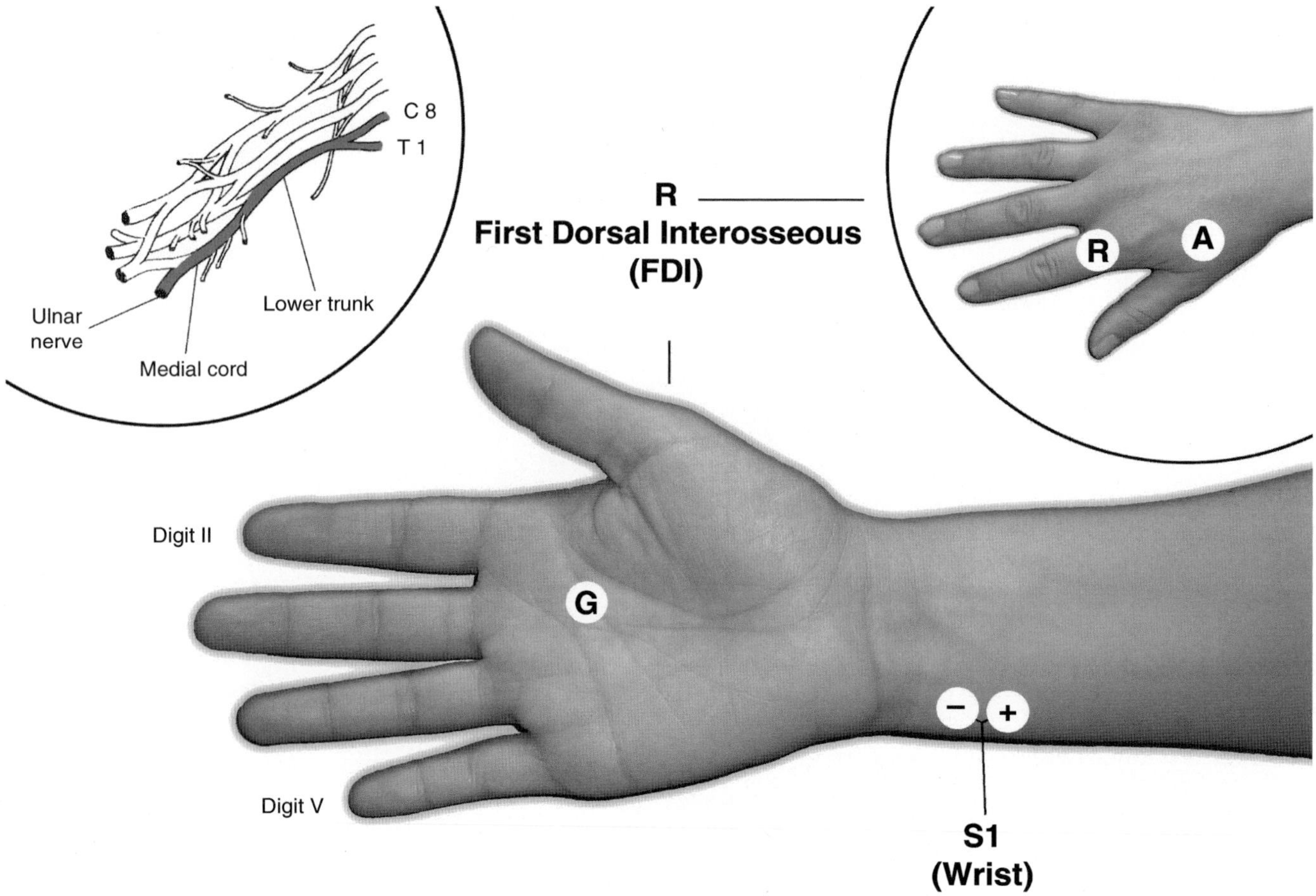

Typical waveform (wrist, below elbow, above elbow – FDI muscle):

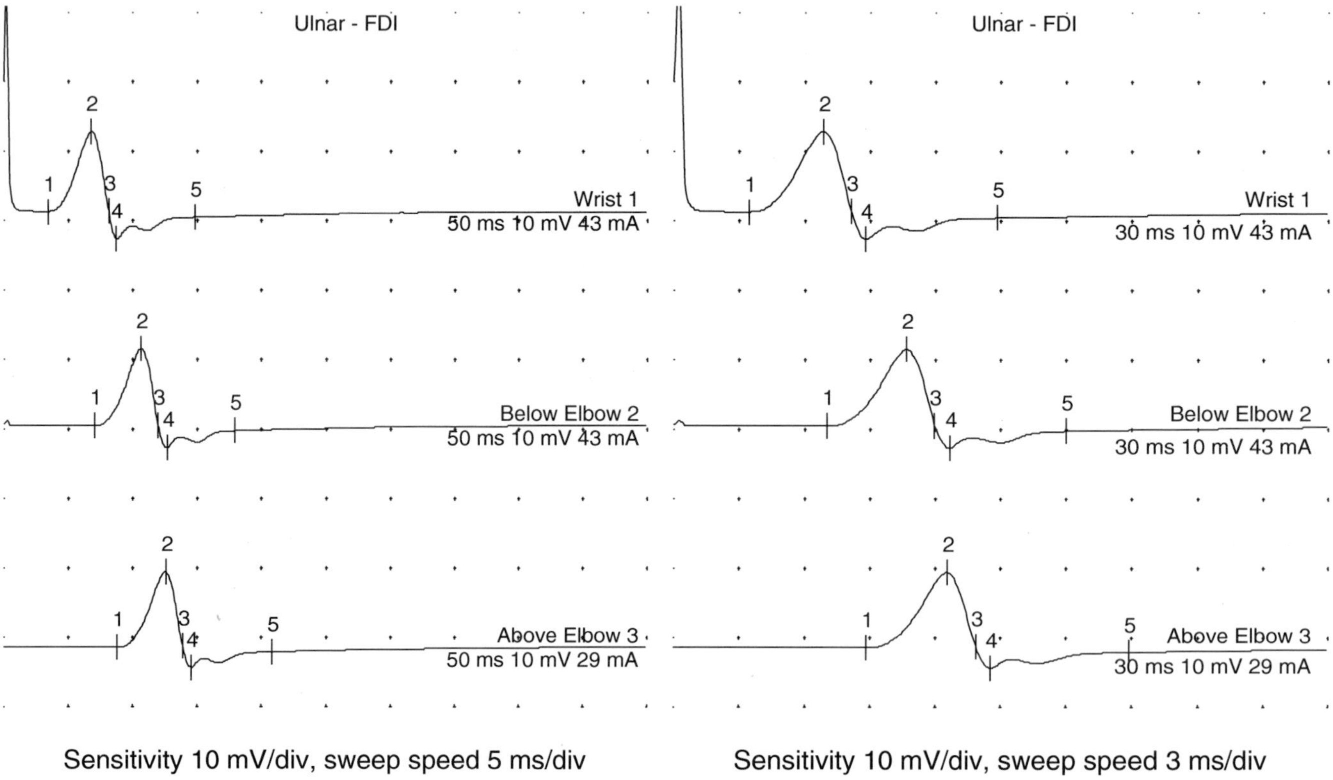

Fig. 3 Compound muscle action potentials (*CMAPs*) recorded at the hand from the FDI muscle, stimulation of the wrist (*upper trace*), below the elbow (*middle trace*), and above the elbow (*lower trace*)

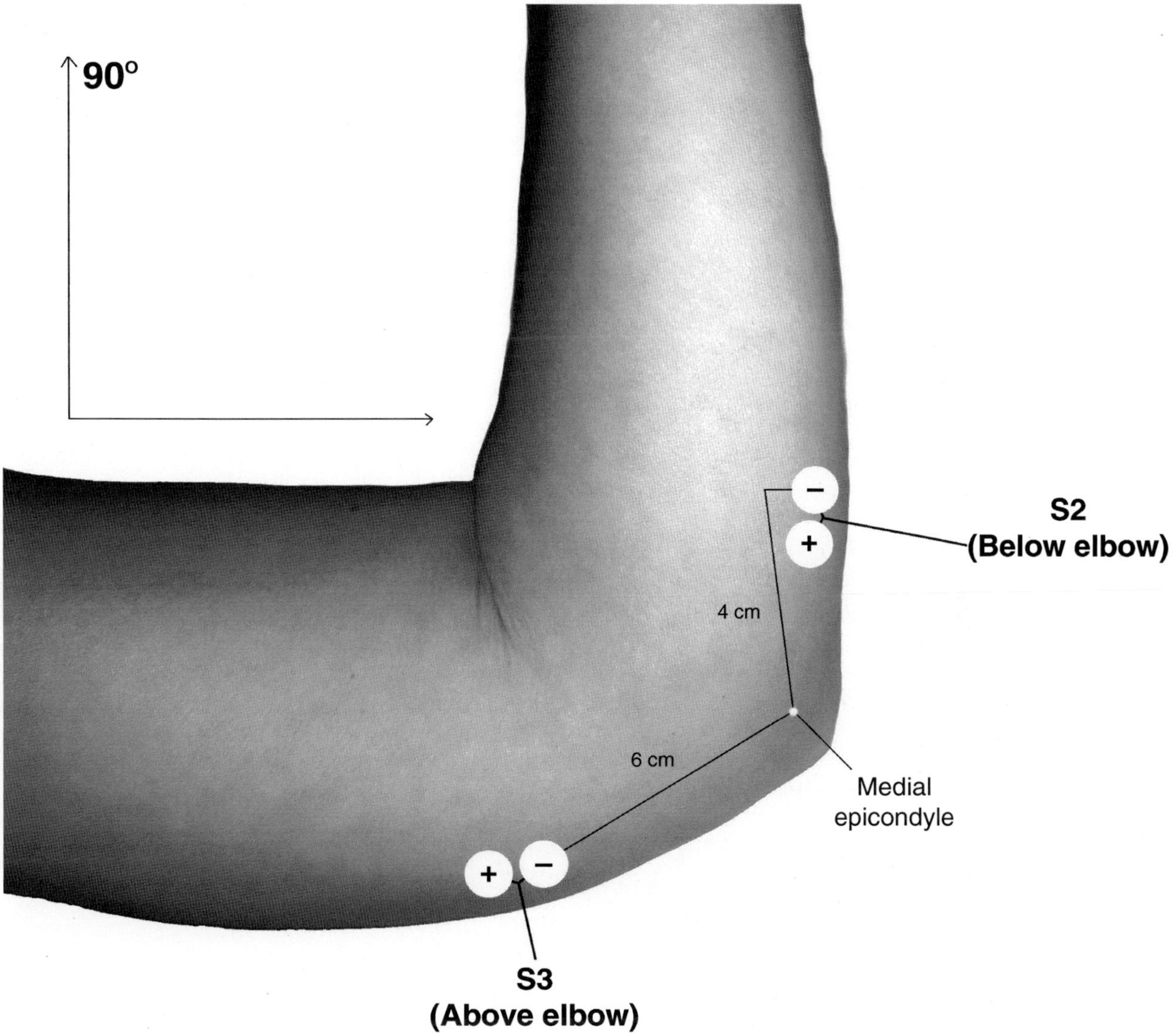

Fig. 4 Proximal stimulation points (S2, below elbow; S3, above elbow) with the elbow flexed to 90°

Table 1 Pathological values – Group 1 [1]

	Mean	Range	Cutoff
Wrist–ADM, onset latency (ms)	3.1	3.0–3.3	2.37
Wrist–FDI, onset latency (ms)	4.3	3.2–5.2	3.46
Above elbow–FCU, onset latency (ms)	6.6	3.7–9.5	2.98
Forearm MNCV (m/s)	46.6	40.5–53.6	60.7
Across-elbow MNCV (m/s)	26.5	17.8–33.3	56.2

Table 2 Pathological values – Group 2 [1]

	Mean	Range	Cutoff
Wrist–ADM, onset latency (ms)	3.0	2.1–6.5	2.37
Wrist–FDI, onset latency (ms)	4.1	2.6–8.6	3.46
Above elbow–FCU, onset latency (ms)	4.65	3.0–9.9	2.98
Forearm MNCV (m/s)	48.0	20.7–64.4	60.7
Across-elbow MNCV (m/s)	31.7	13.6–47.6	56.2

Comment

Uchida and Sugioka [1] observed dispersion of the CMAP recording from the FCU muscle in 2 patients and in the ADM and FDI muscles in 4 patients (Group 1); in the same group, the CMAP latency for the FCU muscle was prolonged in proportion to the delay in the MNCV across the elbow, the latencies for the ADM and FDI CMAPs were slightly prolonged, and the MNCV at the forearm segment was slightly lower compared to the normal value (60.7 m/s). In Group 2, the authors found dispersion of the CMAP recording from the FCU muscle in 5 upper arms, and for the ADM and FDI muscles, it was found in 20 extremities. The CMAP latency of the FCU muscle was prolonged in proportion to the delay of the MNCV across the elbow, and the amplitude of the FCU CMAP was reduced in proportion to the delay of the CMAP latency. The MNCV at the forearm segment was slightly reduced, and the CMAP latencies for both the ADM and FDI muscles were slightly prolonged compared to normal values. In their results [1], the FCU muscle was spared relative to the intrinsic muscles. When the CMAP from the ADM and FDI muscles were not evoked, the CMAP from the FCU muscle could be evoked in many cases, and the temporal dispersion was frequently observed in both ADM and FDI muscles rather than in the FCU muscle. However, the authors observed that in advanced cases of UNE, in which the CMAP from the ADM or FDI muscles could not be evoked or Wallerian degeneration developed in the forearm segment and the forearm MNCV decreases or the latency of The CMAP of the intrinsic muscles increases; the usual measurement of MNCV at the elbow segment is not topodiagnostic. Moreover, in the case of mild UNE in which the SNAP can be evoked, they found the low diagnostic value of the electrophysiological examination of the FCU muscle, and, in these cases, some authors found sensory recordings more useful [2, 4]. Pathological waveform and values in a case of ulnar neuropathy at the elbow (UNE) are reported (Figs. 5 and 6).

Pathological waveform (above elbow – FCU muscle):

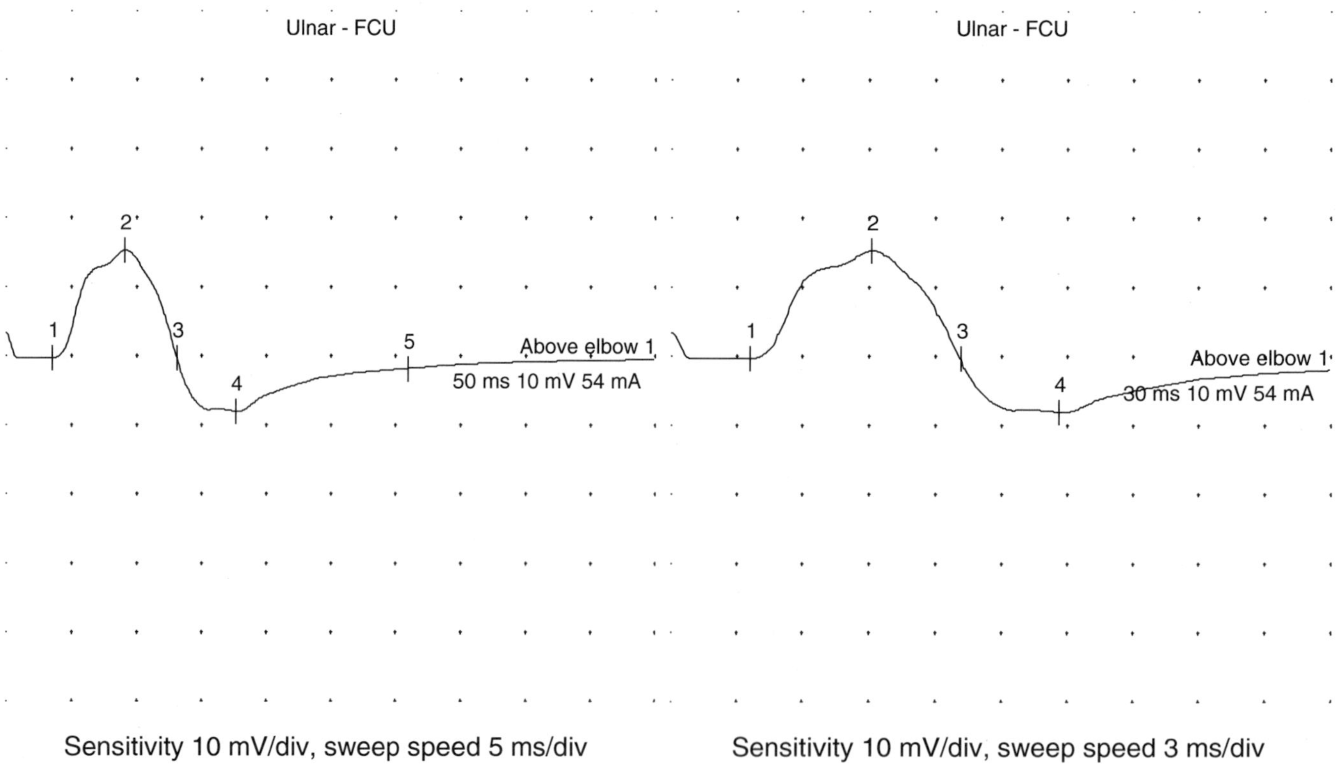

Onset latency (above elbow – FCU): 3.55 ms; **Peak latency** (above elbow – FCU): 9.20 ms; **Onset to peak amplitude** (above elbow – FCU): 15.7 mV; **Peak to peak amplitude** (above elbow – FCU): 23.5 mV; **MNCV** (above elbow – FCU, 14 cm): 39.4 m/s

Fig. 5 Compound muscle action potentials (*CMAPs*) recorded at the hand from the FCU muscle in UNE, stimulation above elbow

Pathological waveform (wrist, below elbow, above elbow – ADM muscle):

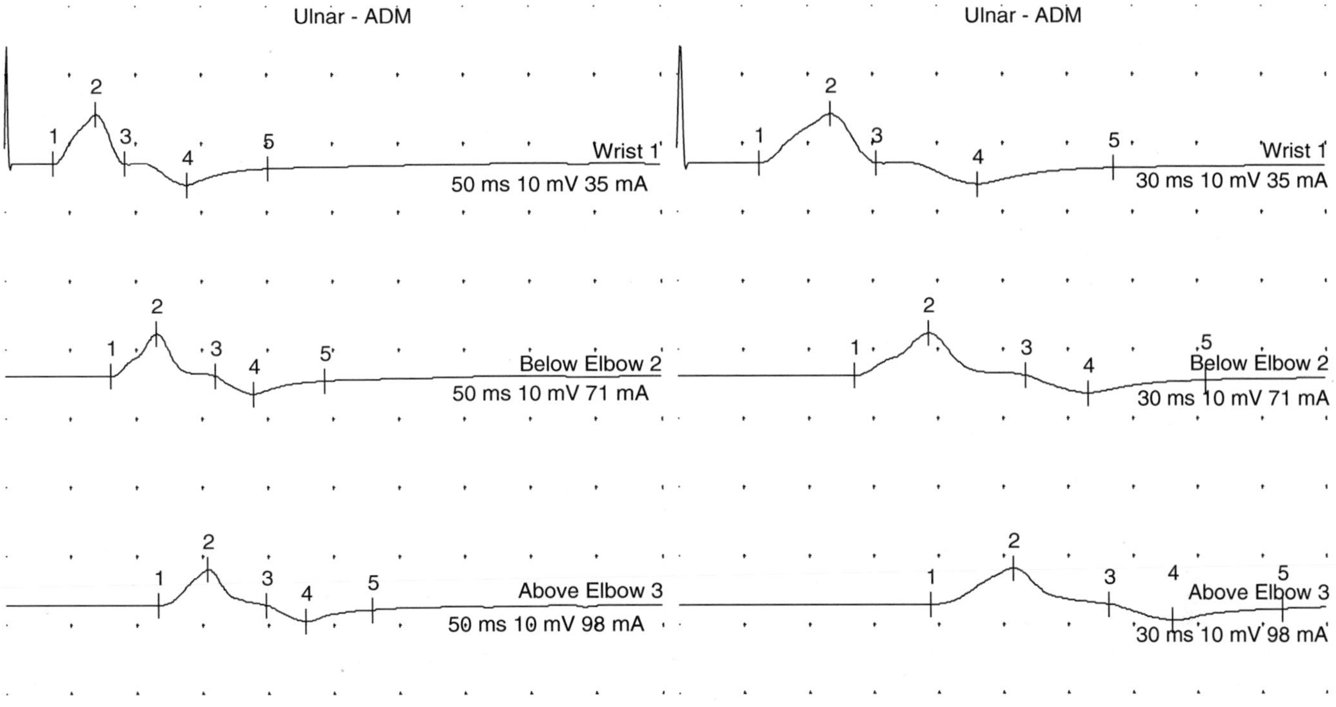

Sensitivity 10 mV/div, sweep speed 5 ms/div

Sensitivity 10 mV/div, sweep speed 3 ms/div

Onset latency (wrist – ADM): 3.75 ms; **Onset latency** (below elbow – ADM): 8.10 ms; **Onset latency** (above elbow – ADM): 11.60 ms; **Peak latency** (wrist – ADM): 7.25 ms; **Peak latency** (below elbow – ADM): 11.70 ms; **Peak latency** (above elbow – ADM): 15.25 ms; **Onset to peak amplitude** (wrist – ADM): 7.2 mV; **Onset to peak amplitude** (below elbow – ADM): 6.1 mV; **Onset to peak amplitude** (above elbow – ADM): 5.3 mV; **Peak to peak amplitude** (wrist – ADM): 10.4 mV; **Peak to peak amplitude** (below elbow – ADM): 8.9 mV; **Peak to peak amplitude** (above elbow – ADM): 7.7 mV; **MNCV** (wrist – below elbow, 23 cm): 52.9 m/s; **MNCV** (below elbow – above elbow, 10 cm): 28.6 m/s

Fig. 6 Compound muscle action potentials (*CMAPs*) recorded at the hand from the ADM muscle in UNE, stimulation at the wrist (*upper trace*), below elbow (*middle trace*), and above elbow (*lower trace*)

References

1. Uchida Y, Sugioka Y (1993) The value of electrophysiological examination of the flexor carpi ulnaris muscle in the diagnosis of cubital tunnel syndrome. Electromyogr Clin Neurophysiol 33:369–373
2. Felsenthal G, Freed MJ, Kalafut R et al (1989) Across elbow ulnar nerve sensory conduction technique. Arch Phys Med Rehabil 70:668–672
3. Nelson RM (1980) Effects of elbow position on motor conduction velocity of the ulnar nerve. Phys Ther 60:780–783
4. Tackmann W, Vogel P, Kaeser HE et al (1985) Sensitivity and localizing significance of motor and sensory electroneurographic parameters in the diagnosis of ulnar nerve lesions at the elbow. J Neurol 231:204–211

Above Wrist to Below Wrist – Hand

M36

Surface Recording Technique, Study from *First Dorsal Interosseous* (FDI) Muscle

Original Settings Sensitivity, sweep speed, low-frequency filter, high-frequency filter, duration of pulse, and the machine used were not specified.

Position The position was not specified in the text; we performed the study with the forearm supine and the elbow fully extended.

Recording Using surface electrodes [1], short-segment incremental study (SSIS) across the wrist was performed recording from the *first dorsal interosseous* (FDI) muscle (Fig. 1). The authors also performed standard electrodiagnostic techniques [2] recording the compound muscle action potentials (CMAPs) from the *abductor digiti minimi* (ADM) muscle, median-innervated second lumbrical (2 L) muscle, and ulnar-innervated palmar *interosseous* (INT) after stimulation at the wrist. The ground (G) electrode position was not specified in the text; the figure shows ground placed on the palm.

Stimulation SSIS started 3 cm proximal to the distal wrist crease (DWC) and then along the ulnar nerve at seven subsequent 1 cm intervals (Fig. 1), ending at 4 cm distal to the DWC [1].

Measurements The authors [1] performed SSIS in ten healthy subjects (sample data were not reported), and it showed that the maximal change in latency (ms) over each of the seven 1 cm segments was 0.4 ms/cm, which was essentially equal to the mean + 2SD value of 0.43. Thus, a change in latency of 0.5 ms/cm or greater was considered abnormal. The maximal change in amplitude over all segments was 117 %/cm, which was essentially equal to the mean + 2SD value of 115 %/cm. Therefore, a change in amplitude of 120 %/cm or greater was considered abnormal.

Comment

McIntosh et al. [1] studied two patients with clinical symptoms consistent with ulnar neuropathy at the wrist (UNW). In addition to standardized electrodiagnostic techniques, they performed SSIS across the wrist, recording from the FDI muscle in an attempt to localize the lesion. In both patients, the routine nerve conduction findings were consistent with, but not diagnostic of, UNW. In patient 1, using the SSIS, no potential was recordable at 2 cm proximal to the DWC, and at 1 cm proximal to the DWC, a low-amplitude CMAP with a prolonged distal latency was recorded. A 491 % increase in amplitude and a decrease in latency of 1.8 ms were found between 1 cm proximal to the DWC and the DWC. Distal to the DWC, the authors [1] observed normal amplitudes and latencies. In patient 2, low-amplitude CMAPs recording from the FDI muscle with normal latency were recorded proximal to 1 cm distal to the DWC. A 128 % increase in amplitude and a 0.5 ms change in latency were observed between the DWC and 1 cm distal to the DWC, and a 200 % increase in amplitude and a 0.9 ms change in latency were seen between 1 and 2 cm distal to the DWC (the main site of compression). Furthermore, between 2 and 3 cm distal to the DWC, there was no increase in amplitude but a 0.8 ms change in latency, and a 134 % increase in amplitude without a significant change in latency was observed between 3 and 4 cm distal to the DWC. SSIS precisely localized the ulnar nerve lesions in both patients, demonstrating normal CMAP amplitude after distal stimulation but markedly reduced amplitude after

G. Gentili, M. Di Napoli, *The Ulnar Nerve*, DOI 10.1007/978-3-319-43292-2_60

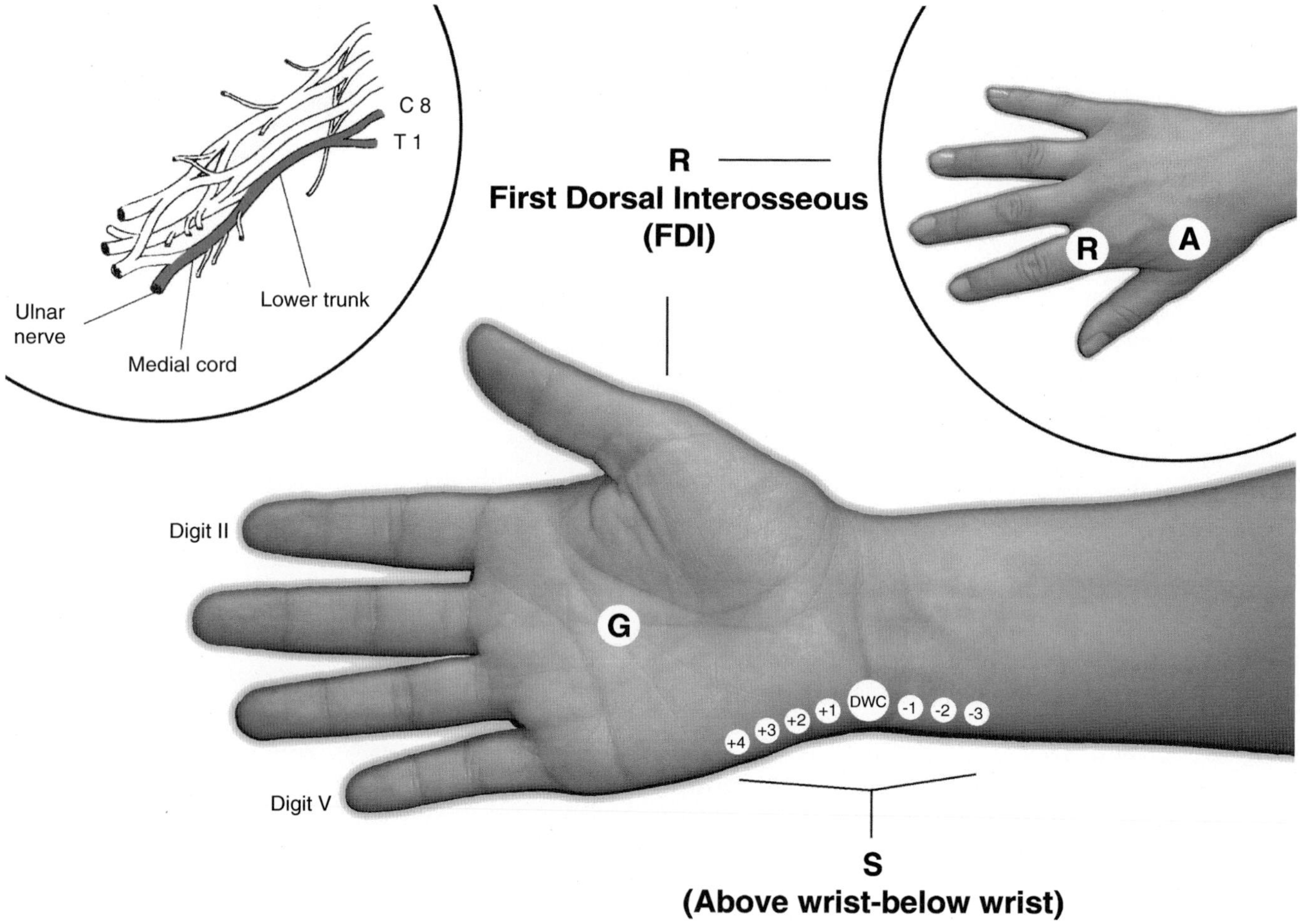

Typical waveform (above wrist to below wrist – FDI muscle):

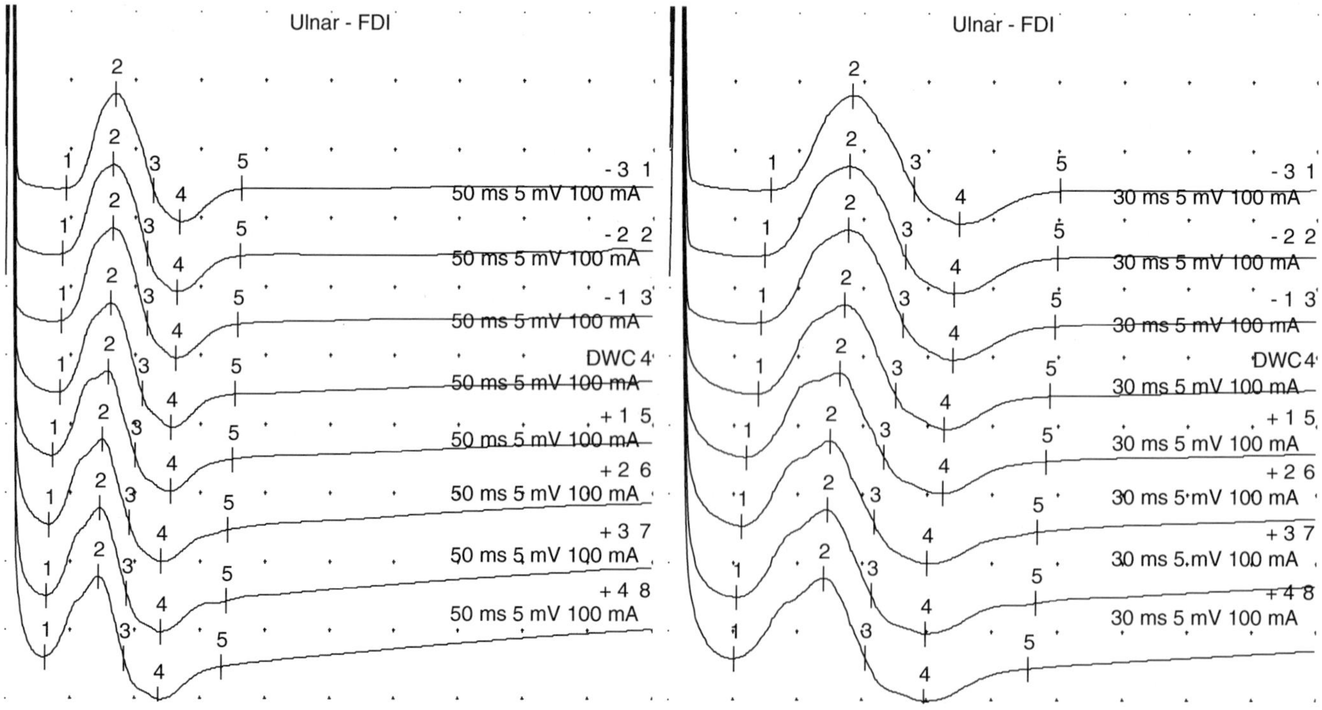

Sensitivity 5 mV/div, sweep speed 5 ms/div

Sensitivity 5 mV/div, sweep speed 3 ms/div

Fig. 1 Compound muscle action potentials (*CMAPs*) recorded at the hand from the FDI muscle, *short-segment incremental stimulation* (SSIS) from the above-wrist to the below-wrist sites proximal stimulation, indicating a significant demyelinating component to each lesion characterized by a conduction slowing over most of the nerve segments that demonstrated amplitude change, and confirming the presence of segmental demyelination. For the authors [1], the motor conduction studies from the FDI muscle may increase the electrodiagnosis of UNW over the motor conduction studies from the ADM muscle because more distal UNW lesions spare the ulnar motor branch to the ADM muscle. They observed conventional FDI studies that use only one stimulation site at the wrist to have a longer conduction distance than the ADM muscle studies and, in patients with focal demyelination, this may tend to normalize FDI distal latencies. McIntosh et al. [1] suggested to use the short-segment incremental study (SSIS) in such cases, since it eliminates this problem and is, therefore, a superior technique for evaluating UNW, because it maximizes the opportunity to detect focal demyelination over nearly the entire course of the ulnar nerve at the wrist. Pathological waveforms and values in patients with UNW – Type I, in whom the lesion occurred just above or within the proximal end of Guyon's canal involving both the deep and superficial branches of the ulnar nerve – are reported (Figs. 2 and 3).

Padua et al. [3, 4] performed the same SSIS method [1] on a 46-year-old man with clinical symptoms suggesting UNW. The patient complained of marked weakness in his left hand, which started 1 month earlier following 6 hours of cycling; he presented deficit of hand muscles innervated by the ulnar nerve, and no sensory deficit was observed by the authors. Conventional motor conduction studies to the ADM and FDI muscles revealed an increase of the distal motor latency (DML) to both segments (4.3 ms and 5.5 ms, respectively) and low CMAP amplitude (0.9 mV and 1.1 mV, respectively). The SSIS study showed a neuroapraxic block with a reduced rate slowing of conduction at the wrist–palm segment (29 m/s registering from the ADM muscle and 35 m/s from the FDI muscle). The orthodromic sensory nerve conduction study of the fifth digit–wrist segment and of the dorsal cutaneous branch of the ulnar nerve appeared normal. One month after the first observation, the patient underwent decompression surgery, and 1 month after surgery, the authors [3, 4] assessed ulnar nerve functionality performing conventional motor conduction and SSIS studies. All tests showed an almost complete resolution of the neuroapraxic block with normalization of conduction velocity. The SSIS enabled them to identify neuroapraxic block and localize the precise site of lesion. Their findings were in agreement with McIntosh et al. [1], and they found SSIS a powerful tool for diagnosis, precise localization, and prognosis of UNW.

Pathological waveform (above wrist to below wrist – FDI muscle):

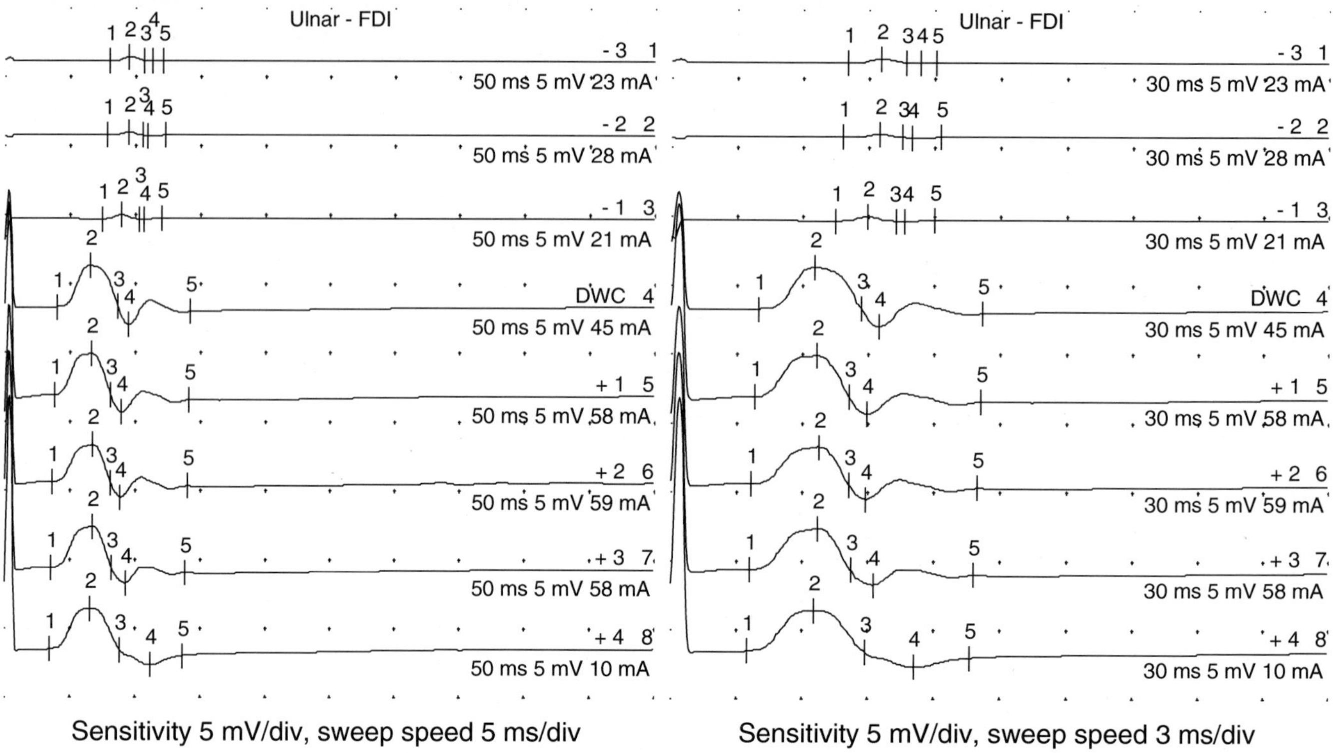

Onset latency (above wrist – FDI, -3): 8.0 ms; **Onset latency** (above wrist – FDI, -2): 7.80 ms; **Onset latency** (above wrist – FDI, -1): 7.45 ms; **Onset latency** (DWC, level 0): 3.95 ms; **Onset latency** (below wrist – FDI, +1): 3.80 ms; **Onset latency** (below wrist – FDI, +2): 3.65 ms; **Onset latency** (below wrist – FDI, +3): 3.55 ms; **Onset latency** (below wrist – FDI, +4): 3.45 ms; **Onset to peak amplitude** (above wrist – FDI, -3): 0.40 ms; **Onset to peak amplitude** (above wrist – FDI, -2): 0.30 mV; **Onset to peak amplitude** (above wrist – FDI, -1): 0.4 mV; **Onset to peak amplitude** (DWC, level 0): 3.10 mV; **Onset to peak amplitude** (below wrist – FDI, +1): 3.0 mV; **Onset to peak amplitude** (below wrist – FDI, +2): 2.70 mV; **Onset to peak amplitude** (below wrist – FDI, +3): 3.0 mV; **Onset to peak amplitude** (below wrist – FDI, +4): 3.0 mV; **Peak to peak amplitude** (above wrist – FDI, -3): 0.30 mV; **Peak to peak amplitude** (above wrist – FDI, -2): 0.30 mV; **Peak to peak amplitude** (above wrist – FDI, -1): 0.40 mV; **Peak to peak amplitude** (DWC, level 0): 4.40 mV; **Peak to peak amplitude** (below wrist – FDI, +1): 4.30 mV; **Peak to peak amplitude** (below wrist – FDI, +2): 3.90 mV; **Peak to peak amplitude** (below wrist – FDI, +3): 4.20 mV; **Peak to peak amplitude** (below wrist – FDI, +4): 4.20 mV

Fig. 2 Compound muscle action potentials (CMAPs) recorded at the hand from the FDI muscle in a patient with UNW (complete focal Type I lesion with a combination of axon loss and conduction block of the motor fibers innervating the FDI muscle) and severe UNE, *short-segment incremental stimulation* (SSIS) from the above-wrist to the below-wrist sites

Pathological waveform (above wrist to below wrist – FDI muscle):

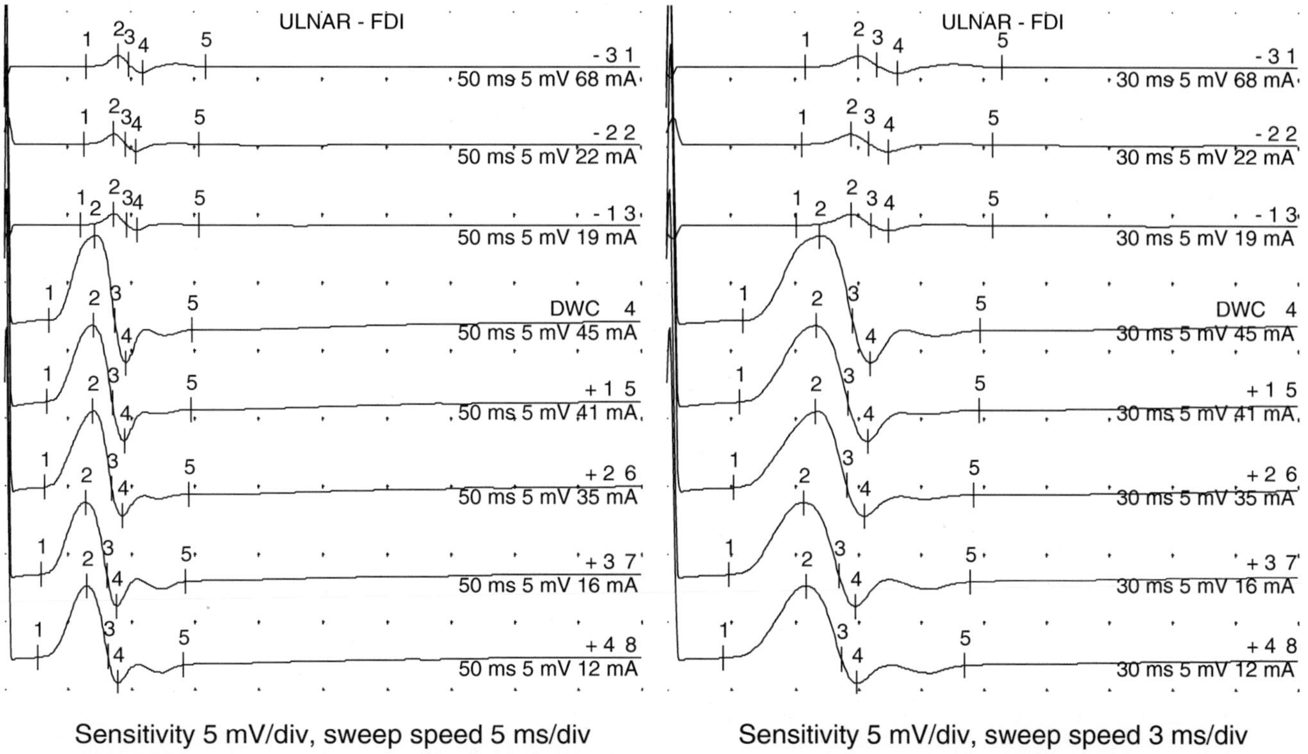

Onset latency (above wrist – FDI, -3): 6.50 ms; **Onset latency** (above wrist – FDI, -2): 6.30 ms; **Onset latency** (above wrist – FDI, -1): 6.05 ms; **Onset latency** (DWC, level 0): 3.55 ms; **Onset latency** (below wrist – FDI, +1): 3.35 ms; **Onset latency** (below wrist – FDI, +2): 3.10 ms; **Onset latency** (below wrist – FDI, +3): 2.85 ms; **Onset latency** (below wrist – FDI, +4): 2.60 ms; **Onset to peak amplitude** (above wrist – FDI, -3): 0.80 ms; **Onset to peak amplitude** (above wrist – FDI, -2): 0.80 mV; **Onset to peak amplitude** (above wrist – FDI, -1): 0.9 mV; **Onset to peak amplitude** (DWC, level 0): 6.30 mV; **Onset to peak amplitude** (below wrist – FDI, +1): 5.70 mV; **Onset to peak amplitude** (below wrist – FDI, +2): 5.70 mV; **Onset to peak amplitude** (below wrist – FDI, +3): 5.40 mV; **Onset to peak amplitude** (below wrist – FDI, +4): 5.40 mV; **Peak to peak amplitude** (above wrist – FDI, -3): 1.40 mV; **Peak to peak amplitude** (above wrist – FDI, -2): 1.40 mV; **Peak to peak amplitude** (above wrist – FDI, -1): 1.30 mV; **Peak to peak amplitude** (DWC, level 0): 9.40 mV; **Peak to peak amplitude** (below wrist – FDI, +1): 8.60 mV; **Peak to peak amplitude** (below wrist – FDI, +2): 7.80 mV; **Peak to peak amplitude** (below wrist – FDI, +3): 7.80 mV; **Peak to peak amplitude** (below wrist – FDI, +4): 7.20 mV

Fig. 3 Compound muscle action potentials (CMAPs) recorded at the hand from the FDI muscle in a patient with UNW (complete focal Type I lesion with a combination of axon loss and conduction block of motor fibers innervating FDI muscle), *short-segment incremental stimulation* (SSIS) from the above-wrist to the below-wrist sites

References

1. McIntosh KA, Preston DC, Logigian EL (1998) Short-segment incremental studies to localize ulnar nerve entrapment at the wrist. Neurology 50:303–306
2. Kothari MJ, Preston DC, Logigian EL (1996) Lumbrical-interossei motor studies localize ulnar neuropathy at the elbow. Muscle Nerve 19:170–174
3. Padua L, Insola A, Lo Monaco M et al (1998) A case of Guyon syndrome with neuroapraxic block resolved after surgical decompression. Electroencephalogr Clin Neurophysiol 109:191–193
4. Padua L, Insola A, Padua R et al (1999) Short-segment incremental studies to localize ulnar nerve entrapment at the wrist. Neurology 52:220–221

Palm, Wrist, Above Elbow, Axilla, Erb's Point – Hand

M37

Surface Recording Technique, Study from *First Dorsal Interosseous* (FDI) Muscle

Original Settings Sensitivity, sweep speed, low-frequency filter, high-frequency filter, duration of pulse, and the machine used were not reported.

Position The subjects were tested with the elbow fully extended and the hand supinated.

Recording Compound muscle action potentials (CMAPs) were recorded from the *first dorsal interosseous* (FDI) muscle with monopolar surface electrodes on the dorsal aspect of the hand [1]. We performed this method using bipolar electrodes recording (Fig. 1), as described by Seror et al. [2]. The ground (G) electrode position was not specified in the text; the figure shows ground electrode placed on the palm.

Stimulation The ulnar nerve was stimulated with bipolar surface electrodes (Figs. 1 and 2) on the palm (S1), at the wrist (S2), above the elbow (S3), axilla (S4), and at the Erb's point (S5) – which is at the angle formed by the clavicle and the posterolateral fibers of the sternocleidomastoid muscle. The anode was proximal in all stimulations [1].

Measurements Distal motor latency (ms) was measured from the stimulus artifact to the onset of the initial deflection of the compound muscle action potential (CMAP). The amplitude (mV) of the CMAP was measured from the baseline to the peak of the negative deflection (negative-peak amplitude). No fixed distances were used between recording and stimulating electrodes. The author studied two cases of suspected ulnar motor deep branch (UMDB) pathology (case 1–43 years, case 2–52 years) in which the site of ulnar nerve lesion (UNL) was not localized clearly despite numerous electrodiagnostic tests. Only the study of the UMDB confirmed the lesion site by demonstrating the very distal conduction block.

Comment

Both cases of UMDB lesions studied by Seror [1] benefited from an early diagnosis because they were referred to the EMG laboratory for a suspected carpal tunnel syndrome (CTS) and special electrodiagnostic techniques (i.e., median–ulnar antidromic comparative study to the digit IV, orthodromic to digit III, and inching test) demonstrated mild median nerve lesions at the wrist. Electrodiagnosis allows the classification of ulnar nerve lesion at the wrist in three [3, 4] or 5 [5] different types, according to the exact location of the nerve lesion. Type 1 is a mixed motor and sensory neuropathy with the lesion located just outside or within the proximal end of the Guyon canal. Type 2 is a pure sensory neuropathy and involves the superficial branch. Type 3 is a pure motor neuropathy caused by a lesion of the UMDB just distal to the superficial branch but proximal to the branch that runs to the hypothenar muscles. Type 4 is a pure motor neuropathy with sparing of hypothenar muscles; this lesion occurs in the UMDB distal to the origin of the superficial branch and distal to the branch going to the hypothenar muscles. Type 5 is also a pure motor neuropathy in which the lesion occurs just proximal to the branches going to the *first dorsal interosseous* (FDI) and *adductor pollicis* (AP) muscles. The conventional distal motor latency from the *abductor digiti minimi* (ADM) muscle and the digit V–wrist sensory conduction are not always sufficient to assess UNL at the wrist when impairment predominates in the UMDB. For this reason, some authors [6–8] proposed several motor and sensory median–ulnar latency difference methods as useful tools in UNL Type 1, 2, 3, 4, and 5 diagnoses. At

G. Gentili, M. Di Napoli, *The Ulnar Nerve*, DOI 10.1007/978-3-319-43292-2_61

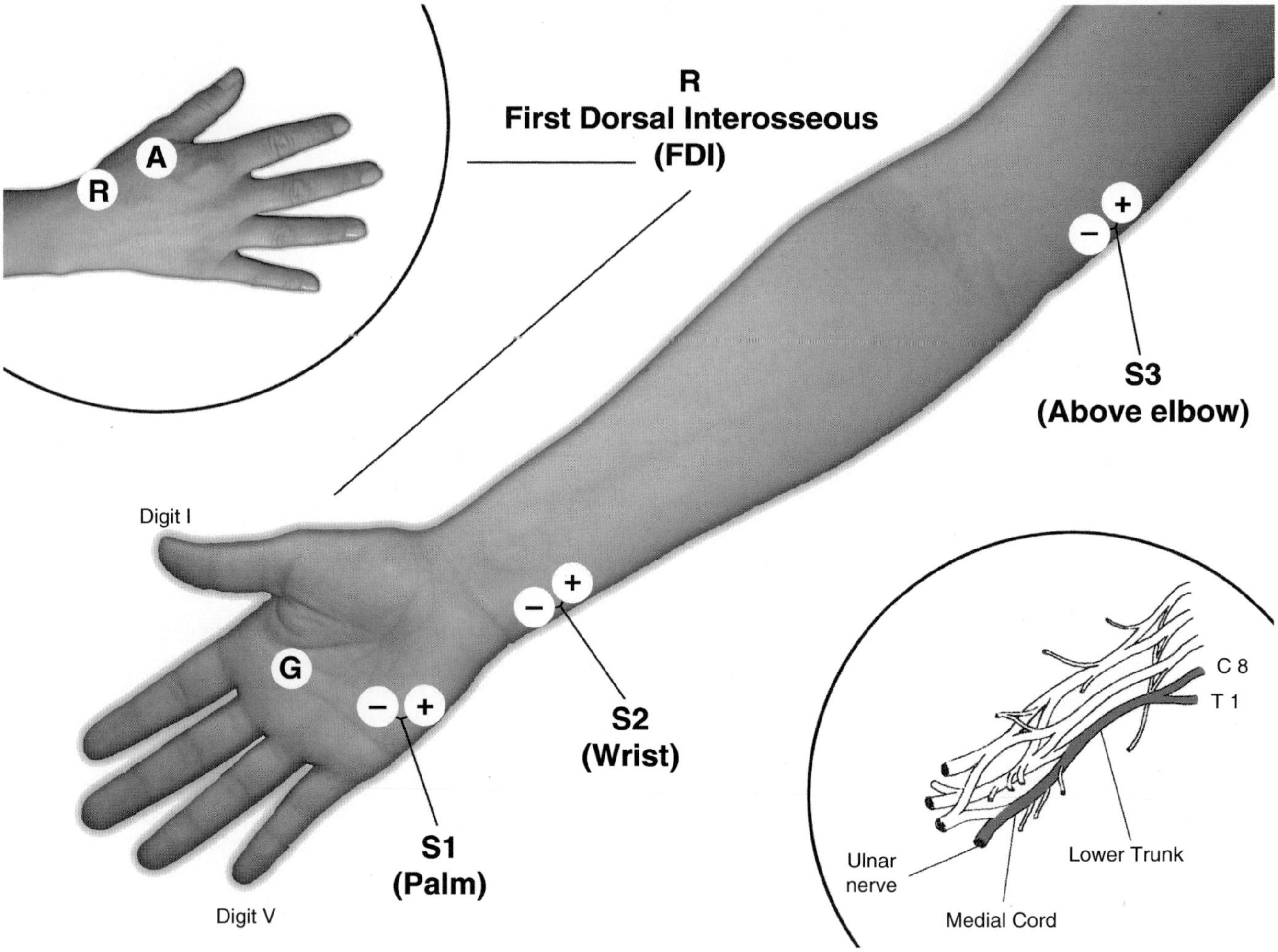

Typical waveform (palm,wrist, above elbow, axilla, Erb's point – FDI muscle):

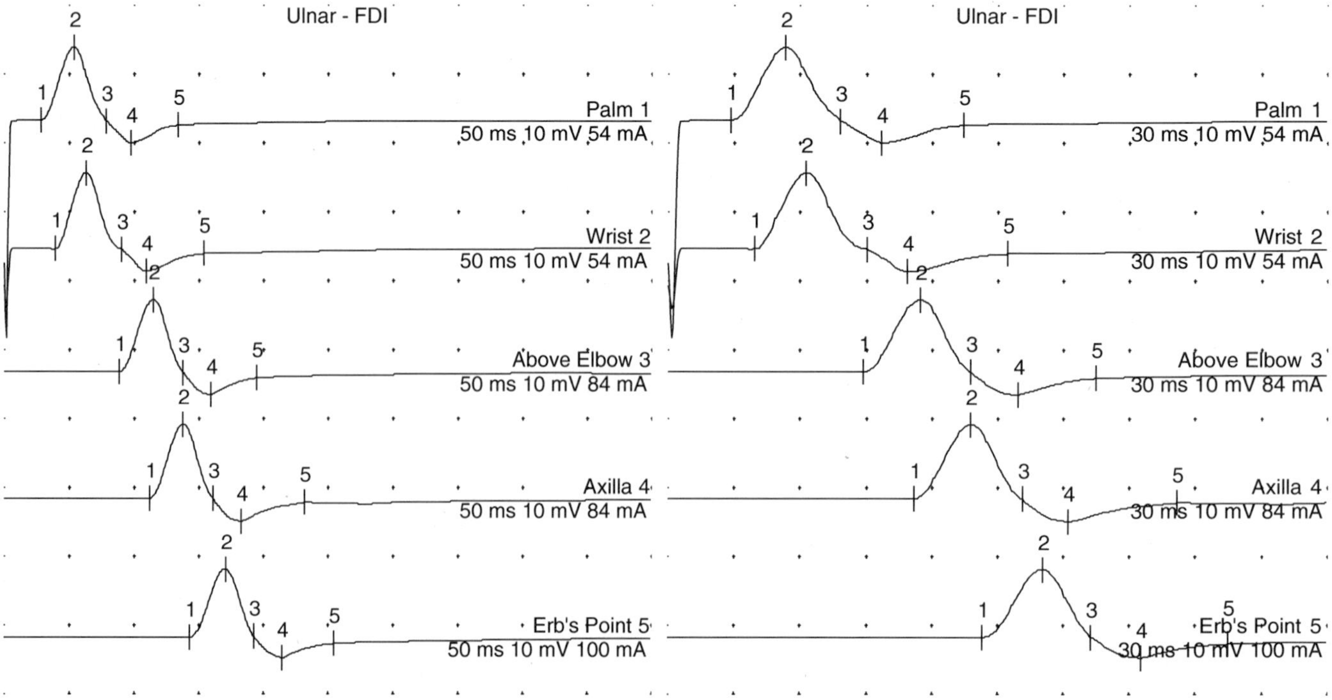

Sensitivity 10 mV/div, sweep speed 5 ms/div   Sensitivity 10 mV/div, sweep speed 3 ms/div

Fig. 1 Compound muscle action potentials (*CMAPs*) recorded at the hand from the FDI muscle, stimulation on the palm, at the wrist, above elbow, axilla, and at the Erb's point

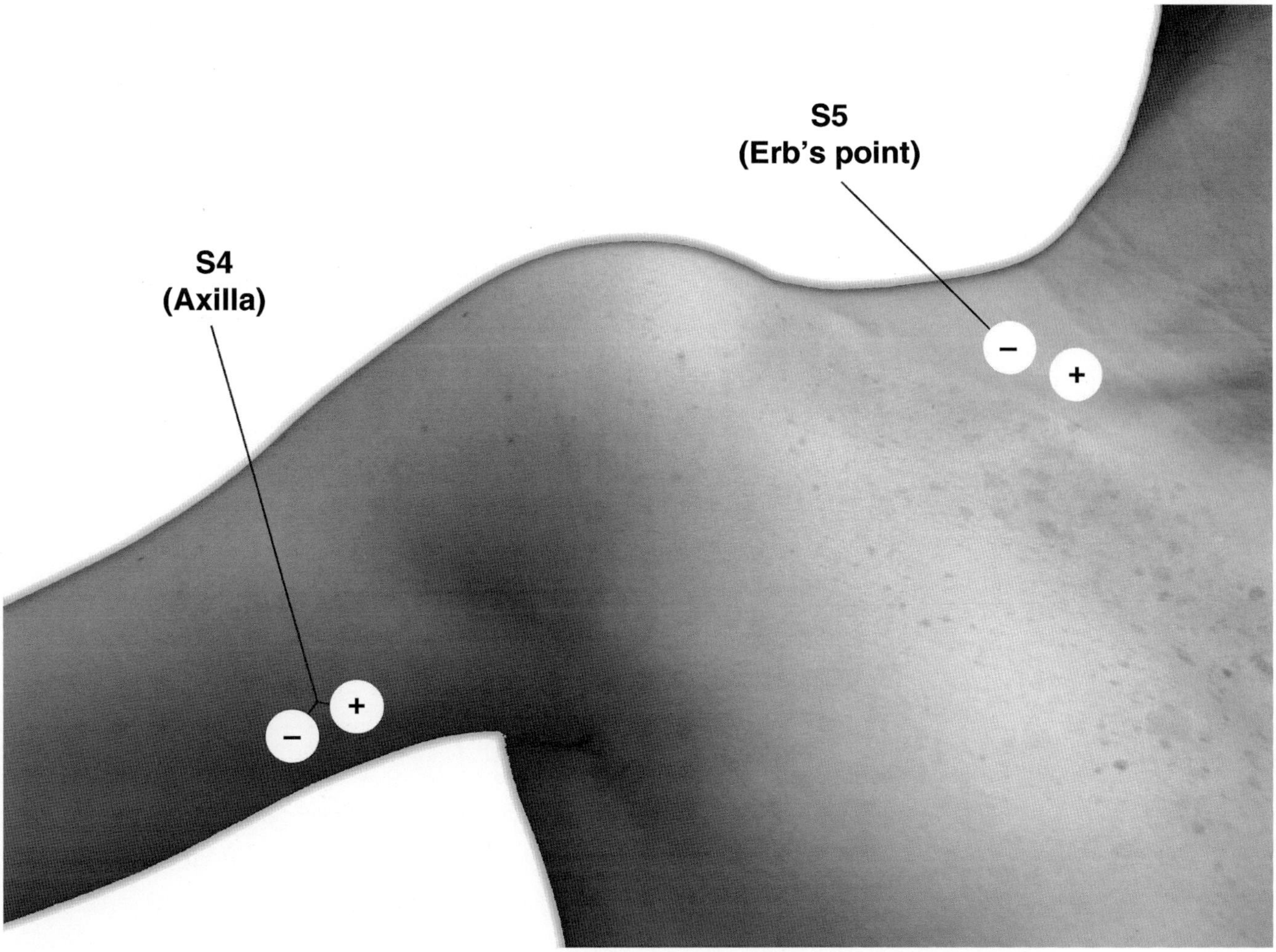

Fig. 2 Proximal stimulation points (S4 – axilla, S5 – Erb's point)

the early UNL stage, routine sensory and motor latency difference tests may be normal or difficult to interpret because the nerve lesions are recent and acute. For the author [1], only the detection of a conduction block can definitively locate the ulnar nerve lesion, and the FDI recording, adding palmar stimulation, can be reliably performed since the FDI is usually the most involved muscle innervated by the UMDB.

Recently, Seror [9] proposed a modified protocol recording from the FDI muscle and stimulating the ulnar nerve with a bar electrode at the palm, wrist, below the elbow, above the elbow, and at the axilla (Fig. 3). Between January 1997 and July 2011, he prospectively recorded clinical and electrophysiological data for 53 cases of UNW; the results were established according to normal data from 90 healthy subjects and compared with data from 100 prospectively studied cases of UNE (Tables 1, 2, and 3).

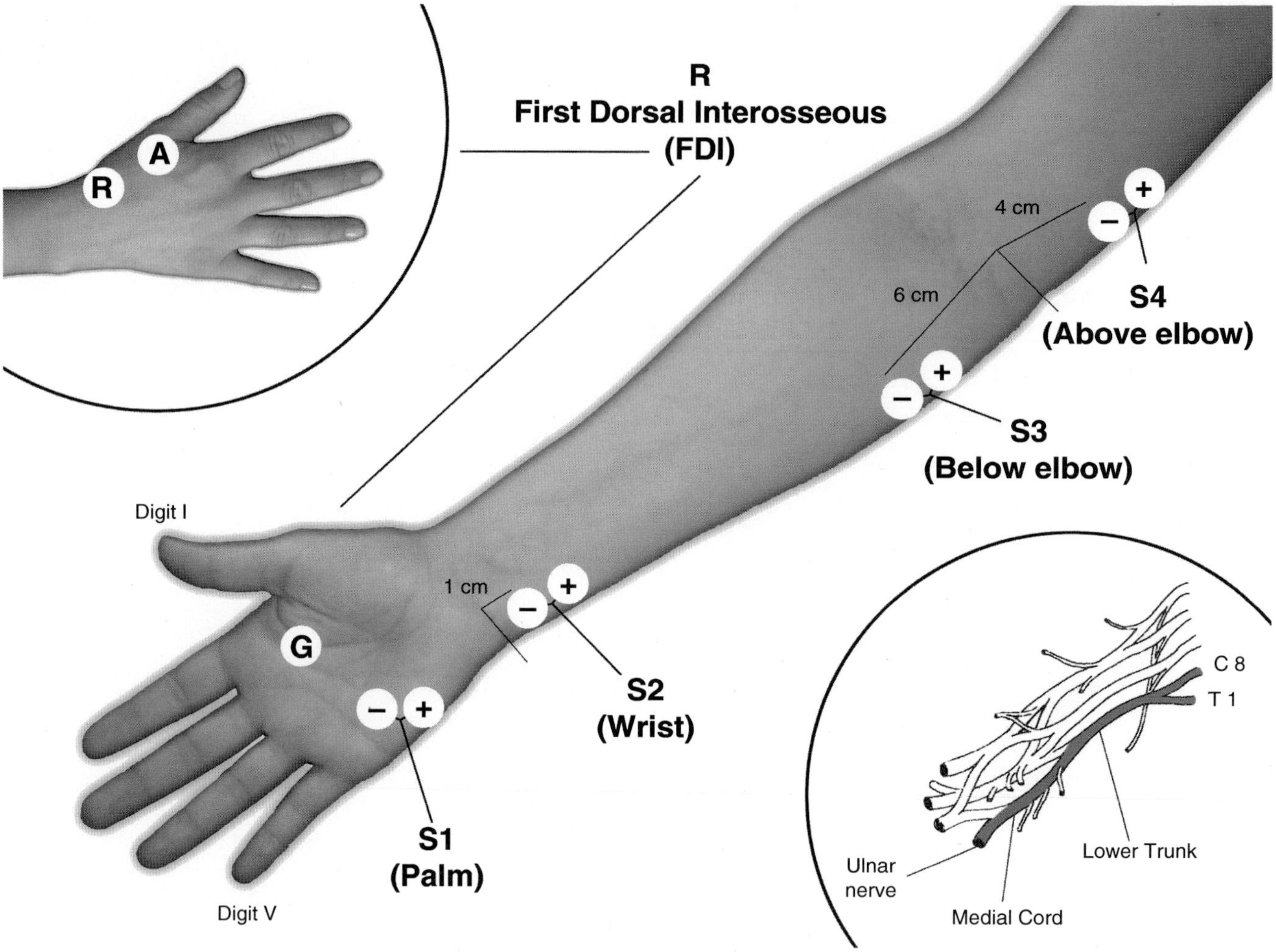

Fig. 3 Modified protocol by Seror [9], stimulation on the palm (S1), at the wrist (S2), below the elbow (S3), and above the elbow (S4)

Table 1 Normal values [9]

	Mean ± SD	Range	Cutoff
Wrist–ADM, distal motor latency (ms)	2.34 ± 0.26		>3.1
Wrist–ADM, CMAP amplitude (mV)	11.2 ± 2.1		< 6.1
Wrist–FDI, distal motor latency (ms)	2.64 ± 0.26		> 3.4
Wrist–FDI, CMAP amplitude (mV)	14.7 ± 3.3		< 6.2
Wrist–FDI – wrist–ADM, latency difference (ms)		−0.2 to 0.9	> 0.9
Wrist–FDI – wrist–APB, latency difference (ms)		−1.0 to 0.3	> 0.3
Wrist–FDI, side-to-side latency difference (ms)		0–0.5	> 0.5
Above elbow–below elbow, MNCV (m/s)	56 ± 6.2		< 47

Table 2 Pathological values, patients with UNW [9]

	Mean ± SD	Range
Wrist–ADM, distal motor latency (ms)	3.33 ± 1.1	
Wrist–ADM, CMAP amplitude (mV)	5.8 ± 3.4	
Wrist–FDI, distal motor latency (ms)	4.47 ± 1.8	
Wrist–FDI, CMAP amplitude (mV)	3.26 ± 3.3	
Wrist–FDI – wrist–ADM, latency difference (ms)		−0.6 to 5.7
Wrist–FDI – wrist–APB, latency difference (ms)		−1.5 to 7.6
Wrist–FDI, side-to-side latency difference (ms)		0–8.4
Above elbow–below elbow, MNCV (m/s)	52.5 ± 7.3	

Table 3 Pathological values, patients with UNE [9]

	Mean ± SD	Range
Wrist–ADM, distal motor latency (ms)	2.61 ± 0.25	
Wrist–ADM, CMAP amplitude (mV)	10.6 ± 2.4	
Wrist–FDI, distal motor latency (ms)	2.94 ± 0.3	
Wrist–FDI, CMAP amplitude (mV)	11.2 ± 3.3	
Wrist–FDI – wrist–ADM, latency difference (ms)		−0.1 to 0.8
Wrist–FDI – wrist–APB, latency difference (ms)		−1.0 to 0.1
Wrist–FDI, side-to-side latency difference (ms)		0–0.5
Above elbow–below elbow, MNCV (m/s)	37.3 ± 6.6	

Comment

Electrophysiological data found by Seror [9] are in general similar to those found in a previous study of 20 cases of UNW [10], demonstrating an high frequency of conduction block (CB) across the wrist (60 % versus 57 %) and a rarely associated UNE (10 % versus 8 %). Ulnar nerve conduction study across the wrist with palmar stimulation increases the diagnostic sensitivity from 66 to 90 %. CB at the wrist was the most definitive electrodiagnostic criteria that allowed locating the lesion at the wrist in 19 cases: (a) in 10 cases, CB was the only abnormal conduction test result (DML to FDI < 3.4 ms); (b) 7 cases showed mild DML to FDI impairment (DML range 3.4–4.0 ms, low CMAP amplitude ≤ 2 mV); and (c) 3 cases showed concomitant UNE and simply finding CB at the wrist definitively located the ulnar nerve lesion at the wrist. CB at the wrist in UNW was 2.7 times more common than at the elbow in UNE and was the only abnormal conduction test in 17 % of cases. For the author [9], considering CB presence or absence may help to determine the best treatment. Pathological waveforms and values in a patient with both ulnar neuropathy at the wrist (UNW) and a traumatic injury at the proximal arm and a patient with multifocal motor neuropathy (MMN) are reported (Figs. 4 and 5, respectively).

Pathological waveform (palm,wrist, above elbow, axilla, Erb's point – FDI muscle):

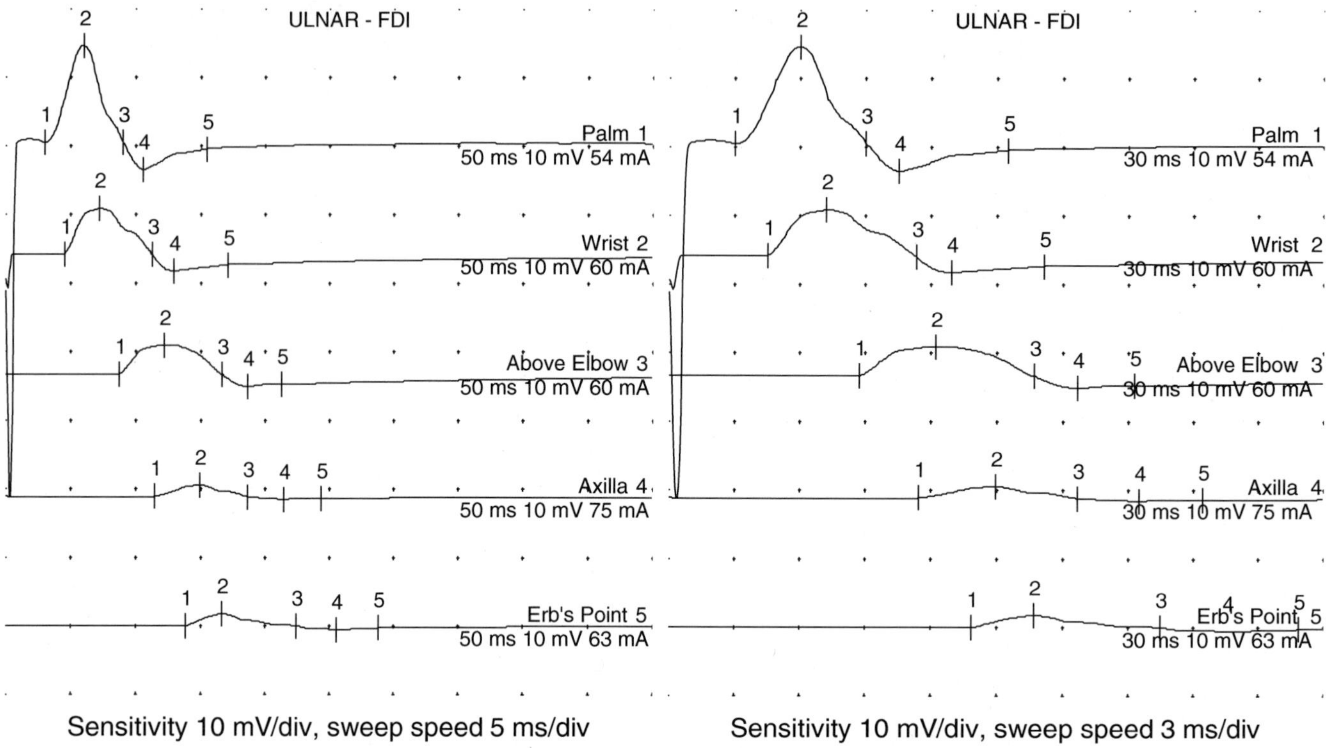

Onset latency (palm – FDI): 3.15 ms; **Onset latency** (wrist – FDI): 5.85 ms; **Onset latency** (above elbow – FDI): 10.70 ms; **Onset latency** (axilla elbow – FDI): 16.90 ms; **Onset latency** (Erb's point – FDI): 19.35 ms; **Peak latency** (palm – FDI): 5.85 ms; **Peak latency** (wrist – FDI): 8.90 ms; **Peak latency** (above elbow – FDI): 13.65 ms; **Peak latency** (axilla – FDI): 20.90 ms; **Peak latency** (Erb's point – FDI): 23.20 ms; **Onset to peak amplitude** (palm – FDI): 0.8 mV; **Onset to peak amplitude** (wrist – FDI): 0.8 mV; **Onset to peak amplitude** (above elbow – FDI): 1.0 mV; **Onset to peak amplitude** (axilla – FDI): 0.9 mV; **Onset to peak amplitude** (Erb's point – FDI): 1.0 mV; **Peak to peak amplitude** (palm – FDI): 1.2 mV; **Peak to peak amplitude** (wrist – FDI): 1.3 mV; **Peak to peak amplitude** (above elbow – FDI): 1.3 mV; **Peak to peak amplitude** (axilla – FDI): 1.1 mV; **Peak to peak amplitude** (Erb's point – FDI): 1.1 mV; **MNCV** (palm – wrist, 7 cm): 25.9 m/s; **MNCV** (wrist – above elbow, 19 cm): 39.2 m/s; **MNCV** (above elbow – axilla, 14 cm): 22.6 m/s; **MNCV** (axilla – Erb's point, 12 cm): 49.0 m/s

Fig. 4 Compound muscle action potentials (*CMAPs*) recorded at the hand from the FDI muscle in a patient with UNW and a traumatic injury at the proximal arm; *top to bottom*: stimulation on the palm, at the wrist, above elbow, axilla, and at the Erb's point

Pathological waveform (palm,wrist, above elbow, axilla, Erb's point – FDI muscle):

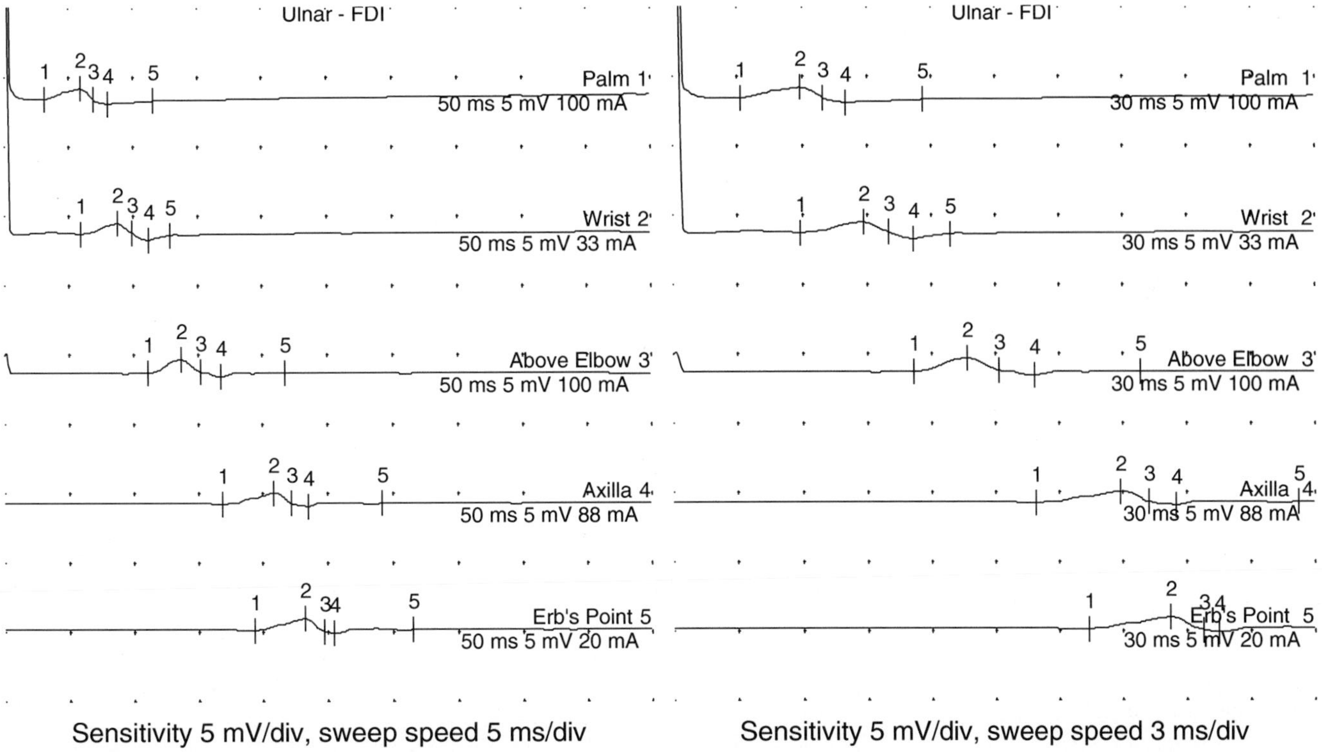

Onset latency (palm – FDI): 2.45 ms; **Onset latency** (wrist – FDI): 5.0 ms; **Onset latency** (above elbow – FDI): 8.50ms; **Onset latency** (axilla elbow – FDI): 10.50 ms; **Onset latency** (Erb's point – FDI): 8.50 ms; **Peak latency** (palm – FDI): 5.80 ms; **Peak latency** (wrist – FDI): 6.60 ms; **Peak latency** (above elbow – FDI): 10.65 ms; **Peak latency** (axilla–FDI): 12.90 ms; **Peak latency** (Erb's point – FDI): 12.90 ms; **Onset to peak amplitude** (palm – FDI): 3.0 mV; **Onset to peak amplitude** (wrist –FDI): 1.2 mV; **Onset to peak amplitude** (above elbow – FDI): 0.9 mV; **Onset to peak amplitude** (axilla – FDI): 0.9 mV; **Onset to peak amplitude** (Erb's point – FDI): 0.9 mV; **Peak to peak amplitude** (palm – FDI): 4.5 mV; **Peak to peak amplitude** (wrist – FDI): 1.8 mV; **Peak to peak amplitude** (above elbow – FDI): 1.3 mV; **Peak to peak amplitude** (axilla – FDI): 1.5 mV; **Peak to peak amplitude** (Erb's point – FDI): 1.5 mV; **MNCV** (palm – wrist, 7 cm): 27.5 m/s; **MNCV** (wrist – above elbow, 17 cm): 48.6 m/s; **MNCV** (above elbow – axilla, 10 cm): 50.0 m/s; **MNCV** (axilla – Erb's point, 10 cm): 50.0 m/s

Fig. 5 Compound muscle action potentials (*CMAPs*) recorded at the hand from the FDI muscle in a patient with a multifocal motor neuropathy (MMN); *top to bottom*: stimulation on the palm, at the wrist, above elbow, axilla, and at the Erb's point

References

1. Seror P (1999) Ulnar conduction block at the wrist. Arch Phys Med Rehabil 80:1346–1348
2. Seror P, Maisonobe T, Bouche P (2011) A new electrode placement for recording the compound motor action potential of the first dorsal interosseous muscle. Clin Neurophysiol 41:173–180
3. Seror P (1988) Comparison of the distal motor latency of the first dorsal interosseous with abductor pollicis brevis (report of 200 cases). Electromyogr Clin Neurophysiol 28:341–345
4. Shea JD, McClain EJR (1969) Ulnar nerve compression syndromes at and below the wrist. J Bone Joint Surg 51A:1095–1103
5. Wu JS, Morris JD, Hogan GR (1985) Ulnar neuropathy at the wrist: case report and review of literature. Arch Phys Med Rehabil 66:785–788
6. Kothari MJ, Preston DC, Logigian EL (1996) Lumbrical-interossei motor studies localize ulnar neuropathy at the wrist. Muscle Nerve 19:170–174
7. Seror P (1995) The value of the special motor and sensory tests for the diagnosis of benign and minor median nerve lesion at the wrist. Arch Phys Med Rehabil 74:124–129
8. Uncini A, Di Muzio A, Awad J et al (1993) Sensitivity of three median to ulnar comparative studies in diagnosis of mild ulnar carpal tunnel syndrome. Muscle Nerve 16:1366–1373
9. Seror P (2013) Electrophysiological pattern of 53 cases of ulnar nerve lesion at the wrist. Clin Neurophysiol 43:95–103
10. Cowdery SR, Preston DC, Herrmann DN et al (2002) Electrodiagnosis of ulnar neuropathy at the wrist. Conduction block versus traditional tests. Neurology 59:420–427

M38

Palm, Wrist – Hand; Palm, Wrist, Below Elbow, Above Elbow – Hand

Surface Recording Technique, Study from *First Dorsal Interosseous* (FDI) Muscle

Original Settings Sensitivity, sweep speed, low-frequency filter, high-frequency filter, duration of pulse, and the machine used were a NICOLET Viking IV and a DANTEC Keypoint.

Position All subjects were tested with the arm and hand resting comfortably on a pillow with the elbow bent at 90°.

Recording Using surface electrodes [1], a modified short-segment incremental study (SSIS) [2] across the wrist and a complete motor conduction study from the palm to the above-elbow site were performed recording the compound muscle action potentials (CMAPs) from the *first dorsal interosseous* (FDI) muscle (Fig. 1). The active recording electrode was placed over the belly of the FDI muscle and a reference electrode on the interphalangeal joint of the digit I. They placed the reference electrode on the digit I to avoid an initial positive deflection that may be seen when the reference electrode is placed on the digit II [3]. Since the first electrophysiologic studies of the ulnar neuropathy at the wrist (UNW), it has been clear that recordings from the FDI muscle are required for optimal electrodiagnostic localization of ulnar nerve lesions at or distal to the wrist and that routine nerve conduction studies to the *abductor digiti minimi* (ADM) muscle are often inadequate for this purpose. The ulnar nerve enters the distal ulnar tunnel about 2 cm proximal to the pisiform and then divides into a superficial sensory branch and a deep motor branch about 1 cm distal to the pisiform. More distally, the deep motor branch typically exits the distal ulnar tunnel 1–1.5 cm beyond the bifurcation by first passing under the fibrous arch of the hypothenar muscles, then giving off motor branches to the hypothenar muscles as it passes through or around them, and finally curving around the hook of the hamate leaving the remaining deep motor fibers to innervate the seven *interossei*, two lumbricals, the *adductor pollicis* (AP), and *flexor pollicis brevis* (FPB) muscles [4, 5]. The ground (G) electrode position was not specified in the text; the figure shows ground placed on the palm.

Stimulation The authors [1] performed a simplified version of the SSIS FDI study previously proposed by McIntosh et al. [2]; the ulnar nerve was stimulated at only two sites, one distal (palm) and one proximal (wrist) to the distal ulnar tunnel (Fig. 1). One stimulus site was approximately 4–5 cm distal to the distal wrist crease, just distal to the palpable hook of hamate, and the other was approximately 2–3 cm proximal to the same wrist crease. They chose a 4–5 cm distance to the distal wrist crease for the palm stimulation because anatomic studies have identified the distal ulnar tunnel to be 4–4.5 cm long, beginning at the proximal edge of the palmar carpal ligament just proximal to the pisiform bone and very near the distal wrist crease and ending at the point where the deep motor branch of the ulnar nerve exits the tunnel in the region of the hook of the hamate bone [6]. No movement of the arm or hand was observed, and costimulation of the median nerve at the wrist or the palm was avoided by assuring that the muscle twitch evoked by ulnar nerve stimulation at these sites did not have thenar component. The ulnar nerve was also studied along its course in the forearm by stimulation at four several sites [1]: on the palm, at the wrist, below the elbow, and above the elbow (Figs. 1 and 2).

Measurements Distal latency (ms) was measured from the stimulus onset to the onset of the initial deflection of the compound muscle action potential (CMAP). The amplitude (mV) of the CMAP was measured from the baseline to the peak of the negative deflection (negative-peak amplitude). By dividing the distance by time, the motor nerve conduction velocity (MNCV) along each nerve segment can be determined and measured in meter per second (m/s), and it

G. Gentili, M. Di Napoli, *The Ulnar Nerve*, DOI 10.1007/978-3-319-43292-2_62

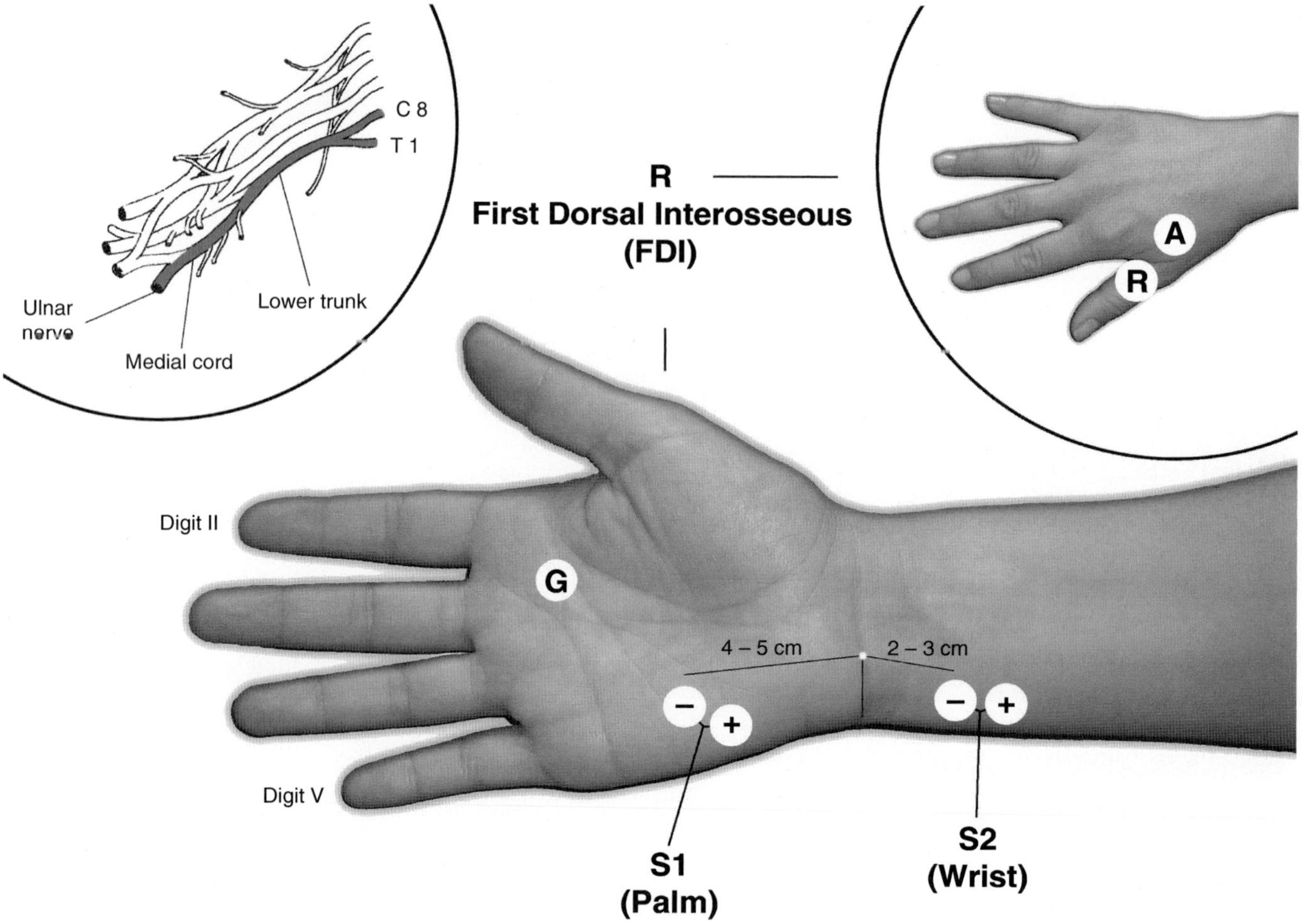

Typical waveform (palm, wrist – FDI muscle):

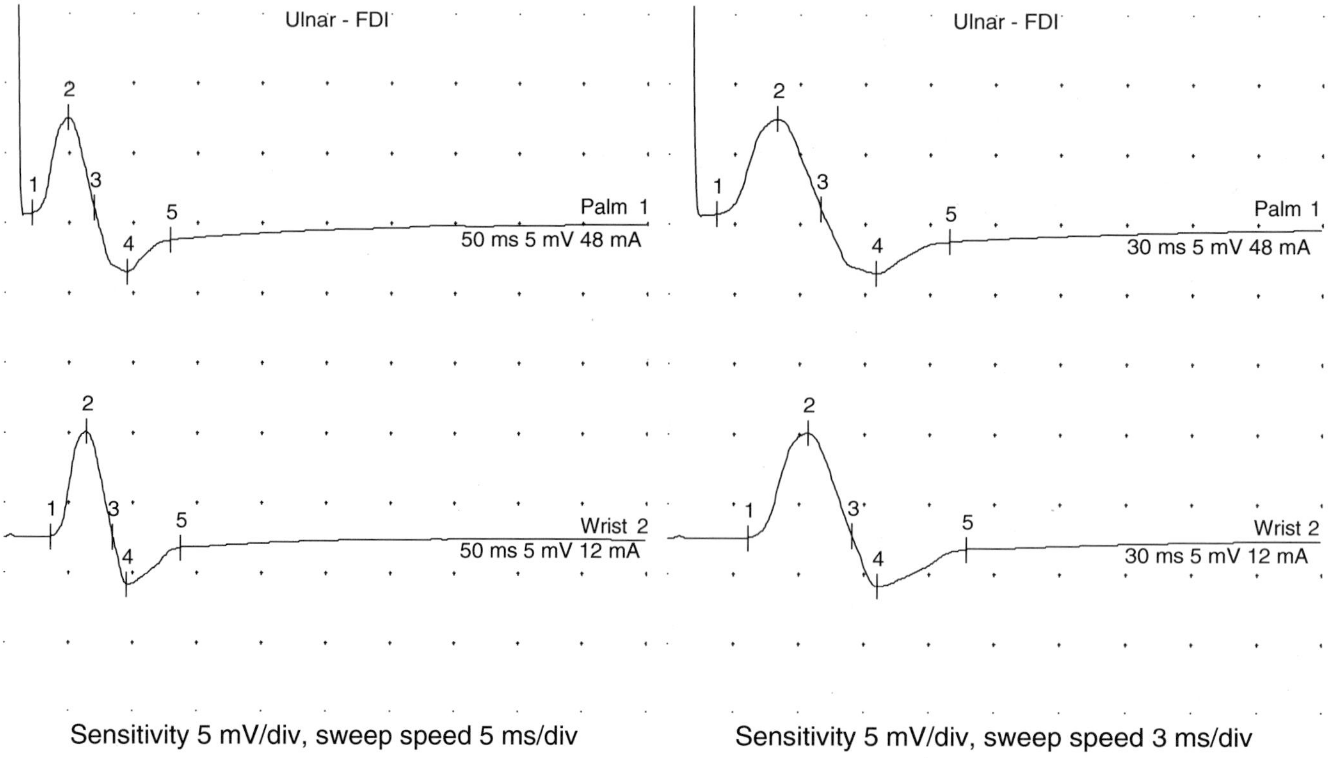

Fig. 1 Compound muscle action potentials (*CMAPs*) recorded at the hand from the FDI muscle, stimulation on the palm, and at the wrist

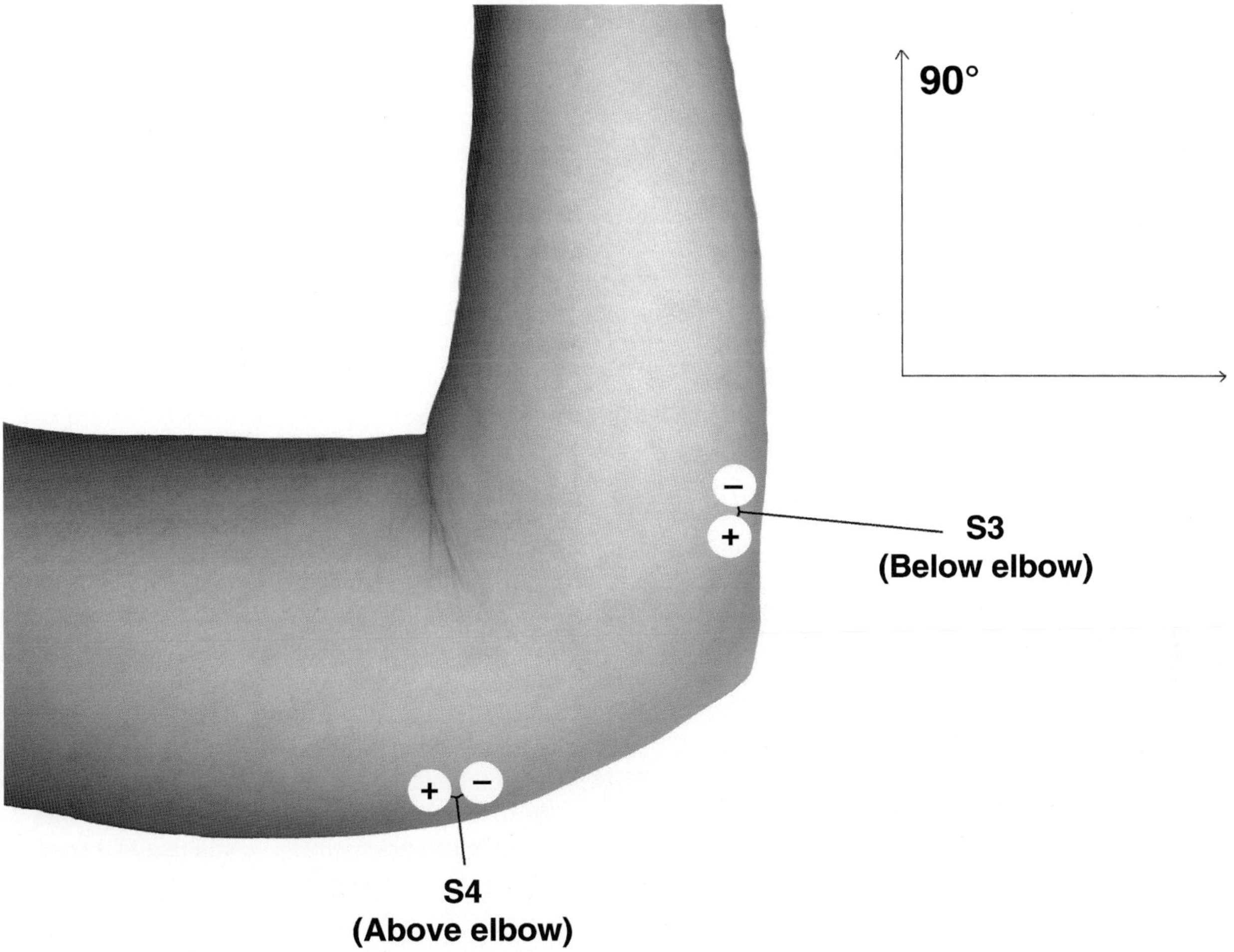

Typical waveform (palm, wrist, below elbow, above elbow – FDI muscle):

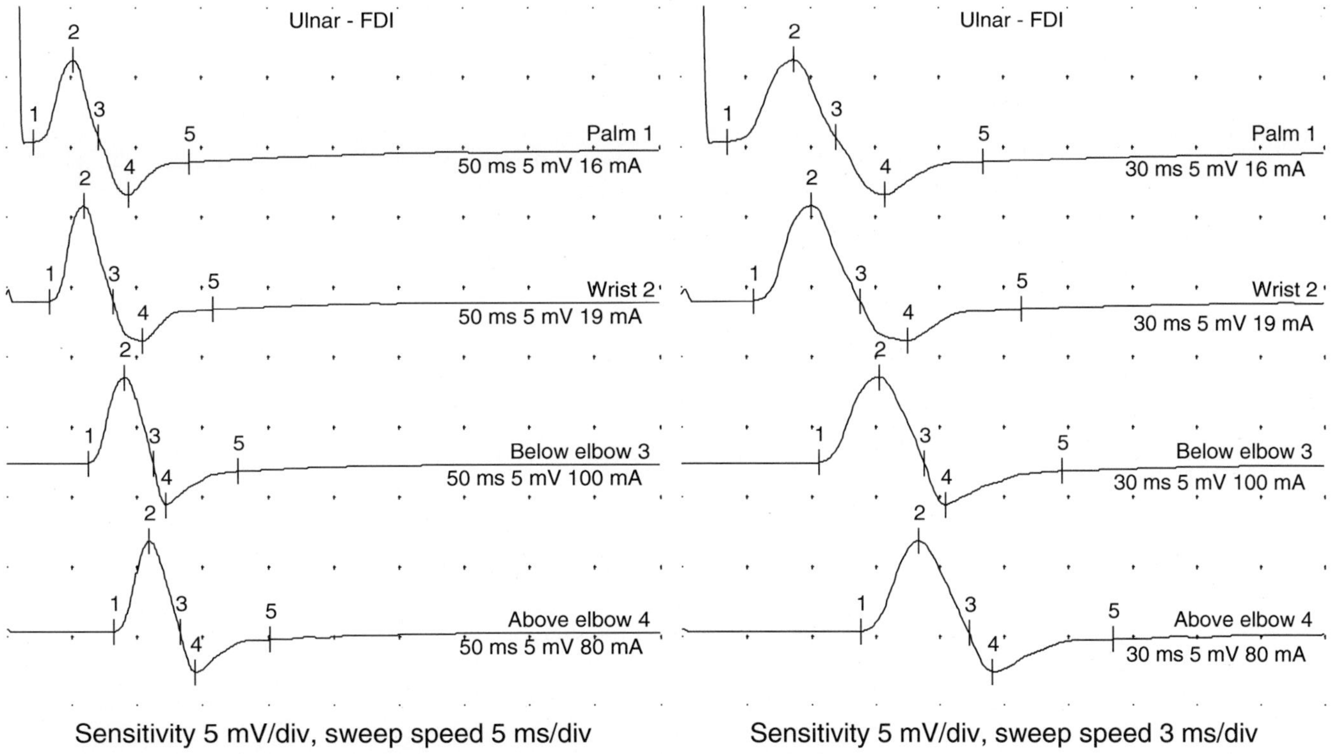

Fig. 2 Compound muscle action potentials (*CMAPs*) recorded at the hand from the FDI muscle, stimulation on the palm, at the wrist, below elbow, and above elbow

was calculated for palm–wrist, wrist–below elbow, and below-elbow–above-elbow segments. The addition of a single palmar stimulation site allowed the determination of the MNCV of FDI fibers that run through the distal ulnar tunnel, as well as the detection of conduction block across this ulnar nerve segment. The authors [1] observed that in the calculation of the palm–wrist MNCV, a linear course of the ulnar nerve is assumed, and the palm–wrist distance measurement is short (mean 7 cm in their study). The ulnar nerve makes a turn into the palm just distal to the hook of the hamate, and therefore if the nerve was stimulated beyond this point, the linear distance measurement would underestimate the curvilinear course of the nerve resulting in a spuriously slow calculated MNCV. For this reason, they perform all palm stimulations just distal to the hamate. The skin and room temperatures were not reported. Ulnar conduction block (CB) across the wrist was defined as per the American Association of Neuromuscular Electrodiagnostic Medicine (AANEM) consensus criteria for the across-elbow ulnar nerve segment, assuming <30 % temporal dispersion and a distal CMAP amplitude at least 20 % lower limit of normal [7]. Definite CB required a 50 % or greater drop in FDI CMAP amplitude and area across the wrist. Probable CB required a 40–49 % drop in amplitude with a 30–39 % drop in area. Clinical criteria for UNW were as follows: (1) ulnar hand muscle weakness with sparing of the deep flexors; (2) normal sensation, or sensory loss confined to digit V, or the ulnar aspect of digit IV, sparing the dorsum and palmar aspect of the hand; (3) an antecedent history of repetitive or unitary wrist trauma, MRI, or radiographic findings localizing an ulnar nerve at the wrist; and (4) no signs or symptoms of a generalized polyneuropathy, lower cervical radiculopathy, or lower trunk brachial plexopathy. Inclusion criteria for UNE were: (1a) ulnar hand muscle weakness sparing the deep flexors with ulnar distribution sensory loss or (1b) ulnar hand muscle weakness sparing the deep flexors but with sensory loss in the ulnar distribution including the dorsum of the hand; (2) absence of an antecedent history of repetitive or unitary wrist trauma or focal pain in the region of the ulnar nerve at the wrist; (3) presence of segmental slowing of >10 m/s of ulnar MNCV across the bent elbow; (4) no signs or symptoms of a generalized polyneuropathy; and (5) no clinical or electromyographic evidence of lower cervical radiculopathy or lower trunk brachial plexopathy. Patients with UNE were classified according to the severity of ulnar motor axon loss as determined by the amplitude of the CMAP evoked from the FDI muscle after stimulation on the palm. Patients with UNE with distal FDI CMAP amplitude below the normal range were classified as severe UNE, whereas those with an FDI CMAP amplitude within the normal range were classified as mild UNE. An effort was made by the authors to estimate the blocked versus degenerated FDI motor axons in our patients with UNW. Precise quantification of FDI motor axon loss and conduction block of FDI motor fibers in acute or subacute UNW would use the asymptomatic FDI CMAP as an indirect estimate of the normal number of FDI conducting axons [8]. However, the authors observed that an indirect estimate of the normal number of FDI conducting axons requires proof that the FDI CMAP amplitude and area are symmetric from side to side in normal subjects, that the contralateral limb is asymptomatic in patients with UNW, and that FDI conduction studies are routinely performed bilaterally. Since these requirements were not met in their patients with UNW and controls, they estimated the fraction of conducting, degenerated, and blocked axons in the UNW group as a whole using the mean FDI CMAP amplitude evoked from palm stimulation in the normal subjects as an indirect measure of presymptomatic FDI motor axon number in their patients with UNW. The following estimates were made by Cowdery et al. [1] who made the estimate of the fraction of blocked axons (FBA), fraction of degenerated axons (FDA), and fraction of conducting axons (FCA):

$$\text{FBA} = \frac{\text{mean UNW FDI CMAP amplitude (palm stimulation)} - \text{mean UNW FDI CMAP amplitude (wrist stimulation)}}{\text{mean normal FDI CMAP amplitude (palm stimulation)}}$$

$$\text{FDA} = \frac{\text{mean normal FDI CMAP amplitude (palm stimulation)} - \text{mean UNW FDI CMAP amplitude (palm stimulation)}}{\text{mean normal FDI CMAP amplitude (palm stimulation)}}$$

$$\text{FCA} = \frac{\text{mean UNW FDI CMAP amplitude (wrist stimulation)}}{\text{mean normal FDI CMAP amplitude (palm stimulation)}}$$

Normal values (Table 1) were obtained from 30 healthy subjects (20 men and 10 women, age range 21–80 years, mean age 51 years), and pathological values were obtained from 20 (Table 2) patients (15 men and 5 women, age range 22–77 years, mean age 54 years) with ulnar neuropathy at the wrist (UNW), 10 (Table 3) patients (6 men and 4 women, age range 32–70 years, mean age 46 years) with mild ulnar neuropathy at the elbow (UNE), and 10 (Table 4) patients (5 men and 5 women, age range 45–83 years, mean age 59 years) with severe ulnar neuropathy at the elbow (UNE). After confirming a normal distribution, the upper limits were defined as the mean+2.5SD, and the lower limits as the mean - 2.5SD. The authors [1] also performed a routine electrophysiologic evaluation in all normal controls and patients: (1) ulnar motor study recording from the *abductor digiti minimi* (ADM) muscle, with stimulation at the wrist and below- and above-elbow sites in the elbow flexed position; (2) median motor study recording from the *abductor pollicis brevis* (APB) muscle; (3) median–ulnar latency difference recording from second *lumbrical* (2 L) and palmar *interosseous* (INT) muscles [9]; (4) ulnar sensory antidromic study to digit V and median sensory antidromic study to digit II; (5) ulnar dorsal sensory conduction study; and (6) concentric needle examination.

Table 1 Normal values [1]

	Mean	Range
Palm–FDI, negative-peak amplitude (mV)	14.7	7.8–23.7
Wrist–FDI, negative-peak amplitude (mV)	14.2	7.0–22.9
FDI palm–wrist amplitude/palm amplitude	0.0	−0.1 to 0.1
Palm–wrist segment, MNCV (m/s)	58	42–73
Wrist–elbow segment, MNCV (m/s)	60	49–71
Palm–wrist, MNCV drop (%)	1.7	−17 to 20
Wrist–FDI, distal latency (ms)	3.7	3.1–4.4

Table 2 Pathological values [1]

	Mean	Range
Palm–FDI, negative-peak amplitude (mV)	7.2	0.8–16.2
Palm–FDI, negative-peak amplitude – % of degenerated motor axons	51	
Wrist–FDI, negative-peak amplitude (mV)	3.5	0.4–14.4
Wrist–FDI, negative-peak amplitude – % of degenerated motor axons	24	
Wrist–FDI, negative-peak amplitude – % of degenerated motor axons	25	
FDI–palm–wrist amplitude/palm amplitude	0.5	0.0–0.8
Palm–wrist segment, MNCV (m/s)	32	14–60
Wrist–elbow segment, MNCV (m/s)	54	40–72
Palm–wrist, MNCV drop (%)	22	−4 to 40
Wrist–FDI, distal latency (ms)	5.1	3.1–8.7

Table 3 Pathological values [1]

	Mean	Range
Palm–FDI, negative-peak amplitude (mV)	14.2	8.2–20.4
Wrist–FDI, negative-peak amplitude (mV)	13.8	8.2–19.5
FDI–palm–wrist amplitude/palm amplitude	0.0	0.0–0.1
Palm–wrist segment, MNCV (m/s)	57	44–75
Wrist–elbow segment, MNCV (m/s)	53	40–62
Palm–wrist, MNCV drop (%)	−4	−15 to 15
Wrist–FDI, distal latency (ms)	3.6	3.0–5.0

Table 4 Pathological values [1]

	Mean	Range
Palm–FDI, negative-peak amplitude (mV)	3.5	0.6–7.8
Wrist–FDI, negative-peak amplitude (mV)	3.5	0.5–7.8
FDI–palm–wrist amplitude/palm amplitude	0.0	−0.1 to 0.17
Palm–wrist segment, MNCV (m/s)	50	39–75
Wrist–elbow segment, MNCV (m/s)	51	37–67
Palm–wrist, MNCV drop (%)	3	−14 to 20
Wrist–FDI, distal latency (ms)	4.4	3.4–5.1

Comment

Cowdery et al. [1] found definite or probable conduction block (CB) at the wrist in 14 (70 %) patients with UNW, and slow wrist–palm FDI MNCV (<37 m/s) was found in 16 (80 %) patients with UNW. Overall, 19 (95 %) of the 20 patients with UNW had either CB (definite or probable) or slow MNCV (<37 m/s) of FDI motor fibers across the wrist. In patients with UNE, the authors observed neither CB nor greatly slowed FDI NCV (<37 m/s) across the wrist, and forearm MNCV (wrist–below-elbow segment) was greater than hand MNCV (wrist–palm segment) in patients with UNW, but not in patients with UNE or in normal subjects, in whom forearm and hand MNCV were not different. Although slow FDI MNCV (<37 m/s) and CB across the wrist were not observed in the patients with UNE, prolonged distal motor latency to FDI and ADM muscles (with low CMAP amplitudes) and slightly low forearm (wrist–below-elbow segment) and hand (wrist–palm) MNCVs were relatively common in severe UNE; the amplitude-dependent slowing of motor conduction was, for authors, likely due to drop out of fast conducting motor axons. The utility of recording from the FDI muscle in their study was the technical ease of palmar stimulation when recording from the relatively distant FDI muscle rather than from the adjacent ADM muscle and the resulting opportunity to demonstrate CB and slow MNCV of the FDI motor fibers across the lesion in the wrist–palm segment of the ulnar nerve. Moreover, adding a palmar stimulation site increased the diagnostic yield for UNW from 60 % (prolonged FDI distal motor latency) to 95 % (markedly slow FDI MNCV or CB across the wrist). There are technical limitations of palmar stimulation of the ulnar nerve that should be considered when performing their study. The determination of CB in UNW requires supramaximal stimulation of the ulnar nerve at the palm and wrist, and the intensities at the palm may be quite high especially for patients with callous formation. In such conditions, there is a risk of costimulating the median nerve at the palm and corecording a thenar CMAP superimposed an FDI CMAP resulting in a higher-amplitude response with palm than with wrist stimulation, simulating a CB. They [1] avoided this problem by ensuring that the cathodal stimulation point was medial in the palm at a distance from the thenar motor branch, that the muscle twitch evoked by the palm stimulation was of ulnar and not the median origin, and that the evoked CMAP did not have an initial positive deflection.

Seror et al. [10] proposed a modified protocol recording from the FDI muscle and stimulating the ulnar nerve with a bar electrode at the palm – 4–7 cm distal to the wrist stimulation (S1) – and at the wrist, 1 cm proximal to the palmar wrist crease (S2). Supramaximal stimuli were delivered (duration 0.2 ms) with a bar electrode (interelectrode distance 2.3 cm) until a CMAP of the same shape and amplitude was obtained. The intensity of stimulation at the wrist was rarely higher than 25 mA but was much higher in the palm because of the thickness of the hypothenar muscles and palmar skin. The recording was made with a pair of disposable surface electrodes (1.9 cm × 1.5 cm); the active recording electrode was placed dorsally over the belly of the FDI muscle at an equal distance to the metacarpophalangeal joints of digit I and digit II, with the reference electrode on the dorsal aspect of the wrist over the trapezium bone – 4.5 cm from the active recording electrode (Fig. 3). The authors determined the reference electrode placement after searching the location that provided the highest CMAP amplitude in 20 wrist controls (they compared three different interelectrode distances: 2, 3.5, and 4.5 cm). Normal values were obtained from 180 hands (Tables 5 and 6) of 90 healthy subjects (53 women and 37 men; mean age 43.3 years; age range 17–84 years, < 30 years (32 subjects), 30–39 years (44 subjects), 40–49 years (44 subjects), 50–59 years (38 subjects), 60–69 years (16 subjects), > 70 years (6 subjects)).

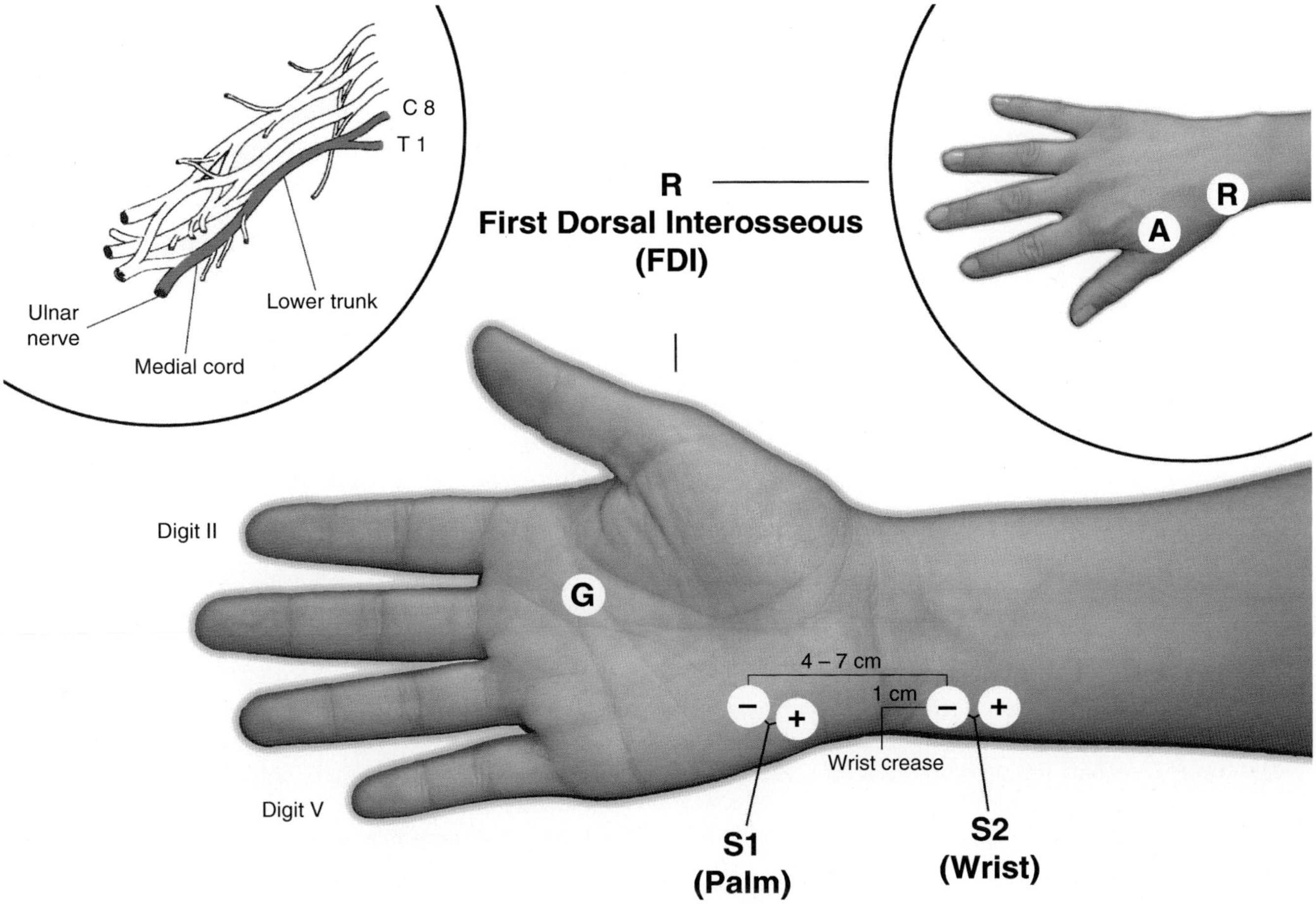

Fig. 3 Modified protocol by Seror [10], stimulation on the palm (S1) and at the wrist (S2)

Table 5 Normal values [10]

	Mean ± SD	Range
Wrist–FDI, distal motor latency (ms) – all subjects	2.65 ± 0.26	2.1–3.4
Wrist–FDI, CMAP amplitude (mV) – all subjects	14.7 ± 3.3	6.2–25.5
Wrist–FDI, side-to-side latency difference (ms) – all subjects	0.11 ± 0.12	0–0.5
Palm–FDI, distal motor latency (ms) – all subjects	1.83 ± 0.24	1.6–2.2
Palm–FDI, CMAP amplitude (mV) – all subjects	14.9 ± 3.4	6.3–26.1

Table 6 Normal values [10]

	Mean ± SD
Wrist–FDI, distal motor latency (ms) – < 30 years	2.68 ± 0.24
Wrist–FDI, CMAP amplitude (mV) – < 30 years	16.38–3.6
Wrist–FDI, distal motor latency (ms) – 30–39 years	2.56 ± 0.26
Wrist–FDI, CMAP amplitude (mV) – 30–39 years	15.45 ± 2.9
Wrist–FDI, distal motor latency (ms) – 40–49 years	2.60 ± 0.26
Wrist–FDI, CMAP amplitude (mV) – 40–49 years	14.05 ± 3.1
Wrist–FDI, distal motor latency (ms) – 50–59 years	2.76 ± 0.27
Wrist–FDI, CMAP amplitude (mV) – 50–59 years	14.77 ± 2.7
Wrist–FDI, distal motor latency (ms) – 60–69 years	2.68 ± 0.26
Wrist–FDI, CMAP amplitude (mV) – 60–69 years	12.81 ± 1.9
Wrist–FDI, distal motor latency (ms) – > 70 years	2.65 ± 0.28
Wrist–FDI, CMAP amplitude (mV) – > 70 years	9.16 ± 2.5

Comment

Seror et al. [10] compared three methods for FDI recording: their new electrode placement (with the reference electrode on the dorsal aspect of the wrist over the trapezium bone – 4.5 cm from the active recording electrode over the FDI muscle), the digit I reference method proposed by Cowdery et al. [1], and the digit V reference method by Prahlow and Buschbacher [11]. They demonstrated that the DML was significantly shorter with the new method (2.6 ± 0.26 ms) than with the second (3.34 ± 0.34 ms) and third (2.9 ± 0.38 ms). The authors also observed, using their method, the smallest SD and the absence of positive initial deflection at the onset of the FDI CMAP. The small SD was one of the most important contributions of their method, because the less the SD, the more homogeneous are the results and the more reproducible is the method; the absence of a positive initial deflection is the other major contribution provided by their method. In fact, the occurrence of a positive initial deflection frequently occurs when the reference electrode is fixed at the base of the digit II – on the second metacarpophalangeal joint [12, 13] – and using digit II provides a short distal motor latency when the latency is measured at the onset of the positive deflection [13]. Cowdery et al. [1] observed that when the reference electrode is fixed on digit I – on the first metacarpophalangeal joint – the CMAP appears not to have a positive initial deflection but has a longer distal motor latency caused, for Seror et al. [10], by a small positive initial deflection which can be seen only with a gain of 0.5 mV/division (35 of 40 hands). Based on their work [10], and with the strict respect of the location of the pair of recording electrodes, the authors recommended that: (1) the distal motor latency to the FDI muscle can be considered abnormal when it exceeds 3.4 ms (mean + 3SD) if the CMAP amplitude is normal (specificity 100 %); (2) the conduction of the ulnar deep motor branch at the wrist is abnormal with distal motor latency to the FDI muscle >0.3 ms than distal motor latency to the APB, > 1 ms than distal motor latency to the ADM muscle, and >0.5 ms that of the contralateral distal motor latency to the FDI muscle, if the CMAP amplitude is normal (specificity 100 %); and (3) FDI CMAP amplitude <6.2 mV is abnormal (specificity 98 %). But this finding has no specific significance concerning the location of the nerve lesion. Pathological waveforms and values in a patient with UNW (Type I, Fig. 4), in which the lesion occurred just above or within the proximal end of Guyon's canal involving both deep and superficial branches of the ulnar nerve, and in a patient with UNW (Type III, Fig. 5), in which the lesion mainly involved the deep branch of the ulnar nerve, and in a patient with severe UNE (Fig. 6), are reported.

Pathological waveform (palm, wrist, below elbow, above elbow – FDI muscle):

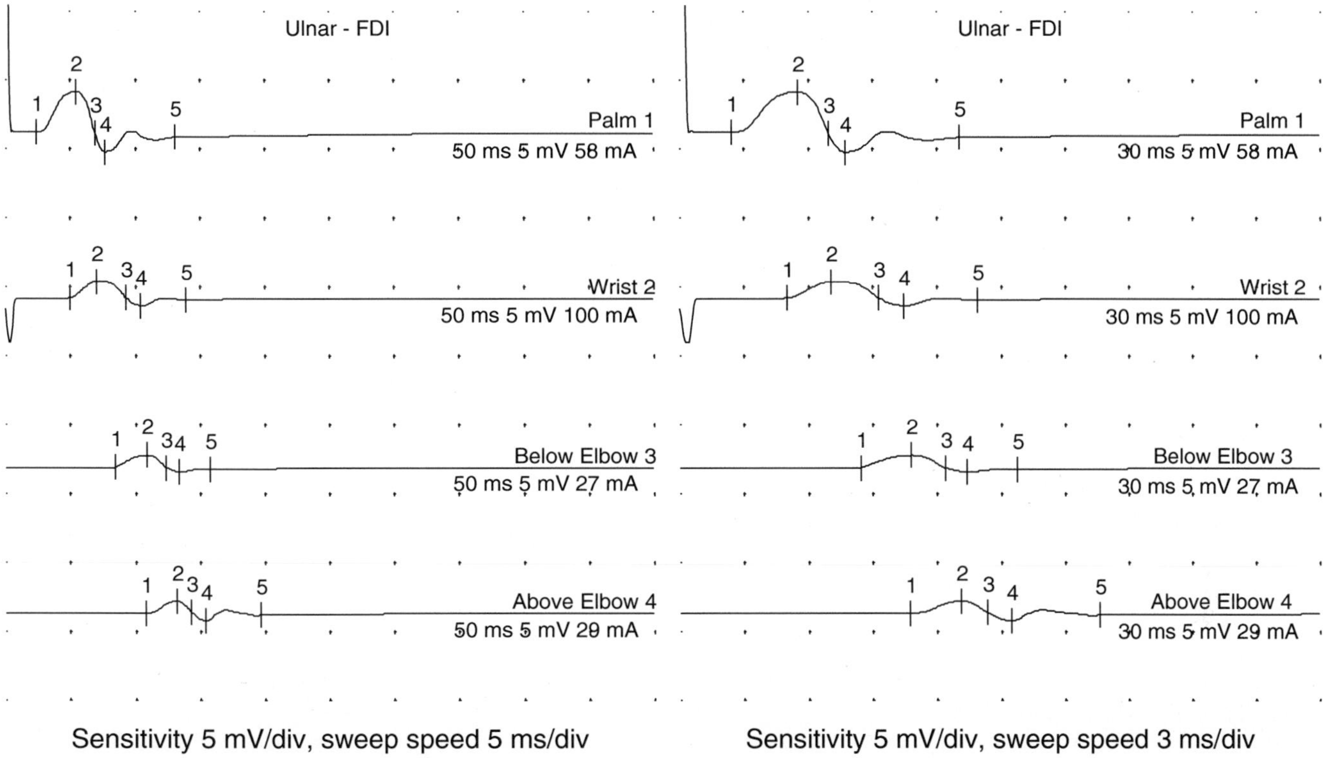

Onset latency (palm – FDI): 2.45 ms; **Onset latency** (wrist – FDI): 5.0 ms; **Onset latency** (below elbow – FDI): 8.50 ms; **Onset latency** (above elbow – FDI): 10.50 ms; **Peak latency** (palm – FDI): 5.80 ms; **Peak latency** (wrist – FDI): 6.60 ms; **Peak latency** (below elbow – FDI): 10.65 ms; **Peak latency** (above elbow – FDI): 12.90 ms; **Onset to peak amplitude** (palm – FDI): 3.0 mV; **Onset to peak amplitude** (wrist – FDI): 1.2 mV; **Onset to peak amplitude** (below elbow – FDI): 0.9 mV; **Onset to peak amplitude** (above elbow – FDI): 0.9 mV; **Peak to peak amplitude** (palm – FDI): 4.5 mV; **Peak to peak amplitude** (wrist – FDI): 1.8 mV; **Peak to peak amplitude** (below elbow – FDI): 1.3 mV; **Peak to peak amplitude** (above elbow – FDI): 1.5 mV; **MNCV** (palm – wrist, 7 cm): 27.5 m/s; **MNCV** (wrist – below elbow, 17 cm): 48.6 m/s; **MNCV** (below elbow – above elbow, 10 cm): 50.0 m/s

Fig. 4 Compound muscle action potentials (*CMAPs*) recorded at the hand from the FDI muscle in a patient with UNW (complete focal Type I lesion with a combination of axon loss and conduction block of the motor fibers innervating the FDI muscle); *top to bottom*: stimulation on the palm, at the wrist, below the elbow, and above the elbow

Pathological waveform (palm, wrist – FDI muscle):

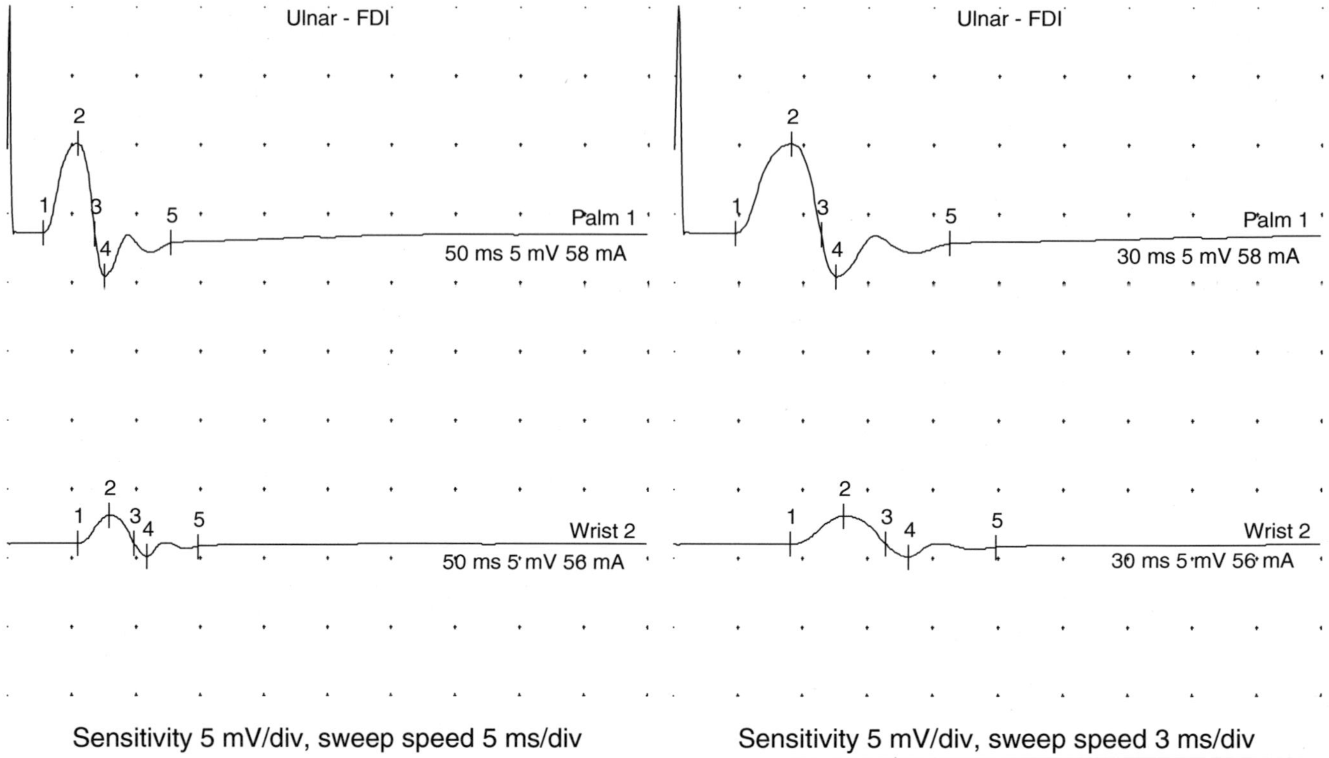

Onset latency (palm – FDI): 2.80 ms; **Onset latency** (wrist – FDI): 5.45 ms; **Peak latency** (palm – FDI): 5.55 ms; **Peak latency** (wrist – FDI): 7.60 ms; **Onset to peak amplitude** (palm – FDI): 6.5 mV; **Onset to peak amplitude** (wrist – FDI): 2.1 mV; **Peak to peak amplitude** (palm – FDI): 9.7 mV; **Peak to peak amplitude** (wrist – FDI): 3.1 mV; **MNCV** (palm – wrist, 8 cm): 30.2 m/s

Fig. 5 Compound muscle action potentials (*CMAPs*) recorded at the hand from the FDI muscle in a patient with UNW (complete focal Type I lesion with a combination of axon loss and conduction block of the motor fibers innervating the FDI muscle), stimulation on the palm (*upper trace*), and of the wrist (*lower trace*)

Pathological waveform (palm, wrist, below elbow, above elbow – FDI muscle):

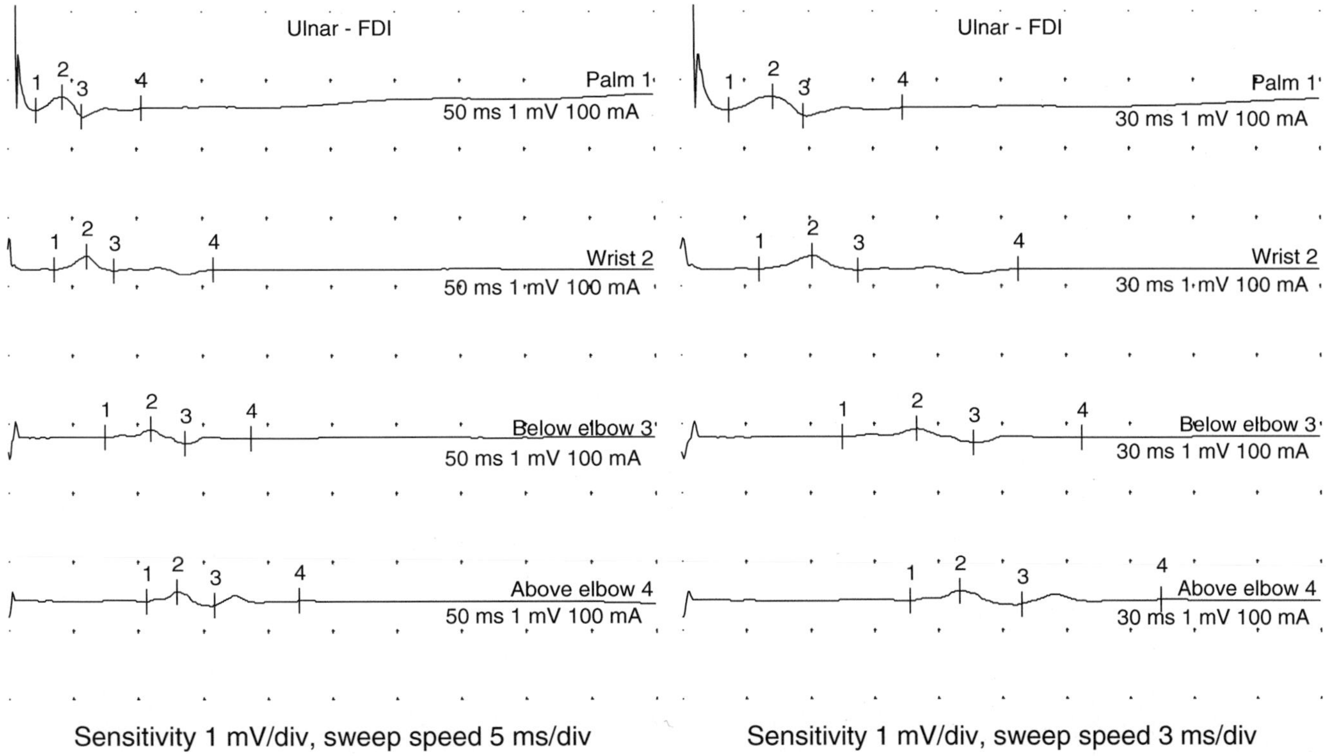

Onset latency (palm – FDI): 2.25 ms; **Onset latency** (wrist – FDI): 3.65 ms; **Onset latency** (below elbow – FDI): 7.50 ms; **Onset latency** (above elbow – FDI): 10.65 ms; **Peak latency** (palm – FDI): 4.50 ms; **Peak latency** (wrist – FDI): 6.20 ms; **Peak latency** (below elbow – FDI): 11.75 ms; **Peak latency** (above elbow – FDI): 13.05 ms; **Onset to peak amplitude** (palm – FDI): 0.2 mV; **Onset to peak amplitude** (wrist – FDI): 0.2 mV; **Onset to peak amplitude** (below elbow – FDI): 0.1 mV; **Onset to peak amplitude** (above elbow – FDI): 0.1 mV; **Peak to peak amplitude** (palm – FDI): 0.2 mV; **Peak to peak amplitude** (wrist – FDI): 0.2 mV; **Peak to peak amplitude** (below elbow – FDI): 0.2 mV; **Peak to peak amplitude** (above elbow – FDI): 0.1 mV; **MNCV** (palm – wrist, 7 cm): 50.0 m/s; **MNCV** (wrist – below elbow, 20 cm): 51.9 m/s; **MNCV** (below elbow – above elbow, 10 cm): 31.7 m/s

Fig. 6 Compound muscle action potentials (*CMAPs*) recorded at the hand from the FDI muscle in a patient with severe UNE; *top to bottom*: stimulation on the palm, of the wrist, below the elbow, and above the elbow

References

1. Cowdery SR, Preston DC, Herrmann DN et al (2002) Electrodiagnosis of ulnar neuropathy at the wrist. Conduction block versus traditional tests. Neurology 59:420–427
2. McIntosh KA, Preston DC, Logigian EL (1998) Short-segment incremental studies to localize ulnar nerve entrapment at the wrist. Neurology 50:303–306
3. Preston DC, Shapiro BE (1998) Electromyography and neuromuscular disorders: clinical electrophysiologic correlations. Butterworth-Heinemann, Newton
4. Simpson JA (1956) Electrical signs in the diagnosis of carpal tunnel and related syndromes. J Neurol Neurosurg Psychiatry 19:275–280
5. Ebeling P, Gilliatt RW, Thomas PK (1960) A clinical and electrical study of ulnar nerve lesions in the hand. J Neurol Neurosurg Psychiatry 23:1–9
6. Gross MS, Gelberman RH (1985) The anatomy of the distal ulnar tunnel. Clin Orthop 196:238–247
7. Olney RK (1999) Consensus criteria for the diagnosis of partial conduction block. Muscle Nerve 22:S225–S229
8. Watson BV, Brown WF (1992) Quantitation of axon loss and conduction block in acute radial nerve palsies. Muscle Nerve 15:768–773
9. Kothari MJ, Preston DC, Logigian EL (1996) Lumbrical-interossei motor studies localize ulnar neuropathy at the wrist. Muscle Nerve 19:170–174
10. Seror P, Maisonobe T, Bouche P (2011) A new electrode placement for recording the compound motor action potential of the first dorsal interosseous muscle. Clin Neurophysiol 41:173–180
11. Prahlow ND, Buschbacher RM (2006) Ulnar nerve motor conduction to the first dorsal interosseous muscle. J Long Term Eff Med Implants 16:333–340
12. Bhala RP, Goodgold J (1968) Motor conduction in the deep palmar branch of the ulnar nerve. Arch Phys Med Rehabil 49:460–466
13. Olney RK, Wilbourn AJ (1985) Ulnar nerve conduction study of the first dorsal interosseous muscle. Arch Phys Med Rehabil 66:16–18

Below Elbow to Above Elbow – Forearm; Below Elbow to Above Elbow – Hand

M39

Surface Recording Technique, Study from *Flexor Carpi Ulnaris* (FCU) and *Abductor Digiti Minimi* (ADM) Muscles

Original Settings Sensitivity was 2–5 mV/division, sweep speed was 3 ms/division, low-frequency filter was 10 Hz, high-frequency filter was 2 kHz, duration of pulse was 0.3 ms, rate of pulse was 1/s, and the machine used was a DANTEC Counterpoint or Keypoint electromyography (EMG) machine.

Position Following the text, short-segment incremental studies were performed with the elbow flexed to 70° (full extension: 0°).

Recording Using surface electrodes, compound muscle action potentials (CMAPs) were recorded from the *flexor carpi ulnaris* (FCU) and *abductor digiti minimi* (ADM) muscles [1]. For FCU muscle recordings, the active electrode (A) was placed over the belly of the muscle – 7 cm distal to station 0 (the line joining the olecranon and medial epicondyle), and the reference electrode was placed distally over the bony prominence of the ulnar head (Fig. 1). Occasionally, the active recording electrode position was shifted 1 cm proximally or distally to achieve optimal FCU CMAP recordings with a distinctly negative waveform. For ADM muscle recordings, the active electrode (A) was placed over the belly of the muscle, and the reference electrode was placed distally at the base of the digit V (Fig. 2). The ground (G) electrode position was not specified in the text; the figure shows ground placed on the forearm (Fig. 1).

Stimulation Short-segment incremental studies (SSIS) were performed using segments of 1 cm at nine different positions distal and proximal to station 0 (Fig. 1). The respective stations were –4, –3, –2, –1 (distal to elbow), 0, +1, +2, +3, +4 (proximal to elbow). Supramaximal stimulation was achieved with single pulses, 0.3 ms in duration. Special care was taken to exclude CMAPs obtained from overstimulation (virtual cathode effect); only CMAPs with similar morphology and distinct onset were accepted to facilitate onset latency determination. The nine stimulation points, set at the 1 cm interval, were used for both FCU and ADM muscle recordings. Stimulation intensities were not significantly different between studies involving FCU and ADM muscles.

Measurements Latency (ms) difference between adjacent stations was measured, and care was taken to obtain optimal CMAP waveform. The authors performed SSIS recording from the FCU muscle, and SSIS was repeated with ADM muscle recording in all subjects. For 25 control subjects, the respective upper normal limits in latency difference for each inching segment (mean + 2SD) were 0.4, 0.4, 0.32, 0.35, 0.38, 0.39, 0.28, and 0.31 ms for the FCU (distal to proximal segments). For the ADM, the values were 0.33, 0.27, 0.34, 0.40, 0.42, 0.39, 0.36, and 0.31 ms, respectively. Latency change over 1 cm segment ≥0.50 ms was considered abnormal (mean + 2SD) for both muscles.

Comment

Lo et al. [1] evaluated 16 consecutive patients (12 men and 4 women, 17 hands) with symptoms of ulnar neuropathy at the elbow (UNE), and they found conduction abnormalities between the +2 and –2 stations. Using the clinical etiology as the reference standard, 13 of 14 hands were localized with SSIS recording from the FCU muscle (sensitivity 93 %), compared with 10 of 14 hands recording from the ADM muscle (sensitivity 71.4 %). For the authors, in the three hands with abnormal SSIS values from FCU recordings only, isolated nerve fascicles innervating the ADM muscle may be spared, whereas those innervating the FCU were specifically affected. In the FCU study, distance between stimulation points and active recording electrode was shorter than ADM study, and they found SSIS from FCU muscle more sensitive than ADM muscle.

G. Gentili, M. Di Napoli, *The Ulnar Nerve*, DOI 10.1007/978-3-319-43292-2_63

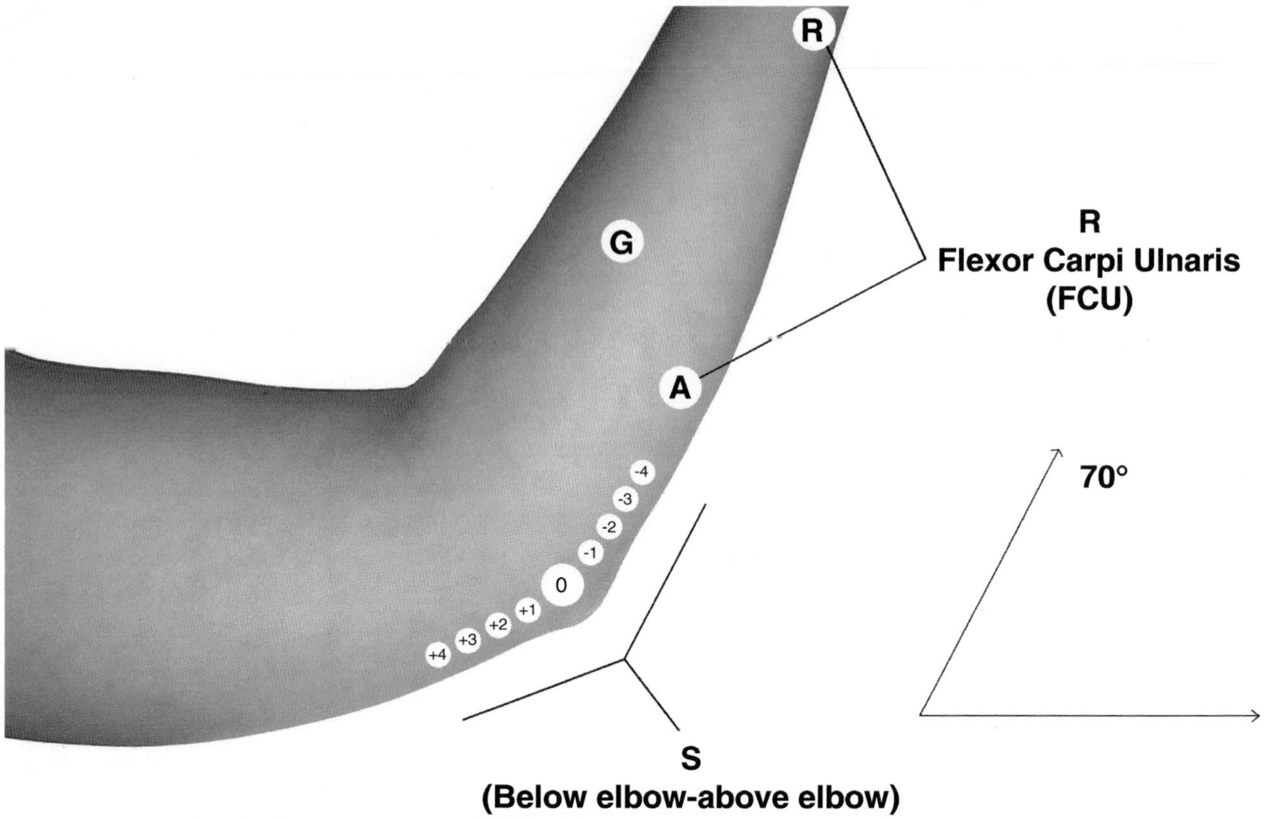

Typical waveform (below elbow to above elbow – FCU muscle):

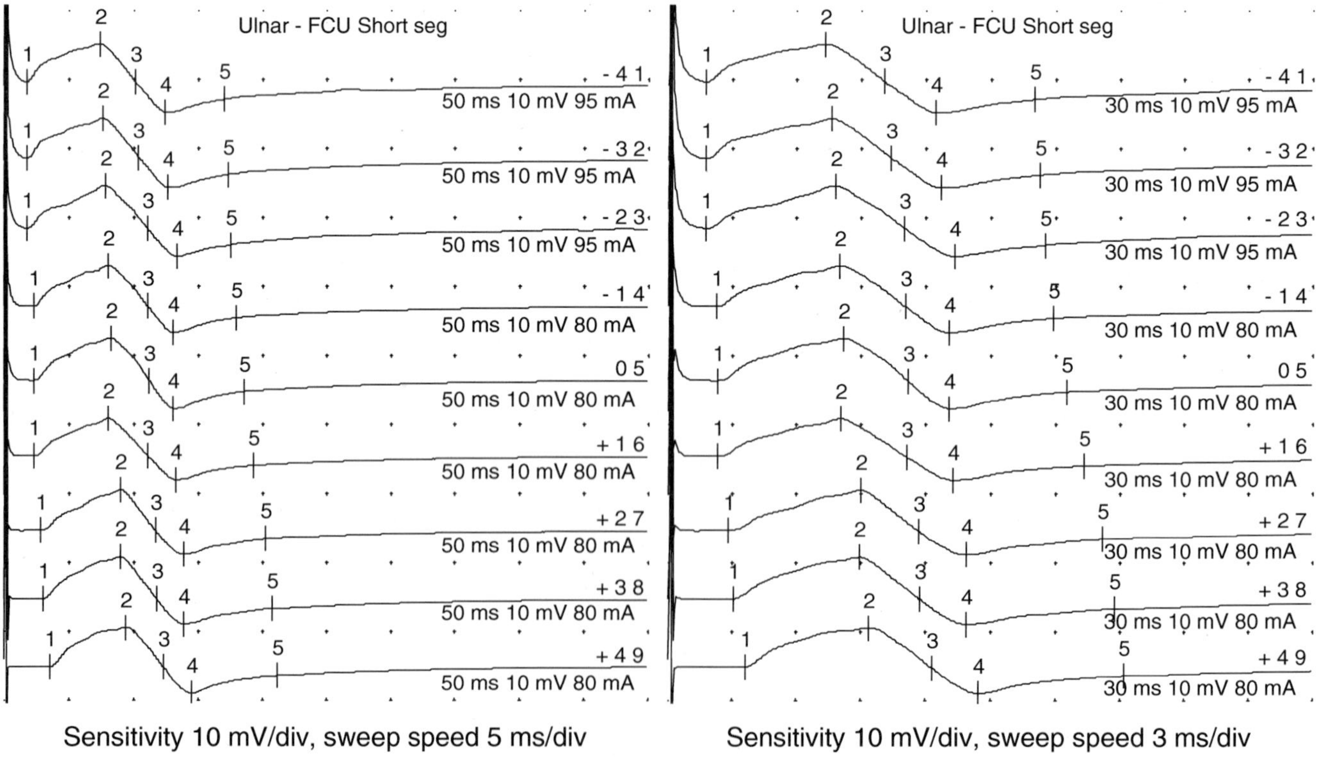

Fig. 1 Compound muscle action potentials (*CMAPs*) recorded at the forearm from the FCU muscle, *short-segment incremental stimulation* (SSIS) from the below elbow to the above elbow sites

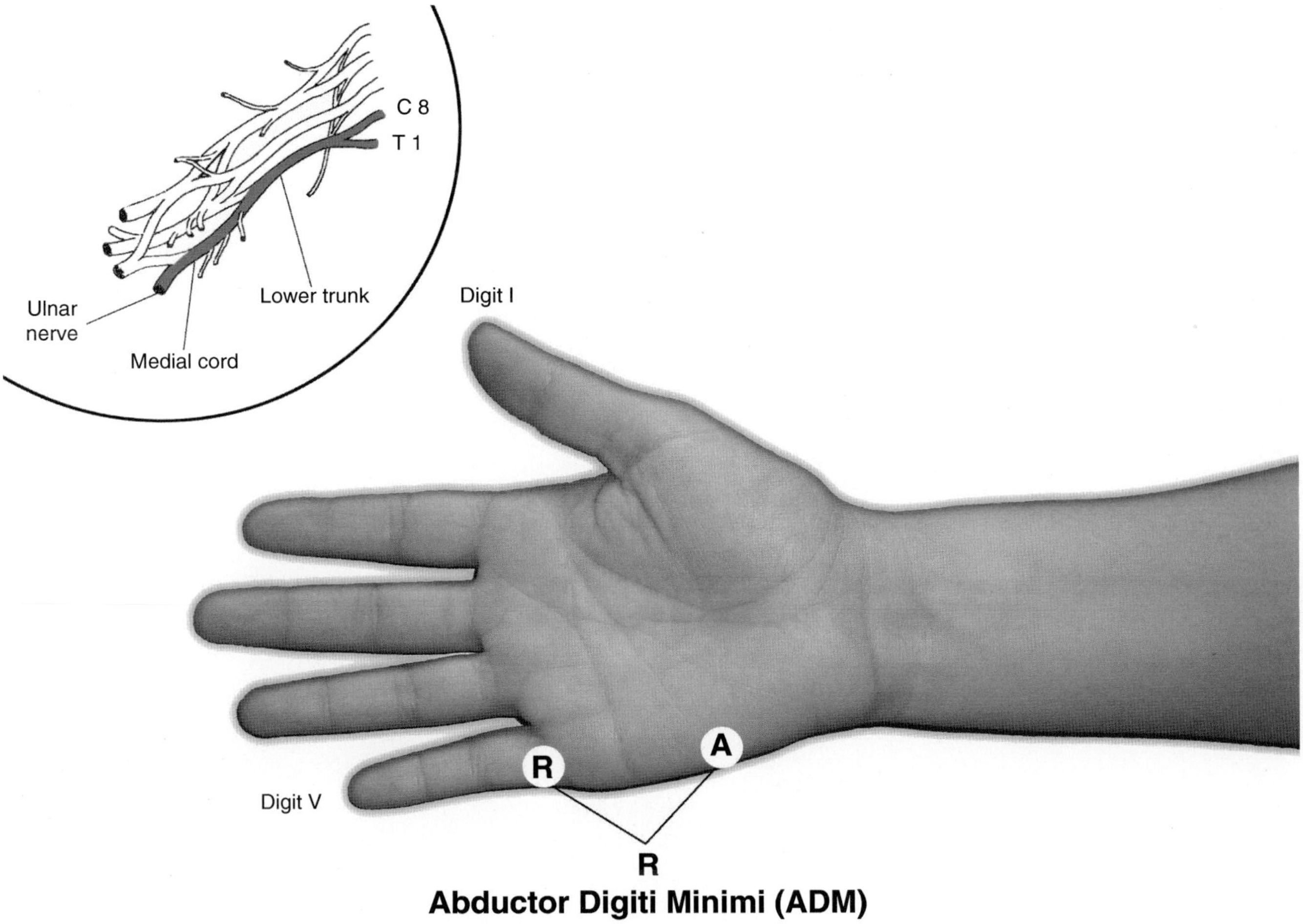

Fig. 2 Distal recording site from the ADM muscle

Reference

1. Lo YL, Leoh TH, Xu LQ et al (2005) Short segment nerve conduction studies in the localization of ulnar neuropathy at the elbow: use of flexor carpi ulnaris recordings. Muscle Nerve 31:633–636

Wrist – Hand

M40

Surface Recording Technique, Study from *First Dorsal Interosseous* (FDI) and *Abductor Digiti Minimi* (ADM) Muscles

Original Settings Sensitivity was 5 μV/division, sweep speed was 2 ms/division, low-frequency filter was 2–3 Hz, high-frequency filter was 10 kHz, and the machines used were an XLTEK Neuromax or a TECA Sapphire 2 M; duration of pulse and rate of pulse were not specified.

Position The arm was positioned slightly abducted with the elbow extended and the forearm supinated.

Recording For the *first dorsal interosseous* (FDI) muscle recording, the active recording electrode (A) was placed on the dorsum of the first web space in the center of the triangle formed by the first carpometacarpal joint, the first metacarpophalangeal joint, and the second metacarpophalangeal joint. The reference electrode (R) was distal to the fifth metacarpophalangeal joint (Fig. 1). For the *abductor digiti minimi* (ADM) muscle, the active recording electrode (A) was placed on the ulnar surface of the hypothenar eminence, halfway between the pisiform bone and the fifth metacarpophalangeal joint. The reference electrode (R) was placed at the same location as for the FDI muscle (Fig. 1). The ground electrode was placed on the dorsum of the hand for both studies. The FDI and ADM muscle recordings were performed separately [1].

Stimulation Supramaximal stimulation was performed at the wrist for both studies [1], on the course of the ulnar nerve (S), with the cathode 8 cm proximal to the active recording electrode for the ADM study (Fig. 1), on a line measured to slightly radial to the tendon of the *flexor carpi ulnaris* (FCU) muscle.

Measurements Distal motor latency (ms) was measured from the baseline to the onset of the first negative deflection of the compound muscle action potential (CMAP), for both ADM and FDI muscle recordings. The technique by Prahlow and Buschbacher [1], with a new electrode placement for motor conduction to the FDI muscle, reliably gave an initial negative deflection, as opposed to placing the reference at the second metacarpophalangeal joint (digit II), such as in the previously described methods by Bhala and Goodgold [2] and Olney and Wilbourn [3]. In 13 subjects, an initial negative deflection was not obtainable (8 left hands, 5 right hands). Negative peak amplitude (mV) was measured from the baseline to the negative peak of the CMAP for both ADM and FDI muscle recordings. Duration (ms) of the negative peak and area (μVs) under the negative peak were also recorded. Normal values for FDI muscle recordings (Table 1) were obtained on 388 ulnar nerves of 199 healthy subjects, 153 were white, 44 were black, and 2 were of other races (105 women and 94 men, women age mean 45.2 ± 15 years, men age mean 45.7 ± 14.5 years, women height mean 162 ± 7 cm, men height mean 176 ± 7 cm, women weight mean 153 ± 32 lbs, men weight mean 184 ± 31 lbs, women hand width – 7.4 ± 0.4 cm right and 7.4 ± 0.5 cm left, men hand width – 8.3 ± 0.6 cm right and 8.3 ± 0.6 cm left). The upper limit of normal (ULN) was taken at the 97th percentile of observed values; the lower limit of normal (LLN) was taken at the 3rd percentile of observed values. Normal values for ADM muscle recordings (Table 2) were obtained from 248 healthy subjects, 55 % were women and 45 % were men, 75 % were white, 22 % were black, and 3 % were of other races (women mean age 40.8 ± 14.8 years, men mean age 44.7 ± 14.7 years, women mean height 163 ± 7 cm, men mean height 176 ± 7 cm, women mean weight 154 ± 33 lbs, men mean weight 182 ± 30 lbs). Normal data for motor conduction to the ADM muscle have been previously published [4]. In all subjects, the skin temperature was maintained at or above 32 °C.

G. Gentili, M. Di Napoli, *The Ulnar Nerve*, DOI 10.1007/978-3-319-43292-2_64

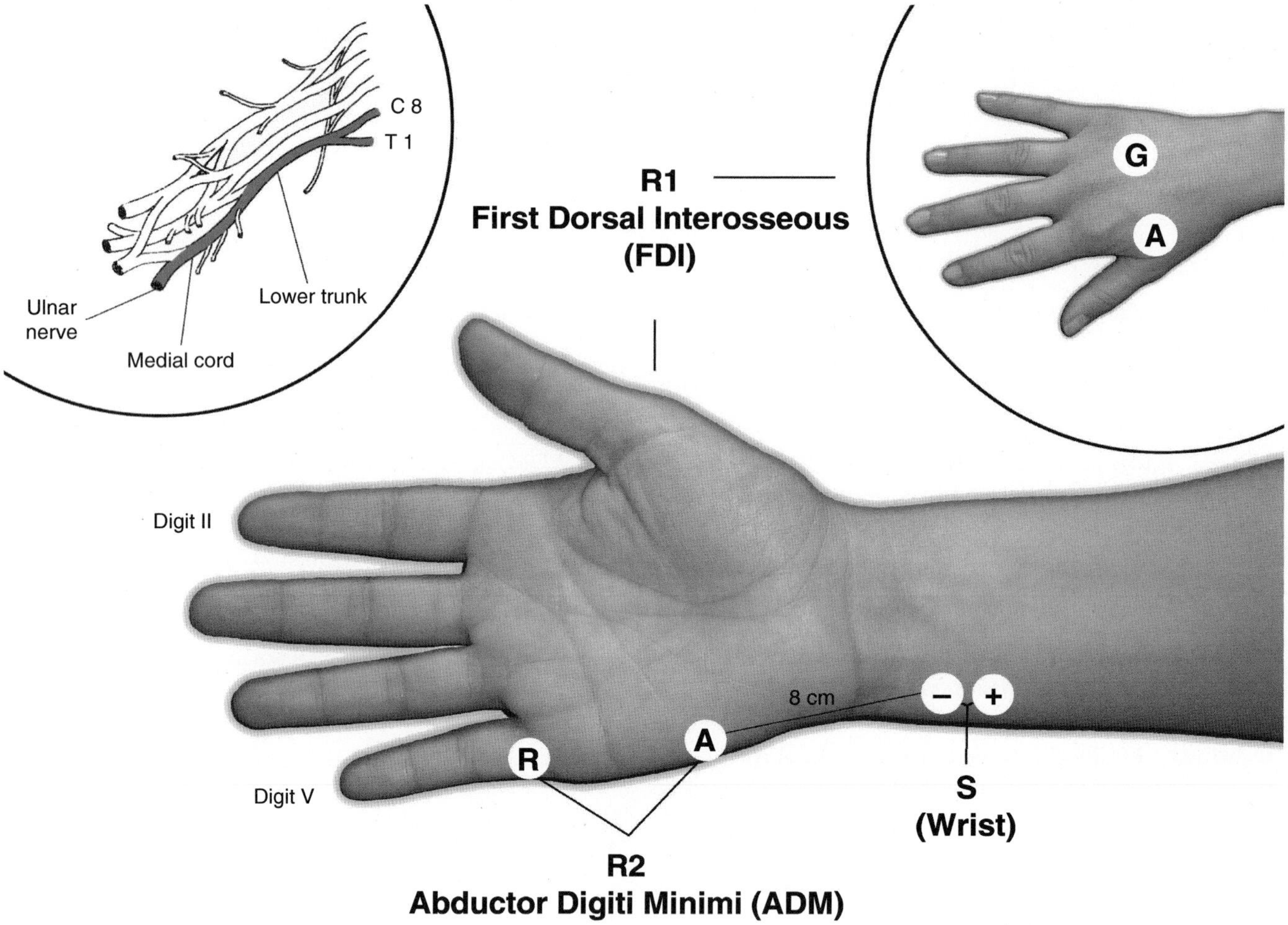

Typical waveform (wrist – FDI muscle, wrist – ADM muscle):

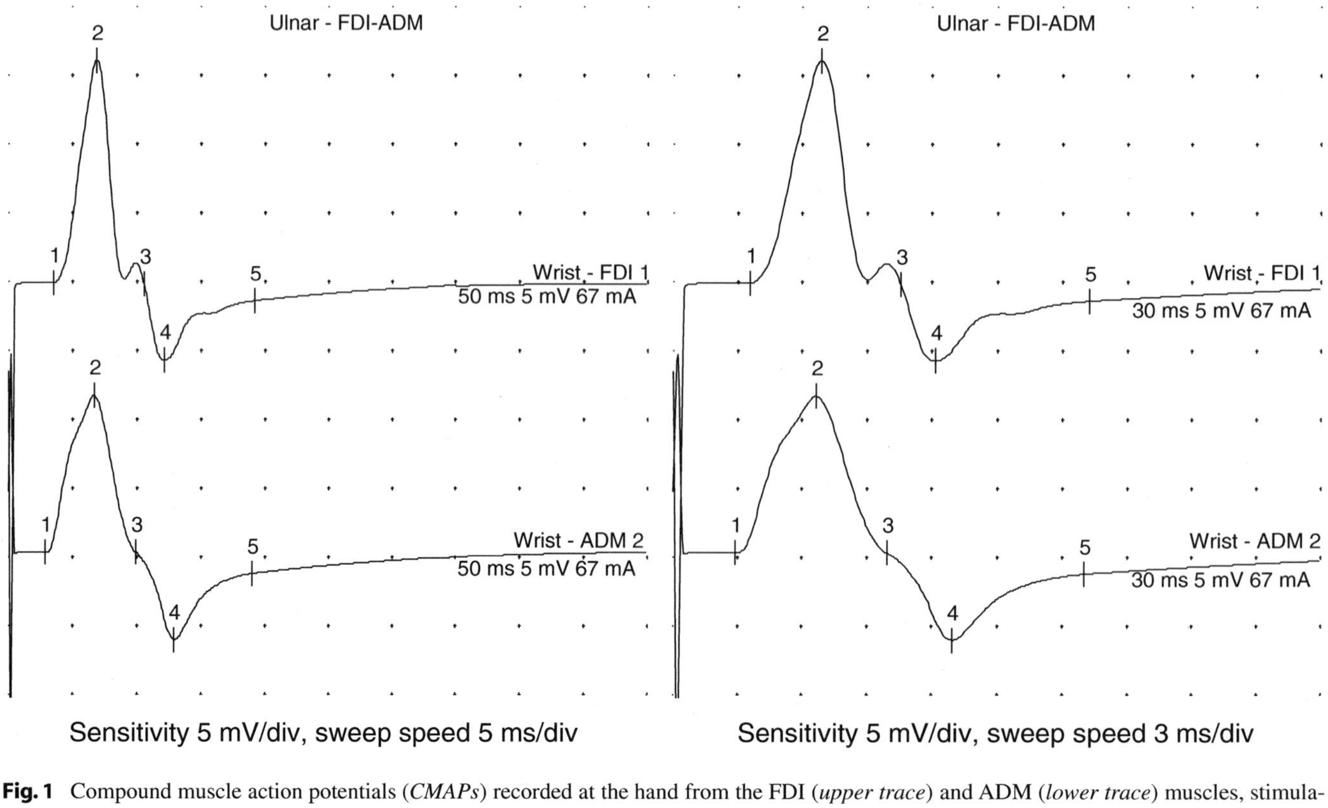

Fig. 1 Compound muscle action potentials (*CMAPs*) recorded at the hand from the FDI (*upper trace*) and ADM (*lower trace*) muscles, stimulation of the ulnar nerve at the wrist

Table 1 Normal values [1]

	Mean ± SD	Range	ULN	LLN
Wrist–FDI, latency (ms) – men	3.8 ± 0.5	3.1–4.9	4.7	
Wrist–FDI, latency (ms) – women	3.4 ± 0.5	2.5–4.6	4.4	
Wrist–FDI, latency (ms) – all subjects	3.6 ± 0.5	2.5–4.9	4.6	
Wrist–FDI – wrist–ADM, latency diff (ms) – all subjects	0.6 ± 0.4		1.3	
Wrist–FDI, amplitude (mV) – all subjects	15.8 ± 4.9	2.7–30.1		5.1
Wrist–FDI, area (μVs) – all subjects	37.6 ± 12.3	4.3–65.9		12.3
Wrist–FDI, duration (ms) – < 180 cm subjects	5.1 ± 0.8	3.3–7.3	6.7	
Wrist–FDI, duration (ms) – ≥ 180 cm subjects	5.5 ± 1.0	4.3–9.3	8.2	
Wrist–FDI, duration (ms) – all subjects	5.2 ± 0.8	3.3–9.3	6.7	
Wrist–FDI, latency side-to-side difference (ms) – all subjects	–0.1 ± 0.4		0.8	
Wrist–FDI, duration side-to-side difference (ms) – all subjects	–0.1 ± 0.8		1.5	
Wrist–FDI, amplitude side-to-side difference (mV) – all subjects	1.0 ± 5.6			–52 %
Wrist–FDI, area side-to-side difference (μVs) – all subjects	1.5 ± 5.6			–49 %

Table 2 Normal values [4]

	Mean ± SD	Mean ± 2SD
Wrist–ADM, distal latency (ms)	3.0 ± 0.3	3.6
Wrist–ADM, duration (ms)	6.0 ± 0.9	7.8
Wrist–ADM, amplitude (mV)	11.6 ± 2.1	7.4
Wrist–ADM, area (μVs)	35.9 ± 7.1	21.7

Comment

With their study, Prahlow and Buschbacher [1] provided a large database of normal ranges for the deep ulnar branch to the FDI muscle. The latency to the FDI muscle was analyzed with respect to height and hand width, and neither was associated with differing results. In fact, hand size as an independent variable did not correlate with latency, and the same was true for height. In addition to the typical measures of latency and amplitude, data were also collected for area and duration. Side-to-side CMAP latency, amplitude, duration, and area differences for the FDI muscle and the FDI-ADM CMAP latency difference in the same limb were reported. The mean latency to the FDI muscle and an upper limit of normal (ULN) were lower for women than for men; duration was longer for taller individuals. For the authors [1], side-to-side differences in latency may be useful in diagnosing some peripheral ulnar neuropathies, and they found a lower ULN value for side-to-side difference (0.8 ms) than ULN value (1.3 ms) reported by Olney and Wilbourn [3]. Bhala and Goodgold [2] reported that the mean FDI-ADM CMAP latency difference was 1.2 ± 0.48 ms (ULN 2.4 ms); Olney and Wilbourn [3] found the mean FDI–ADM CMAP latency difference to be 0.9 ms (ULN 2 ms); Prahlow and Buschbacher [1] found a mean FDI–ADM CMAP latency difference of 0.6 ± 0.4 ms (ULN 1.3 ms). Side-to-side differences in amplitude may be useful in diagnosing some peripheral ulnar neuropathies and radiculopathies, and in their study [1], the ULN for side-to-side decrease in amplitude was 52 %, which was in close agreement with others [5, 6].

Seror et al. [7] compared three methods for FDI recording: the digit V reference method by Prahlow and Buschbacher [1], the digit I reference method proposed by Cowdery et al. [8], and their new electrode placement – with the reference electrode on the dorsal aspect of the wrist over the trapezium bone – 4.5 cm from the active recording electrode over the FDI muscle. The recording was made with a pair of disposable surface electrodes (1.9 cm × 1.5 cm); the active recording electrode was placed dorsally over the belly of the FDI muscle at an equal distance to the metacarpophalangeal joints of digit I and digit II, with the reference electrode on the dorsal aspect of the wrist over the trapezium bone – 4.5 cm from the active recording electrode (Fig. 2). They demonstrated that the DML was significantly shorter with the new method (2.6±0.26 ms) than with the first and the second. The authors also observed, using their method, the smallest SD and the absence of positive initial deflection at the onset of the FDI CMAP. The small SD was one of the most important contributions of their method, because the less the SD, the more homogeneous are the results and the more reproducible is the method; the absence of a positive initial deflection is the other major contribution provided by their method. In fact, the occurrence of a positive initial deflection frequently occurs when the reference electrode is fixed at the base of the digit II – on the second metacarpophalangeal joint [2, 3] – and using digit II provides a short distal motor latency when the latency is measured at the onset of the positive deflection [3].

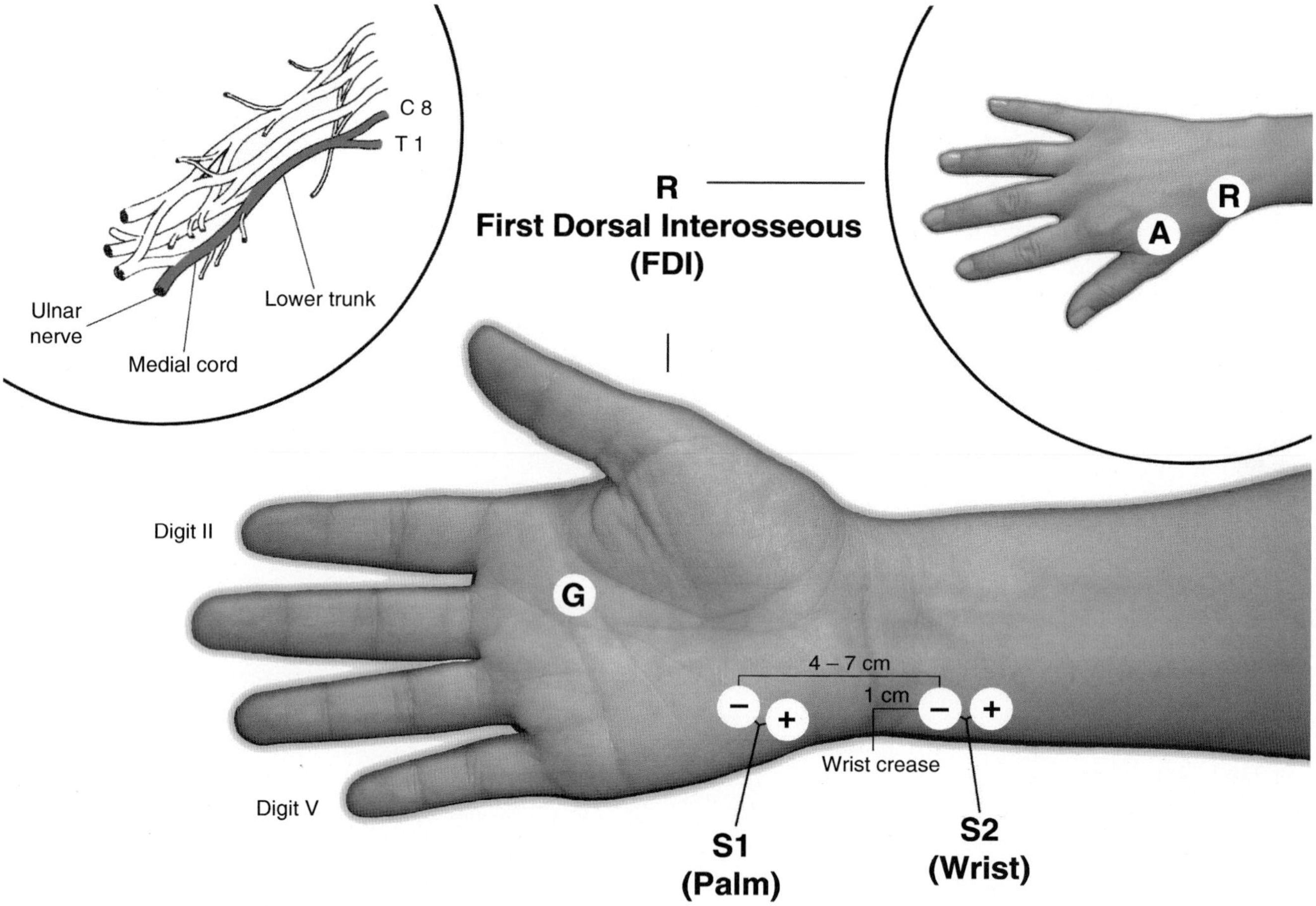

Fig. 2 FDI CMAP recording method by Seror [7], reference electrode over the trapezium bone – 4.5 cm from the active recording electrode

Cowdery et al. [8] observed that when the reference electrode is fixed on digit I – on the first metacarpophalangeal joint – the CMAP appears not to have a positive initial deflection but has a longer distal motor latency caused, for Seror et al. [7], by a small positive initial deflection which can be seen only with a gain of 0.5 mV/division (35 of 40 hands). With the reference electrode fixed on the fifth metacarpophalangeal joint (digit V), following the Prahlow and Buschbacher's method [1], a positive initial deflection was found in 13 of 199 healthy subjects studied by the original authors and in 13 of 40 hands in Seror's study [7] using a gain of 0.5 mV/division. The difference between the two methods is highly significant for the latency and the occurrence of initial positive deflection, but not for the FDI CMAP amplitude. For Seror et al. [7], the method by Prahlow and Buschbacher [1] does not respect the classic belly-tendon method [9], and the recorded CMAP is not specific to the FDI muscle but contains contributions from all intrinsic ulnar-innervated muscles located between the pair of electrodes. The FDI–ADM derivation (belly of the FDI muscle to tendon of the ADM muscle) certainly explains why this complex CMAP may be of such low amplitude [1] in controls (2.7–30.1 mV) or have a polyphasic configuration because of phase cancelation of the CMAPs of the different interosseous muscles. Based on report by Seror et al. [7], and with the strict respect of the location of the pair of recording electrodes, the authors recommended that: (1) the distal motor latency to the FDI muscle can be considered abnormal when it exceeds 3.4 ms (mean + 3SD) if the CMAP amplitude is normal (specificity 100 %); (2) the conduction of the ulnar deep motor branch at the wrist is abnormal with distal motor latency to the FDI muscle > 0.3 ms than distal motor latency to the APB, > 1 ms than distal motor latency to the ADM muscle, and > 0.5 ms that of the contralateral distal motor latency to the FDI muscle, if the CMAP amplitude is normal (specificity 100 %); and (3) FDI CMAP amplitude < 6.2 mV is abnormal (specificity 98 %). But this finding has no specific significance concerning the location of the nerve lesion. Pathological waveform and values in a patient with ulnar neuropathy at the wrist (UNW) are reported (Fig. 3).

Pathological waveform (wrist – FDI muscle, wrist – ADM muscle):

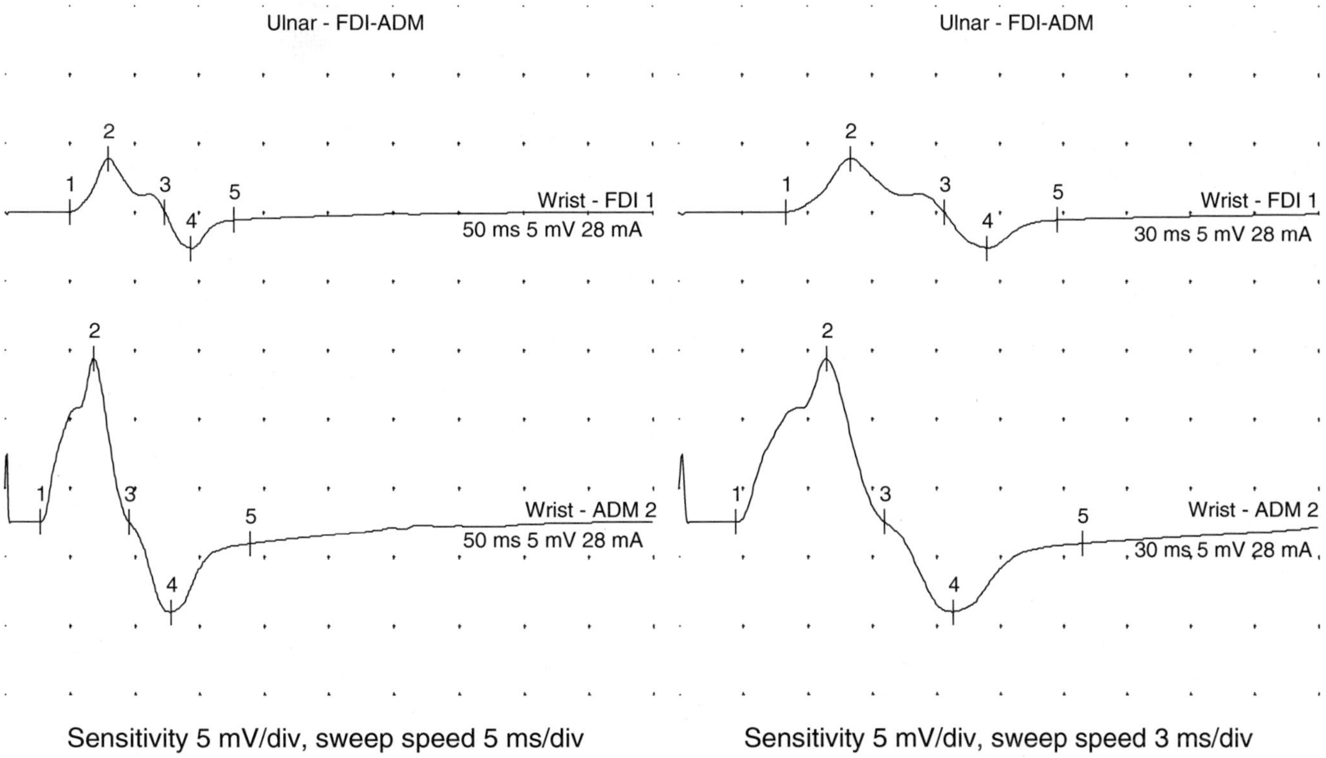

Onset latency (wrist – FDI): 5.0 ms; **Onset latency** (wrist – ADM): 2.65 ms; **Onset latency difference** (wrist – FDI – wrist – ADM): 2.35 ms; **Peak latency** (wrist – FDI): 7.75 ms; **Peak latency** (wrist – ADM): 6.90 ms; **Onset to peak amplitude** (wrist – FDI): 3.9 mV; **Onset to peak amplitude** (wrist – ADM): 11.9 mV; **Peak to peak amplitude** (wrist – FDI): 6.6 mV; **Peak to peak amplitude** (wrist – ADM): 18.5 mV

Fig. 3 Compound muscle action potentials (*CMAPs*) recorded at the hand from the FDI (*upper trace*) and ADM (*lower trace*) muscles in a patient with UNW, stimulation of the ulnar nerve at the wrist

References

1. Prahlow ND, Buschbacher RM (2006) Ulnar nerve motor conduction to the first dorsal interosseous muscle. J Long Term Eff Med Implants 16:333–340
2. Bhala RP, Goodgold J (1968) Motor conduction in the deep palmar branch of the ulnar nerve. Arch Phys Med Rehabil 49: 460–466
3. Olney RK, Wilbourn AJ (1985) Ulnar nerve conduction study of the first dorsal interosseous muscle. Arch Phys Med Rehabil 66: 16–18
4. Buschbacher RM (1999) Ulnar nerve conduction to the abductor digiti minimi. Am J Phys Med Rehabil 78:S9–S14
5. Felsenthal G (1982) An overview of the clinical application of electromyographic and nerve conduction techniques: part III. Prognostic and treatment implications. Md State Med J 31:60–63
6. Eisen A (1985) Electrodiagnosis of radiculopathies. Neurol Clin 3:495–510
7. Seror P, Maisonobe T, Bouche P (2011) A new electrode placement for recording the compound motor action potential of the first dorsal interosseous muscle. Clin Neurophysiol 41:173–180
8. Cowdery SR, Preston DC, Herrmann DN et al (2002) Electrodiagnosis of ulnar neuropathy at the wrist. Conduction block versus traditional tests. Neurology 59:420–427
9. Kimura J (1989) Electrodiagnosis in diseases of nerve and muscle: principles and practice, vol 35, 2nd edn. Davis, Philadelphia, p 709

Wrist, Below Elbow, Above Elbow – Hand

M41

Surface Recording Technique, Study from *Abductor Digiti Minimi* (ADM), and *First Dorsal Interosseous* (FDI) Muscles

Original Settings Sensitivity was 5 mV/division, sweep speed was 2 ms/division, low-frequency filter was 10 Hz, high-frequency filter was 10 kHz, duration of pulse was 0.1 ms, and the machine used was a 2-channel Medtronic SW7 (version 5) EMG machine.

Position All subjects were examined in the supine position with the shoulder externally rotated and 90° abducted, the elbow flexed 135°, and the wrist in the neutral position.

Recording For the *abductor digiti minimi* (ADM) muscle (R1), the active electrode was placed over the motor point of the ADM muscle (Fig. 1), on the muscle's bulk between the pisiform osseous prominence and the 5th metacarpophalangeal joint; the reference (R) was placed on the distal portion of the 2nd metacarpophalangeal joint [1]. For the *first dorsal interosseous* (FDI) muscle recording (R2), the active electrode (A) was placed over the belly of the FDI muscle (Fig. 1), between the first and second metacarpal bones (Fig. 1); the reference (R) was placed distal to the second metacarpophalangeal joint (over the tendon of the FDI muscle). R2 recording was made on the dorsum of the hand. The reference electrode was the same for ADM and FDI recordings, and the compound muscle action potentials (CMAPs) for both recordings were made simultaneously using the same disk electrode type.

The ground (G) electrode was located between the sites of stimulation and recording; the figure shows the ground electrode placed on the palm.

Stimulation The ulnar nerve was stimulated percutaneously at the wrist (WR), 8 cm proximal to the active recording electrode on the ADM muscle; below the elbow (BE), 4 cm below the medial epicondyle and above the elbow (AE), and 6 cm above the medial epicondyle (Figs. 1 and 2). The stimulatory electrode was bipolar and superficial, and the distance between anode and cathode was 3 cm. Each stimulation was supramaximal, which means it was 25 % more than the intensity creating maximal CMAP amplitude.

Measurements Distal latency (ms) was measured from the stimulus onset to the onset of the initial deflection of the compound muscle action potential (CMAP). By dividing the distance by time, the motor nerve conduction velocity (MNCV) along each nerve segment can be determined and measured in meter per second (m/s), and it was calculated for the wrist–below elbow (forearm) and the below-elbow–above-elbow (elbow) segments. Normal values (Table 1) were obtained bilaterally from 50 healthy volunteers (36 men and 14 women, mean age 36.8 ± 6.6 years, age range 16–50 years). Skin temperature was controlled, and it was maintained at or above 32 °C using, if necessary, an infrared device.

Comment

For Azma et al. [1], the results of their study have shown that the minimal ulnar MNCV at the elbow, recorded from the FDI muscle (45.65 m/s), was less than the one recorded from the ADM muscle (47.39 m/s). Besides, it has been evidenced that the minimal normal range of the difference between the MNCV at the elbow and forearm, recorded from the FDI muscle (18.11 m/s), was more than the one recorded from the ADM muscle (16.4 m/s). The authors found that the reduction of MNCV across the elbow was more prominent when recorded from the FDI muscle than recorded from the ADM muscle; they explained the difference in MNCV normal values between FDI and ADM muscle recordings with differences in the type of nerve fibers in terms of speed of conduction or due to the mild and repeated damages to the nerve fascicles of FDI by time with daily activities, as they are more superficial. Pathological waveform and values in a patient with ulnar neuropathy at the elbow (UNE) are reported (Fig. 3).

G. Gentili, M. Di Napoli, *The Ulnar Nerve,* DOI 10.1007/978-3-319-43292-2_65

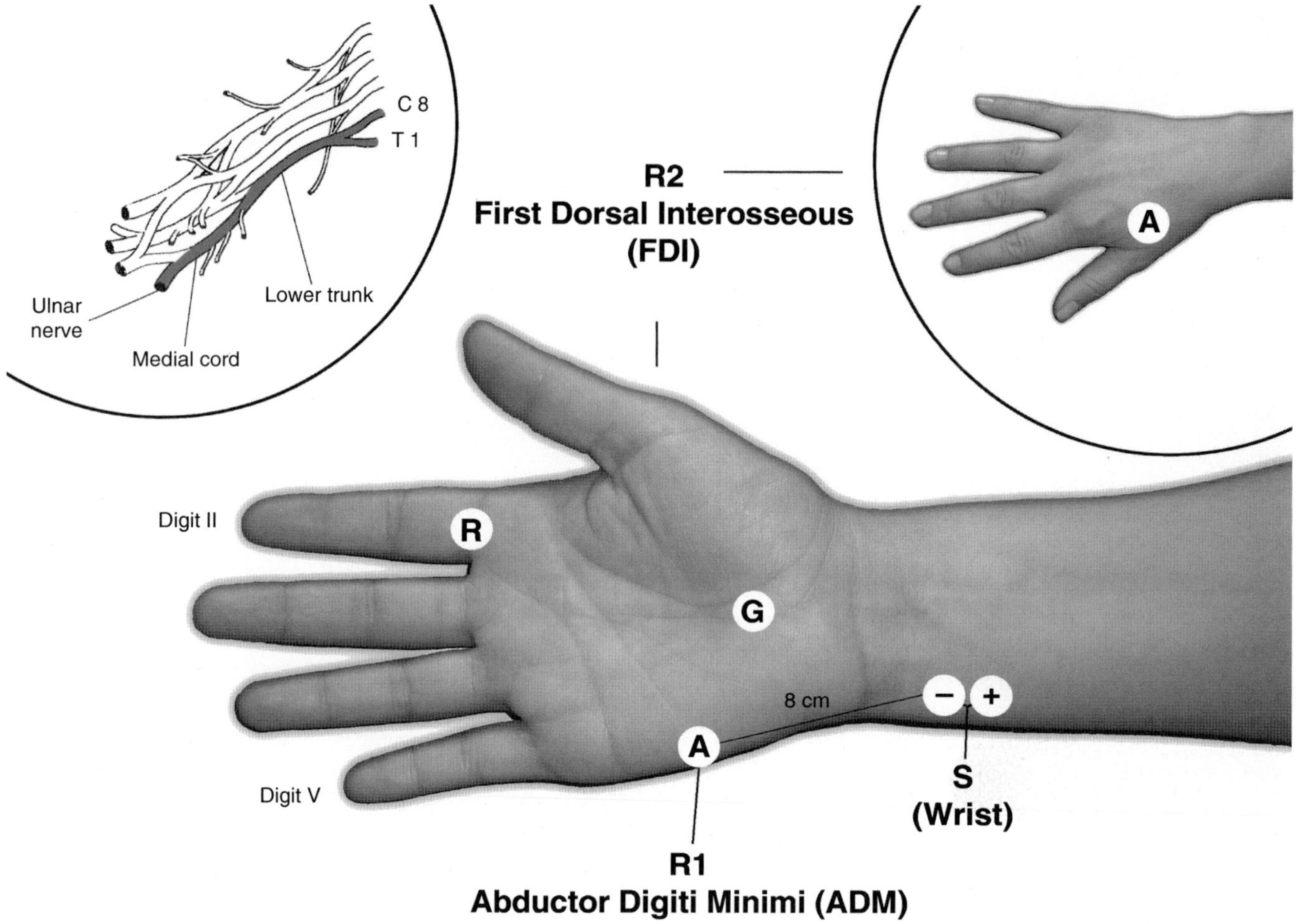

Typical waveform (wrist, below elbow, above elbow – ADM and FDI muscles):

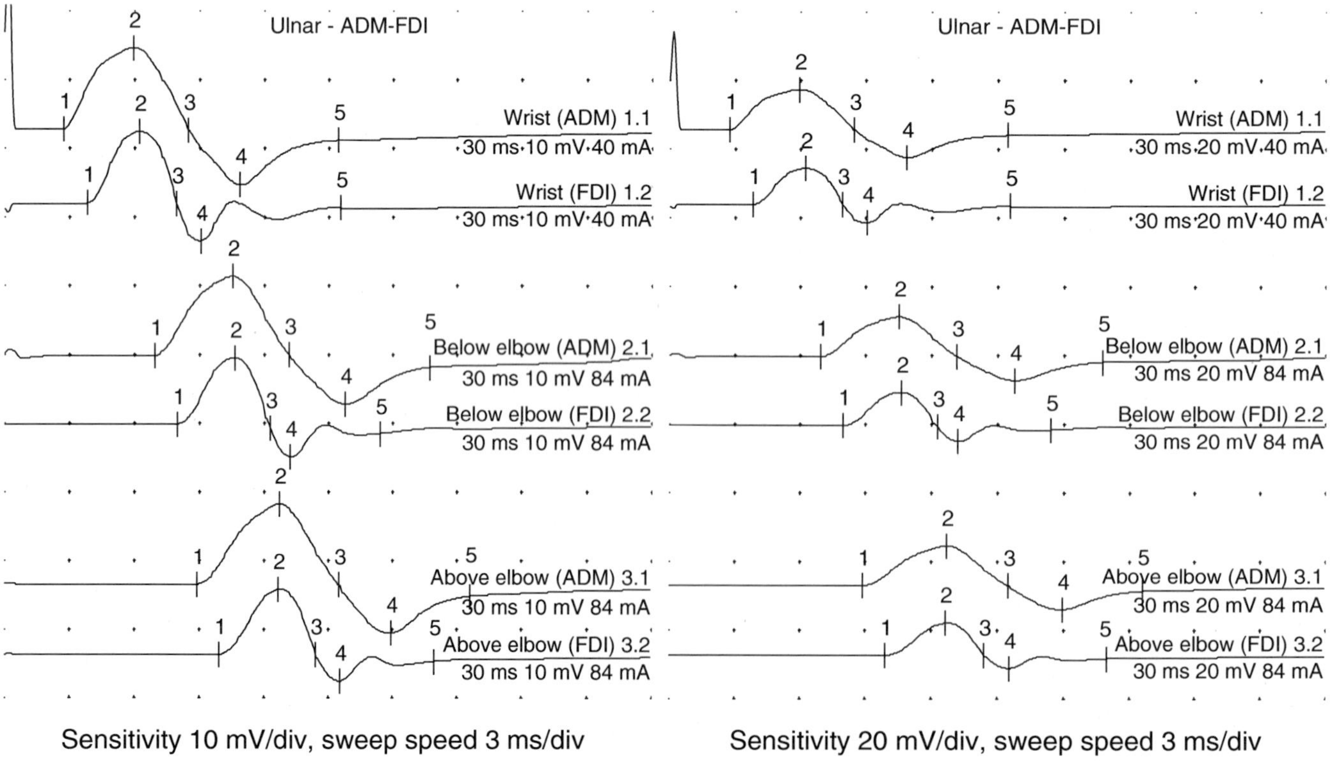

Fig. 1 Compound muscle action potentials (*CMAPs*) recorded at the hand from the ADM and FDI muscles, stimulation of the ulnar nerve at the wrist (*upper traces*), below elbow (*middle traces*), and above elbow (*lower traces*)

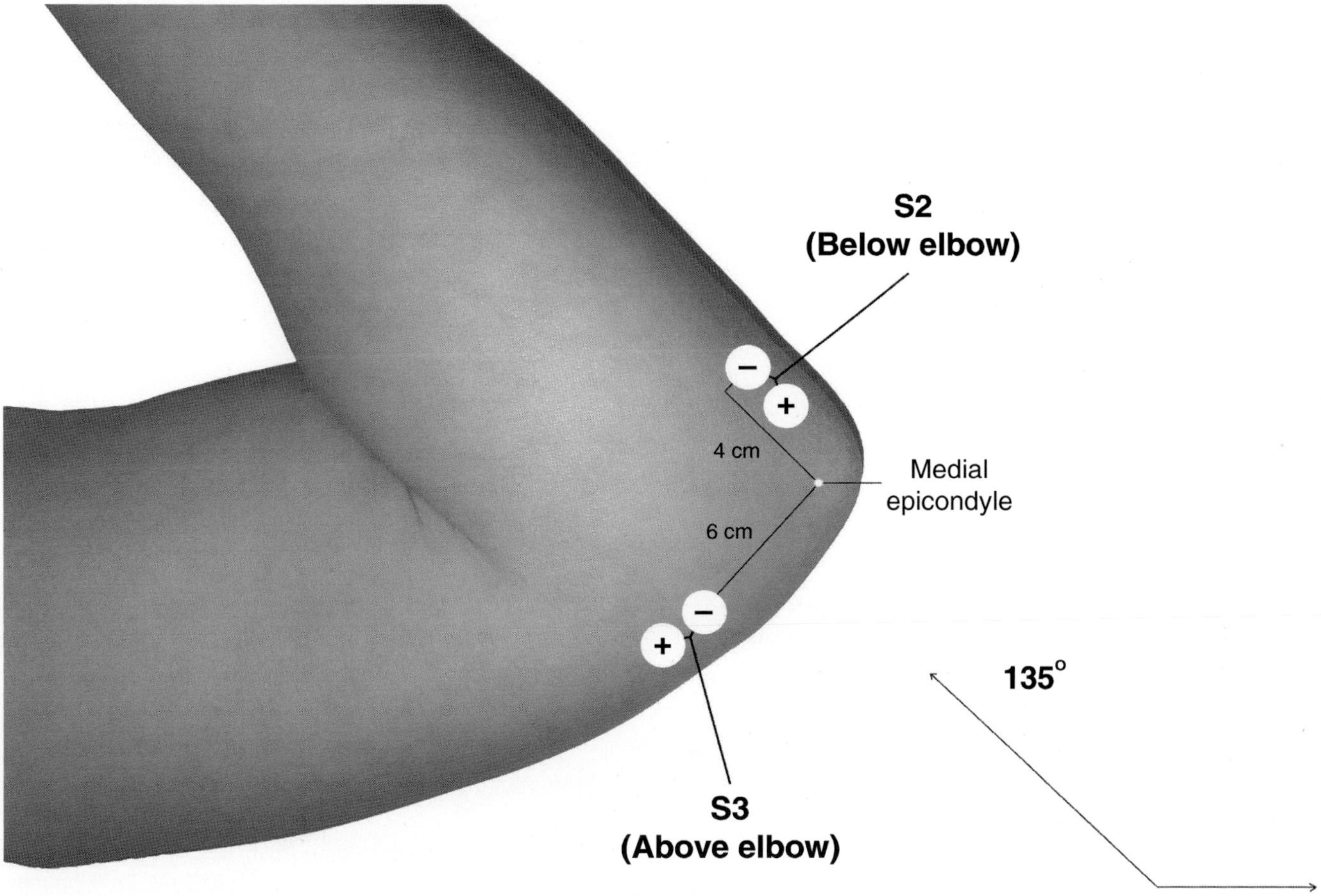

Fig. 2 Proximal stimulation points (S2, *below elbow*; S3, *above elbow*) with the elbow flexed to 135°

Table 1 Normal values [1]

	Mean ± SD
Below elbow–above elbow, elbow MNCV (m/s) – ADM recordings	62.65 ± 7.62
Below elbow–above elbow, elbow MNCV (m/s) – FDI recordings	60.49 ± 7.42
Forearm–elbow MNCV difference (m/s) – ADM recordings	−1.19 ± 7.62
Forearm–elbow MNCV difference (m/s) – FDI recordings	−2.39 ± 7.85

Pathological waveform (wrist, below elbow, above elbow – ADM and FDI muscles):

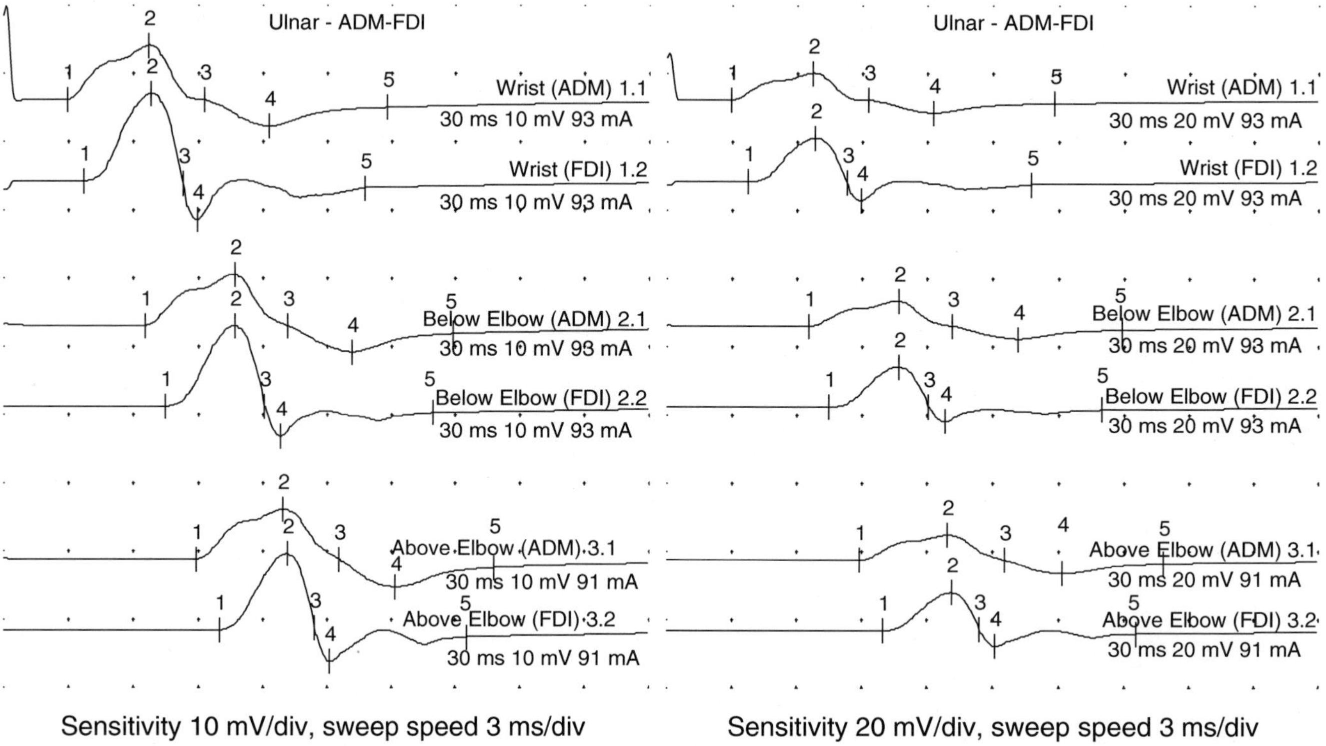

Onset latency (wrist – ADM): 2.90 ms; **Onset latency** (wrist – FDI): 3.70 ms; **Onset latency** (below elbow – ADM): 6.50 ms; **Onset latency** (below elbow – FDI): 7.45 ms; **Onset latency** (above elbow – ADM): 8.85 ms; **Onset latency** (above elbow – FDI):9.95 ms; **Peak latency** (wrist – ADM): 6.25 ms; **Peak latency** (wrist – FDI): 6.30 ms; **Peak latency** (below elbow – ADM): 10.65 ms; **Peak latency** (below elbow – FDI): 10.65 ms; **Peak latency** (above elbow – ADM): 12.80 ms; **Peak latency** (above elbow – FDI): 13.00 ms; **Onset to peak amplitude** (wrist – ADM): 8.1 mV; **Onset to peak amplitude** (wrist – FDI): 12.8 mV; **Onset to peak amplitude** (below elbow – ADM): 7.6 mV; **Onset to peak amplitude** (below elbow – FDI): 11.8 mV; **Onset to peak amplitude** (above elbow – ADM): 7.4 mV; **Onset to peak amplitude** (above elbow – FDI): 11.1 mV; **Peak to peak amplitude** (wrist – ADM): 12.1 mV; **Peak to peak amplitude** (wrist – FDI): 18.6 mV; **Peak to peak amplitude** (below elbow – ADM): 11.6 mV; **Peak to peak amplitude** (below elbow – FDI): 16.3 mV; **Peak to peak amplitude** (above elbow – ADM): 11.5 mV; **Peak to peak amplitude** (above elbow – FDI): 16.0 mV; **MNCV** (wrist – below elbow, ADM, 21 cm): 58.3 m/s; **MNCV** (wrist – below elbow, FDI, 21 cm): 56.0 m/s; **MNCV** (below elbow – above elbow, ADM, 10 cm): 42.6 m/s; **MNCV** (below elbow – above elbow, FDI, 10 cm): 40.0 m/s

Fig. 3 Compound muscle action potentials (*CMAPs*) recorded at the hand from the ADM and FDI muscles in a patient with UNE, stimulation of the ulnar nerve at the wrist (*upper traces*), below elbow (*middle traces*), and above elbow (*lower traces*)

Reference

1. Azma K, Bahmanteimoury K, Tavana B et al (2007) Two measurement methods of motor ulnar nerve conduction velocity at the elbow: a comparative study. Neurol India 55:145–147

Wrist, Below Elbow, Above Elbow – Hand

M42

Surface Recording Technique, Study from *Abductor Digiti Minimi* (ADM) and *First Dorsal Interosseous* (FDI) Muscles

Original Settings Sensitivity, sweep speed, low-frequency filter, high-frequency filter, duration, rate of pulse, and the machine used were not specified.

Position This study was performed with moderate flexion (70–90° from horizontal) of the elbow.

Recording The authors [1] performed ulnar nerve conduction study recording from the motor point of the *abductor digiti minimi* (ADM) and *first dorsal interosseous* (FDI) muscles, using surface electrodes. For the ADM muscle recordings, the active electrode was placed on the *hypothenar* eminence, on a point midway between the distal wrist crease and the crease at the base of the digit V (little finger), and at the junction of dorsal and palmar skin; the reference electrode was placed medially at the base of the digit V (Fig. 1). For the FDI muscle recording, the active electrode was placed over the belly of the FDI muscle (Fig. 1), between the first and second metacarpal bones (Fig. 1); the reference was placed distal to the second metacarpophalangeal joint (over the tendon of the FDI muscle). FDI muscle recording was made on the dorsum of the hand. The ADM and FDI muscle recordings were performed separately. The ground (G) electrode position was not specified in the text; the figures show ground electrode placed on the palm (Figs. 1 and 2).

Stimulation For both ADM and FDI muscle recordings [1], the ulnar nerve was stimulated by surface electrodes at the three points along its course (Figs 1, 2, and 3): at the proximal wrist crease, 3 cm distal to the medial epicondyle (below – BE) and 6–7 cm above this level (above elbow–AE).

Measurements The latency (ms) of the compound muscle action potential (CMAP) was measured from the onset of the stimulus artifact to the onset of the negative deflection from the isoelectric line. By dividing the distance by time, the motor nerve conduction velocity (MNCV) along the nerve segment can be determined and measured in meter per second (m/s), and it was calculated for wrist–below-elbow (BE-wrist – forearm) and below-elbow–above-elbow (AE-BE – across elbow) segments. An AE-BE MNCV more than 10 m/s slower than the BE–wrist MNCV was considered pathological [2]. Skin temperature was checked and maintained above 31 °C by an infrared lamp. All motor conduction studies were performed with moderate flexion (70°–90° from horizontal) of the elbow; this position was chosen because it provides the greatest correlation between surface skin measurement and true nerve length [2–5]. The elbow position was maintained the same when recording from both the ADM and FDI muscles. Kothari et al. [6] noted that sensitivity for detecting MNCV slowing across the elbow (AE-BE segment) was higher when recording from the FDI muscle than from ADM muscle, due to differential fascicular involvement. In their study, Azma et al. [7] found that the reduction of MNCV across the elbow was more prominent when recorded from the FDI muscle than recorded from the ADM muscle; they explained the difference in MNCV normal values between FDI and ADM muscle recordings with differences in the type of nerve fibers in terms of speed of conduction or due to the mild and repeated damages to the nerve fascicles of FDI by time with daily activities, as they are more superficial. Caliandro et al. [1] hypothesized that if in patients with ulnar neuropathy at the elbow (UNE) the fibers to the FDI muscle are more damaged than fibers to the ADM muscle, a relative slowing of MNCV at the segment across the elbow (AE-BE segment) recording from FDI (FDI-MNCV) versus MNCV at the same segment recording from ADM (ADM-MNCV) would occur. The authors [1] called this ratio the "intra-nerve ratio" (IN-RATIO). FDI-MNCV in the AE-BE segment was considered pathological when it was <43 m/s; ADM-MNCV, in the same segment, was considered pathological when it was <40 m/s. To verify this hypothesis, they calculated the ratio between FDI-MNCV and ADM-MNCV in a control group composed of 40 control subjects (25 women and 15 men, mean age 51 ± 18.09 years) and 16 patients with clinical, neurophysiological, and radiological diagnosed lower

G. Gentili, M. Di Napoli, *The Ulnar Nerve*, DOI 10.1007/978-3-319-43292-2_66

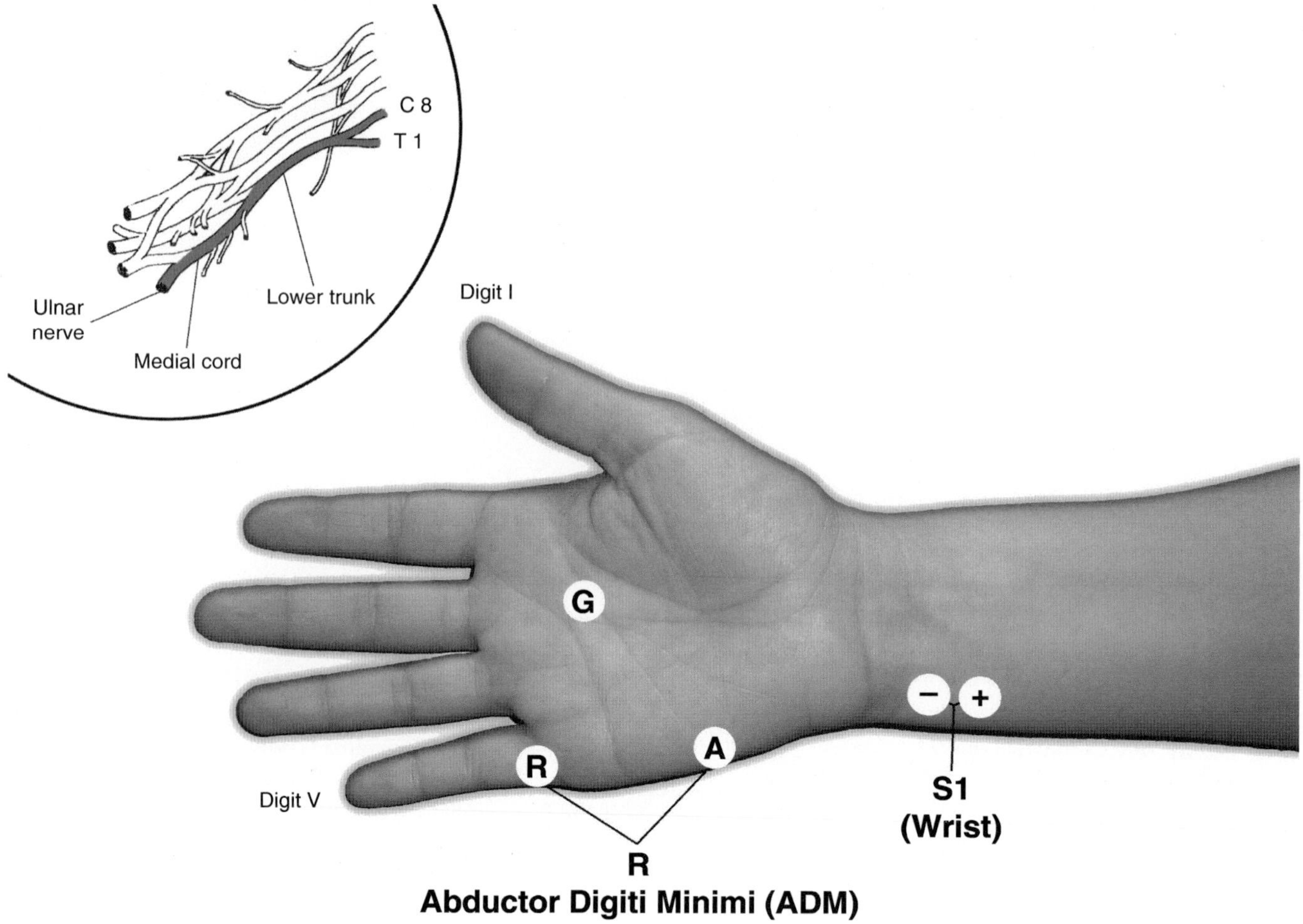

Typical waveform (wrist, below elbow, above elbow – ADM muscle):

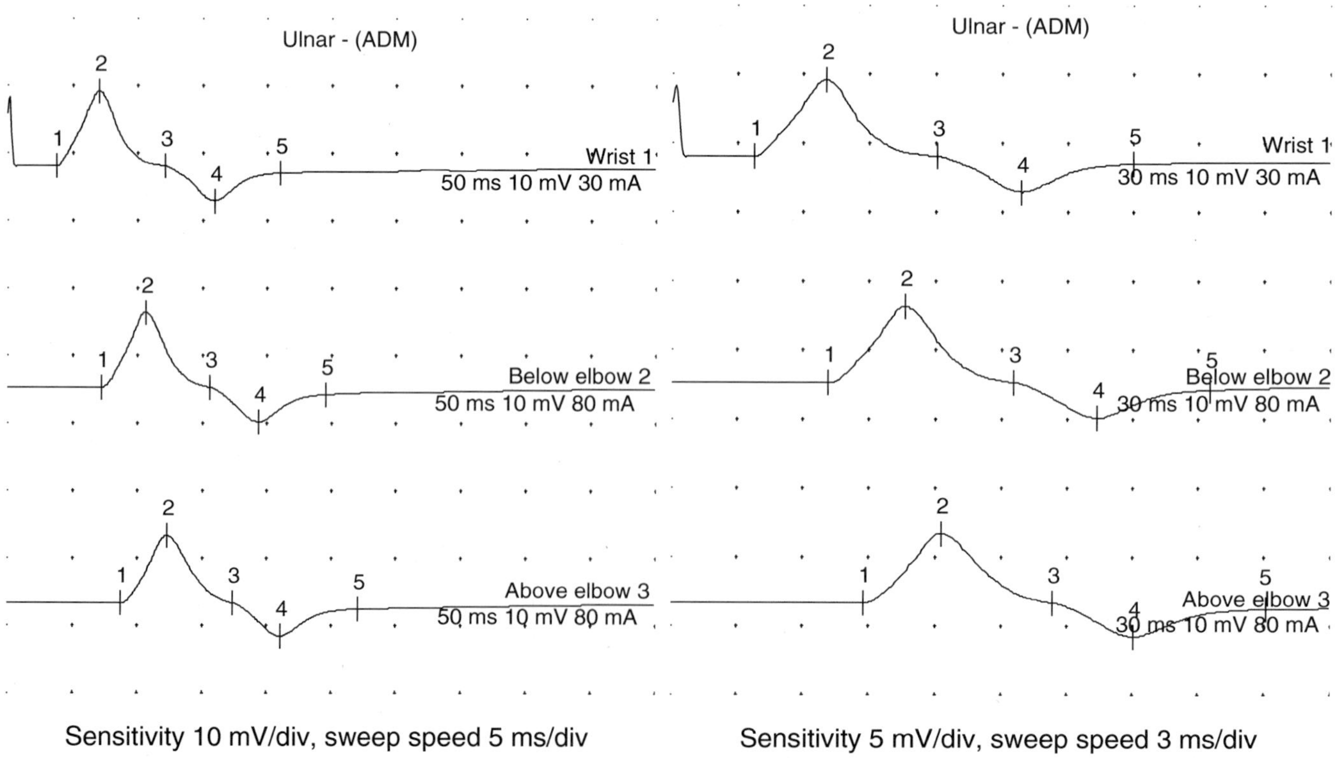

Fig. 1 Compound muscle action potentials (*CMAPs*) recorded at the hand from the ADM muscle; stimulation of the wrist (*upper trace*), below elbow (*middle trace*), and above elbow (*lower trace*)

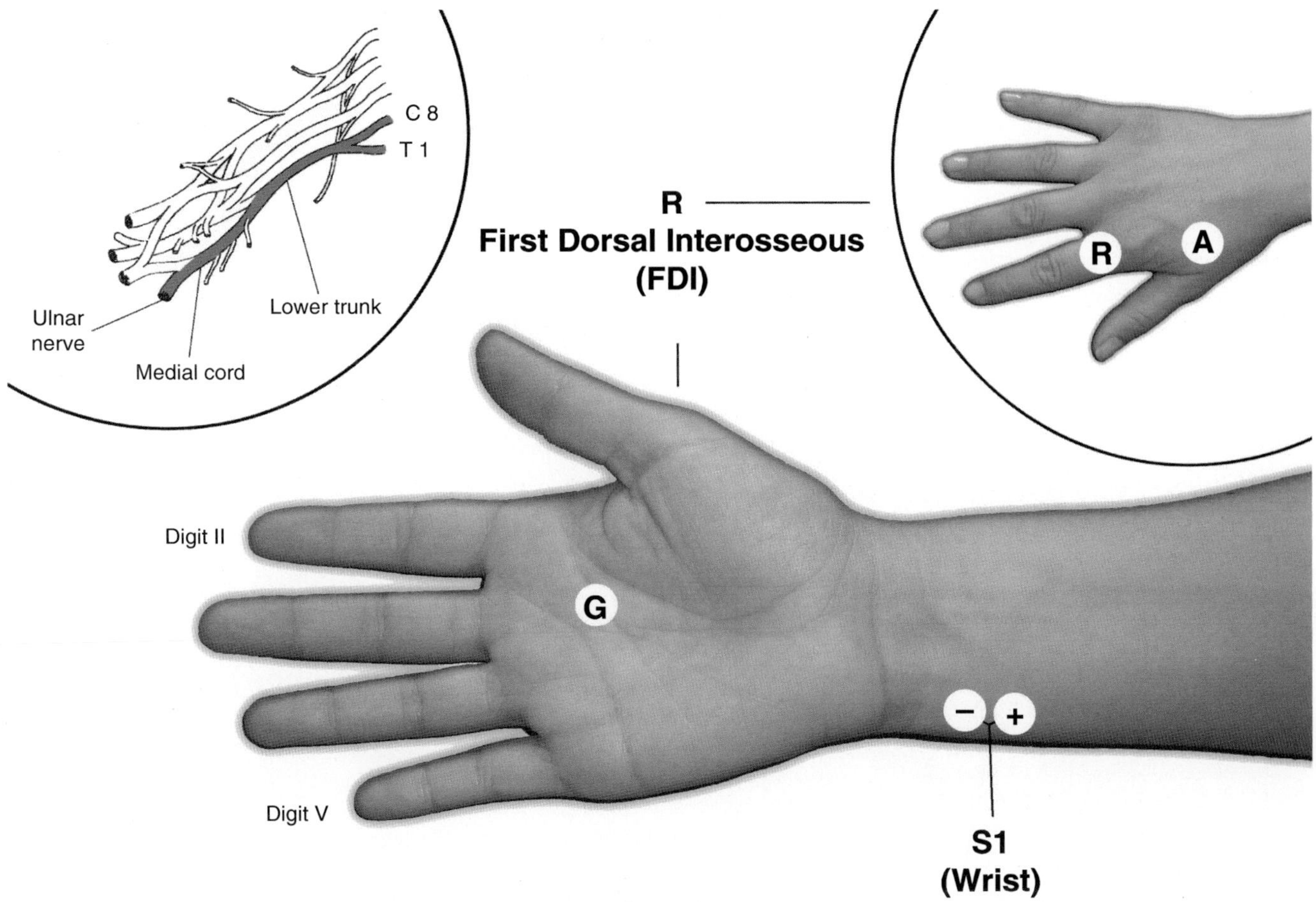

Typical waveform (wrist, below elbow, above elbow – FDI muscle):

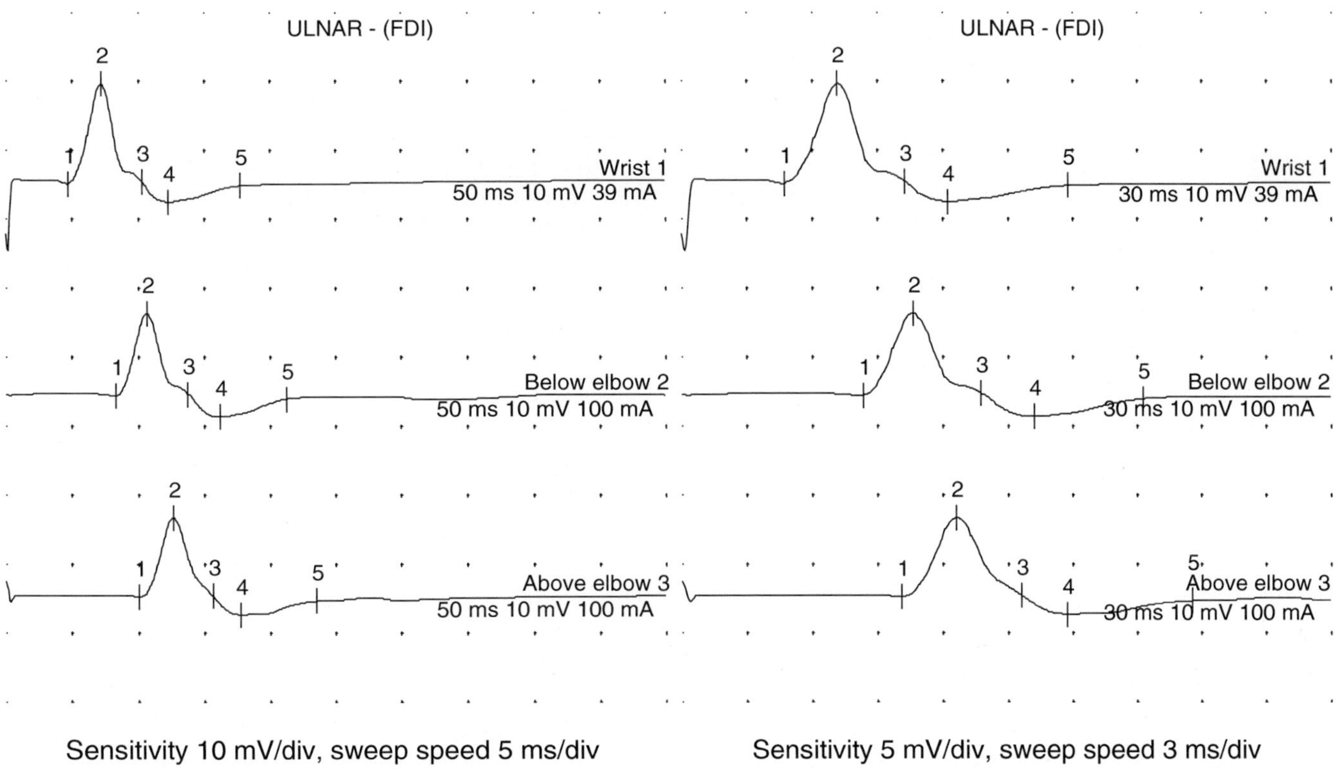

Fig. 2 Compound muscle action potentials (*CMAPs*) recorded at the hand from the FDI muscle; stimulation of the wrist (*upper trace*), below elbow (*middle trace*), and above elbow (*lower trace*)

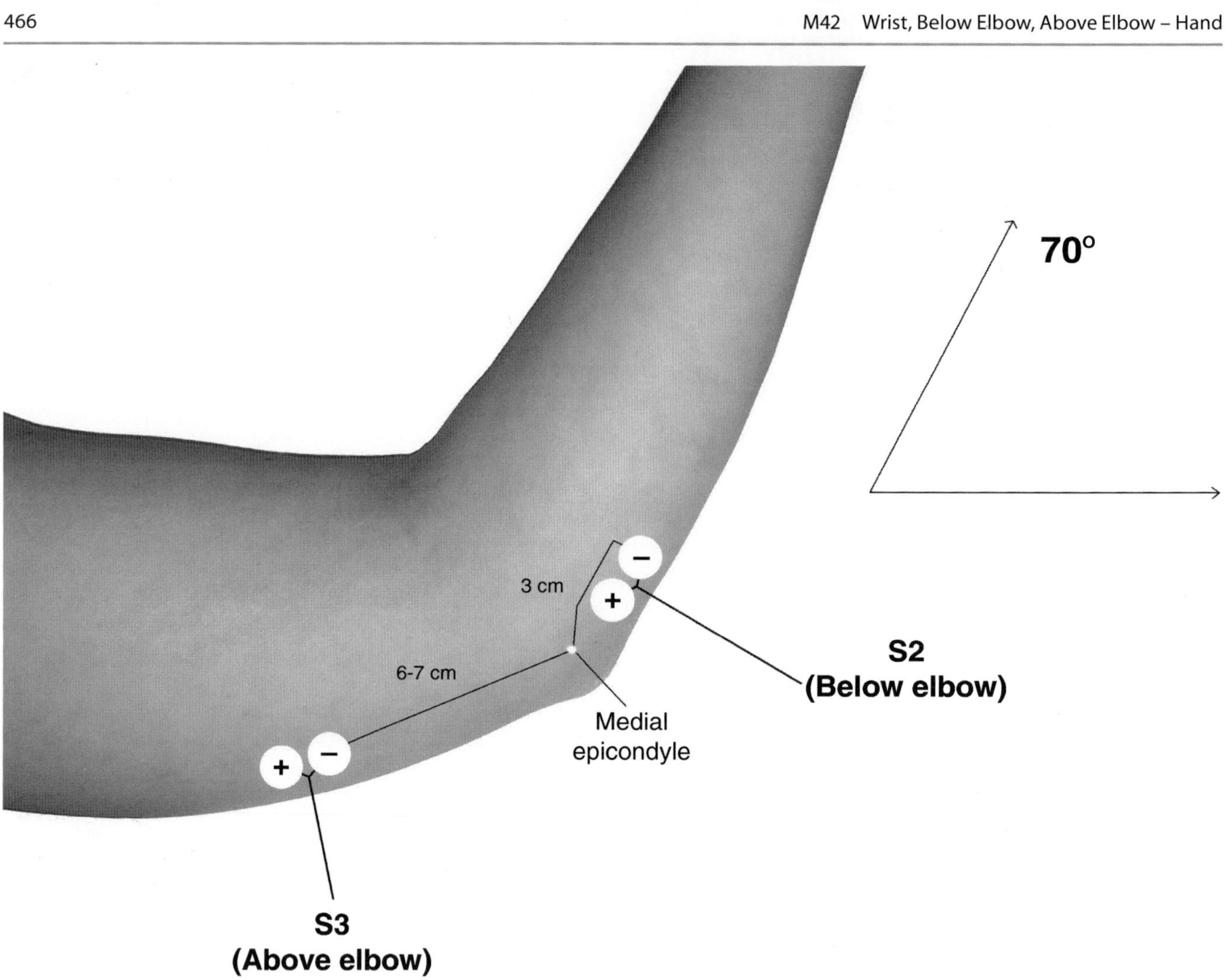

Fig. 3 Below elbow and above elbow stimulation points (elbow flexed to 70°)

cervical radiculopathy (11 women and 5 men, mean age 55 ± 12.35 years) and in a sample of 60 consecutive clinically diagnosed UNE patients (36 men and 24 women, mean age 53 ± 15.52 years). The UNE patients were divided into two groups: (1) patients with clinical UNE and pathological MNCV across the elbow (UNE NF+) and (2) patients with clinical UNE but without pathological MNCV across the elbow (UNE NF-). The patients were considered as having neurophysiological evidence of UNE when either FDI-MNCV or ADM-MNCV across the elbow was pathological, and they were also graded according to a previously published neurophysiological classification [8]: 10 arms were mild UNE, 17 were moderate UNE, 7 were severe UNE, and none showed extreme involvement. In seven arms, the authors found conduction block associated with absolute slowing across the elbow recording from both FDI and ADM muscles, and when in addition to the absolute reduction in FDI-MNCV and ADM-MNCV they used the AAEM criteria (drop in MNCV between AE-BE and BE-wrist segments), the overall sensitivity of the test was increased to 73 % considering all patients (UNE NF+ and UNE NF-) with a specificity of 82 % considering all normal subjects (control subjects and patients with lower cervical radiculopathy). Slower FDI-MNCV and ADM-MNCV in the AE-BE segment were observed in all patients with concomitant sensory and motor symptoms, and both FDI-MNCV and ADM-MNCV were significantly slower in the UNE NF+ sample compared with the UNE NF- patients, control subjects, and the patients with lower cervical radiculopathy. In the UNE NF- patients, Caliandro et al. [1] observed lower IN-RATIO values than they found in control subjects and patients with lower cervical radiculopathy. They calculated the IN-RATIO cutoff value by comparing the UNE NF- and all normal subjects (control subjects and patients with lower cervical radiculopathy) and found that if IN-RATIO was ≤0.97, then its sensitivity in diagnosing UNE NF- patients was 68 % and specificity was 91 %. When in addition to the absolute reduction in FDI-MNCV and ADM-MNCV they used the IN-RATIO, the overall sensitivity of the test was increased to 86 % considering all patients (UNE NF+ and UNE NF-) with a specificity of 91 % considering all normal subjects (control subjects and patients with lower cervical radiculopathy). The authors observed that in normal subjects, FDI-MNCV was faster than ADM-MNCV, and their possible explanation was that fastest fibers to FDI muscle have a larger diameter than the fastest fibers to ADM muscle. In their patients, fibers to FDI muscle were more impaired than fibers to ADM muscle, and this finding was mainly manifest when FDI-MNCV and ADM-MNCV in the AE-BE segment were within the normal range. For the authors, a possible explanation of this finding is that for somatotopic characteristics of ulnar nerve, fibers to the FDI muscle lie adjacent to bone and therefore are more susceptible to damage; when the nerve damage is greater, a larger number of fibers are of interest and from these also fibers to the ADM muscle. IN-RATIO allows to compare the function of two different fascicles of nervous fibers of a nerve within the same segment, eliminating all the variability typical of segmental and comparative techniques in which two different segments of one nerve or two different nerves are compared. Moreover, although the accuracy of latency is very important in the evaluation of IN-RATIO, this is independent of the absolute values of FDI-MNCV and ADM-MNCV, and therefore each laboratory may use its own normal values to diagnose UNE. They found IN-RATIO a useful neurophysiological parameter to increase sensitivity in patients with clinical UNE, and when IN-RATIO is positive and FDI-MNCV and ADM-MNCV in the AE-BE segment are within the normal range, they suggested to classify the patient as having a "minimal UNE." Pathological waveform and values in patients with ulnar neuropathy at the elbow (UNE) are reported (Figs. 4, 5, 6, and 7):

$$\text{IN RATIO (case 1)} = \frac{\text{FDI} - \text{MNCV (m/s)}}{\text{ADM} - \text{MNCV (m/s)}} = \frac{33.9\,\text{m/s}}{37.0\,\text{m/s}} = 0.92$$

$$\text{IN RATIO (case 2)} = \frac{\text{FDI} - \text{MNCV (m/s)}}{\text{ADM} - \text{MNCV (m/s)}} = \frac{39.2\,\text{m/s}}{44.4\,\text{m/s}} = 0.88$$

Pathological waveform (wrist, below elbow, above elbow – FDI muscle), case 1:

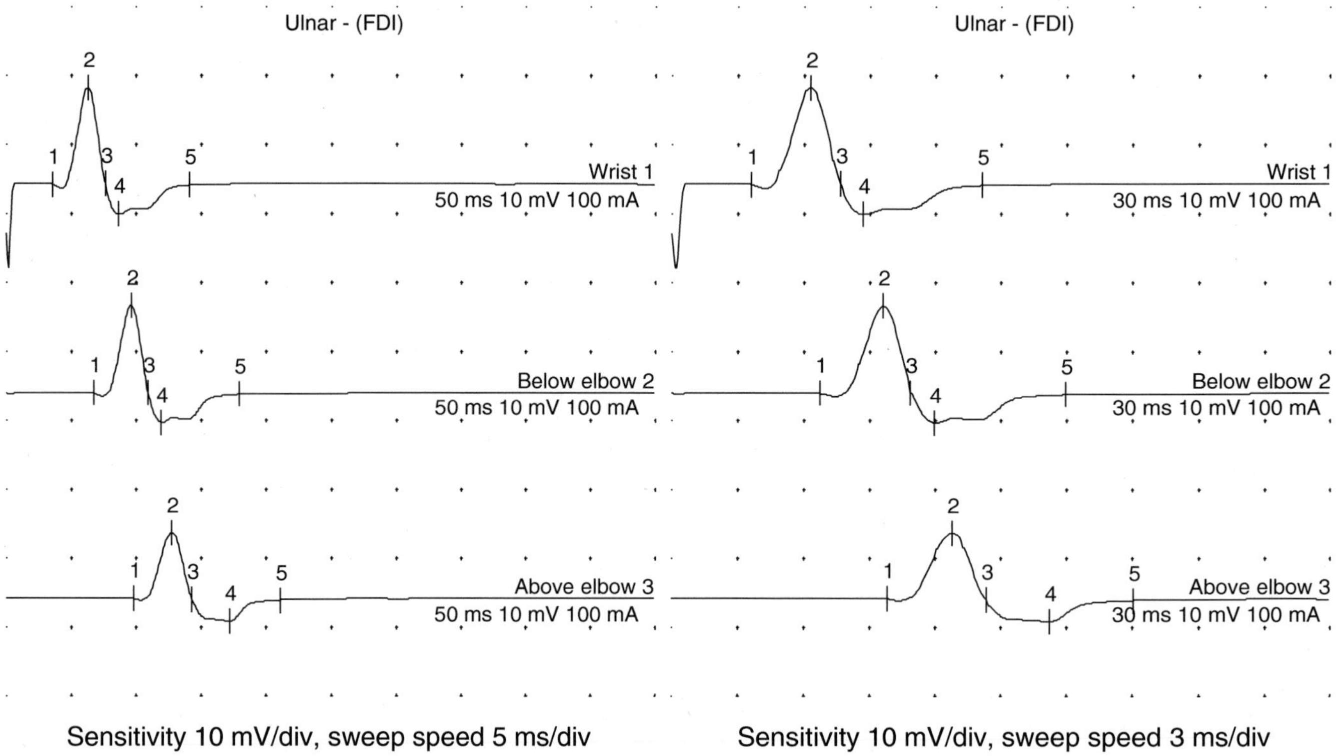

Onset latency (wrist – FDI): 4.20 ms; **Onset latency** (below elbow – FDI): 7.40 ms; **Onset latency** (above elbow – FDI): 10.35 ms; **Peak latency** (wrist – FDI): 6.20 ms; **Peak latency** (below elbow – FDI): 9.40 ms; **Peak latency** (above elbow – FDI): 12.35 ms; **Onset to peak amplitude** (wrist – FDI): 14.70 mV; **Onset to peak amplitude** (below elbow – FDI): 13.20 mV; **Onset to peak amplitude** (above elbow – FDI): 10.0 mV; **Peak to peak amplitude** (wrist – FDI): 18.5 mV; **Peak to peak amplitude** (below elbow – FDI): 17.1 mV; **Peak to peak amplitude** (above elbow – FDI): 13.1 mV; **MNCV** (wrist – below elbow, 18 cm): 56.3 m/s; **MNCV** (below elbow – above elbow, 10 cm): 33.9 m/s

Fig. 4 Compound muscle action potentials (*CMAPs*) recorded at the hand from the FDI muscle in UNE; stimulation at the wrist (*upper trace*), below elbow (*middle trace*), and above elbow (*lower trace*)

Pathological waveform (wrist, below elbow, above elbow – ADM muscle), case 1:

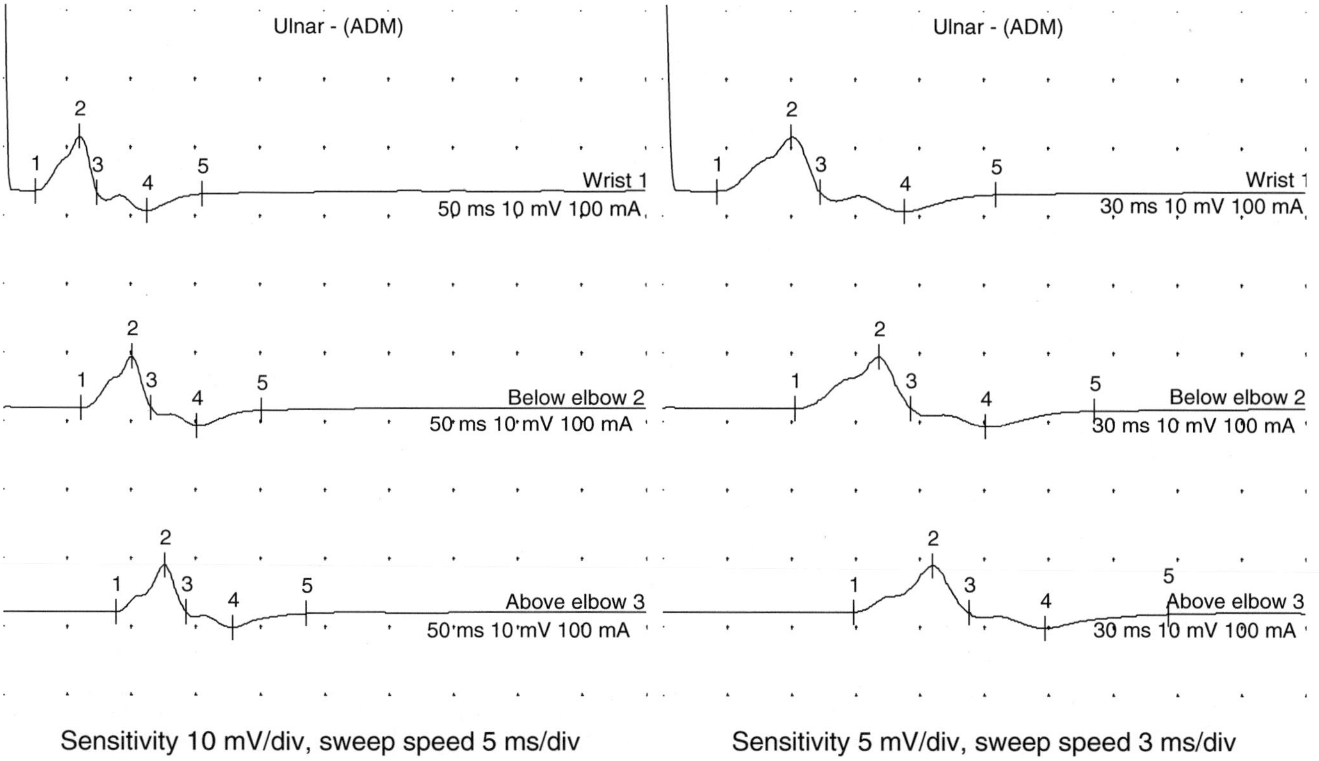

Onset latency (wrist – ADM): 2.55 ms; **Onset latency** (below elbow – ADM): 6.15 ms; **Onset latency** (above elbow – ADM): 8.85 ms; **Peak latency** (wrist – ADM): 6.0 ms; **Peak latency** (below elbow – ADM): 10.05 ms; **Peak latency** (above elbow – ADM): 12.45 ms; **On set to peak amplitude** (wrist – ADM): 8.2 mV; **Onset to peak amplitude** (below elbow – ADM): 7.6 mV; **Onset to peak amplitude** (above elbow – ADM): 6.9 mV; **Peak to peak amplitude** (wrist – ADM): 11.1 mV; **Peak to peak amplitude** (below elbow – ADM): 10.4 mV; **Peak to peak amplitude** (above elbow – ADM): 9.3 mV; **MNCV** (wrist – below elbow, 18 cm): 50.0 m/s; **MNCV** (below elbow – above elbow, 10 cm): 37.0 m/s

Fig. 5 Compound muscle action potentials (*CMAPs*) recorded at the hand from the ADM muscle in UNE; stimulation at the wrist (*upper trace*), below elbow (*middle trace*), and above elbow (*lower trace*)

Pathological waveform (wrist, below elbow, above elbow – FDI muscle), case 2:

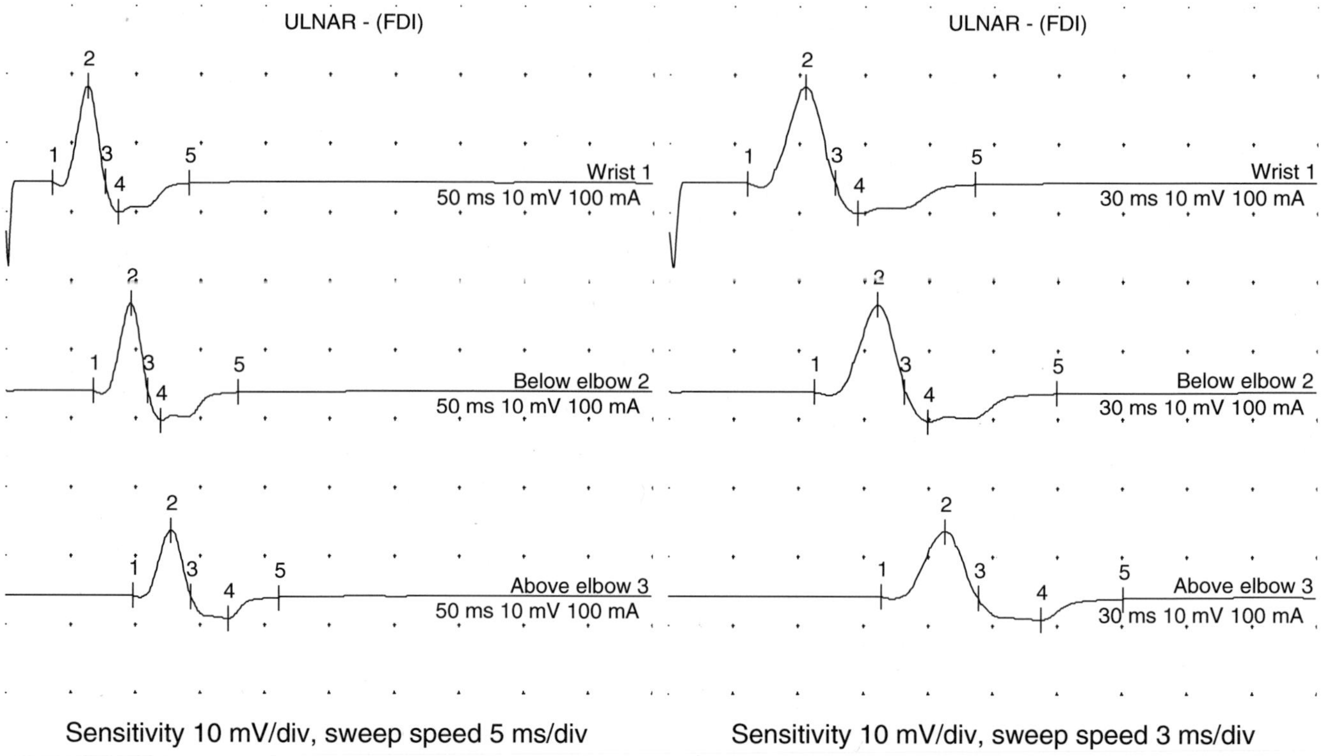

Onset latency (wrist – FDI): 4.45 ms; **Onset latency** (below elbow – FDI): 8.40 ms; **Onset latency** (above elbow – FDI): 10.95 ms; **Peak latency** (wrist – FDI): 6.15 ms; **Peak latency** (below elbow – FDI): 9.30 ms; **Peak latency** (above elbow – FDI): 12.40 ms; **Onset to peak amplitude** (wrist – FDI): 11.0 mV; **Onset to peak amplitude** (below elbow – FDI): 8.10 mV; **Onset to peak amplitude** (above elbow – FDI): 8.10 mV; **Peak to peak amplitude** (wrist – FDI): 14.2 mV; **Peak to peak amplitude** (below elbow– FDI): 10.6 mV; **Peak to peak amplitude** (above elbow – FDI): 10.8 mV; **MNCV** (wrist – below elbow, 22 cm): 55.7 m/s; **MNCV** (below elbow – above elbow, 10 cm): 39.2 m/s

Fig. 6 Compound muscle action potentials (*CMAPs*) recorded at the hand from the FDI muscle in UNE; stimulation at the wrist (*upper trace*), below elbow (*middle trace*), and above elbow (*lower trace*)

Pathological waveform (wrist, below elbow, above elbow – ADM muscle), case 2:

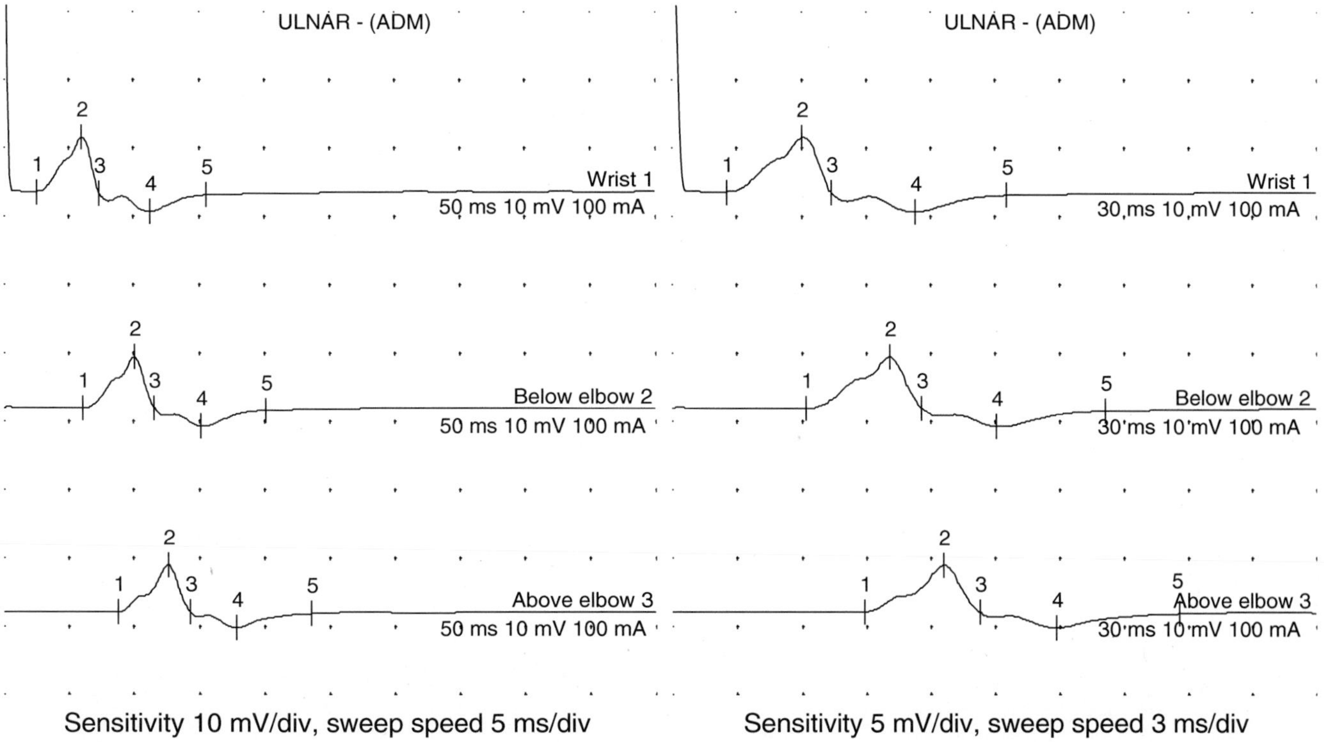

Onset latency (wrist – ADM): 2.70 ms; **Onset latency** (below elbow – ADM): 6.95 ms; **Onset latency** (above elbow – ADM): 9.20 ms; **Peak latency** (wrist – ADM): 5.95 ms; **Peak latency** (below elbow – ADM): 10.0 ms; **Peak latency** (above elbow – ADM): 12.35 ms; **Onset to peak amplitude** (wrist – ADM): 8.7 mV; **Onset to peak amplitude** (below elbow – ADM): 7.3 mV; **Onset to peak amplitude**(above elbow – ADM): 6.6 mV; **Peak to peak amplitude** (wrist – ADM): 11.8 mV; **Peak to peak amplitude** (below elbow–ADM): 10.4 mV; **Peak to peak amplitude** (above elbow – ADM): 9.3 mV; **MNCV** (wrist – below elbow, 22 cm): 51.8 m/s; **MNCV** (below elbow – above elbow, 10 cm): 44.4 m/s

Fig. 7 Compound muscle action potentials (*CMAPs*) recorded at the hand from the ADM muscle in UNE; stimulation at the wrist (*upper trace*), below elbow (*middle trace*), and above elbow (*lower trace*)

References

1. Caliandro P, Foschini M, Pazzaglia C et al (2008) IN-RATIO: a new test to increase diagnostic sensitivity in ulnar nerve entrapment at elbow. Muscle Nerve 119:1600–1606
2. AAEM Quality Assurance Committee: practice parameter for electrodiagnostic studies in ulnar neuropathy at the elbow: summary statement (1999) Muscle Nerve 22:408–411
3. Checkles NS, Russakov AD, Piero DL (1971) Ulnar nerve conduction velocities: effect of elbow position on measurement. Arch Phys Med Rehabil 52:362–365
4. Kincaid JC, Phillips LH, Daube JR (1986) The evaluation of suspected ulnar neuropathy at the elbow. Normal conduction study values. Arch Neurol 43:44–47
5. Kothari MJ, Preston DC (1995) Comparison of the flexed and extended elbow positions in localizing ulnar neuropathy at the elbow. Muscle Nerve 18:336–340
6. Kothari MJ, Heistand M, Rutkove SB (1998) Three ulnar nerve conduction studies in patients with ulnar neuropathy at the elbow. Arch Phys Med Rehabil 79:87–89
7. Azma K, Bahmanteimoury K, Tavana B et al (2007) Two measurement methods of motor ulnar nerve conduction velocity at the elbow: a comparative study. Neurol India 55:145–147
8. Padua L, Aprile I, Mazza O et al (2001) Neurophysiological classification of ulnar entrapment across the elbow. Neurol Sci 22:11–16

Below Elbow, Above Elbow – Hand

M43

Surface Recording Technique, Study from *Abductor Digiti Minimi* (ADM) Muscle

Original Settings Sensitivity was 5 mV/division, sweep speed was 2 mV/division, low-frequency filter was 20 Hz, high-frequency filter was 10 kHz, duration was 0.1 ms, and the machine used was a TECA Synergy.

Position In a supine position, with the shoulder and upper arm externally rotated, the elbow is flexed to 90° and the forearm supinated.

Recording The recording electrodes were placed over the *abductor digiti minimi* (ADM) muscle. The active (A) recording electrode was placed over the belly of the ADM muscle; the reference (R) electrode was placed on the distal tendinous insertion (Fig. 1). The ground (G) electrode was placed over the volar forearm (Fig. 2).

Stimulation The compound muscle action potentials (CMAPs) were recorded following stimulations from the below-elbow and above-elbow sites. The authors used a two-straight-line method instead of conventional curvilinear around the elbow measurement (Fig. 2). First, a straight line was drawn from the medial epicondyle to the olecranon process. The below-elbow distal stimulation point (first stimulation site) was at a fixed 3 cm distance from a point one third the distance from the medial epicondyle to the olecranon process, and the above-elbow proximal stimulation point (second stimulation point) was at a proximal fixed 3 cm distance from a point one third the distance from the medial epicondyle to the olecranon process. If a submaximal stimulation of 10–15 mV at 0.1 ms duration over the proximal point did not produce any CMAPs, the stimulator was moved, in an arc-like fashion, to a point where it did, maintaining the 3 cm distance. The authors used a caliper set at 3 cm to reduce measurement error. CMAPs were elicited using supramaximal stimulation on each site.

Measurements The absolute 6 cm conduction time across the elbow (CTE) was defined as the difference in CMAP onset latency between proximal and distal stimulation points. The CTE was calculated on both arms and conduction time across the elbow (CTE-diff) was defined as the latency difference between the CTE of the opposing limbs. The skin temperature was evaluated with a surface probe placed on the skin overlying the elbow, and the limb was warmed by convection heating, if necessary, to maintain a skin temperature above 32°. Normal values (Table 1) were obtained from 72 healthy subjects (age range 27–85 years) divided into three groups in according to age: Group A (20–40 years, mean age 33.2 years), 24 subjects; Group B (40–60 years, mean age 48.6 years), 24 subjects; and Group C (>60 years, mean age 70.3 years), 24 subjects.

Comment

Landau and Campbell [1] found 95 % of all subjects had CTE-diff lower than 0.40 ms and 99 % of the subjects had CTE-diff lower than 0.50 ms. They recommended that a CTE-diff >0.45 ms could be considered abnormal, attaining a specificity of 97 %. They believed this method a useful tool to the evaluation of ulnar neuropathy at the elbow (UNE) since the internal control for patients with unilateral symptoms would be the contralateral limb (biologically and technically more similar than comparing the across-elbow to the forearm segment). A distance of 6 cm could potentially increase the sensitivity for lesion detection, but distance measurements traced over a nonlinear segment were associated with a marked increase in variability, and a motor nerve conduction velocity (MNCV) decrease of >10 m/s from the forearm to the across-elbow segment could occur from distance measurement alone [2]. The straight-line distances measured with a caliper would decrease measurement variability and should improve specificity. For the authors [1], reducing experimental error to improve specificity is certainly significant, but the techniques used for diagnosis must also have adequate sensitivity. Conduction

G. Gentili, M. Di Napoli, *The Ulnar Nerve*, DOI 10.1007/978-3-319-43292-2_67

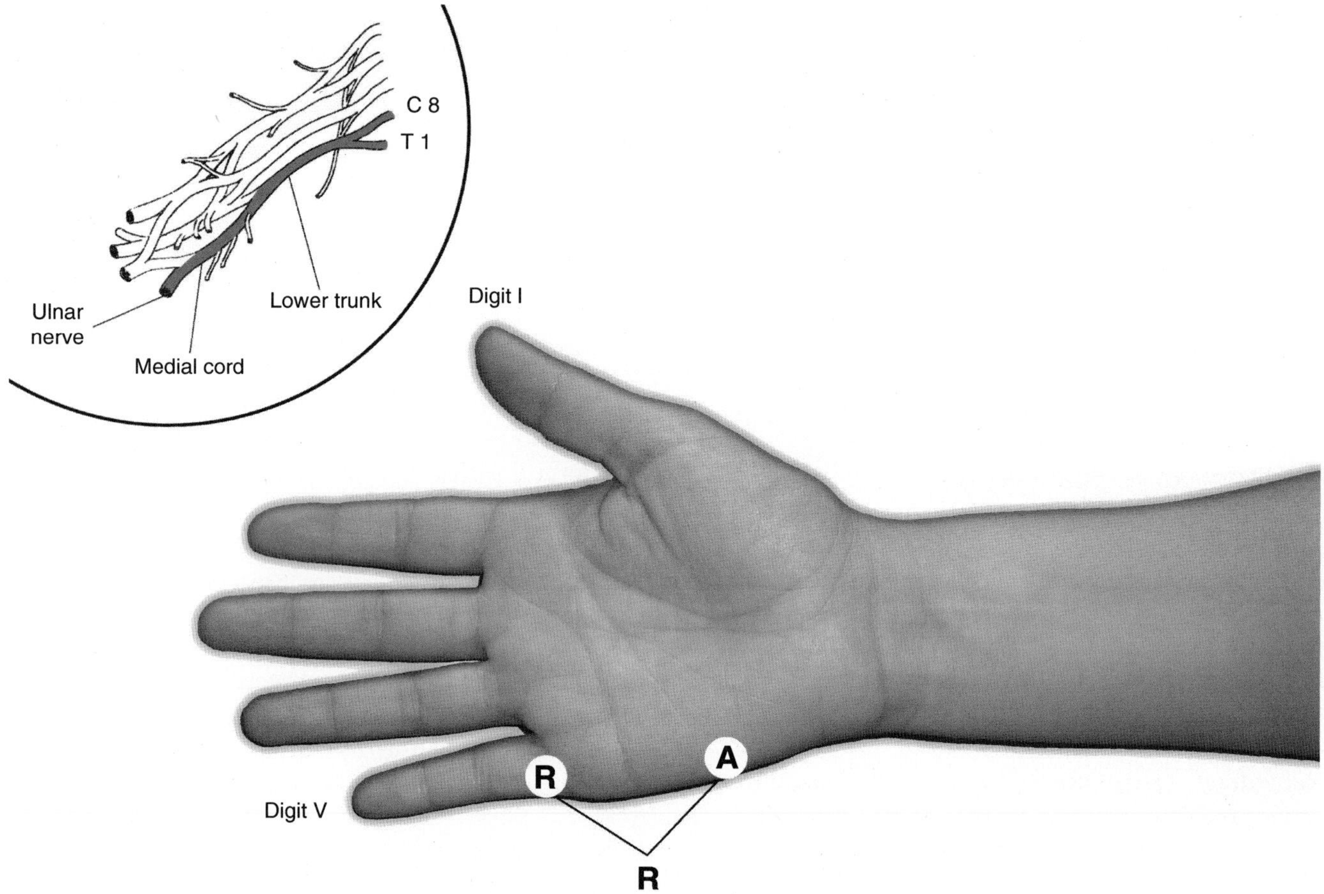

Abductor Digiti Minimi (ADM)

Typical waveform (below elbow, above elbow – ADM muscle):

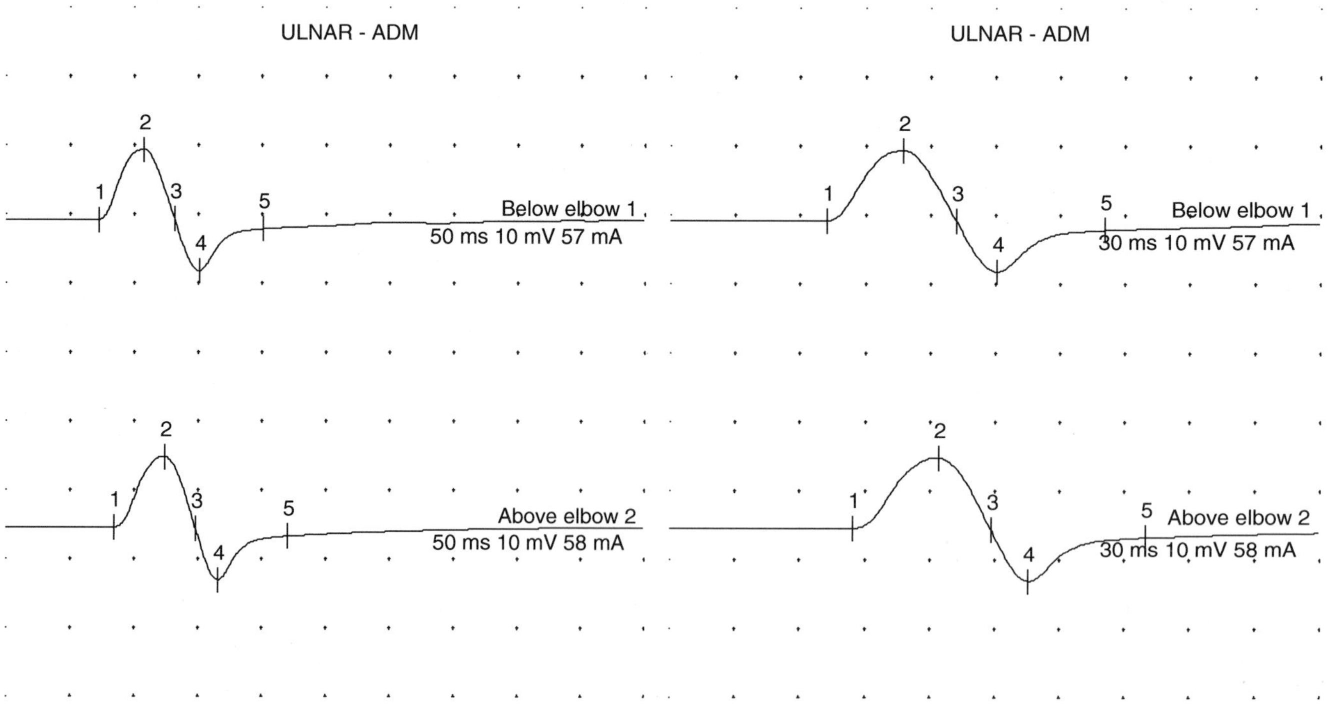

Sensitivity 10 mV/div, sweep speed 5 ms/div

Sensitivity 10 mV/div, sweep speed 3 ms/div

Fig. 1 Compound muscle action potentials (*CMAPs*) recorded at the hand from the ADM muscle; stimulation below elbow (*upper trace*) and above elbow (*lower trace*)

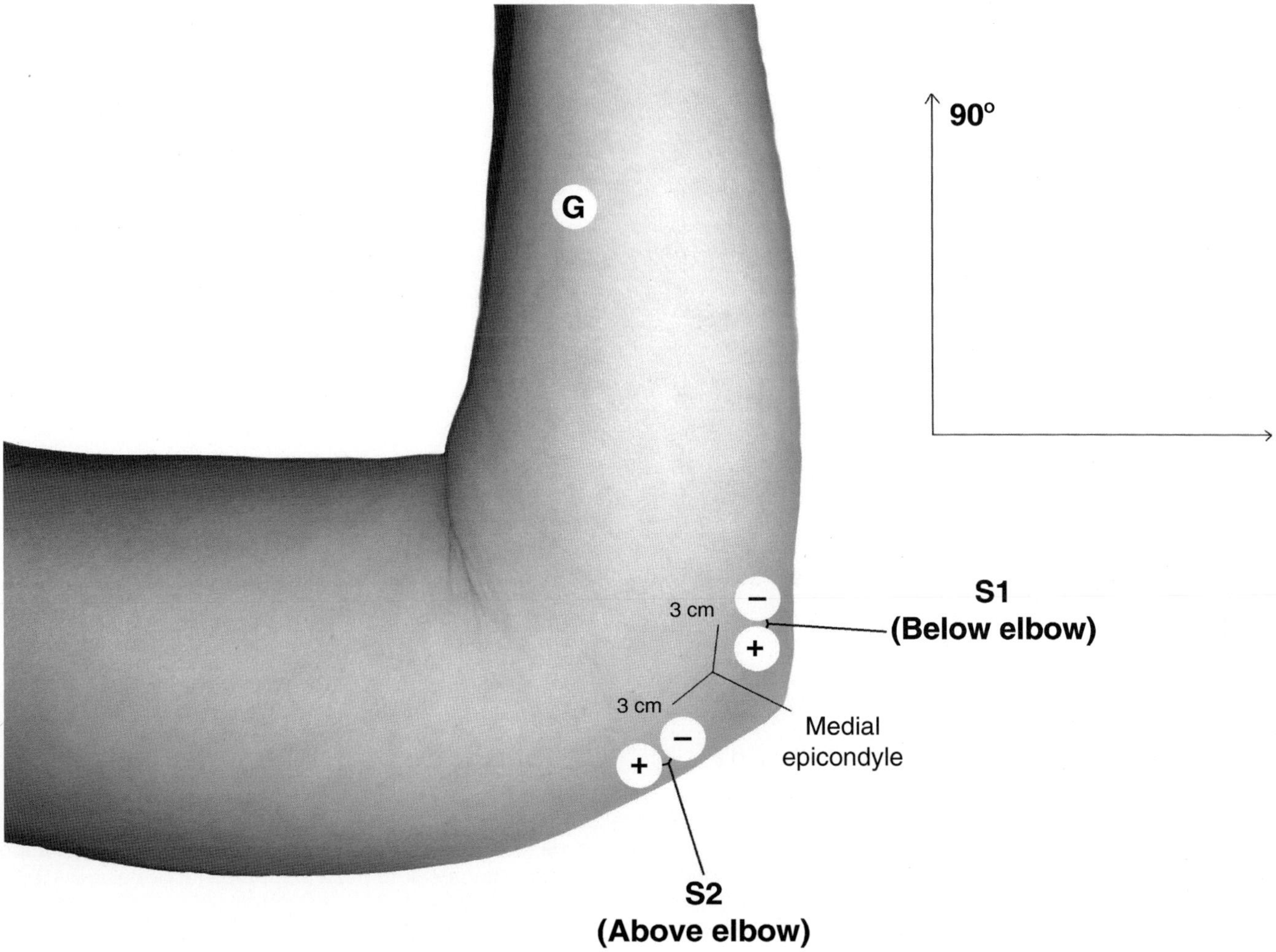

Fig. 2 Below-elbow (S1) and above-elbow (S2) stimulation points (6 cm distance) with the elbow flexed to 90°

Table 1 Normal values [1]

	Mean ± SD	Range	Upper limit
CTE (ms), Group A – 20–40 years	1.16 ± 0.16	0–0.50	1.50
CTE (ms), Group B – 40–60 years	1.23 ± 0.18	0–0.45	1.70
CTE (ms), Group C – >60 years	1.33 ± 0.24	0–0.55	1.80

time across the elbow (CTE) will likely underestimate the true conduction time as the straight-line measurement underestimate the true length of the curved nerve segment (the ulnar nerve in the elbow region curves when the elbow is flexed). Because focal nerve injuries typically occur over a 1 cm segment, when longer nerve distances are used in MNCV calculation, the sensitivity for detecting that lesion declines due to a "dilution" effect of the normally conducting segments [3]. The optimal recording distance to detect focal lesions should balance decreased sensitivity at longer distances and decreased specificity at shorter distances. Some authors [4, 5] demonstrated that distance measurements of 5–6 cm provided the most favorable balance between false-positive and false-negative errors, and the best distance for screening for UNE was found to be 6–8 cm [4]. The two most common locations for UNE are the level of the retroepicondylar groove and at the humeroulnar arcade which lies about 2–3 cm distal to the medial epicondyle [3], and a nerve conduction study needs to accommodate both of these sites. Considering sensitivity, specificity, and the entrapment locations of ulnar nerve, Landau and Campbell [1] developed an electrodiagnostic method to optimally detect UNE using a fixed distance of 6 cm (6 cm test). Pathological waveform and values in a patient with ulnar neuropathy at the elbow (UNE) are reported (Fig. 3).

Pathological waveform (below elbow, above elbow – ADM muscle):

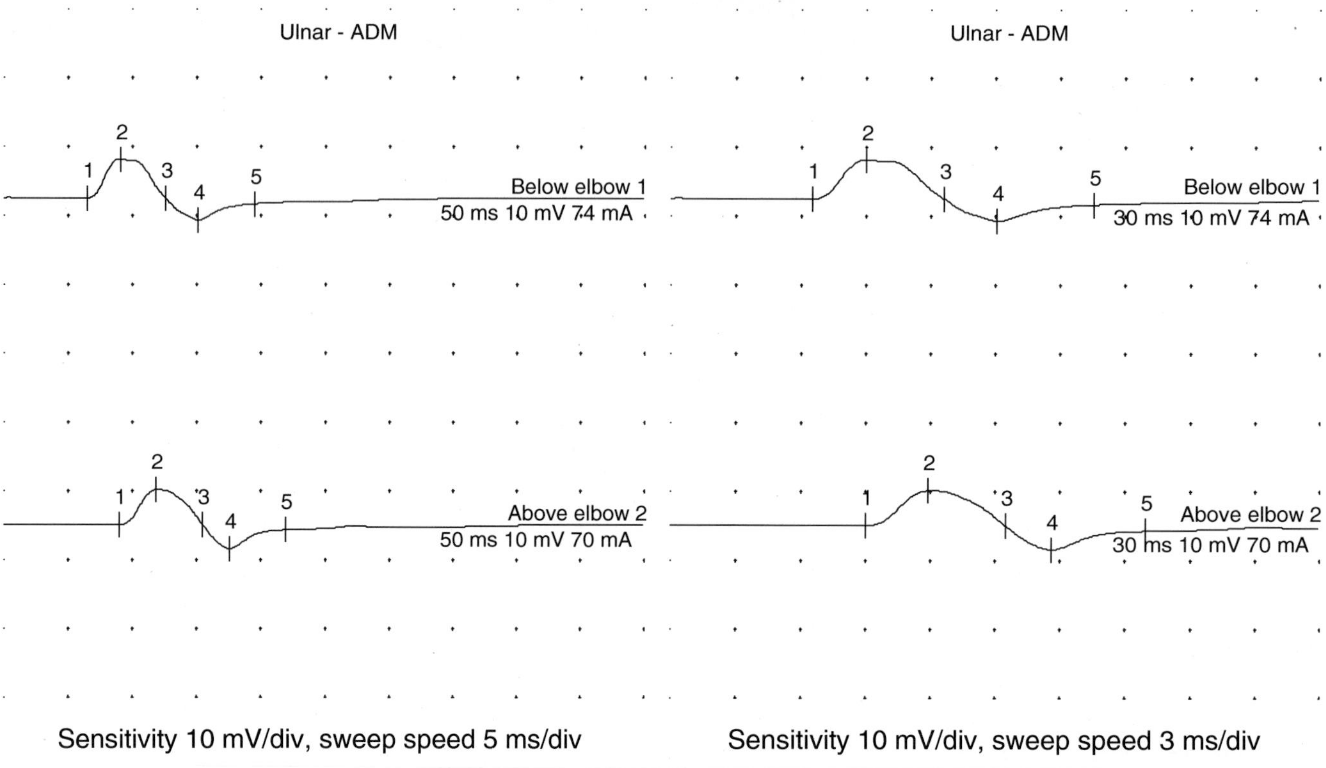

Onset latency (below elbow - ADM): 6.50 ms; **Onset latency** (above elbow - ADM): 9.0 ms; **Peak latency** (below elbow - ADM): 9.0 ms; **Peak latency** (above elbow-ADM): 11.90 ms; **Onset to peak amplitude** (below elbow-ADM): 5.7 mV; **Onset to peak amplitude** (above elbow-ADM): 5.1 mV; **Peak to peak amplitude** (below elbow-ADM): 9.0 mV; **Peak to peak amplitude** (above elbow-ADM): 8.7 mV; **MNCV** (below elbow-above elbow, 6 cm distance): 24.0 m/s

Fig. 3 Compound muscle action potentials (*CMAPs*) recorded at the hand from the ADM muscle in a patient with UNE; stimulation below elbow (*upper trace*) and above elbow (*lower trace*)

References

1. Landau ME, Campbell WW (2009) Conduction time for a 6-cm segment of the ulnar nerve across the elbow: reference values for the 6-cm conduction time test. J Clin Neurophysiol 26:198–200
2. Landau ME, Diaz MI, Barner KC et al (2002) Changes in nerve conduction velocity across the elbow due to experimental error. Muscle Nerve 26:838–840
3. Campbell WW, Pridgeon RM, Sahni KS (1992) Short segment incremental studies in the evaluation of ulnar neuropathy at the elbow. Muscle Nerve 15:1050–1054
4. Landau ME, Diaz MI, Barner KC et al (2003) Optimal distance for segmental nerve conduction studies revisited. Muscle Nerve 27:570–574
5. Van Djik JG, Meulstee J, Zwarts MJ et al (2001) What is the best way to assess focal slowing of the ulnar nerve? Clin Neurophysiol 112:286–293

Wrist – Hand

M44

Surface Recording Technique, Study from *Abductor Digiti Minimi* (ADM) and *Abductor Pollicis Brevis* (APB) Muscles

Original Settings Sensitivity was 5 mV/division (motor waveform and M-wave) and 500 mV/division (F-wave), sweep speed was 2 ms/division, low-frequency filter was 10 Hz, high-frequency filter was 10 kHz, and the machine used was a Medelec Oxford Synergy electromyography. Duration of pulse and rate of pulse were not specified.

Position This study was performed with the subject lying on a bed with the arm and the forearm flexed to 35°–45°; the figures show elbow fully extended.

Recording For the ulnar nerve conduction studies, the active electrode (A) was placed over the belly (motor point) of the *abductor digiti minimi* (ADM) muscle (Fig. 1); the active (A) electrode was placed on the motor point of the ADM muscle, and the reference (R) was placed distally near its tendon of insertion [1]. For the median nerve conduction studies, the active electrode (A) was placed over the belly (motor point) of the *abductor pollicis brevis* (APB) muscle (Fig. 2); the active (A) electrode was placed on the motor point of the APB muscle, and the reference (R) was placed distally to the distal interphalangeal joint of the digit I [1]. The ground (G) position was not mentioned in the text; the figures show ground positioned on the forearm.

Stimulation F-wave recordings of ulnar and median nerves [1] were carried out by stimulating each nerve on its course at the wrist and recording the motor and late responses from the ADM and APB muscles, respectively (Figs. 1 and 2). Routine ulnar motor conduction studies were performed by stimulating it on the wrist and below and above the medial epicondyle and recording the compound muscle action potentials (CMAPs) from the ADM muscle. Conventional median motor conduction studies were performed by stimulating it on the wrist and antecubital fossa and recording CMAPs over the belly of APB muscle. All nerve conduction studies were performed using standard techniques of supramaximal percutaneous stimulation with a constant current stimulator and surface electrode recording. The author used a supramaximal stimulus intensity (increased by 25 %). The anode was proximal, and the cathode was distal for conventional motor conduction studies; the cathode was proximal, and the anode was distal for F-wave studies. No fixed distances, between stimulating and recording electrodes, were reported in the article.

Measurements Distal latency (ms) of the compound muscle action potential (CMAP) was measured from the stimulus artifact to the onset of the negative deflection of CMAP. Motor nerve conduction velocity (MNCV) was calculated by dividing the distance by the difference of proximal and distal onset latencies, and it was measured in meter per second (m/s). To detect the minimum F-wave latency, a series of 16 F-waves was provoked and the shortest latency was derived. The difference of minimum F-wave latency of ulnar nerve to the minimum F-wave latency of ipsilateral median nerve was calculated bilaterally in each subject. The studies were performed in a warm room, and the temperature was maintained at 26°–28°; the limb was warmed if necessary. The author [1] studied 34 ulnar nerves from 17 patients (11 men and 6 women, age range 23–79 years, mean age 36 ± 13.8 years) with left ulnar neuropathy at the elbow (UNE). All the patients had left-sided UNE, and data of healthy right arms were used as control values (Table 1). The author used data of healthy contralateral median and ulnar nerves of the same patient as control values to avoid the interference due to physical characteristics of the individuals (age, height, etc.) which may limit the utility of the reference values. Pathological values were carried out from 17 left arms with clinically and electrophysiologically diagnosed UNE (Table 2).

G. Gentili, M. Di Napoli, *The Ulnar Nerve*, DOI 10.1007/978-3-319-43292-2_68

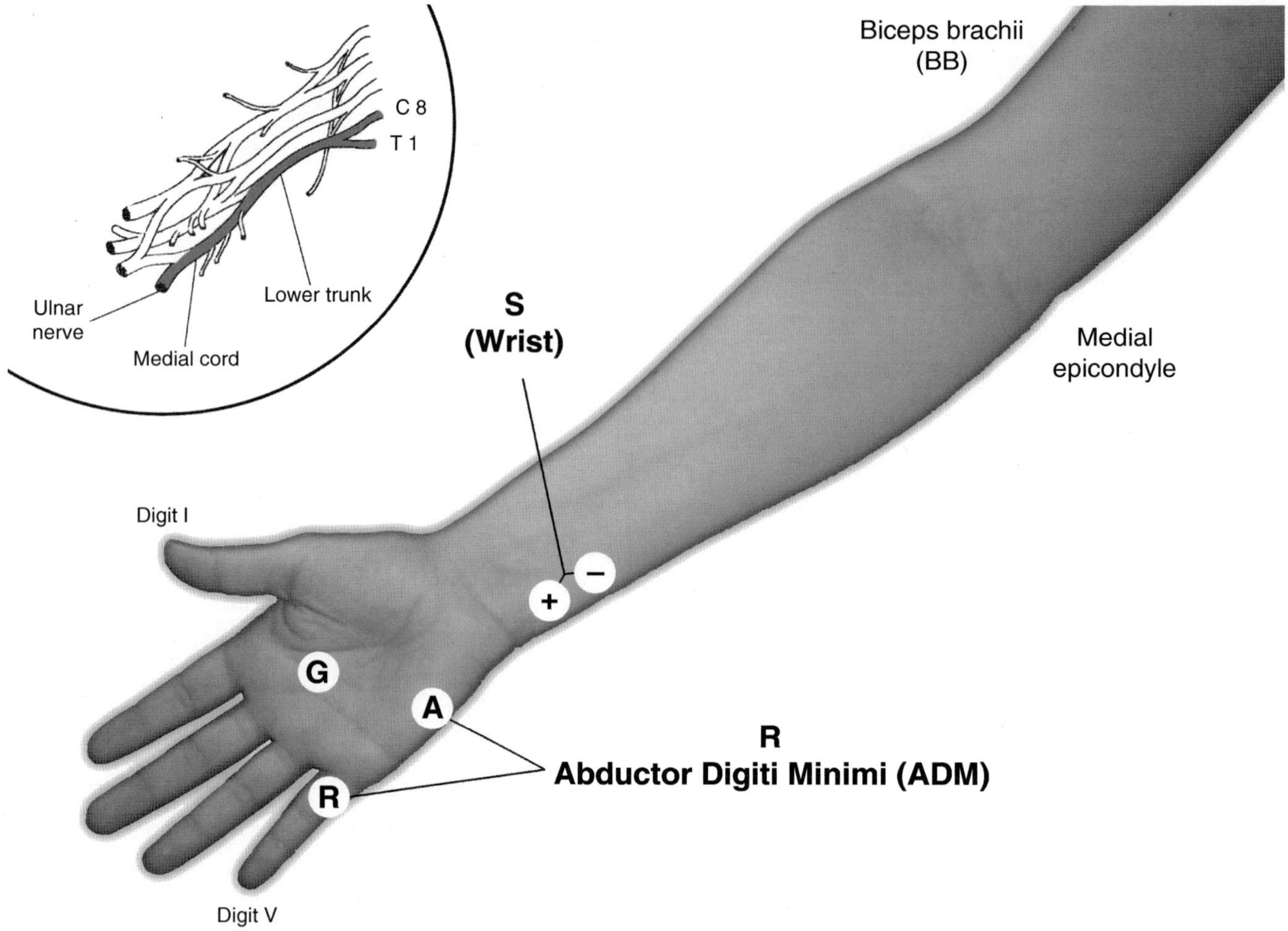

Typical waveform (wrist – ADM muscle):

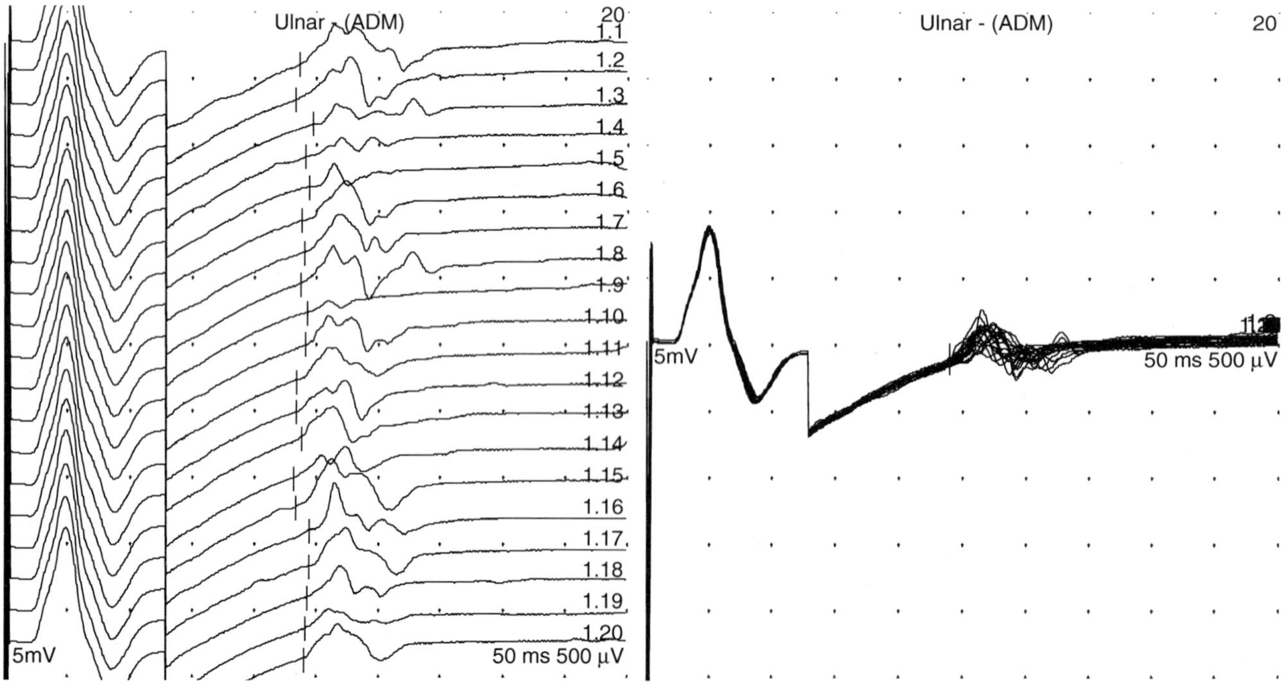

Sensitivity 500 μV/div (CMAP 5 mV/div), sweep speed 5 ms/div

Fig. 1 F-waves recorded at the hand from the ADM muscle, stimulation of the wrist; raster mode, 20 stimuli (*left*); superimposed mode, 20 stimuli (*right*)

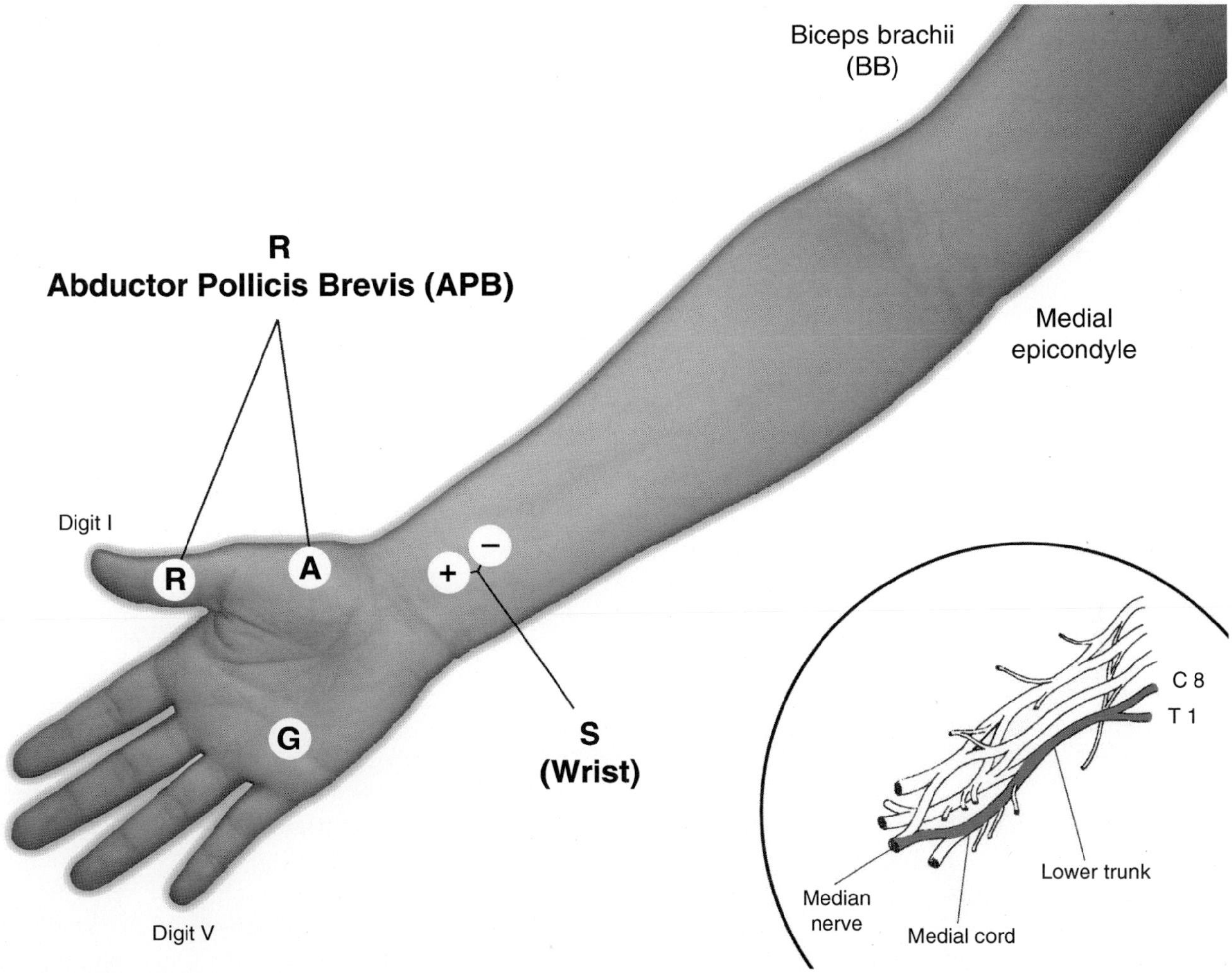

Typical waveform (wrist – APB muscle):

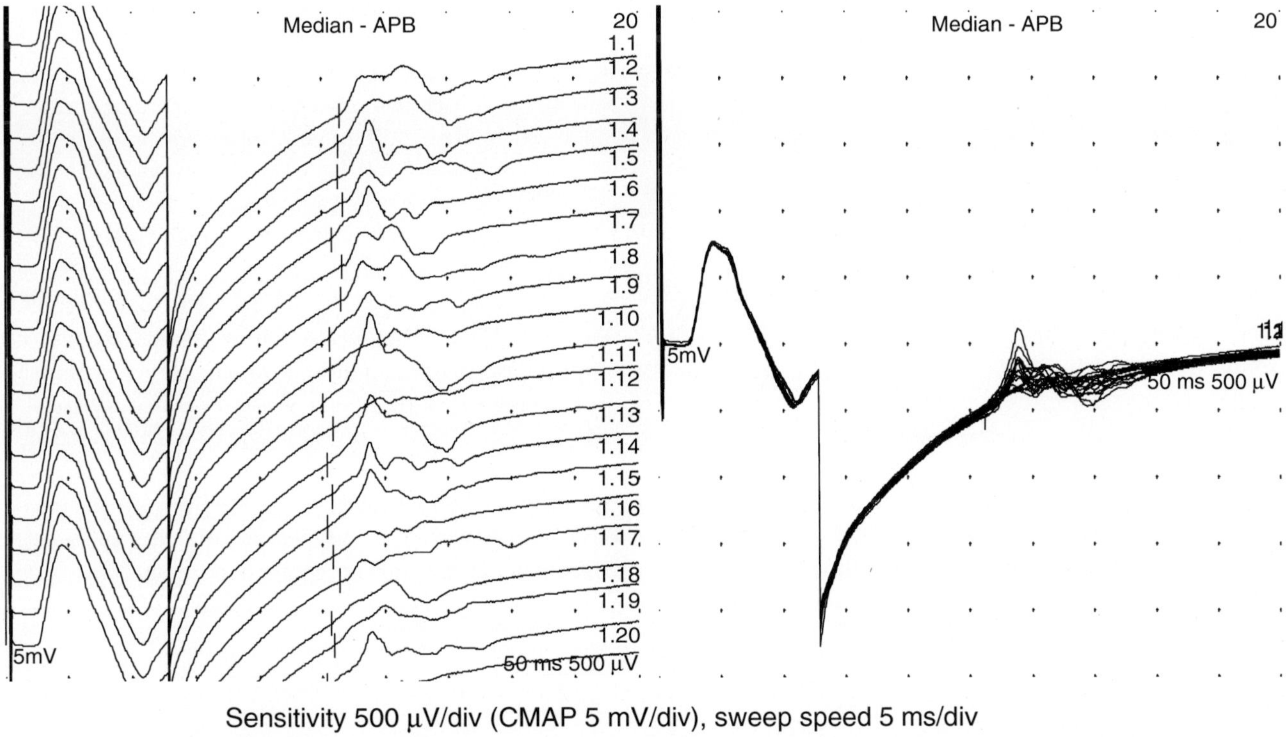

Fig. 2 F-waves recorded at the hand from the APB muscle, stimulation of the wrist; raster mode, 20 stimuli (*left*); superimposed mode, 20 stimuli (*right*)

Table 1 Normal values [1]

	Mean ± SD
Ulnar nerve, wrist–ADM distal motor latency (ms)	2.58 ± 0.30
Median nerve, wrist–APB distal motor latency (ms)	3.51 ± 0.36
Ulnar nerve, forearm MNCV (m/s)	59.26 ± 5.44
Median nerve, forearm MNCV (m/s)	59.17 ± 3.19
Ulnar nerve, across-elbow MNCV (m/s)	51.74 ± 8.56
Ulnar nerve, F-wave minimum latency (ms)	26.60 ± 2.24
Median nerve, F-wave minimum latency (ms)	25.24 ± 1.80
Ulnar–median nerves, F-wave latency difference, FWLD (ms)	0.76 ± 1.03

Table 2 Pathological values [1]

	Mean ± SD
Ulnar nerve, wrist–ADM distal motor latency (ms)	2.78 ± 0.38
Median nerve, wrist–APB distal motor latency (ms)	3.54 ± 0.27
Ulnar nerve, forearm MNCV (m/s)	58.27 ± 5.28
Median nerve, forearm MNCV (m/s)	59.47 ± 4.40
Ulnar nerve, across-elbow MNCV (m/s)	35.12 ± 9.15
Ulnar nerve, F-wave minimum latency (ms)	29.10 ± 3.28
Median nerve, F-wave minimum latency (ms)	25.59 ± 1.69
Ulnar–median nerves, F-wave latency difference, FWLD (ms)	3.05 ± 2.55

Comment

Alemdar [1] decided to conduct a study researching the utility of comparing the F-wave latencies of ipsilateral ulnar nerve (recording from the ADM muscle) to median nerve (recording from the APB muscle) in confirmation of UNE. Sander et al. [2] first described the usefulness of F-wave latencies in entrapment neuropathies, such as the carpal tunnel syndrome (CTS). They revealed that median to ulnar nerve minimum F-wave latency difference (FWLD) could be useful in the diagnosis of CTS. The author [1] applied the FWLD method in the diagnosis of UNE, because of the common pathway (medial cord of the brachial plexus, C8–T1 roots) shared by the ulnar and median nerves. Any pathologic conditions involving C8 and/or T1 roots or medial cord of brachial plexus is expected to prolong F-wave latencies of both nerves, whereas UNE is expected to prolong F-wave latencies of ulnar nerve solely. He performed several conduction techniques for both median and ulnar nerves, such as routine motor conduction study, F-wave latency, and minimum median–ulnar FWLD. In all patients, Alemdar [1] found no statistical differences between healthy right arms and left arm with UNE. There was a significant difference in MNCV across the elbow since the mean ulnar nerve across-elbow MNCV in affected arms (35.12 ± 9.15 m/s) was lower than one in healthy arms (51.74 ± 8.56 m/s). In all affected arms, the minimum F-wave latency of ulnar nerve was longer than one of the median nerves (minimum difference – 1.05 ms, maximum difference 10.90 ms), and the minimum F-wave latency difference (FWLD) in affected arms was greater (3.05 ± 2.55 ms) than ones in healthy arms (0.76 ± 1.03 ms). In 16 affected arms (94.1 %), the minimum FWLD was more than 2.20 ms; the author [1] found, using a 2.20 ms cutoff for ipsilateral ulnar–median minimum FWLD, high sensitivity (94.1 %) and specificity (94.1 %) in the detection of UNE. However, the high rate of carpal tunnel syndrome (CTS) coexistence limits its usage in the diagnosis of UNE as a sole technique; for this reason, Alemdar [1] suggested to use the median–ulnar minimum FLWD method in the confirmation of the results obtained from routine ulnar motor conduction study (MNCV in the across-elbow segment) and other tests like inching technique, which are independent studies from the median nerve conduction measures. Pathological waveform and values in a case of ulnar neuropathy at the elbow (UNE) are reported (Fig. 3).

Pathological waveform (wrist – ADM muscle, wrist – APB muscle):

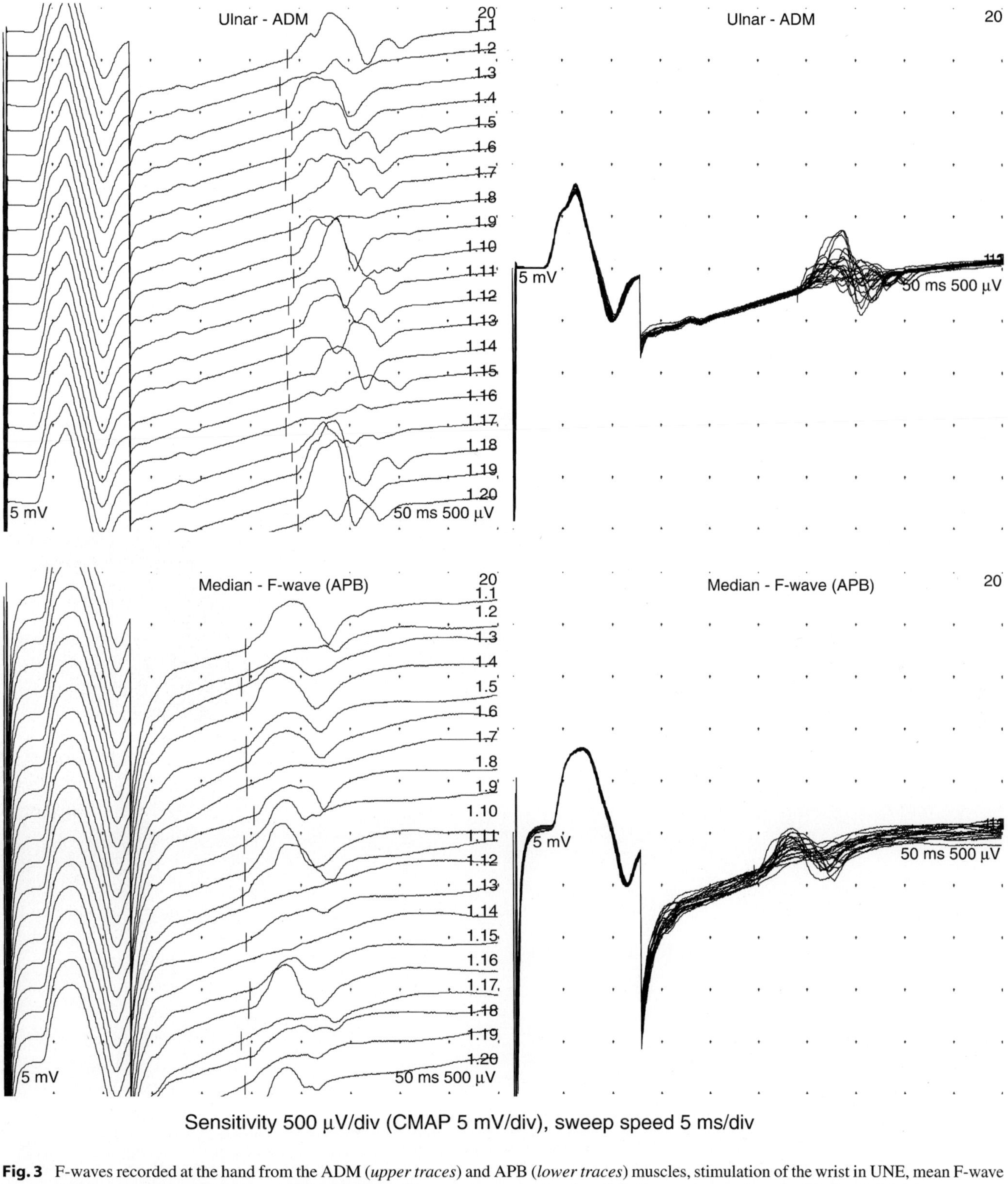

Sensitivity 500 µV/div (CMAP 5 mV/div), sweep speed 5 ms/div

Fig. 3 F-waves recorded at the hand from the ADM (*upper traces*) and APB (*lower traces*) muscles, stimulation of the wrist in UNE, mean F-wave latency 29.20 ms (ADM) and 24.60 ms (APB), FWLD 4.60 ms; raster mode, 20 stimuli (*left*); superimposed mode, 20 stimuli (*right*)

References

1. Alemdar M (2013) Ulnar to median minimum F-wave latency difference in confirmation of ulnar neuropathy at elbow. J Clin Neurophysiol 30:411–414
2. Sander HW, Quinto C, Saadeh PB et al (1999) Sensitive median-ulnar motor comparative techniques in carpal tunnel syndrome. Muscle Nerve 22:88–98

Glossary

Amplitude When referring to an action potential, the amplitude is the maximum *voltage* difference between the two points, usually *baseline* to peak or peak to peak. By convention, the amplitude of potentials which have an initial negative deflection from the *baseline* such as the *compound muscle action potential (CMAP)* and the *antidromic sensory nerve action potential (SNAP)* is measured from the *baseline* to the most negative peak (also called "negative-peak amplitude"), and the amplitude measured from the most positive peak to the most negative peak is called "peak-to-peak amplitude" (Fig. 1).

Amplitude Decay The percentage change in the amplitude of a *compound muscle action potential (CMAP)* or a *compound sensory nerve action potential (SNAP)* between two different stimulation points along the nerve (Fig. 2).

$$\text{Decay} = 100\,\frac{\text{Distal amplitude} - \text{proximal amplitude}}{\text{Distal amplitude}}$$

Antidromic It is the propagation of a nerve impulse in the direction opposite to physiological (Fig. 3). Most sensory *nerve conduction studies* are performed antidromically because of the higher amplitude of the *sensory nerve action potentials (SNAPs)* than the orthodromic studies, with the contrary of *orthodromic* (Fig. 5).

Artifact A voltage change generated by a biological or nonbiological source other than the ones of interest, which may precede or overlap the activity of interest. The stimulus *artifact*, also called "shock *artifact*," represents a cutaneous spread of a stimulating current to the *recording electrode*.

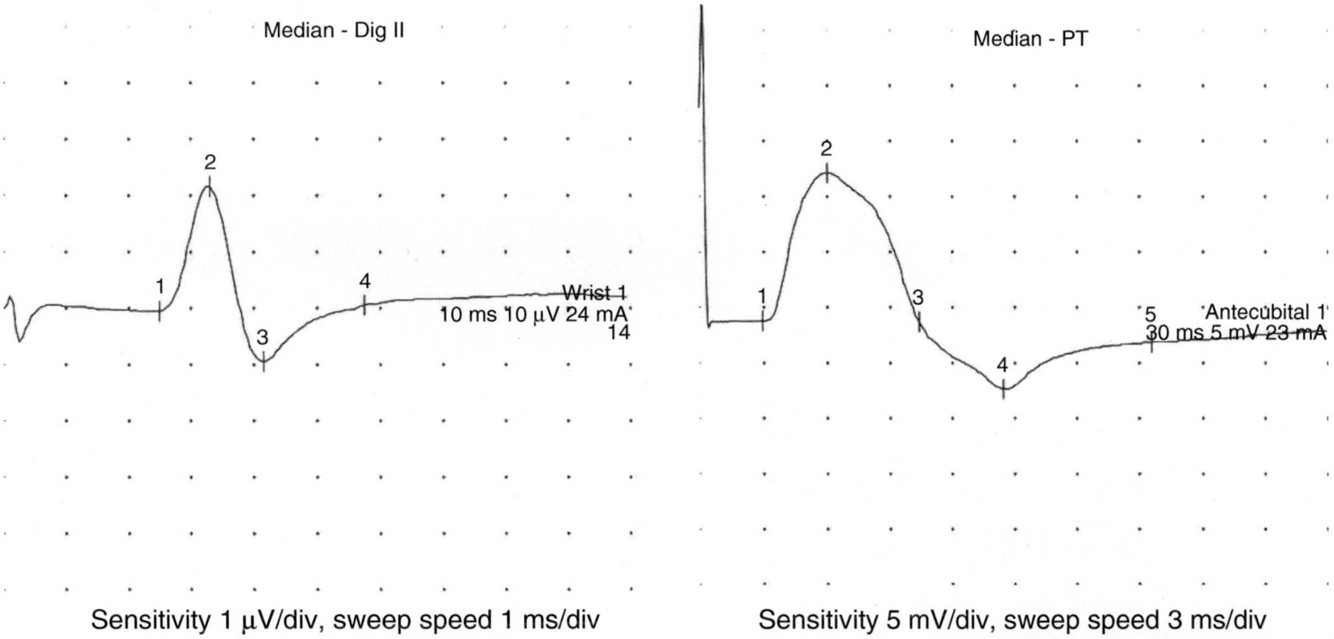

Fig. 1 Antidromic sensory nerve action potential (SNAP) recorded to the digit II, *left*, and compound muscle action potential (CMAP) recorded from the *pronator teres* (PT) muscle, *right*. SNAP: markers 1–2, negative-peak amplitude (µV); markers 2–3, peak-to-peak amplitude (µV). CMAP: markers 1–2, negative-peak amplitude (mV); markers 2–4, peak-to-peak amplitude (mV) (G. Gentili, M. Di Napoli, *The Median Nerve: Motor Conduction Studies*, DOI 10.1007/978-3-319-10473-7, © Springer International Publishing Switzerland 2015)

G. Gentili, M. Di Napoli, *The Ulnar Nerve*, DOI 10.1007/978-3-319-43292-2

$$\text{Decay} = 100\ \frac{\text{Distal amplitude} - \text{proximal amplitude}}{\text{Distal amplitude}}$$

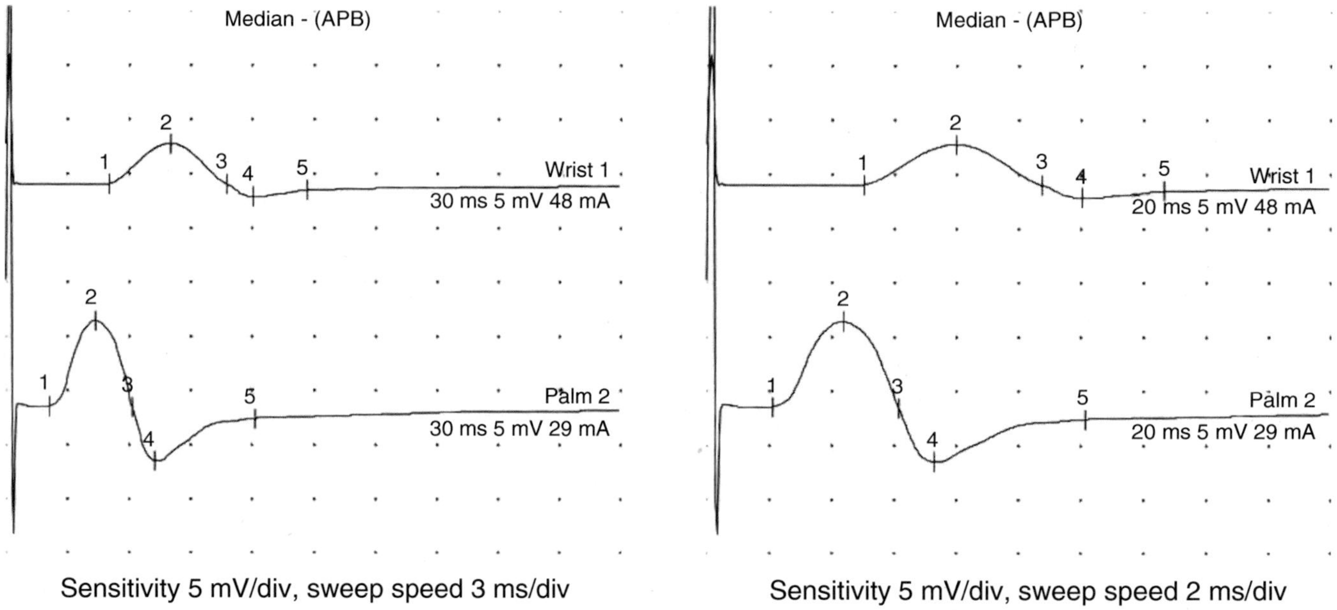

Fig. 2 Compound muscle action potentials (CMAPs) recorded at the hand from the *abductor pollicis brevis* (APB) muscle in a CTS patient, stimulation at the wrist –50 % amplitude decay (*upper trace*) and on the palm (*lower trace*)

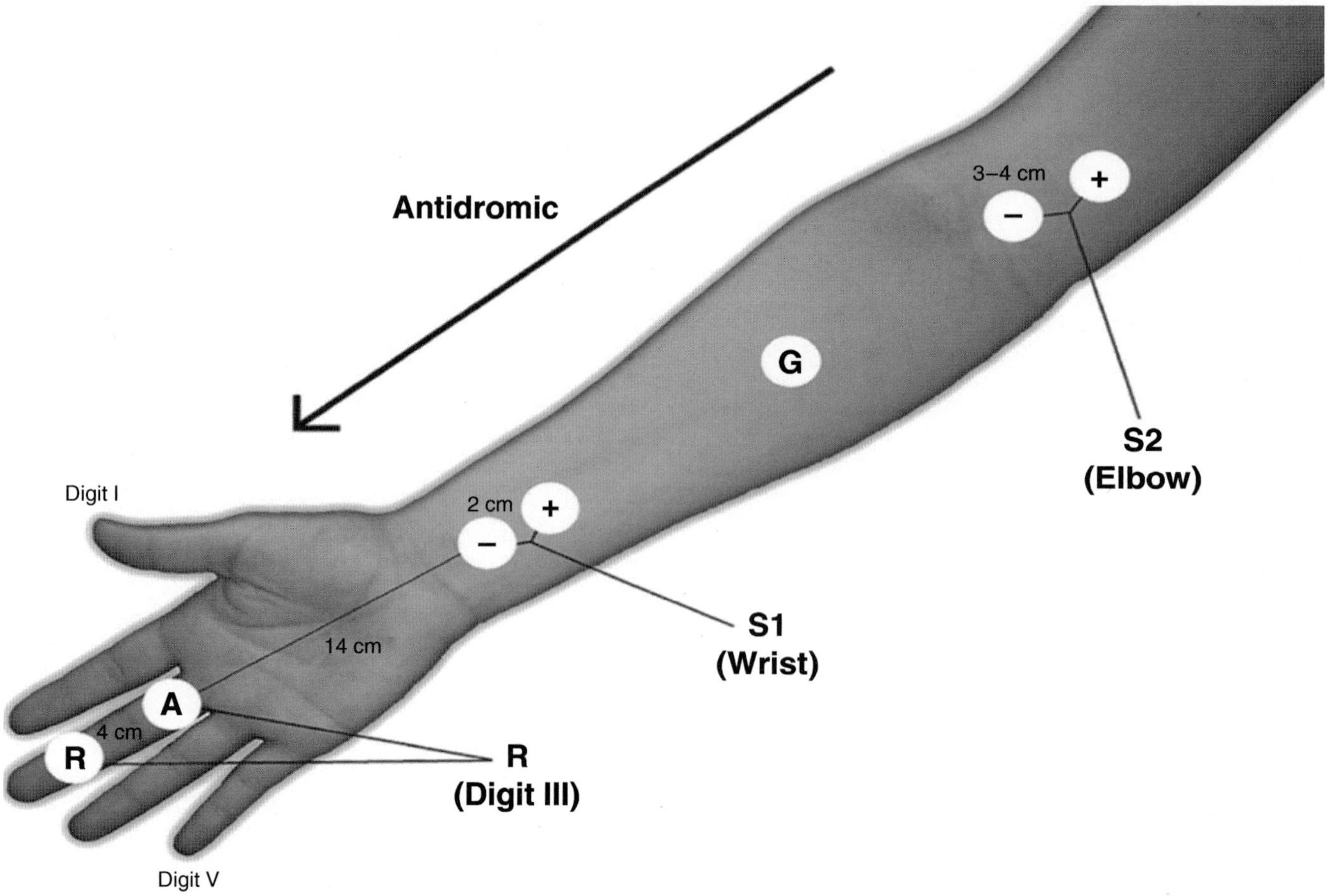

Fig. 3 Sensory antidromic conduction recording to the digit III, stimulation at the wrist (S1) and at the elbow (S2)

Axonal Degeneration It is the degeneration of the segment of the nerve distal to the cell body with preferential distal pathology.

Axonotmesis It is a disruption of nerve cell axons resulting in *axonal degeneration* distal to the injury site. Compare *neurapraxia* and *neurotmesis*.

Baseline In an electrophysiological recording, the *baseline* is the display of zero potential differences between the two input terminals of the recording instrument. It is also called isoelectric line.

Brachial Plexus An anatomical structure composed of roots, trunks, divisions, cords, and terminal nerves. It is formed by the spinal roots from C5 to T1, traverses the shoulder region, and culminates in the named peripheral nerves in the arm (Fig. 4).

Carpal Tunnel Syndrome (CTS) It is a mononeuropathy affecting the median nerve at the wrist. The nerve is subject to compression as it passes through the carpal tunnel, a space bounded dorsally by the bones of the wrist, laterally by the forearm tendons, and volarly by the transverse carpal ligament. Any of these structures can compress the median nerve, and repetitive hand and wrist movement is thought to contribute to the compression.

Chronic Inflammatory Demyelinating Polyradiculoneuropathy (CIDP) A polyneuropathy or polyradiculoneuropathy characterized by a generalized *demyelination* of the peripheral nervous system. In most cases, there is also a component of axonal degeneration. It is distinguished from an acute inflammatory neuropathy or *Guillain–Barré syndrome*.

CMAP It is an abbreviation for *compound muscle action potential*.

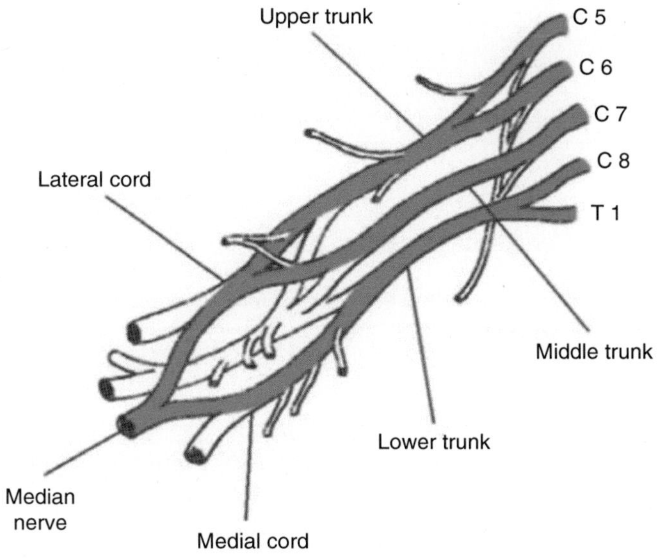

Fig. 4 Brachial plexus, median nerve: spinal roots C5-T1, upper-middle-lower trunks, lateral and medial cords

Collision The interaction of the *two action potentials* propagated toward each other from opposite directions on the same nerve fiber during a *nerve conduction study* so that the refractory periods of the two potentials prevented propagation past each other.

Compound Muscle Action Potential (CMAP) The summation of a nearly synchronous muscle fiber action potential recorded from a muscle commonly produced by stimulation of the nerve supplying the muscle either directly or indirectly. In each compound muscle action potential (CMAP), some parameters as *baseline-to-peak amplitude*, *latency*, and *duration* can be measured (Figs. 5, 6, and 7).

Conduction Block It is the failure of an action potential to propagate past a particular peripheral nerve point, whereas conduction is possible below the point of the block. It is documented by a demonstration of a reduction in the area of a compound muscle action potential greater than that normally seen with stimulation at two different points on a nerve trunk.

Conduction Velocity It is the speed of propagation of an action potential (sensory, mixed, or motor) along a nerve. For a nerve trunk, the maximum conduction velocity is calculated from the *latency* of the evoked potential at a maximal or supramaximal intensity of stimulation at two different points. The distance between the two points (conduction distance measured in mm) is divided by the difference between the corresponding latencies (conduction time measured in ms). The calculated result is the conduction velocity of the fastest fibers and is usually expressed as meters per second (m/s). For the sensory and mixed *nerve conduction studies*, the conduction velocity *(SNCV)* can be calculated using a single site of stimulation and measuring the distance between the stimulating and recording electrodes, while for the motor *nerve conduction studies*, two different sites of stimulation are used, and only dividing the distance between the sites by the difference in *latency* represents the conduction velocity *(MNCV)* of the portion of the nerve which does not include the neuromuscular junction or the muscle (Fig. 8).

Demyelination It represents the disease process affecting the myelin sheath of central or peripheral nerve fibers, manifested by *conduction velocity* slowing, *conduction block*, or both.

Distoproximal Ratio The calculation of the distoproximal ratio allows comparison of a sensory nerve action velocity (SNCV) in the distal median nerve segment (SNCV digit III–wrist) with that of the proximal segment (SNCV palm–wrist) of the same nerve. The distoproximal ratio is calculated comparing digit–palm and palm–wrist sensory nerve conduction velocities (SNCV) as follows:

$$\text{Disto proximal ratio} = \frac{\text{Digit} - \text{palm (m/s)}}{\text{Palm} - \text{wrist (m/s)}}$$

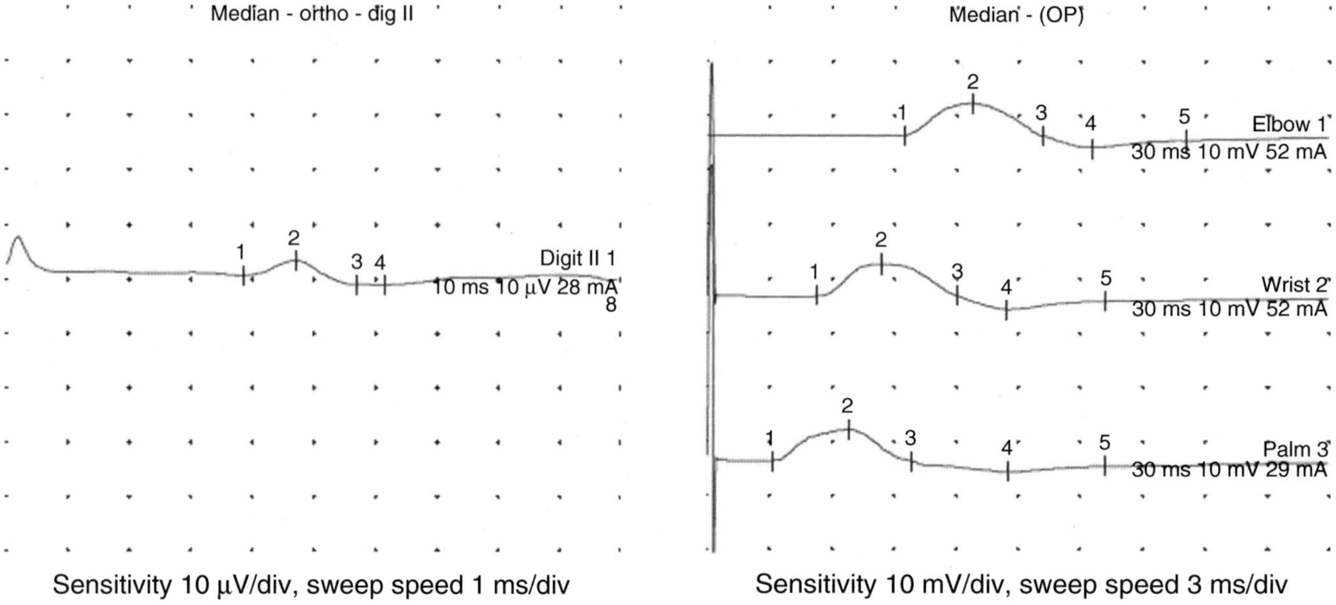

Fig. 5 Orthodromic sensory nerve action potential (SNAP) recorded at the wrist in a CTS patient (stimulation to the digit II), *left*; compound muscle action potentials (CMAPs) recorded at the hand from the *abductor pollicis brevis* (APB) muscle in a CTS patient, stimulation at the elbow (*upper trace*), at the wrist (*middle trace*), and on the palm (*lower trace*), *right*

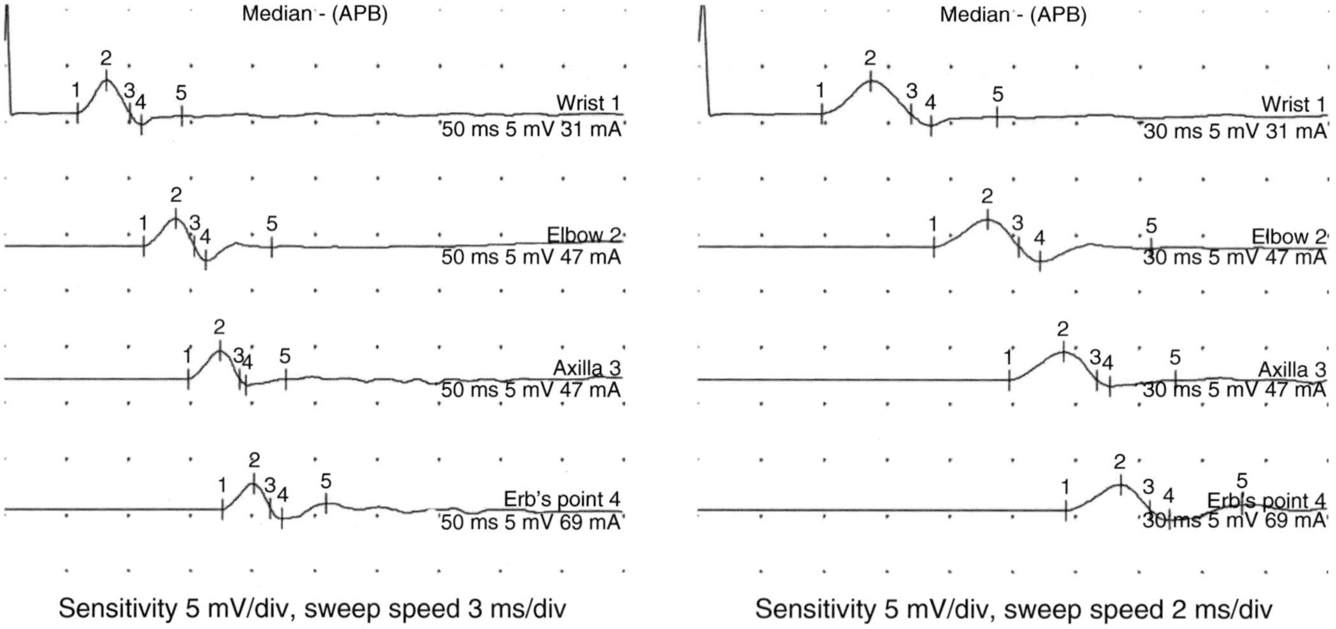

Fig. 6 Compound muscle action potentials (CMAPs) recorded at the hand from the *abductor pollicis brevis* (APB) muscle in a patient with CTS and associated CIDP, stimulation on the Erb's point, at the axilla, at the elbow, and at the wrist (*bottom to top*)

In normal hands, the SNCV in the proximal segment (palm–wrist) was higher than the SNCV of the distal segment (digit–palm), and it is always <1.0. This ratio is reversed (>1) in a patient with CTS.

Duration The interval from the beginning of the first deflection from the *baseline* to its final return to the *baseline* of an action potential or *waveform* unless otherwise specified. For example, the duration of a *sensory nerve action potential (SNAP)* or a *compound muscle action potential (CMAP)* may refer to the interval from the deflection of the first negative phase from the baseline to its return to the *baseline*. Motor potentials have longer duration than sensory potentials (Fig. 9).

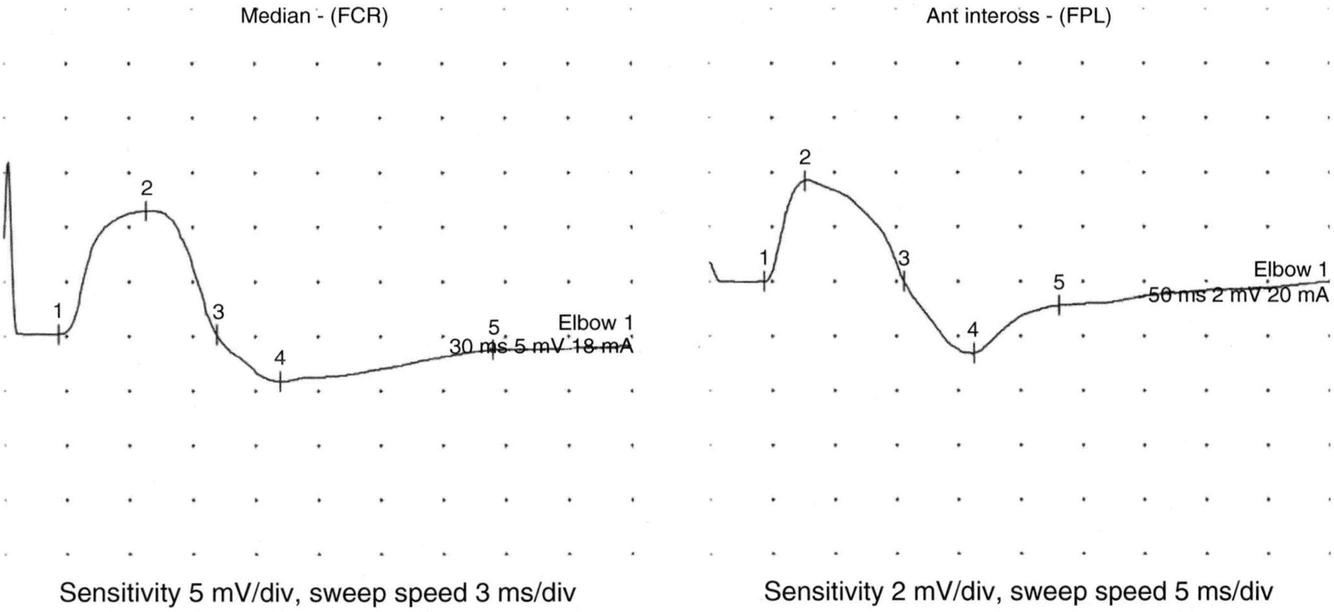

Fig. 7 Compound muscle action potentials (CMAPs) recorded from the *flexor carpi radialis* (FCR) muscle (*left*) and from the *flexor pollicis longus* (FPL) muscle (*right*)

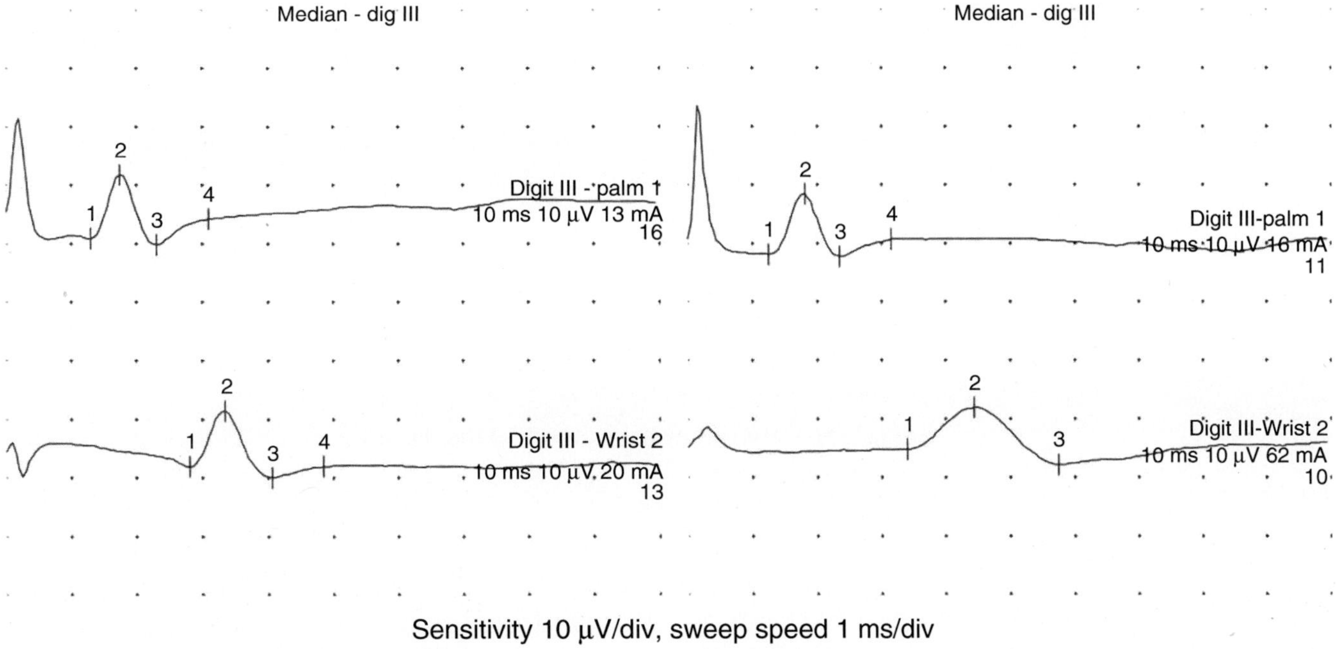

Fig. 8 Orthodromic sensory nerve action potentials (SNAPs) recorded on the palm (*upper trace*) and at the wrist (*lower trace*), digit III stimulation: in a normal subject, *left*, and in a CTS patient (distoproximal ratio 1.50), *right*

Electrode A conducting device used to record an electric potential (*recording electrode*) or to deliver an electric current (*stimulating electrode*). In addition to the single *ground electrode* used in clinical recordings, two electrodes (bipolar configuration) are always required either to record an electrical potential or to deliver an electrical stimulus.

Electromyography (EMG) It is the recording and study of insertion and spontaneous and voluntary activity of muscle with a coaxial needle recording electrode (Fig. 10).

Electroneurography (ENG) It is the recording and study of the action potentials of a peripheral nerve. It is synonymous with *nerve conduction studies*.

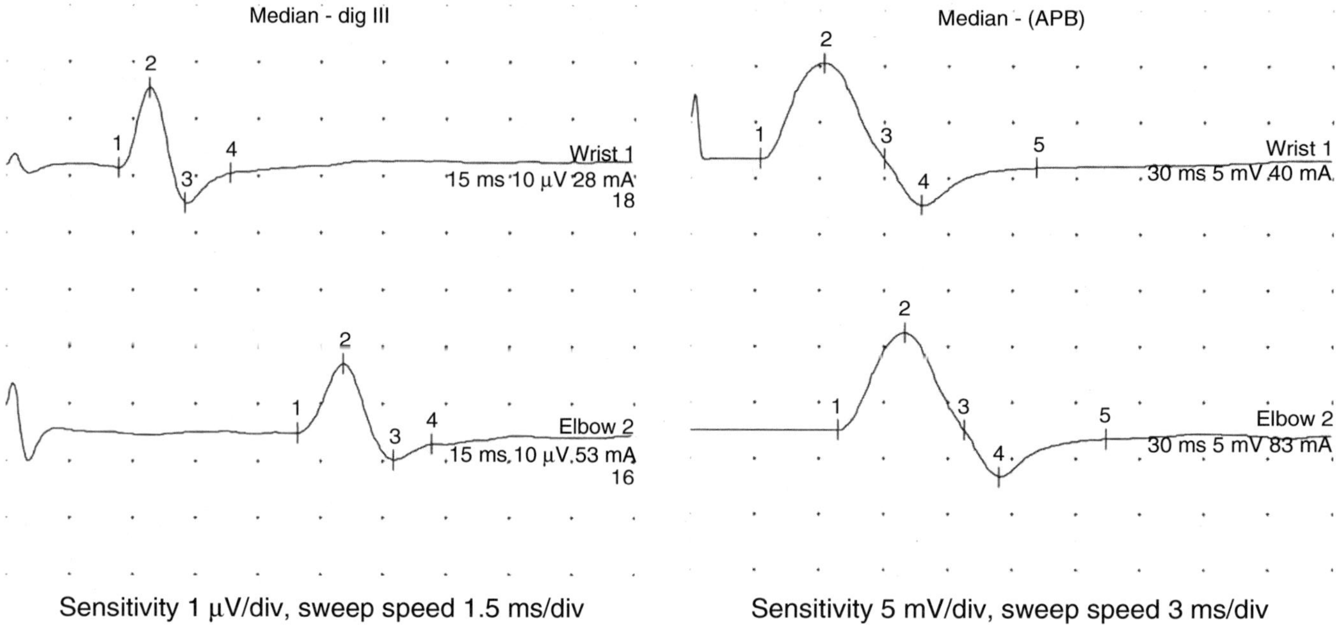

Fig. 9 Antidromic sensory nerve action potentials (SNAPs) recorded to the digit III (*left*); compound muscle action potentials (CMAPs) recorded from the *abductor pollicis brevis* (APB) muscle (*right*); stimulation at the wrist (*upper trace*) and at the elbow (*lower trace*). SNAPs (markers 1–4, duration (ms)), CMAPs (markers 1–5, duration (ms))

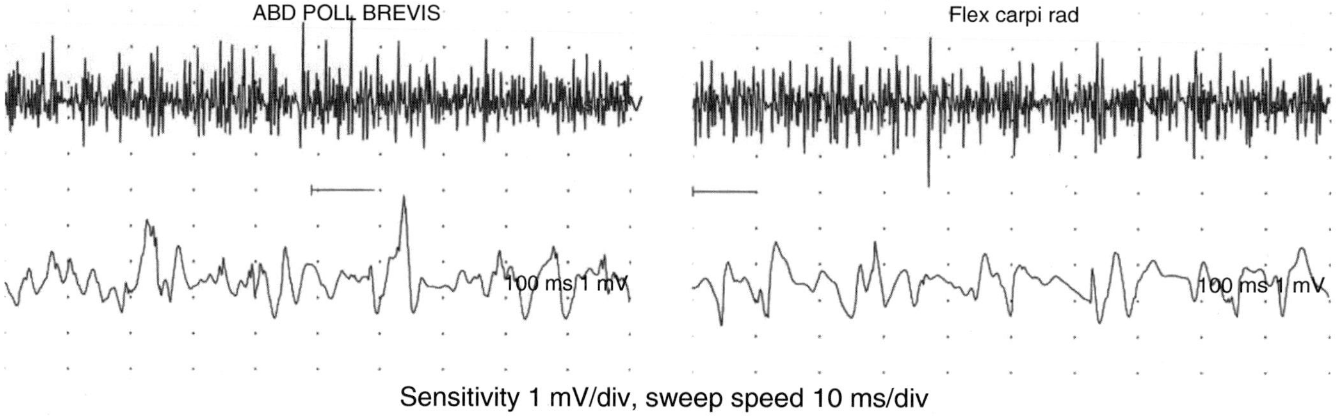

Fig. 10 Voluntary EMG activity recorded from the *abductor pollicis brevis* (APB) muscle (*left*) and from the *flexor carpi radialis* (FCR) muscle (*right*)

Electroneuromyography (ENG) It represents the combined studies of *electromyography* and *electroneurography*. It is synonymous with *clinical electromyography*, a term which describes the scientific methods of recording and analysis of biological electrical potentials from the human peripheral nerve and muscle.

EMG An abbreviation for *electromyography.*

ENG An abbreviation for *electroneurography.*

ENMG An abbreviation for *electroneuromyography.*

Entrapment Neuropathy In general, it is a mononeuropathy caused by the compression of a nerve as it passes through an area of anatomical narrowing (e.g., *carpal tunnel syndrome*).

F-Wave An *action potential* evoked intermittently from a muscle by a supramaximal electrical *stimulus* to the nerve due to an *antidromic* activation of motor neurons. When compared with the maximal *amplitude* of the *M-wave*, it is smaller (1–5 % of the *M-wave*) and has a variable configuration. Its latency is longer and more variable than the *M-wave*. It can be evoked in many muscles of the upper (e.g., the abductor pollicis brevis muscle) and lower extremities, and the latency is longer with more distal sites of stimulation (Figs. 11, 12, and 13). Named "F-wave" by Maglader and McDougal in 1950, it was first recorded from the foot muscles. It is one of the late responses (i.e., *H-wave*, *M-wave*).

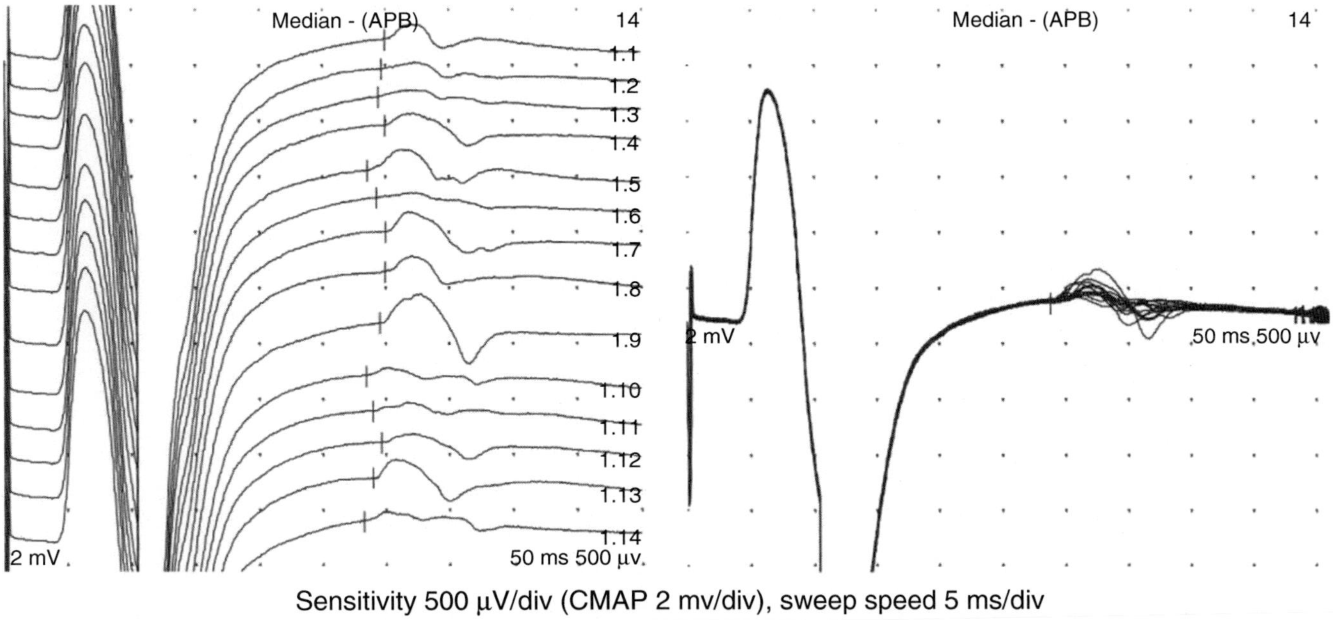

Fig. 11 Normal F-waves (mean latency 28.40 ms) recorded at the hand from the *abductor pollicis brevis* (APB) muscle in a normal subject, stimulation at the wrist; raster mode, 14 stimuli (*left*); superimposed mode, 14 stimuli (*right*)

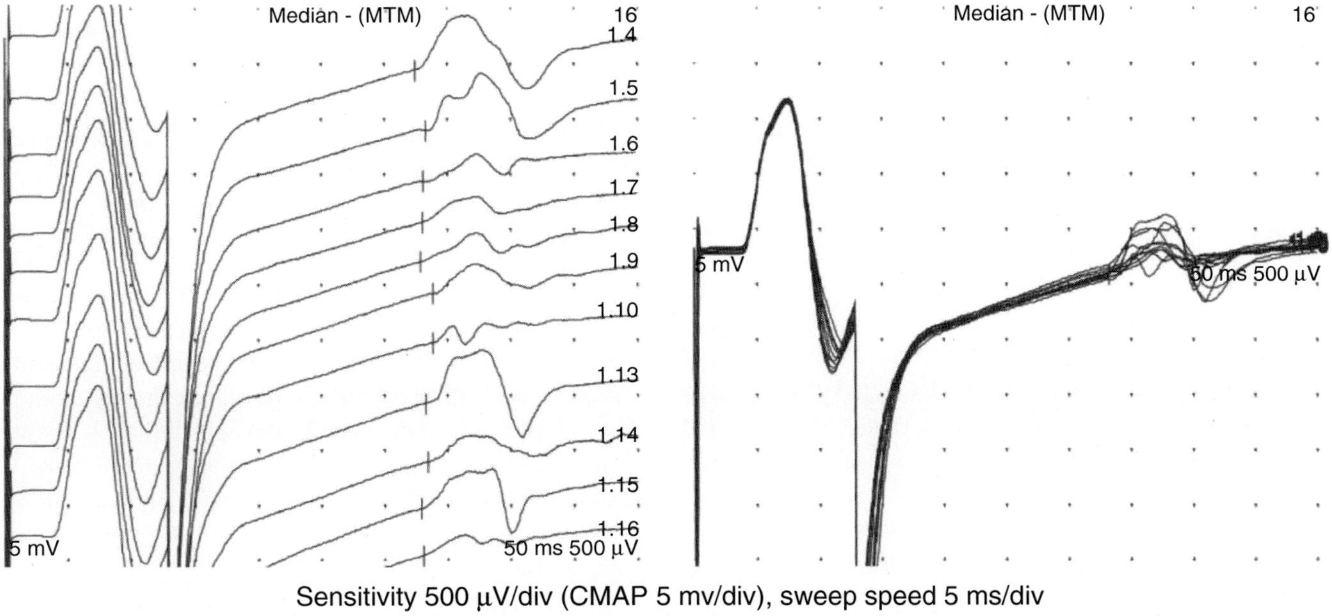

Fig. 12 Abnormal F-waves (mean latency 33.20 ms) recorded at the hand from the *abductor pollicis brevis* (APB) muscle in a CTS patient, stimulation at the wrist; raster mode, 11 stimuli (*left*); superimposed mode, 11 stimuli (*right*)

Far Field It is a region of electrical potentials where the isopotential voltage lines associated with a current source change slowly over a short distance. The term *far-field potential* is sometimes used to designate a potential that does not change in latency, amplitude, or polarity over infinite distances. Compare with *near field*.

Filters Filters are electronic circuits that perform the function of processing the body signals removing unwanted electrical *noise*. Electrodiagnostic studies use low-frequency (high-pass) and high-frequency (low-pass) filters to exclude high- and low-frequency electrical *noise* to reproduce the signal of interest. Specific filter ranges have been selected to ensure exact reproduction of the responses, and in practice the frequency range (according to the type of conduction test) should be set as recommended because a change in the filter range will affect the

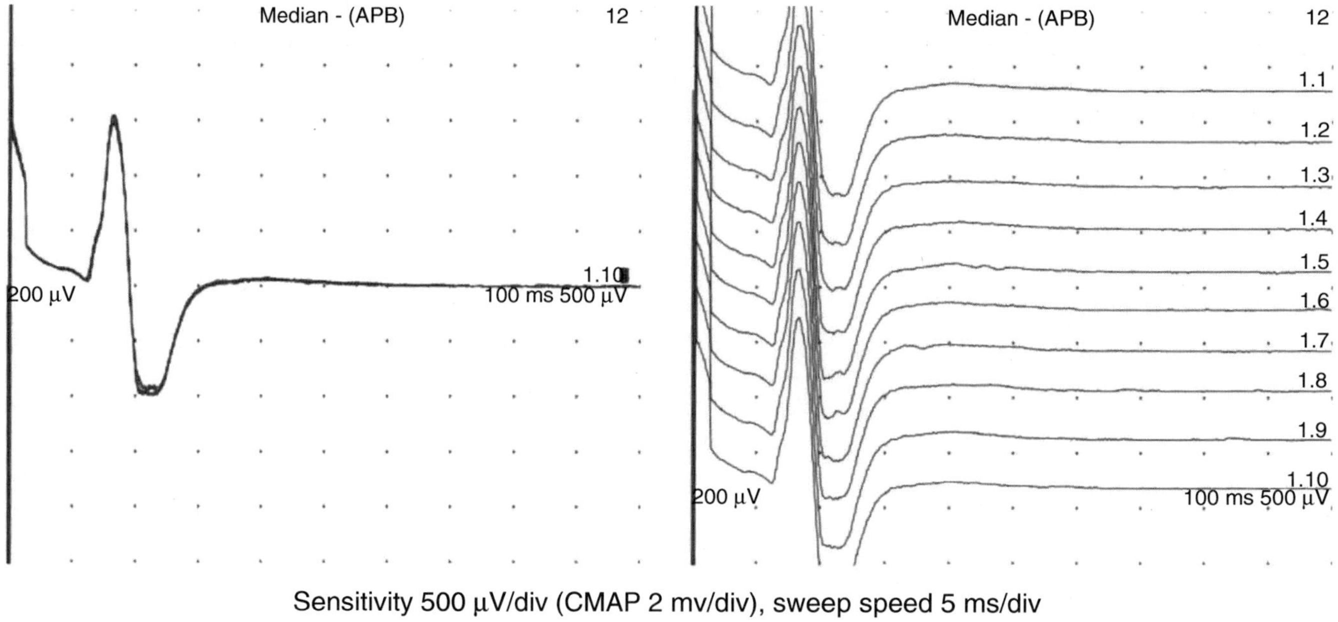

Fig. 13 Absent F-waves recorded at the hand from the *abductor pollicis brevis* (APB) muscle, stimulation at the wrist; raster mode, 10 stimuli (*left*); superimposed mode, 10 stimuli (*right*) in a Guillain–Barré syndrome (1 week of onset of symptoms)

amplitude and *latency* of the potentials. Several differences have been observed between the motor and sensory responses: the *amplitude* is affected more by narrowing the lower-frequency limit than the higher-frequency limit in the former (because of the longer duration of the motor potentials), conversely in the latter, in which the high-frequency limit is more important because of the shorter duration of the sensory potentials. In the motor *nerve conduction study*, the recommended filter range is 2 Hz (high-pass) to 10 kHz (low-pass), while in the sensory *nerve conduction study*, the filter range to be used is 20 Hz (high-pass) to 2–3 kHz (low-pass). For the *electromyography (EMG)*, the suggested filters are 20 Hz for the low-frequency filter and 10 Hz for the high-frequency filter.

Frequency The number of complete cycles of a repetitive waveform in 1 s measured in Hertz (Hz).

Ground Electrode It is a connection from the patient to the earth, used as a common return for an electric circuit and as an arbitrary zero potential reference point. In general, to avoid shock artifact due to current stimulation spread, the ground electrode should be placed between stimulation and recording sites.

Guillain–Barré Syndrome (GBS) It is an eponym for an acute, monophasic polyneuropathy. It is characterized by a time course of progression to a maximum deficit within 4 weeks of onset of symptoms (Fig. 14). The most common clinical presentation is an ascending sensory-motor *neuropathy*. Electrodiagnostic studies most commonly reveal evidence for *demyelination*, but *axonal degeneration* also occurs. Eponym for an acute inflammatory neuropathy. Distinguish from the *chronic inflammatory demyelinating polyradiculoneuropathy (CIDP)*.

H-Wave It is a *compound muscle action potential (CMAP)* with a consistent *latency* recorded from the muscles after stimulation of the nerve. It compares later than the *M-wave*. Stimulus intensity sufficient to elicit a maximal amplitude *M-wave* reduces or abolishes the H-wave. It is most reliably elicited with a stimulus of a long duration (500–1,000 **µ**s). It is due to a spinal reflex, with an electrical stimulation of afferent fibers in the mixed nerve and activation of motor neurons to the muscle mainly through a monosynaptic connection in the spinal cord. It is synonymous with *H-reflex* (Fig. 15). Compare the *F-wave*.

Inching A nerve conduction study technique consisting of applying stimuli at multiple short-distance increments along the course of a nerve. This technique is used to localize an area of focal slowing or a *conduction block*, and it can be performed using *orthodromic* or *antidromic* methods. Following the *antidromic* method, the *sensory nerve action potentials* (*SNAPs*) can be recorded to the digit II (index finger), stimulating the median nerve percutaneously at 12 points between the midpalm and the distal forearm in 1 cm increments (Figs. 16 and 17).

Interpotential Interval It is the time between two different potentials. Measurement should be made between the corresponding parts of each *waveform* (e.g., onset latency, peak latency). The difference between the peak *latencies* of two components of a *waveform* is called *interpeak interval* (Fig. 18).

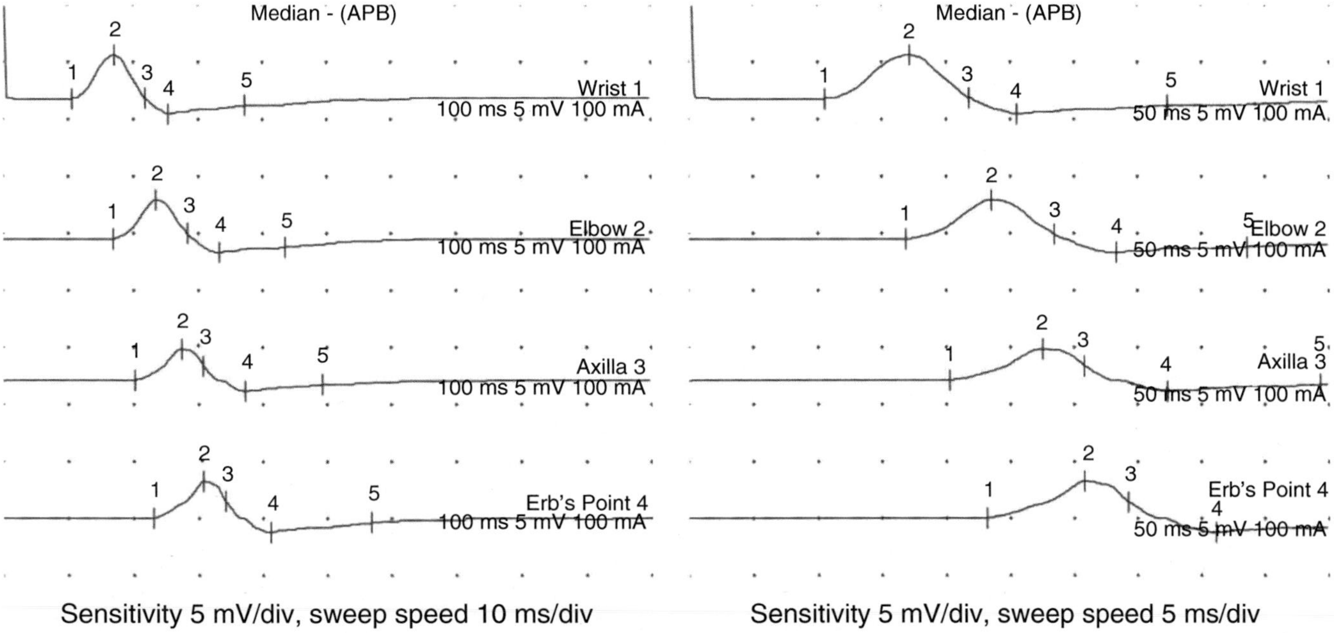

Fig. 14 Compound muscle action potentials (CMAPs) recorded at the hand from the *abductor pollicis brevis* (APB) muscle in a patient with Guillain–Barré syndrome (2 weeks of onset of symptoms) and associated CTS, stimulation on the Erb's point, at the axilla, at the elbow, and at the wrist (*bottom to top*)

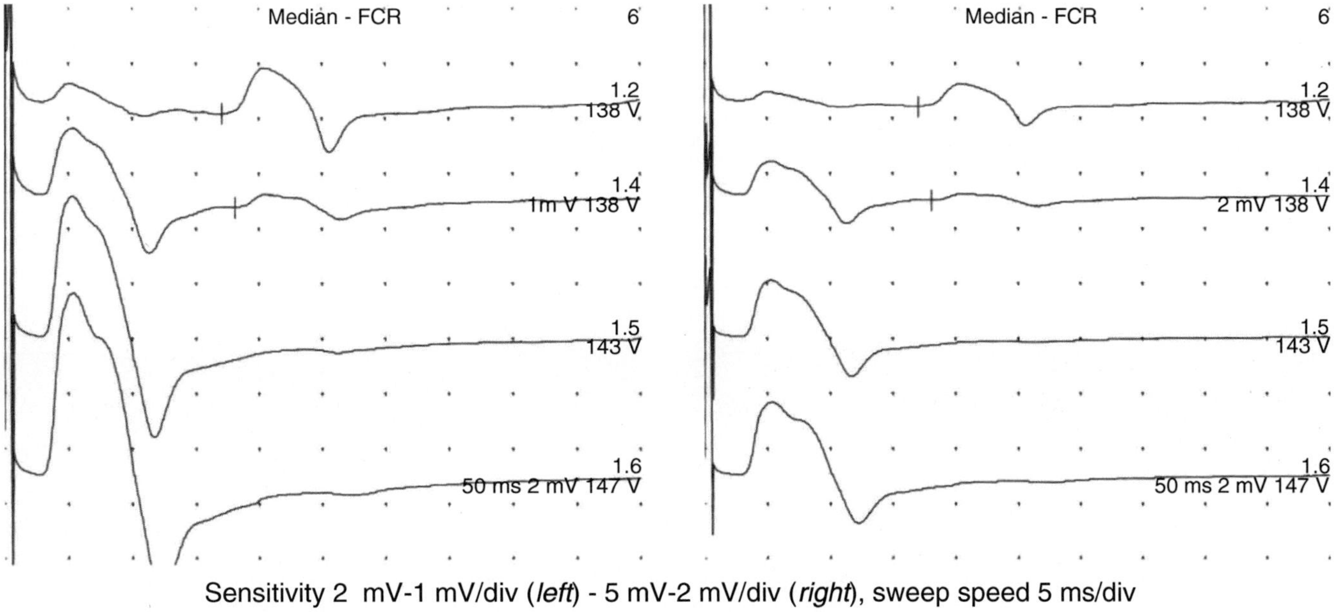

Fig. 15 H-reflex recorded at the forearm from the *flexor carpi radialis* (FCR) muscle, stimulation at the elbow

Latency It is an interval between a *stimulus* and a response. The onset *latency* is the interval between the onset of a *stimulus* and the onset (usually the negative phase) of the evoked potential. The peak *latency* is the interval between the onset of a *stimulus* and a specific peak (usually the negative peak of the negative phase) of the evoked potential. The sensory onset *latency* represents the time for conduction through the largest and fastest cutaneous sensory fibers, while the sensory peak *latency* represents a mix of large and small fibers. The motor distal *latency* (MDL) represents the conduction time from the stimulation site to the neuromuscular junctions, the time delay across the neuromuscular junction, and the depolarization time across the muscle (Fig. 19).

Mixed Nerve Action Potential A compound nerve action potential is recorded from a mixed nerve when an electrical

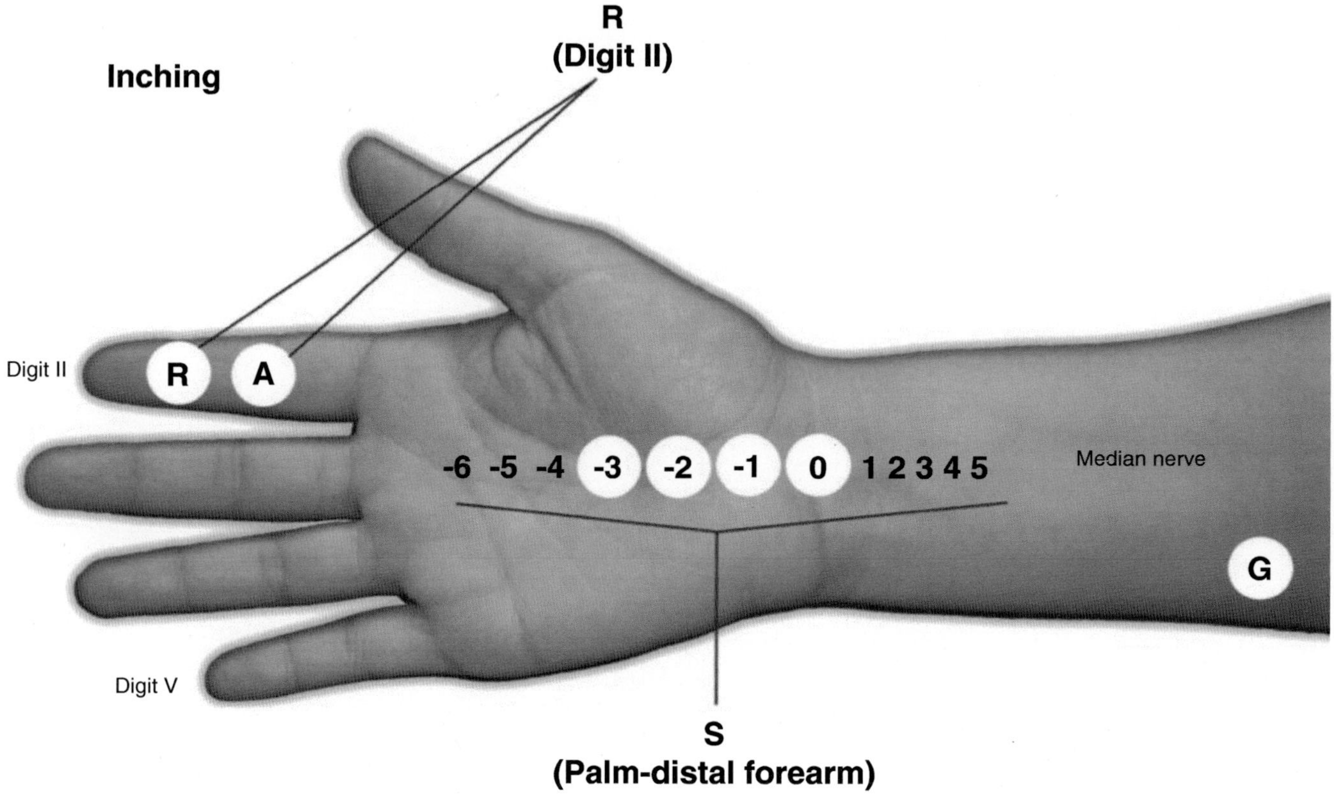

Fig. 16 Antidromic sensory technique (inching) to the digit II, serial stimulation from the palm to the distal forearm

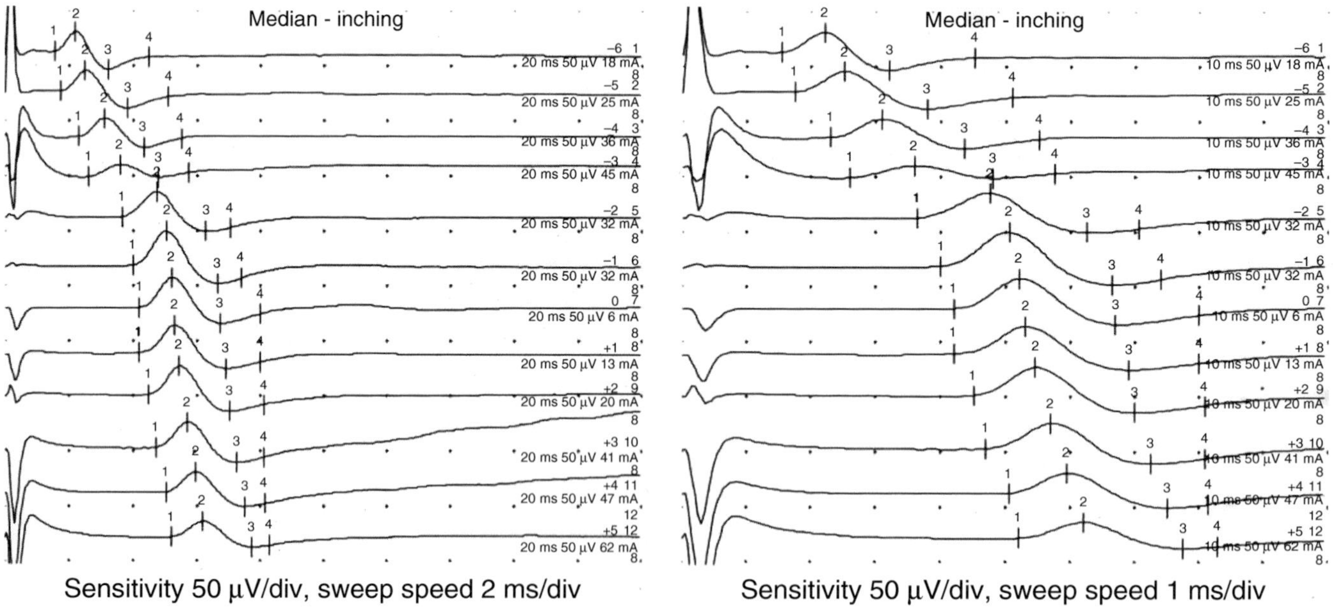

Fig. 17 Antidromic sensory nerve action potentials (SNAPs) recorded to the digit II (inching) in a CTS patient (1.10 ms delay between positions "–3" and "–2"), serial stimulation from the palm to the distal forearm

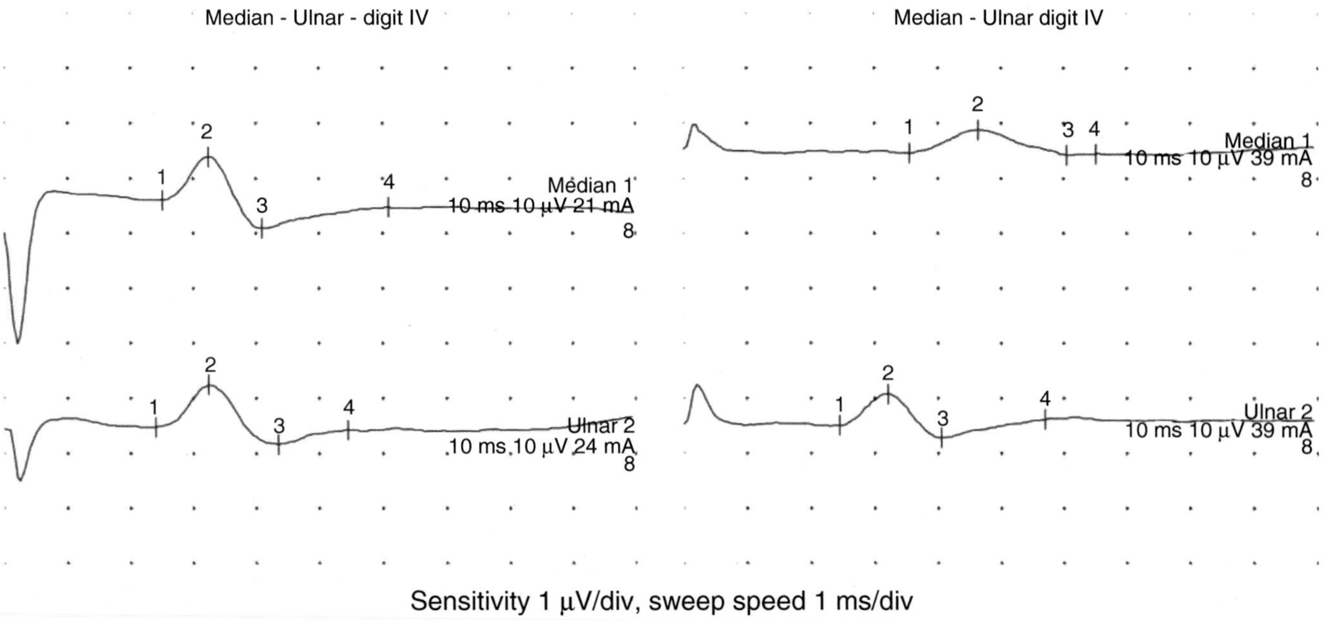

Fig. 18 Median–ulnar nerves comparative sensory conduction study, antidromic sensory nerve action potentials (SNAPs) recorded to the digit IV, stimulation at the wrist: median nerve (*upper trace*) and ulnar nerve (*lower trace*); a normal subject (*left*), a CTS patient – interpotential interval-onset latency 1.10 ms (*right*)

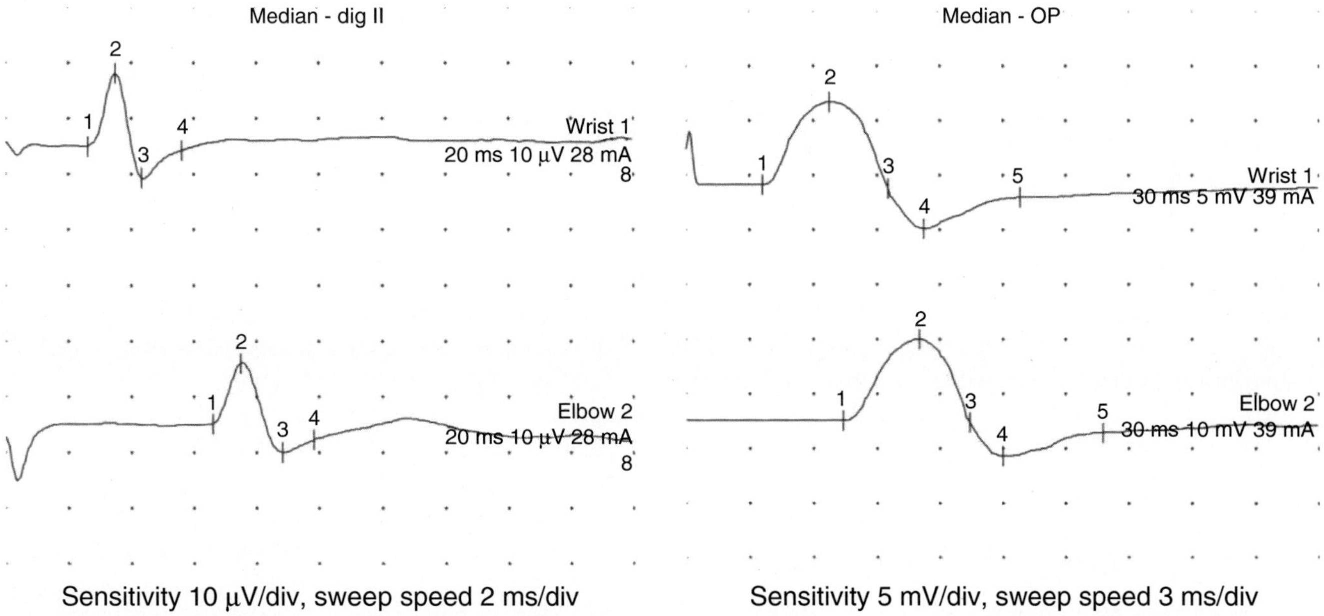

Fig. 19 Orthodromic sensory nerve action potential (SNAP) recorded at the wrist in a CTS patient, (stimulation to the digit II), *left*; compound muscle action potentials (CMAPs) recorded at the hand from the *abductor pollicis brevis* (APB) muscle in a CTS patient, stimulation at the elbow (*upper trace*), at the wrist (*middle trace*), and on the palm (*lower trace*), *right*. Marker 1, onset latency (ms); marker 2, peak latency (ms)

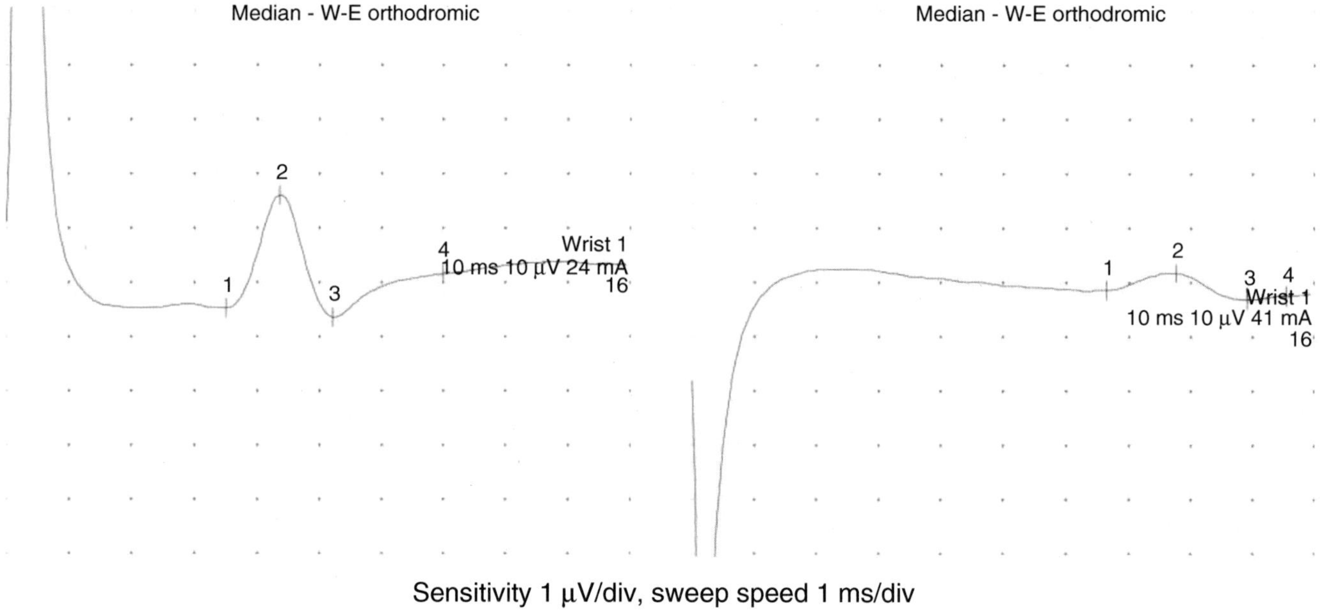

Fig. 20 Orthodromic mixed median nerve responses recorded at the elbow, stimulation at the wrist: a normal subject (*left*), a patient with diabetic polyneuropathy (*right*)

stimulus is applied to a segment of the nerve that contains both afferent (sensory) and efferent (motor) fibers. Some parameters, like *amplitude*, *latency*, *duration*, and number of phases, should be noted for every mixed nerve action potential (Fig. 20).

M-Wave It is a *compound muscle action potential (CMAP)* evoked from a muscle by an electrical stimulus to its motor nerve. By convention, the M-wave elicited by a supramaximal *stimulus* is used for motor *nerve conduction studies*. Ideally, the *recording electrodes* should be placed so that the initial deflection of the *evoked potential* from the *baseline* is negative. The measurement commonly includes (1) the *latency*, commonly called the motor latency, which is the time from stimulation (ms) to the onset of the first phase (positive or negative) of the M-wave; (2) the *amplitude* (mV), which is the baseline-to-peak amplitude of the first negative phase; (3) the *duration* (ms), which refers to the duration of the first negative phase. It is also referred to as the motor response (Figs. 21 and 22).

MNCV Abbreviation for *motor nerve conduction velocity* (m/s). See *conduction velocity*.

Motor Point Also called the "belly point," it is the place where muscle depolarization first occurs. The *motor point* is very important in motor *nerve conduction studies* because if the recording electrode is not placed here, nerve conduction studies can be artificially abnormal because the *CMAP amplitude* may appear artificially reduced. It can be useful to ask the patient an appropriate voluntary muscle contraction. If an initial positive deflection (which makes the onset *latency* difficult to accurately measure) is observed, the *recording electrode* can be moved slightly in order to obtain a clear negative deflection of the *CMAP*.

Multifocal Motor Neuropathy A disease characterized by a selective focal block of motor nerve conduction in multiple nerves. Motor *nerve conduction studies* may permit the identification and localization of the segments of nerve affected by the underlying pathology.

NCS An abbreviation for *nerve conduction study*.

NCV An abbreviation for *nerve conduction velocity*. See *conduction velocity*.

Near Field It is a region of electrical activity where the isopotential *voltage* lines associated with a current source change rapidly over a short distance. The terms *near field* and *far field* are arbitrary designations as there are no agreed-upon criteria defining where the *near field* ends and the *far field* begins. Compare with *far field*.

Needle Electrode It is an electrical device used for recording or stimulating that is positioned near the tissue of interest by the penetration of the skin. *Needle electrodes* are used in *electroneurography* (near-nerve recording technique). The coaxial *needle electrode* is indeed the electrical device that is positioned inside the muscle, and it measures an electrical potential difference between a centrally insulated wire and the cannula of the needle through which it runs. It is synonymous with a concentric *needle electrode*, and it is used mainly in *electromyography*.

Nerve Conduction Study (NCS) It is a recording and analysis of electrical *waveforms* of biological origin elicited in response to electrical or physiological stimuli (sensory or mixed nerve action potentials). It is possible

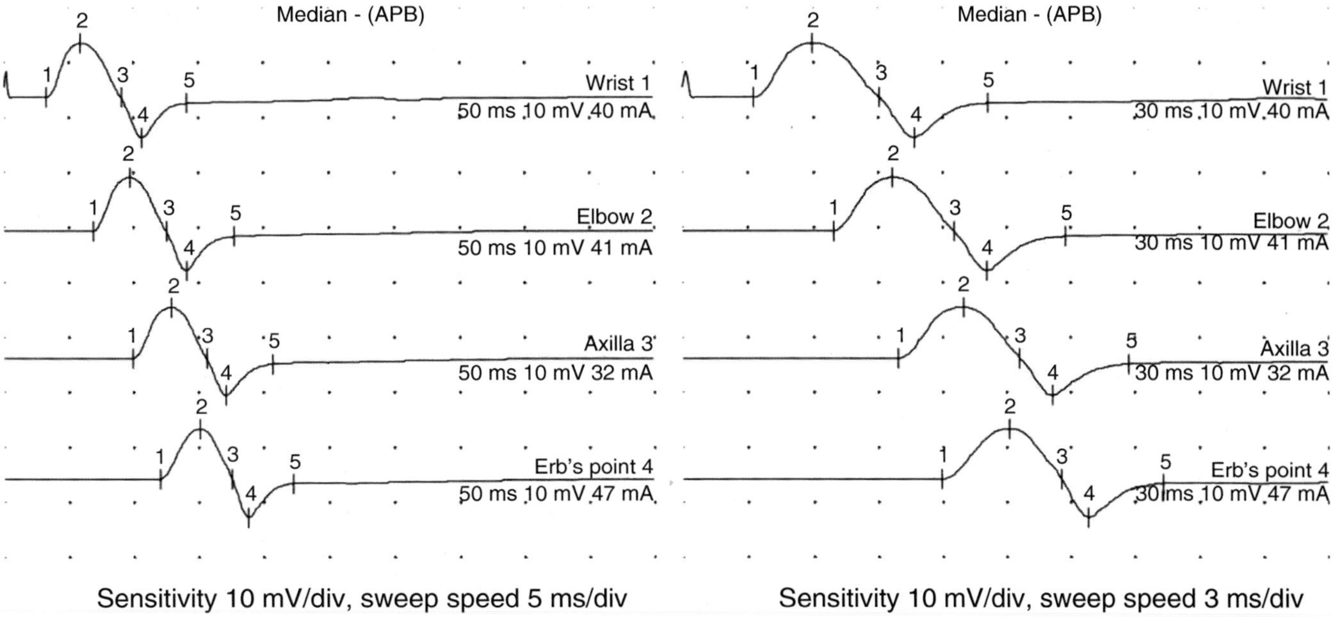

Fig. 21 Compound muscle action potentials (CMAPs) recorded at the hand from the *abductor pollicis brevis* (APB) muscle, stimulation on the Erb's point, at the axilla, at the elbow, and at the wrist (*bottom to top*)

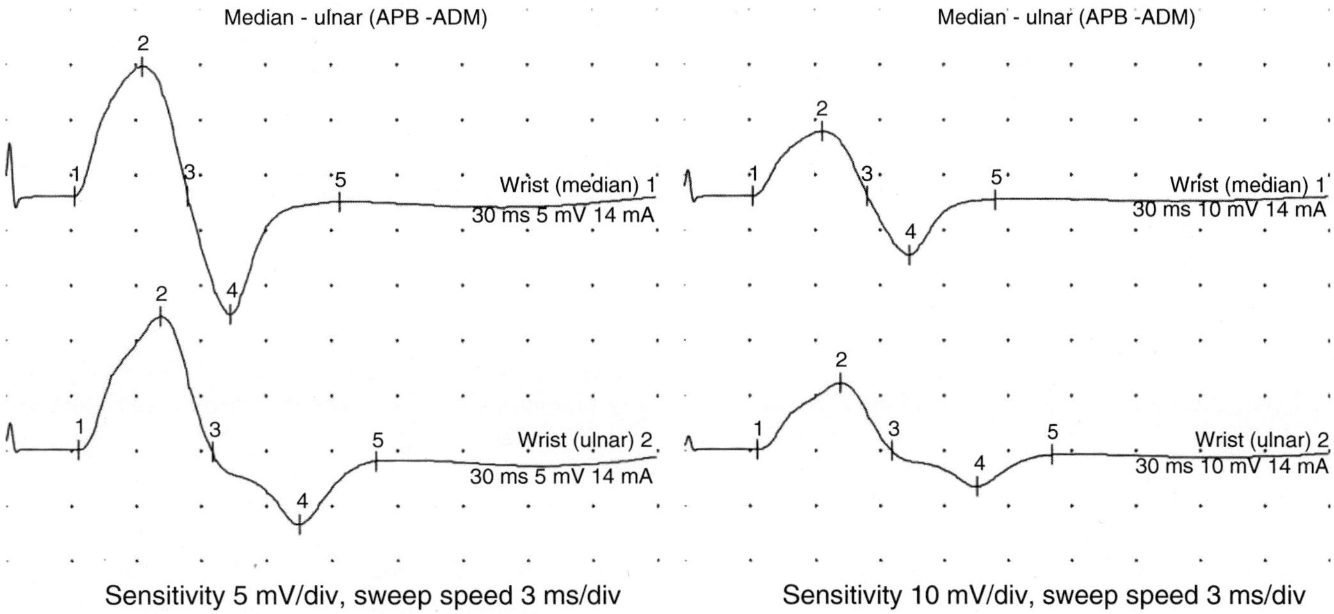

Fig. 22 Median–ulnar nerves comparative motor conduction study, compound muscle action potentials (CMAPs) recorded at the hand from the *abductor pollicis brevis* (APB) muscle (*upper trace*) and the abductor digiti minimi (ADM) muscle (*lower trace*), stimulation of the median and ulnar nerves at the wrist

under standardized conditions to establish normal ranges for amplitude, latency, and duration of the *waveforms* and to calculate the maximum *conduction velocity* of sensory nerves. Synonymous with *electroneurography*.

Nerve Conduction Velocity (NCV) The speed of action potential propagation along a nerve fiber or nerve trunk. Generally assumed to refer to the maximum speed of propagation, unless otherwise specified. See *conduction velocity*.

Neurapraxia A clinical term used to describe the reversible motor and sensory deficits produced by focal compressive or traction lesions of large myelinated nerve fibers. It is due to a *conduction block* most often caused by focal *demyelination* but, when very short lived, presumably caused by focal ischemia. The axon is not injured at the lesion site. Compare with *axonotmesis* and *neurotmesis*.

Neuropathy It is a disorder of the peripheral nerves and is classified according to the anatomical structure of the nerve most affected by the disease: cell body (neuronopathy), axon (axonopathy), or myelin sheath (demyelinating neuropathy). It may selectively affect motor or sensory nerves or both simultaneously. The etiology may be hereditary, metabolic, inflammatory, toxic, or unknown.

Neurotmesis It is a partial or complete nerve severance, including the axons, associated myelin sheaths, and supporting connective tissues, resulting in axonal degeneration distal to the injury site. Compare with *axonotmesis* and *neurapraxia*.

Noise It is the electrical activity not related to the signal of interest (i.e., all *waveforms* generated by electrodes, cables, amplifiers, or storage media and unrelated to potentials of biological origin). The signal-to-noise ratio is the ratio of the desired signal power to the background noise signal power. A common background *noise* is 50–60 Hz (depending on the countries) from electrical devices in the surrounding environment. Since the signals recorded during *nerve conduction studies* (NCS) are based on the differences between the active and references electrodes, making sure that the two electrodes have the same impedance (which is the opposition to current flow), and placing the ground electrode between the stimulator and recording sites will decrease the background *noise*. A good signal-to-noise ratio can be obtained using electrodes of the same type and having intact wires, good connections, and the underlying skin clean and intact.

Orthodromic It is the propagation of a nerve impulse in the same direction as a physiological conduction. The orthodromic nerve conduction studies were performed using a stimulus directed in the way the nerve normally depolarizes, for example, stimulating a sensory nerve away from the sensory receptor (Fig. 23). In clinical practice, sensory *antidromic* nerve conduction studies are the preferred methods because of higher *sensory nerve action potentials (SNAPs)* amplitude than orthodromic sensory nerve conduction studies (Fig. 24). Motor *nerve conduction studies* are orthodromic since the stimulus is always directed toward the neuromuscular junctions and muscles. In contrast with *antidromic*.

Radiculopathy It is any disease of the spinal nerves and roots. The resulting clinical syndrome may include pain, sensory loss, paresthesia, weakness, fasciculations, and muscle atrophy. If more than one spinal root is involved, the term *polyradiculopathy* may be used as a descriptor.

Recording Electrode In a bipolar configuration, two electrodes were used to record the electrical potential difference. The electrode close to the source of the electrical activity to be recorded is the *active electrode*, and the other recording electrode is the *reference electrode* (Fig. 25).

Residual Latency (RL) The calculated time difference between the measured *distal latency* of a sensory nerve and the expected latency, calculated by dividing the distance between the stimulating *cathode* and the active *recording electrode* by the maximum *conduction velocity* measured in a more proximal segment of the nerve.

Sensory Nerve Action Potential (SNAP) A compound nerve action potential recorded from the afferent fibers of a sensory nerve. It can also be recorded from a sensory branch of a mixed nerve or in response to stimulation of a sensory nerve. It may also be elicited when an adequate *stimulus* is applied synchronously to sensory receptors. Commonly, in *electrodiagnostic medicine*, the compound sensory nerve action potentials (SNAPs) are recorded distally to the digits after proximal stimulation at the wrist (the antidromic method); conversely, sensory potentials are recorded at the wrist after stimulation is applied to the digits (the orthodromic method). The *amplitude*, *latency*, *duration*, and configuration should be noted. Generally, the *amplitude* is measured as the maximum peak-to-peak *voltage* (peak-to-peak amplitude) when there is an initial positive deflection or from the baseline to the peak (negative-peak amplitude) when there is an initial negative deflection. The *latency* is measured as either the time to the initial deflection or the negative peak and the *duration* as the interval from the first deflection of the *waveform* from the *baseline* to its final return to the *baseline* (Fig. 26).

Stimulating Electrode A device used to deliver an electrical current. All electrical stimulation requires two *electrodes*; the negative (−) terminal of the stimulator is the *cathode* (where the current flows out), and the positive (+) terminal of the stimulator is the *anode* (where the current flows in). Electrical stimulation for *nerve conduction studies* generally requires application of the *cathode* in the vicinity of the neural tissue to produce depolarization (Fig. 27). Depolarization of a nerve occurs under the cathode (the negative terminal), and the depolarization proceeds from the cathode (both orthodromically and antidromically). The cathode should be placed closer to the active recording electrode than the anode because the anode has the potential to hyperpolarize the nerve and block the depolarization caused by the cathode. This could cause the recorded potential to be reduced or absent. By convention, the stimulating electrodes are called bipolar if they are encased or attached together (e.g., pad electrodes fixed on a plastic bar) and are called monopolar if they are not.

Stimulus In clinical nerve conduction studies, *stimulus* is an electrical stimulus applied to a nerve. It is described

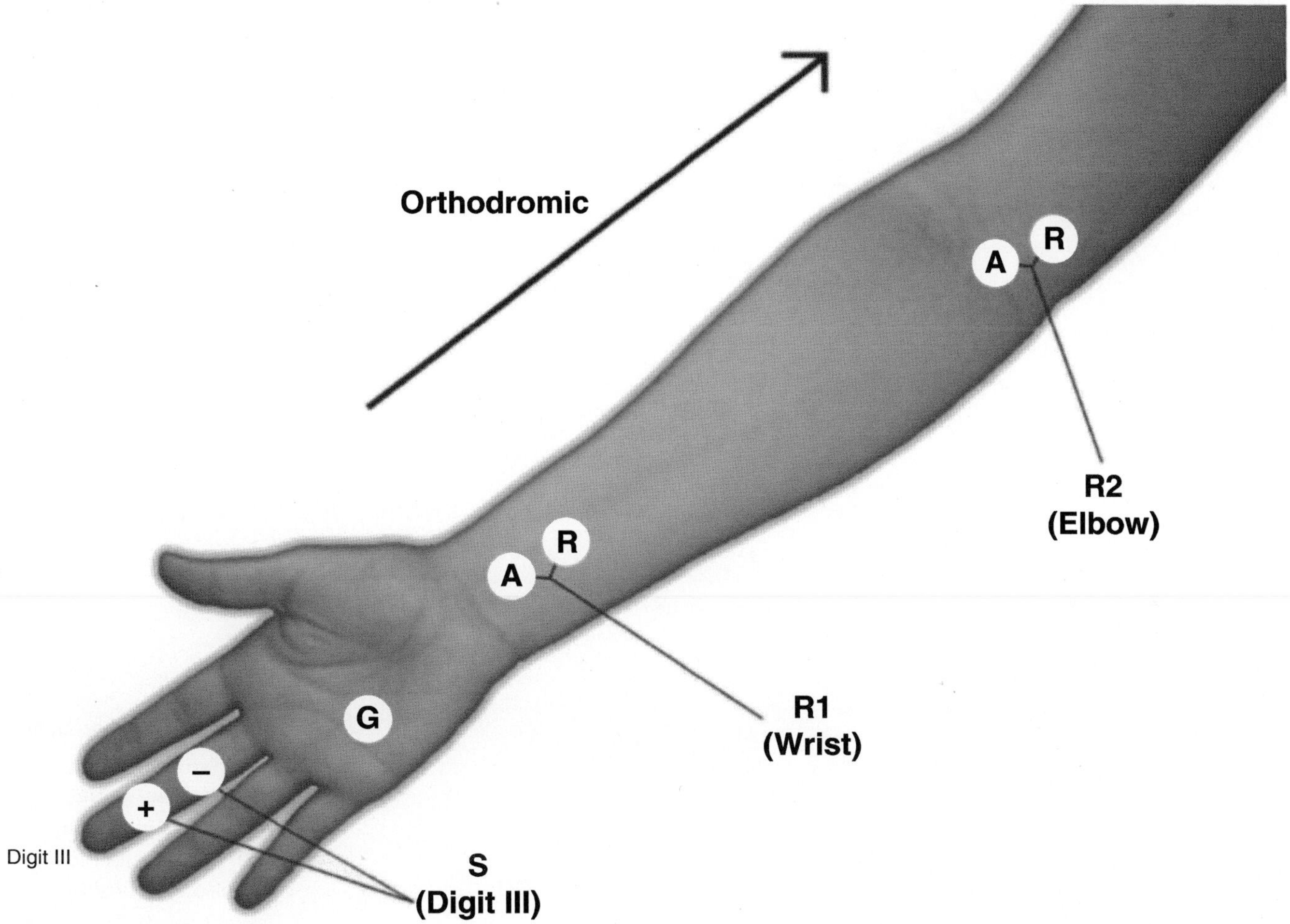

Fig. 23 Sensory orthodromic conduction, recording at the wrist (R1) and at the elbow (R2), stimulation to the digit III

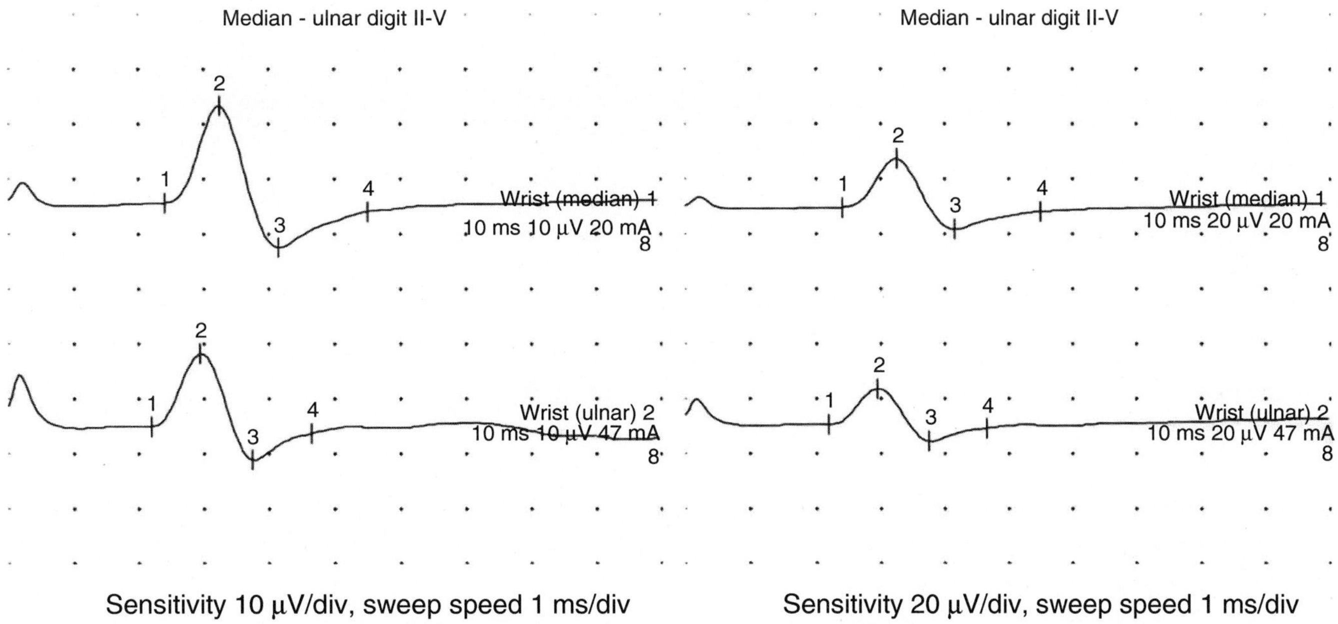

Fig. 24 Triphasic waves. Median–ulnar nerves comparative sensory conduction study, antidromic sensory nerve action potentials (SNAPs) recorded to the digit II and to the digit V, stimulation at the wrist: median nerve (*upper trace*) and ulnar nerve (*lower trace*)

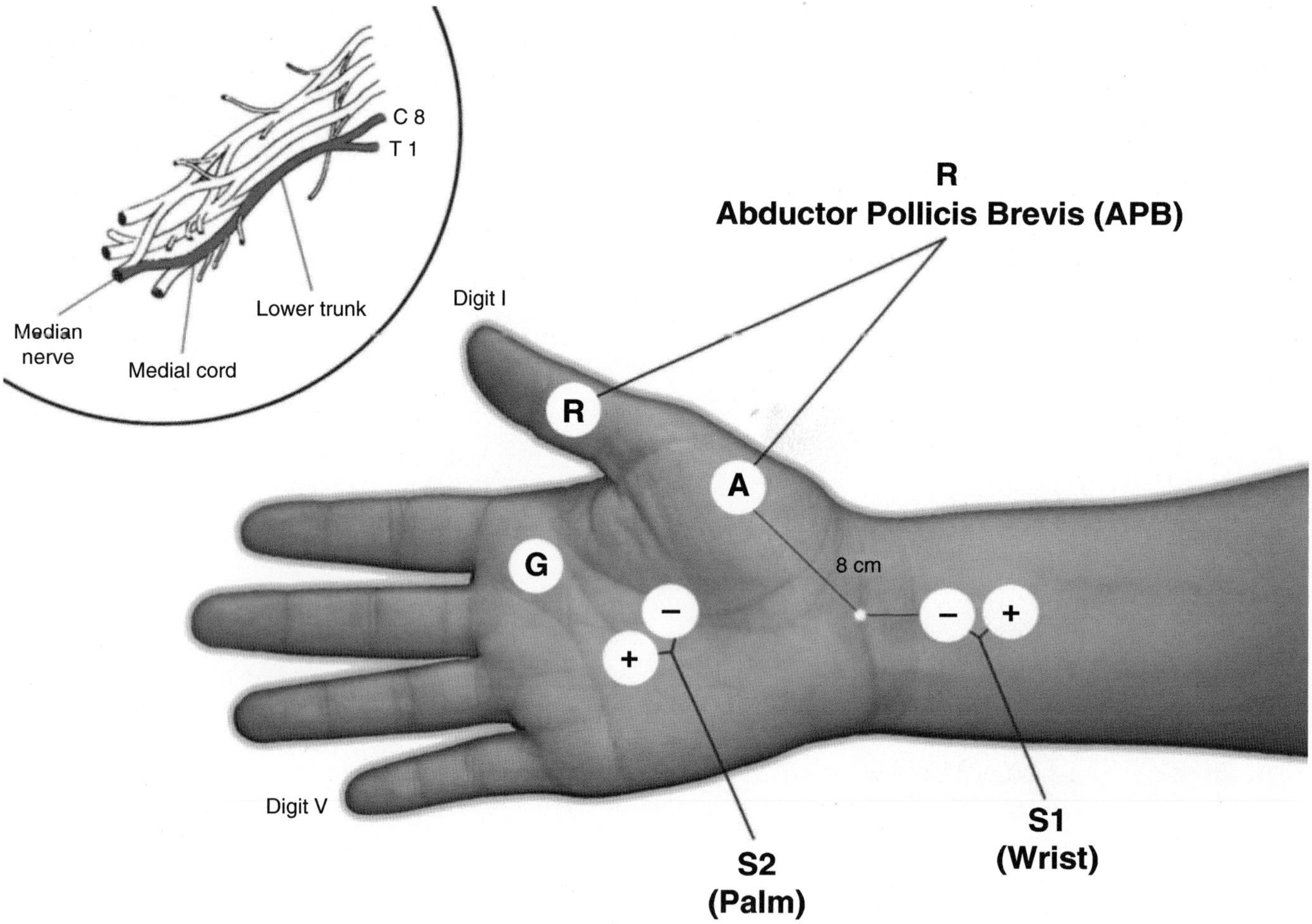

Fig. 25 Recording technique from the *abductor pollicis brevis* (APB) muscle, bipolar configuration: A, *active electrode* (on the motor point of APB muscle); R, *reference electrode* (on the tendinous insertion of APB muscle)

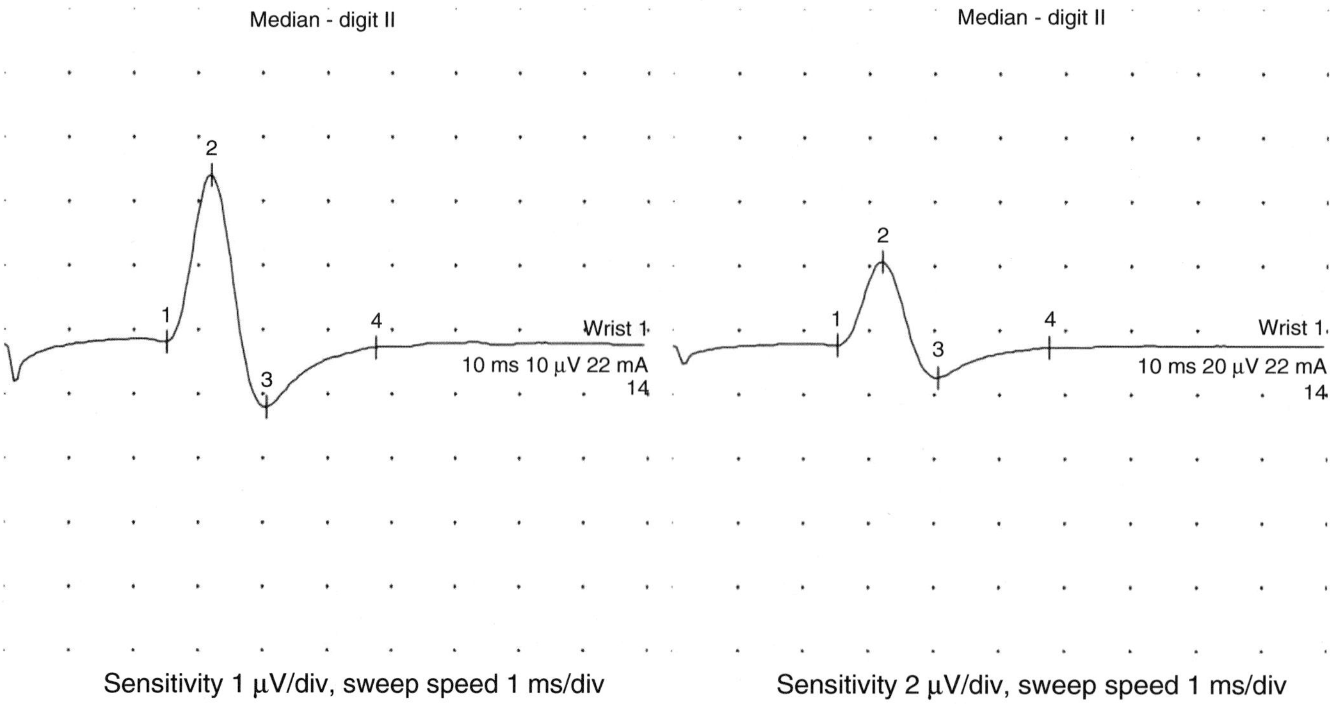

Fig. 26 Antidromic sensory nerve action potential (SNAP) recorded to the digit II (wrist stimulation) marker 1, onset latency (ms); marker 1–2, negative-peak amplitude (µV); markers 2–3, peak-to-peak amplitude (µV); markers 1–4, duration (ms)

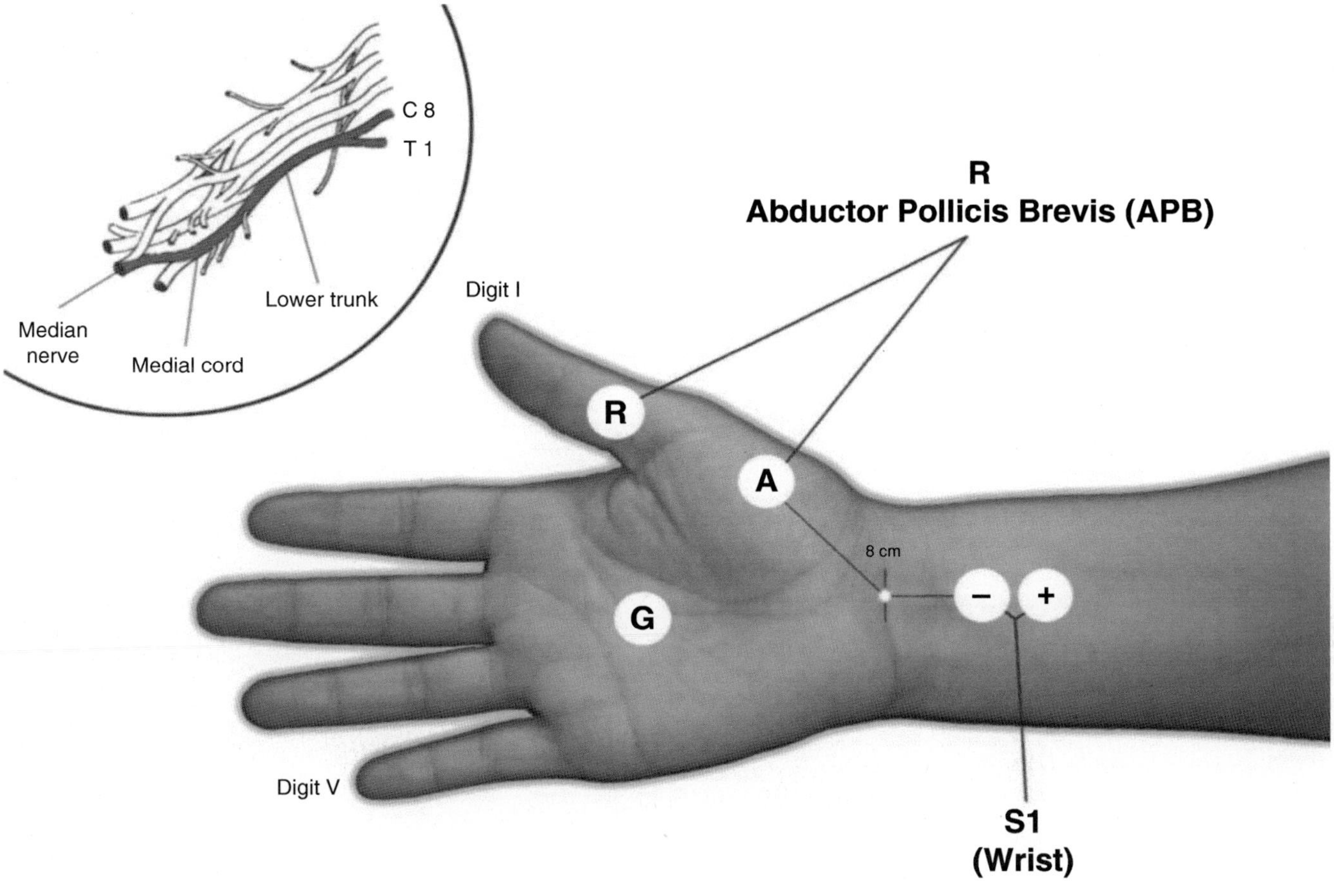

Fig. 27 Recording technique from the *abductor pollicis brevis* (APB) muscle, stimulation of the wrist (S1) and on the palm (S2): negative (–) terminal, *anode*; positive (+) terminal, *cathode*

in absolute terms or with respect to the nerve evoked potential. In absolute terms, it is defined by *duration* (ms), *waveform* (square, exponential, linear, etc.), and strength or intensity measured as *voltage* (V) or current (mA). With respect to the evoked potential, the *stimulus* is graded as subthreshold, threshold, submaximal, maximal, or supramaximal. A threshold *stimulus* is one just sufficient to produce a detectable response. Stimuli less than the threshold *stimulus* are termed subthreshold. The maximal *stimulus* is the stimulus intensity after which a further increase in intensity causes no increase in the *amplitude* of the evoked potential. Stimuli of intensity below this level but above threshold are submaximal. Stimuli of intensity greater than the maximal *stimulus* are defined supramaximal. Ordinarily, supramaximal stimuli are used for *nerve conduction studies*. By convention, an electrical stimulus of approximately 20 % greater voltage/current than required for the maximal *stimulus* is used for supramaximal stimulation. The *frequency*, number, and *duration* of a series of stimuli should be specified.

Sweep Time Also called "time base," it represents the linear time scale with which the potentials are displayed. In clinical practice, the sweep speed usually ranges from 2 to 5 ms/division for motor *nerve conduction studies* and from 1 to 2 ms/division for sensory *nerve conduction studies*. For *late responses*, commonly a sweep speed of 5–10 ms/division is used for F-waves or 10 ms/division for *H-reflex* study. In *electromyography,* a higher sweep rate (10–100 ms/division) is used than in *electroneurography* in order to analyze muscle activity at rest and during minimal and maximal contraction (Fig. 28).

Voltage A potential difference between two recording sites. Expressed in volts (V) or millivolts (mV).

Waveform A transient change in *voltage* represented as a line of differing directions over time, describing the shape of a wave.

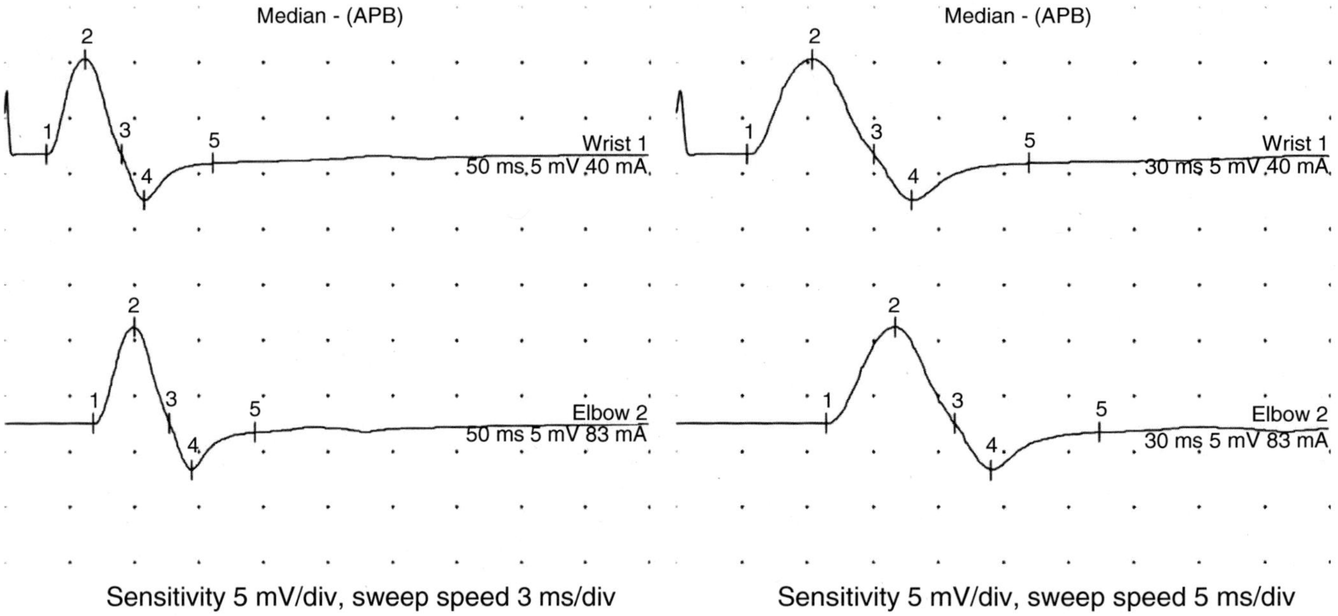

Fig. 28 Compound muscle action potentials (CMAPs) recorded at the hand from the *abductor pollicis brevis* (APB) muscle, displayed at different sweep speed (3 ms/div, *left*; 5 ms/div, *right*) and identical sensitivity, stimulation at the wrist (*upper trace*) and at the elbow (*lower trace*)

Index

G. Gentili, M. Di Napoli, *The Ulnar Nerve*, DOI 10.1007/978-3-319-43292-2

R

S

T

U

Printed by Printforce, the Netherlands